The Princeton Guide to Evolution

The Princeton Guide to Evolution

EDITOR IN CHIEF Jonathan B. Losos
Harvard University

EDITORS

David A. Baum
University of Wisconsin, Madison

Douglas J. Futuyma
Stony Brook University

Hopi E. Hoekstra
Harvard University

Richard E. Lenski
Michigan State University

Allen J. Moore
University of Georgia

Catherine L. Peichel
Fred Hutchinson Cancer Research Center

Dolph Schluter
University of British Columbia

Michael C. Whitlock
University of British Columbia

ADVISORS

Michael J. Donoghue

Simon A. Levin

Trudy F. C. Mackay

Loren Rieseberg

Joseph Travis

Gregory A. Wray

PRINCETON UNIVERSITY PRESS
Princeton & Oxford

Published by Princeton University Press, 41 William Street, Princeton, New Jersey 08540

In the United Kingdom: Princeton University Press, 6 Oxford Street, Woodstock, Oxfordshire OX20 1TW

press.princeton.edu

Library of Congress Cataloging-in-Publication Data

The Princeton guide to evolution / Jonathan B. Losos, Harvard University, editor in chief; David A. Baum, University of Wisconsin, Madison, Douglas J. Futuyma, Stony Brook University, Hopi E. Hoekstra, Harvard University, Richard E. Lenski, Michigan State University, Allen J. Moore, University of Georgia, Catherine L. Peichel, Hutchinson Cancer Institute, Seattle, Dolph Schluter, University of British Columbia, Michael C. Whitlock, University of British Columbia, editors.

pages cm

Includes bibliographical references and index.

ISBN 978-0-691-14977-6 (hardcover)

1. Evolution (Biology) I. Losos, Jonathan B.

QH367.P85 2014

576.8—dc23

2013022360

British Library Cataloging-in-Publication Data is available

This book has been composed in Sabon and Din

Printed on acid-free paper. ∞

Printed in the United States of America

10 9 8 7 6 5 4 3 2 1

Contents

Preface

For more than 150 years, since the publication of *On the Origin of Species*, biologists have focused on understanding the evolutionary chronicle of diversification and extinction, and the underlying evolutionary processes that have produced it. Although progress in evolutionary biology has been steady since Darwin's time, developments in the last 20 years have ushered in a golden era of evolutionary study in which biologists are on the brink of answering many of the fundamental questions in the field.

These advances have come from a confluence of technological and conceptual innovations. In the laboratory, the rapid and inexpensive sequencing of large amounts of DNA is producing a wealth of data on the genomes of many species; comparisons of these genomes are allowing scientists to pinpoint the specific genetic changes that have occurred over the course of evolution. In parallel, spectacular fossil discoveries have filled many of the most critical gaps in our documentation of the evolutionary pageant, detailing how whales evolved from land-living animals, snakes from their four-legged lizard forebears, and humans from our primate ancestors. In addition, providing the data that Darwin could only imagine, field biologists are now tracking populations, directly documenting natural selection as it occurs, and monitoring the resulting evolutionary changes that occur from one generation to the next.

At the same time, evolutionary biology is making an impact throughout human society. Many current issues—such as the rise of new diseases, the increased resistance of pests and microorganisms to efforts to control them, and the effect of changing environmental conditions on natural populations—revolve around aspects of natural selection and evolutionary change. Many disparate areas of modern life—medicine, the legal system, computing—increasingly employ evolutionary thinking and use methods developed in evolutionary biology. Paradoxically, even as our understanding of evolution and its importance to society has never been greater, substantial proportions of the population in a number of countries—most notably the United States and Turkey—dispute the scientific findings of evolutionary biologists and resist the teaching of evolution in schools.

This volume follows on the success of *The Princeton Guide to Ecology*, edited by Simon Levin. Published in 2009, the ecology guide has proven valuable to a wide range of readers, from professional ecologists and graduate students to land planners, economists, and social scientists. With this model in mind, we set out to produce a guide that would be accessible and useful to students and scientists in evolutionary biology and related disciplines, as well as to anyone with a serious interest in evolution. What makes this volume stand out is the breadth and depth of our 107 chapters, each written by authorities in their respective field. In addition, the articles balance accessibility with depth of analysis, making the *Guide* a valuable reference for a broad audience. Certainly, some articles are more technical than others, but readers can easily select chapters appropriate for their interests and expertise.

The *Guide* is divided into eight sections. The introductory section includes four chapters covering the basics of evolution: what it is, the history of its study, the evidence for its occurrence, and a basic primer of genetic and phenotypic variation. The following seven sections cover the major areas of evolutionary biology, each beginning with a synoptic overview by the section editor. Section II: Phylogenetics and the History of Life, covers the history of life and how it is studied. It includes chapters on the evolution of each of the major forms of life, as well as on the study of life's history through the examination of the fossil record and the construction of phylogenetic trees that detail the relationships among species and higher taxa. Section III: Selection and Adaptation, moves to evolutionary processes, focusing on natural selection, the presumed primary driver of evolutionary change. Section IV: Evolutionary Processes, covers gene flow, genetic drift, and nonrandom mating. Section V: Genes, Genomes, Phenotypes, examines the link between genes and phenotypes and how they evolve, focusing on the rapid growth of knowledge and continuing research in genetics and developmental biology and the relationships of these fields to evolutionary biology. Section VI: Speciation and Macroevolution, moves the focus to the species level and above, emphasizing the origin of species—that is, speciation—and evolutionary change that drives large-scale changes in the history of life through time, such as the rise of particular taxa and the extinction of others. Section VII: Evolution of Behavior, Society, and Humans, focuses on behavioral and social interactions that occur within species, including competition for mating success (referred to as sexual selection) and the evolution of traits such as parental care, communication, and altruism. Although chapters in this section are broad in taxonomic scope, many have particular relevance to human biology. Finally, Section VIII: Evolution and Modern Society, addresses how evolutionary biology directly affects the health and welfare of humans today.

This volume could not have been possible without the efforts of the editors and authors, whose work was instrumental to such a wide-ranging and authoritative work. Development of the guide also benefited immensely from the wisdom of our advisors, Michael Donoghue, Simon Levin, Trudy Mackay, Loren Rieseberg, Joseph Travis, and Greg Wray. In addition, the editorial staff at Princeton University Press was indispensable. The entire project was skillfully overseen by executive editor Anne Savarese, and day-to-day management moved smoothly and efficiently under the watchful eyes of editorial assistants Diana Goovaerts and Sarah David, and production editor Karen Carter.

We mourn the loss and gratefully acknowledge the contribution of our distinguished colleague Farish Jenkins, who died on November 11, 2012.

Jonathan B. Losos

Contributors

Aneil F. Agrawal, *Department of Ecology and Evolutionary Biology, University of Toronto*
IV.5 GENETIC LOAD

Michael E. Alfaro, *Department of Ecology and Evolutionary Biology, University of California, Los Angeles*
VI.15 KEY EVOLUTIONARY INNOVATIONS

Garland E. Allen, *Department of Biology, Washington University in St. Louis*
I.2 THE HISTORY OF EVOLUTIONARY THOUGHT

Dan I. Andersson, *Department of Medical Biochemistry and Microbiology, Uppsala University*
VIII.3 EVOLUTION OF ANTIBIOTIC RESISTANCE

Michael J. Angilletta Jr., *School of Life Sciences, Arizona State University*
III.13 BIOCHEMICAL AND PHYSIOLOGICAL ADAPTATIONS

Charles F. Aquadro, *Department of Molecular Biology and Genetics, Cornell University*
V.1 MOLECULAR EVOLUTION

Jonathan W. Atwell, *Department of Biology, Indiana University*
VII.2 EVOLUTION OF HORMONES AND BEHAVIOR

Francisco J. Ayala, *Department of Ecology and Evolutionary Biology, University of California, Irvine*
VIII.13 EVOLUTION AND RELIGION

Doris Bachtrog, *Department of Integrative Biology, University of California, Berkeley*
V.4 EVOLUTION OF SEX CHROMOSOMES

Charles F. Baer, *Department of Biology, University of Florida*
IV.2 MUTATION

Nathan W. Bailey, *School of Biology, University of St. Andrews*
VII.15 EVOLUTION OF APPARENTLY NONADAPTIVE BEHAVIOR

Timothy G. Barraclough, *Division of Ecology and Evolution, Imperial College London*
VI.2 SPECIATION PATTERNS

Spencer C. H. Barrett, *Department of Ecology and Evolutionary Biology, University of Toronto*
IV.8 EVOLUTION OF MATING SYSTEMS: OUTCROSSING VERSUS SELFING

N. H. Barton, *Institute of Science and Technology Austria*
IV.4 RECOMBINATION AND SEX

David A. Baum, *Department of Botany, University of Wisconsin, Madison*
II PHYLOGENETICS AND THE HISTORY OF LIFE

Graham Bell, *Department of Biology, McGill University*
III.6 RESPONSES TO SELECTION: EXPERIMENTAL POPULATIONS

Yehuda Ben-Shahar, *Department of Biology, Washington University in St. Louis*
VII.1 GENES, BRAINS, AND BEHAVIOR

Michael J. Benton, *School of Earth Sciences, University of Bristol*
VI.13 CAUSES AND CONSEQUENCES OF EXTINCTION

Janette W. Boughman, *Department of Zoology, Michigan State University*
VI.5 SPECIATION AND SEXUAL SELECTION

Paul M. Brakefield, *Department of Zoology, University of Cambridge*
V.10 EVOLUTION AND DEVELOPMENT: ORGANISMS

Edmund D. Brodie III, *Department of Biology, University of Virginia*
III.5 PHENOTYPIC SELECTION ON QUANTITATIVE TRAITS

C. Alex Buerkle, *Department of Botany and Program in Ecology, University of Wyoming*
VI.6 GENE FLOW, HYBRIDIZATION, AND SPECIATION

Michael A. Cant, *Biosciences, University of Exeter*
VII.10 COOPERATIVE BREEDING

Paulyn Cartwright, *Department of Ecology and Evolutionary Biology, University of Kansas*
II.15 ORIGIN AND EARLY EVOLUTION OF ANIMALS

Amy Cavanaugh, *Department of Biological Sciences, University of Wisconsin, Rock County*
VIII.5 DOMESTICATION AND THE EVOLUTION OF AGRICULTURE

Michel Chapuisat, *Department of Ecology and Evolution, University of Lausanne*
VII.13 EVOLUTION OF EUSOCIALITY

Deborah Charlesworth, *School of Biological Sciences, University of Edinburgh*
IV.6 INBREEDING

Julia Clarke, *Jackson School of Geosciences, University of Texas at Austin*
II.8 TAXONOMY IN A PHYLOGENETIC FRAMEWORK

Peter R. Crane, *School of Forestry and Environmental Studies, Yale University*
II.13 MAJOR EVENTS IN THE EVOLUTION OF LAND PLANTS

Cameron R. Currie, *Department of Bacteriology, University of Wisconsin, Madison*
VIII.5 DOMESTICATION AND THE EVOLUTION OF AGRICULTURE

David Deamer, *Department of Biomolecular Engineering, University of California, Santa Cruz*
II.10 THE ORIGIN OF LIFE

Michael J. Donoghue, *Department of Ecology and Evolutionary Biology, Yale University*
II.4 HISTORICAL BIOGEOGRAPHY

Dieter Ebert, *Zoological Institute, Universität Basel*
VIII.2 EVOLUTION OF PARASITE VIRULENCE

Scott P. Egan, *Department of Biological Sciences, University of Notre Dame*
VI.9 SPECIATION AND GENOME EVOLUTION

Andrew D. Ellington, *Department of Chemistry and Biochemistry, University of Texas at Austin*
VIII.7 DIRECTED EVOLUTION

Jeffrey Feder, *Department of Biological Sciences, University of Notre Dame*
VI.9 SPECIATION AND GENOME EVOLUTION

Lila Fishman, *Division of Biological Sciences, University of Montana*
IV.7 SELFISH GENETIC ELEMENTS AND GENETIC CONFLICT

Douglas J. Futuyma, *Department of Ecology and Evolution, Stony Brook University*
III NATURAL SELECTION AND ADAPTATION

Dana H. Geary, *Department of Geoscience, University of Wisconsin, Madison*
II.9 THE FOSSIL RECORD

J. Peter Gogarten, *Department of Molecular and Cell Biology, University of Connecticut*
II.11 EVOLUTION IN THE PROKARYOTIC GRADE

Emma E. Goldberg, *Biological Sciences, University of Illinois, Chicago*
VI.14 SPECIES SELECTION

Peter R. Grant, *Department of Ecology and Evolutionary Biology, Princeton University*
VI.10 ADAPTIVE RADIATION

Michael D. Greenfield, *Institut de Recherche sur la Biologie de l'Insecte, Université de Tours*
VII.7 EVOLUTION OF COMMUNICATION

Elizabeth Hannon, *Department of History and Philosophy of Science, University of Cambridge*
VIII.10 CULTURAL EVOLUTION

Sara J. Hanson, *Department of Biology and Program in Genetics, University of Iowa*
V.2 GENOME EVOLUTION

Luke J. Harmon, *Department of Biological Sciences, University of Idaho*
VI.11 MACROEVOLUTIONARY RATES

Richard G. Harrison, *Department of Ecology and Evolutionary Biology, Cornell University*
VI.1 SPECIES AND SPECIATION

Marc D. Hauser, *Independent Scholar*
VII.14 COGNITION: PHYLOGENY, ADAPTATION, AND BY-PRODUCTS

John Hawks, *Department of Anthropology, University of Wisconsin, Madison*
II.18 HUMAN EVOLUTION

Philip Hedrick, *School of Life Sciences, Arizona State University*
IV.1 GENETIC DRIFT

Noel A. Heim, *Department of Geology and Evolutionary Science, Stanford University*
II.9 THE FOSSIL RECORD

Michael E. Hellberg, *Department of Biological Sciences, Louisiana State University*
II.5 PHYLOGEOGRAPHY

David S. Hibbett, *Department of Biology, Clark University*
II.14 MAJOR EVENTS IN THE EVOLUTION OF FUNGI

Hopi E. Hoekstra, *Department of Organismic and Evolutionary Biology, Harvard University*
V GENES, GENOMES, PHENOTYPES

Ary Hoffmann, *Department of Genetics and Zoology, University of Melbourne*
III.8 LIMITS AND CONSTRAINTS

Mark Holder, *Department of Ecology and Evolutionary Biology, University of Kansas*
II.2 PHYLOGENETIC INFERENCE

Kent E. Holsinger, *Department of Ecology and Evolutionary Biology, University of Connecticut*
III.3 THEORY OF SELECTION IN POPULATIONS

Robert D. Holt, *Department of Ecology, University of Florida*
III.14 EVOLUTION OF THE ECOLOGICAL NICHE

Robin Hopkins, *Department of Integrative Biology, University of Texas at Austin*
VI.4 SPECIATION AND NATURAL SELECTION

Gene Hunt, *Department of Paleobiology Smithsonian Institution, National Museum of Natural History*
VI.12 MACROEVOLUTIONARY TRENDS

John Jaenike, *Department of Biology, University of Rochester*
IV.7 SELFISH GENETIC ELEMENTS AND GENETIC CONFLICT

Farish A. Jenkins Jr., *Late Professor of Biology, Harvard University*
II.17 MAJOR FEATURES OF TETRAPOD EVOLUTION

Michael D. Jennions, *Research School of Biology, Australian National University*
VII.6 SEXUAL SELECTION: MATE CHOICE

Laura A. Katz, *Department of Biological Sciences, Smith College*
II.12 ORIGIN AND DIVERSIFICATION OF EUKARYOTES

Paul Keim, *Department of Biology, Northern Arizona University*
VIII.4 EVOLUTION AND MICROBIAL FORENSICS

Laurent Keller, *Department of Ecology and Evolution, University of Lausanne*
VII.13 EVOLUTION OF EUSOCIALITY

Ellen D. Ketterson, *Department of Biology, Indiana University*
VII.2 EVOLUTION OF HORMONES AND BEHAVIOR

Joel G. Kingsolver, *Department of Biology, University of North Carolina, Chapel Hill*
III.7 RESPONSES TO SELECTION: NATURAL POPULATIONS

Hanna Kokko, *Research School of Biology, Australian National University*
VII.6 SEXUAL SELECTION: MATE CHOICE

Mathias Kölliker, *Department of Environmental Sciences, University of Basel*
VII.8 EVOLUTION OF PARENTAL CARE

Allan Larson, *Department of Biology, Washington University in St. Louis*
II.6 CONCEPTS IN CHARACTER MACROEVOLUTION: ADAPTATION, HOMOLOGY, AND EVOLVABILITY

Richard E. Lenski, *Departments of Microbiology & Molecular Genetics, Zoology, and Crop and Soil Sciences, Michigan State University*
VIII EVOLUTION AND MODERN SOCIETY

Andrew B. Leslie, *School of Forestry and Environmental Studies, Yale University*
II.13 MAJOR EVENTS IN THE EVOLUTION OF LAND PLANTS

Tim Lewens, *Department of History and Philosophy of Science, University of Cambridge*
VIII.10 CULTURAL EVOLUTION

John M. Logsdon Jr., *Department of Biology and Program in Genetics, University of Iowa*
V.2 GENOME EVOLUTION

Manyuan Long, *Department of Ecology and Evolution, University of Chicago*
V.6 EVOLUTION OF NEW GENES

Jonathan B. Losos, *Department of Organismic and Evolutionary Biology, Harvard University*
I.1 WHAT IS EVOLUTION?

David B. Lowry, *Department of Integrative Biology, University of Texas at Austin*
VI.4 SPECIATION AND NATURAL SELECTION

Virpi Lummaa, *Department of Animal and Plant Sciences, University of Sheffield*
VII.11 HUMAN BEHAVIORAL ECOLOGY

Florian Maderspacher, *Current Biology, Elsevier, Inc.*
V.8 EPIGENETICS

Gregory C. Mayer, *Department of Biological Sciences, University of Wisconsin–Parkside*
I.3 THE EVIDENCE FOR EVOLUTION

Joel W. McGlothlin, *Department of Biological Sciences, Virginia Polytechnic Institute and State University*
VII.2 EVOLUTION OF HORMONES AND BEHAVIOR

Daniel McNabney, *Department of Biology, University of Rochester*
VI.8 GENETICS OF SPECIATION

John M. McNamara, *School of Mathematics, University of Bristol*
VII.3 GAME THEORY AND BEHAVIOR

Mark A. McPeek, *Department of Biological Sciences, Dartmouth College*
VI.16 EVOLUTION OF COMMUNITIES

Christine W. Miller, *Department of Entomology and Nematology, University of Florida*
VII.5 SEXUAL SELECTION: MALE-MALE COMPETITION

Antónia Monteiro, *Department of Ecology and Evolutionary Biology, Yale University*
V.11 EVOLUTION AND DEVELOPMENT: MOLECULES

Jacob A. Moorad, *Department of Biology, Duke University*
VII.16 AGING AND MENOPAUSE

Allen J. Moore, *Department of Genetics, University of Georgia*
VII EVOLUTION OF BEHAVIOR, SOCIETY, AND HUMANS

Patrik Nosil, *Department of Ecology and Evolutionary Biology, University of Colorado, Boulder*
VI.9 SPECIATION AND GENOME EVOLUTION

Samir Okasha, *Department of Philosophy, University of Bristol*
III.2 UNITS AND LEVELS OF SELECTION

Lorraine Olendzenski, *Department of Biology, St. Lawrence University*
II.11 EVOLUTION IN THE PROKARYOTIC GRADE

Kevin E. Omland, *Department of Biological Sciences, University of Maryland, Baltimore County*
II.1 INTERPRETATION OF PHYLOGENETIC TREES

H. Allen Orr, *Department of Biology, University of Rochester*
VI.8 GENETICS OF SPECIATION

Sarah P. Otto, *Department of Zoology, University of British Columbia*
III.9 EVOLUTION OF MODIFIER GENES AND BIOLOGICAL SYSTEMS

Mark Pagel, *School of Biological Sciences, University of Reading*
VIII.9 LINGUISTICS AND THE EVOLUTION OF HUMAN LANGUAGE

Laura Wegener Parfrey, *Department of Chemistry and Biochemistry, University of Colorado, Boulder*
II.12 ORIGIN AND DIVERSIFICATION OF EUKARYOTES

Bret A. Payseur, *Laboratory of Genetics, University of Wisconsin, Madison*
V.13 DISSECTION OF COMPLEX TRAIT EVOLUTION

Talima Pearson, *Department of Biological Sciences, Northern Arizona University*
VIII.4 EVOLUTION AND MICROBIAL FORENSICS

Catherine L. Peichel, *Fred Hutchinson Cancer Research Center, Seattle*
V GENES, GENOMES, PHENOTYPES; V.12 GENETICS OF PHENOTYPIC EVOLUTION

Robert T. Pennock, *Lyman Briggs College and Departments of Philosophy and Computer Science & Engineering, Michigan State University*
VIII.8 EVOLUTION AND COMPUTING

Dmitri A. Petrov, *Department of Biology, Stanford University*
V.14 SEARCHING FOR ADAPTATION IN THE GENOME

David W. Pfennig, *Department of Biology, University of North Carolina, Chapel Hill*
III.7 RESPONSES TO SELECTION: NATURAL POPULATIONS

Albert Phillimore, *Institute of Evolutionary Biology, University of Edinburgh*
VI.3 GEOGRAPHY, RANGE EVOLUTION, AND SPECIATION

Daniel E. L. Promislow, *Department of Pathology, University of Washington*
VII.16 AGING AND MENOPAUSE

Erik Quandt, *Department of Chemistry and Biochemistry, University of Texas at Austin*
VIII.7 DIRECTED EVOLUTION

David C. Queller, *Department of Biology, Washington University in St. Louis*
III.4 KIN SELECTION AND INCLUSIVE FITNESS; VII.9 COOPERATION AND CONFLICT: MICROBES TO HUMANS

Bruce Rannala, *Department of Evolution and Ecology, University of California, Davis*
II.3 MOLECULAR CLOCK DATING

Richard Ree, *Botany Department, Field Museum of Natural History*
II.7 USING PHYLOGENIES TO STUDY PHENOTYPIC EVOLUTION: COMPARATIVE METHODS AND TESTS OF ADAPTATION

David Reznick, *Department of Biology, University of California, Riverside*
III.11 EVOLUTION OF LIFE HISTORIES

Robert C. Richardson, *Department of Philosophy, University of Cincinnati*
VII.12 EVOLUTIONARY PSYCHOLOGY

Ophélie Ronce, *Institut des Sciences de l'Evolution, Université Montpellier 2, Centre National de la Recherche Scientifique*
IV.3 GEOGRAPHIC VARIATION, POPULATION STRUCTURE, AND MIGRATION

Nick J. Royle, *Department of Biosciences, University of Exeter*
VII.8 EVOLUTION OF PARENTAL CARE

Dolph Schluter, *Department of Zoology, University of British Columbia*
VI SPECIATION AND MACROEVOLUTION

Eugenie C. Scott, *National Center for Science Education, Inc.*
VIII.14 CREATIONISM AND INTELLIGENT DESIGN

H. Bradley Shaffer, *Department of Ecology and Evolutionary Biology, University of California, Los Angeles*
VIII.6 EVOLUTION AND CONSERVATION

Beth Shapiro, *Department of Ecology and Evolutionary Biology, University of California, Santa Cruz*
V.15 ANCIENT DNA

Mark L. Siegal, *Department of Biology, New York University*
V.9 EVOLUTION OF MOLECULAR NETWORKS

Per T. Smiseth, *Institute of Evolutionary Biology, University of Edinburgh*
VII.8 EVOLUTION OF PARENTAL CARE

Rhonda R. Snook, *Department of Animal and Plant Sciences, University of Sheffield*
VII.4 SEXUAL SELECTION AND ITS IMPACT ON MATING SYSTEMS

Jason E. Stajich, *Department of Plant Pathology and Microbiology, University of California, Riverside*
V.3 COMPARATIVE GENOMICS

Stephen C. Stearns, *Department of Ecology and Evolutionary Biology, Yale University*
III.1 NATURAL SELECTION, ADAPTATION, AND FITNESS: OVERVIEW; III.10 EVOLUTION OF REACTION NORMS

Joan E. Strassmann, *Department of Biology, Washington University in St. Louis*
III.4 KIN SELECTION AND INCLUSIVE FITNESS; VII.9 COOPERATION AND CONFLICT: MICROBES TO HUMANS

Sharon Y. Strauss, *Department of Ecology and Evolution, University of California, Davis*
III.15 ADAPTATION TO THE BIOTIC ENVIRONMENT

Alan R. Templeton, *Department of Biology, Washington University in St. Louis, Department of Evolutionary and Environmental Biology, University of Haifa*
VIII.11 EVOLUTION AND NOTIONS OF HUMAN RACE; VIII.12 THE FUTURE OF HUMAN EVOLUTION

John N. Thompson, *Department of Ecology and Evolutionary Biology, University of California, Santa Cruz*
VI.7 COEVOLUTION AND SPECIATION

Michelle D. Trautwein, *Biodiversity Laboratory, North Carolina Museum of Natural Sciences*
II.16 MAJOR EVENTS IN THE EVOLUTION OF ARTHROPODS

Paul E. Turner, *Department of Ecology and Evolutionary Biology, Yale University*
VIII.1 EVOLUTIONARY MEDICINE

Peter C. Wainwright, *Department of Evolution and Ecology, University of California, Davis*
III.12 EVOLUTION OF FORM AND FUNCTION

Michael C. Whitlock, *Department of Zoology, University of British Columbia*
I.4 FROM DNA TO PHENOTYPES; IV EVOLUTIONARY PROCESSES

Brian M. Wiegmann, *Department of Entomology, North Carolina State University*
II.16 MAJOR EVENTS IN THE EVOLUTION OF ARTHROPODS

Patricia J. Wittkopp, *Department of Ecology and Evolutionary Biology, University of Michigan*
V.7 EVOLUTION OF GENE EXPRESSION

Ziheng Yang, *Department of Genetics, Evolution, and Environment, University College, London*
II.3 MOLECULAR CLOCK DATING

Jianzhi Zhang, *Department of Ecology and Evolutionary Biology, University of Michigan*
V.5 GENE DUPLICATION

Carl Zimmer, *Environmental Studies Program, Yale University*
VIII.15 EVOLUTION AND THE MEDIA

I

Introduction

I.1

What Is Evolution?

Jonathan Losos

OUTLINE

1. What is evolution?
2. Evolution: Pattern versus process
3. Evolution: More than changes in the gene pool
4. In the light of evolution
5. Critiques and the evidence for evolution
6. The pace of evolution
7. Evolution, humans, and society

Evolution refers to change through time as species become modified and diverge to produce multiple descendant species. *Evolution* and *natural selection* are often conflated, but evolution is the historical occurrence of change, and natural selection is one mechanism—in most cases the most important—that can cause it. Recent years have seen a flowering in the field of evolutionary biology, and much has been learned about the causes and consequences of evolution. The two main pillars of our knowledge of evolution come from knowledge of the historical record of evolutionary change, deduced directly from the fossil record and inferred from examination of phylogeny, and from study of the process of evolutionary change, particularly the effect of natural selection. It is now apparent that when selection is strong, evolution can proceed considerably more rapidly than was generally envisioned by Darwin. As a result, scientists are realizing that it is possible to conduct evolutionary experiments in real time. Recent developments in many areas, including molecular and developmental biology, have greatly expanded our knowledge and reaffirmed evolution's central place in the understanding of biological diversity.

GLOSSARY

Evolution. Descent with modification; transformation of species through time, including both changes that occur within species, as well as the origin of new species.

Natural Selection. The process in which individuals with a particular trait tend to leave more offspring in the next generation than do individuals with a different trait.

Approximately 375 million years ago, a large and vaguely salamander-like creature plodded from its aquatic home and began the vertebrate invasion of land, setting forth the chain of evolutionary events that led to the birds that fill our skies, the beasts that walk our soil, me writing this chapter, and you reading it. This was, of course, just one episode in life's saga: millions of years earlier, plants had come ashore, followed soon thereafter—or perhaps simultaneously—by arthropods. We could go back much earlier, 4 billion years or so, to that fateful day when the first molecule replicated itself, an important milestone in the origin of life and the beginning of the evolutionary pageant. Moving forward, the last few hundred million years have also had their highs and lows: the origins of frogs and trees, the end-Permian extinction when 90 percent of all species perished, and the rise and fall of the dinosaurs.

These vignettes are a few of many waypoints in the evolutionary chronicle of life on earth. Evolutionary biologists try to understand this history, explaining how and why life has taken its particular path. But the study of evolution involves more than looking backward to try to understand the past. Evolution is an ongoing process, one possibly operating at a faster rate now than in times past in this human-dominated world. Consequently, evolutionary biology is also forward looking: it includes the study of evolutionary processes in action today—how they operate, what they produce—as well as investigation of how evolution is likely to proceed in the future. Moreover, evolutionary biology is not solely an academic matter; evolution affects humans in many ways, from coping with the emergence of agricultural pests and disease-causing organisms to understanding the workings of our own genome. Indeed, evolutionary science has broad relevance, playing an important role in advances in many areas, from computer programming to medicine to engineering.

1. WHAT IS EVOLUTION?

Look up the word "evolution" in the online version of the *Oxford English Dictionary*, and you will find 11 definitions and numerous subdefinitions, ranging from mathematical ("the successive transformation of a curve by the alteration of the conditions which define it") to chemical ("the emission or release of gas, heat, light, etc.") to military ("a manoeuvre executed by troops or ships to adopt a different tactical formation"). Even with reference to biology, there are several definitions, including "emergence or release from an envelope or enclosing structure; (also) protrusion, evagination," not to mention "rare" and "historical" usage related to the concept of preformation of embryos. Even among evolutionary biologists, evolution is defined in different ways. For example, one widely read textbook refers to evolution as "changes in the properties of groups of organisms over the course of generations" (Futuyma 2005), whereas another defines it as "changes in allele frequencies over time" (Freeman and Herron 2007).

One might think that—as in so many other areas of evolutionary biology—we could look to Darwin for clarity. But in the first edition of *On the Origin of Species*, the term "evolution" never appears (though the last word of the book is "evolved"); not until the sixth edition does Darwin use "evolution." Rather, Darwin's term of choice is "descent with modification," a simple phrase that captures the essence of what evolutionary biology is all about: the study of the transformation of species through time, including both changes that occur within species, as well as the origin of new species.

2. EVOLUTION: PATTERN VERSUS PROCESS

Many people—sometimes even biologists—equate evolution with natural selection, but the two are not the same. Natural selection is one process that can cause evolutionary change, but natural selection can occur without producing evolutionary change. Conversely, processes other than natural selection can lead to evolution.

Natural selection within populations refers to the situation in which individuals with one variant of a trait (say, blue eyes) tend to leave more offspring that are healthy and fertile in the next generation than do individuals with an alternative variant of the trait. Such selection can occur in many ways, for example, if the variant leads to greater longevity, greater attractiveness to members of the other sex, or greater number of offspring per breeding event. The logic behind natural selection is unassailable. If some trait variant is causally related to greater reproductive success, then more members of the population will have that variant in the next generation; continued over many generations, such selection can greatly change the constitution of a population.

But there is a catch. Natural selection can occur without leading to evolution if differences among individuals are not genetically based. For natural selection to cause evolutionary change, trait variants must be transmitted from parent to offspring; if that is the case, then offspring will resemble their parents and the trait variants possessed by the parents that produce the most offspring will increase in frequency in the next generation.

However, offspring do not always resemble their parents. In some cases, individuals vary phenotypically not because they are different genetically, but because they experienced different environments during growth (this is the "nurture" part of the nature versus nurture debate; see chapters III.10 and VII.1). If, in fact, variation in a population is not genetically based, then selection will have no evolutionary consequence; individuals surviving and producing many offspring will not differ genetically from those that fail to prosper, and as a result, the gene pool of the population will not change. Nonetheless, much of the phenotypic variation within a population is, in fact, genetically based; consequently, natural selection often does lead to evolutionary change.

But that does not mean that the occurrence of evolutionary change necessarily implies the action of natural selection. Other processes—especially mutation, genetic drift, and immigration of individuals with different genetic constitutions—also can cause a change in the genetic makeup of a population from one generation to the next (see Section IV: Evolutionary Processes). In other words, natural selection can cause adaptive evolutionary change, but not all evolution is adaptive.

These caveats notwithstanding, 150 years of research have made clear that natural selection is a powerful force responsible for much of the significant evolutionary change that has occurred over the history of life. As the chapters in Section II: Phylogenetics and the History of Life, and Section III: Natural Selection and Adaptation, demonstrate, natural selection can operate in many ways, and scientists have correspondingly devised many methods to detect it, both through studies of the phenotype and of DNA itself (see also chapter V.14).

3. EVOLUTION: MORE THAN CHANGES IN THE GENE POOL

During the heyday of population genetics in the middle decades of the last century, many biologists equated evolution with changes from one generation to the next in gene frequencies (*gene frequency* refers to the frequencies of different alleles of a gene; for background on genetic variation, see chapter I.4). The "Modern Synthesis" of the 1930s and 1940s led to several decades in which the field was primarily concerned with the genetics of populations with an emphasis on natural selection (see chapter I.2). This focus was sharpened by the advent of molecular approaches to studying evolution. Starting in 1960 with the application of enzyme electrophoresis techniques, biologists could, for the first time, directly assess the extent of genetic variation within populations. To everyone's surprise, populations were found to contain much more variation than expected. This finding both challenged the view that natural selection was the dominant force guiding evolutionary change (see discussion of "neutralists" in chapters I.2 and V.1), yet further directed attention to the genetics of populations. With more advanced molecular techniques available today, the situation has not changed. There is much more variation than we first suspected.

The last 35 years have seen a broadening of evolutionary inquiry as the field has recognized that there is more to understanding evolutionary change than studying what happens to genes within populations—though this area remains a critically important part of evolutionary inquiry. Three aspects of expansion in evolutionary thinking are particularly important.

First, phenotypic evolution results from evolutionary change in the developmental process that transforms a single-celled fertilized egg into an adult organism. Although under genetic control, development is an intricate process that cannot be understood by examination of DNA sequences alone. Rather, understanding how phenotypes evolve, and the extent to which developmental systems constrain and direct evolutionary change, requires detailed molecular and embryological knowledge (see chapters V.10 and V.11).

Second, history is integral to understanding evolution (see introduction to Section II: Phylogenetics and the History of Life). The study of fossils—paleontology—provides the primary, almost exclusive, direct evidence of life in the past. Somewhat moribund in the middle of the last century, paleontology has experienced a resurgence in recent decades owing to both dramatic new discoveries stemming from an upsurge in paleontological exploration, and new ideas about evolution inspired by and primarily testable with fossil data, such as theories concerning punctuated equilibrium and stasis, species selection, and mass extinction. Initially critical in the development and acceptance of evolutionary theory, paleontology has once again become an important and vibrant part of evolutionary biology (see chapter II.9 and others in Section II).

Concurrently, a more fundamental revolution emphasizing the historical perspective has taken place over the last 30 years with the realization that information on phylogenetic relationships—that is, the *tree of life*, the pattern of descent and relationship among species—is critical in interpreting all aspects of evolution above the population level. Beginning with a transformation in the field of systematics concerning how phylogenetic relationships are inferred, this "tree-thinking" approach now guides study not only of all aspects of macroevolution but also of many population-level phenomena.

Finally, life is hierarchically organized. Genes are located within individuals, individuals within populations, populations within species, and species within clades (a *clade* consists of an ancestral species and all its descendants). Population genetics concerns what happens among individuals within a population, but evolutionary change can occur at all levels. For example, why are there more than 2000 species of rodents but only 3 species of monotremes (the platypus and echidnas), a much older clade of mammals? One cannot look at questions concerning natural selection within a population to answer this question. Rather, one must inquire about properties of entire species. Is there some attribute of rodents that makes them particularly prone to speciate or to avoid extinction? Similarly, why is there so much seemingly useless noncoding DNA in the genomes of many species (see chapter V.2)? One possibility is that some genes are particularly adept at mutating to multiply the number of copies of that gene within a genome; such DNA might increase in frequency in the genome even if such multiplication has no benefit to the individual in whose body the DNA resides. Just as selection among individual organisms on heritable traits can lead to evolutionary change within populations, selection among entities at other levels (species, genes) can also lead to evolutionary change, as long as those entities have traits that are transmitted to their offspring (be they descendant species or genes) and affect the number of descendants they produce. The upshot is that evolution occurs at multiple levels of the hierarchy of life; to understand its rich complexity we must study evolution at these distinct levels as well as the interactions among them. What happens, for example, when a trait that benefits an individual within a population (perhaps cannibalism—more food, fewer competitors!) has detrimental effects at the level of species?

Although evolutionary biology has expanded in scope, genetic change is still its fundamental foundation. Nonetheless, in recent years attention has focused on variation that is not genetically based. Phenotypic plasticity—the ability of a single genotype to produce different phenotypes when exposed to different environments—may itself be adaptive (see chapter III.10). If individuals in a population are likely to experience different conditions as they develop, then the evolution of a genotype that could produce appropriate phenotypes depending on circumstances would be advantageous. Although selection on these different phenotypes would not lead to evolutionary change, the degree of plasticity itself can evolve if differences in extent of plasticity lead to differences in the number of surviving offspring. Indeed, an open question is, why don't populations evolve to become infinitely malleable, capable of producing the appropriate phenotype for any environment? Presumably, plasticity has an associated cost such that adaptation to different environments often occurs by genetic differentiation rather than by the evolution of a single genotype that can produce different phenotypes. Such costs, however, have proven difficult to demonstrate.

Differences observed among populations may also reflect plastic responses to different environmental conditions and thus may not reflect genetic differentiation. However, if consistently transmitted from one generation to the next, such nongenetic differences may lead to divergent selective pressures on traits that are genetically determined, thus promoting evolutionary divergence between the populations. One particular example concerns behavior, which is highly variable in response to the environment—an extreme manifestation of plasticity (see chapter VIII.10). Learned behaviors that are transmitted from one generation to the next—often called *traditions* or *culture*—occur not only in humans but in other animals, not only our near relatives the apes but also cetaceans, birds, and others. Such behavioral differences among populations would not reflect genetic differentiation, but they might set the stage for genetic divergence in traits relating to the behaviors. One can easily envision, for example, how chimpanzee populations that use different tools—such as delicate twigs to probe termite mounds, or heavy stones to pound nuts—might evolve different morphological features to enhance the effectiveness of these behaviors. A concrete example involves human populations that tend cattle—surely a nongenetically based behavior—and have evolved genetic changes to permit the digestion of milk in adults.

4. IN THE LIGHT OF EVOLUTION

In a 1964 address to the American Society of Zoologists, the distinguished Russian-born biologist Theodosius Dobzhansky proclaimed "nothing makes sense in biology except in the light of evolution." Ever since, evolutionary biologists have trotted out this phrase (or some permutation of it) to emphasize the centrality of evolution in understanding the biological world. Nonetheless, for much of the twentieth century, the pervasive importance of an evolutionary perspective was not at all obvious to many biologists, some of whom considered Dobzhansky's claim to be self-serving hype. One could argue, for example, that the enormous growth in our understanding of molecular biology from 1950 to 2000 was made with little involvement or insight from evolutionary biology. Indeed, to the practicing molecular biologist in the 1980s and 1990s, evolutionary biology was mostly irrelevant.

Now, nothing could be further from the truth. When results of the human genome sequencing project first appeared in 2000, many initially believed that a thorough understanding of human biology would soon follow, answering questions about the genetic basis of human diseases and phenotypic variation among individuals. These hopes were quickly dashed—the genetic code, after all, is nothing more than a long list of letters (A, C, G, and T, the abbreviations of the four nucleotide building blocks of DNA). Much of the genome of many species seems to have no function and is just, in some sense, functionless filler; as a result, picking out where the genes lie in this 4 billion–long string of alphabet spaghetti, much less figuring out how these genes function, is not easy.

So where did molecular biologists turn? To the field of evolutionary biology! Genomicists soon realized that the best way to understand the human genome was to study it in the context of its evolutionary history, by comparing human sequences with those of other species in a phylogenetic framework. One method for locating genes, for example, is to examine comparable parts of the genome of different species. The underlying rationale is that genes evolve more slowly than other parts of the genome. Specifically, nonfunctioning stretches of DNA tend to evolve differences through time as random mutations become established (the process of genetic drift; see chapter IV.1), but functioning genes tend to diverge less, because natural selection removes deleterious mutations when they arise, keeping the DNA sequence similar among species. As a result, examination of the amount of divergence between two species relative to the amount of time since they shared a common ancestor can pinpoint stretches of DNA where evolution has occurred slowly, thus identifying the position of functional genes. Moreover, how a gene functions can often be deduced by comparing its function with that of homologous genes in other species and using a phylogeny to reconstruct the gene's evolutionary history (see chapter V.14).

And thus was born the effort to sequence the genomes of other species (see chapter V.3). At first, the nascent field of comparative genomics focused on primates and model laboratory species such as mice and fruit flies, the former to permit comparisons of the human genome with that of our close evolutionary relatives, the latter to take advantage of the great understanding of the genomic systems of well-studied species.

More recently, the phylogenetic scope has broadened as it has become evident that useful knowledge can be gained by examining genomes across the tree of life—knowledge of the genetic causes of Parkinson's disease in humans, for example, can be gained from studying the comparable gene in fruit flies, and much of relevance to humans can be learned from understanding the genetic basis of differences among dog breeds.

Dobzhansky would not have been surprised. Evolutionary biology turns out to be integral to understanding the workings of DNA and the genome, just as it is key to understanding so many other aspects of our biological world (see chapter I.3).

5. CRITIQUES AND THE EVIDENCE FOR EVOLUTION

Unique among the sciences, evolutionary biology's foundation—that species evolve through time—is not accepted by a considerable number of nonscientists, especially in the United States, Turkey, and a few other countries. Public opinion polls repeatedly reveal that most Americans are either unsure about or do not believe in evolution. One yearly poll conducted for more than 30 years, for example, consistently finds that about 40 percent of the US population believes that God created humans in their present form in the recent past.

Yet, the scientific data for evolution is overwhelming (summarized in chapter I.3). Just like the composition and structure of genomes, many other biological phenomena are explicable only in an evolutionary context. Why, if evolution had not occurred, would whales have tiny vestiges of a pelvis buried deep within their blubber? Why would cave fish and crickets have eyes that are missing some parts and could not function even if there were light? Why do human fetuses develop, and then lose, fur and a tail? All these, and many other phenomena, are easily understood as a result of the evolutionary heritage of species but are inexplicable in the absence of evolution.

The case for evolution is built on two additional pillars. First is the fossil record, which documents both the major and minor transitions in the history of life (see chapters II.9–II.18); each year, exciting new discoveries further narrow the gaps in our understanding of life's chronicle. Second is our understanding of evolutionary process, in particular, natural selection, the primary driver of evolutionary divergence. Studies in the laboratory and in human-directed selective breeding clearly demonstrate the efficacy of selection in driving substantial genetic and phenotypic divergence; one need look no further than the enormous diversity of dog breeds to appreciate the power of sustained selection. Moreover, scientists are increasingly documenting the occurrence of natural selection in nature and its ability to transform species, sometimes over quite short periods of time.

The public debate is ironic given that manifestation of evolution has so many important societal consequences (see chapter VIII.1). Evolutionary adaptation of disease-causing organisms has rendered many drugs ineffective, leading to a huge public health toll as diseases thought to have been vanquished have reemerged as deadly scourges (see chapter VIII.3). A recent example is the evolution of resistance to antibiotics in the bacterium *Staphylococcus aureus*, which leads to more than 100,000 infections and 19,000 fatalities a year in the United States. A similar story exists about insect pest species that devour our crops and spread diseases. In the United States alone, the evolution of pesticide resistance results in agricultural losses totaling between $3 billion and $8 billion per year. Perhaps most scary is the realization that the human population is an enormous resource to many organisms and that natural selection continually pushes these species to become more adept at making use of this potential bonanza. Ebola, AIDS, influenza—all are diseases caused by viruses that adapt to take advantage of us; a particularly worrisome concern is that some form of avian flu could evolve to become more virulent to or transmissible between humans, with the potential to produce a pandemic that could kill millions (see chapter VIII.2). All these problems are the result of evolutionary phenomena, and all are studied using the tools of evolutionary biology.

6. THE PACE OF EVOLUTION

For more than a century after the publication of *On the Origin of Species*, biologists thought that evolution usually proceeded slowly. To a large extent, this thinking was a result of Darwin's writing—"We see nothing of these slow changes in progress, until the hand of time has marked the long lapse of ages" (*On the Origin of Species*, chap. 4, 1859). Darwin was, after all, right about so many things, big and small, from accurately deducing the manner in which coral atolls form to correctly predicting the existence of an unknown moth with a 12-inch proboscis from the morphology of a Malagasy orchid. Hence, biologists have learned that it doesn't generally pay to disagree with what Darwin said.

Nonetheless, Darwin was not right about everything. One major mistake was the mechanism of heredity, not surprising, since Mendel's work was unknown to him, and the discovery that DNA is the genetic material was still a century in the future. A second error concerned the pace at which evolution occurs. Darwin expected that natural selection would be weak and consequently that evolutionary change would happen slowly, taking many thousands or millions of years to cause detectable change. Of course, in his day there were no actual data underlying this conclusion; rather, this expectation sprang from Darwin's appreciation of the view promulgated by his mentor, the geologist Charles Lyell, that the slow accumulation of changes caused by weak forces would lead in the fullness of geologic time to major changes, a position in agreement with the prevailing Victorian wisdom about the slow and gradual manner in which change occurs—or should occur—in both nature and human civilization.

Darwin's view influenced evolutionary biologists for more than a century—well into the 1970s, most thought that evolution usually occurred at a snail's pace. Spurred by the results of long-term field studies of natural selection that began in earnest around that time, we now know that Darwin was far off the mark. Many studies now clearly indicate that selection in nature is often strong, and that as a result, evolutionary change often occurs very rapidly (see chapter III.7).

One important consequence of this realization is that we can observe evolution in real time. Pioneered by the study of Galápagos finches by Peter and Rosemary Grant, who documented rapid evolutionary change in these birds (appropriately named after Darwin) from one generation to the next in response to weather-induced environmental changes, the study of real-time evolutionary change in nature has become a cottage industry, with hundreds, or perhaps now thousands, of well-documented examples. This work not only clearly demonstrates the occurrence of evolution but also provides great insights into the processes (usually, but not always, natural selection) that cause it.

Perhaps most exciting, the rapidity by which evolution can occur has opened the door to evolutionary experiments in which researchers can alter environmental conditions and test evolutionary hypotheses over a several-year period. Work at the forefront in this area involved studies on the color of guppies in Trinidad. Observing that the fish were generally much more colorful when they occurred in streams without predators, John Endler moved some fish from streams with predators to nearby areas lacking them; very quickly, the populations evolved exuberant coloration, apparently a result of a female preference for brighter males, which, left unchecked by the absence of predators, led to rapid evolution over 14 generations. Subsequent studies have shown that the guppies freed from predation evolve many other differences, such as in growth and reproductive rates (see chapter III.11). Many similar studies are now ongoing, and it is a safe prediction that field experiments will be an important tool for understanding evolutionary processes in the future.

7. EVOLUTION, HUMANS, AND SOCIETY

Evolution has important implications for humans in a number of ways. Some have already been discussed: humans have used evolutionary principles to alter many species to our own ends (see chapter VIII.5); conversely, wild species are responding to human-caused changes in the environment, adapting to our efforts to control them and responding to new opportunities (see chapter VIII.3). Consequently, it's no surprise that knowledge of evolution is important for efforts to improve artificial selection and combat our evolutionary foes. What is more surprising, perhaps, is the diversity of areas in which an understanding of evolutionary processes is relevant to human society. These include not only medicine (see chapters VIII.1 and VIII.2), conservation (see chapter VIII.6), and criminal forensics (see chapter VIII.4), but also important human pursuits such as creating new molecules in the laboratory (see chapter VIII.7) and devising algorithms to solve analytically intractable problems (see chapter VIII.8).

Beyond purely utilitarian functions, an understanding of evolution can tell us much about ourselves: where we came from and where we may be going, perhaps even shedding light on what it means to be human. In recent years, a series of important fossil discoveries have brought into focus many aspects of the human evolutionary story, from our early primate roots to our recent past. Sequencing of the genomes of humans past and present and of our close primate relatives has complemented these findings in important ways and in some cases has led to unexpected discoveries, such as evidence of lineages, like the Denisovans, for which little fossil data exist (see chapters II.18 and V.15).

But what about our evolutionary future? When I was a boy, the public service television station ran short filler promos speculating that in the future, humans would have a bulbous, brain-packed head with tiny eyes and nostrils. Where this idea came from I have no idea, but it probably represented a mixture of orthogenetic thinking—human evolution has been marked by rapid increase in brain size and so must continue in that direction—with a misguided notion that evolution equals progress, and because intelligence is the hallmark of the human species, it would surely continue to evolve into the future. Even then, I could sense that something was not quite right about this prediction, and today, in fact, many believe that human evolution has ended because selection no longer operates on phenotypic traits: not only has medical care ameliorated the negative consequences of many genetic traits, but human cultural practices such as birth control may have severed the positive link between beneficial traits (e.g., physical strength, intelligence) and reproductive output.

Although these points have validity, they are not absolute. In much of the developing world, selective agents such as malaria can still exert strong selective pressure in the absence of adequate medical care. Moreover, new diseases, such as AIDS, for which, at least initially, no treatment exists, continue to emerge and may impose selection on populations in all parts of the world. Even in the developed world, evidence suggests that some genetically based traits are correlated with survival and reproductive success, and thus that natural selection is still leading to evolutionary change (see chapters VII.11 and VIII.12). Finally, natural selection is only one of several evolutionary processes. Surely, the increased mobility of humans is increasing the homogenizing effects of gene flow and diminishing the diversifying effects of genetic drift that acts in small and isolated populations. Human populations never existed as discretely identifiable genetic "races" (see chapter VIII.11), but ongoing genetic exchange is diminishing the geographic variation that was the result of our past evolutionary history (see chapter VIII.12).

Although selection has been important in shaping human evolution, that does not mean that natural selection can explain all aspects of the human condition. Many human traits—our large brain, altruistic behavior, keen sense of smell—may have evolved as adaptations, but others may represent phenotypic plasticity or may have evolved for nonadaptive reasons. The field of evolutionary psychology focuses particularly on human behavior and is very controversial; some see in most human behavior evidence for adaptation to conditions past or present, but others are more skeptical (see chapter VII.12).

Many look to evolution to help address issues about what it means to be human. Those questions are primarily in the realm of philosophy rather than evolutionary biology and for the most part do not fall within the purview of this volume or this chapter. Nonetheless, I will end with two observations. First, recent advances make clear that plants and animals occupy only a small part of the evolutionary tree of life; a great variety of microbial species constitute most of life's diversity. As a result, the human species is just one of millions of tiny branches on the evolutionary tree, and these microbial species are as well adapted to their ecological niches as we are to ours. It is easy for humans to view life's history anthropocentrically as a great evolutionary progression leading ultimately to us, but microbial species adapted to a great diversity of extreme environments—Yellowstone's hot springs, deep-sea hydrothermal vents—might see things differently. Second, the dinosaurs—members of the class Reptilia—dominated the earth for more than 150 million years. For most of that time, they cohabited with our mammalian ancestors, who were generally small-bodied, minor players in Mesozoic ecosystems. Conventional wisdom has it that our mammal ancestors, thanks to their large brains and warm-blooded physiology, outcompeted dinosaurs, and ultimately would have displaced them. However, evidence for this view is slender; right before the end of their reign, dinosaurs were thriving and showed no evidence of being pushed out by mammals. It is thought provoking to contemplate what the world would be like—where we would be today—had an asteroid not slammed into the earth 65.3 million years ago, wiping out the dinosaurs and clearing the way for the evolutionary diversification of mammals, including our own species.

FURTHER READING

Coyne, J. A. 2009. Why Evolution Is True. New York: Viking.

Darwin, C. 1859. On the Origin of Species by Means of Natural Selection, or the Preservation of Favoured Races in the Struggle for Life. London: John Murray.

Dawkins, R. 2009. The Greatest Show on Earth: The Evidence for Evolution. New York: Free Press.

Futuyma, D. J. 2013. Evolution. 2nd ed. Sunderland, MA: Sinauer.

Grant, P. R., and R. Grant. 2008. How and Why Species Multiply: The Radiation of Darwin's Finches. Princeton, NJ: Princeton University Press.

Reznick, D. N. 2009. The "Origin" Then and Now: An Interpretive Guide to the "Origin of Species." Princeton, NJ: Princeton University Press.

Zimmer, C. 2009. The Tangled Bank: An Introduction to Evolution. Greenwood Village, CO: Roberts & Company.

I.2

The History of Evolutionary Thought

Garland E. Allen

OUTLINE

1. Species and the origins of diversity in the ancient world
2. The eighteenth century and ideas of the transmutation of species
3. The rise of natural history
4. The development of geology in the late eighteenth and nineteenth centuries
5. Ideas of transmutation of species before Darwin
6. Charles Darwin and *On the Origin of Species* (1809–1859)
7. Post-Darwinian controversies and the "eclipse of Darwinism," 1890–1920
8. Heredity and evolution: Mendelism, Darwinism, and the "evolutionary synthesis"
9. Evolutionary theory in the era of molecular biology

GLOSSARY

Catastrophism. Geological theory that the earth's surface was transformed by a series of great catastrophic events in the past, such as floods and massive volcanic eruptions, on a much greater scale than at present.

Evolutionary Synthesis. A term usually applied to developments in evolutionary theory between roughly 1930 and 1950 or 1960, and characterized by the union of Darwinian evolutionary theory with Mendelian genetics (as population genetics), taxonomy, and paleontology.

Population Genetics. The study of populations in terms of gene frequencies and their changes over time, based on mathematical and statistical treatments of interbreeding groups of organisms involving field studies, model building, and experimentation on laboratory populations.

Transformation (Transformism or Development Hypothesis). Older terms for what later (post-1860) became known as evolution, or descent with modification.

Uniformitarianism. Geological term associated with James Hutton and later Charles Lyell to account for the formation of features of the earth's crust by slow, everyday forces such as erosion, and pressure of the oceans against continental boundaries.

1. SPECIES AND THE ORIGINS OF DIVERSITY IN THE ANCIENT WORLD

While philosophers and naturalists in the ancient world did not have a concept of "evolution" in the modern sense, certain traditions or schools of thought in Greece and Rome developed ideas about the origins of biological diversity by natural, as opposed to supernatural, processes. The basic idea that living organisms of one kind can become transformed into living organisms of another kind has its roots in the works of the Greek philosopher Epicurus (341–270 BCE) and his school (the "Epicureans") in Athens. Epicureans were philosophical materialists who believed the world was composed of small particles, or atoms, that were continually in motion in otherwise empty space. By bumping into each other randomly, atoms produced all the physical and chemical processes we observe in the world. One consequence of this general view was that the world was seen to be continually in flux and change; nothing was static or permanent. One of the Epicureans most influential on the development of later European philosophy and natural history was the Roman poet Titus Lucretius Carus (99–54 BCE). In his long poem "De Rerum Naturae" (On the Nature of Things) Lucretius argued forcefully against supernatural explanations of nature and human life, in either the creation of the world or its daily operations. According to Lucretius, nature is purposeless, occurrences are traceable to the random action of atoms, and death is the end of being. Lucretius, along with other Epicureans, was opposed to all forms of established religion. The legacy of the Epicureans is thus a belief in the inevitability of change, a rejection of supernatural causes, and a search for naturalistic explanations. This view would eventually be revived in various forms during the scientific revolution of the seventeenth and eighteenth centuries, and would play an important role in thinking about evolutionary transformation from that period into the nineteenth century.

The materialism of Epicurean philosophy was not the dominant view, however, in either ancient Greece or Rome. Far more pervasive was the idealist (or nonmaterialist) philosophy of Plato (and later Aristotle), which saw the universe and beings within it as creations in the mind of God and had no place for large-scale developmental change. The Platonic universe was based on the distinction between ideal forms, or categories existing in the mind of the Creator, and real, material forms that existed on earth. These categories represented the essence of the forms, which were universal, permanent (immutable), and perfect. For example, the abstract category of "catness" would include all the essential characteristics of what made an animal a cat rather than a dog or horse. While recognizing the existence of variation among real-life cats, Plato's conception focused attention on an idealized world of stable forms or categories (of which "catness" was just one example). The Platonic philosophy is referred to as "idealist" because it was centered on the abstract, nonmaterial world of forms and categories rather than on attention to detailed observations.

Aristotle espoused many of Plato's basic philosophical views, but as a naturalist he paid much closer attention to empirical observations, as his remarkable studies on chick development clearly show. Yet, two aspects of Aristotelian thought provided challenges to later evolutionary thinkers. One was teleological (goal-oriented) thinking: the view that changes in the world are always directed, as in embryonic development, toward a fixed or final goal. Another was the idea of purposefulness, in which the goal also has a purpose, a function, or an adaptation, as in wings as organs of flight. Aristotle also made a distinction between final and efficient causes. Efficient causes are the immediate factors that lead to some particular outcome (in embryonic development, the fertilization of an egg leads to the development of an adult). Final causes are the answers to the long-range questions, why or for what end? The final cause leading to embryonic development is the teleological purpose of forming a completely new individual. By extension, all nature was seen as directed toward purposeful ends. The notions of fixed categories of beings, teleological processes, and purposeful organization of the natural world were thus important legacies of Greece and Rome with which later naturalists had to grapple.

There was little that could be thought of as evolutionary thinking in the period from the fall of Rome in the fifth century through the medieval and early modern (sixteenth to seventeenth) centuries, owing partly to the rise to political power of the Roman Catholic Church and other sects of Christianity that held to an interpretation of the origin of species conforming with the biblical story of creation. In this view, species had been specially created by God and were fixed in their original forms in compliance with the generally pervasive Platonic concept of essences. There matters stood until the consequences of the scientific revolution of the seventeenth century began to be played out in natural history in the eighteenth century.

2. THE EIGHTEENTH CENTURY AND IDEAS OF TRANSMUTATION OF SPECIES

What has been called the *scientific revolution* (roughly 1500–1700) centered on problems of astronomy and dynamics (the science of motion) and grew out of a series of commercial (navigation and calendar reform) and intellectual concerns about the structure and function of the universe, especially the solar system. Building on the work of Nicolaus Copernicus (1473–1543), which centered the universe on the sun rather than the earth, his successors in the seventeenth century, Johannes Kepler (1571–1630), Galileo Galilei (1564–1642), and Isaac Newton (1642–1727), among others, showed that the universe, including events on earth, could be understood as natural "laws" that led to regular and predictable outcomes. God may have started the clockwork of the universe, but it functioned according to its own laws from then on. Organic life, some argued, could also be understood as subject to laws.

A second development, during the middle and later eighteenth century, especially in conjunction with economic developments such as mining, road building, and the growth of industry, was the rapid development of geology. Mining and quarrying for stone (for building), for coal and metal (for a burgeoning weapons industry), and for road and canal construction exposed new layers of the earth's crust, greatly extending the estimates of the earth's age. The discovery of numerous fossils, some similar to and some very different from organisms living on earth today, suggested that life also had a long and varied history. A third development was a result of the economic, social, and political changes surrounding the French Revolution of 1789. This immense upheaval suggested that no established practices, including the supposedly divinely given power of monarchy and the authority of the church, were immutable.

These developments sparked a new materialism and antireligious sentiment, emphasizing that the world in all its aspects was subject to change and flux; nothing was fixed and permanent. These views gained considerable exposure in the works of French philosophes such as Paul-Henri Thiry, Baron d'Holbach (1723–1789), and the encyclopedist Denis Diderot (1713–1784).

One of the English followers of the philosophes was the physician and savant Erasmus Darwin (1731–1802), the grandfather of Charles, whose poetical works such as *Zoonomia; or the Laws of Organic Life* (1794–1796) contained many protoevolutionary ideas: for instance, a constantly changing, dynamic nature, a "world without end." Although none of these developments led to a fully explicit view of evolution in the later nineteenth-century form, they emphasized change and transformation as natural and eternal and thus provided a context for thinking about how species might become transformed by natural causes.

3. THE RISE OF NATURAL HISTORY

Natural history, especially compared with physics and the exact sciences, had been held in low esteem in the sixteenth and seventeenth centuries. But with the expansion of colonial trade, natural history as a source for new crops and commodities began to assume greater importance. As vast numbers of new forms were brought back by commercial voyages, the diversity of the biological world became a matter of great interest and study. New forms from exotic geographic areas were placed in a kind of linear *scala naturae* ("ladder of nature," or "chain of beings," as it was called), ranging from simplest microorganisms to humans. The scala naturae was based on the idea that every organism fit into a place in the natural order, with forms slightly simpler below it and more complex above it. Theoretically, there should be no major gaps in the scala naturae because, following an old principle, nature is a continuity. The scala naturae presented one view of order in the natural world, but it was fixed and static: organisms did not move up or down the ladder, and they did not fall into natural groupings.

A far more useful framework was developed by Swedish botanist Karl von Linné (in Latin, Carolus Linnaeus) (1707–1778). Linnaeus believed that all plants and animals had been created by God and that they could therefore be placed into natural groups that revealed God's divine plan. Whereas previous groupings or classifications had been based on utilitarian principles, Linnaeus sought a system that would group organisms by their natural or structural characteristics, such as numbers and kinds of flower parts, or reproductive organs. Linnaeus's large-scale classification system, including both plants and animals, appeared in the 10 editions of his treatise, *Systema Naturae* (1735–1758). His system involved a hierarchical series of categories—with kingdom as the most inclusive, through phylum, class, order, family, genus, to species as the most specific—that included all organisms of the same kind. He also devised a binomial system for naming organisms by their genus and species names, such as *Canis familiaris* (common dog) or *Acer rubrum* (red maple). The important feature of Linnaeus's system was that organisms grouped by one set of fundamental characteristics would also share other common structural and functional features, at whatever level of the hierarchy one focused. Whether this order originated in God's divine plan or somehow emerged by natural laws (causes) was debatable, but the existence of the Linnaean order was well established by the end of the eighteenth century.

One of the most important concurrent developments in natural history was the rise of comparative anatomy, particularly as espoused by the French anatomist and paleontologist Georges Cuvier (1769–1832). He became familiar with hosts of fossil organisms unearthed during much of the excavation and building projects associated with the Napoleonic era in France and elsewhere. Most of these organisms somewhat resembled modern forms, but many seemed unique. Cuvier studied them, comparing various structures, such as leg bones or skulls. In this manner, Cuvier and others noticed that most organisms in the same Linnaean category showed structures that were modifications of the same basic plan—as in the front limbs of vertebrates—and that were later referred to as *homologies* by the British anatomist Richard Owen (1804–1892) (figure 1). Cuvier also emphasized the close correlation of parts in the anatomy of any organism, arguing that all the parts existed in complete harmony; no part could be changed without affecting all others. Therefore, it seemed impossible to Cuvier that any species could be modified and transformed into another. Despite founding the field of comparative anatomy, Cuvier viewed it as only a reflection of the divine (and immutable) order of the universe.

4. THE DEVELOPMENT OF GEOLOGY IN THE LATE EIGHTEENTH AND NINETEENTH CENTURIES

With the rapid expansion of building, mining, and various forms of excavation between 1790 and 1830, much was being learned about the structure of the earth and (especially) its fossil contents. The succession of strata told a story of continual changes in the earth's surface: submersion, elevation, erosion, further submersion, and so on. Some layers were clearly produced by volcanic action in what appeared to have been

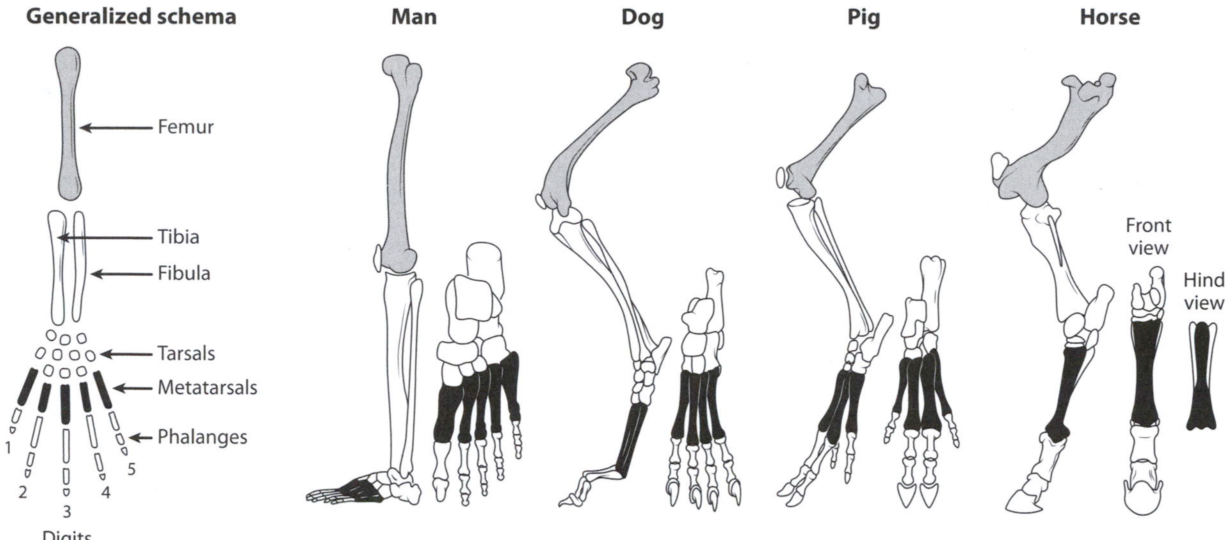

Figure 1. Comparison of the hind appendages of four vertebrates, showing the homologous relationships among the bones. All the femurs are homologous to each other, as are the tibia, fibula, and tarsals; metatarsals; and phalanges. It is evident the same bones have been modified for quite different functions over time. Such comparative homologies provided strong evidence for evolution, as descent with modification from a common ancestor. Special creationists would have to argue that the Creator repeatedly used the same archetypal form, as in the generalized appendage on the far left, for all vertebrate appendages. (From Jeffrey J. W. Baker and Garland E. Allen. 1982. *The Study of Biology*. Reading, MA: Addison-Wesley, 671; used with permission of the authors.)

massive upheavals; others were produced by flooding and underwater deposition of mineral contents. One consequence of these observations was the growing view that the earth was considerably older than the 4004 BCE creation date claimed by Archbishop James Usher. It now appeared that the earth was hundreds of thousands or even possibly millions of years old. One school of thought, *catastrophism*, explained the various upheavals observed in the geological record by a large-scale series of ancient worldwide floods or volcanic eruptions. That these forces were no longer operating to the same extent as in the past fit well with the prevalent theory that the earth had been formed as a spin-off from the sun and had thus originally been very hot but by then had cooled considerably (the "cooling earth theory").

Another school of geology, developed in Britain by James Hutton (1726–1797), emphasized a process called *uniformitarianism*. Hutton claimed that changes in the face of the earth came about in the past by the same sorts of gradual processes observed on the earth at his time and did not require the supposition of vast cataclysms in the past. Hutton's ideas were taken up by a younger geologist in the early nineteenth century, Charles Lyell (1797–1875), who published a three-volume treatise, *The Principles of Geology* (1830–1831), which provided the most influential argument for uniformitarianism. Lyell showed that it was the constant, daily action of wind and water that eroded the land, and the pressure of the sea against the continents that formed and elevated mountains. These processes operated very gradually, almost imperceptibly, yet over long periods of time could have dramatic effects. Uniformitarianism fit well with the positivist philosophy of science, which argued that the best theories required the simplest assumptions and sought the most general and constant laws. Some historians have also argued that uniformitarianism gained its popularity in the wake of the French Revolution, which was claimed to have demonstrated that cataclysmic change was unnatural.

Uniformitarianism posed one problem, however: for vast changes in the earth's surface to have occurred so slowly the age of the earth must be far greater than calculated by earlier estimates. This incongruity raised problems not only with geologists' methods for estimating the age of the earth but also with biblical orthodoxy. The question of the earth's age was to continue to be highly controversial throughout the nineteenth and much of the twentieth centuries.

5. IDEAS OF TRANSMUTATION OF SPECIES BEFORE DARWIN

The idea that species can change in various ways was not original with Darwin but had a variety of incarnations between 1750 and 1850. Erasmus Darwin's idea that species were not fixed was but one of many prevailing theories of transformation in the late eighteenth and early nineteenth centuries. An author more influential with naturalists was Georges Louis Leclerq, Comte de Buffon (1707–1788), whose massive *Histoire naturelle* (Natural History) began to appear in 1749. A committed Newtonian and staunch materialist, Buffon sought natural explanations for all phenomena. Opposed to Linnaeus's view of an abstract, formalistic view of nature as categories, Buffon argued that organisms in nature existed as an array of species and that some individuals could be intermediate in form and function between two species. He developed a theory of degeneration to account for at least the process by which transformation of species could occur within a broad group, such as the family Felidae (cats). The original ancestor could have had the mane of a lion, the stripes of a tiger, and the size of the current large cats. By degeneration, various descendant groups lost one or more of these characters, becoming modern lions (no stripes), tigers (no mane), leopards (no mane, broken stripes), and house cats (small size). Although highly speculative, Buffon's system did promote in a limited way the idea of species transformation by natural processes. *Histoire naturelle* was one of the most influential natural history works of the eighteenth century.

One of the most prolific figures in the period before Darwin was the French naturalist Jean-Baptiste Pierre Antoine de Monet, Chevalier de Lamarck (1744–1829). Lamarck was one of the first naturalists to develop a full-fledged theory of transformation that also provided a mechanism for how organisms could become adapted to their environment. Although Lamarck used many examples in his writings, principally his *Philosophie zoologique* (Zoological Philosophy) of 1809, the development of the giraffe's long neck has been the one most often repeated. Lamarck postulated that as ancestral giraffes ate leaves from lower branches of trees they continually stretched their necks to reach the leaves higher up. The increase in a giraffe's neck during its own lifetime was passed on to its offspring, through the inheritance of acquired characters. Although many people doubted Lamarck's example, the underlying mechanism of the inheritance of acquired characters remained a well-accepted concept throughout the nineteenth (and well into the twentieth) century.

Lamarck's theory of transformation was not what we would recognize as evolution in the modern sense. Although he came to believe in the mutability of species, Lamarck did not maintain that various organisms are related to one another through branching lineages. Rather, he saw simple organisms as continually being generated spontaneously and then gradually becoming transformed by their environment and their own internal drives into more complex forms. For Lamarck, the organism "willed" itself to change and adapt to new conditions and thus was an active participant in its own transformation. More often than not it was this aspect of Lamarck's theory, and not the inheritance of acquired characters, that brought his claims about species transformation into disfavor.

That many naturalists and writers about natural history were beginning to think that species transformation was a reality is evidenced by the immense popularity of a book written in 1844 by the Scottish publisher Robert Chambers (1802–1871) and published anonymously. *The Vestiges of the Natural History of Creation* was a sensational account of the continuous development of life on earth, based on a notion of inevitable progression toward higher, better-adapted forms, including humans. In support of his idea of progressive, goal-directed transformation of species Chambers paid particular attention to the fossil record and what he saw as its parallel in embryonic development. Although he left a place in his process for the Creator (who not only started the whole process but also periodically stepped in to formulate new "laws" of development), Chambers was universally labeled a "materialist," one of the most damning epithets in Victorian culture. Perhaps in part owing to the controversial elements, *Vestiges* was immensely popular and went through numerous editions, to each of which Chambers added new arguments to offset criticisms from both naturalists and theologians. While Chambers in no way presented the idea of evolution in the same vein and with the same attention to detail as Darwin, *Vestiges* did in fact provide an imaginative and fascinating synthesis of geology, paleontology, and natural history that supported the idea of transmutation of species.

6. CHARLES DARWIN AND *ON THE ORIGIN OF SPECIES* (1809–1859)

Early Education and Influences

Charles Robert Darwin (1809–1882) was born into a wealthy upper-middle-class family in Shrewsbury: his father was a successful physician and his mother a daughter of Josiah Wedgwood, of pottery fame. Young Darwin originally attended Edinburgh University to study medicine like his father but was nauseated at dissections and the sight of blood, so he abandoned these plans. Intending to prepare for the ministry, he matriculated at Christ's College, Cambridge, in 1827 but soon realized he did not have the requisite religious convictions for this occupation. Although directionless in the conventional sense, Darwin did show an early interest in natural history. At Edinburgh he met and became friends with the comparative anatomist Robert Grant (1793–1874), who introduced him to current trends in continental comparative anatomy, having studied with Cuvier and Étienne Geoffroy Saint-Hilaire (1772–1844). Grant openly espoused a notion of transmutation of species, which he had gleaned from both Geoffroy and Erasmus Darwin's *Zoonomia*.

At Cambridge, Darwin was strongly influenced first by the writings of William Paley (1743–1805) on natural theology, the view that the study of nature in all its manifestations was a way to understand the glory of the Creator. Paley argued in his book *Natural Theology* (1802) that if one found a watch lying on a heath, it would be clear that it must have had a Creator, and that by studying the watch it would be possible to learn something of the Creator's mind. Similarly, the marvelous adaptations of animals and plants in nature indicated to Paley the existence of a Creator, who produced the most perfect adaptations. Darwin was also strongly influenced by personal contact with two Cambridge professors, the botanist John Stevens Henslow (1796–1861) and the geologist Adam Sedgwick (1785–1873). Both introduced him to organized fieldwork and encouraged him in his natural history inclinations. Darwin considered an extended field trip with Sedgwick to Wales in the summer of 1831 seminal in teaching him how to think scientifically. In these early years Darwin was as much a geologist, in interest and experience, as he was a naturalist. Through his contact with both Henslow and Sedgwick, Darwin began to learn about the vital importance in science of balancing detailed observations with broad-based causal reasoning.

Henslow was to play an even more prominent part in Darwin's life than introducing him to methods of investigation in natural history. Shortly after Darwin graduated in the spring of 1831, Henslow received a request from the British Admiralty to recommend someone to serve as a naturalist and companion to Captain Robert FitzRoy (1805–1865) on the HMS *Beagle*, a hydrographic survey vessel that was to embark on an around-the-world trip later that year. For family reasons Henslow had to decline, and eventually the invitation was extended to Charles. Robert Darwin was at first opposed, considering it dangerous, and thinking it would not lead to any sort of useful career beyond the voyage itself. After intercession by his uncle Josiah Wedgwood, however, Charles gained his father's permission and set off with FitzRoy on the *Beagle* on December 27, 1831. The ship was instructed to collect data about water depths, currents, and climatological and meteorological information as well as to make commercial contacts that would promote British trade throughout South America and the Pacific.

The *Beagle* Voyage

Originally planned to last two years, the voyage extended to five, traversing the Atlantic and Pacific Oceans, and returned to England on October 2, 1836 (figure 2). Darwin took only two books with him initially: the first volume of Lyell's *Principles of Geology* (he had the others sent as they were published) and a new edition of John Milton's *Paradise Lost*, with its highly romanticized engravings by John Martin (1789–1854). Historian David Kohn has noted the significance of this choice in foreshadowing Darwin's own loss of innocence during the voyage about the fixity of species (a metaphoric journey out of Eden) as a result of his experiences during the voyage.

A number of observations of organisms in a variety of geographic settings got Darwin to thinking about the nature of species and their origin. First, reading Lyell in conjunction with his own personal observations of geological formations in South America made Darwin a convinced uniformitarian. Second, in South America he found the fossil of a large, extinct ground sloth (Megatherium) in a chalk cliff while in the forests above he observed present-day living tree sloths; Darwin wondered if the latter could have originated by modification of the former. Third, as the *Beagle* traversed many geographic regions, and as Darwin had frequent opportunities to collect organisms on land, he noted that there were certain patterns in the geographic distribution of species. For example, the fauna and flora of Tristan da Cunha, an island halfway across the southern Atlantic between Africa and South America, were a mixture of those found on the two continents. Darwin also noted that the forms inhabiting islands off the coast of larger land areas were similar to, but modified from, the forms on the mainland. He concluded that migrations had played a significant part in shaping the geographic distribution of life on earth. Fourth, a seminal two weeks spent on the Galápagos Islands, some 700 miles off the coast of Ecuador, provided Darwin with crucial examples of the differences between related species on adjacent islands. Alerted to the fact that the populations of giant tortoises were recognizably different from island to island, Darwin noticed that this phenomenon also held true for other species, such as the finches and mockingbirds. It had become clear to Darwin by this point in the voyage (1835) that the only way to explain these various observations was descent with modification from

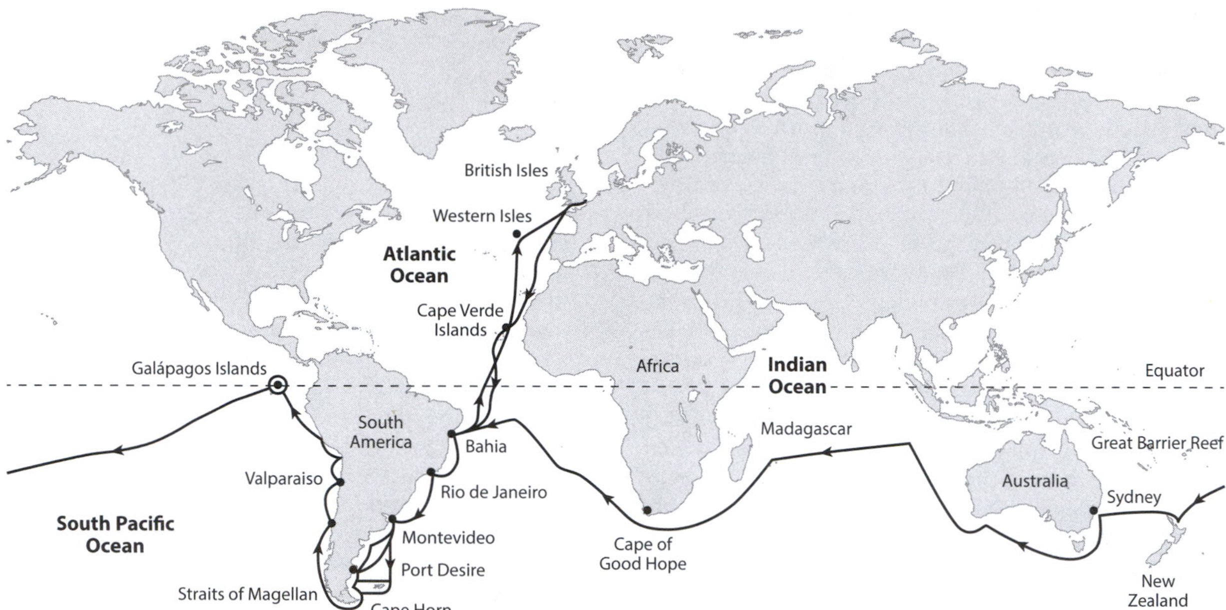

Figure 2. Path of the *Beagle* voyage between December 1831 and October 1836. The voyage completed an extensive around-the-world voyage, exposing Darwin to environments as different as the Canary Islands, the east and west coasts of South America, the Galápagos archipelago, the South Seas, and Australia. (From Jeffrey J. W. Baker and Garland E. Allen. 1982. *The Study of Biology*. Reading, MA: Addison-Wesley, 676; used with permission of the authors.)

common ancestors. At the time, however, Darwin had no mechanism for such modifications to occur.

Genesis of a Theory and Publication of *On the Origin of Species*

Returning home in1836, Darwin began writing up his notes and observations from the *Beagle* and working up his specimens. He became immediately acquainted with a number of leading British naturalists in London, including Lyell and the comparative anatomist Richard Owen (1804–1892), and began to frequent major scientific circles. He was elected a Fellow of the Royal Society in 1839. In 1837 he opened a series of notebooks, two of which (the B and C Notebooks) he labeled "Notebooks on the Transmutation of Species." In these he jotted down his ideas, reading notes, and observations about the possibility of species change. Several lines of evidence were important to Darwin's thinking at this time. One was the ubiquitous nature of variation among organisms of the same species. These variations were often minute, "almost imperceptible," as Darwin called them. He saw these differences as more important than the supposed "essential" nature of a species. The biologist and historian Ernst Mayr (1904–2005) argued that Darwin was at this time beginning to move away from a typological to a population view of species, in which variation, rather than the fixed type, became the focus of the naturalist's attention. A second line of evidence came from animal and plant breeding. For centuries plant and animal breeders had been able to produce widely divergent forms, such as fancy breeds of pigeons (Darwin himself bred pigeons), dogs, cattle, or plant crops by selecting small variations over many generations. Breeders had shown species to be malleable, and to Darwin this suggested that given enough time, full transmutation was possible. It was a gradual process, but as with Lyell's uniformitarianism, great changes could accrue from the buildup of many small events. At this time, however, Darwin knew that without a basic mechanism in nature, analogous to the actions of the animal and plant breeders, his theory lacked plausibility.

Then, in September 1838, Darwin read "for pleasure" Thomas Robert Malthus's (1766–1834) *An Essay on the Principle of Population*, originally published in 1798 and by 1838 in its sixth edition. In this work Malthus, a clergyman and professor of history and political economy at East India Company College, Hertfordshire, put forward his "law" of population: whereas populations grow exponentially, their food supply grows arithmetically, creating constant shortages and thus competition for resources. Written in the immediate wake of the French Revolution, Malthus's essay attempted to show that shortage of resources was not a function of economic and social policies but was inherent in population dynamics. As a naturalist, Darwin immediately recognized that in any species, far more offspring are born than survive to reproductive maturity, thus generating a Malthusian "struggle for existence." The slight variations among members of the same species might give one individual an advantage over another in gaining food or a mate, or avoiding a predator. Those individuals with slight advantages would be expected, on average, to reproduce a little more successfully than others and thus leave more offspring (and vice versa for those individuals with less favorable variations). Assuming, as Darwin did, that many of these variations are inherited, the characteristics of the population would gradually change over time, leading eventually to formation of a new species. Darwin prepared two privately circulated essays, in 1842 and 1844, outlining his overall theory. His hesitancy to publish may have resulted from fear of negative reactions from both scientific colleagues and the public, especially his wife and the church. He also may have wanted to establish his reputation as a solid naturalist by publishing additional technical works, including a theory of coral reef formation and a detailed taxonomic reorganization of the family Cirripedia, the barnacles.

Everything changed when, in the spring of 1858, Darwin received a paper from a young naturalist working in the Malayan archipelago, Alfred Russel Wallace (1823–1913), outlining virtually the same theory of natural selection that Darwin had proposed. It was perhaps no accident that Wallace came up with the same ideas as Darwin, since Wallace, too, had read both Lyell and Malthus and recognized the dramatic role that competition and selection could play in species transformation. Darwin immediately consulted his colleagues Lyell, Thomas Henry Huxley (1825–1895), and Joseph Dalton Hooker (1817–1911), who suggested that one of Darwin's earlier essays and Wallace's paper should be read jointly at a meeting of the Linnaean Society of London that summer. Published in the society's bulletin, these papers established the dual priority of both men, but it also spurred Darwin to publish, on November 24, 1859, his main work, *On the Origin of Species by Means of Natural Selection, or the Preservation of Favoured Races in the Struggle for Life*.

On the Origin of Species: Components of the Darwinian Paradigm

Ernst Mayr argued that Darwin's formulation of the theory of evolution is not a single theory but, rather, a composite of five different theories (thus it functions as a paradigm in Thomas Kuhn's sense). (1) Evolution as Such (descent with modification): As already discussed, by 1859, transmutation was more widely accepted among naturalists, and although controversial, it was not seen as implausible. (2) Common Descent: The idea that current forms have descended from common ancestors by a process of divergence, as in the various finches or mockingbirds of the Galápagos. Common descent as a general principle, found nowhere else in the writings of earlier transmutationists, and one of Darwin's most original ideas, was readily accepted by his contemporaries for its power to explain phenomena such as geographic distribution and homologous structures (those that result from modification of the same basic structures—as in the bones of the hand of the human, the foot of the dog, and the wings of the bat or bird—leading to varied functions). (3) Gradualism: The view that evolution occurs in very slow, small steps and not by sudden, large-scale changes (called "sports" or "monstrosities" in Darwin's day). Gradualism was also one of Darwin's more original concepts. (4) Multiplication of Species: Common descent and divergence lead to an endless multiplication of species over time as new forms replace old ones that eventually become extinct (figure 3). (5) Natural Selection: Factors of the environment select for favorable and against unfavorable variations, in analogy with the work of the practical breeder (see chapter VIII.5). The difference is, of course, that breeders select for variations desirable for their own purposes (such as higher productivity), while natural selection has no ulterior purpose, no goal for improvement. It provides only for how variations are selected for or against in a particular environment at a particular time. Although natural selection was clearly one of Darwin's most important contributions to the transmutation paradigm, the idea had been put forward earlier by several writers in the 1830s, including William Charles Wells (1757–1817), Patrick Matthew (1790–1874), and Edward Blyth (1810–1873), However, none of these writers made a case for the operation of selection as a general process in the way that Darwin did. Natural selection, as the mechanism for how transmutation occurs, was the most controversial component of Darwin's paradigm.

One of the most important problems that Darwin faced, and never properly resolved in his lifetime, centered on the nature of heredity, particularly the origin of variations. Like many of his contemporaries, Darwin believed in blending inheritance, the idea that the contributions of each parent for most traits were blended as an intermediate form in the offspring. This meant that new variations would tend to be diluted and less likely to be maintained in the population unless they were highly selected. Darwin did recognize that some traits were

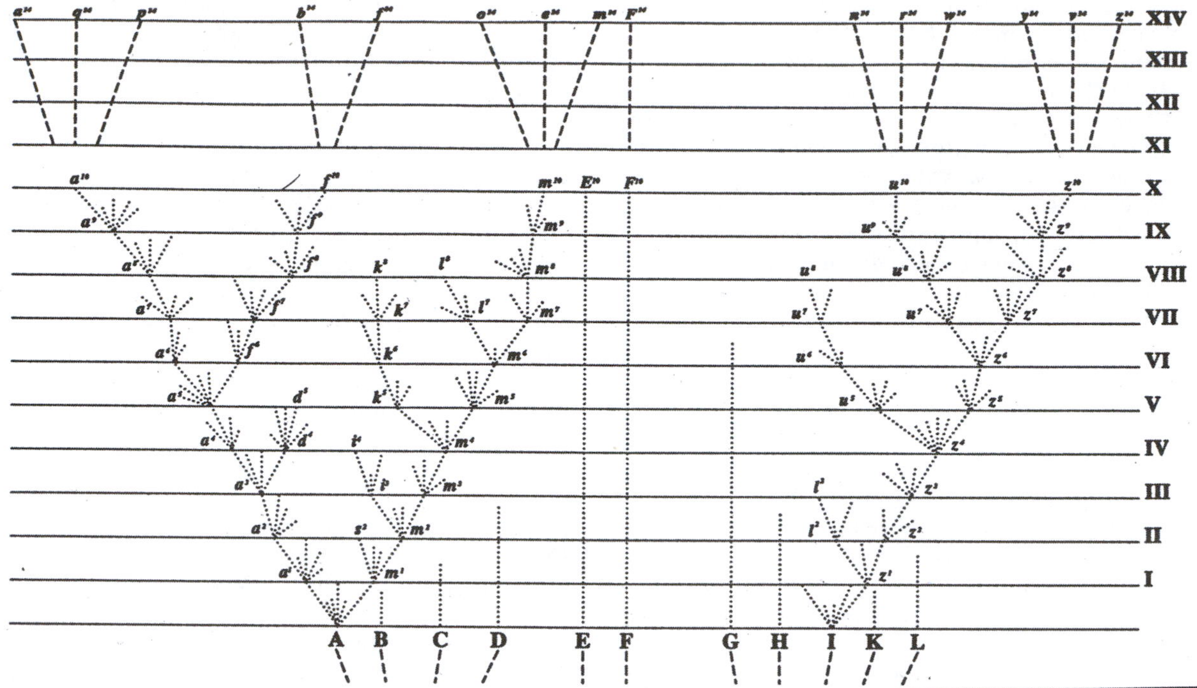

Figure 3. Darwin's diagram in *The Origin*, showing divergence from common ancestors. Horizontal lines represent time periods, from most remote (I) to most recent (XIV). Darwin was aware that most lineages go extinct, as evidenced by the lines that end at a certain point in time. (From Charles Darwin. 1859. *On the Origin of Species by Means of Natural Selection, or the Preservation of Favoured Races in the Struggle for Existence*. London: John Murray.)

inherited in an all-or-nothing way, and these could both skip generations and revert to the original ancestral trait in future generations. On the origin of variations, Darwin held two views. Many variations, he thought, were induced by the environment—for example, food, moisture, temperature—or resulted from the effects of use and disuse, and these could be passed on to the offspring (a kind of Lamarckian inheritance). He also held that many variations occurred by chance, or spontaneously, and were not directly attributable to a specific factor in the environment. Although Gregor Mendel (1822–1884) published his studies on hybridization in 1866, neither Darwin nor most of his contemporaries were aware of this work, which did not come to general scientific attention until after 1900.

The Reception of *On the Origin of Species*

When *The Origin* was released in 1859, all 1250 copies were immediately sold out, either to subscribers or through booksellers. It became the center of much discussion in both scientific and lay circles. Public reaction was largely negative, repelled by Darwin's seemingly materialistic worldview. Theologians were appalled, with one referring to Darwin as "the most dangerous man in England." While some younger scientists, such as Huxley and Hooker, became ardent supporters, the majority of established scientists found the paradigm seriously flawed. (1) A major objection was that as breeding work showed, no new species had ever been produced by selection: although selection had been practiced with dogs for hundreds if not thousands of years, terriers, collies, and Great Danes were all still dogs and could freely interbreed. (2) Blending inheritance meant that new variations, even favorable ones, would hardly ever have a chance to become established except under extremely high levels of selection. (3) While Darwin's theory of slow, incremental change should produce a fossil record with many intermediate forms, the actual strata showed large gaps and discontinuities from one level to another. In most cases there were no intermediates. (4) Estimates of the age of the earth by geologists and physicists, such as William Thompson, Lord Kelvin (1824–1907), claimed that the earth was too young for Darwin's slow, gradual process to have produced all the diversity that currently existed. (5) Although incorrectly attributed to Darwin, the view that humans "came from monkeys" caused considerable controversy by giving humans an animal ancestry that was unacceptable from a secular as well as theological perspective. In the subsequent five editions of *The Origin* Darwin tried to answer some of these suggestions, but it was only well into the twentieth century and the so-called evolutionary synthesis of the 1930s–1950s that many of these original problems with Darwin's paradigm were eventually resolved.

Darwin did have his supporters, however. In England, T. H. Huxley acted as his "bulldog," championing his basic paradigm in both scientific and lay circles. In Germany, August Weismann (1834–1914) and morphologist Ernst Haeckel (1834–1919) both became ardent Darwinians. Weismann's doctrine of the separation of the germ- and somatoplasm contradicted any notion of Lamarckian inheritance. Haeckel more closely followed Darwin in advocating the inheritance of acquired characters and in emphasizing the importance of comparative embryology as evidence for evolution. His "biogenetic law" (that *ontogeny*, or development of the individual, recapitulates *phylogeny*, or development of the evolutionary lineage) was based on the remarkable similarity of the early embryonic stages among related organisms, such as the vertebrates. Although he introduced Darwinism to a wide audience, Haeckel's grandiose ideas gave evolutionary theory a reputation among many biologists for excessive speculation. In the United States, Harvard botanist Asa Gray (1810–1888) championed the basic idea of evolution against his colleague Louis Agassiz (1807–1873) while attempting to reconcile the apparent cruelty and wastefulness of natural selection with Christian theology.

Darwin, Evolution and Society: The Religious Reaction and "Social Darwinism"

The lay reaction to Darwin occurred on both religious and secular grounds. The main religious objection was that evolution as such contradicted the biblical story of creation. Also, the consequence of the view that humans were derived from lower animals was that they lost their place as the special and final creation by God. Darwin had effectively banished God from the everyday universe and, in so doing, had introduced a thoroughly materialistic view of the universe and life, one that lacked purpose (was nonteleological) and thus seemed to lack meaning (see chapters VIII.13 and VIII.14).

On secular grounds, the extension of Darwin's natural history to explain the workings of society flourished in the late nineteenth century. One aspect, known as "social Darwinism" (by its critics), promoted by the English philosopher and sociologist Herbert Spencer (1820–1903), the American political economist William Graham Sumner (1840–1910), and the German Ernst Haeckel, argued that "survival of the fittest" (Spencer's term) was a law for human society as well as nature and thus dictated a laissez-faire economic and social policy. Hard-nosed "Social Darwinists" spoke out against public charity or support for the poor on the grounds that it would encourage the unfit to propagate. Some applied Darwinian models to explain racial differences and the superiority of European races over those in Africa and South America (see chapter VIII.11). A very different extension was promoted by Prince Peter Kropotkin (1842–1921) in Russia, emphasizing cooperation (symbiosis, mutualism) rather than competition as the major characteristic of evolution. The problem with all such extensions of Darwin's (or any other) scientific theory to society is that the theory can be stretched to support almost any a priori political or philosophical views. Supporters from the political left as well as the right claimed the authority of Darwin, partly by drawing on only certain aspects of evolutionary theory (competition and cooperation both exist in the natural world). And since Darwin's metaphor of competition, overproduction, and selection were all processes he initially drew from the social world (i.e., from Malthus), extending these processes to explain society becomes tautological (i.e., circular reasoning).

7. POST-DARWINIAN CONTROVERSIES AND THE "ECLIPSE OF DARWINISM," 1890–1920

Evolutionary debates continued unabated from Darwin's day through the interwar period of the 1920s and 1930s. Of Darwin's five theories, the most generally and quickly accepted (by the 1870s) was descent with modification. The most controversial was the theory of natural selection, which by 1900 most naturalists viewed with considerable skepticism. In addition to voicing earlier objections, some scientists now argued that while selection might occur, it was at best only a negative process: it weeded out the unfit but could not account for the origin of the fit. And of course, Darwin's lack of consistent ideas about heredity was recognized as a major problem. Thus, if the core of Darwinism is evolution by natural selection, by 1900 Darwin's paradigm was, according to one German commentator, "on its deathbed."

In response to the perceived inadequacies of Darwinism, biologists introduced a variety of alternative processes, or mechanisms, by which evolution might occur, among the most important of which were neo-Lamarckism, orthogenesis, and mutationism.

Neo-Lamarckism

Taking their lead from Darwin, a number of naturalists adopted a neo-Lamarckian view to explain the origin of variation from various environmental effects, as well as the use and disuse of parts. Among the strongest supporters were the American paleontologists Alpheus Hyatt (1838–1902), Edward Drinker Cope (1840–1897), Henry Fairfield Osborn (1857–1935), and the Austrian zoologist Paul Kammerer (1880–1926). Neo-Lamarckism solved the problems of adaptation, since variations arose in response to specific environmental demands. It also shortened the time required for evolution to occur, since it did not depend on chance variations. Even in light of Weismann's experiments on mice (see previous discussion) many biologists maintained a neo-Lamarckian position well into the mid-twentieth century, especially in France and the Soviet Union. In the latter, neo-Lamarckism became the basis for agricultural policy after 1948 under the direction of agronomist Trofim D. Lysenko (1898–1976), who claimed that physiological traits such as flowering time could be altered by exposing seeds to varying conditions such as low temperature or artificial light cycles. These effects, Lysenko argued, could be transmitted to future generations.

Orthogenesis

Orthogenesis was the view that evolution inexorably proceeds along certain lines, or directions, and often leads to extinction. The two best-known examples were increase in size of the antlers of the now-extinct Irish elk, and the enormous canine teeth of saber-toothed tigers. A key feature of orthogenesis was the claim that once evolutionary trends were established, even though adaptive initially, they gained momentum and eventually progressed to nonadaptive extremes. Orthogenesis, often accompanied by neo-Lamarckism, was particularly popular among paleontologists because it seemed to explain the various trends in the fossil record such as increase in size or complexity of a trait prior to extinction of the lineage. Orthogenesis also revived a form of Aristotelian teleology in which large-scale trends suggested a direction and purpose in nature, an idea many naturalists were reluctant to give up completely.

Mutationism

Still another alternative to Darwinian evolution was *mutationism*, the claim that large-scale changes could lead to formation of a new species in one generation. Most effectively promoted by Dutch plant physiologist Hugo de Vries (1848–1934) in his two-volume *Die Mutationstheorie* (The mutation theory) in 1904, mutationism gained a significant following because it seemed to be empirically supported and, most of all, was "experimental," meaning that evolution could be investigated in the laboratory. De Vries had observed the evening primrose (*Oenothera lamarckiana*) producing offspring that were not only strikingly different from their parents but also infertile. De Vries thought he had discovered a mechanism by which evolution occurred in short periods of time and could thus be studied experimentally under controlled conditions. As it turned out, *Oenothera* was an unusual plant with very atypical chromosome structures and meiotic processes that accounted for the unusual forms of the offspring. The theory was ultimately abandoned by the 1920s, when other examples failed to materialize and when laboratory genetics revealed a variety of mechanisms for the origin of variations on which selection could act (point mutations, chromosomal rearrangements, recombination).

Isolation and the Role of Geographic Barriers

One of the main problems Darwin had not fully resolved in *The Origin* was that of speciation itself: how one ancestral species diverged into two or more descendant species. While he recognized that geographic separation certainly played a role, he thought it was only one of many factors (self-fertilizing organisms might diverge without being geographically separated). In the decades after Darwin's death this ambiguity led to a controversy between two models of speciation: what would later be called *allopatric* (populations diverge into new species only when separated by a barrier) versus *sympatric* (populations diverge in the same region, not physically isolated from one another). One of the first evolutionists to emphasize the necessity of isolation by means of geographic barriers for speciation to occur was the American missionary-naturalist John Thomas Gulick (1832–1923) in the 1870s and 1880s. Raised by missionary parents in Hawaii, Gulick collected land snails of the family Achatinellidae from adjacent valleys on Oahu. He noted that considerable differences in the shell color and banding patterns existed among specimens from different valleys, which suggested that the populations had diverged as a result of being isolated from one another by the steep valley walls. The same point was made by German naturalist Moritz Wagner (1813–1887) from studies of insects in the Caucasus and Andes Mountains and by American ichthyologist David Starr Jordan (1851–1931) in studies of fish populations in lakes separated by land barriers. The central argument in all these claims was that geographic barriers prevent interbreeding and thus allow each population to accumulate its own unique set of variations, and through natural selection, its adaptation to its particular environment. Isolation came to be regarded as necessary for speciation.

Those who argued for the possibility of some form of sympatric speciation pointed to de Vries's mutation theory and other kinds of "sports" as examples of speciation without isolation. Later, when *polyploidy* (duplication of chromosomes without cell division) was discovered, it, too, was cited as a mechanism for speciation without the necessity for geographic barriers. However, the eventual discrediting of de Vries's mutation theory as a general process and the recognition that polyploidy is restricted to a few species of plants led to increasing acceptance of the role of geographic isolation as a major and necessary component to most speciation in the natural world.

Evolution and the Evangelical Opposition

The somewhat-confused state of evolutionary theory in the period 1910–1930 was not helped by the revival of religious objections, especially in the United States. For a variety of reasons, including suspicion of science as an aspect of "modernism" (the erosion of traditional values due to industrialization and urbanization), the rise of populism, and the association of "social Darwinism" with German aggression in World War I (1914–1918), evolution came under attack by Evangelical Christians and was dramatically highlighted by the trial of teacher John T. Scopes (1900–1970) in Dayton, Tennessee, in 1925. A test case of a bill introduced into the Tennessee legislature earlier that year prohibiting the teaching of evolution in public schools, the trial pitted two highly public figures, as attorneys, against each other: Chicago lawyer Clarence Darrow (1857–1938), who defended Scopes, and two-time presidential candidate and former Secretary of State William Jennings Bryan (1860–1925), who prosecuted the state's position. By the unorthodox strategy of cross-examining his opponent, Darrow was able to turn the trial into a strong plea for the primacy of science over religious ideology, and for rationality over superstition and folk belief. Although Scopes was found guilty and fined a nominal $100, the trial generated worldwide controversy and began a trend to attack evolution in the United States that has persisted, largely in religious circles, to the present (see chapters VIII.13 and VIII.14).

8. HEREDITY AND EVOLUTION: MENDELISM, DARWINISM, AND THE "EVOLUTIONARY SYNTHESIS"

The Problem of Heredity and the Rediscovery of Mendel (1900–1925)

Given Darwin's lack of a hereditary mechanism, it would seem that the rediscovery in 1900 of Gregor Mendel's experiments with hybridization in peas would have immediately resolved many of the problems confronting Darwinian theory (for example, blending inheritance or reversion). This was not the case, however. Many biologists were skeptical of the general applicability of Mendel's work, and a group known as the "biometricians" in England (and their followers elsewhere) claimed that Mendel's work had no bearing on evolution. Led by Darwin's cousin Francis Galton (1822–1911) and his protégé Karl Pearson (1857–1936), biometricians stressed the importance of investigating matters of heredity and variation quantitatively while developing statistical procedures for analyzing data (they introduced methods such as correlation and regression). Committed to Darwin's emphasis on the importance of continuous variations as the raw material for evolution, they rejected Mendelism as a theory of discontinuous variation, allied in many people's minds with de Vries's mutationism. A major battle emerged in England between the biometricians, under Pearson, and the Mendelians, championed by William Bateson (1861–1926), one of the most forceful early promoters of Mendel's work. This controversy, which raged between 1901 and 1908, left the distinct impression that Mendelism, whatever its value might be for plant or animal breeders, had little to contribute to understanding the evolutionary process. Although some Mendelians came to see that small, but discrete, Mendelian variations could serve as the raw material on which selection could act, it took almost two decades for a new generation of investigators, more thoroughly trained in mathematics and statistics than their predecessors, and freed from the earlier hostility to Mendelism, to see the ways in which Mendel's theory could be directly applied to Darwinism. Beginning in the late 1920s and early 1930s, and continuing through the 1940s and 1950s, this period has been labeled the "evolutionary synthesis." It was one of the most important developments in twentieth-century evolutionary biology.

The "Evolutionary Synthesis," 1930–1940

The period of the evolutionary synthesis brought together a number of divergent views and fashioned a comprehensive understanding of the evolutionary process that included Mendelian and quantitative genetics, biometry, classical Darwinian selection theory, taxonomy, biogeography, and paleontology. It was also noted for its role in excluding from evolutionary theory once and for all the older alternative views of neo-Lamarckism, orthogenesis, saltation (macro)mutation, and sympatric speciation (embryology, so central to Darwin's views, was notably absent, leading to its neglect by evolutionary biologists for most of the remainder of the century). By the 1930s, geology and paleontology were beginning to provide a considerably longer time frame for the history of the earth, thus supporting the gradualist views so central to orthodox Darwinism. Incorporating Mendelian genetics into evolutionary theory made it possible to demonstrate that inheritance was not blending but discrete and thus that new variations, even if they were slight and recessive, were fully recoverable in future generations. Moreover, because they were discrete, Mendelian genes could be treated as mathematical entities whose change in frequency from one generation to another provided a quantitative measure of evolution. Combined with Darwin's inherent population thinking, the immediate effect of the synthesis was the emergence of the new field of mathematical population genetics.

For simplicity, the synthesis may conveniently be seen as occurring in three phases. The first (roughly 1918–1940) was dominated by the theoretical population genetics of Ronald A. Fisher (1890–1962), Sewall Wright (1889–1988), and John Burdon Sanderson Haldane (1892–1964) and signified the combining of Mendelism with Darwinian selection theory. This first phase involved the application of the then rather novel methods of mathematical model building to populations of organisms, in which various parameters such as population size, random versus selective mating, alterations in initial gene frequencies, relative fitness, and the effects of selection could be manipulated within the model, and the various outcomes predicted. It was through developments in this phase that the conflict between Mendelians and biometricians was finally resolved. The second phase (roughly 1940–1970) involved the unification of mathematical population genetics with other aspects of evolutionary theory, including paleontology, taxonomy (including the ongoing debate about species definitions), biogeography, field studies of natural variation, population structure, and early studies on the origin of life (abiogenesis). The third phase (roughly 1970–1990s) involved expansion of topics such as extinction (including mass extinctions), evolutionary ecology, experimental evolution (especially using microbial systems), and molecular evolution. All three phases drew on the more general introduction into biology of statistical thinking that had emerged in the physical sciences in the later years of the nineteenth century.

The First Phase: Mathematical Population Genetics

The seminal contribution in the first phase of the synthesis was made by the British mathematician Ronald A. Fisher, who sought to promote a synthesis between orthodox Darwinism and Mendelism and to apply these principles to both agricultural breeding and *eugenics*, or the genetic improvement of the human species. All these interests were brought together in his major book, *The Genetical Theory of Natural Selection* (1930), which aimed to demonstrate the efficacy of natural and artificial selection acting on discrete Mendelian genes. Fisher showed theoretically that, among other things, small selective forces applied over a sufficiently long period could lead to significant genetic change within a population. These changes could be measured quantitatively as shifts in the frequency of certain genes over successive generations: the new, mathematical definition of evolution. Particularly important, Fisher's "fundamental theorem," put forward in the *Genetical Theory*, claimed that the effectiveness of selection was directly related to the total amount of variation in a population: natural or artificial selection could produce change only if there was sufficient variation on which to act. This became an important principle for both the practical breeder and the student of evolutionary mechanisms, as it resolved the old problem of whether selection was a "creative" or only a "negative" force. It was neither: selection required a source of genetic variation to produce inherited change within a population, and both weeded out the unfit and preserved the fit.

Fisher's model was based on populations that were large, *panmictic* (freely interbreeding) groups akin to the physicist's gases in a finite chamber. Like gas molecules, individuals within such a population interacted (mated) randomly, and all combinations were equally possible. A population was defined by its gene frequencies, by what came to be called its *gene pool.* Fisher aimed to treat evolution as lawlike, subject to mathematical formulation like the kinetic theory of gases, according to which organized and predictable outcomes could result from myriads of random events. Despite this oversimplification, there is no doubt that Fisher's approach showed that evolutionary processes could be represented mathematically and that model building was a useful approach to understanding how selection, dominance, recessiveness, and other factors could quantitatively alter the course of evolution in a population.

A mathematical contribution of a different sort was made by American biologist Sewall Wright. A decade of work at the United States Department of Agriculture (USDA) between 1916 and 1925 brought Wright face to face with issues of breeding and the agricultural problems of how to increase the effectiveness of selection at the population level. Wright concluded that the most effective approach was to subdivide a large population (a herd of cattle, for example) into small breeding groups. These would be inbred for several generations and then outcrossed periodically with other subgroups. Inbreeding within the small groups would help fix desirable genotypes, while outbreeding among groups would bring new combinations together that would provide additional raw material for selection. This early work led Wright to recognize that the breeding structure of a population was critical to understanding how evolutionary change most likely occurs in nature. He came to oppose the model of a large panmictic population, promulgated by Fisher, as having little reality in nature. In contrast, Wright claimed that large populations in nature were usually subdivided into smaller subpopulations, which he called *demes*, each occupying its own ecological space ("microniche") within the population's overall geographic range. Two aspects of Wright's model became important innovations in the evolutionary synthesis. One was the shifting balance theory of evolution, and the other, as a consequence, was genetic drift.

Employing the metaphor of a topographical map, Wright's *shifting balance theory* depicted each deme on its own "adaptive peak," separated from other demes by valleys of ecologically less adaptive "terrain." Each deme would thus have its own characteristic distribution of gene frequencies, although through migration, genes could be exchanged between demes within the population at large. The adaptive landscape was not static, however, since climatic or other external conditions, as well as the activities of the organisms themselves, gradually "eroded" the peaks, altering the overall topography. As conditions changed, demes would be forced to move off their adaptive peaks, either "migrating" to another peak where their particular combination of genes might be more adaptive or, failing to find such a peak, becoming extinct. Demes were thus constantly being challenged to undergo a shift in the balance of their gene frequencies owing to the constant interaction of the small subpopulations with their microniches (hence the name "shifting balance theory").

Associated with the shifting balance theory was what Wright referred to as *genetic drift*, the process by which, in small populations, gene frequencies could become fixed, either at 100 percent or 0 percent, simply by the vagaries of random processes such as chance matings, differential survival, and reproductive success. Genetic drift became one of Wright's seminal contributions to evolutionary thinking. However, genetic drift was misunderstood by many of Wright's contemporaries, who thought that he was suggesting natural selection had little or no role in evolution and that most change was due to chance. Nevertheless, Wright always maintained that selection also played a major part in long-range evolutionary processes; his shifting balance theory emphasized only the role of random effects due to small population size.

A third major contribution to mathematical population genetics came from John Burdon Sanderson Haldane (1892–1964), the son of the well-known Oxford physiologist John Scott Haldane (1860–1936). Because of his strong biochemical as well as mathematical background, Haldane emphasized the ways in which genetic variations might act physiologically, for example, in slowing down or speeding up metabolic reactions, and thus provide an understanding of how selection might actually affect gene function. Such changes could also be computed mathematically. More than Fisher or Wright, Haldane also emphasized the importance of gene interactions as targets of selection, and brought embryological considerations into the evolutionary process by pointing out how different selection pressures must come into play at each stage in an organism's life cycle.

The Second Phase: The Genetics of Natural Population

One of the earliest applications of Mendelian genetics to the study of natural populations appeared in the work of Sergei Chetverikov (1880–1959), a Russian butterfly taxonomist who was well acquainted with the problem of variation in natural populations. In 1922, at the Institute of Experimental Biology in Moscow, he was introduced by its director, Nikolai Koltsov (1872–1940), to the laboratory cultures of *Drosophila melanogaster* that H. J. Muller (1890–1967) had just brought from the US laboratory of Thomas Hunt Morgan (1866–1945). Because *Drosophila* was the premier experimental organism for studying genetics, the cultures brought by Muller were highly useful, since they contained point mutations at known positions on the chromosomes. Chetverikov was able to crossbreed these known stocks with flies from natural populations, to uncover variations that were masked in the wild by their dominant alleles. Chetverikov's experiments suggested for the first time that there was considerably more variation in natural populations than had previously been thought. This was important because the rate of evolution, as predicted by Fisher and others on theoretical grounds, depended on the presence of a large amount of variation. Thus, determining how much variation was actually present became an important empirical investigation during the period of the evolutionary synthesis. The Russian school, under Chetverikov and others, contributed significantly to this effort. Unfortunately, owing to World War I (1914–1918), the upheaval of the Bolshevik Revolution (1917), and the fact that most Western biologists did not read Russian, the work of

Chetverikov and his colleagues was not known outside the USSR until the 1940s, when it was brought to light and translated by Theodosius Dobzhansky (1900–1975).

A major figure in the second phase of the synthesis, Theodosius Dobzhansky brought the Russian perspective combining field and laboratory investigations to bear on determining the amount of variation in natural populations and the effect of natural selection on those variations. Dobzhansky's 1937 book, *Genetics and the Origin of Species*, had already outlined the close relationship between genetics, cytogenetics (the microscopic study of chromosome structure and configuration of known genetic strains), and evolution, and became one of the major works outlining the new synthetic approach to evolutionary theory. In a 1947 paper, Dobzhansky reported on a series of innovative field studies on variation and selection, in which he introduced two quite novel approaches. First, he sought to inspect the genotype directly, by observing cytologically the frequency of chromosome variants (inversions) in natural populations, using the methods of cytogenetics he had learned from working in the Morgan laboratory at Columbia and Caltech. Second, he carried the work a step further by developing an experimentally based laboratory approach that allowed him to test under controlled conditions hypotheses about the fitness of different chromosomal variants found in natural populations.

Surveying populations of *Drosophila pseudoobscura* taken at regular geographic intervals across the Southwest from California to Texas, Dobzhansky found that different populations could be characterized by different frequencies of certain chromosomal inversion patterns. These patterns were detected microscopically but had no visible phenotypic effect by which they could be recognized or spotted. More interesting, Dobzhansky found that the frequency of four such arrangements he studied changed within any one population in a regular, cyclic pattern over the course of the year. Dobzhansky hypothesized that these changes were due to changes in selection pressures during different seasons. From an evolutionary point of view, it was clear that selection favored maintaining a range of variations (called *balanced polymorphism*) within a population subjected to regular changing conditions.

Dobzhansky then proceeded to study this phenomenon under controlled conditions in the laboratory. Until the 1940s it had been thought that evolution by selection was too slow a process to be observed by human beings in their lifetime—that is, it was a process whose effects could be seen only in the span of geologic time. Dobzhansky's suggestion that seasonal fluctuation in certain genotypes was produced by changing selection pressures almost seemed preposterous. He pointed out, however, that different chromosomal arrangements very likely had differential adaptive capacities, one type being more adapted to, say, a particular temperature or humidity level. In the lab Dobzhansky exposed flies with different chromosomal inversions to varying conditions and observed that he could alter the frequency of one inversion type over another simply by changing the environment (temperature was the most effective selective agent). He was witnessing evolution right before his eyes.

Dobzhansky's experiments emphasized two important points. The first was that, at last, even the theory of evolution by natural selection could be subject to experimental and quantitative tests. To Dobzhansky and many other biologists at the time it seemed that a problem that had been refractory to experimental study for over a century had at last yielded to rigorous analysis. The second point that Dobzhansky emphasized was that what appeared to be adaptive were not individual genes or their phenotypic traits but, instead, the whole complex of genetic traits carried on a chromosome, a part of a chromosome, or within the entire complex of chromosomes in the population. Selection operates at multiple levels.

The Third Phase: Taxonomy, Paleontology, and Embryology in Relation to Evolution

At the end of *The Origin* Darwin had prophesied that accepting his paradigm of evolution by natural selection would "revolutionize" a variety of fields of natural history, including taxonomy. As far as the latter was concerned, it did not turn out that way at the outset, but by the 1930s and 1940s taxonomists like the German émigré to the United States Ernst Mayr had begun to apply aspects of Darwin's population approach to taxonomic questions, particularly the nature of species and species definitions. Mayr argued in his 1942 book, *Systematics and the Origin of Species*, that the focus of taxonomy should be on the range of variation in populations, not on characterizing the most "typical" member of the group. This was a shift, Mayr noted, from the older "typological" or "essentialist" view of species to a modern, dynamic, or population view. A species was now defined by the variability it displayed, as this was the creative reservoir on which natural selection acted. Taxonomists, he argued, should define their species as active populations in the process of evolving, not simply as the static products of evolution. Mayr called this the "biological species concept" and defined it primarily in terms of reproductive compatibility (see chapter VI.1). This became a highly debated point over the ensuing years, since many biologists, botanists, and microbiologists, for example, felt Mayr's definition was too narrow to encompass the taxa with which they worked (see chapter VI.1).

In the nineteenth century paleontology provided the only hard evidence that evolution had actually occurred, but it was always problematic (see chapter II.9). First, it did not provide the intermediate forms that Darwin's gradualist paradigm required. Also, since fossils are nonliving and are often fragmentary, their life histories are difficult to reconstruct and their taxonomy difficult (many fossils cannot be keyed to the species level with any certainty). Moreover, paleontology was tied professionally to geology, as fossils are often the key to stratigraphy, that is, to correlating strata from one region to another. By the 1930s and 1940s, however, several paleontologists, principally George Gaylord Simpson (1902–1984) and Norman D. Newell (1909–2005), both at the American Museum of Natural History in New York, argued for a closer alliance of paleontology with biology. Simpson had been strongly influenced by Dobzhansky's *Genetics and the Origin of Species* and sought to bring the perspective of genetics, population thinking, and statistics to bear on paleontology. Both Simpson and Newell claimed that paleontologists should think of fossils as "once-living organisms" that existed in populations. In *Tempo and Mode in Evolution* (1944) Simpson argued that paleontology could complement population genetics by revealing large-scale trends and developments over time (tempo, or rates of change, for example) that study of living populations could not. Paleontology could thus contribute to understanding macroevolution, and population genetics could focus on microevolution. Simpson and others were responsible for transforming paleontology into paleobiology and bringing the field into the evolutionary synthesis.

Simpson's challenge was taken up by a younger generation of paleontologists such as Stephen Jay Gould (1941–2002), Thomas J. Schopf (1939–1984), and Niles Eldredge (b. 1943), among others. Gould and Eldredge, for example, published a highly influential paper in 1972, "Punctuated Equilibria: An Alternative to Phyletic Gradualism," that challenged the orthodoxy of slow, gradual change over evolutionary time. Their view of "punctuated equilibrium" characterized macroevolution as a series of rapid transformations followed by long periods of very slow change or stasis (such as the relatively small fossil sample from the pre-Cambrian compared with the Cambrian "explosion," where most of the major body plans of the animal kingdom were laid down). For Gould and Eldredge this sharp change from one series of strata to another was not the result of an imperfect geological record but an actual picture of how evolution works. Punctuated equilibrium was a challenge to the orthodoxy of Darwinian gradualism (figure 4).

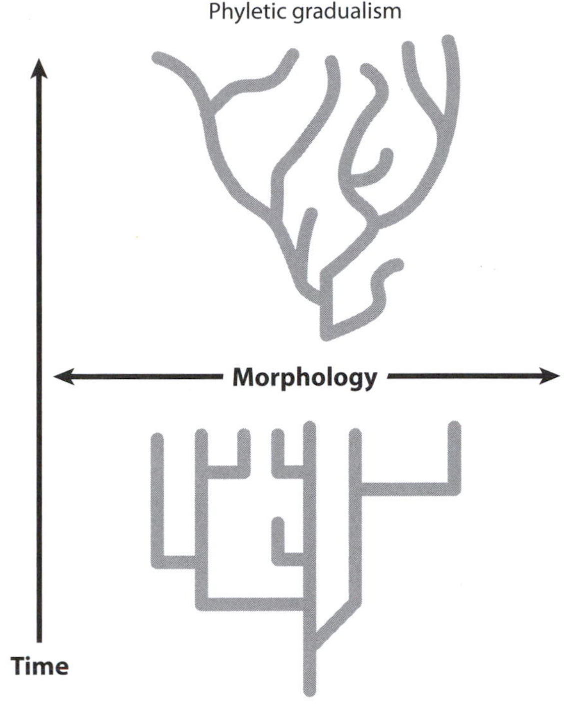

Figure 4. Comparison of evolutionary change according to the theory of phyletic gradualism (top) and the punctuated equilibrium of Gould and Eldredge (bottom). Punctuated equilibrium emphasizes sharp breaks when evolutionary change is rapid, followed by periods of slow change (vertical lines).

Gould, in particular, was also influential in bringing the perspectives of developmental biology back into evolutionary theory after its long hiatus. Gould emphasized that evolutionary lineages were always under developmental constraints that limited the paths they could follow. That is, genetic variations that seriously altered developmental sequences would be selected against in favor of those that worked within the limits of existing embryonic processes. Development, and the constraints it imposed, represented an intermediate set of phenotypes between the genotype of the fertilized egg and the adult phenotype. Gould's emphasis on the importance of development led to the current interest in "evo-devo" (evolutionary-developmental biology; see chapters V.11 and V.12). In recent years, evo-devo has been highly informed by new knowledge in molecular genetics, methods of gene signaling, and genetic control mechanisms (see chapters V.1–V.3 and V.6–V.8).

9. EVOLUTIONARY THEORY IN THE ERA OF MOLECULAR BIOLOGY

With the rapid development of molecular biology after the discovery of the molecular structure of deoxyribonucleic acid (DNA) in 1953, evolutionary biology made great strides in several areas: more precise determination of evolutionary rates, the molecular basis of evolution, molecular taxonomy, comparative genomics of related lineages, and mechanisms of variation in both structural genes (those that code for functional proteins) and control elements (genes and gene products that control the expression of other genes).

Once the structure of DNA was clear, the molecular basis for variation (mutation) could be understood as base-pair substitutions and their effects on the subsequent proteins for which the genes code. With techniques such as chromatography and later electrophoresis (both means of separating molecules with only slightly different structures), it became possible by the 1950s and 1960s to compare molecules (proteins such as the hemoglobins and later DNA itself) from different organisms and work out "molecular phylogenies," that is, the historical lineages of the molecules themselves (see chapter V.1). These often, but not always, confirmed existing lineages based on more conventional morphology. Further, since the rate of base-pair substitution could be determined experimentally, differences in similar proteins or the DNA that produced them yielded a more precise measure of divergence rates and times (what was called the molecular clock; see chapter V.1). Having a calibrated clock aided, for example, in determining more accurately the time of divergence between humans and our pongid ancestry, or between current human geographic populations.

A consequence of accumulating data from molecular genetics was the recognition that many mutations at the DNA level are the result of random drift rather than selective pressures. Such observations led the Japanese geneticist Motoo Kimura (1924–1994) in a 1968 article to propose the neutral theory of molecular evolution, which argued that most mutations in DNA are selectively neutral and remain in the genome because they are not eliminated by natural selection (see chapter V.1). Kimura's idea was given broader coverage by Jack L. King and Thomas H. Jukes in another article, "Non-Darwinian Evolution," in 1969. Kimura emphasized that the neutral theory did not dispense with natural selection as having a key role in the evolution of adaptations at the phenotypic level; but he did point out that the theory aimed to counter the prevailing idea that most mutations were of positive or negative selective value, and thus their frequency in a population was exclusively the result of selection. In many ways Kimura's theory was an extension of Wright's concept of genetic drift. However, it became highly controversial because it seemed to contradict the central role given to selection throughout the evolutionary synthesis. Although the controversy had died down to some degree by the 1990s, it is still often invoked as a null hypothesis in testing the role of variations in evolution. It remains an issue of debate, however, especially in discussions of the relative importance of the role of drift with respect to population size. Especially in taxonomy, comparisons of selected DNA segments have helped delineate monophyletic taxa with a precision not possible with traditional comparative methods. Particularly important has been the elucidation of the many mechanisms by which gene expression is controlled, from transcription of messenger RNA (mRNA), to alternative splicing (varying ways in which the same mRNA transcript can be cut and modified), to posttranslational modification of proteins (including how proteins are folded into their tertiary structure). With the advent of rapid, automated sequencing equipment, the complete genomes of a number of organisms have now been elucidated (mouse, fruit fly, nematodes, yeast, humans, to name a few). Comparative analysis using computers has revealed a remarkable number of conserved sequences of DNA, such as the "homeobox genes" that serve similar or many times quite different functions in widely divergent organisms (the "homeobox," a 180-nucleotide sequence, is found in both *Drosophila* and humans, where it functions in laying down basic patterns of segmentation in the anterior-posterior axis). Even more striking, similarly related sequences have been found in regulatory genes, indicating that evolution can act on existing genetic systems by slight modifications of control elements that can have large effects at the genotypic level (see chapters V.2, V.3, and V.6–V.8). All these discoveries have suggested that evolution may proceed by modular modification of a number of existing genetic elements, just as a contractor can produce a wide variety of houses by arranging and rearranging a few basic room designs. Far fewer genes appear to be needed than previously believed to produce infinite variation for evolutionary change. New mechanisms of variation, their effects on development, and their evolution through time now form the forefront of twenty-first-century evolutionary theory (see chapters V.13 and V.14).

FURTHER READING

Bowler, P. J. 2009. Evolution: The History of an Idea. Berkeley, CA: University of California Press. *The best single treatment of the evolution idea from the seventeenth century through the present day. Well written and clear, this is an often-used textbook in courses on the Darwinian Revolution.*

Browne, J. 1995. Charles Darwin. Vol. 1, Voyaging. London: Jonathan Cape, 2002. Vol. 2, The Power of Place: The Origin and After—the Years of Fame. New York: Alfred A. Knopf. *A fascinating and scrupulously detailed biographical study of Darwin, his influences, and the development of his work.*

Browne, J. 2006. Darwin's Origin of Species: A Biography. *A much-shortened excerpt from the two-volume biography that focuses on the factors and events leading up to the publication of* The Origin, *its reception, and legacy. Beautifully written and an easy, quick introduction to Darwin's major work.*

Darwin, C. 1859. On the Origin of Species by Means of Natural Selection, or the Preservation of Favoured Races in the Struggle for Existence. Facsimile of the first edition, with an introduction by Ernst Mayr. Cambridge, MA: Harvard University Press. *Harvard Press has also issued, more recently,* The Annotated Origin, *with an "Introduction" and explanatory notes by James T. Costa (2009). Both books are facsimile reproductions of the original edition, but the latter has useful explanatory notes on a variety of issues, including updated discussions of Darwin's observations (for example, on ecology), and historical notes identifying people, places, practices.*

Gould, S. J. 2002. The Structure of Evolutionary Theory. Cambridge, MA: Harvard University Press. *This massive publication is both historical and current, in that it covers most of the history of evolutionary theory in the pre-1950 period but also discusses late twentieth-century developments in both micro- and macroevolution.*

Kohn, D., ed. 1985. The Darwinian Heritage. Princeton, NJ: Princeton University Press. *A collection of essays by major historians and philosophers of science, discussing everything from the structure of Darwin's argument to his influence by early nineteenth-century romanticism.*

Mayr, E. 1982. The Growth of Biological Thought: Diversity, Evolution and Inheritance. Cambridge, MA: Harvard University Press. *Not really the growth of biological thought but primarily of evolutionary biological thought, Mayr's tome covers 3000 years of natural history and evolutionary thinking. A great reference and compendium, though, like all the author's work, written from his distinct perspective as a proponent of neo-Darwinism and an architect of the "evolutionary synthesis" of the 1930s–1960s.*

Mayr, E. 1985. Darwin's five theories of evolution. Chap. 25 in D. Kohn, ed., The Darwinian Heritage. Princeton, NJ: Princeton University Press. *Lays out the five components of Darwin's theory or paradigm: Evolution as Such, Common Descent, Gradualism, Multiplication of Species, and Natural Selection.*

Provine, W. 1971. The Origins of Theoretical Population Genetics. Chicago, IL: University of Chicago Press. *Despite its age, this remains one of the most concise and readable accounts of the first phase (mathematical population genetics) of the evolutionary synthesis.*

Ruse, M. 1997. From Monad to Man: The Concept of Progress in Evolutionary Biology. Cambridge, MA: Harvard University Press. *An interesting sweep of the history of evolutionary thinking, both pre- and post-Darwin, and the central role that notions of progress have played to the present day (includes modern evolutionary biologists such as E. O. Wilson, Stephen J. Gould, Richard Lewontin, Richard Dawkins, and John Maynard-Smith).*

I.3

The Evidence for Evolution

Gregory C. Mayer

OUTLINE

1. The fossil record
2. Comparative biology
3. Biogeography
4. Evolution in action
5. Evolution as fact and theory

The evidence for evolution was first comprehensively assembled by Charles Darwin, who succeeded in convincing essentially all his scientific contemporaries of the fact of descent with modification. A signal factor in Darwin's achievement was that he was able to weave together numerous strands of natural history—paleontology, systematics, embryology, morphology, biogeography—into a coherent framework. Since Darwin, genetics has joined this synthesis, and, in a development that might have surprised Darwin, evolution in natural populations has proven to occur sufficiently rapidly that it can be observed on human timescales. The most direct evidence of evolution comes from the fossil record, in which the dynamic changes of life over time are recorded, including many transitions between major taxa. A host of phenomena in comparative biology (e.g., systematics, morphology, embryology, genomics) and biogeography that otherwise appear inexplicable or anomalous are readily explained under the hypothesis of descent with modification. In addition, direct observation of natural and artificial populations shows the process of evolutionary change in action. Together, these sources of evidence lead to a "consilience of inductions" that makes the fact of evolution one of the most securely established generalizations in science.

GLOSSARY

Adaptation. A feature of an organism that fits it to its conditions of existence, giving rise to similarity among organisms leading the same or similar ways of life.

Homology. The correspondence, determined by their relative positions and connections, of organs in different organisms, which is indicative of affinity; the cause of this correspondence is inheritance from a common ancestor.

Oceanic Island. An island that has never been connected to a mainland and thus has received its fauna and flora over water by occasional means of transport.

Phylogenetic Tree. A representation of the history of life, with branching indicating the splitting of lineages, and the connection of the branches indicating the passage of genetic information and materials from one generation to the next.

Progression. The pattern in the fossil record in which earlier forms of life differ from later forms, with major groups first appearing in the record in a generalized form and later as more diversified members of the same group. Some of the earlier forms may become extinct, and later forms may more closely resemble modern forms. Not to be confused with progress, a different concept, according to which evolution proceeds toward some externally defined goal.

Speciation. The splitting of a lineage into two or more daughter lineages reproductively isolated and evolutionarily independent from other lineages.

Tetrapods. The group of four-limbed vertebrates comprising amphibians, reptiles, mammals, and birds; includes species that have secondarily lost their limbs, such as snakes.

Unity of Type. Similarities among organisms leading different ways of life under diverse conditions of existence, going beyond any functional need for similarity.

The evidence for evolution, it has been remarked, is not the result of some crucial experiment but something more like the contents of the American Museum of Natural History. And so it is: the evidence for evolution comes from a plethora of biological and geological subdisciplines—systematics, paleontology, stratigraphy, geochronology, biogeography, morphology, botany, zoology, embryology, genetics—many of which find their objects of study in the vast and varied collections of natural history museums.

It is the diversity of these sources of evidence, all leading to the conclusion that life on earth has undergone a long history of descent with modification, that is the great strength, and the most striking aspect, of the evidence for evolution. The varied sources of evidence are all brought into the unified explanatory scheme of evolution, forming what the philosopher William Whewell (1794–1866) called a "consilience of inductions," each piece of evidence reinforcing the whole. Charles Darwin (1809–1882) in *On the Origin of Species* used precisely such a form of argumentation, marshaling the disparate facts of geology, systematics, morphology, embryology, and biogeography to support his theory of descent with modification.

Progression, Unity of Type, and Adaptation

By the time Darwin began his scientific career, it was already well established that the earth was old and that the fossil record was progressive, that is, that earlier forms of life differed from later forms, that some of the earlier forms had become extinct, and that later forms more closely resembled modern forms. It was also becoming clear that in this progression not only were older forms replaced by newer ones but later forms were in some way related to earlier ones. Thus, a major group would appear in the fossil record in a generalized form and would be succeeded by more diversified members of the same group.

In addition to progression, two other great, but unexplained, classes of phenomena occupied biologists at this time: unity of type and adaptation. *Unity of type* refers to the similarities among organisms living different ways of life, similarities that extend far beyond any functional needs. The same basic skeletal plan of the forelimb—a humerus, then radius and ulna, then carpals, then metacarpals, then phalanges—occurs in all tetrapods, even though the limbs might appear quite different externally and be used for very different functions (figure 1). Such similarities extend to embryological features as well: all tetrapod embryos, for example, have four limb buds, even if the adults (e.g., whales, snakes) lack one or two sets of limbs.

Adaptation refers to those features of organisms that suit them to their conditions of existence (see chapter III.1; for a nuanced discussion of terminology, see chapter II.6), which may be shared by organisms with similar ways of life. Thus, sharks and whales share flattened tail flukes and dorsal fins, both features being of obvious functional importance for aquatic organisms. Despite these adaptive similarities, sharks and whales differ in many features, such as their respiratory, circulatory, and reproductive systems, that mark them as belonging to different major groups of organisms—fish and mammals, respectively.

Darwin provided in his theory of descent with modification a unified explanation of the geological (progression), structural (unity of type), and functional (adaptation) phenomena. Unity of type reflects inheritance of features from common ancestors, and the changes from the common ancestor are due to modification. The most important means of modification, natural selection, leads to adaptation. And when played out over geological time, descent with modification leads to progression. Once admitted, descent with modification and common ancestry would also account for the hierarchical relationships revealed by systematics and for the distributions of organisms across the surface of the globe. It was Darwin's triumph to combine geological, morphological, embryological, systematic, and biogeographic evidence into a single explanatory theory.

This chapter considers the classes of evidence adduced by Darwin to support his theory of descent with modification, including examples from post-Darwinian disciplines such as genetics, and adds as an additional class of evidence observations of evolution in action, a class that was largely unavailable to Darwin.

1. THE FOSSIL RECORD

Progression

The earliest fossils are of simple (prokaryotic), single-celled, photosynthetic bacteria that lived in mats called stromatolites 3.5 billion years ago (see chapter II.11). One and a half to 2 billion years ago, more complex, organelle-bearing (eukaryotic) but still single-celled forms appear, and then multicellular forms. The origins of eukaryotes and multicellularity are not well documented in the record, and molecular data suggest that they may have occurred considerably earlier than the record shows (see chapter II.12). In the last part of the Precambrian, about 600 million years ago, diverse forms of marine invertebrates appear, and then in the Cambrian many more invertebrate groups arise, as well as the first vertebrates (see chapter II.15).

The earliest vertebrates, such as the recently discovered *Myllokunmingia* and *Haikouichthys*, which look remarkably like previously hypothesized generalized vertebrates, were soft-bodied jawless forms that lived about 525 million years ago. Diverse jawless fishes with mineralized hard tissues followed. The first jawed fishes appear in the Late Ordovician, about 445 million years ago. The bony fishes, or Osteichthyes, the most

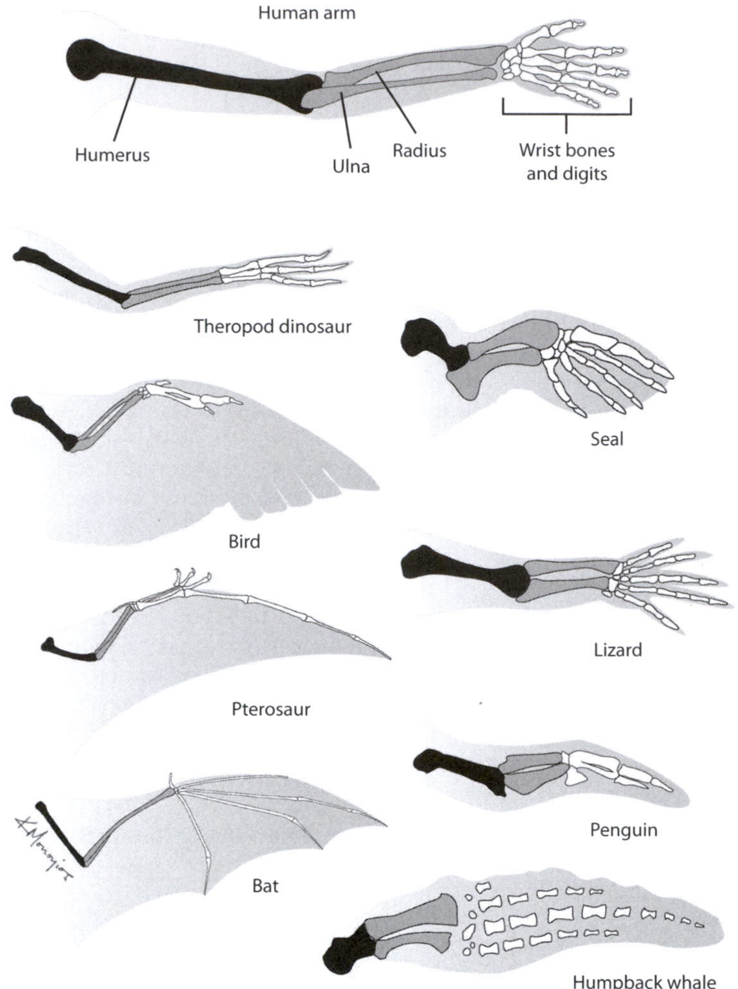

Figure 1. Unity of type and adaptation, illustrated by the forelimbs of tetrapods. The pattern of one bone (humerus), two bones (radius and ulna), many bones (wrist and digits) is present in all, even though the size, shape, and relative proportions of the elements have been modified for varied ways of life, including walking, flying, grasping, and swimming. (Copyright Kalliopi Monoyios.)

diverse living group of jawed fishes, are first found in the Late Silurian, about 420 million years ago. The first four-legged vertebrates (tetrapods) are the approximately 365 million-year-old amphibians *Acanthostega* and *Ichthyostega* from the Devonian (see chapter II.17). Reptiles, the first tetrapods to be independent of water (amniotes), make their appearance in the Pennsylvanian, about 315 million years ago. In the Mesozoic era, the last two major tetrapod groups arise: the first mammals are known from the Triassic/Jurassic boundary (about 200 million years ago), and the first bird is from the Late Jurassic, about 150 million years ago.

The fossil record of vertebrates, then, exemplifies the sequential origin and diversification of the major groups of organisms through geological time. Moreover, as one moves toward the present, many taxa go extinct and others dwindle in diversity, and the array of major groups comes to progressively resemble that of today.

Transitions

Darwin did not have closely spaced transitional forms that could demonstrate evolutionary continuity between major groups; he attributed their absence to the imperfections of the geological record. The record is indeed imperfect: only hard parts of organisms are readily fossilized, only certain sedimentary environments are conducive to fossilization, fossils must survive erosion and metamorphism and, finally, must be exposed on the surface and discovered. As a consequence, fewer than 1 percent of all species that ever lived are represented in the fossil record, and the record is spotty temporally, geographically, and taxonomically (see chapter II.9). For example, only about 30 localities have yielded important fossil faunas of tetrapods from the 60 million years of the Carboniferous, when the first major radiation of amphibians occurred.

Despite the imperfections, the transitional forms that Darwin hoped might be found soon started turning up. In 1861, *Archaeopteryx* was discovered and became the first of the proverbial "missing links" to be found. Clothed in the feathers of a bird with broad wings, it nonetheless had the long bony tail, toothed jaws, and free, clawed fingers of a reptile. T. H. Huxley (1825–1895) saw it clearly as intermediate between reptiles and birds and suggested a relationship to dinosaurs, a suggestion now well documented. Extraordinarily well-preserved fossils show not only the close skeletal resemblance of *Archaeopteryx* and early birds to theropod dinosaurs but also that feathers are not peculiar to birds; they were present on quite a few nonflying dinosaurs as well.

A few years later, Richard Owen (1804–1892) and E. D. Cope (1840–1897) both recognized that certain ancient reptiles (synapsids) bore an affinity to mammals. The origin of mammals from the synapsids is now known in exquisite detail (figure 2). Reptiles have several bones in the lower jaw, and one of the more posterior ones, the articular, forms the jaw joint with the quadrate bone of the skull. Mammals have a single bone in the lower jaw, the dentary (also present in reptiles), which articulates with the squamosal bone of the skull. A long series of fossils, beginning with fully reptilian forms in the Pennsylvanian, lead gradually to the mammals at the Triassic/Jurassic boundary. The dentary enlarges, becoming the largest bone of the lower jaw. The quadrate and articular become smaller. Eventually, the dentary contacts the squamosal, leading to several forms, such as *Probainognathus*, that have a dual jaw joint—both the old reptilian one and the new mammalian one. The transition is so gradual that it becomes a matter of convention to decide which form is the first "mammal": *Morganucodon* is often so regarded, but it, too, has a dual jaw joint. In later forms the quadrate and articular detach from the jaw and become two of the three mammalian ear bones. Many other features that change during this transition—for example, the dentition becomes cusped, a secondary palate forms, the ilium becomes rod shaped—are likewise documented in the fossil record.

Another well-documented transition is that between the lobe-finned osteolepiform fishes and tetrapods (figure 3). Osteolepiforms were "typical" fish, with dorsal and anal fins, rounded heads with short snouts, and their shoulder girdles connected to their heads. But their pectoral and pelvic fins extended from the body in fleshy lobes, and within the lobe was a skeleton that, starting at the base (the end near the body), had a pattern of "one

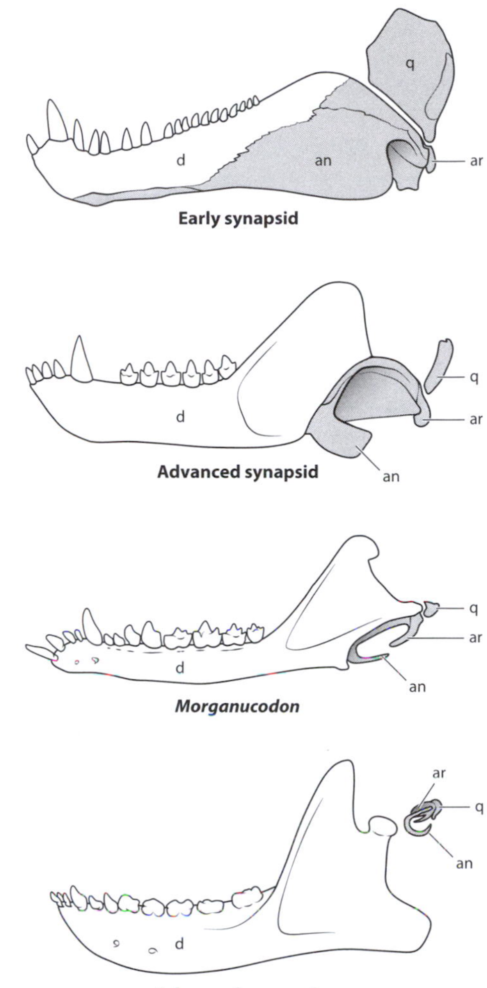

Figure 2. Homology of jaw bones of reptiles and ear bones of mammals. In early synapsids the lower jaw comprises the tooth-bearing dentary (d) and several postdentary bones, including the angular (an) and articular (ar). The latter articulates with the quadrate (q), forming the reptilian jaw joint. In advanced synapsids the latter bones are reduced, while the dentary enlarges. In the earliest mammals (*Morganucodon*) these bones are reduced further, and the dentary makes contact with the upper jaw, forming the mammalian jaw joint. In advanced mammals the angular, articular, and quadrate detach from the jaw entirely, becoming the tympanic, malleus, and incus of the mammalian ear. (After D. Davis 1991, K. Kardong 2012, and R. Carroll 1988.)

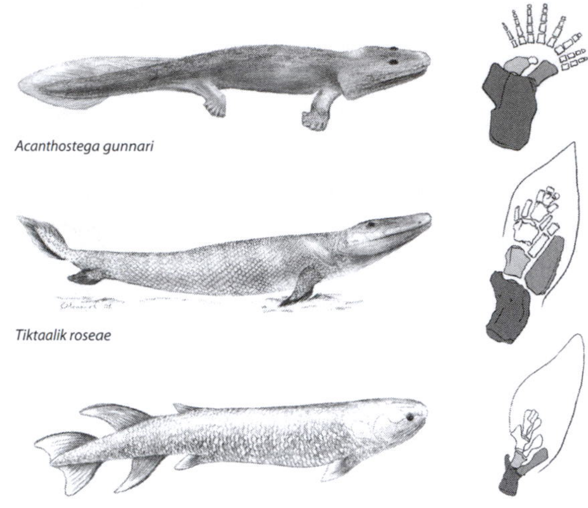

Figure 3. Transition from lobe-finned fish to tetrapods. The lobe-finned osteolepiform fish *Eusthenopteron* has the tetrapod-like "one bone, two bones, many bones" pattern in its fin. *Tiktaalik*, the "fish-apod," has lost the dorsal and anal fins, the head is free of the shoulder girdle, the snout is elongated with the eyes directed upward, and there is a wrist joint in the forelimb. *Acanthostega*, one of the first tetrapods, has digits but retains the caudal fin of its fish ancestors. The bones of the left forelimb shown in gray are, from darkest to lightest, the humerus, the radius, and the ulna, respectively; more distal elements are unshaded. (Copyright Kalliopi Monoyios.)

bone, two bones, many bones." This is the same pattern as in tetrapods: the plan that exemplifies the unity of type of the tetrapod forelimb extends to osteolepiform fishes. In *Panderichthys*, from about 380 million years ago, the head and body are flattened, the snout is elongated, and the eyes are on top of the head; the anal and dorsal fins have been lost. In the remarkable *Tiktaalik* from 375 million years ago, the shoulder girdle (equivalent to our collarbone and shoulder blades) has been freed from the skull—*Tiktaalik* had a neck; and there is a joint within the "many bones" of the forelimb—it also had a wrist. Ten million years later we have the first actual tetrapods, *Acanthostega* and *Ichthyostega*, which have legs with toes but retain some of the gill-cover bones and the caudal fin of their fish ancestors.

The origins of birds, mammals, and tetrapods represent fairly large changes in morphology and way of life, and the intermediate forms bridge the differences between these major taxa. The much smaller transitions that occur as one species evolves into another are also documented in the fossil record, but continuous sedimentary deposition is necessary to record such fine-scale temporal events. Such conditions are not common but occur most often in the fossil record of shelled marine planktonic protists, such as foraminifera and radiolarians, that can be recovered by extracting cores from the seabed. The shells of dead individuals rain down onto the ocean floor, forming an essentially continuous sedimentary record. In these organisms, such as the foraminiferans in the genera *Globorotalia* and *Contusotruncana*, and the radiolarian genus *Eucyrtidium*, such fine-scale changes can be observed, and in the latter even the split into two species from an original one has been recorded.

There are many other examples of transitional forms in the fossil record—such as ancestral whales and snakes with hindlegs—and more are being discovered every year.

2. COMPARATIVE BIOLOGY

Unity of Type

While the fossil record provides direct evidence of change of life over time, comparisons among organisms, either living or extinct, give evidence that the link between the different forms at different times is genealogical. Primary among these comparisons are the observed similarities in fundamental structure among organisms referred to as "unity of type." These similarities, which extend from morphology to development to the genome, are homologies, that is, similarities due to inheritance from a common ancestor (see chapter II.6).

Famous homologies are the "one bone, two bones, many bones" pattern of the tetrapod limb (figure 1), and the jaw and ear bones of reptiles and mammals, both mentioned previously (figure 2). In the case of the limbs, the strong similarity of the same bone among the various tetrapods allows the homologous bones to be easily recognized: for example, humerus, radius, ulna. The similarity in plan is not accounted for by the functional requirements of the limbs but by inheritance from a common ancestor. In the jaw and ear, the homologies are traced with the aid of fossil and embryological (see the section Markers of History) evidence. The angular and articular bones of the reptilian lower jaw are homologues of the tympanic and malleus bones of the mammalian ear, while the quadrate of the reptilian upper jaw is the incus of the mammalian middle ear.

These patterns can be seen not only in the skeleton but in the genome itself at the cellular level. The human genome contains 23 pairs of chromosomes, whereas the genome of great apes has 24 pairs. The difference of one pair can readily be accounted for: human chromosome 2 is homologous to two ape chromosomes, which have become fused in the human lineage. The homology has

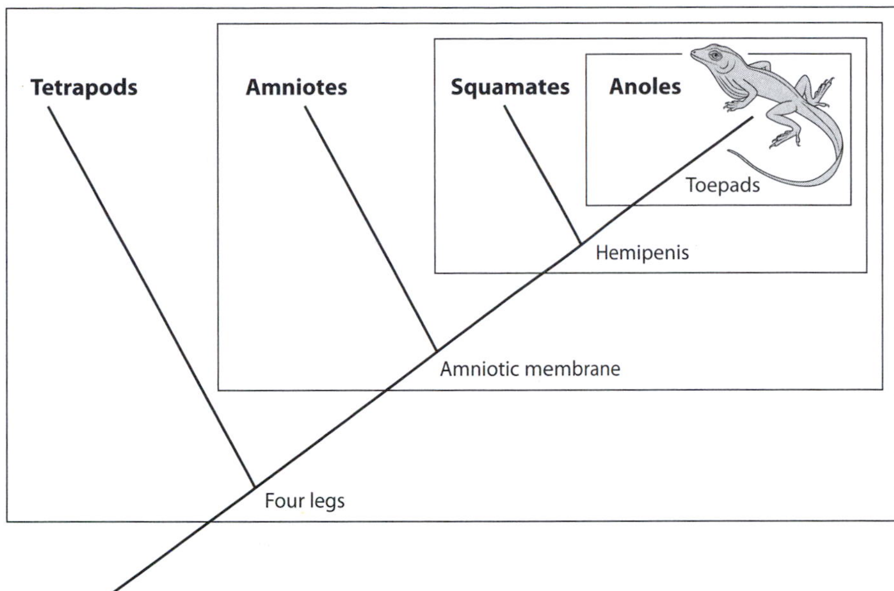

Figure 4. Phylogenetic tree of anole lizards showing nested homologies. The common ancestor of all tetrapods possessed four legs, a trait that, with modifications, was passed on to all of its descendants. One of these descendants evolved the amniotic membrane, thus nesting the amniotes (reptiles, birds, and mammals) within the tetrapods. Squamates (snakes and lizards), which evolved hemipenes, in turn are nested within amniotes. And finally, anoles are characterized by the possession of expanded toepads. Anoles thus possess toepads, hemipenes, amniotic membranes, and four legs, each trait inherited from a successively more distant ancestor, and each marking a more inclusive group. The nested boxes in the figure indicate the named clades.

been confirmed by the discovery of remnants of the central (centromere) and end (telomere) portions of the ape chromosomes within the human chromosome, showing that the latter arose from what were originally two chromosomes.

At perhaps the most basic level, the near universality of the genetic code is another homology that argues for the common ancestry of all life. The genetic code associates each amino acid with a codon of three nucleotides, which carries the information for the making of proteins. But the identity of the three nucleotides is arbitrary, so the code's universality cannot be due to functional constraint but arises as a legacy of the code established in distant progenitors.

Common Ancestry

Although the common ancestry indicated by the genetic code embraces all (or nearly all) living beings, homologies do not generally have such a wide distribution. Rather, homologies characterize smaller groups, and these groups are nested within larger homology-characterized groups, which in turn are nested within yet larger such groups, and so on. This nesting of homologies results from the branching history of life, represented by the phylogenetic tree (see chapter II.1). When a lineage divides, giving rise to a new branch in the tree of life, the characteristics of the splitting lineage are passed on to its descendants. Modifications may occur in a descendant lineage, which will in turn be passed on to its, and only its, descendants. Each homology is the origin of a new feature, shared with descendants but not with collateral relatives. These nested sets of homologies, which are exactly what one would expect from a process of descent with modification, are powerful evidence for evolution.

For example, possession of a toepad, composed of laterally expanded scales at the end of the digit, characterizes lizards of the genus *Anolis* (figure 4). Anoles also have a hemipenis (oddly named, as this means that they have two penises, rather than half a penis, as the word might imply) as the male intromittent organ, but this organ characterizes a larger group, the squamates (lizards plus snakes), within which anoles are nested. The squamates, in turn, are nested within a yet-larger group characterized by the presence of an amniotic membrane around their embryos. The amniotes (reptiles, birds, and mammals) also have four legs, a trait shared with amphibians as well, which together make up the tetrapods.

The tetrapods, also in turn, share features of the limb with certain fishes, forming again a more inclusive group. Continuing in this manner, the bony fishes, vertebrates, chordates, and deuterostomes constitute successively larger groups within which anoles are nested, and this nesting is indicative of the history of common ancestry.

Eventually all, or almost all, life can be subsumed within the nested tree. For organisms as divergent as insects and vertebrates it is hard to recognize morphological homologies, but genetic data show that homologies of gene and genome structure can be recognized in, for example, the presence of homologous sets of developmental regulatory genes (*Hox* genes) in both arthropods and vertebrates. While nestedness holds true for eukaryotes, there is some question whether at the base of the tree of life transfer of genetic material between prokaryotes may be so common as to obscure or efface the nested pattern (see chapter II.11).

Markers of History

Among the most striking evidences of evolution are the features of organisms that appear to reflect a constraint: organisms inherit from their ancestors a developmental system, and evolutionary modifications must take that inherited system as a starting point. Indeed, organismal features give every indication of having arisen by "tinkering" with this inherited preexisting developmental system, and make sense only as a result of descent with modification.

This constraint is evident in the similarities among the embryos of jawed vertebrates. The embryos go through a stage in which they resemble one another in the possession of pharyngeal arches, limb buds, tails, and other traits. After this stage, embryos diverge, developing into the varied forms they will become as adults. What perhaps is most striking is that structures present in the embryo are not always present in the eventual adult. Thus, humans and apes have an embryonic tail that largely disappears in the adult; only a bony vestige remains internally. Humans develop a coat of fine fur, the lanugo, which is lost just before or shortly after birth. All four limb buds develop in tetrapods, such as whales and snakes, that in the adult lack one or both pairs of limbs. In some snakes, a vestigial leg is still visible externally (figure 5). In each case, the eventual adult form develops from a shared state and subsequently passes through stages shared with smaller nested groups until finally arriving at its own specific state.

Many structures show the traces of ancestry. In fetal mammals, the bones that eventually become the bones of the middle ear begin their development in positions along the jaw corresponding to those occupied by the homologous jaw bones of reptiles. The "thumb" of pandas (figure 6) is not homologous to other mammals' inner digits, nor is it even a digit: it is a modified radial sesamoid bone, pressed into service as a makeshift digit, in an exquisite example of tinkering—that is, the modification of available structures, rather than an engineering ideal. In the Australian lungfish (*Neoceratodus*), there is a single dorsal lung, unlike in other lungfish (and tetrapods), in which the lungs are paired (figure 7). In all lungfish (and tetrapods), the lung attaches to the ventral part of the esophagus. In Australian lungfish, the pneumatic duct travels up alongside the right side of the esophagus to the dorsally positioned lung, seeming to trace the course of its morphological modification. Confirming this movement, the left pulmonary artery, instead of going directly to the dorsal lung, travels down the left side of the esophagus, curls under the esophagus, and follows the pneumatic duct up to the lung.

Such examples are not limited to morphology: pseudogenes, nonfunctional versions of genes that are functional in related species or even in other copies in the same organism, occur commonly in organism's genomes (see chapters V.3 and V.5). For example, in most mammals, vitamin C is synthesized by a battery of enzymes. In primates, which get vitamin C in their diet, a gene for one of the necessary enzymes is present as a nonfunctional pseudogene, so that the vitamin is not synthesized. The broken gene is a relict from earlier mammals that do use it in their synthesis of the vitamin. In primates, the gene has been disabled by a mutation, but the now-inactive gene remains as a marker of primates' forebears.

3. BIOGEOGRAPHY

Historically, the biogeographic evidence for evolution was crucial, because it was the evidence that convinced Darwin himself. Many features of the distribution of organisms that are anomalous or merely curious under a hypothesis of special creation became explicable and expected under the hypothesis of descent with modification. Under the latter hypothesis, related species should occur in geographically connected areas or in areas that could have been reached by a common ancestor of the related species. The connectedness and "reachability" of areas are determined chiefly by the distance and geographic barriers between them, and the ability of organisms to cross such barriers. Geographic conditions that either present barriers to dispersal (e.g., the ocean around islands) or facilitate it (e.g., the continuous land of continents), and organisms' abilities to overcome or utilize these geographic

Figure 5. External hindlimb of a snake, the ball python (*Python regius*). Located just lateral to the anal scale, the keratinous claw is here shown slightly pushed away from the body by a probe. The claw is underlain internally by rudiments of the femur and, deeper in the body, the pelvis. (Photo by G. C. Mayer.)

barriers and bridges, are thus key determinants of the distribution of life on earth.

Islands

Bermuda is a small group of coral islands in the North Atlantic, 1000 km to the east of North America. Sitting atop a long-extinct volcanic platform and surrounded by waters of abyssal depth, islands such as Bermuda are called *oceanic*—they have never been connected by land to a mainland. The native non-marine vertebrate inhabitants of Bermuda are few—several land birds, migratory bats, a lizard, a terrapin—and show closest affinity to forms from the North American mainland. Some are identical or nearly so to the North American forms, whereas others are distinct endemic species found nowhere else. Several major groups common on the mainland are lacking entirely—there are, for example, no nonflying terrestrial mammals and no amphibians. Subsequent to human colonization, there have been many successful introductions of vertebrates, including land birds, reptiles, amphibians, and land mammals (figure 8). In its features, the fauna of Bermuda is typical of oceanic islands.

All these characteristics are readily explained by the hypothesis of descent with modification. As a geologically oceanic island, Bermuda lacks organisms that cannot cross 1000 km of ocean (land mammals and amphibians) and is inhabited by descendants of a limited number of successful colonists, whose characteristics, such as the ability to fly, have permitted cross-oceanic dispersal. The nearest relatives of these colonists reside in the adjacent landmass of North America, because the islands are most accessible from there by "occasional means of transport" (as Darwin called them). Some of the colonists have been isolated sufficiently long to have diverged into endemic forms (e.g., the lizard), whereas others more recently arrived have not so diverged (e.g., the terrapin). The success of invasive introduced forms shows that the island fauna was, as Darwin put it, insufficiently stocked

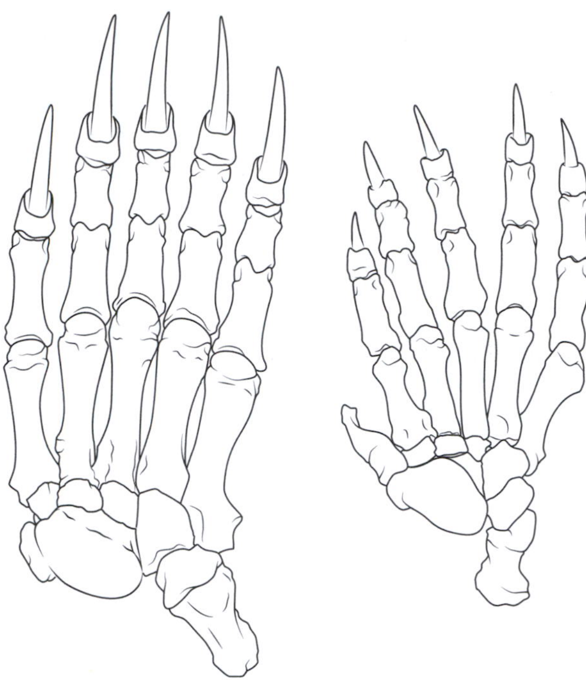

Figure 6. The right manus in the brown bear (*Ursus arctos*; left) and the giant panda (*Ailuropoda melanoleuca*; right). The panda's thumb is not a true digit but a modified radial sesamoid bone. (After D. D. Davis 1964.)

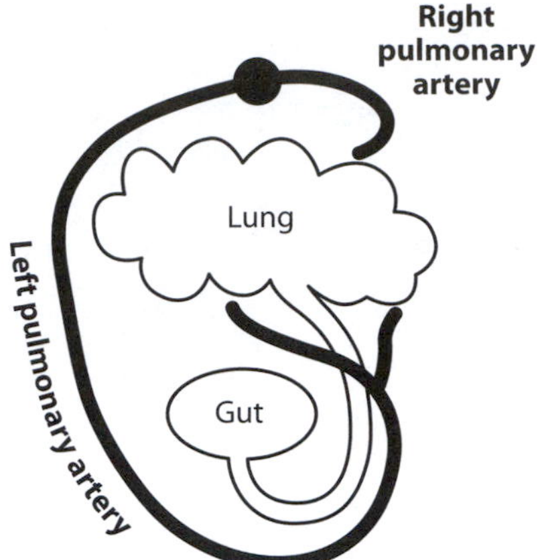

Figure 7. The lung and its arterial blood supply in the Australian lungfish (*Neoceratodus*). Note that the pneumatic duct swings around the gut to make the lung dorsal, and the left pulmonary artery, after branching from the dorsal aorta, follows it down and around (instead of going straight to the lung). (After E. S. Goodrich 1909.)

by the "creative force": it was difficulties of dispersal, not habitat suitability, that caused the fauna to be restricted in species richness and taxonomic diversity.

Continents

On continents, the accessibility of adjacent areas leads, in general, to related forms being found throughout. Darwin noted this for the mammals of South America. The current mammals of the varied South American habitats —tropical, temperate, alpine—are all related to one another and not to mammals of similar habitats in distant places. Although organisms can sometimes disperse over large distances and across barriers (as in the case of many oceanic islands), dispersal occurs more readily between nearby areas, so that the biota of nearby areas, such as the different habitats of South America, have similar compositions. Similarly, the fossil South American mammals Darwin found were related to the ones now present. Thus, related organisms live in places they can reach by dispersal, and descend from the previous inhabitants.

This pattern is also well illustrated by the mammal fauna of Australia, a continent that became isolated from the other continents tens of millions of years ago. Most of the mammals of Australia are marsupials, a group now found elsewhere only in the Americas, where they are much less diverse. In Australia, the marsupials occupy all the major habitats and have a rich fossil record going back tens of millions of years. Having been isolated in Australia, the marsupials have prevailed and diversified into most ecological roles: herbivores (kangaroos), carnivores (thylacine), burrowers (wombats), climbers (koala), gliders (phalangers), and many more. The only native placental mammals are bats and murid rodents. Bats can fly, and rodents are able to colonize across water barriers (although not ones as wide as those that can be crossed by bats).

Geographic distributions that might seem anomalous under the hypothesis of descent with modification are explicable when the movement of the continents is taken into account. For example, *Cynognathus*, a terrestrial synapsid of the Triassic, about 240 million years ago, has been found in South America, Africa, and Antarctica, all now widely separated by seas unlikely to be crossed by a stout, meter-long animal like *Cynoganthus*. But in the Triassic all these continents were connected in a single landmass, Gondwana. Far from being a problematic case, *Cynognathus* shows again the importance of historical connectedness, both genealogical and geological, in the distribution of organisms.

4. EVOLUTION IN ACTION

Changes within Populations

Although fossils record descent with modification over billions of years, evolution often occurs quickly enough

Figure 8. Endemic and invasive oceanic island species. Left: The endemic Bermuda skink (*Eumeces* [*Plestiodon*] *longirostris*) is the only extant native terrestrial reptile of Bermuda. Its nearest relatives are from the nearest mainland, North America, 1000 km to the west. Right: A Bermudian specimen of *Bufo* (*Rhinella*) *marinus*, native to Middle and South America, and introduced to Bermuda in 1885. It is one of several exotic amphibian and reptile species that have thrived and even become pests there, after the barrier to dispersal was overcome by human agency. (Skink photo by Richard Ground; toad photo by G. C. Mayer.)

to be observed within one or a few human lifetimes, and such cases allow the full panoply of evolutionary mechanisms—mutation, gene flow, genetic drift, and natural selection—to be seen in action.

Darwin's chief examples of observed evolution involved the work of animal and plant breeders in producing and elaborating the features of domesticated organisms (see chapter VIII.5). Studies of such species show the importance of selection in establishing and fixing desired characteristics. The many varieties of domestic dog, differing so much in size, shape, and behavior, have all been produced from the wolf in the last few thousands of years. Corn, one of the most highly modified and economically important organisms ever created by humans, was developed by selective breeding from a wild species of grass.

Rapid evolution has also unintentionally been caused by humans, who changed the environment in ways that prompted evolutionary responses from affected organisms (see chapters VIII.2 and VIII.3). Industrial melanism, the evolution of darker coloration in animals living in environments darkened by pollution, is widespread in a variety of insects in industrialized areas of both Europe and North America. When pollution controls have been enacted, the evolutionary change has been reversed, and the lighter-colored forms have again increased in frequency. Both the spread of melanism and its reversal have been observed in Britain in the moth *Biston betularia* over a period of about 150 years.

The use of pesticides and antibiotics has frequently led to the undesired evolution of resistance in the targeted organisms, including rodents, insects, bacteria, and viruses. Multiple-antibiotic-resistant strains of bacteria have become a major health problem, and one of the greatest difficulties in treating acquired immune deficiency syndrome (AIDS) has been the rapid evolution of resistance by the human immunodeficiency virus (HIV). In the evolution of HIV, high mutation rates, short generation time, large population sizes, and strong selection have all combined to make the virus adapt extremely rapidly to drugs. Drug mixtures, which attack the metabolism of the virus in different ways simultaneously, have proven more effective, as the multiple mutations required for resistance to all the drugs are less likely to occur.

Species introduced into new geographic areas, whether by humans or by natural means, are likely to find themselves in new environments and thus are more likely to undergo evolutionary changes. Such introductions have indeed produced many examples of divergence in features such as size, shape, and coloration. One example is the house sparrow (*Passer domesticus*), introduced into North America from Europe about 1851. By the middle of the twentieth century, the sparrow had geographically differentiated in coloration, size, shape, and physiology in a manner parallel to that of native species.

Natural populations have also been observed evolving in response to natural environmental changes (see chapter III.7). Perhaps the best-studied case is that of Darwin's finches (*Geospiza*) in the Galápagos, where populations have been carefully tracked over decades. During this time span, repeated episodes of morphological evolution have occurred in response to climatically induced variations in the food supply. Further, the genetic basis of these changes has been demonstrated by observations in nature. These studies show that natural populations are not evolutionarily static but can adaptively track changes in the environment.

Speciation

Speciation, the origin of new lineages reproductively isolated and evolutionarily independent from other lineages, is a key part of evolution (see Section VI: Speciation and Macroevolution). Without it, evolution might occur within a lineage, but there would be no increases in biodiversity. Divergence of lineages is enhanced by their geographic separation (since it reduces or eliminates the homogenizing force of gene flow). Isolated populations may differentiate independently—including in characteristics affecting reproductive compatibility—to the point that they can no longer exchange genes if brought again into contact. The insensible gradation over space of populations varying from mere geographic isolates to highly differentiated populations approaching genus-level distinction provides evidence for speciation by geographic isolation, and there are many examples, such as in kingfishers of the genera *Tanysiptera* and *Halcyon* on the islands of the southwest Pacific. In most cases though, divergence takes too long to be observed within a human lifetime. Laboratory studies have shown that incipient reproductive isolation can arise in separated populations undergoing differential adaptation, and that reproductive isolation can be enhanced by selection against individuals who hybridize with the "wrong" population.

In some cases, though, speciation occurs sufficiently quickly to be observed in the wild within a human lifetime. Speciation by polyploidy (i.e., the duplication of chromosome sets) is an important mode of splitting in plants (see chapter VI.9). A well-studied case is the recent origin of the British salt marsh grass *Spartina anglica*, a polyploid derived from the natural crossing of the introduced American species *S. alterniflora* (diploid chromosome number $2n = 60$) with the native *S. maritima* ($2n = 62$). *Spartina alterniflora* was accidentally introduced in 1829. In 1870, sterile hybrids with *S. maritima* were first recorded. Sterility of the hybrids, due to mismatching of the parental chromosomes, was overcome by chromosome doubling, and the fertile *S. anglica* ($2n = 122$) was first recorded in 1892. The new species has since spread along coastlines throughout Britain.

5. EVOLUTION AS FACT AND THEORY

It is sometimes noted pejoratively that evolution is a "theory." This pejorative usage confuses a vernacular notion of "theory" as something uncertain or conjectural, with the word's scientific usage. In science, a theory is not a mere conjecture but a connected series of propositions supported by, and explanatory of, many and varied lines of evidence. We refer to the "germ theory of disease" not because we are unsure that microbes cause disease but because the theory is a set of high-level and powerful generalizations that account for and are, in turn, supported by a huge amount of data.

Regarding evolution, it was Darwin who first convincingly assembled the many and varied lines of evidence that could be accounted for by, and provide evidence of, descent with modification. His many successors have carried on his work, and Darwin would have been both pleased and astonished by the further lines of evidence that have been brought to bear. In his own lifetime, transitional fossils began to be found, and we now have them in abundance. Darwin's own attempts at formulating principles of inheritance failed, so he would have been gratified by the explosive growth of genetics and now genomics filling in what he did not know. Crucially, the facts of genetics could have turned out to be incompatible with Darwin's evolutionary views but, instead, his ideas have passed this important test. He would perhaps have been most astonished by the evidence for the rapidity with which evolution can occur, including the formation of new species within a human lifetime.

As briefly reviewed here, all these lines of evidence, all leading to the same conclusion, serve to make descent with modification one of the most securely established high-level generalizations in science and allow us to speak confidently of the fact of evolution.

FURTHER READING

Carroll, R. 2009. The Rise of Amphibians: 365 Million Years of Evolution. Baltimore: Johns Hopkins University Press. *Includes an account of the origins of vertebrates and reptiles, as well as of the origin of amphibians.*

Coyne, J. A. 2009. Why Evolution Is True. New York: Viking Penguin. *An account of the evidence for evolution for a general audience.*

Dawkins, R. 2009. The Greatest Show on Earth. New York: Free Press. *Another account of the evidence for evolution for a general audience.*

Futuyma, D. J. 2013. Evolution. 3rd ed. Sunderland, MA: Sinauer. *The leading textbook of evolutionary biology.*

Grant, P. R., and Grant, B. R. 2008. How and Why Species Multiply: The Radiation of Darwin's Finches. Princeton, NJ: Princeton University Press. *An account of the detailed evolutionary studies conducted by the authors and their associates over four decades, including several episodes of closely observed evolutionary changes.*

Mayr, E. 2001. What Evolution Is. New York: Basic Books. *A summary of the evidence for evolution and its mechanisms by the "Darwin of the twentieth century."*

Prothero, D. R. 2007. Evolution: What the Fossils Say and Why It Matters. New York: Columbia University Press. *A richly illustrated account of the fossil record emphasizing transitions between major groups.*

Shubin, N. 2008. Your Inner Fish. New York: Pantheon Books. *A popular account of the discovery of the fish-amphibian transitional form* Tiktaalik, *and of the traces of common ancestry within vertebrates (including man), and of the common ancestry of vertebrates with other animals.*

Young, D. 2007. The Discovery of Evolution. 2nd ed. Cambridge: Cambridge University Press. *Intended for a general audience, this superb and richly illustrated history serves as a text for evolutionary biology itself, because it introduces and explicates not just the ideas and historical figures but the evidence on which the major discoveries of evolutionary biology are based.*

I.4

From DNA to Phenotypes

Michael C. Whitlock

OUTLINE

1. What is a gene?
2. Descriptions of genetic variation
3. A multiplicity of forms of inheritance

Genetic information is passed between generations in most organisms by DNA. Variation among individuals of the sequence of their DNA is the raw material of evolution. DNA codes for phenotype by sequences specifying proteins and RNAs and by regulatory elements that control when and by how much of each is made. Variation in DNA sequence can translate into variation in phenotype, which may cause fitness differences among individuals on which natural selection can act. This chapter describes the basics of how phenotypes are created from the instructions coded in DNA and introduces some of the descriptions of genetic variation used in evolutionary biology.

GLOSSARY

Allele. One of possibly many versions of DNA sequences at a given locus.

Autosome. A chromosome that is not a sex chromosome.

DNA. Deoxyribonucleic acid; the molecule used by most of life on earth to encode genetic information and to transfer that information from parent to offspring.

Gene. A region of genetic material that encodes a functional unit, like a protein or RNA.

Genetic Variance. Variance among individuals in some quantity measured on their genotypes. Genetic variance can refer to variance among individuals in the number of copies of a particular allele or to variance in the effects of those alleles on phenotypes, depending on context.

Genotype. The genetic code of an organism. Genotype can refer to the whole genome or to the alleles at a specific locus or loci.

Hardy-Weinberg Equilibrium (HWE). The frequencies of genotypes at a locus assuming independent pairing of alleles from maternal and paternal copies.

Heritability. The fraction of phenotypic variance for a trait that can be attributed to genetic effects inherited by offspring from their parents.

Heterozygote. An individual that has two different copies of DNA from each parent at a locus.

Heterozygosity. The frequency of individuals in a population at a locus that are heterozygotes. Often used to refer to the expected heterozygosity, which is the frequency of heterozygotes expected at Hardy-Weinberg equilibrium for the allele frequencies in that population.

Homozygote. An individual that has two identical copies of DNA from each parent at a locus.

Linkage Disequilibrium. An association between alleles at different loci, where the particular alleles appear together in gametes either more or less often than expected by chance.

Locus. Often synonymous with "gene," a location in the DNA sequence.

Messenger RNA (mRNA). An RNA made by copying from DNA, used as a template in translation to produce a protein.

Phenotype. Any observable characteristic of an organism, including, for example, its morphology, behavior, and physiological or developmental processes.

Pleiotropic. Describing a locus that has effects on more than one phenotypic trait.

Protein. A covalently linked series of amino acids. Proteins are responsible for most biological functions.

Sex-Linked Locus. A locus that is located on one of the sex chromosomes.

Deer mice in the species *Peromyscus polionotus* are not always the same color. Some are very light brown, and some are almost black. On the beaches of Florida, where the sand is almost white in color, most mice have very light brown fur, whereas in the nearby fields the soil is dark and loamy, and the mice are darker to match. These differences in color allow the mice to escape predation by visual predators (like coyotes and hawks), because the mice are harder to spot when they are colored like their background.

These color differences are largely controlled by genetic differences between the mice. A large part of what makes a beach mouse pale is that it has a mutation at one particular place in its DNA. This region is responsible for coding the information necessary to make a protein called the *melanocortin 1 receptor*, abbreviated to Mc1r. This protein in the cell membrane can be activated to signal pigment-producing cells to produce the dark-brown pigment called *eumelanin*. Mice with more eumelanin have darker fur; mice with less have lighter brown fur. The difference can determine whether they live or die.

This chapter discusses how individuals of the same species differ from one another in their DNA, their proteins, their morphology, their physiology, their behavior, and ultimately, in fitness. We begin by reviewing, very briefly, how the phenotype of an individual is affected by its DNA.

1. WHAT IS A GENE?

The word *gene* refers to a region of DNA that codes for the instructions for a particular protein or RNA. (RNA is another nucleic acid, and various RNAs have important functions in biology. For example, messenger RNA [mRNA] is a copy of the information from the coding region of a protein-encoding gene, while ribosomal and transfer RNAs both help translate the instructions in mRNA into the correct sequence of amino acids to make a particular protein.) DNA is composed of just four units called *nucleotides*: thymine (T), adenine (A), cytosine (C), and guanine (G). The order in which these four nucleotides occur in the DNA determines its meaning.

If the gene codes for a protein, it contains information describing the order of amino acids that will be linked together to make that protein. A protein is one or more *polypeptides*, that is, covalently linked series of amino acids. There are 20 amino acids commonly used by living organisms. Each amino acid has different biochemical properties that, together with the other amino acids in the protein and other proteins in the organism, determine the biochemical functionality of the protein. Each amino acid has one or more sequences of three nucleotides that code for it; these three-nucleotide sequences are called *codons*. The list of conversions between the particular codons and their associated amino acids is called the *genetic code* (figure 1). There are also three codons that tell the protein translation machinery to stop, marking the end of the coding sequence.

For example, the stretch of DNA that codes for the Mc1r protein is called the *Mc1r* gene. This gene consists of a *coding region*, which specifies the sequence of amino acids that make Mc1r itself, as well as a series of other DNA sequences called *regulatory regions* that interpret signals from the cell to determine when to create Mc1r protein and when to not. Both the coding region and the regulatory region together are called the gene.

In a diploid organism like humans or mice, each cell has two copies of most genes, one each inherited from the mother and father. Many organisms are haploid, carrying only one copy of each gene. Most microorganisms are haploid.

Every copy of a gene may not contain exactly the same DNA sequence, however. If the DNA sequence at a gene is different between two copies, we say that these are different *alleles* of that gene. In the beach mice, for example, one common allele for *Mc1r* codes for the version of the protein that causes the mice to have light brown fur, and another allele codes for Mc1r that causes mice to be darker. In this case, the two alleles differ at only one place in their DNA sequence—a difference that affects the 293rd nucleotide, changing a cytosine nucleotide in the darker allele to a thymine in the lighter allele. This DNA difference causes the 65th amino acid in the resulting Mc1r protein to be the amino acid cysteine rather than arginine (figure 2). The different biochemical properties of the proteins coded by the alleles—the arginine is large and charged, whereas the cysteine is smaller and uncharged—are what cause the fur color phenotype to differ. A difference in the DNA sequence of one nucleotide can cause a difference of one amino acid, which causes a difference in the shape or chemistry of a protein, which causes a difference in the fur phenotype, which in turn causes a difference in fitness between carriers of different alleles.

Not all changes in phenotype result from changes in the coding region. Near each coding region are sections of the DNA that regulate the expression of that gene; that is, these regulatory regions can control when, in what tissues, and in what concentration the protein is actually made. For example, the wildflower in *Phlox drummondii* typically has light blue flowers, as does the related species *P. cuspidata*. Where the two species overlap geographically, however, *P. drummondii* has dark-red flowers. (As a result of this difference in flower color, pollinators move less pollen between the two species, and therefore fewer unfit hybrid offspring are produced.) The dark-red flowers have much more of the pigments cyanidin and malvidin, which in part is a result of an increased expression of the transcription factor MYB. The mutation that increases the expression of MYB is not a change to its

Codon	*Amino acid*	*Codon*	*Amino acid*	*Codon*	*Amino acid*	*Codon*	*Amino acid*
TTT	Phenylalanine (Phe)	TCT	Serine (Ser)	TAT	Tyrosine (Tyr)	TGT	Cysteine (Cys)
TTC		TCC		TAC		TGC	
TTA	Leucine (Leu)	TCA		TAA	Stop	TGA	Stop
TTG		TCG		TAG		TGG	Tryptophan (Trp)
CTT		CCT	Proline (Pro)	CAT	Histidine (His)	CGT	Arginine (Arg)
CTC		CCC		CAC		CGC	
CTA		CCA		CAA	Glutamine (Gln)	CGA	
CTG		CCG		CAG		CGG	
ATT	Isoleucine (Ile)	ACT	Threonine (Thr)	AAT	Asparagine (Asn)	AGT	Serine (Ser)
ATC		ACC		AAC		AGC	
ATA		ACA		AAA	Lysine (Lys)	AGA	Arginine (Arg)
ATG	Methionine (Met)	ACG		AAG		AGG	
GTT	Valine (Val)	GCT	Alanine (Ala)	GAT	Aspartic acid (Asp)	GGT	Glycine (Gly)
GTC		GCC		GAC		GGC	
GTA		GCA		GAA	Glutamic acid (Glu)	GGA	
GTG		GCG		GAG		GGG	

Figure 1. The genetic code

coding region, though, but a change in its gene's regulatory element. Important phenotypic effects based on changes to gene regulation like this are very common.

The Central Dogma

The so-called central dogma of molecular biology is that, for most organisms, DNA is the genetic material that transmits information between generations, and mRNA is copied from genes on the DNA to carry that information to parts of the cell to produce proteins, which do most of the work of the cell. This "dogma" is largely true, but all its steps can have exceptions. Some organisms use RNA as the genetic material that is passed between generations to convey the information encoded therein. Many genes in DNA code for RNAs that themselves have physical function (and do not code for proteins at all). Some information is transmitted between generations by modifications of the DNA, such as methylation, such that genetic transmission of information to subsequent generations is not entirely composed of data in the sequence of the DNA. Parents also modify the environment of their offspring in many cases, affecting the nature of subsequent generations in nongenetic ways. Such enriching exceptions aside, however, the central dogma explains much of the nature of how information is transmitted and used by living organisms.

Variation in DNA Sequence

Differences between alleles, like those causing fur color differences in mice or flower color differences in *Phlox*, are the raw material of evolution. Evolution by natural selection requires heritable differences between individuals, which occur when different individuals carry alleles that produce different phenotypes. A population that contains more than one allele can have *genetic variation*, meaning that all individuals are not genetically identical. This variation at the DNA level can mean that there is variation at the protein level, which may translate into variation in phenotype, which can translate into variation among individuals in fitness. Only when fitness varies among individuals, and when that fitness variation has a genetic basis, can a population evolve by natural selection.

Evolutionary biologists use a variety of techniques to study this genetic variation, ranging from examination at the DNA sequence level to investigation of the genetic basis of differences in phenotypes between individuals. DNA sequencing now allows relatively inexpensive reading of the genome itself; sequencing all or part of the genome in multiple individuals within a species allows the genetic variability of that species to be measured. DNA sequences at a locus can differ in a variety of ways. Most simply, the DNA can differ among individuals at a particular nucleotide site. A *single nucleotide polymorphism* (SNP—pronounced "snip") means that some individuals have one nucleotide while others have a different nucleotide at the same place in their DNA. If this difference occurs in a coding region and causes a change in the amino acid sequence of the resulting protein, it is called a *replacement* or *nonsynonymous* SNP, because the meaning of the DNA is not the same among individuals (figure 2). If a SNP occurs in a coding region but does not change the amino acid sequence, it is called a *synonymous* SNP. (Synonymous changes are possible

Mc1r, dark allele

DNA sequence: . . .ACCAAAAACCGCAACCTGCAC. . .

Amino acid sequence: Thr Lys Asn Arg Asn Leu His

O^{-} / O — C — CH — NH_3^+ | CH_2 | CH_2 | CH_2 | H_2N / H_2N — C — NH

Mc1r, light allele

DNA sequence: . . .ACCAAAAAC**T**GCAACCTGCAC. . .

Amino acid sequence: Thr Lys Asn **Cys** Asn Leu His

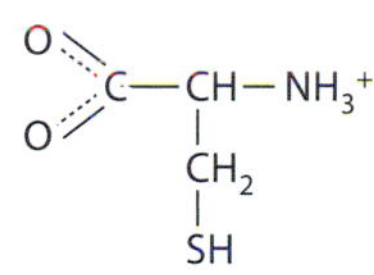

Mc1r, synonymous change

DNA sequence: . . .ACCAAAAACCG**A**AACCTGCAC. . .

Amino acid sequence: Thr Lys Asn Arg Asn Leu His

Figure 2. The coat coloration in beach mice is caused by one change (C to T) in the DNA sequence of *Mc1r*, which causes the codon CGC to turn into TGC, changing it from coding for arginine (Arg) into coding for cysteine (Cys). The chemical structures for Arg and Cys are shown; these two amino acids have very different chemical properties that translate into divergent functions of the protein, causing differences in the coat color of the mice. This is a nonsynonymous change, because the resulting protein sequence differs between the two alleles. In contrast, a change from CGC to CGA at that same codon, shown in the bottom panel, causes no change in the amino acid sequence, because both these codons code for the same amino acid. This is a synonymous change.

because most amino acids are coded for by more than one codon—refer to the genetic code in figure 1.)

Synonymous variation is particularly useful for the study of evolution, precisely because it is unlikely to have large direct effects on phenotype. As a result, synonymous variation is often assumed to be selectively neutral, and as such often allows a "control" for the effects of selection on the genome.

DNA sequence can also vary between alleles by insertions or deletions of an extra set of nucleotides. Deletions may have little effect on phenotype, or they may remove a regulatory element or part of a coding region (or even a whole gene or multiple genes) and potentially have large phenotypic effects. Sometimes, entire genes are duplicated, potentially allowing either that protein to be made faster or for the two copies to diverge to different functions. (See chapter IV.2 for a broader description of the types of genetic variation.)

2. DESCRIPTIONS OF GENETIC VARIATION

The most basic unit in the description of the genetic variation is the *allele frequency*, which is the fraction of all alleles at a locus in the population of interest that have a particular sequence. (Population geneticists often, but not always, use the letters p or q to denote the allele frequency of a particular gene under study.) For example, assume that a diploid population of mice has a total of 500 individuals, which means there are 1000 copies of the *Mc1r* gene in that population, because each individual carries two copies of each autosomal locus. If 30 of those copies code for the light-colored allele, then the frequency of the light-colored allele in that population is $p = 30/1000 = 0.03$. Assuming that all other copies of *Mc1r* in that population are the dark-colored allele, its frequency would then be $q = 1-0.03$, or 0.97.

Every individual of a diploid species carries two copies of each autosomal locus, one from each of its parents. The genotype of the individual at that locus therefore must be described by keeping track of both alleles that it carries. If there are only two alleles in the population, there are three possible genotypes: individuals that carry two copies of one allele, two of the other allele, or one of each. If there are more alleles at a locus, then the number of possible genotypes is higher. If both copies at a locus are the same allele, then we say that the individual is *homozygous* at that locus; if the two copies differ, it is *heterozygous*.

What frequencies might we expect to see of the different genotypes? Genotype frequencies can be predicted from the allele frequencies, provided a long list of evolutionary assumptions are true. If we assume that no selection, mutation, or migration affects the frequencies of alleles at a locus, that the population being studied is extremely large, and all individuals in the population are equally likely to mate with all other individuals, then the genotype frequencies can be predicted from the *Hardy-Weinberg equilibrium* (HWE). (In reality, the HWE is very predictive even with some selection, mutation, and migration, but its usefulness is very sensitive to that last assumption of random mating.) With HWE conditions, two alleles are paired independently to produce a diploid individual for the next generation, and as a result, the probability of a particular genotype is simply the probability that it receives one of its alleles from a random draw from the population (which will occur with a probability equal to the frequency of that allele) times the probability for its other allele. Thus, at HWE, the probability that an individual is homozygous for an allele that occurs at frequency $p = 0.3$ is $p^2 = 0.3^2 = 0.09$. The probability that an individual is heterozygous for two alleles that occur at frequencies 0.3 and 0.6, respectively, is $2\,(0.3)\,(0.6) = 0.36$. (The 2 in that equation represents the two ways of making any given heterozygote, with a particular allele coming from either the father or mother; in either case the genotype of the offspring is the same.) If we sum the expected frequencies of all the possible heterozygotes in a population, we get the *expected heterozygosity*, which is a common measure of genetic diversity. Expected heterozygosity increases with larger numbers of alleles, and it is greater if the frequencies of those alleles are similar.

Genotype frequencies can deviate from HWE, especially if mating individuals are related to each other. If individuals that mate to produce offspring are more closely related than randomly chosen members of the species, they are likely to share alleles, and their offspring are more likely to be homozygous than expected by HWE.

Moreover, allele frequencies can vary substantially over space. If mating occurs locally rather than at random over the range of the species (as is in fact common), then the allele frequencies in one area may evolve to be different than in another area. Such differences can evolve either owing to chance (i.e., genetic drift; see chapters IV.1 and IV.3) or because selection may favor different alleles in different places, especially if the environment also varies spatially. For example, the *Peromyscus* mice described at the beginning of this chapter have very different allele frequencies of *Mc1r* for populations on white sand versus on dark soil.

When more than one genetic locus is examined simultaneously, we may ask whether alleles at one locus are independent of the alleles at another locus. If there is no correlation between alleles at different loci, we say those loci are in *linkage equilibrium*. If, however, two alleles at different genetic loci appear together more often than expected by chance, we say the loci are in *linkage disequilibrium*. Two loci do not have to be physically linked to be in linkage disequilibrium, but physical linkage allows disequilibrium to persist longer. Disequilibrium can be created by chance (if a particular two-locus genotype happens to leave more offspring than expected) or by selection (if two alleles at the different loci work particularly well in combination, for example). Conversely, recombination between the loci tends to reduce linkage disequilibrium.

There are many measures of linkage disequilibrium, but the most basic is D, which measures the excess of gametes in the population of a particular two-locus allele combination. If the allele frequencies of specific alleles at the first and second loci are p_A and p_B, respectively, then the frequency of gametes bearing both of those alleles will be $p_A p_B + D$. If the two alleles were drawn at random and independently, gametes with both would appear with frequency $p_A p_B$, that is, D would be zero. Therefore D measures the deviation in gamete frequencies from independent assortment of alleles at two loci.

Quantitative Genetic Variation

Evolution by natural selection requires that DNA differences also have phenotypic effects. If DNA differences do not affect the morphological, physiological, or behavioral phenotypes of their organisms, then long-term evolution of those phenotypes is impossible. Sometimes, we know the link between differences in DNA sequences and phenotypes (as in the beach mouse *Mc1r* story); far more often, we do not. Moreover, the link between genotype and phenotype is usually far more complicated than shown in the preceding examples. Nearly all traits are affected by multiple loci—sometimes interacting in complex ways—and the environment in which the organism develops almost always affects the traits. In addition, we usually do not know which genes matter for a particular trait.

If a trait is largely controlled by variation at a single locus, and if there are only two common alleles at that locus, then the phenotypes in the population may fall into two or three discrete categories (such as the light-blue or dark-red flowers in the *Phlox* example). If the genetic basis of the trait is more complex, or if the environment plays a stronger role, trait values are likely to vary more continuously. In such a case it becomes very difficult to determine the effects of individual genes from looking at phenotypic variation alone. For example, the height of a human being is strongly affected by that person's genes, but no one gene controls more than about 3 percent of the variation in height. Instead, hundreds of genes and environmental factors like nutrition affect human height. In such cases, to determine how much genetic variation is available for the evolution of a particular trait, we can measure the genetic variance or heritability of that trait. These measures describe the contribution of genetics to the variance among individuals in a given trait.

The phenotypic variance of a trait is simply the variance among individuals in a population of each individual's values of that trait. (Variance is defined in the same way as it is used in statistics, as a measure of the variation in a population.) This phenotypic variance is potentially due to both genetic differences among individuals—either at one or many genetic loci—and variation in the traits among individuals caused by differences in their environments. The relative importance of genes and the environment varies widely from trait to trait and from population to population.

From an evolutionary perspective, to describe the effects of genetic variation on phenotypes, the important issue is whether the results of selection in one generation can be inherited by subsequent generations. Therefore, we wish to describe the phenotypic variation among individuals that can be inherited by their offspring. We call such heritable variation the *additive genetic variance*. It can be less than the phenotypic variance and, indeed, usually is.

Often, the environment in which an organism develops affects the values of its phenotypic traits as much as or more than does its genotype. The variance among individuals in a trait caused by differences in their environment is called the *environmental variance* for that trait. By definition, the heritability of a trait in a population is the fraction of the total phenotypic variance that is additive genetic variance. The heritability is greater if the additive genetic variance is larger, but smaller if there are many strong environmental sources of variation for that trait. For example, within North America and Europe the heritability of human height is approximately 70 to 90 percent, meaning that the majority of variation in height in these populations is caused by genetic variation. In general, the larger the heritability of a trait in a population, the faster that selection can cause an evolutionary change in that trait.

Sometimes, the environment can produce unpredictable variation in a trait, but in other cases, particular environments can cause foreseeable changes in phenotype. For example, when humans grow up with better diets, they tend to be taller. Environmentally induced changes in phenotypes can also be adaptive; for example, when conditions are dry, plants tend to have a greater proportion of their biomass as roots, which increases their ability to capture valuable water. Changes in the phenotype that result from differences in the developmental environment are called *phenotypic plasticity*.

3. A MULTIPLICITY OF FORMS OF INHERITANCE

There are many ways in which DNA sequence variation can affect the phenotypic variance. At one extreme, some changes to DNA sequence probably have no effect at all on the phenotype; at the other extreme, changes to a single base pair can be the difference between life and death. Some alleles have strong effects regardless of the other alleles present in the same individual, whereas other alleles strongly depend on the presence of particular alleles at other loci to produce their effects. This latter interaction between genes is called *epistasis*.

Some alleles produce their effects regardless of the particular alleles with which they are paired at the same locus; such alleles are *dominant*. Other alleles must be present in a homozygous state to produce their effects; we call these alleles *recessive*. Some other alleles show intermediate effects in heterozygotes; such alleles are called *codominant*. The most fit allele can be either dominant or recessive, depending on the locus and the environment.

Each locus may affect many traits. For example, one allele at the gene *tb1* in corn, which was very important during its evolution to domestication under ancient artificial selection, causes the plant to have many branches rather than a single stalk, with more leaves, smaller leaves, smaller ears, and more tillers. When a single locus has effects on multiple traits, we say that it shows *pleiotropy*. Moreover, the dominance and epistasis of alleles may depend on which trait we are considering, because the interactions of alleles within and among loci may differ among traits. Genes are regulated in myriad ways, sometimes by sites close to the start of the coding sequence, sometimes by sites embedded in the introns (DNA between bits of coding sequence), and sometimes by genetic elements far away from the gene, even on different chromosomes. Regulation may depend on details such as whether a stretch of DNA is methylated, which can reflect epigenetic effects transmitted among generations.

Genetic effects can depend on the interaction of sequence differences at multiple sites, either within the same gene or between different genes. These sequence differences can be transmitted together if they are close to each other on the same chromosome, or break apart by segregation and recombination if they are farther apart. All genes are subject to mutation, and the rate of mutation and recombination can vary greatly across the genome. Nearly all genetic effects have the potential of being masked by changes in the environment.

All traits are different in the details of their genetic control. Sometimes this control is relatively straightforward, as when a single base-pair change has large and constant effects. Sometimes the factors affecting the development of a trait are very complex, as when hundreds or thousands of genes contribute small, interacting effects in addition to the multifarious effects of the environment. Sometimes the details matter to the study of their evolution, and sometimes they do not matter so much. Variation is both the fuel of evolution and its fascinating product.

FURTHER READING

Golnick, L., and M. Wheelis. 2008. Cartoon Guide to Genetics. New York: HarperCollins.

Griffiths, A. J., S. R. Wessler, R. C. Lewontin, and S. B. Carroll. 2007. Introduction to Genetic Analysis. 9th ed. New York: W. H. Freeman.

Hartl, D. L., and A. G. Clark. 2008. Principles of Population Genetics. Sunderland, MA: Sinauer.

Hoekstra, H. E. 2010. From mice to molecules: The genetic basis of color adaptation. In J. B. Losos, ed., In the Light of Evolution: Essays from the Laboratory and Field. Greenwood Village, CO: Roberts & Company.

Hopkins, R., and M. D. Rausher. 2011. Identification of two genes causing reinforcement in the Texas wildflower *Phlox drummondii*. Nature 469: 411–414.

II

Phylogenetics and the History of Life

David A. Baum

As laid out by Charles Darwin, evolutionary theory is built around two key postulates (see Section I: Introduction). First, features of living species were acquired over time by evolution along lineages that have branched to form the evolutionary tree of life. Second, the good fit between organisms and their current way of life is explained by natural selection (and variants such as sexual and group selection). Although evolutionary biology has grown significantly in the century and a half since the publication of *On the Origin of Species*, these two points still constitute the central canon of evolutionary biology. They are overwhelmingly supported by empirical data, and are both essential aspects of biological literacy. Mechanisms of evolution, including natural selection, will be covered in the remaining sections of the book. In this section, however, we focus specifically on common ancestry and evolutionary trees, the first of Darwin's postulates.

To motivate this section, and explain why it is placed so early in the book, consider the importance of common ancestry in the historical development of evolutionary theory. During his voyage on HMS *Beagle*, Charles Darwin was struck by changes in the living and fossil organisms he encountered in different parts of the world. These observations, combined with his already excellent knowledge of biological diversity, led him to "see" the pattern of evolutionary descent. He observed numerous patterns that would not be expected under the hypothesis of special creation, but could be explained readily if different species share descent from common ancestors (see chapter I.2). The discovery of *common ancestry*, with its implication that abundant evolution has happened, posed the question, What mechanism could explain such evolution while leading to a fit between organisms and their ways of life? This puzzle occupied Darwin for much of his scientific career, during which he also amassed further evidence for descent from common ancestry. However, it is fair to say that Darwin would have had no reason to begin his studies of pigeons, or to experiment on seed dispersal, and so on, if he had not first discovered the historical fact of evolution. In a book presenting our current understanding of evolutionary biology, it seems fitting, therefore, to begin by covering common ancestry and the history of life on earth, with subsequent chapters dealing with processes of evolution.

The first nine chapters in this section examine the methods by which scientists reconstruct and make sense of evolutionary history, whereas the last nine summarize what we now know about the evolutionary history of different parts of the tree of life.

A recurring motif throughout the chapters in this section is the study of phylogeny. *Phylogenies* are formal representations of Darwin's metaphorical tree of life. Chapter II.1 introduces phylogenies and how to interpret them. This is important, because for all their utility in evolutionary biology, phylogenetic trees are notoriously easy to misunderstand. This chapter introduces phylogenetic terminology and explains what information can and cannot be extracted from a tree diagram.

Chapters II.2 and II.3 introduce the methods used by scientists to obtain phylogenetic trees. The first of these chapters provides an overview of methods that aim to determine the correct tree topology, the ordering of lineage branching. It focuses on phylogenetic inference based on DNA sequence data, far and away the most widely used data source, and introduces the rigorous statistical methods that are now de rigueur in the field. Chapter II.4 extends this discussion yet further, focusing on the challenge of combining information from molecular data with that from geology or paleontology to infer the dates of branch points within a phylogeny. Such time-calibrated trees are especially useful for studies of the migration of lineages around the globe and those aiming to quantify the rates at which different lineages have diversified.

Chapters II.4–II.7 are concerned with inferences about evolutionary history that can be made once phylogenetic trees have been determined. Chapter II.4 deals with historical biogeography, which entails using phylogenetic information to study the mechanisms (e.g., range expansion, range contraction, or long-distance dispersal) by which taxonomic groups came to have their current geographical distributions. The chapter provides an overview of the history of biogeography, a summary of some well-established biogeographic phenomena, and perspectives on controversies and future directions. Chapter II.5 follows nicely by considering phylogeography, a field that merges some aspects of biogeography with insights from population genetics (discussed in Section III: Natural Selection and Adaptation, and Section IV: Evolutionary Processes) to understand the spatial distribution of genetic variation within and among populations. Phylogeography is especially effective at studying recent migration patterns and patterns of gene flow among contemporary populations.

Chapters II.6 and II.7 take two different perspectives on the study of character evolution in the context of phylogenetic information. Chapter II.6 focuses on statistical analyses that can be used to reconstruct ancestral states or to study the dynamics of trait evolution. Such statistical tests are important for, among other things, determining whether pairs of traits show correlated evolution or whether certain traits have altered the rate of speciation and extinction. Chapter II.7 takes a more conceptual approach, clarifying the macroevolutionary patterns that may be seen when looking across phylogenetic trees and providing a guide to the terminology used by biologists to describe these patterns, including such important terms as *adaptation* and *homology*.

Chapter II.8 considers another use of phylogenies: for organizing knowledge of diversity in the form of a taxonomy. The chapter charts the historical shift from taxonomy as a system for capturing the mind of the Creator to taxonomy as a reflection of evolutionary history. It also provides an introduction to phylogenetic nomenclature, an approach to connecting names to biological taxa, and explores some of the recent controversy this approach has engendered.

The aforementioned chapters in the section focus on the effort to make historical inferences by studying living organisms, largely overlooking paleontology (with the exception of some mentions of fossils in the context of molecular dating and historical biogeography). Chapter II.9, therefore, summarizes what can be learned by careful and critical analysis of the fossil record. These insights into paleontological methods and the geological time line provide the last critical pieces of information needed to delve into the actual history of life on earth, as outlined in chapters II.10–II.18.

The nine chapters on evolution of life on earth cover many of the most remarkable evolutionary transitions and describe the history of several of the most familiar and successful groups of living organisms. These chapters begin with the origin of life (see chapter II.10) and end with the origin of humans (see chapter II.18); however, this ordering should not be taken as endorsing the misguided view that humans are inherently more advanced than other living groups. While we are telling the story from the vantage point of humans by exploring events and groups in order of decreasing phylogenetic distance from the human lineage, this telling is not evolutionarily privileged. We could equally validly reorder the narrative around a particular group of bacteria, or plants, or snails, or whatever. That we did not do this reflects our sense that humans naturally tend toward linear stories, and furthermore stories that end at things they consider most important, in this case, humans themselves.

The first chapter in the section, chapter II.10, reviews current knowledge of the origin of life. While there is no consensus view, we now know a lot about the context and timing of life's emergence, and increasingly sophisticated and compelling chemical models have been developed over the last couple of decades.

Chapter II.11 examines the prokaryotic grade, composed of bacteria and archaea, and explores their mode of evolution, including their propensity for lateral gene transfer. Often underemphasized because of their diminutive size, prokaryotic organisms play diverse, important ecological roles and encompass tremendous biochemical diversity. Similar themes emerge in chapter II.12, which examines the origin of eukaryotes and diversification of protists, which is to say eukaryotes that are not animals, fungi, or plants. The remarkable diversity and ecological significance of protists deserves much greater attention than is typical in biological education.

Chapter II.13 explores the origin and diversity of land plants. The invasion of land by green plants was a momentous event in the history of life on earth, eclipsed only by the origin of oxygenic photosynthesis in cyanobacteria (which established an oxygen-rich atmosphere). The radiation of land plants not only permitted the subsequent diversification of animals on land but also allowed for the evolution of diverse interactions with animals and fungi, including those involved in pollination and dispersal. Similarly, fungi, covered in chapter II.14, have played an important role in terrestrial life, especially as decomposers and through biotic interactions, ranging from pathogenic to mutualistic, such as root associations called *mycorrhizae*. Fungal phylogeny is now well understood, including the insight that fungi are closely related to animals, but there is evidence that much of fungal diversity remains to be characterized.

Chapters II.15–II.18 all deal with animals, the most apparent of the major branches of the tree of life. Chapter II.15 sets the stage by exploring the origin of animals from single-celled ancestors and the broad sweep of their diversification into major lineages. A central focus is on the dramatic, yet incomplete, progress that has been made in resolving relationships among the major animal groups.

Chapters II.16 and II.17 delve into two major *clades* (i.e., evolutionary lineages) that have been particularly successful and influential. Arthropods are numerically the most diverse animal phylum, including as they do arachnids, crustaceans, and above all, insects, yet until quite recently there was great uncertainty as to relationships among its major subgroups. As shown in chapter II.17, this contrasts with the situation with tetrapods, whose evolutionary history has been well understood for some time thanks to a rich fossil record and abundant, careful morphological work. Nonetheless, even in tetrapods, modern molecular data have been influential in shaping our understanding of the connection between different lineages and the mechanisms by which certain interesting traits evolved. Finally, chapter II.18 lays out the current state of play in the dynamic field of human evolution, showing how assorted lines of evidence, especially paleoanthropology and genomics, are converging to provide an ever richer and more complete understanding of how evolution came to yield a species that would develop the intellectual capacity to ponder its own evolutionary origin.

The diversity chapters in this section certainly fall far short of communicating all that we now know about the history of life on earth; nonetheless, taken together, they powerfully show how the use of methods described in the first half of the section can help us gain solid insights into the origins of the remarkable diversity of living organism that have existed and do exist on this planet.

II.1

Interpretation of Phylogenetic Trees

Kevin E. Omland

OUTLINE

1. Introduction to phylogenetic trees
2. Misreading trees with species-poor lineages
3. Reading trees correctly: Ancestral state reconstruction
4. Understanding the process of evolution: We are all cousins

All organisms on earth share common ancestry; we are related to every species that has ever existed. Evolutionary biologists since Darwin have sought to infer a "tree of life," a phylogenetic tree showing how all species are related to one another. The concept of phylogeny as the evolutionary history of organisms—and phylogenetic trees as a depiction of that history—is central to evolutionary biology. Phylogenies form the basis for our understanding of relationships among organisms, and they are key tools of modern evolutionary research; however, phylogenetic trees are frequently misinterpreted because of fundamental misconceptions about what trees can and cannot tell us. In particular, people frequently misinterpret trees by reading trees "laterally" from one extant species to the next. This tendency results partly from mistakenly thinking of evolution as a "ladder of progress." Ultimately, a well-informed interpretation of phylogenetic trees goes hand in hand with a clear understanding of the process of evolution.

GLOSSARY

Ancestral State Reconstruction. A procedure that uses a phylogeny as well as character data from extant species to infer likely ancestral character states.

Character State. Alternative states for a given biological character (e.g., brown, blue, hazel are character states of the character eye color; A, C, G, or T are character states of the character corresponding to base position 12 in the actin gene).

Chronogram. A phylogenetic tree with branch lengths scaled to represent time (related to *phylogram*, a tree in which branch lengths are scaled to the amount of character evolution).

Cladogram. An unscaled phylogenetic tree that shows relationships among organisms, but in which branch lengths are meaningless. The key information retained in a cladogram is *topology*, which refers to the composition of clades/monophyletic groups and how they are related to one another.

Derived. Generally the opposite of ancestral; a derived character state is one that has evolved recently relative to an ancestral character state (e.g., scaly skin is an ancestral character state for reptiles, whereas feathers in avian reptiles represent a derived character state).

Monophyletic Group (Clade). A group of species including a common ancestor and all its descendants; a "natural group" with all members more closely related to each other than to any other species (compare with *paraphyletic*).

Paraphyletic Group. A nonmonophyletic group that includes a common ancestor but leaves out some descendants (e.g., "reptiles" leaving out birds). Sometimes contrasted with polyphyletic groups, which include two or more different ancestors and their descendants, for example, bats plus birds without their common ancestor.

Phylogenetic Tree. A branching diagram showing relationships among organisms (e.g., frequently among different species, or among different individuals).

Phylogeny. The evolutionary history of a group of species (or any set of taxa, genes, or tips).

Root Node. The ancestral node at the base of a tree, representing the most recent common ancestor of all species included in that tree.

Shared Derived Character State (Synapomorphy). A character state that defines a monophyletic group (e.g., the presence of mammary glands for mammals).

Species Tree. Phylogenetic tree showing relationships among species, generally based on multiple independent genes (distinguished from *gene tree*, a tree showing relationships based on one gene from multiple individuals, or showing relationships among multiple paralogous genes).

1. INTRODUCTION TO PHYLOGENETIC TREES

Phylogeny is a fundamental concept, and phylogenetic trees are important tools in evolutionary biology; thus, it is crucial to read phylogenetic trees in a way that aligns well with an accurate understanding of the overall process of evolution. Although the next two chapters focus on how to construct trees, this chapter is about what those trees represent and how to interpret them. The challenge of interpreting trees is present even if we know the correct tree (e.g., for animal breeds), or when we have inferred a well-supported uncontroversial tree. Most biologists need not know how various phylogenetic algorithms work, but all biologists should know what trees represent and how trees inform our understanding of the process of evolution.

What does a phylogenetic tree represent? What can a phylogeny tell us? Figure 1A depicts a series of hypothetical speciation events. The arrow at the left labeled X represents a mouse species that lived 4 million years ago. The mouse populations evolved as a result of mutation, genetic drift, and selection for 1 million years as time moved from left to right. At 3 million years ago, an evolutionary fork in the road occurred, some geological change such as a rise in sea level divided the ancestral species into two descendant species, species Y and Z. These species evolved for another 1 million years before a mountain chain formed in the range of species Y, leading to the formation of species A and B. Later, a river formed in the range of species Z, leading to species C and D. Thus, four species are present today. The drawing in figure 1A includes extra information about ancestral species and speciation events, but basically it is a phylogenetic tree. In this tree, the branch lengths are drawn to scale based on time, so this is a scaled tree, specifically a *chronogram*.

Frequently, phylogenetic trees focus on living species. In figure 1B, only species A, B, C, and D are shown, and the branch lengths are not drawn to scale. Such unscaled trees are known as *cladograms*, which focus on the most important information, the evolutionary relationships or tree *topology*. There are essentially only two such pieces of information in this tree. First, A and B are more closely related to each other than either is to C or D. Second, C and D are more closely related to each other than either is to A or B. Crucially, in the language of phylogenetic systematics, "more closely related" means "shares a more recent common ancestor." Cladograms depict the tree topology, which shows the monophyletic groups or clades. Clades are composed of organisms that are more closely related to each other than to any organisms outside the clade, Two clades are shown in figure 1B: (A, B) and (C, D). (The tree in figure 1B also shows that A, B, C, and D all share a common ancestor, but this third piece of information is trivial: we already know that all species on earth share a common ancestor.) Thus, there are only two useful data points on this tree: the two monophyletic groups. Anything else that one thinks can be inferred from this cladogram is overinterpretation.

Many botanical terms used for real trees are used metaphorically for phylogenetic trees. Figure 1C shows the same information as figure 1B, although now as a slanted vertical tree. The "root" is at the base of the tree, representing the furthest point in the *past*. The path of evolution can then be traced along "branches" ("internodes"), always moving away from the root toward the *present*. Branches split at "nodes" ("internal nodes") that represent speciation events. Finally, terminal branches lead to "tips" ("leaves" or "terminal nodes"), which frequently represent extant species that exist in the present (species A–D in this case). The tips of the tree generally represent a single species or a species as a representative of a larger group; however, tips sometimes represent an individual organism or an allele in a gene tree. The trees in figure 1 show each ancestral node with two descendant branches—these trees are bifurcating trees. (Note that trees can have nodes with three or more descendant branches; such nodes are called *unresolved*.)

Evolutionary Trees Have No Trunks

One key botanical term is missing from phylogenetic trees: there is never a trunk, never a main stem, never a main branch. To look at a phylogeny and see a "main path" of evolution is to be misled. Understanding that from every node there are two descendant branches, both of which lead to continued evolutionary change, is key to understanding phylogenetic trees. More important, understanding that there is no trunk is crucial for a clear understanding of the very process of evolution.

Other aspects of the tree metaphor can also lead to misinterpretation, especially the idea that some extant species might be "lower down on the tree," either within a branch, or at the tip of a branch that terminates low down, closer to the root. It is important to trace history starting from the root of the tree and moving toward the branch tips; thus, the lineages are evolving as time moves from the past toward the present. For example, on a vertical tree such as figure 1C, evolution moves from the bottom to the top along the *y*-axis. Critically, the axis

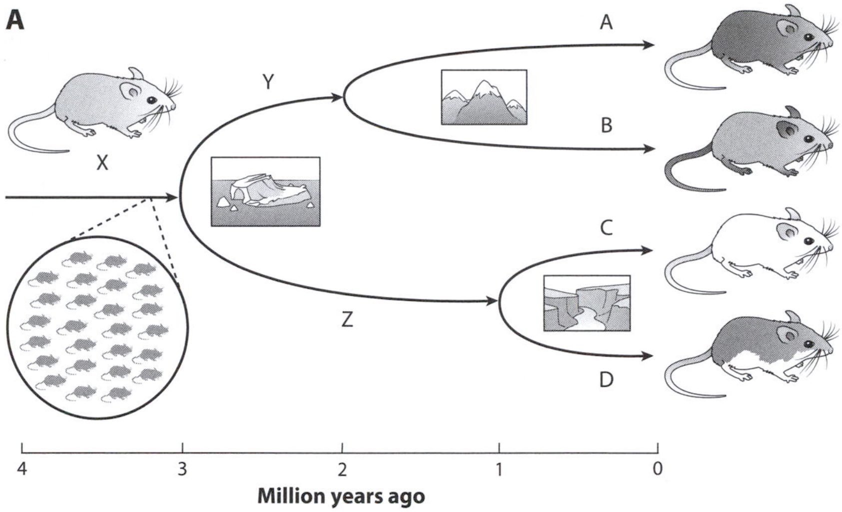

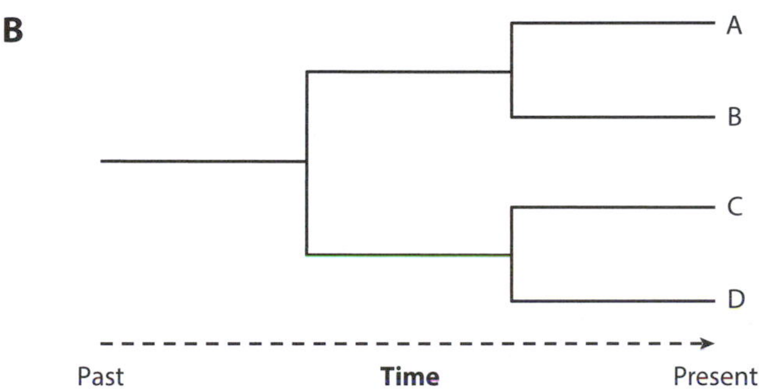

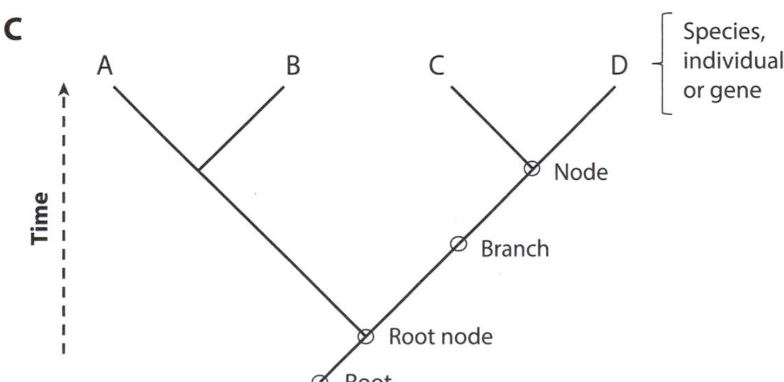

Figure 1. (A) Hypothetical speciation scenario showing one ancestral mouse population repeatedly diverging, leading to four extant species. Arrows represent the paths of evolution from the common ancestor to the descendant species. These arrows form a phylogenetic tree showing relationships among extant species A–D. (B) Relationships among species A–D shown as a horizontal cladogram, an unscaled tree with no information about time of divergence. (C) Vertical slanted cladogram of the same species showing key botanical terms used to label phylogenetic trees.

perpendicular to time has no meaning whatsoever (the apparent "*x*-axis" left to right on a vertical tree). That this axis perpendicular to time has no meaning and that there is no trunk are best illustrated by the concept of branch rotation.

Branches Rotate around Nodes

One of the most important points to understand about phylogenetic trees is that branches can be rotated around any node. David Baum and colleagues (2005) published a perspective article in *Science* titled "The Tree-Thinking Challenge" that made this very point about branch rotation. The simple figures of Baum et al. illustrate how even the most basic information in trees can be misinterpreted (redrawn in figures 2A and 2B). In figure 2A, frogs are next to sharks, so one might think that frogs are more closely related to sharks than to humans; however, figure 2B shows the exact same evolutionary relationships, yet frogs are now furthest from the shark, illustrating that proximity on the page has no meaning when reading evolutionary trees. Clearly, frogs share a more recent common ancestor with other four-limbed animals (tetrapods) than they do with sharks.

In figure 2A, one might think the straight line going from the root to humans would represent a trunk or a main branch; however, because the branches have been rotated in figure 2B, there is no straight line, making it more apparent that there is no trunk. There is no way to determine where evolution is heading. Many people incorrectly think of humans as the end point of evolution, but clear tree thinking can help dispel this notion. In this tree, humans could go in any one of the five positions left to right. Think of trees as being like a mobile, with each of the species able to spin around and appear in any position, yet their connections to each other remain the same. Branch rotation emphasizes the fact that there is no way to find a "main path" of evolution (whether leading to humans or to any other favorite organism) because there is no main path.

2. MISREADING TREES WITH SPECIES-POOR LINEAGES

Phylogenetic trees are most confusing when the trees are "unbalanced," when one side of the tree has few species, but the other side has many species. For example, in figure 2A, the shark branch to the left of the root node is depicted as leading to one descendant species, whereas the branch to the right of the root node is shown leading to four descendant species. Many people mistakenly associate the "species-poor lineage" (e.g., the shark at the left in figure 2A) with the root node (also known as the basal node). This misconception occurs largely because no nodes are *depicted* that separate this basal node from the extant species. A group of biologists centered in Canberra, Australia, published a series of papers explaining the problem of this "basal fallacy." Frank Krell and Peter Cranston (2004) asked, "Which Side of the Tree Is More Basal?" The answer is neither: all extant species are equally distant from the base of the tree, so no extant species should ever be considered basal. For example, with reference to figure 2, the great white shark and humans shared a common ancestor approximately 450 million years ago. We should remember that both lineages have been evolving for the same amount of time since that common ancestor.

Mike Crisp and Lyn Cook (2005) went on to highlight the most serious problem with this way of misreading a phylogenetic tree (e.g., left to right on vertical trees). Because many people associate the species-poor lineage with the base of the tree, they incorrectly assume that these species retain more ancestral characteristics (e.g., the species on the left in figure 2). Crisp and Cook focused on these species-poor lineages that may *seem* to "branch off early." They asked, "Do early branching lineages signify ancestral traits?" They answered *no*—both lineages that descend from the ancestral node continue to evolve by mutation, drift, and selection. As both lineages continue to encounter new environments, food sources, predators, pathogens, etc., they will continue to evolve adaptations—no lineage of organisms stops evolving. Every species is a mix of retained ancestral, shared derived, and uniquely derived characteristics. As an example, Crisp and Cook highlight the platypus, which retains the ancestral tetrapod trait of egg laying, but has evolved many derived character states, including the flattened bill with electrosensory capabilities and a complicated system of 10 pairs of sex chromosomes. In contrast, humans and other placental mammals retain many ancestral tetrapod traits, including several unspecialized jaw characteristics and a single pair of sex chromosomes.

Misreading Trees as Ladders of Progress

Omland, Cook, and Crisp (2008) pointed out that deep-seated biases in the way people think of evolution feed into the misreading of phylogenetic trees. Many people think of evolution as "progressing" linearly from "simple organisms" to more "complex species," then eventually to "the most advanced species," which, unsurprisingly, humans consider to be humans. This flawed view of evolution as a "ladder of progress" can be traced back to Aristotle and Linnaeus, but it is especially apparent in work of the early evolutionary biologist Ernst Haeckel. Although he coined the word *phylogeny*, and he helped bring trees to the study of evolution,

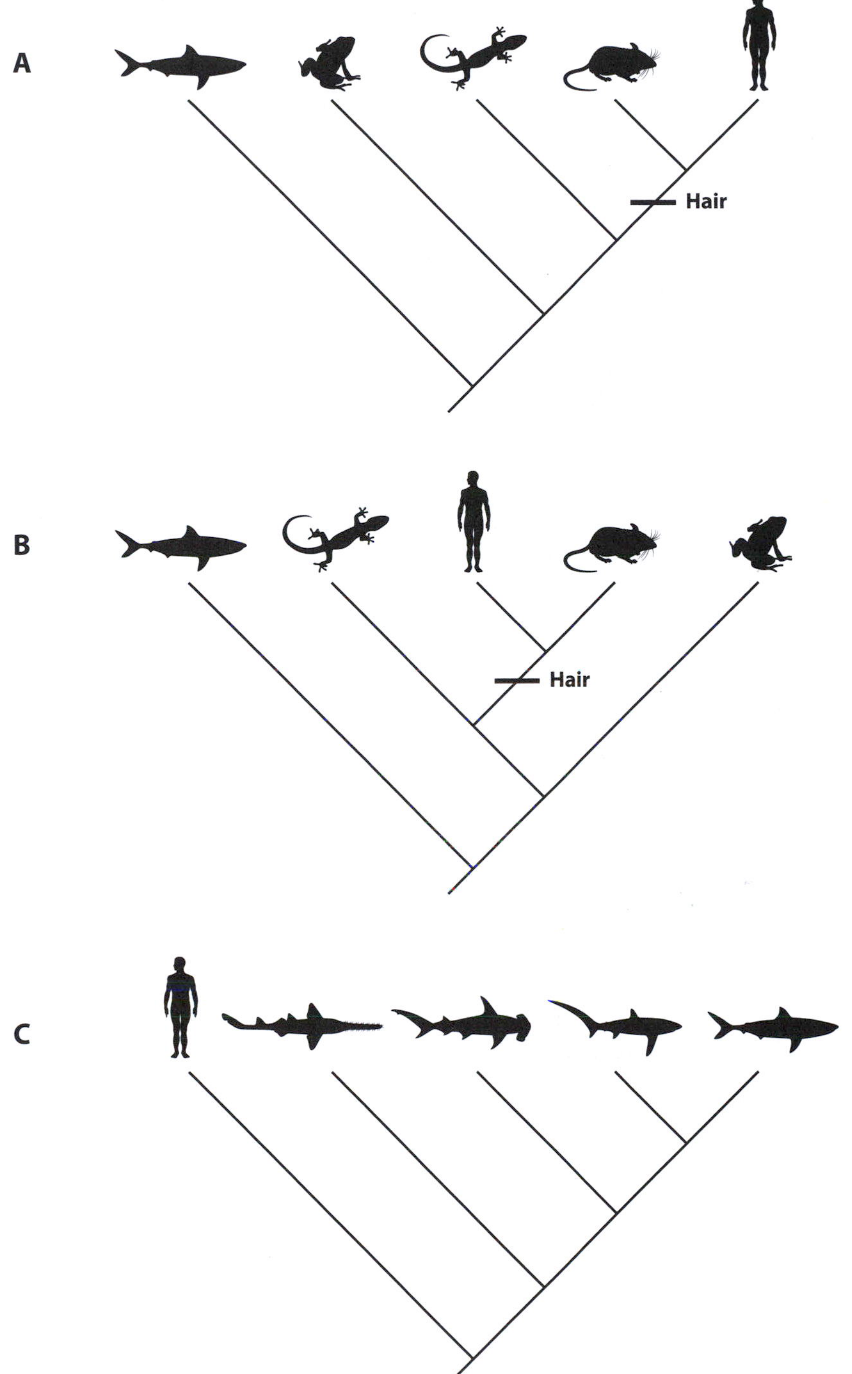

Figure 2. (A) Cladogram showing relationships among five vertebrate lineages: human (*Homo sapiens*), mouse (*Mus musculus*), lizard (*Anolis carolinensis*), frog (*Rana pipiens*), and great white shark (*Carcharodon carcharias*). (B) Rotated cladogram showing the same relationships among the same species, but with branches rotated around each node. (C) Cladogram showing relationships among five vertebrates, but this time the focus is on sharks, with humans included simply as an outgroup. Sharks depicted left to right are sawshark (*Pristiophorus cirratus*), hammerhead (*Sphyrna zygaena*), thresher shark (*Alopias vulpinus*), and great white shark (*Carcharodon carcharias*). (Note that the great white shark lineage and the human lineage are equally old: since they last shared a common ancestor approximately 450 million years ago, both lineages have continued to evolve.)

his "Pedigree of Man" (1866) furthers the mistaken viewpoint that humans are the end point of evolution. Haeckel's tree has a trunk containing extant taxa depicted as ancestral to other extant taxa: "Monera" ancestral to "Amoebae" at the bottom, later with "Amphibia" ancestral to "Pouched Animals" then "Semi-Apes" (lemurs), eventually leading to "MAN" at the very top.

This ladder-of-progress view of evolution is still prominent in biology; species that are mistakenly considered "primitive" are generally shown at the left of vertical trees, whereas species thought of as "advanced" are shown at the right. A major reason species are considered primitive is that they are members of species-poor lineages: monotremes such as the platypus in mammals, ratites such as the ostrich in birds, tuataras relative to lizards and snakes, and mosses among land plants.

Modern genome comparisons and studies of molecular evolution have revealed that our perceptions of which species are "primitive" versus "advanced" (frequently based on a small number of morphological characters) are not borne out at the level of the genome. For example, despite the tendency to assume that humans are "advanced" and chimps are "primitive," a comparison of 14,000 genes revealed that the chimp genome has substantially more genes with evidence of positive natural selection. Likewise, classic work by Alan Wilson, John Avise, and others found no evidence that species considered to have slow morphological evolution (e.g., horseshoe crabs) have slower rates of molecular evolution. Furthermore, no overall genome measures (e.g., total genome size) correspond well to typical human perceptions of which species are considered "more advanced," "more evolved," or "more complex." Many of these words or concepts are also hard to define or are value laden. The word *primitive* carries a strong connotation of inferior, and *advanced* carries the opposite connotation. The words *ancestral* versus *derived* are related terms that can be used for *individual character states*. But as Crisp and Cook (2005) pointed out, even ancestral and derived should not be applied to an *extant species* for two related reasons.

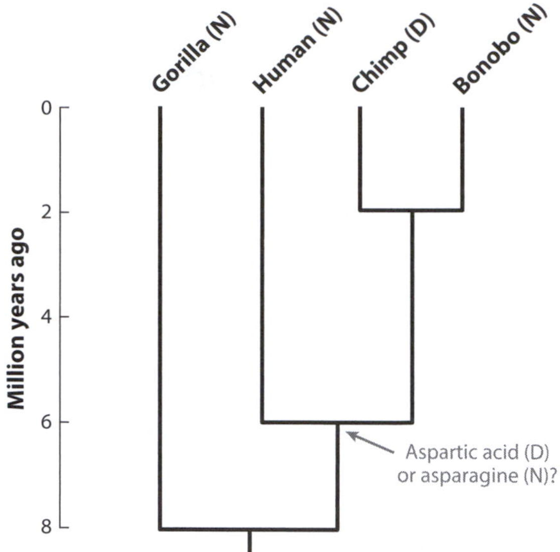

Figure 3. Chronogram showing rough divergence times of the four main lineages of African apes. Scientific names left to right: *Gorilla gorilla*, *Homo sapiens*, *Pan troglodytes*, and *Pan paniscus*. Also shown, amino acids at position 71 in the green sensitive opsin: N = asparagine, D = aspartic acid. (To practice tree thinking, replace the apes in figure 3 with members of the present generation of your family—siblings and cousins. In particular, in place of the gorilla, put one of your second cousins—your parent's cousin's offspring. Can you tell from the tree whether your second cousin is any more "primitive," older, or more ancestral than you are? This analogy emphasizes how flawed it is to even ask such questions about extant species based on a tree.)

Phylogenies Do Not Indicate Which Extant Species Are "Ancestral" or "Older"

Generally it is best to assume that no extant species is ancestral to another extant species. For example, a common misconception is that chimpanzees are ancestral to humans. The human and chimp lineage shared a common ancestor roughly 6 million years ago (figure 3), but that common ancestor was neither a chimp nor a human (as recent fossil finds of the extinct hominid *Ardipithecus* strikingly demonstrate). Thinking that chimps are ancestral to humans assumes that the chimp lineage stopped evolving either before or right after the human/chimp split. In fact, at least as measured by number of genes subject to selection, the chimp lineage has evolved more than the human lineage since their split.

For similar reasons, phylogenies do not indicate which species are "older" versus "younger." Perhaps one might think the age of a species can be inferred by determining when it last shared a common ancestor with another species. Species at the left of typical vertical trees could *seem* "older," and species at the right with other close relatives on the tree could *seem* "younger." A dated chronogram of African apes shows the chimpanzee/bonobo split at the far right at approximately 2 million years ago, and the human/chimpanzee lineage split at approximately 6 million years ago (figure 3; see chapter II.4 for the way such dates are inferred). Does that deeper split mean that humans must be "older" than the other two apes? Does that deep split mean that the species *Homo sapiens* is 6 million years old? Furthermore, if bonobos went extinct, would that suddenly make common chimpanzees several million years older? The answer to all these questions is no—one can tell very little about the "age of species" from a phylogenetic tree. Evolution is a continuous process—allele frequencies are constantly changing, and new character states are evolving—so it would be difficult to determine when one species ends and another begins. Even if one were somehow able to directly observe the process, the origins of livings species would be hard to associate with a specific moment in time.

Taxon Sampling, Tree Focus, and Extinction Affect Which Species *Seem* "Primitive" or "Older"

Which species are sampled ("taxon sampling") can dramatically affect a tree's appearance. For example, a systematist might include only a few species from a species-rich but difficult to sample genus from central Africa. Because of incomplete taxon sampling, this lineage would appear to be species poor and might thus be shown on the left side of a vertical tree; hence, it might *seem* to be "earlier branching" or "older." In fact, every tree ever drawn suffers from incomplete taxon sampling, so there are always missing tips and missing nodes. For a complete tree, at the very least you would want to include every one of the millions of extant species on the planet.

During phylogenetic inference (see chapter II.2), the position of the root of a tree is frequently inferred by comparing the focal group (the in-group) to a few closely related taxa (the out-group). As a result, out-groups frequently appear at the left side of vertical trees, and might therefore *seem* "primitive," "simple," or "ancestral." But as Omland et al. (2008) detail, out-groups are not chosen because they are "ancestral" in some way. Out-groups are chosen because they are closely related to, but not within, the focal in-group.

Returning to figure 2, we can compare a tree focused on humans to a tree focused on sharks. For a systematist interested in relationships among tetrapods, the in-group includes two mammals, a lizard, and an amphibian, with a great white shark as an out-group (figure 2A); thus, the shark would appear at the left and might in some way *seem* "ancestral to tetrapods." However, a marine biologist interested in shark relationships would include multiple sharks in the in-group, and humans could serve as a valid out-group (figure 2C). Thus, in a tree focused on sharks, *Homo sapiens* would appear at the left, which by the same incorrect logic might make humans *seem* "primitive" or "ancestral to sharks." As in this example, humans will be on the species-poor side of many trees; knowing which lineage is species poor does not indicate how "old" or "primitive" the species are in that lineage.

Over time, the process of evolution can lead to changes in which one of two sister lineages has more species. Rates of speciation can speed up and slow down, and extinction can dramatically reduce the number of species in a lineage. Extinction can cause a lineage that once was species rich to become the species-poor lineage millions of years later. Thus, a species-poor sister group can at a later time be the more species-rich lineage. To summarize, left to right order on a vertical tree has no meaning because branches can be rotated around nodes, and because taxon sampling and extinction strongly influence which lineage appears species poor. More generally, the number of species in a lineage does not indicate whether a given characteristic of a species will be ancestral versus derived.

3. READING TREES CORRECTLY: ANCESTRAL STATE RECONSTRUCTION

Given that phylogenetic trees cannot indicate which species are "primitive" versus "advanced," ancestral versus derived, or older versus younger, we can now focus on what it is that trees *can* tell us. Evolutionary trees are tremendously powerful tools for helping us understand evolutionary history, but the focus must be on learning about *characters*. Although trees cannot tell us which *species* are ancestral versus derived, trees can be used to infer which *character states* are ancestral versus derived.

Consider the amino acid sequence of a visual protein, green-sensitive opsin. In amino acid position 71, humans have asparagine (N), whereas chimps have aspartic acid (D) (figure 3). Clearly there has been an evolutionary change somewhere along either the chimp or the human lineage, but did the common ancestor have N or D? Using outdated ladder-of-progress thinking, we might think that chimps would retain the ancestral state, so that the chimp/human ancestor would have had D. Using a phylogenetic approach, we know gorillas are the sister to the human/chimp clade, and gorillas have N. This observation makes it more likely that N is the ancestral state—one change in the chimp lineage is probably more likely than two independent changes in the human and gorilla lineages. This inference is further supported because bonobos and other apes also have N at that site.

Figure 2A illustrates another example of ancestral state reconstruction involving the presence or absence of hair. To be precise, the "character" is body covering, and the "character states" are hair versus no hair. Three of the species have no hair, whereas both mice and humans have hair. On this tree, mice and humans are each other's closest relatives (they form a monophyletic group), so it seems reasonable to infer that hair evolved in the common ancestor of the two mammal species (as indicated by the horizontal line). This reconstruction is more *parsimonious* than the alternative hypothesis—it requires fewer evolutionary changes (only one change).

In many cases, evolutionary biologists use the principle of parsimony to infer ancestral characteristics. Parsimony, which is based on the principle known as *Occam's razor*, assumes that the simplest explanation is the one most likely to be true. It is possible that the two mammal lineages evolved hair independently, but such a reconstruction would be less parsimonious because it would require two evolutionary changes. Data from thousands of other species and from fossils support parsimony in this simple case of mammalian hair. It is worth noting that the ancestral state is directly determined not by which state is the most common but by the distribution of states across the tips (see chapter II.8 for other ways of studying ancestral states).

A useful exercise is to look for patterns of nesting of taxa with certain character states. In figure 2A, the species with hair form a monophyletic group nested within a paraphyletic assemblage of vertebrates without hair. Frequently, the ancestral state can be inferred when character state A is shared by a series of species-poor lineages, with a nested focal clade having character state B. That mammals are nested within nonhairy tetrapods may be even more apparent when the nodes are rotated in a zigzag fashion (figure 2B).

Shared Derived versus Shared Ancestral Traits

Hair is a good example of a trait that is a shared derived character state (*synapomorphy*) of mammals. Hair marks a clade, which means that all organisms with hair are closely related to each other; however, a trait can lead one astray if it is a shared ancestral trait (*symplesiomorphy*). Scaly skin is an ancestral characteristic shared by lizards, turtles, and crocodiles. This trait contributed to misleading taxonomy for hundreds of years. Data from morphology, fossils, and DNA sequences now strongly indicate that "scaly reptiles" do not form a clade; rather, crocodiles are more closely related to birds than they are to lizards and snakes, for example. The data show that birds are clearly part of the reptile clade. This example illustrates that shared derived states provide evidence of evolutionary kinship, whereas shared ancestral states do not. (Figure 2A shows the series of lineages on the left with the shared ancestral trait of "no hair.")

It is important to emphasize that one lineage at the "left" of a vertical tree does not necessarily retain the ancestral state for any given character. For example, in figure 2A, sharks have a cartilaginous skeleton, whereas all the other lineages have bony skeletons. But that fact alone does not mean that the common ancestor at the root node had a cartilaginous skeleton. In fact, considering the bony skeleton of other living and extinct chordate species indicates that the common ancestor had a bony skeleton that became cartilaginous in the shark and ray lineage. Again, it is crucial to avoid the fallacy of thinking of the species at the left as "basal" or "primitive."

Figure 2C shows another example with a species that lacks a tail—human—as the out-group; the four shark species all clearly have a tail. But that tree does not indicate that the common ancestor of these five vertebrates lacked a tail. Presence or absence of a tail at the root node are two equally parsimonious reconstructions (each requires one change of character state). In fact, in this case we know from fossils and additional taxon sampling that the common ancestor had a tail, which was recently lost in the lineage leading to humans and other apes.

4. UNDERSTANDING THE PROCESS OF EVOLUTION: WE ARE ALL COUSINS

In conclusion, phylogenetic trees have a wide range of uses in evolutionary biology and systematics, and subsequent chapters focus on many of these topics, including molecular dating, biogeography, and phylogenetic comparative methods (see chapters II.4–II.8). A central goal of phylogenetics is to define monophyletic groups and sister group relationships, which can be used by taxonomists to develop stable and predictive classifications (chapter II.9 and subsequent chapters); however, knowing how to interpret the basic structure of trees and how to avoid the many widespread misconceptions about trees is a necessary first step for all of these applications.

More generally, knowing how to interpret trees leads to a better understanding of the process of evolution. One might incorrectly think that evolution is goal directed—that natural selection is "advancing" evolution in some way toward "more evolved" characteristics. Understanding phylogenetic trees provides an excellent opportunity to understand that evolution is not proceeding from one extant species to the next toward species with human-like characteristics. Phylogenetic trees are frequently depicted with species unlike humans off on the left side, with humans on the far right. By understanding *tree thinking*, we know that node rotation and taxon sampling can make species appear almost anywhere left to right on a vertical tree; however, some lineages really are species poor; such species-poor sister groups will always be off on the side. But, it is crucial to emphasize that such side lineages do not represent "primitive" early evolutionary stages; side lineages need not retain the ancestral state for any given character.

Focusing on the extant species at the tips of the branches, it is important to remember that relationships among extant species are cousin relationships, not ancestor-descendant relationships. Thus, phylogenetic trees of extant species emphasize relationships among cousins. Our closest cousins are chimpanzees and bonobos, followed by gorillas, then other primates, other mammals, etc. Chimpanzees are our cousins, not our ancestors. Natural selection has molded adaptations that have enabled each of our cousin species to persist. Each extant species depicted at the tips of phylogenetic trees has a unique combination of ancestral and derived characteristics. Knowing how to read the tree of life—which Darwin first conceived of more than 150 years ago—enables us to better understand evolutionary history and better appreciate our common ancestry with all life on earth.

FURTHER READING

Avise, J. C. 2004. Molecular Markers, Natural History and Evolution. 2nd ed. Sunderland, MA: Sinauer.

Baum, D. A., S. D. Smith, and S. S. Donovan. 2005. The tree-thinking challenge. Science 310: 979–980.

Baum, D. A., and S. D. Smith. 2013. Tree Thinking: An Introduction to Phylogenetic Biology. Denver, CO: Roberts & Company.

Brooks, D. R., and D. A. McLennan. 1991. Phylogeny, Ecology, Behavior: A Research Program in Comparative Biology. Chicago: University of Chicago Press.

Crisp, M. D., and L. G. Cook. 2005. Do early branching lineages signify ancestral traits? Trends in Ecology and Evolution 128.

Maddison, D. R., and W. P. Maddison. 1992. MacClade: Analysis of Phylogeny and Character Evolution. Version 3.0 (user's manual). Sunderland, MA: Sinauer.

Omland, K. E., L. G. Cook, and M. D. Crisp. 2008. Tree thinking for all biology: The problem with reading phylogenies as ladders of progress. BioEssays 30: 854–867.

II.2

Phylogenetic Inference

Mark Holder

OUTLINE

1. Logical and statistical inference
2. The parsimony approach
3. Likelihood-based approaches
4. Distance-based approaches
5. Computational aspects of tree estimation
6. Statistical support for clades
7. Bayesian inference

A phylogeny describes the genealogical relationships between different species. In the early 1960s, many biologists were openly skeptical about the prospects of inferring reliable phylogenies. The last 50 years have produced a rich variety of statistical approaches for estimating evolutionary relationships and quantifying the degree of statistical support for different aspects of phylogenetic hypotheses. Current methods use powerful models of biological characters changing over evolutionary time to tease apart historical signals from similarities due to convergence. Today, phylogenetic inference is a routine part of many evolutionary studies, and phylogenies often provide a crucial framework for testing hypotheses.

GLOSSARY

Alignment. The process of adding gaps to DNA sequence data such that each column of the data matrix contains DNA bases that are homologous to each other (all derived from the same base in the common ancestor of the sequences). The aligned data matrix can be referred to as an *alignment*.

Character and Character State. In phylogenetic inference, a *character* refers to a comparable trait that can be studied in multiple species. Characters are hypothesized to be *homologous* (inherited from a common ancestral species). *Character state* refers to the specific form of the character observed in a species. For example, if the character is "number of limbs," the character state for horses would be 4.

Clade. A subtree in a phylogeny. A group of species delimited by an ancestral species and all its descendants.

Data Pattern. The pattern of character states for a set of species. Two different characters that display the same pattern will contain the same phylogenetic signal.

Homoplasy. The creation of the same character state more than once over evolutionary history. Homoplasy can mislead phylogenetic inference because it results in a similarity that is not evidence of a close evolutionary relationship.

1. LOGICAL AND STATISTICAL INFERENCE

A *phylogenetic tree* is a representation of the ancestor-descendant relationships between different species (see chapter II.1). We can collect data from extinct and extant species, and we can directly observe the genealogical relationships of individuals within a population, but phylogenetic relationships must be inferred.

Inference procedures can be classified as logical versus statistical. Logical inference has an appealing property: if our input "premises" are correct and our inferential rule is valid, then our logical conclusion must be correct. As a result, it is tempting to try to cast phylogenetic inference problems into the realm of logical inference; for example, one possible inferential rule might be the following:

> Rule #1: Any two species sharing a homologous attribute (a "character state") must be more closely related to each other than either one is to a species not sharing this character state.

Homologous, in this context, refers to attributes in different organisms that are similar to each other because they were inherited from a common ancestor (see chapter II.7).

One could collect data and arrange them into a matrix in which various columns represented distinct, heritable traits (such as the number of digits on a forelimb), with each row corresponding to a different species. By examining this data matrix and repeatedly applying Rule #1, we could build up a phylogenetic tree piece by piece. Our rule allows us to learn about a piece of the tree by looking for shared character states.

Consider a character corresponding to the concept of "number of limbs" and rows for a human, a lizard, and a snake. Humans and lizards share a trait (presence of four limbs). We are reasonably certain that the most recent common ancestor of humans and lizards had four comparable limbs; thus, the attributes we are scoring are homologous character states (determining which similar attributes represent homologous character states is not a trivial problem; see chapter II.7). It appears we should be able to use our rule to infer that humans are more closely related to lizards than either is to snakes; unfortunately, this result is incorrect. Analysis of other characters (such as the structure of the male reproductive organs, the hemipenes) immediately leads to a conflicting conclusion that lizards and snakes are more closely related to each other than either are to mammals. Clearly, Rule #1 is too simplistic.

Willi Hennig, a German entomologist, dramatically clarified the logic of phylogenetic inference by demonstrating that Rule #1 cannot provide a firm foundation for reconstructing a phylogeny. Hennig pointed out that if we modify the rule to use only character states corresponding to evolutionary novelties ("apomorphies" in his terminology), we ought to arrive at correct inferences. In the context of the human/lizard/snake example, the novel character state is the lack of limbs in snakes. Presence of limbs is not a novelty, because the most recent common ancestor of all three groups had limbs. So, this similarity in homologous traits is not a reason to posit a close relationship between humans and lizards; however, the presence of evertible hemipenes in lizards and snakes is an evolutionary innovation (relative to the most recent common ancestor of humans, lizards, and snakes). Thus, Hennig's approach groups snakes and lizards as more closely related to each other.

Yet even in this simple example we can see some difficulties. Under Hennig's rule, we need a data matrix that makes statements about which character states are homologous and which are ancestral rather than derived (the "polarity" of characters). Before knowing the phylogenetic tree for a group, it is hard to imagine being certain of the attributes of an ancestral species. While we can use information from developmental biology, detailed structural analysis, paleontology, and comparisons to more distantly related organisms to inform polarity decisions, there is no way to avoid all mistakes. Indeed, when we look at real data sets, we almost invariably find conflict between characters. This would not happen if all our homology and polarity diagnoses were correct. When we are not 100 percent certain of our premises, logical inference cannot guarantee the correctness of conclusions. In fact, trying to conduct a logical analysis with conflicting premises will usually lead to no conclusion at all; instead, we must move to the realm of statistical inference.

In statistical estimation the observed data are used to determine the best-fitting value (or range of values) of an unknown quantity (a parameter). Statistical methods provide estimation procedures that account for the possibility of random errors, chance phenomena that might cause the observed data to be somewhat different than expected. For example, a statistical method would have to allow that flipping a fair coin could once in a while yield five consecutive heads. The value for the parameter that is considered optimal is referred to as the *estimate* of the parameter. The formula or procedure used to create the estimate is referred to as the *estimator*. In phylogenetic estimation, the tree and the lengths of its branches are the parameters of interest. In general, statistical estimators work by finding parameter values tending to produce data that is similar to the data that we have observed. This match between parameters and data can be assessed in a number of ways, leading to a wide variety of frameworks for developing estimators.

To move from a general description of statistical estimators to a concrete estimation procedure, we must describe the relationship between parameter and data in the case of phylogenetic inference. We must explicitly describe some sort of error model. Hennigian arguments predict that two closely related species will share derived character states that neither shares with more distantly related organisms. The prediction makes sense, because any novelties that evolve along a lineage should be inherited by the descendant lineages, and they should distinguish those descendant species from all other species; however, multiple changes in the same character can occur, and these will obscure the historical signal. For a full error model, we must also describe how data patterns other than "perfect Hennigian" characters can evolve.

An example using primate phylogeny. Consider an alignment of the entire mitochondrial genome sequence for a human, an orangutan, a baboon, and a squirrel monkey (a small portion of the alignment is shown in table 1). Based on other data, we are confident in the correct phylogeny for this group of species. A wide variety of morphological and biogeographic evidence support the hypothesis that, among these four species,

Table 1. Characters (also referred to as "sites") 21–35 of an alignment of mitochondrial genomes of four species of primates

	Character #														
Species	21	22	23	24	25	26	27	28	29	30	31	32	33	34	35
Human	C	T	C	A	A	A	G	C	A	A	T	A	C	A	C
Orangutan	T	C	C	A	A	A	G	C	A	A	T	A	C	A	C
Baboon	C	C	C	A	A	A	G	C	A	A	G	A	C	A	C
Squirrel monkey	T	T	T	A	A	A	G	C	A	A	G	A	C	A	C

human and orangutan are sister to each other, and the Old World primates (human, orangutan, and baboon) are more closely related to each other than they are to the New World squirrel monkey. It is possible for the topology of a gene tree for a specific locus to differ from the species tree for a variety of reasons, but for very good reasons (that we need not go into here) we can be confident that the true mitochondrial gene tree resembles the species phylogeny.

Despite our confidence in the genealogy of the mitochondrial genome before even examining the sequence data, we can see a wide variety of data patterns. In fact, ignoring sites with gaps that arise as a result of insertion or deletion events, there are 256 possible data patterns, derived from the number of nucleotides (4) raised to the number of species (4). Many sites will show the same nucleotide for all four species; they will appear as "constant" sites in the matrix (such as the last four sites in the matrix shown in table 1). Many other sites will show one DNA base in one of the species, and another base in the other three species. Because the squirrel monkey is the *out-group* (the most distantly related species) in this tree, its sequence has been on a distinct evolutionary trajectory for a longer period of time than those of the other species. Thus, we might expect more sites in which it differs from the other three species (e.g., site 23 in table 1). As the Hennigian logic suggests, we would expect to observe sites in which human and orangutan share a state with each other, while the squirrel monkey and baboon share a different nucleotide.

If positions in the mitochondrial genome behaved like "perfect" phylogenetic characters, then the same site would never change more than once, meaning that we could easily analyze the data using Hennigian logic to obtain the correct phylogeny. In reality, there is a substantial probability that some sites will experience multiple mutational events over the course of this evolutionary history (spanning more than 30 million years of evolution); indeed, some sites in the actual mitochondrial alignment display all four nucleotides, indicating that at least three changes must have occurred. Thus, despite the fact that we expect a relatively high proportion of sites (such as site 31 in table 1) that support the grouping of human with orangutan, we should expect some sites to show a conflicting signal. In this example, the total alignment length is 16,767 sites. Of these, 654 sites show data patterns that favor grouping human and orangutan, 320 favor grouping orangutan and baboon (e.g., site 22 in table 1), and 275 favor a tree that places human and baboon together (e.g., site 21 in table 1). From the perspective of logical analysis, conflicting signals in the data indicate that we cannot treat each character as an inerrant indicator of phylogenetic history. From the statistical perspective, we can still make progress if we can model the connection between various parameters (trees in this case) and the data. This modeling could be derived from a mechanistic understanding of the process by which different data patterns can arise on a tree, or the model could simply predict properties of the data without trying to derive them from first principles about the processes of character evolution.

2. THE PARSIMONY APPROACH

Using a different model of the evolution of characters leads to different estimation procedures. If we think that changes to any character are rare, then we expect to find very few examples of *homoplasy* (see glossary). If we try to find the tree that implies the least amount of homoplasy, we will be led to a parsimony criterion: we prefer the tree that requires the fewest changes in character state to explain the data. Calculating the smallest number of changes required to explain the data sounds daunting. The character states for the ancestral species are not observed, so we cannot simply count the number of changes that occur on each branch of the tree. Fortunately, very efficient algorithms developed in the 1970s can calculate the minimum number of steps required to explain the data (the parsimony score) in one sweep down the tree.

Computer simulation studies have been widely used to study the behavior of tree estimators. Because the true tree for a simulation is known, the accuracy of inference methods can be tested. Even though the parsimony procedure for estimating trees appears to rely on the rarity of changes, many simulation studies have found that parsimony can accurately estimate trees in which large numbers of character changes have occurred. Having a low number of changes *per branch* seems a more important predictor of when a parsimony estimate of the phylogeny will be reliable. If taxon sampling is very dense, parsimony may be able to accurately reconstruct trees even when the total amount of evolution is large.

Despite the fact that parsimony often yields reliable estimates, it has been shown to be sensitive to unequal branch lengths on the tree (as was pointed out by Felsenstein in 1978). Long branches on a tree can be caused by long periods of time between speciation events, or a high rate of character change, or a combination of these factors. Long branches can cause the conflicting signal in our data to be concentrated in specific misleading ways. If two branches on the tree experience a large number of character changes, the probability can be relatively high that they will converge on the same character state independently. Parsimony assigns a penalty of one step to each required change, regardless of where on the tree the change occurs; thus, parsimony will not account for the existence of long branches. When homoplasy is concentrated on a few branches of the true evolutionary tree, parsimony will often erroneously place some of the "long-branch" taxa together. The problem is particularly severe for character types displaying a very limited set of states (such as the four nucleotides in DNA sequence information). Felsenstein (1978) showed that there are cases in which parsimony will reconstruct the tree incorrectly, even if given an *unlimited* supply of data.

3. LIKELIHOOD-BASED APPROACHES

Developing estimators that account for unequal branch lengths requires us to treat the lengths of branches as parameters. Rather than just estimating the tree topology, we must introduce branch lengths into the estimation machinery. The true branch lengths are unknown, but we can assign them values most compatible with the data. The goal of maximum likelihood (ML) tree estimation procedure is to find the combination of tree topology and branch lengths that maximizes a likelihood score.

A fully specified model is a specific set of parameters —a point in "parameter space." If the observed data are what would be expected to arise under a model, then that model fits the data well. The likelihood is a way to assess this fit. The probability that a model with a given set of parameter values would generate a data set identical to the observed data is referred to as the *likelihood of the model*. Note that the "likelihood of a tree" in the statistical sense is not the same as the "probability that a tree is correct." In everyday usage, *probability* and *likelihood* are used synonymously, but in statistical inference the likelihood of a tree is a probability statement about the *data* if we assume that the tree is correct.

Given a particular data matrix, we can calculate the likelihood of any tree model: the probability that it would have given rise to exactly these data. If a model states that the observed data are impossible, then the model will have a likelihood of 0 and be rejected. In phylogenetic analyses, however, a tree model will never completely rule out any data set; nonetheless, some data sets have a very low probability of being generated under a particular tree. If such a data set is observed, then that tree is a poor estimate. In contrast, a tree with a high likelihood is a better estimate, and the tree that results in the maximum likelihood value among all trees is viewed as the best estimate of the tree.

To infer trees using ML, we must be able to assign probabilities to any data pattern that can occur. The probability statements for a data pattern can be constructed by considering the probability of all possible character state changes across a single branch in the tree (the *transition probabilities* for the branch). By assuming that evolutionary events on different branches are independent, one can combine per-branch transition probabilities into probabilities for the evolutionary history of a character across the entire tree. It is often convenient to consider the *rates* at which different changes could occur. Mathematical transformations allow us to extrapolate the effects of an evolutionary process occurring at a certain rate over any timescale of interest.

After a description of character evolution is formulated in terms of rates of change, it is possible to calculate a likelihood for any combination of tree topology and branch lengths. Felsenstein's (1981) pruning algorithm makes the probability calculations feasible. However, the need to maximize the likelihood score over a large number of parameters still makes ML much slower than parsimony.

By maximizing the likelihood we can find the parameter values that match the data most closely. This provides estimates of the tree and of branch lengths. One implication of treating branch lengths as unknown parameters is that every character we observe in a data matrix provides information. An entirely constant character will not "directly" prefer one tree over another, but it will provide evidence that branch lengths are short. This could have an indirect effect on which tree best fits the data, so even constant characters can alter tree inference.

The inclusion of an explicit model of character evolution is both a strength of ML methods and a target of criticism. ML methods can use all the available data to pick up on fairly subtle patterns. ML is much more resistant to branch length inequality than parsimony; however, misspecification of the model can lead to incorrect tree inference. Our models of character evolution are dramatic oversimplifications of the real evolutionary processes. Fortunately, numerous computer simulations have demonstrated that ML tree estimation is fairly robust against violations in the details of the model of character evolution, as long as the dominant aspects of evolution (e.g., unequal branch lengths, different rates of evolution for different characters) are incorporated into the models.

4. DISTANCE-BASED APPROACHES

Calculating the likelihood of a tree involves considering the probabilities associated with a huge number of possible evolutionary scenarios that could have led to the observed data. Furthermore, ML inference must consider a huge parameter space of branch lengths and rates of character change (in addition to the space of all trees). Distance-based methods for tree reconstruction simplify the tree inference problem by trying to explain only the observed divergences between the tips of the tree.

A tree makes a prediction about the *evolutionary distance* (the number of character state changes that have occurred) between each pair of tips of the tree. We can observe pairwise divergence in the characters that we study, so we have an empirical estimate of the tip-to-tip distance. From the character matrix, we can calculate a divergence between each taxon to every other taxon and summarize these calculations in a taxon-by-taxon distance matrix. The combination of tree topology and branch lengths with tip-to-tip divergences closest to the observed distance matrix is judged to be the best estimate of phylogeny.

Distance-based approaches treat the distance matrix as if it were the only data relevant to tree inference. Because distance methods do not have to "map" evolutionary events on the tree for each character, they can be very fast. The price paid for this computational benefit is unclear. Condensing a character matrix into a distance matrix implies a loss of information. When we compare characters between two taxa (tips on a tree) the number of differences represents a *minimum* number of evolutionary events that must have occurred. This minimum will usually underestimate the actual number of events. Thus, the observed pairwise distance matrix is not an error-free representation of the evolutionary distance between tips. Models can be used to correct the pairwise distance estimates for repeated changes at the same position (the "multiple hits" problem). But even a corrected pairwise distance is often an imprecise estimate of the true number of evolutionary events. Relying on a summary of the data (the distance matrix) rather than the full data should make distance methods less powerful than character-based methods. In general, no compelling statistical reasons have been advanced for preferring distance-based methods over likelihood-based approaches. Distance-based approaches continue to be widely used, however, because they provide reasonable estimates of the tree very quickly even for very large data sets.

5. COMPUTATIONAL ASPECTS OF TREE ESTIMATION

The preceding sections have focused on the statistical basis of estimating a tree, specifically on the correspondence between various estimation methods and different ways of assessing the fit between a phylogenetic hypothesis and the observed data. Developing the computational machinery to conduct phylogenetic inference is a complementary, and very active, area of research. Whether we use an ML score, a parsimony score, or a distance-based score, we must still find the tree that produces the optimal score. The number of possible trees is enormous, so scoring every possible tree is not feasible. In general, software for phylogenetic estimation works by generating a rough initial solution, then trying to improve the estimate by looking at similar trees.

A procedure called *stepwise addition* builds up a tree estimate by adding taxa to a growing tree one at a time. At each step the attachment point with the best score for a taxon is chosen. The procedure is not guaranteed to produce the best-scoring tree. Placements made in early steps may be suboptimal when new data are added to the tree. The initial approximation of the tree obtained by stepwise addition can be perturbed by rearranging a few of the relationships while keeping most of the tree's structure intact. If the perturbation results in a tree with a better score, we have improved our solution, and we can continue searching for a better tree. If we try a large number of perturbations and fail to find a tree with a better score, we can terminate the search. The final tree will be a good approximation of the tree with the optimal score even if we cannot guarantee that our search found the best tree. Repeatedly performing searches from different starting points can reveal whether this type of hill-climbing approach appears to be working on a particular data set. If each starting point yields a different final tree, then the landscape of tree scores is very complex and there is a good chance that none of the searches identified the "global" optimum.

Many variations of this general strategy of tree searching have been studied, and it is now feasible to reconstruct trees of hundreds and even thousands of taxa. When dealing with large data sets, one can rarely be confident that the optimal tree has been found; however, it is unlikely that one could reconstruct a huge tree with no error at all. The crucial question becomes, "What aspects of the tree are strongly supported by the data?" (see below). Strongly supported branches in a tree are usually easy to find during tree searching, so most phylogenetic analyses are limited more by the amount of information in the data rather than by the efficiency of tree-searching software.

6. STATISTICAL SUPPORT FOR CLADES

With enough computational resources, we can be confident that a given phylogeny represents the best estimate of evolutionary relationships that can be obtained from our data. Explicit criteria can help us choose among a set of alternative families of models, and formal tests of model adequacy can identify cases in which our inferential models are clearly unrealistic. Even with a satisfactory model of the evolutionary processes generating the data, we still must acknowledge the possibility that limitations in our data can lead to an incorrect estimate. Our estimates are based on a finite sample of data.

Given a clade in our estimate of phylogeny, such as the grouping of human with orangutan, we would like to know whether the grouping could simply be the result of sampling error rather than a true evolutionary signal. *Sampling error* refers to the mistakes in estimation caused by a small sample of data. A common approach in statistics is to calculate a *P* value to evaluate the strength of evidence about a proposition. Roughly speaking, a *P* value is the probability of seeing at least as much evidence against a proposition even if the proposition is true. It helps us assess whether it is plausible to discount our result as merely an artifact of sampling error.

To calculate a *P* value for a clade within a phylogeny, we quantify the support for the group in a numerical statistic. The most appealing choice is the difference in score between the best-scoring tree (our estimate) and the best tree that does *not* contain the clade of interest. For example, in the primate mitochondrial genome example, the tree with the best parsimony score grouped human and orangutan; this tree required 7990 changes to explain the data. The best alternative tree places orangutan with the baboon; that tree required 8324 steps to explain the sequence data; thus, we can say that the human + orangutan tree is 334 changes better than the next-best tree. Obtaining a large difference in scores clearly implies that we have more compelling evidence. This same style of analysis could have been conducted using ML scores or distance-based approaches to scoring trees.

We might want to calculate a *P* value for the hypothesis that human and orangutan are *not* close relatives. To do this we must answer the question, "If human and orangutan were *not* a true group on this phylogeny and we randomly sampled 16,767 sites, what is the probability that we would obtain a data set that yields a score difference of at least 334 steps in favor of an incorrect tree?" This is not a trivial question to answer. Phylogenetic trees are difficult parameters to deal with, and we do not have the convenience of calculating a simple number for a tree and looking it up in a standard statistical table; nevertheless, we can still apply the core insights of statistical testing and identify groupings in a phylogenetic estimate that are weakly supported and likely to be overturned by subsequent analyses. If we think our data are very "clean" (unlikely to generate patterns that support spurious groupings), then it seems very unlikely that we would see a score difference of 334 entirely from sampling error. If our data appear to have lots of homoplasy, we might see this much-erroneous signal. Fortunately, our actual data set gives a hint about how "messy" the signal was. We can look at the number of sites that supported different groupings in the original data set to see how variable the score would be as a result of resampling. Alternatively, we could use a computer to simulate the effect of sampling error by generating many artificial data sets. By counting the proportion of simulated data sets that display at least 334 steps of support for spurious groupings, we can approximate a *P* value.

Bootstrapping is the most common way to assess the effect of sampling error on tree inference. In bootstrapping, we create many "pseudoreplicates" of our original data by randomly sampling from the pool of characters we observed. In each pseudoreplicate, a different set of characters will be overrepresented and another set will be excluded. By conducting phylogenetic analysis on each of these pseudoreplicates, we can discover which groupings in the tree are sensitive to sampling error. The bootstrap proportion for a group is the proportion of pseudoreplicate analyses that supported the group. Bootstrapping is computationally demanding, because hundreds of tree searches must be performed, and it does not directly yield a *P* value; however, it does provide a useful summary of clades that are well supported.

7. BAYESIAN INFERENCE

Bayesian statistics is a major branch of statistical theory, and it has had a large impact on phylogenetic inference. Bayesian approaches formalize the ways we can use data to update our beliefs about the world. We start by considering all parameter values, assigning each parameter value a prior probability. This prior probability represents our degree of belief before examining the data. For instance, in the case of the primate tree, a person who had never studied anthropology or mammalogy might be completely uncertain about whether the correct grouping for the Old World primates would be human + orangutan, human + baboon, or orangutan + baboon. In such a case, the person might assign each tree a prior probability of 1/3 to reflect the fact that he thinks each scenario is equally likely (and the sum of probabilities must be 1). Technically speaking, this prior probability for the tree topology is actually an integral of probability densities over all the possible branch-length combinations for the tree. Before looking at the data, few people would be confident about specifying a reasonable branch length, but some combinations might seem implausible. For example, we might be surprised if the branch leading to human were thousands of times longer than the branch leading to its closest relative; by assigning such combinations low prior probability, one is able to bring previously learned insights to bear on an analysis. There is an element of subjectivity or arbitrariness in prior specification.

The next step is to update prior beliefs into posterior probability statements. This step is not subjective at all; in fact, it is simply an exercise in applying the rules of probability. Specifically, Bayes' theorem states that the probability associated with a parameter value in light of the data is proportional to the parameter value's prior probability multiplied by the likelihood of that parameter value; thus, Bayesian inference is closely tied to exactly the same likelihood function that forms the basis of ML inference.

One very attractive feature of Bayesian inference is its ability to produce a single-best estimate of a parameter (for instance, the phylogeny with the highest posterior probability), but also an easily interpreted statement of support. If the posterior probability for a tree or a clade is close to 1, then those aspects of evolutionary history are strongly supported. If the tree with the highest posterior probability has a probability of 0.4, then we immediately know it is a questionable inference.

The result of a Bayesian analysis is a summary that blends any prior knowledge with information from the data. In practice, we often have only vague knowledge of the model before seeing the data, so the effect of the likelihood dominates the inference. In such cases, the results of Bayesian point estimation are usually similar to ML point estimates. If we do have strong biological knowledge about some aspect of character evolution, then we can express that as a prior probability statement and use that information in a Bayesian framework.

Because Bayesian techniques aim to describe the probability of all possible parameter values, the computational tools required differ considerably from the tree-searching tools used to find parsimony or ML tree estimates. In the vast majority of cases, Bayesian phylogenetic inference is conducted by a computer-simulated walk through the space of all parameter values. We can design the rules for the simulation in a very specific way so the walk tends to avoid parameter values with low likelihood (or lower prior probability). Running these simulations for a large number of iterations provides a set of parameters that are sampled in proportion to their posterior probability. This Markov chain Monte Carlo simulation approach is an elegant solution to the difficulties of exploring a large parameter space, but it requires considerable care to ensure it provides reliable results.

FURTHER READING

Felsenstein, J. 2003. Inferring Phylogenies. Sunderland, MA: Sinauer. *A comprehensive, authoritative text on phylogenetic estimation.*

Giribet, G. 2007. Efficient tree searches with available algorithms. Evolutionary Bioinformatics Online 3: 341–356. Available online at www.ncbi.nlm.nih.gov/pmc/articles/PMC2684131/. *A discussion of methods for searching for an optimal tree.*

Holder, M. T., and P. O. Lewis. 2003. Phylogeny estimation: Traditional and Bayesian approaches. Nature Reviews Genetics 4: 275–284. *A review that focuses on the distinction between Bayesian methods and ML approaches to tree estimation.*

Lemey, P., M. Salemi, and A. Vandamme, eds. 2009. The Phylogenetic Handbook: A Practical Approach to Phylogenetic Analysis and Hypothesis Testing. Cambridge: Cambridge University Press. *An advanced book covering theory and descriptions of software for tree estimation.*

Lewis, P. O. 1998. Maximum likelihood as an alternative to parsimony for inferring phylogeny using nucleotide sequence data. In D. E. Soltis et al., eds., Molecular Systematics of Plants II. Boston: Kluwer, 132–163. *An excellent resource for learning how ML is used in phylogenetic estimation.*

Ronquist, F., and A. R. Deans. 2010. Bayesian phylogenetics and its influence on insect systematics. Annual Review of Entomology 55: 189–206.

II.3

Molecular Clock Dating

Bruce Rannala and Ziheng Yang

OUTLINE

1. The molecular evolutionary clock
2. Molecular clock dating
3. Testing the molecular clock
4. Statistical methods for divergence time estimation
5. Maximum likelihood estimation of divergence times
6. Bayesian estimation of divergence times
7. Fossil calibrations
8. Relaxed clocks and prior model of rate drift
9. Perspectives

This chapter reviews the history of the molecular clock, its impact on molecular evolution, and the controversies surrounding mechanisms of evolutionary rate variation and the application of the clock to date species divergences. We review current molecular clock dating methods, including maximum likelihood and Bayesian methods, with an emphasis on relaxing the clock and on incorporating uncertainties into fossil calibrations.

GLOSSARY

Fossil Calibrations. The use of the fossil record to specify the ages of nodes (divergence events) on the phylogenetic tree. In the simplest case, an interior node on the tree is assigned a fixed age, and a molecular clock is then applied in an analysis of the sequence data to estimate the absolute ages of the remaining nodes. More sophisticated calibration methods use Bayesian methodology to accommodate uncertainties in the fossil record, by specifying a distribution for a node age (instead of a fixed constant).

Fossil/Sequence Information Plot. A regression-based method for determining how much remaining uncertainty for node ages is due to uncertainties in fossil calibration times (or lack thereof) and how much to insufficient sequence data.

Molecular Clock. The hypothesis (or observation) that DNA (or amino acid) sequences accumulate changes at a constant rate through time (and among species). A "relaxed" clock model allows rates to vary across lineages in an orderly way; there may be a "local clock" with constant rates in subsets of species (in a likelihood analysis), or there may be lineage-specific rates that are either independent observations from a common distribution or correlated between ancestral and descendant species (in a Bayesian analysis).

Nonparametric Rate-Smoothing Method. One of the first methods for modeling sequence substitution rate evolution among lineages (a relaxed molecular clock). This early heuristic procedure penalizes changes in rate between ancestral and descendant branches while maximizing the probability of the data (i.e., the likelihood), this was referred to as a penalized likelihood.

1. THE MOLECULAR EVOLUTIONARY CLOCK

In the early 1960s, it was observed that the amino acid differences between aligned hemoglobin or cytochrome *c* sequences from different species were roughly proportional to the times of divergence between the species (according to the fossil record). These observations led Emile Zuckerkandl and Linus Pauling to propose the hypothesis of a *molecular evolutionary clock* in 1965. The clock was envisaged as a stochastic one, with "ticks" corresponding to nucleotide or amino acid substitutions, which occur at random time intervals. Although particular substitutions occur at random times, the rate at which substitutions occur is assumed to be constant or "clocklike" through time and across lineages. The process is analogous to the way in which the random decay of isotopes can be used to construct an atomic clock. Furthermore, much the way that different isotopes have a characteristic rate of radioactive decay, different proteins can have different evolutionary rates, meaning that their molecular clocks tick at different rates.

The molecular clock hypothesis had an immediate and profound impact on the emerging field of molecular evolution, greatly expanding the role of molecular analysis in studies of phylogeny and the timing of significant evolutionary events; nonetheless, the molecular clock hypothesis has been a focus of controversy throughout the five decades of its history. The reliability of the clock and its implications for the mechanism of molecular evolution were a focus of immediate controversy. The molecular clock hypothesis was proposed at a time when the neo-Darwinian theory of evolution was generally accepted by evolutionary biologists, according to which the evolutionary process is dominated by natural selection. A constant rate of evolution among species as different as mice and monkeys was incompatible with that theory. Species living in different habitats, with different life histories and generation times, must be under very different regimes of selection (and therefore should have different substitution rates). When the *neutral theory of molecular evolution* was first proposed (by Motoo Kimura in 1968 and by Jack King and Thomas Jukes in 1969), the observed clocklike behavior of molecular evolution was considered major supporting evidence.

The neutral theory emphasizes random fixation of neutral or nearly neutral mutations (see chapter V.1). Under such a model, the rate of substitution is equal to the neutral mutation rate, independent of factors such as environmental change and population size variation. If the mutation rate is similar and the function of a protein remains the same across species (so that the same proportion of mutations are neutral), a constant substitution rate is expected. Rate differences among proteins are explained by the presupposition that different proteins are under different functional constraints, with a different proportion of amino acids experiencing neutral mutations.

The neutral theory is not the only mechanism compatible with clocklike evolution; neither does the neutral theory always predict a molecular clock. For example, the efficiency of DNA repair mechanisms may vary among lineages leading to differences in the rate of neutral mutations and a violation of the clock (but not of the neutral theory). Controversies also exist concerning whether the neutral theory predicts rate constancy over generations or over calendar time, or whether the clock applies only to silent (synonymous) DNA changes, or instead to protein evolution as well.

Since the 1980s, DNA sequences have accumulated rapidly, replacing the protein sequences predominantly used in earlier studies. DNA sequences have now been used to conduct extensive tests of the clock and to estimate evolutionary rates in different groups of organisms. An interesting early observation was that primates have lower rates of DNA substitution than rodents, and that humans have lower rates than other apes and monkeys—characterized as the *primate slowdown* and *hominoid slowdown*, respectively. Two major factors that could account for such between-species rate differences are generation time (with a shorter generation time causing more germ-line cell divisions per calendar year and a higher substitution rate) and DNA repair mechanism (with less reliable repair mechanisms associated with higher mutation [and substitution] rates). Perhaps because of the generation time effect or other correlated life history variables, for example metabolic rate, substitution rates tend to be negatively related to body size, with high rates in rodents, intermediate rates in primates, and slow rates in whales. Species with small body sizes tend to have shorter generation times and higher metabolic rates. The negative correlation between substitution rate and body size has been supported in some studies but questioned in others. The disagreements do not appear to have been resolved.

2. MOLECULAR CLOCK DATING

The molecular clock hypothesis provides a simple yet powerful way of dating evolutionary events. Under the clock assumption, the expected distance between sequences increases linearly with time of divergence. When external information about the geological ages of one or more divergence events on a phylogeny is available, based on the fossil record or certain geological events, the distances between sequences or the branch lengths on the tree can be converted into absolute geological times. This is known as *molecular clock dating*.

The earliest application of the clock to estimate divergence times was by Zuckerkandl and Pauling in 1962, who used an approximate clock to date duplication events among α, β, γ, and δ globins of the hemoglobin family. The molecular clock has since been used widely to date species divergences. The outcomes of molecular clock analyses have often produced controversies, usually because the molecular dates are at odds with the fossil record. One controversy concerns the origin of the major animal forms. Fossil forms of metazoan phyla appear as an "explosion" around 540 million years ago in the early Cambrian, but most molecular estimates of the ages of these divergence events have been much older, sometimes twice as old. Another controversy surrounds the origins and divergences of modern mammals and birds following the demise of the dinosaurs about 65 million years ago at the Cretaceous-Tertiary boundary (the KT boundary). Molecules again generated much older dates than expected by paleontologists.

Part of the discrepancy between molecular and fossil data is due to the incompleteness of the fossil record. Fossils provide information concerning the date by which a newly diverging lineage had developed diagnostic morphological characters. There may be a lag between the time that a lineage arose and the age of the first fossil with the derived traits of the descendants. Molecular dating, in contrast, infers ages of nodes (divergence events among ancestral lineages) in a phylogenetic tree. Fossil-based dates therefore tend to be younger than those derived from molecular data. Another source of discrepancy can be inaccuracies and deficiencies in molecular time estimation. Despite sometimes acrimonious controversies, the interactions between molecules and fossils have been a driving force in this research area, since they have prompted reinterpretations of fossils, critical evaluations of molecular dating techniques, and the development of more advanced analytical methods.

Our focus in this chapter is on statistical methods for testing the clock hypothesis, and on likelihood and Bayesian methods for dating species divergence events under global and local clock models. In such analyses, fossils are used to calibrate the clock, that is, to translate sequence distances into absolute geological times and substitution rates. A special case of molecular dating applies to viral genes, which evolve so fast that DNA substitutions may be observed over a few years (rather than thousands of millennia as with eukaryotes). One can use the dates at which particular viruses were isolated to calibrate the clock and to estimate divergence times, using essentially the same techniques as discussed here. Indeed, such dated viral sequences are sometimes referred to as "fossil sequences," although most such samples were isolated during the last 100 years and are not true fossils.

3. TESTING THE MOLECULAR CLOCK

Several statistical tests have been developed to examine whether the rate of molecular evolution is constant over time. The simplest, known as the *relative rate test*, examines whether two species *a* and *b* evolve at the same rate by using a third out-group species *o* (figure 1). As species *a* and *b* share the same common ancestor *y*, the distance from *y* to *a* should equal the distance from *y* to *b* if the hypothesis of the molecular clock is true: $d_{ya} = d_{yb}$ (figure 1A). Equivalently, one can formulate the clock hypothesis relative to the out-group as $d_{ao} = d_{bo}$ and test whether the difference between the two calculated distances $d = d_{ao} - d_{bo}$ is significantly different from 0. The sequence distances and their variances can be calculated under any model of nucleotide or amino acid substitution, and the calculated *d* and its standard error can be used to construct a test based on the normal distribution.

It is also possible to conduct this relative-rate test using a likelihood ratio test. The null model assumes the clock and involves two parameters (t_1 and t_2 in figure 1A). The more general model does not assume the clock. The general model is unable to identify the root of the tree, so that the parameters in the model are the three branch lengths in the unrooted tree (b_1, b_2, b_3 in figure 1B). Note that the test is applied to sequence data alone, without knowledge of absolute times and rates, so that both the *t*s of figure 1A and *b*s of figure 1B are measured by distance, the expected number of changes per site. Using maximum likelihood analysis (see chapter II.2), one calculates the optimized log likelihood values under the null (clock) and alternative models (nonclock), ℓ_0 and ℓ_1, and then compares $2\Delta\ell = 2(\ell_1 - \ell_0)$ against a chi square distribution with one degree of freedom to decide whether the clock (the null model) should be rejected.

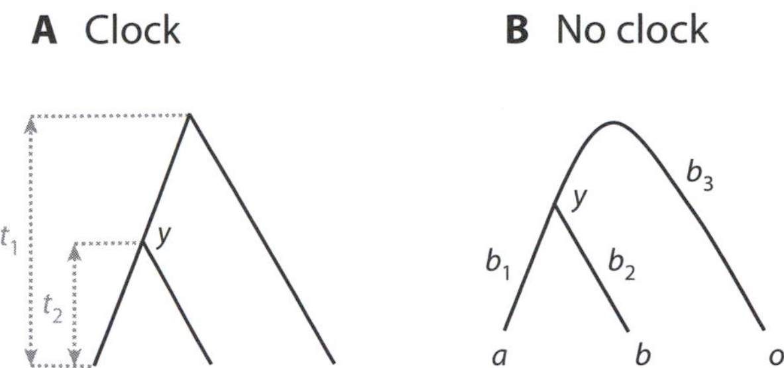

Figure 1. The relative rate test compares the rates of evolution in two species (*a* and *b*) using a third species *o* as the out-group.

The likelihood ratio test may be applied to a tree of arbitrary size. Under the null hypothesis of the clock, there are $s - 1$ parameters corresponding to the ages of the $s - 1$ interior nodes on the rooted tree with *s* species. The more general nonclock model allows every branch on the unrooted tree to have its own rate, meaning there are $2s - 3$ free parameters for the $2s - 3$ branch lengths. Twice the log likelihood difference between the two models, $2\Delta\ell = 2(\ell_1 - \ell_0)$, can be compared with the χ^2 distribution with $(2s - 3) - (s - 1) = s - 2$ degrees of freedom to decide whether the clock is rejected.

Several caveats about these molecular clock tests should be noted. Although a constant rate implies the equality $d_{ya} = d_{yb}$ in figure 1, the inverse is not necessarily true; the distances can be equal without a clock. For example, if the rate of evolution has been accelerating or decelerating over time, but the rate change affects all lineages in the same way, the tree will look clocklike, judged by the distances, even though the clock is violated. Information on absolute times of divergences is needed to detect such violations of the clock. Also, failure to reject the clock may simply be due to lack of information in the data or lack of power of the test. In general, the likelihood ratio test applied to multiple species has far more power than the relative-rate test applied to only three species.

Whether the molecular clock holds in empirical data sets depends on the level of species divergences. In general, the more ancient the divergences among the groups being studied, the less likely that a molecular clock hypothesis will be valid. For example, the molecular clock generally holds among the hominoids. Among primates, the clock may be acceptable for nuclear genes but is often rejected for faster-evolving mitochondrial genes. Among various orders of mammals, the clock is most often rejected even for nuclear data. Beyond vertebrates, the clock typically provides a very poor description of the evolutionary process.

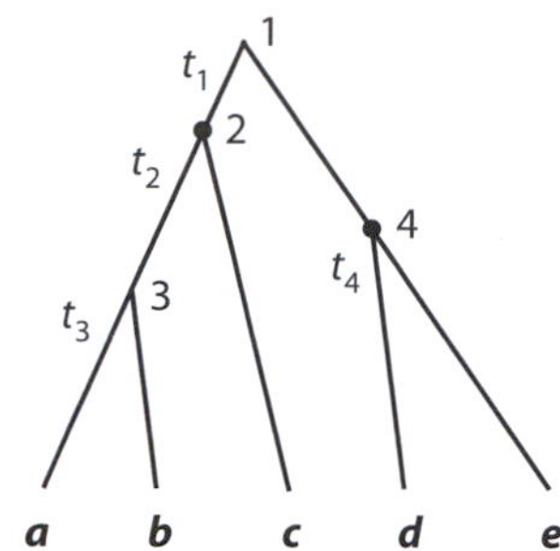

Figure 2. A tree of five species to explain the maximum likelihood and Bayesian methods for dating species divergences. Nodes 2 and 4 are the calibration nodes, while nodes 1 and 3 are the noncalibration nodes.

4. STATISTICAL METHODS FOR DIVERGENCE TIME ESTIMATION

In recent years, more sophisticated statistical methods for estimating divergence times using both multiple fossil calibrations and sequence data have been developed. Both distance methods (based on calculations of pairwise distances) and likelihood methods (based on a simultaneous analysis of multiple sequences on a phylogenetic tree; see chapter II.2) can be used to estimate the distances from the internal nodes to the present time. The assumed substitution model may be important, as a simplistic model may not correct for multiple hits properly and may underestimate distances. Often the underestimation is more serious for large distances than for small ones, and the nonproportional underestimation may generate systematic biases in estimates of divergence time.

A rooted tree topology representing the ancestor-descendant relationships among lineages is typically assumed to be known in molecular clock dating, although some methods simultaneously estimate the tree and the divergence times. Uncertainties in the tree may (or may not) be important to the estimation of divergence times, for example depending on whether the uncertainties affect the placement of the fossil calibrations, and depending on the number and location of the calibration nodes. The use of several alternative, fully resolved phylogenetic tree topologies in a dating analysis may provide an assessment of the robustness of time estimation to uncertainties in the tree topology.

Besides possible errors of the substitution model and the tree topology, two additional problems that may arise are violations of the molecular clock and uncertainties in the fossil calibrations. In the past few years, considerable effort has been expended in dealing with these two problems in the likelihood and Bayesian frameworks. Below we discuss the likelihood and Bayesian methods of divergence time estimation, with an emphasis on the Bayesian method. The latter can incorporate uncertainties in fossil calibrations by specifying a prior distribution on divergence times, and it can deal with a violation of the clock through a prior model that allows substitution rate to vary across evolutionary lineages.

5. MAXIMUM LIKELIHOOD ESTIMATION OF DIVERGENCE TIMES

As mentioned above, a rooted tree of s species comprises $s-1$ ancestral nodes. Suppose that the ages of c ancestral nodes are known without error, determined from fossil data. The model then involves $s-c$ parameters: the substitution rate and the ages of the $s-1-c$ nodes that are not calibration points. For example, the tree shown in figure 2 has $s = 5$ species, with four interior node ages: t_1, t_2, t_3, and t_4. Suppose nodes of ages t_2 and t_4 are fixed according to the fossil record. Then three parameters are estimated under the model: μ, t_1, and t_3. Given those rate and time parameters, each branch length (in units of expected substitutions) is simply the product of the rate and the time duration of the branch. For example, the length of the branch from nodes 2 to 3 in figure 2 is $\mu(t_2-t_3)$. The likelihood function, that is to say the probability of the sequence data given the branch lengths on the tree, can be calculated using standard algorithms (see chapter II.2). Times and rates are estimated by maximizing the likelihood function.

The description above assumes the molecular clock. What if a clock model is rejected? One possible solution is to remove some species so that the clock approximately holds for the remaining species. This may be useful if one or two lineages with grossly different rates can be identified and removed, but awkward if the rate variation is more complex. Another approach is to take explicit account of among-lineage rate variation when estimating divergence times. Considering the tree of figure 2, for example, one may assign one rate for all branches to the left of the root, and another for those to the right. This approach is known as the *local-clock* method. The implementation is very similar to that described for the strict molecular clock discussed above. The only difference is that, under a local-clock model with k rates of evolution, one estimates $k - 1$ extra rate parameters. The local-clock method may be straightforward to use if biological considerations allow us to assign branches to rate classes; however, in general, too much arbitrariness is involved in applying such a model.

Another method for accommodating among-lineage substitution rate variation in divergence date estimation, developed in the late 1990s; is Michael Sanderson's nonparametric rate-smoothing (NPRS) method. This approach allows that the rate of substitution may evolve more slowly than the rate of lineage branching, so that closely related lineages will tend to share similar rates. One implementation of this approach, called *penalized likelihood*, penalizes changes in rate between ancestral and descendant branches while maximizing the probability of the data (i.e., the likelihood), thus allowing estimation of both rates and times. A smoothing parameter, λ, estimated through a cross-validation procedure, determines the importance of penalizing rate changes relative to the likelihood. Both the likelihood calculation and rate smoothing are achieved through heuristic search procedures. If a probabilistic model of rate change (see below) is instead adopted there is no need for either a rate-smoothing parameter or cross-validation. The NPRS method has the advantage that it can deal with uncertainties in the fossil calibrations, implemented by placing constraints on the ages of calibrated nodes ($t_L < t < t_U$); however, the NPRS method is identifiable (a necessary condition for reasonable results that depend on the data) only if at least one node age is known without error; thus the method does not provide a solution to the general problem that all fossil calibrations have some error associated with them.

6. BAYESIAN ESTIMATION OF DIVERGENCE TIMES

The Bayesian method is currently the only framework that can simultaneously incorporate multilocus sequence information, prior information on substitution rates, prior information on rates of cladogenesis, and so on, as well as fossil calibration uncertainties, to estimate divergence times. In a Bayesian analysis, one assigns prior distributions on evolutionary rates and nodal ages, and the analysis of the sequence data then generates the posterior distribution of rates and ages, on which all inference is based. Computation in Bayesian molecular dating is achieved through Markov chain Monte Carlo (MCMC) algorithms, which generate samples from the posterior distribution (see chapter II.2).

A Bayesian MCMC dating method was developed in the late 1990s by Jeff Thorne, Hiro Kishino, and Ian Painter. A model describing substitution rate change over time is used to specify the prior probability on rates, while fossil calibrations are incorporated as minimum and maximum bounds on node ages in the tree. This approach has formed the basis for several later extensions. Here we describe the general structure of these models.

Let x be the sequence data, $\mathbf{t}$ the $s - 1$ divergence times (nodal ages) and $\mathbf{r}$ the lineage-specific rates. Bayesian inference is based on the posterior probability of $\mathbf{r}$, $\mathbf{t}$, and other parameters (θ):

$$f(\mathbf{t}, \mathbf{r}, \theta|\mathbf{x}) \propto f(\mathbf{x} \mid \mathbf{t}, \mathbf{r}, \theta) f(\mathbf{r} \mid \mathbf{t}, \theta) f(\mathbf{t} \mid \theta) f(\theta). \quad (1)$$

Here $f(\mathbf{t})$ is the prior probability distribution on times and $f(\mathbf{r} \mid \mathbf{t}, \theta)$ is the prior on rates given the divergence times and model parameters, θ, while $f(\mathbf{x} \mid \mathbf{t}, \mathbf{r}, \theta)$ is the likelihood function of the sequence data, $\mathbf{x}$. The MCMC algorithm generates samples from the joint posterior probability distribution of times ($\mathbf{t}$), rates ($\mathbf{r}$) and model parameters (θ).

It should be noted that Bayesian estimation of species divergence times differs from a conventional Bayesian estimation problem, in that the errors in the posterior estimates do not approach zero when the amount of sequence data approaches infinity; indeed, theory developed by Yang and Rannala in 2006 specifies the limiting distribution of times and rates when the length of sequence approaches infinity. The theory predicts that the posterior distribution of times and rates condenses to a one-dimensional distribution as the amount of sequence data tends to infinity. Essentially there is only one free variable, and each divergence time is completely determined given the value of this variable; the variable encapsulates all the information jointly available from all the fossil calibrations. Any specific divergence time is obtained as a particular transformation of this single free variable, and the divergence time estimates are completely correlated across nodes. By examining the fossil/sequence information plot (figure 3), which is a regression of the width of the credible interval for the divergence time against the posterior mean of the divergence time, one can evaluate how closely the sequence data approach this limit. This can be used to determine whether the remaining uncertainties in the posterior time estimates are due mostly to the lack of precision in fossil calibrations or to the limited amount of sequence data. If the correlation coefficient of the regression is near 1, then little improvement in divergence dates can be gained by sequencing additional genes. This method thus allows a decision to be made as to whether digging for fossils or doing additional sequencing, or both, would be a better investment of effort. The theory

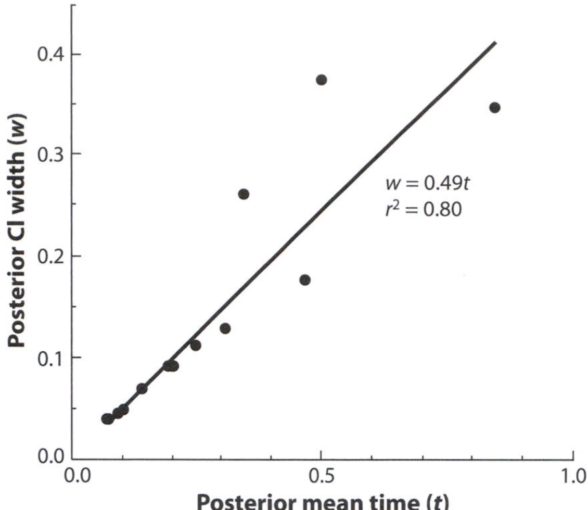

Figure 3. The fossil/sequence information plot for a Bayesian analysis of primate divergence times. Two large nuclear loci are analyzed, using two fossil calibrations derived from a Bayesian analysis of primate fossil occurrence data. The rooted tree has 15 species, with 14 internal nodes. The posterior means of the ages for the 14 internal nodes are plotted against the 95% posterior credibility intervals. The correlation $r = 0.9$ indicates that the sequences are informative, but improvement is likely with more sequence data. The slope of the regression $b = 0.49$ reflects the precision of the fossil calibrations: every 100 million years of divergence adds 49 million years to the Bayesian credibility interval. (Wilkinson et al. 2011.)

highlights the critical importance of reliable and precise fossil calibrations in molecular clock dating.

7. FOSSIL CALIBRATIONS

Fossil calibrations are incorporated into a Bayesian analysis through the prior probability distribution placed on divergence times (node ages). Thorne and colleagues allowed minimum and maximum age bounds on node ages, implemented in the MCMC algorithm by not proposing new divergence times that violate such bounds. The prior for the ages of the noncalibration nodes assumes that the tree is the result of a random cladogenesis (speciation) process (a Yule pure-birth process), possibly with extinction (a birth-death process).

The bounds on node ages were "hard," since they assign zero probability for any ages outside the interval. Such priors represent strong conviction on the part of the biologist and may not always be appropriate. In particular, fossils often provide good minimum bounds but rarely provide good maximum bounds. As a result, the researcher may be forced to use an unrealistically large maximum age bound to avoid precluding an unlikely (but not impossible) ancient age for the node. Such a "safe" approach may be problematic because the bounds may greatly influence posterior time estimation. On the other hand, failing to use a sufficiently old maximum bound can also have a pathological outcome. For example, if the true age of a fossil is larger than a hard maximum bound used in an analysis, the molecular data may conflict strongly with the fossil-based prior, resulting in overinflated confidence in the ages of other nodes.

Yang and Rannala (2006) subsequently developed more flexible distributions to mathematically describe fossil calibration uncertainties. These distributions use so-called soft bounds and assign low (but nonzero) probabilities over the whole positive half-line ($t > 0$). A few examples are shown in figure 4. The basic model used is a birth-death process, generalized to account for species sampling with fossil calibration information incorporated into the probability distribution by multiplying the probabilities for the branching process conditioned on the calibration ages and the probability distribution on calibration ages based on fossil information alone. A subsequent Bayesian approach to this problem by Ho and colleagues, implemented in the program Beast, did not use the "conditional" birth-death prior described above, instead multiplying unconditional probabilities, which is incorrect according to the rules of the probability calculus. The effects of this error on inferences obtained using the Beast program is difficult to judge, and the results should therefore be interpreted with caution.

Considerable effort has been spent on developing objective priors that best summarize the fossil record to represent our state of knowledge concerning the ages of calibration nodes. Studies of fossil preservation and discovery, errors in fossil dating techniques, and morphological character evolution in fossils and modern species may all contribute to this goal.

8. RELAXED CLOCKS AND PRIOR MODEL OF RATE DRIFT

Thorne and colleagues implemented a Bayesian "relaxed clock" in which substitution rates may vary across species. In their model, the rate at each node is specified by conditioning on the rate at its ancestral node. Specifically, given the rate r_A at the ancestral node, the rate r at the current node has a lognormal distribution. This means that the logarithm of the rate "drifts" according to a *Brownian motion* process, while the rate itself drifts according to a *geometric Brownian motion* process (figure 5). Parameter σ^2 in the Thorne et al. model controls how rapidly the rate drifts, which is to say how clocklike the tree is. A large σ^2 means that the rates vary rapidly over time or among branches and the clock is seriously violated, while a small σ^2 means that the clock roughly holds.

An alternative model of rate variation assuming independent rates was independently implemented in the

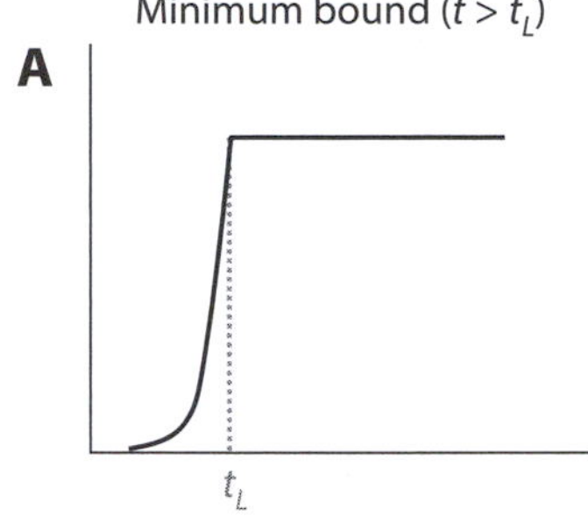

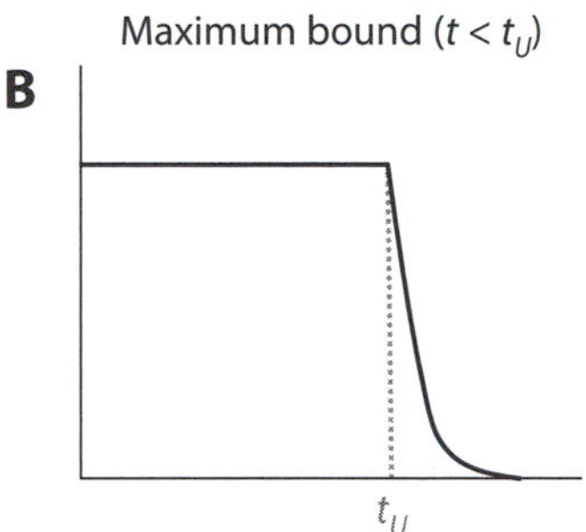

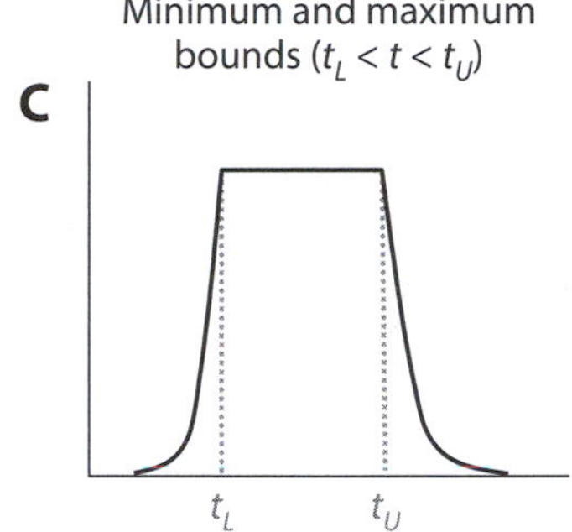

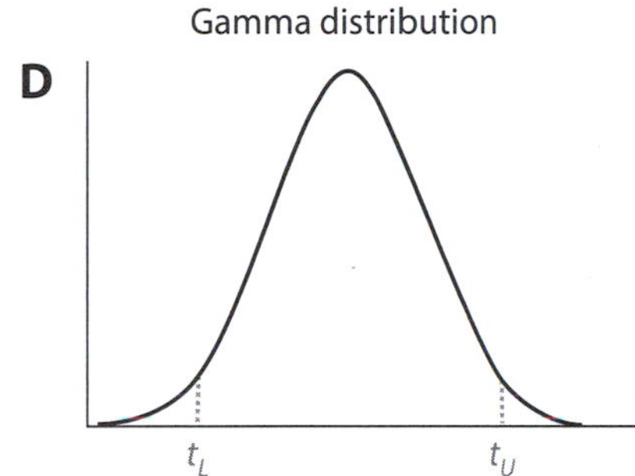

Figure 4. Probability densities used to describe the likely age of a node based on the fossil record. (Redrawn according to Yang and Rannala 2006.)

late 2000s by Rannala and Yang, and by Alexei Drummond and colleagues. In this model, the rate for a branch is a random variable drawn from a common probability distribution such as the lognormal or the gamma. The rates effectively evolve independently on each lineage, but the extent of rate variation has some form of evolutionary constraint (imposed by the prior distribution on rates).

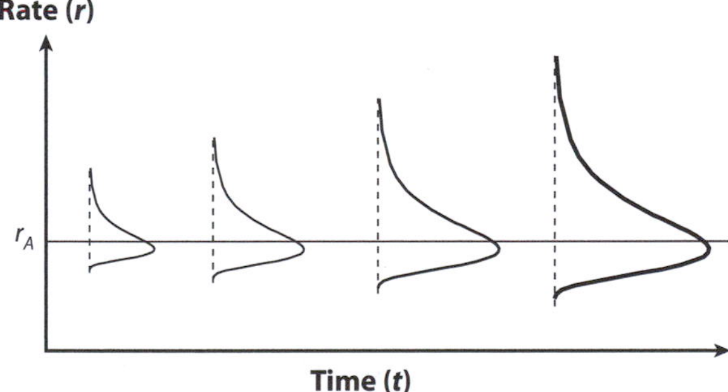

Figure 5. The geometric Brownian motion model of rate drift. Given the ancestral rate r_A time t ago, the current rate r has a lognormal distribution centered around r_A, with the variance being greater the larger t is. In other words, the logarithm of the rate $y = \log(r)$ drifts according to a Brownian motion process: given the ancestral log rate $y_A = \log(r_A)$ time t ago, the current log rate $y = \log(r)$ has a normal distribution with variance $t\sigma^2$. Parameter σ^2 measures the degree of variability of the evolutionary rate.

9. PERSPECTIVES

Bayesian statistics is currently the only framework that can integrate information and uncertainties from different sources in order to obtain reasonable statistical estimates of nodal ages. In particular, it can deal with violation of the molecular clock through its use of the prior model of evolutionary rate change, and it can incorporate uncertainties in the fossil calibrations by specifying prior distributions on divergence times. In contrast, attempts to achieve those two objectives in the maximum likelihood framework have been unsuccessful; nevertheless, a number of challenging problems remain in Bayesian molecular clock dating. First, use of multiple fossil calibrations in a Bayesian analysis may impose significant computational challenges. This is suggested by the observation that different dating programs may produce very different priors and thus different posterior time estimates. Second, fossil calibrations in a molecular dating analysis should be a statistical summary of the relevant part of the fossil record; thus, to generate good calibrations for a molecular dating analysis, probabilistic modeling and statistical analysis of fossil data (in particular, fossil occurrences and morphological measurements) will be necessary. Methods for molecular dating are currently the subject of intensive research and can be expected to change dramatically over the next decade. With improvements in sequence and fossil data sets, as well as more refined analytical methods, the degree of conflict between fossils and molecular data is gradually diminishing.

FURTHER READING

Bromham, L., and D. Penny. 2003. The modern molecular clock. Nature Reviews Genetics 4: 216–224. *A discussion of the clock in relation to theories of molecular evolution.*

Felsenstein, J. 1981. Evolutionary trees from DNA sequences: A maximum likelihood approach. Journal of Molecular Evolution 17: 368–376. *Original description of the likelihood ratio test for a molecular clock.*

Morgan, G. J. 1998. Emile Zuckerkandl, Linus Pauling, and the molecular evolutionary clock. Journal of the History of Biology 31: 155–178. *A history of the molecular clock.*

Smith, A. B., and K. J. Peterson. 2002. Dating the time of origin of major clades: Molecular clocks and the fossil record. Annual Review of Earth and Planetary Sciences 30: 65–88. *A discussion of conflicts between sequence and fossil data concerning divergences of mammals at the KT boundary and Cambrian origins of major animal phyla.*

Wilkinson, R. D., M. E. Steiper, C. Soligo, R. D. Martin, Z. Yang, and S. Tavare. 2011. Dating primate divergences through an integrated analysis of palaeontological and molecular data. Systematic Biology 60: 16–31. *A Bayesian integrated analysis of paleontological and sequence data.*

Yang, Z. 2006. Computational Molecular Evolution. Oxford: Oxford University Press, chapters 4 and 7. *Descriptions of statistical methods for estimating divergence times with clock and relaxed-clock models.*

Yang, Z., and B. Rannala. 2006. Bayesian estimation of species divergence times under a molecular clock using multiple fossil calibrations with soft bounds. Molecular Biology and Evolution 23: 212–226. *Description of soft bounds, calibration densities, the infinite-sites theory, and fossil/sequence information plot.*

Zuckerkandl, E., and L. Pauling. 1965. Evolutionary divergence and convergence in proteins. In V. Bryson and H. J. Vogel, eds., Evolving Genes and Proteins. New York: Academic, 97–166. *Early application of the molecular clock to analyze divergence of protein sequences among mammals.*

II.4

Historical Biogeography

Michael J. Donoghue

OUTLINE

1. Early developments
2. Cladistic biogeography
3. Inferring ancestral areas
4. A fresh look at old patterns
5. Beyond the standoff

Historical biogeographers try to understand how life and the earth have evolved together, accounting for current geographic distribution patterns in terms of past events. They try to understand where lineages originated, and how, when, and why they have spread, adapted to new environments, and diversified. These spatially oriented questions are as central to evolutionary biology today as they were at the time of Darwin and Wallace. Although we are still analyzing patterns that were noted long ago, the landscape of ideas and methods has changed dramatically over the years. A key recent period, beginning in the 1970s, saw the rise of cladistic biogeography in its various forms. Although this period of conceptual and methodological turmoil served to clarify fundamental issues, in hindsight it appears that an overly narrow view of the permissible questions and admissible evidence resulted in relatively little progress in understanding empirical patterns. Today, the emphasis has shifted to more integrative approaches, especially in regard to inferring ancestral areas using methods capable of accommodating information on the ages of lineage-splitting events and on the relative likelihood of geographic movements at different times in the past. Consequently, we are now taking a fresh look at a number of long-recognized biogeographic patterns. The major challenge ahead is to fill the wide gap still separating those focused on general patterns and the relationships among areas of endemism from those developing and using methods to infer ancestral areas within particular lineages.

GLOSSARY

Ancestral Area/Center of Origin. Geographic area where a clade (or multiple clades) originated and began to diversify before spreading to other areas; this may or may not be a center of diversity.

Area of Endemism. A geographic area that harbors multiple *endemics*.

Clade. A monophyletic group. An entire branch of a phylogenetic tree, including an ancestor (e.g., an ancestral species) and all its descendants.

Cladogram. A branching diagram that depicts hypothesized relationships among the terminals. Includes *area cladogram*, with geographic areas inhabited by terminal taxa replacing the names of those taxa (often species; figure 1), and *general area cladogram*, a summary tree for a set of area cladograms for different groups of organisms occupying the same areas of endemism. The term *phylogeny* (or *phylogenetic tree*) is commonly used when the intention is to convey information about the inferred evolutionary history of a group of organisms.

Dispersal. The movement of organisms resulting in the expansion of the geographic range of a species; long-distance dispersal generally refers to movement well outside the current range across a natural dispersal barrier (a mountain range, ocean, etc.).

Endemic. A species or clade that is naturally restricted in its distribution to a particular geographic area.

Integrative Biogeography. The attempt to incorporate multiple sources of relevant evidence in biogeographic inferences; moving beyond area cladograms, this might include information on the timing of lineage-splitting events and on the likelihood of movements between areas at different times in the past.

Niche Conservatism. The tendency for related species (possibly even large clades) to retain ancestral ecological characteristics. In historical biogeography, this is manifested in the phenomenon of *habitat tracking*, where lineages spread and contract within retained environments; niche conservatism ultimately underlies disjunct distributions.

Pseudocongruence. When area cladograms for different groups of organisms are the same even though the groups diversified at different times and in response to different causal events (figure 1).

Vicariance/Cladistic Biogeography. An approach to the identification and analysis of biogeographic patterns in which area cladograms are the underlying source of evidence and general area cladograms provide the basis for identifying common causes (often vicariance events related to earth history).

Vicariance/Vicariance Event. The splitting of an ancestral geographic range (of a species or clade) or of an entire biota by the formation of a barrier, often geological in nature, such as drifting continents, mountain building, or climate change.

Historical biogeography, to paraphrase Leon Croizat (1964), is the study of how life and the earth have evolved together. The fundamental aim is to account for the current geographic distributions of species and clades in historical terms. From this standpoint it is important to infer where lineages originated, how, when, and why they spread to other areas, and how such movements influenced genetic variation, adaptive evolution, speciation, and extinction. This problem can just as well be viewed from the standpoint of particular biotas, asking how these were assembled through time and, therefore, how long the component lineages have been interacting. And from yet another angle, studies of patterns of endemism and disjunction can help us understand the history of the earth and its changing climates.

Historical biogeography forms a natural bridge between evolutionary biology and the earth sciences, including paleontology and climatology. It also connects directly to the study of ecological processes (perhaps especially to life history and reproductive biology), and to genetic studies of population histories within species (*phylogeography*; see chapter II.5). It is clear that historical biogeography is, and must always be, very broadly integrative, and this necessity has presented challenges for the development of a unified methodology (Morrone 2009; Lomolino et al. 2010). There has been a tendency to default to a narrative mode of explanation, as opposed to a hypothesis-testing mode, and it has proven difficult to move beyond individual case studies to draw very general conclusions (Crisp et al. 2011). Not surprisingly, the history of historical biogeography has been marked at intervals by heated debate over proper methodology and relevant evidence.

1. EARLY DEVELOPMENTS

Historical biogeography has deep pre-evolutionary roots (Lomolino et al. 2004). The recognition of many biogeographic patterns and the origin of several general conceptual approaches go back to the eighteenth century. For example, Georges-Louis Buffon (1707–1788) made the fundamental observation (now known as Buffon's law) that widely separated but environmentally similar regions are typically inhabited by quite different assemblages of species. Likewise, the recognition that floristic belts and species diversity tend to shift in parallel along latitudinal and altitudinal gradients traces to Alexander von Humboldt (1769–1859).

From the beginnings of evolutionary thought in the early to mid-nineteenth century, geographic patterns figured prominently both as challenges to evolutionary explanation and by providing support for descent with modification. The ideas of Charles Darwin (1809–1882) and Alfred Russel Wallace (1923–1913) were to a significant extent inspired by their direct experiences with geographic patterns in nature (for Darwin, think Galápagos; for Wallace, the Malay Archipelago). Both used biogeographic patterns as key evidence of shared ancestry, but also recognized the need to explain oddball disjunctions that might otherwise be seen as favoring separate creation. Darwin devoted two chapters of *The Origin* to biogeography, and these were largely designed to defend the continuity of evolution: "so in space, it certainly is the general rule that the area inhabited by a single species, or by a group of species, is continuous, and the exceptions, which are not rare, may, as I have attempted to show, be accounted for by former migrations under different circumstances, or through occasional means of transport, or by the species having become extinct in the intermediate tracts."

As but one example, Darwin and Wallace both devoted considerable energy to explaining (invoking two different mechanisms) bipolar distributions (lineages distributed at both high northern and high southern latitudes, but absent in between), which are now perfectly understandable in evolutionary terms. Under the circumstances at the time, it is hardly surprising that they focused special attention on the likelihood of movements (including chance long-distance dispersal over existing geographic barriers) over long stretches of geological time.

Wallace, more so than Darwin, devoted his energies in the late 1800s to historical biogeography, and is frequently identified as "the father of zoogeography." He wrote three major books on the subject—*The Malay Archipelago* (1869), *The Geographical Distribution of Animals* (1876), and *Island Life* (1880)—and famously identified the biogeographic discontinuity between the

Australian and Oriental faunas now known as Wallace's line. Other key figures at that time included the British botanist Joseph Dalton Hooker (1817–1911), who, in contrast to Wallace and Darwin, favored the rise and fall of land bridges ("extensionism") as an explanation for major intercontinental disjunctions, and the American botanist Asa Gray (1810–1888), who focused special attention on disjunctions in temperate forests around the Northern Hemisphere. The British ornithologist Phillip Sclater (1829–1913) first circumscribed the world's major terrestrial biogeographic regions, and Edward Forbes (1815–1854) likewise identified marine biogeographic realms.

In the first half of the twentieth century, William Diller Matthew (1871–1930) emphasized "centers of origin," proposing that primitive forms would occupy more inaccessible areas. John Willis (1868–1958) formulated the "age and area" hypothesis, whereby the geographic range of a taxon depended on its age. Stanley A. Cain (1902–1995) provided a badly needed critique of the center of origin concept, and the criteria by which these "centers" were recognized in practice.

George Gaylord Simpson (1902–1984) and Phillip Darlington (1904–1983) were prominent among zoological biogeographers in the Wallace-Darwin-Matthew tradition. They drew special attention to migration routes of different types ("corridors," "sweepstakes routes"), envisioning movements over a more or less fixed landscape. At first they dismissed continental drift as a factor, although a dynamic earth clearly provided a completely new interpretation for patterns at the level of whole biotas. Later they reconsidered this stance, though Darlington, in particular, cautioned that drifting continents might be too ancient to be relevant to many modern geographic patterns.

Beginning with E. V. Wulff (1885–1941), phytogeographers connected plant disjunction patterns to past continental movements, and this oriented Peter Raven and Daniel Axelrod's masterful synthesis in 1974. In the meantime, led by Sherwin Carlquist, botanists continued to focus on dispersal biology and the colonization of oceanic islands. A significant development during this period was Leon Croizat's "panbiogeography" (see Croizat 1964). Importantly, Croizat emphasized the search for general patterns ("generalized tracks"), vicariance explanations ("baselines"), and the merging of biotas ("nodes"). To this day Croizat's approach has attracted a small but vocal following, though many fault his approach for shunning phylogenetic information.

2. CLADISTIC BIOGEOGRAPHY

With the rise of phylogenetic systematics came the use of phylogenetic trees in historical biogeography. Willi Hennig (1966) featured the placement of trees on maps, and the interpretation in terms of lineage movements (e.g., the "progression rule" to identify a center of origin). Another dipterist and contemporary, Lars Brundin, had the more profound influence through his empirical work on the chironomid midges of the Southern Hemisphere. Brundin compared multiple phylogenies of midge lineages overlaid on a map of the southern continents, seeking general patterns caused, perhaps, by drifting continents. Brundin's work, combined with Croizat's outlook, spawned the development by Gareth Nelson and colleagues of what was initially known as vicariance biogeography or, later, cladistic biogeography (e.g., Platnick and Nelson 1978). This reflected a strong negative reaction to the one-off dispersal scenarios for individual groups that had come to dominate the biogeographic literature. Instead, the focus was squarely on vicariance-causing events (especially continental movements) that could impact multiple lineages, and on inferring relationships among areas of endemism as a way to elucidate earth history.

Methodologically, vicariance biogeography focused almost exclusively on the comparison of cladograms for different groups of organisms, but with the names of the terminal taxa replaced by the areas of endemism they occupy (so-called area cladograms). The goal was to derive a general area cladogram that best summarized the geographic relations among the areas in an underlying set of area cladograms (e.g., Wiley 1988). This was easy in the simple three-area cases used repeatedly to illustrate the basic approach, but proved far more difficult in practice, especially when species were widespread (found in more than one area), when an area appeared in more than one place in an underlying area cladogram, when particular areas were missing from any of the area cladograms, or with a combination of these factors. One set of methods focused on tallying shared geographic "components," employing various assumptions about widespread taxa, etc. In recent years, this general approach has spawned several others, including "three-area analysis" and "paralogy-free subtree analysis," which variously focus on subsets of the underlying area cladograms. Another popular method, known as Brook's parsimony analysis (BPA), decomposed each area cladogram into a set of binary characters representing the area connections, and compiled these into a matrix that was analyzed by parsimony to derive a single tree showing area relationships.

Much of the emphasis in the cladistic biogeography literature was on philosophical and methodological issues, which were reviewed repeatedly, but with few breakthroughs. Real-world cases were typically complicated enough that area cladograms alone lacked the power to confidently resolve area relationships. Among

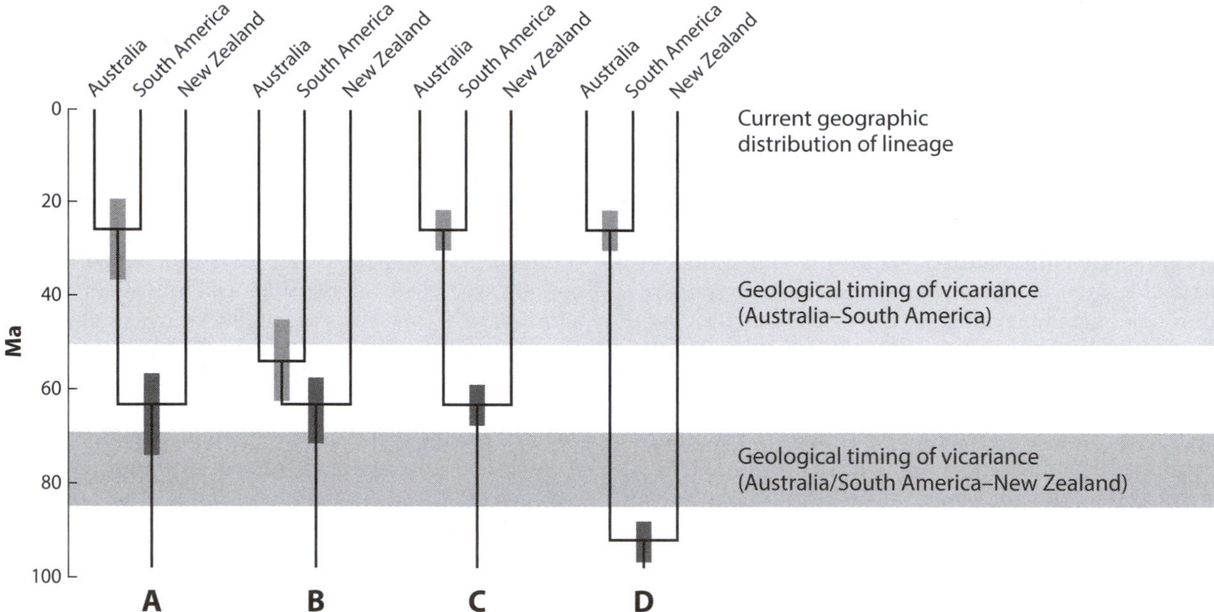

Figure 1. Four clades showing the same area cladogram: New Zealand (Australia + South Africa). The inferred ages for the divergences in Clades A and B (confidence limits indicated by bars at the nodes) are consistent with vicariance caused first by the separation of New Zealand from Australia + South America, and then by the separation of Australia and South America. In contrast, the inferred divergence times in clades C and D are inconsistent with these geological causes; these clades illustrate instances of pseudocongruence. (Modified from *Trends in Ecology and Evolution*, 26/2, Michael D. Crisp, Steven A. Trewick, and Lyn G. Cook, "Hypothesis Testing in Biogeography," p. 69 [2011], with permission from Elsevier.)

other problems, cases of pseudocongruence came to light, in which events that took place at different times in the past yielded the same area cladogram (Donoghue and Moore 2003; figure 1). By the mid-1990s even the leaders of the vicariance movement had become frustrated by the lack of progress. Unfortunately, however, positions hardened. For example, although information on the timing of lineage-splitting events has been widely viewed as potentially useful in sorting out competing biogeographic explanations (e.g., these might be far too old or too young in relation to a particular earth history event; figure 1), such information has been dismissed by some cladistic biogeographers as being too questionable or distracting from the mission of identifying general patterns in area cladograms.

3. INFERRING ANCESTRAL AREAS

One of the hallmarks of the development of cladistic biogeography was the dismissal of the problem of inferring areas of origin for individual lineages or centers of origin for entire biotas. Such pursuits had been central in the history of biogeography, but were denigrated by cladistic biogeographers as fundamentally unscientific. Given such strong objections, it came as a pleasant surprise when the issue resurfaced in the 1990s, now in a phylogenetic framework. This development began with an inference that the common ancestor of the plant clade Asteraceae had probably lived in South America and, following a healthy debate, culminated in the development of a new optimization procedure designed specifically for biogeographic problems (Ronquist 1997). This procedure entailed the key realization that reconstructing ancestral geographic areas was not the same as reconstructing ancestral character states. Area reconstruction needed to place values (or costs) on the events that specifically mattered to historical biogeography, namely speciation, extinction, dispersal, and vicariance. The resulting "event-based" approach, implemented in the software package DIVA, was easily applied to real-world data sets and quickly became the method of choice for biogeographic studies focused on individual clades.

As this discussion proceeded, it was recognized that the problem of reconstructing the geographic evolution of clades was parallel in many respects to the problem of deciphering gene trees within species trees, or of parasites on their hosts (e.g., Page and Charleston 1998). All three of these problems can be conceived as involving trees of different sorts that track one another with greater or lesser fidelity (e.g., in historical biogeography, trees of geographic areas and trees of the taxa in those areas), and it was natural to identify parallel processes in the different systems. For example, vicariant speciation could be equated with codiversification in the case of gene and species trees, speciation within an area to gene duplication, extinction to lineage sorting, and dispersal to lateral gene transfer. The hope was that methods developed in one area could be ported to problems in the other areas, and that a general theory would emerge; however, despite the similarities, there are also disanalogies. For example, in biogeography the repeated movement, at different times, often among a limited set of areas via the same corridor (e.g., iterated movement of lineages through the Bering Land Bridge), might yield many more instances of pseudocongruence than in the other systems. Nevertheless, it is noteworthy that maximum likelihood and Bayesian approaches began to be applied to the gene-tree and host-parasite cases, primarily to evaluate more complex models specifically including time as a variable.

The development of maximum likelihood methods for the inference of ancestral areas got under way with Ree et al. (2005), who developed what was later dubbed the dispersal-extinction-cladogenesis (DEC) model. The parsimony-based approach in DIVA was unable to take into account the timing of splitting events in the underlying lineage tree, or changes in the likelihood of movements among areas through time (e.g., the existence of a corridor during some period but not in others, or the increasing distance between two continents). The underlying model of diversification (how daughter species inherit geographic ranges) is another important concern, and the method developed by Ree et al. (2005) specifically provided for maintenance of a widespread species through a speciation event within a subregion. Initially, likelihood calculations were made using simulations, but an analytical solution (calculating probabilities of dispersal and extinction as functions of time using a rate matrix) was soon implemented in the software package Lagrange. Such approaches are open ended, and recent efforts have incorporated additional relevant information such as the physical sizes of geographic regions and the distances that separate them.

4. A FRESH LOOK AT OLD PATTERNS

Over the past decade, considerable progress has been made in understanding several classical biogeographic patterns. To give the flavor of such findings, I will briefly highlight below just three intercontinental disjunction patterns; similar advances have been made within several continents and biogeographic regions (e.g., the Mediterranean, Australia, and southern Africa), and in some island systems (e.g., Hawaii and the Canary Islands). In general, new dating information—both on lineages and on geological and climatic events—is having an important impact on our interpretations, and concerted movements and long-distance dispersal are once again being considered as possible causes alongside continental movements (de Queiroz 2005).

Aside from insights into intercontinental disjunction patterns, historical biogeography has increasingly entered discussions of global biodiversity patterns and conservation. A good example is provided by the latitudinal species-richness gradient. Many ecological explanations have been put forward for this gradient, but it is possible that it is largely explained by the initial diversification of many lineages in the tropical climates that were widespread in the Paleocene and Eocene, followed by more recent movements into temperate, boreal, and arid biomes with global cooling since the Eocene (e.g., Wiens and Donoghue 2004). A similar explanation may also hold for the equivalent latitudinal gradient seen in marine organisms.

Laurasian connections. Disjunctions have long been evident between eastern Asia and eastern North America, and these are embedded in a broader Laurasian distribution pattern that also involves endemics in Europe and western North America. Dated phylogenies imply that there have been multiple movements of lineages around the Northern Hemisphere, and that many of these were movements through Beringia (as opposed to the North Atlantic Land Bridge) at times when climates were accommodating. It appears that Asia has been a source area for a number of lineages, but movement has occurred in both directions. Fossil evidence indicates that a number of lineages now confined to Asia were more widespread around the Northern Hemisphere in the Eocene and Oligocene, and that ranges have become restricted with the cooling and/or drying of climates in some regions (Europe and western North America, in particular). Collectively, these patterns provide evidence of niche conservatism. Many Northern Hemisphere plant lineages, for example, diversified solely within temperate forests throughout the Cenozoic, though a few radiated into drier regions as these have spread since the Miocene in western North America and around the Mediterranean basin.

Gondwanan connections. Gondwanan disjunction patterns have long attracted attention, with an emphasis on the role of past continental movements. Although there are disjunctions that may well have been caused by the drifting of the southern continents, new information on divergence times cautions against the universal application of such an explanation. Indeed, many Africa/South America plant disjunctions that were once thought to be the result of the breakup of Gondwanaland, are being reinterpreted in light of evidence that their divergences are far younger than 100 million years ago. Disjunctions dated to the Eocene are now often interpreted as reflecting movements through the Northern Hemisphere when climates were warmer and connections across southern Laurasia more continuous. Other recent studies have featured "west-wind drift" as an explanation for significant connections between Australia and New Zealand; these may have been established through repeated dispersal with the prevailing winds after the breakup of the southern continents. Finally, contrary to earlier theories emphasizing the southward migration of groups that originated in the north, a number of studies have argued for a Southern Hemisphere origin of some major, now-cosmopolitan, groups, including birds and the sunflowers and their relatives (Asteraceae).

New World connections. It has long been appreciated that South America existed in relative geographic isolation through much of the Cenozoic, and that regular north-south connections were established only more recently (especially with the formation of the Isthmus of Panama), setting off "the great American interchange." Recent studies have reported a wider than expected range of dates for disjunctions between North and South America, with plant lineages in particular showing a number of older splits, perhaps reflecting more frequent long-distance dispersal than in terrestrial vertebrates. At the same time, however, new geological evidence is suggesting the possibility of earlier corridors between the two landmasses (e.g., in the Eocene-Oligocene some 35–33 Ma), and that the docking of the continents may have taken place considerably earlier than the previously accepted time frame of about 3 million years ago. Regarding directionalities, recent studies have largely substantiated the view that more tropical plant clades moved north from South America, while more high-elevation clades moved from north to south. A number of studies have documented the movement of alpine elements into the Andes from the north, accompanied by greatly elevated rates of diversification. In mammals, where it has been possible to integrate an extensive fossil record, it appears that more lineages moved from North America to South America than in the opposite direction, and the northern elements tended to radiate more extensively in the south (e.g., deer, canids, sigmodontine rodents) than did the southern elements in the north (e.g., sloths, agoutis, and opossums).

5. BEYOND THE STANDOFF

Over the last several decades a counterproductive chasm opened between two broad camps focused on different aspects of historical biogeography. One group has attempted, primarily using area cladograms as evidence, to discern general patterns in disjunctions across multiple lineages, with the aim of assessing past connections among areas of endemism. Another group has attempted to infer ancestral areas and pathways of movement within particular lineages using phylogenetic trees and methods that incorporate varying amounts of other relevant information, such as the absolute timing of splitting events. Both these objectives are not only perfectly legitimate, but also, within limits imposed by the data, attainable. Most importantly, they are complementary and jointly necessary to fully understand historical biogeography. If nothing else, they could directly benefit one another. On the one hand, the search for general patterns would benefit from information on the timing of events. For example, this information would help weed out instances of pseudocongruence that plague the analysis of area cladograms. On the other hand, the inference of past biogeographic movements within particular lineages would benefit greatly from knowledge of patterns and processes in other groups of organisms. Methodologically, it should be possible to bring all this information together in a likelihood or Bayesian framework, and we might even realize the long-imagined feedbacks between biogeography and the earth sciences. For example, information on the history of areas could be incorporated into estimations of divergence times within clades, or biogeographic information could be used in choosing among competing geological models.

Fortunately, a new generation of historical biogeographers has the opportunity for a fresh start, identifying and synthesizing the best ideas and methods that have been developed. The resulting integrative historical biogeography will yield a far richer understanding of the spatial history of lineages and of the earth itself, and will add even greater value to the study of ecology, evolutionary biology, and biodiversity conservation.

FURTHER READING

Crisp, M. D., S. A. Treweck, and L. G. Cook. 2011. Hypothesis testing in biogeography. Trends in Ecology & Evolution 211: 66–72. *A recent critique of the field, with an emphasis on framing testable hypotheses.*

Croizat, L. 1964. Space, Time, Form: The Biological Synthesis. Published by the author. Caracas, Venezuela. *One of several prodigious self-published volumes, this provides an insight into the iconoclastic views of Croizat, the founder of "panbiogeography."*

de Queiroz, A. 2005. The resurrection of oceanic dispersal in historical biogeography. Trends in Ecology & Evolution 20: 68–73. *An argument that the importance of long-distance dispersal across oceans was underestimated in the era of vicariance biogeography.*

Donoghue, M. J., and B. R. Moore. 2003. Toward an integrative historical biogeography. Integrative and Comparative Biology 43: 261–270. *An argument for integrating the timing of lineage-splitting events into biogeographic analyses, especially to identify instances of pseudocongruence.*

Hennig, W. 1966. Phylogenetic Systematics. Urbana: University of Illinois Press. *The founding document of phylogenetic biology, which includes examples of the use of phylogenetic trees in relation to biogeography.*

Lomolino, M. V., B. R. Riddle, R. J. Whittaker, and J. H. Brown. 2010. Biogeography. 4th ed. Sunderland, MA: Sinauer. *A textbook treatment of the entire discipline of biogeography.*

Lomolino, M. V., D. F. Sax, and J. S. Brown. 2004. Foundations of Biogeography. Classic Papers and Commentaries. Chicago: University of Chicago Press. *A compilation and analysis of key historical writings on biogeography; an excellent starting point for understanding the history of the discipline.*

Morrone, J. J. 2009. *Evolutionary Biogeography: An Integrative Approach with Case Studies.* New York: Columbia University Press. *A recent overview of the wide variety of methods in use in historical biogeography.*

Page, R.D.M., and M. A. Charleston. 1998. Trees within trees: Phylogeny and historical associations. Trends in Ecology & Evolution 13: 356–359. *A comparison of historical biogeography with the gene-tree/species-tree problem and the host-parasite codiversification problem.*

Platnick, N. I., and G. Nelson. 1978. A method of analysis for historical biogeography. Systematic Zoology 27: 1–16. *A key early paper on the vicariance/cladistic biogeographic approach.*

Ree, R. H., B. R. Moore, C. O. Webb, and M. J. Donoghue. 2005. A likelihood framework for inferring the evolution of geographic range on phylogenetic trees. Evolution 59: 2299–2311. *The introduction of a maximum likelihood method for historical biogeography, taking into account the timing of lineage-splitting events and changes in the likelihood of geographic movements through time.*

Ronquist, F. 1997. Dispersal-vicariance analysis: A new approach to the quantification of historical biogeography. Systematic Biology 46: 195–203. *The introduction of a parsimony-based approach to the inference of ancestral geographic ranges that minimizes the number of implied dispersal and extinction events.*

Wiens, J. J., and M. J. Donoghue. 2004. Historical biogeography, ecology, and species richness. Trends in Ecology & Evolution 19: 639–644. *An argument for the reintegration of historical biogeography and ecology in explaining global distribution patterns such as the latitudinal species-richness gradient.*

Wiley, E. O. 1988. Vicariance biogeography. Annual Review of Ecology and Systematics 19: 513–542. *An overview of the concepts and methods of vicariance biogeography, after the first decade, with simple explanations of component analysis and Brooks parsimony analysis.*

II.5

Phylogeography

Michael E. Hellberg

OUTLINE

1. Direct interpretation of single-locus gene genealogies
2. Comparative phylogeography
3. Lineage sorting and the coalescent
4. Multilocus gene genealogies
5. Testing models of population history

Phylogeography is the study of the history of populations within species. These studies emerged from analysis of mitochondrial DNA sampled from multiple populations, providing a genealogical perspective within species. Early studies helped identify geographical barriers that separated differentiated populations and suggest where recent range expansions had occurred. The development of coalescent theory led to analyses that could not only discern whether populations were isolated, and, if so, for how long, but also identify demographic changes since that point and past gene flow between populations. An emerging multilocus, model-testing framework places greater emphasis on identifying key factors in shaping population history than on estimating parameter values.

GLOSSARY

Coalescent. Genealogy of alleles tracing back in time to a common ancestral sequence.

Haplotype. Stretch of DNA passed across generations without recombination.

Lineage Sorting. Process during which the genealogies of genes within isolated populations move toward becoming reciprocally monophyletic by the differential loss and replication of gene lineages that were shared at the time of isolation.

Mismatch Distribution. Frequency plot of all pairwise sequence differences between sampled haplotypes from a population.

Reciprocal Monophyly. When the members of two defined groups are all more closely related to other members of their own group than to those of the other.

Phylogeography draws its conclusions from the genealogical analysis of genetic data sampled from populations. These conclusions can reveal the history of populations. Phylogeography thus provides insights into the state of populations during the process of their divergence toward forming new species, allowing us to answer such questions as: How large were the populations? When did they diverge? How much migration occurred after the initial population divergence? Phylogeography can also reveal how populations responded to past climatic changes, thereby suggesting possible impacts of future change. Phylogeography can identify long-isolated populations that are not sustained by demographic connections to other regions, and thus may inform conservation and management programs. Phylogeography can also help reveal our human past, via genetic analysis of our own lineage (By what routes did our ancestors populate the world? Did *Homo sapiens* exchange genes with other hominids?), and through studies of the evolution of pathogens that have helped shape our recent history.

The term *phylogeography* was coined to mark the incorporation of tree thinking (phylogeny) into the study of genetic variation within species. Phylogeography was born of the study of mitochondrial DNA (mtDNA) variation among animal populations. Analysis of these data resulted in a genealogy of mtDNA haplotypes. The ability to discern relationships among these mitochondrial haplotypes, combined with this molecule's uniparental inheritance, appeared to provide the means to trace the splitting and coalescing of populations back through time.

This promise stimulated the collection of millions of base pairs of mtDNA sequence data. Spreading PCR

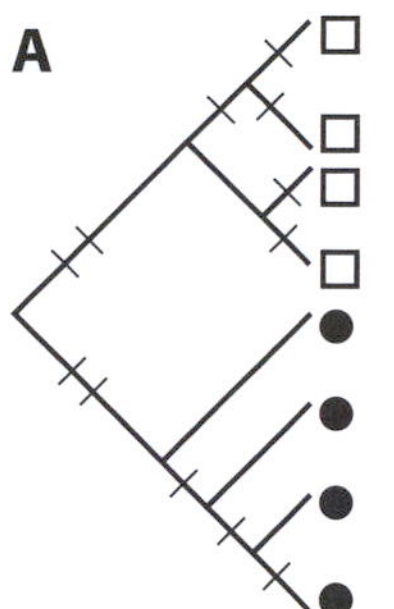

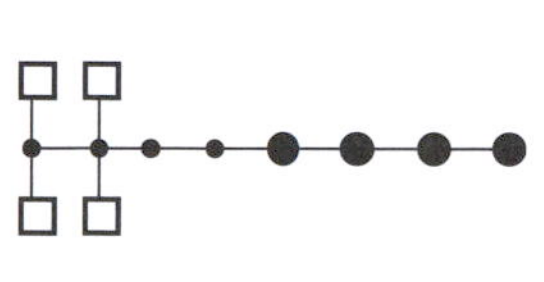

Figure 1. Phylogeny of alleles (A) taken from individuals sampled from either side of a barrier (B). (C) The same alleles shown in a haplotype network, based on the locations of mutations indicated on tree A.

technology, automated DNA sequencing machines, and small set of primers that amplified short (350–800 bp) gene regions across a wide phylogenetic range facilitated this explosion. Possessing a dog-eared photocopy of *The Simple Fool's Guide to PCR*, put together by Steve Palumbi's lab in Hawaii in 1989, marked one as a giddy member of a growing club of explorers. Broad geographical sampling commonly revealed distinct mtDNA clades within species, and these were (sometimes fecklessly) associated with present barriers and past events.

The promises of mtDNA phylogeography were eventually exposed, however, not as a lie so much as a tease. For elucidating recent population isolation or for discerning the details of historical events, trees drawn from mtDNA trees were insufficient. For example, the ability to detect population mixing was limited because traditional phylogenetics rarely considered mixing among its units of study (species), even though such mixing (migration) is central to population genetics. The history of phylogeography has been the continuing reconciliation of the perspectives of phylogenetic systematics and populations genetics, combined with the realization that many of the inferences we thought we could make directly from gene trees based on one locus require more complex analytical tools and new sources of data. Since those initial days, the field has been engaged in a search for new analyses and sources of data that allow us to sate our hunger for the possibilities that mtDNA data first suggested but ultimately could not satisfy.

1. DIRECT INTERPRETATION OF SINGLE-LOCUS GENE GENEALOGIES

The typical phylogeographic study entails first acquiring genetic data from an orthologous region of DNA from multiple individuals sampled from different geographic locales, then inferring the genealogy of alleles from those individuals to make inferences about the degree to which populations have been isolated or connected. Initially, the history of mitochondrial haplotypes was equated with that of the populations from which they were drawn. While the analyses used to infer phylogenetic relationships among alleles were often taken straight from those developed by systematists, the connections of these genealogical patterns to geography were far less sophisticated, often amounting to simply placing a tree of haplotypes over a map of their origins.

In a simple example (figure 1), mtDNA haplotypes are determined for samples taken from across a species' geographic range. Phylogenetic analysis reveals that they fall into two reciprocally monophyletic clades. When this genealogy of haplotypes is mapped on a physical map of their origins, the phylogenetic gap between the genes corresponds to a potential geographic barrier to movement between populations, in this case a mountain range.

Many studies followed this same approach, often revealing phylogeographic breaks corresponding with known biogeographic breaks or with geographic features such as rivers (for terrestrial animals) or strong currents (for marine populations). But geography did not always rule. Most remarkable were instances in which animal behavior trumped all else, such as the distinct mtDNA from American and European eels: both species migrate to the Sargasso Sea to spawn before their larval offspring eventually return to their respective ancestral home waters.

Not every study revealed such patterns, however. In some instances, mtDNA haplotypes were identical across multiple populations, sometimes owing to slow mtDNA substitution rates, as with most plants and some animals, including corals and sponges. For these taxa, chloroplast DNA (like mtDNA, usually with uniparental inheritance) and the internal transcribed spacer (ITS) region between ribosomal RNA genes (whose many copies may be homogenized within a genome) sometimes served as alternative markers for phylogeography. In other cases, some portion of the geographic range was fixed for a single haplotype. In the Northern Hemisphere, the poleward edge of the range of many species showed little genetic diversity (figure 2), a pattern consistent with a recent recolonization following the glacial retreat. The range over which mtDNA diversity was low often

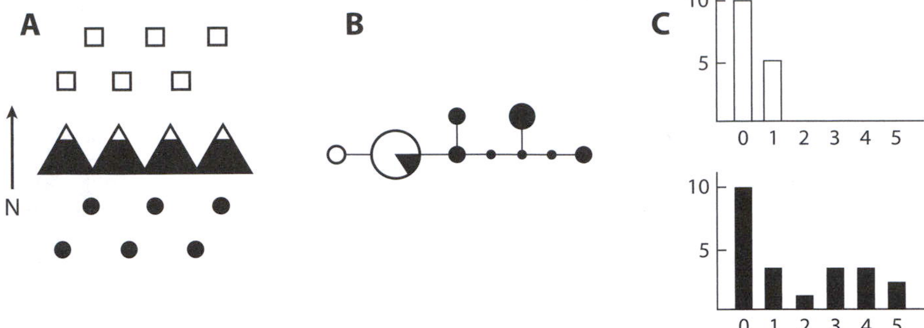

Figure 2. (A) Two populations, the northern poleward of a barrier. (B) Haplotype network of the alleles sampled, with area of circles proportional to the frequency of the haplotype. (C) Mismatch distributions for the two populations, with the top pattern consistent with a recent population expansion following a bottleneck.

coincided with habitat made unsuitable by the equatorial advance of glaciers and cool temperatures during Pleistocene glacial maxima.

Data accumulating around this time (the late 1980s) pointed to other instances of population expansion revealed by mtDNA, most notably the "mitochondrial Eve" data for humans. A realization grew: genealogies of haplotypes within species are not strictly the same as those above the species level, and they should not be analyzed the same way. A first step was to realize that in contrast to phylogenetic analysis, in which one deals only with descendant taxa (sister groups descended from a common ancestor; chapter II.1), ancestral and descendant haplotypes commonly coexist, making bifurcating trees a misleading way to illustrate their relationships. *Haplotype networks* (figures 1 and 2) arose as a means of visualizing the mutational steps among alleles within a species and their frequencies in different populations.

Another useful tool emerged for combining information on the distribution of genetic differences among haplotypes and the frequency of these haplotypes: *mismatch distributions* (figure 2). These are frequency plots of the number of mutational steps between all possible pairs of sequences in a sample plotted against the frequency of each genetic distance class. Early simulation work showed that these mismatch distributions looked qualitatively different for populations that had maintained a constant size compared with those that had recently undergone a population expansion: while constant-size populations yielded erratic peaks and valleys, growing ones consistently produced a single peak in the mismatch distribution. In the extreme, a population expanding from a population in which all members shared the same haplotype would create a wave in the mismatch distribution over time, beginning with all pairwise differences at zero, then sliding to the right in the graph as mutations accumulated.

The position of the wave's peak, in mutational units, can (if a molecular clock applies) be translated into the time since the population expansion. Lessios et al. (2001) used this approach to distinguish between two alternative histories for the Caribbean long-spined sea urchin (*Diadema antillarum*), an important grazer that suffered a massive die-off in 1983. This mass mortality tipped coral reef communities to higher algal cover at the expense of corals. But were these low-coral reefs a newly damaged state, or a reversion to an old form? One view was that the pre-1983 abundance of *D. antillarum* was a recent condition linked to human activities; an alternative hypothesis was that high urchin densities were the norm even before human impacts. A sample of mtDNA from *D. antillarum* did show a single wave in the mismatch distribution, but its peak was at about 3 mutational steps for the 642 bp surveyed, equating to a time long before humans appeared, even when calibrated with the fastest molecular clock.

As illustrated by the *Diadema* example, the direct interpretation of single-locus gene genealogies can provide unprecedented insights into population history. But there are problems. For one, biologists sometimes hope to draw broad conclusions about the history of regional biotas (say, for the purpose of setting conservation priorities), but different taxa sampled from the same populations do not always give the same results. Parameter estimation based on the genealogy of a single gene region often produces high variances and can easily be swayed by locus-specific selection. Furthermore, biologists came to realize that the history of a single gene need not reflect exactly the history of populations or species from which it was sampled. Finally, it became apparent that the human eye is too prone to seeing patterns: a phylogeographic break dropped just about anywhere in a haplotype tree could inspire a post hoc explanation mechanistically linking it to some physical feature

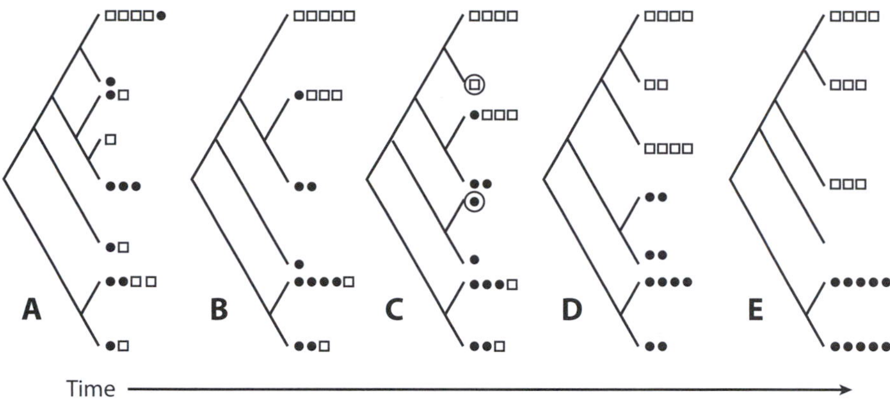

Figure 3. Sorting of alleles into reciprocally monophyletic groups over time.

or past event. A more explicit hypothesis-testing framework was needed.

2. COMPARATIVE PHYLOGEOGRAPHY

Comparative phylogeography examines the population histories of multiple species sampled from the same communities. These studies might aim to identify major barriers or past events common to many species. Regions themselves thus become the focus of study, with resident species serving as replicate recorders of history. Such work is often aimed at recognizing regions whose distinctive histories warrant special consideration in conservation and management. Patterns of differentiation that share the location of phylogeographic breaks are said to be *concordant*. Such geographically coincident genetic breaks may reflect a shared history, as when a strong new geographic barrier emerges, simultaneously splitting populations of many taxa. Whole communities can also respond together to shared changes in their environment, as with the many continental European species experiencing northward range expansions following the most recent glacial interlude.

Shared histories among species are not the rule, however. Co-occurring species often show phylogeographic breaks in different places, or some might show evidence of a range expansion while others have maintained constant population size. For instance, among coastal marine animals, those with intertidal distributions appear to have less stable population histories than do sympatric subtidal species, perhaps because exposure to air makes them more vulnerable to climatic change than species thermally buffered by water. The effects of differences in microhabitats can be explored more quantitatively with *ecological niche modeling*, in which distributions, including past ones, are predicted using a combination of presence-absence data and environmental data such as rainfall and temperature.

Even concordant phylogeographic breaks do not necessarily indicate a common history. Lineages dividing at the same place may have become isolated at different times by that same barrier, or have been in different places before arriving at the common barrier, patterns termed *pseudocongruency*. Such patterns suggest that community members do not always coevolve over sustained periods, and the patterns can also provide insights into how communities are assembled.

3. LINEAGE SORTING AND THE COALESCENT

The reciprocal monophyly of alleles, in which all haplotypes are phylogenetically closer to other members of their own population than to any others, is a sure sign that populations are isolated and experience no gene flow between them. But reciprocal monophyly does not emerge at the onset of isolation; far from it. Consider a well-mixed population that is instantaneously split in two (figure 3A). At first, any differences between the two would be due purely to stochastic effects at initial sampling. As time goes on (figure 3B), two independently evolving populations, both of finite size, will lose some haplotype lineages by genetic drift. Over longer periods (figure 3C), new alleles may arise via mutation, further distinguishing the populations. Eventually, one population will become monophyletic (figure 3D), and after a period of paraphyly, the other will follow, at which point lineage sorting is complete.

As this process of lineage sorting is dictated primarily by genetic drift, the time it takes will be proportional to the effective population size (N_e). For mitochondrial genes, this will take on the order of N_e generations. For diploid nuclear genes with equal numbers of males and females in the population, it will take four times longer.

That such calculations, and others now forming the basis of modern phylogenetic analyses, can be made at all owes to the study of the coalescent: the genealogy of alleles tracing back in time to a common ancestral allele. It is important to recognize that identical alleles still have a nonzero coalescence time, even though we cannot see such coalescences because they are not marked by mutations. Framing things in the retrospective way simplifies calculations, as only the lineages that lead to the haplotypes sampled must be accounted for, not all individuals in the population (as with forward-running simulations).

Coalescent analyses have begun to take full advantage of the richness of information in sequence data from populations. Fundamentally, they provide a null model of what populations that had become recently isolated will look like. From this foundation, additional parameters that impinge on the frequency and phylogenetic relationships of haplotypes can be added, including population size, time of isolation, and levels of migration in each direction (Hey and Nielsen 2004). The statistical validity of adding such parameters can be tested, and their values estimated if addition is warranted.

4. MULTILOCUS GENE GENEALOGIES

While work on the coalescent was motivated by mtDNA data, the large variances that necessarily result from estimating multiple parameters with a single marker led to interest in analyzing multiple loci to infer population history. Multiple markers can improve parameter estimation and increase the power of tests while also allowing for insights that qualitatively exceed anything provided by mtDNA or any single marker.

Multilocus genotypes permit associations among alleles at different loci (linkage disequilibrium) to be the basis for recognizing isolated populations. These clustering analyses have the additional advantage of not requiring that populations or species be defined a priori; identifying these units is part of the analysis. Linkage disequilibrium builds up and breaks down rapidly, allowing populations isolated on the order of dozens of generations to be recognized. Such sensitivity can detect recent phylogeographic breaks not visible to other approaches but, because these analyses are based on genotypes alone and not the relationships among alleles, sensitivity comes at the expense of the ability to infer the timing of the split or to detect hints of any past dispersal between populations.

Different types of loci can also say different things. Because the effective population size of mtDNA is smaller than that of nuclear genes, it is more sensitive to population bottlenecks and recent gene flow. At the same time, bottlenecks recorded by mtDNA may also clear the record of changes in population size that occurred further back in time. Nuclear genes from the roughhead blenny *Acanthemblemaria aspera*, for example, suggest a population expansion about 400,000 years ago, while mtDNA flags a far more recent expansion beginning just 20,000 years ago (Eytan and Hellberg 2010). Gene regions under selection can increase the breadth of past events that can be inferred. Loci under strong balancing selection, such as the MHC loci in vertebrates or S-alleles in self-incompatible plants, can provide a lower limit for the size of past bottlenecks. The use of sequence data to detect past bouts of locus-specific selection does not usually fall within the purview of phylogeography, but suffice it to say that the combination of tests for selection with reconstructions of population history holds much promise.

Critically, work with the coalescent has revealed that replicate genes experiencing the same population histories can show high levels of stochastic variation among their gene genealogies. Thus, from first principles, the gene genealogy of any single gene is unlikely to reflect the genealogy of the species or population from which it was sampled (figure 4). In fact, it has been shown theoretically that, under certain conditions of branch length variation, the most commonly occurring gene tree will *not* be the species tree (i.e., the tree of population splitting). The potential for disagreement between gene trees and species trees is greatest when population sizes are large and when divergence was recent and rapid (Maddison 1997), conditions that match many events in phylogeography. Theory suggest that reciprocal monophyly of nuclear genes is not the most likely outcome for a sample of loci from isolated populations until 1.66 N_e generations after their split (Rosenberg 2003).

Analyses that can identify population isolation despite high interlocus variation and long times to monophyly are thus appropriate for the timescales at which most phylogeographic studies are directed. The number of such analyses is growing, with the present pool divided among faster programs requiring resolved gene trees as input, and more accurate ones that take uncertainty in gene trees into account. The use of such approaches generally requires that terminals (either populations or species) be defined a priori, which has turned attention to the difficult problem of genetically delimiting taxa when few, if any, genes are expected to have coalesced.

5. TESTING MODELS OF POPULATION HISTORY

As more and more complex analyses have offered the potential to estimate an increasing number of parameters, the data generated by empirical phylogeography have not kept up with the needs of these data-hungry approaches. Fortunately, however, the crux of a question

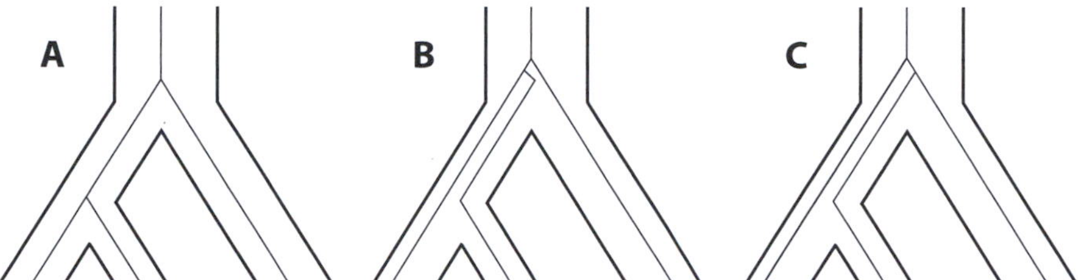

Figure 4. Agreement and disagreement between gene trees and species trees. In (A), the gene tree within the species tree perfectly reflects the species phylogeny in terms of both order and timing of splitting events. In (B), the gene topology matches the species tree, but divergence between the sister species' gene greatly precedes that of the species. In (C), the gene tree disagrees with the species tree.

may hang not on the values that a parameter takes, but rather on alternative hypotheses that can be tested in more tractable ways.

The first steps toward such tests arose from initial inspection of gene trees. Montane grasshoppers live in the northern Rockies, high on mountains in habitat that would have been uninhabitable during recent glacial periods. Knowles (2001) tested two alternative hypotheses for the history of these now-differentiated populations: one in which all sky island populations arose from a single isolation event (having previously been one undifferentiated population), the other in which regional subdivision existed prior to fragmentation. By comparing her mtDNA data against distributions of gene trees generated under conditions matching the two alternatives, she ruled out the null model of a single fragmentation event.

Testing of alternative models is appropriate when prior information suggests a few specific alternatives. A more general approach, and one suitable for a first look at populations with no data on their history, is to test whether particular parameters are needed at all to sufficiently explain the data. For example, one could ask whether isolation alone (and the resultant population-specific coalescent processes taking place) suffices to explain patterns of shared sequence variation in two recently formed populations, or whether migration parameters are also needed. This can again be done by comparing the data to expectations generated by simulation. A conceptually similar approach can be used to ask whether species pairs whose distributions presently are divided by the same barrier were split at the same time. Hickerson et al. (2006) used approximate Bayesian computation (ABC) to estimate the number of different divergence times needed to model splits between eight pairs of urchin species residing on opposite sides of the Isthmus of Panama. Seven of these species pairs, presumed to have been sundered about 3 million years ago, diverged simultaneously, but a final pair appears to have split more recently.

But historical sciences like phylogeography often require many alternative hypotheses. Information theory can enable the simultaneous testing of many alternatives (Carstens et al. 2009). Under this approach, alternative hypotheses are ranked by their power to explain the data, enabling the researcher to identify the particular set of parameters (the model) that best reduces uncertainty about past processes while using the fewest parameters. For example, for two populations that have diverged, likelihoods calculated for a full model that includes ancestral and descendant population sizes and migration in each direction can be compared against submodels from which some of the parameters have been removed. From evidence ratio scores, the information theory metric used to rank hypotheses, the relative strength of each model can be compared. If all the best models have migration parameters set to zero, then the history of the populations is likely one of allopatry. Thus, even when no single model can be identified as best, parameters common to all top models can be found, and a smaller subset of models (now alternative hypotheses) that future work must distinguish can be identified.

FURTHER READING

Carstens, B. C., H. N. Stoute, and N. M. Reid. 2009. An information theoretic approach to phylogeography. Molecular Ecology 18: 4270-4282. *Introduces a way to simultaneously evaluate the fit of multiple alternative models to a phylogeographic data set for a salamander found in the northwestern United States.*

Eytan, R. I., and M. E. Hellberg. 2010. Nuclear and mitochondrial sequence data reveal and conceal different demographic histories and population genetic processes in Caribbean reef fishes. Evolution 64: 3380–3397. *A phylogeographic survey of two hole-dwelling blennies infers different times of past population expansions from mtDNA and nDNA.*

Hey, J., and R. Nielsen. 2004. Multilocus methods for estimating population sizes, migration rates and divergence time, with applications to the divergence of *Drosophila pseudoobscura* and *D. persimilis*. Genetics 167: 747–760. *An updated and empirical application of the influential IM model.*

Hickerson, M. J., B. C. Carstens, J. Cavender-Bares, K. A. Crandall, C. H. Graham, J. B. Johnson, L. Rissler, P. F. Victoriano, and A. D. Yoder. 2010. Phylogeography's past, present, and future: 10 years after Avise, 2000. Molecular Phylogenetics and Evolution 54: 291–301. *Reviews the rise of model-based phylogeographic analyses and points to an emerging synthesis of ecological niche models, studies of natural selection, and community assembly.*

Hickerson, M. J., E. A. Stahl, and H. A. Lessios. 2006. Test for simultaneous divergence using approximate Bayesian computation. Evolution 60: 2435–2453. *Employs ABC to test whether multiple pairs of sea urchins divided by the Central American Isthmus were sundered simultaneously.*

Knowles, L. L. 2001. Did the Pleistocene glaciations promote divergence? Tests of explicit refugial models in montane grasshoppers. Molecular Ecology 10: 691–701. *Early model-based approach to evaluating alternative hypotheses for the divergence of grasshoppers living on Rocky Mountain "sky islands."*

Lessios, H. A., M. J. Garrido, and B. D. Kessing. 2001. Demographic history of *Diadema antillarum*, a keystone herbivore on Caribbean reefs. Proceedings of the Royal Society B 268: 2347–2353. *Employs mismatch distributions of mtDNA to distinguish alternative ecological histories for a Caribbean urchin.*

Maddison, W. P. 1997. Gene trees in species trees. Systematic Biology 46: 523–536. *Classic discussion of the relationship between these two kinds of phylogenetic trees.*

Rosenberg, N. 2003. The shapes of neutral gene genealogies in two species: Probabilities of monophyly, paraphyly, and polyphyly in a coalescent model. Evolution 57: 1465–1477. *Eye-opening exploration of the number of generations it takes for samples of nuclear genes from two recently formed species to become reciprocally monophyletic.*

II.6

Concepts in Character Macroevolution: Adaptation, Homology, and Evolvability

Allan Larson

OUTLINE

1. Darwinism and character variation
2. Evolutionary analysis of character homology
3. Testing hypotheses of character adaptation
4. Character evolvability

Evolutionary analysis requires deconstructing an organism into separately measurable parts that we call characters. This operation succeeds if the characters have biological validity, representing semiautonomous units of evolutionary change within the context of the organism as a whole. All empirical tests of Darwinian evolutionary theory rely on the biological validity of the characters constructed to test it. This article presents a critical analysis of the evolutionary character concepts used to test Darwinian evolutionary theory in a macroevolutionary framework (comparisons among species encompassing millions of years of evolutionary time). The term *macroevolution* often carries connotations of a rejection of Darwinian evolutionary theory or at least a perception that Darwinian theory is inadequate to explain major features of evolution. I briefly review the major conjectures of Darwinian evolutionary theory and the ways in which construction of characters and analysis of their variation can test these conjectures.

GLOSSARY

Adaptation (as a Process). Evolution of a population by natural selection in which hereditary variants most favorable to organismal survival and reproduction are accumulated and less advantageous forms discarded; includes *character adaptation* and *exaptation*.

Character Adaptation. A character that evolved gradually by natural selection for a particular biological role; character adaptation contrasts with *disaptation*, *exaptation*, and *nonaptation*.

Convergence. Evolution of superficially similar characters by different developmental means in different population lineages. In diagnostic tests of homology, it passes the conjunction test but fails tests of similarity and congruence.

Deep Homology. Relationship between similar morphological characters that evolved in parallel by separate evolutionary activations of homologous developmental pathways.

Developmental Constraint. A bias in the morphological forms that a population can express, caused by the mechanisms and limitations of organismal growth and morphogenesis.

Disaptation. A character that reduces fitness relative to contrasting conditions evident in a population's evolutionary history. A primary disaptation is disadvantageous within the populational context in which it first appears; a secondary disaptation acquires a selective liability not present at its origin as a consequence of environmental or evolutionary change.

Evolvability. Ability of a population to produce new morphological characters by mutation or genetic recombination, often by activating latent developmental modules.

Exaptation. Co-option of a character by natural selection for a biological role other than one through which the character evolved by natural selection.

Function. Biological role through which an adaptive character evolved by natural selection.

Genetic Assimilation. By experimentally selecting individuals most susceptible to an environmentally induced change of development, a formerly latent developmental module comes to be expressed even without the environmental treatment formerly needed to activate it.

Gradualism. Accumulation of individually small quantitative changes in organismal form in a population leads over many generations to qualitative change in organismal structure; contrasts with *saltation*, in which a single genetic change induces a large qualitative change in organismal structure.

Homology. Two characters are homologous if they derive, with or without some modification, from an equivalent character of a common ancestor. Diagnostic tests of character homology include similarity (physical resemblance), conjunction (alternative states do not occur together in the same organism at the same developmental stage), and congruence (sharing of homologies among species forms a nested hierarchy of groups within groups that can be summarized as a cladogram). Homology contrasts with serial homology (fails conjunction test), *parallelism* (fails congruence test), and *convergence* (fails similarity and congruence).

Modularity. As applied to development, a process of pattern formation or morphogenesis that is semiautonomous with respect to other aspects of organismal development, and which produces a characteristic arrangement of morphological substructures in the adult body. Developmental modules often feature characteristic patterns of gene expression. Ectopic expression of a module during organismal development can lead to evolution of new structures.

Nonaptation. A character selectively indistinguishable from contrasting conditions present in a population's evolutionary history.

Orthology. Homology relationship between DNA sequences whose genealogies coalesce to a common ancestral molecule with no intragenomic gene duplication (intragenomic here referring to a haploid genome) and no horizontal transfer between genomes of different organisms.

Parallelism. Origins of similar characters independently in two different population lineages, usually because these lineages share homologous developmental constraints that channel production of morphological variation in similar directions. It is diagnosed by failure of the congruence test of homology but passing tests of similarity and conjunction; reversal of a derived character to an ancestral condition is a special case of parallelism using this diagnosis.

Paralogy. Homology relationship between DNA sequences whose genealogical coalescence to a common ancestral molecule includes at least one intragenomic gene duplication (intragenomic here referring to a haploid genome) but no horizontal transfer between genomes of different organisms.

Saltation. Evolution of a large, qualitative change in phenotype in a single mutational step; contrasts with *gradualism*. Also, genetic assimilation of a qualitative change in organismal structure initially caused by an environmental treatment.

Xenology. Homology relationship between DNA sequences whose genealogical coalescence features at least one horizontal transfer between genomes of different organisms.

1. DARWINISM AND CHARACTER VARIATION

My favorite concise account of Darwinian evolutionary theory is by Mayr (1985). He consolidates the many connotations acquired by the term *Darwinism* into five principal theories testable by measuring character variation.

The first and most fundamental theory is that life has a long history of irreversible change with hereditary continuity from past to present life. Mayr (1985) calls this theory *Evolution as Such*. Cambrian organisms of the Burgess Shale, for example, would not be mistaken for any organisms alive today because their characters contrast with those of living forms, yet those characters reveal homologies critical for establishing historical continuity between extinct and living forms.

Darwin's "second" theory states that all past and present forms descend from a shared common ancestor of life on earth (called *Common Descent* by Mayr). Life's history thus takes the form of a branching tree of population lineages. This theory makes the prediction that sharing of characters among species forms a nested hierarchy of groups within groups. Common descent of species was firmly established during Darwin's lifetime by studies of morphological characters, and we now measure its details with great precision using character variation revealed by DNA sequence data.

Multiplication of Species denotes the spatial dimension of evolution in Darwinian theory, geographical processes by which population lineages branch to form two or more descendant lineages. A lineage is an unbranched series of ancestor-descendant populations through time. Geographic isolation of two populations typically precedes evolution of genetic differences that prevent them from merging should they make secondary geographic contact. Character variation across geographic space is the primary means by which species lineages are diagnosed, based on Darwin's principle of divergence of character among geographically isolated populations.

The remaining two Darwinian theories pertain specifically to populational processes of evolution, typically measured as change in organismal morphology. *Gradualism* states that quantitative change in organismal characters leads to qualitative change; by accumulating, over many generations, hereditary variants that individually have very small effects on organismal appearance, diverging populations eventually acquire sharply contrasting morphological characters. This is Darwin's theory of gradualism, whose alternative, traditionally called *saltation,* is that sharply contrasting characters arise as such within a generation. Saltation is compatible with but not identical to *punctuated equilibrium*, which postulates that morphological change occurs in geologically brief (fewer than 1 million years) events of branching speciation, followed by morphological evolutionary stasis within species maintained over a much longer interval of evolutionary time. Saltation associated with branching speciation is a special case of punctuated equilibrium, as would be phenotypically continuous divergence that leads to qualitative change at the formation of a new species. A saltation occurring within a species and not associated with branching speciation does *not* constitute punctuated equilibrium. *Punctuated phyletic evolution* denotes a punctuated pattern of character evolution not associated with branching speciation.

Darwin's fifth theory, *Natural Selection*, is itself a composite of many subtheories. New variation occurs at random with respect to its potential utility to an organism that possesses it. Variation is "heritable," in the sense that organisms resemble their parents more closely than they do individuals drawn at random from their population. Variant forms that enhance their possessors' fitness are thereby transmitted to the next generation at a higher rate than are contrasting characters less conducive to survival and reproduction. By accumulating many such changes across many generations, a population can gradually construct a new character qualitatively different from ancestral conditions.

If a new character arises by saltation, natural selection might increase its frequency in the population if the new character enhances fitness, but natural selection does not formally explain the character's origin. The mechanistic explanation lies alternatively in specifying how a genetic change and its interaction with environmental conditions during development produced a discontinuous phenotype. *Developmental constraint* denotes the hypothesis that the structure of organismal development, particularly specific interactions between proliferating cells at critical stages, makes some morphological forms more accessible to a population than are various conceivable alternatives.

Developmental constraint is compatible with gradualism if its main consequence is to make certain directions of continuous character change more accessible than others. Hypotheses of developmental constraint nonetheless often include an argument that disparate morphologies often lack developmentally accessible intermediate conditions. If the physical processes of cell proliferation and differentiation during development preclude a gradual transition between disparate states of a character, then the developmental-constraint hypothesis is saltational.

Darwin and many of his followers have argued that single variants of large phenotypic effect are inevitably detrimental and that natural selection would eliminate them. This assumption underlies Darwin's commitment to gradualism, but numerous discoveries in evolutionary developmental biology now challenge this assumption.

2. EVOLUTIONARY ANALYSIS OF CHARACTER HOMOLOGY

Darwin's theory of common descent is the foundation for testing the biological validity of character constructs through the concept of evolutionary *homology*: two characters are homologous if they derive, with or without some modification, from an equivalent character of a common ancestor.

Evolutionary homology was applied first to organismal anatomy and form. The forelimbs of a human and an orangutan are homologous as vertebrate forelimbs because they descend, with much modification, from the forelimbs of a common ancestral form. Wagner (1989) elaborated in his concept of biological homology the properties that we expect of homologous organismal structures. First, homologies are historically unique; they arise in a particular population lineage at a particular place and time and occur only in the descendants of that lineage. Second, they have evolutionary continuity; two characters are homologous to each other only if there is an unbroken chain of lineal descent connecting them to each other and to their common ancestral origin. Third, homologies are individuated; they exist as semiautonomous components within the context of the organism as a whole. A vertebrate forelimb, for example, has an individual evolutionary history and semiautonomous developmental dynamic within an organism.

We construct characters and test hypotheses of their homology also at the cellular level. Among the many cellular-level characteristics used in evolutionary studies are the detailed structures of chromosomes as they appear during cell division. Chromosomes are homologous to each other if they descend with some modification from a common ancestral chromosome. Chromosomes of an orangutan each have homologous chromosomes in human cells despite some minor rearrangements of chromosomal contents. Perhaps most effective for testing precise hypotheses of common descent of species is homology at the level of DNA sequences. For example, the gene encoding hemoglobin β in humans is homologous to the gene encoding hemoglobin β in orangutans; these DNA segments descend with modification from an ancestral gene encoding hemoglobin β.

Having defined the concept of homology, we must establish principles for testing hypotheses of character homology, evaluating whether the general principle of homology explains our comparisons for a particular set of characters. There are three diagnostic tests of homology that can be applied at the organismal, cellular, and molecular levels (Patterson 1988). Whether a set of structures passes or fails these tests separates homology from contrasting explanations for character variation and resemblance.

The first diagnostic test is that of character *similarity*. This test is the definitive one at the molecular level. If one compares the DNA base sequences of two pieces of DNA and finds greater than 70 percent sequence similarity, then those sequences undoubtedly trace to a common ancestral DNA sequence. The hereditary pathway that connects them to their common ancestral sequence is nonetheless sometimes a contorted one, as revealed by the second and third diagnostic tests.

For characters of organismal morphology, similarity testing implies a nontrivial structural correspondence that transcends differences in exact form. For example, the forelimbs of humans, horses, bats, birds, lizards, and frogs are all homologous to each other as vertebrate forelimbs despite enormous dissimilarities in overall form; the homologies reveal themselves in the major bones present and the patterns by which the bones connect to each other and to the rest of the body; specific homologies of the limb bones and their developmental genetic origins are evident across all four-limbed vertebrates and even among bony fishes. Note that although the forelimbs of bats and birds are homologous as vertebrate forelimbs, they are not homologous as wings. The most recent common ancestor of birds and bats had forelimbs that did not form wings, and the structural modifications that make a bird's forelimb a wing are very different from those that make the bat's forelimb a wing. Bird wings and bat wings thus pass similarity testing as vertebrate forelimbs, but they fail similarity testing as wings.

The second diagnostic test for homology is the *conjunction* test; if two organismal structures are hypothesized to be homologous to each other, a single organism should not express both structures at the same stage of development. The hypothesis that human arms and bird wings are homologous as vertebrate forelimbs would be rejected were we to find angels as often depicted in Italian Renaissance artwork; no living or fossil forms, however, have the characteristic arms of humans and wings of birds present in the same organism.

Duplication of body parts is nonetheless common in evolutionary history. Vertebrate forelimbs and hindlimbs have sufficient similarity in skeletal structures that they would perhaps pass similarity testing, but they clearly fail conjunction testing. We apply the term *serial homology* (or *homonomy*) to structurally similar features that pass similarity testing but fail conjunction testing. Intraorganismal duplication of structures separates serially homologous structures from homologous ones. In tracing the evolutionary history of serially homologous structures, one must traverse at least one event of intraorganismal duplication of the structure. For example, multiple intraorganismal duplications separate a neck vertebra from a trunk vertebra.

For molecular characters, the unit of comparison is not the organism but the haploid genome as transmitted through a sperm or egg. DNA segments passing the similarity test must be present one time only per haploid genome to pass the conjunction test. The conjunction test separates molecular sequence homology into the contrasting subcategories of *orthology* and *paralogy*. A pair of sequences is orthologous if their coalescence to a common ancestral molecule features no intragenomic gene duplication (intragenomic here referring to a haploid genome). A pair of sequences is paralogous if its path of coalescence to a common ancestral molecule includes at least one event of intragenomic duplication.

The third diagnostic criterion is the *congruence* test (figure 1). A homology has by definition only a single origin on the tree of life, and we hypothesize that a homology is transmitted from its lineage of origin to all descendant lineages (unless secondarily lost). If we assume that a new homology spreads throughout its population shortly after arising and is not subsequently lost, the homology should characterize all and only the descendants of that ancestral population. Hence, a homology should characterize a particular clade of the phylogenetic tree. Different homologies often characterize different branches of the phylogenetic tree of species, but the tree structure restricts the relationships between the set of species having homology A and the set of species having homology X to one of three possible relationships: (1) sets A and X are identical, (2) one set is nested within the other one, or (3) sets A and X are mutually exclusive. Each of these conditions passes the congruence test. If the two sets intersect but do not meet conditions (1) or (2), that is, if they partially overlap in a non-nested way, then at least one character, A or X, fails the congruence test.

For example, the hypothetical homology of bird wings and bat wings failed similarity testing but passed conjunction testing. The hypothesis that vertebrate wings are homologous fails congruence testing with at least three very strong morphological homologies: feathers, hair, and mammary glands. Feathers characterize all living birds and some fossil forms more closely related to birds than to other living vertebrates. Mammary glands

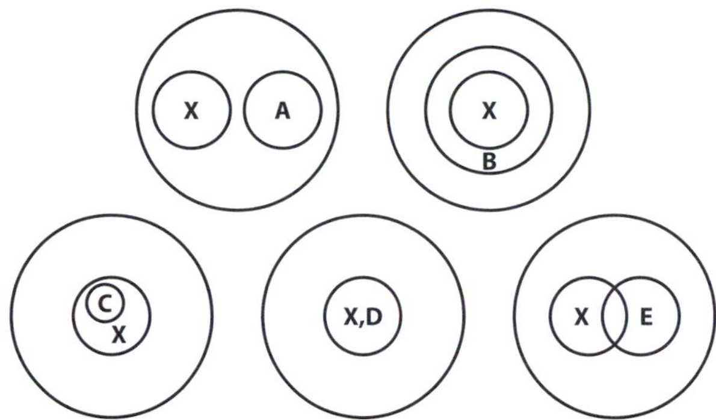

Figure 1. Congruence test of homology as applied to morphological characters. Circles in the Venn diagram denote sets of organisms or species possessing a particular homology or hypothetical homology. Assume that the outer circle denotes all organisms having vertebrae and that X denotes organisms possessing hair or fur. Character X is tested for congruence with characters A–E. Character X passes the test if it forms with others a nested hierarchy of groups within groups. Character A, presence of feathers, illustrates one way to pass congruence testing: the two hypothetical homologies characterize mutually exclusive subgroups of the larger group. Character B, presence of an amniotic membrane in the egg, illustrates another way to pass congruence testing: X characterizes a nested subset of amniotic vertebrates. Character C, presence of a placenta, identifies a nested subset of the organisms that have character X, again passing the congruence test. Character D, presence of mammary glands, identifies an equivalent set of organisms to character X, also passing the congruence test. Character E, presence of wings, fails congruence testing because it partially overlaps with organisms possessing X (bats), but it is not a nested subset of X because organisms possessing feathers (A) also have wings and are outside the group possessing X. Character E must be separated into two different homologies, E_1 = bat wings and E_2 = bird wings. As noted in the text, character E "wings" also fails similarity testing; although bat and bird wings are homologous as vertebrate forelimbs, the forelimb modifications that make them wings are very different. (Modified from Patterson.)

and hair characterize mammals but no other vertebrates. Wings characterize all birds and a subset of mammals (bats), thus failing the congruence test. If we treat bird wings and bat wings as separate homologous characters not homologous to each other, then each one passes congruence testing with each other and with feathers, mammary glands, and hair.

Note that the nested hierarchy of groups within groups corresponds to the pattern of a branching tree of common descent of the species being studied. A tree of life, termed a *cladogram* or *phylogenetic tree*, is a way to depict results of congruence testing of many characters (see chapter II.1). Phylogenetic analysis in all its complexities is fundamentally a congruence test of homologies.

In DNA sequence comparisons that pass the similarity test, failure of the congruence test implies horizontal transfer of DNA between organisms that typically are not familial relatives and often not of the same species. Such transfer may involve viral transmission of DNA from one organism to another. Domestic cats and their closest wild relatives have in their genomes a viral gene absent from other felids and carnivores but that has strong sequence similarity to viral genes in rat genomes. A hypothesis that cats and their closest wild relatives are more closely related to rats than to other felids and carnivores is contradicted by many strong homologies. A better hypothesis is that an ancestor of cats and their closest wild relatives acquired the rat viral gene while feeding on rats, and that the gene inserted into the germ line, became fixed in the ancestral population, and was passed to the descendants of that lineage.

Morphological characters that fail the congruence test but pass the other two (conjunction and similarity) are *parallelisms*. A major discovery of evolutionary developmental biology is that species often share a latent developmental potential to produce characteristics that are not actively expressed in the species at a given time (Carroll 2008). A developmental switch, controlled by interactions among proliferating cells at a critical developmental stage, may determine which pathway is taken. Interactions between genetic and environmental factors influence whether a genetic switch follows one pathway or a contrasting one.

Conrad H. Waddington demonstrated one such developmental switch in the fruit fly *Drosophila melanogaster*. Flies are unusual among insects in having a pair of halteres, also called balancers, in the position where most insects have a pair of hindwings. Although physically very different from hindwings, halteres are homologous to hindwings and represent an evolutionary change that occurred in a common ancestor of all flies more than 200 million years ago. Expression of a pair of hindwings has not been a normal condition in flies since that time, yet a latent developmental potential to produce hindwings remains and can be activated either by genetic mutation (*bithorax* and related mutations) or by treating the developing eggs with ether at a critical stage. Waddington experimentally selected for increased susceptibility of genetically normal flies to respond to ether treatment of the egg by developing hindwings rather than halteres. Many generations of selection produced lineages that developed hindwings at a high rate even without ether treatment or the *bithorax* mutation. He called this phenomenon *genetic assimilation*. A trait originally induced by environmental means (ether treatment) had been stabilized by selective accumulation of the genetic variants most conducive to expressing a developmental pathway that had been latent for more than 200 million years.

The hindwings of Waddington's experimental flies are not homologous to hindwings of other insects, but the developmental *potential* to express hindwings is a separate character likely homologous among all winged insects. Because species share latent developmental pathways, parallelisms for complex characters are more likely than Darwin and his early followers could have anticipated. Because the developmental pathway to produce hindwings, although latent in flies, was presumably constructed by natural selection in an ancient common ancestor of winged insects, activation of this pathway in fly development produces a highly ordered structure with potential utility. This observation negates the traditional Darwinian assumption that mutations of large effect (saltations) are inevitably harmful and eliminated by natural selection.

Evolutionary loss of a homologous body part in different evolutionary lineages is a particularly common parallelism. For example, loss of limbs occurred in parallel in a common ancestor of snakes and in some legless lizard groups, including most amphisbaenians. Thus, absence of limbs is not homologous between snakes and amphisbaenians, as revealed by failure of congruence testing of limblessness with other skeletal homologies. The same is true for the loss of limbs in whales (mammals) and caecilians (amphibians).

Note that parallelism as defined here includes evolutionary reversal as a special case; reversal implies a transformation from an evolutionarily derived character state to an ancestral state, whereas parallelism in a stricter sense features two separate origins of the same derived state. Parallelism is the primary source of error in species taxonomies derived from morphological characters. It is a biologically interesting phenomenon because it usually reveals *deep homology* of developmental pathways that species share despite variation among them in whether the shared pathway is active or latent during typical organismal development.

Convergence denotes cases where hypothetically homologous morphological structures fail both similarity and congruence testing, but pass conjunction testing. These are typically contrasting states of a more inclusive homology, illustrated above by convergent evolution of wings of bats and birds by very different modifications of their homologous forelimbs. In contrast to parallelisms, convergently evolved characters do not imply deep homology of developmental pathways.

Evolution of new homologies often occurs by serial duplication of structures followed by fusion and modification of the repeated parts to form a new structure. This process is perhaps best illustrated by evolution of insects in the animal phylum Arthropoda. Serial repetition of many nearly identical body segments is most evident in millipedes and centipedes, fellow arthropods that lack structures equivalent or homologous to an insect thorax. The insect thorax is hypothesized to have arisen by a fusion of three originally identical segments, each one bearing a pair of legs, as observed in centipedes and millipedes. Fusion of the three segments produced the insect thorax. Individual segments of the insect thorax are serially homologous to each other and individually homologous to particular segments of other arthropods; however, the thorax produced by developmental fusion and elaboration of these segments has no homologue in millipedes and centipedes.

There is no necessary one-to-one correspondence between homology of morphological structures and homology of the genes whose expression contributes to their development. The product of a particular gene typically contributes to multiple, nonhomologous morphological structures. Although shared patterns of gene expression often characterize homologous structures, and shared patterns of gene expression play a role in similarity testing of hypothetically homologous structures, gene expression is not a sure guide to homology. Homologous morphological structures can undergo evolutionary changes in the developmental genetic processes underlying their formation while maintaining evolutionary continuity and thus homology in the fully formed structure. For example, digit formation in salamanders follows a fundamentally different developmental pathway from the one it follows in other terrestrial vertebrates despite clear homology of the digits among terrestrial vertebrates; following the evolutionary origins of digits, a lineage ancestral to living salamanders evolved a novel developmental mechanism for forming the digits without destroying their homology to digits of other vertebrates. This change is evident in contrasting developmental constraints on loss of digits in salamanders versus those in other terrestrial vertebrates (figure 2): parallelisms for loss of digits typically start with loss of the fifth digit in hindlimbs of salamanders, whereas such loss typically begins with loss of the first digit in other tetrapods.

Experimental study of deep homology reveals *modularity* in patterns of development and associated gene expression. Modularity features coordinated changes in separately measurable components of morphology arising from shared developmental processes and patterns of gene expression (see Wagner et al. 2007 for a thorough discussion of modularity). Genetic and developmental modules are often called toolkits. A given developmental module presumably evolved by natural selection in the context of one morphological structure, but this does not prevent its later being activated at a new location in the body and at an atypical time during development to generate a new structure. Most biologists would not consider these new structures homologous to the original structure despite their using the same module. For example, patterns of expression of homeobox genes in vertebrate limbs reveal modules homologous to

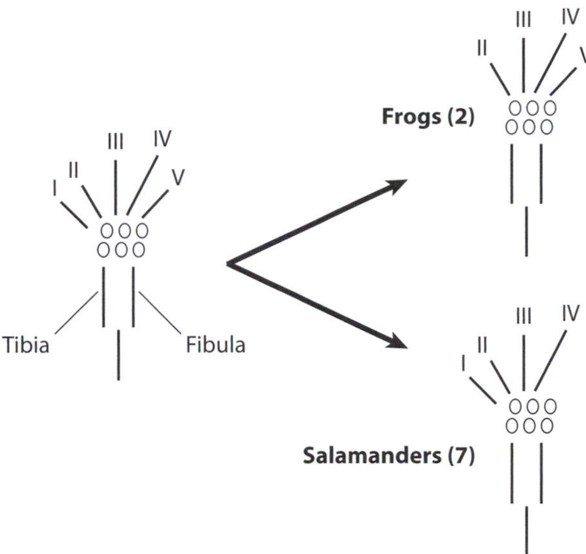

Figure 2. Empirical evidence for contrasting developmental constraints on loss of digits from the hindlimbs of frogs and salamanders. The inferred ancestral condition in both frogs and salamanders is to have five digits on the hind foot. Digits are numbered I–V based on their characteristic appearance and ordering relative to the fibula and tibia. The evolutionary history of frogs features two evolutionary losses of a single digit; in both cases, digit I is lost. Experimental treatment of frog limb buds with a mitotic inhibitor can cause species that normally have five digits to lose one, always the first digit. In contrast, in salamanders, seven independent evolutionary reductions in digit number from V to IV feature loss of digit V, which is also the digit lost on treatment of *Ambystoma mexicanum* with a mitotic inhibitor (see figure 4B). The developmental constraint seen in frogs also characterizes other tetrapod vertebrates, whereas the contrasting constraint in salamanders is a uniquely derived condition in that group. (Modified from Alberch 1991.)

ones used in the development of the caudal axial skeleton of the same organism. The best evolutionary hypothesis is that the module evolved originally as a means for forming the caudal skeleton, followed by a separate expression of the module in the pectoral and pelvic body regions to induce formation there of bony, paired appendages. Nonhomologous characters thus can share homologous modules of gene expression.

3. TESTING HYPOTHESES OF CHARACTER ADAPTATION

Adaptation, the observation that the form of a character serves its biological utility, was at one time thought to contradict biological evolution. The apparent correspondence between form and function implied teleology in the origin of characters, that a preexisting purpose guided their design. Darwin's theory showed, however, that by accumulating the most favorable of randomly produced variants while discarding less favorable ones, a population would evolve useful structures without a predetermined goal. Since the acceptance of Darwin's theory, the term *adaptation* has come to denote a character that evolved gradually by natural selection because successively accumulated variations contributed incrementally to a biological role evident in the resulting composite structure. Adaptation is not necessarily the most parsimonious explanation for all characters, and it requires a sequential testing of relevant hypotheses.

Investigating the role of natural selection in evolution requires that we expand our testing of homology to include not only presence of the character in question but also its utility to organisms possessing it. Assessing the biological role of a character requires identifying the environmental contexts in which the character occurs and inferring environmental contexts of its past evolutionary history.

To argue empirically that a character evolved by natural selection, one must first reject the null hypothesis that the character in question and its historically antecedent condition are equivalent in their utility to the organisms bearing them (Baum and Larson 1991). Adaptation is always relative. Darwin argued that the characters constructed by natural selection are not expected to be perfect in any role, only slightly better than the alternatives against which they have had to compete in their evolutionary history. Depending on the character whose utility is being studied, appropriate tests might exploit the existence of alternative forms segregating in the same population, for instance the sickle-cell allele of hemoglobin β and the contrasting allele called hemoglobin A in various human populations. In other cases, the test might involve comparing different living species that share similar morphological and ecological characteristics except for the character in question; for example, populations of *Xiphophorus* fishes that have a sword on the caudal fins of males might be compared with those retaining the ancestral condition of lacking the sword. In yet other cases, physical manipulation and experimental analysis or biomechanical modeling might be used to assess the relative utility of ancestral and derived characters. In all such tests, phylogenetic analysis of character origins is a critical tool for identifying the contrasting conditions to be compared (Baum and Larson 1991).

Gould and Vrba (1982) introduced the term *nonaptation* to denote a character that has no detectable utility, one for which the null hypothesis of the preceding paragraph cannot be rejected. The contrasting term *aptation* denotes any character that enhances survival or reproduction relative to its antecedent condition. They divided aptation into two subcategories, adaptation and *exaptation*. An adaptation is defined as an aptation that originated by natural selection for a biological role that the character continues to serve. An exaptation, in contrast, is an aptation that serves a role other than one for which the character might be inferred to have evolved by natural selection. Gould and Vrba (1982) use as an example the utility of feathers for flight in birds. Phylogenetic analysis of birds and their fossil relatives shows that the origin of feathers preceded the origin of flight, thus rejecting the hypothesis that feathers could have originated by natural selection for enhanced flight. Feathers arose coincidentally with evolution of homeothermy in birds, favoring an explanation that feathers are an adaptation for thermoregulation but an exaptation for flight.

Baum and Larson (1991) expanded this terminology to include the possibility that a character is deleterious relative to its antecedent condition. A primary disaptation is one whose lowered utility pertains to the environmental conditions under which the character arose, whereas a secondary disaptation is one that became relatively unfavorable following a subsequent change in environmental conditions. A primary disaptation usually signals processes analogous to natural selection that operate at the level of gene replication and transmission. *Selfish DNA* is the hypothesis that some DNA sequences, and also possibly their consequences on organismal morphology, become prominent in a population because *selection at the genic level* favors their proliferation. For example, occurrence of a tailless condition in some mouse populations is associated with a genetic allele that produces tailless mice in the heterozygous condition and lethality when homozygous. The morphological condition is undoubtedly detrimental relative to the antecedent condition (normal development of the tail) in relevant populations. Nonetheless, during spermatogenesis in heterozygous individuals, gametes containing the tailless allele physically destroy those containing the contrasting allele (reviewed by Burt and Trivers 2006). Most offspring of heterozygous males receive the tailless allele. Occurrence of the tailless allele thus represents an interaction between natural selection, which acts to decrease the frequency of the allele, and selfish DNA, which acts to increase its frequency in the population.

4. CHARACTER EVOLVABILITY

Evolutionary biologists sometimes try to evaluate the potential of a species for undergoing further evolutionary change and diversification, including production of new species and new characters. Darwinian theory and its population-genetic models make clear that large geographic distributions and large amounts of genetic variation enhance the opportunities for a species lineage to give rise to new species and new characters. Because genetic variation in a population often stabilizes organismal development rather than expressing itself as greater variation in organismal morphology, a species that is relatively uniform in organismal morphology nonetheless can have the potential to produce a great range of organismal morphologies. If genetic variation is reordered, as might occur in the founding of a new population by a small number of individuals drawn from the ancestral one (Carson and Templeton 1984), organismal development can be destabilized to reveal alternative morphologies whose developmental pathways were latent in the ancestral population. Changes in genetic variation in this case act analogously to the environmental challenges (ether treatment of fruit fly eggs) that shifted development in Waddington's genetic-assimilation experiments discussed in part 2.

Evolvability has emerged over the past 20 years as a concept that encapsulates the potential of a species to produce new organismal characters. Pigliucci (2008) notes that this term carries disparate connotations among authors, a major contrast being whether the term pertains strictly to expressed morphological variation or the potential to produce unexpressed characters by new mutation or recombination: "Variation is a measure of the realized differences within a population, whereas variability is the propensity of characters to vary (whether or not they actually do) and depends on the input of new genetic variation through mutation or recombination." Evolvability should pertain to "variability" and not just to "variation" as Pigliucci (2008) defines these terms. Mutagenesis experiments, perturbing development by environmental treatments, and founding laboratory strains by small numbers of wild-caught individuals are means of measuring variability. Such experiments potentially reveal the diversity of latent developmental modules whose expression is inducible by genetic change.

Conceptual diagrams used by Alberch (1991) illustrate the concept of evolvability and the ways in which two hypothetical species can differ in evolvability. Figure 3A shows a series of six alternative organismal morphologies (A–F) whose expression depends on interactions between quantitative values of two parameters, X_1 and X_2. The parameters might be the amount of a gene product, variation in the multilocus genotype of a quantitative developmental character, or environmental conditions, such as temperature, that influence cellular growth or proliferation. An organism can express only one of the six alternative developmental outcomes, depending on the specific inductive interactions that occur among cells during a critical moment in development. Continuous changes in genetic or environmental conditions can switch development from one pathway to another as shown. Species 1 expresses

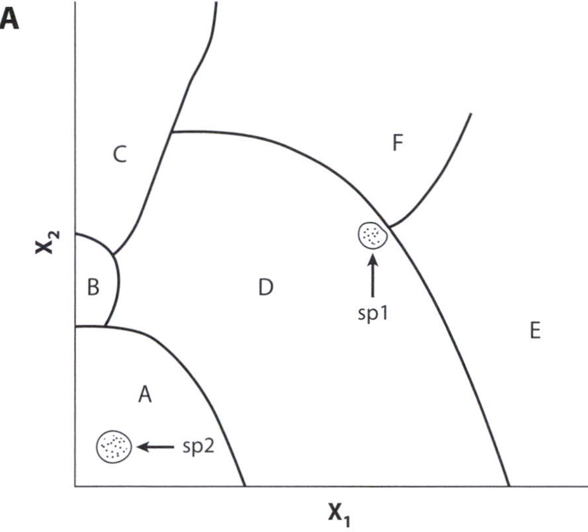

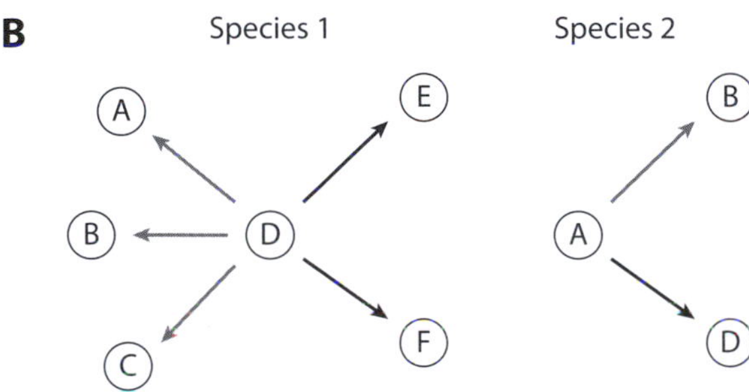

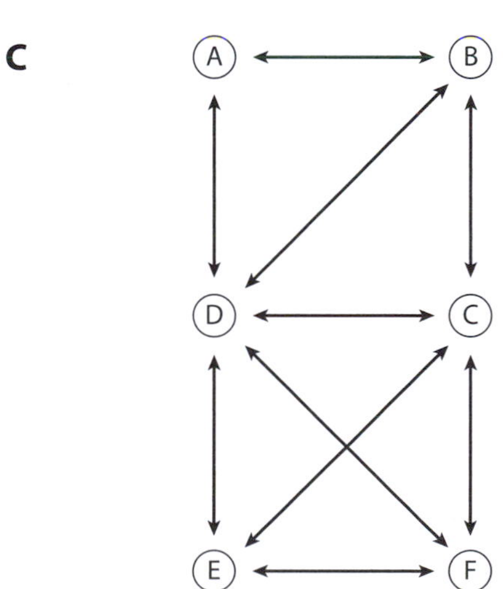

Figure 3. Alberch's (1991) conceptual diagram of character evolvability for two species. (A) Parameters X_1 and X_2 denote continuously varying genetic or environmental variables that influence embryonic induction to specify one of five contrasting developmental pathways (A–F). These alternative pathways could be, for example, those specifying contrasting hindlimb structures as illustrated in figure 4. Species 1 has parameter values specifying developmental pathway D, although it is near the critical thresholds for expressing pathways E and F. Species 2 has parameter values well within the range that specifies pathway A. (B) Evolvability of species 1 and 2 contrasted according to whether alternative developmental pathways are directly accessible by changes in parameters X_1 and X_2. Thick arrows denote transformational changes more accessible than those depicted by thin arrows. (C) Transformational diagram corresponding to the parameter space in part A. On the basis of direct accessibility of alternative developmental pathways by changes in parameters X_1 and X_2, species expressing pathway D have the highest evolvability (five connections) and those expressing pathway A have the lowest evolvability (two connections), with pathway C having four connections and the remaining pathways each having three.

pathway D, which borders in parameter space all the other contrasting pathways. Species 2 expresses pathway A, which borders only pathways B and D in parameter space. Species 1 has a higher evolvability than does species 2 in this diagram because continuous changes in parameters X_1 and/or X_2 access a greater number of contrasting pathways than would comparable changes in species 2 (figure 3B). Figure 3C summarizes the relative evolvabilities of species expressing each of the contrasting developmental pathways. Relative evolvabilities of these pathways in decreasing order are D (five connections), C (four connections), B = E = F (three connections), and A (two connections). Note that although the parameter space is continuous, differences between the organismal morphologies represented by pathways A–F can be discontinuous (figure 3C).

An example from Alberch's work on foot morphology in salamanders illustrates morphological outcomes of different developmental pathways for the hind foot (figure 4). Figure 4A shows the skeletal structure of the hind foot characteristic of *Ambystoma mexicanum*. Figures 4B and 4C show alternative morphologies produced experimentally by treating the developing limb buds with the mitotic inhibitor colchicine, which alters inductive interactions in the developing limb. Both experimentally produced abnormal limbs match the normal limbs of distantly related salamander species. *Hemidactylium scutatum* (figure 4D) normally expresses the developmental pathway produced experimentally in *A. mexicanum* shown in figure 4B. *Proteus anguinus* normally expresses the developmental pathway induced in *A. mexicanum* shown in figure 4E. These results show that *Ambystoma mexicanum* shares with each of the other species as a "deep homology" a developmental pathway normally expressed in the other species but latent in *A. mexicanum*.

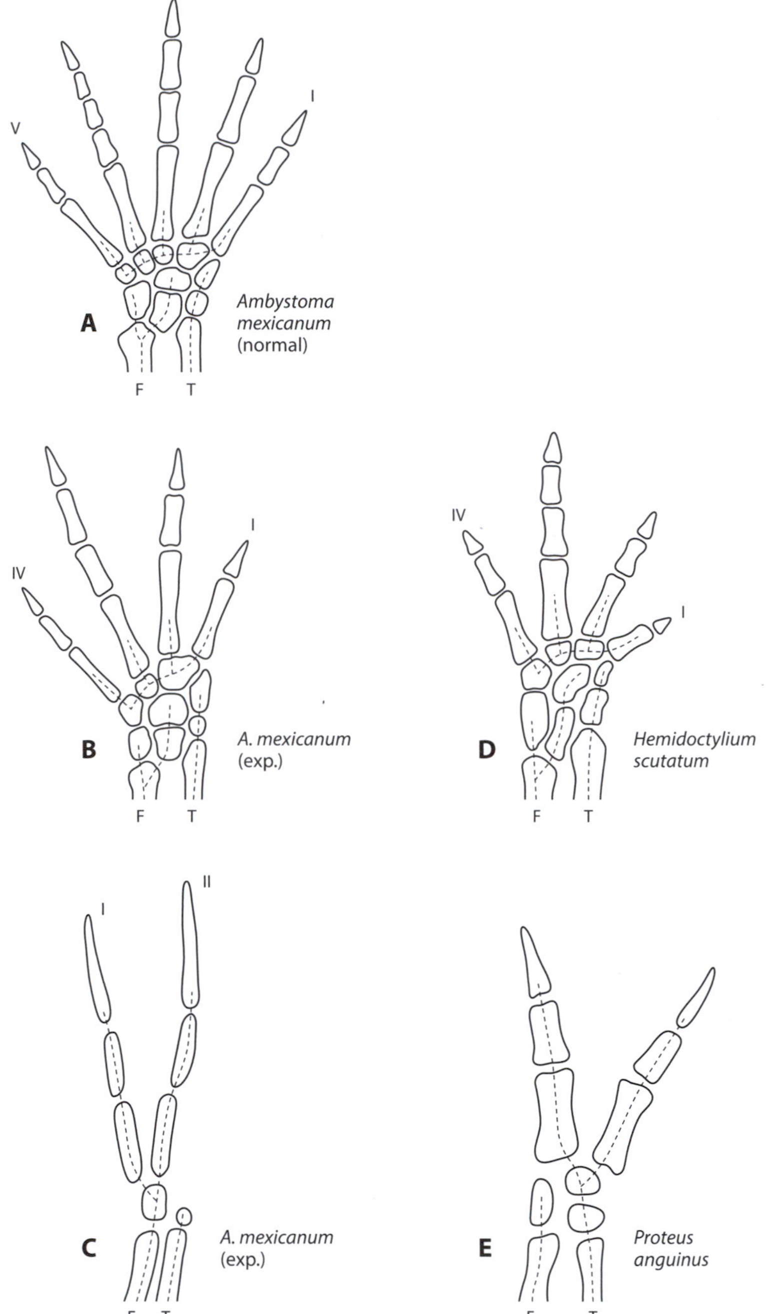

Figure 4. Morphological outcomes of three alternative developmental pathways for the structure of the hind foot in salamanders. Morphology A characterizes adult *Ambystoma mexicanum*. Morphologies B and C represent two alternative pathways activated by treating developing limb buds with the mitotic inhibitor colchicine. Morphology B resembles the normal condition expressed in the species *Hemidactylium scutatum* (morphology D). Morphology C resembles the normal condition expressed in the species *Proteus anguinus* (morphology E). The contrasting developmental pathways expressed by *Hemidactylium scutatum* and *Proteus anguinus* thus occur in latent form in *Ambystoma mexicanum*, indicating shared deep homology of developmental pathways among these species. Should one of these species have access by genetic mutation or recombination to a larger number of contrasting developmental pathways for foot skeletal morphology than do the other species, that species would have greater evolvability for this character. If all three species share the same set of accessible developmental pathways, the one capable of activating alternative conditions with the smallest amount of genetic change would be judged to have greater evolvability for this character. F=fibula; T=tibia (Modified from Alberch 1989.)

It is possible, although not yet empirically demonstrated, that the various salamander species in figure 4 differ from each other with respect to evolvability of the skeletal structure of the hind foot. If genetic modification of *A. mexicanum* can more easily access a larger number of alternative developmental pathways than are available to either *H. scutatum* or *P. anguinus*, then *A. mexicanum* would have higher evolvability for this character. Quantitative comparisons of species for evolvability of characters is a new research endeavor that promises to reveal the developmental and genetic factors underlying evolution of new homologies and parallelisms in character evolution.

FURTHER READING

Alberch, P. 1989. The logic of monsters: Evidence for internal constraint in development and evolution. Geobios: Mémoire spécial 12: 21–57. *A good summary of the theory and phenomenology underlying the hypothesis of developmental constraint in evolution.*

Alberch, P. 1991. From genes to phenotype: Dynamical systems and evolvability. Genetica 84: 5–11. *Explains the conceptual transition from developmental constraint to evolvability.*

Baum, D. A., and A. Larson. 1991. Adaptation reviewed: A phylogenetic methodology for studying character macroevolution. Systematic Zoology 40: 1–18. *A thorough discussion of empirical tests of character adaptation, disaptation, exaptation, and nonaptation.*

Burt, A., and R. Trivers. 2006. Genes in Conflict: The Biology of Selfish Genetic Elements. Cambridge, MA: Harvard University Press. *Selfish genetic elements provide a mechanism by which a primary disaptation can persist in a population despite natural selection against it.*

Carroll, S. B. 2008. Evo-devo and an expanding evolutionary synthesis: A genetic theory of morphological evolution. Cell 134: 25–36. *Explains how evolution of genetic regulatory systems can stabilize developmental pathways.*

Carson, H. L., and A. R. Templeton. 1984. Genetic revolutions in relation to speciation phenomena: The founding of new populations. Annual Review of Ecology and Systematics 15: 97–131. *Explains how founding of new populations by a small number of individuals can reveal previously unexpressed phenotypes, thereby providing a means of measuring evolvability.*

Gould, S. J., and E. S. Vrba. 1982. Exaptation: A missing term in the science of form. Paleobiology 8: 4–15. *A classic work providing precise terminology for describing the role of natural selection in character evolution.*

Larson, A. 2009. Adaptation. In S. Levin, ed., The Princeton Guide to Ecology. Princeton, NJ: Princeton University Press. *Discusses testing hypotheses of character adaptation, disaptation, exaptation, and nonaptation in a microevolutionary context.*

Mayr, E. 1985. Darwin's five theories of evolution. In D. Kohn, ed., The Darwinian Heritage. Princeton, NJ: Princeton University Press. *An excellent concise summary of Darwinian evolutionary theory.*

Patterson, C. 1988. Homology in classical and molecular biology. Molecular Biology and Evolution 5: 603–625. *Compares empirical tests for character homology at the levels of organismal morphology and DNA sequences.*

Pigliucci, M. 2008. Is evolvability evolvable? Nature Reviews Genetics 9: 75–82. *A detailed discussion of the concept of evolvability and how it transcends traditional Darwinian theory.*

Wagner, G. P. 1989. The origin of morphological characters and the biological basis of homology. Evolution 43: 1157–1171. *A thorough discussion of the biological basis of morphological homology.*

Wagner, G. P., M. Pavlicev, and J. M. Cheverud. 2007. The road to modularity. Nature Reviews Genetics 8: 921–931. *Explains developmental modules as well as modularity in patterns of character covariation and function.*

II.7

Using Phylogenies to Study Phenotypic Evolution: Comparative Methods and Tests of Adaptation

Richard Ree

OUTLINE

1. Phylogeny and the comparative method
2. Ancestral state reconstruction
3. Model-based inferences of trait evolution
4. Analysis of multiple traits: Correlated evolution and phylogenetic tests of adaptation
5. Trait evolution and lineage diversification
6. Accuracy and confidence in ancestral inferences
7. Future directions for comparative methods

Phylogeny, in describing the genealogy of species, provides a historical framework for understanding the evolution of phenotypic diversity. Modern comparative biology uses the analysis of trait variation across species to infer the history of organic evolution and to elucidate evolutionary principles and processes. Comparative methods account for the fact that species are not entirely independent, but instead share evolutionary history by virtue of common ancestry. These methods commonly employ statistical models of trait evolution that can be used to estimate ancestral states, rates of change, directional trends, and correlations between traits. They can also be used to study the links between phenotypic evolution and patterns of lineage diversity, including rates of speciation and extinction.

GLOSSARY

Ancestral State. The phenotype or trait value of an ancestral species, usually inferred from the states of extant species given a phylogenetic tree.

Character (Trait). Any distinct, observable feature of an individual (e.g., aspects of morphology or behavior, gene sequences) or an emergent property of groups of individuals (e.g., ecological niche, geographic range, sexual dimorphism, mating system).

Likelihood. An optimality criterion based on probability distributions defined by a statistical model in which the preferred parameters are those that maximize the probability of the observed data. Given a stochastic model of trait evolution on the branches of a phylogenetic tree, likelihood can be used to assess ancestral character states.

Model of Trait Evolution. A statistical description of changes in states or values of a trait occurring stochastically through time, based on instantaneous rates that parameterize the underlying probability distributions.

Optimality Criterion. A set of principles or rules that measure the fit of data (e.g., observed character states) to a given hypothesis (e.g., a phylogenetic tree); in comparative biology, the most commonly used optimality criteria are *parsimony* and *likelihood*.

Parsimony. Also known as *Occam's razor*, an optimality criterion based on the principle that the simplest explanation is most likely correct. For example, in ancestral state reconstruction, parsimony means choosing ancestral states that minimize the amount of change required to explain all the states observed at the terminal nodes of a phylogenetic tree.

1. PHYLOGENY AND THE COMPARATIVE METHOD

Charles Darwin's theory of "descent with modification"—that species arise and diverge naturally from common ancestors—kick-started a revolution not only in the way biologists classify the living world (systematics), but also in the way they interpret its phenotypic diversity (comparative biology). Patterns of descent are described by phylogenetic trees, the availability of which continue to increase dramatically as a result of advances in genetic sequence acquisition, inference algorithms, and computational resources (see chapter II.2). Similarly, the use of phylogeny as a comparative framework for studying the history of "modification"—the sequence, tempo, and mode of evolutionary change in morphological characters, behaviors, and other traits of species—has seen much progress. Comparative methods facilitate analyses that are explicitly historical (e.g., what was the diet of the original Darwin's finch?), or not amenable to experiments (e.g., do larger-bodied predator species require larger home ranges?). The following sections review the theory and methods commonly employed in making inferences about evolutionary history and processes from comparative phylogenetic data.

Some Terminology

In comparative biology, a *trait*, or character, can be thought of generally as any distinct, observable feature of a species (see chapter II.1). To understand how comparative methods work, one needs to know some terms for the component parts of phylogenetic trees. The tips of branches that typically represent extant species are called *terminal* (or *leaf*) *nodes*; points where branches split and diverge are called *internal nodes*. Individual branch segments connecting nodes to each other are called *internodes*, or simply branches, and represent single lineages (species). Branches can have associated length values, which measure the evolutionary distance between nodes, for example, in units of time or genetic divergence. It is generally assumed that phylogenies are rooted, meaning they have an explicit temporal orientation, with each branch connecting an ancestral node (earlier in time) to a derived one (descendant, later in time). An internal node or branch and all its descendants is called a *clade*; the terminal nodes in a clade represent a *monophyletic group*. A bifurcating split in a branch yields sister clades, by definition the same age. The deepest internal node in a tree—the most recent common ancestor of all its leaf nodes—is called the root node.

2. ANCESTRAL STATE RECONSTRUCTION

Which came first, the chicken or the egg? In this age-old dilemma lies a question about the evolutionary sequence of ancestral states. Oviparity (egg laying) is a trait shared by all birds, as well as crocodilians, lizards, snakes, and turtles. Since these taxa form a monophyletic group, it is likely their common ancestor was also oviparous (see chapter II.1)—an inference corroborated by the fossil record. Thus, comparative data clearly show that the egg came before the chicken.

Questions involving ancestral states are ubiquitous in comparative studies. In principle, the most direct and accurate means of inferring the state of an ancestor is to examine fossils of the ancestor as it existed in the past; however, this is often impractical, as the organisms and/or traits of interest may not preserve well. Moreover, even if such fossils are available, it is often difficult to be confident that they represent the true ancestral species, rather than a divergent and extinct side branch. Comparative methods for ancestral state reconstruction generally focus on the hypothetical common ancestors represented by the internal nodes of a phylogeny. The general problem to be solved is, What ancestral states at those nodes make the most sense, in light of the observed data—species' traits arrayed across the tips of a phylogeny?

Parsimony

One approach to the answer appeals to the idea of simplicity: optimal ancestral states are those requiring the least change along the branches of the tree. In the case of a discrete character (e.g., red versus blue petals in a clade of flowering plants), this means the fewest transitions between states (colors). This is the principle of parsimony, also known as Occam's razor. As illustrated by the chicken-and-egg example, if all species in a monophyletic group share the same state, it is parsimonious to infer that their common ancestors—the internal nodes all the way down the tree—were also the same. Otherwise, when species vary in their character states, algorithms are needed to find the ancestral values that fulfill the criterion of minimal change.

The parsimony criterion begs the question, How is change quantified? The answer requires assumptions about the relative "cost" of state transitions. For discrete states, for example, red petals versus blue, an unbiased view would assume equal costs of change in both directions, from red to blue and vice versa; however, equality may not always be preferred. For example, if it were known that red pigments in plants require an extensive and complex biosynthesis pathway involving many genes, in any of which a simple knockout mutation would disrupt the production of necessary precursors and result in blue petals, the cost of red-to-blue transitions might be down weighted. Under such weighting, inferring many changes from red to blue could be more parsimonious than inferring a few changes from blue to red. In general, assumptions about transition costs between n discrete states can be expressed as an $n \times n$ array of values, known as a step matrix. However, while the preference for an asymmetrical step matrix may be empirically grounded, in practice it may be difficult to objectively justify any specific choice of relative weights.

For continuous traits, such as body size, or petal color recorded as wavelength of reflected light, the parsimony criterion generally posits that the cost of change along a branch is proportional to the squared difference of the ancestral and descendant values—so-called squared-change parsimony. The optimal solution is that set of ancestral states that minimizes the sum of squared differences over all branches of the phylogeny. If these differences are weighted inversely by branch length, following the reasoning that more change is expected on longer branches, the inferred ancestral states are exactly equivalent to estimates under the assumption of Brownian motion evolution (see below).

Likelihood

With statistical comparative methods, the optimality criterion is based on likelihood rather than parsimony. The question shifts from "What is the least amount of change required to explain the observed states?" to "What is the probability of the observed states having evolved, given a model specifying ancestor-descendant probabilities of change?" The key difference lies in modeling evolution as a stochastic process governed by probability distributions, with the probability of change being a function of time. In phylogenetic terms, *time* refers to the length of the branch between ancestral and descendant nodes. This contrasts with parsimony, in which branch lengths are generally ignored, and change thus tends to be underestimated on longer branches. Note that the use of stochastic models of evolution does not imply that changes are themselves random—that is, nonadaptive. Stochastic models can also describe the unpredictable effects of natural selection. For both discrete and continuous characters, specifying a model allows ancestral states to be estimated by maximum likelihood methods.

In Markov models of discrete characters, a common assumption is exponentially distributed waiting times between transition events. In such models, the expected waiting time is dictated by the instantaneous rate of change of the character. Analogous to step matrices, transition rates between n states are commonly given in an $n \times n$ rate matrix, usually denoted Q. For a single branch of length t, probabilities of all pairs of ancestor and descendant states are easily computed as $P(t) = \exp(Qt)$. This approach integrates over all possible paths of change along a branch to calculate the probability that the descendant had state 0, 1, ... or n given that the ancestor had, say, state 0. The likelihood of character data having evolved on a given phylogeny is obtained by recursively calculating these probabilities from a tips-to-root traversal of the tree's branches.

For continuous characters, the most widely used model of evolution is Brownian motion, in which a trait's value changes stochastically, in small positive or negative increments, at a constant rate. This process can be thought of as a random walk (often compared to the staggering of a drunken sailor). It is named for the random fluctuations in the position of pollen grains under the microscope, as first seen by Robert Brown in 1827. Brownian motion generally predicts that the trait values of a descendant will fit a normal distribution, centered on the value of the trait in the ancestor, with variance proportional to the intrinsic rate of change and the time separating the ancestor and descendant. Alternatives to Brownian motion include directional random walks, in which the mean of expected outcomes is shifted, and constrained random walks, such as the Ornstein-Uhlenbeck model, in which a "rubber band" parameter pulls values toward an optimum. The latter can be applied to questions such as whether phenotypes in different clades have evolved toward distinct adaptive peaks.

3. MODEL-BASED INFERENCES OF TRAIT EVOLUTION

Analysis of Single Traits

Stochastic models of trait evolution define the probability of observed states at the tips of a phylogenetic tree. Their parameters (e.g., instantaneous rates of change) can be estimated by maximum likelihood. As an alternative, one can apply Bayesian statistical methods, which assume some prior knowledge (a prior probability density) and then use likelihoods to estimate a probability that any ancestral trait value or combination of trait values is true (a posterior probability density).

In some cases, the model parameters are of greater interest than ancestral states, the latter being regarded as nuisance variables. In that case it is normal to integrate over all possible values of the ancestral states, measuring their individual contributions to the total probability. Parameter estimates may be sought because they can shed light on past evolutionary dynamics; moreover, one can test evolutionary hypotheses framed in terms of competing models. For example, is there a directional trend in the evolution of flower color, such as a higher rate of change from red to blue than vice versa? Likelihoods obtained using a model that constrains the "forward" and "reverse" rates to be equal can be compared to those from a model in which the rates are free to vary. Various statistical methods, including likelihood-ratio tests, Bayes' factors, and other information content criteria, can be brought to bear on whether the observed data support one model or the other. In this case, support for the two-rate model would lend credence to the hypothesis of a directional trend.

By using phylogenies in which branch lengths are in units of absolute time, as can be obtained from fossil-calibrated *molecular clock* analysis, absolute rates of trait evolution can be estimated from comparative data. The *felsen* is a recently proposed measure of evolution that corresponds to an increase in one unit of variance per million years, calculated under the assumption of Brownian motion, for natural log-transformed trait values.

Rates of change are not the only parameters of interest in evolutionary models. In some cases, parameters that transform the tree itself may be invoked to test hypotheses about the tempo and mode of change. For example, the question of whether change in a trait has been gradual along branches, or punctuated (concentrated at cladogenesis events), can be framed as whether the likelihood is increased by scaling all branch lengths by a common power, κ. If κ is significantly less than 1, corresponding to branches having rates of trait evolution that are more equal than expected given their branch lengths, the punctuated hypothesis is supported. Other transformations have been designed to detect the signature of Ornstein-Uhlenbeck evolution as well as accelerating or decelerating Brownian motion evolution.

Phylogenetic Signal

Given the null expectation of similarity by descent, a basic question in comparative analysis is, To what degree does a trait actually covary with phylogeny? In other words, how much "phylogenetic signal" does a trait have? The question is often raised in the context of niche conservatism, that is, the tendency for closely related species to share ecological traits, and in studies of why some traits are more labile in evolution than others. Various tests of phylogenetic signal have been proposed. A common theme is the calculation of a test statistic that measures the fit of the data to the tree (e.g., as defined by a stochastic model of evolution), with the significance of the test statistic being judged against a null distribution generated from random permutations of the data across the tips of the tree. Another common strategy is to measure the fit of the data while transforming the tree's branch lengths to be increasingly starlike (i.e., such that all its terminal branches appear to radiate from a single ancestral node). These tests can accommodate both discrete and continuous characters, and a range of evolutionary models. They can be useful for ascertaining whether an individual trait does or does not exhibit signal; however, measurement of phylogenetic signal as a quantity directly comparable across traits and trees is a more challenging problem, formally defined for continuous characters only in the context of Brownian motion evolution. Simon Blomberg and colleagues devised a statistic, K, that is greater than 1 if close relatives are more similar than expected from Brownian motion, and less than 1 ($0 < K < 1$) if they are less similar. This statistic is commonly used to compare the strength of phylogenetic signal across different combinations of traits and trees.

4. ANALYSIS OF MULTIPLE TRAITS: CORRELATED EVOLUTION AND PHYLOGENETIC TESTS OF ADAPTATION

Independent Contrasts and the Phylogenetic Regression

A common goal in comparative biology is to study correlated change in different traits, or correlations between phenotype and environment, as correlations can reveal evidence of adaptive, functional, or genetic constraints. As should be clear from the previous sections, the central problem facing such analyses is that species' traits are not statistically independent data points, owing to similarities inherited from shared ancestry. A great deal of research has focused on ways to account for this nonindependence. A seminal advance in this area was Joseph Felsenstein's method of phylogenetically independent contrasts, allowing measurement of the correlation between two continuous traits while taking account of nonindependence due to shared ancestry. The method assumes that both traits evolved according to Brownian motion, and that the branch lengths of the phylogeny correspond to expected variances in trait values. It calculates contrasts (differences in trait values) for each trait at all pairs of sister nodes on the phylogeny, using a recursive algorithm that proceeds from the tips of the tree toward the root, assigning weighted averages of trait values to the common ancestor of each sister pair. These contrasts are independent of phylogeny and can thus be studied using standard bivariate techniques for correlation and regression.

Independent contrasts sparked a cascade of theory that further explored, in mathematical terms, the covariance of species arising from phylogeny and the detection of trait correlations. These investigations have drawn heavily from statistical techniques based on matrix algebra, in which the phylogeny is transformed into a variance-covariance matrix specifying the shared and independent histories of species. A significant outcome of this work was a more general framework for the phylogenetic regression of traits based on generalized least squares (GLS). With GLS, the phylogenetic variance-covariance matrix can be constructed using arbitrary models of trait evolution, relaxing the need to assume Brownian motion. The framework has spawned a wide variety of related methods for parameter estimation, hypothesis testing, and ancestral state reconstruction from multivariate data sets of both continuous and discrete traits.

Discrete Markov Models of Correlated Evolution

Tests for correlated evolution are often motivated by hypotheses of adaptation. For example, are transitions to C4 photosynthesis favored in plant lineages that occupy arid environments? For discrete characters, a popular method for studying the correlated evolution of discrete traits uses Markov models, and extends the general framework described previously for univariate hypothesis testing. A correlated Markov model for two binary-valued traits has four discrete "states," corresponding to the four combinations traits that a lineage can have (00, 01, 10, 11). The four states define a 4×4 rate matrix, Q, of which the 12 off-diagonal entries represent instantaneous rates of change. However, elements are set to zero where they correspond to transitions involving two changes (i.e., $00 \rightarrow 11$ and $01 \rightarrow 10$), reflecting the assumption that only one trait can change in an instant of time. Thus, the rate matrix allows up to eight free parameters describing rates of change between each pair of character states: $q_{00\rightarrow01}$, $q_{01\rightarrow00}$, $q_{10\rightarrow11}$, $q_{11\rightarrow10}$, $q_{00\rightarrow10}$, $q_{10\rightarrow00}$, $q_{01\rightarrow11}$, and $q_{11\rightarrow01}$. These can be used to test specific hypotheses. For example, if the independent trait is mesic versus arid habitat preference, and the dependent trait is C3 versus C4 photosynthesis, the hypothesis that evolutionary "gains" of C4 are concentrated in arid-inhabiting lineages could be formulated as a model in which $q_{10\rightarrow11} >> q_{11\rightarrow10}$, and possibly also that $q_{01\rightarrow00} >> q_{00\rightarrow01}$. This model can be tested by comparing its likelihood versus a model in which these rates are equal.

5. TRAIT EVOLUTION AND LINEAGE DIVERSIFICATION

A common theme in the preceding discussion is that comparative methods accept a given phylogeny as fixed, and account for its topology and branch lengths in making historical inferences, but otherwise treat the evolution of traits as independent of the processes that shaped the tree itself—namely, speciation (cladogenesis) and extinction. With this perspective, one can imagine the branches of the tree as static structures, their lengths and topology unaffected by the traits of the species they represent; however, abundant evidence has been found that traits do affect speciation and extinction. Moreover, a large body of theoretical and empirical work has focused on estimating the birth and death rates of lineages (including where these rates have shifted) from phylogenetic trees. What are the connections between these lines of inquiry?

A major branch of comparative biology studies the links between species' traits and lineage diversification. For example, a pervasive idea is that certain traits represent evolutionary "key innovations" that played a role in the success of unusually large clades. In parametric terms, this amounts to asking whether rates of diversification are state dependent. To address such questions, comparative analyses often initially use a combination of methods that separately infer the history of trait evolution (e.g., where on the phylogeny did the putative innovation arise?) and the history of lineage diversification (where did rates of diversification change?), and subsequently associate the results. For example, bilaterally symmetrical flowers are thought to represent an adaptation for specialized animal pollination, which in turn may enhance the potential for reproductive isolation and the origin of new plant species. This hypothesis would be supported if bilateral symmetry is associated with higher rates of diversification. Independent evolutionary transitions in flower symmetry on the phylogeny of plants thus represent naturally replicated experiments that can be brought to bear on the question. In fact, it has been shown that bilateral clades are larger than their radially symmetrical sister groups more often than can be attributed to chance. Comparative analyses thus support the idea that bilateral flowers are key innovations in plants.

The sister-clade approach is appealing in its simplicity. By definition, sister clades are the same age, so their relative sizes directly reflect differences in net diversification (speciation minus extinction); however, the sister-group method relies on confident inferences of ancestral states, and a sufficient number of transitions that yield replicated sister-clade contrasts. Meeting these criteria can be difficult in practice. In particular, ancestral-state reconstructions can be positively misled if assumptions about the model of evolution are violated—including the assumption that diversification is not state dependent! The heart of the problem is that asymmetry in the direction of character evolution and inequality in state-dependent rates of speciation and extinction can each yield similar phylogenetic distributions of states (Maddison et al. 2007). If diversification rates are highly state dependent and unequal, character reconstructions that apply a standard Markov model might erroneously infer asymmetrical transition rates; conversely, if rates of trait evolution are asymmetrical, tests for state-dependent diversification might be falsely positive. In both cases, estimates of ancestral states will likely be inaccurate.

To solve this problem, joint models of trait evolution and diversification have recently been developed that incorporate parameters for the rate and direction of trait change as well as state-dependent rates of speciation and extinction. With such models, the likelihood function generally cannot be solved analytically, so parameter estimation requires numerical integration over all trees with extinct branches that are otherwise consistent with the observed tree. Analyses accounting for interactions between trait evolution and lineage proliferation are now becoming quite common. For example, a recent study of the nightshade family of plants used a joint model to show that self-incompatibility (which is frequently lost, but rarely, if ever, regained in this clade) is associated with higher net diversification relative to self-compatible lineages, demonstrating species selection (see chapter VI.14) for obligate outcrossing.

6. ACCURACY AND CONFIDENCE IN ANCESTRAL INFERENCES

How reliable are the basic tools of comparative biology—the phylogenetic relationships of species and theoretical models of trait evolution—for inferring patterns and processes in evolutionary history? The answer depends on a number of factors. Of primary concern is whether the tree is accurate, and whether the models are valid descriptors of the evolution of traits of interest. For example, ancestral state estimates are more likely to be accurate and unambiguous if the rate of evolution is low relative to the rate of lineage proliferation. Conversely, if a trait evolves quickly and exhibits rampant homoplasy (convergent and/or parallel evolution), ancestral states will tend to be more uncertain. Both parsimony and likelihood methods can be led astray, yielding positive support for erroneous conclusions, if their underlying assumptions are not met. Unfortunately, in the absence of independent lines of evidence (e.g., fossils), it is not always easy to determine whether and how these assumptions are violated. For example, even if one is able to demonstrate that a character's phylogenetic distribution is consistent with Brownian motion evolution, it is more difficult to confidently establish that it actually evolved according to that process.

7. FUTURE DIRECTIONS FOR COMPARATIVE METHODS

Opportunities for evolutionary insight from comparative analysis will continue to grow with the accumulation of phylogenies and expanding knowledge of species' traits. Research on improving the utility and power of comparative methods is important and ongoing. A continuing trend is that increasingly sophisticated comparative methods enhance the potential for statistical inferences of ancestral states and evolutionary processes. In particular, joint models of trait evolution and lineage diversification represent a significant step toward a unified framework for exploring the reciprocal interactions between these two processes; however, many challenges remain. For example, methods are generally lacking for multivariate analyses, and most are ill equipped to deal with inconstant rates of evolution or non-Markovian processes, such as the influence of density dependence or species interactions on trait evolution. Integration of trait data from fossils deserves greater attention, as do models in which trait change can be associated directly, as either cause or consequence, with speciation (cladogenesis). The latter are important because comparative methods generally assume that the state of an ancestor is inherited identically by both daughter species at divergence. Traits that violate this assumption include geographic ranges, which can be subdivided at speciation, and traits that underlie ecological speciation or otherwise directly promote reproductive isolation (such as host associations, habitat preferences, mate selection, etc.).

FURTHER READING

Felsenstein, J. 1985. Phylogenies and the comparative method. American Naturalist 125(1): 1–15. *A seminal paper describing the method of independent contrasts.*

Freckleton, R. P. 2009. The seven deadly sins of comparative analysis. Journal of Evolutionary Biology 22(7): 1367–1375. *A practical guide to recognizing and avoiding some common pitfalls in comparative phylogenetic inference.*

Maddison, W. P., P. E. Midford, and S. P. Otto. 2007. Estimating a binary character's effect on speciation and extinction. Systematic Biology 56(5): 701–710. *This paper introduces an important new method for jointly modeling trait evolution and lineage diversification.*

Martins, E. P., and T. F. Hansen. 1997. Phylogenies and the comparative method: A general approach to incorporating phylogenetic information into the analysis of interspecific data. American Naturalist 149(4): 646–667. *This paper was important in establishing generalized least squares analysis in comparative biology.*

Pagel, M. 1999. Inferring the historical patterns of biological evolution. Nature 401: 877–884. *A nice review of statistical comparative methods, with several interesting examples of their application.*

Price, T. 1997. Correlated evolution and independent contrasts. Philosophical Transactions of the Royal Society B 352: 519–529. *An interesting counterpoint to the theory underlying independent contrasts, inspired by the process of adaptive radiation.*

II.8

Taxonomy in a Phylogenetic Framework

Julia Clarke

OUTLINE

1. Taxonomy in historical context
2. Incorporating an evolutionary perspective
3. Species in a phylogenetic framework
4. Concerns about and misunderstanding of phylogenetic nomenclature
5. The future of phylogenetic nomenclature

In biology, naming of groups of organisms is a separate but linked enterprise to determining relationships among them. Although, historically, an array of properties of interest were considered relevant for clustering organisms and applying names, today most biologists are interested specifically in discovering and naming phylogenetic groups: organisms related by virtue of descent from common ancestry. Because named groups of organisms, or taxa, figure prominently in our evolutionary theories, many biologists are deeply invested in how names are applied. A major focus has been on naming clades, all the descendants of a common ancestor and that ancestor. Some workers have focused on adapting the ranked Linnean system of taxonomy, while others have proposed a new phylogenetic system. Names for taxa defined phylogenetically utilize specimen or species specifiers and refer explicitly to clades of the tree of life. A taxon name may be tied explicitly to a clade (see chapter II.1) through definitions of three basic forms: node-based, stem-based, or apomorphy-based. These phylogenetic definitions for clade names can be described algorithmically, which may help address informatics needs in the face of increasingly dense taxonomic sampling and assembly of larger and larger sections of the tree of life. Debate concerning the format for the definitions of species names is linked to ongoing controversy over the reality and nature of species.

GLOSSARY

Binomen. A two-part species name comprising a generic name and a specific name (ICZN) or epithet (ICNB, ICBN) under the rank-based codes, or of a praenomen and a species name under the *PhyloCode*.

Clade. A monophyletic group; a group of organisms including an ancestor and all its descendants.

Phylogenetic System of Nomenclature. An integrated set of rules and principles governing the naming of taxa and application of taxon names that is based on the principle of common descent and formalized in the *PhyloCode*.

Specifier. A species, specimen, or apomorphy in a phylogenetic definition of a name that serves to specify the clade to which the name is applied. An *internal specifier* is a part of the clade to be named, and an *external specifier* is outside that clade and used in stem-based definitions of names.

Taxon (Pl., Taxa). A named group(s) of organisms.

Taxonomic Definition. A statement specifying the meaning of a name (i.e., the taxon to which the name refers).

1. TAXONOMY IN HISTORICAL CONTEXT

The discipline of taxonomy is concerned with identifying significant groups of organisms, or *taxa*, and giving them scientific names that can be used to facilitate communication about these organisms and their features. Whether in evolutionary theory, public policy, or conservation, there is no doubt that what groups of organisms we recognize as taxa matters. Taxa are routinely discussed not only with reference to conservation status but also in patents for biotic compounds or in the assessment and governance of public health risks. Assessment of phylogenetic diversity is important to developing conservation priorities. This is one of the main reasons taxonomy now strives to ensure that the groups recognized as taxa are those united by evolutionary relatedness—but this was not always the case.

In the eighteenth and nineteenth centuries, especially in Europe, there was a penchant for ordering (and reordering) the natural world as a way to organize knowledge of living organisms, many of which were newly known to Western science. The resulting classification systems emphasized particular attributes or ecological factors used to determine taxon membership. At this time, taxa were viewed as static and divinely determined groups of organisms. It was in this environment that Carl Linnaeus (1707–1778) proposed and refined an all-encompassing system of classification or taxonomy. Linnaeus's taxonomy was composed of named groups of organisms, *taxa* (singular = *taxon*), arranged in relation to one another. In addition to proposing a taxonomic system (the organization of organisms into ranked categories by different character systems), Linnaeus also developed a system of nomenclature: a system of rules governing taxon names.

Linnaeus recognized five ranks of taxa from species up to class and brought into broad use a binomial (two-part) name for species, consisting of a genus name combined with a species epithet. The composition of taxa was determined by the presence of characteristics considered to define the taxon. Different character systems were thought to naturally distinguish different categories of taxa at distinct ranks. The hope was that, by using a limited number of key features or clusters of features, it would be possible to classify all known life, at that time a few thousand described species.

The taxonomic endeavor started by Linnaeus quickly took hold, becoming the focal point of natural history for at least 200 years, although his particular taxonomic scheme was largely revised. The nomenclatural system he initiated was accepted but greatly expanded to accommodate the great diversity of species found by explorers in the nineteenth century. Separate nomenclatural codes eventually formalized the rules and governance of taxonomic systems for the naming of bacteria (BC), plants (ICBN), and animals (ICZN). All took the Linnaean system as their base and were thus focused on ways to name ranked taxa whose membership was determined by defining characters.

2. INCORPORATING AN EVOLUTIONARY PERSPECTIVE

The publication of *The Origin* by Charles Darwin in 1859 signaled a profound revolution in natural history with the realization that living organisms are linked by descent from common ancestry. Darwin articulated the view that while taxa may be identified based on their distinctive characteristics, the taxa we wish to discuss share characteristics by virtue of evolutionary history. Two members of a taxon should be more closely related to each other than to any organism not a part of that taxon. The centrality of defining characters present in all members of a class is incompatible with the mutability fundamental to evolution; however, a shift in nomenclatural approach did not immediately occur.

The hierarchical aspects of the Linnaean system as a whole seemed to fit well with the nested relationships implied by a single tree of life; however, the Linnaean system was built on taxonomic rank and the idea that distinct kinds of characters (e.g., reproductive, locomotory) characterize different ranks. Subsequent naturalists determined that taxa at each rank in the Linnaean hierarchy do not share any essential properties that could allow ranks to be recognized as natural entities. A family of plants, for example, does not share special family-category properties with a family of birds, or even with other families of plants.

In the twentieth century, evolutionary taxonomy, which arose with the modern synthesis of the 1940s, advocated the application of names with consideration of shared history but also an emphasis on certain characteristics. By contrast, Willi Hennig (1913–1976) and others working around the same time emphasized that discovering and naming of monophyletic groups (= clades) of organisms should supersede the emphasis on characteristics. Such a perspective resulted from an interest in discovering and communicating about groups of organisms related by virtue of common descent (monophyletic groups or clades).

As discussed in chapter II.1, monophyletic groups have the property that its members share a more recent common ancestor with each other than with any organism outside the group. The phylogenetic approach contrasts with evolutionary taxonomy and older approaches that would allow for the recognition of groups unified by collections of characters not due to shared ancestry (*polyphyletic* groups) or taxa that exclude descendants that have lost or transformed particular features (*paraphyletic* groups). For example, evolutionary taxonomists accepted the utility of a concept of a taxon Reptilia that included crocodiles, lizards, snakes, and extinct apparently "reptile-like" dinosaurs but excluded birds. While this concept may seem intuitive, it actually communicates less about the natural world than a taxon Reptilia that includes one complete branch of the tree of life, a monophyletic group, rather than artificially excluding birds. Recognizing that birds are nested within Reptilia, specifically as most closely related to Crocodylia, has explanatory power and is useful for identifying biological questions of interest. For example, this relationship was recognized in large part based on bony characteristics (e.g., aspects of the skull such as an antorbital fenestra). The later recognition that both crocodilians and birds share parental care among Reptilia, could have been anticipated by a taxonomy that reflected monophyletic groups. Specifically knowing that dinosaurs include birds, and that crocodilians

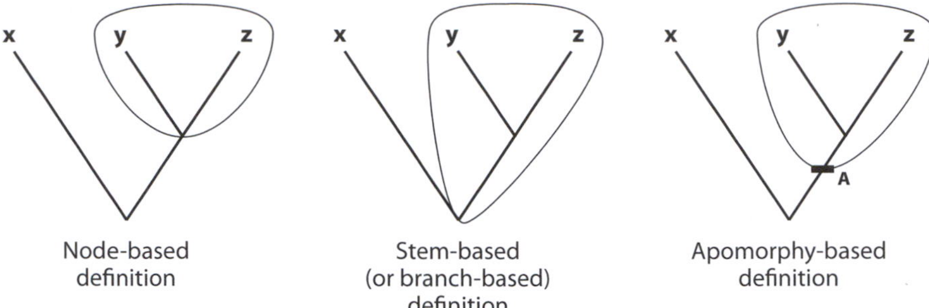

Figure 1. The three basic forms of a clade name definition in a phylogenetic frame (after De Queiroz and Gauthier 1992). Species or specimens X–Z and apomorphy A are specifiers in the (left to right) node-based, stem-based (or branch-based), and apomorphy-based definitions of the taxon names.

are most closely related to that clade, makes the discovery of parental care in dinosaurs not a surprise but predicted.

Over the decades following Willi Hennig's seminal 1966 book, the importance of naming monophyletic groups of organisms (clades) became largely the consensus view. At first, the nested hierarchy of the Linnaean taxonomic system was thought to be easily translatable into a nested hierarchy, a taxonomy, communicating a particular phylogenetic hypothesis represented by more or less inclusive clades with ranks; however, such rank-based systems (or variants that used numerical annotations, or indented lists) did not allow a set of taxon names to be unambiguously adjusted in response to a new hypothesis of relationships. Also, there remained misleading nonequivalency of ranks and a lack of sufficient ranks to represent the tree of life. There were taxonomies based on phylogenies that used ranked names for clades of organisms but no system of nomenclature that was explicitly built on a phylogenetic framework.

In recent years a number of systematists have argued that the rank-based, Linnaean system of nomenclature has undesirable features if we equate taxa with monophyletic groups. The primary goal of biological nomenclature is to allow names to be assigned to taxa in such a way as to minimize ambiguity about content of the taxon and maximize stability over time. However, finding that one taxon is embedded within another of the same rank would, for example, require a change in name (at least the suffix used to indicate rank in the Linnaean system) of one or both taxa. Also, Linnaean systems require that new species be placed in sets of higher taxa regardless of the actual degree of known phylogenetic resolution. For example, perhaps new species L is known only to be a member of a large clade M previously identified as a class, but its specific relationships to subclades of M is unresolved based on the data available. Regardless, in order to avoid creating paraphyletic groups, L would need to be placed in an existing order, family, etc., or new taxa at these ranks would need to be named, all of which would imply more phylogenetic structure than the data supported.

3. SPECIES IN A PHYLOGENETIC FRAMEWORK

Although there were important precursors in preceding decades, proposal of a formal phylogenetic system of taxonomy dates to the early 1990s. De Queiroz and Gauthier proposed that what we want to name in the tree of life are clades, and that definitions of taxon names should explicitly reference ancestor-descendant relationships. To aid this enterprise they argued that the definitions of the names of taxa should be phylogenetic, proposing three main kinds of phylogenetic definitions of taxon names: node-based, stem-based, and apomorphy-based. All required that certain tips of the tree of life be *specifiers*, species or specimens that serve as referents in the definition of a name. For example, tips X-Y-Z (figure 1) are specifiers. *Node-based* definitions of clade names take the form: the most recent common ancestor of specifiers Y and Z and all of its descendants. *Stem-based* or *branch-based* definitions take the form: all taxa more closely related to Y (or Y and Z) than to X. *Apomorphy-based* definitions take the form: the most recent common ancestor that shares apomorphy A with Y and Z and all its descendants. For example, one proposed node-based phylogenetic definition of the taxon name Aves linking it to the most recent common ancestor of all extant birds and all of its descendants takes the general form: Aves is the name for the most recent common ancestor of carefully chosen species specifiers (the Andean Condor, *Vultur gryphus*; Great Tinamou, *Tinamus major*; and Ostrich, *Struthio camelus*) and all its descendants.

The *PhyloCode* (ICPN: International Code of Phylogenetic Nomenclature), under development since the late 1990s, is a code that formalizes the rules and recommendations of phylogenetic nomenclature and establishes an organizational structure overseeing the implementation of this practice. As of this writing, it is not yet formally in effect. The fundamental form of definitions for the names of taxa first outlined by De Queiroz and Gauthier is retained in the *PhyloCode*. As with other nomenclatural codes, the *PhyloCode* is not expected to dictate taxonomic practice, which groups should be recognized as taxa, but to provide rules to govern the names of taxa that are recognized. Under the *PhyloCode*, a practitioner can name any group, even those that are nonmonophyletic. Furthermore, ranks can be used in conjunction with names defined phylogenetically, although unlike the traditional codes, the ranks are not part of the definitions of those names.

The *PhyloCode* includes rules governing publication of new names, conversion of existing names that were previously established (i.e., validly published and named) under the rank-based codes, as well as rules for priority (which of two names for the same taxon is correct) and synonymy (when two names apply to the same taxon). Other fundamentals of the system include requiring the formal registration of names and their associated phylogenetic definitions in a database, *RegNum*. A current draft of the code is downloadable from www.ohiou.edu/phylocode/ (or from phylocode.org).

Beginning in the 1990s, questions were raised about addressing the Linnaean species binomen, or two-part species name (e.g., *Homo sapiens*) in a phylogenetic system of nomenclature given that the first part of the binomen, genus, is a taxon of rank. Early drafts of the *PhyloCode* covered only clades and did not address species. It was noted that some species taxa are recognized by utilizing a criterion of monophyly, making them equivalent to clades; however, there was little consensus on how, or even if, species should be incorporated into the code, but it was generally recognized that a complete system of nomenclature based on phylogenetic principles would be expected to formally address species. Species figure centrally in the languages of evolutionary theory and public policy. They are the most numerous named taxa and commonly employed in many metrics of standing biodiversity. A diverse public is accessing knowledge about species daily, whereas it is often only specialists who are invested in the names defined for its major subclades.

The debate over the form that species names and their definitions should take in a phylogenetic system is linked to the extensive debate over the nature and importance of taxa recognized as species. Within any community of biologists there is always heterogeneity with respect to the concept of species (see chapter VI.1). Some taxonomists working in a phylogenetic framework have wanted to consider, discuss, and name only clades, and remove all discussion of species. Several of these authors consider that retention of species in a phylogenetic framework would conflict with the rank-free aspects of the system. Others have proposed that instead of "species," we should focus on the least-inclusive taxonomic unit or LITU, the smallest clades below which there is no phylogenetic structure. Yet others simply wish to discuss clades that might approximate the content of traditional species, without according those clades any special status or named rank. The latter authors prefer to use Linnaean species epithets for small clades and name them using node- or branch-based definitions.

The community has debated how to convert species names under the *PhyloCode*, given that epithets or specific names (i.e., the second part of the binomen; e.g., *sapiens*) were never required to be unique in the rank-based codes. Consequently, species epithets have been used repeatedly in distinct clades of organisms, leading to concerns about homonymy (the same name being used for different taxa). The sheer number of existing species names, more than a million named under the rank-based codes, also presents a challenge if species names are to take a different form in a phylogenetic system, because all these names would have to be converted and registered in *RegNum*. Such conversion was ultimately deemed nonpracticable and undesirable given the large number of named species.

Based on this reasoning, it was established in a *PhyloCode* article that all new species names will be required to be validly established under the appropriate existing bacterial, botanical, or zoological rank-based codes; however, this *PhyloCode* article interprets an established species name within the context of a phylogenetic system. Under the approach adopted, the first part of a species binomen (called in the *PhyloCode*, the praenomen) is recommended to be a converted clade name. It does not need to be a converted genus name after the species name is validly established compliant with the appropriate rank-based code. After the establishing first publication, clade names can be used in combination with species names (the second part of the binomen) that are not genera in rank-based codes, or a species name can be used alone with further recommended identifying information (e.g., author and publication date).

Named species are heterogeneous entities recognized by an array of criteria, and in many cases they are not clades. The *PhyloCode* article interprets established species names in a phylogenetic system and provides additional recommendations for increased explicitness in taxonomic practice. Species are recognized as distinct biological entities from clades that can be identified by a broader array of criteria than monophyly alone. While this position does place limits on biologists who equate species only with clades and/or want to apply species epithets to unranked clades without discussing species, this route was adopted in the face of lack of consensus concerning a unified way to accommodate all the diverse interpretations of species within the *PhyloCode*. Debates are predictably ongoing concerning the equivalency of species taxa, their nature as biological entities (e.g., as taxa or functional units), their boundaries, and how they can be appropriately named.

4. CONCERNS ABOUT AND MISUNDERSTANDING OF PHYLOGENETIC NOMENCLATURE

Concerns about phylogenetic nomenclature have been diverse, including some based on a misunderstanding of the system, perhaps confused by changes in the system from its earliest forms to its ultimate articulation in the *PhyloCode*. Some authors appear to have been confused by the intentions of the system and its implications for taxon names established under other codes; however, the *PhyloCode* does not require that all existing taxon names be replaced with new names. Likewise, it does not enforce particular taxon concepts (e.g., require monophyly) or disallow ranks. Other critics have maintained that the *PhyloCode* intends to replace the rank-based codes. Although at some point in the distant future the phylogenetic community could decide this is the right decision, the *PhyloCode* is presently designed to function alongside the rank-based codes; indeed, it explicitly requires valid establishment of species names, which are also the most broadly accessed taxon names, under these codes.

Other objections to phylogenetic taxonomy and the *PhyloCode* have been philosophical. One argument in favor of rank-based codes is that the imprecision of the definitions of taxon names under these codes yields flexibility. By precisely tying a name to a particular clade with particular specifiers, the argument goes, we may discover that we have not applied a widely used name to the particular biological entity we most wish to discuss. Other phylogeneticists value the stability of knowing that a name will always refer to one specific set of ancestor-descendant relationships, even if the list of species it contains changes with new data.

5. THE FUTURE OF PHYLOGENETIC NOMENCLATURE

In some systematic communities, phylogenetic taxonomy is in broad use. These communities, however, are heterogeneous in the way they tend to deploy phylogenetic nomenclature. Some systematists prefer not to use apomorphy-based definitions; some reject the recommendation in the *PhyloCode* of applying widely used names to the living members (the "crown" group) of major clades. To some, larger questions remain contentious; for example, are complete definitions possible for biological entities that may have, by their nature, imprecise edges and boundaries? There has been much debate over how recognition of a proposed temporal framework for biological kinds may affect their properties.

While the formal publication of the *PhyloCode* and its start date are not yet firmly set, the community of phylogenetic taxonomists continues to increase. Some authors have noted the fit between this system of phylogenetic nomenclature and computer-based methods for tracking biodiversity (*phyloinformatics*). A given set of specifiers is sufficient for a computer to apply the definition of a name unambiguously with the input of a current estimate of phylogenetic relationships. To some in the systematic community, these properties are desirable; to others, the flexibility/imprecision of the rank-based codes is preferable, even though it means no way to automate taxonomic practice. It will be interesting to see how the differences of opinion are resolved in the future. The one thing we can be sure of is that systems of taxonomy and nomenclature will also continue to evolve as phylogenetic methods and the scope of the questions asked with them continue to expand.

FURTHER READING

Baum, D. A., and S. D. Smith. 2013. Tree Thinking: An Introduction to Phylogenetic Biology. Denver, CO: Roberts & Company. *A recent systematic textbook that includes a chapter on taxonomy and nomenclature.*

Barkley, T. M., P. DePriest, V. Funk, R. W. Kiger, W. J. Kress, and G. Moore. 2004. Linnaean nomenclature in the 21st century: A report from a workshop on integrating traditional nomenclature and phylogenetic classification. Taxon 53: 153–158. *A recent attempt at reconciling phylogenetic systematics with rank-based codes (see further discussion in Laurin 2008)*

Bryant, H. N., and P. D. Cantino. 2002. A review of criticisms of phylogenetic nomenclature: Is taxonomic freedom the fundamental issue? Biological Reviews of the Cambridge Philosophical Society 77: 39–55. *A response to criticisms of phylogenetic nomenclature.*

Cantino, P. D., and K. De Queiroz. 2010. International Code of Phylogenetic Nomenclature, version 4c. Downloadable from www.ohiou.edu/phylocode/. *The current draft of the* PhyloCode, *with a preface describing its development and glossary.*

De Queiroz, K., and J. Gauthier. 1992. Phylogenetic taxonomy. Annual Reviews of Ecology and Systematics 23: 449-480. *The first in-depth address of the motivations for the proposal of a phylogenetic system of nomenclature and description of what such a system would look like.*

Ereshefsky, M. 2002. The Poverty of the Linnaean Hierarchy. A Philosophical Study of Biological Taxonomy. Cambridge:

Cambridge University Press. *A nuanced yet accessible treatment of the philosophical issues with Linnaean taxonomy.*

Hennig, W. 1966. Phylogenetic Systematics. Urbana: University of Illinois Press. *A landmark contribution to systematics discussing the necessary centrality of phylogenetics in taxonomy.*

Laurin, M. 2008. The splendid isolation of biological nomenclature. Zoologica Scripta 37: 223–233. *A response to critiques of phylogenetic nomenclature.*

Pleijel, F., and G. W. Rouse. 2003. Ceci n'est pas une pipe: Names, clades and phylogenetic nomenclature. Journal of Systematics and Evolutionary Research 41: 162–174. *A discussion of phylogenetic nomenclature and one early proposal that species not be recognized but only named clades (includes LITU concept).*

Rieppel, O. 2006. The PhyloCode: A critical discussion of its theoretical foundation. Cladistics 22: 186–197. *A critical view of phylogenetic nomenclature from a philosophical perspective.*

II.9

The Fossil Record

Noel A. Heim and Dana H. Geary

OUTLINE

1. Fossilization and taphonomy
2. The nature of the fossil record
3. Marine diversity in the Phanerozoic
4. The value of the fossil record

The fossil record documents the history of life over the course of the past 3.5 billion years, demonstrates that evolution has occurred, and provides otherwise inaccessible insights into the evolutionary process. This chapter outlines briefly how the fossil record has been formed, and explores the nature of the fossil record in relation to its central role in understanding evolution. Evolutionary biology is a historical science, and the process of evolution is often played out over intervals of time much too long for direct observation. Thus the fossil record provides the dimension of time that is essential for a complete understanding of the process that unites all of biology.

GLOSSARY

Body Fossil. The fossilized remains of a once-living organism. Body fossils represent the actual organism and are distinct from trace fossils.

Stratigraphy. The study of how and why rocks are deposited in their observed vertical and lateral successions. A key stratigraphic concept is that of superposition, where vertical successions of rock are ordered with the oldest *strata* (layers) at the bottom and the youngest at the top. Although younger rocks are always deposited on top of older rocks, primary "layer cake" stratigraphic successions can be altered through tectonic folding and faulting.

Taphonomy. The study of the ways in which the dead remains of once-living organisms become preserved as fossils. Much of taphonomy is concerned with understanding which biological information is preserved by fossils and which is lost.

Time Averaging. The mixing of noncontemporaneous individuals into a single sedimentary or fossil assemblage.

Trace Fossil. The record of behavior preserved in the sedimentary record. Trace fossils include footprints, burrows, feeding traces, coprolites, and insect leaf damage. Trace fossils are frequently not attributable to a specific species.

Unconformity. A surface separating two stratigraphic units that represents "missing" geologic time. Unconformities are formed through the erosion of previously deposited sediments, prolonged intervals of nondeposition, or a combination of the two.

Uniformitarianism. The idea that the processes that are observable and operating today, and only those processes, can be used to explain the geological and biological evolution of the earth as preserved in the geological and fossil records.

The value of the fossil record is perhaps most clearly illuminated if we try to imagine our knowledge of evolution without it. Plenty of evidence would indicate that evolution had occurred, but the rich history of life would simply be a matter for conjecture. Fossils reveal our history in amazing and often-unpredictable ways: giant ground sloths and saber-toothed cats roaming the landscapes of our present-day cities just a few millennia ago, diverse genera of camels, horses, and rhinos not long before that, and earlier still, flightless birds that preyed on tiny horses and flying reptiles the size of airplanes, to name just a few. We would know that whales are mammals, and we could use molecular evidence to determine their closest living relatives, but how much richer the story becomes when we can actually find their ancestors on the ancient shores of a warm tropical ocean and touch the diminutive leg bones of a 15 m *Basilosaurus*. We would likely suppose that great calamities had occurred, but who could imagine a global deep freeze complete with tropical glaciers, followed by the most remarkable blossoming of life in history?

The fossil record, however, is not only a catalog of wonderful organisms. It also demonstrates without a doubt that evolution happened. It provides us with transitional organisms between major groups that demonstrate clearly how one evolved from the other (e.g., land vertebrates from fishes, birds from dinosaurs, mammals from reptiles, and many more). The fossil record gives us the ability to determine the actual time frame of evolutionary change, and importantly, it continues to provide valuable information about how evolution occurs.

Evolution can be slow on human timescales. Thus the sequence of evolutionary events preserved in the fossil record is essential for testing the validity of existing evolutionary theory and modifying or expanding our set of ideas on how the process works. For example, the fossil record demonstrates that rates of evolution need be neither slow nor constant, even within lineages. Paleontologists have also shown that evolutionary history has been marked by various types of extinction events (e.g., mass extinctions, mass depletions, turnover events; see chapter VI.13). The field of macroevolution, the study of evolution at or above the species level, is deeply grounded in the fossil record (see chapters VI.12 and VI.13).

1. FOSSILIZATION AND TAPHONOMY

Fossils are the remains or traces of ancient organisms preserved in the rock record. Fossils are found in sedimentary rocks, including sandstones, siltstones, and limestones—rocks that have formed by the accumulation of particles of sediment and/or the skeletal remains of organisms. Sedimentary rocks form in surficial or near-surface environments, especially aquatic ones such as lake and ocean bottoms, and thus record the occurrence and activities of living organisms for the past 3.5 billion years.

The study of the ways in which freshly dead organisms are incorporated into the fossil record is called *taphonomy*. Taphonomy explores a broad variety of processes, including decay, postmortem transport, fossilization, time averaging, and postburial alteration. Taphonomic studies can reveal not only how certain organisms become preserved as fossils but also why certain other organisms are not readily fossilizable.

The chance of an individual organism becoming a fossil is vanishingly small, yet museums and outcrops abound with fossil remains. In large part, our abundant fossil record results from the fact that life has such a rich and deep history—countless trillions of organisms have lived and died on our earth. Aside from a good dose of luck, several reasonably predictable factors are key in determining the likelihood of an individual becoming a fossil. The path to becoming a fossil is logically divided into two phases: those processes that happen between death and burial in sediment (collectively termed *biostratinomy*) and those that occur after burial (collectively termed *diagenesis*).

Biostratinomic processes are easily visualized because most people have witnessed at least some of them. The death of an organism may be followed by partial or complete consumption by a predator, an interval of scavenging and decay (scavenging on a small scale), mechanical abrasion, chemical dissolution, and bioerosion, as the remains wash back and forth in the surf, roll along in a river, or rest on the ground or the seafloor. Naturally then, a significant fraction of organisms disappear forever in these ways, their atoms eventually recycled in a multitude of other organisms but leaving no remains to fossilize. Biostratinomic processes are not entirely destructive, however; they also control the nature of many fossil deposits through the winnowing, sorting, and concentrating of hard parts by normal sediment transport in water or wind.

It is intuitively obvious that hard parts such as shells, bones, and teeth will withstand postmortem degradation better than soft tissue. It is not surprising, then, that clams have a better fossil record than do slugs, or that more is known about the teeth of ancient mammals than about their livers. Even hard parts have an organic matrix, however, and the ratio of mineral to organic material can be very important in preservation. The cuticle of trilobites, for example, contained a higher proportion of calcium carbonate than does that of a lobster or a crab, which explains why trilobites have a much better fossil record.

Perhaps the single most critical factor in becoming a fossil is having the good fortune to be buried as quickly as possible in sediment of some kind. Rapid burial avoids the ravages imposed by physical and biological processes at the surface. The habitat occupied by an organism has a very important effect on its potential for burial. In general, most terrestrial habitats are areas of net sediment erosion, whereas oceans, lakes, and some river systems are more likely to be areas of net sediment accumulation (although it is geological [tectonic and sea level] factors that govern how much sediment can accumulate, not the habitat per se).

Once the remains of an organism become permanently buried, a variety of biological, chemical, and physical processes may predominate. Enormous spans of time may be involved, of course, and the burial history of an object may be complex. Naturally there are many ways for buried remains to be rendered forever unrecognizable: from complete dissolution in the shallow subsurface to destruction by extreme heat and pressure at depth. Here we focus on those processes that result in fossilization.

Waters that percolate through a deposit play a critical role in determining the fate of the buried remains. These fluids may dissolve skeletal material away or may precipitate minerals, or both. When mineral-laden waters deposit precipitates in preexisting spaces in skeletal material, the result is *permineralization*. Wood and bone, with their abundant natural pores, are often fossilized in this way, sometimes with fine structural details preserved. The minerals deposited are typically silicates, pyrite, or carbonates. *Petrifaction* is a closely related process in which organic matter is completely replaced by precipitating minerals. In other cases, both the skeletal and soft tissue components of an organism may be replaced by precipitating minerals (*replacement*). The dissolution of tissues or hard parts and subsequent precipitation of minerals in their place may happen virtually simultaneously, thereby preserving fine detail, or these events may be widely spaced in time and record only the basic shape of the organism. Fossils may form in more unusual ways as well, including freezing, desiccation, or entrapment in sap (with eventual modification into amber). Under particular conditions (including low oxygen and pH), tissues may be preserved as phosphate, occasionally with spectacular results (e.g., cellular preservation in the Doushantuo Formation [590–565 Ma], arthropod larvae in the Orsten Formation [501–488 Ma]).

The fossilization processes described above, in which various parts of ancient organisms become preserved, generate what are called *body fossils*. *Trace fossils*, on the other hand, record the activities of ancient organisms and may include footprints, feeding traces, borings, burrows, and even fossilized feces (*coprolites*). Trace fossils often provide a wealth of information that would be unavailable from even the best of body fossils, including such important things as feeding behavior, diet, and locomotion.

Taphonomic processes have acted and continue to act as a filter for the preservation of organic remains in the geological record; not all living individuals, species, or higher taxa have an equal chance of preservation. Paleontologists sometimes estimate "paleontological completeness" as the probability of sampling any member of a given taxon within its stratigraphic range (the total amount of geologic time between the oldest- and youngest-known fossils of that taxon). For example, in an interval of about 5 million years, estimates of the probability of finding any genus belonging to a particular group range from 5 percent for polychaetes to 90 percent for brachiopods, trilobites, graptolites, conodonts, and cephalopods (Foote and Sepkoski 1999). Other well-known marine animal taxa fall between, including sponges, corals, crinoids, bivalves, and gastropods, all with probabilities of approximately 40–50 percent.

Fortunately, the taphonomic filter is imperfect and the geological record is scattered with deposits of exquisitely preserved fossils from organisms that would typically be destroyed before fossilization. *Lagerstätten* is the term generally reserved for fossil deposits with exceptionally well-preserved soft tissues. The importance of these relatively few deposits for understanding the history of life has been profound. Certainly the most well-known lagerstätten is the Middle Cambrian Burgess Shale, discovered in the Canadian Rockies by C. D. Walcott in 1909. The Burgess Shale deserves its fame; it provides a window on an entire community of organisms, including a spectacular variety of soft-bodied forms, from a critical interval near the early expansion of animal life. Of the more than 100 species described from the Burgess Shale, probably fewer than 15 percent have hard parts that would be preserved under more typical conditions of fossilization. Our knowledge of the early history of animal life has been dramatically enhanced in recent years by discovery of other lagerstätten, each containing otherwise-unknown organisms, and many taking the origins of important groups further back in time.

Lagerstätten are not restricted to any particular time interval. The Messel Oil Shale near Frankfurt, Germany, preserves animals and plants from an Eocene lake and its surroundings, including numerous fish, birds, mammals (e.g., primates, bats, pygmy horses, and hedgehogs), and insects (the latter with distinct coloration still present). Other lagerstätten include the Hunsruck Slate (Devonian of Germany), which has revealed much about the nature of trilobite limbs and other soft parts, the Rhynie Chert (Devonian of Scotland) with its early vascular plants, and various Cretaceous localities in China known for their feathered dinosaurs and early flowering plants.

Aside from their exceptional preservation, lagerstätten do not share a particular mode of fossilization. Each deposit differs in the mineralogy of its specimens and in the particular suite of physicochemical conditions that led to their preservation. The general themes that prevail are the following: (1) postmortem exposure to aerobic (oxygenated) conditions is simply not possible: decay processes that occur in the presence of oxygen destroy organic remains very quickly; (2) rapid burial is usually important, although environments with low sedimentation rates may allow for preservation if bottom waters lack oxygen; (3) the chemistry of ambient fluids is key to the mineralization of remains (e.g., pyritization requires low oxygen and high iron concentrations).

Fossil material with well-preserved DNA is not impossible, but it is more rare than one would imagine, given the coverage it receives in the popular press. Most confirmed cases of DNA preservation involve frozen or desiccated organisms less than 1 million years old. Other chemical signatures of life are much more durable, however. Sponge-specific sterane biomarkers from Precambrian sediments of the Arabian Peninsula, for example, indicate that basal metazoans first evolved nearly 100 million years before their earliest known body fossils.

2. THE NATURE OF THE FOSSIL RECORD

Many important questions about evolution can be addressed only by the fossil record. For most of these questions, the issue of temporal resolution is fundamental. How long do individual species persist? How rapidly do they replace one another? (We note that paleontologists use a morphological species concept, which can be supported by data on paleoecology, the geographic and temporal ranges of particular forms, and quantitative methods to objectively differentiate morphological groups.) What is the pace of morphological change within a lineage? How rapid and how simultaneous are mass extinctions? How quickly can communities recover from a major environmental catastrophe? What is the relationship between the timing of a particular group's first appearance and the timing of major geological events such as continental fragmentation? Before addressing these and other basic questions, a paleontologist must establish the basic temporal parameters of the collections with which he or she is working.

The amount of time represented by a given sample is of critical importance, and a wide spectrum of possibilities exists. A set of fossils found in geographic and stratigraphic proximity may represent a true ecological snapshot; that is, it may include only organisms that lived at the same time and in the same place. The Miocene Ashfall Fossil Beds in Nebraska come close to these criteria; this deposit preserves complete skeletons of multiple species of rhinos, horses, camels, dogs, birds, turtles, and many more organisms that apparently fell victim to the effects of a sudden influx of volcanic ash. All of the animals preserved likely died within a few days or weeks of one another. At the opposite end of the spectrum, a fossil sample may contain individuals that lived at widely separated time intervals. For instance, many modern coastlines include rocky outcrops, often bearing fossils that are Pleistocene or older in age. As these fossils weather out of the nearshore rocks (or unconsolidated sediment), they mingle with the remains of recently dead organisms; the resultant mixture of shells that lived thousands or even millions of years apart may eventually become a fossil assemblage. This mixing of noncontemporaneous material into a single fossil deposit is known as *time averaging*. Aside from erosion and redeposition (as just described), time averaging may be caused by sediment winnowing or low sedimentation rates, coupled with the accumulation of skeletal material. Once remains are buried, they are still subject to time averaging through *bioturbation*, the mixing of the upper layers of sediment by burrowing organisms.

Time averaging has been quantified in modern marine settings, primarily on collections of mollusks. Radiocarbon and amino acid racemization dating techniques have shown that most shells are fewer than 3000 years old in modern nearshore settings and fewer than 10,000 years old on the continental shelves.

Although time averaging may seem like a hindrance to understanding ancient ecology and evolution, it has its advantages, too. A fossil collection that contains individuals accumulated over a few thousand years is certainly not an ecological snapshot, but the noise caused by short-term environmental and other fluctuations is reduced. Time averaging in the fossil record permits evolutionary change to be examined without conflating responses to ephemeral or short-period phenomena (e.g., seasonality) with more lasting evolutionary change. For example, two samples of living gastropods collected 100 m apart in the same bay are likely to be more different in species composition and relative abundances than the death assemblages collected at the same locations. This is because localized faunas change rapidly over time through stochasticity in birth, death, recruitment, and locomotion; thus, fewer fossil samples are needed to adequately capture the overall composition of an area when time averaging has occurred.

The fidelity of the fossil record can also be assessed via *live-dead comparisons*, in which the relative abundance of species in living communities is compared with the relative abundances in nearby accumulations of remains (that, given time, might become a fossil collection). Studies by taphonomists have shown that life assemblages are generally well represented by their corresponding death assemblages in a variety of marine settings (Kidwell and Flessa 1996). These studies have focused primarily on mollusks, which are readily fossilizable and ecologically important, but the results are very likely applicable to a broad spectrum of taxonomic groups. Live-dead comparisons have also demonstrated that most fossil accumulations only very rarely involve long-distance transport of skeletal material, and when they do there is abundant independent evidence that transport has occurred, including size-sorting, fragmentation, imbrication, and/or sedimentological context.

The resolution of a paleontological study is determined by a number of factors: the mode of fossilization, time averaging, and the amount of geologic time missing between fossil collections. Careful examination of modern ecosystems, lagerstätten, and modern sedimentary environments can provide a wealth of information on the first two factors. But the issue of stratigraphic resolution, or missing time, is one of the most commonly criticized aspects of paleontological studies, beginning with Charles Darwin himself.

Darwin famously devoted a chapter in *The Origin* to the imperfection of the fossil record. This chapter ended with a metaphor:

> ...The geological record as a history of the world imperfectly kept, and written in a changing dialect; of this history we possess the last volume alone, relating only to two or three countries. Of this volume, only here and there a short chapter has been preserved; and of each page, only here and there a few lines. Each word of the slowly-changing language, more or less different in the successive chapters, may represent the forms of life, which are entombed in our consecutive formations, and which falsely appear to have been abruptly introduced. (Darwin 1872, chapter 10)

Darwin's main concern with geological completeness, or the lack thereof, was his conviction that natural selection required species to evolve slowly and continuously, leading to an expectation of multiple intermediate forms between ancestral and descendant species. Instead, just as with living species, Darwin found significant morphological gaps between closely related fossil species. The conclusion was inevitable; the fossil record is incomplete.

Darwin was, of course, not a naive geologist, having studied geology at the University of Edinburgh and the University of Cambridge. It was common geological knowledge then as now that no single place records the entirety of earth history. Darwin's extreme view of geological incompleteness, however, was not entirely based on geological observations but rather colored by the failure of the fossil record to conform to his theory: "But I do not pretend that I should ever have suspected how poor was the record in the best preserved geological sections, had not the absence of innumerable transitional links between the species which lived at the commencement and close of each [geological] formation, pressed so hardly on my theory" (Darwin 1872, chapter 10). Darwin faced a choice when the geological record did not produce the innumerable intermediate forms between species his theory had predicted: reject his theory of evolution by natural selection or attribute the failure of his hypothesis to the poor quality of the fossil record. He chose the latter.

Darwin's choice was based on strict adherence to the principle of *uniformitarianism*, which was first proposed by James Hutton in 1788 and fully developed by Charles Lyell in *The Principles of Geology*. One of the key tenets of uniformitarianism to which both Lyell and Darwin adhered was that rates of natural phenomena have not changed over geologic time—volcanoes erupt with unvarying frequency, lakes fill with sediment at a constant rate, within-lineage speciation rates are invariant, etc. Darwin's worldview was informed by a strict adherence to constant rates and precluded him from seriously considering the possibility that distinct species appeared rapidly in geologic time.

The geologic timescale was understood in a general way before Darwin wrote *The Origin*; the Paleozoic, Mesozoic, and Cenozoic eras and nearly all the geological periods were named in the first half of the nineteenth century. Geologists have been working diligently ever since to refine these subdivisions, which form the basis for correlations of rocks of different ages across the globe. The most significant development since Darwin's day, however, has been the application of radiometric dating to provide numerical ages for the timescale (plate 1). Darwin could only guess at the age of the earth or the amount of time represented by any given sequence of sedimentary rocks (or the gaps contained therein). Our ability to quantify these time intervals now, combined with the wealth of modern paleontological and biological studies of species-level change, has changed our expectations of the way evolution should appear in the fossil record.

We now appreciate that stratigraphically adjacent paleontological samples will commonly be separated in time by tens of thousands of years. It is clear from considerable biological work that such time spans are more than enough time for distinctive new species to be generated. Recognizing these facts, paleontologists no longer have an expectation of geologically gradual sequences: transitions from ancestral to descendant species that play out continuously over hundreds of thousands or even millions of years. Such changes are occasionally preserved, but it is more common that a given stratigraphic sequence provides a snapshot of morphology every 30,000–50,000 years, and that the changes representing the evolution of a new species have occurred between adjacent samples. A new species will therefore appear to have arisen "suddenly," but our understanding of the time elapsed defines this "geological instant" as, typically, some tens of thousands of years. This sort of rapidity is what Stephen Jay Gould and Niles Eldredge refer to in their theory of punctuated equilibrium, first proposed in the early 1970s. As they argue, the apparently rapid changes in fossil sequences should be expected, given our knowledge of the absolute timing of the accumulation of sediments forming the rock record and the time intervals typically required for species-level change.

Although Darwin's bleak assessment of the fossil record is now recognized as overly pessimistic, paleontologists take care to approach the fossil record realistically, knowing that stratigraphic gaps can alter our perceptions of evolutionary patterns and processes (Kidwell and

Holland 2002). The recognition of unconformities in the field is therefore critical to the interpretation of fossil sequences. Fortunately, recognition of these gaps between sedimentary strata has occupied the time of a great many stratigraphers who have shown that unconformities can be readily identified and commonly fall at predictable positions within stratigraphic successions.

Thus, the fossil record is neither a complete nor unbiased archive of past life; however, the nature of the fossil record is fairly well understood, and it is possible to design meaningful evolutionary studies using fossil data that are not dominated by the fossil record's inherently nonrandom structure.

3. MARINE DIVERSITY IN THE PHANEROZOIC

A major focus of paleobiological research over the past 35 years has been the documentation and explanation of biodiversity changes over time. Precambrian life was chemically diverse and often locally abundant, but it generally would not have been much to look at with the naked eye. It is sometimes referred to as the Age of Slime; microbes were dominant, multicellular organisms evolved only after billions of years and for the most part remained tiny; therefore, most paleobiologists interested in the geological history of biodiversity focus on the Phanerozoic eon (the past 542 million years). The name *Phanerozoic* means obvious life, and indeed the interval is represented by voluminous, widely distributed, fossiliferous sedimentary rocks. The Cambrian explosion occurred in the eponymous earliest period of the Phanerozoic and marks the geologically sudden appearance of large metazoans with readily fossilizable hard parts. Thus, although Cambrian diversity was far from zero, it does mark the Phanerozoic low and provides a good temporal starting point for understanding biodiversity dynamics over evolutionary timescales.

John Phillips, an English geologist of the Victorian era, published one of the first estimates of marine diversity over the Phanerozoic. Although his compilation reflected only taxa known in Great Britain and the diversity axis of his seminal figure bears no numbers, his plot shares many features of more recent compilations. Specifically, Phillips's figure (figure 1) shows increasing diversity from the earliest Paleozoic to the Holocene, with major drops in diversity at the end-Permian and the end-Cretaceous (recognized today as two of the great mass extinctions, although Phillips did not use that term). These extinction events not only record sharp drops in diversity, but they mark major shifts in the composition of global faunas. It is no accident that these events mark the most fundamental divisions of the Phanerozoic: the Paleozoic, Mesozoic and Cenozoic eras.

In the late twentieth century, new compilations of global marine diversity, particularly by the late J. Jack Sepkoski Jr., were used to explore biodiversity dynamics. Sepkoski trolled the paleontological literature to compile a data set containing the times of first and last appearance of 2800 marine fossil families (later updated to include >36,000 marine genera). The data were global in scope, although with a bias toward North America and Western Europe, where fossil collecting has been most intense. Sepkoski's family- and genus-level compilations showed very similar diversity trends: an initial slow increase through the Cambrian, a rapid Ordovician radiation, a plateau in diversity through the remainder of the Paleozoic, followed by the major end-Permian mass extinction, with subsequent apparently exponential increase toward the Holocene. Twice during the Paleozoic plateau and twice during the Mesozoic-Cenozoic increase, mass extinctions removed 11–14 percent of the families, including the well-known event at the end of the Cretaceous that defined the extinction of the large terrestrial dinosaurs. Sepkoski's iconic plot of Phanerozoic diversity (figure 1) is probably the single most frequently used figure in paleontology presentations at national meetings.

Sepkoski did not simply tabulate Phanerozoic marine diversity, he also proposed an evolutionary model to explain the observed patterns. Sepkoski identified three evolutionary faunas: the Cambrian, Paleozoic, and Modern. Each of these three statistically defined evolutionary faunas includes taxa with similar diversity histories and broadly similar ecologies; the Cambrian Fauna is dominated by trilobites and a variety of other small grazers or deposit feeders, the Paleozoic Fauna is characterized by brachiopods, crinoids, and certain types of corals and bryozoans, most of which lived on the sediment surface and fed on suspended material in the water column, and the Modern Fauna (dominant ever since the global ecosystem recovered from the end-Permian mass extinction) is characterized by snails and clams, many of which live below the sediment surface, and a host of predators (including crabs, lobsters, fish, mammals, and reptiles) capable of feeding on organisms with hard skeletons.

Sepkoski used a coupled logistic model of diversification to explain these diversity patterns. In this model, each evolutionary fauna undergoes *logistic (density-dependent) growth*, meaning that diversification rates are initially exponential but then decline in response to the accumulating diversity. The coupled logistic growth model accounts well for many features of the empirical diversity curve, suggesting that global diversity may move toward equilibrium and that cohorts of higher taxa share common macroevolutionary histories.

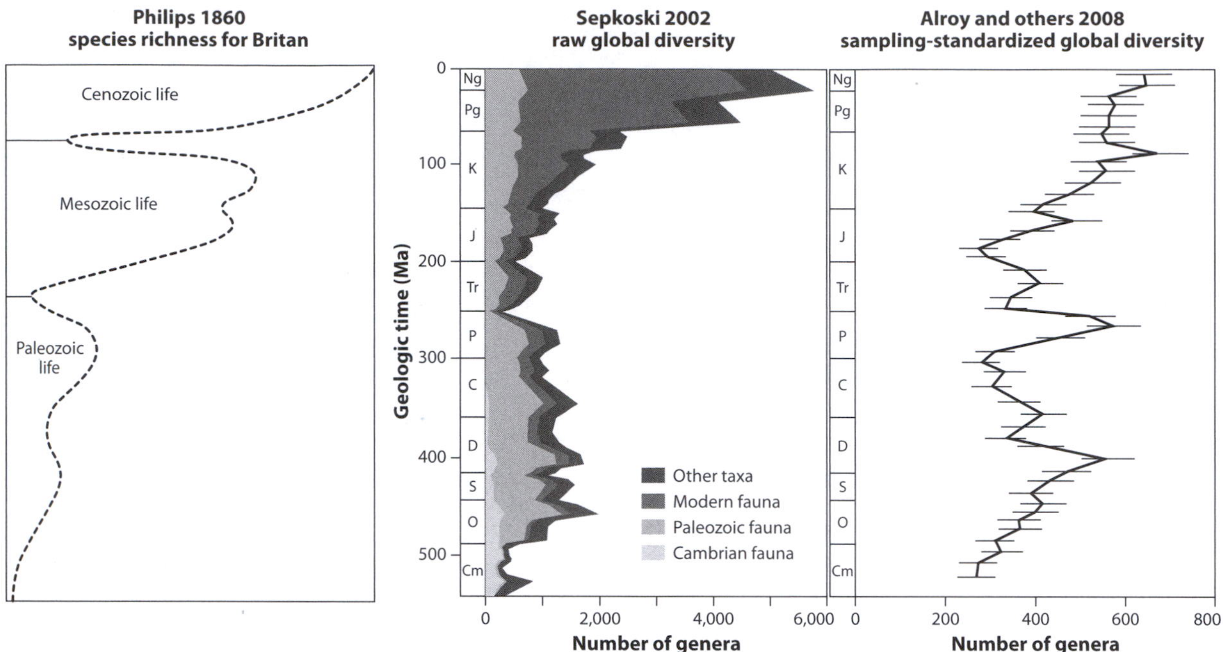

Figure 1. Modeled diversity curves showing the number of marine animal taxa through the Phanerozoic. Phillips's diversity curve is an estimate of the number of species based on observed fossils and the number of outcrops of a given age within Great Britain. Sepkoski's curve is a global estimate of the number of genera based on times of first and last occurrences. Alroy's curve is a sampling-standardized global estimate of the number of genera based on the number of genera actually sampled in each time interval; the total ranges of genera are not considered.

Currently, the most controversial aspect of the Sepkoski diversity curve is the shape and magnitude of the Mesozoic-Cenozoic radiation. Sepkoski's compendium suggests that it is exponential, implying that if the coupled logistic model is correct, the Modern Fauna has not yet reached equilibrium. Others, most notably John Alroy of Macquarie University, have argued that the exponential increase toward the Holocene is a sampling bias. Their primary argument is that the age distribution of sedimentary rocks at the earth's surface is nonrandom and nonuniform, and that paleontological sampling intensity is dictated by this distribution. (In other words, diversity simply appears to be higher in intervals that have more rocks exposed at the surface.) To test this hypothesis, John Alroy and Charles Marshall created the Paleobiology Database (PaleoDB; paleodb.org) to catalog fossil occurrences from around the globe. Thanks to an ongoing community-wide effort, the PaleoDB currently includes more than 110,000 individual fossil collections comprising more than 950,000 fossil occurrences. The main difference between the PaleoDB and Sepkoski's data is that Sepkoski reported only the time intervals of a taxon's first and last appearance, while the PaleoDB records all individual occurrences with the total stratigraphic range (as defined by the oldest and youngest occurrences). Paleontological sampling intensity within each interval can therefore be accounted for in the PaleoDB. The PaleoDB diversity curve (figure 1) shares many of the broad features of the Sepkoski curve, but the rise toward the recent is not exponential and appears to have reached an equilibrium. Although compelling, the PaleoDB diversity curve is certainly not the final "answer." The nuances of Phanerozoic marine diversity will continue to emerge as the PaleoDB grows, new subsampling techniques are found, and paleoenvironmental data become more comprehensive; however, it seems clear that our overall understanding of past marine diversity is generally good and that logistic (density-dependent) growth models consistently provide the most accurate description of diversification trends. Furthermore, each of the diversity curves produced by Phillips, Sepkoski, and Alroy suggest the possibility that the equilibrium state of diversity has changed over the Phanerozoic.

The other great macroevolutionary advance to come from Sepkoski's global compendium was recognition of the "big five" mass extinctions and a myriad of smaller extinctions. The earth's biosphere experienced major mass extinctions at or near the end of the Ordovician, Devonian, Permian, Triassic, and Cretaceous periods. These events, which are covered in more detail in chapter VI.13, were relatively rapid, taxonomically devastating, and global in extent.

4. THE VALUE OF THE FOSSIL RECORD

This chapter offers a very brief introduction to the nature of the fossil record and its importance for understanding the history of life and the evolutionary processes that have shaped that history. Molecular genetics, developmental biology, comparative morphology, and the fossil record all provide powerful and independent evidence for evolution. The beauty of the evidence offered by the fossil record is its accessibility. No matter where you are, a fossiliferous outcrop is likely not too far away. A fossil hunter cannot expect museum-quality specimens at every roadcut, but pulling from the earth a small part of a very ancient trilobite or tree fern or turtle can stir the imagination and awaken a connection to our deep ancestry.

FURTHER READING

Ager, D. V. 1993. The Nature of the Stratigraphical Record. 3rd ed. New York: John Wiley & Sons. *An excellent introduction to stratigraphy, including issues of completeness.*

Alroy, J. 2010. Geographical, environmental and intrinsic biotic controls on Phanerozoic marine diversification. Palaeontology 53: 1211–1235. *The most recent summary of PaleoDB diversity and methods.*

Foote, M. 2010. The geological history of biodiversity. In M. A. Bell et al., eds., Evolution since Darwin: The First 150 Years. Sunderland, MA: Sinauer. *A review of Phanerozoic marine animal diversity focusing on geological completeness and diversity dependence.*

Foote, M., and A. I. Miller. 2007. Principals of Paleontology. 3rd ed. New York: W. H. Freeman. *A recent revision of a classic paleobiology textbook.*

Gould, S. J. 1989. Wonderful Life: The Burgess Shale and the Nature of History. New York: W. W. Norton. *An enjoyable account of the Burgess Shale, with emphasis on evolution as a historical science. Some ideas presented here sparked a controversy over just how strange these organisms were.*

Gould, S. J. 2002. The Structure of Evolutionary Theory. Cambridge, MA: Harvard University Press. *Gould's final book. Remarkably long, but with much of value.*

Kidwell, S. M., and K. W. Flessa. 1996. The quality of the fossil record: Populations, species, and communities. Annual Reviews in Earth and Planetary Sciences 24: 433–464. *A review of the fidelity of the fossil record.*

Kidwell, S. M., and S. M. Holland. 2002. The quality of the fossil record: Implications for evolutionary analyses. Annual Review of Ecology and Systematics 33: 561–588. *A review of the ways in which stratigraphy influences perceptions of the fossil record.*

Knoll, A. H. 2003. Life on a Young Planet: The First Three Billion Years of Evolution on Earth. Princeton, NJ: Princeton University Press. *The finest popular book on the fossil record and early life.*

Miller, W., III, ed. 2007. Trace Fossils: Concepts, Problems, Prospects. New York: Elsevier. *An excellent edited volume on the history, methods, and importance of ichnology, the study of trace fossils.*

Raup, D. M. 1979. Biases in the fossil record of species and genera. Bulletin of the Carnegie Museum of Natural History 13: 85–91. *A summary of the biases in the fossil record written by one of the most important paleontologists of the last 50 years.*

Sepkoski, D., and M. Ruse, eds. 2009. The Paleobiological Revolution: Essays on the Growth of Modern Paleontology. Chicago: University of Chicago Press. *Includes some very thoughtful pieces on the development of paleobiology.*

Smith, J. M. 1984. Paleontology at the high table. Nature 309: 401–402. *An important comment on paleontology by a nonpaleontologist.*

II.10

The Origin of Life

David Deamer

> Nothing in biology makes sense except in the light of evolution. —Theodosius Dobzhansky

OUTLINE

1. Defining life in evolutionary terms
2. Plausible sites for the origin of life
3. Conditions required for life's origin
4. Self-assembly of boundary membranes and compartments
5. Prebiotic polymerization reactions
6. How could evolution begin?
7. Evolution in the laboratory

The prebiotic environment had a variety of simple carbon compounds and energy sources that could drive chemical reactions. These reactions produced ever more complex carbon compounds, some of which could assemble into membranous compartments, while others could link up to make polymeric chains. The polymers became encapsulated in the compartments, producing vast numbers of protocells. The variable protocellular compartments are a microscopic version of what is now called *combinatorial chemistry*; that is, each protocell contained a different mix of polymers and monomers, and represented a natural experiment. Some of the polymers happened to have potential catalytic abilities, while others could in some way replicate. A rare few of the cellular compartments contained both catalysts and replicating molecules in which the catalysts could speed up replication, and the replicating polymers could carry a kind of genetic information that coded for the monomer sequence in the catalysts. Biological evolution began with compartmented systems of molecules that could grow and reproduce. Within the populations of primitive microorganisms there would be variations in their abilities to compete for resources, and to withstand stress. These variations became selective factors that initiated subsequent stages of biological evolution.

GLOSSARY

Amphiphile. A molecule having both hydrophilic and hydrophobic components. Typical hydrophilic groups are carboxylate, phosphate, and sulfate, and the hydrophobic portion is a hydrocarbon. Examples of amphiphiles include most lipids, fatty acids, sterols, and detergents.

Combinatorial Chemistry. A way to perform multiple chemical reactions in parallel. Each combination can be different, so information about reaction conditions is obtained much more rapidly than by performing a series of experiments with different conditions one after another.

Lipid Bilayer. All cell membranes use a bimolecular lipid layer as a barrier to free diffusion of ionic and polar solutes.

Liquid Crystal. An organized structure formed by amphiphilic molecules, often in the form of semicrystalline layers (smectic phase) or rods (nematic phase), in which the molecules are not fixed in place but instead move by diffusion within the layer or rod.

Multilamellar Matrix. The nanoscale organization of many phospholipids and soaps that associate as layers of molecules.

Protocell. A microscopic membrane-bounded system of interacting polymers that represents an evolutionary stage on the pathway to the origin of life.

RNA World Hypothesis. The conjecture that life passed through a phase in which RNA functioned both as a catalyst and to store genetic information. The RNA life-forms were later replaced by living systems using RNA, DNA, and proteins.

Vesicle. Spherical structures bounded by a lipid bilayer, sometimes referred to as liposomes.

Dobzhansky was among the great pioneering geneticists of the last century, and the quote at the beginning of the chapter is the title of an essay he wrote in 1973, late in his career. The quote is often used to challenge biology teachers and professional biologists to think more deeply about evolution, and it will be used here in a slightly different context: Does the origin of life make sense in the light of evolution?

Although Charles Darwin knew virtually nothing about biochemistry and molecular biology, he would have agreed that evolution began with the origin of life. In *On the Origin of Species*, he asks the question we are addressing here: "Looking at the first dawn of life, when all organic beings, as we may believe, presented the simplest structure, how, it has been asked, could the first steps in the advancement or differentiation of parts have arisen?"

What would Darwin have written if he understood that life began 4 billion years ago, when the systems of replicating molecules were much simpler than bacteria today? He would likely propose that evolution by natural selection could not begin with a single molecule, but there must have been a way to generate large numbers of primitive systems of molecules in the prebiotic environment. Furthermore, there would be considerable variation in the properties of such systems. The requirement for variation within a population means that the first life-forms capable of evolution were not simply a mixture of reproducing molecules, but instead consisted of microscopic systems of interacting molecules encapsulated in some sort of boundary structure, referred to here as *protocells*. Each protocell was a kind of natural experiment, and the primary hurdle they needed to overcome was to capture available sources of energy in order to grow by polymerization of nutrient monomers, then to reproduce. Heterotrophic life today does this by accumulating simple molecules from the environment. Chemical energy is then used during metabolism to activate them so they can be linked into polymers such as proteins and nucleic acids. The earliest cells also needed to store genetic information and replicate it when they reproduced, so their properties were passed along to the next generation. A certain amount of error in this process was inevitable, what we now call *mutations*. The imperfections in replication were important, because they meant that life could explore different niches and begin the long trek to cellular life as we know it today.

The next question concerns a process by which large numbers of natural experiments could be generated by organic carbon compounds in the prebiotic environment. Three things are necessary: a way to produce the microscopic equivalent of test tubes, a suitable source of energy, and a way to synthesize polymeric molecules. If the polymers can be encapsulated as protocells and provided with an energy source, the system has the potential to become more complex. Because life somehow emerged from the complex environment of the prebiotic earth, and did so soon after liquid water first condensed on the planet, it seems possible that its origin can be reproduced in the laboratory under just the right set of conditions and components.

1. DEFINING LIFE IN EVOLUTIONARY TERMS

At some point in the near future, a claim will likely be made that artificial life has been fabricated in the laboratory. For this claim to be convincing, it will be necessary to show that the system has properties that fall within an accepted definition of life. The problem is that no definition is generally accepted by biologists. Even the simplest microorganisms are extraordinarily complex, and dictionary-style definitions don't easily encompass such complexity. Because life is a complex phenomenon, one approach to a definition is to describe a minimal set of properties associated with the living state. What follows is a single paragraph that incorporates properties of terrestrial life as we know it; taken together, the properties exclude anything that is not alive:

> The machinery of life is composed of polymers, very long molecules composed of subunits called monomers. The primary polymers of life are nucleic acids and proteins, often called *biopolymers* by definition. The polymers interact within a membranous boundary that has three primary functions: containment, transport of nutrients, and energy transduction. Biopolymers are synthesized in the container by linking together monomers—amino acids and nucleotides—using energy available in the environment. Polymer synthesis is the fundamental process leading to growth of a living system. Nucleic acids have a unique ability to store and transmit genetic information, and proteins called enzymes have a unique ability to act as catalysts that increase the rates of metabolic reactions. The genetic and catalytic polymers are incorporated into a cyclic feedback-controlled system in which information in the genetic polymers is used to direct the synthesis of the catalytic polymers, and the catalytic polymers take part in the synthesis of the genetic polymers. During growth, the cyclic system of polymers reproduces itself, and the cellular compartment divides. Reproduction is not perfect, so that variations arise, resulting in differences between cells in a population. Because different cells have varying capacities to grow and survive in a given environment, individual cells undergo selection according to their ability to compete for nutrients and energy. As a result, populations of cells have the capacity for evolution.

There is no doubt that a claim of synthetic artificial life would be convincing if the system incorporated all the above properties; however, if the properties are deleted one by one, the definition becomes blurred and the claim weaker. Suppose the system reproduced perfectly so that evolution could not occur. Would it still be considered alive? Most would say yes, so the ability to evolve might not be a fundamental property of life. But consider another system in which all the nutrients required for growth were present in the medium so that no metabolism was required. This system would resemble a virus that requires the cytoplasm of living cells to reproduce; viruses, however, can evolve, so they seem to exist in the border between life and nonlife.

2. PLAUSIBLE SITES FOR THE ORIGIN OF LIFE

We can now briefly describe a few examples of sites and conditions proposed as being conducive to the origin of life. Each of these is characterized by one or more properties believed to have promoted certain chemical or physical processes conceivably related to steps involved in the pathway to life. For instance, mineral surfaces such as clay were suggested many years ago by Desmond Bernal and promoted by Graham Cairns-Smith, who thought a genetic takeover might have occurred as organic compounds were adsorbed to and organized by the clay: "We have, as it were, identified the organization responsible for the 'crime against common sense,' the origin of life. And it is true that the proposition that our ultimate ancestors were mineral crystals was not widely anticipated." There is some experimental support for the idea that clays were involved in life's origins. James Ferris and coworkers have made an extensive study of montmorillonite clay and demonstrated that chemically activated mononucleotides in the form of imidazole esters do in fact adsorb to the mineral surface. When concentrated as near neighbors, polymerization into oligomers up to 15 or more nucleotides in length can occur.

Another surface reaction has been suggested by Gunther Wächtershäuser, who proposed that life could begin as two-dimensional synthetic chemistry on a special mineral surface called pyrite, a crystalline mineral composed of iron sulfide. According to Wächtershäuser's idea, pyrite has two special properties. The first is that it has a positive surface charge; it therefore adsorbs negatively charged solutes such as carbonate and phosphate. Furthermore, when hydrogen sulfide reacts with iron in solution to produce iron pyrite, the reaction can potentially donate electrons to the bound compounds and drive a series of energetically uphill chemical reactions that otherwise could not occur in solution. Wächtershäuser sees these reactions as the beginning of metabolism, occurring on a flat mineral surface rather than in the volume of a cell. He refers to this stage of life's history as the "Iron-Sulfur World." After metabolic processes were initiated in this way, the reaction pathways would become encapsulated in membranes to produce the more familiar forms of cellular life.

Yet another site was proposed by Jeffrey Bada and Stanley Miller, who suggested that the early earth may have been covered by a global ice sheet. Under these conditions, organic compounds are preserved for much longer time intervals, and they speculated that occasional melting produced by impact events would release the organics and initiate chemical reactions necessary for life to begin.

All these proposals involve reactions of relatively simple compounds; however, for cellular evolution to begin, there must have been a point at which complex interacting systems of polymeric molecules were encapsulated within boundary membranes, and none of these proposals addresses this requirement. The rest of this article will describe a process by which the first protocells could emerge and be exposed to selective processes.

3. CONDITIONS REQUIRED FOR LIFE'S ORIGIN

We can start by describing conditions likely necessary for life to begin on the early earth 4 billion years ago, and then see whether a plausible site exists encompassing all the conditions. Certainly liquid water was required, a dilute solution of potential monomers. The primary monomers of life today are amino acids, nucleobases, pentose sugars (ribose and deoxyribose), and phosphate, all of which have been synthesized in simulated prebiotic conditions or demonstrated to be present in carbonaceous meteorites. There must also have been an energy source capable of driving polymerization reactions, involving a chemical reaction in which monomers could be linked into random polymers, some of which would have weak catalytic activity. A process would be required to concentrate dilute solutions of monomers to the point they could react; furthermore, the polymeric products of the reaction must be confined in some sort of compartment so they can interact with one another rather than being dispersed. The result will be large numbers of compartmented systems of polymers that are exposed to selective conditions so that evolution can begin. The next section will describe a plausible site and experimental systems in which membranous boundary structures self-assemble from amphiphilic molecules. Within the structure, polymers can be synthesized from monomers by an energy source that was ubiquitous in the prebiotic environment, followed by encapsulation of the polymers to produce protocells.

4. SELF-ASSEMBLY OF BOUNDARY MEMBRANES AND COMPARTMENTS

It has long been known that a phospholipid called lecithin can be extracted from egg yolks. Experiments conducted in the 1960s showed that if lecithin is dried on a microscope slide and then exposed to water, long wormlike structures called myelin figures grow (figure 1). Alec Bangham and his coworkers at the Animal Physiology Institute at Babraham, Cambridge, UK, added a dilute salt solution to egg lecithin in a test tube and found a milky suspension was produced consisting of immense numbers of cell-sized spherical globules. Using an early version of electron microscopy, they observed that the globules were multilamellar structures composed of many lipid bilayers. Furthermore, the globules could be dispersed into vesicles now called *liposomes*, and the *membranous lipid bilayer* is now understood to be the primary structural component of all cell membranes.

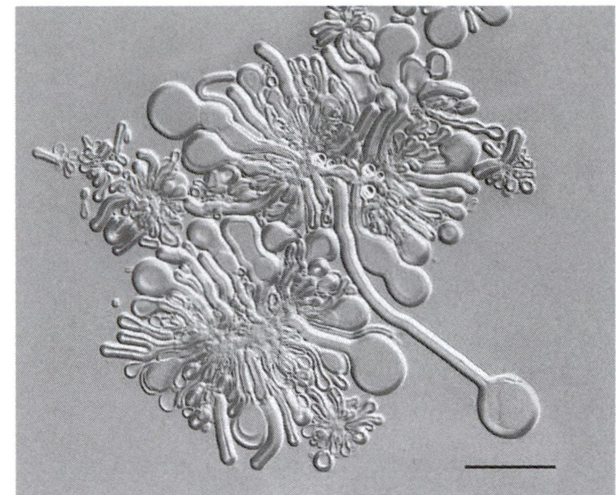

Figure 1. When phospholipid on a microscope slide is exposed to a dilute salt solution, the dry material begins to absorb water and grow out into tubular structures called myelin figures. These are unstable and ultimately break up into vesicles called liposomes. The tubules and vesicle boundaries are composed of lipid molecules that self-assemble into multilamellar bilayers in aqueous phases. A reasonable assumption is that the earliest forms of cellular life also used lipid bilayer membranes as boundary structures, but composed of lipid-like molecules such as fatty acids and alcohols available in the prebiotic environment. Bar shows 25 micrometers.

Could similar self-assembled compartments have been present on the prebiotic earth? This question has been addressed by investigating organic compounds in carbonaceous meteorites. One such meteorite fell near Murchison, Australia, in September 1969, and more than 100 kg of scattered fragments were collected and distributed to interested scientists. In 1970, Keith Kvenvolden and a group of researchers at NASA Ames analyzed a sample of the meteorite and convincingly demonstrated that amino acids, one of the essential organic compounds composing all life on earth, were present in the meteorite. This study established that amino acids, the fundamental building blocks of proteins, can be synthesized by a nonbiological process that occurred in the asteroid parent body of the meteorite. It seems reasonable to think that amino acids would have been synthesized on the prebiotic earth by similar reactions.

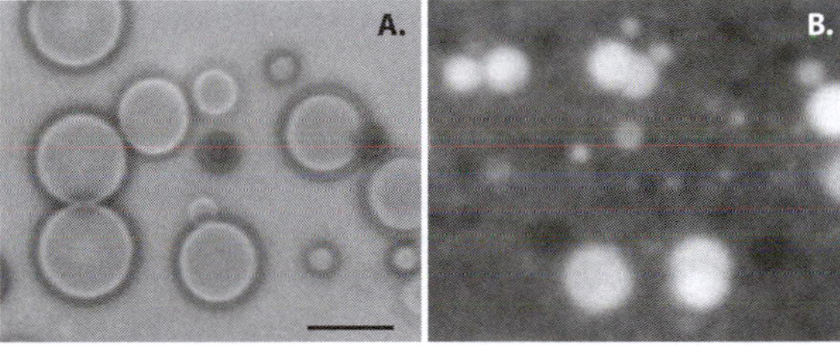

Figure 2. Carbonaceous meteorites like the Murchison chondrite contain long-chain monocarboxylic acids that can assemble into microscopic vesicles (A). These are true membranes capable of encapsulating a fluorescent dye such as pyranine (B). Bar shows 25 micrometers.

But what about membrane-forming compounds? In 1985, samples of the Murchison meteorite were extracted with an organic solvent, and a drop of the solution was dried on a microscope slide. When a dilute salt solution was added, amphiphilic compounds in the extract assembled into cell-sized membranous vesicles (figure 2), suggesting that similar cellular compartments were likely to be present when the first liquid water appeared on the earth more than 4 billion years ago.

The next question concerns how polymers can be encapsulated in the empty membranous compartments to produce protocells. It is known that lipid membranes fuse into multilamellar structures when dried, so one possibility is that wet-dry cycles in hydrothermal sites associated with volcanic activity on the early earth could carry out such a process. In early studies it was found that liposomes dried in the presence of nucleic acids or proteins trapped the polymers between the layers. When water was added back to the dry film, the lipid layers formed vesicles again, but now with up to half the large molecules trapped inside (figure 3). This seems a very plausible process by which primitive protocellular systems of polymeric molecules could be produced on the early earth.

5. PREBIOTIC POLYMERIZATION REACTIONS

Given cellular compartments and a way to encapsulate biopolymers, how could prebiotic polymers have been synthesized? This question has not yet been answered, but several possibilities have been experimentally

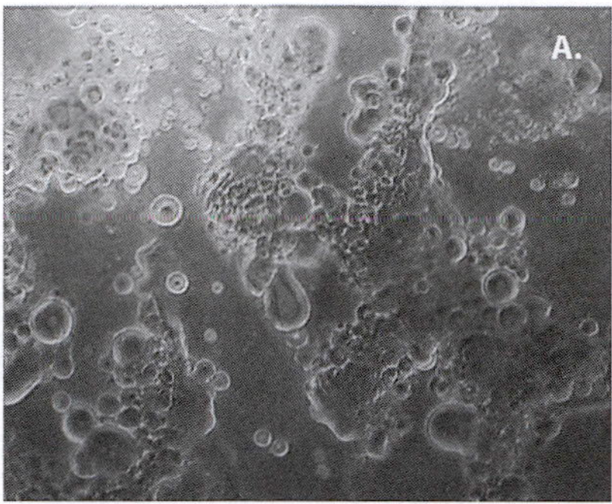

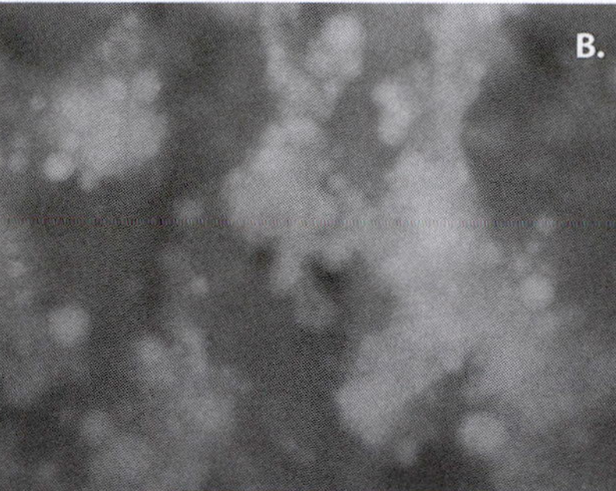

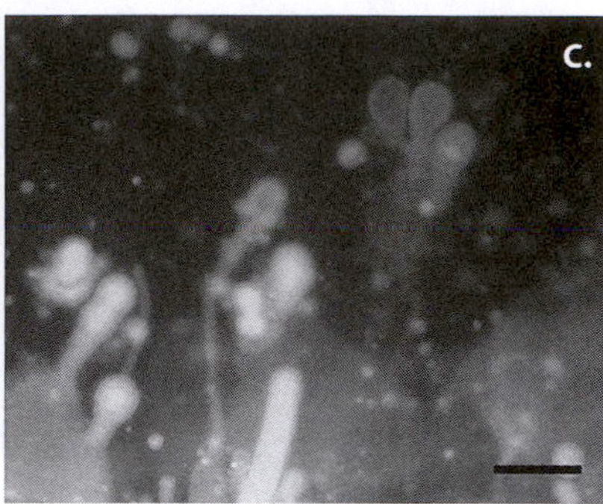

Figure 3. Phospholipid vesicles can be dried in the presence of a solute, in this case short strands of duplex DNA. When water is added to the dry material, multilamellar vesicles assemble (A), trapping the DNA between lipid bilayers. Here the DNA has been labeled with a fluorescent dye so that it can be visualized in association with the vesicles (B). After a few minutes, the multilamellar structures begin to form myelin figures that then break up into smaller vesicles containing the DNA (C). Cycles of wetting and drying would be common in the prebiotic environment, and represent a simple process by which cellular compartments could form containing encapsulated reactants and polymers. Bar shows 50 micrometers.

tested. In one approach, dispersions of lipid vesicles were prepared and mononucleotides, the monomers of RNA, were added to make a ratio of about one nucleotide per lipid in the solution. The mixture was warmed to 85 °C while being dried with a gentle stream of carbon dioxide to simulate the prebiotic atmosphere. A small amount of water was then added, the mixture was stirred for a few seconds to disperse the lipid vesicles, and the wet-dry cycle was repeated up to seven times. The idea was to simulate the conditions of a volcanic hydrothermal area on the early earth in which a continuous drying and wetting process occurred at the edges of pools. The water would be fairly hot (80–90 °C) and weakly acidic.

When the solution was analyzed for the presence of polymers, it was found that RNA-like molecules had been synthesized, ranging from 20 to 100 nucleotides in length. The yields were low by the standards of organic synthesis: fewer than 0.1 percent of the nucleotides had been linked into longer polymers, representing a few micrograms of product from the milligram quantities of nucleotides present in the mixture; however, there is no reason to think that high-yield polymerization reactions occurred in the prebiotic environment, in which mixtures of hundreds of different organic compounds would be present. The reactions leading to early biopolymers were almost certainly very low yield.

An important outcome of these experiments was that when the last cycle of hydration was completed by adding water, the lipid captured the RNA in vesicles. Such *protocells* represent a first step toward cellular life in an RNA world, that is, microscopic membrane-bounded compartments containing complex mixtures of polymers with the potential to be both catalysts (ribozymes) and carriers of information.

6. HOW COULD EVOLUTION BEGIN?

To summarize what has been said so far, the origin of life can be understood as an emergent phenomenon that occurs when water, mineral surfaces, and atmospheric gases interact with organic compounds and a source of energy. Hydrothermal conditions and processes act in concert to "pump" a random assemblage of simple organic compounds away from equilibrium toward increased complexity. The primary conditions are cyclic processes driven by a suitable input of energy, and capture of small amounts of the mixture in compartments that permit a natural version of combinatorial chemistry, resulting in vast numbers of microscopic molecular systems, each a kind of natural experiment.

We can now consider the physical and chemical conditions prevailing on the prebiotic earth that could drive the first steps of evolution. The early earth had oceans, volcanic landmasses, and an atmosphere of carbon dioxide and nitrogen gas. The most plausible site for the origin of life was not the open ocean or dry land; instead, there is reason to think that the most conducive conditions for life to begin existed in places where liquid water and the early atmosphere formed an interface with mineral surfaces such as volcanic rocks. Interfaces have special properties, because they allow three essential processes to occur that happen nowhere else: wet-dry cycles, concentration and dilution, formation of compartments, and combinatorial chemistry.

Cycles: The fluctuating environment required to provide cycles most likely took the form of pools in volcanic sites where hot water constantly underwent wetting and drying. The pools contained complex mixtures of dilute organic compounds from a variety of sources, including extraterrestrial material delivered during the last stages of the earth's formation, and other compounds produced by chemical reactions associated with volcanoes and atmospheric reactions. Because of the fluctuating environment, the compounds underwent cycles in which they were dried and concentrated, then diluted on rewetting.

Compartments: During the drying cycle, the dilute mixtures would form very thin films on mineral surfaces, a process necessary for chemical reactions to occur. Not only would the compounds react with one another under these conditions but the products of the reactions would also become encapsulated in microscopic compartments by membranes that self-assembled from soap-like compounds. This process produced vast numbers of protocells that appeared all over the early earth, wherever water solutions were undergoing wet-dry cycles in volcanic environments similar to today's Hawaii or Iceland.

Combinatorial chemistry: The protocells represented compartmented systems of molecules, each different in composition from the next, and each representing a kind of microscopic natural experiment. Most of the protocells remained inert, but a few happened to have the capacity to capture energy and smaller molecules from outside the encapsulated volume. As smaller molecules were transported into the internal compartment, energy was used to link them into long polymeric chains. Polymers have emergent properties that far exceed what monomers can do; for instance, both the primary biopolymers of today's life—proteins and nucleic acids—can act as catalysts, and nucleic acids carry and transmit genetic information, yet individual amino acids and nucleotides lack these properties.

Life began when a few of the immense numbers of protocells found a way not just to grow but also to incorporate a cycle involving catalytic functions and genetic information. According to this hypothesis, cellular systems of molecules, not individual molecules, were the first forms of life.

7. EVOLUTION IN THE LABORATORY

We can now address a simple question central to our understanding of the origin of life: Can nonliving molecular systems evolve? Can genetic information really appear out of nowhere? An answer to that question was provided by Andrew Ellington and Jack Szostak in 1990, then elaborated by David Bartel and Szostak in 1993. Their goal was to determine whether a completely random system of molecules could undergo selection in such a way that defined species of molecules emerged with specific properties. Bartel and Szostak began by synthesizing many trillions of different RNA molecules about 300 nucleotides long, all present as random sequences of nucleotides. They reasoned that buried in those trillions were a few catalytic RNA molecules called ribozymes that happened to weakly catalyze a ligation reaction, in which one strand of RNA is linked to a second strand. The RNA strands to be ligated were attached to small beads on a column, then exposed to the trillions of random sequences simply by flushing them through the column. This process could fish out any RNA molecules with even a weak ability to catalyze the reaction. They then amplified those molecules in an enzyme-catalyzed process and put them through for a second cycle, repeating the process for 10 rounds.

The results were astonishing. After only four rounds of selection and amplification, an increase in catalytic activity was seen, and after 10 rounds, the ligation rate was 7 million times faster than the uncatalyzed rate. It was even possible to watch the RNA evolve. Nucleic acids can be separated and visualized by a technique called *gel electrophoresis*. At the start of the reaction, nothing could be seen, but with each cycle new bands appeared. Some came to dominate the reaction, while others went extinct.

Bartel and Szostak's results demonstrate fundamental principles of evolution at the molecular level. At the start of the experiment, each molecule of RNA was different from all the rest. There were no species, just a mixture of trillions of different molecules, but then a selective hurdle was imposed in the form of a ligation reaction that allowed only certain molecules to survive and be reproduced enzymatically. After a few generations, groups of molecules began to appear that displayed ever-increasing catalytic function. In other words, in a mixture that initially contained completely random RNA molecules, species of molecules appeared in an evolutionary process closely reflecting the natural selection outlined by Darwin for populations of living organisms. These RNA molecules were defined by the sequences of bases in their structures, which caused them to fold into specific conformations that had catalytic properties. The sequences are analogous to genes, because the information they contained was passed between generations during the amplification process.

The inescapable conclusion is that genetic information can emerge in random mixtures, as long as there are populations containing large numbers of polymeric molecules with variable sequences of monomers, and a way to select and amplify a specific property. A similar process must have occurred on the prebiotic earth to bring the first forms of life into existence. The origin of life is best understood as a metaphor of combinatorial chemistry, but at a level far beyond what is possible in the laboratory. Will we ever discover the combination of ingredients that gave rise to life? There is reason to be optimistic. We need to apply what we know about the chemistry and physics of living systems to develop plausible hypotheses, then be brave enough to test them experimentally.

FURTHER READING

This article was adapted in part from First Life *(Berkeley: University of California Press, 2011), which is a more detailed account of the research briefly described here.*

Bartel, D. P., and J. W. Szostak. 1993. Isolation of new ribozymes from a large pool of random sequences. Science 261: 1411–1418. *The authors convincingly demonstrate how a catalytic function can emerge from a completely random system of RNA molecules undergoing selection and amplification.*

Deamer, D. W., and J. Szostak, eds. 2010. Origins of Life. Cold Spring Harbor, NY: Cold Spring Harbor Laboratory Press. *Each chapter of this multiauthored book provides an expert analysis of processes leading to the origin of life.*

Hazen, R. M. 2007. Genesis: The Scientific Quest for Life's Origin. *Washington, DC:* National Academies Press. *Robert Hazen, a research scientist at the Carnegie Institution of Washington, has written an engaging first-hand account of what it is like to investigate the origin of life.*

Sullivan, W. T., and J. Baross, eds. 2007. Planets and Life: The Emerging Science of Astrobiology. Cambridge: Cambridge University Press. *The editors of this book invited experts to write chapters about their specialty, using language that would be appropriate for undergraduate courses.*

ONLINE RESOURCES

Astrobiology: Life in the Universe. National Aeronautics and Space Administration.
astrobiology.nasa.gov/
NASA's Astrobiology website introduces concepts and news related to the origin and distribution of life in the universe.

Exploring Life's Origins: A Virtual Exhibit.
exploringOrigins.org/
This website, featuring molecular animations by Janet Izawa, visualizes the results of origins of life research for broad audiences.

II.11

Evolution in the Prokaryotic Grade

J. Peter Gogarten and Lorraine Olendzenski

OUTLINE

1. What is a prokaryote?
2. Archaea and Bacteria
3. Rooting the tree of life
4. Symbiosis, syntrophy, and eukaryotic origins
5. Horizontal gene transfer in the evolution of prokaryotes
6. Darwin's coral of life
7. Biased gene transfer
8. Sex, recombination, and procreation
9. Transfer of genes within and between groups
10. Biochemical innovation as a result of horizontal gene transfer

Prokaryotes are defined as organisms that lack a double membrane-bounded nucleus, but comprise two separate evolutionary lineages, the Archaea (or Archaebacteria) and the Bacteria (or Eubacteria). In prokaryotes and in many single-celled eukaryotes, genes can be transferred between related and unrelated organisms. As a consequence, genes coexisting in a genome have different histories from one another, and organisms can acquire new traits not only through gradual modification of ancestral traits, but also through transfer of genetic material from unrelated organisms.

GLOSSARY

Aminoacyl tRNA Synthetases. Enzymes that charge the tRNA with their cognate amino acid.

Archaea. One of the three domains of life, distinguished based on ribosomal RNA sequence, RNA structure, ether-linked lipids in the cell membrane, and the absence of peptidoglycan in their cell walls. Also known as Archaebacteria.

ATPase/ATP Synthase. Multi-subunit enzymes that use an electrochemical transmembrane gradient of protons to synthesize ATP from ADP and inorganic phosphate. The reaction is reversible, and in some organisms sodium ions are used instead of protons.

Bacteria. One of the three domains of life, characterized by distinct ribosomal structure and rRNA sequence, and cell walls (if present) containing peptidoglycan.

Conjugation. Process of DNA transfer from one cell to another in which bacteria are joined by pili or other structures.

CRISPR Elements (Clustered Regularly Interspaced Short Palindromic Repeats). Regions found in some bacterial genomes and most archaeal genomes that confer immunity against exogenous genetic elements such as phages, plasmids, and other invading genetic elements.

Duplicated Genes. Sets of homologous genes that arose by evolution from each other in an individual or lineage. Also called paralogous genes.

Flagellins. Proteins that make up the bacterial and archaeal flagella.

Gene Transfer Agent (GTA). Particles evolved from phages that transfer host genetic material between prokaryotes and that lost the activity to propagate independently of the host genome.

Halophile. Organism that thrives at high salt concentrations.

Horizontal Gene Transfer (HGT). A process by which genes or gene fragments are transferred among closely or distantly related organisms that are not in a direct ancestor-descendant relationship. It differs from vertical inheritance, whereby offspring acquire genes from their parents. Also known as lateral gene transfer.

Hyperthermophile. Organisms that have optimal growth temperatures from 80 °C to up to 121 °C.

Inteins. Protein introns; intervening sequences similar to introns that are transcribed and translated together with the host protein and that splice out only after translation.

Phage. Virus that attacks Bacteria. Phages are able to introduce foreign genes into a cell.

Plasmids. Circular extrachromosomal DNA elements present in prokaryotes that can replicate independently of the main chromosome. Plasmids carry few nonessential genes.

Prokaryotes. Cellular organisms whose genetic material is not surrounded by a membrane to form a nucleus. They include two domains of life: Archaea and Bacteria.

Recombination. A process in which new DNA is incorporated into a genome. Incorporation happens at regions of similarity between the existing genomic DNA and the incoming DNA. Recombination can be reciprocal, where the introduced DNA replaces an already-existing gene, or illegitimate, whereby newly introduced DNA does not replace existing DNA.

RNA Polymerase. An enzyme that synthesizes RNA using DNA as template.

Rooted Tree of Life. A phylogenetic tree that shows the relationships among all major lineages and includes at its root the most recent common ancestor (MRCA) of all living organisms (also known as last universal common ancestor, or LUCA). The root can be determined using a gene family whose history includes an ancient gene duplication. The phylogeny of such a family including homologues from a variety of different organisms can be rooted by using the ancient paralogues as an out-group.

Thermoacidophile. An organism that is both a thermo- and an acidophile. Many thermoacidophiles thrive at temperatures between 60 and 90 °C and at pH <3. All currently known thermoacidophiles are Archaea.

Thermophile. Organism that grows optimally at temperatures between 50 and 80 °C.

Transformation. Uptake of free DNA from the environment and integration into the genome of a cell.

1. WHAT IS A PROKARYOTE?

Prokaryotes are microorganisms defined by the absence of a nucleus. In eukaryotes (domain Eukarya), the nucleus is created by a double membrane that forms a compartment separating most of the genetic material from the rest of the cell. This compartment exists for most of the cell cycle. The nucleus can be considered a character that evolved in the lineage leading to the eukaryotes. Thus the presence of a nucleus is a derived characteristic of the eukaryotes. The difference in cellular structure between prokaryotes and eukaryotes formed a long-standing perceived dichotomy in the tree of life. However, recognition of the Archaea as a lineage separate from the Bacteria by Carl Woese, George Fox, Otto Kandler, and colleagues forced the realization that prokaryotes comprise at least two major lineages that are superficially similar in cellular morphology, but that do not constitute a monophyletic group. Because the absence of the nucleus reflects the ancestral state, and a group defined by an ancestral state is considered paraphyletic (see chapter II.1), in a cladistic system of classification there is no domain "Prokaryota" (cf. below and the section on taxonomy). The term *prokaryote* thus describes an organizational level or a grade of evolution; it is useful for describing a group of organisms with similar features but should not be regarded as a valid taxonomic group in a cladistics system.

Prokaryotes are ubiquitous in nature, inhabiting almost all possible niches on earth. It is estimated that there are 10^{30} prokaryotes on our planet. A single human carries about 10^{14} prokaryotes, 10 times more than the number of eukaryotic cells in the human body. Prokaryotes are the oldest forms of life on earth; evidence for their existence in the form of laminated rocks called stromatolites stretches back to 3.5 billion years BP, and in the form of graphite-rich sediments to 3.8 billion years BP. They are abundant in ocean water and soil and often predominate in extreme environments where eukaryotes are rare, such as hydrothermal vents, hot springs, hypersaline basins, and the deep subsurface.

Typical unicellular prokaryotes usually range in size from 0.2 µm to about 2.0 µm in diameter. Among the largest is the bacterium *Thiomargarita namibiensis*, found in the deep sea, whose cells can reach up to 1 mm in diameter. *Epulopiscium fishelsoni*, a bacterium found in the gut of the surgeon fish, was originally classified as eukaryotic because its cells can be 500 µm long. The morphology of prokaryotes is typically limited to rods, spheres, and filaments; however, some, such as the caulobacters, cyanobacteria, myxococci, and actinobacteria, have evolved complex developmental stages, including multicellularity and cell differentiation. When filamentous forms occur, they are composed of flattened cells stacked together. Resting stages may occur in the form of spores and cysts.

Although the morphological diversity of prokaryotes is relatively limited, their metabolic diversity is a defining characteristic. Prokaryotes carry out unique biochemical processes not found elsewhere in the biosphere, for example, nitrogen fixation, methane production, ammonia oxidation, and various unusual forms of anaerobic respiration, including sulfate, nitrate, and nitrite reduction, as well as the reduction of a variety of metals including iron. Heterotrophic prokaryotes are able to utilize a vast array of organic carbon sources and contribute to the breakdown of complex organic compounds. The nutritional mode of heterotrophic prokaryotes is typically absorptive: they hydrolyze substrates externally and transport small molecules into their cells. Autotrophic prokaryotes are able to fix CO_2 into cellular matter using either light energy (*phototrophy*) or redox energy obtained from reduced inorganic chemicals in a process termed *chemoautotrophy*. Phototrophy has been found in at least six different bacterial phyla, and all plastids found in eukaryotes can trace their ancestry back to an endosymbiotic event involving a primary endosymbiosis of a eukaryotic cell with a cyanobacterium. Chemoautotrophy is found in diverse phyla of both Bacteria and Archaea.

As a whole, prokaryotes have evolved the ability to couple the oxidation of almost any possible electron donor to the reduction of a huge variety of electron acceptors. In this way, they are able to drive the geochemical transformations of the major biologically reactive elements that make up the biogeochemical cycles of earth.

2. ARCHAEA AND BACTERIA

Initially designated as Eubacteria and Archaebacteria, the two prokaryote groups have been more recently renamed Bacteria and Archaea. Characters that support the division of prokaryotes into two separate lineages include ribosomal RNA sequences, promoter structure, membrane lipid and cell wall composition, antibiotic sensitivity, and the composition of flagellins, the proteins that make up prokaryotic flagella. Phylogenetic trees generated by using molecular sequences of ribosomal proteins, RNA polymerases, tRNAs, proton-pumping ATPases/ATP synthases, elongation factors, and aminoacyl tRNA synthetases typically show Bacteria and Archaea as two separate lineages, influenced by instances of gene transfer.

Archaea

The Archaea are dominated by two well-established phyla, the Euryarchaeota and the Crenarchaeota. Many of the known Euryarchaeota are found in extreme environments and include the halophiles and methanogens. Methanogens gain energy from converting hydrogen (H_2) and carbon dioxide into methane. Thermophilic Euryarchaeota are those that grow optimally at temperatures greater than 50 °C and include the thermophilic Archaeoglobi, the thermoacidophilic Thermoplasmata, and the hyperthermophilic Thermococci. The halophilic Archaea, often referred to as haloarchaea, grow best in salt concentrations of 3.5–4.5 molar. Officially their class is still called Halobacteria because they were named before the Archaea were recognized as a distinct group and long before this domain was labeled Archaea.

The first Crenarchaeota were characterized as sulfur-dependent hyperthermophiles, but they have since been found in many other environments. Some theories on the origin of eukaryotes suggest the crenarchaeotes as ancestors of the eukaryotes. In these theories the Crenarchaeota are labeled Eocytes.

Nanoarchaea are extremely small hyperthermophilic cells that are symbiotic with the crenarcheaote *Igniococcus* and initially were suggested to represent a basal archaeal phylum, the Nanoarchaeota; however, more recent analyses suggest that they are a divergent group within the Euryarchaeota.

The mesophilic Thaumarchaeota, which include marine ammonia oxidizers, were originally seen as a lineage branching within the thermophilic Crenarchaeota. The Thaumarchaeota have been suggested to represent a third major phylum of the Archaea. Thaumarchaeota may possess several ancestral features of the Archaea that have diverged in Crenarchaeota and Euryarchaeota.

The Korarchaeota are another candidate phylum. At present they contain a group of uncultured thermophilic Archaea whose genes in molecular phylogenies often either branch with the Crenarchaeota, or constitute a branch deeper than the split between Crenarchaeota and Euryarchaeota.

Bacteria

More than 50 phyla belonging to the domain Bacteria have been recognized. Approximately half these lineages have no cultured representatives and are recognized solely on the basis of rRNA sequence data obtained from natural environments. Lineages that contain no cultured representatives are termed Candidate Phyla or Candidate Divisions. The Taxonomic Outline of the second edition of *Bergey's Manual of Systematic Bacteriology* (Garrity et al. 2004) describes 24 phyla, each with cultured representatives. In August 2011, 30 phyla of Bacteria were recognized in the List of Prokaryotic Names with Standing in Nomenclature. While the phylogenetic analyses of bacterial phyla yields branching patterns among lineages that are unresolved or shrub-like, suggestive of a large simultaneous radiation at the base of the bacterial tree, the placement of individuals into phyla usually is unambiguous.

Many bacterial phylogenies contain two deep branching lineages, Thermotogae and Aquificae, that include mainly thermophiles (organisms that grow optimally at temperatures higher than 50 °C) and hyperthermophiles (organisms that grow optimally from 80 °C up to 121 °C). Other major bacterial lineages include Cyanobacteria, Firmicutes, Actinobacteria, Proteobacteria, Nitrospirae, Bacteriodetes, Chlorobi, Spirochetes, Chlamydiae, Planctomycetes, and Verrucomicrobia. Cyanobacteria are ubiquitous in marine and terrestrial habitats, can have complex morphologies (including filamentous forms with cell specialization), and generate oxygen during photosynthesis (oxygenic photosynthesis). They can be symbiotic with eukaryotic hosts and are the ancestors of eukaryotic plastids. The Firmicutes and Actinobacteria have a specialized cell wall structure characterized by multiple peptidoglycan layers. The major antibiotic-producing bacteria are found within this lineage. The Firmicutes include the endospore-forming *Bacillus* and *Clostridia.* Species of *Clostridia* cause tetanus and botulism. The nonspore-forming *Lactobacillus* found in fermented foods such as yogurt, pickles, and cheese, as well as probiotic supplements, are also members of the Firmicutes. The Actinobacteria include the filamentous actinomycetes, and nonfilamentous forms such as *Mycobacterium*, the genus that includes the causative agent of tuberculosis, *Propionibacterium,* which includes bacteria responsible for the holes in Swiss cheese and acne, and *Bifidobacterium*, species of which inhabit intestinal tracts and are used in probiotic supplements. One of the best-characterized lineages of Bacteria is Proteobacteria, which is divided into five classes: Alphaproteobacteria (containing nitrogen-fixing endosymbiotic rhizobia, plant-tumor-inducing *Agrobacterium*, and parasitic *Rickettsia*), Betaproteobacteria (some photosynthetic bacteria, nitrifying bacteria, and pathogens such as *Neisseria*, the causative agent of gonorrhea), Gammaproteobacteria (purple sulfur and iron phototrophic bacteria, sulfur-oxidizing bacteria, and the Enterobacteriaceae, such as *E. coli* and many other bacteria found in the human digestive tract), Deltaproteobacteria (including sulfur- and iron-reducing bacteria and the multicellular Myxobacteria), and Epsilonproteobacteria (*Helicobacter*, the bacteria that cause ulcers, and some marine sulfur-oxidizing bacteria). The heterotrophic Bacteroidetes and green sulfur-oxidizing phototrophic Chlorobi are species-rich phyla of obligate anaerobes found in many environments. Members of the Nitrospirae are nitrite- and iron-oxidizing bacteria. Spirochetes are responsible for syphilis, Lyme disease, and yaws, and are characterized by unique, sheathed spiral cells with internalized flagella although nonspiral members of this phylum have been characterized. The Chlamydiae are unusual obligate intracellular parasites and form a common sexually transmitted infection in humans, while the Planctomycetes are found in a variety of aquatic environments and exhibit internal membrane compartments reminiscent of nuclei. The Verrucomicrobia are poorly understood microbes that have cell projections containing tubulin, a protein normally found in eukaryotes and believed to have been acquired by these bacteria through horizontal transfer from a eukaryote.

3. ROOTING THE TREE OF LIFE

With the discovery that the Archaea are a separate lineage, a three-domain model of the tree of life emerged. The evolutionary relationship among these three groups has become one of the most important questions in studies of the early evolution of life. One question debated by some is the location of the root of the tree of life, corresponding to the Most Recent Common Ancestor (MRCA) of all extant organisms. While the existence of the Archaea was confirmed by comparing sequences of ribosomal RNA molecules, rRNA sequence analysis alone cannot be used to identify the root of the tree of life since no out-group exists when rRNAs from across all three domains are compared (see chapter II.2). However, sequences from protein families that have undergone ancient gene duplications can be used to root the tree of life, with each paralogue acting as an out-group for the other, producing a reciprocal rooting of the tree. These analyses generally support placing the root on the bacterial branch or within the bacterial domain; however, analyses of some proteins are interpreted to support a rooting along the eukaryal branch or within the Archaea.

Analyses of most ancient duplicated genes and of the amino acid compositional bias created through the early expansion of the genetic code place the root on the bacterial branch and reveal the Archaea as the lineage most closely related to the eukaryotic nucleocytoplasm (figure 1A). The recognition of the Archaea as a distinct group specifically related to the eukaryotic nucleocytoplasm is widely accepted based on similarities of the transcription machinery, presence of N-linked glycoproteins, lack of formylmethionine, shared resistance or sensitivity to various antibiotics, presence of tRNA introns, histones in some Archaea and most eukaryotes, and similarity between the eukaryotic endomembrane energizing vacuolar ATPase and the archaeal ATP synthase.

Two major hypotheses for the topology of the tree of life are currently seriously debated. The "classical" three-domain model corresponds to a monophyletic Archaea (figure 1A), whereas the alternative "eocyte" model contains a paraphyletic archaeal grouping in which a subset of the Archaea (the Crenarchaeota, sometimes referred to as eocytes) is more closely related to the eukaryotes (figure 1B). Under the classical model (figure 1A), eukaryotes might have accumulated eukaryotic characteristics, such as the cytoskeleton, long before uptake of the mitochondria. The accumulation of numerous derived characters in the eukaryotic lineage, including evolution of the nuclear, cytoskeletal, and spliceosomal machinery, before the endosymbiotic event leading to the mitochondria, provides support for a deep-branching protoeukaryote lineage. However, if some of these derived eukaryotic features, such

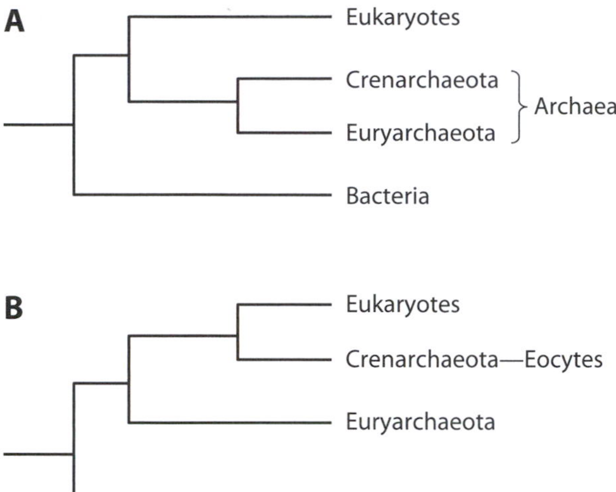

Figure 1. Possible relationships between the three domains of life. (A) The three-domain model shows a monophyletic archaeal domain (Crenarchaeota and Euryarchaeota). (B) In the eocyte version of the tree of life, the Archaea form a paraphyletic group, with only the Crenarchaeota/Eocytes forming the sister group to the eukaryotic nucleocytoplasm.

as the emergence of spliceosomal introns and the nuclear envelope, were triggered by the mitochondrial endosymbiont, or if the premitochondrial derivation of protoeukaryotes was rapid, then a more recent crenarchaeal origin of the Eukarya is plausible, consistent with a tree of life depicting a paraphyletic Archaea (figure 1B).

4. SYMBIOSIS, SYNTROPHY, AND EUKARYOTIC ORIGINS

Many prokaryotes live inside or on a larger host as mutualists, commensals, or pathogens. Such close symbiotic relationships of prokaryotes with larger organisms led to the acquisition of novel metabolic capabilities by the host, for example, nitrogen fixation in the root nodules of plants or chemoautotrophy in animals such as the giant tube worms and clams found at hydrothermal vents. Symbiosis with prokaryotes can lead to morphological evolution of the host; structures such as the cecum, colon, and rumen have evolved in different mammalian lineages to accommodate the prokaryotic communities that contribute to food digestion. Some deep-sea fish signal mates or attract prey using specialized structures that evolved solely to house luminescent bacteria. Some squid use luminscent bacteria contained in ventral light organs for counterillumination against light radiating from above, making the squid less visible to predators when viewed from below.

Prokaryotes also live in close associations with other prokaryotic members of their communities and engage in a type of metabolic interdependence known as syntrophy. Common in anaerobic environments, syntrophic prokaryotes take advantage of the metabolic abilities of their partners to break down compounds they cannot digest by themselves and overcome energy barriers, allowing them to survive using metabolisms that yield very little free energy for growth. *Desulfovibrio vulgaris*, which grows well when provided with sulfate, can ferment lactate in the absence of sulfate if the methanogen *Methanococcus paludis* is also present. When grown together, *Desulfovibrio* produces acetate, carbon dioxide, and hydrogen from the fermentation of lactate. The methanogen in turn uses the hydrogen produced by *Desulfovibrio* to reduce carbon dioxide and yield methane. The use of hydrogen by the methanogen drives the overall reaction, which otherwise would not yield energy because of the accumulation of hydrogen.

Syntrophic, endosymbiotic relationships are postulated to have played an important role in the origin of eukaryotes. All extant eukaryotes are the product of lineage fusion that took place at least once early in their evolution, when a bacterium belonging to the Alphaproteobacteria became an endosymbiont in the ancestor of all known eukaryotes. In other words, eukaryotes are derived from at least two distinct ancestors: an organism related to the Archaea was ancestor of the nucleocytoplasm, and an Alphaproteobacterium was the ancestor of the endosymbiont that evolved into the eukaryotic cell organelle known as the mitochondrion. Even eukaryotes that possess no functioning mitochondria have been shown to have evolved from ancestors that already had the alphaproteobacterial endosymbiont (see chapter II.12). Additional bacterial lineages likely contributed to the origin and early evolution of the eukaryotes; however, the selective advantages that led to the formation and success of these symbioses are still open questions.

5. HORIZONTAL GENE TRANSFER IN THE EVOLUTION OF PROKARYOTES

While the tree of life depicts bifurcating lineages as the standard representation of species evolution, we now understand that evolution is not solely a steadily bifurcating process; lineages also exchange genetic material or fuse with one another to form a new line of descent. The comparison of whole genome sequences from a great variety of prokaryotes has underscored the importance of horizontal gene transfer (HGT) in their evolution.

Phylogenetic trees constructed using single genes can have radically different topologies from one another, depending on what gene is being analyzed. This pattern is consistent with the transfer of genes between unrelated lineages; as a result, prokaryotic genomes are mosaics: different parts of a microbial genome have differing evolutionary histories, largely as a result of horizontal gene transfer. Because of this, phylogenetic reconstruction of organismal evolution is more complicated than in animals, for example, where HGT is rare. For example, in the deeply branching bacterial lineages Aquificae and Thermotogae, the ribosomal protein and RNA gene trees place these lineages as sister to the rest of the Bacteria; however, the majority of genes in the Aquificae group them with the Epsilonproteobacteria, whereas most Thermotogae genes group them among the clostridia, members of the Firmicutes. This suggests large-scale HGT between particular bacterial lineages sometime in the past.

6. DARWIN'S CORAL OF LIFE

The reconstruction of prokaryotic evolution is based mainly on the study of extant organisms. In most molecular phylogenies, only the lineages leading to extant organisms are included. In one of his notebooks, Darwin contemplated that the image of a tree was not appropriate to depict evolution, because in a botanical tree the whole tree is alive, whereas in the tree of life only the top layer represents living organisms. This layer of extant life rests on extinct ancestors, a fact that Darwin noted would be more appropriately captured by the image of a coral. One misconception resulting from the tree image is that the Most Recent Common Ancestor (MRCA) of all extant organisms appears as a singular organism without coexisting lineages. If we take extinct lineages into consideration, it is clear that the MRCA was surrounded by many other lineages. Furthermore, a consequence of gene transfer is that different genes trace their history back to different MRCAs; that is, the MRCA of all extant ribosomes probably existed in a different organism and at a different time than the MRCA of all ATP-synthases, or the MRCA of all extant RNA polymerases.

7. BIASED GENE TRANSFER

Although all genes are subject to horizontal gene transfer, some are more frequently transferred than others, and not all groups of organisms experience gene transfer to the same extent. Such differential gene transfer may contribute to maintaining coherent groups such as species. More closely related organisms often exchange genes more frequently, thus creating a phylogenetic relationship or signal that mimics or reinforces that seen with vertical inheritance. Genes are transferred mainly within major bacterial groups with a transfer bias toward partners more closely related to each other, as judged by rRNA phylogeny. Both shared ancestry (vertical inheritance) and biased HGT seem to cause the observed phylogenetic tendencies observed in genome analyses. Horizontal transfers between divergent species may be draped over a phylogenetic backbone of vertical descent and do not necessarily affect the resolution attainable for a treelike vertical inheritance process of organismal evolution. In prokaryotic lineages, distinguishing the extent to which biased gene transfer contributes to a particular phylogenetic pattern as compared with that of vertical shared ancestry is ongoing.

8. SEX, RECOMBINATION, AND PROCREATION

Sex in most eukaryotes involves the pairing of homologous chromosomes that can then undergo reciprocal crossing-over during meiosis. Daughter cells of a set of meiotic cell divisions receive single copies of each chromosome; overall, the collection of daughter cells contains the full complement of genes in the parent cell, redistributed reciprocally among each homologous chromosome. A generally acknowledged advantage of sexual recombination is that beneficial mutations that arose in different individuals in a population can come together in a single individual in future generations. Without recombination, in order for two separately arising beneficial mutations to come together in the same individual, they would need to evolve successively in the same lineage; however, it must be noted that in many eukaryotic lineages sex is not required for reproduction. Many single-celled eukaryotes, for example, yeast, propagate mainly through mitosis. Even in animals, where reproduction is linked primarily to sex, asexual lineages occur where asexual propagation (parthenogenesis) is the norm (e.g., bdelloid rotifers), with sex occurring only when food becomes scarce (e.g., in the planktonic crustacean *Daphnia*). In many instances, eukaryotes that propagate asexually have been shown to facilitate recombination by mechanisms similar to those found in prokaryotes.

Recombination in prokaryotes uses different mechanisms from those found in meiotic recombination. DNA can be introduced to a genome via various mobile genetic elements, including plasmids, transposons, integrons, and integrative conjugative elements (ICEs) such as conjugative transposons. Transformation occurs when DNA is taken up from the environment by competent recipient cells. In conjugation, DNA is transferred between two cells that become physically joined by pili. Conjugation can lead to the transfer of either single genes, plasmids, or a substantial portion of the genome. In the cell fusion observed in haloarchaea, two complete genomes are brought together in a single cell, and recombination between the two genomes can occur. Additionally, genetic material can be packaged into phages, viruses that attack bacteria and can be delivered to a recipient cell via a process called transduction. Transduction by phage or virus results in the transfer of foreign DNA into a prokaryotic cell. Gene transfer agents (GTAs) first discovered in some marine Alphaproteobacteria seem to be derived from phage, but they function only in the transfer of genomic DNA between cells and no longer propagate independently from their host. Mechanisms for transferring genes in nature are abundant, but the frequency with which transferred DNAs overcome barriers to attain successful long-term integration into new genomes needs further elucidation.

Gene transfer in prokaryotes is usually unidirectional and not reciprocal. Either a new gene or gene fragment is added to the genome (illegitimate recombination), or it replaces a homologous piece of DNA previously present in the genome (homologous recombination). Usually the replaced fragment is not transferred back to the donor; however, the above-mentioned case of cell fusion in haloarchaea shows that reciprocal recombination in prokaryotes does occur.

Many of the mechanisms for gene transfer may have evolved for purposes unrelated to the acceleration of adaptive evolution through recombination: the immediate selective advantage of taking up DNA from the environment might have been for food; phages, viruses, and conjugative plasmids can thus be considered parasites that have been selected for their own propagation. However, once these mechanisms were in existence, they also allowed for recombination and gene flow within and between populations. Interestingly, virulence resistance via CRISPR elements (Clustered Regularly Interspaced Short Palindromic Repeats) is a trait that greatly decreases the interaction between the bacterial genome, phages, and other extrachromosomal elements; ironically, however, the CRISPR loci themselves are often acquired through the transfer of plasmids.

9. TRANSFER OF GENES WITHIN AND BETWEEN GROUPS

Gene transfer has many advantages for the evolution of groups of organisms. Homologous recombination allows beneficial mutations that occurred in different organisms to come together in a single individual. Similarly, a slightly deleterious mutation can be unlinked through recombination from a beneficial mutation that has occurred in the same individual. However, acquisition of DNA also can be detrimental to the recipient, because the acquired DNA can encode molecular parasites, such as transposons, self-splicing introns, or inteins (protein introns, intervening sequences similar to introns that are transcribed and translated together with the host protein and that splice out only after translation).

The probability of acquiring a molecular parasite is smaller if genes are traded only with close relatives followed by homologous recombination of only small pieces of transferred DNA. The specificity of phages and conjugative plasmids, and the use of recognition sequences by the machinery that imports naked DNA from the environment, decrease the probability of incorporating DNA from an unrelated organism; however, the specificity for within-group transfer is not strict. For example, GTAs can transfer genes to divergent recipients, even to members of different bacterial phyla. *Agrobacterium tumefaciens* uses its conjugation machinery to transfer tumor-inducing DNA into a plant genome, and many bacteria import DNA without the use of a species-specific recognition sequence.

Prokaryotes possess restriction endonucleases, proteins that act as a kind of immune system to cleave foreign DNA from a virus or phage that may invade a host cell. The smaller fragments created by restriction endonuclease from the foreign DNA can still undergo homologous recombination, but their digestion into smaller pieces decreases the probability of incorporating a complete parasitic genetic element. Methylated DNA is not cleaved by the restriction endonucleases and is thus recognized as belonging to the host. While restriction endonucleases protect against invasion by molecular parasites, the restriction endonuclease and protecting DNA methylase can be considered selfish mobile genetic elements. The protecting DNA methylase has a higher turnover than the restriction endonuclease, and an organism that loses the genes for the restriction/modification system will lose protection faster than the endonuclease activity decays. Under these conditions, the remaining endonuclease activity will attack the genomic DNA of the organism.

Small deletions occur frequently in prokaryotic genomes. If such a deletion inactivates a beneficial gene, the deletion will be negatively selected and removed from the population; however, if the deletion occurs in a parasitic genetic element, the element is inactivated and may be deleted through further deletions. The deletion pressure in prokaryotic genomes tends to inactivate genes that are not under purifying selection in the host; however, this piecewise deletion is circumvented by those selfish genes that lead to addiction, such as the methylating enzyme–restriction enzyme pairs described above.

Sex in eukaryotes restricts recombination to partners belonging to the same species. In contrast, prokaryotes are more promiscuous, frequently incorporating genes from very divergent donors. Despite being more selective in choosing with whom to exchange genes, eukaryotic genomes contain many more remnants of parasitic elements in their genomes than bacteria and archaea. The likely reason for this is the difference in population size. In the case of eukaryotes, the smaller population size allows for less efficient weeding out of mutations with a slight selective disadvantage. The remnant of a molecular parasite increases the amount of DNA and possibly protein that need to be synthesized by a tiny fraction that is too small to be subject to purifying selection in a small population, but that could be weeded out in the very large populations that characterize many prokaryotes.

10. BIOCHEMICAL INNOVATION AS A RESULT OF HORIZONTAL GENE TRANSFER

Horizontal gene transfer plays an important role in the creation of novel biochemical pathways. An existing pathway can be transferred between divergent organisms, providing an adaptive advantage to the recipient; however, in addition, the transfer of genes encoding individual enzymes can create novel pathways that did not previously exist in either the donor or the recipient. Pathways created through gene transfer have greatly impacted the biosphere. One example is acetoclastic methanogenesis in *Methanosarcina*, which arose through the acquisition of two enzymes from cellulolytic clostridia. These two enzymes in clostridia lead to the production of acetate. In *Methanosarcina*, these enzymes work in the opposite direction, producing acetate and introducing it into the methanogenic pathway. Today methanogenesis from acetate represents the major pathway for biological methane production, most of it occurring in members of the genus *Methanosarcina*. Oxygenic photosynthesis is another important process whose assembly is believed to have involved horizontal gene transfer. Chlorophyll-based photosynthesis occurs in several bacterial phyla. In all cases except cyanobacteria, light capture uses only a single photosystem and entails relatively reduced electron donors such as hydrogen sulfide. The photosystems in different phyla are very divergent and can be distinguished on the basis of their sequence and their primary electron acceptor. Only when two of these divergent photosystems came together in the ancestor of the cyanobacteria did it become possible for these photosystems to work in series—generating enough energy to raise electrons from the redox potential of water to that of nicotinamide adenine dinucleotide phosphate (NADP). This allowed cyanobacteria to use water, an effectively unlimited resource, as an electron source, and to generate molecular oxygen as a by-product. Thus, it is fair to say that the oxygen-rich atmosphere we enjoy today would not have arisen without HGT.

FURTHER READING

Bapteste, E., M. A. O'Malley, et al. 2009. Prokaryotic evolution and the tree of life are two different things. Biology Direct 4: 34. *Provides the philosophical, scientific, and epistemological perspectives on the reasons a strictly bifurcating phylogenetic tree does not adequately describe evolution in the prokaryotic domains because of extensive chimerism and horizontal gene transfer.*

Boekels-Gogarten, M., J. P. Gogarten, and L. Olendzenski, eds. 2009. Horizontal Gene Transfer: Genomes in Flux. Methods in Molecular Biology: 532. New York: Humana Press. *Chapters summarizing definitions, philosophy, mechanisms, case studies, and analysis methods used in research on horizontal gene transfer and the impact of HGT on microbial, animal, and plant evolution.*

Doolittle, W. F. 1999. Phylogenetic classification and the universal tree. Science 284: 2124–2129. *Reviews the reasons a tree-based classification system may be unsuitable for microorganisms.*

Dworkin, M., S. Falkow, E. Rosenberg, K.-H. Schleifer, and E. Stackebrandt, eds. 2006. The Prokaryotes: A Handbook on the Biology of Bacteria. 3rd ed. Vol. 1–7. New York: Springer. *Organismal approach to prokaryotic diversity, providing information on habitat, culture techniques, biochemistry, and unique attributes of each group of prokaryotes.*

Fournier, G. P., A. A. Dick, D. Williams, and J. P. Gogarten. 2011. Evolution of the Archaea: Emerging views on origins and phylogeny. Research in Microbiology 162: 92–98. *Discusses the origin of eukaryotes and their relation to the archaeal phyla.*

Fournier, G. P., and J. P. Gogarten. 2010. Rooting the ribosomal tree of life. Molecular Biology and Evolution 27: 1792–1801. *Using a phylogenetic tree of concatenated ribosomal proteins, this analysis of amino acid compositional bias detects a strong and unique signal associated with the early expansion of the genetic code, placing the root of the translation machinery along the bacterial branch.*

Garrett, R. A., and H.-P. Klenk, eds. 2007. Archaea: Evolution, Physiology, and Molecular Biology. Malden, MA: Blackwell. *Documents the early development of the field of study of Archaea by researchers active in their discovery.*

Garrity, G. M., et al., eds. 2001–2012. Bergey's Manual of Systematic Bacteriology. 2nd ed. Vol. 1–5. New York: Springer. *Survey of bacterial and archaeal phyla, including information on phylogeny, morphology, ecology, and physiology of cultured species.*

Gogarten, J. P., W. F. Doolittle, et al. 2002. Prokaryotic evolution in light of gene transfer. Molecular Biology and Evolution 19: 2226–2238. *Discusses biased gene transfer as a force creating exchange groups whose members appear similar to one another, regardless of their shared ancestry.*

Margulis, L. 1993. Symbiosis in Cell Evolution: Microbial Communities in the Archean and Proterozoic Eons. 2nd ed. New York: W. H. Freeman. *A classic text on the role of endosymbiosis in the evolution of eukaryotic cells.*

Oren, A., and R. T. Papke, eds. 2010. Molecular Phylogeny of Microorganisms. Norfolk, UK: Caister Academic Press. *Methods and concepts of phylogeny and their usefulness in reconstructing evolutionary relationships of microorganisms, including rooting of the tree of life and impact of horizontal gene transfer on formation of microbial groups.*

Pace, N. R. 2009. Mapping the tree of life: Progress and prospects. Microbiology and Molecular Biology Reviews 73: 565–576. *Describes current understanding of the structure of the Bacteria and Archaea phylogenetic tree constructed using ribosomal RNA sequences.*

Woese, C. R. 1987. Bacterial evolution. Microbiological Reviews 51: 221–271. *Summarizes the shortcomings of prior taxonomies and gives an overview of the use of ribosomal RNAs for constructing phylogenetic relationships among Bacteria and Archaea. It also includes an overview of the major groups known at the time.*

II.12

Origin and Diversification of Eukaryotes

Laura A. Katz and Laura Wegener Parfrey

OUTLINE

1. Origin of eukaryotes
2. Timing of the origin and diversification of eukaryotes
3. A brief history of eukaryotic classification
4. Major clades of eukaryotes
5. Distribution of photosynthesis in eukaryotes
6. Extant symbioses
7. Genome diversity in microbial eukaryotes
8. Origins of multicellularity

Eukaryotes are marked by tremendous diversity in terms of size, shape, ecology, and genome structure. Eukaryotes are defined by two evolutionary innovations: the nucleus and cytoskeleton. Although named for the presence of a nucleus (*eu* = true, *karyon* = kernel or seed), it is the cytoskeleton and related proteins that allowed for the dramatic variation in morphology (i.e., shape and size) among eukaryotes. As with Archaea and Bacteria, the other two domains of life, eukaryotes are predominantly single-celled microbes, with plants, animals, and fungi representing just three of approximately 75 major lineages. This chapter discusses the origin of eukaryotes and current views of the relationships among eukaryotic lineages, and then highlights the diversity of eukaryotes by exploring the distribution of several characters (e.g., photosynthesis and multicellularity) across these lineages.

GLOSSARY

Algae (Sing., Alga). A descriptive term for any photosynthetic (i.e., plastid-containing) eukaryote. Algae are broadly distributed across the eukaryotic tree of life and are not a monophyletic group. The term can refer to taxa that have primary, secondary, or tertiary plastids.

Amoebae. A descriptive term for microbial eukaryotes that move using cytoplasmic projections called *pseudopodia*. Amoeboid organisms are found in many different lineages of eukaryotes.

Cytoskeleton. The cellular scaffolding that is made out of proteins and gives eukaryotes their shape, enables cellular movement, and participates in many subcellular processes.

Endosymbiosis. The intimate associate of two organisms where one (endosymbiont) lives within the other (host). Ancient endosymbiotic events led to the acquisition of mitochondria and plastids, both of which have played a large role in shaping the evolutionary history of eukaryotes.

Lateral Gene Transfer. Transfer of genetic material between distantly related organisms, by nonsexual means, in contrast to vertical inheritance of genes from parent to offspring. Lateral gene transfer obscures the structure of the tree of life.

Nucleus. A double-membrane-bound organelle that contains the genome, is the site of transcription, and is present in all eukaryotic cells.

Plastid. An organelle derived from the endosymbiosis of an algal cell (either cyanobacterium or eukaryotic alga). Plastids are generally involved in photosynthesis and include the chloroplasts of plants.

Protist. A descriptive term for eukaryotes that are not plants, animals, or fungi, used most commonly for microbial taxa. Protists do not constitute a monophyletic clade, and early classifications that lumped protists together in groups such as Protista or Protozoa are invalid.

Slime Molds. A diverse collection of organisms found in at least five of the major clades of eukaryotes originally thought to be fungi, because they form a multicellular fruiting body and release spores at one stage in their life cycle.

1. ORIGIN OF EUKARYOTES

The events that led to the origin of eukaryotes remain one of the outstanding questions in biology. Hypotheses put forth to explain the origin of eukaryotes must account for the presence of three features of eukaryotic cells: nucleus, cytoskeleton, and mitochondria. The nucleus and cytoskeleton are the defining features of eukaryotes, and both are absent in Bacteria and Archaea. These features must, therefore, have been present in the ancestral eukaryote, though their origins remain unclear. Research over the past two decades has demonstrated that mitochondria were also present in the last common ancestor of all extant eukaryotes, although they subsequently evolved into reduced mitochondria-related organelles in numerous lineages. In contrast to the nucleus and cytoskeleton, the mechanisms for the origin of mitochondria has been robustly established: they were acquired through endosymbiosis of an alphaproteobacterium. We discuss the origin of these defining features of eukaryotes in relation to hypotheses on the origin of the eukaryotic domain. Such hypotheses can generally be divided into those that invoke endosymbiosis and those that assume an autogenous mechanism (evolution within a single lineage).

The origin of the eukaryotic cytoskeleton, which gives eukaryotes their distinct morphologies and motilities, remains a mystery. The renowned biologist Lynn Margulis argued that the eukaryotic cytoskeleton resulted from endosymbiosis between an archaeon and a spirochete (a bacterium); however, no eukaryotic cytoskeletal proteins appear to have evolved specifically from within the spirochetes. Although homologues of some cytoskeletal proteins have been found in bacteria (e.g., *FtsZ* and *MreB* are bacterial homologues of tubulin and actin, respectively), evidence is lacking that the eukaryotic proteins derive from any specific bacterial lineage. Moreover, the bulk of eukaryotic cytoskeletal proteins lack clear homologues in either of the other domains; hence, it is unclear how the many proteins underlying the cytoskeleton arose.

Similarly, few data or convincing models exist to explain the origin of the eukaryotic nucleus and the associated endoplasmic reticulum. Many hypotheses argue that the nucleus resulted from an endosymbiotic event between a bacterium and an archaeon, two different bacteria, or even between an archaeon and a virus; however, few proteins within the nucleus show convincing affinities with any specific lineage of bacteria or archaea. Other theories posit an autogenous origin of the nucleus, driven by a selection pressure such as separation of transcription and translation or protection from viruses. The nuclear envelope provides spatial separation of transcription and translation in eukaryotes, whereas these processes occur in close proximity in Bacteria and Archaea. This has led to the suggestion that one advantage of the nuclear envelope is to allow for processing of pre-mRNAs (e.g., removal of introns) prior to translation. Other hypotheses posit that the nucleus arose as an autogenous product of the endosymbiotic event that gave rise to mitochondria. Additional data are needed to support or reject these hypotheses.

One of the few certainties in the origin and diversification of eukaryotes is that mitochondria, found in many but not all eukaryotes, are derived from an endosymbiotic alphaproteobacterium. Evidence for this symbiotic origin includes the membrane structure of mitochondria, its mode of division, and the presence of a bacterial-derived genome whose genes have close affinity to homologous genes in the Alphaproteobacteria. Moreover, the preponderance of evidence indicates that the acquisition of mitochondria occurred prior to the evolution of the last common ancestor of all extant eukaryotes. This evidence includes the broad phylogenetic distribution of mitochondria on the eukaryotic tree of life. While some eukaryotic lineages without mitochondria do exist, such as *Trichomonas, Giardia,* some ciliates, and some fungi, their nested relationships amongmitochondrial-containing taxa demonstrates that their ancestors had mitochondria. In fact, eukaryotes originally thought to completely lack mitochondria turn out to have double-membrane-bound organelles derived from mitochondria. These remnant mitochondria are alternatively termed *hydrogenosomes* if they are anaerobic and hydrogen-producing, or *mitosomes* if they are highly reduced.

2. TIMING OF THE ORIGIN AND DIVERSIFICATION OF EUKARYOTES

Eukaryotes likely arose in the Paleoproterozoic, 2.1–1.7 billion years ago. This estimate emerges from the first appearance of putative eukaryotes in the fossil record as well as from molecular clock analyses (see chapter II.3 for a discussion of molecular clock dating). The earliest fossils of eukaryotes are found around 1800 million years ago (Ma) and cannot be assigned to any extant clade. These fossils are identified as eukaryotes because they have complex structures that require a cytoskeleton to build, or inferred behaviors (e.g., budding) found only in eukaryotes. For example, fossils have been found from 1500 Ma with a cell wall composed of hexagons (akin to the patterning of a soccer ball) and others have been found covered with cylindrical processes that extend symmetrically from the cell. During the early evolution of eukaryotes, earth was very different from today: oceans were predominantly anoxic and sulfidic, and complex multicellular life was absent.

Fossils that can be assigned to modern clades of eukaryotes begin to appear around 1200 Ma with the appearance of the red algal fossil *Bangiomorpha*. Molecular clock analyses suggest that all the major clades of eukaryotes appeared prior to 1000 Ma and were present for hundreds of millions of years before leaving fossil evidence. Beginning 800 Ma, the diversity and abundance of eukaryotic fossils increases markedly: there are testate (shelled) amoebae and green algal fossils as well as biomarkers (fossilized lipids that indicate the presence of particular groups of organisms) that suggest the presence of ciliates and other taxa. The radiation of many eukaryotic lineages around 800 Ma coincides with a shift in the chemistry of the oceans toward conditions beginning to resemble modern oceans in that they became more fully oxygenated and were no longer sulfidic.

3. A BRIEF HISTORY OF EUKARYOTIC CLASSIFICATION

The relationships among eukaryotes have been subject to much debate and revision, as have the definitions of the groups themselves. Early classification schemes such as those of Carl Linneaus divided the living world into plants and animals, so that all photosynthetic eukaryotes (both multicellular and microbial) were considered plants (i.e., "algae") and motile microbes were considered animals (i.e., "protozoa," including ciliates, flagellates, and amoebae). This worldview is problematic from a modern perspective because it does not entail naming of natural (i.e., monophyletic; see chapter II.8) groups. Moreover, many microbial taxa fall into both categories at the same time, such as the photosynthetic amoeba *Paulinella* (see section 5 below).

Following Charles Darwin's publication of *On the Origin of Species* in 1859, representations of biodiversity were transformed to capture the concepts of ancestors and descendants. For example, Ernst Haeckel's 1866 depiction of the "Tree of Life" divided living things into Plantae, Animalia, and Protista. The Protista included a grab bag of problematic organisms such as amoebae, flagellates, ciliates, fungi, some animals (e.g., sponges), and bacteria. The concept of these basic divisions persisted into the 1970s with the development of the five-kingdom system. Four of these five kingdoms of life are eukaryotic: plants, animals, fungi, and protists; however, from an evolutionary perspective, these divisions are inappropriate, because "protists" are nonmonophyletic—animals, plants, and fungi all evolve from within microbial lineages. Thus, labels such as protist and alga remain useful as descriptive terms, but have no phylogenetic meaning.

In recent decades, molecular data have redrawn the fundamental division of living organisms into three domains (bacteria, archaea, and eukaryotes), all of which are predominately microbial. This classification was first created based on the evolutionary history of a single gene that is part of the ribosome of all living beings: the small subunit ribosomal RNA. Additional genes and a few differences in cell structure also support the three-domain concept of life; however, extensive lateral gene transfer among many lineages complicates the concept of the tree of life as evolution has occurred through a combination of vertical and lateral descent (chapter II.11). For example, the acquisitions of mitochondria and plastids represent dramatic examples of lateral transfer in which an entire genome from one lineage is captured by another through endosymbiosis.

In the past decades, we have gained a much better understanding of relationships among eukaryotes from analyses of molecular data combined with insights from ultrastructure—details of subcellular structures that are revealed by electron microscopy. Beginning in the 1960s it was found that protists comprise 70+ lineages that can be distinguished on the basis of ultrastructural features. We now understand that animals, plants, and fungi emerge out of these microbial lineages, indicating that the distinction between macroscopic and microscopic eukaryotes is a false one, driven by an excessive focus on the world we can physically see. Molecular data have confirmed the monophyly of the majority of the lineages defined by ultrastructure. In recent years, most of these lineages have been characterized with molecular data from multiple genes (which yield more robust results than single gene analyses) and have been grouped into four major clades: Opisthokonta, Amoebozoa, Excavata, and SAR (Stramenopiles + Alveolates + Rhizaria). For example, more than 30 of the 70+ lineages defined by ultrastructural identities have been shown to fall within the Rhizaria. Thus, the phylogenetic relationships among eukaryotes are stabilizing, although the placement of some photosynthetic lineages and other "orphan lineages" is still subject to debate. We discuss each of these major clades below.

4. MAJOR CLADES OF EUKARYOTES

The Opisthokonta unite the animals and fungi along with their microbial relatives. The affinity of animals with choanoflagellates was first noted more than a century ago; specialized sponge cells termed choanocytes have similar structure as choanoflagellates (see chapter II.15). The Opisthokonta are the best supported of the major clades in molecular studies, and also share the morphological trait of a single posterior flagellum (in clades with flagella) and the molecular synapomorphy of an insertion into the *EF-1*α gene (see chapters II.1 and II.8 for definitions of phylogenetic terms). Other members of the group include the Ichthyosporea (parasites of fish) and an enigmatic slime mold called *Fonticula*. Thus, multicellularity evolved many times within the Opisthokonta (see section 8 below for a discussion of multicellularity in eukaryotes, and chapter II.14 for fungi).

The Amoebozoa include many of the organisms typically called "amoebae," including the star of high school biology *Amoeba proteus*, two clades of slime molds, and *Entamoeba histolytica*, the causative agent of amoebic dysentery; however, this clade does not contain all amoebae, since amoeboid lineages can be found in virtually every major clade of eukaryotes. Amoebozoa emerged out of molecular phylogenetic analyses and are generally recovered in multigene molecular analyses, but there is no defining synapomorphy for Amoebozoa. The largest clade within the Amoebozoa, the Tubulinea, does have a morphological synapomorphy: cylindrical pseudopodia in which the cytoplasm streams in just one direction (monoaxial streaming). Examples of Tubulinea include the lobose testate (shelled) amoebae whose fossilized remains have been so useful in reconstructing paleoclimate and our old friend *Amoeba*.

The Excavata were originally defined on the basis of a morphological and ultrastructural character, an "excavated" ventral feeding groove. The clade is only sometimes recovered, and then only in molecular analyses with many genes and dense taxonomic sampling. The most familiar Excavata are the causative agents of parasitic diseases. These include *Giardia*, a major cause of diarrhea worldwide that exacerbates malnutrition in children in the developing world. Other parasitic excavates include trypanosomes that cause sleeping sickness and *Trichomonas vaginalis*, which causes the sexually transmitted disease trichomoniasis. Members of Excavata are heterotrophic with the exception of one lineage of euglenids (a close relative of the trypanosomes) that acquired photosynthetic ability by engulfing a green alga endosymbiont hundreds of millions of years ago (see section 5). Many taxa, and all the parasites, within the Excavata are anaerobic and have highly reduced mitochondria, either hydrogenosomes or mitosomes. Some of these parasites also have rapid rates of evolution that artificially pulled them to the base of early ribosomal DNA trees (see chapter II.2 for a discussion of long-branch attraction artifacts). These two observations led to the now-disproven hypothesis that members of the Excavata represented early diverging eukaryotes that had branched off the eukaryotic lineage prior to the acquisition of mitochondria.

The final major clade of eukaryotes that is moderately well supported, SAR, is the amalgamation of the three other monophyletic clades, the stramenopiles, alveolates, and rhizarians. The stramenopiles contain diatoms (algae with beautiful silica shells), kelps, and the causative agent of the Irish potato famine (*Phytophthora*). The alveolates include the morphologically diverse ciliates, the dinoflagellates critical to the survival of coral reefs, and the apicomplexa, which include the malaria parasite. Both the stramenopiles and alveolates are defined by ultrastructural synapomorphies. The stramenopiles have specific hairs on one of their flagella and the alveolates have sacs (alveoli) underlying their cell membrane that lend rigidity. In contrast, Rhizaria is a large, heterogeneous collection of amoebae, flagellates, and parasitic lineages that lacks diagnostic ultrastructural features. The amoeboid members of Rhizaria tend to have filose (fine) or reticulating networks of pseudopodia, a morphological feature that generally (but not always) distinguishes Rhizaria from members of the Amoebozoa. The SAR clade also contains the largest number of named species among the microbial eukaryotes, as it contains the diatoms, ciliates, and foraminifera, each encompassing thousands of described species.

The phylogenetic position of lineages that are wholly or predominantly photosynthetic is much less resolved. This instability is likely driven by gene transfer from the algal symbionts to the host nucleus, complicating phylogenetic reconstruction. The most popular hypothesis is that all the primary photosynthetic lineages (green algae, red algae, and glaucophytes) form a monophyletic clade, and that a single endosymbiotic event in the ancestor of this clade gave rise to all plastids (see section 5 for more information on plastids).

5. DISTRIBUTION OF PHOTOSYNTHESIS IN EUKARYOTES

The shift from historical classification systems that lumped together all photosynthetic organisms toward a system emphasizing monophyletic groups has led to the realization that photosynthesis is patchily distributed across eukaryotes (figure 1). Much of the energy that powers ecosystems—terrestrial, marine, and freshwater—is driven by photosynthetic eukaryotes making up the diverse clades of algae (including land plants, a lineage of green algae). Photosynthesis in eukaryotes was likely acquired through endosymbiosis of a single cyanobacterium more than a billion years ago, much like the endosymbiosis of an alphaproteobacterium that led to mitochondria. Like mitochondria, plastids have a small, circular genome with roughly 100 genes. As the cyanobacterium was reduced to a plastid, many of the genes in the cyanobacterial genome were lost, and others were transferred to the host nucleus. Today the chloroplasts of plants contain about 110–120 genes, and the plant nucleus contains many times that number: generally around 10 percent of a plant's nuclear genome is derived from cyanobacteria.

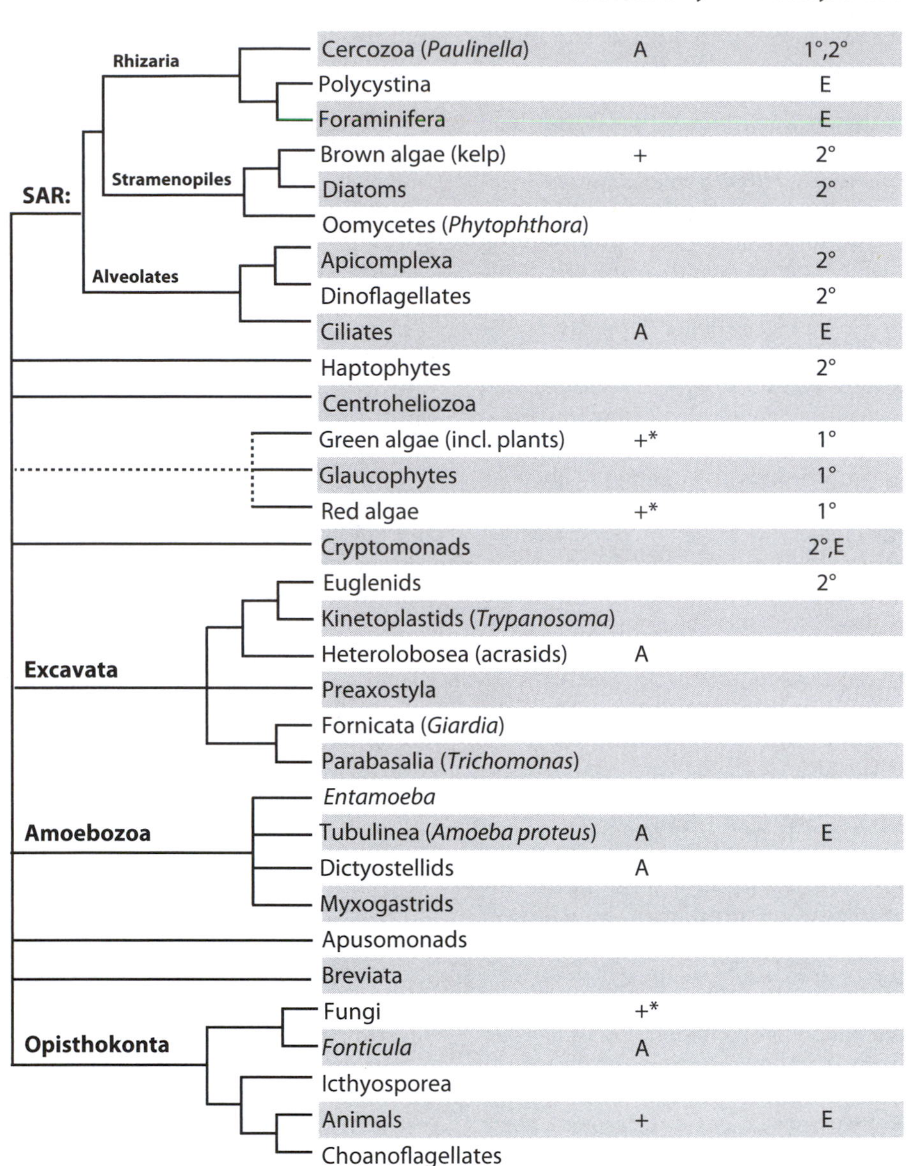

Figure 1. Cartoon tree of eukaryotes that highlights the lineages discussed in the text. The dashed line connecting red algae, green algae, and glaucophytes reflects the remaining uncertainty about this relationship. Symbols indicate the trait is found in at least one member of the clade. Multicellularity: + = multicellular lineages; +* = multiple origins of multicellularity; A = aggregative multicellularity such as slime molds. Photosynthesis: 1° = primarily photosynthetic lineages; 2° = secondarily photosynthetic. E = Extant symbioses with algae.

The prevailing hypothesis is that the initial acquisition of plastids occurred only once in the ancestor of all primary photosynthetic lineages. Evidence in support of this hypothesis includes the similarity among all plastids in the transport machinery that moves macromolecules from the host to and from the plastid, and the similarity of the plastid genes themselves according to phylogenetic analyses; however, the possibility that plastids were acquired multiple times cannot be ruled out at this time, mainly because the three lineages thought to be direct descendants of that single acquisition, the primary photosynthetic lineages (red algae, green algae, and glaucophytes), are not monophyletic in many phylogenetic analyses. Moreover, cyanobacteria appear to have gone through substantial extinction and radiation events since the time of the endosymbiosis, making inferences about the ancestry of plastids more difficult.

After the initial establishment of photosynthesis in eukaryotes, plastids spread through the eukaryotic tree of life by secondary, tertiary, and even quaternary endosymbiosis. *Secondary endosymbiosis* is the engulfment of a unicellular alga by a nonphotosynthetic eukaryote. In all known cases, secondary endosymbiosis involves red or green algae (glaucophytes are not known to have become secondary plastids). Rather than being fully digested, the alga is retained, reduced, and over many generations, fully integrated into the host cell as a secondary plastid. In most lineages, the reduction process has resulted in the loss of all the components of the symbiotic algal cell except for the plastid; however, two distantly related lineages, the cryptomonads and chlorarachniophytes, still have a remnant nucleus from the red and green algal symbiont, respectively. A now-defunct hypothesis placed the six or more lineages (including dinoflagellates and photosynthetic stramenopiles) with secondary red algal plastids into a clade called chromalveolates; however, it is now clear that secondary endosymbiosis involving red algae has occurred multiple times across the eukaryotic tree of life.

There are also photosynthetic endosymbioses that are established but have not (yet) proceeded to the point where the retained alga can be considered a plastid. One such case is a second example of a primary endosymbiosis of a cyanobacterium by the amoeba *Paulinella* (within Rhizaria). Several photosynthetic lineages of *Paulinella* have been isolated from around the world and it is estimated that this endosymbiosis began about 65 Ma.

Other lineages of eukaryotes take advantage of photosynthesis despite their lack of fully integrated plastids. Instead they rely on plastids that are retained from food organisms (termed *kleptoplastids*) or on extant symbioses with algal lineages. The process of kleptoplasty (retaining stolen plastids) gives us green, photosynthetic animals like the sea slug *Elysia*. Similarly, Foraminifera and Radiolaria (large amoebae that fall within the Rhizaria) include many members that reside in the upper layers of the oceans and farm algae, generally dinoflagellates. During the day these amoebae expose the symbiotic algae to sunlight by extending them in their extensive pseudopodial network to photosynthesize, and at night the algae are drawn back into the amoeba cell body and sugars are harvested.

6. EXTANT SYMBIOSES

Extant eukaryotes are involved in many important symbioses, where they can serve as either hosts or symbionts, or both. For example, corals are an association between animals (host) and photosynthetic dinoflagellates (symbionts) in which the dinoflagellates provide the cnidarian host with sugars and amino acids and the coral provides nutrients and protection. When this symbiosis goes awry, the corals can suffer, as is the case in coral bleaching, which arises when dinoflagellates leave in response to high temperatures or stress. The "green animals" and "farming amoebae" discussed above harbor algal symbionts within their tissues, presumably to take advantage of the energy produced by photosynthesis.

Symbioses involving heterotrophic eukaryotes are also rampant; for example, ciliates and parabasalid flagellates aid in the digestion of cellulose in ruminants (e.g., cows) and the hindguts of termites. Microbial eukaryotes can also play host to symbiotic bacteria, archaea, or other eukaryotes, and some of these symbioses play a role in human health. Some amoebae within the Amoebozoa are able to serve as Trojan horses, harboring pathogenic bacteria that can eventually emerge and cause disease. In 1976, for example, numerous people were sickened at an American Legion convention in Philadelphia by bacteria (later assigned to the genus *Legionella*) that were harbored inside amoebae. Since then, considerable effort has been spent documenting the numerous bacteria that can live within microbial eukaryotes. Associations between eukaryotes and bacterial endosymbionts can also alter the pathogenicity of human parasites such as *Trichomonas* and *Entamoeba*. For example, in the presence of pathogenic *Escherichia* and *Salmonella* bacteria, *Entamoeba* becomes much more invasive and causes more tissue damage.

7. GENOME DIVERSITY IN MICROBIAL EUKARYOTES

Textbooks generally depict genomes as stable entities that are passed from generation to generation with minimal change. While this is generally true, there is a surprising amount of variation across eukaryotes in the genome content both within individual organisms during their life cycle and among individuals belonging to the same species. The most common forms of genome variation are ploidy-level variation (shifts in copy number of the whole genome beyond haploid and diploid) and differential amplification of portions of the genome. The extensive variation in genome content within individuals during their life cycle suggests that eukaryotes have the ability to distinguish between the genome that will be inherited in the next generation (i.e., germ line genome) and the somatic genome that accumulates variation during the life of a cell.

In many lineages, the germ line and somatic genomes are segregated into separate nuclei (in the same or different cells). Animals are the most familiar case of segregated genomes, with germ line cells sequestered early in development. In some animals, including copepods, nematodes, and hagfish, the somatic genome is extensively modified from the zygote, as chromosomes are fragmented and a portion of the genome is eliminated. Ciliates are also characterized by the presence of two types of nuclei that contain genetically different genomes, the micronucleus (germ line) and the macronucleus (somatic), though both these genomes exist within a single cell. The micronuclear genome behaves like a typical eukaryotic genome in that it goes through meiosis and mitosis and has a few long chromosomes; in contrast, the genome of the macronucleus is fragmented, sometimes extensively, yielding hundreds to thousands of tiny somatic chromosomes. Foraminifera are the third lineage with genetically distinct somatic and germ line genomes, although this feature is present only in a subset of genera. Here one of the nuclei in the multinucleate life cycle stage expands and is transcriptionally active while the remaining generative nuclei are quiescent until meiosis.

8. ORIGINS OF MULTICELLULARITY

Contrary to the popular belief that multicellularity is rare, involving only plants, animals, and fungi, there have been many origins of multicellularity among eukaryotes (figure 1). For example, many lineages of algae have become multicellular. Several of these multicellular algae are familiar, including giant kelps and the red algae used in making sushi. There are also multiple origins of multicellularity among the green algae, among whose descendants are the land plants. Further, there have been several origins of multicellularity in terrestrial environments where organisms have evolved to produce multicellular fruiting bodies that likely aid in dispersal. Such organisms constitute the numerous lineages of slime molds, which include the dictyostelids and acrasids as well as isolated genera scattered across the eukaryotic tree (figure 1). Comparative analyses that include these lesser-known origins of multicellularity may be helpful in clarifying the selective pressures and developmental mechanisms underlying the origin of multicellularity.

Synthesis

Eukaryotes, cells defined by the presence of both a nucleus and a cytoskeleton, have inhabited the earth for nearly 2 billion years. During this time, eukaryotes have evolved into a tremendous diversity of forms found in all major ecosystems. The bulk of eukaryotic diversity is microbial, though we are most familiar with the macroscopic lineages: plants, animals, and fungi. While the combination of powerful microscopes and molecular data have begun to transform our understanding of the origin and diversification of eukaryotes, there is still much to be discovered about this incredible branch on the tree of life.

FURTHER READING

Archibald, J. M. 2009. The puzzle of plastid evolution. Current Biology 19: R81–R88. *This article reviews the evolutionary history of plastids and discusses the hypotheses on the acquisition and spread of plastids across the eukaryotic tree.*

Bonner, J. T. 1998. The origins of multicellularity. Integrative Biology 1: 27–36. *Contains a comprehensive list of multicelled eukaryotes, focusing on protist groups. Presents evolution of multicellularity in a phylogenetic context.*

Embley, T. M., and W. Martin. 2006. Eukaryotic evolution, changes and challenges. Nature 440: 623–630. *This article reviews the evidence for the presence of mitochondria in the last common ancestor of eukaryotes and discusses hypotheses for the origin of eukaryotes and the timing of this event.*

Hjort, K., A. V. Goldberg, A. D. Tsaousis, R. P. Hirt, and T. M. Embley. 2010. Diversity and reductive evolution of mitochondria among microbial eukaryotes. Philosophical Transactions of the Royal Society B 365: 713–727. *A review of the recent findings on the origin and evolution of mitochondria in eukaryotes.*

Katz, L. A. 1999. The tangled web: Gene genealogies and the origin of eukaryotes. American Naturalist 154: S137–S145. *A synthetic paper on the impact of lateral events in the history of eukaryotes.*

Knoll, A. H, E. J. Javaux, D. Hewitt, and P. Cohen. 2006. Eukaryotic organisms in Proterozoic oceans. Philosophical Transactions of the Royal Society B 361: 1023–1038. *A review of the early history of eukaryotes as elucidated from the fossil record.*

Parfrey, L. W., D.J.G. Lahr, and L. A. Katz. 2008. The dynamic nature of eukaryotic genomes. Molecular Biology and Evolution 25: 787–794. *A synthesis of the common types of genomic variability in eukaryotes with examples from lineages across the tree of eukaryotes.*

Patterson, D. J. 1999. The diversity of eukaryotes. American Naturalist 154: S96–S124. *Outstanding review of eukaryotic diversity. Presents hypotheses of relationships among eukaryotic groups as well as a long list of taxa of unknown affinities.*

ONLINE RESOURCES

Micro*scope.
www.mbl.edu/microscope
*Maintained at the astrobiology institute of the Marine Biological Laboratory, Woods Hole, MA, with oversight from David J. Patterson (University of Sydney, Australia). Micro*scope aims to provide a comprehensive listing of microorganisms, with links to images, DNA sequences, educational materials, information about cellular biology of organisms, and other relevant web pages. This valuable resource serves as a starting point to access online information about protists.*

II.13

Major Events in the Evolution of Land Plants

Peter R. Crane and Andrew B. Leslie

OUTLINE

1. Phylogenetic framework
2. Origin and diversification of early land plants
3. Origin and diversification of vascular plants
4. Origin and diversification of seed plants
5. Origin and diversification of angiosperms
6. Innovation in the land plant body
7. Innovation in land plant reproduction
8. Coevolution with animals
9. Patterns of extinction

Although land plants represent merely one branch in the eukaryotic tree of life, they are essential to the energetics and functioning of terrestrial ecosystems. Land plants appear to have arisen from a single colonization of the land surface around 450 million years ago. In the early phases of this colonization, plant innovations centered on the elaboration of a new kind of plant body capable of withstanding the rigors of life on land and exploiting the new opportunities that terrestrial existence provided. Subsequently, this phase of vegetative innovation was followed by successive transformations of the reproductive system, for example, resulting in seeds and flowers, which facilitated increasingly efficient reproduction and dispersal. The unfolding of land plant diversification from a single origin makes it possible to understand all living groups in a relatively simple phylogenetic framework, in which increasingly less inclusive groups are characterized by successive innovations in plant structure and biology.

GLOSSARY

Alternation of Generations. A type of life cycle with multicellular organisms in both the haploid and diploid phases. The diploid phase (sporophyte) produces haploid spores by meiosis that later germinate and develop into the haploid phase (gametophyte), which produces gametes by mitosis. Gamete fusion (fertilization) gives rise to a diploid zygote and marks the transition from the haploid to diploid phase.

Heterospory. The earliest land plants, as well as many living "pteridophytes," were *homosporous*, meaning that their sporophytes produced haploid spores of a single size that could each germinate to produce bisexual gametophytes. In contrast, several extinct and extant lineages are *heterosporous*, meaning they produce two types of spores: larger megaspores, which develop into gametophytes (megagametophytes) that produce only female gametes, and smaller microspores, which develop into gametophytes (microgametophytes) producing only male gametes. Typically, these two types of spores are produced in separate sporangia (megasporangia and microsporangia, respectively).

Seed. The seed can be considered an extreme form of heterospory in which a single megaspore develops within a megasporangium that never opens to release the spore. In seed plants, this megasporangium is further enveloped by a protective covering called the *integument*, and the whole structure is termed an *ovule*. The ovule becomes a seed on fertilization, and at maturity it contains an embryo, an accompanying food source, and a protective seed coat.

Vascular Tissue. Vascular tissues are specialized cells that transport water and nutrients through the plant body. The *xylem*, or water-conducting tissue, is composed principally of cells called *tracheids* (or similar but modified cells called *fibers* or *vessels*) that are dead at maturity and have walls that are typically thickened by the deposition of the complex biopolymer lignin. The *phloem*, which conducts sugar and nutrients

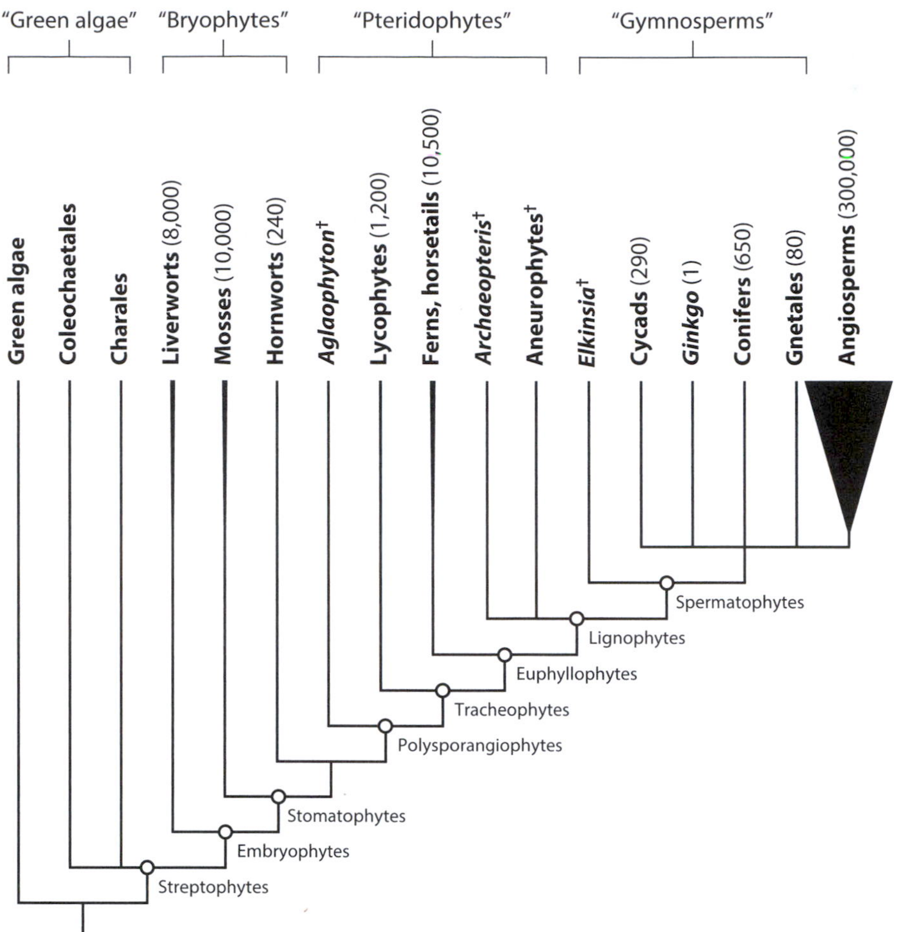

Figure 1. Phylogenetic framework for the evolution of land plants, including the 11 major living land plant groups and several extinct groups (indicated by a dagger) discussed in the text. The general topology is based on results from analyses of morphological data, which are broadly supported by results from molecular data sets. Major clades are indicated by a hollow circle; important clades discussed in the text include embryophytes (land plants), tracheophytes (vascular plants), and spermatophytes (seed plants). Paraphyletic groups, such as "bryophytes," are indicated in quotes. The relationships among the "bryophyte" groups have been controversial, and this topology reflects the current hypothesis based on analyses of several large molecular data sets. Likewise, relationships among the five living groups of seed plants remain controversial, with different molecular data sets generating different topologies, especially in regard to the position of the Gnetales. Here these relationships have been left unresolved. The approximate number of species in each group of extant land plants is given after the group name. The thickness of the line for each living plant group is scaled to reflect its proportion of total land plant species diversity (groups with fewer than 1000 species are depicted with the same line thickness). Note the overwhelming diversity of angiosperms (flowering plants) compared with all other groups; approximately 90% of all living terrestrial plant species are angiosperms.

through the plant, contains living transport cells called *sieve cells* as well as other cell types.

1. PHYLOGENETIC FRAMEWORK

A phylogenetic framework in which to understand the 11 major groups of living land plants was originally developed from structural data and has been confirmed by analyses based on molecular sequences (figure 1). Plants are a monophyletic subgroup of eukaryotes defined by the presence of the photosynthetic pigments chlorophylls *a* and *b*, the storage of starch in plastids (chloroplasts), and several other features. In addition to land plants, this definition includes freshwater and marine organisms often grouped together as "green algae" and contrasts with treatments in which plants are defined more narrowly as synonymous with land plants. Molecular and structural data, including characteristics of cell division, support the placement of certain freshwater "charophycean green algae" (*Coleochaetae*, *Chara*) as the closest relatives of land plants; these groups together form a clade called the streptophytes.

The land plant clade, or embryophytes, is composed of several major monophyletic groups that are widely used in discussions of plant evolution (figure 1), including vascular plants (tracheophytes), seed plants (spermatophytes), and flowering plants (angiosperms). There are also corresponding paraphyletic groups (see chapter II.1) at each level. For example, "green algae" are those plants that are not land plants, "bryophytes" are those land plants that are not vascular plants, "pteridophytes" are those vascular plants that are not seed plants, and "gymnosperms" are those seed plants that are not flowering plants. Analyses of molecular data suggest that living "gymnosperms" (conifers, cycads, *Ginkgo*, Gnetales) may make up a monophyletic group, although when extinct seed plants are included, "gymnosperms" are clearly paraphyletic. Within flowering plants, phylogenetic analyses support monocots (such as grasses and palm trees) as a monophyletic group, but "dicots" are now known to be paraphyletic. Two major monophyletic groups, eudicots and eumagnoliids, together account for almost all species previously treated as dicots.

2. ORIGIN AND DIVERSIFICATION OF EARLY LAND PLANTS

Land plants are distinguished from "green algae" by a suite of reproductive and vegetative innovations, many of which are important in overcoming shortages of water. In land plants, the egg is formed and fertilized in a specialized flask-like structure called the *archegonium* within which the young embryo begins its development. Land plants also have desiccation-resistant spore walls impregnated with the complex biopolymer sporopollenin; this enables their reproductive progagules to survive drying and disperse among suitable environments. The land plant body is covered in a waxy cuticle that prevents water loss, and all land plant groups except liverworts also have *stomata*, pores in the cuticle that are used to regulate the exchange of gases and water vapor with the atmosphere.

Spore-like microfossils that may have been produced by plants or other eukaryotes are known from as early as the middle Cambrian (ca. 510 Ma), although their exact affinities are unclear. Spores that were most likely produced by embryophytes appear around the Middle Ordovician (ca. 465 Ma), with forms broadly similar to those of some living "bryophytes." Beginning around the same time, and continuing through much of the Silurian (ca. 443–416 Ma), the fossil record preserves a variety of enigmatic organic sheets and tubes suggestive of land plant cuticle and simple water-conducting tissues. Such structures provide further evidence that the diversification of early land plants was under way, even though our knowledge of the plants involved is extremely fragmentary (see chapter II.10).

Macroscopic evidence of terrestrial plants is scarce until the middle to late Silurian (ca. 420 Ma), when small branching axes bearing multiple sporangia appear in the fossil record. These fossils represent the first-known appearance of a major clade, the polysporangiophytes, which exhibit an important innovation in the vegetative plant body. While living "bryophyte" groups (hornworts, liverworts, mosses) produce a morphologically diverse array of sporophytes, they are all small, unbranched structures bearing a single sporangium at their tip. In contrast, the branched sporophytes of polysporangiophytes can produce multiple sporangia from a single embryo. Unlike the situation in "bryophytes," the larger sporophytes of at least some early polysporangiophytes were also probably nutritionally independent of their corresponding gametophyte. Polysporangiophyte fossils become increasingly common in the latest part of the Silurian (ca. 416 Ma) and into the Early Devonian, where a glimpse into early polysporangiate diversity is provided by the classic earliest Devonian Rhynie Chert locality in Scotland (ca. 413 Ma). This assemblage preserves an early terrestrial ecosystem in exquisite detail, including arthropods, fungi, polysporangiophytes, and early representatives of a polysporangiophyte subgroup, the vascular plants, which were to become the most important constituent of later terrestrial ecosystems.

3. ORIGIN AND DIVERSIFICATION OF VASCULAR PLANTS

Living vascular plants are distinguished from hornworts, liverworts, and mosses by their specialized vascular tissues (see glossary) and their larger and more elaborate sporophyte. Rhynie Chert polysporangiophytes exhibit a mixture of features that appear to reflect various stages in the evolution of these tissues and structures. The gametophyte and sporophyte of at least some Rhynie Chert plants were similar in size and morphological complexity, which is unlike the life cycle of any living terrestrial plant and may reflect an intermediate stage between the small sporophytes of "bryophytes" and the earliest polysporangiophytes (some *Cooksonia* species) and the more elaborate sporophytes of true vascular plants. Some Rhynie Chert polysporangiophytes (e.g., *Aglaophyton*, *Horneophyton*) contain water-conducting tissues that are similar to the unspecialized conducting cells found in some living mosses, while others (*Rhynia*) have water-conducting cells with pronounced internal thickenings but lack the extensive deposition of the resistant biopolymer lignin characteristic of modern tracheids. Paleobotanists once grouped plants like *Aglaophyton*, *Horneophyton*, and *Rhynia* together as Rhyniophytes, but their structural features suggest they are most likely of diverse relationships: some are likely on the stem lineage leading to true vascular plants (tracheophytes), while others may represent early branches within the tracheophytes.

Asteroxylon, a vascular plant from the Rhynie Chert with modern-type tracheids, appears to be related to living lycophytes, an early branch of vascular plant evolution distinguished by their kidney-shaped sporangia and characteristic pattern of vascular tissue development. The Early and Middle Devonian fossil record (ca. 416–385 Ma) shows that lycophytes diversified quickly into two main groups: zosterophylls, which were diverse during the earliest phases of vascular plant evolution but became extinct by the end of the Devonian, and lycopods, which diversified through the Devonian and include about 1200 living species that occur in a wide range of modern biomes. One extinct lycopod subgroup included large trees (greater than 40 m tall in some cases) that dominated the well-known Pennsylvanian tropical coal swamps of Eurasia and North America. While most of these arborescent lycopods were extinct by the end of the Permian (ca. 251 Ma), some members of this lineage survived into the Mesozoic. The only living representative of this group is the genus *Isoetes*, a small rosette plant of wet places.

Lycophytes shared the Early and Middle Devonian landscape with "trimerophytes," a poorly understood but important assemblage of plants with naked, variously branched photosynthetic axes bearing elongated terminal sporangia. In basic vegetative construction, "trimerophytes" were similar to Rhynie Chert plants such as *Aglaophyton*, *Horneophyton*, and *Rhynia*, but they often had more complex branching patterns and include forms such as the Early Devonian plant *Pertica,* which had a well-defined central axis and lateral branches. While the morphology and anatomy of some "trimerophytes" is quite well characterized, their relationships to each other and to living groups are poorly understood; nevertheless, these plants are thought to represent the early evolution of the second major lineage of vascular plants (the euphyllophytes; lycophytes being the other) that ultimately gave rise to several groups of ferns, living and extinct horsetails (relatives of living *Equisetum*), and seed plants.

The exact pattern of relationships among these groups is controversial; molecular data suggest that living ferns, horsetails, and Psilotales (a group of simple, fernlike plants) form a monophyletic group to the exclusion of seed plants while analyses of morphological data incorporating fossil taxa do not always support this relationship. Whatever the precise pattern of relationships, indisputable horsetails, with their characteristic jointed stems and whorled leaves, are first known from the Late Devonian (ca. 385–359 Ma), and from then on this lineage is a ubiquitous feature of the plant fossil record. As in lycopod evolution, some groups (called calamites) attained the stature of large trees during the Mississippian and Pennsylvanian; however, despite the long evolutionary history of this group, their morphological and ecological diversity has been relatively limited; both extinct and extant forms tend to favor wet environments.

Ferns have a complicated evolutionary history characterized by several successive waves of diversification. There are diverse possible fern relatives in the Middle and Late Devonian and a number of probable early fern groups in the Mississippian, although none of these plants can be readily assigned to living groups. However, ferns closely related to living Marattiales were prominent components of terrestrial ecosystems in the Pennsylvanian and formed a major component of many coal-swamp environments.

Most groups of modern ferns (termed leptosporangiate ferns or Filicales) belong to a clade with a characteristic pattern of sporangial development in which a single initial cell gives rise to the sporangium. A number of Carboniferous fern groups have been included in this lineage based on their sporangial morphology and stem anatomy, although their exact phylogenetic relationships are unclear. Early diverging extant Filicales (e.g., the living family Osmundaceae) first appear in the late Permian (ca. 251–244 Ma), and plant assemblages during the Mesozoic often include representatives of other early diverging extant filicalean groups such as Dicksoniaceae, Dipteridaceae, Gleicheniaceae, Matoniaceae, Marsileaceae, and Schizaeaceae. While these groups contain relatively few species today, they were important components of Jurassic and Cretaceous plant communities. Over the Late Cretaceous and Cenozoic, certain subgroups of filicalean ferns (the polypod ferns) have undergone a further major radiation that accounts for most extant fern diversity.

One additional group of Middle and Late Devonian free-sporing plants developed an important structural innovation that had major consequences for the subsequent history of life on land. These "progymnosperms" (e.g., *Aneurophyton*, *Archaeopteris*) had a layer of actively dividing cells in their axes capable of producing a cylinder of secondary xylem (wood) to the inside and cylinder of secondary phloem to the outside. The presence of this feature, called a bifacial cambium, defines the lignophyte clade that also includes all extinct and living seed plants. Although many groups of vascular plants have developed some form of secondary growth, a bifacial cambium allowed lignophytes to increase the girth of their axes (through the production of wood) while maintaining connections for the transport of nutrients throughout the plant (through the secondary phloem). Large, long-lived sporophytes that could potentially grow for an indeterminate amount of time were one result of this innovation, and by the Late Devonian (ca. 380 Ma), "progymnosperms" such as *Archaeopteris* had sporophytes similar in basic structure to modern seed plant trees such as living conifers.

4. ORIGIN AND DIVERSIFICATION OF SEED PLANTS

While early lignophytes like *Archaeopteris* exhibited vegetative features similar to those of seed plants, their reproductive biology was fundamentally the same as that of "bryophytes," lycopods, ferns, and horsetails. All these groups reproduce by spores dispersed through the air which then germinate to produce a free-living haploid gametophyte (see glossary). Fertilization occurs when motile male gametes released by gametophytes swim through films of water in the environment and fertilize eggs in the archegonia on the same or nearby gametophytes. Seed plants, however, have a very different kind of reproductive biology in which female gametophytes develop on the parent plant inside structures called ovules (see glossary). Seed plant reproductive biology therefore includes a unique process called pollination where small male gametophytes (contained within pollen grains) are transported to the ovules by wind, water, or animals. After pollen germination, the male gametophyte grows and ultimately produces gametes that will fertilize an egg cell in the female gametophyte. After fertilization, the embryo begins to grow inside the ovule, which is now termed a seed. This radically new kind of reproductive biology, seen in five living groups (conifers, cycads, *Ginkgo*, Gnetales, and angiosperms), created new opportunities for maternal investment in offspring, gave sporophytes greater control over the reproductive process by minimizing the influence of the outside environment, and allowed seed plants to colonize a much greater variety of terrestrial habitats by reducing dependence on external films of water for fertilization and enclosing embryos within a protective coat. These advantages have likely contributed to the ecological dominance of seed plants in most terrestrial ecosystems since the Permian.

The earliest fossils that can be recognized unambiguously as seed plants (e.g., *Archaeosperma*, *Elkinsia*, *Moresnettia*) date from the latest Devonian (ca. 365 Ma) and are generally grouped together within a potentially paraphyletic group, the "hydraspermans," characterized by particular type of pollination biology. The diversity of hydrasperman and other early seed plant groups increases through the Mississippian and Pennsylvanian, although many of these plants had not yet acquired the full suite of features seen in living seed plant groups; for example, their pollen grains germinated in a manner similar to that of "pteridophyte" spores. Seed plants with fully modern seed plant reproduction, including probable early conifers and cycads as well as several now-extinct groups, do not appear until the Pennsylvanian and Permian.

Seed plants continued their taxonomic and ecological radiation through the Mesozoic. In addition to familiar groups such as cycads, *Ginkgo*, and several families of conifers, Mesozoic fossil assemblages contain a wide diversity of extinct groups that were ecologically important at different times and in different parts of the world (e.g., Bennettitales, Caytoniales, Corystospermales). The phylogenetic relationships of many of these extinct seed plant groups are uncertain; for example, analyses based on morphological data usually resolve the groups Bennettitales, Erdtmanithecales, and Gnetales as closely related to each other, and in turn closely related to flowering plants. However, molecular data from the limited set of extant seed plants suggest instead that Gnetales are more closely related to conifers, and that cycads and *Ginkgo* also belong to this group to the exclusion of angiosperms.

5. ORIGIN AND DIVERSIFICATION OF ANGIOSPERMS

Nearly all seed plant diversity, and therefore nearly all terrestrial plant diversity as a whole, comprises flowering seed plants, or angiosperms (figure 1). While the phylogenetic position, and hence the origin, of angiosperms relative to other seed plant groups remains uncertain, knowledge of phylogenetic patterns within the group has been clarified substantially by advances in molecular phylogenetics. Angiosperms are distinguished from other seed plants by a suite of innovations in both vegetative and reproductive biology that favor efficiency and speed in functionality, growth, and development. Angiosperm leaves are characterized by multiple, reticulate vein orders, and the stems of most species contain specialized xylem elements called *vessels*. Both these features allow angiosperms to rapidly and efficiently move water through the plant body, although they have both apparently originated convergently in a few other groups of living and extinct seed plants (for example, in living *Gnetum* and extinct gigantopterids).

It is in reproductive biology, however, that the differences between angiosperms and other seed plants are most pronounced. Angiosperm ovules and seeds are borne inside a closed structure called the *carpel* so that unlike other groups of seed plants, angiosperm pollen grains land and germinate away from the ovules on a specialized receptive tissue called a *stigma*. Angiosperm ovules also typically have two integuments (see glossary) rather than one as in most other seed plants. The female gametophyte of angiosperms is very simple and presumably highly reduced (typically seven cells with eight nuclei) and fertilization is also unique: in most species one male gamete fuses with the egg, while another fuses with a diploid nucleus in the female gametophyte. The triploid nucleus resulting from this second fertilization event develops to form the nutritive tissues of the seed, the endosperm. Through this process of double fertilization the development of the nutritive tissue is linked directly to successful formation of the zygote.

Unequivocal angiosperm flowers, with a carpel or ovary surrounded by characteristic pollen-producing organs (stamens) and a perianth (petals or similar structures), are first recorded in the fossil record around the middle of the Early Cretaceous (ca. 125 Ma). Unambiguous angiosperm pollen grains with complex pollen walls structurally similar to those of living taxa occur somewhat earlier (ca. 135 Ma). In the Late Cretaceous and continuing through the Cenozoic, the diversity of angiosperm species and their ecological importance has increased explosively. Angiosperms are by far the most diverse group of terrestrial plants, and they ultimately gave rise to forms as diverse as grasses, palms, water lilies, orchids, oaks, and sunflowers. Angiosperms are the autotrophic foundation of modern terrestrial ecosystems; through their intricate and complex interactions with microbes, fungi, animal herbivores, pollinators, and seed dispersers, their diversification has been of fundamental importance in the origin of terrestrial biodiversity as a whole. The diversification of angiosperms may have even made possible the occupation of new kinds of habitats and the origin of new kinds of terrestrial biomes. For example, the increased water-cycling rates made possible by angiosperm leaves may have contributed to the development of extensive rain forests in the tropics.

6. INNOVATION IN THE LAND PLANT BODY

In very broad terms, the history of innovation in land plants can be divided into two phases: a period of intense vegetative innovation from about the Late Ordovician to the Late Devonian (450–360 Ma) and a period of reproductive innovation that began around the Late Devonian and still continues. The first phase is marked by a suite of adaptations relating to the acquisition, retention, and transport of water and other fluids and also includes a strong trend toward increasing the size and stature of the sporophyte. This elaboration of the sporophyte was most likely driven by the advantages of large size for increased reproductive output and by competition for light as the most favorable ecological sites on land became more crowded. One manifestation of this trend can be seen in the rooting structures of Silurian and Devonian plants, which become increasingly elaborate and more efficient at gathering water and nutrients from the soil compared with those of the earliest land plants and modern "bryophytes," in which delicate root hairs (called rhizoids) collect water but do not deeply penetrate the soil. Increasing root penetration may also have significantly increased terrestrial weathering rates, potentially leading to a drawdown of atmospheric CO_2 and global cooling in the Late Devonian and Carboniferous.

These increasingly complex rooting structures served to support increasingly large and complex aerial plant bodies. Much morphological and anatomical evolution between the late Silurian and Late Devonian consisted of various groups developing larger, more ramified, and more specialized photosynthetic axes, resulting in plants such as Middle and Late Devonian *Rhacophyton* and *Pseudosporochnus* (generally considered relatives of early ferns) with complex, three-dimensional arrays of photosynthetic branches that in some ways mimic large compound leaves. From the Late Devonian through the Pennsylvanian, multiple lineages (horsetails, ferns, "progymnosperms," seed plants) independently developed laminar leaves with specialized photosynthetic tissue between the leaf veins. Many of the same lineages also developed large arborescent forms on which these new photosynthetic organs were arrayed. While there were certainly important innovations in vegetative structure through the late Paleozoic, Mesozoic, and Cenozoic (for example, the origin of vessels and high vein density in angiosperms, or the reorganization of the plant body seen in monocotyledons), the basic vegetative structure of modern terrestrial plants was already established by the Late Devonian.

7. INNOVATION IN LAND PLANT REPRODUCTION

Following the basic innovation of the branching sporophyte by the late Silurian, the reproductive biology of terrestrial plants appears to have changed little over the Devonian, even though their vegetative structure became increasingly complex. While several lineages developed heterospory (see glossary and further reading), reproduction remained fundamentally based on the simple release of wind- or water-dispersed spores; however, beginning in the Late Devonian and continuing through the Mississippian and Pennsylvanian, there was a diversification of reproductive morphology associated with the emergence, radiation, and rise to ecological dominance of seed plants. In seed plants, reproduction is more complicated than the dispersal of spores; reproductive success depends on pollen output, the number of ovules, the efficiency with which pollen can reach the ovules, and the efficacy of subsequent fertilization, embryo development, and seed dispersal. The result was a diverse array of morphologies linked to the dispersal of gametes (accomplished through pollination) and the dispersal of embryos (contained in the seeds).

Reproductive innovation increased still further after the Pennsylvanian, and the evolutionary history of seed plants (especially angiosperms) can be thought of as a nearly limitless exploration of ways to more efficiently invest reproductive resources, move pollen, disperse seeds, and ensure the survival of offspring. Some aspects of this evolutionary history reflect basic advances in the mechanics of terrestrial reproduction. For example, the pollen grains of early seed plants and some living gymnosperms such as cycads and *Ginkgo* produce motile sperm that swim through liquid inside ovules, similar to the way motile male gametes of ferns swim through films of water on the forest floor; however, in groups such as angiosperms, conifers, and Gnetales, the motile phase is lost entirely and male gametes are delivered directly to eggs by tubes that develop from the pollen grains. Resource allocation in seeds has also become generally more efficient through time; early groups of seed plants such as medullosan seed ferns from the Pennsylvanian developed nearly fully provisioned seeds (technically ovules) before fertilization, and perhaps before pollination, had occurred. In contrast, most flowering plants make almost no investment in the growth of ovules or seeds until after fertilization.

Other aspects of seed plant reproductive evolution reflect responses to constantly changing ecological interactions with pollinators and seed dispersers. For example, fleshy cone tissues that attract vertebrate seed dispersers, usually birds, have evolved independently in several groups of living conifers. Similarly, "flowerlike" reproductive structures occur in insect-pollinated living "gymnosperms" (e.g., *Welwitschia* in the Gnetales) and in extinct "gymnosperms" that were most likely insect pollinated (most Bennettitales) as well as in flowering plants. In angiosperms, structural innovations such as the bisexual flower and the carpel function in conjunction with complex pollination and seed dispersal interactions involving insects, birds, and mammals, and such interactions have long been thought to be centrally important in generating the enormous diversity in the group.

8. COEVOLUTION WITH ANIMALS

Plant evolution has unfolded in an ecological theater that was constantly being rebuilt by the changing positions of oceans and continents, as well as by major fluctuations in climate. In addition, there have been major changes in the reciprocal interactions between animals, plants, and microbes as all three groups have evolved. For example, the earliest terrestrial ecosystems appear to have been very different from those we would recognize today; fossil evidence suggests plants were largely free from animal herbivory, and the earliest Silurian and Devonian insects seem to have been primarily detritivores or carnivores. Similarly, the earliest terrestrial vertebrates in the Devonian, Mississippian, and Pennsylvanian show no obvious adaptation in dentition or body size for the direct consumption of plants (see chapters II.17 and II.18). This implies that the flow of energy in Devonian ecosystems was very different from that in modern ecosystems; fungi that actively decomposed plants and insects that fed on this decomposed plant matter probably occupied a central place in a food web with few direct links between animals and plants. Direct insect herbivory, however, was well established by the Pennsylvanian, and the intensity and complexity of plant-herbivore interactions has continued to increase over time. The first unequivocal vertebrate herbivores are recorded in the early Permian and have likewise increased in diversity and ecological importance through time, leading ultimately to the extremely large dinosaur herbivores of the Jurassic and Cretaceous and the large mammalian herbivores of the Cenozoic that replaced them (see chapter II.18). The evolution of large grazing mammals together with changing climates during the later Cenozoic may also have helped create the open habitats that have been exploited very effectively, and relatively recently, by derived groups of herbaceous angiosperms (e.g., grasses, Asteraceae).

Early indications of potential interactions between plants and pollinators come from pollen found in coprolites and gut contents of Mesozoic insects. At the same time, the aggregation of ovules and pollen-producing organs into a single flowerlike structure, as seen, for example, in extinct Bennettitales, may have created opportunities for insects to deposit and receive pollen during the same visit; however, direct evidence of plant-pollinator interactions is sparse until the rise of angiosperms, and it is only within this group that animal pollination has been exploited to its fullest potential and highest degree of specialization. While Early and mid-Cretaceous flowers are not highly specialized compared with those of many modern taxa, most were clearly pollinated by insects (probably beetles and flies). Beginning in the Late Cretaceous, and increasing through the Cenozoic, a dramatic radiation of floral morphology has been occurring as increasingly intricate interactions have developed between angiosperms and a huge range of insect pollinators such as bees, beetles, flies, and butterflies as well as birds, bats, rodents, and even primates.

The history of coevolution in relation to seed dispersal broadly parallels that of pollination. Evidence has been found of diffuse interactions between animal dispersers and many Mesozoic plants; for example, the seeds of fossil cycads, *Ginkgo*, and *Caytonia* were surrounded by fleshy tissues that may have been important in attracting animal dispersers, as they are today in many living "gymnosperms." Among angiosperms, indications of relatively generalized interactions with animal dispersers first appear during the Early Cretaceous. Many Cretaceous angiosperms produced small fleshy fruits that suggest some kind of animal dispersal; however, in the early Cenozoic, there is an abrupt increase in both fruit and seed size that appears to be linked with the availability of bird and mammal dispersers as well as with a change from relatively open plant communities to more dense vegetation and closed forest canopies. Such environments may have favored larger seeds with more stored nutrients for the establishment of seedlings. In general, the great variety of fruit and seed morphology seen among living angiosperms is linked to an equally great variety of interactions with animal seed dispersers.

9. PATTERNS OF EXTINCTION

The fossil record provides ample evidence of extinction in the history of plants on land, just as it does in the history of terrestrial animals, but successive episodes of mass extinction seem to have had much less influence on plant evolution. A few significant perturbations in the history of plant life do coincide with major extinction events in the animal record; for example, the apparent loss of the ecologically important plant group glossopterids at the Permian-Triassic boundary, and the significant loss of species diversity in North America at the Cretaceous-Paleogene boundary. These episodes often appear linked to regional climatic changes, however, rather than to fundamental shifts in the composition of global vegetation. There are also equally large perturbations that do not coincide with major extinctions in the animal world, for example, the loss of coal swamps and their characteristic plants toward the end of the Paleozoic, which seems to have been caused by increasingly arid climates. It also appears that the great diversity in some groups of plants, for example, angiosperms and certain groups of filicalean ferns, may reflect low extinction rates as well as high speciation rates. In the broadest sense, the major changes in terrestrial vegetation through time appear to reflect a more gradual pattern of displacement by competition among plant groups rather than repeated resetting of the evolutionary clock by mass extinction. Large-scale evolutionary patterns among plants seem to predominantly reflect successive biological innovations that resulted in increased vegetative and reproductive efficiency against a background of changing ecological conditions.

FURTHER READING

Algeo, T. J., and S. E. Scheckler. 1998. Terrestrial-marine teleconnections in the Devonian: Links between the evolution of land plants, weathering processes, and marine anoxic events. Philosophical Transactions of the Royal Society B 353: 113–130. *Discusses increases in terrestrial weathering rates that may have accompanied the innovation of deeply rooted plants in the Devonian and provides an example of the broad effects of plant structural evolution on terrestrial ecosystems.*

Bateman, R. M., and W. A. DiMichele. 1994. Heterospory: The most iterative key innovation in the evolutionary history of the plant kingdom. Biological Reviews 69: 345–417. *Summarizes the evolution of heterospory in different vascular plant lineages and offers potential explanations for its frequent occurrence.*

Boyce, C. K. 2008. How green was *Cooksonia*? The importance of size in understanding the early evolution of physiology in the vascular plant lineage. Paleobiology 34: 179–194. *Offers a new interpretation of the physiology of some of the earliest polysporangiophytes and discusses the evolution of the sporophyte in early terrestrial plants.*

Boyce, C. K., T. J. Brodribb, T. S. Feild, and M. A. Zwieniecki. 2009. Angiosperm leaf vein evolution was physiologically and environmentally transformative. Proceedings of the Royal Society B 276: 1771–1776. *This study suggests that angiosperm vegetative innovations, in particular their high leaf vein density relative to other groups of vascular plants, may have altered terrestrial ecosystems by increasing the rate of water cycling and therefore contributing to the formation of extensive tropical rain forest biomes.*

Friis, E. M., K. R. Pedersen, and P. R. Crane. 2011. Early Flowers and Angiosperm Evolution. Cambridge: Cambridge University Press. *A summary and synthesis of the known information about early angiosperm ecology, biology, and floral morphology based on Cretaceous fossils and their living relatives.*

Kenrick, P., and P. R. Crane. 1997. The Origin and Early Diversification of Land Plants: A Cladistic Study. Washington, DC: Smithsonian Institution Press. *A summary, synthesis, and analysis of the morphology, anatomy, and systematics of early vascular plants and their living relatives.*

Mathews, S. 2009. Phylogenetic relationships among seed plants: Persistent questions and the limits of molecular data.

American Journal of Botany 96: 228–236. *Summarizes the various hypotheses of relationships among living seed plant groups based on molecular and morphological data sets and discusses some potential reasons for the variety of results.*

Pryer, K. M., E. Schuettpelz, P. G. Wolf, H. Schneider, A. R. Smith, and R. Cranfil. 2004. Phylogeny and evolution of ferns (monilophytes) with a focus on the early leptosporangiate divergences. American Journal of Botany 91: 1582–1598. *This study proposes a general hypothesis for relationships among living "pteridophytes" based on a large molecular data set.*

Shaw, A. J., P. Szövényi, and B. Shaw. 2011. Bryophyte diversity and evolution: Windows into the early evolution of land plants. American Journal of Botany 98: 352–369. *Summarizes the current state of research on the diversity and systematics of bryophytes, especially in relation to the early evolution of land plants.*

II.14

Major Events in the Evolution of Fungi

David S. Hibbett

OUTLINE

1. Fungi in the tree of life
2. Losses of flagella and diversity of the "basal fungal lineages"
3. Evolution of the dikaryon and multicellular fruiting bodies
4. Evolution of decayers and plant pathogens
5. Evolution of mycorrhizae, lichens, and endophytes
6. Evolution of animal pathogens and mutualists
7. The age of Fungi

Fungi represent one of the few clades of eukaryotes that evolved complex multicellular forms and diversified extensively on land, the others being plants and animals. In the process, Fungi have become integral to the functioning of ecosystems. They are the master decayers of plant biomass, playing a pivotal role in the global carbon cycle. As parasites and pathogens, they attack plants, animals, and each other, but they have also evolved intricate mutualistic symbioses, such as mycorrhizae and associations with the fungus gardening leaf-cutter ants. The number of extant species of Fungi is a matter of conjecture; one commonly cited estimate suggests 1.5 million fungal species, but only about 100,000 species have been described, and fungal molecular ecologists routinely detect species and even major clades that have no match in DNA sequence databases. Much progress has been made in reconstructing the phylogenetic relationships of the Fungi, but there are still unresolved branches in the fungal evolutionary tree. Adaptation in Fungi is largely biochemical in nature, involving the diversification of enzymes, effectors, and secondary metabolites that allow these anatomically simple organisms to exploit diverse substrates and hosts.

GLOSSARY

AFTOL. Assembling the Fungal Tree of Life, a multilaboratory collaboration supported by the US National Science Foundation that sought to reconstruct the phylogeny of Fungi and construct a higher-level classification for the group.

Biotroph. Fungus that obtains nutrition by association with a living host, whether as a pathogen, parasite, or mutualist.

Fruiting Body. Macroscopic multicellular structure that produces spores, including mushrooms.

Hyphae. Fungal filaments.

Mold. Asexual spore-bearing filamentous fungus (mostly Ascomycota). Also an informal term for certain fungus-like organisms (water molds, slime molds).

Mycelium. Network of hyphae; a fungal thallus.

Mycorrhiza. Symbiotic association of fungal hyphae and plant roots.

Saprotroph. Fungus that obtains nutrition by decaying dead organic matter. Necrotrophs (Fungi that kill and then decay a living host) blur the boundaries between saprotrophs and biotrophs (see above).

Yeast. Unicellular, nonflagellated fungus.

1. FUNGI IN THE TREE OF LIFE

Linnaeus considered Fungi to be members of the kingdom Plantae, and ever since mycology has been one of the traditional disciplines of botany; however, unlike plants and algae, Fungi have cell walls composed of chitin (as opposed to cellulose) and they are heterotrophic, with an absorptive mode of nutrition. The "five kingdom" system of the late 1960s was based largely on cellular, ultrastructural, and biochemical characters, and it classified Fungi as a separate kingdom of eukaryotes, distinct from the plants and animals, as well as the paraphyletic "protists." Today, it is known from DNA sequence data that the Fungi form a clade in the eukaryotic supergroup Unikonta ("*unikonts*"), which also includes animals and several protist lineages.

Fungus-like taxa outside the unikonts include the Oomycetes, which have a filamentous habit and absorptive heterotrophy like true Fungi, but are actually members of the Heterokonts, a group that includes kelps and diatoms. Oomycetes include some of the most destructive plant pathogens, such as *Phytophthora infestans* (infamous as the cause of the Irish potato famine), as well as "water molds" that plague aquarium fish (some species also attack mammals, including humans). The suffix *-mycetes* and the term *mold* reflect the historical classification of these organisms as Fungi.

Another group of eukaryotes formerly classified as Fungi are the *slime molds*, which are in a clade within the unikonts termed the Mycetozoa. Slime molds include cellular forms (e.g., *Dictyostelium discoideum*), which forage as amoeboid unicells but can also aggregate to form a pseudomulticellular structure capable of gliding motility, and plasmodial forms (e.g., *Physarum polycephalum*), which alternate between amoeboid and uniflagellated swarm cells and eventually form a coenocytic, often brightly colored, multinucleate "supercell" that scavenges for bacteria along the forest floor. The closest relatives of Fungi within unikonts are the nucleariids, which are unicellular amoebae. Based on comparison with nucleariids and other clades of unikonts, such as choanoflagellates, it is likely that the common ancestor of the Fungi was a unicellular, possibly aquatic heterotroph, with a single posterior flagellum.

2. LOSSES OF FLAGELLA AND DIVERSITY OF THE "BASAL FUNGAL LINEAGES"

Reconstructing the deepest branching events in the fungal phylogeny is a work in progress. The topology presented here (figure 1) is based on a study that analyzed six genes in almost 200 species and that is reflected in the AFTOL (see glossary) classification of the Fungi. Other studies have resolved relationships differently, however, and this picture may change as more complete fungal genomes are included in phylogenetic analyses. Some of the most problematic aspects of the fungal phylogeny concern the "basal fungal lineages," a paraphyletic assemblage containing diverse chytrids and zygomycetes (these are informal, purely descriptive terms, defined below).

Chytrids are Fungi that produce swimming zoospores, and in this regard they resemble the putative aquatic protist ancestors of the Fungi and other unikonts. Chytrids occur in both freshwater and marine environments, but they have also adapted to terrestrial habitats, including soil and the bodies of terrestrial organisms. Some chytrids have developed mycelial (see glossary) growth or produce rootlike extensions (rhizoids) that anchor the cells on the substrate and facilitate absorption of nutrients. Three independent clades of chytrids were included as formal taxa in the AFTOL classification: the Chytridiomycota, Neocallimastigomycota, and Blastocladiomycota. In addition, the recently recognized and informally named clade "Rozellida" or "Cryptomycota" is based largely on environmental sequences from aquatic habitats; the only formally named taxon in the group is the chytrid genus *Rozella*. Finally, *Olpidium* is a group of plant and algal parasites of uncertain placement that may represent another independent chytrid lineage.

Most chytrids are saprotrophs and can often be isolated by "baiting" with organic substrates containing cellulose or keratin (hemp seeds and snake skin are often used), but others have evolved biotrophic nutritional modes involving diverse hosts. For example, *Rozella allomycis* (Rozellida/Cryptomycota) is an intracellular parasite that inhabits another chytrid, the filamentous *Allomyces* (Blastocladiomycota). Other members of Rozellida/Cryptomycota appear to be phagotrophic or intracellular parasites, based on observations of cells harvested from water and sediments and visualized using fluorescently labeled DNA probes. Some chytrids have evolved plant parasitic lifestyles, such as *Synchytrium endobioticum* (Chytridiomycota), which causes black wart disease of potatoes, and *Olpidium brassicae* (which is of uncertain placement; see below), which causes club root diseases of cabbages and other Cruciferae. Animals are not immune to attack by chytrids; *Catenaria anguillulae* (Chytridiomycota) infects and kills nematodes, while *Batrachochytrium dendrobatidis* (also Chytridiomycota) is a devastating pathogen implicated in the ongoing global decline of frogs and other amphibians. More benign associations are manifested in the Neocallimastigomycota, which includes anaerobic rumen Fungi that benefit their herbivore hosts by digesting plant fibers.

Like chytrids, zygomycetes are united only by shared primitive characters (symplesiomorphies), specifically, the development of a mycelium lacking regular septa, and the absence of complex multicellular fruiting bodies. Beyond that, zygomycetes are wildly diverse in form and ecology. Zygomycetes are distributed across about five independent clades, the Mucoromycotina, Kickxellomycotina, Zoopagomycotina, Entomophthoromycotina (collectively, these four groups have often been classified as the Zygomycota, a name still in common use), and Glomeromycota.

Just about every imaginable fungal nutritional mode has evolved in the zygomycetes. The most familiar

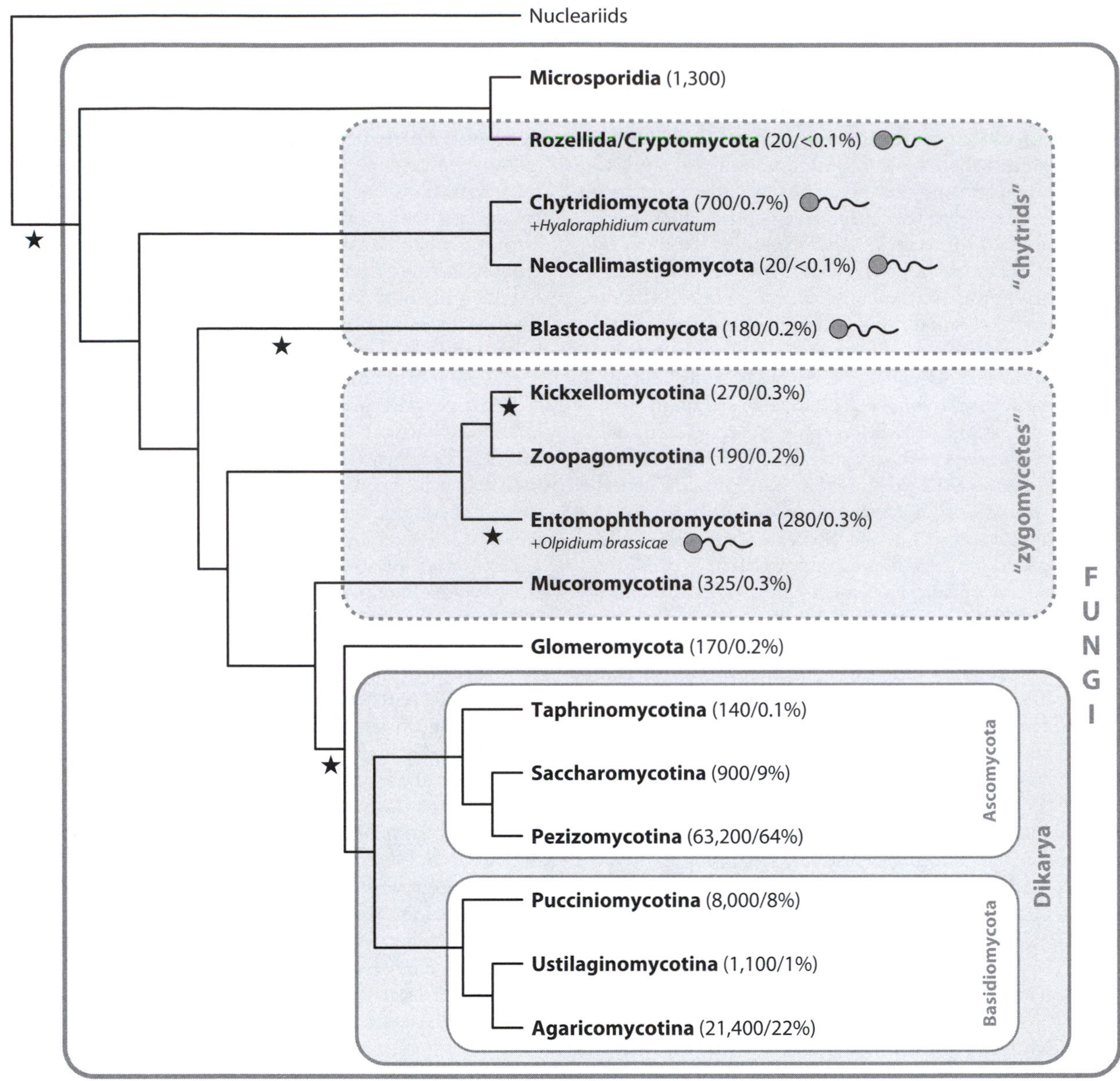

Figure 1. Phylogenetic relationships of Fungi. Numbers of described species and proportion of all described Fungi in each group are indicated in parentheses following taxon names. Dashed lines delimit paraphyletic groups, not formal taxa. Branch lengths are not proportional to time. Relationships among chytrids and zygomycetes are particularly controversial.

group, the Mucorales (Mucoromycotina), includes saprotrophic molds that are common agents of food spoilage, some of which are used to make fermented foods such as tempeh (*Rhizopus oryzae*), but members of this group also cause an extremely dangerous (and thankfully rare) fungal infection of humans called mucormycosis, while others parasitize mushrooms. Biotrophic associations with animals are widespread in other groups of zygomycetes as well. For example, members of the Entomophthoromycotina include insect associates, such as the fly pathogen *Entomophthora muscae*, which have potential applications in biocontrol (saprotrophs also occur in this group). Kickxellomycotina include a fascinating array of "Trichomycetes," which live in the hindgut of aquatic arthropods, such as mosquito larvae and isopods, and Zoopagomycotina contains species that can attack and kill nematodes, rotifers, amoebae, and other Fungi.

Mycorrhizal associations have evolved in two groups of zygomycetes, the Endogonales (Mucoromycotina) and the Glomeromycota. Of these, the arbuscular-mycorrhizal (AM) associations of Glomeromycota are the most widely documented, and have been estimated to occur in 80 percent of plant species. Fossils of putative Glomeromycota spores have been found in the roots of early land plants from the Rhynie Chert, which has contributed to the view that the establishment of AM symbioses was a key event facilitating the colonization of land by plants. Members of the Glomeromycota and Endogonales form associations with extant nonvascular plants (bryophytes) as well as vascular plants, and both groups have been proposed to represent the oldest lineage of mycorrhizal Fungi. Both groups also contain simple fruiting bodies, including the "pea truffles" of the Endogonales and aggregations of spores termed sporocarps in certain Glomeromycota.

One of the most controversial issues in fungal evolution concerns the number of losses of the eukaryotic flagellum. In addition to the zygomycete groups discussed above, two other clades of nonflagellated Fungi are of relevance to this problem. One is an obscure planktonic fungus called *Hyaloraphidium curvatum*, which is nested within the Chytridiomycota. The other is a clade called Microsporidia, which are obligate intracellular parasites of diverse animals that are notable for their highly reduced genomes (some are fewer than 3 million base pairs, smaller than many bacteria), vestigial mitochondria lacking a genome (*mitosomes*), and a unique infection mechanism, the polar tube, through which the microsporidial cytoplasm is injected into the host cell.

The placement of Microsporidia within Fungi is not resolved; they could be the sister group to the rest of the Fungi, or they could occupy several positions closely related to different groups of chytrids or zygomycetes, with different implications for the number of losses of the flagellum. Finally, uncertainty in the placement of the chytrid *Olpidium brassicae* also makes it difficult to estimate the number of losses of the flagellum. In the six-gene study mentioned above, *O. brassicae* is closely related to certain zygomycetes (Entomophthoromycotina), and the Microsporidia are the sister group of the Rozellida/Cryptomycota clade (represented in that study by *R. allomycis*). This topology implies five independent losses of the flagellum within the Fungi, but alternative trees requiring fewer losses cannot be rejected.

3. EVOLUTION OF THE DIKARYON AND MULTICELLULAR FRUITING BODIES

About 98 percent of the described species of Fungi are in a strongly supported group called the Dikarya. The synapomorphy for which the group is named, the *dikaryon*, is formed through the cytoplasmic fusion of mating compatible haploid cells (monokaryons), but without the immediate fusion of the haploid nuclei, yielding a stable binucleate condition. Nuclear fusion typically occurs immediately prior to meiosis in specialized meiosporangia (cells producing haploid spores via meiosis), which in most Dikarya are the only truly diploid cells in the life cycle.

The Dikarya is composed of two sister clades, the Ascomycota and Basidiomycota, each of which is divided into three major groups, the Saccharomycotina, Taphrinomycotina, and Pezizomycotina (Ascomycota), and the Pucciniomycotina, Ustilaginomycotina, and Agaricomycotina (Basidiomycota). In Ascomycota, the meiosporangium is a saclike ascus, which produces spores internally, whereas in Basidiomycota the meiosporangium is a pedestal-shaped basidium, which produces spores externally on minute stalks. Both Ascomycota and Basidiomycota also contain species with asexual reproductive modes. Most common molds, including those used to make blue cheeses, and the fungus that produced the first antibiotic, penicillin, are asexual members of Ascomycota. In some cases, connections between asexual and sexual forms have been confirmed through cultural or molecular studies, but for many asexual forms the corresponding sexual stage, if it exists, has not been observed.

Many Basidiomycota produce hook-shaped projections called *clamp connections* at cell junctures in dikaryotic hyphae, which serve to apportion the individual haploid nuclei to daughter cells in mitosis. Ascomycota produce putatively homologous structures, termed croziers, but these are formed only at the bases of asci. Although many Ascomycota and Basidiomycota have filamentous growth, some have a yeast form (or alternate between the two forms), and two groups of Ascomycota, the Saccharomycotina and Taphrinomycotina, are composed almost entirely of yeasts. Clamp connections and croziers are features of hyphal forms, implying that the common ancestor of Dikarya was a mycelial fungus and that yeast forms in both Ascomycota and Basidiomycota have been derived by parallel simplification.

Multicellular fruiting bodies have evolved only within the Dikarya (excluding the minute sporocarps of certain Endogonales and Glomeromycota). Some fruiting bodies are inconspicuous and relatively simple crusts, but others are massive and may have complex, developmentally integrated structures. Examples include the gigantic bracket Fungi that grow on tree trunks, mushrooms with multiple layers of veil tissues (like the iconic "fly agaric," *Amanita muscaria*), or the diverse gasteromycetes (Fungi with spores produced inside the closed fruiting body), such as stinkhorns, puffballs, and bird's nest Fungi. There has been extensive convergent evolution of fungal fruiting bodies, with forms such as gilled mushrooms, coral Fungi, polypores, and puffballs all having been derived repeatedly. The majority of macroscopic fungal fruiting bodies, including all the preceding examples, are in the Agaricomycotina (Basidiomycota), but some Pezizomycotina (Ascomycota) have also evolved fruiting bodies, including various cup Fungi and the edible morels (*Morchella* spp.). One group of Taphrinomycotina, the genus *Neolecta*, produces simple club-shaped fruiting bodies; it is not clear whether these represent an independent origin of fruiting bodies or a retained plesiomorphic condition. Some Dikarya produce fruiting bodies underground, including the prized truffles in the genus *Tuber* (Pezizomycotina). Such hypogeous Fungi have fully adapted to the terrestrial habit, releasing their spores directly into soil or exploiting small mammals and other animal vectors for spore dispersal. At the same time, other lineages of Dikarya, both Ascomycota and Basidiomycota, have reverted to aquatic habitats. Lacking flagellated swimming cells, marine and freshwater Dikarya often have spores with elongate projections that aid in dispersal. Subsequent sections of this chapter highlight ecological innovations within the Dikarya, many of which are also echoed in the various chytrid and zygomycete lineages described previously.

4. EVOLUTION OF DECAYERS AND PLANT PATHOGENS

Dikarya play a key role in the global carbon cycle by decomposing plant cell walls, which are composed of the carbohydrates cellulose and hemicellulose, as well as lignin, a heterogeneous polymer highly resistant to microbial attack that gives woody tissues rigidity. Saprotrophic Dikarya attack diverse plant-derived substrates, including leaf litter, organic matter in soils, herbivore dung, and wood. The diversity and evolution of the enzymatic apparatus that enables Dikarya to cause decay is being revealed through biochemical and comparative genomic research. Many different kinds of carbohydrate-active enzymes involved in dismantling the cellulose and hemicellulose components of plant cell walls have diversified in both Ascomycota and Basidiomycota, and both groups include numerous taxa that decay nonwoody tissues. Selected economically important examples include the edible button mushroom *Agaricus bisporus* (Basidiomycota, Agaricomycotina), which has been adapted for cultivation on manure-straw compost mixture, and the mold *Trichoderma reesei* (Ascomycota, Pezizomycotina), a source of industrial cellulase enzymes. In contrast, efficient decay of massive woody substrates is limited to certain Agaricomycotina (although some Ascomycota do decay wood to an extent). Two principal forms of wood decay occur in Agaricomycotina: white rot, in which the carbohydrate and lignin portions of wood are degraded, and brown rot, in which the lignin fraction is chemically modified but not appreciably degraded. Lignin decay appears to involve the action of enzymes called class II fungal peroxidases, which are of interest for their potential application in bioremediation and other "green" technologies. The brown rot mode of decay has evolved independently in multiple lineages of Agaricomycotina with associated losses of genes encoding lignin-degrading peroxidases.

Plant pathogens have evolved in numerous lineages of Dikarya and have huge impacts on natural ecosystems and human agriculture. Selected examples in Ascomycota include the chestnut blight fungus (*Cryphonectria parasitica)*, which devastated one of the dominant and most prized timber trees of North America, the American chestnut, and the ergot fungus (*Claviceps purpurea*), which infects rye and produces a mixture of alkaloids that cause a horrifying suite of symptoms in humans who ingest the grain, including convulsions, hallucinations, and impaired blood circulation leading to gangrene.

Plant pathogens also occur throughout Basidiomycota, with the largest concentrations among the "smuts" (Ustilaginomycotina), such as the corn smut fungus (*Ustilago maydis)*, and the "rusts" (Pucciniomycotina), which include economically important pathogens of diverse crops that have complex life cycles involving multiple host plants. An example is the barberry-wheat rust (*Puccinia graminis*). The reliance of wheat rust on barberries (*Berberis*) has been recognized for centuries, leading to campaigns to control the pathogen by eliminating the alternate host in the area of grain fields.

Plant-pathogenic and saprotrophic Dikarya use many of the same families of carbohydrate-active enzymes to degrade plant cell walls. In addition, an active area of current research concerns the diversity and function of *fungal effectors*, small secreted proteins produced by plant-pathogenic Dikarya that suppress the immune response of host plants. Studies on fungal effectors are being conducted in Ascomycota, such as the rice blast fungus (*Magnaporthe grisea*), a widespread pathogen of one of the world's most important cereal crops, and Basidiomycota such as *Melampsora lini*, a rust that infects flax. Comparisons of genes encoding fungal effectors often suggest strong diversifying selection, reflecting an evolutionary arms race between host and pathogen.

5. EVOLUTION OF MYCORRHIZAE, LICHENS, AND ENDOPHYTES

Mutualistic biotrophic associations of Dikarya and photosynthetic organisms include mycorrhizae, lichens, and endophytes. In each case, the fungus obtains its carbon nutrition in the form of sugars produced by the photosynthetic host. Mycorrhizal Dikarya aid their hosts by facilitating uptake of mineral nutrients, while some endophytes may confer enhanced drought tolerance, resistance to herbivores, or other benefits. Nonetheless, these associations may best be regarded as reciprocal parasitisms, with selection driving each partner to maximize its fitness at the expense of the other. An extreme outcome of such instability is manifested in mycoheterotrophic plants such as Indian pipe (*Monotropa uniflora*), a nonphotosynthetic angiosperm that actually draws sugars out of its fungal partners (mushrooms in the Agaricomycotina), thus reversing the normal flow of carbon in mycorrhizal symbiosis.

Most of the mycorrhizal associations of Dikarya are ectomycorrhizae (ECM), in which the hyphae of the fungus ensheathe and penetrate the roots of woody plants. ECM hosts include pines, oaks, hickories, birches, willows, dipterocarps, eucalypts, and certain legumes. These are often the dominant trees of extensive forests, so ECM have a large impact on terrestrial ecosystems, even if they do not involve as many plant species as the AM of Glomeromycota. ECM have evolved in certain Pezizomycotina (Ascomycota), but they are most diverse in the Agaricomycotina (Basidiomycota) (ECM also occur in the Endogonales). Within the Agaricomycotina, the ECM habit has clearly evolved repeatedly, but the precise number of origins is not well resolved. One lineage of mostly ECM mushrooms, the Boletales, also contains several mycoparasites and decayers, suggesting that reversals from the ECM habit to other nutritional modes are possible. ECM-forming Dikarya produce some of the most prized edibles, such as truffles, porcini mushrooms, and chanterelles, which are expensive, in part because they must be collected in forests containing their tree hosts.

Lichens and endophytes have also evolved in both Ascomycota and Basidiomycota, but they are much more numerous in Ascomycota (Pezizomycotina). Lichens, which involve associations with eukaryotic green algae as well as cyanobacteria, are able to colonize truly harsh environments, such as rock surfaces and tree trunks, and they play important ecological roles as primary colonizers. Endophytes are cryptic Fungi that produce no visible symptoms in the plants they inhabit. The diversity of endophytes has been revealed through direct culturing approaches as well as molecular surveys, which have found them in every plant species investigated so far. Phylogenetic studies in Ascomycota have revealed complex patterns of transitions between lichenized, endophytic, pathogenic, and saprotrophic lifestyles. Intriguingly, an "endolichenic" habit, in which Fungi inhabit lichen thalli, appears to have been an evolutionary precursor to the endophytic habit in some lineages of Ascomycota.

6. EVOLUTION OF ANIMAL PATHOGENS AND MUTUALISTS

Dikarya have evolved biotrophic associations with diverse animals, including humans, both as pathogens and as mutualists. Many human-associated Fungi occur as commensals or produce relatively mild infections, such as the yeast *Candida albicans*, which causes genital and oral yeast infections, as well as dermatophytes (*Trichophyton* and other genera), which are able to digest the keratin in skin, hair, and nails, causing annoying maladies known as ringworm and athelete's foot. Other fungal pathogens are much more serious, such as *Coccidioides immitis*, the causal agent of coccidioidomycosis (San Joaquin Valley fever), which initiates as a lung infection but can become a deadly systemic infection. Another lung-borne pathogen is the unicellular *Pneumocystis* spp., which was thought to be a protist until it was shown with ribosomal RNA gene sequences to be a fungus. Like many fungal pathogens, *Pneumocystis* is common as a commensal in healthy individuals but can cause a serious opportunistic infection, particularly in immunocompromised individuals.

Diverse Dikarya attack insects, with perhaps the most dramatic being members of the genus *Cordyceps*, which produce elongate, stalked fruiting structures arising from the host's body. *Cordyceps* species attack diverse arthropods, and one lineage has even switched from the underground larvae of cicadas to the underground fruiting bodies of certain truffles (an interkingdom host jump). Moreover, *Cordyceps* is closely related to the plant pathogen *Claviceps purpurea*, mentioned previously, further highlighting the evolutionary lability of fungal nutritional modes. Other nonhuman animals attacked by Fungi include nematodes, which are trapped by various "predatory" Fungi, whose mycelia are equipped with constricting or nonconstricting rings, and adhesive knobs, networks, and hyphal branches.

All of the preceding examples are in the Ascomycota, but Basidiomycota have also evolved diverse associations with animals, both benign and antagonistic. Examples include human pathogens, such as the yeast *Filobasidiella (Cryptoccus) neoformans*, which can cause fungal meningitis, or the "dandruff fungus" *Malassezia globosa*. One of the best-studied mutualistic associations of Fungi and animals is that of the neotropical leaf-cutter ants, which harvest large volumes of plant biomass to feed to several lineages of Agaricomycotina that are closely related to free-living saprotrophic mushrooms. In a striking case of convergent evolution, certain Old World termites, another group of social insects, cultivate a different group of saprotrophic Agaricomycotina in the genus *Termitomyces*.

7. THE AGE OF FUNGI

Fungi, with their ephemeral and often microscopic forms, are not well represented in the fossil record. Nonetheless, there are branching hyphae from the Silurian period (ca. 430 Ma), which may be the oldest fungal fossils, and forms attributed to Glomeromycota, Chytridiomycota, and Ascomycota occur alongside the oldest land plant fossils in the Devonian Rhynie Chert (ca. 410 Ma). Molecular clock studies have estimated the Fungi to be much older than the fossils alone would suggest, albeit with significant variance in age estimates. Various molecular clock analyses have suggested that the most recent common ancestor of the Fungi as a whole existed anywhere from 850 million to 1.5 billion years ago; the Dikarya are probably at least 500 million years old, and perhaps much older. While the exact timing and sequence of branching events deep in the fungal phylogeny are not resolved with confidence, it is clear that Fungi, through their activities as saprotrophs, pathogens, and mutualistic symbionts, have played a major role in shaping the evolution of terrestrial ecosystems.

FURTHER READING

Alexopoulos, C. J., C. W. Mims, and M. Blackwell. 1996. Introductory Mycology. 4th ed. New York: Wiley. *This textbook contains a wealth of information on fungal forms, ecology, and impacts on humans.*

Arnold, A. E., J. Miadlikowska, K. L. Higgins, S. D. Sarvate, P. Gugger, A. Way, V. Hofstetter, F. Kauff, and F. Lutzoni. 2009. A phylogenetic estimation of trophic transition networks for ascomycetous Fungi: Are lichens cradles of symbiotrophic fungal diversification? Systematic Biology 58: 283–297. *Technical article reconstructing the historical pattern of switches between endophytic, endolichenic, and plant-pathogenic lifestyles.*

Barron, G. L. 1977. The Nematode-Destroying Fungi. Topics in Mycobiology No. 1. Guelph, Ontario: Canadian Biological Publications. *A classic text describing the surprisingly diverse Fungi and fungus-like organisms that trap and kill nematodes.*

Berbee, M. L., and J. W. Taylor. 2010. Dating the molecular clock in fungi—how close are we? Fungal Biology Reviews 24: 1–16. *Review article describing the opportunities and pitfalls of molecular clock dating in Fungi, including the integration of biogeography and fossils.*

Floudas, D., M. Binder, R. Riley, K. Barry, R. A. Blanchette, B. Henrissat, A. T. Martinez, et al. 2012. The Paleozoic origin of enzymatic lignin decomposition reconstructed from 31 fungal genomes. Science 336: 1715–1719. *Comparative genomic study describing the evolution of wood decay, and suggesting a fungal role in the decline of coal formation at the end of the Carboniferous.*

Hibbett, D. S., M. Binder, J. F. Bischoff, M. Blackwell, P. F. Cannon, O. E. Eriksson, S. Huhndorf, et al. 2007. A higher-level phylogenetic classification of the Fungi. Mycological Research 111: 509–547. *Current classification of Fungi with references to the phylogenetic analyses that form the basis of the taxonomy.*

James, T. Y., F. Kauff, C. Schoch, P. B. Matheny, V. Hofstetter, C. Cox, G. Celio, et al. 2006. Reconstructing the early evolution of the fungi using a six gene phylogeny. Nature 443: 818–822. *Comprehensive 6-gene, 200-species phylogenetic analysis of all major groups of Fungi.*

Jones, M.D.M., I. Forn, C. Gadelha, M. J. Egan, D. Bass, R. Massana, and T. A. Richards. 2011. Discovery of novel intermediate forms redefines the fungal tree of life. Nature 474: 200–203. *This technical article describes a major clade of Fungi known almost entirely from environmental DNA sequences.*

Kirk, P. M., P. F. Cannon, D. W. Minter, and J. A. Stalpers. 2008. Ainsworth & Bisby's Dictionary of the Fungi. 10th ed. Oxon, UK: CAB International. *Essential desk reference for fungal biologists, with counts of described species for genera and more inclusive groups.*

McLaughlin, D. J., D. S. Hibbett, F. Lutzoni, J. Spatafora, and R. Vilgalys. 2009. The search for the fungal tree of life. Trends in Microbiology doi:10:1016/j.tim.2009.08.001. *Review article describing progress in fungal systematics from a historical perspective.*

Rodriguez, R. J., J. F. White Jr., A. E. Arnold, and R. S. Redman. 2009. Fungal endophytes: Diversity and functional roles. New Phytologist 182: 314–330. *Useful overview of the biology of endophytic Fungi.*

Spatafora, J. W., K. W. Hughes, and M. Blackwell, eds. 2006. A phylogeny for kingdom Fungi. Mycologia 98: 829–1103. *This special issue of* Mycologia *contains phylogenetic studies on all major groups of Fungi.*

Staijch, J., M. L. Berbee, M. Blackwell, D. S. Hibbett, T. Y. James, J. Spatafora, and J. W. Taylor. 2009. The Fungi. Current Biology 19: R840–R845. *Concise overview of fungal phylogeny emphasizing selected topics of interest to molecular biologists, and others.*

Stergiopoulos, I., and P.J.G.M. de Wit. 2009. Fungal effector proteins. Annual Reviews of Phytopathology. 47: 233–263. *Authoritative review of fungal effectors in selected model systems.*

Taylor, T. N., E. L. Taylor, and M. Krings. 2009. *Paleobotany.* 2nd ed. Amsterdam: Elsevier. *This is a lavishly illustrated compendium that presents the fossil record of Fungi in a broad botanical context.*

II.15

Origin and Early Evolution of Animals

Paulyn Cartwright

OUTLINE

1. The Cambrian explosion and the origin of animal phyla
2. Animal phylogeny
3. Multicellularity and the origin of sponges (phylum Porifera)
4. The origin of the nervous system and the evolution of sensory structures in Cnidaria
5. The origins of Bilateria and the phylogenetic placement of Ctenophora, Acoela, Myxozoa, and Placozoa
6. Animal diversity

Around 600 million years ago, members of the animal clade were present but distinct from animals seen on earth today. The origin of most living animal lineages occurred relatively suddenly during the Cambrian period (543–510 Ma). The rapid appearance of animal phyla in the fossil record is referred to as the *Cambrian explosion*. Understanding the evolutionary relationships by reconstructing the animal tree of life is fundamental for unraveling key transitions in animal evolution. While much progress has been made, the phylogenetic position of many major animal lineages remains uncertain. Through the study of animals' closest living relatives, the choanoflagellates, and early diverging animal lineages, such as sponges and cnidarians, we can begin to understand how major innovations such as the evolution of multicellularity, the nervous system and sense organs, and bilateral symmetry evolved. Future investigation into the phylogenetic placement of several enigmatic taxa will increase our understanding of when and how major transitions in animal evolution occurred. Whole genome sequencing of early diverging animals has revealed that the ancestral animal genome was remarkably complex and that the ancestral animals had a molecular toolkit permitting the development of diverse and complex animal body plans.

GLOSSARY

Bilateria. A clade of animals that display bilateral symmetry and are traditionally divided into two major groups, the protostomes and the deuterostomes. Includes most major animal phyla except sponges, cnidarians, and ctenophores.

Cambrian Explosion. Refers to an interval in the history of life from 543 to 510 million years ago when most animal phyla suddenly appeared in the fossil record.

Choanoflagellates. A clade of single-celled or colonial eukaryotes that are the closest living relatives to animals.

Early Diverging Animals. An informal term for animals that are not members of Bilateria. Includes sponges, cnidarians, and ctenophores. (Also sometimes called "basal animals" or "basal metazoans"; see chapter II.1 for a critique of such terms.)

Metazoa. Another name for multicellular animals; eukaryotic organisms that are generally motile and possess an embryonic stage that undergoes gastrulation.

Molecular Toolkit. The set of genes and gene pathways present in a genome that can be co-opted to facilitate body plan evolution. Usually refers to signaling molecules and transcriptional factors that are important for regulating and specifying key aspects of animal development.

1. THE CAMBRIAN EXPLOSION AND THE ORIGIN OF ANIMAL PHYLA

The origin and diversification of major animal lineages represents a key episode in the history of life. To understand these critical events, we must turn to the fossil record, which provides the only tangible evidence for the origin of animal life. The earliest fossils thought to be animals are a diverse assemblage of macroscopic fossils known as the Ediacaran fauna, appearing in the

Vendian period (610–550 Ma). Many of these fossils bear a superficial resemblance to jellyfish, crustaceans, and worms, but none of the fossils can be definitively assigned to modern groups of animals. Many scientists dispute their affinity to animals at all. For example, paleontologist Adolf Seilacher claims that ediacarans may have been an independent experiment in multicellular life that subsequently went extinct.

The next glimpse at possible animal life is found in the Doushantuo fossil formation in China, which dates back to 580 million years ago. This formation holds a diverse set of microscopic fossils that bear a striking resemblance to embryos of cnidarians (jellyfish and sea anemones) and embryos of bilaterian animals such as worms. Detailed cellular structure of these microscopic fossils has been preserved by a rare form of preservation known as phosphatic fossilization. As with the Ediacaran fossils, assignment to modern animal lineages is controversial.

Most animal phyla bearing unmistakable characteristics of modern-day animals made their first appearance in the early Cambrian period, between 510 and 543 million years ago. During a window of slightly more than 30 million years, most modern animal phyla, including mollusks, annelids, arthropods, and chordates, suddenly appeared in the fossil record. This time period, while long in human terms, is incredibly brief compared to the entire history of life, which spans more than 3.5 billion years. As a result, this key event is often referred to as the *Cambrian explosion*.

Two famous fossil formations, the Chenjiang in the Yunning Province of China (525 Ma) and the Burgess Shale in British Columbia, Canada (500 Ma), provide a remarkable glimpse of Cambrian animal diversity. These fossil formations contain preservations of animals that lack hard parts, providing detailed preservations of soft-bodied forms in addition to a diversity of skeletonized fossils. In these fossil formations we find representatives from sponge, ctenophore, arthropod, annelid, priapulid, and chordate lineages. Another fossil formation similar to the Burgess Shale, the Marjum Formation in Utah (500 Ma), has yielded a remarkable diversity of jellyfish fossils (phylum Cnidaria). Table 1 provides names and dates of some of the earliest animal fossil representatives of major phyla.

Differing explanations have been given for the Cambrian explosion. Some argue that a long hidden history of animal evolution dates well into the Precambrian, with glimpses of this diversity provided by the Doushantuo phosphatized embryos, but that Precambrian conditions were not conducive to preservation of macroscopic animals. Others have argued that the Cambrian radiation represents a true explosion of animal diversification. The timing of this explosion could be due to changes in the abiotic environment, including fragmentation of a supercontinent that provided more opportunities for geographic isolation and speciation. Studies of trilobites show that at the time of their first appearance in the fossil record 525 million years ago, they already showed significant biogeographic differentiation; their origins must therefore have occurred well before the Cambrian (Lieberman 2003). In addition, a warming climate with reduced ice cover and increasing levels of atmospheric oxygen provided conditions more conducive to the survival of macroscopic animals. Changes in ocean chemistry could have facilitated biomineralization used to build skeletons as protection from predators. The evolution of skeletal armor and hardened feeding structures (e.g., teeth) may have triggered an arms race that further increased diversification. Another explanation is that the genetic toolkit present in single-celled ancestors happened to have features that enabled rapid diversification of metazoan body plans. No single hypothesis can adequately explain this significant event in the history of life. Evidence from fossils, phylogenies, biogeography, paleoecology, genomics, and developmental biology, when interpreted together, provide the best explanation for the Cambrian explosion.

Table 1. Earliest animal fossil representatives of some major phyla

Phylum	*Name*	*Date in millions of years (fossil formation)*
Porifera (sponges)	*Palaeophragmodictya*	560 (Ediacara)
Ctenophora (comb jellies)	*Maotianoascus*	540 (Meishucun)
Arthropoda	*Anomalocaris*	530 (Burgess Shale)
Chordate	*Yunnanozoon, Haikouichthys*	525 (Chengjiang)
Brachiopoda	Many	525 (Chengjiang)
Urochordate	*Shankouclava*	525 (Chengjiang
Mollusca	*Fordilla*	514 (Greenwich)
Cnidaria (jellyfish)	Unnamed	500 (Marjum)

2. ANIMAL PHYLOGENY

Reconstructing the animal tree of life remains one of the grand challenges in evolutionary biology and is essential to understanding key transitions in animal evolution. The challenges arise in part because of the rapid divergence and ancient origin of the major animal lineages, which date back more than 500 million years, as discussed above. Most of our current understanding of

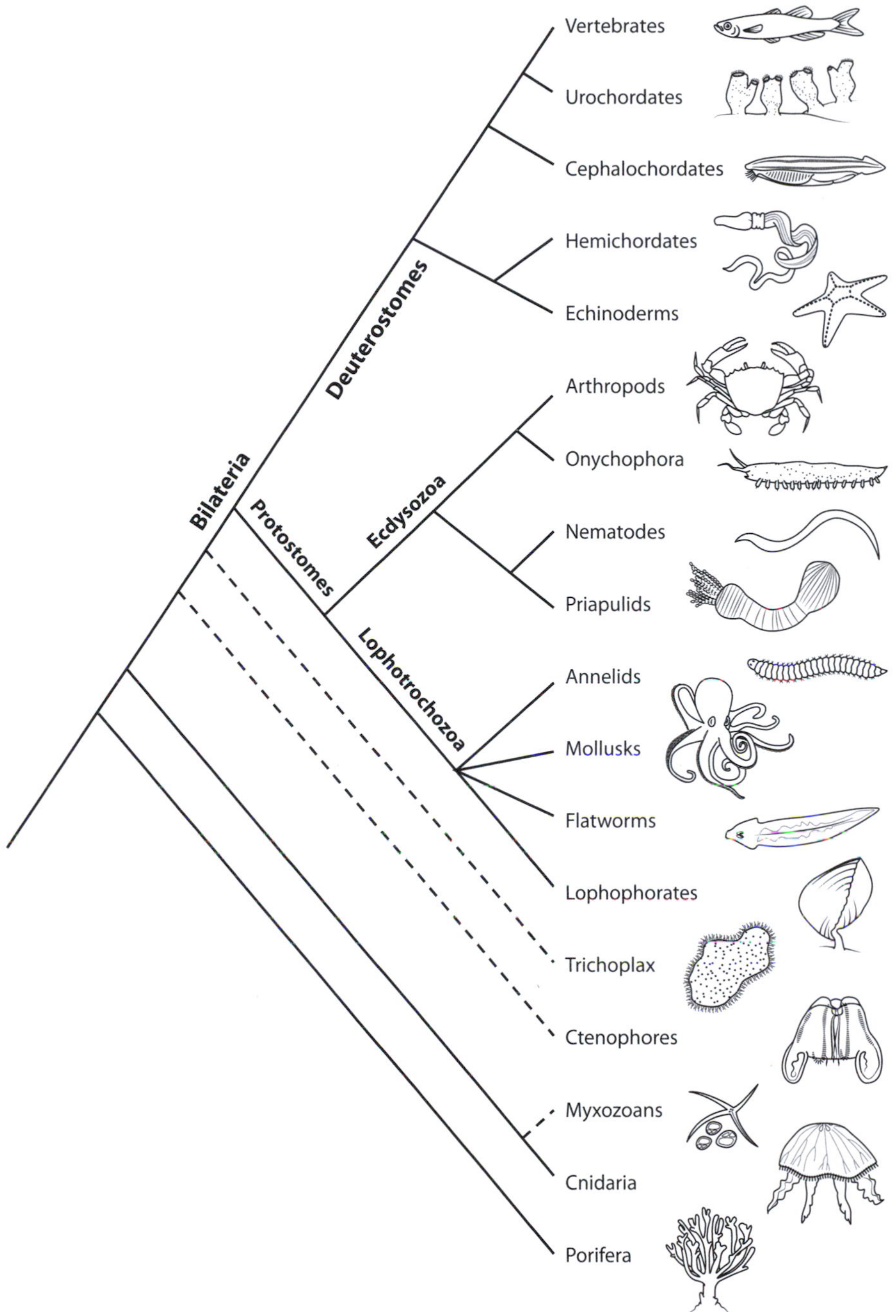

Figure 1. Evolutionary relationships of major animal phyla.

animal relationships comes from recent studies using large data sets of DNA sequences. Figure 1 summarizes our current understanding of the evolutionary relationships of most major animal phyla.

The closest living relatives to animals are the cho anoflagellates. Choanoflagellates exist as single cells or small colonies. The name *choanoflagellate* comes from a distinct collar structure, made up of closely packed microvilli, that surrounds their flagellum. The phylum Porifera (sponges), which is outside the major clade that includes all other known lineages of animals, also possess cells, called *choanocytes*, with flagellated collar cells. Sponges display a diversity of forms in their adult stage, but have long been thought to be united by the presence of choanocytes and a distinctive body plan: benthic filter feeders with chambers that circulate water and filter food; however, recent molecular phylogenetic studies suggest that Porifera may not possess a single common ancestor and may instead comprise two to four separate lineages (Calcarea, Demospongiae, Hexactinellida, and Homoscleromorpha). Separate sponge lineages imply that a sessile, filter-feeding existence is not a unique feature uniting sponges but was, instead, present in the last common ancestor of all animals.

The cnidarians, which include corals, sea anemones, and jellyfish, diverged from the main animal lineage following the divergence of sponges. All cnidarians possess stinging cells that house complex intracellular, venom-containing structures called *nematocysts*. Cnidarians are radially or biradially symmetrical, comprise just a few cell types, are diploblastic, meaning they lack organs and consist of an outer ectodermal cell layer and an inner gastrodermal cell layer, but lack a third layer, the mesoderm. Despite this simple construction, cnidarians comprise a huge diversity of forms and habitats, including, for example, the simple freshwater *Hydra* polyp, intertidal anemones, expansive coral reefs, benthic colonial hydrozoans, and pelagic jellyfish.

Ctenophores are commonly referred to as comb jellies because they have ciliated rows, called *combs*, arranged longitudinally along their bodies that are used for locomotion. Ctenophores superficially resemble jellyfish, and traditionally ctenophores and cnidarians were classified in a group called Coelenterata; however, morphological and molecular data do not support this hypothesis, although the exact phylogenetic placement of ctenophores remains controversial (see below).

The remaining animal lineages belong to the clade Bilateria. Bilaterian animals, as the name implies, display *bilateral symmetry*, in conjunction with *triploblasty*: possession of a third, middle cell layer, the *mesoderm*. Given their bisymmetrical organization, early wormlike bilaterians could move in a forward motion with their sensory and feeding structures located at the anterior ends. With the acquisition of a mesoderm, they could also form organs and organ systems. The simplest bilaterian organization is exemplified by the small wormlike animals in Acoelomorpha (acoels and nematodermatids). These small worms lack a through-gut and instead have one opening for eating and excreting, similar to cnidarians. They also lack a body cavity called a *coelom* that is found in almost all other bilaterians. Although it has long been thought that Acoelomorpha are an early diverging relative to the main group of bilaterians, their phylogenetic placement has recently come into question (see below).

The majority of bilaterian animals are divided into two major groups, Protostomia and Deuterostomia. These clades are named based on the fate of the *blastopore* (an opening in the early embryo); in protostomes, the blastopore generally develops into a mouth, whereas in deuterostomes, it forms an anus; however, in many animals the blastopore is an ephemeral feature and the opening to the gut develops anew. Thus, although Protostomia and Deuterostomia are indeed separate lineages, as evidenced by molecular data, the actual names should not be overinterpreted.

Recent work has revealed that Protostomia comprises two main clades, the Lophotrochozoa and the Ecdysozoa (Halanych 2004). Members of Lophotrochozoa include annelids, mollusks, bryozoans, phoronids, and brachiopods and are characterized by either a trochophore larva (annelids and mollusks) or a lophophore feeding structure in the adult (bryozoans, phoronids, and brachiopods). *Spiral cleavage*, a particular pattern of cell division in early embryogenesis, was likely ancestral for Lophotrochozoa but subsequently modified in the lophophore-bearing taxa. Relationships between major lophotrochozoan lineages remain uncertain.

Ecdysozoa are a group of animals united by having a cuticular exoskeleton that is shed through molting. Ecdysozoa includes arthropods, onychophorans, nematodes, nematomorphs, kinorhynchs, priapulids, and tardigrades. Relationships between major ecdysozoan lineages, especially with respect to the placement of tardigrades, remain uncertain.

Deuterostomia is a large group of animals supported by numerous molecular phylogenetic analyses. Morphological characters that unite deuterostomes are the development of the coelom through pinching off of the gut (*enterocoely*), and a coelom divided into three separate sections (*tripartite*). Deuterostomes comprise two main clades, Ambulacraria, which includes xenoturbellids, echinoderms, and hemichordates, and Chordata, which includes tunicates, cephalochordates, and craniates (vertebrates and their relatives). Molecular phylogenetic studies have revealed the inclusion of the small worm *Xenoturbella* in Deuterostomia, the placement of hemichordates as the closest relative to echinoderms, and the possible placement of tunicates, and not cephalochordates, as the closest relatives to vertebrates. These findings have required the reevaluation of the evolution of many deuterostome-specific characters. For example, xenoturbellids lack all of the traditional, diagnostic deuterostome traits, in addition to lacking a through-gut and a coelom, suggesting extensive loss of characters in this lineage. Likewise, hemichordates and chordates have gill slits, which implies that gill slits might be ancestral for deuterostomes, with subsequent losses in echinoderms and xenoturbellids. The placement of tunicates relative to craniates is controversial. Classically, tunicates have been considered the closest relative to craniates + cephalochordates, and this hypothesis has been supported by some molecular phylogenetic studies with ribosomal RNA genes (rDNA). Other molecular phylogenetic studies combining morphology with rDNA genes, or employing multiple protein markers, place tunicates as the closest relative to craniates, with cephalochordates falling outside of tunicates + craniates. This latter hypothesis is supported by the presence of migratory neural crest cells found in vertebrates and also reported present in tunicates but not cephalochordates. This resolution also implies that metameric segmentation, classically used to unite craniates + cephalochordates, is ancestral for Chordata and was secondarily lost in tunicates.

3. MULTICELLULARITY AND THE ORIGIN OF SPONGES (PHYLUM PORIFERA)

Animals evolved from a single-celled eukaryotic ancestor; the evolution of multicellularity is thus a key event in the evolution of animals. Choanoflagellates represent the closest living relative to animals. They exist as free-living single cells or as small colonies. As noted above, choanoflagellates display a distinctive morphology that includes a collar-shaped structure surrounding their flagellum. Sponges possess a cell type, the *choanocyte*, that is similar to an individual choanoflagellate. In sponges, the choanocytes line interior cavities and function to circulate water and filter food particles. Given that sponges represent either a grade (i.e., distinct lineages united by ancestral characters) or a clade sister to all other animals, and choanoflagellates are the closest relatives to animals, it is likely that the ancestor of animals resembled a modern-day choanoflagellate. Insights into the transition from a single-celled organism to a complex multicellular animal can therefore be gleaned from the study of choanoflagellates.

Two primary functions are required for multicellularity in animals: the ability of cells to adhere to one another, and the ability of cells to communicate to one another. These functions are necessary precursors for cellular specialization and coordination required for proper animal development and function. Recent research has shown that single-celled choanoflagellates possess a large array of genes that in animals function in cell signaling and cell adhesion, including C-type lectins, cadherins, components of the extracellular matrix, and participants in protein kinase signaling pathways (King 2004). Given the surprising complexity of the choanoflagellate genome, it is certain that many of the molecules necessary for multicellularity were already present in the animal ancestor. Although the role of these genes in choanoflagellates has not been sorted out, it is thought they may function in adhering cells to surfaces, catching prey, mating, and responding to environmental cues. Later in evolution, these same gene families were likely co-opted for cell-to-cell communication in multicellular animals.

Some choanoflagellates form small colonies, providing a glimpse of how the first multicellular animal may have been organized. Colonies of the choanoflagellate *Proterospongia*, for example, show signs of limited functional specialization and cellular differentiation. *Proterospongia* has two types of cells; the outer flagellated collar cells propel the colony through the water, whereas the inner amoeboid cells divide to enlarge the colony. The cell signaling pathways likely function in the coordination among cell types within these small colonies.

Further cellular specialization and coordination between cells is evident in sponges. As noted above, sponges likely do not form a single lineage but instead represent a grade. They are multicellular, but lack organized tissues, muscles, nerves, or a gut. The adult sponge is organized around chambers that circulate water and filter food, which is ingested by specialized amoeboid cells. Although adult sponges are markedly different from all other animals, sponge embryos possess some of the hallmarks of animal development. Sponges undergo a process akin to *gastrulation* (reorganization of cells into layers), although sponges do not form true epithelia as found in other animals. This is followed by a ciliated larval stage. Gastrulation, which is lacking on choanoflagellates, is critical for setting up the adult body plan in animals, in that it provides spatial organization for the differentiation of specific cell types. It certainly evolved in the ancestor of all animals and marks the beginning of the evolution of animal body plans.

4. THE ORIGIN OF THE NERVOUS SYSTEM AND THE EVOLUTION OF SENSORY STRUCTURES IN CNIDARIA

Sponges lack nerve cells; however, they do possess the ability to sense and respond to their environment. Recent whole-genome sequencing of the sponge *Amphimedon queenslandia* revealed that sponges have genes that code for components of the nervous system, even though they lack nerve cells. Although their functions are unknown, these protoneuronal components may have served as the building blocks for the first nerve cell to evolve in animals.

Cnidarians display clear signs of organized tissues, including muscle cells and nerve cells. Comparative genome studies reveal that cnidarians possess a large array of genes involved in neural development and signaling that is nearly as complex as seen in bilaterians, despite the lack of a centralized nervous system or brain in Cnidaria. Instead, cnidarians possess a diffuse nerve net, dispersed among epithelial cells. Cnidarians also possess a type of neural cell not found in other phyla: stinging cells called nematocysts, which possess mechanosensory capabilities. This type of decentralized nervous system likely represents an ancestral condition for animals. Even among cnidarians, there are various degrees of complexity and specialization of the nervous system. Although they lack a central nervous system, many cnidarians have areas where nerve cells form a plexus or longitudinal track.

The nerve cells of cnidarians express various types of neural peptides that often show spatially structured expression patterns. For example, RFamide-positive neurons are expressed around the mouth in hydrozoan polyps and near epithelial muscles in hydromedusae. The pelagic colonial hydrozoan *Aglantha digitale* has elaborate nerve-ring systems that control complex behavior among the different polyps and medusae in the colony, including directional swimming and food capture. These nerve cell complexes and nerve rings are signs of the centralization and coordination of a nervous system as needed for complex behaviors.

Cnidarians possess the ability to sense and respond to light, chemicals, and touch. These abilities are derived from nerve cells, which can communicate with nonneural cells to elicit behavioral responses through signaling mechanisms. For example, cnidarians possess seven-pass transmembrane G protein–coupled receptors (GPCRs) that can respond to mechanical or chemical stimuli through ion channel signaling. The most developed sense organs occur in medusozoan cnidarians, which include Scyphozoa, the true jellyfish, Cubozoa, the box jellyfish, and Hydrozoa, which includes colonial hydroids and hydromedusae. Scyphozoans and cubozoans organize their sense organs in structures called *rhopalia*, located on the bell of the medusa. Rhopalia house statocysts that serve as balance organs, and eyes for responding to visual cues. The cubozoans possess complex eyes that include lenses, retinas, and corneas. Cubozoans express the developmental regulatory gene *PaxB* in their developing rhopalia (Kozmik et al. 2003). This gene is a homolog to *Pax* genes in bilateria that are involved in eye and ear development. Likewise, a number of regulatory genes involved in sensory structures in bilaterians are also expressed in the sensory structures of the jellyfish *Aurelia*. The genomic and developmental evidence suggest that the molecular and cellular precursors to complex centralized nervous systems and sense organs were present in the common ancestor of cnidarians and Bilateria (with some present before the divergence of sponges) and were modified and elaborated in individual animal lineages.

5. THE ORIGINS OF BILATERIA AND THE PHYLOGENETIC PLACEMENT OF CTENOPHORA, ACOELA, MYXOZOA, AND PLACOZOA

The phylogenetic placement of four enigmatic animal groups—ctenophores, acoels, placozoans (*Trichoplax*), and myxozoans—has been controversial. Understanding their position in the animal tree of life is critical for elucidating the evolutionary transitions that occurred between major animal lineages, as many of these taxa display traits that might be seen as intermediate forms. Key evolutionary innovations such as the origin of mesoderm, a through-gut, bilateral symmetry, and a centralized nervous system all occurred sometime between the divergence of cnidarians and the last common ancestor of Bilateria. Many classic and recent molecular studies have attached the four enigmatic animal groups to the stem lineage of Bilateria; however, other studies have placed ctenophores and placozoans at the base of Metazoa (diverging before Cnidaria), myxozoans within Cnidaria, and acoels as derived deuterostomes. Given the important implications of these phylogenetic hypotheses, they are worth exploring in more detail.

Recent work has provided evidence that ctenophores have some degree of bilateral symmetry, a developed nervous system, muscles, and evidence of mesodermal tissue, suggesting they are closely allied to bilaterians; however a phylogenetic analysis using a large molecular data set positioned ctenophores as sister to all other animals, suggesting that the group's bilaterian features evolved independently in the ctenophore lineage (Dunn et al. 2008). This study, however, has been questioned, and subsequent studies have refuted this claim. Resolution of the phylogenetic placement of Ctenophora has important implications for our understanding of the evolution of bilaterian symmetry and various organ systems.

Acoels are small wormlike animals that, like cnidarians, lack a coelom and a through-gut; however, unlike the diploblastic cnidarians, they are triploblasts, possessing a mesoderm. Hence it appears, at least superficially, that acoels represent a transitional form between diploblasts like cnidarians and bilaterians. Although originally placed in the phylum Platyhelminthes, which is now known to be part of the lophotrochozoan clade in Protostomia, numerous molecular phylogenetic studies have supported the placement of acoels as sister to other Bilateria, consistent with morphological evidence that acoels possess some but not all bilaterian features. A more recent but still controversial study has suggested that acoels are not early diverging bilaterians but are derived deuterostomes (Philippe et al. 2011); specifically, this study placed acoels as the closest relatives of xenoturbellids (in a clade with echinoderms). This phylogenetic placement, if corroborated, would suggest that acoels are not an example of a transition to bilaterians but instead resulted from secondary losses of derived bilaterian structures. Clarification of the position of acoels in the animal tree of life awaits further study.

Myxozoans are a diverse group of minute freshwater and marine organisms that alternate their parasite life cycle between invertebrate and fish hosts. Historically, myxozoans were thought to be protists; however, a number of characteristics led many to consider the possibility that they might be animals, including a multicellular stage in their life cycle and the possession of intracellular polar capsules used to attach to a host. These polar capsules bear remarkable similarity to nematocysts in the stinging cells of cnidarians, leading to the suggestion that myxozoans could be animals related to cnidarians. Molecular phylogenetic studies using nuclear ribosomal genes have placed myxozoans within bilateria, but outside its major subclades (Protostomia and Deuterostomia); however, other studies, combining nuclear ribosomal genes with morphology or using protein-coding genes, place myxozoans with cnidarians. The failure to come to a consensus in the placement of myxozoans with molecular data is likely due to their highly aberrant DNA sequences (Evans et al. 2010); however, the remarkable similarities between polar capsules and nematocysts of cnidarians, along with some molecular support, suggest that myxozoans are probably cnidarians and that the polar capsules and nematocysts represent a single evolutionary origin.

Placozoans are microscopic animals that lack a mouth, gut, organs, nervous system, bilateral symmetry, or even an anterior-posterior axis. The only described species in this phylum is *Trichoplax adhaerens*. These organisms have just a dorsal and ventral cell layer; they digest food particles intracellularly with their ventral cells. The phylogenetic position of placozoans is controversial. Their simple construction suggests they could be the sister group to all other animals. This hypothesis is supported by the analysis of the structure of the *Trichoplax* mitochondrial genome, which is more protist-like than animal-like (Dellaporta et al. 2006); however, the presence of cell junctions, which are lacking in sponges, suggests that placozoans are more derived, and analyses of rDNA support a divergence from other animals after sponges but before cnidarians. This conclusion has been confirmed by analysis of multiple genes, although other hypotheses of relationships have also been reported. Although it is tempting to view the simplicity of the *Trichoplax* as an ancestral condition to all other animals, this conclusion should perhaps be viewed as tentative, pending further studies of the animal tree of life.

6. ANIMAL DIVERSITY

The ecological, molecular, and evolutionary processes that occurred at the dawn of animal evolution were fundamental for setting the stage for the diversity of animal life that we see on earth today. All major animal lineages evolved in the oceans, but subsequent diversification has resulted in numerous groups that have invaded freshwater. In addition, after land was colonized by plants (see chapter II.14), multiple animal lineages invaded land, most famously the tetrapods (Vertebrata) and insects (Arthopoda), but also including arachnids (Arthropoda), isopods (Arthropoda), gastropod snails (Mollusca), and earthworms (Annelida). In addition to these major ecological transitions, animal diversification on both land and water has been accompanied by shifts into different ecological niches. In these transitions, animals had to overcome major challenges associated with respiration, reproduction, and osmoregulation. Animals have evolved parasitic (e.g., tapeworms), mutualistic (e.g., anemones and anemone fish), and highly cooperative (e.g., bee colonies) lifestyles. Through the course of evolution, animals have generally increased in overall complexity, although animals have also reevolved simpler body plans. The most extreme example of the latter, the reevolution of a single-celled organism, is found in the canine transmissible venereal tumor (CTVT), whose ancestor was a dog. The past 500 million years have been witness to evolutionary processes that have resulted in the amazing diversity of animals we observe on earth today, which display an unimaginable diversity of forms, ecological habitats, and lifestyles.

FURTHER READING

Cartwright, P., S. L. Halgedahl, J. R. Hendricks, R. D. Jarrard, A. C. Marques, A. G. Collins, and B. S. Lieberman. 2007. Exceptionally preserved jellyfishes from the Middle Cambrian. PLoS ONE 2(10): e1121, 1–9. *This study documents the first exquisitely preserved true jellyfish from the Cambrian.*

Dellaporta, S. L., A. Xu, S. Sagasser, W. Jakob, M. A. Moreno, L. W. Buss, and B. Schierwater. 2006. Mitochondrial genome of *Trichoplax adhaerens* supports Placozoa as the basal lower metazoan phylum. Proceedings of the National Academy of Sciences, USA 103(23): 8751–8756. *Provides molecular evidence that the enigmatic invertebrate* Trichoplax *was an early diverging animal.*

Dunn, C. W., A. Hejnol, D. Q. Matus, K. Pang, W. E. Browne, S. A. Smith, E. Seaver, et al. 2008. Broad phylogenomic sampling improves resolution of the animal tree of life. Nature 452: 745–749. *Provides a hypothesis about the relationship of major animal phyla using sequences from multiple protein-coding genes.*

Evans, N. M., M. T. Holder, M. S. Barbeitos, B. Okamura, and P. Cartwright. 2010. The phylogenetic position of Myxozoa: Exploring conflicting signals in phylogenomic and ribosomal datasets. Molecular Biology Evolution 50(3): 456–472. *Discusses the controversies surrounding the phylogenetic placement of Myxozoa.*

Gould, S. J. 1989. Wonderful Life: The Burgess Shale and the Nature of History. New York: W. W. Norton. *The book that introduced the Burgess Shale fossil fauna to a broader audience.*

Halanych, K. M. 2004. A new view of animal phylogeny. Annual Review of Ecology and Systematics 35: 229–256. *A comprehensive review of major issues surrounding animal phylogenetics.*

King, N. 2004. The unicellular ancestry of animal development. Developmental Cell 7: 313–325. *Provides evidence that choanoflagellates share a similar genetic toolkit with animals.*

Kozmik, Z., M. Daube, E. Frei, B. Norman, L. Kos, L. J. Dishaw, M. Noll, and J. Piatigorsky. 2003. Role of *Pax* genes in eye evolution: A cnidarian *PaxB* gene uniting *Pax2* and *Pax6* functions. Developmental Cell 5: 773–785. *Provides molecular evidence uniting light-sensing organs in all animals.*

Lieberman, B. S. 2003. Taking the pulse of the Cambrian radiation. Journal of Integrative and Comparative Biology 43: 229–237. *Documents the role played by geological change in the Cambrian radiation.*

Minelli, A. 2009. Perspectives in Animal Phylogeny and Evolution. New York: Oxford University Press. *Provides a comprehensive overview of major features considered important in animal evolution.*

Philippe, H., H. Brinkmann, R. R. Copley, L. L. Moroz, H. Nakano, A. J. Poustka, A. Wallberg, K. J. Peterson, and M. J. Telford. 2011. Acoelomorph flatworms are deuterostomes related to *Xenoturbella*. Nature 470: 255–258. *Provides molecular evidence that acoelomophs are deuterstomes.*

II.16

Major Events in the Evolution of Arthropods

Brian M. Wiegmann and Michelle D. Trautwein

OUTLINE

1. Arthropod origins
2. Phylogenetic framework
3. Colonization of land
4. Evolution of flight
5. Complete metamorphosis
6. Life history specializations

The animal phylum Arthropoda is by almost any measure the most successful clade of metazoan life. Its members occupy and often dominate all ecosystems on earth and have had a long and rich history of diversification beginning in or just prior to the Cambrian (546 million years ago, Ma). They can boast more species than any other animal group and include familiar and abundant forms—spiders, crustaceans, centipedes and millipedes, and insects, along with numerous unique, less well-known, or extinct types—trilobites, sea spiders, horseshoe crabs. Adaptations for exploiting nearly any food source make them a major component of the living world. The enormous abundance of arthropods in oceans (e.g., krill), tropical rain forests (beetles, ants, termites), arctic and alpine zones (flies), and deserts (flies, bees, wasps) make them critically important to sustaining these environments. They are pollinators, parasites, decomposers, predators, and disease vectors. The evolutionary success of arthropods has often been attributed to aspects of their segmented and compartmentalized body plan, resilient and replaceable exoskeleton, developmentally modifiable segmented appendages, and to the opportunities that these adaptations have afforded them in occupying novel environments like dry land, fresh water, and the air.

GLOSSARY

Metamorphosis. The episodic growth process common in many arthropod groups involving major changes in body form and life history through discrete life stages.

Sclerotization. The chemical hardening or "tanning" of the arthropod cuticle that allows for strength in a protective covering as well as flexibility and growth during the molting process. Arthropod cuticles contain *chitin*, a complex polysaccharide, embedded in a protein matrix (sclerotin).

Tagma. The specialized body regions of arthropods; these regions are variously modified within the major groups, for example, head-thorax-abdomen in insects versus cephalothorax-opisthosoma in spiders. The differential development of tagma (tagmosis) of arthropods is regulated by the *Hox* family of regulatory genes.

1. ARTHROPOD ORIGINS

Arthropods originated in marine environments in the Cambrian (or perhaps even earlier, in the Precambrian), an era considered remarkable for the diversity of body plans and anatomical adaptations that first appear in the fossil record. From about 550 to 520 Ma, the earliest fossils of all the major "types" (phyla) of complex multicellular animals emerged. The rapid origin of new body plans and the diversification of many marine animals at this time is often called the Cambrian explosion. This remarkable moment in earth history may have been the result of a substantial increase in dissolved oxygen in the oceans, together with significant melting of glaciers. These conditions would have provided abundant food and new ecological niches for increased survival, as well as offering a chemical environment more conducive to the production of rigid, and readily fossilizable, protective structures, skeletons, and hardened teeth (see chapter II.10) . During this time it seems that animals experienced dramatic alterations to the gene regulatory networks controlling the blueprint for the development of animal body plans.

All arthropods share certain unique features: segmented bodies, jointed legs that are often modified to perform various tasks, and a sclerortized exoskeleton made of chitinous cuticle that is periodically molted for growth. The Cambrian fossil record, especially from China (Chengjiang, 520 Ma) and Canada (Burgess Shale, 505 Ma) reveal that these key arthropod features were not gained all at once but were successively acquired in an increasingly specialized modern arthropod form. Impressive fossils have been brought to light that provide a view of how anatomical diversity is assembled over time, and how unique features, or combinations of characteristics unknown in extant groups, evolved in animal groups that once flourished in Cambrian seas. Some of these fossils cannot be easily assigned to extant groups of arthropods and are considered "stem-group" taxa, or unique, early diverging lineages that left no modern descendants. Examples of these "protoarthropods" include *Opabinia* from the Burgess Shale, which had a long, single anterior appendage or proboscis presumed for the purpose of feeding on worms in tubes, or the large, segmented predator anamolocaridids like *Kerygmachela* (510 Ma, Greenland) or *Anomalocaris* (505 Ma, Burgess Shale, Canada; Chengjiang, China) that share certain features with modern arthropods, but also lack key features like the fully sclerotized cuticle found in all living groups. The astounding morphological diversity found in the Cambrian fossil record continues to provide new insights and significant reinterpretation of the origin and evolution of taxa, appendages, body plans, and ecological habits within the arthropod radiation.

Despite their early morphological variation, arthropods are now well known as animals with compartmentalized body parts, or *tagma*, with specialized functions, for example, the insect head, thorax, and abdomen. The genes controlling the specification of these body regions in early embryonic development can be expressed at different times or in slightly different areas to direct the organization of the alternative combinations of tagma that differentiate spiders from millipedes and crabs from insects.

2. PHYLOGENETIC FRAMEWORK

Our understanding of the evolutionary relationships of major arthropod lineages has undergone dramatic change in recent years, resulting from a growing contribution of new, large-scale molecular data to traditional, morphological evidence. Though controversies remain, there is a growing consensus on the history of arthropod relationships that is quite altered from previous long-standing views (figure 1). Arthropods, with their successively repeating segmented body plan, were long considered close relatives of annelids (ringed worms, such as earthworms) that also have a segmented body structure. Now, however, it is widely accepted that arthropods are more closely related to nematodes (roundworms, such as pinworms and hookworms) as part of Ecdysozoa—a group defined by molting or ecdysis (see chapter II.15).

The closest relatives of arthropods are two small, yet compelling phyla, Tardigrada (waterbears) and Onychophora (velvet worms), both known from Cambrian fossils, that together with arthropods form the clade Panarthropoda. Within Panarthropoda, there is genomic evidence (e.g., from micro RNAs) that velvet worms are the sister group to Arthropoda. A contrary view held by some is that tardigrades are not part of Panarthropoda, but instead are the closest relatives to nematodes. Tardigrades are microscopic (up to 1 mm) animals known for their ability to survive extreme conditions (desiccation, temperature extremes, vacuum of space). Onychophorans, on the other hand, are soft-bodied caterpillar-like creatures that tend to live in moist forest environments and emit slime to capture prey. Understanding the exact evolutionary arrangement of these closely related phyla is key to understanding the ancestral or original features of the very first, ancient arthropods.

Arthropods have all descended from a single common ancestor that probably lived in the early Cambrian (or very late Precambrian). For several decades, prominent hypotheses suggested that arthropods might be a polyphyletic lineage, yet there is now overwhelming molecular and morphological evidence of their monophyly. Within Arthropoda (also called Euarthropoda), there are three main extant lineages: Chelicerata (sea spiders, horseshoe crabs, and arachnids), Myriapoda (millipedes and centipedes), and the most species rich by far—Pancrustacea (crustaceans and hexapods, the latter including insects). Trilobites, marine organisms with an extensive fossil record that lived from the Cambrian to the end of the Permian, are the most prominent extinct arthropod lineage.

Some of the earliest-known arthropod fossils are considered ancestors of chelicerates from close to the end of the Cambrian (>500 Ma). Chelicerata, named for their chelicerae, appendages near the mouth that function, for example, as envenomating fangs in spiders (Araneae), is the sister group to the rest of Arthropoda. Their monophyly is well supported by various types of evidence, though it is still debated whether sea spiders

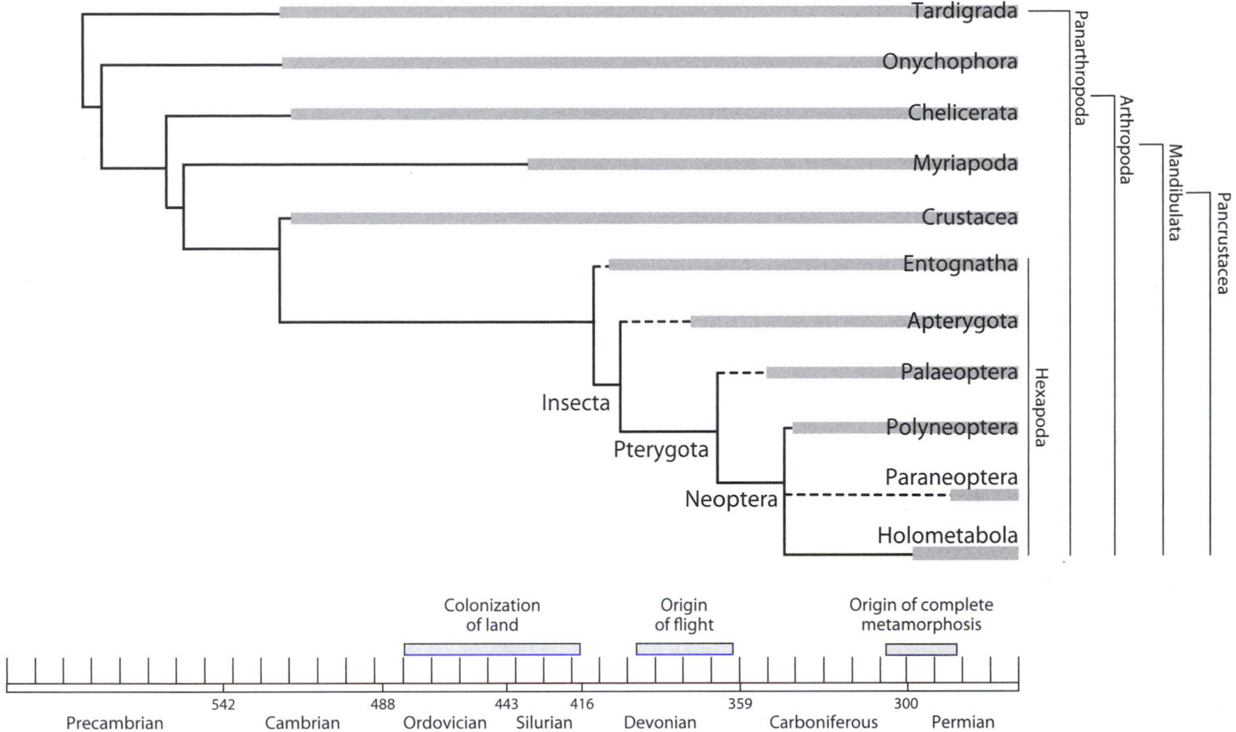

Figure 1. The timing and pattern of arthropod evolution shown here is a reflection of the emerging consensus by both morphological and molecular analyses. Gray bars represent the earliest fossils known for each group. Range bars indicate the hypothesized time intervals in which major milestones in arthropod evolution may have taken place.

(Pycnogonida) are a lineage that diverged earlier than chelicerates and is sister to all other arthropods. Also, the resolution of the relationships among arachnids remains a major challenge, including the controversies regarding the monophyly of the mites (Acari) and the placement of some major arachnid lineages such as scorpions, harvestmen, and pseudoscorpions (among others). Though extant chelicerates comprise primarily terrestrial arachnids today (spiders, with 35,000 known species and mites with ca. 30,000), they originated in marine Cambrian environments. Sea scorpions (Eurypterida) are early chelicerates (now extinct) well known from the fossil record. They were predatory ocean dwellers and one of several major lineages of extremely large-bodied arthropods (up to several meters in size) that disappeared by the end of the Permian. Ancient chelicerates that are still represented by modern groups are marine sea spiders (Pycnogonida), a stilt-legged group from the Cambrian, and the marine horseshoe crab, a creature often referred to as a "living fossil" because of its relatively unchanged appearance compared with Silurian fossils (445 Ma).

Among chelicerate groups, spiders are one of the most ubiquitous and cosmopolitan today. These predators are recognized by their four pairs of legs and two body regions (with the head and thorax fused into a cephalothorax or prosoma), yet the feature that has most likely defined their success is the less visible pair of silk-generating spinnerets. Spiders appeared in the Devonian. The earliest lineages are not thought to have been web makers but instead lived in silk-lined burrows. Most of living spider diversity is found in the Araneomorphae, including commonly known groups such as orb weavers, running spiders (wolf spiders), and jumping spiders, a lineage that first appeared in the Jurassic.

Mites and ticks are as diverse as spiders, yet because of their often-microscopic size, they are much less well known. They also can be found in a greater diversity of habitats (thousands of species are aquatic) and exhibit a range of feeding habits—with many mites acting as ectoparasites and endoparasites of animals (including mollusks, other arthropods, and every class of vertebrate). Coevolution between mites and their specific hosts has been established in groups including birds, marsupials, bees, and many others. Mites are known from Devonian fossils, with ticks appearing in the Cretaceous.

Myriapoda, the many-legged millipedes (generally herbivorous, with two pairs of legs per segment) and centipedes (often carnivorous, with one pair of legs per segment), is a smaller, terrestrial arthropod group whose placement has been central to several controversial hypotheses of arthropod relationships. Unlike chelicerates, myriapods are unknown from Cambrian fossils, and make their first appearance as millipede fossils from the Silurian. Today, Myriapoda includes about 13,000 species; most of these are millipedes (8000 species), followed by centipedes (3000 species) and two common yet less familiar and less speciose groups, the minute soil-dwelling lineages Symphyla (200 species) and Pauropoda (700 species). Arthropleurids, a notable, extinct lineage of early millipedes, are another example of gigantism in arthropod history. They grew to more than 2 m in length yet went extinct in the Permian along with many other groups.

Opinion regarding the placement of Myriapoda within arthropods varies, with even the monophyly of the group occasionally questioned. Traditional morphological hypotheses have long considered Myriapoda the closest relatives to insects. Both groups are primarily terrestrial and share morphological features, such as a tracheal system for respiration on land, and malphigian tubules—organs used like kidneys for excretion of nitrogenous waste. Yet now that molecular evidence has convincingly demonstrated that the insects are actually a crustacean lineage, scientists are reinterpreting these traits as independently evolved, but similarly modified, adaptations for living on land. The consensus now is that myriapods are most likely the closest relatives to Pancrustacea, and together the two form Mandibulata, a group defined by the possession of chewing mandibles, or jaws. Large-scale molecular studies do conflict on Mandibulata, with myriapods sometimes placed as the closest relatives to the chelicerates in a group lightly known as Paradoxipoda. Yet Mandibulata, the more traditional organization, now has extensive morphological and molecular support.

The very earliest fossils that can be identified as extant arthropods are crustaceans from the Cambrian. The traditional definition of Crustacea is a diverse and species-rich group (37, 000 species) that has long been suspected to be paraphyletic (see chapter II.1) because few morphological characters are shared by all its members. Molecular studies have now shown that, indeed, crustaceans are not all descended from a recent common ancestor and form a monophyly only when grouped together with Hexapoda (insects); thus insects themselves, the most species rich of all Arthropoda, are actually a crustacean lineage and together with crustaceans are known as Pancrustacea. Though several crustacean groups have adopted a terrestrial lifestyle (woodlice, land crabs), crustaceans are almost entirely aquatic and particularly abundant in marine environments. It appears that insects, which dominate both land and air but do not inhabit the oceans, emerged as the primary land-invading branch of the largely marine pancrustacean clade. Crustaceans play a major role in marine ecosystems analogous to the role played by insects in terrestrial ecosystems. Perhaps most significantly, tiny crustaceans and larvae are a primary component of plankton, a foundation of the food web, including an Antarctic species of krill considered to have the highest biomass on the planet

Within Crustacea, classes and the relationships among these higher-level divisions are not well established. Particularly, the specific lineages that are the closest relatives to hexapods are still uncertain. Alternate hypotheses consider the hexapod sister group to be either the Branchiopoda (fairy shrimp and water fleas), the species-rich Malacostraca (crabs, shrimp, krill) or a clade consisting of two very small marine crustacean classes, Remipedia and Cephalocarida. Molecular data and morphological features of the brain and the eye all return conflicting results regarding these relationships. Other major crustacean lineages are Ostracoda and Maxillopoda (barnacles).

Among crustaceans, a curious phylogenetic enigma that was recently resolved relates to a group known as tongue worms, or Pentastomida. These elongate creatures are vertebrate parasites, found primarily in the respiratory tract of reptiles and other intermediate hosts, as well as humans (although with limited success). As internal parasites, their morphology is very simplified; as a result, their placement in the tree of life was almost entirely unknown. Different hypotheses placed them within invertebrate groups as diverse as Nematoda, Annelida, Tardigrada, and Onychophora. Recently, new molecular data have firmly placed these unusual parasites as an early diverging crustacean lineage.

Hexapods (including insects), defined by six legs and three body regions, are the most successful arthropod lineage in terms of species richness, diversity, and biomass (>1 million species). Though hexapod monophyly has been hotly debated, recent molecular evidence firmly places the three relatively small groups of noninsect hexapods (Collembola, Protura, and Diplura) as the closest relatives to the remarkably large clade, Insecta. These common, yet unfamiliar hexapods are minute wingless soil-dwelling animals with a fossil record that extends back to the Devonian (>400 Ma).Within Insecta, there are two primitively wingless ground-dwelling groups, the bristletails and the silverfish (sometimes referred to as Apterygota), with the latter being the closest relatives to all winged insects (Pterygota). There is much debate over the relationships of the earliest winged insects. The wings of dragonflies and mayflies are unusual in that they cannot be folded flat across the back and have thus been dubbed "old wings" or Paleoptera, in contrast to the rest of insects grouped in Neoptera (or new wings). Yet exactly how dragonflies and mayflies relate to each other and to the remaining winged insects is unclear.

The origin of flight likely accounts for much of the great success of insect lineages in dominating terrestrial ecosystems for the past 300 million years. Although this pattern of success is seen in the rapid diversification of the early neopteran lineages (Polyneoptera) such as grasshoppers, earwigs, and walking sticks that emerged in such rapid succession over the course of 50 million years, little morphological or genetic trace of their relationships remains. In contrast, the relationships among insects that undergo complete metamorphosis, known as Holometabola, are now well known. Analyses of the genomes of insect model organisms identify Hymenoptera (130,000 species of wasps, bees, and ants) as the sister group to all other holometabolan lineages. The other major radiations of insects (beetles: 400,000 species; moths: 140,000 species; flies: 250,000 species) all emerged within 50 million years of each other during the late Permian (260–220 Ma). This explosion of insect diversity likely corresponds with the evolution of flowering plants that provided many new niches for insect species to exploit.

3. COLONIZATION OF LAND

Existing fossil evidence indicates that arthropods were the first animals to transition from aquatic to terrestrial life, perhaps even before or coincident with the emergence of terrestrial plants. Convincing trace fossils of arthropod tracks in coastal environments show that terrestrial life may have begun by the early to mid-Ordovician, yet fossils of actual arthropods don't appear until tens of millions of years later, in the late Silurian. Early chelicerates were the first arthropods on land, yet four major independent colonizations have since occurred: arachnids, myriapods, crustacean isopods, and hexapods.

Though arthropods originated in marine environments, all major lineages, including panarthropods-onychophorans and tardigrades, have colonized land at some point. Indeed, paleontological evidence shows that the colonization of terrestrial environments by arthropods is best viewed as an ongoing process, with many lineages independently capitalizing on the opportunity to colonize semiaquatic, shoreline, or dry land many times throughout their multimillion-year history. For example, a Caribbean crab lineage has been shown to have moved inland as recently as 4 million years ago.

Many barriers to survival on land, such as withstanding desiccation, enabling gas exchange from the atmosphere throughout the body, excreting waste, and finding food and mates in a less predictable environment, have been overcome numerous times in arthropods, demonstrating the remarkable power of natural selection to produce evolutionary adaptations required for life in challenging environments (especially when those environments provide sufficient untapped resources).

4. EVOLUTION OF FLIGHT

Possibly the most significant event in the evolutionary history of arthropods was the origin of wings, and thus powered flight, more than 350 million years ago. Winged insects were the very first animals to achieve flight, and for almost 100 million years they were the only animal inhabitants of the air, with no competition from the flying vertebrate predators they have today (i.e., birds, bats). A putative winged insect fossil, identified by Grimaldi and Engel in 2004, places the origin of flight sometime within the Late Devonian and thus allows for the many different varieties of winged insects fossils found by the late Carboniferous (including dragonflies, mayflies, and cockroaches). During the Carboniferous, some winged insects, like previously mentioned arthropod groups, grew to spectacularly large sizes. Most impressively, some dragonflies from the family Meganeuridae had a wingspan of 75 cm. These winged giants disappeared by the end of the Permian, possibly because of the appearance of new winged predators such as pterosaurs, or because of the end of a period of high oxygen concentration in the atmosphere.

The morphological origins of wings remain a matter of great debate. Competing hypotheses propose that wings are modified thoracic lobes once used by insects for gliding, or that they developed from gills (that can still be found on the abdomens of mayfly nymphs). Yet the earliest insects are clearly terrestrial and substrate dwelling, and there are no known transitional fossils to offer insight into the origin of insect flight. Aside from gliding or respiration, it has been proposed that preflight wings served as protective structures, locomotory paddles for skimming on the water surface, spiracle covers, or thermoregulators, or played a role in sexual selection displays. Recent behavioral and comparative studies of jumping and gliding in terrestrial insects, such as tree-dwelling bristletails (Thysanura), by R. Dudley and colleagues, provide compelling evidence for the origins of insect flight from jumping and through directed gliding to fully powered flight. This idea corresponds well with the origin of flight in the Devonian, when vegetation was gaining height, becoming more tree- or shrub-like.

Wings provided insects the freedom to disperse widely, to escape predation, and to colonize diverse habitats. Throughout millions of years of insect evolution, wings have been modified to serve many alternate purposes. In beetles, forewings (called elytra) are hardened and function as a protective cover, while in flies, the hindwings have been greatly reduced to paddle-like halteres that function as gyroscopes. Wings have also been secondarily lost many times, in many lineages (walking sticks, grasshoppers, roaches, wasps, beetles, moths, flies). In some cases, this wing loss is conspicuous and exhibited by every species in an order—such as fleas and lice. In other groups, such as flies, wings have been lost many times by individual, often-unfamiliar, lineages (interestingly, some wingless fly groups, such as sheep keds and bat flies, have blood-feeding, ectoparasitic lifestyles that mimic those of fleas and lice). Wingedness, in combination with the evolution of complete metamorphosis, set the stage for the hyperdiversity of insects that we see on earth today.

5. COMPLETE METAMORPHOSIS

The evolutionary innovation of three distinct life stages within some groups of insects, known as *complete metamorphosis*, is another significant milestone in the history of arthropod evolution. Though various types of metamorphosis take place across Metazoa (tunicates, amphibians, fish), and even within arthropods (crustaceans such as crabs), complete metamorphosis within insects appears to be a unique Paleozoic innovation that enabled diversification within major insect lineages in the Mesozoic and Cenozoic. Insects that undergo complete metamorphosis, Holometabola, are the single lineage making Arthropoda so exceptionally diverse in terms of numbers of species. Taken together, these groups account for most of known life on earth. Holometabola include the four major superradiations of insects that make up most of their species diversity: beetles, moths, wasps (as well as bees and ants), and flies.

The spectacular diversity of holometabolous insects must be due, in part, to the selective advantages conveyed by complete metamorphosis. One evident advantage of this form of development is that it allows insect larvae to avoid direct competition with their adult forms in terms of both habitat and diet. In addition, holometabolan larvae experience internal wing development, allowing them to live and feed within substrates (plant matter, wood, bodies of other organisms) that, structurally, would be unsuitable for their adult forms. Another benefit of complete metamorphosis is the relative brevity of the life cycle; thus the increased risk of predation due to the restricted mobility of the wingless larval stage is effectively reduced. There are, however, exceptions to this rule, such as several types of beetles, flies, and others that have multiyear life cycles.

In *Evolution of the Insects*, Grimaldi proposes that it is not necessarily complete metamorphosis or the shortening of life cycles in general that has made holometabolan lineages so successful, but that holometabolan lineages appear to have "more effective control over their development." This flexibility has allowed particular lineages to finely tune their larval development and thus provide unique opportunities for their evolutionary success. For example, cyclorraphan flies (>50,000 species, e.g., *Drosophila*, houseflies), one of the largest animal radiations in the Cenozoic, have benefited from a modified form of metamorphosis in that they encase the pupa in an impervious puparium made from the last larval skin. This allows these flies to develop in a wide variety of substrates (e.g., low-oxygen decaying organic muck, petroleum, salty environments, vertebrate bodies) that would otherwise be toxic to a pupa with a more permeable membrane. Many other holometabolan lineages feature similar key innovations or life history specializations that paved the way for their extreme species richness.

6. LIFE HISTORY SPECIALIZATIONS

Arthropod evolutionary history is replete with amazing examples of specialization in life history traits and biological associations that have allowed groups with key adaptations to exploit ready resources. Arthropods exhibit almost every possible feeding strategy, including nectar/pollen feeders (bees and flies), decomposers (flies, crabs), predators (centipedes, spiders, mantis shrimp), plant feeders (beetles, moths), blood feeders (flies, fleas) and parasites (lice, wasps, flies). In addition to the diversity of their life histories, it is complex ways that arthropods interact and accomplish communication, competition, dispersal, aggregation, and escape that make their evolutionary story so compelling.

Arthropod specializations include examples that are both familiar and surprising. The use of silk by common orb-weaving spiders (Araneomorphae) to catch prey and construct shelters is just one example of a remarkable coordinated set of morphological and behavioral adaptations that has allowed this diverse clade to flourish since the Devonian (340 Ma) as major terrestrial predators. The evolution of insect societies (frequently encountered in our homes and elsewhere) is another major arthropod achievement. Eusociality involves behavioral, physiological, developmental, and morphological adaptations that enable survival and reproduction through group living and a division of labor. Eusociality has emerged independently in termites, ants, paper wasps, and honey bees—each developing unique systems to achieve the benefit of cooperative interaction (see chapter VII.13). Examples of coevolution, meaning reciprocal adaptations in closely associated organisms, are common within arthropods—from the strange mites that move among flowers on the beaks of hummingbirds to moths with mouthparts adapted to utilize the nectar of specific flowers they also pollinate.

There are many thousands of examples of arthropod species that have evolved specific, highly specialized features that allow them to use specific plants, animals, or fungi as food, to communicate, to mimic other organisms, or to hide from them, and to rapidly colonize newly opened niches. Viewed from almost any angle, arthropods provide innumerable demonstrations of the "tangled bank" so eloquently described by Darwin in the closing passage of *On the Origin of Species* to depict the interconnected evolutionary processes and patterns that make up the history of life on earth.

FURTHER READING

Budd, G. E., and M. J. Telford. 2009. The origin and evolution of arthropods. Nature 457: 812–817. *A good summary of the current consensus on arthropod evolution and the implications of fossil discoveries and phylogenetic analyses on interpretation of arthropod head development.*

Edgecomb, G. D. 2010. Arthropod phylogeny: An overview from the perspective of morphology, molecular data and the fossil record. Arthropod Structure and Development 39: 74–87. *An authoritative review of the current evidence from fossils, comparative anatomy, and molecular sequences of the classification and phylogeny of arthropods.*

Grimaldi, D., and M. S. Engel. 2005. Evolution of the Insects. New York: Cambridge University Press. *A beautifully illustrated, encyclopedic treatment of insect evolutionary history, fossil record, and biology.*

Meusemann, K., et al. 2010. A phylogenomic approach to resolve the tree of life. Molecular Biology and Evolution 11: 2451–2464. *Evidence from large-scale expressed gene samples of the phylogenetic history of animals, including a large sample of arthropod taxa.*

Regier, J. C., et al. 2010. Arthropod relationships revealed by phylogenomic analysis of nuclear protein-coding sequences. Nature 463: 1079–1084. *A new phylogenetic estimate of arthropod relationships based on nuclear gene sequences that firmly establishes the position of Hexapoda within the Pancrustacea.*

II.17

Major Features of Tetrapod Evolution

Farish A. Jenkins Jr.

OUTLINE

1. Tetrapod ancestry
2. The fish-tetrapod transition
3. Amniote origins
4. Synapsids
5. Diapsids: Lepidosaurs and their relatives
6. Diapsids: Archosaurs

Tetrapod evolution spans the last 375 million years, beginning with the origin of limbed terrestrial vertebrates from finned fishes. The earliest tetrapods were amphibians that radiated during the Late Paleozoic; only three groups of highly specialized amphibians survive today. The development of an amniotic egg provided novel mechanisms to enhance embryonic growth and allowed eggs to be laid on land. Amniotes came to dominate Mesozoic and Cenozoic terrestrial faunas worldwide, generated powered aerial flight three times, reverted to aquatic habitats and the marine realm repeatedly, created endothermic temperature controls, evolved limbs for running, digging, and climbing, flew in water with fins converted from limbs, many times lost limbs altogether, forged armor, concocted venoms and other weapons, adapted to eat almost everything macroscopic of biological origin, and attained body sizes across four orders of magnitude. A skeletal outline of tetrapod evolution and diversity is presented here.

GLOSSARY

Amniote. Tetrapods that reproduce by means of an amniotic egg, comprising an eggshell membrane and extraembryonic structures (including an amnion and chorioallantoic membrane) that provide a protective environment, nutrition, waste storage, and respiratory exchange.

Anapsid. The earliest recognized of three clades of amniotes that arose in the Late Paleozoic. Anapsids are characterized by skulls in which the chamber housing major jaw-closing muscles is completely enclosed. The term *anapsid* signifies lack of any external openings (fenestrae) that otherwise occur in diapsids and synapsids.

Diapsid. Amniotes in which the chamber housing jaw-closing muscles possesses two fenestrae. Diapsids include numerous extinct Mesozoic taxa such as dinosaurs, pterosaurs, and ichthyosaurs. Lizards, snakes, crocodilians, and birds are extant diapsids.

Synapsid. Amniotes that possess a single fenestra in the chamber housing jaw-closing muscles. Synapsids include extinct Late Paleozoic pelycosaurs and Permo-Triassic mammal-like reptiles (therapsids). Mammals are the only extant synapsids.

Tetrapod. Literally, four feet. Tetrapods are vertebrates with limbs instead of fins. The fish-tetrapod transition and the emergence of vertebrates onto land was initiated during the Late Devonian, some 375 million years ago.

1. TETRAPOD ANCESTRY

The ancestry of tetrapods may be traced to bony fishes (Osteichthyes) that first appear in the fossil record during the Silurian (439–408 million years ago, Ma). By Devonian times (408–362 Ma), bony fishes had diverged into two distinct clades, ray-finned fishes (Actinopterygii) and fleshy-finned fishes (Sarcopterygii). In their pectoral and pelvic fins, both possess elongate, slender, rodlike scales (*lepidotrichia*) and a skeletal base for fins within the trunk. In actinopterygians, lepidotrichia radiating from the skeletal base constitute the fin's sole skeletal framework, with fin movements controlled by musculature within the body wall. Sarcopterygians are distinctive for the muscles and robust skeleton present within the fin; in this group, lepidotrichia are largely confined to the fin margins.

Sarcopterygians diversified into three distinctive assemblages that were major components of Devonian fish faunas: lungfish (Dipnoi), coelacanths (Actinistia), and a

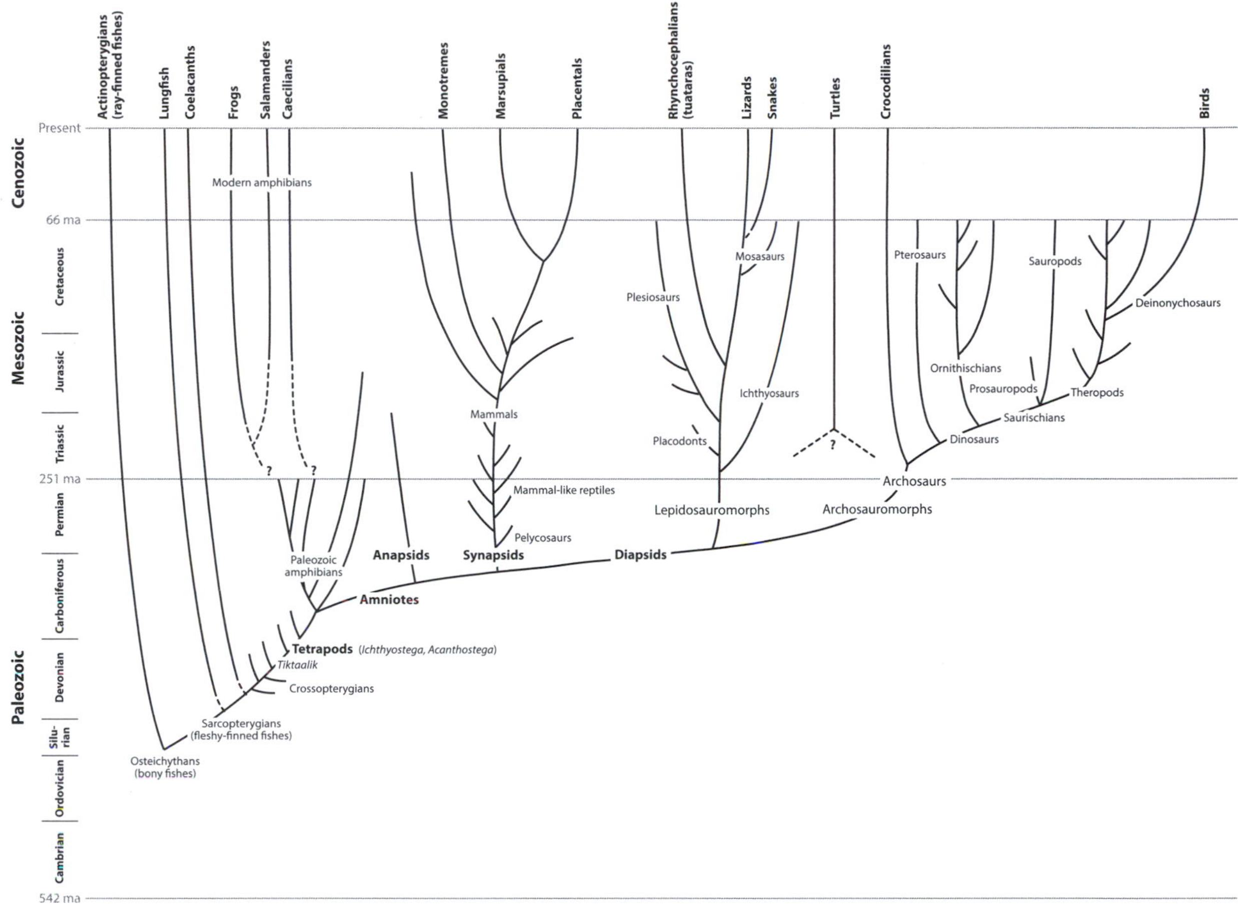

Figure 1. Overview of tetrapod evolution in the context of geologic time.

variety of so-called crossopterygians (figure 1). Lungfishes and coelacanths have few living representatives, and neither is generally considered closely related to tetrapods. Rather, tetrapods arose from a particular crossopterygian group (the elpistostegalids) that did not survive the Paleozoic in their piscine form. Lungs were probably present in Devonian sarcopterygians as accessory respiratory organs, for they are present in extant lungfish and, in modified form, in living coelacanths. Lungs were thus not a novel development among emergent tetrapods. Although the respiratory function of gills was eventually abandoned, tetrapods retained and adapted the skeletal elements of the gill system for feeding and other functions.

Tetrapods carried into a terrestrial sphere many crossopterygian features, including the pattern of skull bones persisting in modified shapes throughout tetrapod evolution. A distinctive tooth structure in which the enamel and dentine are deeply infolded into the tooth's interior (*labyrinthodonty*) was also retained. The tetrapod vertebral pattern likewise originated among crossopterygians; vertebrae comprised a single midline element (*neural arch*) with three crescentic, supporting ossicles (paired pleurocentra, and an intercentrum) surrounding a large notochord (Benton 2005). Although the proximal homologues of the bones in the tetrapod forelimb (humerus, radius, and ulna) and hindlimb (femur, tibia, and fibula) can be confidently identified in crossopterygian fins, the distal elements as precursors to tetrapod feet have long remained problematic. Nonetheless, extant lungfishes and the coelacanths both paddle with alternating strokes of contralateral fins, much as tetrapod limbs are used in walking.

2. THE FISH-TETRAPOD TRANSITION

Elpistostegalid sarcopterygians, initially recognized from incomplete fossil remains of two taxa, *Elpistostege* and *Panderichthys*, exhibit features suggesting a proto-tetrapod condition. The discovery by Daeschler and colleagues (2006) of *Tiktaalik roseae*, an elpistostegalian from the Late Devonian Fram Formation (375 Ma) in the Canadian Arctic, documents many anatomical details of an amphibious transitional stage in the emergence of tetrapods (figure 1).

Tiktaalik possesses eyes elevated above the skull roof, a feature shared with various living vertebrates that gaze just above the air-water interface (notably the mudskipper *Periophthalmus*, frogs, crocodilians, and hippopotamus). The broadly triangular skull and rather flat body bear greater resemblance to those of a Paleozoic amphibian than to a fish. Most surprisingly, *Tiktaalik* possesses a neck. Typically, a fish skull is connected to the shoulder girdle through a series of intermediate bones (operculo-gular and extrascapular series), but in *Tiktaalik* these are lost, allowing the head independent mobility. Fish feeding in water can readily position the entire body to direct the mouth toward prey. Such maneuvers are more difficult on land, where a mobile head is distinctly advantageous for a predator.

Shubin and colleagues (2006) described the forefin skeleton, which is homologous to the vertebrate forelimb and includes precursors to the tetrapod wrist and possibly digits. In addition to a humerus, radius, and ulna, there are wrist bones (an ulnare and intermedium) and a robust distal skeleton of small, short bones (radials). The eight rows of radials making up the bony margin of the *Tiktaalik* fin are possible precursors to the polydactylous feet of early tetrapods. The geometry of the shoulder, elbow, and distal joints provides evidence that the pectoral fins could assume both finlike and limblike postures.

The interpretation that the fins of *Tiktaalik* supported the body on land is corroborated by other aspects of skeletal architecture. The expanded, overlapping ribs recall those in the earliest-known tetrapods, which augmented rigidity for the trunk and axial support for the body. Such support was critical as elpistostegalid fishes left the buoyancy of water and encountered gravitational forces on land. This specialization for terrestrial existence, once believed to have evolved as an adaptation in the earliest-known tetrapods, in fact developed in amphibious fishes that were behaving like tetrapods. Although *Tiktaalik* is clearly piscine in retaining a well-developed branchial skeleton to support gills, a scaly body covering, and lepidotrichia, this amphibious sarcopterygian could move from shallow water margins onto shore.

What selective forces may have promoted a land invasion? The open seas of the Devonian were populated by large, predaceous fish—such as sharks and placoderms. Elpistostegalids, with flattened head and body and no dorsal fins, were adapted for nearshore shallow waters; *Tiktaalik* was recovered from deposits created by freshwater rivers overrunning their banks. In the shallowest habitats, elpistostegalids entered a natural refugium free of large predators and provisioned with abundant, unexploited invertebrate faunas on which to feed.

The earliest and most primitive tetrapods known, *Ichthyostega* and *Acanthostega* of Late Devonian age, retain many crossopterygian features, including labyrinthodonty, the pattern of skull bones, the location of lateral line sensory canals, the configuration of vertebral components, a finned tail, and gills. Relative to *Tiktaalik*, major alterations appear in the vertebral column and limbs. In *Ichthyostega,* the vertebral column is robustly constructed to support gravitational loading of the trunk. The closely abutted neural spines of the vertebrae are arranged to reinforce the back against sagging. Zygapophyses, processes that interconnect adjacent neural arches, are developed to reinforce spinal stiffness and regulate intervertebral movements. A specialized vertebra, the sacrum, anchors the hindlimb to the spine. The limbs are robust, with numerous small bones that form flexible wrists and ankles; fore- and hind feet are polydactylous. Tetrapod anatomy underwent relatively modest changes during the initial radiation of amphibians through the remainder of the Paleozoic and early Mesozoic. By comparison, extant amphibians (frogs, salamanders, and caecilians) are morphologically quite specialized, yet most remain dependent on an aqueous environment for the development of externally laid eggs (Clack 2002; Carroll 2009).

Vertebrate development from a fertilized ovum requires a protected environment, and supplies of oxygen and nutrients. Extant amphibians (anamniotes) and amniotes exhibit divergent solutions to providing these necessities. An amphibian ovum is surrounded by a vitelline membrane and successive "jelly" (mucopolysaccharide) capsules. Water absorbed through the membrane and capsules enlarges the space for embryonic growth and provides oxygen for the embryo. Although the capsular layers have diverse physiological and protective functions, they limit oxygen diffusion, which in turn limits the size of the hatchling. No longer bound to aqueous or moist environments, the amniotic egg overcame these constraints.

3. AMNIOTE ORIGINS

Turtles, snakes, lizards, tuataras, crocodiles, birds, and mammals present a rich array of living tetrapod diversity. Each group has evolved such distinctive anatomical, physiological, and behavioral characteristics that they cannot be confused. Yet each lineage converges on a singular event in its common ancestry: the development of an amniotic egg. The eggshell membrane and extraembryonic structures form an elaborate support system, including a protective casing (shell membrane, shell), a watery environment (amnion-enclosing amniotic fluid), a nutritional source (yolk sac), waste storage (allantois), and a respiratory exchanger (chorioallantoic membrane). The system's origin remains obscure. The only feature of egg development shared by both amphibians and amniotes is the derivation of a maternally derived envelope in the oviduct. The Paleozoic fossil record has yet to reveal any evidence of amniotic eggs (Sumida and Martin 1997).

Amniote evolution can be traced only by a suite of skeletal features inherited and further transformed by amniote lineages that survive to the present day. The earliest-known amniotes, entrapped within hollow, upright stumps of giant lycopods (see chapter II.3) , exhibit distinctive features shared with later amniotes. Some represent losses from the common amphibian pattern, such as loss of palatal fangs and labyrinthodonty. Others are novel, including additional skull components. On the palate, enlarged fang-like teeth are developed, and a prominent transverse flange of the pterygoid is tooth bearing. In the postcranial skeleton, amniotes fuse three separate bones in the amphibian ankle—the intermedium, tibiale, and proximal centrale—into a single bone, the astragalus.

By the Late Carboniferous, three types of amniotes are known on the basis of skull roof patterns (figure 2). In anapsids, the chamber behind the eye from which jaw-closing muscles originate is completely enclosed by external skull bones. Numerous extinct Late Paleozoic taxa exhibit an anapsid lack of fenestration. Synapsids are defined by a single lateral opening (temporal fenestra) exposing the muscle chamber, whereas diapsids have two fenestrae, one above the other. In contrast to anapsids, synapsids and diapsids underwent extensive Mesozoic and Cenozoic radiations, ultimately yielding the vast variety of amniote life (Benton 2005). An anapsid ancestry for turtles has been suggested, but not unequivocally demonstrated. Most recently, molecular and morphological analyses have suggested that turtles are diapsids that have secondarily achieved an anapsid condition, but it is unclear to which diapsid group (lepidosauromorphs or archosaurs) they are most closely related (figure 1). Turtle origins remain an unresolved problem in tetrapod evolution.

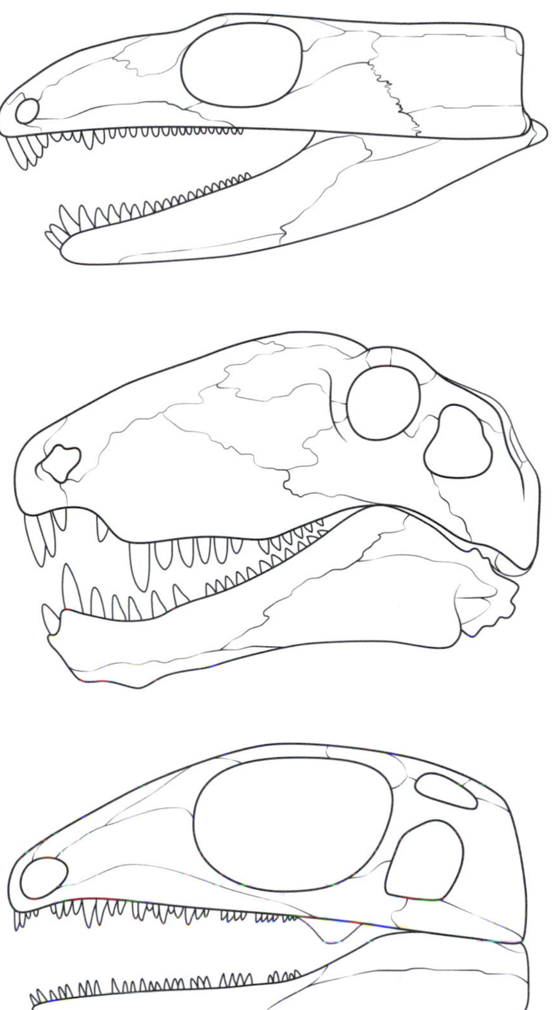

Figure 2. Representative skulls of the three major groups of amniotes. Top: The anapsid condition, in which there is no opening (fenestra) behind the orbit. Middle: The synapsid condition, with a single fenestra to the chamber from which jaw-closing muscles arise. Bottom: The diapsid condition, exhibiting two fenestrae of the chamber. (Top: *Eocaptorhinus laticeps* modified from M. J. Heaton. 1979. Cranial Anatomy of Primitive Captorhinid Reptiles from the Late Pennsylvanian and Early Permian, Oklahoma and Texas. Oklahoma Geological Survey, Bulletin 127, fig. 1; middle: Pelycosaur *Dimetrodon limbatus* modified from A. S. Romer and L. W. Price. 1940. Review of the Pelycosauria. Geological Society of America Special Papers, 28, fig. 62; bottom: *Petrolacosaurus kansensis* modified from R. R. Reisz. 1981. A Diapsid Reptile from the Pennsylvanian of Kansas. Special Publication of the Museum of Natural History, University of Kansas, 7, fig. 1.)

4. SYNAPSIDS

Living mammals are synapsids possessing such distinctive skeletal characteristics as a jaw joint between the dentary and squamosal bones, a secondary (bony) palate separating the nasal cavity from the mouth, and three middle ear bones (malleus, incus, and stapes). Although none of these skeletal configurations was present when synapsid evolution began, the development of these and other mammalian features is now traceable through 300 million years of the geological record with considerable precision.

Synapsids evolved through three successive radiations: pelycosaurs, therapsids (so-called mammal-like reptiles), and mammals. The earliest-known undoubted synapsid, the pelycosaur *Archaeothyris florensis*, occurs in Middle Pennsylvanian deposits in Nova Scotia. Pelycosaurs, the largest tetrapods in Late Pennsylvanian faunas, diversified in the Early Permian to occupy both carnivorous and herbivorous niches, with some forms developing greatly elongated neural spines supporting a dorsal "sail."

Therapsids supplanted pelycosaurs during the Late Permian. Substantially more varied than pelycosaurs, therapsids ranged from large, heavy-bodied, robust-limbed Dinocephalia to relatively gracile Theriodontia. During the Permian, therapsids underwent considerable cranial diversification; dental specializations were accompanied by substantial increases in the size of temporal fenestrae and the mass of jaw-closing muscles.

Only two groups of therapsids survived the widespread extinctions during the Permo-Triassic transition. The most abundant Triassic therapsids were the anomodonts, robust herbivores that replaced teeth with a keratinous, slicing mechanism. Most important in terms of mammalian origins were cynodonts. Members of certain families of cynodonts (Thrinaxodontidae, Probainognathidae) exhibited dental, cranial, and postcranial features that represent the most advanced level of therapsid organization from which mammals were derived.

Approximately 5000 diverse species of living mammals testify to the huge scope of synapsid evolution since the Early Jurassic. Throughout this radiation, the defining mammalian characters in the skull and postcranial skeleton have largely persisted. These characters did not arise as an integrated complex, but rather as a mosaic initiated at different times in synapsid history. Whereas brain enlargement and diphyodonty (deciduous, or "milk," teeth followed by the adult dentition) are novel attributes of early mammals, the development of other features can be traced to earlier stages. Carnivorous pelycosaurs possessed elongate, paired caniniform teeth set in an otherwise homogeneous dentition. Various therapsids subsequently developed prominent single canine teeth, together with modest differentiation of incisors and postcanines. Distinctive cusps on anterior and posterior postcanine teeth are a cynodont feature, presaging the differentiation of mammalian premolars and molars as well as tooth-on-tooth occlusion.

In early cynodonts the size of the temporal muscle substantially increased, and an incipient coronoid process on the jaw (a lever for the temporal muscle) and a masseter muscle first appeared. In pelycosaurs, a dentary (the tooth-bearing bone) and eight postdentary bones composed the lower jaw. Throughout synapsid evolution, postdentary bones diminished as the dentary enlarged. In the most advanced cynodonts, a mammalian jaw joint between the dentary and squamosal evolved alongside the primitive quadrate-articular joint. Although most postdentary bones were ultimately lost, several persisted to serve novel functions. The angular (with a distinctive flange, the reflected lamina) persisted as the tympanic bone supporting an eardrum. The articular and quadrate were retained as the malleus (adherent to the inner surface of the tympanic membrane), and incus (linking the malleus and stapes). In therapsids the close association of the stapes, the original bone of the middle ear, with the quadrate set the stage for the quadrate (incus) and articular (malleus) to associate as a miniaturized, sound-conducting bony chain between the tympanic membrane and inner ear.

Major advances toward a mammalian grade of postcranial organization occurred among cynodonts. The joints between the skull and first cervical vertebra, and the first and second cervicals, became specialized for flexion-extension and rotation of the head, respectively, which are movements important in feeding, defense, and other behaviors. Cervical, thoracic, and lumbar regions became distinctly differentiated, contributing to the wide range of movements and postures utilized by mammals. A small tuberosity on a major ankle bone, the calcaneum, was the forerunner of an elongate lever, the calcaneal tuberosity, into which flexor muscles insert to power mammalian walking, running, and jumping. But not all mammalian characteristics had their beginnings among cynodonts; numerous features emerged later among Jurassic and Cretaceous mammals.

Despite the rich synapsid fossil record, the origins of such quintessential mammalian attributes as hair, endothermy, and mammary glands remain obscure. A furry coat in Early Cretaceous mammals is the earliest persuasive evidence of endothermy. Given the plausible thermoregulatory function of Permian pelycosaurs' extensive dorsal sails, which might have served to gather radiant heat, synapsids conceivably could have had a long history of temperature regulation.

5. DIAPSIDS: LEPIDOSAURS AND THEIR RELATIVES

Extant diapsids are represented by lepidosaurs (the squamates: tuataras, lizards, and snakes) and certain archosaurs (crocodilians and birds). As diverse as this assemblage may seem, diapsid evolution during the Mesozoic yielded an even more extraordinary array of tetrapods. Some invaded the oceans (placodonts, plesiosaurs, ichthyosaurs, mosasaurs), others became the first powered fliers (pterosaurs), and still others dominated terrestrial herbivorous and carnivorous niches (ornithischian and sauropodomorph dinosaurs, and theropod dinosaurs, respectively). A Late Pennsylvanian precursor to this diversity, *Petrolacosaurus kansensis*, is clearly diapsid by virtue of the double fenestration of the skull (figure 2), but it lacked many of the specialized features found in lepidosaurs and archosaurs.

Lizards (>4,000 species), snakes (~3,000 species), and tuataras (1–2 species; technically, rhynchocephalians) share numerous derived features indicative of common ancestry. The epidermis is shed entirely at discrete intervals. The cloaca, an opening for the digestive, urinary, and reproductive tracts, is a transverse slit. A fracture plane within each caudal vertebra allows the tail to break away (autotomy), facilitating escape.

Although rhynchocephalians were moderately diverse during the Mesozoic, and included herbivorous, possibly venomous, and even aquatic forms, today they are represented only by two relict species of the genus *Sphenodon* that survive on coastal islands off New Zealand. Tuataras differ from true lizards in a number of features, including the presence of a premaxillary process, or "beak," in place of premaxillary teeth. The upper dentition is a double row of teeth between which the lower teeth precisely occlude. Tuataras lack an external ear.

The Early Jurassic *Gephyrosaurus bridensis* is the earliest-known well-documented lizard with such defining features as fusion of the primitively paired frontals and parietals, fracture planes for caudal autotomy, and an astragalocalcaneum. Lizards that are possibly related to modern groups appear during the Middle and Late Jurassic. Limb reduction or loss is widespread. Iguania (including iguanids, agamids, and chamaeleonids) is generally considered the most generalized assemblage of extant lizards. Most are territorial ambush predators; the fleshy tongues of iguanids and agamids snatch prey with sticky saliva. Scleroglossa is a diverse assemblage of mostly active, foraging predators without established home territories. The highly protrusible tongues of many of these lizards serve both prehensive and chemosensory functions (Pianka and Vitt 1989).

Snakes were derived from lizards, and are likely most closely related to monitors (Varanidae) and alligator lizards (Anguidae). Fossil snakes are rare and not known before the Late Cretaceous. *Najash rionegrina*, a Late Cretaceous terrestrial snake, possessed a pelvis with diminutive hindlimbs. The positioning of the pelvis and definitive sacral vertebrae suggest external hindlimbs, in contrast to hindlimb rudiments in some modern snakes that lie within the body wall. Some interpret this as evidence that snakes derived from a terrestrial, burrowing ancestor, but for lack of adequate evidence there is no consensus on snake origins. In modern snakes, as in other elongate squamates, only the right lung is functional. In some taxa, the respiratory surface is limited to the anterior part of the very elongate lung; posteriorly, the surface is nonrespiratory, but the bag serves as an air reservoir during the prolonged period of prey ingestion when the glottis is blocked. In one feature, however, snakes are unique among tetrapods: the right aortic arch predominates, rather than the left. Also, in the most specialized taxa, almost all bones save for those encasing the brain (frontal, parietal) are involved in highly mobile articulations and capable of displacement. Unfettered from the constraints of the primitive amniote cranial plan, serpents suspend from the protective braincase a startlingly flexible and expansive feeding apparatus (Greene 1997).

During the Mesozoic, numerous lepidosaurs and their relatives turned to the sea. Aquatic varanoid lizards gave rise to the large, open-water, predatory Late Cretaceous mosasaurs, some exceeding 10 m in length, with paddling limbs and sculling tails. Triassic placodonts, armored and superficially turtle-like, developed broad, flat teeth for crushing shelled invertebrates in coastal waters. Pachypleurosaurs, likely propelled by undulation, were small, fish-eating forms (ca. 0.5 m in body length) with long necks. Nothosaurs were similar to pachypleurosaurs, but larger (1–4 m body length); these anguilliform (eellike) swimmers were likely ancestral to plesiosaurs.

Plesiosaurs, appearing in the Jurassic and persisting almost to the end of the Cretaceous, diverged into two distinctive morphotypes. Pliosauroids, with huge heads, relatively short necks, and robust teeth, were adapted for large prey. Plesiosauroids, with slender teeth indicative of smaller prey, bore undersized heads on enormously elongated necks. The expansion of ventral elements in the shoulder and pelvic girdles, together with limb shape, are evidence that plesiosaurs engaged in subaqueous flight, much as do extant sea turtles, seals, and penguins.

Ichthyosaurs are the epitome of aquatic diapsids. Early Triassic ichthyosaurs were anguilliform swimmers; their backbone of cylindrical centra permitted a wide range of undulations. Jurassic and Cretaceous ichthyosaurs developed a thunniform (tuna-like) body shape, a pronounced dorsal fin and crescentic tail, and deep discoidal centra that contributed to body rigidity—all features of powerful swimmers. The disproportionately large eyes of later ichthyosaurs, with diameters greater than 200 mm in *Ophthalmosaurus* and 260 mm in *Stenopterygiius*, were evidently an adaptation for visual foraging at great depths in near darkness.

6. DIAPSIDS: ARCHOSAURS

Archosaurs, a diverse assemblage of diapsids that dominated terrestrial faunas throughout much of the Mesozoic, are represented today only by crocodiles and birds (figure 1). Their origins are to be found among Early Triassic archosauromorphs; distinctive features, such as socket-implanted (thecodont) teeth were largely retained in later archosaurs. True archosaurs also appear in the Early Triassic. They are distinguished from other diapsids by openings in front of the orbit (antorbital fenestra) and in the lateral side of the jaw (mandibular fenestra). The morphological diversity among the ankle joints of early archosauromorphs and later archosaurs reflects shifts in stance and gait, and defines major archosaur groups. In Ornithodira, which includes pterosaurs and dinosaurs, the astragalus and calcaneum are immobile, fixed to the distal ends of the tibia and fibula, respectively. Ankle flexion and extension occurs at a mesotarsal joint between these fixed elements and the next row of anklebones, the distal tarsals. In Crurotarsi, which includes crocodilians and various extinct Triassic taxa, the ankle joint is developed between the astragalus and calcaneum. The calcaneum (and the remainder of the distal elements of the foot) flexes and extends on the immobile astragalus (Benton 2005).

The relatively flattened skulls, fused parietals, and small temporal fenestrae of primitive Late Triassic crocodilians are features that persist in modern crocodiles. Whether bipedal or quadrupedal, the gracility and length of their limbs indicate that early crocodilians were agile, perhaps even cursorial, predators. Early Jurassic crocodilians, initiating a trend toward separating the nasal and oropharyngeal passages, formed a complete secondary bony palate with internal nasal openings (internal choanae, linking the mouth and nasal cavities) halfway along its length. There is no evidence that these early forms were particularly aquatic, but later in the Jurassic mesosuchian crocodiles occupied aquatic habitats. Mesosuchian skulls elongated and flattened; the internal choanae displaced posteriorly, invading the pterygoids at the back of the palate. In some mesosuchians the limbs were relatively small and the tail was prolonged for sculling. Eusuchians appeared in the Cretaceous and have changed relatively little since. A major eusuchian specialization was the completion of the posterior migration of the internal choanae, now completely surrounded by the pterygoids. Crocodilians thus achieved a complete separation of the nasal and oropharyngeal passages, an adaptation uniquely useful to semiaquatic ambush predators (Ross and Garnett 1989).

Pterosaurs, the first vertebrates to have evolved powered flight, present one of the most spectacular radiations in Mesozoic history, encompassing a great range of size, diverse skull and tooth shapes, and an array of diets. The principal wing membrane (brachiopatagium) arose from the trunk and forelimb and extended to the distal tip of a hyperelongated fourth digit. Exceptionally well-preserved pterosaurs provide unequivocal evidence of a furry pelage on the body and wing membranes, a strong indicator that pterosaurs were partly, if not fully, endothermic. Braincase endocasts reveal that pterosaurs possessed relatively large brains, although not as large as those of birds of comparable body mass. The immense size of the flocculus, where reflexes are coordinated that maintain a fixed gaze under dynamic conditions, together with very long semicircular canals in the inner ear, indicate that pterosaurs possessed a highly refined sense of equilibrium.

Beaks, cranial crests, body size, and dental specializations are the sources of pterosaur diversity. Beaks varied from short and deep to long and attenuated, and from downturned and prehensive to upturned and probing. Crests occurred as bony ridges on beaks, or as soft tissue or bony adornments on the head. Extreme dental specializations ranged from bristlelike teeth in filter feeders to batteries of blunt teeth fused to the jaws of shellfish crushers. Sizes ranged from the robin-sized *Pterodactylus elegans* (wingspan 25 cm) to *Quetzalcoatlus northropi*, the largest-known flying vertebrate, with a wingspan of 11–12 m. Found in Late Cretaceous Texas floodplain deposits 400 km from the coastline, *Quetzalcoatlus* possessed an unusually long neck and a robust (but toothless) beak, evidence that these huge pterodactyls might have been either predatory or carrion feeders (Wellnhofer 1991).

Relatively few features unite dinosaurs as a group. Among them are loss of the postfrontal in the skull and the tendency to reduce manual digits V and IV. Structures associated with the insertions of locomotor muscles are accentuated; the deltopectoral crest of the humerus is elongate, and a cnemial crest (for insertion of the quadriceps femoris) appears on the tibia. More than two vertebrae contribute to the sacrum.

By Late Triassic time, two major lineages of dinosaurs are distinguishable on the basis of pelvic structure. Ornithischians, ultimately comprising a great diversity of herbivorous forms, had a pubis with two processes: a prepubic process projecting anteriorly, and a postpubic process lying posteriorly near the ischium. The postpubic process superficially resembles the pubis of birds, and thus originated the term Ornithischia ("bird hip"), although pubic configurations in birds and ornithischians were derived independently. The Saurischia, comprising herbivorous sauropodomorphs and carnivorous theropods, had pubes that retained the primitive anteroventral orientation and were widely separated from the ischium. In both groups the acetabulum (hip socket) was widely fenestrated (Weishampel et al. 2004).

Ornithischian dinosaurs developed several masticatory specializations to process plant material. A toothless premaxilla and a neomorphic predentary bone in the lower jaw supported a cropping, presumably keratinous beak. Lanceolate (leaf-shaped) tooth crowns bore multicusped edges; wear facets developed through tooth-on-tooth occlusion. In many forms the enamel was asymmetrically distributed to promote self-sharpening. A robust, elevated coronoid process was a mechanical lever for powerful jaw-closing muscles. But unlike many mammalian herbivores, few major radiations of ornithischians seem to have employed rapid escape as an antipredatory tactic. Rather, as defenses against contemporary carnivores, they variously developed large body size (iguanodontids, hadrosaurids), armor and tail clubs (ankylosaurids, nodosaurids), spikes and plates (stegosaurids), shields and horns (ceratopsians), and thickened skulls for head ramming (pachycephalosaurs).

Saurischian dinosaurs comprise two divergent radiations: herbivorous sauropodomorphs (prosauropods and gigantic sauropods), and predaceous theropods. By the Late Triassic, prosauropods were well established as the dominant herbivores in practically all known terrestrial faunas. Typically, prosauropods had small skulls, robust bodies, and long tails. They range in size from the diminutive *Thecodontosaurus* (0.5 m) to the bulky melanorosaurid *Riojasaurus* (8–9 m). Prosauropods persisted into the Jurassic, but their niche as large-bodied browsers was subsequently occupied by sauropods, huge animals with very small heads mounted on elongate necks, shortened trunks, and massive, straight (graviportal) limbs. The huge size attained by sauropods represented, in part, a defensive strategy, but for some, long necks and legs also permitted foraging at canopy levels, a resource unavailable to any contemporary ornithischian.

Theropods are far less well represented in the fossil record; the biomass of carnivores invariably is far smaller than that of their prey. The earliest group, the coelophysians, was widely distributed and present from the Late Triassic through the Early Jurassic. *Coelophysis* (Late Triassic), bipedal and gracile, is a representative early form with features of a generalized theropod. The limb bones have extensive marrow cavities, giving them a hollow appearance when fossilized. The forefoot retains four digits, although the fourth is commonly lost in later, more derived forms. The fourth digit on the hind foot is sufficiently reduced in size that the foot is functionally tridactyl. Tridactyl trackways are a common trace fossil signature of theropods.

Later theropods developed larger heads and deeper mandibles than those of coelophysians. Allosaurids, ranging up to 12 m in length, were dominant predators from the Late Jurassic through the Early Cretaceous. Tyrannosaurids were the largest predators of Late Mesozoic ecosystems. Relative to an allosaurid skull, the tyrannosaurid skull was longer, deeper and more massively built, with a very wide snout. In cross-section, the extremely robust teeth were as wide as long, and thus thicker than allosaurid teeth. Bearing such a heavy head, the neck and trunk are shortened. The very slender, small shoulder girdle bore tiny, short forelimbs with only two divergent digits. But not all theropod groups depended on a powerful dental battery. Both the ornithomimosaurs and the oviraptosaurs were toothless, the latter bearing a cutting beak.

The Deinonychosauria, comprising two families (Dromaeosauridae, Troodontidae), share a few skull characters (such as loss of the prefrontal bone bordering the orbit) that unite the group and are shared with birds. Postcranial evidence for a deinonychosaurian-avian relationship is stronger. Both modern birds and the dromaeosaurid *Deinonychus* possess tubercles (epipophyses) for muscular attachment on the second and third cervical vertebrae; these are well developed in the primitive bird *Archaeopteryx*. The length of the forefoot in deinonychosaurs and *Archaeopteryx* equals or exceeds the length of the hind foot. The pubic orientation in *Deinonychus* is almost vertical, approaching the backwardly reflected avian position, in contrast to the primitive theropod anteroventral orientation.

Representatives of several Early Cretaceous theropod families possessed a covering of protofeathers; another theropod (*Caudipteryx*) was fully feathered but flightless nonetheless. Together this evidence indicates that feathers likely evolved initially for thermoregulation rather than for flight. Their development was a critical stage in the evolution of powered flight. Two theories have offered seemingly contrasting interpretations of the origin of powered flight. In the "cursorial" or "from the ground up" scenario, incipiently developed feathers on the forelimbs of cursorial theropods provided sufficient lift to launch a runner into the air. In the "arboreal" or "from the trees down" scenario, primitive feathers provided sufficient lift to support gliding. In both cases, the development of an aerofoil generating even the most incipient aerodynamic lift (even if insufficient for sustained, powered flight) represents the structural innovation from which avian flight evolved. Although long represented as competing theories, until recently there was little evidence to support either theory, but two studies now provide evidence that both gliding and running with protowings are realistic precursors to powered flight and are not mutually exclusive. From their initial radiations in the Mesozoic (Chiappe and Witmer 2002), birds now constitute more than 10,000 living species, a major component of tetrapod diversity, and testimony to the vast scope of evolutionary change within archosaurs.

FURTHER READING

Benton, M. J. 2005. Vertebrate Palaeontology. 3rd ed. New Jersey: Blackwell. *An updated edition of a synoptic textbook encompassing the relationships, morphology, and evolution of major vertebrate clades.*

Carroll, R. A. 2009. The Rise of Amphibians: 365 Million Years of Evolution. Baltimore: Johns Hopkins University Press. *A leading paleontologist's summary of discoveries and interpretations of the evolution of Paleozoic and modern amphibians.*

Chiappe, L. M., and L. M. Witmer, eds. 2002. Mesozoic Birds: Above the Heads of Dinosaurs. Berkeley: University of California Press. *Well-illustrated research reports and subject summaries on avian origins, avian-theropod relationships, the anatomy and systematics of Mesozoic birds, and locomotor evolution.*

Clack, J. A. 2002. Gaining Ground: The Origin and Evolution of Tetrapods. Bloomington: Indiana University Press. *A compendium of geological, paleoecological, anatomical, paleobiological, and taxonomic data bearing on the origin and early evolutionary history of tetrapods.*

Daeschler, E. B., N. H. Shubin, and F. A. Jenkins Jr. 2006. A Devonian tetrapod-like fish and the evolution of the tetrapod body plan. Nature 440: 757–763.

Greene, H. W. 1997. Snakes: The Evolution of Mystery in Nature. Berkeley: University of California Press. *An extravagantly illustrated survey of the diversity and biology of snakes, with perspectives on evolution and biogeography.*

Pianka, E. R., and L. J. Vitt. 2003. Lizards: Windows on the Evolution of Diversity. Berkeley: University of California Press. *An authoritative and well-illustrated summary of the adaptations, life history, and diversity of lizards.*

Ross, C. A., and S. Garnett, eds. 1989. Crocodiles and Alligators. New York: Facts on File. *A superbly illustrated account of the evolution, diversity, life histories, and conservation of crocodilians.*

Shubin, N. H., E. B. Daeschler, and F. A. Jenkins Jr. 2006. The pectoral fin of *Tiktaalik roseae* and the origin of the tetrapod limb. Nature 440: 764–771.

Sumida, S. S., and K.L.M. Martin, eds. 1997. Amniote Origins: Completing the Transition to Land. New York: Academic. *Research summaries of the phylogeny, biogeography, feeding, reproductive biology, functional anatomy, and physiology relating to amniote origins.*

Weishampel, D. B., P. Dodson, and H. Osmólska, eds. 2004. The Dinosauria. 2nd ed. Berkeley: University of California Press. *The sourcebook for current interpretations of the evolution and major features of all dinosaurian taxa, including biogeography, taphonomy, paleoecology, extinction, and the early history of birds.*

Wellnhofer, P. 1991. The Illustrated Encyclopedia of Pterosaurs. London: Salamander Books. *A compendium on the origin, evolution, and diversity of pterosaurs, including detailed illustrations of diverse taxa as well as life reconstructions.*

II.18

Human Evolution

John Hawks

OUTLINE

1. Origin of the hominins
2. Early *Homo*
3. Neanderthals and the origin of modern humans
4. Recent human evolution

Living humans are the sole living representatives of a lineage, the hominins, which diverged from other living apes 5 to 7 million years ago. Hominins remained limited to Africa for two-thirds of their history. With chimpanzee-sized bodies and brains, early hominins diversified into several lineages with different dietary strategies. One of these found a path toward technology, food sharing, and hunting and gathering, giving rise to our genus, *Homo*, approximately 2 million years ago. As populations of *Homo* spread throughout the world, they gave rise to regional populations with their own anatomical and genetic distinctiveness. Within the last 100,000 years, a massive dispersal of humans from Africa absorbed and replaced these preexisting populations. In the time since this latest emergence from Africa, humans have continued to disperse, interact, and evolve. The rise of agricultural subsistence shifted human ecology, fueling evolution.

GLOSSARY

Acheulean. A style of stone tool manufacture associated with early humans (*Homo erectus*) during the Lower Stone Age era across Africa and Eurasia. Acheulean technology is derived from the older, Oldowan technology and is a progenitor of the more complex stone tools that characterize the Middle Stone Age.

Australopithecines. Members of the hominin clade with the bipedal gait and dentition of modern humans, but lacking the enlarged brains of the genus *Homo*. Australopithecine species have been assigned to a diversity of genera, but most are now included within *Australopithecus*.

Hominins. Modern humans and extinct species more closely related to humans than to chimpanzees or gorillas.

Oldowan. The earliest stone tool industry, which emerged about 2.6 million years ago and persisted until about 1.6 million years ago, when it was replaced by the more sophisticated Acheulean technology.

Orthograde. An upright posture associated with a bipedal gait, such as occurs in modern humans.

Pronograde. The posture of holding the body parallel to the ground, such as is typical of most quadrupedal vertebrates.

While Darwin avoided discussion of the evolution of humans in *On the Origin of Species*, he soon tackled the issue in *The Descent of Man*, which defined the starting point for modern evolutionary anthropology. In the nineteenth and early twentieth centuries, the main theme of anthropology was a perceived lack of fossil progenitors, prompting a much-hyped search for a "missing link." Gradually this concern diminished as paleoanthropologists, especially over the last half century, succeeded in uncovering thousands of fossil specimens, representing diverse human ancestors and collateral relatives. While questions still remain, these fossil data provide a rich history of the origin of many of humanity's distinctive physical traits. Furthermore, archaeological finds have provided information on the behavior of hominins during the last half of human evolution, giving details about diet and social organization. Today, geneticists can add evidence from whole-genome comparisons of living humans, other primates, and some ancient hominins. Through all these lines of evidence a remarkably clear picture of human evolution is now emerging.

We can roughly consider human evolution in three parts. The first, running from 7 million up to around

4 million years ago, saw the origination of the hominin lineage and the initial appearance of our bipedal pattern of locomotion. The second, from 4 million up to around 1.8 million years ago, was the age of the australopithecines. This group of species had a stable set of adaptations in body size and locomotion, while showing substantial dietary and geographic diversity. Our own genus, *Homo,* arose about 1.8 million years ago from the australopithecines. The spread of *Homo* throughout the world, along with many later dispersals and population expansions, laid the foundation for today's human populations.

1. ORIGIN OF THE HOMININS

Chimpanzees and bonobos are our closest relatives among living primates. Whole genome comparisons suggest that our common ancestors with these apes lived between 4 million and 7 million years ago. Our common ancestors with gorillas lived a bit earlier, within the last 10 million years, and with orangutans even earlier, before 12 million years ago. Hence it is during the period between 10 million and 4 million years ago that paleontologists look for the immediate precursors of the hominin lineage.

A rich record of fossil apes has been recovered from the Miocene geological epoch, which lasted from 23 million to 5.2 million years ago. Before 15 million years ago, all known apes lived in Afro-Arabia. Early in the Middle Miocene, some apes dispersed into Asia and Europe, including the Asian ancestors of orangutans. Miocene apes ranged extensively in body size and adaptive niche, and evolved a diversity of locomotor strategies. Many were pronograde quadrupeds, essentially like living Old World monkeys such as macaques and baboons. A few had shoulders and vertebral columns, indicating an orthograde posture or climbing, but no early apes are known to have had the long arms and below-branch suspensory capability of today's great apes. Vertical, orthograde posture was once thought to be an ancestral feature of all apes (including humans); however, some anthropologists now believe that this suspensory body plan evolved convergently in the African and Asian apes.

Living humans are obligate bipeds, with pelvis, foot, and vertebral adaptations that impede effective quadrupedal gait and climbing. The origin of hominins is entangled with this unique adaptation, but the earliest members of our lineage surely did not have the full package of bipedal adaptations found in later hominins. All living apes can move bipedally, and some Miocene apes such as *Oreopithecus* may have specialized in terrestrial bipedality. Recognizing the beginnings of the hominin adaptation to bipedality has been central to identifying early hominins, whose identity remains subject to debate.

The earliest candidate fossils for being hominins share a suite of dental resemblances with later members of our lineage, including small canine teeth, low-crowned molar teeth, and thick molar enamel. Some paleoanthropologists suggest that such dental traits are shared much more broadly with other Miocene lineages, and may not indicate hominin affinities. Skeletal adaptations to upright posture and bipedal stance provide strong evidence that later fossils, after 4.2 million years ago, are human relatives. For earlier fossils, evidence of posture and stance is more equivocal. *Sahelanthropus tchadensis* from north central Africa is the earliest known, at around 7 million years ago. Represented by a nearly complete skull and jaw, it shows an orthograde placement of the skull atop the spinal column. *Orrorin tugenensis*, from western Kenya dating to 6 million years ago, also has a femur consistent with bipedal weight bearing. *Ardipithecus kadabba*, 5.5 million years old from Ethiopia, combines the aforementioned hominin dental features with a toe bone, suggesting that the toe generated force during bipedal walking, as occurs in modern humans. It remains unclear whether these fossil taxa lived before or after the divergence of the human and chimpanzee lineages, and if after, whether they are on the human or chimpanzee side of this evolutionary split.

Ardipithecus ramidus, dating to 4.4 million years ago from Ethiopia, comprises a large fossil sample including one nearly complete skeleton. From its limb proportions, grasping feet, and apelike hands, *Ardipithecus* was a habitual quadruped that also had good climbing abilities. But several of its features are similar to those of hominins, including a shortened pelvis and an upright posture. The teeth and jaws of *A. ramidus*, like those of earlier *A. kadabba*, are among its most hominin-like features. It is often interpreted as the earliest well-documented member of our lineage. However, the data do not rule out the possibility that it is an early member of the chimpanzee or gorilla lineages.

Australopithecines

The first fossils to show clear evidence of a commitment to terrestrial bipedal locomotion are assigned to *Australopithecus anamensis*. Between 4.2 and 3.9 million years ago, this species existed in East Africa. After this time, the same region was inhabited by *Australopithecus afarensis*, which is present in more than a dozen fossil-bearing localities representing hundreds of known specimens, all dated between 3.9 and 2.9 million years ago. The teeth of these two closely similar species show several temporal trends, toward larger postcanine teeth and functional changes in the canine-premolar cutting anatomy. Because of these trends, most paleoanthropologists regard *A. anamensis* and *A. afarensis* as successive members of a single evolving lineage.

Other lineages of hominins may have been present at the same time, including *Kenyanthropus platyops* from Kenya and *Australopithecus bahrelghazali* from Chad, both between 3.5 and 3.3 million years ago. These are possibly distinct from *A. afarensis* because of cranial and dental peculiarities, but in each case the single specimen is fragmentary. Likewise, a partial foot skeleton from Woranso-Mille, Ethiopia, may represent yet another lineage with a distinct locomotor strategy, possibly a direct descendant of earlier *Ardipithecus*.

A diversity of contemporaneous forms is much clearer among the hominins near the Plio-Pleistocene boundary. From approximately 2.8 to 2.3 million years ago, South Africa was the home of *Australopithecus africanus*, also represented by large fossil samples and in most respects similar in cranial anatomy and teeth to *A. afarensis*. Additionally, by 2.5 million years ago, the robust australopithecines appeared in East and later in South Africa. *Robust* refers to the chewing mechanics of these hominins, which combined powerful jaw muscles with extraordinarily large molar and premolar teeth. The robust australopithecines had approximately the same body size as other australopithecines but clearly had a different diet, featuring many more leaves and hard seeds. *Australopithecus robustus* was a South African form, apparently descended from *A. africanus*. *Australopithecus boisei* was the apex of this trend toward plant dietary specialization, and constitutes the majority of hominin fossils from East Africa between 2.5 and 1.5 million years ago.

The australopithecines were obligate bipeds, meaning their skeletal adaptations to bipedality precluded effective quadrupedal movement. Their feet had a first toe aligned with the other toes, minimal opposability or grasping ability, and arches similar to the feet of living people. Their knees were angled to promote effective weight support in a bipedal stance, and did not rotate to facilitate grasping with the feet. In contrast to nonhuman apes, humans and australopithecines have short hipbones that make a broad, bowl-shaped structure to support the viscera when upright. In addition, the broader hip and shorter ischium enabled effective muscle control of the lower limbs during bipedal walking and running. Our bipedal form of locomotion is not as fast as chimpanzee or gorilla knuckle walking, but it is highly energetically efficient.

However, despite their clear bipedality, australopithecines had relatively long, heavily muscled arms, curved toes and finger bones, and a long clavicle and apelike shoulder blade, all suggesting that climbing remained important to *A. afarensis* and *A. africanus*, even as these hominins moved into more open grassland settings. Still, with hands and legs ill suited for suspension or above-branch quadrupedal walking, early hominins must have climbed in a manner analogous to recent humans. With female masses around 35 kg and males up to 50 kg, they approximated living chimpanzees in body size. The most complete skeletal individuals, such as the "Lucy" skeleton of *A. afarensis*, had statures of 100 to 140 cm, much shorter than the average of any recent human population. Australopithecines had small brains, approximately 450 ml on average, which contrasts with the 1350 ml brains of living humans.

2. EARLY *HOMO*

By 1.8 million years ago, a very different kind of hominin had emerged and spread into Eurasia. *Homo erectus* was the size and stature of recent human hunter-gatherer populations, bigger than any known australopithecine. The skulls of *H. erectus* also contained disproportionately larger brains than australopithecines, initially between 600 and 900 ml, and relatively small teeth. The earliest clear fossil evidence of *H. erectus* occurs at Dmanisi in the Republic of Georgia and Modjokerto, Java, with additional fossil discoveries in East and South Africa. In each of these areas, remains of *H. erectus* existed along with evidence of stone tool manufacture and transport of stone. The evidence indicates that *H. erectus* relied on a higher-quality diet including meat, which imposed greater demands on technical abilities and social organization, but created opportunities for dispersal and range expansion, explaining the species' extra-African distribution.

At present, identifying the population that gave rise to *H. erectus* is one of the most engaging problems in the study of human evolution. Stone tools are known from several sites in Ethiopia and Kenya before 2.5 million years ago, and cut marks on animal bone indicate that these tools were often used for butchering animals. Between these earliest stone tools and the appearance of clear examples of *Homo* fossils lie nearly 700,000 years of time, during which the distinctive features of *H. erectus* must have been evolving, but the identity of the organisms themselves remains mysterious. Between 1.9 and 1.5 million years ago, a number of fossil crania and a handful of partial skeletons may represent a species known as *Homo habilis*. The crania have larger brains than typical for australopithecines, ranging from 500 to 750 ml, and their teeth and jaws are smaller than earlier australopithecines. Still, it remains unclear whether *H. habilis* was ancestral to *H. habilis*, and scholars disagree about how many species these early specimens represent.

Australopithecus sediba is an exceptionally interesting sample, dating to 2 million years ago from Malapa, South Africa. Two very complete skeletons of this species combine *Homo*-like teeth and hands with the body proportions, brain size, and possible arboreal adaptations of earlier hominins. Whether this species could be ancestral to *H. erectus* or *H. habilis* or both could be influenced by analysis of a handful of fossil fragments from East Africa. These have, in the past, been assigned to *Homo*, but until more is known about their anatomy, it will be difficult to test hypotheses about their relationships.

The expansion of brain size from *Australopithecus* to *Homo* is correlated with many aspects of life history and behavior. Neural tissue imposes a high metabolic cost, which humans met by adopting dietary and behavioral strategies that provide high caloric returns. The first postnatal year of human brain development includes a rapid expansion of brain size and concomitant shape changes, in contrast to developmental trajectories of other primates. Neural development in humans extends across a long childhood, with late sexual maturation and an adolescent growth spurt. These ontogenetic patterns appeared in concert with increasing brain size in Pleistocene humans. An increased dependence in hunting and meat scavenging compared to other primates yielded a net increase in diet quality, but imposed several risks, such as competition with large carnivores, unreliability of game, and long training necessary for skill development. Modern humans mitigate these risks by food sharing, sexual division of labor, and gathering of plant foods and animal resources including honey. Hunter-gatherer social groups are relatively egalitarian, with decision making regulated by a coalition of many group members. In this setting, learning of social rules and communication about social norms are fundamental determinants of survival and reproduction. This social environment is thought to be a major selective driver of larger brains that allowed for more sophisticated communication and inferences about the intentions of other social actors. Whereas australopithecines had vocal tracts similar in form to those of chimpanzees and gorillas, early *Homo* had both vocal and auditory traits that could have supported humanlike sound production and reception.

After its origin, *Homo* diversified into regional populations with some morphological differences. In East Asia, *Homo erectus* occupied a range from north China to Java, which was connected to the Asian mainland during periods of low sea level. Across this range, populations developed regional variation in the shape of the browridge and forehead, extent of muscle development of the jaw and neck, and shape of the teeth. Some of these people made a deepwater crossing to the island of Flores by 1 million years ago, where later they may have evolved into a late-surviving isolated dwarf population called *Homo floresiensis*.

In Africa, the fossil record is sparser but supports the idea that *Homo* increased in variability in the period after 1.2 million years ago. The West and South Asian archaeological records show that these regions were also occupied by early human populations, but scant fossils remain. Europe was inhabited by 1.2 million years ago, but the skeletal record represents chiefly the last 800,000 years.

Everywhere they lived, humans used stone tools. The basics of production involved the procurement of stone raw material either from rocky outcrops or from rounded cobbles in streambeds. People were selective about material, choosing fine-grained stone with predictable fracture dynamics, which they sought and transported over kilometers. Removing a sharp flake by itself yields a reliable cutting edge; removing several flakes from a rock, or "core," can shape an edge suitable for chopping or piercing bone. This basic technological pattern is called Oldowan. After 1.6 million years ago, however, mainly in Africa and later in Europe and West Asia, people shaped core tools into symmetrical tools with long edges, called hand axes. The resulting Acheulean stone industry persisted for some 1.3 million years. Along with stone, archaeologists know that Pleistocene humans often used fire, wooden spears and other implements, and sometimes tools made of bone.

By 300,000 years ago, brain size had increased the range of *Homo* to between 800 and 1300 ml. Most paleoanthropologists refer these later remains to species other than *H. erectus*. In Africa and Europe, they are often called *Homo heidelbergensis*, while many scientists call them "archaic *Homo sapiens*." Whatever they are called, these people began to experiment with different technical forms, including a process of stone tool manufacture known as a *prepared core* technique. The result was a greater control over the shape of end products, sometimes yielding blades and points that were attached (hafted) onto spears as compound tools. These stone industries are called Middle Stone Age (MSA).

3. NEANDERTHALS AND THE ORIGIN OF MODERN HUMANS

Genetic evidence has greatly clarified our understanding of the human populations of the last 250,000 years. Archaeology and skeletal remains help to complete the story, adding perspective on the causes and timing of the key events. This was a time of vast migrations and mixture of distant populations with each other.

By 250,000 years ago, MSA people had developed regional tool industries with little evidence of interregional movement or exchange. A small skeletal sample represents these MSA populations from across Africa. These represent the earliest humans with modern anatomical characteristics, including a high forehead, face tucked beneath the front of the braincase, and a rounded cranial vault. The functional import of these changes is not yet understood, but they seem to reflect a basic shift in developmental patterning.

The later MSA peoples, after 120,000 years ago, became regionally differentiated. In both southern Africa and the Maghreb, people collected shells and marked objects, for example, with natural pigments and ostrich eggshells. In Mozambique, people gathered large stores of wild grains; in Ethiopia, they transported obsidian over hundreds of kilometers.

Some African population dispersed into West Asia by 105,000 years ago, taking with it a subset of the genetic variation present in Africa. In western Asia and Europe they encountered the Neanderthals, whose remains are dated between 200,000 and 300,000 years ago. Beginning from a common anatomical background with modern humans, Neanderthals evolved a number of traits that appeared nowhere else: long, barrel-shaped skulls with a rearward projection called an "occipital bun," thick curving long bones with large joints, and at least in the European part of their range, body proportions now associated with inhabitants of very cold environments. Neanderthals were a small population dispersed over a large space, and even more than their contemporaries in Africa, depended heavily on meat from large prey animals. The Neanderthals were probably a minor component of the overall Pleistocene human population, but their skeletal and archaeological remains are numerous, so we understand their lifeways better than other populations. Additionally, it has proved possible to obtain a partial genome sequence from Neanderthals, which has shed great light on the genetic ancestry of modern humans outside of Africa.

A sign of the weaknesses of the skeletal record is the Denisova genome, from the Altai Mountains of southern Siberia. This genome represents a population living at the same time, but to the east of the Neanderthals, but substantially distinct from the known Neanderthals genetic sample. Living people in Australia and New Guinea derive around 5 percent of their ancestry from a population similar to the Denisova individual. Neanderthals themselves contributed between 2 and 4 percent of the ancestry of present populations throughout the world (including Australasia), except within Africa itself. These genetic results may help to explain morphological features that imply some degree of regional continuity of human populations in Europe, East Asia, and Australasia; however, the spread of Africans within the last 100,000 years accounts for more than 90 percent of the ancestry of living people, but a small multiregional component of ancestry has remained in the face of this and subsequent migrations.

4. RECENT HUMAN EVOLUTION

After modern human populations became established throughout the world, evolution continued to shape our biology. Early human populations in Europe and northeast Asia likely found themselves poorly suited for the low temperature and insolation of these regions. The tropical regions of Asia had a similar physical geography but very different floral and faunal communities than Africa. Watercraft allowed people to colonize Australia, Melanesia, and other island regions, and facilitated the migration of people from the Bering Land Bridge into the southern parts of the Americas before 14,000 years ago. Rapid evolution by natural selection in all these novel environments was inevitable.

As humans dispersed throughout the world, they also increased vastly in numbers. At the end of the last glaciation, people expanded their dietary breadth to a greater number of plant and animal species, a process called the Broad Spectrum Revolution. Some experimented with planting and keeping seed crops; others began managing herd animals more intensively. Over many generations, these processes led to domestication of former wild species, settlement of many human groups into villages and cities, and the rise of political and economic elites. Pastoralists sustained large populations on formerly less hospitable plains and steppes, sometimes migrating over long distances. Civilization was one result of this agricultural revolution; warfare and serfdom were others.

Human skeletal traits (and by inference genes) have changed during the last 20,000 years at a rate unmatched by earlier periods. Humans became more gracile as cranial muscle attachments and structures such as the browridge became lighter. After the introduction of agriculture, smaller teeth and jaws became common, and a higher proportion of individuals failed to develop third molars, or "wisdom teeth," entirely. Along with such evolutionary changes, skeletal samples document the catastrophic health effects resulting from agriculture and village life.

Pathogens have been among the most obvious causes of recent human evolution. For example, more than 20 different alleles that protect to some extent from falciparum malaria are known from different human populations, many of which have arisen within the last few thousand years. Diet is another important cause of recent evolutionary changes, as some human groups have specific genetic adaptations to starchy grains and milk consumption. The physical environment has exerted its own selection on populations at high altitude, with selection affecting oxygen transport, and at high latitude, with recent strong selection on genes associated with pigmentation.

Industrial populations of the last 200 years have undergone further radical changes in longevity, residence patterns, and family size. Nevertheless, selection and evolution of modern human populations is ongoing, with documented selection on quantitative traits of medical and biometric interest. The future direction of human evolution cannot be predicted from our past history (see chapter VIII.12), but the pace of recent evolution suggests that our species may have many more changes ahead.

FURTHER READING

Aiello, L. C., and J.C.K. Wells. 2002. Energetics and the evolution of the genus *Homo*. Annual Review of Anthropology 31: 323–338. *A summary of the impacts of a transition to an energy-dense (high-meat) diet and its significance to some of the main differences between* Homo *and its progenitors.*

Antón, S. C., W. R. Leonard, and M. L. Robertson. 2002. An ecomorphological model of the initial hominid dispersal from Africa. Journal of Human Evolution 43: 773–785. *Analysis and discussion of the dispersal dynamics of* Homo.

Boehm, C. 1993. Egalitarian behavior and reverse dominance hierarchy. Current Anthropology 34: 227–254. Available from www.jstor.org/stable/2743665. *An interesting discussion of the conditions under which egalitarian human societies are expected to arise, with implications for the social structures in archaic humans.*

Brumm, A., G. M. Jensen, G. D. van den Bergh, M. J. Morwood, I. Kurniawan, F. Aziz, and M. Storey. 2010. Hominins on Flores, Indonesia, by one million years ago. Nature 464: 748–752. Available from dx.doi.org/10.1038/nature08844. *Description of a diminutive fossil hominin from the Indonesian island of Flores.*

Dunbar, R.I.M. 2003. The social brain: Mind, language, and society in evolutionary perspective. Annual Review of Anthropology 32: 163–181. *A review of the hypothesis that the unusually large brains of* Homo *evolved in response to the cognitive demands of living in social groups with complex bonds.*

Green, R. E., J. Krause, A. W. Briggs, T. Maricic, U. Stenzel, M. Kircher, N. Patterson, et al. 2010. A draft sequence of the Neandertal genome. Science 328: 710–722. Available from dx.doi.org/10.1126/science.1188021. *Description of the genome sequence of Neanderthals and a comparison with modern human variation.*

J. Hawks, E. T. Wang, G. Cochran, H. C. Harpending, and R. K. Moyzis. 2007. Recent acceleration of human adaptive evolution. Proceedings of the National Academy of Sciences USA 104: 20753–20758. Available from dx.doi.org/10.1073/pnas.0707650104. *An analysis showing that contrary to common perception, humans have experienced recent, rapid adaptive evolution that is likely ongoing and accelerating.*

Leakey, M. G., F. Spoor, F. H. Brown, P. N. Gathogo, C. Kiarie, L. N. Leakey, and I. McDougall. 2001. New hominin genus from eastern Africa shows diverse middle Pliocene lineages. Nature 410: 433–440. *Description of Kenyanthropus and discussion of its relationship to* Australopithecus.

McBrearty, S., and A. S. Brooks. 2000. The revolution that wasn't: A new interpretation of the origin of modern human behavior. Journal of Human Evolution 39: 453–563. *A discussion of the origins of many modern human traits, such as sophisticated Middle Stone Age tools and other artifacts; argues that these traits emerged gradually within Africa but then rapidly spread to the rest of the world 40,000–50,000 years ago.*

McHenry, H. M., and K. Coffing. 2000. *Australopithecus* to *Homo*: Transformations in body and mind. Annual Review of Anthropology 29: 125–146. *A review providing useful perspectives on the origin of the genus* Homo.

Patterson, N., D. J. Richter, S. Gnerre, E. S. Lander, and D. Reich. 2006. Genetic evidence for complex speciation of humans and chimpanzees. Nature 441: 1103–1108. *A genomic analysis that dates the human-chimpanzee divergence and argues for gene flow between the two lineages after their initial split.*

Reich, D., R. E. Green, M. Kircher, J. Krause, N. Patterson, E. Y. Durand, B. Viola, et al. 2010. Genetic history of an archaic hominin group from Denisova Cave in Siberia. Nature 468: 1053–1060. Available from dx.doi.org/10.1038/nature09710. *Description of the genome sequencing of archaic human remains that were contemporaneous with Neanderthals, but geographically separated.*

White, T. D., B. Asfaw, Y. Beyene, Y. Haile-Selassie, O. C. Lovejoy, G. Suwa, and G. Wolde. 2009. *Ardipithecus ramidus* and the paleobiology of early hominids. Science 326: 75–86. Available from dx.doi.org/10.1126/science.1175802. *Description of* Ardipithecus ramidus *and discussion of its significance for understanding of human origins.*

Wood, B., and T. Harrison. 2011. The evolutionary context of the first hominins. Nature 470: 347–352. Available from dx.doi.org/10.1038/nature09709. *A review of the challenge of identifying fossil hominins and determining whether they belong to the lineage leading to modern humans.*

III

Natural Selection and Adaptation

Douglas J. Futuyma

Natural selection is the centerpiece of Darwin's great book, and is prominent in its title: *On the Origin of Species by Means of Natural Selection, or the Preservation of Favoured Races in the Struggle for Life*. In fact, this book was a hastily written "abstract" of a book he had started, and intended to be much larger, titled simply *Natural Selection*. To be sure, Alfred Russel Wallace independently conceived the idea, but it was Darwin who deduced its many implications and followed its ramifications in detail, and with whom the concept is usually, and rightly, associated. Natural selection is the most important of Darwin's many original ideas. It is one of the most important ideas in the history of the world.

Why? Because for the first time, there existed a purely scientific explanation of the most powerfully impressive examples of design and purpose in nature: the features of organisms that equip them most exquisitely for survival and reproduction. The adaptations of organisms—including humans—had long been attributed to the Creator's beneficent design, indeed were among the most important arguments for the existence of such a supernatural Creator. As an alternative explanation lacking the slightest supernatural tinge, natural selection made biology a science. Philosopher Daniel Dennett (1995) has nominated natural selection as the best idea in history—and has also termed it "Darwin's dangerous idea," for it threatened the theological substructure of much of Western philosophy and has many ramifications outside biology.

Natural selection occurs whenever there is a consistent, average difference in fitness (reproductive success) among sets of "individuals" that differ in some respect that we may refer to as phenotype (see chapter III.1). Most (but not all) evolutionary biologists would add that the phenotypic difference is at least partly inherited. If this is the case, the difference in reproductive success may result in one phenotype increasing in frequency while another decreases. The eventual outcome may be complete replacement of one phenotype by the other. Throughout *On the Origin of Species* and most subsequent evolutionary discourse, the "individuals" are individual organisms, such as dark versus pale moths or people with different blood types. However, the theory also applies, mutatis mutandis, to other kinds of biological "individuals" that produce more such "individuals": genes, genotypes, cell types within an organism, different populations of the same species, or different species. Thus, selection can act at different levels of biological organization (see chapter III.2).

In order to establish that there exists an average, nonrandom difference in fitness, we must estimate reproductive success of a number of individuals in each class, for we cannot tell whether a difference between two different individuals is caused by, or even correlated with, their phenotypes; one may have just been luckier. Luck—the random, unpredictable element in frequency changes due to sampling error—is termed *genetic drift* (see chapter IV.1). Natural selection, in contrast, is the nonrandom component of variation in reproductive success. A higher proportion of yellow than of brown land snails in a pasture might be trampled by cattle, just by chance, but if the grass is mostly yellow rather than brown, we can predict with considerable confidence that brown snails will suffer higher predation from thrushes, which use color vision to find prey. The thrushes, but not the cattle, act as an agent of natural selection, which is the antithesis of chance.

It is extremely important to recognize that natural selection is not an agent, and certainly not remotely similar to a rational agent; casual talk of "Mother Nature" or simply of "nature" as a personified entity (as in "nature tends to . . . ," or "nature selects" or "nature has found a solution") is misleading. It is even misleading to say, "selection acts at different levels of biological organization" (as in the previous paragraph!), because selection does not "act." Natural selection really is no more than a statistical difference in reproductive rate, often owing to some mechanistic relationship between a property of the phenotype and some feature of the environment. A common result of the difference in reproductive rate is that one type replaces others. That's all there is to it. When we personify natural selection (as almost all biologists do at times, including Darwin, for the sake of comfortable, less stilted discourse), we can easily slip into describing selection as if it could plan, as if it had the species' welfare in mind, as if it were beneficent—or cruel. Such modulation from one concept to another has resulted in frequent misunderstanding and misrepresentation of natural selection and evolution. For example, many writers have referred to natural selection as if it had a goal, such as perpetuation of the species or "evolutionary progress"; but selection cannot have a goal of any kind. Natural selection results in the evolution of characteristics that enhance the survival of the individual organism (or of its genes), characteristics that we might metaphorically call "selfish"; indeed, characteristics that promote "cooperation" among organisms call for special explanations, such as *kin selection*, in which cooperation with or aid to a related individual increases the frequency of their shared genes, including the genes that underlie the propensity to cooperate or aid (see chapter III.4, also chapters VII.8–VII.10). But natural selection itself is not selfish—or cruel, or kind—except in a purely metaphorical sense. Above all, natural selection is not a normative "law of nature," prescribing right or ethical conduct. The behavior that may have evolved by natural selection holds no prescription for moral or ethical human conduct.

Among the levels at which natural selection can occur is the level of the individual gene or DNA sequence (see chapter IV.7). This occurs whenever different sequences make different numbers of copies that are transmitted to the next generation. For example, transposable elements proliferate within the genome at higher rates than "normal" genes. Natural selection can also occur at the level of species, for certain characteristics enhance the rate of origin of new species or diminish the likelihood of species extinction (see chapter VI.14). For instance, the number of species in lineages of herbivorous insects has generally increased faster than in closely related lineages that have other feeding habits. Neither gene selection nor species selection has molded the advantageous characteristics of individual organisms; rather, they have affected properties at the gene level or at the species level. But individual selection, selection among individual organisms within populations, is at the center of evolutionary theory. It is at this level that selection explains most of the adaptive features of organisms. An alteration of a bird's beak caused by a gene mutation is a feature of an individual bird, not a feature of the gene or of the entire species, except insofar as more or fewer individuals of the species possess it. And it is the process of individuals' birth and death rates that alters the frequency of the mutated gene, on the one hand, and the character of the entire species, on the other.

Understanding the dynamics and consequences of natural selection requires familiarity with theory, as framed in terms of population genetics. Most population genetic models of selection consist of (1) a postulated relationship (*mapping*) between genotype and phenotype, (2) a frequency distribution of genotypes, and (3) a postulated relationship between phenotype and fitness. Together, these determine the frequency distribution of genotypes and phenotypes in subsequent generations. For example, if body size differs among the three genotypes (A_1A_1, A_1A_2, and A_2A_2) at a locus with two alleles in a sexually reproducing population, the outcome of selection on body size depends, first, on whether or not the heterozygote (A_1A_2) is intermediate in size between the two homozygotes (i.e., on the mapping between genotype and phenotype). The rapidity of change depends on the frequency distribution of the phenotypes; all else being equal, evolution under selection is faster if all the genotypes are common than if one homozygote makes up most of the population. And the final outcome depends on the "mode" of selection, the relationship between fitness and phenotype. If largest individuals are most fit, the outcome is fixation of the A_1A_1 genotype if the heterozygote is intermediate in size, but stable maintenance of variation (polymorphism) if the heterozygote is largest. Some relationships among genotype, phenotype, and fitness are more complex, resulting, for example, in diverse, historically contingent outcomes that depend on initial conditions.

Much of this theory is cast in terms of frequencies of genotypes and alleles (see chapter III.3), and is readily applied to genetic data (e.g., DNA sequence variation) and to phenotypic data when the phenotypes are distinct categories (e.g., black and white); however, the variation in most phenotypic traits (e.g., human stature) is continuous, or gradual, owing to variation at many gene loci, as well as influences of environmental variables on growth and gene expression. In this case, the effects of individual genes are difficult to discern, so the frequencies of alleles and genotypes are very difficult to measure. Evolution of such traits is most often analyzed by the statistical tools of evolutionary quantitative genetics, which are founded on population genetics (see chapter III.3). These statistical tools are mostly variances, covariances, and correlations, which are used to partition variation in a character into its genetic and nongenetic components, and to characterize the degree to which different characteristics are inherited together. The genetic variance in a character plays a major role in its "response" to natural selection (see chapters III.5

and III.6), and the correlations among characters affect the extent to which characters can evolve independently. Consequently, the theory and analytical methods of population genetics and quantitative genetics are often indispensable both for predicting and studying evolutionary changes caused by selection, and for understanding the limits of natural selection—the ways in which species may fail to adapt (see chapter III.8).

The genetic theory of natural selection is often useful—although not always indispensable or easily applied—for analyzing and understanding *adaptations*, which many biologists would define as features of organisms that have evolved by natural selection. It is the adaptations of organisms, often so exquisitely suited to particular tasks or ways of life, that "so justly excite our admiration," as Darwin wrote, and that constituted the argument for supernatural design before Darwin demolished it. The study of adaptations pervades much of biology and is at the core of many chapters throughout this book. For this section, we have assembled a set of chapters treating the adaptive evolution of several classes of characteristics that have been the focus of extensive evolutionary research. The evolution of *biological systems* (especially modes of reproduction and inheritance; see chapter III.9) concerns such questions as why some organisms reproduce sexually and others asexually. A *reaction norm* (see chapter III.10) is the variety of phenotypes that a genotype may have, depending on environmental conditions (e.g., slender or obese as a function of diet). Often the reaction norm describes *phenotypic plasticity*, the capacity for nongenetic, advantageous modifications of the phenotype. Why characteristics are or are not phenotypically plastic is a major focus of this chapter. Chapter III.11 treats the evolution of the great array of *life histories* of organisms, focusing on such features as reproduction (why do some species have so few, and others so many, offspring?), generation time (why do humans, periodical cicadas, and century plants take so long to reach reproductive age, compared with most other species?), and life span (why does maximum life span differ among species? Why do we grow senescent and die?).

Much of biology is concerned with analyzing the *function of phenotypic traits* such as anatomical or cellular structures, a study closely related to the physiological and biochemical processes inherent in living systems. The raison d'être of morphological and physiological traits, their adaptive advantages, is a traditional field of evolutionary biology, although the questions and research methods continue to change (see chapters III.12 and III.13). In particular, traditional analyses of the function of an anatomical or physiological characteristic may be joined with studies of the ways in which variation in the trait affects fitness in natural populations, or with genetic analyses of variation in the trait, or with paleontological and phylogenetic studies of its history of evolutionary change.

Many phenotypic traits are adaptations to environmental factors, including other organisms. Evolutionary biology has been intimately associated with ecology ever since Darwin, who may as justly be called the first great ecologist as the first great evolutionary biologist. This section's chapter on the evolution of *ecological niches* (see chapter III.14) concerns the conditions under which species might evolve broad or narrow (specialized) tolerance of abiotic environmental conditions (e.g., temperature), habitat use, or diet. The species' ecological niche, in turn, influences its interactions with other species—its *biotic environment* (see chapter III.15). Adaptation to other species, such as competitors, prey, predators, parasites, and mutualists, has greatly shaped many of the features of species and been of paramount importance in the evolution of biological diversity. These chapters provide a foundation for several other chapters in the guide, especially in Section VI: Speciation and Macroevolution.

FURTHER READING

Dennett, D. C. 1995. Darwin's Dangerous Idea. New York: Simon and Schuster.

III.1

Natural Selection, Adaptation, and Fitness: Overview

Stephen C. Stearns

OUTLINE

1. Natural selection explains adaptation
2. Concepts are tools
3. Definitions and complications
4. Fitness and units of selection
5. Connecting selection to fitness in hierarchies
6. Polished adaptations or rough history
7. Adaptationist storytelling
8. How to recognize selection and adaptation
9. Can we do without these concepts? Absolutely not.

This chapter defines natural selection, adaptation, and fitness, the core concepts in the process driving an important part of evolutionary change, discusses how they relate to each other, and comments on their appropriate and inappropriate use.

GLOSSARY

Density-Dependent Selection. Selection that favors different genotypes or phenotypes at different population densities.

Frequency-Dependent Selection. A mode of natural selection in which either rare types (negative frequency-dependent selection) or common types (positive frequency-dependent selection) are favored.

Genotype. The information stored in the genes of one individual; it can refer to anything from one gene to all the genes, depending on context.

Group Selection. Selection generated by variation in the reproductive success of groups.

Individual Selection. Selection generated by variation in the reproductive success of individual organisms.

Interactor. The organism in its ecological role, in which it develops, grows, acquires food and survives, mates, and reproduces.

Kin Selection. Adaptive evolution of genes caused by relatedness; an allele causing an individual to act to benefit relatives will increase in frequency if that allele is also found in the relatives and if the benefit to the relatives more than compensates the fitness cost to the individual.

Life History Traits. Traits directly associated with reproduction and survival, including size at birth, growth rate, age and size at maturity, number of offspring, frequency of reproduction, and life span.

Maladaptation. A state of a trait that leads to demonstrably lower reproductive success than an alternative existing state.

Phenotype. The material organism, or some aspect of it, as contrasted with the information in the *genotype* providing the blueprint for the organism.

Replicator. The organism in its role as information copier, the mechanism that copies the DNA sequence of the parent and passes it to the offspring.

Trade-off. An evolutionary change that increases fitness in one trait or context but causes a decrease in fitness in another trait or context.

1. NATURAL SELECTION EXPLAINS ADAPTATION

The astonishing precision and elegance with which complex biological structures efficiently function calls out for explanation. The human eye can detect a single photon, allowing it to see a match lit at a distance of 10 miles on a dark night. The nostrils of a migrating salmon can discriminate a difference in concentration of the molecules characteristic of its home stream corresponding to one molecule more in one nostril than the other. The ears and brain of an echo-locating bat can detect a returning echo less than a millionth as loud as the cry it emitted milliseconds earlier and from that information decipher the position, movement, and surface texture of a rapidly and erratically flying insect. Darwin's greatest idea, natural selection, explains the evolution of the myriad of such astonishing adaptations through the operation of simple mechanisms that can be readily observed. A triumph of human thought, it organizes and explains much of biology, strongly contributes to the impact of biology on other disciplines, and as a major scientific principle not contained in chemistry or physics, elevates biology to their rank in its power to explain the natural world.

2. CONCEPTS ARE TOOLS

Natural selection, adaptation, and fitness are concepts used by biologists to organize their understanding of evolutionary processes and to facilitate their communication; like all scientific concepts, they are tools invented to describe aspects of reality. Here I first define selection, adaptation, and fitness. I then note complications introduced by sex, traits, age structure, relatives, and frequency dependence; the complications call attention to important issues. After comparing definitions of fitness and the units of selection to which they apply, I discuss things on which selection can operate in principle but so inefficiently that it produces rough history rather than polished adaptation. I recount the controversy over adaptationist storytelling; and then discuss how to attenuate that controversy by demonstrating selection and adaptation. I conclude by reemphasizing the stature of these major ideas.

3. DEFINITIONS AND COMPLICATIONS

Natural selection is a process of sorting by reproductive success that occurs in populations of replicating units, whether those units are molecules, cells, organisms, or larger units. Four conditions—all necessary and together sufficient—state when natural selection on a trait will occur and elicit a response:

1. The units expressing the trait must vary in their reproductive success.
2. The trait must vary among the units in the population.
3. The correlation between the trait and reproductive success must be nonzero.
4. For a response to occur, the trait must be heritable.

When these conditions hold, the differential reproductive success of the units expressing a heritable trait correlated with reproductive success will change the frequencies of the states of the trait in the population from one generation to the next. Those positively correlated with reproductive success will increase, and those negatively correlated will decrease in frequency. Natural selection can work in populations of anything that satisfies the conditions, not just populations of organisms. Among other units that have been proposed to experience natural selection are groups, species, words, and ideas.

Note that the word *selection* is misleading, for nothing actively selects. The action of "selection" is caused by whatever contributes to the correlation of a trait with reproductive success, a set of many things that have inspired much research.

These four conditions can be more simply expressed as heritable variation + differential reproductive success = natural selection. While its simplicity is attractive, that shorter definition misses two important points. First, it neglects the key role played by the correlation between the trait and reproductive success. If that correlation is zero, the trait (or gene) drifts aimlessly; it does not change systematically. Thus, the simpler definition misses the point that the conditions for selection and random drift are quite similar, differing only in the value of the correlation with reproductive success. Second, the simpler definition ignores the important distinction between selection and the response to selection. Drawing that distinction calls attention to the two main steps in the process, the two things that need to be measured to establish that selection is occurring and eliciting a response: the correlation of the trait with reproductive success, and its heritability.

Adaptation is both a process and a state. As a process, adaptation describes the portion of evolutionary change in a trait that is driven by natural selection. (There are other reasons for inherited evolutionary change, genetic drift being an important one.) As a state, adaptation describes that aspect of the current condition of a trait that can be reliably ascribed to the past action of natural selection. Although traits are often loosely referred to as "adaptations," what is meant more precisely is not the entire trait, which usually has a long complex history and is a mosaic product of several processes, but that aspect of the trait that has been produced by natural selection. Identifying that aspect is one aim of the adaptationist research program.

Fitness is a word meant to capture a measure of relative reproductive performance that can be used to predict long-term dynamics. If heritable, traits that confer greater fitness increase in a population subject to natural selection; those that confer less fitness decrease. The meaning of fitness is context specific in ways that illuminate several major issues in conceptualizing evolutionary change. Among the factors defining those contexts are sex, traits, age structure, spatial structure, relationships, and hierarchies.

First Complication: Sex

In an idealized population of asexual organisms reproducing by binary fission, fitness is relatively easy to define. Because every reproductive act results in two daughter cells, there is no variation in the population in number of offspring per reproductive cycle. What can vary are two components of fitness—the time it takes a cell to divide, and the probability that a daughter cell will survive—and the genetic constitutions of the daughter cells, which differ only through mutations. If survival is fixed, then fitness reduces to a measure of cell cycle time; if cell cycle time is fixed, then fitness reduces to probability of survival per cell cycle; a combined composite measure allows projection of the frequencies of performance variants. Here the unit of replication is the entire asexual genome; the unit of interaction is the entire asexual cell; and any changes in performance introduced by a mutation are linked to the entire genome and expressed in the entire cell.

Sex introduces the complication of recombination, which in eukaryotes is caused both by the independent segregation of chromosomes and by crossing-over within chromosomes (see chapter III.10). Sex has two important consequences for fitness: mutations in one gene are no longer linked to the entire genome but to a local region within a chromosome whose size diminishes through generations of crossing-over, and effects on performance are not expressed in one genetic background but in the many genetic backgrounds of the recombinant descendants. Sex thus uncouples the unit of replication—the gene—from the unit of interaction—the organism. It also associates the consequences of a genetic change—a mutation—with average effects on recombinant descendants rather than with cloned daughters that inherit the rest of the genome with its gene-gene interactions intact. This is one reason population geneticists chose to conceptualize evolution as changes in gene frequencies and to associate selection coefficients with alleles. Population genetics is in part an analytic reaction to the existence of sex.

Second Complication: Traits

Organisms can be analyzed into traits produced by a developmental map linking genotypes to phenotypes. Focusing on traits rather than whole organisms unveils a rich structure demanding special analysis. From the perspective of the gene, one gene often affects two or more traits, a pattern called *pleiotropy*; from the perspective of the trait, one trait is often influenced by more than one gene, a pattern called *epistasis*; each captures a different aspect of the web of connections between genes and traits. From the perspective of the whole organism, traits are connected in relationships that result in *trade-offs*, a term that includes the genetic effects of epistasis, pleiotropy, and linkage and adds to them physiological effects mediated by hormones, energy allocation, and signaling conflicts. Trade-offs represent the functional relationships among traits that result in a particular class of correlated responses to selection; a trade-off exists when an evolutionary change that increases fitness in one trait or context causes a decrease in fitness in another trait or context. In one widely used analytical framework that neatly expresses the complications introduced by traits, the linkages among traits are represented as the off-diagonal elements of genetic and phenotypic variance-covariance matrices, and the correlations of the traits with fitness are represented as a vector, the selection gradient (see chapter III.5).

The objective existence of traits is supported by the genetic observation that pleiotropy is not homogeneously distributed over the genome but is organized into sets of traits, that is, modules within which genes have strong pleiotropic effects and among which they have weaker or even no such effects.

Traits whose combination yields a fitness measure are called *components of fitness* or *life history traits*; the most important and frequently used components of fitness are age-specific birth and survival rates. All traits are related to the components of fitness, both mechanistically through the functional connections of genetics, development, morphology, and physiology, and statistically through their correlational impact on survival and reproduction, a level at which ecology and behavior also play a role. The key trade-off between reproduction and survival that shapes the evolution of life span and aging is also called *the cost of reproduction* (see chapter III.11).

Thus a focus on traits introduces three essential concepts with many consequences: first, the idea that all traits are involved in trade-offs, implying that evolutionary improvements in one trait are linked to evolutionary costs in others; second, the idea that fitness is a composite measure that summarizes the contributions of many traits to reproductive performance; and third, the idea that combinations of traits can yield synergistic fitness effects; for example, light bones and wings are of much less use by themselves than they are in combination.

Third Complication: Ages, Stages, and Sites

There is good reason to introduce the complications of age-, stage-, and site-specific birth and death rates to the analysis of fitness and adaptation: organisms reproducing or surviving at various ages and stages and at different sites can differ in their contributions to future generations, with two very important consequences: aging and maladaptation, which are, in this view, analogous (see chapter III.11).

In any population that is growing or stable, reproduction earlier in life contributes more to future generations than does reproduction later in life, for there is always a risk of dying between the earlier and the later reproductive events, and surviving offspring produced earlier in life will start contributing to future generations through their own reproduction before those produced later in life can do so. That idea, combined with the recognition of trade-offs between reproduction, growth, and survival, yields the theory of the evolution of life histories and aging and with it our understanding of why all living things must grow old and die, why some organisms are large and others small, why some mature earlier and others later, and why some have many offspring and others only a few (see chapter III.11).

Similar insights come from the recognition of spatial heterogeneity in populations. In any population that is distributed among geographical sites, some sites are usually *sources* and others, *sinks*. Conditions in sources support excess reproduction that generates emigrants, whereas conditions in sinks do not permit reproduction adequate to maintain the population within the sink, which continues to exist only because of immigration from sources. Organisms will be adapted to sources and maladapted to sinks; this has important implications for both basic science and conservation biology (see chapter III.14).

Just as aging is a by-product of selection for reproductive success early in life, so maladaptation of organisms in sinks is a by-product of selection for reproductive success in sources.

Fourth Complication: Relatives (Kin Selection and Inclusive Fitness)

Because both sexual and asexual organisms have relatives, both genetic systems share a basic property: they encounter situations where actions taken by the focal organism have consequences for the probability that a relative will contribute genes to future generations. Since the same gene can be transmitted to the next generation either through the reproduction of the focal organism or through the reproduction of its relatives, how relatives behave toward each other is one path to reproductive success. Recognition of this fact, and development of its consequences, led to the ideas of kin selection and inclusive fitness, which have yielded many insights into altruism, cooperation, and conflict (see chapter VII.9).

Consider a focal organism, one of its relatives, and an act with fitness consequences for both. A new mutation that affects the probability of that act will invade the population if the benefits gained by the relative (b: its increase in reproductive success as a consequence of the act), multiplied by the coefficient of relationship between the focal organism and the relative (r: the probability that the relative has a copy of the gene that is identical by descent from a shared ancestor), is greater than the cost to the focal organism (c: its decrease in reproductive success as a consequence of the act): that is, if $b \times r > c$. This can also be written as $b \times r > c \times 1$; that is, the benefit to the recipient, weighted by the coefficient of relationship to the donor, must be greater than the cost to the donor, weighted by its coefficient of relationship to itself, which is 1.

Fifth Complication: Frequency and Density Dependence

Often the fitness of genetic or phenotypic variants depends on the frequencies of the other variants present in the population. Here the tool of choice is game theory, and the questions to ask are, under what conditions will a variant increase when rare, thus invading the population, and under what conditions will a common variant resist invasion by all other variants, persisting in a state of evolutionary stability? (See chapters III.3 and III.5.)

Two examples suggest the importance and scope of frequency dependence. First, what is the stable ratio of males and females in a randomly mating monogamous population? If the population consists mostly of females, males are favored, for each of the rare males will have several mates, whereas each female will have at most one. If the population consists mostly of males, females are favored, for each of the rare females will find a mate, but some males will not mate at all. If offspring sex ratio is heritable, the evolutionary equilibrium will be equal frequencies of males and females: a 50:50 sex ratio. Second, what is the stable frequency of a host genetic variant that confers resistance to a specific pathogen genotype? When that variant is rare, it will increase in frequency, for the commoner variants suffer from infection by the pathogen. As the host variant becomes common, pathogen variants are selected that improve the pathogen's ability to infect that host genotype, and its spread is halted or reversed. Here being rare is advantageous, and being common is costly. The prediction is that the number of genetic variants conferring resistance will increase until there are so many of them that even the commonest is effectively rare.

Negative frequency dependence—the advantage of being rare and the cost of being common—efficiently maintains variation in populations. Here there is no single best solution: many rare variants persist stably.

Often the fitness of genetic or phenotypic variants depends on population density. Many things change with population density, among them the importance of traits mediating intraspecific competition, of traits conferring resistance to predators and diseases, and of traits that allow communication with potential mates over short versus long distances. When population densities fluctuate regularly from low to high, traits can experience alternating and sometimes conflicting selection pressures. At low densities many organisms tend to grow rapidly to large size, maturing early and having many offspring. At high densities many organisms tend to grow slowly to smaller size, maturing later and having fewer offspring.

Frequency and density dependence are widespread. Their consequences for population dynamics and community ecology are being analyzed by the growing field of adaptive dynamics, which explicitly considers the effects of population dynamics on fitness and the feedbacks between ecological and evolutionary processes.

4. FITNESS AND UNITS OF SELECTION

If life had remained completely asexual, genes—the replicators—would have remained consistently associated with organisms—the interactors—and there would be little occasion to wonder whether genes or organisms are the units on which selection operates. The evolution of sexual reproduction, which enabled genes to move horizontally into many lines of descent, produced the conditions under which kin selection can be most readily detected. While kin selection does operate in asexual lines of descent, for example, selecting for programmed cell death in the development of multicellular organisms, this was seen only retrospectively after its effects had been recognized in situations where degrees of relationship differed more dramatically.

The realization that it was useful to conceptualize the evolutionary process as driven by differences in the fitness of genes led to ideas like the selfish gene, with support coming from successful analyses of altruism and cooperation in terms of kin selection. Two things need to be said about that. First, that selection acts to increase the copy number of genes in populations rather than promoting the survival of organisms is strongly supported by the theory of the evolution of aging and by the many experiments that have confirmed it. The soma is disposable; it is the genes that persist. This body of evidence is just as strong as or stronger than the behavioral evidence supporting the kin selection argument. Second, it is not necessary to decide whether selection acts on genes or on organisms because both levels are involved: selection acts through the differential reproductive success of organisms—the interactors—to generate differences in populations in the number of copies of genes—the replicators.

5. CONNECTING SELECTION TO FITNESS IN HIERARCHIES

The controversy over whether cooperation and altruism are produced by selection acting on groups, on kin, or on individuals has now continued for nearly fifty years and occupied thousands of pages. Many feel it has been settled in favor of kin selection; others continue to argue for group selection. The space available here will not do the controversy justice, but two brief remarks are in order (see chapter III.2).

First, it helps to distinguish between selection acting on groups and selection acting on genes to increase the fitness of individuals as mediated by their social interactions within groups. Attention to this point reveals that some, but not all, of the evidence produced in favor of group selection is simply evidence that selection has acted to improve the fitness of individuals or genes in the social context of the group.

Second, the conditions that determine the potential strength of selection at any level in a hierarchy are easy to state (although they may be hard to measure). Consider a hierarchy with just two levels—individuals and groups—and an extreme case used here to make a point clearly. If a population of individuals is organized into a set of groups, and the genetic variation in the population is distributed in such a manner that there is no variation within the groups, while each group differs genetically from every other group, then all the potential for a response to selection consists of the differences among groups. And if, in such a group-structured population, there are no differences in the reproductive success of any of the individuals within groups, and large differences in the reproductive success of the groups, then the strength of selection acting on individuals will be zero, and the strength of selection acting on groups will be strong. Any population that fulfills these two conditions will unquestionably experience group selection. Relaxation of these extreme conditions leads to intermediate cases in which selection is acting at both levels of the hierarchy.

The question is how often conditions favoring group selection are fulfilled in reality. There are good reasons to think they are infrequent. In most populations, there is more genetic variation among individuals within groups than there is among groups. That distribution has been well measured, for example, in humans, where about 85 percent of genetic variation is among individuals within groups. Furthermore, the births and deaths of individuals that create variation in individual reproductive success are much more numerous and occur on a much shorter interval than do the splitting and local extinction of groups. During the time it takes for one episode of group selection to happen, there has usually been opportunity for millions of events of individual selection to take place. Under such circumstances, group selection would have to be very strong indeed to overcome modest but consistent differences in individual reproductive success.

Those who favor kin and individual selection point out that the conditions under which group selection works—high degrees of relationship within groups, low degrees of relationship among groups—are also the conditions under which kin selection most efficiently selects for individual traits that benefit group cohesion. Those who favor group selection express the opposite perspective on the same situation: namely, that conditions that enable kin selection also favor group selection. A compromise view is that selection can act at any level of a hierarchy, that its efficiency in eliciting a response is determined by the distribution of genetic variation and the variation in reproductive success among units at that level, and that in principle it can act simultaneously at all levels. It usually does so, however, much more efficiently at the levels of genes and individuals than at that of groups.

6. POLISHED ADAPTATIONS OR ROUGH HISTORY

The realization that selection acts simultaneously at many levels, but usually much more frequently, and much faster, at lower levels than at higher ones, suggests two points. First, we will see the appearance of polish and design associated with adaptations only when those traits have experienced a vast number of selective events, for the thoroughness and efficiency with which the space of phenotypic alternatives is explored depend on the number of selective events and the rate at which they occur. For example, the ability of a bat to fly rapidly through a network of closely spaced branches in complete darkness can be explained only by a vast number of selective events. Second, selective events at higher levels that occur infrequently, such as differences in the splitting and extinction rates of species and higher clades, cannot create polished adaptations, for they occur too infrequently to do so, but they can generate the rough and arbitrary look that we attribute to history. For example, the current dominance of mammals and birds in the world fauna has as much to do with the disappearance of the many groups that were eliminated in the end-Cretaceous extinction as with any adaptations characteristic of mammals and birds, both of which originated millions of years before that extinction event (see chapter VI.14). Occasionally selective events at higher levels can produce broad patterns, such as the distribution of asexual reproduction up at the tips of phylogenetic trees regularly associated with sexual ancestors, a pattern that suggests that asexual lineages go extinct more rapidly than do sexual lineages.

7. ADAPTATIONIST STORYTELLING

The power of natural selection to generate complex adaptations from mutations whose effects are random with respect to the needs of the organism strikes some with the force of an epiphany. Such converts are then inclined to explain most of what they see as the product of selection. They usually find it easy to posit a scenario in which selection could have produced the trait in question by invoking some set of conditions tailor-made for the purpose, and they can get carried away and claim that the states of certain traits are adaptations without producing evidence sufficient to exclude alternatives, despite the fact that there are always at least the alternatives that the trait attained that state through genetic drift, or that it emerged as a by-product of selection on other traits, or that it exists in that state because it is constrained and cannot be otherwise. That is why loose thinking within the adaptationist tradition elicited a strong critique, a critique so strong that it produced a temporary overreaction during which any claim of adaptation was suspect. The pendulum has recently swung back to a balanced position that admits that adaptations are frequent but demands evidence to support the claim.

8. HOW TO RECOGNIZE SELECTION AND ADAPTATION

The problem with a claim of adaptation arises when one has not observed the evolutionary process that produced the state of the trait in question. There are at least three ways to support the claim of adaptation. If one has observed heritable changes in the trait that resulted from the correlation of trait state with reproductive success, then the *change* in the trait is an adaptation. If one can perturb the trait and demonstrate, with credible controls, that the original state has higher fitness than the perturbed states, then the original state was an adaptation relative to the perturbed states. If a developmental change in a phenotype that improves reproductive success relative to the unchanged state occurs only in response to a specific environmental signal, and the change is not expressed without the signal under circumstances where it would decrease fitness if it were expressed, then the *change* in the trait is an adaptation.

While it is true that adaptation is asserted more frequently than such tests are applied, such tests have nevertheless often been done. For example, altering widowbird tail lengths by cutting and pasting showed that longer tails enhance male reproductive success. And raising mosquitofish in fresh and brackish water demonstrated that those living in freshwater were maladapted to that environment, genetic analysis revealing that the freshwater population was swamped by gene flow from the nearby, much larger population living in brackish water. Both outcomes are possible, and both are accepted.

9. CAN WE DO WITHOUT THESE CONCEPTS? ABSOLUTELY NOT

Selection, adaptation, and fitness are well-tested ideas that have proven their usefulness in organizing information, communicating it efficiently, and motivating research for 150 years (selection) or more (adaptation). They combine to explain the millions of cases in biology of precision and complexity—the pervasive matches of structure to function—that cannot be explained by any other set of concepts. Because of their power and scope in making sense of universal features of biological systems, they belong to the very small set of elite concepts that explain all of biology.

FURTHER READING

Bell, G. 2008. Selection: The Mechanism of Evolution. 2nd ed. New York: Oxford University Press. *An up-to-date account of selection as the principal agent of evolution.*

Endler, J. A. 1986. Natural Selection in the Wild. Princeton, NJ: Princeton University Press. *A summary, now somewhat dated, of the prevalence of selection in nature.*

Rose, M. R., and G. V. Lauder, eds. 1996. Adaptation. San Diego: Academic. *A multiauthor rehabilitation of the concept of adaptation.*

Sober, E., ed. 2006. Conceptual Issues in Evolutionary Biology. Cambridge, MA: MIT Press. *A collection of papers taking sides on controversies over adaptation, the units of selection, and the definition of fitness.*

III.2

Units and Levels of Selection

Samir Okasha

OUTLINE

1. The group selection controversy
2. Kin selection, inclusive fitness, and the gene's-eye view
3. Species selection
4. Major evolutionary transitions

The levels-of-selection question asks at which level or levels of the biological hierarchy the process of natural selection takes place. After a brief historical introduction, this chapter sketches some of the more important positions in the levels-of-selection debate. Topics discussed include group selection and kin selection, the gene's-eye view of evolution, species selection, and the major evolutionary transitions.

GLOSSARY

Altruism. A behavior that is costly for the individual that performs it but beneficial to others, where the costs and benefits are measured in units of Darwinian fitness: expected number of offspring.

Genic Selection. Sometimes used to describe any selection process resulting in gene frequency change; also used more narrowly to refer to selection between the genes within the genome of a single organism.

Group Selection. The idea that natural selection can operate on whole groups of organisms, favoring some types of group, or group traits, over others.

Kin Selection. The selection of a behavior because of the behavior's effect on the reproductive success of the relatives of the organism performing it.

Level of Selection. A level in the biological hierarchy at which natural selection takes place; thus, if selection favors some types of individual in a population over others, selection occurs "at the level of the individual," and so on.

Multilevel Selection. The idea that natural selection can operate simultaneously at more than one level of the biological hierarchy, for example, at the individual level and at the group level.

Species Selection. Selection acting at the level of species, favoring those species best able to avoid extinction and/or to leave daughter species. Sometimes thought to play an important role in macroevolution.

Unit of Selection. Used by some authors as a synonym for *level of selection*, or the entity on which natural selection acts, but by others to mean a unit transmitted intact across the generations, also called a *replicator*.

The *levels-of-selection question* is a fundamental one in evolutionary biology, for it arises directly from the logic of Darwinian theory. In *On the Origin of Species*, Darwin argued that if the organisms in a population vary, and if some variants are more successful than others in the "struggle for life," and if offspring tend to resemble their parents, then evolution by natural selection will occur—over time, the fittest variants will eventually supplant the less fit. It is easy to see that in principle, Darwin's argument could apply to biological entities other than individual organisms, for example, genes, cells, groups, colonies, or species. Any entity exhibiting variation, differential fitness, and heritability (or parent-offspring resemblance) could in theory be subject to evolution by natural selection, no matter the level of the biological hierarchy occupied by that entity. This possibility is what gives rise to the levels-of-selection question.

The question of levels of selection is intimately linked with the paradoxical problem of altruism, both historically and conceptually. *Altruism* in biology refers to a behavior that is costly to the individual that performs it but beneficial to others, where the costs and benefits are measured in terms of fitness—that is, survival and reproduction. At first glance, the existence of altruism seems hard to square with Darwinian principles: surely natural selection should lead individuals to behave in ways that benefit themselves, not others? Yet altruism is quite common in nature, particularly among social animals (and has even been found among microbes). For example, sterile workers in social insect colonies devote their entire lives to assisting the reproductive efforts of the queen, foregoing personal reproduction. One possible solution to this paradox, first suggested by Darwin himself, is to invoke selection at the group or colony level, rather than at the individual level. Groups containing many altruists, all working for the common good, might have an advantage over other groups, thus leading to evolution of altruism. Darwin suggested that self-sacrificial behaviors in early hominids might have evolved by this mechanism, an idea still discussed.

The phenomenon of altruism suggests that selection may sometimes operate at levels *above* that of the individual organism. There is also evidence that selection can operate at levels below the individual, for example, on cells and genes. As early as 1903, August Weismann argued that selection could operate on the hereditary particles within the "germplasm" of an individual, a process he called "germinal selection." Today, it is quite common to think of mammalian cancer as a form of intraindividual selection, in which some somatic cell lines gain a short-term replicative advantage over others, to the detriment of the whole organism. More importantly, abundant evidence has been found that selection can occur among the genes within a single individual's genome. This arises because, in sexual species, the genes within an individual are not transmitted en masse to its offspring. So, for example, some genes are able to subvert the rules of fair meiosis and gain access to more than half the host organism's gametes, a phenomenon known as *meiotic drive*. Selection of this sort, sometimes called *genic selection*, leads to a conflict of interest between the genes within a single organism, known as *intragenomic conflict* (see chapter IV.7).

A note on terminology: the expressions *level of selection* and *unit of selection* are sometimes used interchangeably. With this usage, if selection operates at the colony level, for example, it follows that the colony is the unit of selection, and vice versa; however, other authors have distinguished levels from units of selection, using the latter to mean, roughly, entities transmitted intact from one generation to another, or *replicators*, as they are sometimes called. On this usage, the unit of selection is almost always the gene or allele. To avoid confusion over terminology, the expression "unit of selection" is not used in this article.

1. THE GROUP SELECTION CONTROVERSY

The issue of group-level selection has long been a source of controversy in biology. Darwin himself discussed selection primarily at the level of the individual organism, though he did countenance the possibility of colony- or group-level selection in a few cases. Similarly, the founders of the neo-Darwinian synthesis—R. A. Fisher, J.B.S. Haldane, and Sewall Wright—were concerned primarily with selection acting on individual organisms (however, Wright's "shifting balance" theory can be interpreted as involving a limited form of intergroup selection). This focus on individual-level selection is easy to understand, for the bulk of the adaptations we see in nature appear designed to benefit individual organisms, not their groups.

Fisher and Haldane stressed that natural selection acting on individuals would not necessarily lead to adaptations beneficial to entities at higher levels. Thus, Haldane gave examples of adaptations that were "biologically advantageous for the individual, but ultimately disastrous for the species," while Fisher wrote that it was "entirely open" whether selection between individuals would on aggregate benefit or harm the whole species. Nonetheless, by the mid-twentieth century, many biologists routinely assumed that evolution would produce adaptations for the "good of species," or the "good of the group," even though the evolutionary mechanism they had in mind was apparently individual-level selection. Thus, for example, the ethologist Konrad Lorenz argued that the submissive displays of weaker animals were a feature designed to benefit the species, by preventing costly conflicts. From a modern perspective the fallacy in Lorenz's reasoning is obvious, but this is thanks to the considerable wisdom of hindsight.

The "good of the group" tradition came under severe attack in the 1960s and 1970s by biologists such as George C. Williams, John Maynard Smith, and Richard Dawkins. In *Adaptation and Natural Selection*, Williams (1966) argued that group selection, while logically possible, was unlikely to have been a potent evolutionary force, and would usually be trumped by individual selection, owing to individuals' having a faster rate of turnover than groups. He also argued that the hypothesis of group selection was unnecessary—the empirical facts could be explained without it. A similar argument by Maynard Smith, that altruism was better explained by invoking kin selection and inclusive fitness theory (see chapter III.4) than group selection, was bolstered by early mathematical models. They seemed to show that group selection would have significant effects only for a limited range of parameter values; more recent work suggests the situation is somewhat more complicated than this, however.

Williams also made an important distinction between *group adaptation* and what he called *fortuitous group benefit*. The former refers to a feature of a group that benefits the group and evolved for that reason, while the latter refers to a feature of a group that happens to benefit it but did not evolve for that reason. For example, it is conceivable that sexual reproduction is beneficial for a species, as it increases the amount of genetic variation in the species (compared with parthenogenesis), thus allowing rapid evolutionary response to environmental change. This may well be true, but it is unlikely the reason sexual reproduction evolved in the first place—more likely, the benefit is an incidental side effect. If so, then sexual reproduction does not count as a group adaptation, by Williams's light, as it did not evolve *because* of the benefit it confers on the whole species.

As a result of the work of Williams, Maynard Smith, and others, the concept of group selection fell into widespread disrepute in evolutionary biology, where it remained for decades. More recently, this situation has begun to change somewhat, partly because of theoretical developments suggesting that a "trimmed down" version of group selection, often called *multilevel selection*, might be effective in certain contexts; partly because of experimental work showing an unexpected potency to group selection in the laboratory; partly because of conceptual developments suggesting that group and kin selection are really two sides of the same coin (see below); partly because of recent work on the "major evolutionary transitions" suggesting that group selection may have played a role (see below); and partly because there is some evidence to suggest that selection between competing hominid groups may have been an important factor in the evolution of our own species. The issue continues to be debated, but all parties agree that the naive "good of the group" tradition of the 1950s and 1960s was a flawed way of understanding evolution.

2. KIN SELECTION, INCLUSIVE FITNESS, AND THE GENE'S-EYE VIEW

Kin selection theory, and the associated concept of inclusive fitness, emerged from W. D. Hamilton's seminal work on the evolution of altruism in the 1960s and 1970s. The basic idea is simple. As noted above, altruism poses a prima facie challenge to Darwinism, because altruists sacrifice their own fitness to help others; so it seems that altruism, and the genes that cause it, should be disfavored by natural selection. Hamilton realized that the logic of this argument breaks down if altruists can direct their benefits toward relatives, rather than toward unrelated members of the population. Relatives share genes, so there is a certain probability that the beneficiary of the altruistic action will *itself* be an altruist; if so, then the gene for altruism may spread. This idea is encapsulated in *Hamilton's rule*, which says that a gene causing an altruistic action will spread so long as $b > c / r$, where c is the cost to the donor, b the benefit to the recipient, and r the *coefficient of relationship* between them (roughly, the probability that donor and recipient both inherited the gene from a common ancestor). In other words, altruism can spread by natural selection, so long as the cost to the donor is offset by a sufficient amount of benefit to sufficiently closely related relatives (see chapters VII.10 and VII.13).

The main empirical prediction of kin selection theory is that individuals should behave more altruistically toward relatives than nonrelatives. This broad qualitative prediction has been amply confirmed in diverse species, and in some cases kin selection models enjoy a close quantitative fit with the data. It seems likely that kin selection played a major role in the evolution of the highly cooperative social insect colonies, like honey bees, as the relatedness of the insects in such colonies to each other and to the queen is typically quite high (see chapters VII.10 and VII.13). Though kin selection theory has its detractors, most evolutionists regard it as a major part of the explanation of mechanisms by which social behavior evolved.

How does kin selection theory relate to the traditional levels-of-selection issue? According to one view, kin selection provides an explanation for a way in which altruism can evolve without appealing to the group selection concept, thus undermining the main motivation for the latter. This was the dominant view for many years; however, many recent theorists regard kin selection and modern versions of group selection as essentially equivalent. This view is underpinned by mathematical results showing that it is often possible to "translate" between kin selection and (at least some) group selection models. Thus, in social insect colonies, for example, one can interpret the foraging behavior of sterile workers as an adaptation for helping relatives, or as an adaptation designed to boost the whole colony's fitness. These explanations may sound different, but mathematically they amount to essentially the same thing. This is intuitive, because a colony is composed mostly of relatives.

Kin selection is often associated with the *gene's-eye view of evolution*, which sees phenotypic adaptations as "strategies" designed by genes to help gain an advantage over their alleles in the competition for increased representation in the gene pool (Dawkins 1976, 1982). The gene's-eye view is a useful way to understand kin selection, and was employed by Hamilton himself in his early papers. Altruism seems anomalous from the individual organism's viewpoint, but from the gene's own viewpoint, it makes good sense. A gene is under selection to maximize the number of copies of itself found in the next generation; one way of doing this is to cause its host organism to behave altruistically toward other bearers of the gene. But interestingly, Hamilton showed that kin selection can also be understood from the organism's point of view. Though altruistic behavior reduces an organism's personal fitness, it may increase its *inclusive fitness* (see chapter III.4), defined as an organism's personal fitness plus the sum of its weighted effects on the fitness of every other organism in the population, the weights determined by the coefficient of relationship, *r*. Given this definition, natural selection will act to maximize the inclusive fitness of individuals in the population.

In *The Selfish Gene*, Dawkins argues that all adaptations, not just social behaviors, should be regarded as "for the good of the gene," since genes are the ultimate beneficiaries of the evolutionary process. This is sometimes summarized in the slogan "the gene is the unit of selection." Though heuristically valuable, it is important to see that viewing selection in this "genic" way does not resolve the traditional levels of selection question. Whether natural selection occurs at the individual level, the group level, or some other level, the net result will be the spread of one gene at the expense of its alleles—so it is always possible to take a gene's-eye view of the selection process. Therefore, it does not make sense to oppose the gene's-eye view to group selection or to individual selection, as Dawkins himself recognized in his later work.

However, there is a quite different sense in which selection sometimes occurs at the genic level. In cases of intragenomic conflict, there may be selection between the genes within the genome of a single individual (see chapter IV.7). The phenomenon of meiotic drive, discussed earlier, illustrates this. In meiotic drive, selection takes place between the two alleles at a single locus in a heterozygote, leading the organism's gametes to contain one of the alleles in greater proportion than the other. Genes that bias fair meiosis in their favor, and more generally genes that profit at the expense of their host organism, are known as *selfish genetic elements*. These genes spread by a selection process that can be called genic or gene-level selection; in this sense, the genic level *is* a distinct level of selection that can be contrasted with individual- and group-level selection. The important point to note is that gene-level selection in this sense is relatively rare; whether it occurs in any particular case is a matter of empirical fact, not perspective.

3. SPECIES SELECTION

It is sometimes suggested by macroevolutionary theorists that natural selection can occur at the species level (see chapter VI.14). Since species leave daughter species, the Darwinian notion of fitness, the expected number of offspring, is applicable to whole species. Conceivably, natural selection might favor some species over others, depending on their species-level characteristics. For example, it has been argued that mollusk species with a large geographic range have a survival advantage over those with smaller ranges, and that geographic range is a heritable trait, passed on from parent species to their offspring. This may explain why the fossil record appears to indicate that average geographic range, in certain mollusk clades, has increased over time. Similarly, it has been proposed that species selection sometimes favors ecological generalists over ecological specialists within the same clade, as the former are less prone to extinction (see chapter VI.14).

That species selection (and lineage-level selection more generally) is a logical possibility is clear, but opinions differ over its empirical importance. The original proponents of species selection, such as Stephen Jay Gould and Niles Eldredge, tried to establish the "autonomy" of macroevolution from microevolution; large-scale evolutionary patterns and trends cannot be understood as the long-term consequences of the within-population evolutionary changes studied by neo-Darwinians, they argued. Rather, macroevolution was governed by irreducible dynamic processes of its own, such as species selection; however, opponents of this "autonomy" thesis have argued that in the most commonly cited examples of species selection, lower-level causal processes, for example, individual selection, *are* in fact responsible; the differential survival/reproduction of the species is simply a side effect, so selection is not in fact acting at the species level at all. (This point is closely linked to G. C. Williams's distinction between "group adaptation" and "fortuitous group benefit," discussed above.) This debate continues today.

Species selection is not simply a higher-level analogue of group selection. In most group selection models, the fitness of a group is defined as the average or total fitness of the individuals in the group. This is because such models have been concerned with explaining the spread of an *individual* trait, often a prosocial behavior, in a population subdivided into groups. (So it is not assumed that groups must "beget" other groups.) In species selection theory, by contrast, the fitness of a species is defined differently, as the expected number of offspring *species* that it leaves, a quantity that bears no necessary relation to the average fitness of its constituent organisms. This is because the point of species selection is to explain the changing proportions of different types of *species* in a clade, not different types of individuals. Hence species selection and group selection, as usually understood, are of different logical types.

4. MAJOR EVOLUTIONARY TRANSITIONS

In the last twenty years, many biologists have become interested in major transitions in evolution, or "evolutionary transitions in individuality" (Buss 1987; Maynard Smith and Szathmáry 1995; Michod 1999). Such transitions occur when a number of free-living biological entities, originally capable of surviving and reproducing alone, become integrated into a cohesive whole, giving rise to a new higher-level entity, and thus an increase in hierarchical complexity. Evolutionary transitions of this sort have occurred numerous times in the history of life; they include the following: individual replicators → networks of replicators; genes → chromosomes; prokaryotic cells → eukaryotic cells with organelles; single-celled organisms → multicelled organisms; solitary organisms → integrated colonies. The challenge is to understand such transitions in Darwinian terms. Why was it advantageous for the lower-level units to come together, sacrifice their individuality, and form themselves into a corporate body? What prevented "cheaters" from selfishly pursuing their own interests and undermining the integrity of the whole?

During an evolutionary transition, the potential exists for selection to act at more than one hierarchical level. Thus, for example, in the transition to multicellularity, the potential exists for selection to act at two levels: between cells within the emerging proto-organisms or cell groups, and between the cell groups themselves. Moreover, in order for cell groups to emerge as genuine organisms, or "evolutionary individuals," it is necessary for the higher level of selection to dominate; otherwise, selfish cells trying to replicate as fast as possible will undermine the functionality of the cell group. Thus, according to one theory, successful evolutionary transitions require "conflict suppression" or "policing" mechanisms, to ensure the good behavior of the lower-level units, and to align their reproductive interests with those of the whole. In essence, what is required is that the lower-level entities cease to behave as individuals in their own right, and become parts of a larger, integrated whole.

The literature on evolutionary transitions has subtly transformed the original levels-of-selection debate. In the original debate, the existence of the biological hierarchy was taken for granted; the question was about selection and adaptation at preexisting hierarchical levels. By contrast, theorists of the evolutionary transitions are aiming to explain the origin of hierarchical organization itself; that is, why it is that the biological entities we see today form a nested hierarchy. Despite this difference, many themes in the original discussion have proven relevant for understanding evolutionary transitions, including the tension between cooperation and conflict; the importance of kinship in permitting the spread of altruism; the pulling of individual and group selection in different directions; and the suppression of cheaters as a means to promote group welfare. Each of these principles plays a role in recent work on the evolution of multicellularity, for example. The cells within a typical multicelled organism are clonally related, and thus have identical evolutionary interests; sophisticated policing mechanisms exist to prevent rogue cells from undermining the organism's integrity. This illustrates how ideas originally formulated to explain facets of animal social behavior are applicable much more generally, at many levels of the biological hierarchy.

Finally, the literature on evolutionary transitions shows that the traditional levels-of-selection debate raised issues of real importance. Many biologists regarded the group selection debates of the 1960s and 1970s as rather overblown, arguing that, in practice, selection on individual organisms is the preeminent evolutionary mechanism, whether about other theoretical possibilities. But in light of the evolutionary transitions, this attitude is hard to defend. What we call an "individual organism" is itself a highly cooperative group of cells, each specialized in a different task. Moreover, a eukaryotic cell is itself a multispecies assemblage, as it was formed by the union of two prokaryotic cells, and in addition contains numerous organelles with their own genes, whose evolutionary interests are not always fully aligned with those of their host. Today's "individual organisms" did not always exist, and were not always the cohesive and integrated entities that they (mostly) are today. Thus "individuality" is a derived trait, something whose evolution we need to explain, not something we can take for granted. Most likely, selection acting at multiple hierarchical levels will constitute an important part of the explanation.

FURTHER READING

Buss, L. 1987. The Evolution of Individuality. Princeton, NJ: Princeton University Press. *A landmark study of the evolution of multicellularity that launched the modern approach to evolutionary transitions. Buss defends a "hierarchical" view of evolution, involving multiple levels of selection, and argues that many assumptions of traditional neo-Darwinism are less generally applicable than is often thought.*

Dawkins, R. 1976. The Selfish Gene. Oxford: Oxford University Press. *The original statement of the "gene's-eye view" of evolution. Dawkins argues that the true "unit of selection" is the "germ-line replicator" transmitted intact down the generations. The book elaborates the implications, practical and theoretical, of this "genic" approach to Darwinian evolution.*

Dawkins, R. 1982. The Extended Phenotype. Oxford: Oxford University Press. *A further elaboration and refinement of the gene's-eye view of evolution. Dawkins makes the important distinction between "replicators" and "vehicles of selection," arguing that the traditional group selection debate is about whether groups are vehicles, or replicators.*

Hamilton, W. D. 1996. Narrow Roads of Gene Land. Vol. 1, Evolution of Social Behavior. New York: Freeman. *The first volume of Hamilton's collected papers, containing his famous papers from 1963 and 1964 on the evolution of altruism that launched the theory of inclusive fitness. Also relevant is a 1975 paper in which Hamilton shows that the evolution of altruism can be equally understood in terms of multilevel selection.*

Keller, L., and H. K. Reeve. 1999. Levels of Selection in Evolution. Princeton, NJ: Princeton University Press. *Useful collection of papers on various aspects of the levels-of-selection issue, with an emphasis on conflict and co-operation between levels, particularly in relation to social behavior. The editors' introduction offers an interesting perspective on ways in which the levels-of-selection question should be approached.*

Lewontin, R. C. 1970. The levels of selection. Annual Review of Ecology and Systematics 1: 1–18. *Classic paper that lays out very clearly the logic of Darwinian explanation. Lewontin argues that selection will operate wherever there is "heritable variation in fitness," and shows that these conditions can in principle be met at many different levels of the biological hierarchy.*

Maynard Smith, J., and E. Szathmáry. 1995. The Major Transitions in Evolution. Oxford: Oxford University Press. *Important and wide-ranging study of evolutionary transitions, from the very earliest stages of life on earth up to the present, that set the stage for much subsequent work. Maynard Smith and Szathmáry argue that evolutionary transitions involve a change in the way information is transmitted across generations. The levels-of-selection issue is in the background of many of these transitions.*

Michod, R. 1999. Darwinian Dynamics: Evolutionary Transitions in Fitness and Individuality. Princeton, NJ: Princeton University Press. *A key theoretical study of the ways in which "evolutionary transitions in individuality" can arise through natural selection, with a particular emphasis on the transition from single-celled to multi-celled organisms. Michod adopts a multilevel selection framework, and also outlines a "philosophy of fitness."*

Okasha, S. 1996. Evolution and the Levels of Selection. Oxford: Oxford University Press. *A study of the levels-of-selection problem from the perspective of philosophy of science, with a focus on foundational and conceptual issues. Particular attention is paid to causality as it relates to the levels of selection.*

Sober, E., and D. S. Wilson. 1998. Unto Others: The Evolution and Psychology of Unselfish Behavior. Cambridge, MA: Harvard University Press. *An interesting though controversial take on the levels-of-selection question. Sober and Wilson present a sustained defense of "group selection" against its detractors, arguing that group selection is in fact implicit in many evolutionary theories purporting to do without it, and offer an interesting revisionist history of the debate.*

Williams, G. C. 1966. Adaptation and Natural Selection. Princeton, NJ: Princeton University Press. *A classic of modern evolutionary biology. Williams calls for more "discipline" in the study of adaptation, and argues persuasively against the "good of the group" approach to evolution prevalent at the time. Also contains an embryonic version of the gene's-eye view of evolution.*

III.3

Theory of Selection in Populations

Kent E. Holsinger

OUTLINE

1. An example of natural selection
2. Fisher's fundamental theorem of natural selection
3. Patterns of selection
4. Components of selection
5. Maintenance of polymorphism
6. Selection and other processes
7. Synthesis and conclusions

This chapter provides an introduction to the genetics of natural selection. It focuses on selection associated with differences in probability of survival determined by alternative alleles at a single locus, but it also illustrates some of the properties associated with natural selection when selection arises at other stages in the life cycle, when selection varies in space or time, when selection interacts with other evolutionary processes (like mutation, migration, and genetic drift), and when fitness depends on the genotype at more than one locus.

GLOSSARY

Absolute Viability. The probability of survival from zygote to adult.

Bateman's Principle. In most species, females invest more heavily in offspring than males and remate less quickly, leading to greater variation in reproductive success among males than females.

Cline. A gradual change in allele frequency along a geographical transect.

Directional Selection. A mode of natural selection in which one of the homozygous genotypes has the highest fitness, the heterozygote is intermediate, and the other homozygote has the lowest fitness.

Disruptive Selection. A mode of natural selection in which the heterozygous genotype has a lower fitness than either of the the homozygous genotypes.

Equilibrium. A state of a population in which its allele frequency does not change from one generation to the next. It may be a *monomorphic equilibrium* in which only one allele is present, or a *polymorphic equilibrium* in which various evolutionary forces are precisely balanced and the allele frequency remains constant.

Fertility Selection. Natural selection associated with differences in the number of offspring produced. These differences may depend on the genotype of both male and female partners or they may be related only to differences among females, *fecundity selection*, or only to differences among males, *virility selection*.

Fitness. The probability of survival from zygote to adult. More generally, the performance of a genotype in survival and reproduction.

Fixation. A population is fixed for an allele if that allele is the only one present. The population could also be called *monomorphic* for that allele.

Mean Fitness. The average probability of survival from zygote to adult, calculated as the product of genotype frequency and genotype viability and summed across all genotypes.

Mode of Natural Selection. The type of natural selection as determined by which genotype has the highest fitness, which is intermediate, and which has the lowest fitness.

Monomorphic Equilibrium. An equilbrium in which only one allele is present.

Monomorphism. A population is monomorphic at a particular locus if only one allele is present in the entire population. The population could also be said to be fixed for that allele.

Overdominance. A mode of natural selection in which the fitness of heterozygotes is greater than that of homozygotes.

Polymorphic Equilibrium. An equilibrium in which more than one allele is present.

Polymorphism. A population is *polymorphic* at a particular locus if more than one allele is present at that locus in the population.

Relative Viability. The viability of a genotype relative to another genotype, specifically the absolute viability of one genotype divided by the absolute viability of another.

Selection Coefficient. The difference in relative fitness between a particular genotype and the genotype with a relative fitness of 1.

Stabilizing Selection. A mode of natural selection in which the heterozygous genotype has a higher fitness than either of the homozygous genotypes.

Viability Selection. Natural selection associated with differences in the probability of survival.

1. AN EXAMPLE OF NATURAL SELECTION

The basics of natural selection are easy to understand. To illustrate them we'll use a numerical example based on data from *Drosophila pseudoobscura* collelcted by Theodosius Dobzhansky nearly 70 years ago. This species has chromosome inversion polymorphisms. Although the inversions contain many genes, they are inherited as if they were alternative alleles at a single Mendelian locus, so we can treat them as single-locus genotypes and study their evolutionary dynamics. We'll be considering two inversion types: the Standard inversion type, *ST*, and the Chiricahua inversion type, *CH*.

Dobzhansky counted the number of each of the three genotypes both at the egg stage and at the adult stage. He then calculated the fraction of each genotype that survived, its fitness. Data from one of his experiments are shown in table 1. As you can see, the genotypes differ in fitness. In fact, as you can also see from the table, the frequency of heterozygotes increased and the frequency of homozygotes decreased within this generation. That is not an evolutionary response, since there has been no transmission of genetic information from one generation to the next. But the differences are heritable, so the genotype frequencies will change in response to natural selection from one generation to the next.

Of course, we'd like to be able to predict how these frequencies will change over time. To do that, we need to build an algebraic model that allows us to describe how genotype and allele frequencies change in response to natural selection. We will use the notation in table 2 throughout our development of this model. Notice that if we know the frequency of each genotype in eggs and the total number of eggs, we can calculate the number of individuals as the product of the number of eggs and the genotype frequency. For example, if the number of eggs is N and the frequency of the *ST/ST* homozygote is x_{11}, then the number of *ST/ST* homozygotes is Nx_{11}. If we also know the probability that each genotype will survive from egg to adult (its fitness), we can calculate the number of adults as the product of the number of zygotes and their fitnesses. For *ST/ST* that's $Nx_{11}w_{11}$. Putting this all together, we can calculate the frequency of the *ST* chromosome in adults as

Table 1. Data from Dobzhansky's experiment on *Drosophila pseudoobscura*.

	ST/ST	*ST/CH*	*CH/CH*
Number in eggs	41	82	27
Number in adults	25	74	12
Fitness	0.61	0.90	0.44
Frequency in eggs	0.27	0.55	0.18
Frequency in adults	0.23	0.67	0.11

Note: The frequencies in adults do not sum to 1 because of rounding.

Table 2. Notation used to describe natural selection.

Symbol	*Definition*
N	total number of eggs
x_{11}	frequency of *ST/ST*
x_{12}	frequency of *ST/CH*
x_{22}	frequency of *CH/CH*
w_{11}	fitness of *ST/SH*
w_{12}	fitness of *ST/CH*
w_{22}	fitness of *CH/CH*

$$p' = \frac{Nx_{11}w_{11} + Nx_{12}w_{12}/2}{Nx_{11}w_{11} + Nx_{12}w_{12} + Nx_{22}w_{22}} = \frac{x_{11}w_{11} + x_{12}w_{12}/2}{x_{11}w_{11} + x_{12}w_{12} + x_{22}w_{22}}$$

We need to assume that these differences in survival are the only differences relevant for natural selection, that no mutation or migration is occurring, and that the population is large enough that we can ignore genetic drift. That's a lot of assumptions, but if we make them, the Hardy-Weinberg principle guarantees two things: (1) the frequency of the *ST* chromosomes in newly formed zygotes of the next generation will be the same as it is in adults of this generation, and (2) the genotype frequencies in those zygotes will be in Hardy-Weinberg proportions. As a result, we can make that formula a little simpler:

$$p' = \frac{p^2w_{11} + pqw_{12}}{p^2w_{11} + 2pqw_{12} + q^2w_{22}} \tag{1}$$

It probably doesn't look like it, but we can do a lot with that formula thanks to a theorem proven by Sir Ronald Fisher.

2. FISHER'S FUNDAMENTAL THEOREM OF NATURAL SELECTION

To start, let's take a look at the denominator of equation (1): p^2 is the probability that a randomly chosen egg is *ST/ST* and w_{11} is the probability that it survives to adulthood, and the other two terms, $2pq$ and q_2, are the same probabilities for *ST/CH* and *CH/CH*, respectively. So the denominator is the probability that a randomly chosen egg survives to adulthood or, equivalently, the average probability of survival. For convenience we often refer to the probability of survival as *fitness*, and statisticians often use the word *mean* to refer to averages. So the name Fisher gave to this quantity is *mean fitness*, and we typically symbolize it with the symbol $\bar{w}$.

The mean fitness is a property of a population, not of any one individual in it, and it depends not only on the fitness of each genotype but also on their frequencies. So it might be more appropriate to write it as $\overline{w(p)}$ to emphasize that the mean fitness of a population depends on its allele frequency. In fact, if the genotype fitnesses remain constant, the only way the mean fitness of the population can change over time is if its allele frequencies change.

This is where Fisher's fundamental theorem of natural selection comes in. It tells us that allele frequencies will change in response to natural selection in such a way that the mean fitness of the population increases from one generation to the next. Mathematically, we say that

$$\overline{w(p^i)} \geq \overline{w(p)},$$

with equality holding only when the allele frequency has reached a point where $\overline{w(p)}$ is at a maximum. Recall that $\overline{w(p)}$ is the average probability of survival, so another way of stating Fisher's theorem is to say that natural selection increases adaptation in the sense that it increases the average probability of survival in the population.

3. PATTERNS OF SELECTION

Armed with Fisher's theorem, we're now ready to make progress in understanding how populations will respond to natural selection. The key is to understand how the shape of $\overline{w(p)}$ depends on the *mode of natural selection* (figure 1). The mode of natural selection is determined by which of the three genotypes has the highest fitness, which has an intermediate fitness, and which has the lowest fitness. There are three modes of natural selection: directional selection, disruptive selection, and stabilizing selection. Throughout the discussions that follows, we'll assume that p refers to the frequency of an allele we'll label A_1 and that $1 - p$ is the frequency of the alternative allele A_2.

Directional Selection

Directional selection occurs when one of the homozygotes has the highest fitness, the heterozygote has an intermediate fitness, and the other homozygote has the lowest fitness. In figure 1A the fitnesses are $w_{11} = 1.0$, $w_{12} = 0.95$, and $w_{22} = 0.80$. In figure 1B the fitnesses are $w_{11} = 0.80$, $w_{12} = 0.95$, and $w_{22} = 1.0$. In Figure 1A, Fisher's theorem tells us that natural selection will cause the frequency of A_1 to increase in every generation. Only if p^i is greater than p, will $\overline{w(p^i)}$ be greater than $\overline{w(p)}$ as required by Fisher's theorem. Moreover, the frequency of A_1 will continue increasing until it equals 1, meaning that all the alleles in the population are A_1 and that allele A_2 has been lost. Under these conditions we say that allele A_1 is *fixed* in the population and that the population is *monomorphic*. Similarly, natural selection like that in figure 1B will lead to fixation of allele A_2.

In short, directional selection will eventually cause a population to become monomorphic for the homozygous genotype with the highest fitness.

Disruptive Selection

Disruptive selection occurs when the heterozygote has a lower fitness than either of the homozygotes (figure 1C). In this case, the outcome of natural selection depends on the initial allele frequency. If the initial frequency of allele A_1 is smaller than the value of p associated with the dashed line in figure 1C, its frequency will decline from one generation to the next until the population is fixed for allele A_2. If, on the other hand, its initial frequency is larger than the value of p associated with the dashed line, its frequency will increase from one generation to the next until the population is fixed for allele A_1. If it happened that the initial frequency were exactly equal to the value of p associated with the dashed line, it wouldn't change from one generation to the next. The population would be in *equilibrium*.

In this case, the equilibrium is not important biologically, because if the allele frequency departs ever so slightly from the equilibrium value, natural selection will push it farther and farther away. In short, disruptive selection will eventually cause a population to become monomorphic for one of the alleles, but which allele becomes fixed depends on the initial frequency of the

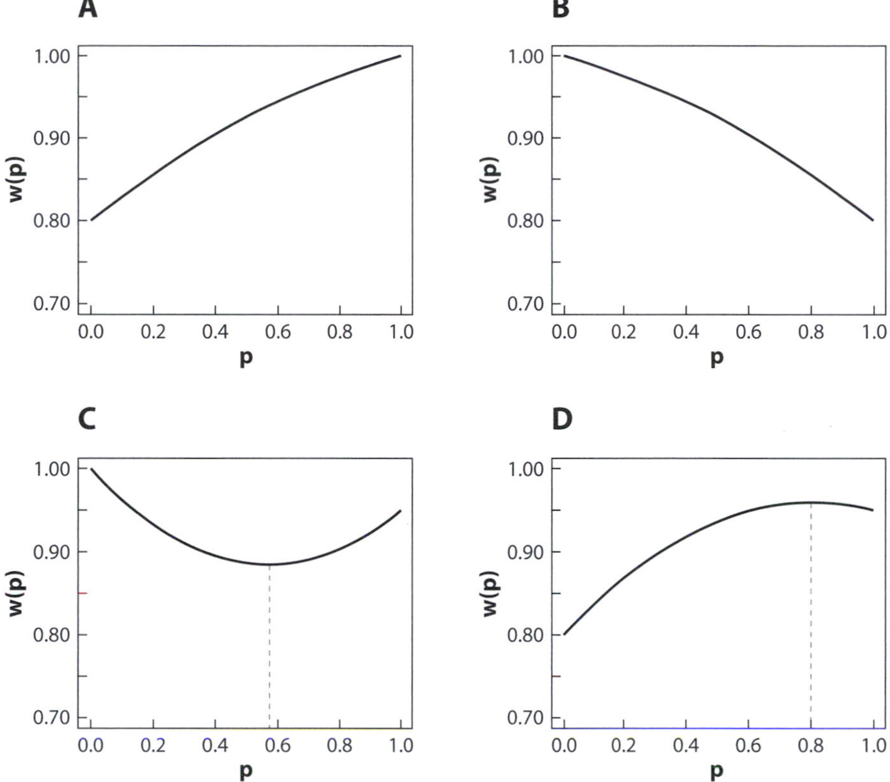

Figure 1. Patterns of natural selection at one locus with two alleles. (A) Directional: $w_{11} > w_{12} > w_{22}$. (B) Directional: $w_{11} < w_{12} < w_{22}$. (C) Disruptive: $w_{11} > w_{12}$, $w_{22} > w_{12}$. (D) Stabilizing: $w_{11} < w_{12}$, $w_{22} < w_{12}$. The dashed vertical lines in (C) and (D) indicate the allele frequency corresponding to a polymorphic equilibrium. There is no polymorphic equilibrium in (A) or (B).

allele. Two populations subject to identical disruptive selection pressures will diverge even further if their initial allele frequencies are different enough.

Stabilizing Selection

Stabilizing selection occurs when the heterozygote has a higher fitness than either of the homozygotes (figure 1D). In this case, the population always evolves to an intermediate allele frequency. If the initial frequency of allele A_1 is smaller than the value of p associated with the dashed line in figure 1D, its frequency will increase from one generation to the next until it reaches a value corresponding to that dashed line. If, on the other hand, the initial frequency of allele A_1 is larger than the value of p associated with the dashed line in figure 1D, its frequency will decrease from one generation to the next until it reaches a value corresponding to that dashed line.

Under stabilizing selection, the *polymorphic equilibrium* is important biologically, because even if the allele frequency happens to depart from the equilibrium value, natural selection will pull it back. Unlike the situation under directional or disruptive selection, in which natural selection acts to eliminate variation, under stabilizing selection natural selection acts to preserve it. Moreover, if two populations are subject to the same stabilizing selection pressures, they will converge on the same allele frequency no matter how different they were initially.

Returning to Our Example

If we now return to our initial example, we recognize that we have an example of stabilizing selection. The fitness of the *ST/ST* karyotype is 0.61, that of the *ST/CH* karyotype is 0.90, and that of the *CH/CH* karyotype is 0.44. The heterozygous karyotype is the most fit, and we therefore predict that natural selection will lead to maintenance of both inversion types in the population. We would even predict that if we start an experimental population with a low frequency of *ST* its frequency will increase, and that if we start with a high frequency of *ST* it frequency will decrease. That's precisely what Dobzhansky's experiments showed.

Table 3. The number of offspring produced by singly inseminated females of *Drosophila melanogaster* as a function of mating type in Clark and Feldman's experiment (simplified for this presentation).

	ey^2/ey^2	$ey^2/+$	$+/+$
ey^2/ey^2	56	61	55
$ey^2/+$	115	115	99
$+/+$	78	74	57

Note: Female genotypes are in rows; male genotypes are in columns.

4. COMPONENTS OF SELECTION

So far we have discussed natural selection as if it happens only as a result of differences in the probability of survival, but selection can happen at any life stage, and when it does, the results can be quite different from those we have discussed so far. The specific type of natural selection we have been discussing so far is *viability selection.* Some other important types of selection are *fertility selection*, *sexual selection*, and *gametic selection.* Any one or all of these types of natural selection can influence how allele frequencies change from one generation to the next.

Fertility Selection

In its most general form, fertility selection occurs when the number of offspring produced from a mating depends on both male and female genotypes. For example, Andrew Clark and Marcus Feldman studied the number of offspring produced by *Drosophila melanogaster* in experimental crosses (table 3). Their results show not only that the number of offspring produced may depend on the genotypes involved, but also that the same pair of genotypes can produce different numbers of offspring depending on which genotype is male and which is female.

For example, a cross in which $ey^2/+$ is female and ey^2/ey^2 is male produces 115 offspring, but one in which the genotypes are reversed produces only 61. Perhaps even more surprisingly, the number of offspring a +/+ female produces depends on whether she mates with a +/+ male (57 offspring), an $ey^2/+$ male (74 offspring), or an ey^2/ey^2 male (78 offspring). Even though we don't know why these differences exist, they are reproducible, so we do know they will lead to changes in genotype and allele frequencies over time, even if the three genotypes have equal probabilities of survival.

Sexual Selection

Sexual selection occurs when genotypes differ in their probability of mating. In most species females invest more in offspring than males, and they are not able to remate as quickly as males. As a result, there is more competition for mates among males than females, and females are less likely to go unmated than males. This is known as *Bateman's principle.* As a result, sexual selection often takes one of two forms: male-male competition, in which males compete for access to females, and female choice, in which females select the males with whom they will mate. In the Clark and Feldman experiment, for example, female *Drosophila melanogaster* preferred wild-type, +/+, males to either heterozygotes or homozygotes for ey^2 regardless of their own genotype.

Sexual selection favors traits that enhance the probability of attracting mates, like the enormous, colorful train on a peacock or the elaborate display in the vicinity of a male bowerbird nest. These traits may reduce the probability of survival, leading to a conflict between viability selection and sexual selection. The outcome will represent a compromise between these competing forces.

Gametic Selection

Gametic selection occurs when gametes differ in their probability of accomplishing fertilization (see chapter III.2). In flowering plants, genes expressed in pollen are likely to influence the rate at which pollen tubes grow down the style, and allelic differences in these genes may be associated with differences in fertilization probability. Similarly, in animals, many genes are expressed in sperm, and sperm competition can also cause allelic differences in those genes to be associated with differences in fertilization probability. In perhaps the most famous example of gametic selection, 90 percent or more functional sperm in heterozygotes for the *t* allele in house mice carry the *t* allele. Sperm carrying the wild-type allele are functionally inactivated by their *t* partner. Thus, gametic selection strongly favors the *t* allele.

Just as with sexual selection, however, alleles favored by gametic selection may be disfavored by selection at other stages in the life cycle. While gametic selection strongly favors the *t* allele in house mice, for example, homozygotes for the *t* allele are either inviable or male sterile. So viability selection strongly favors the wild-type allele. As with the conflict between viability selection and fertility selection, the outcome represents a compromise between the competing forces of gametic selection and viability selection (see chapter IV.7).

5. MAINTENANCE OF POLYMORPHISM

We have already seen that viability selection will maintain both alleles in a population when heterozygotes are most fit. Although this simple property is not universal when more than two alleles are present at a locus or when other forms of selection are acting, broadly speaking if heterozygotes are more fit than homozygotes, then selection will tend to maintain polymorphisms within populations. But heterozygote advantage, or *overdominance*, is not the only mechanism by which natural selection can maintain multiple alleles in populations.

Frequency-Dependent Selection

Our discussion of natural selection has so far assumed that the fitness of different genotypes remains constant over time. But it may be that the fitness of a genotype depends on the frequency of other genotypes in the population. For example, in many flowering plants, self-fertilization is prevented by genes that prevent pollen from germinating on the style of plants that share an allele with the pollen grain. As a result, even outcross pollen can fail to germinate if it happens to land on the stigma of a plant with which it shares an incompatibility allele. In this kind of system, rare genotypes will consistently be more successful in mating than common ones, because their pollen will rarely be deposited on an incompatible stigma. This is an example of negative frequency dependencee in which the fitness of an allele or a genotype increases as its frequency decreases. When fitnesses vary in this way, selection may maintain a large variety of different alleles within a population.

Spatial Variation

The fitnesses of genotypes can also depend on the particular place in which they are found. Plant genotypes favored in a sunny meadow, for example, may be different from those favored in a shady forest. If offspring from any individual can be dispersed among different habitats, the differences among them may lead to maintenance of genotypes that do well in each habitat, although the conditions under which spatial heterogeneity promotes polymorphism are complex.

For example, imagine a plant population growing in an area with sunny and shady habitats, and suppose that offspring from any one plant might be dispersed into either habitat. In many plant populations, the number of seeds produced is relatively independent of the number of seeds from which the population started, because only a certain amount of biomass can be produced in a given area, and the number of seeds produced is often proportional to biomass. In that case, the seed output from each habitat will not be influenced by the genotype composition within it. Under these conditions, selection will maintain polymorphism so long as different genotypes are favored in the two habitats.

But suppose that seed output from each habitat *does* depend on the genotype composition within it. Then selection will maintain a polymorphism only so long as the fitness of the heterozygote exceeds that of both homozygotes, where fitnesses are calculated as the average fitness within each habitat weighted by habitat frequency. When dispersal between habitats is limited, the condition that determines whether selection will maintain variation represents a compromise between these two extreme scenarios. In general, while spatial heterogeneity in selection can make it more likely that genetic variation is maintained, spatial heterogeneity alone is not sufficient to ensure that populations remain genetically variable.

Temporal Variation

With frequency-dependent selection, fitness varies over time, but it varies predictably as a function of the frequency of different genotypes or alleles. The fitness of genotypes may also vary over time because the environmental conditions change in ways unrelated to the genotype frequency. If one homozygous genotype is favored under some circumstances and the other is favored under different circumstances, the differences over time may lead to the maintenance of a polymorphism, but as with spatial variation, the conditions under which temporal variation promotes polymorphism are complex. In particular, if the fitness of heterozygotes varies more over time than the fitness of homozygotes, natural selection may eliminate genetic variation even when heterozgotes are more fit on average.

6. SELECTION AND OTHER PROCESSES

In our derivation of equation (1), we assumed that the only fitness differences relevant for natural selection were viability differences and that the viabilities remained the same over time. We've now seen that relaxing those assumptions makes predicting the response to selection complex. It shouldn't come as a surprise that the interaction of selection with mutation, migration, and drift can also be quite complex.

Mutation

We've already seen that in the absence of mutation, directional selection will lead to fixation of the allele associated with the highest fitness. But suppose that in every generation, the deleterious allele arises anew by mutation in each generation, and so persists. The population will reach an equilibrium in which loss of the allele associated with selection will be balanced by gain of the allele through mutation.

Consider the simplest case, when the deleterious allele is completely recessive, so that the relative fitnesses of the genotypes are 1, 1, and $1 - s$ for the homozygote for the advantageous allele, the heterozygote, and the homozygote for the deleterious allele, respectively, where s is the *selection coefficient*. If μ is the mutation rate from the advantageous allele to the deleterious allele, the equilibrium frequency of the deleterious allele is approximately $\sqrt{\mu/s}$. If the deleterious allele is only partially recessive so that the relative fitness of the heterozygote is $1 - hs$, then the equilibrium frequency of the deleterious allele is approximately $\mu/(hs)$ (see chapter IV.3).

Because completely recessive alleles are "hidden" from natural selection in heterozygotes, they may occur in a much higher frequency than a partially recessive allele in which the heterozygotes have the same relative fitness (figure 2). On the other hand, because the deleterious phenotype is expressed only in recessive homozygotes, the mean fitness in a population with a pure recessive is a little higher than that in a population with a partial recessive (see chapter IV.5).

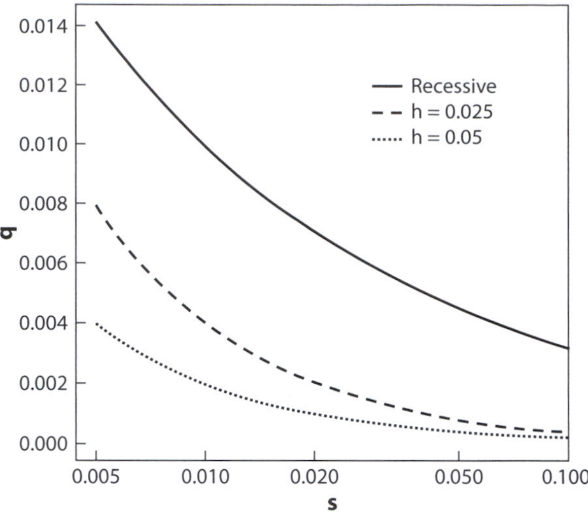

Figure 2. The equilibrium frequency of a deleterious allele with different selection coefficients and different degrees of dominance. $\mu = 10^{-6}$ in all cases.

Gene Flow and Migration

To most people the word *migration* conjures up images of birds heading south for the winter (or north for the winter in the Southern Hemisphere). To a population geneticist, the word *migration* means the movement of an organism from the population in which it was born to another in which it reproduces (see chapter IV.3). If a robin returns to the place where it was born to reproduce, to a population geneticist there has been no migration. Only if that robin reproduces in a place different from where it was born would a population geneticist say that migration has occurred—and the migration would be from where it was born to where it reproduced, not from where it was born to where it spent the winter. *Migration* is often used interchangeably with the phrase *gene flow*, although gene flow is sometime used to include the movement of genes into a new population founded through colonization as well as the movement of genes into an existing population.

We already learned that if the fitness of genotypes differs from one habitat to another and dispersal among habitats is limited, natural selection may lead to the maintenance of a genetic polymorphism. But suppose that the habitats, instead of being intermixed as discrete units, have a sharp boundary as you move from one part of a region to another. For example, some plant species have genotypes that are able to survive on soils with high levels of heavy metals like those found on mine tailings. These genotypes usually have lower fitness than nontolerant genotypes when growing on soil with low levels of heavy metals. In the absence of migration, we would expect all plants that grow on mine tailings to be heavy-metal tolerant and all plants that grow elsewhere to be nontolerant. But both pollen and seed move from one habitat to the other. If the area of tailings is very large, nearly every plant in the center of the tailings habitat will be a resistant genotype. As you move closer to the boundary, however, the frequency of nonresistant genotypes increases. Similarly, resistant genotypes are found out of the area of the tailings, but their frequency decreases as the distance from the mine tailings increases. The gradual change in allele frequencies along a geographical transect is referred to as a *cline*.

The allele frequency at any position along a cline represents a balance between natural selection favoring the genotype best suited to local circumstances and the introduction of other alleles through gene flow or migration. The width of a cline is related to the strength of selection in each habitat and the extent of dispersal between habitats. The stronger the selection, the narrower the cline, and the more the gene flow, the broader the cline.

Genetic Drift

In our discussion of natural selection so far, we've ignored the fact that in real populations, allele frequencies can change simply because some individuals are lucky and some are unfortunate. Some individuals may just happen to have a large number of offspring while others have few or none, and these differences may be completely unrelated to genotypic differences at the locus we happen to be studying. When such changes occur, it is referred to as *genetic drift* (see chapter IV.1). In large

Table 4. A hypothetical set of relative fitnesses corresponding to directional selection for allele A_1 with $s = 0.01$.

	A_1A_1	A_1A_2	A_2A_2
Fitness	1	$1 - s = 0.99$	$1 - 2s = 0.98$

populations, we can usually ignore the influence of such random changes, but when populations are small, these types of changes can have a larger influence on the way allele frequencies change than natural selection. Although the details vary depending on the mode of selection, roughly speaking, random changes in allele frequency, genetic drift, predominate when the product of effective population size, N_e and the selection coefficient, s is less than 1. Natural selection predominates when $N_es > 1$. Thus, whether natural selection has an important influence on allele frequency changes depends on both the strength of the selection and the size of the population.

Consider the set of relative fitnesses in table 4. If the effective size of the population is 50, $N_es = 0.5 < 1$, so allele frequency changes will primarily reflect the random effects of genetic drift, even though there is strong directional selection in favor of allele A_1. If, on the other hand, the effective size of the population is 200, $N_es = 2 > 1$, so allele frequency changes will primarily reflect the effects of natural selection and cause the frequency of A_1 to increase in every generation. When $N_es < 1$, allelic differences are effectively neutral, meaning that even though genotypes have different fitnesses, genetic drift rather than natural selection is the predominant influence on allele frequency changes in the population.

If we could ignore genetic drift, then every advantageous mutation that arises would be guaranteed to increase in frequency as a result of natural selection. But when a new mutation arises, it is usually unique. There will be only one copy in the entire population. As a result, there's a good chance it will be lost even if it is highly advantageous. In fact, J.B.S. Haldane showed that if the selection coefficient in favor a favorable allele is s in heterozygotes and $2s$ in homozygotes, the probability that it will be fixed by selection is only $2s$. In other words, an allele providing a 20 percent fitness advantage to those homozygous for it has an 80 percent chance of being lost as a result of genetic drift. Most newly arisen mutations are lost, even if they are highly favorable.

7. SYNTHESIS AND CONCLUSIONS

Population geneticists have explored the genetic consequences of natural selection using mathematical models and experiments since the early 1920s, amassing a deep understanding of many complex phenomena, only a few of which have been described here. In spite of the many complexities and subtleties this work has revealed, several broad principles apply:

- Natural selection usually increases the adaptedness of individuals, making them better suited to the biotic and abiotic environments in which they exist.
- Natural selection will eliminate variation and lead to a population composed of a single genotype, if there is a single homozygous genotype that is more fit than all others. It will preserve variation, when the most fit genotype is a heterozygote. Natural selection will also eliminate variation if heterozygotes are less fit than homozygotes, but the single genotype that remains will depend on the genotype frequencies from which the population started.
- While natural selection tends to increase the level of individual adaptation, several evolutionary processes may result in compromises that reduce individual fitness. If mutation regularly introduces deleterious mutations, the frequency of the alleles will represent a balance between the mutation rate and the strength of selection. Similarly, if migration or gene flow among populations introduces genotypes that are not optimal within local populations, the fitness of individuals will be smaller than in the absence of migration.
- While differences in fitness usually lead to predictable changes in genotype frequency, in small populations, changes in allele frequency may be largely random even when genotypes differ in fitness. In particular, the chance that advantageous alleles become more common depends not only on how strongly they are favored but also on their frequency. As a result, most new mutations are lost, even if they are highly advantageous.
- Natural selection can occur simultaneously at several different life history stages: survival from zygote to reproductive adult, finding of mates, production of gametes, and production of offspring. Traits that enhance fitness at one life history stage may reduce it at another, for example, a peacock's enormous train may enhance reproductive success of males while reducing their probability of survival.
- Corresponding with the different life history stages at which natural selection can operate are several different hierarchical levels at which it can operate: gametic, individual, and mated pair. Other chapters in this guide (see chapters III.2

and VII.9) illustrate that natural selection can even operate at the level of groups of kin and maybe even at the level of groups of unrelated individuals that share certain traits.

In short, the theory of evolution by natural selection provides a richly textured framework by which to understand an enormous diversity of evolutionary phenomena. Yet underneath all of this diversity lies Darwin's fundamental insight: Heritable characteristics that enhance the probability of survival and reproduction will tend to become more common, and those that reduce fitness will tend to be eliminated.

FURTHER READING

Clark, A. G., and M. W. Feldman. 1981. Density-dependent fertility selection in experimental populations of *Drosophila melanogaster*. Genetics 98: 849–869.

Endler, J. A. 1986. Natural Selection in the Wild. Princeton, NJ: Princeton University Press.

Frank, S. A. 2012. Natural selection. V. How to read the fundamental equations of evolutionary change in terms of information theory. Journal of Evolutionary Biology 25: 2377–2396.

Hurst, L. D. 2009. Genetics and the understanding of selection. Nature Reviews Genetics 10: 83–93.

Travis, J. 1989. The role of optimizing selection in natural populations. Annual Review of Ecology and Systematics 20: 279–296.

III.4

Kin Selection and Inclusive Fitness

David C. Queller and Joan E. Strassmann

OUTLINE

1. The problem of altruism
2. Inclusive fitness and Hamilton's rule
3. Kinds of social selection
4. Comparative evidence in social insects
5. Experimental evidence in microbes
6. Kin recognition
7. Challenges to kin selection

Kin selection is selection that operates via effects on relatives. It is the explanation for altruistic behavior, where an actor gives up fitness in order to help other individuals, because the trait can spread only through possession of the altruism gene by beneficiaries. It also applies to other forms of behavior toward relatives, including selfish behavior. Kin selection, and the associated but broader concept of inclusive fitness, is supported by many theoretical models and empirical studies.

GLOSSARY

Altruism. A behavior that is costly to its performer but that aids other individuals.

Eusocial. Societies with overlapping generations, cooperative care of young, and reproductive division of labor; usually applied to insects though some vertebrates and shrimp meet these criteria.

Hamilton's Rule. A formulation that states that a trait is favored when it raises average inclusive fitness; for an altruistic behavior it is $-c + rb > 0$, where c is cost to self, b is benefit to partner, and r is their relatedness.

Haplodiploid Hypothesis. Hamilton's hypothesis that the unusually high sister relatedness of ants, bees, and wasps was critical in the evolution of their eusociality.

Inclusive Fitness. The sum of all an individual's fitness effects, on self and others, each multiplied by relatedness; it is a quantity that is maximized by selection.

Kin Recognition. A mechanism by which individuals identify their kin.

Kin Selection. Selection on genes causing behavior in one individual through the effects on fitness of other individuals who share the genes.

Phenotype Matching. Kin recognition mechanisms involving comparison of unknown individuals to a template, often learned from known kin.

Relatedness. Genetic similarity above the level expected by chance, usually due to pedigree ties.

1. THE PROBLEM OF ALTRUISM

Eusocial insect colonies are marvels of cooperation. Ants and termites follow chemical trails to good food sources, while honey bees communicate food locations via dance. Leaf-cutter ants can strip large trees bare of leaves, while blind army ants subdue prey many times their size and can even dismember vertebrates. Food is carried back to nests and distributed to growing young, often after storage, processing, and even cultivation. Nest structures can be built from leaves, mud, paper, and even, in army ants, their own bodies. Many ants excavate elaborate tunnel systems reaching far below ground, while some termites erect earthen towers stretching far above it. These colonies are carefully protected. Guards swarm out of the entrance to attack intruders with bites, stings, and venom, or occasionally employ more innovative defenses, such as termites that rupture their bodies to release glue to entangle enemies. Others simply block the entrance, sometimes using specially shaped heads as shields. Not one of these functions—foraging, building, feeding, or defense—is carried out in isolation; instead each is a regulated and finely coordinated group activity. For each of them, one could ask the same questions one would ask of any complex adaptation. Where do the right variations come from? How are the traits built by step-by-step selection?

But an additional, more special question arises. Beneath all this complex cooperation is a somewhat more subtle form that has been even more challenging to conventional evolutionary theory. All the tasks described above are carried out by the colony's workers, a class of individuals that normally does not reproduce (see chapter VII.13). Their actions benefit the reproduction of one or a few queens, along with their mates (the latter existing only as the queen's stored sperm in ants, but still present in termites). How are these adaptations inherited, given that the actors have no offspring? And going one step further, how could they evolve to have no offspring in the first place?

Darwin recognized this problem and viewed it as the "one special difficulty, which at first appeared to me insuperable, and actually fatal to my whole theory." His solution was selection at the level of the family. He noted that animal breeders can select for tasty steaks, despite the death required for tasting, by breeding from the same stock, which would have the same characteristics.

2. INCLUSIVE FITNESS AND HAMILTON'S RULE

Darwin's idea would work straightforwardly for clonal organisms. If we could breed from a clonemate of the cow who died to produce the desirable steak, it would be genetically just like breeding from the tasty cow herself. Indeed, we are familiar with this kind of explanation for the altruistic behavior at the lower level of cells in a body. A cow consists of a clone of cells, so though a cow's liver cell or brain cell never reaches the next generation, the cells' traits are inherited via the identical germ-line cells. At the level of the organism, individual cows do not have clonal relatives, but they do have siblings and more distant relatives. These too share genes with the tasty cow, but to a lesser degree.

Darwin did not know about genes, but in the 1960s W. D. Hamilton quantified the idea of gene sharing or relatedness, leading to the idea we now call *kin selection*. Hamilton's math led to a useful new concept, inclusive fitness, and a quite simple result, which we call Hamilton's rule. *Inclusive fitness* is the sum of all an individual's fitness effects, on self and others, each multiplied by the individual's relatedness to each party. Because relatedness to self is one, in the absence of effects on relatives, this reduces to the individual's effect on its own fitness. Hamilton's rule essentially says that a behavior will evolve when its average inclusive fitness is positive: $\Sigma sr > 0$, where s is a fitness change caused by the actor, and r is the relatedness of the actor to the individual who experienced the fitness change. Returning to the cow, we would like to apply selection of intensity s to the tasty cows themselves, but since they are already dead, we apply selection intensity s to their full siblings and will get a response that is half as strong, because full siblings are related by one-half.

The relatedness that matters for kin selection is genetic similarity above random levels in the relevant population. Cows and pigs share some genes, but breeding a pig will not give us tastier cow steaks; the pig is not relevant to evolution in the cow population. More subtly, our cow shares genes with all other cows, but breeding a random cow will still not give us tastier steaks. Likewise, helping random individuals in the population does not increase the frequency of the helping gene. The main reason that genes are identical above random levels is pedigree kinship, although other possibilities can also be addressed using inclusive fitness.

3. KINDS OF SOCIAL SELECTION

Kin selection and inclusive fitness have been most famously applied to altruistic behavior, such as that of worker social insects. Altruists sacrifice personal fitness but increase the fitness of another. In this case, Hamilton's rule says altruism will evolve if $-c + rb > 0$, where c is the fitness cost to the altruist, b the benefit to the beneficiary, and r their relatedness (the relatedness of 1 of the altruist to itself is omitted). There can be multiple rb terms if there are multiple beneficiaries. The equation makes it clear that altruism cannot evolve if it benefits only nonrelatives, and that altruism can evolve with some suitable combination of high relatedness, high benefits to relatives, and low costs to self.

It is important to remember, however, that kin selection and inclusive fitness apply not only to altruism but also to any effects on relatives. These effects are often organized into four classes according to the signs of the effects on self versus relatives. Altruism involves negative effects on self and positive effects on relatives. A positive effect on both parties is called *mutual benefit*. Selfish behavior involves a positive effect on self and a negative one on the partner. Note that relatedness reduces selfishness, but does not preclude it: one can evolve to harm relatives if $b_{\text{self}} - rc_{\text{partner}} > 0$. The last category, spite, which involves harm both to self and others, should not normally evolve. However, harm to self and to some partners can evolve if it sufficiently benefits other partners, as in the case of bacteriocins. These are bacterial poison-antidote systems, where release of the poison—sometimes suicidally by cell bursting—kills neighbors who lack the system, removing competitors of those who possess it.

For relatives that are not genetically identical, kin selection predicts the possibility of conflict. For example, when two or more honey bee queens hatch out in a queenless colony, they usually fight to the death, despite being sisters in a species that otherwise has extraordinary altruism. The reason is that neither can survive on her own, and each would do better genetically by taking over the colony herself than by letting her sister have it. Similar logic explains parent-offspring conflict, for example, over timing of weaning.

4. COMPARATIVE EVIDENCE IN SOCIAL INSECTS

In seeking evidence of kin selection, most attention has focused on the role of relatedness, because that was the most novel aspect of the theory. In addition, much of the early attention centered on a peculiarity of relatedness that Hamilton noticed. He knew that most social insects were ants, bees, and wasps of the order Hymenoptera, and that this order has a peculiar genetic system called haplodiploidy. Here the females are typical diploids and pass on a random half of their genome to each offspring; however, males are haploid, and they pass on their full genome, but only to daughters. The system is perpetuated by whether a mother uses stored sperm to fertilize her egg or not. If she does not fertilize the egg, the offspring is a haploid male, who is fatherless. If she does, the offspring is a diploid female. Full sisters—those who share the same father—will be unusually highly related because their paternally inherited parts are completely identical. When averaged with the relatedness of one-half for the maternally inherited parts, full sister relatedness turns out to be three-quarters, well higher than the corresponding relatedness of one-half in diploid organisms. Hamilton believed this extra dose of relatedness explained why sociality had evolved so much more often in the haplodiploid Hymenoptera than in diploids, and also why workers are always female in the Hymenoptera.

However, both the prediction and the evidence for this haplodiploid hypothesis are more complicated. Haplodiploid females are more related to their full sisters but worker females normally also rear brothers, to whom they are related by only one-quarter. Their average relatedness to an equal mixture of full sisters and brothers is one-half, exactly like diploids. They can still gain an advantage if they rear more sisters than brothers, but the advantage is considerably smaller than that first envisioned by Hamilton. On the empirical side, there are issues of what should be compared. Parental care is an important preadaptation for the allo-parental care seen in social insects. The Hymenoptera are the most parental of insect orders, so perhaps the ability to provide benefits, rather than haplodiploidy, is the real reason for the preponderance of social insects in this group. Moreover, hymenopteran parental care is nearly exclusively female, so it is females and not males that are preadapted for allo-parental care. This does not mean that kin selection was not involved; it simply means that the most relevant part of Hamilton's rule for explaining helping by Hymenoptera females is their ability to provide benefits. In agreement with this, helping to rear siblings has evolved many times in diploid vertebrates, which also have parental care as a preadaptation (see chapter VII.10). The specific benefits provided in different groups may vary (see chapter VII.13).

This could mean that the extra dose of sister relatedness was not the main factor favoring eusociality, but it does not mean that kin selection and relatedness are unimportant. A central lesson of this history is that benefits and costs are just as important to kin selection as relatedness. But the importance of relatedness remains clear. Social insect colonies are essentially always groups of relatives. Phylogenetic analysis shows that eusocial groups always arose in monogamous settings, and never in situations where multiple fathers would lower relatedness (although this sometimes evolves secondarily in advanced social insects, where workers are so specialized that they would be very ineffective as reproductives).

The best evidence for the importance of kin selection in social insects comes from sex investment ratios. As noted above, female workers would pass on more genes by investing in full sisters related by three-quarters than in brothers related by one-quarter. They do just that by feeding sisters more, or even by killing brothers, so sex ratios are usually female biased. In species where different colonies have different relatedness structures, for example, with a singly mated queen versus a multiply mated queen, the colonies with more full sisters (singly mated) specialize in rearing females while the others tend to specialize in males.

5. EXPERIMENTAL EVIDENCE IN MICROBES

Kin selection is not just for relatively smart animals, such as social insects and vertebrates. The process applies to any organism that interacts with kin, and even microbes turn out to be very social (see chapter VII.9). The bacteriocin poisons that are secreted to harm competitors have already been mentioned. Bacteria also secrete many beneficial products, particularly for food digestion, and these products become available to neighbors, who may be kin. Microbes also cooperate for motility or dispersal. Even the primary energy pathways can have a social dimension because, compared to fermentation, respiration is slow but very efficient. Therefore, in a competitive environment, respiration can involve an energy cost to the individual but, by being less wasteful, leaves more carbon resources for neighbors.

Microbes bring two great advantages to the study of kin selection. First, the genes underlying social traits are more often known and easier to manipulate. Second, one can manipulate their population structure and follow selection over many generations (see chapter III.6). Many microbial studies have documented the importance of relatedness in favoring cooperation or altruism genes. For example, the bacterium *Pseudomonas aeruginosa* produces and secretes molecules that bind iron in way that allows the cell, or other cells, to take it up. High relatedness structures favor secretion over nonsecretion, because some of the benefit of secretion goes to neighbors.

Another example comes from the social amoeba *Dictyostelium discoideum.* These amoebas collectively produce a fruiting body in which 20 percent of the cells die to form a stalk, promoting the dispersal of the other 80 percent, which become spores. Genetic selection experiments have isolated numerous cheater mutants that produce excess spores, for example, by shirking on stalk production. When populations are maintained under low-relatedness conditions, it leads to the spread of mutants that cheat but cannot fruit on their own, even though the result is sharply declining spore production of the population. Under very high relatedness, however, groups are either nearly all wild-type cooperators, which do well in the absence of cheaters, or nearly all cheaters, which cannot produce spores without having cooperators to exploit.

6. KIN RECOGNITION

Kin selection requires some method to make kin the specific targets of behavior. The simplest mechanisms involve proximity. For example, birds growing up in a nest can generally count on each other being kin. In more continuous populations, limited dispersal can make neighbors close kin. This mechanism depends on the extent to which neighbors are also each other's closest competitors, which is determined by factors like the timing of altruism and competition during the life cycle. This increased competition can sometimes cancel out the kinship effect.

The advantages potentially available through kin selection suggest that organisms should often have evolved mechanisms to explicitly recognize kin. This could be said to be one of the great predictions of kin selection theory, something that essentially started a new field. The prediction has been confirmed repeatedly across the tree of life, from microbes to mammals. Learning is often an important component of these mechanisms. A bird might learn who its nest mates are by proximity and then remember that information for later use after leaving the nest. In addition to remembering individuals, a more general mechanism of phenotype matching can be employed. Characteristic cues of known relatives (which could include self) are learned and remembered. Later individuals are compared against this stored template and scored as kin according to the degree of match. This mechanism can allow the identification of unfamiliar relatives, which is important in large social insect colonies where all the individuals cannot be learned individually. Social insects generally use learned colony odors to separate friend and foe, though some small-colony wasps recognize individuals by color patterns.

Some phenotype matching systems rely on genetic cues, either learned as in social insects, or innate, as in social amoebas. A significant puzzle, known as Crozier's paradox, is how the required variability is maintained at these cue loci. In theory, the most common cue allele receives more benefits from those who match, or less harm from those who do not. This advantage drives the common cue to fixation, rendering the locus useless for discrimination. One possible solution is that genetic kin recognition systems rely on cues whose variability is maintained for other reasons, such has host-pathogen selection at immune-system genes.

Kin recognition can be challenging. Errors in recognition can have costs, both if nonrelatives are favored, and if relatives are disfavored. Thus we would predict that recognition would be most effective in organisms where cooperating was most important. In an insightful study, Griffin and West explored the relationship between the importance of helping in social vertebrates, and the extent to which it was preferentially directed toward relatives. Species like kookaburra and superb fairy-wrens, where helping was not particularly beneficial, were less likely to discriminate kin. Seychelles warblers and pied kingfishers show greater benefits of helping and had a correspondingly higher preference for kin.

7. CHALLENGES TO KIN SELECTION

Kin selection theory has revolutionized our understanding of cooperative social interactions. Ever more advanced models, phylogenetic comparative studies, and experimentation, including experimental evolution, support it. However, this does not mean kin selection has not faced challenges or that it can explain all forms of cooperation.

Kin selection is not refuted by benefits that go to nonrelatives. Some beneficial effects on others are simply by-products of self-interested behavior. For example, when one parent benefits its own fitness by caring for its young, it also enhances the fitness of the other parent, who is typically unrelated. Secretion by a bacterium of products that help unrelated neighbors could still be favored provided the bacterium itself gets a net gain.

Other beneficial effects are neither by-products nor kin-selected benefits. Much human cooperation occurs between nonrelatives who do not share breeding interests (see chapter VII.11). This cooperation can be explained as *reciprocal altruism*, which requires direct or indirect payoffs to the actors. The kinds of accounting and retaliation necessary for reciprocal altruism to work make it unlikely in most animals, but there are still other ways of cooperating for immediate direct benefits. For example, all individuals in a herd may benefit by saturating local predators. Huddling for warmth can help all involved. Another major kind of cooperation is mutualism. Even members of different kingdoms can benefit in mutualisms. For example, in a lichen mutualism, the fungus provides structure and protection, while the alga provides carbon. Cleaner shrimp eat parasites they pick off the fish they clean. There are many more such examples throughout the tree of life (see chapter VII.9).

The key difference between all these other forms of cooperation and kin-selection cooperation is that the former all require direct benefits to the actor. Only kin selection can lead to true altruism, where there is a fitness cost to the actor that is not repaid. Since inclusive fitness includes not just kin benefits but also direct benefits to self, it can account for all these forms of cooperation.

There is a kind of true altruism that involves some individuals helping others with no evident selfish gain to their soma, or to their genes. This is the kind of thing that kin selection predicts should not be favored by selection, so if it were pervasive, it would mean trouble for kin selection theory. Unicoloniality is a feature of sociality in a small number of ant species that appears to break all rules. There is a great deal of movement from one nest to another, resulting in workers rearing brood to which they are unrelated. In some species, such as the Argentine ant, single cooperative colonies can extend for hundreds of kilometers.

Unicoloniality appears to be the result of family recognition gone astray, through genetic bottlenecks and loss of kin recognition alleles during invasion. Kin selection does not preclude the existence of this kind of trait, but it does predict that the trait is maladaptive. The lack of relatedness is predicted to result in the loss of selection on adaptive worker behaviors, since genes for effective worker behavior will not be favored by selection and should degrade through mutation and drift. Unicoloniality is therefore expected to be evolutionarily short-lived and this appears to be true. The trait is scattered through the twigs of the ant phylogeny and does not give rise to successful clades, according to an analysis by Heikki Helantera and colleagues.

8. INCLUSIVE FITNESS AND OTHER APPROACHES

Inclusive fitness is a very easy way to think about kin selection. It is simple and yet quite general. It provides a quantity maximized by selection and therefore allows us to continue the very useful practice of imagining individuals as fitness-maximizing agents (now inclusive-fitness maximizing) that act as if they had goals. However, the process of kin selection can be modeled by a variety of approaches from population genetics, game theory, evolutionary dynamics, and quantitative genetics. The result of such models is generally a version of Hamilton's rule, though the models may not be formulated to make this immediately apparent.

Indirect genetic-effects models extend the quantitative genetic tradition to social traits. They are becoming increasingly popular because they (like inclusive fitness) start with phenotypic traits for which the underlying genetics may be unknown, which is true for most traits biologists want to study. These models work with selection coefficients and heritabilities, and show that relatedness can be thought of as a ratio of heritabilities. For example, a trait that increases the fitness of full siblings is half as heritable through those effects as it is when it increases the fitness of self—exactly the same as relatedness of self to full siblings.

A related approach that yields compatible results is group selection (see chapter III.2). Here selection is decomposed into selection between groups and selection within groups. Selfish behavior is usually favored within groups, but cooperation may be favored when it sufficiently increases group reproduction. Relatedness among group members falls out as a crucial parameter here too. Older ideas of group selection and adaptation were lacking primarily in that they paid scant attention to relatedness or equivalently to the relative heritabilities of group and individual effects. Current group selection models do pay attention to this and give results compatible with kin selection.

To sum up, kin selection is an empirically well-supported form of selection that occurs through effects on relatives. It is most easily understood via inclusive fitness thinking, but can also be modeled by a variety of other approaches.

FURTHER READING

Bourke, A.F.G. 2011. Principles of Social Evolution. Oxford: Oxford University Press.

Frank, S. A. 1998. Foundations of Social Evolution. Princeton, NJ: Princeton University Press.

Griffin, A. S., and S. A. West. 2003. Kin discrimination and the benefit of helping in cooperatively breeding vertebrates. Science 302: 634–636.

Hamilton, W. D. 1996. Narrow Roads of Geneland. Vol. 1, The Evolution of Social Behavior. Oxford: Freeman.

Helanterä, H., J. Strassmann, J. Carrillo, and D. Queller. 2009. Unicolonial ants: Where do they come from, what are they, and where are they going? Trends in Ecology & Evolution 24: 341–349.

Hughes, W., B. Oldroyd, M. Beekman, and F. Ratnieks. 2008. Ancestral monogamy shows kin selection is key to the evolution of eusociality. Science 320: 1213–1216.

Ratnieks, F., K. R. Foster, and T. Wenseleers. 2006. Conflict resolution in insect societies. Annual Review of Entomology 51: 581–608.

Strassmann, J., and D. Queller. 2007. Insect societies as divided organisms: The complexities of purpose and cross-purpose. Proceedings of the National Academy of Sciences USA 104: 8619–8626.

Tsutsui, N. D. 2004. Scents of self: The expression component of self/nonself recognition systems. Annales Zoologici Fennici 41: 713–727.

West, S. A., S. P. Diggle, A. Buckling, A. Gardner, and A. S. Griffins. 2007. The social lives of microbes. Annual Review of Ecology and Systematics 38: 53–77.

III.5

Phenotypic Selection on Quantitative Traits

Edmund D. Brodie III

OUTLINE

1. How selection works
2. Selection is a statistical process
3. The genetic response to selection
4. Modes of selection
5. The multidimensional phenotype
6. Indirect selection and misleading covariances
7. Genetic correlations and correlated response to selection

Natural selection is the primary driver of adaptive evolution. Despite its power as a force of change over time, it is a remarkably simple process with only a few basic criteria. Whenever variation in a trait is associated with differences in reproductive success, selection occurs. This fundamental relationship between phenotype and fitness is measured as a covariance, and allows for a quantitative assessment of the strength of selection. Selection can occur in different modes that have distinct effects on the distributions of traits. While directional selection that changes the average character is most commonly considered, stabilizing and disruptive selection can directly impact the variance of a trait. Hundreds of studies of selection suggest that each of these modes is common in natural populations, and that directional selection tends to be strongest in magnitude. Correlational selection simultaneously affects two or more traits and can result in integration of characters within an organism. Relatively few studies have attempted to examine correlational selection in the wild. Genetic architecture determines whether and how any of these forms of selection are transmitted across generations. Selection can cause lasting changes only when traits are heritable and offspring resemble their parents. When traits are inherited together, selection on one trait can drag other traits along in a correlated response. Such indirect effects of correlations limit the independent evolution of single traits and can constrain evolution of the phenotype as a whole.

GLOSSARY

Correlated Response to Selection. The change in a trait across generations due to genetic correlations with another trait experiencing selection.

Covariance. A statistical measure of the degree to which two characters vary together; when standardized it is known as the *correlation*.

Fitness. How much an individual contributes to future generations, usually measured empirically as total lifetime reproductive success of an individual.

Frequency Distribution. Describes the probability of observing a particular value of a trait in a population. Usually assumed to be bell shaped or "normal," with the mean near the peak of the bell and the width of the curve measured by the variance.

Heritability. The resemblance of parents and offspring measured as the proportion of phenotypic variance in a trait due to the additive effects of genes.

Indirect Selection. Selection experienced by a trait not because it is causally related to fitness but because it is correlated with another trait that is.

Multivariate. Comprising more than one character or dimension; for example, a multivariate phenotype exhibits many traits at once.

Phenotype. Any observable characteristic of an organism; collectively, all its traits and their patterns of integration.

Population. A collection of sexually reproducing individuals with the potential to interbreed. For asexual organisms, a group of individuals of a single species in a local area.

Variance. The statistical spread of a distribution measured as the average squared deviation from the mean of the distribution; distributions with high variance are wider than those with low variance.

When Darwin and Wallace independently conceived of natural selection as the primary driver of evolutionary change, they imagined a process operating on all forms of life to gradually modify them to be better suited to the environmental challenges facing them. The forms better suited to survival and reproduction leave more copies of themselves and replace the alternatives in a population, leading over time to changes in forms from the ancestral condition, an idea that became known as *descent with modification*. This simple idea transformed human thinking about life on earth and remains one of the most important intellectual contributions in history.

Although Darwin and Wallace wrote specifically about "natural selection," we now understand the same general process of selection to apply across the spectrum of biological systems. Sexual selection is a subcategory of natural selection that operates through differences in the ability of organisms to acquire mates. Humans practice artificial selection to improve crops and livestock and to change the qualities of cultivated forms of plants and pet animals. Accidental selection resulting from the application of antibiotics and other biocides often leads to the evolution of resistance in pathogens and pests, posing serious problems in pest control. All these forms of selection operate in the same way and have similar impacts on populations, natural or managed.

1. HOW SELECTION WORKS

The power of selection, as both a process and a concept, lies in its simplicity. Only three conditions are necessary, and sufficient, for selection to cause change over time. When these three requisites are met, populations will change because of selection, and adaptive evolution will proceed, as follows:

1. *Variation.* Differences among individuals are the essential substrate on which selection works. Without such differences, some forms that are more successful than others could not exist. The source of the variation is not important for this criterion. The phenotypic differences might arise through influences of environment, different abilities to acquire resources, or fundamental genetic differences. Traits may vary quantitatively, as in differences in body mass, the length of a wing, or the number of flowers on an inflorescence, or variation might be qualitative, as in song forms of a bird, the presence of horns, or the color of flowers. So long as the traits of one individual are not identical to those of others, selection can occur.
2. *Heritability.* In order for the effects of selection to have lasting consequences across generations, offspring must resemble their parents. Heritable variation is essential as the mechanism of transmission of selection in one generation to the distribution of traits in the next. Darwin did not have a clear understanding of how inheritance worked when he described selection, but he nonetheless emphasized that resemblance among relatives was essential to the process. With the rediscovery of Mendel's work on particulate inheritance in the early 1900s, biologists began to understand the mechanism by which offspring reliably expressed traits similar to their parents. We now usually think of heritability as having a genetic basis, but some consistent environmental influences such as cultural transmission can satisfy the same criterion.
3. *Differential Reproduction.* The crux of the selection argument is that some individuals are more successful than others. Success is measured in terms of the number of copies of oneself left in the subsequent generation. This success in representation in the next generation is known to evolutionary biologists as fitness and is most easily measured as the number of offspring produced in a lifetime. It is fitness relative to others within a population that matters to selection, not the absolute amount of reproductive success. Selection essentially compares variants within a population, so a variant needs only to perform better than alternatives to be successful.

Whenever these three criteria are met, it is unavoidable that selection will cause changes to the mean and/or variance of traits in a population. Those variants that are associated with higher relative fitness will leave more offspring in subsequent generations, and heritability ensures that those offspring resemble their parents. Whether the differences in reproduction come about through the struggle for existence in the wild, or through humans selecting the breeding stock for a domestic improvement plan, the process is the same.

The two-part nature of evolution by natural selection becomes clear from this dissection. Phenotypic selection works within a generation whenever some trait values are associated with differential reproductive success. Inheritance then mediates the effects of phenotypic selection to determine evolutionary change. This separation of the force and response to the force allows us to understand in detail the contributions of each component of the process.

2. SELECTION IS A STATISTICAL PROCESS

The language typically used to discuss selection—active terms like the "force" of selection, identifying "agents" and "targets" of selection, and labeling some traits as "favored" by selection—obscures how straightforward the process is. In fact, selection requires no consciousness, no end goal, and no long-term direction. Even the idea that selection is a "process" is a bit misleading.

In fact, phenotypic selection is nothing more than a simple statistical by-product. To understand why, it helps to think of the traits of organisms in terms of distributions. For any variable trait in a population of organisms, say, body size or running speed, we can measure the population in terms of its *frequency distribution*. Usually that distribution is shaped something like a normal or "bell" curve, with the mean or average value of the trait near the peak of the bell, and the spread of the curve measured by the variance. Fitness or reproductive success has a similar kind of distribution with its own mean and variance.

Selection as a statistical process becomes apparent when we consider these distributions together in a bivariate form. On an x-y plot we can represent each individual in the population in terms of its trait value (x) and its reproductive success (y). This bivariate distribution determines whether a trait experiences selection or not. If some values of the trait are associated with higher values of reproductive success, then selection occurs.

This association is measured as the *covariance*. The covariance is the key to understanding selection on quantitative traits because it determines the strength of the relationship between fitness and a trait. If the covariance is positive, large values of a trait are disproportionately reproduced, whereas smaller values of a trait are overproduced if the covariance is negative. If the covariance is zero, then all values are equally reproduced on average and no selection occurs.

What we are trying to understand with selection is whether and how the average trait changes over time—will body mass, for example, get larger or smaller because of differences in reproductive success? In the early 1970s, George Price demonstrated that this change was neatly and fully predicted by the covariance between fitness and phenotype. Without going into Price's full mathematical proof, the simplest form of the Price equation shows that the change in the average trait, $\Delta\bar{z}$, can be expressed in terms of the covariance of fitness, w, and that trait, z:

$$\Delta\bar{z} = COV(w, z)$$

The Price equation underscores the statistical nature of selection. Any trait that covaries with fitness, whether the association is causal or not, will experience selection. Whether the covariance obtains because some lizards are fast enough to outrun predators and others are not, or because humans choose to plant seeds from only the sweetest watermelons, the inequality in reproductive success is what drives the selection. The covariance is easy to measure and provides a metric for the strength of total selection on a trait known as the *phenotypic selection differential, s*. The variance in fitness itself is known as the *opportunity for selection* because it places a limit on the strength of selection on any character. Without variance in fitness, there could be no covariance between fitness and phenotype.

3. THE GENETIC RESPONSE TO SELECTION

The type of selection described above is a purely phenotypic process. It happens in one generation, and it changes distributions of traits within that generation. In order for the phenotypic selection to have evolutionary consequences, changes that take place in one generation must be transmitted to the next; this is where inheritance comes into play.

It is easiest to understand how this works by considering the simple form of truncation selection, in which every individual above a threshold value survives and breeds; those below die and leave no offspring. Truncation selection is often practiced in agricultural contexts to improve the quality of crops, but it can occur in nature too. In the interaction between Japanese weevils and *Camellia* fruits, a female weevil uses her long beak to chew through a protective fruit coat into the seed to lay her eggs. *Camellia* fruits with coats thicker than the beak length of weevils survive, but those with thinner coats are parasitized and eaten by weevil larvae. This form of truncation selection results in an increase in the mean fruit coat thickness of surviving fruits compared to the population as a whole.

To determine whether fruit coat thickness evolves, we have to know whether the trait is heritable. In the most general sense, a trait is heritable if offspring resemble their parents more than they do other adults. More specifically, we are interested in the portion of resemblance that is likely to be passed along generation after generation, which is determined by the proportion of total phenotypic variation attributable to additive genetic effects. Additive genetic effects have the same

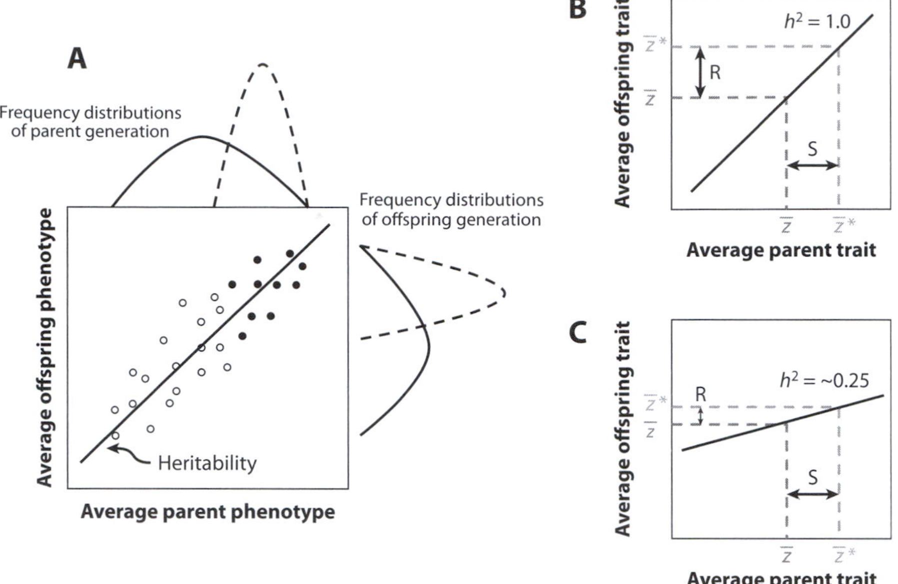

Figure 1. Heritability plot and response to selection. (A) The relationship between the average of a trait in two parents ($\bar{z}$) and the average trait of the offspring they produce. The slope of the relationship is the heritability of the trait. On the margins are the frequency distributions of each generation (solid curves). The means of each generation will be the same in the absence of selection. If truncation selection occurs and only parents with traits above a threshold value reproduce (solid symbols), the distributions change (dashed curves), with new means. At the top of the plot, the mean of the parent generation is increased, and this difference is *s*, the selection differential. (B and C) Response to selection without the distributions or the parent-offspring pairs. The means of the parent generation and the offspring that would be produced are shown as $\bar{z}$. After truncation selection, the mean of the parent distribution becomes $\bar{z}^*$ and the difference between the parent means is the strength of selection, *s*. Heritability determines the average traits of the offspring that would be produced after selection, $\bar{z}^*$. The difference between mean of the offspring produced and those that might have been produced is the response to selection, $R = \bar{z}^* - \bar{z}$. The same strength of selection, *s*, can generate a strong response (B) or a weak response (C) depending on the heritability of a trait.

influences on trait variation regardless of the genetic background, so these are most generally expected to transmit change across many generations.

Graphically, a heritability plot shows a positive relationship between the average traits of parents and their offspring (figure 1A). The stronger the relationship, the greater the proportion of variance explained by additive genetic effects. The slope of this regression is equal to the heritability of a trait. Regardless of heritability, the mean of the trait in the offspring generation is expected to be the same as the mean in the parent generation in the absence of selection. The effects of selection on distribution of the trait in the offspring generation can be seen by noting the parents that survive and reproduce after truncation selection (figure 1). In the case of *Camellia* fruits, only trees that make a fruit coat thicker than the length of the beak of local weevils will successfully reproduce. The difference between the mean of all possible parents and the mean of those that do reproduce is the phenotypic selection differential, *s*, the same value as predicted by Price's equation, $\Delta\bar{z}$, and described above. The difference between the mean offspring trait and the mean that would have been produced if all possible parents had reproduced equally is the response to selection, *R*. It is predicted by the product of the heritability, h^2, and the phenotypic selection differential, *s*.

$$R = h^2 s$$

This relationship is famously known as the *breeder's equation*. The stronger the heritability (which, as a ratio of variances, is bound between 0 and 1), the more faithfully the effects of selection in one generation are translated into phenotypic change across generations. Fruit coat thickness in *Camellia* is known to have a heritability around 0.7. While the strength of selection on *Camellia* varies greatly among populations, it averages around $s = 0.1$, indicating that fruit coat thickness changes around one-tenth of a standard deviation each generation. The fruit coat of *Camellia* differs fivefold across islands in the Pacific, matching the beak length of weevils on each island, suggesting that evolutionary response has occurred.

The breeder's equation is based on statistical principles, and the estimates of heritability and phenotypic selection that go into it are notoriously approximate. Nonetheless, this simple product of selection in one generation and genetic basis of response across generations is remarkably accurate at predicting quantitative evolutionary changes. One particularly impressive example comes from a field study of the alpine skypilot flower in Colorado. Candace Galen bred plants to obtain a heritability estimate for flower size ($h^2 \approx 1.0$) and observed pollinator behavior to determine the strength of the selection differential on the same trait ($s = 0.07 - 0.17$). Combining the estimates using the breeder's equation, she predicted that flower size should increase 4–17 percent per generation because of the preferences of bumble bees for larger flowers. By following the population across generations, she was able to detect an observed evolutionary increase of 9 percent, firmly within the range of her predictions.

4. MODES OF SELECTION

Selection can take several different statistical forms having very distinct consequences for the evolution of quantitative traits (see chapter III.3). When one thinks of selection that changes form, driving adaptation of the kind associated with classic evolutionary radiations like leg lengths of *Anolis* lizards on Caribbean islands and the beak depth of finches on the Galápagos, this is generally called *directional selection*. Directional selection changes the mean of a trait and is visualizable as a monotonically increasing (or decreasing) function of fitness with phenotype (figure 2A). Truncation selection is an extreme form of directional selection. Sexual selection, such as that documented by Carl Gerhardt and colleagues in which female gray tree frogs prefer males with longer calls, typically generates directional selection.

It is often assumed that most traits of organisms are relatively well adapted to their current environments. This means that neither larger nor smaller values of traits should be associated with higher fitness; instead, it is the intermediate values of traits that have highest relative fitness, so the resulting selection function is nonlinear with an intermediate peak. *Stabilizing* (or optimizing) *selection* acts to reduce the variance in a distribution without changing the average value of a trait (figure 2B). For this reason it is usually regarded as a mechanism of stasis and stability rather than one that generates new adaptations; however, the shape of a phenotypic distribution may be considered adaptive at the population level.

Whereas stabilizing selection occurs because the individuals with the most deviant trait values have lowest fitness, in *disruptive selection* these individuals fare best and the intermediate values fare worst. Disruptive selection is the mathematical opposite of stabilizing selection; the resultant function is convex, whereas that of stabilizing selection is concave (figure 2C). The effect of disruptive selection is to inflate the variance of a distribution, and in this way it acts to promote polymorphism within populations. Most cases of disruptive selection in nature involve relatively distinct forms with alternative strategies, such as those of black-bellied seedcrackers, in which birds with different bill shapes specialize on alternative food types. Field studies conducted by Tom Smith in Cameroon demonstrated that birds with intermediate bill shapes are less efficient at foraging on either food type, thereby suffering lower fitness.

Disruptive selection is expected to be uncommon because it is unstable over time. As the phenotypic distribution of a trait shifts toward one side or the other of the convex function, the prevailing effect of selection in the population becomes more directional. For disruptive selection to persist, temporal fluctuations in selection must occur. The most important of these may be *negative frequency-dependent selection*, in which the fitness associated with a trait value depends on its relative frequency. If rare values always have higher fitness, this has the effect of rocking the convex function back and forth across generations, thus maintaining the average form of disruptive selection through time. Negative frequency-dependent selection has been observed in a variety of species with polymorphic trophic strategies and color patterns. Flower color polymorphisms such as the yellow or purple inflorescences of the orchid *Dactylorhiza sambucina* may be maintained in this fashion. Pollinators of *D. sambucina* receive no reward for visitation, and they learn to avoid the most commonly encountered flower color, thus lending an advantage to rare types.

Phenotypic selection of each mode has been widely documented in nature (see chapter III.7). Joel Kingsolver and his colleagues have reviewed the majority of these studies and revealed that directional selection is the mode most commonly observed. The strength of directional selection, particularly when it results from sexual selection, is stronger than many researchers expect, averaging around $s = 0.15$, or a change of 15 percent of one standard deviation each generation. Despite the presumption that most selection should be stabilizing, it is not as commonly observed as directional selection. In fact, disruptive selection is detected at nearly the same frequency and the same strength as stabilizing selection. These patterns may be influenced by the tendency of researchers to study selection in traits they suspect to be experiencing directional selection, and by the statistical difficulty in detecting nonlinear functions.

5. THE MULTIDIMENSIONAL PHENOTYPE

Each of the above modes of selection assumes that relationships exist between fitness and single traits in

A Directional selection

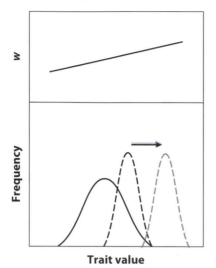

w

Frequency

Trait value

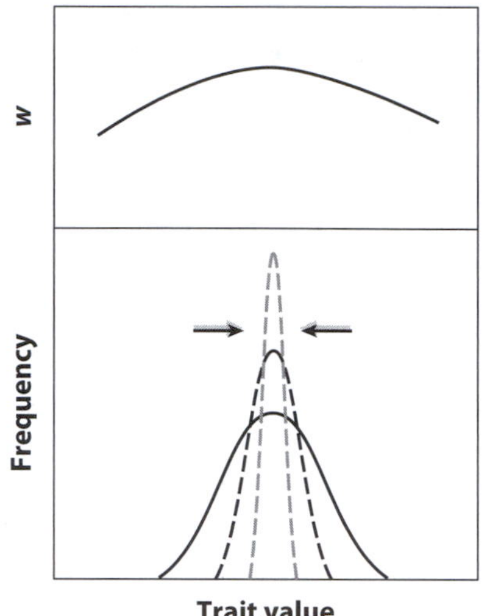

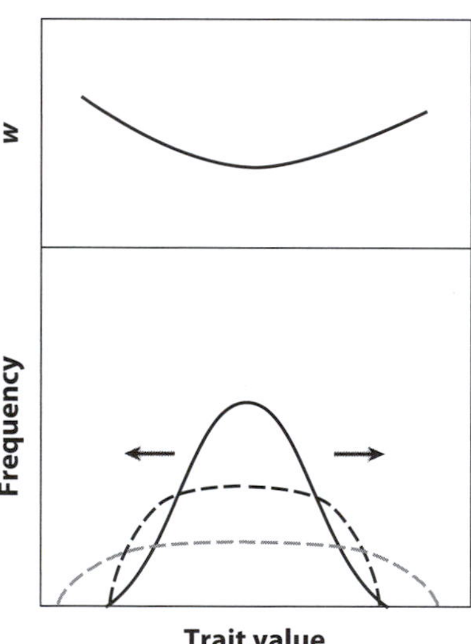

D Correlational selection

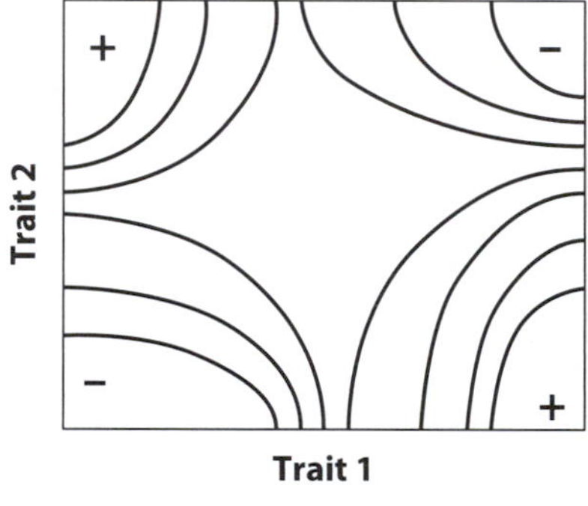

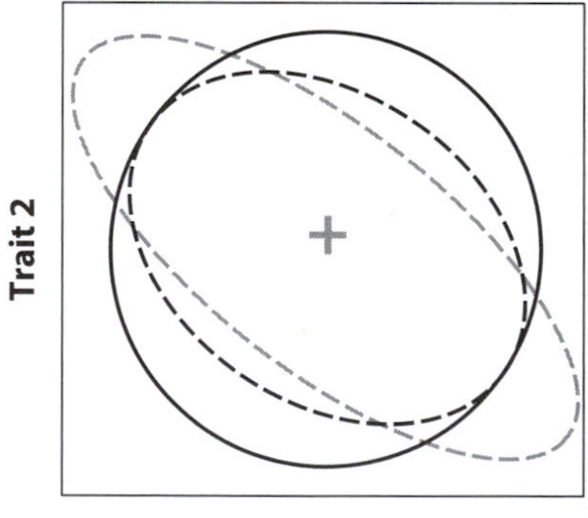

Figure 2. Modes of selection. Selection functions relating fitness (*w*) to trait variation are shown at the top of each plot, with the frequency distributions of the trait below. The general change in the frequency distribution due to selection within a generation is shown with a black dashed line. If heritability results in a response to selection, the long-term effect of selection is shown in the dashed gray distribution. (A) Directional selection; (B) stabilizing selection; (C) disruptive selection; (D) correlational selection (shown as a saddle-shaped contour plot, wherein lines on the surface indicate equal fitness and peaks are denoted with a +; bivariate frequency distributions are shown as ellipses).

isolation. In fact, organisms comprise an uncountable number of traits that both function and are inherited as part of an integrated whole. This fact renders selection and response multivariate problems that must consider whether and to what extent traits are correlated to make accurate predictions about evolutionary change.

A character sometimes does not experience selection alone, but does so in combination with another. Neither individual trait covaries with fitness, nor is there a discernible function relating variance in fitness to the individual trait; when the two traits are considered simultaneously, however, a pattern may emerge, a mode known as *correlational selection*, which describes selection on combinations of traits rather than on individual characters in isolation. The effect of correlational selection is to change the bivariate distribution of traits without necessarily changing the mean of either character (figure 2D). Field studies of the northwestern garter snake showed this mode of selection with respect to escape behavior and color patterns. No covariance between fitness and either the number of reversals during escape or the degree of stripedness was observed, indicating that snakes had an equal probability of survival regardless of these traits, when viewed individually. However, when both traits were examined together, correlational selection was detected. Survival was highest for snakes that had striped patterns and made few reversals, or for those with spotted patterns that performed many reversals. Snakes with the other two possible combinations had lower survival. This form of selection acts to increase the correlation between behavior and color pattern, thus shaping the integration of the multivariate phenotype.

Correlational selection can take several shapes with slightly different effects. The selection on garter snakes described above can be envisioned as a saddle-shaped function in bivariate space (plate 2). Much like disruptive selection, the highest fitness is associated with extreme combinations of traits while the opposite combinations of extremes have the lowest fitness. This form of selection is sometimes called *epistatic* because it is the interaction between traits, rather than the traits themselves, that predicts fitness. Any trait value of one character could have the same fitness, depending on the value of a second character with which it is paired. A second form of correlational selection can be envisioned as a ridge of higher fitness along a line of matching trait values. In this case, there is no particular advantage to having extreme phenotypes, but rather to being equally matched. This ridge form of selection also leads to increased correlations among traits and builds integration across the phenotype. Several studies of birds and insects, including Dolph Schluter's 1994 analysis of survival in song sparrows, have revealed matching correlational selection for body mass and wing size, suggesting that a range of body sizes are equally fit as long as they are paired with appropriate-sized wings. Correlational selection is understudied in nature, owing in part to the logistical challenges of studying multivariate functions.

6. INDIRECT SELECTION AND MISLEADING COVARIANCES

Selection is typically thought of as a causal process, but this interpretation contradicts the covariance approach. To be sure, covariances can arise because of causation, but they do not require functional relationships to occur. Statistical analyses of selection alone can rarely demonstrate causal links; experimental approaches are normally required to disentangle such paths.

Correlated characters make the problem particularly confounding. If two characters are positively correlated, one trait might covary with fitness because it is correlated with another trait that is functionally more important. Flat-tailed horned lizards of the North American Southwest have a crown of bony projections around the skull presumably used in defense against predators. Kevin Young and colleagues used the covariance approach to show that the length of the parietal horns on the back of the head covaried strongly with survival in the face of attack by shrikes; however, the length of parietal horns strongly correlates with the length of squamosal horns on the side of the head. Squamosal horns, too, covary with survival. Both types of horns are positively correlated with overall body size, which also covaries with fitness. Each of these traits experiences phenotypic selection, but that does not imply that each is causally important.

This problem is known as *indirect* (or *correlated*, distinct from "correlational") *selection*. In the example above, it is assumed from behavioral studies that the parietal horns are important in defense because they are used to stab predators in the face during attacks. If the length of parietal horns drives the variation in survival, then squamosal horns might experience indirect selection because they are correlated with the causally important trait. A functional relationship between one trait and fitness causes all correlated traits to experience selection as well. In a landmark paper in the early 1980s, Russell Lande and Steve Arnold described an analytical approach that helped disentangle the multivariate problem. Using multiple regression, it is possible to determine how much of the covariance between a trait and fitness is independent of other correlated traits. This approach has changed the way that selection is studied in the wild, and it ameliorates the problem of confounding trait correlations.

A related problem arises when covariances with fitness arise as a result of environmental factors. This problem is easy to imagine in plant populations. Individuals growing in particularly good soil or local environments may have larger than average leaves, flowers, or other characters. By virtue of being in that rich environment, they might also produce more seeds. Other plants that grow in poor soils have lower reproduction and make smaller leaves and flowers. A selection analysis would reveal a positive covariance suggesting that leaf size and flower size experience directional selection; however, the covariance is not causal, and because the differences in phenotype and fitness are caused by the environment, there would be no response to selection.

7. GENETIC CORRELATIONS AND CORRELATED RESPONSE TO SELECTION

A different set of correlations changes the way selection is transmitted across generations. Genetic correlations describe how traits are inherited together and can arise from a variety of mechanisms. *Pleiotropy* occurs whenever a gene or group of genes influences variation in multiple traits at once. Pleiotropic effects can be generated whenever a single gene product like an enzyme is used in common developmental pathways or influences the expression of more than one character. The endocrine system is a common example, wherein a single product like testosterone determines the expression of behavioral traits including aggression and parental care, morphological traits like body size, and physiological traits like immune function.

Genetic correlations also occur whenever independent genes become associated statistically or physically. When two genes are found close together on a chromosome, they are physically linked, and recombination between them is unlikely. Alleles at these genes form nonrandom combinations known as *linkage disequilibrium* that is measured as a genetic correlation. Even without physical linkage, patterns of assortative mating and selection (such as the correlational selection described above) can lead to combinations of alleles occurring together more often than expected by chance, and likewise generating genetic correlations.

When selection occurs on one trait, all the genetically correlated traits will respond to that selection, even if they experience no selection themselves. This *correlated response to selection* occurs because selection on one trait in the parent generation also changes the distribution of the genetic variance in correlated traits. Just as heritability causes evolution across generations in a single trait, genetic correlations cause correlated evolution in linked traits. This effect is clear when we consider traits expressed in both sexes. *Silene latifolia* is a dioecious plant, with male and female reproductive structures on separate plants. Because flower size in both sexes is controlled by many of the same genes, the flower size of males and females is strongly genetically correlated. Lynda Delph and colleagues have demonstrated that natural selection favors large female flowers, in part because they can make more seeds. At the same time, selection favors many small flowers in males, because that is how they increase their mating success. In an artificial selection experiment, Delph allowed only the smallest female flowers to reproduce, causing an evolutionary reduction in female flower size as predicted. Male flower size, too, responded to this selection with a reduction in size. Even though males experienced no selection directly, the genetic correlation between the sexes was enough to drag along the evolution of a second trait.

This experiment highlights the usual interpretation that genetic correlations can constrain evolutionary response (see chapter III.8). In *Silene*, male and female flowers are selected in opposite directions, creating a conflict in selection. Genetic correlations cause selection on one sex to drag the other along in the same direction in evolutionary time; thus the positive genetic correlation acts as a constraint against selection favoring different flower sizes in the two sexes. More generally, genetic correlations and the correlated responses to selection they cause are expected to constrain phenotypic evolution, because selection on a single trait will cause changes throughout the integrated phenotype.

FURTHER READING

Brodie, E. D., III, A. J. Moore, and F. J. Janzen. 1995. Visualizing and quantifying natural selection. Trends in Ecology & Evolution 10: 313–318. *A review of the analytical tools for measuring and graphically representing selection.*

Endler, J. 1986. Natural Selection in the Wild. Princeton, NJ: Princeton University Press. *The most complete review of studies of the selection in natural populations.*

Falconer, D. S., and T. C. Mackay. 1996. Introduction to Quantitative Genetics. New York: Longman. *Introduces and explains the theory behind heritability, genetic correlations, and the response to selection.*

Grant, P. R., and B. R. Grant. 1995. Predicting microevolutionary response to directional selection on heritable variation. Evolution 49: 241–251. *The classic study comparing predicted and observed rates of evolution in Galápagos finches over two decades.*

Hartl, D. L., and J. K. Conner. 2004. A Primer of Ecological Genetics. Sunderland, MA: Sinauer. *Chapters 4–6 offer a good introduction to the basic concepts of selection and response to selection, with examples from empirical studies.*

Kingsolver, J. G., E. H. Hoekstra, J. M. Hoekstra, D. Berrigan, S. N Vignieri, C. E. Hill, A. Hoang, P. Gilbert, and P. Beerli. 2001. The strength of phenotypic selection in natural populations. American Naturalist 157: 245–261. *A synthetic review of published estimates of selection coefficients comparing the strength of different modes.*

Kingsolver, J. G., and D. W. Pfennig. 2007. Patterns and power of phenotypic selection in nature. Bioscience 57: 561–572.

Young, K. V., E. D. Brodie Jr., and E. D. Brodie III. 2004. How the horned lizard got its horns. Science 304: 65. *A short description of a field study using regression approaches to disentangle natural selection on correlated characters.*

III.6

Responses to Selection: Experimental Populations

Graham Bell

OUTLINE

1. Will adaptation evolve?
2. How fast will adaptation evolve?
3. Does sex accelerate adaptation?
4. Is adaptation gradual or saltational?
5. What is the limit to adaptation?
6. Is adaptation based on gain or loss of function?
7. Is adaptation repeatable?
8. Is adaptation predictable?
9. Is adaptation reversible?
10. How do ancestry, selection, and chance contribute to adaptation?
11. How can selection maintain diversity?
12. What limits the extent of specialization?

Evolutionary biology has been an observational and comparative science for most of its history because "natural selection always acts with extreme slowness" (Darwin 1859, p. 121) and therefore produces adaptation "by minute steps, which, if useful, are augmented in the course of innumerable generations" (Weismann 1909, p. 24). Artificial selection in crop plants and domestic animals was from the first used to justify the general principle of modification, but the deliberate choice of breeding individuals by human agency made it only a simile for natural selection. A century passed before the invention of the chemostat led to the realization that microbial cultures, with their huge populations and rapid turnover, could act as time machines enabling us to investigate evolutionary change through natural selection in real time. One path led to a series of brilliant biochemical studies showing how individual enzymes and whole metabolic pathways could evolve (Mortlock 1984). Another led to investigations of more general evolutionary processes (Dykhuizen 1990), and this research has continued to the present day with experiments of increasing scope and power. The promise of experimental evolution is to provide decisive tests of specific hypotheses about adaptation, the core process of evolution. Questions such as what would happen if the tape of life were replayed or how is sex maintained have elicited endless debate, yet they can now be settled by experiments using laboratory microcosms. In this brief note—by no means a review of the field (see Bell 2008)—I shall list a dozen basic questions about adaptation and describe briefly how they have been illuminated by selection experiments in the laboratory. Most of them involve the very simplest scenario of asexual unicellular microbes growing in homogeneous culture medium—a small, cloudy tube of bacteria, yeast, or algae.

GLOSSARY

Adaptation. In the sense used here, a change in the genetic composition of the population caused by natural selection and resulting in elevated fitness in defined conditions.

Beneficial Mutation. Any genetic change that causes elevated fitness and thereby contributes to adaptation.

Evolutionary Rescue. The survival of a population in conditions lethal to its ancestor as the consequence of adaptation through natural selection.

Experimental Evolution. The study of natural selection under controlled conditions, usually in the laboratory.

Microbe. A unicellular organism, usually a bacterium, yeast, or alga.

Selection Experiment. The repeated propagation of a population in controlled conditions with the object of discovering how it adapts to those conditions through natural selection.

1. WILL ADAPTATION EVOLVE?

Adaptation can be demonstrated in the laboratory simply by subjecting a microbial culture to a stress such as high temperature or a toxic chemical. As a sample of the culture is repeatedly transferred to fresh medium, any variant type that is resistant to the stress will tend to spread, replacing more sensitive types, thus raising the average rate of growth. This process is bound to occur when there is heritable variation in fitness, and is driven by differences in relative fitness among types within the population, whether these types are already present when the stress is applied or arise afterward by mutation and recombination. It is not bound to produce permanent results, however, because the stress may be so severe that even the most resistant types are incapable of sustained growth. The population will then diminish, transfer after transfer, until eventually it becomes extinct.

Hence, we can recognize three levels of stress with different evolutionary outcomes. If only a mild stress is applied, the population will continue to have a positive rate of growth. The population may well evolve, because the new conditions cause shifts in relative fitness, but its persistence does not depend on a higher level of adaptation. On the other hand, if a very severe stress is applied, the population is certain to become extinct sooner or later. At intermediate levels of stress, the population at first declines but may recover later if resistant types with positive rates of growth have spread. The signature of evolutionary rescue is the U-shaped trajectory of abundance as collapse is followed by recovery. The boundary between the zones of recovery and extinction is set by variation in absolute rather than relative fitness, which depends on the presence of genotypes with positive rates of growth, which is more likely in larger populations because these encompass a broader range of variation. This has been demonstrated by culturing yeast populations of different sizes in medium with high concentrations of salt and showing that rescue occurs in those sufficiently large to include one or two resistant cells.

2. HOW FAST WILL ADAPTATION EVOLVE?

When a population is stressed, its mean fitness at first declines and then increases as it becomes better adapted through natural selection. The rate at which it becomes adapted depends solely on the amount of variation in fitness. If this is small, there is little difference between the least fit and the most fit types in the population, so that selection will be weak and adaptation slow. Conversely, a large amount of variation in fitness implies strong selection and rapid adaptation (see chapter III.5). For the population to become permanently modified, this variation must be heritable, of course, leading to the conclusion that the rate of adaptation is equal to the heritable variance of fitness (Fisher 1930).

This classical result is always valid for infinitely large populations. It will not always accurately predict the rate of adaptation of finite populations, however, because the most fit types may be absent, so that adaptation is slow until they are generated by mutation or recombination. In this case, adaptation will occur in two stages: a waiting stage, before the first high-fitness type appears, and an establishment phase, during which it spreads to fixation. The dynamics of the establishment phase are indeed governed entirely by the difference in fitness between the superior type and the average of the population, and are (almost) independent of population size. The length of the waiting period, on the other hand, is inversely proportional to the rate of supply of beneficial mutations, which is higher in larger populations: more beneficial mutations are likely to occur within a given span of time in larger than in smaller populations. Hence, the rate of adaptation should increase with population size. This can be tested experimentally by allowing cultures of different volume to evolve under stress: after a prescribed number of transfers, the larger cultures, containing more cells, are found to be better adapted. The relationship seems to be a power law, with each doubling of population size leading to the same small fractional increment in the rate of adaptation.

3. DOES SEX ACCELERATE ADAPTATION?

The function of sex has long been vigorously disputed by evolutionary biologists (see chapter III.9). The prevailing hypothesis (although by no means the only one) is that the fusion of gametes followed by the recombination of their genomes generates genetic variation, with the effect of accelerating adaptation, a view dating back to Weismann. In the simplest terms, beneficial mutations arising in different lineages can be brought together by fusion and packaged together in the same lineage by recombination. The evolutionary effect of sex is best studied in eukaryotic microbes such as yeasts and unicellular algae, because they all have periods of purely vegetative growth between sexual episodes. In organisms like these, where sex can be controlled by environmental or genetic manipulations, it is possible to set up populations that are equivalent in every way except that some go through a sexual cycle of fusion and recombination from time to time, whereas others are perennially asexual. Experiments like this have shown that when sex is induced in a population growing in a novel and stressful environment, its immediate effects are a drop in average fitness, relative to a comparable asexual line, accompanied by an increase in the variance of fitness. The increased variation generates stronger selection, eventually driving the fitness of the sexual line above that of the asexual line. Over a period of several hundred generations, sexual populations thereby adapt faster than purely asexual populations in stressful environments; moreover, this effect depends on population size, because it is only in large populations that two or more beneficial mutations are likely to occur at the same time, and therefore only in large populations that fusion and recombination are likely to accelerate adaptation. Consequently, sex accelerates adaptation much more effectively in large than in small populations.

4. IS ADAPTATION GRADUAL OR SALTATIONAL?

The classical account of evolution is gradualist: populations become adapted through the substitution of beneficial alleles of small effect at many loci over long periods of time (see chapter V.12). Generally speaking, this is not borne out by laboratory experiments, in which adaptation can occur very rapidly through the substitution of a few alleles of large effect. It is true that most of the beneficial alleles that appear by mutation when a population first experiences stressful conditions have rather small effects on fitness. This can be demonstrated by isolating mutants resistant to an antibiotic and then measuring their fitness in the absence of the antibiotic. This provides the distribution of fitness of new mutations at the time when they first arise. A few of these mutants are fitter than their ancestor, but most are only slightly superior, and very few are much more fit. If such beneficial mutations are allowed to spread, however, and collected only when they have become fixed, a very different picture emerges: the bulk of these fixed beneficial mutations have large effects, often amounting to a doubling of fitness. The reason is that beneficial mutations that increase fitness only slightly, although they may be very numerous, are likely to be overtaken by the much faster spread of mutations that greatly increase fitness, despite their rarity. The rapid spread of large-effect beneficial mutations is often observed in laboratory experiments. This provides a concrete alternative to the gradualist interpretation of adaptation, especially when populations are severely stressed.

5. WHAT IS THE LIMIT TO ADAPTATION?

If conditions remain unchanged, the first few mutations to be fixed may well increase fitness substantially, but the supply of these large-effect mutations is sure to be limited, and as the supply is depleted, only those of more modest effect remain available for selection. Consequently, the rate of adaptation will tend to diminish over time. This pattern has been demonstrated by long-term serial-transfer experiments with *E. coli* in which replicate lines have been maintained in a simple glucose medium for tens of thousands of generations. These increased rapidly in fitness over the first 2000 generations, but the rate of adaptation was clearly decreasing throughout this period, and was much slower in the subsequent 20,000 generations. This illustrates a process whereby adaptation is driven by mutations of smaller and smaller effect as time goes by. Nevertheless, although adaptation decelerated, it did not stop completely: some improvement is still being made after 50,000 generations in culture. Hence, there may be no definite limit to adaptation, but rather a continuously diminishing response, with alleles of diminishing effect being substituted at longer and longer intervals. Moreover, the supply of large-effect mutations, while it will become very meager, may not vanish completely. A very rare mutation conferring the ability to utilize citrate as a carbon substrate appeared in the long-term lines after about 35,000 generations and resulted in a large increase in fitness. It is not safe to conclude that a population has lost all capacity to adapt, even after a very long period of observation. In practice, of course, conditions will seldom remain constant for tens of thousands of generations, so the prolonged improvement of lines in uniform laboratory conditions implies that natural populations may be actively and often rather rapidly adapting most of the time.

6. IS ADAPTATION BASED ON GAIN OR LOSS OF FUNCTION?

Fitness is increasing over time in any population adapting to novel conditions, and in this sense adaptation always involves a gain of function. At a genetic or biochemical level, however, this is often based on a loss of function, in the sense that a particular chemical reaction can no longer be carried out. This may occur because a reaction that promotes growth in most conditions becomes unnecessary in highly simplified laboratory culture, and variants that lose the ability to conduct it gain an advantage because of some economy in time or materials. The uptake of glucose by bacteria, for example, is regulated by a series of genes that switch on transport systems when glucose is available and switch them off when it is not. When experimental lines are cultured in glucose medium, mutations that cause defects in this regulatory system and leave glucose transport permanently switched on are often among the first to spread. This increases fitness, in laboratory conditions where glucose is continuously supplied, through the degradation of a coordinated biochemical pathway. Once these loss-of-function mutations have become established, however, the stage is set for gain-of-function mutations to spread. These are mutations that alter, rather than degrade, a reaction so as to create a new biochemical ability. The experimental evolution of amide utilization illustrates this sequence. The ancestral strain expresses an amidase enzyme enabling it to use acetamide, the simple two-carbon amide responsible for the characteristic odor of mouse cages. The amidase is hardly able to process the four-carbon amide butyramide at all, however, so that if butyramide is supplied as the sole carbon source, growth is very slow. Among the first beneficial mutations to appear are alleles conferring faster growth simply by expressing very large quantities of the amidase, for example, by loss-of-function mutations in the genes that regulate its expression or by increasing the number of copies of the structural gene. Once these have become fixed, further adaptation is based on a gain-of-function mutation in the gene encoding the amidase that alters its structure so that it processes butyramide more efficiently. The ability to utilize larger and more complex amides can subsequently evolve in a similar fashion, through the successive modification of the amidase by gain-of-function mutations to increase its activity on particular substrates. The evolution of new metabolic capabilities by bacteria often begins with exaptation, the use of an inefficient enzyme normally responsible for degrading some other substrate; this is followed by deregulation and amplification to increase the supply of this inefficient enzyme and culminates in the modification of the enzyme to produce a new and more efficient version. The exact course of this exaptation-deregulation-amplification-modification (EDAM) process varies from case to case, but the transition from loss-of-function to gain-of-function mutations during the course of adaptation may be quite general.

7. IS ADAPTATION REPEATABLE?

The spread of loss-of-function mutations affecting glucose uptake in bacterial cultures growing in glucose-limited medium has been repeatedly observed and can be confidently expected whenever such experiments are conducted. The specific mutations involved, however, differ from case to case. There are several genes in the outer and inner membranes of the cell that contribute to the regulation of glucose uptake, and any of them may be affected; moreover, the mutations themselves may be substitutions of single nucleotides, or frameshifts, or short insertions or deletions. There are many other examples of similar changes occurring in response to the same agent of selection. The long-term *E. coli* lines, for example, have consistently lost the ability to utilize ribose as a substrate for growth. In other cases, however, the same agent of selection leads to different outcomes in replicate selection lines. Bacteria cultured on a range of different substrates will often become adapted to each of them, evolving higher rates of growth; at the level of fitness, adaptation is quite highly repeatable. Whether the same underlying genetic changes are responsible in each case can be evaluated by culturing all replicate lines that have adapted to a single particular substrate on each of the other substrates: they will show the same pattern of growth on these exotic substrates if they have acquired the same genetic changes, but different patterns of growth otherwise. In one extensive experiment involving nearly a hundred different substrates, there was little correlation in most cases between lines cultured on the same substrate, indicating a low level of repeatability. The cause may be the historical nature of adaptation, arising from the stochastic appearance of mutations and the interaction between their effects.

The repeatability of adaptation can be assessed at a genomic level by obtaining complete sequences and expression profiles for replicate lines under uniform selection. In the long-term *E. coli* lines, similar changes in expression evolved in a limited suite of about 50 genes, although the details often differed between lines, for example, by the production of similar changes in gene expression by different regulatory mutations. Likewise, a small group of genes consistently had modified sequences in the lines, although the particular mutations that had occurred varied from line to line. At present, it seems that microbial adaptation in uniform laboratory conditions often involves a few themes and many variations. The few themes are the major genes where beneficial mutations can occur; the variations are the many possible alleles of these genes.

8. IS ADAPTATION PREDICTABLE?

The spread of loss-of-function mutations in glucose transport systems is not only repeatable but also predictable from first principles. In other cases, such as loss of ribose metabolism in the long-term *E. coli* lines, the event is repeatable, but the reason for it is not understood. Where genetic changes are tightly coupled to fitness, the course of adaptation is usually rather highly predictable. Lactose metabolism in *E. coli* depends on three processes acting in succession: diffusion through pores in the cell wall, active transport across the cell membrane by a permease, and hydrolysis by β-galactosidase in the cytoplasm to split the molecule into glucose and galactose. The effect of altering the activity of the permease or β-galactosidase on the overall flux through this pathway can be calculated from biochemical principles and compared with the observed relative fitness of a series of mutants in lactose-limited medium. Since the flux through the pathway is equivalent to fitness when lactose is the sole carbon source, the biochemical and evolutionary estimates should coincide, and experiments show that they are indeed very highly correlated.

A more extensive attempt was made to predict evolutionary change in the genome of phage T4. This small genome is better understood than any other, with each of its 288 genes being well characterized at a molecular level; hence, it should be possible to predict the genetic basis of adaptation to a stress such as high temperature with a high degree of confidence. When this was tested by experiment, about one-half of all changes were correctly predicted, at least to some extent, such as the gene or region of the gene in which they should occur, whereas the other half were unexpected despite our deep functional knowledge of the genome. This can be seen as a glass half full: the demonstrable possibility of predicting a large proportion of evolutionary changes in this small genome clearly foreshadows eventual success in predicting the response of larger genomes in cellular organisms.

9. IS ADAPTATION REVERSIBLE?

A population that becomes adapted to a novel environment will usually recover a high level of adaptedness to ancestral conditions once these are again imposed. Any change in the morphological, physiological, and developmental features that contribute to adaptation, however, may not be as readily reversible (see chapter VI.12). This follows from the historical nature of evolutionary change, with each step in the adaptive walk arising stochastically and depending for its success in some degree on the modifications that have evolved previously. The simplest and most dramatic experiment is to delete an essential gene and investigate how a population can recover the ability to grow. If the structural gene encoding β-galactosidase is deleted in *E. coli,* the population is unable to grow on medium with lactose as the sole carbon source. Supplementing normal growth medium with lactose, however, permits the cells to grow while creating strong selection for any mutants able to utilize lactose once all other substrates have been used up. This procedure eventually leads to the evolution of populations able to grow slowly on lactose alone. This is not attributable to the restoration of the original β-galactosidase gene, of course, which would involve a very long and improbable series of mutations; rather, a different gene, present in the ancestor and producing an enzyme able to hydrolyze lactose at a very low rate, has become modified and appropriately regulated through a series of five beneficial mutations that together confer the ability to grow on lactose alone.

The irreversibility of evolution is not only an academic issue. It has often been suggested that the resistance that often evolves in bacterial populations to any given antibiotic would disappear if the antibiotic were withdrawn completely, because of the resurgence of the original susceptible type. Experiments have shown, however, that although resistant populations readily become adapted to an antibiotic-free environment, they usually do not lose their resistance, often because compensatory mutations, occurring at loci not concerned with resistance, arise and spread by restoring normal rates of growth without affecting resistance to the antibiotic. An expanded program of laboratory selection experiments to investigate the evolutionary dynamics of antibiotic resistance might make an important contribution to public health.

10. HOW DO ANCESTRY, SELECTION, AND CHANCE CONTRIBUTE TO ADAPTATION?

The issues of repeatability, predictability, and reversibility are tied to the contributions of three processes to overall evolutionary change. The first is phylogenetic: a particular feature was inherited from more or less remote ancestors. The second is adaptive: it evolved through natural selection because it confers higher fitness. The third is neutral: once having arisen by chance, it persists because it has no appreciable effect on fitness. The contribution made by each of these three factors to biological diversity has been strenuously debated, for example, between those who believe that almost all features, from the most fundamental aspects of development to the most trivial details of morphology, have been precisely sculpted by natural selection and those who maintain that chance and history are often responsible. All three may affect any particular feature, although it is usually difficult to evaluate their relative contributions; in experimental situations, however, evolutionary change can be unambiguously partitioned between history, selection, and chance. In one very elegant experiment, *E. coli* lines that had become adapted to glucose medium were switched to medium in which maltose was the only carbon source. Some were able to grow well, whereas others grew only feebly. After several hundred generations their evolved ability to grow on maltose could be partitioned among the three fundamental processes. Selection will cause a general increase in growth in all the lines; if history has an effect, however, the differences among the ancestral lines will be retained among their descendants; and finally, the divergence of replicate lines founded from the same ancestral line is attributable to chance. In this case, almost all the variation among the maltose lines was attributable to selection, with only very small contributions from chance and history. This finding is consistent with those of most experiments in which exposure to a new environment elicits strong selection, leading to a rapid and consistent increase in fitness that largely effaces any historical signal. This does not necessarily apply to features other than fitness. In the maltose experiment, for example, cell size also changed over time, but history, selection. and chance contributed more or less equally to the final state of the lines. It is likely generally true that selection is almost exclusively responsible for overall adaptedness, whereas the morphological and physiological features that underlie adaptation have more complex roots.

11. HOW CAN SELECTION MAINTAIN DIVERSITY?

Selection drives the spread of highly adapted types and is therefore likely to lead to the fixation of the best type and the elimination of all others. In practice, populations are often rather diverse, and there has been much debate about whether diversity is attributable to forces such as mutation and recombination running counter to selection, or whether selection itself often acts to preserve diversity (see chapter III.3). In general, selection has the property of preserving diversity when rare types consistently have an advantage that is lost once they become more abundant. Although this may seem an unusual and onerous requirement, it is likely to occur whenever a range of growth conditions is available, as in most natural habitats. Simple media with a single limiting substrate such as glucose are laboratory artifacts: bacteria outside the lab grow in very complex media with many carbon sources, none of which is limiting. This creates an opportunity for specialization, because a genotype able to utilize any substrate with exceptional efficiency will tend to spread. If it depletes this substrate as it grows, however, the genotype necessarily limits the extent of its own spread. It will have high fitness when it is rare, when its preferred substrate is present at relatively high concentration by virtue of the scarcity of types that can consume it efficiently, but low fitness once it has become abundant, when its own consumption has removed the substrate from the medium. Hence, bacterial populations cultured in complex media become more diverse than comparable lines cultured in simpler media. Hence, populations growing in complex medium do not become dominated by a single generalist type with maximal growth on all substrates; nor do they consist of narrow specialists, each restricted to a single substrate. Rather, the outcome of selection in complex media is a set of imperfect overlapping generalists, each superior on a restricted range of substrates. This may be a plausible explanation of the broad but not unlimited metabolic diversity of natural bacterial communities.

The term *substrate* can be extended to denote any aspect of the conditions in which an organism is growing. Even the simplest laboratory microcosm then substantiates some irreducible level of complexity. A simple glass vial, for example, does not supply a perfectly homogeneous environment, because conditions at the surface will differ from those below. A culture of the soil bacterium *Pseudomonas fluorescens* in a glass vial that is left to stand on the bench for a day or two develops a thick mat at the surface, where the cells are stuck together by cellulose-like fibrils. These cells have evolved from the normal cells growing in the body of the medium through loss-of-function mutations in the operon responsible for the synthesis of the polymer. They have an advantage because they monopolize the supply of oxygen diffusing into the medium at the surface, but this advantage is balanced by the lower availability of nutrients in this zone. A third type may also appear at the bottom of the vial as an extreme low-oxygen specialist. By isolating these types and competing them against one another, it can be shown that each type is superior so long as it is rare, but loses this advantage once it has become sufficiently abundant. Hence, each is able to invade a culture dominated by the other but cannot completely replace it, and diversity is actively maintained by selection. This divergent selection is ultimately based on the oxygen gradient initially set up by purely physical forces, and if this gradient is destroyed by shaking the vial, diversity does not evolve. Nevertheless, the initial gradient is greatly exaggerated by the growth of the mat-forming type, so the environmental complexity sustaining diversity is itself reinforced by the evolution of specialized types.

12. WHAT LIMITS THE EXTENT OF SPECIALIZATION?

Selection in complex environments can maintain a range of specialized types, but not an unlimited range. The extent of an experimental adaptive radiation is governed by two features of adaptation: functional interference, and degradation through disuse. Functional interference arises when adaptation to one way of life necessarily leads to loss of adaptation to another. The long-term *E. coli* lines maintained on glucose, for example, lost the ability to utilize a range of other carbon sources. In simple growth media, a universal source of interference is the contrast between rapid but wasteful use of resources and slower but more efficient use. Laboratory cultures are often taken over at first by types that process the limiting resource very rapidly, because in removing it from the medium they deny it to their competitors. They do so at the expense of discarding incompletely metabolized molecules that can be used as substrates by more frugal types as these substances accumulate in the medium. This "cross-feeding" arising from the biological modification of an initially simple medium often leads to the evolution of complex bacterial communities in laboratory microcosms. It is ultimately based on the irreconcilable demands of different metabolic processes, especially the conflict between fermentation and respiration as energy-producing pathways.

A specialized type restricted to a particular way of life derives no benefit from being adapted to others. Loss-of-function mutations in the genes that are required only for these other ways of life are then neutral and will tend to accumulate without limit (see chapter III.14). A population that has become adapted to a novel way of life may then be severely impaired if ancestral conditions of growth are restored. For example, green algae possess a carbon-concentrating mechanism that increases the efficiency of photosynthesis by transporting carbon dioxide to the site in the chloroplast where the initial reactions of photosynthesis occur. This is an expensive process that is switched on only when the external concentration of carbon dioxide is low. If lines are grown at elevated concentrations of carbon dioxide, diffusion alone is sufficient to maintain high internal concentrations and the carbon-concentrating mechanism is unnecessary. Consequently, when these lines have been propagated for a few hundred generations, their carbon-concentrating mechanism becomes degraded, and they are unable to function normally when grown at normal atmospheric levels of carbon dioxide.

Adaptation to environments that vary in time is quite different: any type unable to survive conditions of growth that recur from time to time will become extinct, and more generally selection will favor generalists able to grow moderately well in all conditions. If specialization is limited by interference between different specialized functions, then generalists will be less well adapted to given conditions than the corresponding specialist, whereas degradation through disuse should be halted by the recurrent change in conditions, and broad generalists with no impairment of function may evolve. When bacterial populations are cultured with fluctuating temperature, or algae are exposed to alternating light and dark, the usual outcome is the evolution of generalists, often with fitness in either environment comparable with that of the corresponding specialists, suggesting that degradation through disuse is a frequent consequence of specialization.

This article outlines only a few of the simplest issues tackled by experimental evolution without referring to more complex themes such as metabolic pathways, social behavior, multispecies communities, host-pathogen dynamics, sexual selection, speciation, and multicellular organisms. With the aid of the appropriate model system, there are few fundamental questions in evolutionary biology that cannot be investigated by selection experiments. It may be objected that studying highly simplified laboratory systems may not be relevant to the behavior of populations embedded in highly diverse communities of interacting organisms living in a complex and shifting environment. It is certainly true that particular adaptations, at all scales, from photosynthesis to webbed feet, will require particular explanations, which in many cases cannot be tested experimentally. The justification for experimental evolution, however, is that a distinctive category of evolutionary mechanisms is operating on all characters in all organisms—indeed, in all self-replicating entities—and that it can be elucidated by experiment in much the same way the very different mechanisms governing physiological or developmental processes are being elucidated. We have now acquired a useful understanding of the way these mechanisms operate and the outcomes they produce in simple laboratory systems. The study of more complex systems is only just beginning, and field experiments have seldom even been attempted. Nevertheless, it is clear that the future of the experimental research program lies in greater realism, so that a steadily broader range of evolutionary phenomena will become explicable in terms of clear, testable mechanisms.

FURTHER READING

Bell, G. 2008. Selection: The Mechanism of Evolution. 2nd ed. Oxford: Oxford University Press. *Broad review of evolutionary biology from a primarily experimental perspective.*

Bell, G., and A. Gonzalez. 2009. Evolutionary rescue can prevent extinction following environmental change. Ecology Letters 12: 942–948. *Simple demonstration of evolutionary rescue.*

Buckling, A., R. C. Maclean, M. A. Brockhurst, and N. Colegrave. 2009. The *Beagle* in a bottle. Nature 457: 824–829. *Review of microbial evolution experiments, mostly after 1990.*

Darwin, C. R. 2011. The Annotated *Origin*: A Facsimile of the First Edition of *On the Origin of Species*, ed. J. T. Costa. Cambridge, MA: Harvard University Press. *The first edition, with a modern line-by-line commentary.*

Dykhuizen, D. E. 1990. Experimental studies of natural selection in bacteria. Annual Review of Ecology and Systematics 21: 373–398. *Review of microbial evolution experiments before 1990.*

Elena, S. F., and R. E. Lenski. 2003. Evolution experiments with microorganisms: The dynamics and genetic bases of adaptation. Nature Reviews Genetics 4: 457-469. *Review concentrating on the long-term* E. coli *lines.*

Fisher, R. A. 1930. The Genetical Theory of Natural Selection. Oxford: Oxford University Press. *The classic post-Mendelian synthesis of evolutionary biology.*

Kassen, R. 2002. The experimental evolution of specialists, generalists, and the maintenance of diversity. Journal of Evolutionary Biology 15: 173–190. *Review of experimental studies of balancing and frequency-dependent selection.*

Kawecki, T. J., R. E. Lenski, D. Ebert, B. Hollis, I. Olivieri, and M. C. Whitlock. 2012. Experimental evolution. Trends in Ecology & Evolution 27: 547–560. *Most recent review of experimental evolution for nonspecialists.*

Mortlock, R. P., ed. 1984. Microorganisms as Model Systems for Studying Evolution. New York: Plenum. *Collection of articles on biochemical evolution in the laboratory.*

Rainey, P. B., and M. Travisano. 1998. Adaptive radiation in a heterogeneous environment. Nature 394: 69–72. *Seminal paper on experimental diversification.*

Travisano, M., J. A. Mongold, A. F. Bennett, and R. E. Lenski. 1995. Experimental tests of the roles of adaptation, chance, and history in evolution. Science 267: 87–90. *Ingenious partition of processes responsible for genetic change in experimental populations.*

Weismann, A. 1909. The selection theory. In A. C. Seward, ed., Darwin and Modern Science. Cambridge: Cambridge University Press, 18–65. *Written 50 years after* On the Origin of Species *by the most prominent evolutionary biologist of the day.*

III.7

Responses to Selection: Natural Populations

Joel G. Kingsolver and David W. Pfennig

OUTLINE

1. Measuring selection in natural populations
2. Strength and patterns of phenotypic selection
3. Microevolution in natural populations
4. Local adaptation and population divergence
5. Limits to selection and evolutionary responses

GLOSSARY

Allopatry. A geographical distribution in which populations (or species) occur in different locations or habitats (contrast with *sympatry*).

Divergent Selection. The situation in which selection acts in contrasting directions in two populations.

Local Adaptation. The evolution of features in separate populations that render the members of each such population better able to survive and reproduce in its particular habitat.

Microevolution. Generally refers to inherited change in the characteristics of a group of organisms across generations that occurs within populations and species. Often contrasted with *macroevolution*, generally regarded as large-scale evolutionary change, ranging from the origin of species and major new features (e.g., novel traits or even new body plans) to long-term evolutionary trends.

Natural Selection. Variation in reproductive success that is correlated with variation in phenotypic traits among individuals. Natural selection can produce evolutionary change when such trait variation is inherited.

Phenotypic Selection. A form of selection that occurs when individuals with particular phenotypes survive and produce offspring at higher rates than do individuals with other phenotypes within a population.

Selection Gradient. A measure of the strength of selection acting on quantitative traits. For selection on a single trait, it is equal to the slope of the best-fit regression line in a scatterplot showing relative fitness as a function of phenotype. For selection acting on multiple traits, it is equal to the slope of the partial regression in a scatterplot showing relative fitness as a function of all phenotypic traits.

Sexual Selection. A form of natural selection that arises from variation in fitness resulting from either within-sex competition for reproduction or between-sex choice of mates.

Sympatry. A geographical distribution in which populations (or species) occur in the same location and habitat (contrast with *allopatry*).

1. MEASURING SELECTION IN NATURAL POPULATIONS

In *On the Origin of Species*, Darwin proposed a new mechanism that drives evolution and generates adaptation: natural selection (see chapter III.1). Yet despite the centrality of natural selection to his theory, Darwin never actually attempted to measure selection in nature. Furthermore, in the century following the publication of *The Origin*, selection was generally regarded as too weak, and evolutionary change too slow, to be observed directly in natural populations.

Research in the past four decades has demonstrated that selection and evolution in natural populations can be faster and more dynamic than Darwin and other early evolutionary biologists thought possible. Selection has now been detected in hundreds of populations in nature; moreover, numerous examples of rapid evolutionary change—microevolution on a timescale of 1 to 100 years—have also been reported.

What have we learned from these studies of selection and evolution in natural populations? Here we focus on phenotypic selection, because natural selection acts on the phenotypes of individual organisms. We describe how the strength of selection can be quantified and the patterns of phenotypic selection observed in nature. We explore the conditions that have promoted rapid evolution in nature and how such evolution can lead to local adaptation. Finally, we consider some of the limits to selection that can slow or alter evolutionary change.

Phenotypic selection occurs when individuals with particular phenotypes survive and produce offspring at rates higher than those of individuals with other phenotypes within a population. Phenotypic selection requires phenotypic variation, whereby individuals differ in some characteristics, and differential reproduction, whereby some individuals have more surviving offspring than others because of their distinctive characteristics. Thus, phenotypic selection results from differences in relative fitness (i.e., relative to the mean fitness or reproductive success) of individuals with different phenotypes.

To determine whether selection is acting on some trait in a population, we must first estimate the fitness associated with various trait values (see chapter III.5). Typically, this is done by measuring trait values for a sample of individuals of similar age (e.g., hatchlings), marking each individual, then following them over time to determine their survival and reproductive success. Ideally, these data can be used to determine the relationship between the total fitness and trait values of each individual; in practice, however, most investigators measure only individual components of fitness: survival, mating success, or fecundity.

Once we estimate the fitness associated with different trait values, we then plot fitness against trait value and fit a regression line (i.e., the best-fit line) through the data points. From the slope and shape of this regression line, we can determine the strength and mode of selection acting on the focal trait. When this fitness function is monotonic (always increasing or always decreasing, indicating *directional selection*; see chapter III.4), the fitness (w) of the trait (z) can be estimated by the simple linear regression equation:

$$w = \alpha + \beta z,$$

where α is the y-intercept of the fitness function and β is the fitness function's slope. In this case, β measures the strength of directional selection (figure 1A). By contrast, when the fitness function has curvature (indicating stabilizing and disruptive selection; see figures 1B and 1C), quadratic regression is required to estimate the strength of selection. Here, fitness is estimated by:

$$w = \alpha + \beta z + (\gamma/2)z^2.$$

Here γ measures the amount of curvature in the fitness function—that is, the strength of quadratic selection. When $\beta = 0$ and γ is significantly negative (i.e., when fitness is highest at an intermediate phenotypic value), we conclude that *stabilizing selection* is acting on the trait of interest (figure 1B). By contrast, when $\beta = 0$ and γ is significantly positive (i.e., when the fitness function contains an intermediate performance minimum), we conclude that *disruptive selection* is acting (figure 1C).

The coefficients β and γ (called the directional selection gradient and quadratic selection gradient, respectively) provide useful measures of the strength of phenotypic selection in a population. To allow comparisons among different types of traits and organisms, we can standardize selection gradients by the amount of variation in the trait (e.g., by the standard deviation) to obtain standardized measures of selection, β_s and γ_s.

2. STRENGTH AND PATTERNS OF PHENOTYPIC SELECTION

Scores of studies have quantified phenotypic selection in natural populations using this approach over the past three decades; thousands of estimates of the strength of phenotypic selection are now available, especially for terrestrial plants, birds, lizards, frogs, and insects. Several general patterns emerge from these studies. First, there is abundant evidence for directional selection on morphological and life history traits in many study systems. The strength of such selection varies widely among species and traits, but it is often sufficiently strong to generate rapid evolution change (assuming that genetic variation is available; see below). Moreover, the magnitude of directional selection is not always consistent over time; indeed, in any given population, directional selection can vary in magnitude and even direction over time (i.e., reversals in the direction of selection are sometimes detected). Another common pattern associated with directional selection is that it often acts on body size in diverse taxa. Indeed, in most natural populations studied, directional selection favors increasing body size; that is, larger size tends to be associated with higher survival, fecundity, and mating success.

Studies have also found that the strength of directional selection depends on the component of fitness (e.g., survival, fecundity, mating success) involved. For example, directional selection through mating success and fecundity is generally greater, and more consistent in direction over time, than directional selection through survival. This basic pattern holds both among and within

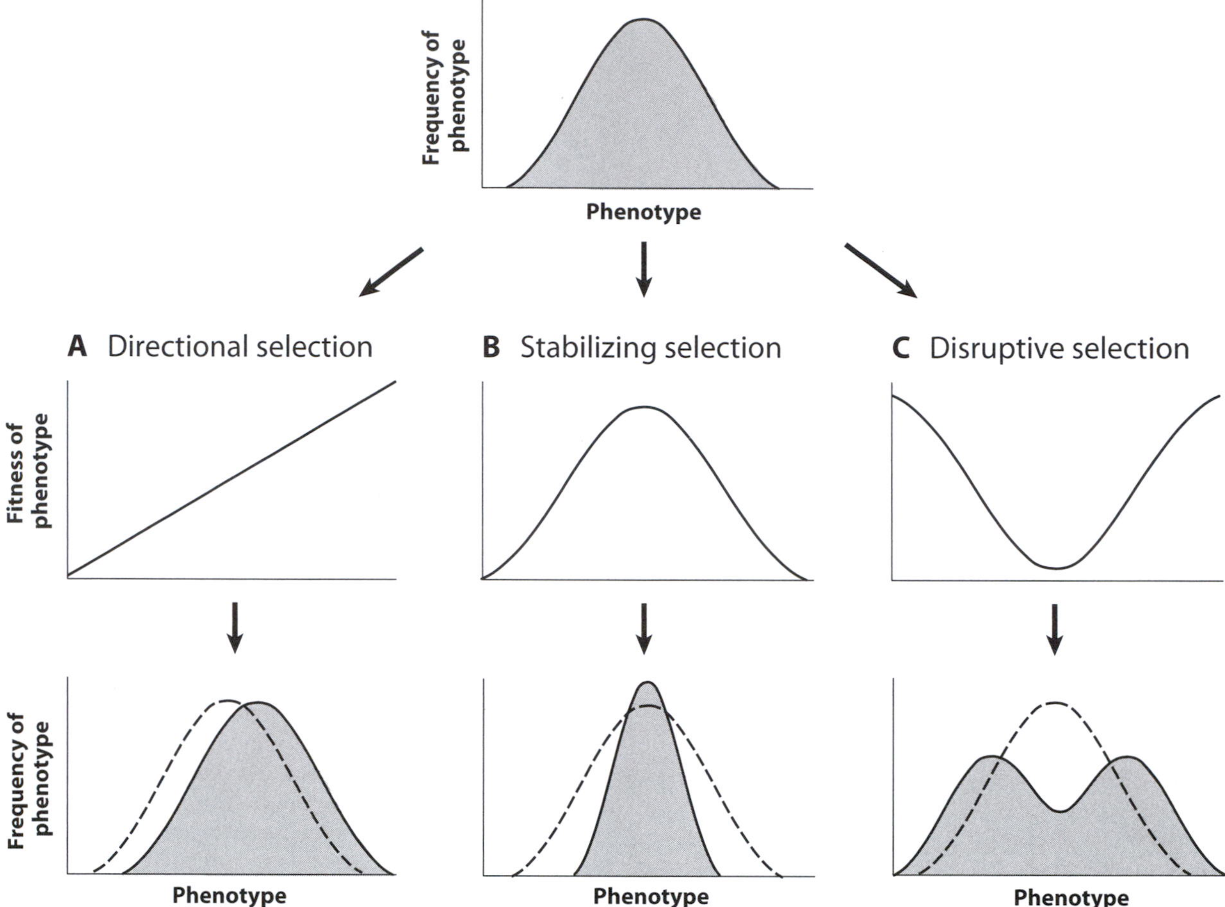

Figure 1. Three different modes of selection can act on a quantitative trait: (A) directional selection, in which extreme phenotypes on one end of the phenotype distribution have the highest fitness and those on the other end have the lowest; (B) stabilizing selection, in which intermediate phenotypes have the highest fitness and extreme phenotypes on either end have the lowest; and (C) disruptive selection, in which extreme phenotypes on both ends of the phenotype distribution have the highest fitness and intermediate phenotypes have the lowest. The graph on the top row shows the distribution of phenotypes in a hypothetical population before selection, the graphs in the middle row show fitness associated with different phenotypes during each of three different modes of selection, and the graphs in the bottom row show the distribution of phenotypes following selection (in each panel, the dashed line shows the distribution of phenotypes before selection).

studies, as well as for different types of traits and organisms. These results suggest that sexual selection (selection due to differences in mating success) is frequently stronger than viability selection (selection due to differences in survival). In this sense, the struggle for existence may be less intense than the struggle for mates.

If populations of organisms are well adapted to current environments, then we would expect stabilizing selection (figure 1B) to be common; however, field studies of phenotypic selection in natural populations provide little evidence for significant stabilizing selection in most study systems. In particular, current estimates of quadratic selection gradients suggest that stabilizing selection is no more common than disruptive selection (figure 1C).

This surprising result can be explained in several ways. First, there may be trade-offs among various components of fitness, such that (for example) a trait value that increases survival may also decrease mating success or fecundity (see chapter III.12). Consequently, net selection on the trait may be less than directional selection via each fitness component. Second, phenotypic and genetic correlations between traits may cause indirect, correlated selection (see chapter III.4); as a result, direct selection on a trait may be balanced by opposing indirect selection on a correlated trait. Third, directional selection on a trait may alternate in direction in time or space, reducing the cumulative effects of selection. However, for most traits in most populations, none of these hypotheses is strongly supported by the available data. It is likely that experimental manipulations of traits or environments will be needed to reliably detect stabilizing selection in the field.

3. MICROEVOLUTION IN NATURAL POPULATIONS

If directional selection is common in natural populations, does this lead to detectable microevolutionary changes? For convenience, we can arbitrarily define rapid, contemporary evolution as detectable changes in the mean phenotype of a population in a few human generations—for example, within the 150 years or so since Darwin's *On the Origin of Species*. Rapid microevolution has now been reported for both morphological and life history traits in numerous field populations of microbes, plants, and animals. The rates of evolutionary change vary widely, with many slower rates and a long tail of rapid rates.

What ecological conditions promote rapid microevolution? A common theme is colonization of new geographic regions and environments, leading to newly adapted populations. *Drosophila subobscura* provides an elegant example of rapid, repeatable microevolution following colonization. In its native range from northern Africa to Scandinavia, *D. subobscura* exhibits a strong geographic cline in wing and body size. During the late 1970s, *D. subobscura* was introduced independently into both South (by 1978) and North (by 1982) America, and rapidly expanded its geographic range on each continent. Studies in the mid-1980s showed that when reared under the same temperature conditions, there was no significant geographic differentiation in size among populations within either North or South America. However, by 2000, population divergence had generated size clines in both North and South America that paralleled the European cline. Substantial evolutionary increases in size at higher latitude populations in North and South America were particularly important in these clines. Rapid evolutionary changes in size and age of reproductive maturity have also been detected in newly introduced populations of salmon, and other animal and plant species.

Colonizing species can also generate selection and microevolution in native species. For example, the soapberry bug (*Jadera haematoloma*) is a plant-feeding insect native to the southeastern United States. It uses its long beak to feed on the fruit capsules of its host plants. Prior to 1925, soapberry bug populations in Florida fed exclusively on fruits of the native balloon vine. Starting in 1926, flat-podded golden rain tree, an Asian relative of balloon vine with flatter and narrower fruit capsules, was introduced and widely used by gardeners in Florida as an ornamental. By the 1960s, many soapberry bug populations had evolved a shorter beak, enabling them to feed more effectively on flat-podded golden rain tree. Comparable rates of evolutionary change in native species in response to colonizing competitors or predators have also been reported.

A second major cause of rapid evolution is recent environmental change due to human activities. Local adaptation of populations in response to herbicides, heavy metals, insecticides, and soil pH has been detected in many plants and insects; resistance to antibiotics and other antimicrobial agents has occurred in numerous human pathogens; and evolutionary change in response to recent global climate change has been detected. For example, evolutionary changes in the timing of flowering in annual plants, and in the seasonal cues that control active development in mosquitoes, have been detected in association with the recent elevation in mean environmental temperatures. The frequency and rates of microevolution in nature may increase as human activities increasingly alter climate and other major components of our physical, chemical, and biological environments.

4. LOCAL ADAPTATION AND POPULATION DIVERGENCE

Selection does not always act in a similar way on all populations of any given species; indeed, the mode (figure 1), magnitude, and even direction of selection might differ in different populations. Regarding the latter, when directional selection acts in opposing directions in different populations, it is referred to as *divergent selection* (figures 2A and 2B). Divergent selection is important, for two main reasons. First, divergent selection can promote local adaptation, in which different populations evolve different adaptive trait values or even different adaptive traits altogether (see chapter IV.3). Second, divergent selection may ultimately favor the evolution of barriers to genetic exchange between populations and thereby lead to speciation (see chapter VI.4). Here, we briefly review the causes of divergent selection before discussing its role in local adaptation and speciation.

There are three main agents of divergent selection. First, divergent selection can arise because of differences between populations in their abiotic environments. For instance, populations might diverge from one another as a result of experiencing different soil chemistries, climates, or resources. As an example, populations of plants that grow on serpentine soils must adapt to extreme soil conditions, including the absence of essential nutrients and the presence of heavy metals. Such populations may diverge from conspecific populations residing on nonserpentine soil to such a degree that they eventually become separate species (indeed, serpentine soils are characterized by high

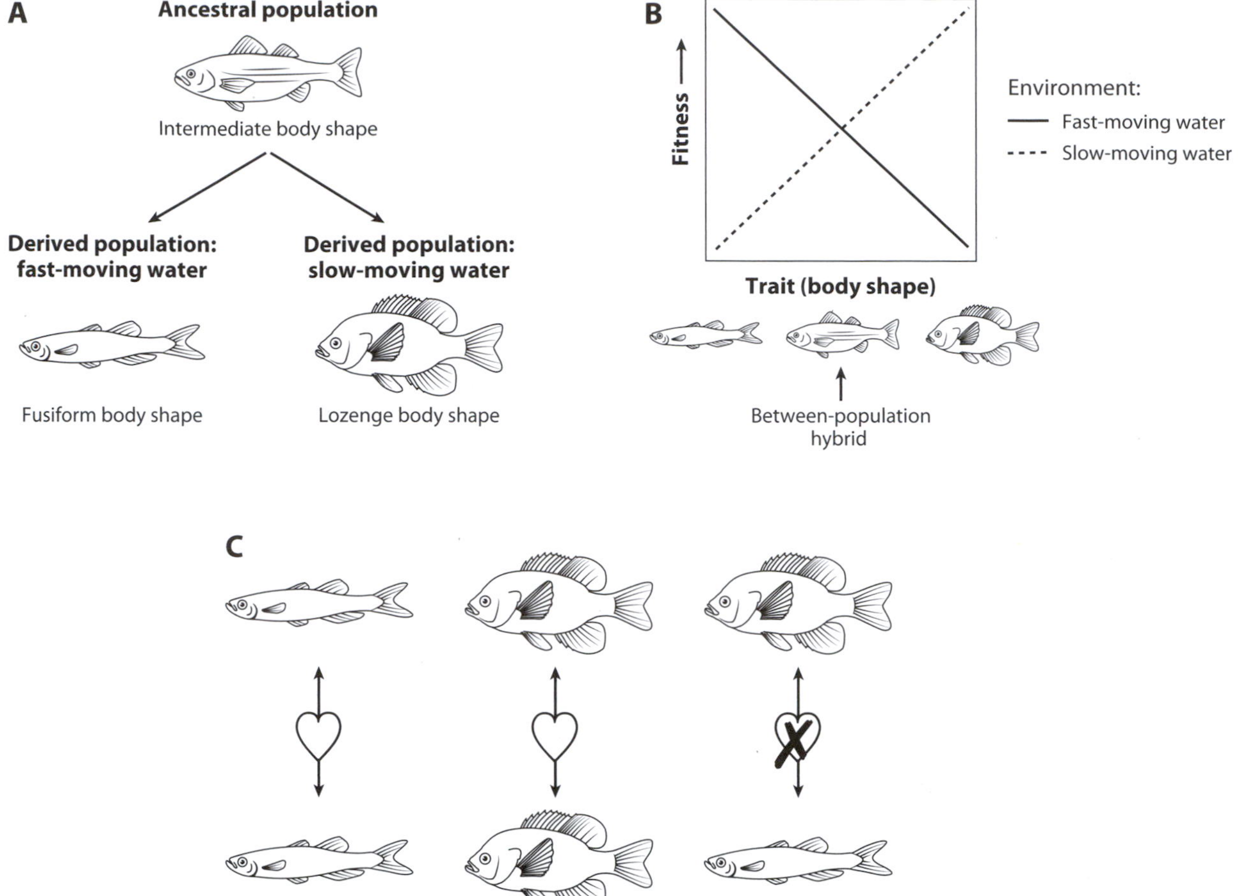

Figure 2. Selection is divergent when it acts in contrasting directions in different populations. (A) For example, an ancestral population of fish that splits into two separate populations that come to occupy two different environments–fast-moving water and slow-moving water–might experience divergent selection on body shape if (B) different body shapes are favored in different water-flow regimes. Moreover, such divergent selection might even promote the evolution of reproductive isolation and, possibly, speciation. Specifically, (B) if offspring produced by matings between populations perform poorly in both environments, postzygotic isolating mechanisms might prevent gene flow between populations on secondary contact; (C) additionally, if individuals prefer mates with a similar body shape, prezygotic isolating mechanisms might also prevent gene flow.

levels of endemism). Generally, differences in the abiotic environment can be potent agents of divergent selection.

Second, divergent selection can arise when populations differ in their interactions with other species, especially predators, parasites, or competitors (see chapter III.5 and III.15). For example, when conspecific populations differ in their exposure to a heterospecific competitor—such that some populations co-occur with the heterospecific and others do not—those populations in sympatry with the heterospecific will experience a different selective environment from that experienced by conspecifics in allopatry. Consequently, populations might diverge from one another in traits associated with resource use, reproduction, or both. In particular, the sympatric population might experience selection for traits that minimize resource competition and/or costly reproductive interactions with the heterospecific (e.g., mismatings or signal interference), and thereby undergo a form of divergent trait evolution known as *character displacement*.

The classic example of character displacement comes from finches from the Galápagos Islands. As Lack first observed, different species of finches typically differ in beak morphology more in areas where they are sympatric than where they are allopatric. Such divergence likely reflects selection to minimize competition between species (beak size and shape correlates with the types of seeds on which each species feeds, so differing in beak size reduces overlap in diet and thereby competition for food). More generally, character displacement has been detected in numerous species, and it has been shown to affect traits involved in both resource use and reproduction.

Third, divergent selection can arise when the direction or strength of sexual selection differs between different populations. For instance, populations might diverge in display traits, sensory systems, and/or mate preferences owing to differences in either abiotic or biotic environments. As an example, two species of threespine sticklebacks (of the *Gasterosteus aculeatus* complex) have undergone character displacement in certain small lakes in southwestern Canada, which has resulted in one species expressing a distinctive "benthic" ecomorph and the other a distinctive "limnetic" ecomorph. The benthic ecomorph inhabits the heavily vegetated littoral zone of lakes, whereas the limnetic ecomorph occupies open water. As a result of occupying different habitats, the light environments experienced by these two ecotypes also differ. In the littoral zone, where the benthic ecomorph forages and mates, red coloration is more difficult to detect. By contrast, in open water, where the limnetic ecomorph forages and mates, red coloration is more discernible. Interestingly, benthic females are less sensitive to variation in red than are limnetic females, and, unlike limnetic females, benthic females do not tend to prefer redder males. Male red coloration, in turn, is "tuned" to female perception of red color: males are redder in populations where females are actually sensitive to, and thus prefer, redder males. There are a number of other examples in which display traits, sensory systems, and/or mate preferences have diverged between populations experiencing different selective regimes.

As the examples above illustrate, an important consequence of divergent selection is that it can promote local adaptation. But how frequently does divergent selection lead to local adaptation, and what are the patterns it produces? A standard means for assessing local adaptation is reciprocal transplant experiments, in which samples of individuals (and genotypes) from different populations are reared together in a "common garden" at each site. These experiments can determine whether genotypes from the native or "home" population have higher relative fitness at their native site than genotypes from populations at other ("away") sites: that is, whether local genotypes are locally adapted to their own sites.

A recent review (Hereford 2009) of more than 70 reciprocal transplant studies in field populations indicates an overall frequency of local adaptation of 71 percent and an average fitness advantage of native genotypes at their native site of 45 percent. The magnitude of local adaptation was greater when there were larger environmental differences among sites, as we would expect if these environmental differences create the divergent selection that caused the local adaptation. Interestingly, trade-offs in relative fitness across sites were weak, suggesting that strong local adaptation to one site does not necessarily produce low fitness at other sites. Overall, these results suggest that local adaptation in natural populations is widespread. The phenotypic basis for local adaptation is not always known, but may frequently result from population differences in many traits.

Local adaptation is not the only significant consequence of divergent selection, however. As noted above, divergent selection can even promote speciation. Indeed, differences that arise between populations owing to divergent selection can reduce or prevent gene flow when populations come into secondary contact, thereby reproductively isolating them. This route to speciation, dubbed *ecological speciation*, occurs when barriers to gene exchange between populations evolve as a consequence of ecologically based divergent selection (see chapter VI.4).

To illustrate how divergent selection can promote speciation, imagine that different populations of a fish invade two different environments: one containing only slow-moving water and one containing only fast-moving water (figure 2A). These two different populations would likely experience divergent selection pressures (figure 2B) and might therefore evolve different body shapes. Consequently, hybrids produced by any matings that might occur between such populations (if they were ever to come into secondary contact) would likely be disfavored because of ecologically based divergent selection; in this case, because these hybrids would likely have an intermediate body shape, which would perform poorly in either parental environment (see figure 2B). Such reduced fitness of between-population hybrids might act as a postzygotic isolating mechanism preventing gene flow between ancestral and derived populations. Additionally, if individuals prefer mates with a similar body shape (figure 2C), prezygotic isolating mechanisms might also prevent gene flow between these populations, thereby potentially completing the speciation process. Generally, ecological speciation becomes more likely if a strong association exists between the gene(s) conferring local adaptation and the gene(s) conferring reproductive isolation.

As support for ecological speciation, numerous studies have found that divergent selection has promoted differences between populations in traits such as body shape, size, or coloration that influence mate preferences. The ecological speciation model is also supported by laboratory experimental evolution studies, comparative studies, and by instances of parallel speciation, in which

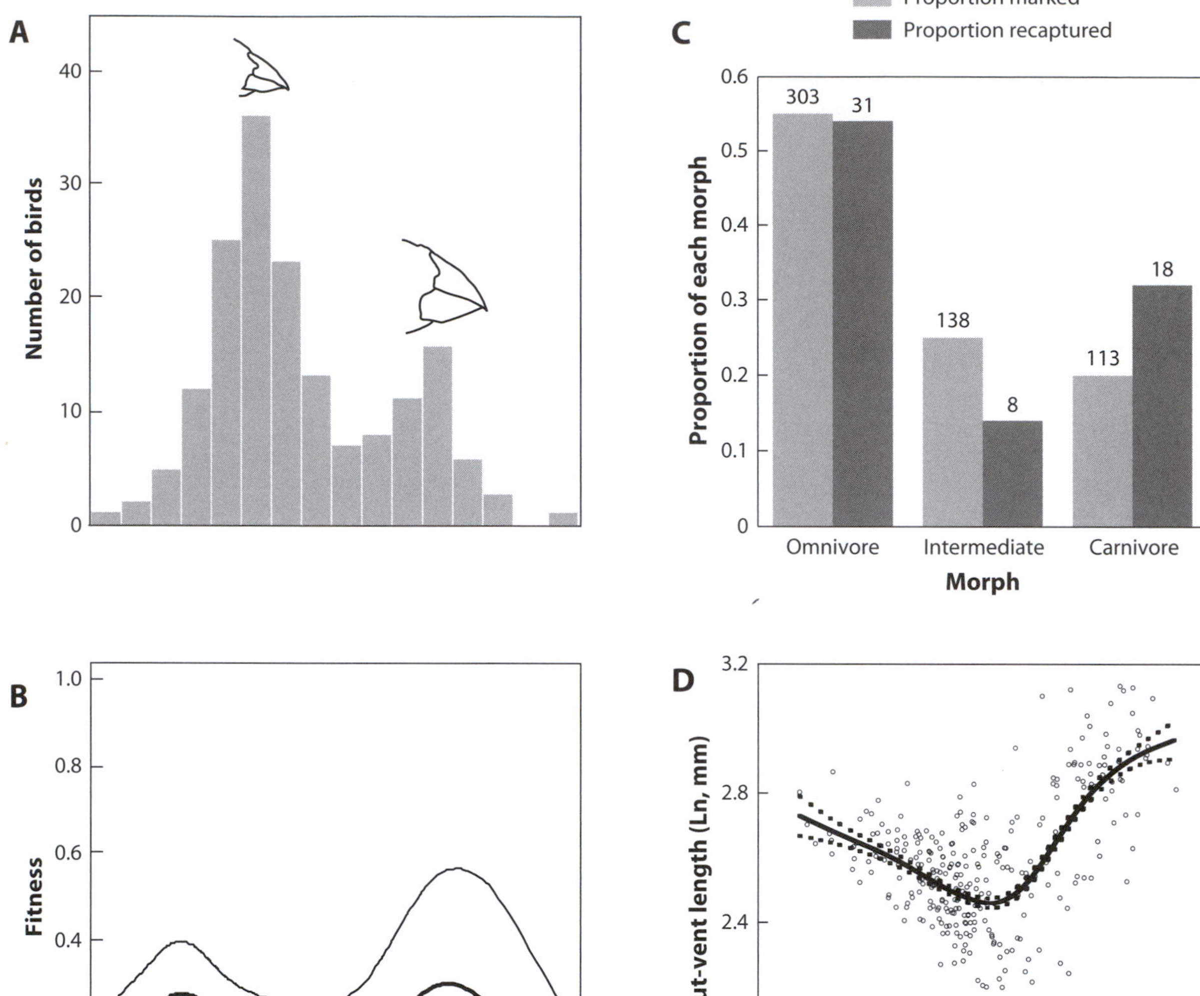

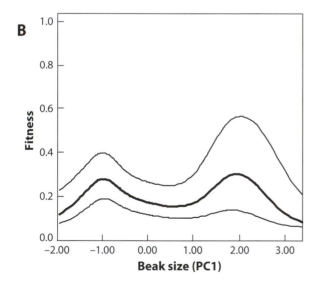

Figure 3. Two examples of disruptive selection in the wild. (A) The adult population of Darwin's finches (*Geospiza fortis*) at El Garrapatero (Santa Cruz Island, Galápagos archipelago) consists of a small-beaked morph and a large-beaked morph. The data come from a study by Podos and colleagues. (B) This population experiences disruptive selection, in which individuals with small and large beaks have higher fitness than those with intermediate-sized beaks. Selection on beak morphology is depicted as a cubic spline (heavy line) with 95% confidence intervals (lighter lines). These data are from work by Hendry and colleagues. (C) Similarly, disruptive selection disfavors individuals with intermediate trophic phenotypes in Mexican spadefoot toad tadpoles (*Spea multiplicata*). This graph shows the probability of survival for individuals expressing different ecomorphs—omnivores, carnivores, and intermediates—based on a mark-recapture experiment in a natural pond in Arizona. Individuals expressing an intermediate phenotype had lowest survival (numbers above each bar show sample sizes), demonstrating that this population experiences disruptive selection. (D) Disruptive selection can also be visualized by a cubic-spline estimate of body size (a fitness proxy) on a composite shape variable of trophic morphology (morphological index). The cubic spline (solid line) is bracketed by 95% confidence intervals (dashed lines). As in panel (B), the presence of an intermediate fitness minimum suggests that disruptive selection acts on trophic morphology. The data in panels (C) and (D) are from work by Martin and Pfennig.

reproductive isolation has evolved independently and repeatedly between populations adapting to contrasting environments.

Finally, broadly defined, divergent selection includes the special case of disruptive selection (figure 1C). As noted above, disruptive selection occurs when two or more modal phenotypes in a population have higher fitness than the intermediate phenotypes between them. Recent empirical studies suggest that disruptive selection might be more widespread than formerly presumed; indeed, several studies of natural populations have documented disruptive selection; two examples are shown in figure 3. Disruptive selection might play a general and important role in maintaining diversity within species, especially when such selection leads to the evolution of a mating polymorphism or a resource polymorphism (as in the two examples depicted in figure 3).

5. LIMITS TO SELECTION AND EVOLUTIONARY RESPONSES

In this article, we have emphasized that directional selection and local adaptation are common in nature. We also described how colonization events and anthropogenic environmental changes have generated microevolutionary changes in many populations during the past century, but directional selection does not inevitably lead to evolutionary change and adaptation. First, evolution also requires appropriate genetic variation in the traits under selection (see chapter III.4). Although genetic variation has been documented for numerous phenotypic traits for many natural populations, lack of genetic variation can still limit the response of some populations to environmental change. For example, tropical species of *Drosophila* with narrow geographical distributions have little genetic variability in resistance to cold and to desiccation. This lack of genetic variation limits their potential to adapt to cooler or drier environmental conditions beyond their current geographic ranges—and their capacity for evolutionary responses to future climate changes. Limits to genetic variance and other factors might constrain evolutionary responses to selection in a variety of ways (see chapter III.8).

Additionally, environmental change not only causes selection but can also alter patterns of phenotypic and genetic variation in important ways. Long-term population studies with mammals and birds illustrate this point. Yellow-bellied marmot populations in the Rocky Mountains in the western United States hibernate as adults during the long, snowy winter months. Adult mass at the start of hibernation has increased markedly over the past 35 years, largely as a result of three factors related to weather: earlier seasonal emergence from hibernation, earlier weaning of young, and a longer growing season. The longer growing season and larger adult size at hibernation have also reduced adult mortality, causing large increases in population density in recent years. Selection and adaptive evolution have likely played little direct role in these phenotypic changes.

Environmental change can also affect both selection and genetic variation, however, in ways that alter the rate of microevolution. Higher spring temperatures in the Netherlands during the past 35 years have generated directional selection on the timing of breeding in populations of the great tit, a common bird in northern Europe. Additionally, increasing spring temperature has increased the heritability of breeding time in the population, by altering patterns of phenotypic and genetic variation. The increase in both selection and heritability has accelerated the rate of microevolution of breeding time in this population.

These examples illustrate how ecology and evolution are intimately interconnected in the responses of populations to environmental change. The interactions among genetic variation (see chapter III.4), phenotypic plasticity (chapter III.10), life history (chapter III.11), and selection are essential for a full understanding of the evolutionary responses and adaptation of natural populations in a changing world.

FURTHER READING

Boughman, J. W. 2001. Divergent sexual selection enhances reproductive isolation in sticklebacks. Nature 411: 944–948. *The paper presents an empirical example in which divergent sexual selection has apparently favored the evolution of reproductive isolation in natural populations.*

Hereford, J. 2009. A quantitative survey of local adaptation and fitness trade-offs. American Naturalist 173: 579–588. *This paper presents a meta-analysis of field experimental studies, and it documents that local adaptation is widespread in natural populations.*

Hoffmann, A. A., and C. M. Sgrò. 2011. Climate change and evolutionary adaptation. Nature 470: 479–485. *This paper illustrates how lack of genetic variation in key phenotypic traits can limit the adaptive responses of some tropical species to climate change.*

Kingsolver, J. G., and S. E. Diamond. 2011. Phenotypic selection in natural populations: What limits directional selection? American Naturalist 177: 346–357. *This synthetic review explores an important paradox: the abundant evidence for directional selection, and the limited evidence for stabilizing selection.*

Kingsolver, J. G., and D. W. Pfennig. 2004. Individual-level selection as a cause of Cope's rule of phyletic size increase. Evolution 58: 1608–1612. *In this paper, the authors present the results of a meta-analysis that suggests that selection favors larger body size in many species.*

Nosil, P., and H. D. Rundle. 2009. Ecological speciation: Natural selection and the formation of new species. In S. A. Levin, ed., The Princeton Guide to Ecology. Princeton, NJ: Princeton University Press, 134–142. *This book chapter presents a concise overview of the role of ecological factors, especially biotic interactions, in speciation.*

Reznick, D. N., and C. K. Ghalambor. 2001. The population ecology of contemporary adaptations: What do empirical studies reveal about the conditions that promote adaptive evolution. Genetica 112–113: 183–198. *This review documents how anthropogenic environmental change, species invasions, and other ecological factors contribute to rapid adaptive evolution in contemporary populations.*

Siepielski, A. M., J. D. DiBattista, J. A. Evans, and S. M. Carlson. 2011. Differences in the temporal dynamics of phenotypic selection among fitness components in the wild. Proceedings of the Royal Society B 278: 1572–1580. *This synthetic review explores the evidence for temporal variation in the strength and direction of selection in natural populations.*

III.8

Evolutionary Limits and Constraints

Ary Hoffmann

OUTLINE

1. Lack of genetic variation as a limit and constraint
2. Trade-offs
3. Multivariate selection
4. Gene flow in marginal populations limiting range expansion
5. Limits and constraints: biodiversity and conservation

Although evolution is a powerful process that leads to rapid changes in the characteristics of organisms, limits to evolution arise from a lack of genetic variation, a loss of well-adapted genotypes in populations due to gene flow, trait interactions leading to trade-offs, and/or the difficulty of evolving simultaneous changes in a number of traits. Signatures of genetic constraints at the molecular level include a loss of functional genes as a result of mutational decay.

GLOSSARY

DNA Decay. The loss of functionality of genes as a consequence of mutations and a lack of selection to remove these mutations.

Gene Flow. Movement of genes in space, as a consequence of organisms moving and contributing offspring after they have moved. Gene flow can also occur when gametes move in space (as in movement of pollen).

Genetic Correlation. Effects of the same genetic variants on different traits, as a result of pleiotropy or genetic linkage (tendency for genes to be coinherited because they are located nearby on the same chromosome).

Genetic Variation. Arises because there are different forms (alleles) of the same genes affecting traits. Can be measured by looking for DNA variation in functional genes, variation in their protein products, or variation in characteristics inherited across generations (*heritability*).

Heritability. The proportion of phenotypic variation in a trait that is controlled by genetic factors (as opposed to environmental factors). Varies from 0 (all variation due to environmental effects) to 1 (all variation due to genetic factors).

Pleiotropy. The multiple characteristics that can be affected by the same genes.

Although evolution can rapidly change the morphology, physiology, and behavior of populations and species, there is also ample evidence that the effects of evolution can be constrained. The fossil record includes many examples of lineages showing long periods of morphological stasis (see chapter VI.11). Some lineages show remarkably little change in appearance across tens or even hundreds of millions of years; cockroaches, the tuatara, cycads, and horseshoe crabs provide a few well-known examples. Such lineages are often considered to have reached an evolutionary dead end, unable to evolve further and constrained to a narrow ecological niche. Yet the effects of selection can be extremely powerful (see chapters III.6 and III.7), perhaps best exemplified by the power of artificial selection in generating crop plants and domesticated animals with little similarity to their wild relatives, and by the rapid adaptation of many species to anthropogenic stresses, such as the development of antibiotic resistance in microbes and pollution/toxin resistance in many invertebrates, fungi, and plants (see chapter III.7); however, even in cases where selection is intense, evolution is not necessarily an inevitable outcome. Although pesticide resistance is widespread in insect pests, weeds, and fungi that cause plant disease, many agricultural chemicals have remained effective against pests for several decades. In these situations the pest populations lack the ability to evolve resistance, evidently because they are constrained in some way.

These types of observations have led evolutionary biologists to search for the reasons underlying constraints. Is it the case that evolutionary change has reached a fundamental limit, unable to occur even when conditions are conducive to adaptation because the set of genetic changes required for adaptation simply are not possible in a species? Or is there some other reason for an evolutionary constraint? A population or species might have the potential to evolve, but other factors like the movement of genes among populations and/or trait interactions make evolutionary change difficult despite ongoing selection. Both fundamental limits and other forms of constraints can prevent populations and species from adapting to new environments. Constraints restrict species to living under a particular set of environmental conditions. In this way, limits and other constraints drive biodiversity; without them, there might be a few common species adapted to a wide range of conditions, rather than a diversity of species, the majority restricted to a narrow range of ecological conditions.

Explanations for evolutionary constraints can be divided into two categories: those that reflect the nature of genetic variation required for evolution and adaptation, and those that have their origin in the ecological processes to which populations and species are exposed.

1. LACK OF GENETIC VARIATION AS A LIMIT AND CONSTRAINT

Natural selection will not change a trait if the trait lacks genetic variation. In the absence of genetic variation, any increase or decrease in the mean value of a trait after selection will not be passed on to the next generation. A way to think about this issue is to imagine a population of individuals derived from a single clone. Barring mutation, all individuals in the population will then be genetically identical, and there will be no genetic variation for any trait in this population. Because of differences in environmental conditions experienced by individuals, they will still differ somewhat in appearance and performance. However, even if these differences affect the fitness of individuals, the differences will not be passed on to subsequent generations because all individuals are genetically identical.

Although natural populations of plants and animals will be genetically diverse rather than derived from a single clone, they may nevertheless have traits lacking genetic variation. One possibility is that populations and species lack specific genes that are required to adapt to new environmental conditions. For instance, Antarctic marine fish have lost the genes coding for proteins and regulatory mechanisms needed to live in warm environments, preventing colonization of warmer waters. And many species of *Drosophila* vinegar flies lack copies of genes coding for heat shock proteins (HSPs) essential for surviving in hot conditions, where HSPs protect other proteins from degradation. In both these situations, the absence of the appropriate functional genes likely represents a fundamental evolutionary limit. Constraints are also associated with development that influence the body plans and morphological options available to species (see chapters V.11 and V.12).

At the molecular level, this limit can be overcome only by a change in the genome, such as a duplication of another gene or set of genes that evolves a new function and restores the proteins that are required for living in warm water or surviving hot conditions (see chapter V.5). Such changes might then produce an evolutionary lineage consisting of species with the ability to colonize a warmer environment; however, genes are also subjected to DNA decay as mutations accumulate, so they gradually become functionally inactive. DNA decay will occur when there is no selection on a gene to remove mutations that lead to inactivation. The genomes of species contain many genes in a state of decay and on the way to becoming nonfunctional pseudogenes (see chapter V.1). Once DNA decay and eventual gene loss have occurred, function may not easily be restored unless there is a further duplication event and further evolutionary changes to produce a new function.

Because traits are typically affected by a number of genes (and the regulatory mechanisms acting on these genes), genetic variation in a trait may be lost only when there is a cumulative effect of molecular changes at multiple loci, or when a key regulatory gene in a developmental pathway is inactivated. The absence of genetic variation in a trait can be detected through the inability of selection to change the distribution of the trait when artificial or natural selection is imposed in a particular direction (for instance, increased resistance of organisms to warmer or colder conditions, ability of phytophagous insects to use a new host plant, ability of animals to tolerate a disease agent). It can also be detected through a loss of heritable variation (or heritability) in a trait, which is often estimated from family studies (see chapter III.5). Heritability reflects the extent to which variation in a trait is determined by genetic rather than environmental factors, and estimates vary from 0 (all trait variation due to environment rather than genes) to 1 (all variation genetic in origin). A heritability value of 0 points to an evolutionary constraint, resulting in a lack of similarity between parents and their offspring (or between other related individuals). In practice, heritability estimates are prone to large standard errors, meaning that it can be difficult to distinguish between a low heritability and a true value of 0 that might reflect an evolutionary constraint.

Heritability estimates for morphological traits are typically high in outbreeding species (for instance, estimates for height in populations of domestic animals and humans are often around 0.5 to 0.8), while they tend to be lower for behavioral and physiological traits; however, heritability estimates for natural populations of animals and plants can be quite variable, particularly for traits that are important in determining the ecological niche of species. For instance, the distribution of various *Drosophila* species coincides closely with their levels of resistance to desiccation and cold stresses; species sensitive to cold and dry conditions tend to be confined to moist and warm tropical rain forests, whereas others that are widespread have a high level of resistance to these stresses. Comparisons of heritability for these resistance traits across species indicate that the sensitive tropical species tend toward a very low level of heritable variation (Kellermann et al. 2009). This may help explain why such species are restricted in their distribution—an evolutionary limit exists, based on a lack of genetic variation preventing them moving out of their warm and moist habitats. In a rather different context, Bradshaw and coworkers first found that the limited number of plant species growing on old mine tailings in Europe are the same species that exhibit genetic variation for tolerance of toxic contaminants in soil—these species all contain heritable variation for tolerance in populations not exposed to contaminants. In contrast, plant species that have been unable to evolve and colonize the mine tailings lacked this heritable variation in the first place. These types of cross-species comparisons highlight the potential importance of a lack of genetic variation in limiting evolutionary responses.

As mentioned above, one of the mechanisms likely to drive a loss of genetic variation is DNA decay. When a species becomes restricted to an area because it is dependent on a host plant with a restricted distribution for food or for breeding (as in some birds and phytophagous insects) or because it becomes confined by physical barriers (mountain ranges, water bodies, caves, etc.), there can be a loss of purifying selection for particular characteristics and their underlying genes (e.g., the genes to recognize other host plants and detoxify compounds in them, or to tolerate extremes of temperature). In the absence of purifying selection, the genes will start to undergo decay as they accumulate mutations that may become fixed by genetic drift. Eventually, the decay process will decrease the evolutionary potential of a species should its environment change, perhaps forever confining it to living within a particular set of conditions until a further change in its genome, such as a gene duplication process. It is not clear how often (or how quickly) DNA decay acts to limit further evolutionary change (or how easy it is for lineages to escape decay).

Another mechanism that can contribute to a loss of genetic variation in traits is strong selection (see chapter III.7). If directional selection for increased expression of a trait persists for many generations, the alleles favored by selection are expected to increase and eventually go to fixation. Once this occurs across all the loci affecting a trait, genetic variation in the trait is expected to decrease toward zero, preventing any further selection response until the strong directional selection is alleviated, and new mutations can accumulate.

2. TRADE-OFFS

While evolutionary constraints due to DNA decay and loss of genetic variation arise because genes are absent and nonfunctional or lack genetic variation, limits can also arise because of the *pleiotropic effects* of genes, which occur when the same genes affect multiple traits. Genes have an enormous potential for pleiotropic effects because proteins encoded by genes are embedded in networks of interacting biochemical processes, and these networks in turn are likely to influence the expression of multiple traits. Moreover, genes that regulate the expression of other genes can have pleiotropic effects by influencing multiple networks. Because of the complex and indirect ways in which genes influence phenotypes (see chapter V.13), selection for a decrease or increase in a trait will favor a set of underlying allelic changes that simultaneously impact other traits.

Genetically based trade-offs occur through pleiotropy when these simultaneous effects influence traits closely related to fitness but in opposing directions. For instance, in insects there is evidence for a genetic trade-off between development time and reproduction, because alleles promoting fast development lead to early emergence and early reproduction by adults, but a cost is paid in that adults emerge at a smaller body size with reduced reproductive output. Similarly, in plants there is evidence for a genetic trade-off between flowering time and seed production, with early-flowering plants tending to produce smaller flowers and potentially less seed. These trade-offs may result in different traits being favored in different environments. For instance, several studies have shown that insect populations living in cool conditions are under strong selection to complete development within a short growing season, resulting in smaller body size and reduced overall reproductive output. Similarly, populations of an invasive wetland plant in cold latitudes have genetically adapted to flower early at the cost of a smaller flower size, compared with populations of the same species in warmer areas (Colautti et al. 2010). In these cases, a genetically based reduction in reproduction can become an evolutionary constraint if reproductive output is no longer sufficient to sustain a population.

Evolutionary trade-offs can be studied either by considering allelic variants of genes individually or by examining patterns of genetically based correlations among traits. An allele that increases insecticide resistance by boosting a detoxification mechanism will be favored when the chemical is present, but is typically selected against when it is absent, perhaps because the detoxification mechanism is energetically costly. This type of information can point to a mechanistic basis of a trade-off, but might not necessarily provide information about an evolutionary constraint, because other mechanisms of insecticide resistance, such as decreased sensitivity of the target site of the chemical, might be selected instead, and these other mechanisms might not be associated with a trade-off.

For this reason, trade-offs are often characterized by looking at genetic correlations that reflect the effects of the many underlying genes (see chapter III.5). These correlations can be measured through family studies by considering how multiple traits are inherited across generations. Negative genetic correlations for traits affecting fitness in opposing directions, like the association of more rapid development with reduced reproductive output, are then taken as evidence of trade-offs that potentially contribute to evolutionary constraints; however, a negative genetic correlation does not necessarily reflect an evolutionary constraint; unless the negative genetic correlation is particularly strong, the possibility still exists that selection can proceed at least partly independently on the two traits. Some alleles may affect one trait but have no pleiotropic effects on other traits, and might then still be selected. As well as being assessed through family studies, trade-offs can also be investigated by carrying out selection experiments (see chapter III.6), designed to test whether selection on one trait is invariably associated with fitness costs involving a different trait. For instance, many selection experiments in insects, mice, and worms have been undertaken to explore whether an increase in early reproduction is invariably associated with a decrease in longevity as a consequence of allocation of resources to reproduction rather than to maintenance. This trade-off was proposed by George Williams as a way of explaining constraints on the evolution of increasing life span (see chapter III.11 and section VII).

When testing for constraints due to pleiotropy, it is important to distinguish genetic interactions among traits due to chromosomal linkage from those due to pleiotropy. If genes are closely linked along a chromosome, the alleles affecting traits can end up in linkage disequilibrium—an allele that increases development time might by chance end up linked to an allele affecting reproductive output but from a different, closely linked gene. This issue can be particularly important in selection experiments involving a small number of individuals, with the possibility of strong disequilibrium among loci at the start of the experiment. Although pleiotropic effects can be difficult to distinguish from linkage in practice, linkage associations between traits are expected to be lost as recombination takes place among the linked loci, and so need not impose long-lasting constraints.

Experiments aimed at examining trade-offs and constraints do not necessarily have to involve selection for specific traits, but can also involve the process of *experimental evolution*, whereby populations are held for multiple generations in various environments, to which they usually adapt (see chapter III.6). This approach has been widely used in microbes; for instance, Lenski and colleagues placed *E. coli* bacteria at a temperature of 20 °C, near the lower limit at which these bacteria can be maintained, and monitored adaptation after 2000 generations. They found that the bacterial populations improved in their competitive performance at 20 °C by 8 percent, but at a cost to performance at 40 °C of 20 percent. These findings were interpreted as evidence of trade-offs associated with temperature adaptation; *E. coli* can adapt to low temperatures, but this ability comes at a cost that may constrain evolution in an environment where temperatures are fluctuating.

Although most tests of trade-offs have taken place in experimental laboratory conditions, it is possible to test for the role of trade-offs in generating constraints under field conditions. By crossing a native prairie legume under controlled conditions and then transplanting seed to various locations along a temperature and moisture gradient designed to reflect expected conditions under future climate change, Etterson and Shaw were able to measure natural selection under different field conditions and test for a pattern of genetic correlations among traits related to fitness along the gradient. They showed that because of antagonistic associations among traits, the response to selection along the moisture gradient was expected to be much less than predicted from the heritability of the traits when considered alone. For instance, leaf number was selected to increase where moisture stress occurred, but changes in number were antagonistic to changes in leaf thickness, which was also under selection, constraining the extent to which these traits were expected to change.

Evolutionary constraints due to pleiotropy can be distinguished from those due to a lack of genetic variation by examining the presence of genetic variation in each of the correlated traits. In the former case, each trait is expected to show genetic variation. This situation applied to most of the traits under selection in the prairie legume studied by Etterson and Shaw. In selection experiments, the persistence of genetic variation in traits even after a selection limit is reached can point to pleiotropy rather than lack of genetic variation acting as an evolutionary constraint.

3. MULTIVARIATE SELECTION

Although trade-offs due to trait interactions are often regarded as essential for evolutionary constraints, other types of genetic interactions among traits can also prevent selection responses. Evolution is constrained when no genetic variation is available for a population to respond in the direction in which selection acts, and this does not necessarily require trade-offs or negative genetic correlations among traits. It is the way in which multiple traits under selection interact at the genetic level that drives eventual selection limits.

Understanding this type of selection limit requires an understanding of selection on multiple traits at once (*multivariate selection* rather than univariate or bivariate selection; see chapter III.4). When a set of traits is under selection, the outcome of selection can be predicted by considering how the traits interact at the genetic level and how selection acts on combinations of traits. As an example of this approach, Blows and colleagues have investigated the cuticular hydrocarbons from males of a *Drosophila* species. The nine different hydrocarbons are under sexual selection caused by female choice, although they are also under natural selection. In family studies, the level of each hydrocarbon shows genetic variation individually, but when considered together the cuticular hydrocarbons show little variation for the specific combination favored by females. In other words, the genetic interactions among the cuticular hydrocarbons prevent much variation from being expressed in the direction favored by sexual selection. In selection experiments, female choice drives only small changes in hydrocarbon profiles before a constraint is reached. Natural selection is also important in determining the hydrocarbon profile of the males, because in selection experiments a change in hydrocarbons due to female choice can be mostly lost when selection is relaxed for a few generations. The evolutionary constraints in this system can therefore really be understood only by considering the ways in which traits are interacting.

4. GENE FLOW IN MARGINAL POPULATIONS LIMITING RANGE EXPANSION

Gene flow occurs when individuals or propagules move from one population to another and then contribute to the genetic constitution of the other population (see chapter IV.3). This process can both enhance and retard evolutionary adaptation. The former occurs when gene flow increases genetic variation by introducing new genetic variants into a population that can then be selected to increase fitness. On the other hand, when too much gene flow occurs, the effects of selection can be overwhelmed by an influx of nonadapted genotypes. Gene flow can then act as an evolutionary constraint.

The effects of excessive gene flow have been modeled for populations at the geographic margins of species ranges. In some marginal populations, a decreased density of individuals occurs, compared with more centrally located populations. This can result in directional gene flow into the marginal populations, which in some situations may be sufficient to retard adaptation to the conditions experienced by the marginal population. This process might then be sufficient to prevent further expansion of the species.

Although models suggest that directional gene flow can act as a constraint, the empirical data supporting this hypothesis are quite limited. Part of the problem is that marginal populations (particularly for plants) are often just as dense as more centrally located populations, making unidirectional gene flow from high-density central populations to low-density marginal populations unlikely. When gene flow is measured across multiple populations with molecular markers, it often seems to occur in both directions rather than only from central to marginal populations. Perhaps the strongest evidence for this evolutionary constraint comes from transplant experiments suggesting that populations can survive outside their normal range. For instance, annual cocklebur plants moved to north of their range in North America were able to survive and reproduce successfully if induced to reproduce early (Griffiths and Watson 2006), pointing to a role of gene flow in preventing the evolution of early reproduction and subsequent range expansion.

5. LIMITS AND CONSTRAINTS: BIODIVERSITY AND CONSERVATION

Evolutionary constraints are a key determinant of biodiversity. If all species could successfully adapt to changing conditions, it is likely that far fewer species would exist, whereas constraints promote species-level biodiversity. One possible reason tropical areas have a high number of species is that tropical species are more likely to suffer from evolutionary constraints and lack adaptive potential; this seems to be the case for tropical *Drosophila*, which have a low evolutionary potential to adapt to colder and drier conditions when compared to more widespread species. It is possible that many ecologically specialized species lack adaptive potential because of a low level of genetic variation, or the presence of strong pleiotropic interactions that prevent them from easily adapting, or both. This may, for instance, prevent insect herbivores and parasites from expanding their diets to encompass new hosts, and prevent the spread of species from the humid tropics into drier and cooler climate zones. One possibility is that the processes of DNA decay and mutation accumulation are more likely in some regions—perhaps because species become more easily confined to a narrow-range host plant or because of a geographic barrier. However, it is currently not clear whether particular groups of species from some regions lack adaptive potential, or why this might be the case.

Recognizing adaptive constraints is important for conservation of species and ecological communities, because it can help to identify groups of species at potential risk because of an inability to adapt, or even entire communities if common patterns exist across species groups. Unless they are sufficiently phenotypically plastic (see chapter III.10), these species may be particularly prone to extinction due to disease, climate change, or other stressors. If evolutionary constraints arise because of gene flow, it might be possible to alter levels of gene flow to promote adaptive changes in marginal populations.

Evolutionary constraints are also important from an applied perspective. For instance, when pests and weeds are unable to evolve resistance to pesticides, there is the potential to continue using the pesticides. If the mode of action of pesticides and the genetic basis of resistance are understood, it might be possible to predict the likelihood of resistance developing in a particular evolutionary lineage of pests. On the other hand, where evolution is likely, it may be possible to slow the rate of evolution by ensuring ongoing gene flow between susceptible populations and those under selection. This practice has been adopted in the management of resistance to toxins introduced into crop plants, where resistant crops are interspersed with susceptible cultivars to slow adaptation by pests.

FURTHER READING

Bradshaw, A. D. 1991. The Croonian lecture, 1991: Genostasis and the limits to evolution. Philosophical Transactions of the Royal Society B 333: 289–305. *Outlines the notion of* genostasis, *in which a lack of genetic variation leads to inability of a species to adapt to stressful conditions, as illustrated by lack of evolutionary ability of many plant species to colonize soil contaminated by mine tailings and other examples.*

Colautti, R. I., C. G. Eckert, and S. C. Barrett. 2010. Evolutionary constraints on adaptive evolution during range expansion in an invasive plant. Proceedings of the Royal Society B 277: 1799–1806. *Shows how trait interactions can constrain range expansion during an invasion.*

Connallon, T., and A. G. Clark. 2011. The resolution of sexual antagonism by gene duplication. Genetics 187: 919–937. *Indicates how constraints due to antagonistic effects of genes in the different sexes can lead to gene duplication to allow genes to evolve sex-specific functions.*

Etterson, J. R., and R. G. Shaw. 2001. Constraint to adaptive evolution in response to global warming. Science 294: 151–154. *Describes an experimental study in which plants were exposed to environmental changes related to climate change and the response to selection was limited by interactions among traits.*

Futuyma, D. J. 2010. Evolutionary constraint and ecological consequence. Evolution 64: 1865–1884. *Provides a link between stasis in the fossil record and potential explanations of evolutionary constraints and suggests that microevolutionary constraints may not always be responsible for stasis.*

Kellermann, V., B. Van Heerwaarden, C. M. Sgrò, and A. A. Hoffmann. 2009. Fundamental evolutionary limits in ecological traits drive *Drosophila* species distributions. Science 325: 1244–1246. *Outlines an example involving* Drosophila *species in which a lack of genetic variation seems to limit the distribution of species, providing a connection between ecological niches and evolutionary limits.*

Sexton, J. P., P. J. McIntyre, A. L. Angert, and K. J. Rice. 2009. Evolution and ecology of species range limits. Annual Review of Ecology and Systematics 40: 415–436. *Provides an overview of ways in which gene flow can limit range expansion at species margins and constrain evolution, with a discussion of empirical studies.*

Walsh, B., and M. W. Blows. 2009. Abundant genetic variation plus strong selection = multivariate genetic constraints: A geometric view of adaptation. Annual Review of Ecology and Systematics 40: 41–59. *Outlines ways in which a response to selection can be stopped when genetic interactions occur among traits that constrain the expression of variation in the direction of selection; a very thorough review.*

III.9

Evolution of Modifier Genes and Biological Systems

Sarah P. Otto

OUTLINE

1. Evolution of biological systems
2. Evolution of dominance
3. Direct versus indirect selection
4. The evolution of genetic transmission
5. The evolution of the mutation rate
6. The evolution of sex and recombination
7. The evolution of haploidy versus diploidy
8. On evolution and optimization

The features that define how an organism lives and reproduces—how its genes are transmitted over time and space—have been molded by evolution. Scientists study this process by tracking changes over time at genes that alter the biological system (so-called modifier genes). Genes that modify a particular feature evolve in response to both direct and indirect selection, where the former depends on which modifier allele(s) an individual carries, and the latter depends on genetic associations that develop between the modifier and other genes affecting fitness. This chapter reviews the philosophy of modifier models and how they are used to study the evolution of biological systems.

GLOSSARY

Dominance. The degree to which phenotype is affected more by one allele than another at a gene. In population genetics, the phenotype of interest is often fitness. Here, if an allele is dominant, the fitness of a heterozygote is closer to the homozygote carrying this allele than to the fitness of the opposite homozygote.

Genetic Transmission. Processes associated with the inheritance of genes from parents to offspring, including mutation, segregation distortion, and recombination.

Modifier Gene. A gene whose alleles alter a feature of interest in a species, such as its mating system, mutation or recombination rate, or life history characteristics.

Ploidy Level. The number of homologous copies of each chromosome carried by a cell (excluding sex chromosomes). A haploid carries one copy, a diploid two, and a polyploid more than two.

1. EVOLUTION OF BIOLOGICAL SYSTEMS

Arguably the most fundamental equations in evolutionary biology describe the changes that occur within a population under natural selection. For example, as described in chapter III.3, selection favoring one version of a gene (say, the "A" allele) over another version of the gene (the "a" allele) causes the frequency, p, of the A allele to rise over time. This rise can be predicted using mathematical models under a particular set of assumptions, for example, that the population is diploid (with two copies of every gene), completely sexual, and randomly mating, which in turn implies that organisms and the genes they carry are well mixed across the species range. In other words, such models make a series of assumptions about the "biological system": how the organism lives out its life, how it reproduces, and how it moves over space.

Yet one might wonder how the biological system itself evolves. Why do some species live their lives as diploids, whereas other species are predominantly haploid (with one copy of every gene), and yet others alternate between haploid and diploid phases? Why do some species reproduce sexually, whereas other species eschew sex and reproduce clonally? Why do some species move freely over the landscape, while others choose to stay put? That is, how do the features that define how an organism lives and reproduces—how its genes are transmitted over time and space—evolve? Of course,

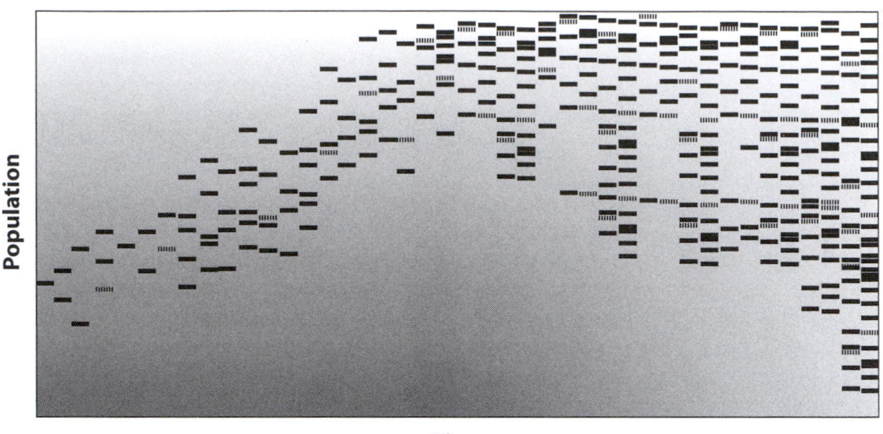

Figure 1. Tracking a modifier gene. In this schematic, each individual carrying a new modifier is represented by a line within a population (vertical axis) at a particular point in time (horizontal axis). Fitness is represented by background shading (lighter implying fitter). By altering the biological system, modifiers can become associated over time with particularly fit genotypes, as shown here. Initially, the modifier appears in an individual of relatively low fitness, but over time it becomes associated with fitter individuals (e.g., if a mutation modifier, a carrier might have produced a new beneficial mutation). Over time, survival and proliferation of the fittest individuals allows the associated modifier to spread throughout the population. Individuals who would have survived but die because of the direct costs of the modifier (e.g., costs of error correction) are represented as dotted lines. Alternatively, had the modifier allele stayed associated with low-fitness individuals, it would have been eliminated with the deaths of those individuals (not shown).

these features may themselves be directly subject to natural selection (e.g., to avoid costs of dispersal), but they may also evolve because they alter the genetic constitution and environmental context of an individual's descendants. Because most introductions to the field of evolutionary biology focus on the direct action of natural selection (see chapter III.3) and sexual selection (see chapter VII.6), assuming a particular biological system, most biologists and laypeople have little idea how evolutionary biologists would address these broader questions about how and why a species lives and reproduces the way it does (see chapters III.3 and VII.4).

This chapter provides an introduction to the evolution of biological systems. The focus is on a body of theory that takes some feature of a biological system and examines how genes that shape this feature are expected to evolve. For example, whether yeast reproduce sexually depends on the activity of a series of genes that regulate entry into meiosis. Variants of these genes (called *modifier genes*) make it more or less likely for the cell to enter meiosis. These variants thus modify the transmission of the parents' genes to offspring, determining whether they are inherited as a direct copy of the parental genome (asexual reproduction) or a mixture of two parental cells (sexual reproduction via meiosis). To study how the mode of reproduction and inheritance might evolve, evolutionary biologists develop mathematical models that track variants at such modifier genes. As described below, a particular modifier allele can rise or decline in frequency either because it directly affects the fitness of its carriers and/or because it alters the array of offspring and future descendants that are produced (figure 1). Modifier theory integrates these effects by tracking a modifier allele across generations to determine how the biological system is expected to evolve over the long term (see examples and links to other chapters in Table 1).

The results of modifier theory are particularly interesting in cases where it is hard to predict ahead of time which of several possible paths evolution is likely to take. For example, take the case of haploid versus diploid life cycles. Diploids have two sets of chromosomes and so have twice the amount of DNA as haploids. As a result, they suffer twice as many new mutations each generation, resulting in diploids suffering twice the deleterious mutation load compared to haploids (i.e., twice the reduction in fitness due to deleterious mutations; see chapter IV.5). Thus, from the perspective of the species, it would be optimal to be haploid. On the other hand, any particular individual is more fit if it is diploid, because the functioning of the second copy of the gene can mask any deleterious mutations the individual happens to carry. Thus, from the perspective of the individual, it would be optimal to be diploid. Which is the right perspective?

Modifier theory sidesteps this question, asking not what is optimal but instead what evolves. As we shall see, models that track genes that modify the alternation of generations between haploid and diploid phases find that haploidy is expected to evolve under some circumstances and diploidy under others. Interestingly,

Table 1. Examples of direct and indirect selection shaping the evolution of biological systems

Feature	*Direct selection*	*Indirect selection*
Mating preferences (chapter VII.4)	Costs involved in searching for a preferred mate or rejecting a mate	Modifier increasing preference becomes associated with sons that are attractive to females with a similar preference
Migration rate (chapter IV.3)	Energetic costs and associated risks of moving, e.g., risk of predation	Modifier increasing migration rate becomes associated with alleles that were selectively favored in other habitats and with reduced competition among relatives
Mutation rate (chapter IV.2)	Costs involved in repairing DNA damage and in failing to do so	Modifier increasing mutation rate becomes associated with novel deleterious and beneficial alleles
Sexual versus asexual reproduction (chapter IV.4)	Transmission advantage of asexuality; offspring inherit 100% of genes from an asexual parent rather than only 50% from a sexual parent	Modifier increasing frequency of sex becomes associated with different genetic combinations
Selfing rate (chapter IV.6)	Transmission advantage of selfing; selfing individuals can gain extra fitness by fertilizing their own ovules	Modifier increasing frequency of selfing becomes associated with more homozygous gene combinations

modifier models can also be used to understand how the direct effects on an individual's fitness and the long-term effects on the mean fitness of a lineage both play a role in the evolution of the biological system.

The goal of this chapter is to illustrate some of the insights provided by modifier models about the evolution of biological systems.

2. EVOLUTION OF DOMINANCE

Working in the 1900s, R. A. Fisher, Sewall Wright, and J.B.S. Haldane developed the mathematical underpinnings of our understanding of evolution. They were also the first to develop and analyze modifier models. Indeed, a modifier model of dominance played a central role in one of the most infamous early debates in evolutionary biology. Data from an increasing number of species indicated that wild-type alleles tend to be dominant over deleterious mutations, partially to fully masking the effects of mutations when both alleles are present in a heterozygote. To explain this phenomenon, Fisher considered a scenario with two genes: one gene under selection with a wild-type allele (A) subject to recurrent mutation to a less fit allele (a), and one gene that alters the degree of dominance, h, of the mutant allele with fitnesses given by:

Genotype:	*AA*	*Aa*	*aa*
Fitness:	1	$1 - h_{ij}\, s$	$1 - s$

where ij describes the genotype at the modifier locus (MM, Mm, or mm). Figure 2 shows how to derive equations for this two-gene model. Fisher then argued that modifier alleles that increased the fitness of heterozygotes (decreasing h_{ij}) would leave more descendants and hence should rise in frequency. Fisher concluded that biological systems would evolve by successive modification to the point where wild-type alleles were more dominant over mutant alleles.

Wright did not dispute Fisher's reasoning; indeed, Wright explored a similar model to reach the opposite conclusion. Wright observed that while the model predicts that modifiers of dominance would spread, the rate at which they spread is extremely low, being proportional to the mutant frequency at the A gene. Wright concluded that such a weak evolutionary force would likely be overwhelmed by other forces, including mutation at the modifier gene, side effects of the modifier gene, or random genetic drift. Instead, Wright argued that the fitness benefits of many genes diminish as their function increases, because the gene product becomes less a limiting resource; consequently, fitness is less affected in heterozygotes that still have one functioning gene copy than in homozygote mutants, whose gene product is much more limiting.

The interchange between Fisher and Wright over the power and efficacy of evolution is fascinating, and it eventually led to the breakdown of communication between them (Provine 1986). Subsequent authors, including Haldane, painted a less black-and-white picture, pointing out that selection on modifiers of dominance becomes strong when both alleles at the A gene are common

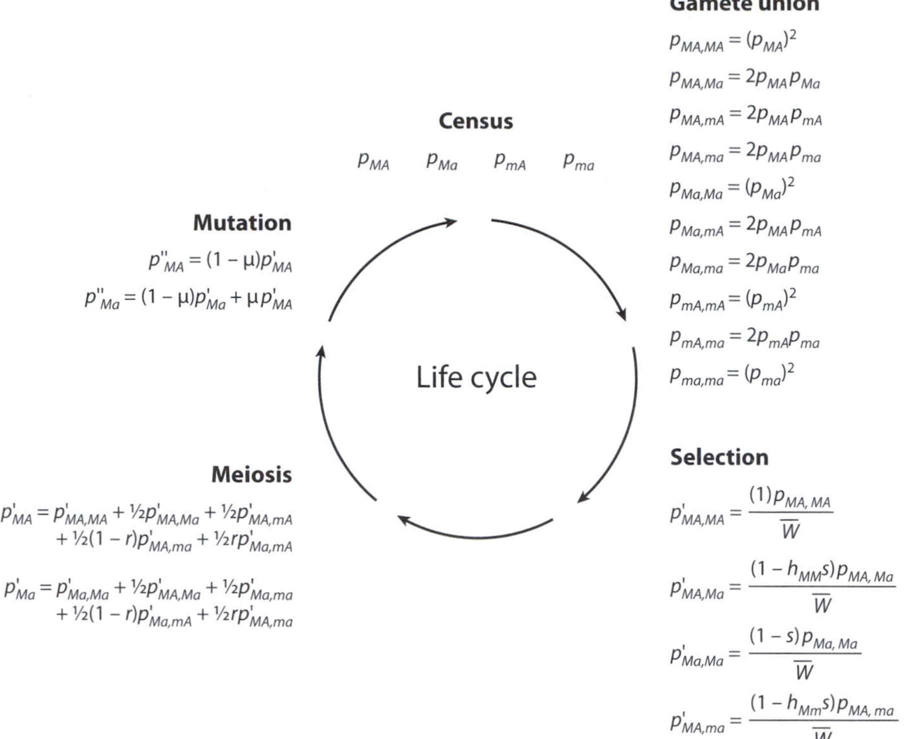

Gamete union
$p_{MA,MA} = (p_{MA})^2$
$p_{MA,Ma} = 2p_{MA}p_{Ma}$
$p_{MA,mA} = 2p_{MA}p_{mA}$
$p_{MA,ma} = 2p_{MA}p_{ma}$
$p_{Ma,Ma} = (p_{Ma})^2$
$p_{Ma,mA} = 2p_{MA}p_{mA}$
$p_{Ma,ma} = 2p_{Ma}p_{ma}$
$p_{mA,mA} = (p_{mA})^2$
$p_{mA,ma} = 2p_{mA}p_{ma}$
$p_{ma,ma} = (p_{ma})^2$
Census
p_{MA} p_{Ma} p_{mA} p_{ma}
Mutation
$p''_{MA} = (1-\mu)p'_{MA}$
$p''_{Ma} = (1-\mu)p'_{Ma} + \mu p'_{MA}$
Life cycle
Meiosis
$p'_{MA} = p'_{MA,MA} + ½p'_{MA,Ma} + ½p'_{MA,mA} + ½(1-r)p'_{MA,ma} + ½rp'_{Ma,mA}$
$p'_{Ma} = p'_{Ma,Ma} + ½p'_{MA,Ma} + ½p'_{Ma,ma} + ½(1-r)p'_{Ma,mA} + ½rp'_{MA,ma}$
Selection
$p'_{MA,MA} = \frac{(1)p_{MA,MA}}{\overline{W}}$
$p'_{MA,Ma} = \frac{(1-h_{MM}s)p_{MA,Ma}}{\overline{W}}$
$p'_{Ma,Ma} = \frac{(1-s)p_{Ma,Ma}}{\overline{W}}$
$p'_{MA,ma} = \frac{(1-h_{Mm}s)p_{MA,ma}}{\overline{W}}$

Figure 2. Model for the evolution of dominance. The evolution of dominance can be explored using a model to track changes at two genes: a modifier (M) and a gene under selection (A). We start (top) by censusing the frequency, p_{kl}, of each gamete (for kl equal to MA, Ma, mA, or ma). These gametes come together to form diploids, which are then subject to selection ($\bar{W}$ represents the mean fitness in the population). The surviving adults undergo meiosis (r is the recombination rate between M and A), with each gamete subject to mutation (at rate μ from A to a), producing the next generation of gametes. This cycle is repeated over and over to determine whether a modifier that alters the degree of dominance rises or falls in frequency. (Only a few examples of the equations for selection, meiosis, and mutation are shown.)

(e.g., during the initial spread of wild-type alleles over previous alleles or when different alleles are favored over different parts of the species' range) or if modifiers can affect dominance at multiple genes simultaneously (e.g., by promoting the degradation of misfolded proteins).

While it is sometimes said that Fisher's modifier theory of dominance ultimately lost out to Wright's physiological theory, this view is misleading. Both sides relied on insights obtained from modifier theory, with Fisher focusing on the direction of selection acting on dominance and Wright focusing on the strength of this selection.

3. DIRECT VERSUS INDIRECT SELECTION

In modifier models of dominance, the fitness of an individual depends on its modifier genotype (through h_{ij}). The same is true of models of epistasis, where modifiers alter the fitness of individuals carrying specific gene combinations at different loci (Liberman and Feldman 2005). In such cases, we say that the modifier gene is under *direct selection.*

Not all modifiers affect fitness directly. They can nevertheless evolve because they alter the types of descendants produced, experiencing *indirect selection* according to the fitness of these descendants. For example, a modifier that alters the frequency of recombination may have little direct effect on the fitness of its carrier, but it will change the genetic makeup of any offspring produced. If these offspring happen to have high fitness, on average, then the modifier allele will rise to higher frequency (figure 1). In essence, the modifier allele hitchhikes along with the successes (and failures) of the descendants it produces. To determine whether a modifier spreads or not requires that we track each descendant carrying the allele and the type of offspring it in turn produces. It can take several generations before the fate of the modifier becomes clear, in which case various mathematical techniques (such as local stability analyses and quasi-linkage equilibrium methods) are used to determine the direction of modifier evolution over longer periods of time.

One of the earliest examples of a modifier experiencing indirect selection was used by Fisher in a verbal account of the widespread observation that approximately equal numbers of males and females are found in many species. Whether an individual produces sons or daughters may have little consequence for the fitness of the parent, especially if parental investment ends after offspring are produced. Thus the sex ratio of an individual is often not under direct selection, yet indirect selection does act. Specifically, if an individual carries a gene that causes it to produce more offspring of the rarer sex, those offspring would make up a larger proportion of the mating pool of that sex and would contribute more genes, on average, to the grandoffspring and the great-grandoffspring, etc. Consequently, modifier alleles promoting the production of the currently rare sex would rise in frequency, causing the sex ratio to evolve toward 50:50. This argument was initially verbal. Indeed, the roots of the idea trace back to Darwin's *The Descent of Man and Selection in Relation to Sex* before genetic inheritance was understood, but it has subsequently been modeled and verified mathematically (e.g., Bodmer and Edwards 1960).

Of course, many biological systems evolve under the influence of both direct and indirect selection, as illustrated in Table 1, so that both must be considered when predicting the evolution of the system. It can be difficult, however, to intuit exactly how direct and indirect selection combine to influence the frequency of modifier alleles as they are transmitted from generation to generation, particularly because we must account for associations that develop between modifier alleles and other genes within the genome. Mathematical models are particularly illuminating in such cases, because they help guide and correct our evolutionary reasoning. These models are developed as illustrated in the case of dominance (figure 2), considering each step within the life cycle of an organism, so that modelers need not guess ahead of time what the net result of direct and indirect selection will be.

We turn next to a few examples of the application of modifier theory and the insights they have provided.

4. THE EVOLUTION OF GENETIC TRANSMISSION

One early and general result from modifier theory investigated the transmission of genes from parents to offspring. Assuming that a single randomly mating population has reached equilibrium under constant selection, modifier alleles that cause perfect transmission—where offspring are accurate copies of their parents—are always able to spread. This result was shown to apply to a number of processes that alter transmission, including mutation, migration, recombination, and segregation distortion, each of which is predicted to evolve toward zero. This general result is known as the *reduction principle* (for further information, see Altenberg and Feldman 1987).

While initially surprising, the reduction principle does make sense: if the world weren't changing and if selection favored certain alleles or combinations of alleles, then biological systems should evolve to exactly replicate parental genotypes, given that these parents survived to reproduce and so have genotypes that work well in the current environment.

The reduction principle tells us that, all else being equal, the simplest form of transmitting genes by copying them perfectly across generations should evolve. The real world is, however, much more complex; all else is not equal. Direct costs of perfect transmission make it impossible to replicate DNA without mutation and, in many species, to segregate chromosomes properly without recombination. Furthermore, the world is changing. Environmental changes, mutation, and drift can all push biological systems away from their equilibrium points and cause the reduction principle to fail. Research in this area thus attempts to figure out exactly when and why organisms transmit their genes the way they do, imperfectly.

5. THE EVOLUTION OF THE MUTATION RATE

All biological systems are subject to errors during the replication of their genetic material; mutations thus represent a universal example of imperfect transmission. The frequency of mutational errors is influenced by a variety of genes, including those involved directly in DNA replication (e.g., polymerases) as well as genes involved in the detection and repair of errors (e.g., excision repair and mismatch repair genes) (see chapter IV.2). Both direct and indirect selection are thought to shape the evolution of the resulting mutation rate.

Direct selection on the mutation rate results from the costs and benefits of detecting and repairing DNA damage and replication errors. These include energetic costs of producing error-correcting proteins and reduced growth rates resulting from high-fidelity replication. The benefits include avoiding mutations that arise during development and directly reduce the fitness of an individual (e.g., cancer-causing mutations).

Indirect selection results from transmitting different numbers of mutations to future generations. On the one hand, alleles increasing the mutation rate produce offspring that suffer a higher load of deleterious mutations. On the other hand, these offspring are also more likely to carry beneficial mutations that improve fitness in the current environment.

What is the net result of these selective forces? The proof of the reduction principle allowed only deleterious mutations and assumed no costs, and so it cannot tell us. Subsequent work has shown, interestingly, that the answer depends on the way the organism reproduces.

If the organism reproduces asexually, clones evolve a mutation rate that maximizes the lineage's long-term mean fitness. Consequently, higher mutation rates are expected to evolve in a changing environment where there is an advantage to producing beneficial mutations, as observed in a number of empirical studies (see review by Sniegowski et al. 2000).

With sex and recombination, however, genes that modify mutation rates evolve toward lower rates than would be optimal for the population. To understand this result, one again must think about the genes that influence the mutation rate and what happens to them from generation to generation. Carriers of a modifier that increases the mutation rate produce descendants that are more likely to carry deleterious as well as advantageous mutations. Deleterious mutations do not, on average, persist for long within the population, killing off their carriers as well as copies of the mutation modifier. Advantageous mutations persist longer, but sex and recombination separate these advantageous mutations from the mutation modifier that initially produced them. Thus, individuals carrying modifiers that increase the mutation rate soon bare the costs of producing more deleterious mutations, but because of sex, they share the benefits of their advantageous mutations with noncarriers. The more sex and recombination within a population, the more diluted the benefits become, hindering the evolution of higher mutation rates. Thus, mutation rates are expected to evolve to a lower level than what would maximize the mean fitness of the species.

6. THE EVOLUTION OF SEX AND RECOMBINATION

One of the most puzzling aspects of biological systems is that the vast majority of species engage in sexual reproduction, at least occasionally. Given that parents have survived to reproduce and so are "tried-and-true," why should a parent shuffle its genome with the genome of another? What makes this question even more puzzling is that sexual reproduction generally entails many direct costs, including the cost of finding a mate, the risk of remaining unmated, and the dangers of disease transmission and predation during mating, not to mention the fact that transmitting only half of one's genes to offspring automatically halves a parent's fitness (the twofold cost of sex).

Theoretical models have confirmed that genomic shuffling is generally a bad idea. Only if the environment changes very rapidly over time (changing direction every two to five generations) or over space is the mean fitness of offspring higher with genomic shuffling than without. Thus, under most circumstances, the short-term effect of a modifier that increases the frequency of sexual reproduction is to produce less fit descendants. The same holds for a modifier that increases the frequency of recombination. Basically, by breaking apart the gene combinations that have been favored by past generations of selection, sex and recombination tend to produce less fit offspring. That is, these offspring suffer from two types of fitness load: a segregation load and a recombination load (see chapter IV.5).

Why then is sex so prevalent? The key is thought to lie not in the mean fitness of offspring but in the variability of their fitness. Genetic variation is the fuel of evolutionary change, and if sex increases genetic variation, the response to selection will be faster among sexually produced offspring, promoting the spread of modifiers that increase the frequency of sex. Put another way, if the fittest individuals of a generation are produced sexually, their survival and reproduction spread modifier genes that enhance the rate of sexual reproduction. Lutz Becks and Aneil Agrawal have recently provided experimental evidence for this phenomenon in rotifers.

There are some big "ifs" in the preceding paragraph, however: "if sex increases genetic variation" and "if the fittest individuals of a generation are produced sexually." These conditions are not necessarily true. The problem is that while a single sexual individual can produce more variable offspring than a single asexual, a group of asexuals can be remarkably diverse, and altogether produce more variable offspring than a group of sexual individuals.

Evolutionary theory has thus sought conditions where (1) genetic variation is lacking, (2) sexual reproduction can increase this variation, and (3) the long-term advantages of increasing genetic variation outweigh the short-term segregation and recombination loads. Two main conditions have been found that satisfy these requirements and allow the spread of genes that increase the rate of sexual reproduction. First, genetic variation may be lacking because of past selection; this explanation requires that individuals at the extremes in fitness (high and low) are less fit, on average, compared with intermediate individuals, which in turn requires negative fitness interactions either within a gene (dominance) or among genes (epistasis). Second, genetic variation may be lacking because populations contain only a finite number of genetic combinations, and it is unlikely that the fittest alleles all occur within the same individual. Indeed, after a period of selection, a finite population tends to become composed of individuals that have similar fitness but carry different mixtures of good and bad genes. Sex and recombination can recover some of this hidden genetic variation by combining genes into different configurations, producing some particularly sick offspring and some particularly healthy offspring. With the survival and spread of the extremely fit individuals, genes promoting sex and recombination hitchhike along in frequency.

The jury is still out regarding whether the puzzle of sex is solved by the need for variation depleted by past selection and/or past drift in finite populations, but there is no doubt that modifier theory has clarified which evolutionary explanations can work and the conditions needed for them to do so.

7. THE EVOLUTION OF HAPLOIDY VERSUS DIPLOIDY

As one final example, let us return to the question of ploidy evolution. All sexual organisms pass through a stage with a reduced genome following meiosis (*haploid*) and a doubled genome following the union of gametes (*diploid*). Some organisms, like ourselves, spend little time in the haploid stage, whereas others, including a variety of fungi, algae, and mosses, spend little time in the diploid stage.

Evolutionary biologists can explore the conditions favoring the evolution of haploidy or diploidy by tracking changes at genes that alter the timing of the life cycle (figure 3A). Genetic changes that promote meiosis soon after gametes unite produce a predominantly haploid species, and vice versa for a predominantly diploid species.

Tracking such modifier genes, we find another interesting result. Haploid life cycles are favored in species that have low effective rates of recombination—for example, when reproduction is typically asexual or often involves inbreeding. In addition, haploid life cycles are favored when heterozygous diploids have a lower fitness than homozygous diploids, on average (i.e., when beneficial alleles are partially recessive and/or deleterious mutations are partially dominant). In contrast, diploidy evolves in highly sexual populations as long as heterozygotes have relatively high fitness (figure 3B). Additionally, in large multicellular organisms, modifiers that promote diploidy are more likely to spread because of the direct benefits of protecting an individual from the cancer-causing mutations that arise during development. Otto and Gerstein (2008) review empirical work testing these predictions.

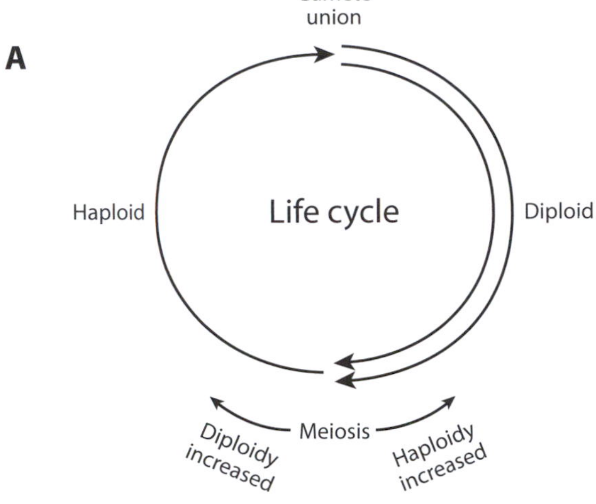

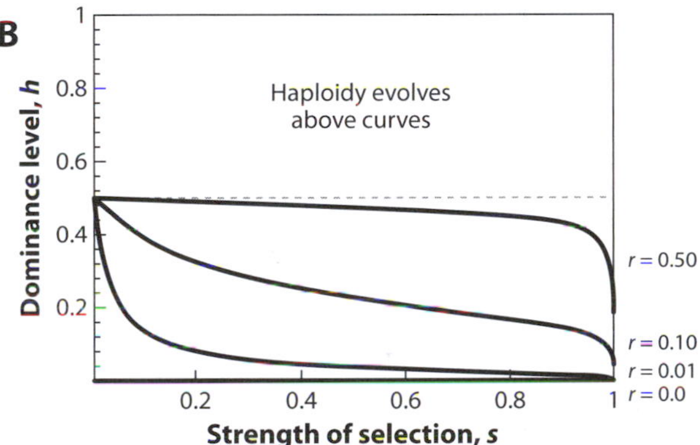

Figure 3. The evolution of haploidy versus diploidy. (A) With an alternation of generations, the extent of the haploid or diploid phase can be altered by genes that control the relative timing of meiosis and gamete union. (B) Modifier alleles that slightly expand the haploid phase are favored above the curves, with diploidy favored below the curves. The size of the region in which haploidy is favored expands when there is less genetic mixing (here shown as less recombination, r, between M and A). Selection acts against mutations with strength $1 - hs$ in heterozygous diploids and strength $1 - s$ in haploids and homozygous diploids.

8. ON EVOLUTION AND OPTIMIZATION

In the introduction, we pointed out that haploidy is optimal from the perspective of a species: haploids carry, on average, half as many mutations, and these mutations are immediately exposed to selection and purged. From the perspective of the individual, however, it is best to be diploid: deleterious mutations carried by a diploid can be masked by the second gene copy, allowing that individual to survive and reproduce. Which perspective is correct?

In some ways, the answer is neither. Considering what is best for the group would predict a world without diploids. Considering what is best for the individual would predict a world in which haploids are rare.

In another sense, however, the answer is that both perspectives provide clues about the evolutionary forces acting on a biological system. Because we track modifier genes over generations, whether those genes spread or disappear is influenced by what is best for the individuals that carry them in any particular generation and also what is best for the lineage (or group) of descendants that inherit the gene.

In many of the examples highlighted in this chapter, the way the biological system evolves depends on the amount of genetic mixing occurring within a species. When reproduction is primarily sexual, modifier alleles are more likely to evolve in ways that are best for the individual (e.g., lower mutation rates even when beneficial mutations are needed, more diploid life cycles even though deleterious mutations accumulate), because genetic mixing via sex prevents genetic differences from building up between the group of descendants that carry a particular modifier gene and those that do not. With little sex, modifier alleles are more likely to evolve in ways that are best for the group (e.g., higher mutation rates when beneficial mutations are needed, more haploid life cycles), because the modifier tends to be inherited for several generations alongside whatever genetic changes it causes, allowing the modifier to reap the long-term benefits of these changes.

What modifier theory allows us to do is to look at the biological world and to change the question from "What is best?" to "What will evolve?" What will evolve is sometimes best for the individual, sometimes best for the species, and sometimes neither. To understand how complex biological systems evolve and how this evolution has led to the remarkable diversity of life requires us to pay attention to how the genes that mold an organism change over time and how this evolution depends on the features of the organism and its environment.

FURTHER READING

Altenberg, L., and M. W. Feldman. 1987. Selection, generalized transmission and the evolution of modifier genes. I. The reduction principle. Genetics 117: 559–572. *A technical proof of the reduction principle, showing when and why evolution favors perfect transmission.*

Barton, N. H. 1995. A general model for the evolution of recombination. Genetical Research 65: 123–145. *A tour de force analysis of the conditions under which increased recombination evolves.*

Bodmer, W. F., and A. W. Edwards. 1960. Natural selection and the sex ratio. Annals of Human Genetics 24: 239–244. *A mathematical treatment demonstrating that the sex ratio evolves toward equal parental investment in daughters and sons.*

Liberman, U., and M. W. Feldman. 2005 On the evolution of epistasis I: Diploids under selection. Theoretical Population Biology 67: 141–160. *A mathematical study exploring, for the first time, the evolution of epistasis using a modifier model.*

Mayo, O., and R. Bürger. 1997. The evolution of dominance: A theory whose time has passed? Biological Reviews 72: 97–110. *An examination of the evolution of dominance, providing a strong historical overview.*

M'Gonigle, L. K., and S. P. Otto. 2011. Ploidy and the evolution of parasitism. Proceedings of the Royal Society B 278: 2814–2822. *An exploration of genetic factors that could influence the transition between free-living and parasitic life cycles.*

Otto, S. P. 2009. The evolutionary enigma of sex. American Naturalist 174: S1–S14. *A review of the evolutionary forces acting on rates of sex and recombination.*

Otto, S. P., and A. C. Gerstein. 2008. The evolution of haploidy and diploidy. Current Biology 18: R1121–R1124. *A primer describing theoretical and empirical results on the evolutionary transitions between haploidy and diploidy, as well as open questions.*

Sniegowski, P. D., P. J. Gerrish, T. Johnson, and A. Shaver. 2000. The evolution of mutation rates: Separating causes from consequences. Bioessays 22: 1057–1066. *An excellent overview of the evolution of the mutation rate.*

III.10

Evolution of Reaction Norms

Stephen C. Stearns

OUTLINE

1. Two major features of the genotype-phenotype map
2. Induced responses: Examples of adaptive plasticity
3. Robust traits: Examples of canalization
4. Reaction norms: Phenotypic plasticity and canalization
5. The evolutionary significance of plasticity and canalization
6. The Baldwin effect and genetic assimilation

Because the genetically identical members of clones can develop different phenotypes when they encounter different environments, we infer that one genotype can produce different phenotypes depending on the environment encountered. What difference does this make to the evolutionary process? To help answer that question, evolutionary biologists use the concepts of phenotypic plasticity, reaction norms, and canalization to describe the patterns observed. This chapter describes how those patterns are thought to evolve, what consequences they have for further evolution, whether they can be predicted and whether they are adaptive, nonadaptive, or maladaptive. It concludes with a discussion of the nature, origin, and evolutionary significance of genetic assimilation, one of the consequences of phenotypic plasticity.

GLOSSARY

Canalization. The limitation of phenotypic variation by developmental mechanisms that buffer it against genetic and environmental variation.

Genetic Assimilation. Environmentally induced phenotypes become genetically fixed and no longer dependent on the original environmental stimulus.

Phenotypic Plasticity. Sensitivity of the phenotype to differences in the environment.

Reaction Norm. A property of a genotype and a particular type of phenotype plasticity that arises when a trait is a continuous function of an environmental variable; it describes the mechanism by which development maps the genotype into the phenotype as a function of the environment.

1. TWO MAJOR FEATURES OF THE GENOTYPE-PHENOTYPE MAP

Throughout this chapter, I assume that the phenotype is a fixed property of an individual that develops once in a lifetime and does not change thereafter. This restriction excludes from the discussion both learned behavior and seasonally cyclic morphological changes, which are reversible and more dynamic than the plasticity discussed here.

What is at stake here is how best to build development into our models of the microevolutionary process. This, one of the major projects currently confronting evolutionary biologists, is often expressed as trying to understand the major features of the genotype-phenotype map, the set of rules linking the information contained in the genome to the material stuff of the organism. Understanding those major features—two of which are phenotypic plasticity and canalization—is thought by many to be a key to future breakthroughs in biology.

2. INDUCED RESPONSES: EXAMPLES OF ADAPTIVE PLASTICITY

Induced responses are classical examples of adaptive phenotypic plasticity. They satisfy one definition of an *adaptation*: a change in a phenotype that occurs in response to a specific environmental signal and improves reproductive success; otherwise, the change does not take place. Some water fleas in the genus *Daphnia* develop helmets and spines that protect them against predators, but only when they detect predators. Predators feed less effectively on spiny, helmeted *Daphnia*, but helmets and spines are costly. Individuals that do not produce them have higher reproductive rates than individuals that do produce them; that is why the spines and helmets are not produced when predators are not present.

Similarly, barnacles of the genus *Chthamalus* react to the presence of a predatory snail, *Acanthina*, by altering their development. If the snail is present, the barnacles grow into a bent-over form that suffers less from predation but pays for it with a lower reproductive rate. If the snail is not present, the barnacles develop into a typical form with normal reproduction. Tollrian and Harvell review many other examples of induced responses.

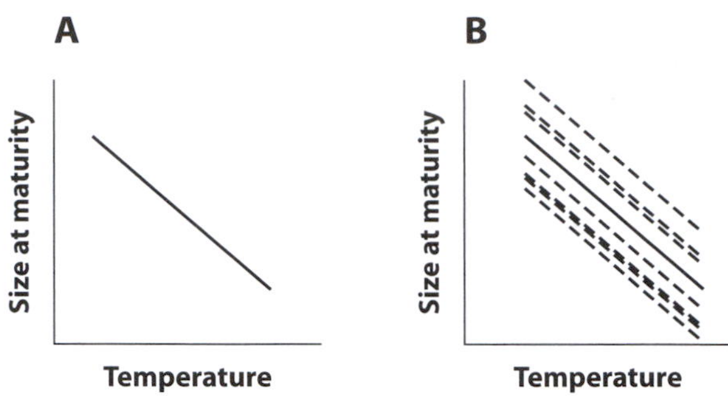

Figure 1. (A) An example of a reaction norm, which is a property of a single genotype: individuals all belonging to a single clone mature at smaller sizes when reared at higher temperatures. (B) An example of a bundle of reaction norms. Each dashed line represents the sensitivity of a single genotype to temperature; the solid line represents the population mean reaction norm.

3. ROBUST TRAITS: EXAMPLES OF CANALIZATION

Traits that exhibit very little phenotypic variation despite considerable environmental and genetic variation are called *canalized* because the phenotypic outcome is kept constant, as though development were confined within a canal that allowed no deviations from its course. When the canalization breaks down, for whatever reason, genetic variation for the hidden trait is revealed, demonstrating that the normal state was genetically canalized.

For example, Rendel found that *Drosophila melanogaster* normally have exactly four scutellar bristles, but in flies homozygous for the mutation *scute*, the number of bristles is reduced to an average of two with some variation. The mutation both reduces the average number of bristles and allows previously hidden variation for bristle number to be expressed. Because this variation responds to selection for fewer or greater number of bristles, we know it is based on genes other than the *scute* locus, and can infer that because of developmental buffering, the phenotypic effect of mutations in genes affecting this canalized trait had been suppressed in wild-type flies.

Many developmentally stable features of the phenotype appear to be canalized, including the four limbs of tetrapods, the six legs of insects, the eight legs of spiders, and the seven cervical vertebrae of almost all mammals, from whales to giraffes, none of which respond developmentally to environmental variation or to the genetic variation normally encountered from one generation to the next by a developmental system in sexually reproducing organisms.

Neither plasticity nor canalization is an absolute property. Some traits are more plastic and some more canalized than others; we detect both patterns through comparisons. We will return to canalization after developing the tools needed to analyze plasticity, the most important of which is the concept of a reaction norm.

4. REACTION NORMS: PHENOTYPIC PLASTICITY AND CANALIZATION

Narrowly defined, a *reaction norm* is a property of one trait, one genotype, and one environmental factor. We can measure it by raising individuals from one clone at different levels of an environmental factor, measuring the trait at each level, and plotting it as a function of the environmental factor. The resulting line (figure 1A) describes how development maps the genotype into the phenotype as a function of the environment. A population of genotypes can be represented as a bundle of reaction norms (figure 1B); the average reaction of the population to the environmental factor is the population mean reaction norm.

Depicting trait variation as a bundle of reaction norms does two important things. First, it shows us at a glance how genes and environments interact to determine the trait. Consider three genotypes (G1, G2, and G3) sampled from a population of parthenogenetic lizards, reared as clones, and raised at three population densities of low, medium, and high (figure 2), and two traits, number of digits per foot and fecundity. Figure 2A depicts the reaction norms of the three genotypes for number of digits per foot. In fact, they would lie on top of one another, for every individual in the entire population has exactly five digits per foot at all population densities; in the figure they are separated to show that three genotypes were measured. These are perfectly flat reaction norms. The trait is insensitive to environmental variation, expresses no genetic variation, and

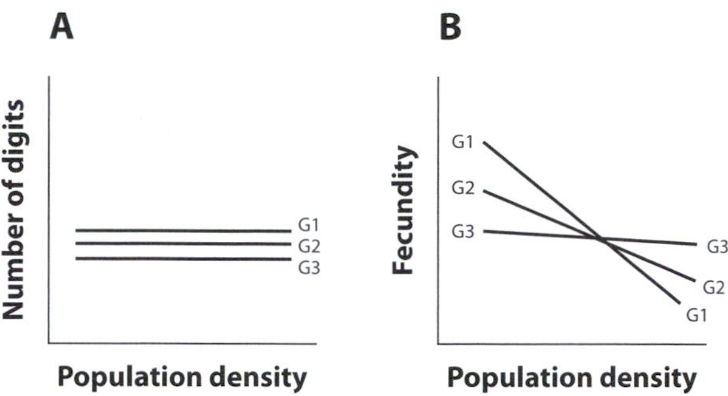

Figure 2. (A) The number of digits in the hand of a lizard is not sensitive to population density and does not vary genetically; it has a flat reaction norm that is both genetically and environmentally canalized. (B) The fecundity of the same three genotypes is sensitive to population density; here, sensitivity to density varies with genotype.

cannot respond to selection. It is both genetically and environmentally canalized.

For fecundity, the situation is quite different (figure 2B). All three genotypes reduce their fecundity at higher population densities, but they differ in sensitivity to changes in density. G1 is quite sensitive. It has the highest fecundity at low population density and the lowest fecundity at high population density. G3 is not very sensitive. It has the lowest fecundity at low population densities and the highest fecundity at high population densities. G2 is intermediate.

Second, figure 2B also gives us a starting point for thinking about how natural selection operates on phenotypic plasticity. Phenotypic plasticity, the sensitivity of a trait to a change in an environmental factor, is the slope of the reaction norm. If there is genetic variation in the slopes of reaction norms, selection can change plasticity by selecting genotypes whose reaction norms have larger or smaller slopes, but it will only do so under the particular circumstances discussed next.

Selection on Plasticity: A Matter of Frequency and Quality of Encounters

Plasticity is a second-order effect, defined by the difference in phenotypic response to two or more environments and measured in two or more individuals. The strength of selection on plasticity depends on the frequency with which environments are encountered in space and time and on the reproductive performance of the genotypes in each environment. Environments rarely encountered have little influence on selection; those encountered frequently influence selection more strongly. Frequency of encounter depends on both the frequency of the different environmental types and the size of the population encountering them. Environments in which survival and reproduction are good more strongly influence selection than environments in which survival and reproduction are poor, for subpopulations in good environments contribute more to population growth and recruitment. Thus both the frequency and the quality of an environment encountered determine the degree to which an evolved reaction norm will be adapted to that environment. If an environment is rarely encountered and of poor quality, it makes little difference how the organism responds to it with phenotypic plasticity, for it will not contribute much to future generations. We can expect all portions of a reaction norm to be adjusted by selection to fit the organism to each environment only if all environments are encountered with roughly equal frequency, and if all environments are of roughly equal quality. While that situation is conceivable, it is probably not often the case, suggesting that some parts of reaction norms will usually be better adapted than others.

The Costs of Plasticity: Usually Environment-Specific and Small

If there were no costs or limits to plasticity, the appropriately plastic organism would outcompete all others, for it could adjust to every environment encountered with maximal reproductive performance. For the reasons given in the previous section, such perfect adjustment of plasticity to all environments is unlikely, and the existence of imperfectly plastic phenotypes is not a puzzle, for they can simply be the by-products of environments that are rarely encountered, of poor quality, or both. The issue of costs of plasticity nevertheless remains interesting, for the costs could further limit evolution, particularly if there is a cost of being plastic per se.

The many attempts to measure the costs of plasticity were recently reviewed by Auld, Agrawal, and Relyea. They found it helpful to distinguish two types of costs of plasticity: production costs, specific to the environment in which the phenotype is produced, and maintenance costs, independent of any specific environment and associated with maintaining the general capacity to respond. In more than 200 sets of paired estimates of the costs of plasticity, detected costs were most frequently environment specific and therefore probably production costs rather than costs of maintenance.

Measured costs of plasticity have usually been small, probably for two associated reasons. The first is that fitness costs are not likely to be large in environments frequently encountered, for those are the environments in which selection has had the greatest opportunity to adjust the reaction norms to produce the optimal phenotype. If fitness costs are measured by comparing two frequently encountered environments, it should be no surprise that the estimates will be small. The second reason is that whenever costs are incurred, mutations that reduce costs—compensatory mutations—will be selected. The opportunity to select such mutations depends on the frequency and quality of the environments encountered; their efficacy also depends on the degree to which their expression must be specific to an environment to shape the reaction norm to fit it. There is, however, another type of cost of plasticity: the fitness cost of adjusting a phenotype to an expected environmental state on the basis of a signal that happens at times to mislead. Such costs could be quite high; they are quite relevant to predicting the consequences of global warming, and they have not yet been measured.

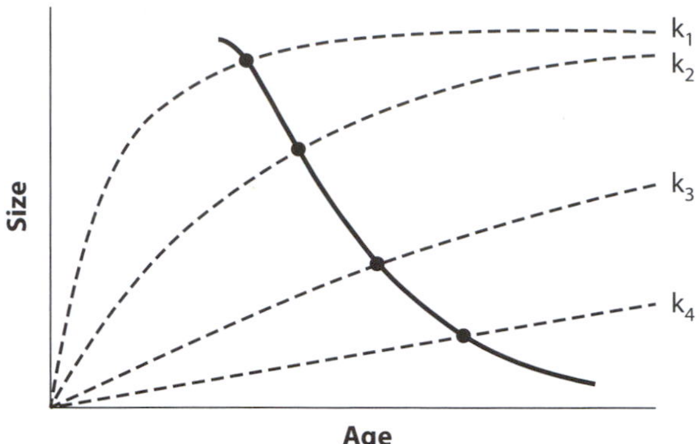

Figure 3. An optimal reaction norm for age and size at maturity (dark line) with predicted maturation events (Xs) for four different growth trajectories (dotted lines).

Is All Plasticity Adaptive? No, But Often Part of It Is

Selection is thus most likely to shape a plastic response that fits the phenotype to the local environment if that environment is frequent, of high quality, and capable of supporting a large population. Those are general evolutionary conditions. Other conditions are physical and chemical. For example, all chemical reactions proceed more slowly at lower temperatures; there is therefore no reason to invoke a past history of selection to explain the observation of slower development at lower temperatures *per se*. Only if the observed response differs from that predicted from chemistry unmodified by evolution might we suspect adaptation. Such was found to be the case by Berven, Gill, and Smith-Gill in frogs maintaining populations at high and low altitude. As would be expected from chemistry alone, frogs developed more slowly at lower temperatures at high altitude; however, in addition, frogs that had evolved at high altitude developed more rapidly at those low temperatures than did frogs from low altitude raised at the same low temperatures. This indicates an evolutionary adjustment of developmental rate; it also shows that part of a plastic response can be adapted while another part is the inevitable consequence of chemistry and physics.

Can the Plastic Response Be Predicted? Yes, for Life History Traits

For traits that are direct components of reproductive success, the relationship between a change in the trait and a change in fitness can be calculated, and from that calculation an optimal reaction norm can be predicted. Such a reaction norm is optimal in the sense that any other response would yield lower fitness given the trade-offs assumed. For traits like age and size at maturity, for example, one trade-off usually assumed is that earlier maturation implies less exposure to the risk of death because of a shorter juvenile period, but also because of less time to grow to a large size that would support the production of many offspring (see chapter III.11). Starting with work by Stearns, Crandall, and Koella, models based on such assumptions have often predicted the evolution of optimal reaction norms for age and size at maturity that embody this rule: if growing fast, mature large and young; if growing slow, mature old and small (figure 3).

The qualitative prediction of figure 3 is often but not always observed; the exceptions suggest ways in which the assumptions of the models may have been violated. Other shapes are predicted if growth rates are correlated with adult or juvenile mortality rates, if growth is determinate rather than indeterminate, if environmental heterogeneity is dominated by spatial structure rather than temporal change, and if explicit account is taken of population dynamics with frequency- and density-dependent effects.

The position and shape of such a reaction norm are seen as genetically determined and shaped by a history of selection. The particular point along the reaction norm at which an individual matures depends on the environment in which that individual has been raised. The reaction norm plot thus reveals how nature and nurture—genes and environment—interact to determine the actual age and size at which an individual matures. That maturation event is thus determined both by the history of selection encountered by the population and by a particular individual's history of developmental interaction with the environment.

Why Canalization Is Not the Opposite of Plasticity

Canalization, the limitation of phenotypic variation by developmental buffering, can act to buffer environmental

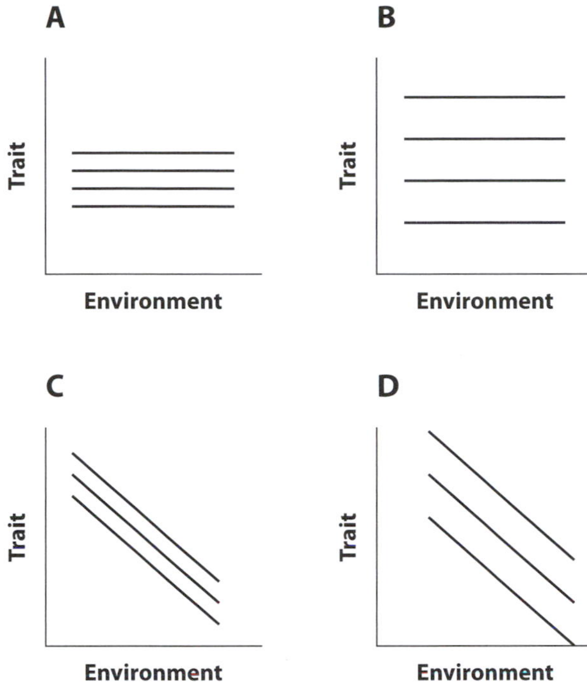

Figure 4. (A) A set of reaction norms suggesting both genetic and environmental canalization when compared with those in B. (C) A set of reaction norms suggesting genetic but not environmental canalization when compared with those in D.

or genetic variation, or both. Again, reaction norm plots are a convenient way to visualize some helpful distinctions.

In figure 4, comparison of A and B suggests that both sets of reaction norms are canalized with respect to environmental variation (they are *environmentally canalized*), because all the reaction norms are flat, but those in A are in a tight bundle, whereas those in B are in a loose bundle. This suggests that those in A may be more *genetically canalized* (i.e., canalized with respect to genetic variation) than those in B. Comparison of C and D indicates that both sets of reaction norms are not canalized with respect to environmental variation, since all the reaction norms have negative slopes; however, those in C, like those in A, are in a tight bundle, whereas those in D, like those in B, are in a loose bundle, suggesting that those in C may be more genetically canalized than those in D. Thus a trait can be genetically canalized but environmentally plastic; a trait can also be environmentally canalized but genetically free to vary. Care must be taken to specify precisely what pattern is under analysis.

Reaction norm plots can show that traits are *not* canalized, but at least two reasons can always be given for why they might appear genetically canalized. The first is that they actually are buffered against variation by developmental mechanisms (e.g., the impact of the *scute* mutation on bristle number in *Drosophila*); the second is that they have been under strong selection that has depleted the genetic variation in the population. Therefore the description of the pattern must be followed by an analysis of the mechanisms that produced it before a conclusion about the causes driving those mechanisms can be drawn. That search for causes has begun.

Selection for Canalization

Several hypotheses for the selection of canalization have been proposed; while they are not mutually exclusive, and therefore all could act at once, the evidence for some is better than others. Schmalhausen (1986; originally published in English in 1949) suggested that canalization is a result of stabilizing selection, for if there is a single optimal phenotype, then any deviation from it has lower fitness, and canalization would buffer the phenotype from such costly deviations, whether caused genetically—by mutation, recombination, or gene flow—or environmentally. Evidence from genetically and environmentally controlled experiments on fruit flies (Stearns et al. 1995) supports Schmalhausen's idea, but, as Wagner et al. (1997) point out, the ability of stabilizing selection to shape canalization depends on the amount of genetic variation present in the population and the degree to which canalizing genes have deleterious pleiotropic effects on other characters or direct effects on the same character. Their arguments suggest, given the amount of genetic variation usually present, that the further evolution of canalization will be quite slow on a microevolutionary timescale and that the evolution of canalization through stabilizing selection is quite unlikely ever to achieve complete fixation of a trait.

Another reason for selection of canalization was suggested by Kawecki (2000), who pointed out that in a fluctuating environment, selection tends to produce the phenotypes that work best in the previous environment, not the one currently encountered. (This resembles the complaint that the generals are always fighting the last war.) In such circumstances canalization reduces variation in fitness, increasing geometric mean fitness.

Siegal and Bergman (2002) proposed a strikingly different reason for the existence of canalization. They investigated selection for robustness in genetic networks required to deliver products reliably or the organism would fail to develop properly; they concentrated on selection for developmental stability rather than for stabilizing selection on the phenotype. They discovered that phenotypic canalization was a by-product of selection on developmental stability, whose strength increases with the complexity of the genetic network. Since then, others have established that this conclusion is independent of the details of their model, and that such selection is stronger in larger as well as more complex genetic networks in which more mutations can have an effect. Selection for canalization as a by-product of developmental stability can also be strong even in a small genetic network if it is perturbed by gene flow from other populations, which can have larger effects than single mutations. Their work now motivates further research to establish how much of canalization can be attributed to selection for developmental stability and how much to selection for a specific, optimal phenotype.

Evidence for Canalization

Some of the best evidence for canalization comes from the study of heat-shock proteins. Heat-shock proteins are molecular chaperones that accompany other proteins to the intracellular sites where they function, protecting them on their journey. When the production of one heat-shock protein, HSP 90, is inhibited, development is altered and hidden genetic variation is released in both fruit flies (*Drosophila*: Rutherford and Lindquist) and wall cress (*Arabidopsis*: Queitsch, Sangster, and Lindquist). These results establish at least one mechanism that causes canalization and extend Rendel's classical results, based on mutation in the *scute* gene, to the molecular level. Note that there is no need to postulate that the effects of HSP 90 evolved because they suppress genetic variation; that consequence could simply be a side effect of its direct chemical function, which is to stabilize proteins in their journey from biosynthesis to cellular action.

Evidence for canalization is also now coming from work inspired by Siegal and Bergman's ideas on selection for developmental stability supported by robust genetic networks. Duplicating a pathway in such a network can ensure proper development when one pathway is knocked out or perturbed; gene duplications with that effect have been found in flower development (Lenser et al. 2009). And when the inactivation of a gene in such a network has no effect on the phenotype, one suspects some buffering mechanism; such genes are common in yeast. Other genes buffer the effects of such genetic inactivation, and it turns out that the buffering genes also confer robustness to environmental and stochastic changes (Lehner 2010). From the local point of view of a genetic control network, this makes good sense, for at this level of intracellular detail, the source of the perturbation is unknown and irrelevant, but the consequences of the perturbation, no matter what the source, are serious. One therefore expects the buffering mechanisms to be general and not specific to the source, whether genetic or environmental.

5. THE EVOLUTIONARY SIGNIFICANCE OF PLASTICITY AND CANALIZATION

Plasticity has many evolutionary consequences, among the most important of which is that it extends the range of conditions under which organisms can survive and reproduce and thus reduces the frequency at which populations go extinct. Another major consequence, genetic assimilation, is discussed in the next section.

One evolutionary consequence of canalization is that it renders invisible and therefore neutral any mutations that could affect a genetically canalized trait, allowing a greater proportion of them to be stored in the population than would otherwise be the case (see chapter IV.1). If canalization breaks down, this variation is released from buffering and expressed in phenotypes, providing additional material on which selection can act. How frequently this happens in nature is at this point uncertain; in a sample consisting mostly of life history traits, there was no consistent pattern (Hoffmann and Parsons 1991).

Another potential evolutionary consequence of canalization is that it plays a role in the origin of long-term constraint. Once a trait has been canalized and its form and function have been fixed, other traits coevolve with the fixed form of the trait. This can embed the canalized trait in a network of interactions with other traits so that continued successful function depends on the canalized trait remaining canalized; the process selects for further buffering of its canalized state. While this may be one source of developmental constraint, evidence supporting the idea remains scarce, and, as Wagner et al. (1997) point out, the mechanism that leads to complete fixation of a trait is unlikely to be stabilizing selection.

6. THE BALDWIN EFFECT AND GENETIC ASSIMILATION

In a novel environment, a plastic response can help an organism survive where a canalized organism would die. Such a response is not an adaptation to that environment, which has never been encountered before; it is a preadaptation that can allow the adaptive process to continue. One result can be a change in the genetic determination of the trait when it is expressed in the new environment.

This process, or something like it, has been called the Baldwin effect (as described by Baldwin), genetic assimilation (by Waddington), genetic accommodation (by West-Eberhard), and stabilizing selection (by Schmalhausen). Each of the labels emphasizes a different part of a complex process; genetic accommodation is the most inclusive.

Lande analyzed this process by framing it as the evolution of reaction norms in a novel environment. He started from a background condition in which the trait was canalized with minimum genetic and phenotypic variation and there was no correlation between reaction norm elevation and slope. He found that in the first generation in the novel environment, mean fitness drops, and the mean phenotype moves toward the new optimum, a change at that point based only on plasticity. Then adaptation occurs in two phases. First, increased plasticity rapidly evolves, allowing phenotypes to approach the new optimum in the novel environment. Then the new phenotype undergoes slow genetic assimilation that reduces plasticity (the slope of the average reaction norm) but compensates with genetic evolution of the elevation of the reaction norm in the original environment.

Whatever one calls it, such a process has several important consequences. It reduces the probability that a population will go extinct when it encounters novel conditions. And it makes clear that the phenotype has an important role in evolutionary change that has been underappreciated, a role described by West-Eberhard as plasticity and behavior taking the lead in evolution. The generality and significance of this important idea are under continuing examination.

FURTHER READING

Auld, J. R., A. A. Agrawal, and R. A. Relyea. 2010. Re-evaluating the costs and limits of adaptive phenotypic plasticity. Proceedings of the Royal Society B 277: 503–511.

Hoffmann, A. A., and P. A. Parsons. 1991. Evolutionary Genetics and Environmental Stress. Oxford: Oxford University Press.

Kawecki, T. J. 2000. The evolution of genetic canalization under fluctuating selection. Evolution 54: 1–12.

Lande, R. 2009. Adaptation to an extraordinary environment by evolution of phenotypic plasticity and genetic assimilation. Journal of Evolutionary Biology 2: 1435–1446.

Lehner, B. 2010. Genes confer similar robustness to environmental, stochastic, and genetic perturbations in yeast. PLoS ONE 5: e9035. doi:10.1371/journal.pone.0009035.

Lenser, T., G. Theissen, and P. Dittrich. 2009. Developmental robustness by obligate interaction of Class B floral homeotic genes and proteins. PLoS Computational Biology 5: e10000264 doi:10.1371/journal.pdbi.1000264.

Pigliucci, M. 2001. Phenotypic Plasticity: Beyond Nature and Nurture. Baltimore, MD: Johns Hopkins University Press.

Schmalhausen, I. I. 1986. Factors of Evolution: The Theory of Stabilizing Selection. Chicago: University of Chicago Press.

Siegal, M. L., and A. Bergman. 2002. Waddington's canalization revisited: Developmental stability and evolution. Proceedings of the National Academy of Sciences USA 99: 10528–10532.

Stearns, S. C., and R. E. Crandall. 1984. Plasticity for age and size at sexual maturity: A life-history response to unavoidable stress. In G. Potts and R. J. Wootton, eds., Fish Reproduction. London: Academic.

Stearns, S. C., M. Kaiser, and T. J. Kawecki. 1995. The differential genetic and environmental canalization of fitness components in *Drosophila melanogaster*. Journal of Evolutionary Biology 8: 539–557.

Stearns, S. C., and J. C. Koella. 1986. The evolution of phenotypic plasticity in life-history traits: Predictions for norms of reaction for age- and size-at-maturity. Evolution 40: 893–913.

Tollrian, R., and C. D. Harvell. 1999. The Ecology and Evolution of Inducible Defenses. Princeton, NJ: Princeton University Press.

Wagner, G. P., G. Booth, and H. Bagheri-chaichian. 1997. A population genetic theory of canalization. Evolution 51: 329–347.

West-Eberhard, M. J. 2003. Developmental Plasticity and Evolution. New York: Oxford University Press.

III.11

Evolution of Life Histories

David Reznick

OUTLINE

1. What is the life history and why is it of interest?
2. The theory of life history evolution: A sampler
3. Other aspects of life history evolution
4. What have we learned?
5. Future research

The life history is a composite of all the variables that contribute to the way in which an organism propagates itself. The most important variables are how old it is when it begins to reproduce, how much it invests in reproduction, as opposed to other activities or structures, and how it allocates resources to offspring (many small versus few large). We are interested in life histories from a theoretical perspective because these variables are closely allied to an organism's fitness, or its ability to contribute offspring to the next generation. We are interested in life histories from a natural history perspective because organisms in the real world display a vast diversity of life histories, so we would like to know why they evolved. Life history theory predicts the optimal allocation of resources to growth, maintenance, storage, and reproduction in response to external features of the environment, such as risk of mortality. The evolutionary optimum is defined as that allocation that results in the largest number of successful descendants. The empirical study of life history evolution has revealed that the different components of the life history evolve in a way that is consistent with assumptions made by theory. For example, there is support for the proposal that trade-offs are made between different components of the life history, such that investing heavily in reproduction early in life is associated with shorter life span. Empirical research also includes the experimental test of life history theory in natural populations and, as such, presents one of the best examples available of the experimental test of evolutionary theory in nature. Specifically, investigators have manipulated the risk of mortality in natural populations and have seen life histories evolve as predicted by theory. One future direction of life history research is to use it as a forum for developing a better understanding of the definition of fitness. Another is to use it as a forum for understanding how ecological and evolutionary processes interact with one another.

GLOSSARY

Adaptation. Any feature of an organism that has been shaped by the process of evolution by natural selection.

Antagonistic Pleiotropy. Pleiotropy means that a single gene influences more than one feature of the phenotype. The gene's effects are described as antagonistic if the gene changes one feature of the phenotype in a way that increases fitness and another feature in a way that decreases fitness.

Fitness. In the context of life history evolution, fitness is the relative success of an individual at contributing offspring to the next generation. If we set the average success of an individual in a population as equal to 1, then we can score the relative fitness of individuals relative to this average as being either greater or less than 1.

Semelparity/Iteroparity. *Parity* refers to giving birth. *Iteroparity* means giving birth repeatedly during the course of an individual's lifetime. *Semelparity* means giving birth only once, then dying.

Trade-off. In the context of life history evolution, a trade-off is a causal linkage between two features of the life history, such that increasing the resources devoted to one feature causes a decline in the amount devoted to the other. Trade-offs can also be dictated by other constraints, such as between the number and size of individual offspring, as shaped by the finite volume that a mother can devote to developing offspring.

1. WHAT IS LIFE HISTORY AND WHY IS IT OF INTEREST?

An individual's life history is the composite of all the variables that contribute to the way it propagates itself. The two generic classes of variables that make up a life history pertain to timing and resource allocation. Timing variables include how old the individual is when it begins to reproduce and how often it reproduces. Allocation variables address how an organism divides up the resources with which it has to work. The categories of allocation include growth, maintenance, reproduction, and storage, such as in the form of fat. Reproductive allocation includes the ways in which these resources are in turn divided among offspring. An organism like an elephant, whale, or human produces only a handful of babies during its lifetime and devotes a very large amount of resources to each of them. An ocean sunfish (*Mola mola*) can produce many millions of eggs each time it reproduces, devotes little to each of them, and casts them to the fates of ocean currents. Allocation variables also include parental investment before or after the fertilization of gametes. For many organisms, all resources are invested in seeds, eggs, or live-born offspring who are on their own from birth. For others, the parents provide continued care, and hence make continued resource investment after birth.

We study life histories for two reasons. One is their theoretical importance. The other is their remarkable diversity.

Life History Theory

Life history theory is really a special application of the more general theory of evolution by natural selection. We use the word *fitness* to describe the differences among individuals in their ability to contribute offspring to the next generation. The key to understanding fitness is that it is first determined by the odds of surviving to maturity, then by the number of offspring that in turn survive to reproduce in the next generation. These features can evolve if they vary among individuals and if the variation is at least partly heritable. Life history characteristics therefore can evolve to maximize fitness and can be considered adaptations. Life histories represent a particular category of adaptations, as detailed in the next section. The reason they stand out as special is that they apply to the actual currency of fitness, which is the production of offspring. For that reason, a body of theoretical and empirical work has grown around explaining how and why life histories have evolved.

Life history theory occupies a special place in evolutionary biology because it contributed to a more generalized concept of fitness. When Darwin presented evolution by natural selection in *The Origin*, he defined fitness as survival. If individuals with a given phenotype are able to live longer, then they have more opportunities to reproduce. In the theory of life history evolution, fitness is defined instead as a composite measure of survival and age-specific reproduction, or all the different variables that make up the life history. One such composite variable is the intrinsic rate of increase (r), or the per capita rate at which a population increases in size. An alternative index of fitness that is sometimes used is R_0, or the expected number of offspring produced during the lifetime of the individual. The rank order of the relative fitness of different phenotypes is equal to the rank order of each phenotype's value for r or R_0.

An important consequence of defining fitness in terms of such composite variables and in the context of life history evolution is that natural selection can cause the evolution of seemingly disadvantageous traits. For example, it can favor the evolution of reduced life span or reduced number of offspring produced during an individual's lifetime, providing there are compensations, or *trade-offs*, in other components of the life history (see chapters III.1 and III.8).

Life History Diversity

Consider some of the extremes of life histories that we see in nature. The seeds of the desert annual plant *Linanthus parryae* lie dormant in the soils of the Mojave Desert for most of the year. They germinate in response to unpredictable winter rains. A single rain can cause some seeds to break dormancy, sprout, grow, and, if there is sufficient moisture, flower and set seed, all within a few months. All that remain at the end of the brief growing season are short, dry stalks and seed pods; otherwise, the species exists as a "seed bank" in the soil, accumulated in those years when there was sufficient rain to support a complete life cycle. Individual seeds remain dormant, possibly for decades, before responding to a promising winter rainstorm. Whole years can pass with no living plants being visible on the surface of the desert.

In contrast, bristlecone pines (genus *Pinus*), which inhabit equally arid, but high-elevation environments dispersed throughout the southwestern United States, replace the speed of the *Linanthus* life cycle with the ability to persist. These are the oldest-known living things, with the known ages of some individuals being close to 5000 years. It takes centuries for them to reach maturity. Their seed-producing cones take more than a year to mature. The large trunk of an old individual may contain only a narrow strip of living tissue connecting the crown of the tree to its roots; the remainder consists of decay-resistant dead wood.

The topic of life history evolution thus presents us with compelling questions and contrasts. If attributes like development time, frequency of reproduction, and number of propagules are so important in defining fitness, why are they also so variable? What features of the environment shape this remarkable diversity of life histories?

2. THE THEORY OF LIFE HISTORY EVOLUTION: A SAMPLER

Think of your life as a pie. The pie represents the available resources enabling you to survive and reproduce. The pie is divided into four slices, one each representing allocations devoted to maintenance, growth, storage, and reproduction.

Maintenance includes the cost of replacing the parts of your body that are constantly wearing out. Skin, blood cells, and the lining of your intestinal tract consist of cells that are in constant need of replacement, so they occupy a significant component of maintenance. Many other components of your body are also constantly being replaced or modified over time, at some cost in terms of consumed energy and resources. For adult humans, maintenance is the largest slice of the pie.

Growth is a major investment early in life, but many organisms either stop growing or dramatically reduce growth rate when they attain maturity, causing the size of this particular slice to decline with age. Mammals and birds, for example, virtually cease growth at maturity, so the investment in growth precedes the investment in reproduction. Fish, amphibians, reptiles, and most plants can grow throughout their lives, so there can be a temporal overlap between investment in growth and investment in reproduction.

Storage can be in the form of energy, primarily fat tissue, but also in the form of limiting nutrients or elements. In birds, calcium is a stored nutrient; it plays an important role in body fluids and bones, but it is also used periodically and in large amounts to produce eggshells. I will concentrate only on energy storage here. You may have difficulty envisioning fat as adaptive, but that is because we are all inclined to make more deposits than withdrawals. For many organisms, fat is a key feature of the life history, because it enables organisms to stockpile resources at a time when they are available, then save them for a later time when resources may be scarce, and it is ideal for producing offspring.

Reproduction is the piece of the pie devoted to making babies. This investment includes the gametes (eggs and sperm), and other forms of care provided by the parents. These other investments in reproduction can include finding, preparing, and defending a nest site, competing with others of the same sex to obtain a mate, or convincing members of the opposite sex to mate. You might be tempted to think of investment in reproduction as what is done with the surplus after all the critical needs for survival are met, but that is not necessarily the case. In some extremes it can displace maintenance. In most species of salmon, for example, reproduction happens just once, and it is like the flight of a ballistic missile. As salmon migrate from the sea into freshwater, they cease to feed. The entirety of the resources required to migrate upstream (sometimes hundreds of miles), prepare a nesting site, compete for mates, and produce gametes are derived from stored resources. Salmon cease investing in vital functions, like their immune systems, as they swim upstream. In their final days, they are depleted of fat reserves and can be covered with festering infections.

A consequence of thinking of life as a pie is realizing that the pie is finite in size. This means that increasing the size of any of these four slices necessarily reduces the size of other slices. This interdependence of the slices defines a central tenet of most life history theory, which is that all organisms must face *trade-offs* in allocating their resources to the various components of the life history, such that a gain in any one function comes at some cost to another function. For example, in a fish, reptile, or amphibian, increasing the current investment in growth might carry the benefit of producing more offspring in the future, if the individual survives, because larger individuals can produce more offspring, but increasing growth now can mean reproducing less now. The goal of life history theory is to predict the optimal allocations of this finite quantity of resources in a given set of circumstances. I offer an introduction to two types of life history theory that illustrate these principles.

r and K selection. This theory was first well articulated by Robert MacArthur and E. O. Wilson in 1967. It does not represent the inception of our interest in life history evolution, but it was the first big idea that captured the imaginations of evolutionary biologists and made the study of life history evolution an important empirical endeavor.

r and K are parameters from the equation used to describe logistic population growth, or population growth in conditions where resources can be limiting:

$$dN/dt = rN(1 - N/K)$$

r = intrinsic rate of increase, or rate of exponential population growth when resources are unlimiting
N = population size
K = carrying capacity, or the maximum number of individuals the environment can sustain

The expression in parentheses can be thought of as a damping factor, or the degree to which the potential growth rate of the population is reduced because the environment is partially occupied. If N is very small, then the population growth rate is close to r. When $N = K$, the population ceases to grow.

In MacArthur and Wilson's original formulation of the theory, they predicted that r-selection favors the evolution of "productivity," or selects for traits that enable an individual to sustain rapid population growth. In addition, r-selected populations are those that are persistently far from their carrying capacity, perhaps because their abundance is frequently reduced by unpredictable events, like storms or droughts, or because they are kept scarce by predators. K-selection is selection in favor of the ability to persist in the face of limiting resources, such as when individuals are members of populations that are persistently close to their carrying capacity, perhaps because they occupy stable environments. MacArthur and Wilson envisioned how r and K selection would come into play when an organism colonizes an island. It will encounter abundant resources and experience an initial phase of r-selection, which would favor genotypes that "have a shorter developmental time, a longer reproductive life, and greater fecundity, in that order of probability" (MacArthur and Wilson 1967). After population expansion, island colonists will then experience K-selection, which "favors efficiency in the conversion of food into offspring" (p. 149).

Eric Pianka elaborated on MacArthur and Wilson's predictions. He argued that r-selection would favor the evolution of early maturity, increased investment in reproduction, and the production of many offspring, each of which receive little investment from the parents, and short life span. All these changes in the life history, save the last, will increase the value of r. Pianka predicted that K-selection would instead favor the evolution of delayed maturity, decreased reproductive investment, the production of few offspring, each receiving high parental investment, and long life span. All these changes will reduce r, but are assumed to increase the ability of an organism to persist when resources are limiting. The concept of trade-offs is implicit in his predictions. Because r-selected organisms invest heavily in reproduction early in life, they invest less in growth and maintenance and have shorter lives. Conversely, because K-selected species invest more in their own growth and maintenance and less in reproduction, they tend to have longer lives.

Demographic Theory: How Mortality Risk Shapes Life History Evolution

There is a prominent feature of life histories that is not included in r- and K-selection. Many populations have age or stage structure, such as juveniles versus adults. Age classes can differ in the way they are affected by selection. Populations of the annual plant *Linanthus parryae* do not have age structure once seeds have germinated because all individuals germinate, mature, reproduce, and die in the same season. Populations of bristlecone pines do have age structure because they grow for centuries before they attain maturity, then live and continue to grow for millennia as mature, reproducing individuals. The probability of survival from one year to the next can change with the age, size, or stage of development of the individual. For example, if predators prefer large prey, then older, larger individuals may experience increased risk of mortality. On the other hand, older, larger individuals may be better at evading predators and defending themselves, so mortality risk may decline with age. The expected number of offspring is zero prior to maturity. The number of offspring produced can increase progressively after maturity if the individual grows, since fecundity is often directly proportional to size. In species with postnatal care, like birds and mammals, reproductive success often increases with age because the parents become more experienced and are better providers.

Since the populations of many organisms have such age structure, a body of theory was developed to address the consequences of differences among age groups in the risk of mortality and reproductive success. The theory deals with risks imposed from the outside, such as those of predators, disease, and competitors.

One common prediction from this body of theory is that when the risk of adult mortality is high, natural selection will favor the evolution of earlier maturity and increased allocation to reproduction. If we think of life as a pie, then this enlargement of the slice devoted to reproduction early in life means that some other slice must be smaller. There could be a reduction in growth, in maintenance, or in storage. Wherever the trade-off occurs, the expectation is that investing more in reproduction now will in some way detract from the future. Another prediction that emerges from some models is that a selective increase in the mortality rate of juveniles can also select for the evolution of delayed maturity and a decrease in the rate of investment in reproduction. The reason for the delay in the context of this model is also a product of the trade-offs associated with the pie of life. Deferring the investment in reproduction means investing more in growth and maintenance. In real life, this shift in investment can in turn mean more quickly outgrowing the size classes that are susceptible to predators, if predation on small individuals is the source of mortality. These alternative life histories resemble those predicted by r- and K-selection, but they can evolve independently of the population's proximity to its carrying capacity. They evolve instead as a consequence of the probability of surviving as a function of age, the expected reproductive success of different age classes, and the sorts of trade-offs that exist between investments in different slices of the pie of life.

Alternative models have been developed that can yield different predictions. For example, Conrad Istock modeled the evolution of complex life cycles, which include larval and adult life stages separated by metamorphosis. In this context, higher mortality during the larval versus adult life stage is predicted to select for more rapid rather than slower larval development because metamorphosis can mean being able to make a transition to a less dangerous environment.

3. OTHER ASPECTS OF LIFE HISTORY EVOLUTION

Many features of life history evolution have emerged as specialized subdisciplines, or even disciplines in their own right. Here I offer a sampling of these spin-offs.

Iteroparity versus Semelparity

One prominent feature of the life histories of diverse organisms is whether they reproduce once, then die (*semelparity*) or are capable of reproducing multiple times (*iteroparity*). Pacific salmon (genus *Oncorhynchus*) are famous among fish for their single, suicidal investment in reproduction. Likewise, agaves are desert plants that may grow for decades before flowering once, then dying. The difference between the alternatives of single versus multiple reproductive events becomes more interesting when these alternatives are expressed in close relatives. Some populations of Atlantic salmon (*Salmo salar*) and American shad (*Alosa sapidissima*) are semelparous, while other populations have many individuals that will return to the sea after breeding, then come back to freshwater rivers to breed a second time. Some closely related species of plants, such as in the genus *Echium*, native to North Africa, Europe, Madeira, and the Canary Islands, can be either semelparous or iteroparous. The existence of such diversity among populations within a species or between closely related species tells us that semelparity is a life history that can evolve from iteroparous ancestors. It thus poses the question, "Why does semelparity evolve"? One simple answer, derived from theory, is that the evolution of semelparity is driven by the cost of producing the first offspring. Envision a salmon that must swim hundreds of miles upstream to reproduce. The cost of producing the first egg is swimming this great distance, facing barriers like waterfalls and rapids, surviving predators like bears, competing for a nesting territory, preparing the nest, then laying the first egg. All subsequent eggs are cheap by comparison and a far better investment, even to the point of death, than facing the risks associated with return to the sea, then once again facing the start-up costs of reproduction.

Offspring Number versus Size

Demographic life history theory generally considers how evolution shapes the size of the slice of the pie of life devoted to reproduction. It often does not address whether that slice is invested in many offspring, each receiving few resources, or few offspring, each receiving abundant resources. A separate body of theory has been developed that deals with the division of reproductive allocation among individual offspring. One benchmark paper in this literature was published by Christopher Smith and Stephen Fretwell. They predicted the optimal number and size of offspring for a parent to produce, given some simple assumptions. First, they assumed that the fitness of the parent increases with the number of offspring produced. Second, they assumed that the probability of survival of the offspring increases with its size at birth. It then becomes possible to define the fitness of the parent as the number of surviving offspring, which is in turn defined by some optimal combination of offspring number versus offspring size. Farrah Bashey found that guppies that live in rivers with abundant predators produce many small offspring, while those that live in headwater streams produce few but large offspring. In rivers with predators, guppies are scarce and food is abundant. In rivers without predators, guppies are abundant and food is scarce. She found that the fitness of small and large offspring is the same when food is abundant, but that large offspring have a substantial fitness advantage when food is scarce. This combination of results suggests that guppies gain by making more and smaller offspring in streams with predators because there is no sacrifice in being small at birth; however, when predators are absent and food is scarce, the advantage of making larger babies offsets the cost of making fewer.

Life Span and Senescence

A related body of theory deals with the evolution of life span and senescence. If there are trade-offs among various components of the life history, then it follows that maturing at an early age and investing heavily in reproduction can evolve only if there is some reduction in the investment in growth, storage, or maintenance. Any such reduction can be expected to take away from future reproduction or survival. George Williams captured this anticipated trade-off by proposing his *antagonistic pleiotropy* model for the evolution of senescence. He proposed that genes that increase reproductive investment early in life would at the same time cause the evolution of an earlier onset of senescence. Senescence is not defined as life span. It is instead defined as an age-specific deterioration in function, such as the ability to reproduce or the number of offspring produced. It is also defined as an acceleration in death rate with age. A consequence of earlier or more rapid senescence is a shorter life span.

It is this extension of life history theory to a consideration of senescence that shows how life history evolution can lead to seemingly counterintuitive responses. Darwin's original concept of evolution by natural selection was that it acted primarily by selecting for the evolution of increased life span, yet here we argue that it can actually cause the evolution of decreased life span. This can happen because the benefit of early maturity and high reproductive investment early in life more than offsets the cost of reduced life span, at least in some circumstances. One of those circumstances is when an organism has a high risk of mortality imposed on it from the outside, such as a high risk of predation. If the chances of living long are small in spite of the amount invested in reproduction early in life, then the cost of high early investment is small, because most individuals will not survive long enough to realize that cost, regardless of what they invest in reproduction.

4. WHAT HAVE WE LEARNED?

Empirical research on the evolution of life histories has progressed in diverse directions. Here I will describe two of them. One productive line of research involves compiling descriptions of the life histories of many species of organisms into a single analysis that probes for statistical relationships among components of the life history, or between the life history and features of the environment. These analyses are capitalizing on the vast number of life history descriptions that have been generated over the past few decades. One pattern that has emerged in many studies is that the life histories of groups of organisms, like mammals or birds, array along what some refer to as a fast-slow continuum. At one end of the gradient are species that mature at an early age, devote abundant resources to reproduction, produce many offspring, and are short-lived. At the other end are organisms that are old at maturity, devote less to reproduction, often by producing few, well-provisioned offspring, and are longer lived. These studies show that the life history of any given species is not a random aggregation of life history components. Life histories instead evolve in an organized fashion, meaning that the way any one feature of the life history evolves is well correlated with the way the other components of the life history evolve. These correlations among different features of the life history are often consistent with the idea of trade-offs.

In one recent study, Peron et al. (2010) analyzed the association between senescence and the early life history. They compiled data from the published literature on 81 free-ranging populations of 72 species of birds and mammals. They evaluated the association between events early in the life history and the age at onset of senescence, defined as the age when an acceleration in mortality rate occurred. They found that the early life history predicted two-thirds of the variation in the age of onset of senescence and, specifically, that higher juvenile mortality, an earlier age at first reproduction, and the production of more offspring early in life combined to predict an earlier onset of senescence. This pattern of association was predicted by Williams and is consistent with antagonistic pleiotropy, or a more general trade-off between investment in reproduction early in life and the future potential to reproduce.

A second empirical approach has been the experimental study of life history evolution, either in the laboratory or in nature (see chapter III.6). This approach can bring us closer to the study of the cause and effect relationship between some feature of the environment that causes life histories to evolve and the way life histories evolve in response. It therefore offers a more direct way of testing and evaluating life history theory. A pioneering effort in experimental evolution was performed by Michael Rose and Brian Charlesworth on *Drosophila melanogaster*. They selected for reproductive success either early or late in life in replicate laboratory populations, by propagating successive generations of some populations from flies that had been the offspring only of young parents, and other populations from the offspring only of old parents. After 20 generations of selection, the early lines laid more eggs early in life, and the late lines laid more eggs late in life. The most interesting outcome was that the flies from the early lines had a higher rate of egg production early in their lives, but also had shorter life spans than flies from the late lines. These responses confirm the life history theory that predicts how life histories will evolve in response to differences in age-specific reproductive success. They show that selection on reproductive success can cause either a decrease or an increase in life span. Many other such experiments on the evolution of senescence in *Drosophila* have supported the predictions of life history theory.

I and my colleagues have pursued similar goals in our study of life history evolution in natural populations of guppies. The rivers in the northern range mountains of Trinidad offer a natural laboratory for studying life history evolution in action because they flow over steep gradients punctuated by waterfalls that separate fish communities. Species diversity decreases as waterfalls block the upstream dispersal of some species. The succession of communities is repeated in many parallel drainages, providing us with natural replicates. In the downstream localities, guppies live with a diversity of predators that prey on adult size classes of guppies (high-predation communities). Waterfalls often exclude predators but not guppies, so guppies found above waterfalls have greatly reduced risks of predation and increased life expectancy (low-predation communities). The only other fish found in these localities rarely preys on guppies, and when it does, preys on small, immature size classes.

We used mark-recapture methods to demonstrate that guppies from high-predation (HP) environments sustain much higher mortality rates than those from low-predation (LP) environments. We also found that, as predicted by life history theory, HP guppies mature at an earlier age, devote more resources to reproduction, produce more offspring per litter, and produce significantly smaller offspring than LP guppies. Laboratory studies confirm that all these life history differences between HP and LP guppies have a genetic basis. Genetic analyses imply that HP guppies invade guppy-free environments and evolve into LP phenotypes, and that some rivers represent independent replicates of this process.

Rivers can be treated like giant test tubes, since fish can be introduced into portions of stream bracketed by waterfalls, creating in situ experiments. We transplanted guppies from a high-predation environment below a barrier waterfall to a previously guppy-free environment above a waterfall, thus reducing their risk of mortality. As predicted by theory, the descendants of these fish evolved delayed maturation and reduced reproductive allocation, with some changes happening in four years or less. We also experimentally increased the risk of mortality by transplanting predators over a barrier waterfall that previously excluded predators but not guppies. These guppies evolved earlier ages at maturity within five years. Our results thus successfully test some predictions from life history theory in a natural setting. One by-product of this work is that it also reveals the potential rate of evolution by natural selection in nature. It can be remarkably fast.

5. FUTURE RESEARCH

Science moves like waves that gather energy, crest, then dissipate as they reach the shore. The cresting of a wave of science may mean that the important questions have been answered, but it may also mean that the easy questions have been answered, leaving behind the hard ones that few want to face, or perhaps that other topics have captured the imagination of the scientific community. The study of life history evolution seems to have crested, yet there remain important questions to answer. The results to date also suggest important new avenues of research. I will discuss one example of each of these venues for future inquiry.

The Elusiveness of Trade-Offs and Condition-Dependent Fitness

Trade-offs play a central role in life history theory and are supported by broad comparative studies and selection experiments; however, some studies appear to contradict the existence of trade-offs. One curious result that defies the expectations of antagonistic pleiotropy was reported for natural populations of water fleas (*Daphnia pulex*). Some genetic lineages grew faster and to larger body sizes, began to produce eggs at an earlier age, and produced eggs at a higher rate throughout their lives than other lineages in the same population. This seems like an impossible combination of attributes occurring as natural variation within a population. The rapidly developing lineages are superior in every way and should displace the other genotypes, yet the slowly developing genotypes persist. Later work suggested that the superiority of the rapidly developing lines is expressed only when food availability is abundant. When food is scarce, such genotypes lose out to those that are smaller, develop more slowly, and produce fewer eggs. These results tell us that our perception of trade-offs can depend on the environment we choose for assessing them. They also tell us that life history evolution represents adaptation to multiple dimensions of the environment. Risk of mortality is important, but so is resource availability.

Interactions between Ecology and Evolution

The rapidity with which life histories evolve raises another question about the factors that shape life history evolution and, more generally, about how ecological and evolutionary processes might relate to one another. The traditional perspective of the relationship between ecology and evolution is that ecological processes happen so much more quickly than evolution that we can treat organisms as constants (meaning that they do not evolve) when modeling ecological processes or studying ecological interactions in the real world. The traditional perspective is also that ecology shapes evolution. Field selection experiments with guppies tell us that evolution happens in a time frame comparable to that of ecology, so perhaps it is better to think of ecology and evolution as processes that interact in real time and reciprocally shape each other. In the case of guppies, the presence and absence of predators is associated with changes in guppy populations and features of the ecosystem that suggest that such interactions are important. When predators are absent, guppies become abundant and certain types of resources, like the invertebrates that guppies feed on, become rare. The individual growth rates of guppies are slower, and in other ways it appears that they have less to eat. Some features of guppy adaptations to low-predation environments suggest that they are also adapting to the way they have changed the environment in the absence of predators. This potential interaction between ecology and evolution represents our current, unfinished research.

There is good cause to think that what we are studying in guppies might be applicable to a diversity of organisms and in a diversity of climates. The study of ecology is replete with examples of organisms that have a major impact on the structure of their ecosystem. Every time we see such examples, we should think about how such impacts cause natural selection and evolution, both of the organism in question and of all others that occupy the ecosystem. A recent perspective by Tom Schoener (2011) offers a good summary of progress in the empirical study of such interactions. A second, by James Estes and others (2011), presents one context, the recent elimination of apex consumers in many ecosystems, that dramatically illustrates how individual species can restructure their ecosystems and hence potentially change the kind of selection that they, and many other members of the ecosystem, experience (see chapter VI.16).

Life history evolution has found practical applications. One is in fisheries biology. When humans harvest fish, they function as predators of the largest-size classes of prey. The field of fisheries management has focused on the population biology of exploited fish populations (meaning their abundance and age structure), but has not traditionally considered whether humans are causing these populations to evolve. A likely response to such predation is the evolution of earlier maturity and smaller body size. Smaller sizes in turn mean having less to harvest as food. We are now seeing such changes in some fish species used as sources of food. The mind-set had originally been that the resources of the ocean are too vast to deplete. As it became obvious that they could be depleted, the perspective of fisheries managers shifted to thinking that they could be actively managed. If they were overharvested, then harvesting could be reduced, with the expectation that exploited populations would rebound to their former state. If, on the other hand, exploited populations of fish have evolved and if their ecosystems have changed in response to their changed abundance and life histories, there may be no easy return to what they were in the past. Fisheries managers are only beginning to integrate evolution and its implications into their thinking about managing the exploitation of natural populations.

More generally, as the study of life history evolution has matured, it has become more integrated into other aspects of the study of ecology and evolution.

FURTHER READING

Estes, J. A., J. Terborgh, et al. 2011. Trophic downgrading of Planet Earth. Science 333(6040): 301–306.

MacArthur, R. H., and E. O. Wilson. 1967. The Theory of Island Biogeography. Princeton, NJ: Princeton University Press.

Pianka, E. R. 1970. On r- and K-selection. American Naturalist 104: 592–597.

Peron, G., O. Giminez, A. Charmantier, J. M. Gaillard, and P. A. Crochet. 2010. Age at the onset of senesence in birds and mammals is predicted by early-life performance. Proceedings of the Royal Society B 277: 2849–2856.

Reznick, D. 2011. Guppies and the empirical study of adaptation. In J. B. Losos, ed., In the Light of Evolution. Greenwood Village, CO: Roberts & Company, 205–232.

Roff, D. A. 1992. The Evolution of Life Histories. New York: Chapman & Hall.

Schoener, T. W. 2011. The newest synthesis: Understanding the interplay of evolutionary and ecological dynamics. Science 331(6016): 426–429.

Stearns, S. C. 1992. The Evolution of Life Histories. Oxford: Oxford University Press.

III.12

Evolution of Form and Function

Peter C. Wainwright

OUTLINE

1. Form and function in organismal design
2. Measuring the evolution of form and function
3. Key features of life's functional systems: Multifunctionality, genes, and complexity
4. General principles of the evolution of complex functional systems

GLOSSARY

Adaptation. A process of genetic change in a population whereby, as a result of natural selection, the average state of a character improves with reference to a specific function, or whereby a population is thought to have become better suited to some feature of its environment. Also, *an* adaptation: a feature that has become prevalent in a population because of a selective advantage conveyed by that feature in the improvement of some function. In this chapter the term is used mostly in the latter sense—as a noun describing a trait that has evolved through this process and helps make the individuals in the population better suited to their habitat.

Fitness. The ability of an individual to survive and reproduce. It can be measured conceptually as the contribution to the gene pool of the next generation.

Functional Morphology. The study of the relationship between form and function in organisms.

Morphospace. Any axis or set of axes that describes the parameters required to illustrate the form of organisms, such as dimensions of parts of the body. By plotting the positions of living forms into a theoretical morphospace it is possible to determine what forms are especially common and which do not exist.

Natural Selection. The process by which organismal traits become more or less common in the population as a function of differential survival and reproduction of the individuals that vary in these traits.

Performance. The ability of an individual to execute tasks important to its daily life. Performance is hierarchical, with fitness being the most integrative, an inclusive measure of performance that sums across many underlying performance traits. Underlying fitness may be running speed, maneuverability, and the ability to avoid detection by not moving (among other performance traits). Underlying running speed are a number of more proximate performance traits, such as the capacity of an individual muscle to generate power, the length of the stride, and the time required for the muscle to relax prior to its next contraction.

1. FORM AND FUNCTION IN ORGANISMAL DESIGN

Form and function are inextricably linked in living creatures because patterns of natural selection reflect the impact of alternative morphologies on the ability of the organism to perform the tasks determining its survival and reproduction. The details of construction of a lizard's limbs and body will determine how fast it can run and how effectively it can evade specific predators. To the extent that limb dimensions affect maneuverability and sprint speed, they can be expected to evolve if a new sort of predator comes on the scene, favoring a different escape strategy; thus, the primary reason that lizard limb dimensions evolve is that they underlie locomotor abilities, which in turn shape the survival and reproductive success of individual lizards. Form evolves mainly because it shapes performance and hence fitness; therefore, a key to understanding how form shapes fitness is to understand how it determines function.

What aspects of form are most often studied? With mobile animals, most of the body is typically concerned with locomotion, feeding, or reproduction, and these systems are often similar in that they involve muscles and linkages of skeletal elements. The mechanical properties of these systems are shaped to a large degree by the sizes of muscles, the leverage of muscles acting across joints, and the mechanical properties of the muscles and skeletal elements themselves. In vertebrate animals, the skeleton is made mostly of bone; in arthropods the skeleton is made of chitin that is variably mineralized, and in a number of animals there is no skeleton per se, but stiff, contracted muscles often playing a similar role. Approaches to analyzing feeding and locomotor systems are very similar, since the task is to work out the mechanisms by which force and movement are transmitted from the muscles driving the behavior through the system and onto the environment. But not all animal performance is about movement; many other performance traits have a morphological basis. Some examples are camouflage, respiration, attractiveness to potential mates, acuity of vision, and the ability to detect sounds or chemicals. The relevant structures may be microscopic, but sensory systems normally have a performance basis in the size and shape of their component parts.

2. MEASURING THE EVOLUTION OF FORM AND FUNCTION

Studies of the evolution of form and function come mostly in two flavors: population-level analyses in which natural selection is measured, and comparative analyses across species, wherein a longer-term view of the evolutionary process is gained.

Studies of Natural Selection Acting on Variation within a Population

Studies of this type often involve some combination of two approaches in which the impacts of form on performance and on fitness are separately estimated. The strength and form of natural selection can be measured on size or shape of structures, using the standard methods of studying natural selection in natural populations (see chapters III.3 and III.4). The typical strategy in this type of study is to capture a large number of individuals in a population, measure the traits of interest in each individual, and mark the individuals with an identifying tag before returning them to their habitat. After sufficient time for some mortality or growth to occur, individuals are recaptured and their identities recorded. The starting form of survivors is compared to the entire original sample to determine whether form affected (or is correlated with) the probability of survival. In a handful of studies, researchers have measured not only form but also performance traits after initially capturing individuals, permitting deeper understanding of the ways in which selection is acting on performance and its underlying traits. In most cases, however, researchers measure selection acting on form without knowing the relationship between form and performance. In these cases they either assume they know how form and function are related, or they simply are not interested.

Understanding the functional significance of form deepens our understanding of adaptation. A standard approach here is to model the functional system, such as the wing or bill of a bird, then to parameterize the model with key measurements from animals to estimate functional properties expected to directly impact performance. Here, measurements of individual size and shape are used to estimate a functional property. A common example with muscle-skeleton systems is to measure the mechanical advantage of a muscle acting across a joint with an input lever and an output lever. This might be done in a jaw by measuring the distance between the attachment of the jaw muscle on the jaw to the jaw joint as the input lever, and the distance from jaw joint to the location on the teeth where the food is held as the output lever. The ratio of the input lever to the output lever gives the mechanical advantage—or the proportion of an input force (produced by the jaw muscle) transmitted through the lever to the food item. Mechanical advantage of levers reflects a trade-off between transmission of force and transmission of movement (usually the movement of a contracting muscle). A lever with a low mechanical advantage transmits less force, but more motion, than a lever with a high mechanical advantage. In fact, there is a one-to-one exchange of force transmission and movement transmission in a lever, so that mechanical advantage modified during evolution to enhance force transmission will transmit less movement. This model is based on principles from physics. Levers are central to mechanical engineering, but other functional systems might be more appropriately modeled by using chemical engineering, optics, or electrical engineering. Whether the system is a lever or some other functional device, the modeling approach is especially good for comparative studies because it allows one to efficiently compare the functional implications of variation in form. The use of models has also been very useful in studies of large numbers of species to relate form and function to habitat use and feeding habits of animals.

Comparative Analyses

The modeling approach is a powerful way to study diversity across species. In this arena an important additional tool is an estimate of the evolutionary relationships among the species being studied, or a *phylogeny* (see chapter II.1). The phylogeny provides a road map to the sequence and pattern of evolutionary modification of functional systems and can be used to identify the sequence of assembly of complex adaptations, as well as patterns of association between form and aspects of the species' ecological niches, such as habitat and feeding habits. By focusing on variation in parameters from functional models, one can draw a tighter, more causal link between form and these ecological patterns.

In a comparative analysis, one infers a past process by evaluating the outcomes of evolution. By studying numerous evolutionarily independent transitions to the same habitat or feeding specialization, one can develop an understanding of whether the modifications to form accompanying the transition always occur in the same way, or alternatively whether multiple solutions to the switch have occurred during evolution. The combination of species values of traits and the phylogeny allows an estimate of sequential changes that have occurred. This is particularly useful if one is interested in how a complex novelty came about during evolution.

One of the classic questions in the evolution of organismal designs concerns whether complex functional systems are rapidly assembled during evolution or are assembled piecemeal over a long period of time. One example of this sort of analysis is the origin of powered flight in birds. Did all the underlying innovations for powered flight occur at one time in the ancestor of birds? The phylogeny of birds and the closely related theropod dinosaurs from which they arose reveals a fascinating story. Both bipedalism (walking and running on two feet) and feathers clearly evolved long before the origin of powered flight. In fact, the surprisingly ancient history of feathers within theropods indicates that feathers, an integral and crucial element of powered flight in birds, most likely evolved for their insulating value and were only incorporated into the suite of adaptations for flight after many millions of years of keeping theropods warm. The evolution of powered flight involved the modification of feathers, both as key structures in the wings and to help produce a smooth body surface facilitating efficient airflow. In addition to modifications of feathers, the muscles and skeleton of the forelimb were modified considerably into a structure that supports the flight feathers and moves powerfully in the pattern used during flight. This is one of many examples in which the phylogenetic context of complex adaptations provides surprising and interesting insights into their evolutionary history.

Perhaps not surprisingly, the dominant pattern that has emerged in studies of the evolution of complex innovations is that key components of the system evolve earlier and are modified for a new function as the system is gradually assembled. The importance of this common pattern is that it underscores the dependence of evolution on available building blocks. Some major, complex innovations have huge effects on the subsequent evolution of the lineages that possess them. Innovations such as powered flight, the origin of endothermy, and air breathing in fishes are major breakthroughs that opened up whole new ways of life. Some major innovations are followed by a burst of diversification as organisms radiated out into the new niches made possible by the innovation (see chapter VI.10). Interestingly, it is much more common for even the best innovations to show little impact on the success of the group for a prolonged period of time, and in some cases there is never a spectacular burst of diversification. This last point is illustrated by a major innovation in the feeding system of fishes called *pharyngognathy*, in which bones and muscles associated with the gill arches are modified into a functionally versatile second set of jaws. All bony fish have this second set of jaws, but in six lineages of ray-finned fishes this system has been independently modified in a similar way, making the jaws powerful and very versatile for processing food items. This condition, pharyngognathy, evolved independently in cichlids and in labrids (wrasses and parrot fish), two of the most successful and ecologically diverse groups of fishes in tropical freshwater systems and coral reefs, respectively. In these groups, pharyngognathy seems an important innovation that contributed to exceptional diversification; however, pharyngognathy also evolved independently in damselfishes, halfbeaks, surfperches, and false scorpion fishes, but these groups have thus far failed to diversify ecologically to any notable degree. The point is that innovations, in and of themselves, do not guarantee spectacular diversification. Lineages possessing valuable innovations must also find themselves in appropriate ecological circumstances that promote realization of the potential provided by the innovation.

3. KEY FEATURES OF LIFE'S FUNCTIONAL SYSTEMS: MULTIFUNCTIONALITY, GENES, AND COMPLEXITY

So, on the surface of the problem it is fairly straightforward to evaluate the relationship between form and function and the adaptive value of size and shape of organisms and their parts—by determining which designs confer higher performance, and which have higher fitness. But several factors, other than the direct linkages between form, performance, and fitness, characterize biological systems and make the study of the evolution of form and function especially interesting.

The first is that functional systems in organisms do not occur in isolation. Most structures participate in multiple performance traits that all compete for the shape and properties of the structure. You might think the wings of an insect would be shaped solely for their role in powering flight, a physically demanding function, but insect wings can also be used to absorb warmth that radiates from the sun and to provide camouflage when the animal is at rest. Ultimately, the wing size and shape evolve that maximize fitness, or the integration of all the underlying performance traits that directly or indirectly determine reproductive output. Because the ideal wing design may differ for different performance traits, wing form reflects trade-offs that balance these different demands. In some cases other functions of insect wings have become so important that the animal has lost the ability to fly. An example is an eastern North American katydid, *Pterophylla camellifolia*, in which the forewings of the male are inflated to form a resonating chamber used in sound production, and the animal cannot fly.

Trade-offs come about when a single structural trait contributes to two or more performance traits. They are thought to be one of the dominant factors explaining the diversity of organismal design because of the inherent constraints they place on maximizing performance. The fact that most parts of an organism participate in more than one performance trait is a major reason that systems can be honed by natural selection but not reach the highest possible performance in any one functional system. It is often possible to evaluate the adequacy of design in a particular system for a particular function, but it may be difficult to identify all the performance trade-offs the organism faces in building these structures.

A second factor is that organismal form is a consequence of genetic programs that normally allow single gene products to contribute to the formation of many structures (i.e., the genes are *pleiotropic*). As a result, we can expect it to be difficult to modify genes to effect specific morphological changes without causing other changes. These genetic correlations can limit adaptation, at least in the short term, because the unintended, correlated changes to form may have a negative consequence for other performance traits. This constraint is alleviated to a considerable degree by trait-specific differences in gene regulation in different parts of the body, but the potential for manifold impacts of changes to widely used genes is a major factor in the evolution of developmental pathways.

A third factor is complexity itself. Perhaps no other intrinsic feature of living functional systems has so much impact on the dynamics of their evolution as the fact that they are inherently complex. Even at its most basic levels, complexity impacts the evolution of functional systems. Potential diversity is fundamentally a function of the number of parameters required to describe a form, or its degree of complexity. In general, the story of the ways in which intrinsic properties of functional systems influence their diversification is a story about the many implications of complexity.

4. GENERAL PRINCIPLES OF THE EVOLUTION OF COMPLEX FUNCTIONAL SYSTEMS

The nature of the relationship between form and function plays a prominent role in shaping evolutionary dynamics of functional systems, and a number of processes have been identified that highlight this interplay. It is not simply the adaptive value of traits that govern their evolution; the way in which they impact function is also important. Let us consider, for example, how trade-offs can change or even be abolished during evolution.

Decoupling Trade-Offs

We observed earlier that structures in organisms normally participate in multiple performance traits and that these competing demands can lead to trade-offs limiting diversification. One key to overcoming this sort of constraint is to break the linkage between the two performance traits. If the structure in question must no longer serve both performance traits, the constraint would be released, perhaps allowing modifications of the structure to underlie the evolution of an enhanced ability in its function. There are many examples of this sort of decoupling of performance traits during evolution. One involves the feeding system of fishes. Primitively, fishes (like most other vertebrates) capture and process prey with their oral jaws. This is the condition seen in cartilaginous fishes and a few lineages of bony fishes, such as sturgeon and lungfish. But prey capture in an aquatic environment, without the benefit of appendages, is dominated by the highly dense and viscous nature of water. As a result, the vast majority of fishes use suction feeding to capture prey. In suction feeding, the mouth is rapidly expanded, drawing in water and prey. Effective suction feeding on quick elusive prey is enhanced by light bones and levers in the jaw that favor the transmission of displacement over force. But once caught, prey is processed by biting; in particular, if the prey has a tough outer shell, robust jaws are needed with levers favoring force transmission over movement. So when the oral jaws are used for both prey capture and processing, trade-offs can limit adaptation in both functions and overall diversification of jaw design.

This trade-off was broken with the origin of teleost fishes, when a second set of jaws evolved, the pharyngeal jaws. These new jaws are located at the back of the oral cavity just in front of the esophagus; following their origin, they took over the role of prey processing, releasing this function from the oral jaws and permitting much more extreme oral specializations for prey capture. The introduction of a new set of jaws that came to specialize in prey processing released a major constraint on the oral jaws.

This is an example of what is called a *functional duplication*: when a novelty arises that can perform a function of an existing structure. The general implication is that performance of the function by the new system releases constraints on the form of the original system. The ultimate consequences of duplication for the original system may be to enhance other performance traits, or even to permit breakthroughs in design that had been prevented by the need to perform both functions. Functional duplication has the potential to increase the overall performance of the organism and is a common evolutionary phenomenon operating at many levels of biological design. For example, gene duplication is thought to be one of the most common mechanisms of gene diversification and is the focus of a huge amount of research. The principles operating in gene duplication are effectively the same as those operating at the level of the organism.

One way two performance traits can become decoupled is for one to be taken up by a second system, through evolutionary modifications. A second way is when the original structural system is duplicated, and a descendant copy is later modified to specialize in a different performance trait. Exactly this history has been rather common among certain body parts and groups of organisms. Decapod crustaceans have body plans based on repeated body segments with homologous parts. Appendages and their function vary from one part of the body to another, and species can differ considerably in the degree of diversity in appendage function along the body. Some of the diversity includes use of appendages as thrust devices in swimming, legs for walking, claws for defense and prey processing, and mouthparts for chewing. A less well-appreciated phenomenon occurs in muscular systems, when muscles become subdivided into two descendant muscles. Typically the two new muscles attach to the same structures as the original muscle, but over the course of time, one may migrate and develop novel attachments and novel functions. This phenomenon is important in the evolutionary history of tetraodontiform fishes—puffer fishes, triggerfishes, and their relatives. Primitively, these fish have two muscles that attach to the jaw and provide the power behind the bite. These muscles have been subdivided numerous times during evolution so that some lineages have five, eight, or even ten jaw muscles. Across tetraodontiforms several examples can be found in which the muscle no longer attaches on the jaw and has evolved an entirely novel function in compressing the oral cavity.

To summarize this section: alterations of form occur by which performance traits are decoupled and constraints on morphology alleviated. This is one major way in which complex organismal systems evolve.

Many-to-One Mapping of Form to Function

Another mechanism exists by which complexity enhances the evolutionary potential of organisms. When performance and functional traits have a complex underlying basis, almost always, many combinations of those underlying parts will give any particular value of the functional trait. As an example, consider the oral jaw biting system of a vertebrate in which the strength of the bite is a function of just two traits: the force-producing capacity of the jaw muscle multiplied by the mechanical advantage of the lever through which it acts on the jaw. Many combinations of muscle force and mechanical advantage will give any particular value of bite strength. This *many-to-one mapping* of form to function is inherent in any performance trait determined by multiple traits.

Many-to-one mapping of form to function has far-reaching implications for the evolution of functional systems. In general, this phenomenon acts to soften the impact of trade-offs, because there is almost always more than one way to modify an existing system to create the functional properties favored by natural selection. There are at least three important evolutionary consequences of many-to-one mapping of form to function. First, and most important, the capacity for multiple forms to have the same functional property can permit the optimization of two or more functional properties shaped by the same structures, even when changes in some components result in a functional trade-off. This highly nonintuitive result comes about because of the potential for alternative values of underlying parameters to produce a single value of one function, while permitting change in the second function. In essence, changes to dimensions of structures may be neutral with respect to change for one function, while producing a potentially adaptive change in a second function. In cases where this phenomenon is looked for, it has virtually always been found. The flexibility conferred by many-to-one mapping on design releases constraints on evolution by providing pathways through morphospace with little or no effect on key functional properties of the organism.

The second consequence of many-to-one mapping is that lack of a one-to-one match of form to functional properties means that there is the potential that morphological and functional diversity will be decoupled. Imagine a group of 12 species of finches that feed on prey ranging from insects to seeds of various sizes and hardness. If we measure the dimensions of the bill and its levers as well as the jaw muscle used to close the bill, we can calculate the diversity among the species in these traits. One common way to measure diversity among species in trait values is to calculate the variance among species in the trait. Total diversity of several traits can be obtained by summing the variances of several traits. We can also measure diversity among species in the capacity to exert a forceful bite. Because many combinations of bill-lever mechanical advantage and strength of jaw muscle can produce a given value of bite force, it is very likely that diversity among species in the morphology will be greater than diversity in bite force. The potentially tight relationship between morphological and functional diversity is weakened by many-to-one mapping. Conformation at the lower level precisely determines functional properties at the level above, but as we have seen, the reverse is not true when many forms have the same functional property. Lack of correlation between diversity among species in form and diversity of functional properties is surprisingly common, and just one of the reasons that knowledge of how an organism works is necessary before the meaning of variation in structures can be interpreted. Some variation is neutral with respect to function, but these emergent properties of complexity make it possible for decoupled diversity even when the form directly underlies the function.

The third consequence is that many-to-one mapping results in a strong signal of evolutionary history in the design of organisms. If only a single combination of muscle force and lever mechanical advantage produces a given value of biting force, then when natural selection favors jaws with this biting strength, it will always produce the same form. But strict convergent evolution (the independent evolution of the same form in response to the same selective pressure in two lineages) is actually surprisingly uncommon. Different lineages typically start at different places in morphospace; when natural selection favors a specific functional property in different lineages, it is more likely that the ancestral form will simply move to the closest form satisfying the function, rather than always converging on the same combination of traits. Natural selection will produce a combination of traits giving the optimal functional property, but not a particular one rather than other possible combinations, unless they have negative consequences for other functional properties. What this means is that the form produced during a bout of evolution in response to a specific selective force will depend in part on the starting form in the population when the response began. In other words, there will be a strong phylogenetic signal during the evolution of morphological systems.

FURTHER READING

Hazen, R. M., P. L. Griffin, J. M. Carothers, and J. W. Szostak. 2007. Functional information and the emergence of biocomplexity. Proceedings of the National Academy of Sciences USA 104: 8574–8581.

Hughes, A., and R. Friedman. 2005. Gene duplication and the properties of biological networks. Journal of Molecular Evolution 61: 758–764.

Koehl, M.A.R. 1996. When does morphology matter? Annual Review of Ecology and Systematics 27: 501–542.

McShea, D. W. 1996. Perspective: Metazoan complexity and evolution: Is there a trend? Evolution 50: 477–492.

Wainwright, P. C. 2007. Functional versus morphological diversity in macroevolution. Annual Review of Ecology and Systematics 38: 381–401.

Walker, J. A. 2007. A general model of functional constraints on phenotypic evolution. American Naturalist 170: 681–689.

III.13

Biochemical and Physiological Adaptations

Michael J. Angilletta Jr.

OUTLINE

1. Physiological diversity
2. How do we know that physiological variation is adaptive?
3. Biochemical mechanisms inform models of physiological adaptation
4. Adaptive variation in tolerance
5. Adaptive variation in regulation
6. Adaptive acclimation
7. Constraints on physiological adaptation
8. Implications for global change biology

Organisms can thrive in diverse environments by evolving biochemical processes to tolerate extreme conditions or to avoid these extremes by regulating internal conditions. Both tolerance and regulation impose costs, often expressed as trade-offs with other traits that affect fitness. By analyzing these trade-offs, one can predict how natural selection will shape physiological strategies in particular environments. Our understanding of physiological adaptation has been tested through comparative analyses of populations along environmental clines and experimental evolution of populations in the laboratory. These approaches have often led to surprising insights, suggesting that we can better understand physiological adaptation by also considering processes such as mutation, drift, and migration. Although biologists still have much to learn about physiological adaptation, our current knowledge has already helped to predict biological impacts of global change.

GLOSSARY

Abiotic Factor. A variable that describes a physical (nonliving) characteristic of the environment, such as temperature, humidity, or pH.

Extreme Environment. An environment in which some abiotic factor approaches a value that limits an organism's survival or reproduction.

Generalist. An organism that tolerates a wide range of environmental conditions.

Heritability. The proportion of phenotypic variation that results from genetic variation among individuals.

Optimality Model. A mathematical model that defines the relationship between a phenotype and fitness in a specified environment. This relationship can be used to find the phenotype that maximizes fitness (the optimal phenotype).

Osmolyte. A soluble compound affecting the osmosis of a cell.

Regulation. The act of maintaining an internal state that differs from the state expected if the organism were to exchange energy or materials passively with its environment.

Specialist. An organism that tolerates only a narrow range of environmental conditions.

Tolerance. The ability to survive and reproduce in a given environment.

Trade-off. A decrease in the quality of one trait stemming from an increase in the quality of another trait.

1. PHYSIOLOGICAL DIVERSITY

Earth provides a home for millions of species, some residing on its surface while others dwell within dark crevices, loose soils, or deep oceans. Each of these species represents a unique way of accomplishing a single goal: persistence. In the process of achieving this goal, organisms must do more than just survive and reproduce; they must also forage, grow, and develop until they have acquired sufficient size, experience, and resources to leave offspring. These functions rely on a suite of biochemical processes that are common to all organisms, as well as specialized processes that have evolved within certain types of organisms (e.g., aerobic respiration, photosynthetic nutrition, neuromuscular communication). The biochemistry of life depends on abiotic factors such as temperature, pressure, and pH. Yet life occurs in just about every environment on our planet, spanning an amazingly broad range of conditions.

Although life occurs everywhere, no species does so. Each species functions under a limited range of conditions, referred to as its *physiological tolerance*. The relationship between a species' tolerance and its distribution is most evident from a global perspective. Tropical waters are known for their great diversity of fish, most of which swim very well at temperatures around 25 °C. Cool a tropical fish by just a few degrees, and its ability to swim also declines; for instance, the zebra fish (*Danio rerio*) cannot swim well at temperature below 15 °C. Yet close to the poles, many species of fish patrol icy waters that rarely exceed 0 °C. Despite their incredibly cold bodies, these fish show no sign of stress. Yet if one were to warm these fish just a few degrees, they would become stressed! Thus, environmental stress is relative; conditions that stress some species enable other species to thrive. Clearly, the different thermal tolerances of tropical and polar fish reflect adaptation to their local environments. Similar patterns of adaptation exist along gradients of humidity, salinity, acidity, toxicity, and pressure.

Species adapt to environmental stress in two ways. As in the case of polar fish, a species can acquire mutations that enable it to tolerate extreme internal conditions (e.g., low temperature). The evolution of physiological tolerance involves changes in the structures or concentrations of proteins, which in turn alter membranes, tissues, and organs. Alternatively, a species can acquire mutations that enable it to regulate its internal conditions within tolerable limits (e.g., thermoregulation). Some of the most effective forms of regulation involve not only physiology but behavior and morphology as well. Depending on the circumstances, natural selection can enhance tolerance, regulation, or both. Although polar fish evolved the ability to tolerate subzero body temperatures, polar mammals evolved the ability to maintain body temperatures that greatly exceed the temperatures of their surroundings. In reality, tolerance and regulation evolve together according to their relative costs and benefits to a species.

This chapter outlines four themes emerging from studies of physiological adaptation. First, optimality models help biologists understand how trade-offs shape physiological diversity (see chapter VII.3). Second, biochemical mechanisms of tolerance and regulation determine the trade-offs during physiological adaptation. Third, physiological variation within and among species often reflects adaptation to local environmental conditions. Finally, observed physiological strategies sometimes differ from optimal physiological strategies because of genetic constraints (see chapter III.8). Although these themes are illustrated through examples, the brevity of this article precludes a detailed treatment of many fascinating cases of physiological adaptation. Interested readers should consult the references that follow this article for additional perspectives on evolutionary physiology.

2. HOW DO WE KNOW THAT PHYSIOLOGICAL VARIATION IS ADAPTIVE?

How do we know that physiological diversity resulted from natural selection rather than some nonadaptive process? To be confident, we must compare the observed patterns with those predicted by theoretical models. Optimality models have been invaluable in this endeavor. Such models tell us the selective pressures on physiological traits given a set of hypothetical constraints, usually referred to as trade-offs. A satisfactory match between a model's predictions and a researcher's observations supports the idea that the physiological variation is adaptive. A substantial mismatch indicates that either the variation is nonadaptive or that our evolutionary model omits some important constraint.

Physiologists have used two approaches to test the predictions of optimality models. The oldest and most widely adopted approach is to compare the traits of species that evolved in different environments (see chapter II.7). In recent decades, comparative analyses have been aided by statistical methods that control for the effects of common descent; since any two species have inherited some phenotypes from a common ancestor, common descent can inflate or mask signals of adaptation. The second approach is to expose experimental populations to controlled environments and then quantify the genetic divergence of traits (see chapter III.6). Both comparative analysis and experimental evolution have advantages and disadvantages. Comparative analysis tells us how physiology has evolved in complex environments, but it cannot disentangle the myriad of factors that covary among environments. Experimental evolution isolates hypothetical selective pressures by manipulating some environmental factors while controlling others; nevertheless, this approach is practical only for studying species with short generations that are easily raised in laboratories. Thus, comparative analysis and experimental evolution are complementary approaches to testing hypotheses about physiological adaptation.

3. BIOCHEMICAL MECHANISMS INFORM MODELS OF PHYSIOLOGICAL ADAPTATION

To model the trade-offs constraining physiological adaptation, we must know the mechanisms by which extreme environments affect biochemical processes. Abiotic conditions can alter the structures of proteins in ways that inhibit chemical reactions. Since a protein's structure depends on weak bonds between amino acids, changes in conditions within cells can disrupt this structure. Yet, modifying the sequence of amino acids that form a protein can improve the protein's function under extreme conditions. As an example, consider the modifications that improve function at extreme temperatures. Amino acids that increase the stability of a protein improve function at high temperatures, whereas those that decrease stability improve function at low temperatures. Thus, biochemical adaptation to one thermal extreme necessarily results in maladaptation to the other. Adaptation to osmotic pressure imposes a similar trade-off. To limit water loss in a hyperosmotic environment, organisms must maintain high concentrations of solutes that are compatible with proteins. In a hypo-osmotic environment, however, high concentrations of solutes would cause cells to swell with water, reducing the concentrations of chemical reactants and potentially causing death.

Can species evolve to tolerate wide ranges of conditions? One way to become a generalist is to produce multiple forms of the same protein, each capable of functioning under different conditions; however, an organism would have to invest the energy required to synthesize additional proteins, which would reduce the energy available for other activities. Another strategy would be to make one type of protein at a time and modify the concentrations of each protein as conditions change. In this case, energy would still be needed to turn over proteins on a regular basis. Given a limited amount of energy, biochemical adaptation to a wide range of conditions would compromise performance under any single condition. In other words, a jack of all environments would be a master of none.

Enhanced regulation imposes an energetic cost that also mediates trade-offs in performance. Regulation of internal conditions involves metabolic processes that rely on the energy stored in chemical bonds, such as the covalent bonds of macromolecules. Mammals and birds regulate their body temperature by catabolizing carbohydrates to generate heat. Many marine fish regulate their osmotic pressure by coupling the catabolism of adenosine triphosphate to the transport of salts across the epithelia of the gills. Terrestrial organisms reduce their rates of water loss by forming a water-resistant cuticle. Each of these forms of regulation involves cellular machinery that requires energy to produce and maintain; therefore, an organism possessing this machinery would gain a physiological advantage in a stressful environment but suffer an energetic disadvantage in a benign environment.

4. ADAPTIVE VARIATION IN TOLERANCE

Consistent with the biochemical mechanisms described above, current optimality models assume that the ability to tolerate one environmental extreme leads to an inability to tolerate another. This hypothetical constraint prevents the evolution of a species that performs extremely well under all conditions. Given this trade-off, two predictions emerge. First, species should evolve to perform best under the conditions that they experience most frequently. Second, species should evolve to perform over the narrowest range of conditions needed to persist in their environment. Thus, constant environments would favor specialists, whereas variable environments would favor generalists (see chapter III.14).

These predictions about physiological adaptation have been tested extensively by comparing species distributed along latitudinal gradients. Since the mean temperature decreases and thermal variation increases from the equator to the poles, we should expect thermal tolerance to vary among species at different latitudes. Consistent with this expectation, tropical species tolerate high temperatures better, but low temperatures worse, than do temperate or polar species. Moreover, in many groups of plants and animals, species from higher latitudes tolerate a wider range of temperatures. Broad geographic patterns of tolerance have also been observed for other abiotic factors, such as moisture, salinity, and pH.

These patterns of environmental tolerance suggest that a species transplanted to a novel environment could not function as well as one that evolved in that environment. In fact, many experiments have been conducted in which (1) organisms were reciprocally transplanted between distinct environments and (2) the performances of native and transplanted individuals were compared. In one of these experiments, Amy Angert and her colleagues moved two species of plants (*Mimulus cardinalis* and *Mimulus lewisii*) that normally occur at different altitudes. Native individuals outperformed transplanted individuals, indicating that adaptation to high altitude resulted in maladaptation to low altitude (and vice versa). Subsequent experiments by Angert and others confirmed that temperature was an important factor. In the laboratory, individuals from high altitude grew better at low temperature than did individuals from low altitude. In a field experiment, genotypes grown at low altitude survived according to their ability to photosynthesize at high temperatures; however, greater photosynthesis came at the expense of cold tolerance, leading to selection against these genotypes at high altitude. Although the biochemical basis of adaptation differs from case to case, the majority of transplant experiments have revealed adaptation of tolerance along abiotic gradients.

Despite the wealth of evidence from comparative analyses, recent insights from experimental evolution have challenged our notions about the adaptation of tolerance. Model organisms—representing species from all kingdoms of life—have been exposed to a multitude of environmental conditions in the laboratory. And in all cases, some degree of physiological adaptation occurred. One of the most widely studied species, *Escherichia coli*, adapts readily to thermal, acidic, and nutritional stresses. Some experimentally evolved populations have been screened for mutations that conferred tolerance. Shaobin Zhong and his colleagues found that adaptation to nutritional stress requires mutations to downregulate proteins that transport the usual source of carbon and upregulate proteins that transport an alternative source. Adaptation to an environment containing only lactulose consistently involved duplication of genes encoding a protein that transports this substrate. By contrast, adaptation to methyl-galactoside involved deletion of a particular region of the genome that, if present, suppresses the expression of a protein that transports methyl-galactoside. Exposure to a mixture of these substrates nearly always caused the evolution of a mixture of specialists, each of which used one of the substrates. This result accords with the common assumption that generalization imposes an energetic cost that should be avoided when possible. Unfortunately, many other cases of experimental evolution conflict with this theoretical view. In most experiments that exposed populations to fluctuating conditions, adaptation led to a population of generalists that could outperform specialists. Once biologists understand the biochemical mechanisms that enable certain genotypes to succeed over a wide range of conditions, they will need to revise current models of optimal tolerance accordingly.

5. ADAPTIVE VARIATION IN REGULATION

When tolerance cannot evolve because of costs or constraints, a species can regulate its internal state to persist in extreme environments. The benefit of physiological regulation depends on an organism's tolerance of environmental conditions. A specialist, which performs well only within a narrow range of conditions, would benefit greatly from regulation. The cost of regulation depends on the time and energy required to maintain an internal state that deviates from the external one. From an optimality perspective, we should expect either a high benefit or a low cost to cause the evolution of effective regulation.

Much evidence of adaptive regulation comes from studies of thermal and hydric states, which often depend on one another. In particular, mammals and birds provide outstanding examples of adaptive regulation in the face of varying costs. In cold environments, these animals rely on metabolic reactions to generate the thermal energy needed to maintain warm bodies (*endothermy*). In hot environments, excess thermal energy can be dissipated through the evaporation of water. For many species, these regulatory processes result in a nearly constant body temperature. Nevertheless, both mammals and birds adjust the intensity of thermoregulation when either energy or water becomes scarce. Experimental manipulations of feeding rate, ambient temperature, and thermal insulation have shown that mammals and birds let their bodies cool considerably when maintaining an elevated temperature becomes energetically costly. Furthermore, these animals let their bodies warm to unusually high temperatures when dehydrated. This tradeoff between balancing thermal and hydric states also occurs in organisms that rely primarily on solar radiation to thermoregulate (*ectothermy*).

As with physiological tolerance, physiological regulation varies adaptively along abiotic gradients. Comparisons of populations within and among species of *Drosophila* have generated a comprehensive view on the regulation of water loss, reinforced by studies of experimental evolution. In general, flies from temperate environments resist desiccation better than do flies from tropical environments. This resistance to desiccation comes from enhanced regulation of water loss rather than enhanced tolerance of dehydration. Allen Gibbs and his colleagues used experimental evolution to discover mechanisms underlying the adaptation of water regulation in *Drosophila melanogaster*. Populations exposed periodically to dry conditions evolved genotypes that develop relatively long chains of hydrocarbons in their cuticles, a biochemical strategy thought to reduce water loss. This example illustrates the complementary nature of comparative and experimental approaches to the study of physiological adaptation.

6. ADAPTIVE ACCLIMATION

Organisms benefit greatly from the ability to adjust their physiology in response to environmental conditions, a process usually referred to as *acclimation*. In a fluctuating environment, acclimation enables an organism to specialize for conditions that will likely occur in the future given conditions of the past. But acclimation involves costs as well as benefits. Energy must be expended to restructure cells and tissues as environmental conditions change. Moreover, tuning one's physiology to match expected conditions involves an element of risk beyond the commitment of energy; if past conditions relate poorly to future conditions, the organism might commit to the wrong strategy! In an unpredictable environment, natural selection would favor generalists that do not acclimate during environmental change.

Since species differ in their ability to acclimate, we can ask whether these differences reflect adaptation to the variability and predictability of their environments. Comparative studies of acclimation are less common than other studies of physiology, but enough data exist to challenge our current notions about optimal acclimation. Since environmental variation increases with increasing latitude, species at high latitudes should evolve a greater capacity for acclimation. Consistent with this expectation, rodents from high latitudes can adjust the length of their intestine more readily than can rodents from low latitudes. The most flexible species of rodents also occupy the widest range of habitats. Still, other comparisons failed to support the view that abilities to acclimate have adapted to local environments. For example, both tropical and temperate genotypes of *Drosophila melanogaster* readily adjust their tolerances of high and low temperatures in the lab, although these genotypes experience very different levels of thermal variation in nature. Likewise, other species of flies adjust their rates of water loss when raised under dry or humid conditions, regardless of whether they come from environments that experience such conditions. The widespread capacity for acclimation in these species could reflect dispersal among distinct environments; dispersal leads to variation in environmental conditions among generations, which strongly selects for genotypes that can tune their physiology to current conditions.

Experimental evolution enables researchers to control or manipulate environmental variation, to see whether adaptation involves a change in the ability to acclimate. In a recent experiment, Brandon Cooper and his colleagues compared populations of *Drosophila melanogaster* exposed to either constant or fluctuating temperatures for more than 30 generations. These populations diverged such that genotypes from the fluctuating environments were better able to adjust their cellular membranes to developmental temperature. The specific adjustment, involving the ratio of two phospholipids, was the very kind expected to confer greater performance in either hot or cold environments. This finding supports the view that environmental variation promotes the evolution of acclimation.

7. CONSTRAINTS ON PHYSIOLOGICAL ADAPTATION

When we use an optimality model to predict physiological variation, we assume that nonadaptive processes have not constrained adaptative ones. Processes such as mutation, genetic drift, and gene flow can slow adaptation in two ways. First, some of these processes can reduce the amount of genetic variation in a population. Second, all of these processes can increase the frequency of maladaptive alleles in a population. Both factors have influenced the evolution of physiology.

How does genetic variation constrain physiological adaptation? Adaptation depends not only on selection but also on heritability. If the physiological variation among individuals was caused by environmental factors rather than genetic factors, selection cannot produce evolutionary change. In general, physiological variation is no less heritable than morphological variation; nevertheless, the type of physiological variation present within populations does not always reflect the type that enables the evolution of optimal phenotypes. For example, consider a species that occurs throughout a wide range of latitudes. Current optimality models predict that adaptation would lead to tropical genotypes that perform best at high temperatures and temperate genotypes that perform best at low temperatures. But how likely is this form of adaptation when one considers genetic constraints? Specializing for function at extreme temperatures might require mutations of hundreds of genes, which would take a very long time to accumulate. Instead, adaptation sometimes takes a more convenient course, as illustrated by the studies of Atlantic silversides (*Menidia menidia*) conducted by David Conover and his colleagues. Northern and southern populations of these fish experience different temperatures in nature, but both populations grow best at the same temperature in the lab. Even more surprising, northern genotypes outgrow southern genotypes over a wide range of temperatures, including those temperatures more common in the south. Why do these apparently superior genotypes not spread throughout the entire range? The answer lies in understanding the cost of their rapid growth. These fish grow rapidly by consuming large amounts of food, which reduces swimming speed and increases predation risk. The benefit of rapid growth outweighs this cost in highly seasonal environments, but the reverse seems true in less seasonal environments; therefore, trade-offs in fitness between environments exist even when trade-offs in growth do not. The adaptation of feeding behavior in silversides likely reflects insufficient genetic variation to adapt biochemical functions to low temperature.

Gene flow between distinct environments also slows the adaptation of physiology by increasing the frequency of maladaptive alleles (see chapter IV.3). This phenomenon can operate even on very small spatial scales. For example, many species of plants and fungi that live in contaminated soils have evolved biochemical mechanisms to regulate the absorption of toxic metals. These species generally exhibit clines in metal tolerance between contaminated and uncontaminated sites. Experiments have shown that alleles conferring greater fitness at contaminated sites reduce fitness at uncontaminated sites. Yet the close proximities of these sites sometimes enable deleterious alleles to persist through gene flow. Gradual clines in metal tolerance have been observed along transects running parallel to seed dispersal, whereas sharp clines have been observed along transects running perpendicular to seed dispersal. Although dispersal can help to establish populations in stressful environments, ultimately this process limits physiological adaptation to these environments.

8. IMPLICATIONS FOR GLOBAL CHANGE BIOLOGY

The adaptation of physiology—in the past, present, and future—has significant consequences for the persistence of species in a changing environment. Given the physiology diversity highlighted in the preceding sections, biologists have become keenly aware of the need to consider how physiological adaptation influences the dynamics of populations and the ranges of species. Most models that incorporate physiological diversity have focused on the responses of species to global climate change. Changes in temperature, humidity, and precipitation have altered and will continue to alter the geographic distributions of species. Importantly, physiological adaptation can ameliorate or exacerbate these effects. For example, a model developed by Michael Kearney and his colleagues indicates that the Australian distribution of the dengue mosquito (*Aedes aegypti*) is limited by moisture in the north and temperature in the south. Their model also shows that adaptation of desiccation resistance during the next few decades could enable this species to spread throughout northern Australia. Predictions of this kind depend not only on the selective pressures created by climate change but also on the constraints that limit physiological adaptation (see section VII). Models that consider physiological adaptation should become increasingly relevant to global change biology.

FURTHER READING

Angilletta, M. J. 2009. Thermal Adaptation: A Theoretical and Empirical Synthesis. Oxford: Oxford University Press.

Chown, S. L., and S. W. Nicolson. 2004. Insect Physiological Ecology: Mechanisms and Patterns. Oxford: Oxford University Press.

Feder, M. E., A. F. Bennett, and R. B. Huey. 2000. Evolutionary physiology. Annual Review of Ecology and Systematics 31: 315–341.

Garland, T. 2001. Phylogenetic comparison and artificial selection: Two approaches in evolutionary physiology. Advances in Experimental Medicine and Biology 502: 107–132.

Garland, T., and P. A. Carter. 1994. Evolutionary physiology. Annual Review of Physiology 56: 579–621.

Hochachka, P. W., and G. N. Somero. 2002. Biochemical Adaptation. Oxford: Oxford University Press.

Karasov, W. H., and C. Martínez del Rio. 2007. Physiological Ecology: How Animals Process Energy, Nutrients, and Toxins. Princeton, NJ: Princeton University Press.

McNab, B. K. 2002. The Physiological Ecology of Vertebrates: A View from Energetics. Ithaca, NY: Cornell University Press.

Natochin, Y. V., and T. V. Chernigovskaya. 1997. Evolutionary physiology: History, principles. Comparative Biochemistry and Physiology A 118: 63–79.

III.14

Evolution of the Ecological Niche

Robert D. Holt

OUTLINE

1. Natural history, niches, and evolution
2. What is an ecological "niche"?
3. Complexities in the niche concept
4. The issue of genetic variation in niches
5. Demographic constraints on niche evolution
6. Niches evolving in communities

Every species and clade has a niche characterizing the range of environments (including abiotic as well as biotic factors) within which it persists, and outside of which it goes extinct. The niche describes how an organism with a particular phenotype performs in its demography (birth and death rates) as a function of environmental conditions. Given genetic variation in these traits, niches can evolve, sometimes quite rapidly, but niches also can show surprising conservatism. To understand niche evolution, one must draw on and integrate many areas of knowledge, ranging from detailed mechanistic understanding of individual performance to the mapping of genes to phenotypes, from life histories, mating systems, and population dynamics to population genetics, community ecology, and the broad spatial and historical perspectives of landscape ecology, biogeography, and paleobiology. Understanding niche evolution and conservatism is important to many basic questions in evolution, ecology, and biogeography, and it is also highly germane to many crucial applied issues.

GLOSSARY

Allee Effect. A positive effect of increasing population size on population growth rate.

Asexual Reproduction. Reproduction by cloning (i.e., making offspring that are genetically identical to the parent).

Dispersal. Movement of individuals across space; immigration is dispersal into a site, emigration is movement away from a site.

Evolutionary Rescue. A population that is declining in numbers toward extinction because the environment has changed, or because it has colonized outside its niche, yet nonetheless persists (because natural selection increases mean fitness sufficiently rapidly to allow the population to rebound from low numbers) is said to have experienced evolutionary rescue.

Extinction. The event marked by the death of the last individual of a population, species, or larger clade.

Hutchinsonian Niche. The range of environmental conditions (both abiotic, such as temperature, and biotic, such as density of a predator) for which the intrinsic growth rate r of a population is positive. If one plots r as a response surface undulating over an abstract space, where the axes are environmental variables, the niche is defined by that subset of this variable space where $r > 0$.

Intrinsic Growth Rate, r. The difference between birth and death rates (per individual, per unit time) when a population is sufficiently scarce that one can ignore competition for resources, interference, and other density-dependent effects.

Source-Sink Dynamics. A mechanism for sustaining some populations of a species outside its Hutchinsonian niche. In a heterogeneous landscape, a source habitat is one with conditions inside a species' niche, where a population persists. This population can export individuals that end up in a habitat with conditions outside the niche, and so maintain a sink population.

1. NATURAL HISTORY, NICHES, AND EVOLUTION

From the air, many landscapes in northern climes such as Yorkshire, England, display lovely mosaics of land and water, tapestries of green vegetation, moors and woodland dotted with seemingly endless small ponds and lakes, reflecting glacial molding of the earth's surface during the Pleistocene. A naturalist out for a Sunday stroll to scan for an elegant but rare butterfly, the small pearl-bordered fritillary (*Boloria selene*)—rumored to occur in a grassland sprinkled with violets, the butterfly's required host plant—might from the corner of one eye see a glint of blue as a kingfisher dives to nab a fish in a pond, even as she hears the call of a swift soaring overhead to catch aerial insects. Each species seems to have its place, or way of life. What all naturalists know in their bones is that the world is intrinsically a highly heterogeneous place, and that to find a particular species, one must seek out habitats that match its conditions for life. These specifications for what a species needs to persist—which can often be quite subtle—constitute its niche.

So species have discernible niches. Each species across all these distinct habitats perused by our Yorkshire naturalist had a common ancestor, possibly a very long time ago, that also had its own niche, and so the current niche differences among these taxa must have emerged during evolution. Like any trait, given genetic variation, the niche requirements of a species can evolve, but sometimes species or clades can be surprisingly constant in their niches and key organismal traits related to niches—a phenomenon called *niche conservatism* (Wiens et al. 2010). In this chapter, I first present some necessary ecological background, including an exposition of the basic concept of the niche, and a brief discussion of some subtleties in the concept. I then turn to the crucial issue of the existence of genetic variation in the niche—which is necessary to fuel niche evolution—and sketch how the demographic context of selection can sometimes constrain niche evolution. I touch on how the community context often modulates niche evolution, and conclude by suggesting that the theme of niche evolution and conservatism is central to a range of vitally important applied questions.

2. WHAT IS AN ECOLOGICAL "NICHE"?

To understand what governs niche evolution, or its absence, it is important to have a crisp understanding of what is meant by the term *niche*. The word has many overlapping meanings in ecology (see Schoener 2009). In everyday English, a "niche" refers to a recess in a wall (e.g., a place that might hold a statue), and so statements about niches seem to be statements about the environment. In ecological usage, however, the word refers more subtly to how organisms relate to the environment. Our focus here is on the basic idea first formalized by the renowned ecologist G. E. Hutchinson (in his 1957 essay "Concluding Remarks," discussed in Hutchinson 1978). Hutchinson suggested that the environment in which an organism lives could be graphically represented in terms of a set of axes, defining, for instance, the ranges of conditions impinging on organismal function (e.g., temperature, pH, toxin concentration), or resource availability (e.g., algal food supply for a zooplankter), or the intensity of different mortality sources (e.g., the abundance of a predatory fish species). The crucial idea is that one considers not just individual survival but more abstractly the dynamics of populations or lineages of reproducing individuals, reflecting the outcome of survival and births over many generations, and how these dynamics depend on environmental conditions. We imagine that a genetically homogeneous group of a few individuals of a given species is introduced into a habitat with a certain set of environmental conditions (including abiotic factors such as temperature, as well as biotic factors such as food availability, abundance of predators, etc.). For simplicity, we assume the species is asexual (i.e., a clone), or at least that mates have no trouble finding each other (for sexual species), and that the environment is constant, so that the genotype of these introduced individuals corresponds to a particular phenotype. We then watch what happens.

These individuals have an expected birth rate and death rate. The difference between birth and death rates is the net growth rate of the population, which for a population at low abundance is called its *intrinsic rate of growth*, denoted r. This concept is closely related to the population genetic concept of absolute fitness of a genotype. Because we are examining what is happening at low density, we assume that density-dependent effects such as crowding or competition for resources are negligible. If the intrinsic growth rate in a given habitat is positive, the population can grow in that environment; if negative, then without immigration or evolution, the population is doomed to extinction. If one now repeats this protocol across a large range of environmental conditions, and measures r for each, one builds up a profile, a kind of abstract landscape describing what is called the *niche response surface* for that particular genotype as a function of its environment (thereby making explicit the dependence of absolute fitness on the environment). Figure 1 shows a schematic example of niche response surfaces (which are curves in this case) for two related species, across a range of values of one environmental factor (e.g., temperature). The shape of the entire niche response surface is of ecological interest, and evolution can sculpt this shape. But a particularly important distinction is provided by the boundary in environment space separating zones of positive and negative population growth; this boundary defines the *Hutchinsonian niche*. This boundary cleaves the environmental states of the world into that set of conditions where a lineage goes extinct ($r < 0$, outside the niche), and another set where it potentially persists ($r > 0$, inside the niche). In figure 1, species 1 has a broader niche than does species 2, and the two species differ in the shape of their niche response and the environment in which growth is maximal. Some environments could potentially harbor both species (assuming they do not strongly compete), other environments just one. The niche of the clade spans

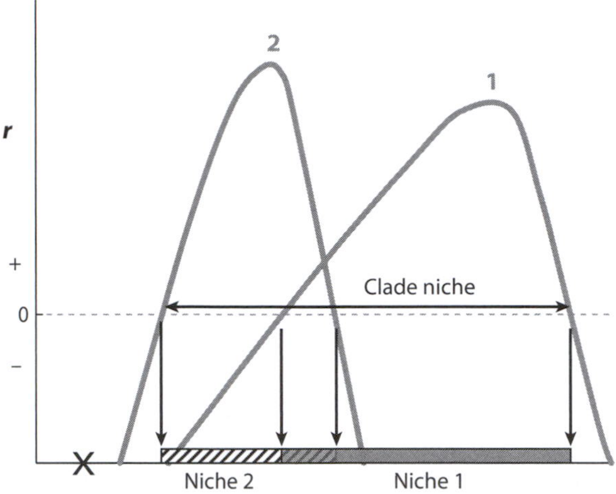

Figure 1. Niche response surfaces along a single environmental niche axis, for two related species. A niche response surface is given by intrinsic growth rate, r, expressed as a function of an environmental variable (e.g., temperature). The Hutchinsonian niche of a species is defined by those environments where a species has a positive, rather than negative, growth rate. Species 1 is expected to go extinct if placed in environments outside the region shown in gray, and species 2 likewise perishes outside the hatched region. The niche of the clade is the union of these two niches. In a habitat with conditions X, neither species can persist.

more environmental space than any single member, but there are some environments (e.g., X in the figure) where neither species can persist.

The niche concept with minor modifications can pertain to individuals, or genotypes, or aggregates of individuals in populations, species, or broader phylogenetic clades. If conditions are outside the niche, individuals are not expected in the long run to have descendants, populations are not expected to persist, species will go extinct, and, finally, phylogenetic clades as a whole can disappear. Understanding niche limits can help explain why species' borders occur where they do along gradients, or what determines the range of hosts that can sustain a parasite, or why phylogenetic clades disappear or proliferate in the fossil record.

To make this idea of the niche more concrete, let us return to our naturalist wandering over the Yorkshire landscape. Were she to dip a bucket in a pond and examine its contents under a microscope, the sample would teem with zooplankters, but with different types in different ponds. One small crustacean, the water flea (*Daphnia magna*) occupies some, but not all, water bodies (schematically depicted in figure 2A). English ecologists hypothesized that this distributional pattern could be explained by this species' niche requirements and carried out lab experiments to test this idea. This species lives and breathes in water, so one niche axis—in versus out of the water—is so blatantly obvious, there is no need to quantify it. The ecologists surmised that more subtle aspects of water chemistry might explain why *D. magna* is absent from many lakes and ponds, even though it is present in others nearby. In particular, pH and calcium concentration should be key niche axes. Maintaining internal ionic balance is important for any organism, and pH influences that. Water fleas shed and replace their exoskeletons at each molt, and so require calcium. Conveniently, the daphnid grows asexually, so a clone was brought into the lab, and replicated into many copies. Small populations were then introduced into containers with different water chemistries and tracked, permitting the genotype's intrinsic growth rate to be assessed across a wide range of combinations of pH and calcium availability (figure 2B).

Almost without exception, water bodies where these abiotic factors predicted negative growth lacked the species (figure 2C); thus understanding abiotic niche requirements by using just two abiotic variables has strong explanatory power for interpreting this species' distribution in Yorkshire. But the niche boundary in this two-dimensional space does not quite explain everything about the species' distribution. Some sites seem suitable, yet lack the species. Maybe other unmeasured niche dimensions (e.g., the presence or absence of a voracious predator) explain these absences. Alternatively, r may be positive but low, making recovery from chance disturbances less likely. Finally, some ponds may simply be hard to reach or newly formed, and so not occupied because of the chance vicissitudes of colonization. Intriguingly, and conversely, a few sites have conditions a little outside the niche, but do have the species. One plausible explanation is that regular immigration from suitable sites ("sources") can sustain populations in what is called a "sink" habitat, where conditions are outside the niche. Another possibility comes from the fact that the niche was quantified for just a single clone, yet *Daphnia magna* harbors considerable genetic variation. Maybe some genetic variants have niche requirements differing from the measured clone. Despite this possibility, it is clear that to an excellent approximation, the pH and calcium requirements describing niche limits of this clone also must describe the niches of a much wider array of genetic types, providing a plausible example of niche conservatism in a clade.

This experimental study helps define this species' niche in the Yorkshire landscape but does not elucidate those aspects of organismal function that actually account for its niche response structure. A full understanding of the latter requires one to delve deeply into the rich mechanistic details of organismal biology, including physiology,

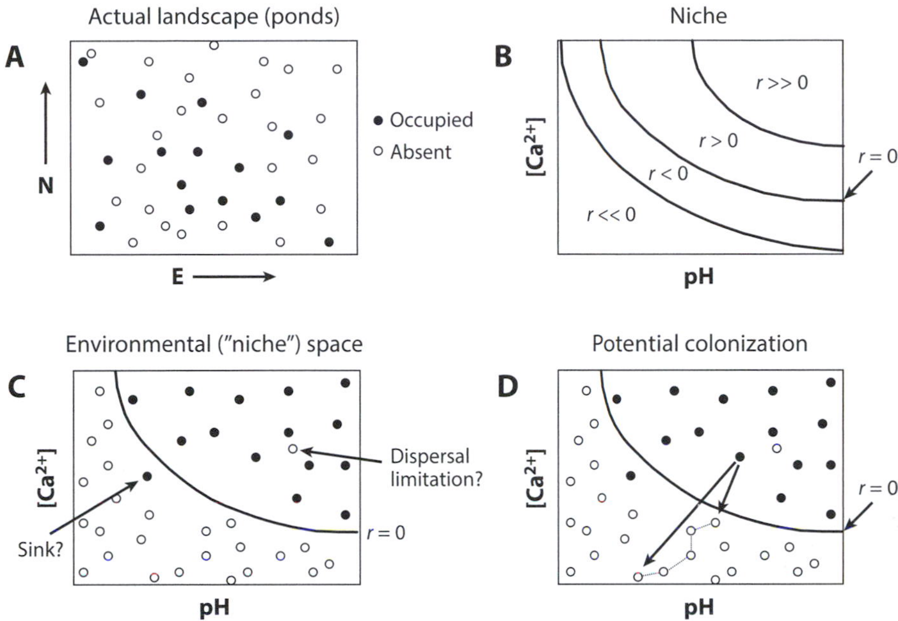

Figure 2. This figure idealizes the study of the ecological niche of *Daphnia magna* reported by Hooper et al. (2008). (A) A map of ponds and lakes in the Yorkshire landscape, some with the zooplankter, and some without. (B) A laboratory study of intrinsic growth rates r of a clone of this zooplankter, grown with abundant food and no other species, as a function of pH and calcium concentration. The contours can be viewed as heights on a "mountain" emerging from the page, describing the growth rate of the daphnid, its niche response surface (to these variables). The Hutchinsonian niche consists of all those combinations of these abiotic variables where $r > 0$. (C) The water chemistries of each lake and pond in A, plotted in this same abstract environment space. In most cases, occupied ponds have conditions within the niche, and unoccupied ponds have conditions outside the niche; there are, however, a few intriguing exceptions (see text for more detail). (D) The long arrow describes a colonization event from an occupied pond with conditions inside the niche, to a pond with conditions well outside, where $r \ll 0$. Evolution is unlikely, because extinction is rapid. If instead there can be a chain of colonization events, or spatially coupled sink populations, where at each shift a much smaller change in the niche is required for persistence (as indicated by the short arrow to just outside the niche, followed by short steps along a chain of habitats linked by thin lines), then gradually the lineage may evolve to include even radically different habitats in its niche. (Figures adapted from Holt and Barfield 2011.)

morphology, behavior, and life history; each species' story is likely to have some unique aspect that must be unraveled to really understand its niche. In effect, a full portrayal of niche evolution and conservatism requires a detailed understanding of the natural history and organismal biology (in its fullest sense) of each species. Covering this rich body of literature is beyond a short article —to do full justice to the theme would be like writing an advanced general biology text! Instead, the remainder of this chapter highlights general conceptual issues, illustrated by examples, which almost always arise when contemplating the evolution of species' niches.

3. COMPLEXITIES IN THE NICHE CONCEPT

There are other dimensions of the niche concept that are important in ecology and evolution, which we barely touch on here. Direct measurements of a species' niche are difficult, and in practice, ecologists often attempt to indirectly quantify the niche by characterizing patterns of resource use (see Schoener 2009). For instance, for an insectivorous lizard species, instead of plotting population growth rate against an environmental variable, as in figure 1, one might plot frequency of consumption of insects as a function of insect size, scaled against insect availability at each size.

The resource utilization niche concept has been particularly important in grappling with the problem of understanding the degree to which two species can be similar and still coexist. Species do not live alone, but instead are found in communities of interacting species. When species compete for resources or otherwise interfere, it can be a challenge to understand their coexistence. Indeed, in laboratory settings where pairs of related protozoa are put together, usually just one species dominates and persists, forcing the other to extinction. We will return briefly to this important issue at the end of this chapter.

Moreover, and crucially, understanding species coexistence requires analysis not just of what species need and can tolerate (as in the Hutchinsonian niche) but also of how they impact their environments via depletion of resources, or augmentation of natural enemies, or even alterations in physical or chemical conditions (what is called "niche construction"). This impact dimension of the niche (Chase and Leibold 2003) depends not just on the species but also on the feedback mechanisms present in the environment itself. As species evolve, these feedbacks may themselves change, altering conditions for persistence and coexistence. Even defining the "environment" can be quite tricky, since organisms can move to select their habitats and otherwise affect their living conditions. The environment in which organic evolution unfolds is itself in part determined by evolutionary processes.

Sometimes the growth rate of a species when rare can be boosted by an increase in its abundance, via what are called Allee effects; for instance, reproduction may be facilitated because mates can more easily find each other when the species is more common, or deaths may be reduced because there is protection in numbers against predation. Because of Allee effects, a species if sufficiently abundant may be able to persist in some environments where it cannot increase when initially rare (i.e., $r < 0$); the population "persistence niche" may exceed the population "establishment niche."

4. THE ISSUE OF GENETIC VARIATION IN NICHES

Leaving aside such complexities, we return to the question of why the zooplankter does not inhabit a wider range of environmental conditions. The Hutchinsonian niche is a kind of abstract landscape (as in figures 1 and 2B), describing how absolute fitness (intrinsic growth rate) varies for a genotype (or species or lineage) over an abstract environmental space. To understand how niches evolve (and when they might not), it turns out one needs to think about two other kinds of landscapes (one abstract, one not), as well. Consider a thought experiment for *Daphnia magna* in Yorkshire. A waterspout sucks an aliquot of a daphnid from an established population and plops it into a pond, with conditions outside the niche of the source population, so the average growth rate of the colonizing population is negative. Without genetic variation, the colonizing clone simply goes extinct.

But given appropriate genetic variation in the source population, or if by chance a favorable mutation arises in the introduced population, evolution may occur that allows the population to persist and become established —and the niche of the clade will have expanded. Evolution by natural selection arises from variation in relative fitnesses among individuals (with a genetic basis) and can lead to evolutionary rescue of a population placed outside its niche (Gomulkiewicz and Holt 1995). A second landscape metaphor often usefully describes selection. The "adaptive landscape" portrays fitness as a function of genotype or a phenotypic measure (such as body size or temperature tolerance) in a given environment (including biotic interactions within and among species). In some (not all) cases, evolution can be described by a hill-climbing metaphor, where selection among alternative genotypes moves a population toward a local optimal fitness. This metaphor breaks down when individuals interact, such that fitness depends strongly on relative frequency; in this case, in effect the hill itself undulates as evolution occurs. But if a population is outside its niche, in general its numbers will be declining, and at low densities, so individuals may not encounter each other very often. This makes frequency dependence in selection less likely, and the adaptive landscape metaphor becomes a reasonably accurate characterization of the way selection occurs. As Charles Darwin noted in *On the Origin of Species*, reflecting on the struggle for existence, "a plant on the edge of a desert is said to struggle for life against the drought"; if fitness is determined largely by the ways in which individuals cope with physical and chemical conditions (i.e., the external environment), selection will straightforwardly favor whichever phenotype best tolerates these abiotic factors. The adaptive landscape describes how variation in phenotypes translates into variation in fitness in a given environment, and hence in the strength and direction of selection. One of the near-magical features of Darwinian evolution is that the effects on genetic composition of populations of even small differences in fitness cumulate and become amplified over time, leading to dramatic transformation within and among populations.

If niches are to evolve, there must be genetic variation among individuals in their phenotypes, leading to a heritable basis for variation in fitness as a function of the environment (i.e., in the niche). This issue requires much more empirical study and has not been addressed in detail in many species; nonetheless, there are some clear examples. At the level of entire species, there is considerable evidence for genetic variation among populations in climatic tolerances (Hoffmann and Sgrò 2011), implying intraspecific variation in ecological niches (see chapter IV.3). For instance, the Canadian tiger swallowtail (*Papilio canadensis*) ranges from Michigan to Alaska. Laboratory experiments suggest that Michigan caterpillars are so intolerant of many Alaskan summer temperatures that were one to move a population from

Michigan to Alaska, it would go extinct. The ecological niche of the entire species is thus larger than that of local populations. In forestry, there are economic incentives to plant seedlings that will successfully mature into adult, log-worthy trees; thus many transplant studies have been carried out. The lodgepole pine, *Pinus contorta*, extends from Colorado to the Yukon. Seedlings often cannot grow and survive when planted at locations across the species' range where thermal conditions are either much warmer or much cooler than their natal habitats. Though the physiological mechanism is not understood, this again suggests the existence of considerable geographic variation in the ecological niche of the lodgepole pine. Were a devastating blight to sweep across the range of the species and lead to mass local extinctions, leaving one remnant population behind, these experiments suggest one could not quickly restore the original range of the lodgepole, using individuals drawn from that sole surviving population.

The ultimate source of genetic variation that can permit niche evolution is of course mutation. Experiments probing the niche limits of clonal organisms have shown that when large populations are placed outside their niches (e.g., thermal tolerance zones for *E. coli*, salt concentrations for yeast), typically they go extinct, but very occasionally novel "Lazarus mutations" arise that can rescue these populations from extinction (see chapter III.6). Quantitative traits in sexual species can be under stabilizing selection, yet the species can maintain a pool of heritable variation in those traits because of recurrent mutation. This pool can provide the raw material to fuel niche evolution. Laboratory selection experiments on *Drosophila* (fruit flies) reveal that there can be substantial standing genetic variation permitting evolution of some niche traits; basically, conditions that are stressful for most individuals in the population may not be stressful for all.

Genetic variation in traits influencing the niche within species thus surely occurs, permitting species to be selected for increased fitness when absolute fitness is low (i.e., when conditions are outside the niche). But there is also increasing evidence that such variation may be lacking for crucial characters, leading to one explanation for niche conservatism for at least some species, along some niche axes (see chapter III.8). For instance, desiccation resistance and upper thermal limits can have little or no genetic variation in *Drosophila* populations. Plant species may be missing from soils with heavy concentrations of toxic metals, even though they reside in other habitats nearby, because they have no discernible genetic variation for resistance to those toxic conditions. Such examples are contrary to the conventional wisdom that genetic variation is ubiquitous for almost any trait and allows evolutionary responses to almost any selective pressure (Futuyma 2010). Even with genetic variation in single traits affecting the niche, genetic correlations among traits may constrain selective responses and hamper niche evolution (see chapter III.8). Leaving aside such genetic explanations for niche conservatism, ecological factors can also at times constrain niche evolution.

5. DEMOGRAPHIC CONSTRAINTS ON NICHE EVOLUTION

The third conceptual landscape pertinent to niche evolution is the "real" landscape, describing the ways in which environments as experienced by a lineage are structured over space and time. If lakes differing substantially in abiotic conditions are closely juxtaposed, our colonizing population of zooplankters is likely to end up in a lake with conditions well beyond its ancestral niche boundary (like the long arrow in figure 2D). Thus, its initial rate of decline will be large, rapidly reducing populations to low numbers and extinction. Theoretical studies suggest that the harsher the environment faced in colonization (as measured by the rate of decline in numbers), the less likely one will observe adaptation rather than extinction. If the geometry of the landscape is such that colonization is sporadic, and into habitats to which a species is so poorly adapted that the habitats lie well outside the niche, one expects evolutionary stasis even over long time horizons.

Reasons for Failed Adaptation in Colonization outside the Niche

Failed invasion outside the niche can reflect both the scarcity of appropriate genetic variation and demographic constraints operating outside the niche. If the colonizing population is initially genetically homogeneous, the potential for adaptation and persistence rests entirely on novel genetic variation, created by mutation—otherwise, the population is doomed. The likelihood of such mutations arising depends on the number of replication events that occur before a population goes extinct. If a population is plummeting rapidly to extinction, there will be scant opportunity for favorable mutations to arise; moreover, mutations of small positive effect on fitness (which arguably are more common than mutations of large effect) may not suffice. By assumption, outside the niche, $r < 0$. For a mutation to be favored by selection, it must have an effect $\delta > 0$ on fitness (i.e., per capita growth rate) giving the mutant a higher relative fitness. But will the mutation be captured by evolution? Maybe not! The absolute growth rate of this mutant is $r + \delta$. If r is negative, and δ is very small, then the net growth of the mutant type, $r + \delta$, will still be negative (i.e., deaths of individuals carrying the mutation will exceed their births), and the lineage generated by the mutation will go extinct (along with the rest of the population). The harsher the environment (i.e., the lower r is), the larger the effect of mutation on fitness must be (i.e., the larger δ must be) for the new mutation to have any chance to persist. If most genetic variants that arise in the colonizing population have a small effect on the phenotype (and thus fitness), most will not lead to persistence. Mathematical models that take into account the inherent stochasticity of mutation and the chance vicissitudes of small population sizes have rigorously shown that the initial step of adaptation in a population suddenly exposed to an unfavorable environment (as can occur during colonization) requires mutations of large positive effect on fitness, and extinction may simply overwhelm the scope for adaptive evolution if such mutations rarely occur.

A comparable argument holds if adaptation depends not on novel variation but instead on variation sampled from a genetically variable source. For an introduction of an asexual species into a habitat to succeed, some individuals in the initial pulse of colonists must have a heritable positive growth rate, even though the average growth rate is negative. Figure 3 schematically shows what is needed. We imagine clonal genetic variation to be present among the colonists, expressed as variation in intrinsic growth rates among individuals in the colonized habitat. The left hump shows a population placed into a quite harsh environment; the right hump describes the same population in a less harsh environment. Both populations have equivalent levels of genetic variation in growth rates (the width of the curves is equivalent); however, in the harsh environment, note that no clones have a positive growth rate, so the population is doomed (without novel, highly favorable mutations). In the less harsh environment, a small number of individuals have a positive growth rate, so there is a chance the population will persist.

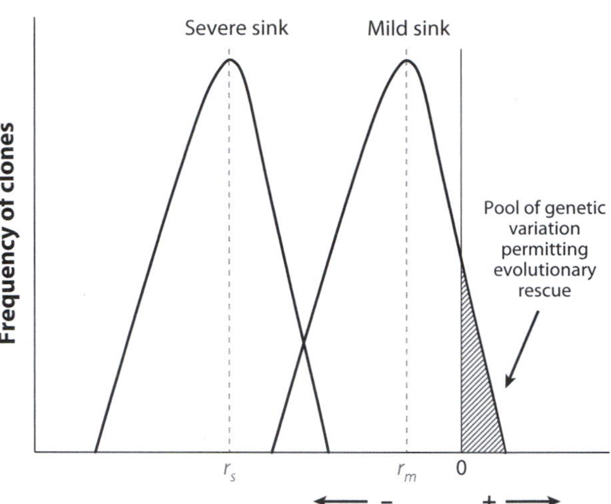

Figure 3. A demographic constraint on evolution during colonization outside the niche. We imagine that a group of colonists of an asexual (clonal) species has been taken from one habitat (a source), and placed in another (a sink). There is genetic variation among clones in their growth rates, so selection will occur. The humped curves depict heritable variation among individuals in their expected growth rates for two possible sinks; the curve on the left is for a severe sink, so the average growth rate is very negative, whereas the curve on the right is for a mild sink, where the average growth rate is only slightly negative. The amount of genetic variation in the growth rate of the colonizing population in each sink habitat is similar (as expressed by the width of these curves); however, for the population to persist without new mutations, some of these variants must have a positive growth rate, so what matters is not so much the mean growth rate but the tail of the distribution that exceeds zero growth. In the severe sink, no genetic variants have a positive growth rate, so extinction is assured. In the mild sink, some genetic variants have a positive growth rate, so there is at least a fighting chance of persistence.

The latter could describe colonization into a habitat only slightly outside the ancestral niche (as in the short arrow of figure 2D); adaptation and thus niche evolution would probably be more likely than they would be for colonization into a sharply different habitat (as in the long arrow of figure 2D). In the former case, the probability of some colonists having a positive growth rate is much higher. Also, with mutations arising in the sink, selection may be able to sort among a larger supply of mutations with rather modest effects on the phenotype and fitness, since only a small change in fitness might permit a positive growth rate in the novel habitat. If the structure of the environment experienced by an evolving lineage consists of gradual transitions between environmental states, rather than abrupt disjunctions, adaptation thus may be more likely to occur, and niches in a phylogenetic lineage will be evolutionarily labile, rather than conserved. Quite similar reasoning pertains to environments varying in time rather than space. Abrupt temporal changes in the environment that greatly lower fitness usually lead to extinction, rather than adaptation; however, if the same change occurs, but spread out in time rather than in a steep step, species may be able to adapt and persist, with an evolving niche tracking small environmental changes.

Extending these arguments, one can reason that the potential supply of favorable variation should increase with the initial number of colonists; larger numbers mean a more generous sample of preexisting variation found in the source, and they also provide a greater opportunity for novel mutations to arise in the sink as the population takes longer to decline to extinction from higher numbers. Experimental studies of adaptive evolution in harsh environments have shown exactly this predicted effect (see chapter III.6). For instance, in lab experiments, a sink habitat was created for an asexual species of yeast by adding salt to growth media. Most of the experimental introductions into the sink habitat went extinct, but some persisted and eventually grew as a result of adaptive evolution. The likelihood of persistence increased with the initial number of colonists, consistent with theoretical expectations. In the absence of absolute constraints on variation, niche evolution can certainly occur during colonization well outside the niche, but it may require a very large number of individuals in each colonization episode.

Reasons for Evolutionary Stasis in Sink Populations Maintained by Immigration

Returning to our spatial scenario, now imagine that some ponds in the Yorkshire landscape are connected by streams, permitting recurrent dispersal. Such recurrent immigration can sustain sink populations in environments outside their niches. This raises the issue of the interplay of gene flow and selection in determining local adaptation. The story is richly complex (Holt and Barfield 2011), but a few highlights are worth noting (see chapter IV.3).

First consider asexual species. The simplest effect is that immigration sustains the population in the first place, ensuring recurrent opportunities for adaptation to occur. If the spatial coupling of the habitats permits many separate attempts at colonization, one eventually might be successful. In general, the length of time required before recurrent immigration by a species succeeds in its adapting and persisting outside the niche increases with increasing harshness of the sink habitat, and with reduced numbers of individuals per colonizing episode. Even rare events, such as the appearance of a mutation with a large positive effect on fitness, are likely if one waits long enough. If density dependence is weak, an increase in immigration (as measured by the number of individuals arriving per colonization bout) can facilitate adaptation. Increased immigration enhances sampling of genetic variation from the source and also boosts local numbers, increasing the potential for local mutational inputs of variation. For a given rate of immigration, by contrast, increasing harshness in the sink (i.e., increasing difference in the environment between the source and sink) makes it harder for adaptation to occur; this reduces local population size, shrinking variation, and also makes mutations with small effect on fitness less likely to be captured by selection (as argued above for single colonization bouts). Lab experiments with the bacterium *Pseudomonas* have demonstrated these effects, using antibiotics to create sink habitats; a single antibiotic made a mild sink, and a cocktail of antibiotics generated a harsh sink. In both cases, increased immigration increased the rate at which the population adapted to the local environment. Moreover, the likelihood of adaptation was reduced when the sink environment was harsher (and adaptation was not seen at all over the timescale of the experiment with low immigration into the harsh sink). Both effects match theoretical expectations.

A quite different effect can arise when one considers sexual species. Given recurrent immigration, if immigrants mate with residents during each generation before selection occurs, a "migrational load" arises, diluting the effectiveness of selection. The reason is that immigrants tend to carry alleles that are maladaptive in the sink; when mixed with better-adapted alleles of residents, these lower the fitness of the offspring of resident individuals who mate with these immigrants. This is particularly likely to occur when immigration occurs into environments strongly differing from the source. In this case, resident numbers may be low, so most residents may mate with immigrants, rather than each other, and the genes flowing into the sink are likely quite maladapted there. The negative effect of gene flow on adaptation in marginal populations, leading to constraints on niche evolution within a species, is theoretically very plausible, but robust examples have been surprisingly hard to demonstrate. Douglas Futuyma (2010) has argued that speciation (defined by reproductive isolation between lineages) is crucial in diversification because it permits local adaptations to be captured by a lineage rather than washed away by gene flow. Speciation can potentially facilitate niche evolution and diversification in a clade, but note that the genes permitting persistence in a local environment must already be present—or the reproductively isolated population will simply disappear! This negative effect of immigration on niche evolution as a result of gene flow constraining local selection is also more likely for some life histories than others. If selection in the sink occurs immediately on immigrants, before they have a chance to mate with residents, then the migrational load imposed on local adaptation by immigration is weakened; the only migrants left will be those that by chance have higher fitnesses locally, and immigration, by boosting genetic variation, should facilitate niche evolution.

Another effect arises when reciprocal movement between source and sink habitats occurs, rather than one-way migration from the source to the sink (as assumed above). In this case, to understand evolution in the niche, one has to grasp that selection in effect averages over all the environments experienced by a lineage, but with differential weightings for different conditions. Because there may be few individuals in the sink, and they have low reproductive value there, selection tends to be automatically weighted toward conditions in the source (in effect "success breeds success"). If there is a trade-off between performance in the sink and that in the source, selection tends to favor the latter. Trade-offs are often simply assumed by evolutionary ecologists, but in practice they have been rather hard to definitively demonstrate. Trade-offs between fitness in habitats within and outside the initial niche tend to produce niche conservatism, particularly given large differences between source and sink. An alternative genetic mechanism for niche conservatism involves deleterious mutations. If a species for whatever reason is abundant in one habitat, and scarce in another, it may lose its ability to utilize the latter, not because of trade-offs in performance, but because selection is ineffective at weeding out deleterious mutations that degrade performance there (see chapter III.8; Holt 1996).

In some circumstances, however, these demographic and genetic constraints can be overcome, and niche evolution will occur. This can at times be dramatic, as in the adaptive radiations found on many oceanic islands. Analyses of the scenarios discussed above help identify circumstances for which niche evolution may be quite rapid. For instance, in source-sink environments, if dispersal is high into the sink, many individuals are forced to experience sink conditions, thus automatically increasing the "weighting" that selection provides such habitats, relative to sources. Transient periods when conditions are favorable in the sink (e.g., because competitors are absent) can also facilitate adaptation to it. Factors that go beyond these demographic models can make niche evolution more likely. For instance, individuals may have plastic responses, permitting them to shift their phenotypes so as to boost fitness in the sink environment. This dampens the rate of decline in the population and can permit adaptation to occur using even genetic variants of modest effect. Some species do seem to have abundant genetic variation that can respond to novel conditions (Hoffmann and Sgrò 2011), and in some cases, stress itself can pump up mutation rates or break down the developmental stabilization (canalization) of characters, which might provide variation for niche evolution.

6. NICHES EVOLVING IN COMMUNITIES

We have focused on how a single species evolves (or not) in a fixed environmental template. But as has been known since the time of Darwin, interactions among species crucially modulate the opportunity for niche evolution. The reason a habitat is a sink for a particular species may be that a superior competitor or voracious predator resides there, keeping r negative. Remove that other species, and the colonizing species may persist; then adaptation to local abiotic conditions can leisurely occur. This scenario helps explain the explosive evolution of adaptive radiations on islands, and can also account for rapid evolution in invasive species occupying novel environments. The spotted knapweed (*Centaurea maculosa*), for instance, was successfully introduced into North American environments at sites where the climate matches that found in its ancestral Eurasian range, but it then rapidly evolved the ability to live in different climatic regimes as it moved through the anthropogenically disturbed landscapes of the West, where disturbance had removed or weakened potential native competitors. Such bouts of rapid evolution ultimately slow, often as a result of intensifying interspecific interactions. Interspecific interactions may be highly significant in governing the likelihood of niche conservatism, versus rapid evolution. As a metaphor for this, consider dancers entering an empty dance floor. At first, some dancers (wallflowers) may stay put because of mysterious internal constraints, but other dancers wander widely and quickly across the entire floor. But as more and more dancers enter the room, it gets harder to move, because the space is preempted. Eventually, in a really crowded room, even though everyone continues to jostle and move locally in time to the music, no one really gets anywhere very fast.

One of the grand themes in the dance of life is a comparable patterning of movement in evolving and diversifying clades, measured against the spatially and temporally shifting template of environmental opportunities we call niche space. Understanding the determinants of the moves and halts in this dance—niche evolution and conservatism—is a crucial dimension of basic evolutionary biology, ranging from adaptive radiations, to biogeographical limits of species ranges, to understanding how ecological communities are structured. It is also increasingly a crucial dimension of applied evolutionary biology, for instance, in understanding species invasions and impacts of climate change (Hoffmann and Sgrò 2011), mitigating the risks of extinction of endangered species, or analyzing the conditions for disease emergence and the evolution of antibiotic resistance. There is the potential for creative application of many of these ideas to urgent applied questions. Consider conservation of an endangered species, which is declining because of environmental change. One hopeful message of models of niche conservatism and evolution is that in altered environments, anything that can be done that can improve the demographic performance of a population—even though the feasible measures that can be applied on their own cannot save the population—can indirectly make it more likely that evolution can help rescue it from extinction. Conversely, an understanding of these issues can help craft management strategies to prevent unwanted niche evolution, such as the evolution of resistance by microbes to antibiotics, or of agricultural pests to control measures. The central unifying theme of niche evolution and conservatism in ecology and evolutionary biology is one that cries out for a much deeper understanding, both empirically and theoretically.

FURTHER READING

Chase, J. M., and M. A. Leibold. 2003. Ecological Niches: Linking Classical and Contemporary Approaches. Chicago: University of Chicago Press. *An accessible text treating ecological perspectives of both the Hutchinsonian niche and the impact dimension of the niche.*

Futuyma, D. J. 2010. Evolutionary constraint and ecological consequences. Evolution 64: 1865–1884. *An overview of how evolution can often be constrained, and how this sheds new light on a range of classical issues in ecology and evolutionary biology.*

Gomulkiewicz, R., and R. D. Holt. 1995. When does evolution by natural selection prevent extinction? Evolution 49: 201–207. *A characterization of the requirements for evolution by natural selection to act sufficiently strongly and rapidly to permit a species to persist in a novel, harsh environment.*

Hoffmann, A. A., and C. M. Sgrò. 2011. Climate change and evolutionary adaptation. Nature 470: 479–485. *A review of examples and theory related to rapid evolution in novel environments.*

Holt, R. D. 1996. Demographic constraints in evolution: Towards unifying the evolutionary theories of senescence and niche conservatism. Evolutionary Ecology 10: 1–11. *An exploration of alternative genetic causes for niche conservatism in source-sink environments, such as trade-offs and deleterious mutational loads.*

Holt, R. D., and M. Barfield. 2011. Theoretical perspectives on the statics and dynamics of species' borders in patchy environments. American Naturalist 178: S6–S25.

Hooper, H. L., R. Connon, A. Callaghan, G. Fryer, S. Yarwood-Buchanan, J. Biggs, S. J. Maund, T. H. Hutchinson, and R. M. Sibly. 2008. The ecological niche of *Daphnia magna* characterized using population growth rate. Ecology 89: 1015–1022. *An excellent empirical study quantifying the ecological niche for a zooplankter.*

Hutchinson, G. E. 1978. An Introduction to Population Ecology. New Haven, CT: Yale University Press. *A classic text discussing the niche concept, with strong ties to community ecology.*

Schoener, T. W. 2009. Ecological niche. In S. Levin, ed., The Princeton Guide to Ecology. Princeton, NJ: Princeton University Press.

Wiens, J. J., D. D. Ackerly, A. P. Allen, B. L. Anacker, L. B. Buckley, H. V. Cornell, E. I. Damschen, et al. 2010. Niche conservatism as an emerging principle in ecology and conservation biology. Ecology Letters 13: 1310–1324. *A synoptic review of niche conservatism, in terms of its existence, the mechanisms that explain it, and its implications.*

III.15

Adaptation to the Biotic Environment

Sharon Y. Strauss

OUTLINE

1. Defining adaptation to the biotic environment
2. Differences between adaptation to biotic versus abiotic environments
3. Factors that influence adaptation to the biotic environment, and our ability to detect them
4. Conflicting selection and community complexity complicate detection of biotic adaptation
5. Lessons from introduced species
6. Changing relative abundances of species may alter selection from the biotic environment

Adaptation to the biotic environment describes the evolutionary response of a population to the web of interactions with other organisms that influence fitness.

GLOSSARY

Conflicting Selection. When selection acts in opposing directions on a trait, often as a result of interactions with different species, changing environmental conditions, or differences in ecology between sexes or life stages within a species.

Conifer. A cone-bearing, typically evergreen tree or shrub, such as pines, firs, etc.

Density Dependent. When the strength of an effect or an interaction is correlated with population size.

Frequency Dependent. When the strength of an effect or interaction is correlated with relative abundance; frequency dependence can act with respect to the relative abundance of genotypes within a population, or the relative abundance of a species within a community.

Mesocosm. Small-scale outdoor, semicontrolled ecosystem, typically in ponds or streams, in which water and natural community members are known or controlled, and in which some species or conditions are experimentally manipulated. Mesocosms lie between ends of the continuum running from laboratory fish-tank studies to large-scale whole lake or watershed manipulations.

Mutualistic. When an association between two organisms is beneficial to both.

Phenotype. An organism's observable characteristics or traits, including morphology, development, biochemical or physiological properties, and behavior; phenotypes are typically influenced both by genes (the organism's genotype) and by the environment.

Phytochemical. A chemical compound, typically bioactive, found in plant tissues and not directly involved in the primary metabolic pathways of photosynthesis or respiration.

Plasticity. The ability of a genotype to change its phenotype in response to changes in the environment.

1. DEFINING ADAPTATION TO THE BIOTIC ENVIRONMENT

Organisms naturally occur in environments characterized by abiotic physical properties such as temperature and water availability, and by biotic properties stemming from the interactions among organisms. The *biotic community* is defined as the collection of interacting or potentially interacting organisms across both local and regional spatial scales. Individuals within communities interact directly, when, for example, one organism eats another, or indirectly, when one organism is affected by another via effects on a third. To illustrate the latter, when a predator eats a prey species A, which is a strong competitor of species B, then species B benefits from the predator of species A, even though it has no direct interaction with that predator. Indirect effects in communities can often be as strong as direct effects.

We could thus define *selection from the biotic environment* as the net selective effect of all organisms with which individuals within a population interact, either directly or indirectly. *Adaptation to the biotic environment* is the evolutionary response of a population to biotic selection. In several other contributions to this volume, aspects of evolution in response to the biotic environment are considered—coevolution, character displacement, niche evolution, and evolution in communities (see chapters III.7, VI.7, and VI.16). In this section, the focus will be on evidence for adaptation to the biotic environment in general.

2. DIFFERENCES BETWEEN ADAPTATION TO BIOTIC VERSUS ABIOTIC ENVIRONMENTS

Douglas Schemske (2009) has argued that a fundamental difference between adaptation to biotic and abiotic environments derives from the dynamic versus static nature of selection from these sources. This argument does not suggest that abiotic conditions like climate aren't changing but rather that if an organism is adapted to temperatures of, for example, –5 °C, that organism remains adapted to that temperature through time. Many temperate tree species have adaptations that prevent water in their cells from freezing and rupturing cell membranes at –5 °C, and many tropical species have been unable to adapt to temperatures at which water is solid. If a tree possesses an adaptation that prevents its freezing solid at –5 °C, that tree will remain adapted to that temperature over its lifetime (barring huge stressors that might compromise its overall functionality). A genetically based adaptation can be passed on to offspring that will also possess resistance to freezing at –5 °C, and so on, for generations. In contrast, a selective agent that is biotic—an agent whose individuals are living, reproducing, recombining, and mutating—can be a moving target. Selection by biotic agents differs from abiotic selection in that the agents can change through time as a result of evolution or coevolution. The degree to which organisms remain adapted to interactions with other organisms may therefore change through time, and it is this feature in particular that differentiates adaptation to biotic and abiotic environments.

3. FACTORS THAT INFLUENCE ADAPTATION TO THE BIOTIC ENVIRONMENT, AND OUR ABILITY TO DETECT ADAPTATION

Several factors might affect the degree to which organisms are adapted to their biotic environment, as well as our ability to detect such adaptation. These include, but are not limited to, the relative generation times of interactors and the nature of selection by community members—specifically the degree to which one or a few selective agents dominate the selective landscape, the degree to which community composition fluctuates through time and space, and the possibility that adaptation in one species might drive another species to extinction.

Differences in generation times among evolving interactors can be important determinants of the scale at which we find biotic adaptation. A tree that lives 500 years, with a generation time of 50 years, will present a relatively similar phenotype to thousands or millions of its tiny-scale insect herbivores that can reproduce every three weeks during the course of the growing season. The scale insects may thus be able to generate huge amounts of genetic variation on which selection can act over the life span of the tree. This asymmetry in generation time and population size may promote adaptation by scale populations to individual trees (which has, in fact, been documented). Trees, on the other hand, having long life spans, are unlikely to show adaptation to specific populations of scale insects, though they might be generally adapted to scale insect feeding. Large asymmetries in generation time, as also in humans and microbes, can generally affect the patterns and scale of adaptation to biotic interactions in these interacting partners.

4. CONFLICTING SELECTION AND COMMUNITY COMPLEXITY COMPLICATE DETECTION OF BIOTIC ADAPTATION

One of the challenges in detecting adaptation to the biotic environment lies in the difficulty of measuring the effects of selection exerted by many interacting species at once. For example, a population might be adapted to its local, diverse biotic environment, but because the community is diverse, a focal species may not be adapted to selection from any one interactor alone. This scenario would be especially true if there were conflicting selection from a variety of interactors. For example, if one finds that a focal species does not display the optimal phenotype for interacting with some key community members, one might conclude that the focal species is not adapted to the biotic environment; however, this conclusion could be in error, if interactors important in shaping selection were not included in the original study. In the complex communities in which species live, including both direct and indirect effects that shape selection, it can often be extremely difficult to identify key selective agents on traits, and to measure their interactive effects in shaping the phenotypes of species in communities.

If conflicting selection poses an obstacle to detecting adaptation to the biotic environment, then one approach might be to explore adaptation to biotic environments in simpler communities, or to explore traits for which we feel very confident that only one or a few interactors play the major selective role.

Simpler Communities

It is recently appreciated that co-occurring populations of interacting species may be locally adapted to one another; these effects have typically been investigated in low-diversity communities. In artificial, decades-old pasture communities planted with only a few species, genotypes of the perennial clover *Trifolium repens* grow best when planted with genotypes of the perennial rye grass *Lolium perenne* from their own field than with *Lolium* from different fields; thus, there appears to be adaptation by clovers to local long-lived genotypes of the competing grass.

Similarly, in experimental mesocosms, Palkovacs and others (2009) manipulated the presence and population of origin of two common Trinidadian fish competitors, the guppy (*Poecilia reticulata*) and a killifish (*Rivulus hartii*). They then measured a number of ecosystem properties in these mesocosms, including the biomass of aquatic invertebrates, prey of these fish. Killifish-guppy coevolution significantly influenced the standing biomass of aquatic invertebrates and other ecosystem properties. Locally coevolved populations more effectively exploited invertebrates in the environment than populations of the same fish species that had no prior history with one another.

David Reznick and others document adaptations by fish prey to the introduction of fish predators in these same relatively simple fish communities (see chapter III.11). In experiments in which predators are introduced to previously predator-free pools, changes in phenotypes of guppy prey fish mirror patterns of phenotypic differences among guppy populations living in naturally predator-free or predator-rich pools. For example, prey guppies experiencing low-predation regimes are present at high densities and have fewer food resources per individual than those from high-predation sites. Even when experimentally reared in the same environment, guppies from low-predation sites mature earlier and at smaller size. These different phenotypes are adaptive; that is, phenotypes from high-predation environments have higher fitness in the presence of predators than those from low-predation environments, and vice versa. Similar kinds of adaptations to predators are now being found in ocean fish stocks overharvested by human predators. Humans are causing evolution in prey fish populations by selecting for earlier age at reproduction and reproduction at smaller sizes. Such selection may exacerbate losses of fishing yields, as fish are smaller, reproduce less, and are less abundant.

Other simple systems also provide evidence for adaptation to predation. On islands, where species diversity is often naturally lower than on mainlands, there is clear evidence for selection on lizard prey limb length by ground-dwelling predators. When predators are experimentally introduced to islands, prey lizards move from primarily ground dwelling to bush dwelling, and undergo changes in limb morphology that are likely to have both plastic and genetic bases.

Rapid adaptation to biotic interactions also is found in simple laboratory communities (see chapter III.6). We can document trait changes and adaptation in algal prey to rotifer predators, in bacterial hosts to viruses and vice versa, in flies to parasites, and in competing flour beetles to each other.

In summary, when strong directional selection is imposed on populations as a result of biotic interactions with one or two interactors, in most cases, we see evolution in traits, and the resultant phenotypes are better adapted to these interactions than are those in previous generations.

Complex Natural Communities

Detecting adaptation to interactions in diverse natural communities where multiple species can have comparable selective impacts is a great challenge. Species often select for opposing trait values, or selection from one species may result in trait changes that affect other species through genetic linkages between traits. Unlike the lizards on islands experiencing predation from one or two ground-dwelling predators, tadpoles living in ponds can be subject to predation from fish, amphibians, and a suite of invertebrates like diving beetles, dragonfly larvae, and bugs. These predators feed on tadpoles in different ways—some lie in wait at the bottom, others chase in the open water, some are fast, others slow. This diversity of predators prevents a single adaptive solution to reduce predation. Conflicting selection from diverse predator assemblages results in tadpoles that have plastic responses to changing predation regimes, and that may not be optimally adapted to any single predator. As mentioned above, in complex species-rich communities, the norm for most communities, it becomes very difficult to distinguish between a lack of adaptation to biotic interactions, and adaptation that integrates across many diverse and often-conflicting selective pressures.

Despite these difficulties, adaptation to biotic interactors in diverse natural communities has been shown, though typically when only one or a few interactors have large effects on an organism's fitness. Good examples of adaptation to the biotic environment in nonisland native communities come from studies of conifer species and the animals that eat seeds borne in conifer cones. Conifer cone shapes and seed number are determined largely by whether the seeds are preyed on by birds, squirrels, or moths, each of which selects for a different suite of cone traits. Lodgepole pine cones adapted to squirrel predation have few seeds and are relatively wide, whereas cones adapted to crossbill bird predators when no squirrels are present have thicker scales and more seeds; crossbills in these squirrel-free areas have also evolved deeper bills and a better ability to handle thick-scaled cones. When humans introduced squirrels to two previously squirrel-free islands, cone crops adapted only to feeding by crossbills were decimated by the squirrels, which drove the crossbills locally extinct. To complicate the selective landscape, in some areas, cone-feeding moths are also present at high densities, and they select for small cones with few seeds. Selection by moths counters selection by crossbills for large cone size and large seed number; cone shapes reflect the dominant cone predators in the local area.

In a related tree species, the limber pine, variation in cone structure is increased by conflicting selection pressures exerted by a mutualistic seed-dispersing bird, Clark's nutcracker, and pine squirrels that eat seeds. Variation in cone structure is twice as great where both agents of selection co-occur than where only the disperser is present, suggesting that spatial variation in community diversity can enhance and maintain variation in cone traits, and that cone traits are evolving in response to these cone-feeding community members.

Adaptation to Species' Interactions Might Result in Extinction

A possible outcome of adaptation to an interaction is the extinction of one or all interacting species. For example, a predator or pathogen might adapt to using a prey species so efficiently that it drives that species to extinction, at least locally. The results of such adaptation are difficult to observe, as it is hard to detect the absence of a species, let alone know that adaptation caused such an absence. We have some indications, however, that adaptation could cause species' extinctions. Many aquatic organisms produce propagules that remain in resting stages in lake sediments and that can be brought out of dormancy, even after many years. Laminated lake sediments provide a temporal sequence of the genotypes present through time for such organisms. Ellen Decaestecker and colleagues used lake sediments to "revive" genotypes of *Daphnia* water flea hosts and genotypes of its microsporidian parasite in a temporal sequence. From the time it was first detected in the sediments, the parasite successively increased the number of spores it made each generation, becoming increasingly more virulent to the *Daphnia* host through time. In this case, both species continued to persist in the system, but one might imagine that such parasite adaptation could result in extinction of the host, or of both species. Thus, adaptation to another species does not necessarily result in coexistence, but extinction as a result of adaptation is hard to detect, as the interaction is transient.

5. LESSONS FROM INTRODUCED SPECIES

Some of the best evidence that organisms are adapted to their biotic environment comes from cases in which new biotic interactors are introduced to an area by humans and have profound ecological and evolutionary effects on existing, evolutionarily naive local populations and species.

Introductions on islands illustrate how the lack of adaptation to a biotic interactor can cause extinction, as reviewed by Fritts and Rodda (1998). Extinctions of local fauna on previously predator-free islands from introduced predators like rats, cats, snakes, and mongoose have been rampant. Most of these native species of birds, mammals, and reptiles lacked the genetic or plastic variation to respond adequately to new potent interactions.

Introductions on the continental scale can also have large evolutionary effects. Some species, when introduced to new areas, experience explosive population growth, even though they might be uncommon in their native habitat. One hypothesis that has received substantial support suggests that organisms introduced to new areas may escape enemies that otherwise hold their populations in check in their native habitats, or that they possess novel weapons that allow them to easily overcome native, evolutionarily naive species. Whatever the underlying mechanism, when such explosive population growth happens, there are often large ecological impacts that, in turn, can generate strong selection from the introduced species on native species, and sometimes, vice versa.

A classic example of evolution in both native and invader species is the introduction of cane toads, originally from Central and South America, to Australia. Cane toads are poisonous, secreting bufotoxin from glands behind their eyes and on their backs, and are also poisonous as tadpoles. Many naive Australian predators that typically eat native frogs and tadpoles have died from eating newly introduced toxic cane toads. Selection from eating cane toads has favored snake individuals with smaller mouths, perhaps because smaller-mouthed predators can eat only small, less toxic toads, and thus have the opportunity to survive and learn from their mistakes. Extremely strong, novel directional selection from a new agent can override historical patterns of selection, presumably exerted at least in part by interactions among native biota. Numerous evolutionary changes, morphological and behavioral, have occurred in both the native Australian fauna and in the toads. Some native predators have adapted by evolving genetically based aversion to the toad, and others by simply learning not to eat them, by bitter experience.

6. CHANGING RELATIVE ABUNDANCES OF SPECIES MAY ALTER SELECTION FROM THE BIOTIC ENVIRONMENT

Species' relative abundances often fluctuate greatly from year to year, and these fluctuations may alter the strength and direction of selection from the biotic environment, simply because the proportion of individuals engaged in the interaction will then vary. In the context of adaptation to biotic interactions, changes in relative abundance or relative frequency of interactors may result in shifting selection pressures, as John Thompson (2005) discusses. Changes in the population size of a species affect two separate, but related, ecological properties: processes that act in a density-dependent fashion, and processes that act in a frequency-dependent manner. Both changes in density and changes in relative frequency of interactions within and among species can lead to cascading changes in other species.

The degree and direction of conflicting selection will thus reflect not only which species are present or absent but also their relative abundances. In a study that manipulated the relative abundance of specialist aphid and generalist mollusk herbivores of black mustard plants, Lankau and Strauss (2008) showed that generalist herbivores, which eat many plant species, imposed strong selection for high levels of a plant phytochemical defense, while specialist herbivores, which eat only plants in the mustard family, were unharmed by these defenses, and actually preferred the larger, more defended, plants —those that had escaped early attack by generalists. Moreover, there was also evidence that producing phytochemical defenses was costly to plants. Thus, in a year with many slug and snail generalists, selection is expected to favor highly defended plants, whereas in a year when these herbivores are less abundant, selection will favor less defended plant genotypes that do not incur the costs of defense. Selection owing to the complexity of communities and shifting relative abundances of species thus suggests that adaptation to biotic environments must reflect adaptation to fluctuating community composition. A related consideration is how selection changes as a function of the relative frequency of interactions between individuals of the same species versus other species.

Up to now, the discussion has focused on adaptation to a biotic environment comprising other species, but populations also exhibit traits adapted to differing densities of their own species. Within-species competition is often very strong, and traits favored at high population density often differ from those favored at low density. In black mustard plants, individuals producing higher levels of the phytochemical sinigrin, a compound that reduces fungal densities in the soil, are good competitors against other species that rely on soil fungi for nutrient acquisition. In contrast, mustard plants themselves do not need these fungi, and the costs of sinigrin production are high; so, in patches of high densities of black mustard plants, individuals producing low levels of sinigrin have an advantage. Thus, sinigrin production is favored in competition with other species, and selected against in competition with individuals of the same species. Moreover, low sinigrin-producing individuals lose in competition to other species. This set of costs creates a cyclic network in which high-sinigrin genotypes lose to low-sinigrin genotypes that lose to other species that, in turn, lose to high-sinigrin genotypes. The genetic diversity within black mustard and the species composition of the community are both maintained, resulting in changing relative frequencies and abundances of these genotypes and species, and also shifting selection on the sinigrin trait. In the absence of evolution in the trait, diversity in the community would not be maintained.

Other examples of adaptation to population density can be found in behavioral and immunological traits. Barnes and colleagues have shown (2011) that the frequency of an allele associated with resistance to pathogens such as tuberculosis and leprosy in humans is greater in older urban settlements. Human populations with a long history of living in towns were more resistant to these infections, suggesting adaptation to high-density living. In another study of how the nature of selection depends on density, *Lolium perenne* grass cultivars differed in their infection rates by a rust fungus in plots where grass densities were high; the same cultivars showed no differences in infection rates when grown in low-density plots, where infection rates were overall much lower. Thus, differences in grass densities among plots alter selection on disease resistance in *L. perenne*.

Animal "personalities" are also traits that can respond to density and frequency, and that appear to have both environmental and heritable components. These traits include aggression against conspecifics and/or heterospecifics, the tendency to disperse and colonize new habitats, and sociality. Each of these behavioral traits may have costs and benefits that are density or frequency dependent. Colonizers reach new habitats with ample resources, but may pay the costs of dispersal in loss of reproduction or increased predation. In the intertidal zone, long-distance dispersing bryozoans (moss animals) colonized more areas, but also had much lower reproductive rates than short-dispersal colonizers. It has been shown in social spiders and other species that more aggressive, bold genotypes are generally better defenders of territories and resources but may suffer costs of aggression when frequencies of aggressive individuals increase. In small mammals and birds that exhibit large fluctuations in population density, numerous studies document changes in behavior, immunological condition, and body size. Some of these changes have a genetic basis, many of them are plastic, and many exhibit genetically based plasticity. Thus an important aspect of adaptation to the biotic environment includes traits that vary in their contribution to fitness at high versus low population densities. As in the black mustard examples, these trait changes can also have impacts on direct and indirect interactions with other species in the community.

In summary, there is good evidence that organisms exhibit adaptation to the biotic communities in which they are embedded. Our best evidence comes from simplified communities where the impacts of one or a few species on the evolution of a focal species, or on each other, are easily detected. As community complexity increases, and as species fluctuate in composition and in their relative densities across both time and space, detection of adaptation to the biotic environment becomes more difficult. Recent studies showing local adaptation by genotypes to other genotypes present in the same community point to the importance of adaptation to biotic interactions even in more complex communities. Ecologically important invasive species enable us to watch evolutionary changes occurring rapidly throughout communities. A current challenge remains to devise experiments and observations to detect the importance and prevalence of adaptation to interactions in complex communities, and to determine its role in ecosystem function.

FURTHER READING

Aarssen, L. W., and R. Turkington. 1985. Biotic specialization between neighboring genotypes in *Lolium perenne* and *Trifolium repens* from a permanent pasture. Journal of Ecology 73: 605–614. *A classic paper that shows adaptation by neighbors to each other.*

Barnes, I., A. Duda, O. G. Pybus, and M. G. Thomas. 2011. Ancient urbanization predicts genetic resistance to tuberculosis. Evolution 65: 842–848.

Bassar, R. D., A. Lopez-Sepulcre, M. R. Walsh, M. M. Turcotte, M. Torres-Mejia, and D. N. Reznick. 2010. Bridging the gap between ecology and evolution: Integrating density regulation and life-history evolution. Year in Evolutionary Biology 1206: 17–34.

Benkman, C. W., T. L. Parchman, and E. T. Mezquida. 2010. Patterns of coevolution in the adaptive radiation of crossbills. Year in Evolutionary Biology 1206: 1–16. *A review of the body of work on coevolution between crossbills, conifers, and other species that interact with cones.*

Bohannan, B.J.M., and R. E. Lenski. 2000. Linking genetic change to community evolution: Insights from studies of bacteria and bacteriophage. Ecology Letters 3: 362–377. *This paper shows how microorganisms in vials can evolve in the laboratory to partition resources and habitats.*

Decaestecker, E., S. Gaba, J.A.M. Raeymaekers, R. Stoks, L. Van Kerckhoven, D. Ebert, and L. De Meester. 2007. Host-parasite "Red Queen" dynamics archived in pond sediment. Nature 450: 870–873. *Study uses a technique to revive ancestral genotypes buried in lake sediments and thus reconstructs past coevolutionary relationships.*

Fritts, T. H., and G. H. Rodda. 1998. The role of introduced species in the degradation of island ecosystems: A case history of Guam. Annual Review of Ecology and Systematics 29: 113–140. *A sobering review of effects of the brown tree snake in Guam.*

Lankau, R. A., and S. Y. Strauss. 2008. Community complexity drives patterns of natural selection on a chemical defense of *Brassica nigra*. American Naturalist 171: 150–161. *An illustration of how different herbivores exert opposing selection on plant defense traits.*

Lankau, R. A., and S. Y. Strauss. 2011. Newly rare or newly common: Evolutionary feedbacks through changes in population density and relative species abundance, and their management implications. Evolutionary Applications 4: 338–353. *A review of how rapid changes in density or frequency of a species can result in strong selection and even maladaptation.*

Schemske, D. W. 2009. Biotic interactions and speciation in the tropics. In R. K. Butlin, J. R. Bridle, and D. Schluter, eds., Speciation and Patterns of Diversity. Cambridge: Cambridge University Press.

Shine, R. 2010. The ecological impact of invasive cane toads (*Bufo marinus*) in Australia. Quarterly Review of Biology 85: 253–291.

Thompson, J. N. 2005. The Geographic Mosaic of Coevolution. Chicago: University of Chicago Press. *A book describing the impacts of shifting community composition across the landscape on trait evolution, and coevolution between species.*

Touchon, J. C., and K. M. Warkentin. 2008. Fish and dragonfly nymph predators induce opposite shifts in color and morphology of tadpoles. Oikos 117: 634–640.

IV

Evolutionary Processes

Michael C. Whitlock

As the great population geneticist (and statistician) Sir Ronald Fisher said in the first sentence of his foundational *Genetical Theory of Evolution*, "Natural selection is not evolution." A population evolves when the frequencies of its genotypes change over time. The most important of these changes are typically caused by natural selection, but selection is not the only mechanism by which evolution occurs. When alleles are passed from one generation to the next, the next generation may by chance not exactly match the generation of its parents, especially if the population size is small. Alleles can mutate to new alleles, and alleles can arrive by migration from genetically diverged populations. DNA sequences may recombine with genetically distinct sequences. The details of mating can matter, because genotype frequencies can change as a result of mating between relatives, mating between similar individuals, or mating with nearby individuals. All these factors—random genetic drift, mutation, migration, recombination, and nonrandom mating—can change the genotype frequencies in a population from one generation to the next; in other words, they can cause *evolution*.

This section discusses these nonselective evolutionary processes and some of their important consequences. We start with a discussion of random genetic drift (see chapter IV.1). In a finite population, chance in part dictates which individuals happen to succeed and leave offspring; therefore, chance can cause the allele frequency to change from one generation to the next. This random process, called *genetic drift*, changes the allele frequency from one generation to the next. Such drift has greatest effect for alleles that have very small effects on fitness. However, most of the genetic variation in a typical genome may be close to selectively neutral, making drift an important part of the evolutionary process for a large fraction of the genome.

New alleles appear in populations as a consequence of mutation (see chapter IV.2). Ultimately, without mutation the evolutionary process would cease, because all genetic variation has its origin in a mutation and all evolution depends on genetic variation. Mutation brings in new variants that may be selectively beneficial, but it also introduces alleles that are deleterious for their carriers. Some mutations may even have little selective effect at all. The net effect on evolution of mutation rests in the balance of these classes of mutants and in the ways mutation interacts with other evolutionary processes.

Dispersal and migration are also sources of new alleles to a population; moreover, gene flow from other places tends to make a population more similar to other populations in the species (see chapter IV.3). Gene flow is the glue holding species together; without dispersal, populations would grow more and more different from each other until they were unrecognizable as the same species. Movement between populations can bring in valuable new alleles that have proven successful elsewhere; on the other hand, an influx of alleles from elsewhere can interfere with adaptation to local conditions.

Migration, mutation, and drift change allele frequencies and genotype frequencies. Other evolutionary processes change the frequencies of combinations of alleles without directly affecting the frequencies of those alleles; for example, recombination mixes alleles into new chromosome combinations, changing the patterns of association of alleles at different loci (see chapter IV.4). If new beneficial alleles appear on different genetic backgrounds, only recombination can bring them together so that all individuals might benefit from both. Most of the other evolutionary processes—including selection, mutation, migration, and drift—can generate associations between alleles at different loci, and recombination whittles those associations away. The benefits of recombination are largely responsible for the evolution of sex, a characteristic feature of many organisms.

Deleterious alleles repeatedly appear in every population by mutation. They are ultimately removed mainly by selection, but this is not instantaneous. As a result, populations usually have extremely large numbers of deleterious alleles, mostly at low frequency, and for some species each individual may carry hundreds or even thousands of alleles that reduce fitness (see chapter IV.5). This causes a reduction in the mean fitness of the population called *mutation load*. Other factors can reduce fitness of a population as well; for example, genetic drift can cause mildly deleterious alleles to reach high frequencies and recombination can break up favorable combinations of alleles that work well together.

Inbreeding and other forms of nonrandom mating affect the combinations of alleles at the same locus in diploid individuals; more inbred individuals are more likely to have two copies of the same allele at a locus than expected in a randomly mated population (see chapter IV.6). As a result, inbred individuals are more likely to be homozygous with increased expression of the effects of recessive alleles. Many of these recessive alleles are deleterious, meaning that inbred individuals can have reduced fitness relative to what is possible for outbred individuals (*inbreeding depression*).

Inbreeding is not a fixed property of a species, however, and the tendency of an organism to inbreed can respond to selection and evolve over time. Many plants (and some animals) are capable of an extreme form of inbreeding called *selfing*, in which a hermaphrodite individual fertilizes its own ovules with its own sperm or pollen. All else being equal, this gives the selfing individual a genetic advantage, because it can transfer two copies of its genes into each offspring. Many factors promote an increase in the rate of selfing; many, including inbreeding depression select against self-mating. Mating system evolution is a fascinating, active area of investigation (see chapter IV.8).

The mechanisms of evolution discussed in this section usually take as a starting place the rules of genetic inheritance as described by Mendel; however, in many cases, evolution can occur because of biases in transmission of genetic elements. "Selfish genetic elements" can transfer from one generation to the next in excess of the proportions expected under Mendelian inheritance (see chapter IV.7). Differences between one allele and another can cause one allele to be passed into offspring differentially; in other words, there can be selection within an individual for which genetic material is transmitted to offspring. This can take the form of genetic parasites like transposable elements or reproductive parasites like *Wolbachia* infections that can sometimes change sex ratios of their hosts' offspring to their own advantage. Moreover, some alleles can differentially be passed to offspring at the expense of the other copy of a gene in the parent. These processes can have dramatic effects, ranging from skewed sex ratios to potential speciation. Evolution can occur even in the seemingly simple process of transmission of genetic material to offspring.

Both selective and nonselective processes can generate evolution. The demographic and genetic processes discussed in this section can have strong influences on each other, and on the mechanisms by which selection operates. The amount of genetic variation in a species is the raw material of evolution, and this variation is determined by a balance between different kinds of selection, genetic drift, mutation, migration, recombination, and patterns of mating. Each of these processes interacts in myriad ways, giving rise to many important features of biological evolution. None of these factors alone—even selection—is sufficient to understand evolution.

In the evolutionary theater, natural selection without doubt is the star of the show, but important roles are played by mutation, migration, recombination, drift, and details of the transmission of genetic material. These other players change the pace and direction of evolution, and without them the outcome of natural selection would be completely different.

IV.1

Genetic Drift

Philip Hedrick

OUTLINE

1. Genetic drift
2. Effective population size
3. Neutral theory
4. Coalescence
5. Future directions

Genetic drift is the chance change in genetic variation resulting from small population size. The effective population size, which can incorporate unequal numbers of male and female parents, variation in progeny number, or variation in numbers over different generations, is a useful concept for understanding genetic drift. The neutral theory incorporates the effects of genetic drift and mutation to understand the amount and pattern of molecular genetic variation. Coalescent approaches provide a way to estimate the past population size and other evolutionary factors.

GLOSSARY

Coalescence. The point or event in the past at which common ancestry occurs for two alleles at a gene because of genetic drift.

Effective Population Size (N_e). An ideal population that incorporates such factors as variation in the sex ratio of breeding individuals, the offspring number per individual, and numbers of breeding individuals in different generations.

Founder Effect. Impact on genetic variation in a population when it grows from a few founder individuals.

Genetic Bottleneck. A period during which only a few individuals survive and become the only ancestors of the future generations of the population.

Genetic Drift. Chance changes in allele frequencies that result from small population size.

Linkage Disequilibrium. Statistical association of alleles at different loci.

Molecular Clock. A constant rate of genetic substitution over time for molecular variants.

Neutral Theory. The theory that states that genetic change is primarily the result of mutation and genetic drift, and that different molecular genotypes are neutral with respect to each other.

Population. A group of interbreeding individuals existing together in time and space.

The primary goals of population genetics are to understand the genetic factors determining evolutionary change and stasis and the amount and pattern of genetic variation within and between populations (Hartl and Clark 2007; Hedrick 2011). In the 1920s and 1930s, shortly after widespread acceptance of Mendelian genetics, the theoretical basis of population genetics was developed by Ronald A. Fisher, J.B.S. Haldane, and Sewall Wright. As they showed, the amount and kind of genetic variation within and between populations are potentially affected by selection, inbreeding, genetic drift, gene flow, mutation, and recombination. Fisher thought that selection was most important and that genetic drift played only a minor role in evolutionary genetics, whereas Wright advocated a central role for genetic drift as well as selection, and in fact, genetic drift was sometimes called the "Sewall Wright effect."

Generally, these evolutionary factors can have particular effects; for example, genetic drift and inbreeding can be considered to always reduce the amount of variation, and mutation to always increase it. Other factors, such as selection and gene flow, can either increase or reduce genetic variation, depending on the particular situation. In addition, the factors other than genetic drift generally have deterministic effects; for example, given certain relative fitness values for genotypes, selection results in a predictable genetic change. Genetic drift is different from these other factors in that it has a nondeterministic or stochastic effect; that is, genetic changes resulting from genetic drift are random in direction.

The development of population molecular data in the late 1960s, DNA sequence data in the 1980s, and genomic data in recent years have revolutionized population genetics and produced many new questions and some answers. Population genetics and its evolutionary interpretations provided a fundamental context in which to interpret this new molecular genetic information. For example, Motoo Kimura in 1968 introduced the *neutral theory of molecular evolution* that assumes that genetic variation results primarily from a combination of mutation-generating variation and its elimination by genetic drift (Kimura 1983). This theory is called neutral because allele and genotype differences at a gene are selectively neutral with respect to each other. This theory is consistent with many observations of molecular genetic variation.

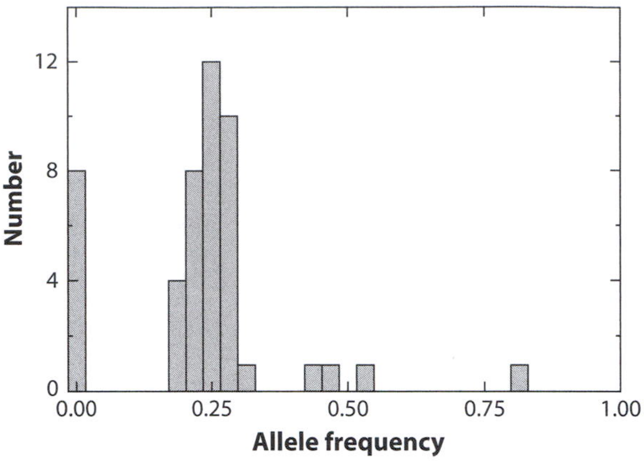

Figure 1. Histogram of the number of the alleles with different frequencies for the 11 microsatellite loci in lake trout from Swan Lake. (After Kalinowski et al. 2010.)

1. GENETIC DRIFT

Before we discuss genetic drift, let us first define the evolutionary or genetic connotation of the term *population*. As a simple ideal, a population is group of interbreeding individuals that exists together in time and space. *Genetic drift* refers to chance changes in allele frequency that result from the sampling of gametes from generation to generation in a population. Since the beginning of population genetics, there has been controversy concerning the importance of genetic drift. Part of this controversy has resulted from the large numbers of individuals observed in many natural populations, large enough to think that chance effects would be small in comparison to the effects of other factors, such as selection and gene flow.

Under certain conditions, a population may be so small that genetic drift is significant even for loci with sizable selective effects, or when there is significant gene flow. For example, some populations may be continuously small for relatively long periods of time because of limited resources in the populated area. In addition, some populations may have intermittent small population sizes. Examples of such episodes are the overwintering loss of population numbers in many invertebrates, and epidemics that periodically decimate populations of both plants and animals. Such population fluctuations generate *genetic bottlenecks*, or periods during which only a few individuals survive and become the only ancestors of the future generations of the population.

Small population size is also important when a population grows from a few founder individuals, a phenomenon termed *founder effect*. For example, many island populations appear to have started from a very small number of individuals. If a single female who was fertilized by a single male founds a population, then only four genomes (assuming a diploid organism), two from the female and two from the male, can start a new population. In plants, a whole population can be initiated from a single seed—only two genomes, if self-fertilization occurs. As a result, populations descended from a small founder group may have low genetic variation, or by chance have a high or low frequency of particular alleles.

Kalinowski et al. (2010) provided an excellent example of a founder effect in lake trout that invaded Swan Lake in Montana in the late 1990s. The number of founders was not observed, but samples taken less than a decade after they invaded provided a genetic signal. First, a limited number of alleles at 11 microsatellite loci, only four or fewer alleles with a frequency greater than 2 percent, were observed in the founders, while samples from the putative source, Flathead Lake, averaged more than 12 alleles per locus. Second, the allele frequencies in Swan Lake sample, clustered around 0.25, 0.5, and 0.75 (one, two, or three out of four copies) (figure 1), whereas many other alleles observed in Flathead Lake were not found. This suggests that the population was founded primarily by only two individuals, only four genomes, and that the chance effects of this founding event are reflected in the allele frequencies.

Another situation in which small population size is of great significance is one in which the population (or species) in question is one of the many threatened or endangered species (Allendorf et al. 2013). For example, all approximately 500 whooping cranes alive today descend from only 20 whooping cranes that were alive in 1920 because only a few had survived hunting and habitat destruction. All 200,000 northern elephant seals alive today descend from as few as 20 that survived nineteenth-century hunting on Isla Guadalupe, Mexico.

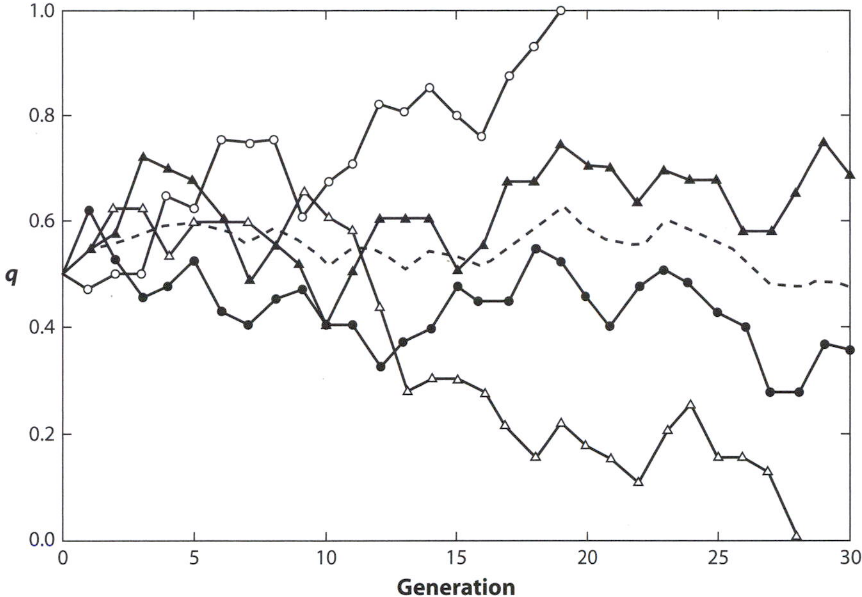

Figure 2. Frequency of allele A_2 over 30 generations for four replicates (solid lines) of a population of size 20. The mean frequency of allele A_2 is indicated by the broken line. (After Hedrick 2011.)

Further, all the living individuals of some species are descended from a few founders that were brought into captivity to establish a protected population, such as Przewalski's horses (13 founders), California condors (13 founders), black-footed ferrets (6 founders), Galápagos tortoises from Española Island (15 founders), and Mexican wolves (7 founders).

All these examples of restricted population size can have the same general genetic consequence: a small population size causes chance alterations in allele frequencies. Genetic drift has the same expected effect on all loci in the genome. In a large population, on average, only a small chance change in the allele frequency will occur as the result of genetic drift. On the other hand, if the population size is small, then the allele frequency can undergo large fluctuations in different generations in a seemingly unpredictable pattern and can result in chance fixation or loss of an allele. These effects describe both the impact of genetic drift over all the different loci (the total genome) in a given population and the impact of genetic drift at a single locus over replicate populations, as discussed below.

Figure 2 illustrates the type of allele frequency change expected in a small diploid population with two alleles, A_1 and A_2 (Hedrick 2011). This example uses Monte Carlo simulation with uniform random numbers to imitate the change in frequency of allele A_2 (q) in four replicate populations. Here the solid lines are the four replicates of a diploid population of size $N = 20$ ($2N = 40$), and the broken line is the mean frequency of A_2 over the four replicates. All the replicates were initiated with the frequency of A_2 equal to 0.5. One of these simulated replicates went to fixation for A_2 in generation 19, and another lost A_2 in generation 28. The other two replicates were still segregating for both alleles at the end of 30 generations. As shown here, genetic drift may cause large and erratic changes in allele frequency in a rather short time.

On the other hand, the mean frequency of A_2 for the four replicates varied much less: it ranged from 0.625 in generation 19 to 0.475 in generation 30 but was generally near the initial frequency of 0.5. If there are enough replicate populations, then there is no expected change in the mean allele frequency from genetic drift, so that

$$\bar{q}_0 = \bar{q}_1 = \bar{q}_2 \ldots \bar{q}_t \ldots \bar{q}_\infty,$$

where $\bar{q}_t$ is the mean frequency of A_2 in generation t over all replicates (0.5 in this example). The constancy of the mean frequency occurs because the increases in allele frequency in some replicates are cancelled by reductions in allele frequency in other replicates.

Individual replicates eventually either go to fixation for A_2 ($q = 1$) or to loss of A_2 ($q = 0$). The proportion of populations expected to go to fixation for a given allele is equal to the initial frequency of that allele (see chapter V.1). In other words, if the initial frequency of A_2 is q_0, and assuming that genetic drift is the only evolutionary factor influencing it, then the *probability of fixation* of that allele, $u(q)$ (proportion of replicate populations eventually fixed for it), is

$$u(q) = q_0.$$

For example, if the initial frequency of A_2 is 0.1, only 10 percent of the time will a population become fixed for that allele. On the other hand, if the initial frequency of A_2 is 0.9, 90 percent of the time it will become fixed. This can be understood intuitively because the amount of change necessary to go from a frequency of 0.1 to 1.0 is much greater than from 0.9 to 1.0. This finding is a fundamental aspect of the neutrality theory used in molecular evolution; that is, without differential selection, the probability of fixation of a given allele is equal to its initial frequency.

Because the mean allele frequency does not change but the distribution of the allele frequencies over replicate populations does, the overall effect of genetic drift is best understood by examining the heterozygosity (or the variance) of the allele frequency over replicate populations (or multiple, independent loci). This is because the heterozygosity decreases as allele frequencies get closer to 0 or 1, a general consequence of genetic drift as shown in figure 2. The simplest approach to understanding the general effect of genetic drift is to examine the relationship between the heterozygosity over time in a small diploid population of size N. The expression giving this relationship is

$$H_t = \left(1 - \frac{1}{2N}\right)^t H_0 \text{ or } \frac{H_t}{H_0} = e^{-t/2N},$$

where t is the number of generations and N is the effective population size (see below). For example, we can predict how much the level of heterozygosity is reduced after 30 generations from genetic drift with an effective population size of 20. In this case, $H_t/H_0 = e^{-30/40} = 0.472$, or the level of heterozygosity is predicted to be reduced by 52.8 percent. Although we had only four replicate populations in the example in figure 2, by generation 30, two replicates had become fixed, reflecting this expectation.

If we go back to the lake trout example of Kalinowski et al. (2010), the average heterozygosities in the source Flathead Lake and in Swan Lake are 0.88 and 0.68, respectively. Let us assume only one generation of genetic drift ($t = 1$), because around 7000 lake trout were already present in Swan Lake only 10 years (two generations) after their discovery there. Therefore, $H_0 = 0.88$ and $H_1 = 0.68$ (assuming $t = 1$), and solving for the equation above for N, then

$$N = \frac{H_0}{2(H_0 - H_1)} = 2.2,$$

again suggesting that there were primarily two founders that established this population.

2. EFFECTIVE POPULATION SIZE

The number of breeding individuals in a population may be much less than the total number of individuals in an area, the census population size, but even the breeding population number might not be indicative of the population size appropriate for evolutionary considerations. For example, other factors, such as variation in the sex ratio of breeding individuals, the offspring number per individual, and numbers of breeding individuals in different generations, may be evolutionarily important. As a result, the *effective population size* (N_e), a theoretical concept that incorporates variation in these factors and others, is quite useful (Charlesworth 2009).

The concept of the effective population size makes it possible to consider an ideal population of size N in which all parents have an equal expectation of being the parents of any progeny individual. In general, the effect of genetic drift in a diploid population is a function of the reciprocal of twice the effective population size, $1/(2N_e)$. If N_e is large, then this value is small and there is little genetic drift influence. Or, if N_e is small, then this value is larger and genetic drift may be important.

A straightforward approach often used to tell the impact of various factors on the effective population size is the ratio of the effective population size to breeding (or sometimes census) population size N, that is, N_e/N. Sometimes, this ratio is only around 0.1 to 0.25, indicating that the effective population size may be much less than the number of breeding individuals (Palstra and Ruzzante 2008).

Assuming there are N individuals in the population, N_f is the number of females, and N_m is the number of males ($N = N_f + N_m$), then the effective population size becomes

$$N_e = \frac{4N_f N_m}{N_f + N_m}.$$

If there are equal numbers of females and males, $N_f = N_m = ½N$, then $N_e = N$; however, in some species, the numbers of females and males are often unequal. Frequently, the number of breeding males is smaller than the number of breeding females ($N_m < N_f$), because some males mate more than once.

Let us assume the most extreme situation possible, one male mates with all the females in a colony or harem, as is thought to occur in some vertebrate populations where males control female harems, such as elephant seals. In this case, the expression above becomes

$$N_e = \frac{4N_f}{N_f + 1}.$$

Note that the maximum value of this expression, when N_f becomes large, is 4.0. In other words, because each sex must contribute half the genes to the progeny, restricting the number of breeding individuals of one sex can greatly reduce the effective population size.

There may be a nonrandom distribution of progeny (gametes) per parent because of genetic, environmental, or accidental factors. For example, some birds have strongly determined numbers of eggs in a clutch, so the variance of egg number in a clutch may be near zero. Or, in some human populations, a relatively uniform number of offspring per parent may lower variation because of efforts to control population growth. On the other hand, if whole clutches or broods survive or perish as a group, then the variance of progeny number may be larger. Even more extreme, in some organisms with very high reproductive potential, a substantial proportion of the progeny may come from only a few highly successful parents.

To examine the impact of variance in the number of offspring, let us assume that the population is not changing in size (the number of progeny per individual is two), then the effective population size is

$$N_e = \frac{4N - 2}{V + 2},$$

where V is the variance in the number of progeny. If $V = 2$ (the variance equals the mean number of progeny of two), then $N_e \approx N$. If $V = 0$, where there are exactly two progeny from each individual, then $N_e \approx 2N$ or $N_e/N \approx 2$. Therefore, if V is kept low, the effects of small population size can be avoided to some extent, and the effective population size may actually be larger than the breeding or census number; often, however, the variance in progeny number is larger than the mean, and as a result, N_e/N is lower than unity. In some organisms, such as many shellfish or fish, there may be very high variance in reproduction, where, in a given year, most of the recruited young may be from a few parents. For example, if $V = 40$, then $N_e/N \approx 0.1$.

When the effective population size varies greatly in size in different generations, it can have a large impact on the overall effective population size. The variation in population size can result from regular cyclic variation in population numbers, periodic decimation of the population because of disease or other factors, or seasonal variation in population numbers. When this occurs, the lowest population numbers determine, to a large extent, the overall effective population size, because after these bottlenecks, all remaining individuals are descendants of the bottleneck survivors.

The effective population size over t generations becomes approximately

$$N_e = \frac{t}{\sum \frac{1}{N_{e.i}}},$$

where $N_{e.i}$ is the effective population size in generation i. For example, assume that the population in three subsequent generations has effective population sizes of 10, 100, and 1,000. Applying the expression above gives the effective population size of 27.0, closest to the lowest of the three populations sizes in different generations, and much smaller than the mean census number of 370 and $N_e/N = 0.073$, a quite low proportion.

The effective population size can be estimated using demographic information such as sex ratios, variance in progeny production, and variance in N_e over time. In addition, N_e can be estimated from observations of the effect of genetic drift on genetic variation over time in a population. For example, in a small population, both the change in allele frequency between generations and loss of heterozygosity are expected to be much higher than in a large population. Another approach is to measure the *linkage disequilibrium*, or the statistical association of alleles at different loci, as an indicator of the effective population size. For large populations, very little association of alleles at different loci is expected (unless they are tightly linked), whereas for small populations, large associations can be generated by chance.

The most comprehensive estimate of N_e using linkage disequilibrium is by Tenesa et al. (2007) who used data from about 1 million SNPs (single nucleotide polymorphisms) in four human samples from Nigeria, Europe, China, and Japan that provided about 20 million closely linked SNP pairs. Figure 3 gives the estimates of N_e from these data for each of the 22 autosomes and the X chromosome. The average N_e estimates over all chromosomes for the Nigerian, European, Chinese, and Japanese samples are 6286, 2772, 2620, and 2517, respectively. The European, Chinese, and Japanese populations have very similar N_e estimates for nearly all the chromosomes, and the overall N_e estimate for the African sample is about 2.4 times as large. This pattern is consistent with the hypothesis that the non-African populations descended from a migrant African population that represented a subset of the variation present in Africa at that time.

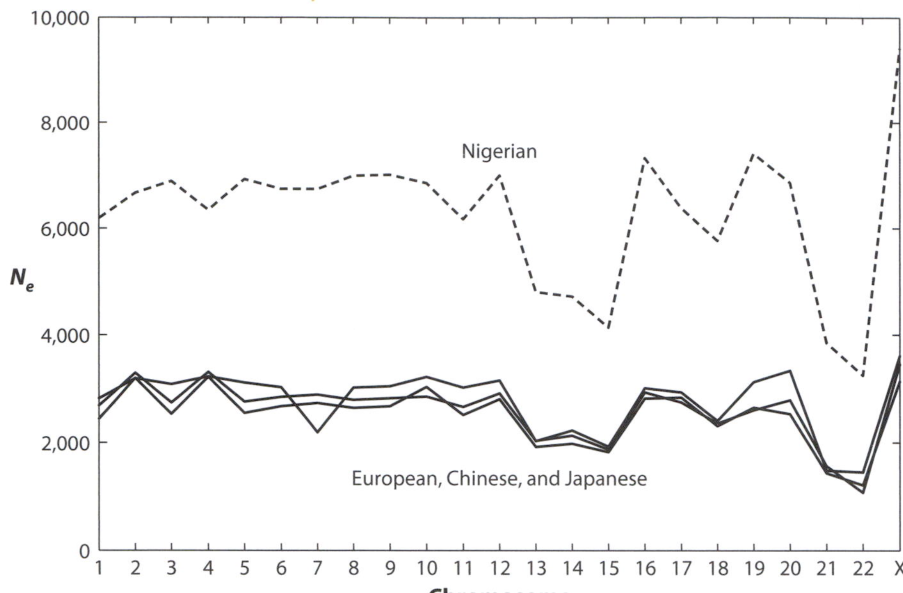

Figure 3. The effective population size for each chromosome estimated from linkage disequilibrium between about 20 million closely linked SNPs in four human populations. (After Tenesa et al. 2007.)

3. NEUTRAL THEORY

Neutral theory generally assumes that selection plays a minor role in determining the maintenance of molecular variants and proposes that different molecular genotypes have almost identical relative fitnesses; that is, they are neutral with respect to each other. The actual definition of selective neutrality depends on whether changes in allele frequency are determined primarily by genetic drift. In a simple example, if s is the selective difference between two alleles at a locus, and if $s < 1/(2N_e)$, the alleles are said to be neutral with respect to each other because the impact of genetic drift is larger than selection.

The neutral theory is also consistent with a *molecular clock*; that is, there is a constant rate of substitution over time for molecular variants (see chapter II.3). To illustrate the mathematical basis of the molecular clock, let us assume that mutation and genetic drift are the determinants of changes in frequencies of molecular variants. Let the mutation rate to a new allele be u so that in an effective population of size $2N$ (we will drop the subscript e in this discussion and just assume $N_e = N$), there are $2Nu$ new mutants per generation. The probability of chance fixation of a new neutral mutant is $1/(2N)$ (the initial frequency of the new mutant). Therefore, the rate of allele substitution k is the product of the number of new mutants per generation and their probability of fixation, or

$$k = 2Nu\left(\frac{1}{2N}\right) = u.$$

In other words, this elegant prediction from the neutral theory is that the rate of substitution is equal to the mutation rate at the locus and constant over time. Note that substitution rate is independent of the effective population size, a fact that may initially be counterintuitive. This independence occurs because in a smaller population there are fewer mutants; that is, $2Nu$ is smaller, but the initial frequency of these mutants is higher, increasing the probability of fixation, $1/(2N)$, by the same magnitude by which the number of mutants is reduced. This simple, mathematical prediction and others from the neutral theory provide the basis for the most important developments in evolutionary genetics in recent decades.

One of the appealing aspects of the neutral theory is that if it is used as a null hypothesis, predictions about the magnitude and pattern of genetic variation are possible. Initially, molecular genetic variation was found consistent with that predicted from neutrality theory. In recent years, examination of neutral theory predictions in DNA sequences has allowed tests of the cumulative effect of many generations of selection and a number of examples of selection on molecular variants have been documented.

If it is assumed that an equilibrium exists between mutation producing new alleles and genetic drift eliminating them, then

$$H_e = \frac{4N_eu}{4N_eu + 1},$$

the equilibrium heterozygosity for the neutral model. Note that for this equilibrium, the allele frequencies,

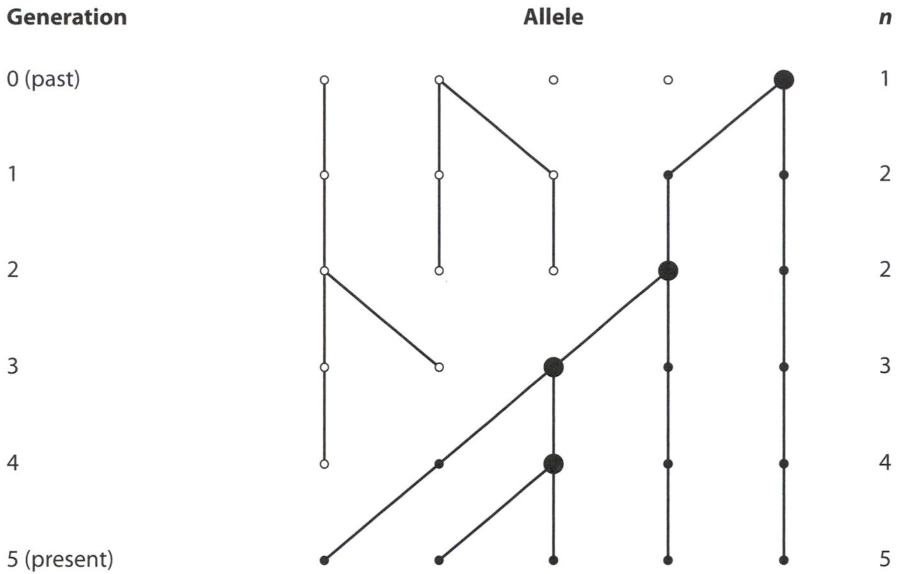

Figure 4. The ancestry of five alleles sampled in the present generation. If we go backward in time, bottom to top, we see the effects of genetic drift in a very small population resulting in coalescence in a single allele in generation 0 (*n* is the number of ancestral alleles in a given generation). (After Hedrick 2011.)

and even the identity of the alleles, are constantly changing as new mutants are generated and old mutants impacted by genetic drift, and it is only the distribution of alleles that remains more or less constant. This equation predicts that as $4N_eu$ increases, the amount of heterozygosity will also increase; therefore, neutrality predicts that an increase in either effective population size or mutation rate will result in an increase in heterozygosity. Surveys of microsatellite loci, which have a high mutation rate, generally have a high heterozygosity, consistent with predictions from the neutral theory; for example, some microsatellite loci have mutation rates of $u = 10^{-4}$, and if $N_e = 10{,}000$, then $H_e = 0.8$, as often found for microsatellite loci.

4. COALESCENCE

Traditionally, population genetics examines the impact of various evolutionary factors on the amount and pattern of genetic variation in a population and how these factors influence the future potential for evolutionary change. Generally, evolution is conceived of as a forward process, examining and predicting the future characteristics of a population; however, rapid accumulation of DNA sequence data over the past two decades has changed the orientation of much of population genetics from a prospective one, investigating the factors involved in observed evolutionary change, to a retrospective one, inferring evolutionary events that have occurred in the past. That is, understanding the evolutionary causes that have influenced the DNA sequence variation in a current sample of individuals, such as the demographic and mutational history of the ancestors of the sample, has become the focus of much population genetics research.

When DNA variation is being determined in a population at a given locus, a sample of alleles is examined. Each one of these alleles can have a different history, ranging from descending from the same ancestral allele, that is, *identical by descent*, in the previous generation to descending from the same ancestral allele many generations before. The point at which this common ancestry for two alleles occurs is called *coalescence*. If one goes back far enough in time in the population, all alleles in the sample will coalesce into a single common ancestral allele. Research using the coalescent approach is the most dynamic area of theoretical population genetics because it is widely used to analyze DNA sequence data in populations and species.

To illustrate the coalescent process, figure 4 gives a hypothetical example of the ancestry of five alleles sampled in the present generation, generation 5. If we go down in figure 4, forward in time, we can see the effect of genetic drift in a very small population: some alleles are lost (such as the middle allele in the first generation because it has no descendants), and some alleles increase in frequency, such as the right-hand allele (it has two descendants in the second generation). After five generations, only the right-hand allele remains, the other four original alleles have been lost. Of course, if the population size were larger, coalescent events from genetic drift would be less frequent and spaced out over many generations.

The theory of coalescence allows us to examine only the alleles ancestral to those sampled in the present generation. If we go up in figure 4, back in time, we see that the five alleles sampled in the present generation 5 are descended from four alleles in generation 4. In other words, there is a coalescent event because two alleles in generation 5 are descended from the same ancestral allele in generation 4, indicated by a larger circle. If we continue up the figure, the number of ancestral alleles declines because of three additional coalescent events, until only one ancestral allele remains in generation 0. Notice that other alleles were present in the past, but they have left no descendants in the present-day sample.

In this example, only genetic drift is assumed to influence the alleles, and mutation is not included. If mutation is included, observed alleles that have a common ancestor may actually have somewhat different DNA sequences. Coalescent theory and molecular data allow estimation of past events; for example, the past effective population size thousands of generations ago can be estimated using contemporary molecular data.

5. FUTURE DIRECTIONS

Genomic data from many individuals in a population or species and theoretical coalescent approaches will provide new insights into the population genetics of many species in coming years. In some cases, ancient samples of organisms will provide a way to validate these predictions about genetic drift, gene flow, mutation, and selection (see chapter V.15).

FURTHER READING

Allendorf, F., G. Luikart, and S. Aiken. 2013. Conservation and the Genetics of Populations. 2nd ed. Oxford: Blackwell. *A recent summary of the application of population genetics to conservation.*

Charlesworth, B. 2009. Effective population size and patterns of molecular evolution and variation. Nature Reviews Genetics 10: 195–205.

Hartl, D., and A. Clark. 2007. Principles of Population Genetics. 4th ed. Sunderland, MA: Sinauer. *A summary of the principles of population genetics.*

Hedrick, P. 2011. Genetics of Populations. 4th ed. Boston: Jones and Bartlett Publishers. *A recent and thorough summary of the principles of population genetics.*

Kalinowski, S., C. Muhfeld, C. Guy, and B. Cox. 2010. Founding population size of an aquatic invasive species. Conservation Genetics 11: 2049–2053.

Kimura, M. 1983. The Neutral Theory of Molecular Evolution. Cambridge: Cambridge University Press. *Summary of the neutral theory from the view of its major architect, Motoo Kimura.*

Palstra, F., and D. Ruzzante. 2008. Genetic estimates of contemporary effective population size: What can they tell us about the importance of genetic stochasticity for wild population persistence? Molecular Ecology 17: 3428–3447.

Tenesa, A., P. Navarro, B. Hayes, D. Duffy, G. Clarke, M. Goddard, and P. Visscher. 2007. Recent human effective population size estimated from linkage disequilibrium. Genome Research 17: 520–526.

IV.2

Mutation

Charles F. Baer

OUTLINE

1. The meaning of *mutation*
2. Types of mutations
3. Causes of mutation
4. Mutation and evolution: Basic principles
5. How random is mutation?
6. Variation in mutation rate: Among taxa
7. Variation in mutation rate: Within the genome
8. The mutational spectrum and mutational bias
9. Mutation, genome size, and genomic complexity
10. Mutation and extinction
11. Mutation and evolution: Other long-term consequences

Evolution depends on genetic differences among individuals, and ultimately all genetic variation has its origins in mutation. There are many ways in which DNA can change in a heritable fashion—from changes in single nucleotides, to rearrangements, to wholesale insertion or deletion of new sequences of DNA—and many causes of these mutations. Mutation rates vary among individuals, among species, among regions of the genome, and according to environmental conditions, and these mutation rates themselves can evolve. It may be energetically expensive to minimize mutation rates, and lineages with low mutation rates may slow their rates of evolution. On the other hand, a large fraction of mutations are deleterious, and on average the typical mutation can harm the function of organisms, even to the point where populations cease to survive.

GLOSSARY

Ectopic Recombination. Recombination between two nonhomologous sites in the genome. Recombination requires sequence similarity, but the similar sequences need not represent the same feature in the genome of the common ancestor.

Gene Conversion. Nonreciprocal exchange of genetic information from one homologous nonsister chromatid to the other in a heterozygous individual during meiotic recombination that results from template-directed repair of double-stranded breaks. The effect is to convert one allele in a heterozygote into the other. Gene conversion is a kind of mutation, but the state of the "mutant" allele depends on the state of the other allele present in the individual.

Mutation. A change in the nucleotide sequence of the genome from the parent to the offspring.

Transposable Element (TE). A genetic element that encodes the information necessary for its own replication, independently of the replication of the "host" individual's genome. The behavior of a TE is analogous to the replication of a parasitic organism; TEs are examples of "selfish" genetic elements.

1. THE MEANING OF *MUTATION*

Many a textbook chapter, research paper, and grant proposal begins with the phrase "mutation is the ultimate source of genetic variation." Without mutation, every locus will ultimately fix, the population will be devoid of genetic variation, and evolution will cease. Ultimately, all life everywhere would be (genetically) identical. This logic has a profound implication: *all* evolutionary innovation (see, e.g., chapter II.6) must ultimately have as its origin a single mutant allele in a single population. The mutant allele must initially increase in frequency in the population by genetic drift when rare, before proceeding to fixation.

The primary focus of this chapter is the ways in which mutation influences evolution; its secondary focus is the mechanisms by which mutation itself evolves (for a fuller exploration of theories on the evolution of mutation rate, see chapter III.9). Before embarking, it is useful to define exactly what is meant by *mutation.* Prior to the rediscovery of Mendel's work, "mutation" referred to a heritable, discontinuous change in the phenotype, a so-called sport. Following the rediscovery of Mendel's work, it was recognized that many, if not most, mutations obeyed Mendel's laws. The Oxford dictionary defines *mutation* as "the changing of the structure of a gene, resulting in a variant form that may be transmitted to subsequent generations, caused by the alteration of single base units in DNA, or the deletion, insertion, or rearrangement of larger sections of genes or chromosomes." From the perspective of evolution, the mutations that matter are *heritable mutations*. In organisms with a distinct germ line (e.g., animals), somatic mutations can harm the individual (e.g., cause cancer), but since they are not passed to the next generation, they have no evolutionary consequence beyond their effect on an individual bearer.

Because faithful transmission of biological information is necessary for life to continue, living organisms have evolved multiple mechanisms to ensure accurate replication of the genome. It is useful to consider the mutational process in terms of "input," that is, damage to the DNA or replication errors, and "output" (i.e., "mutation" per se), which is the fraction of the input that makes it into the next generation after DNA repair. The distinction between input and output matters, because the mutational process can evolve in two ways, either by changing the input (e.g., by avoiding environmental mutagens), or by changing output (e.g., by evolving a proofreading polymerase).

2. TYPES OF MUTATIONS

Three fundamental types of mutation have been identified, each of which comes in several varieties and has a variety of causes, and consequences. Base substitutions occur when one nucleotide is substituted for another at the same homologous position in the genome. A base substitution may be either a transition (purine → purine or pyrimidine → pyrimidine) or a transversion (purine → pyrimidine, or vice versa). Insertions and deletions (collectively, *indels*) occur when an additional sequence is inserted into an existing sequence, or part of an existing sequence is deleted. Genome rearrangements, including inversions and translocations, occur when pieces of chromosomes change position in the genome. Inversion occurs when a piece of a chromosome becomes detached and reattaches in the opposite orientation (e.g., gene order goes from ABCDE to ADCBE, following inversion of the segment BCD. Translocation occurs when a piece of a chromosome becomes detached and is reattached either on a different chromosome or in a different place on the same chromosome. In eukaryotes, rearrangement can result in the suppression of recombination, either from mispairing at synapsis or from lethal recombinant genotypes resulting from improper gene dosage.

Two important classes of indels are short tandem repeats (STRs), and copy-number variants (CNVs). STRs are a particular type of CNV consisting of a short motif repeated one after another ("in tandem") multiple times, and are highly mutable because the DNA polymerase tends to "slip" during replication, with the result that one of the resulting daughter strands contains either one (or more) additional repeat(s) or one (or more) fewer repeat(s). Mutation rates of STRs can be many orders of magnitude greater than base-substitution mutation rates. More generally, CNVs are sequences present in multiple copies in the genome that vary in number between (haploid) genomes; CNVs are generated by any mechanism capable of generating an indel mutation. An important feature of CNVs is their use as substrate for ectopic (nonhomologous) recombination.

An important source of CNVs is transposable elements (TEs), genetic elements that encode their own replication throughout the genome, independent of the host cell's replication machinery; they are "selfish" elements because natural selection operating at the level of the TE will favor an increase in TE copy number, even to the detriment of the fitness of the host organism.

3. CAUSES OF MUTATION

Mutations may have their ultimate cause in factors either endogenous or exogenous to the organism. Replication errors, TEs, and free-radical by-products of metabolism are examples of the former; environmental mutagens are examples of the latter. Each comes with its own implications for evolution. The strategy of an organism wanting to avoid the deleterious consequences of exogenous mutation is simple: "don't go there." For example, to reduce the mutational input from incident UV radiation, an organism might evolve the choice of spending more time in the shade. Similarly, if circumstances favor an increase in metabolic rate, the organism will need to cope with the increased mutational input from metabolic by-products. One way to reduce the mutational output would be to evolve more efficient DNA repair; another possibility might be to increase free-radical "scavenging" mechanisms.

4. MUTATION AND EVOLUTION: BASIC PRINCIPLES

Evolution requires variation. All else equal, the more genetic variation produced by mutation, the more opportunity for evolution, and the faster evolution can proceed; however, that genetic variation comes with a cost: most mutations that are not neutral are deleterious, and only a relatively few mutations are beneficial.

Before turning to the many ways in which deleterious mutation influences evolution, consider neutral mutations, those with no effect on fitness. The simplest case is that of a single, bi-allelic locus (alleles A and a at frequencies p and $q = 1 - p$, respectively) in an infinite population in which A mutates to a with probability μ and a mutates to A with probability ν. At equilibrium, $\hat{q} = \frac{\mu}{\mu+\nu}$. If μ and ν are equal, at equilibrium the allele frequencies will be equal; if, say, μ is 10 times greater than ν, at equilibrium the population will consist of 10 times as many a alleles as A alleles. Less obvious is the timescale. It can be shown that the magnitude of change in allele frequency in one generation due to mutation is on the order of the mutation rate itself, typically a very small number.

In a closed finite population at genetic equilibrium, the amount of "standing" genetic variation present at a single neutral locus, $\hat{H}$ (H represents "heterozygosity"), is proportional to the product of the mutation rate, μ, and the genetic effective population size, N_e (see chapter IV.1); that is, (H)^$(N_1 e\mu)$. This result is intuitive: the higher the mutation rate, the more genetic variation there will be, and similarly, the larger the population, the more genetic variation it can hold. If different groups harbor different amounts of genetic variation, it might be because they differ in N_e (usually the explanation of first resort) or because they differ in mutation rate, or both.

The role of mutation in determining the standing genetic variance for a quantitative trait (e.g., height) is determined jointly by the per locus mutation rate μ, the number of loci that affect the trait, n, and the phenotypic effects of mutant alleles, a. In the simplest case, that of a haploid organism, the genome-wide mutation rate for the trait is $U = \sum_i^n \mu_i$. The genetic variation introduced into the population by mutation each generation (the *mutational variance*, V_M) is equal to the product of U and the expected squared effect of a new mutation, denoted as $E(a^2)$, so that $V_M = UE(a^2)$. The same principles apply to diploids, except that dominance must be taken into consideration.

For a neutral quantitative trait, the standing genetic variance in a population, V_G, is proportional to the product of effective population size (N_e) and the mutational variance, that is, $V_G \propto N_e V_M$. The scaling with N_e and mutation rate (μ) is the same as for single-locus variation, but the standing quantitative variance also depends on the number of loci that affect the trait (n, the mutational "target size") and the average effects of alleles at those loci, $E(a^2)$.

Clearly, not all mutations are neutral; for example, many human genetic disorders are caused by mutations of large effect (usually recessive) at individual loci. For loci under directional selection, mutation adds genetic variation to the population and selection removes it; at equilibrium, referred to as *mutation-selection balance* (MSB), the two forces exactly offset. For a single haploid locus, the equilibrium frequency of the deleterious mutant allele, $\hat{q} \approx \frac{\mu}{s}$, where μ is the mutation rate from wild type to mutant and s is the selection coefficient against the mutant allele. This result is intuitive: the greater the mutation rate and the weaker the strength of selection, the greater the equilibrium frequency of the mutant allele (see chapter IV.5).

Similar reasoning applies to quantitative traits; many traits are not neutral, in which case natural selection would prefer the most-fit allele be fixed at every locus (balancing selection leads to a similar conclusion). At MSB the standing genetic variance, V_G, is established by the counterbalancing effects of the input of genetic variation by mutation (V_M) and the removal of deleterious alleles by natural selection; in many cases $V_G \approx V_M/S$, where S represents the average selection coefficient against a mutant allele.

In finite populations, the effects of selection and N_e become entangled. If the strength of selection (s) acting on an allele is less than (about) $1/2N_e$, the evolutionary dynamics are governed by genetic drift rather than selection; that is, the allele is said to be "nearly neutral" (see chapter IV.1). The *strength* of selection is an inherent property of an allele, whereas the *efficiency* of selection depends on the population size. The consequences of the relationship between efficiency of selection and N_e are profound; we return to this result throughout the chapter.

5. HOW RANDOM IS MUTATION?

Mutation is almost always assumed to be a "random" process, by which is meant that mutations do not occur based on the potential future effect on fitness. The conclusion that mutation is random in this regard stems from the pioneering work of Luria and Delbrück in the 1940s. It was known that when *E. coli* sensitive to the bacteriophage T1 were plated in the presence of T1, initially no colonies would grow on the plate, but eventually colonies would begin to appear, and those colonies consisted of resistant bacteria that bred true. The question was, Are the resistant cells derived from slow-growing mutants that existed in the population prior to plating in the presence of phage, or from mutations that occurred subsequent to exposure to phage? The question cut to the heart of evolutionary biology, because if the resistant mutants occurred only (or much more frequently) after exposure to phage, it would mean the environment directly influences the heritable genome in such a way as to increase the fitness of the affected organism—in which case evolution would be "Lamarckian" (although Darwin himself was Lamarckian in this regard, particularly in later editions of *On the Origin of Species*). Luria and Delbrück and others convincingly demonstrated that preexisting mutants were sufficient to explain the delayed population growth and there was no need to invoke acquired immunity.

In 1988, in a paper provocatively titled "The Origin of Mutants," John Cairns and colleagues reopened the issue of "directed mutation." Cairns et al. employed a different system, in which the selective agent—the ability to utilize lactose as a carbon source—does not kill sensitive (*Lac-*) cells but merely prevents their growth. An excess of *Lac+* mutations occurring subsequent to the onset of selection, and moreover, the selective environment, apparently did not increase the rate of nonadaptive mutations. The basic feature—the appearance of adaptive mutations after the onset of selection unaccompanied by a simultaneous increase in nonadaptive mutations—was subsequently observed in other systems and precipitated a lively controversy. Cairns et al. initially speculated that starvation induced "highly variable" transcription, which, when coupled with reverse transcription, would result in eventual incorporation of the adaptive mutant into the genome. Although Cairns's hypothetical mechanism was not borne out, other mechanisms were proposed. Barry Hall proposed that starvation might induce a transient state of hypermutation in a small subset of cells that would then revert to wild-type mutation rate after starvation was alleviated by an adaptive mutation. Alternatively, Roth and coworkers argue that apparently adaptive mutations can be explained by very slow growth in cells carrying additional copies of the critical genes, providing additional targets for beneficial mutations; the apparent high frequency of adaptive mutations is therefore a product of simple Darwinian selection plus increased mutational target size.

The mutation rate in *E. coli* and other microbes apparently does increase under various physiological stresses, including starvation. In many cases, the mechanism involves recombination induced by double-strand breaks in the DNA, followed by the (mutagenic) action of an inherently error-prone polymerase; however, it remains unclear whether the stress-induced increase in mutation represents an adaptation or is simply a feature of a sick organism functioning at a subpar level; evolutionary orthodoxy suggests the latter.

6. VARIATION IN MUTATION RATE: AMONG TAXA

Mutation rates vary among organisms, among genomic locations, among sequence motifs, and among nucleotides. At the outset, it is important to distinguish exactly what is meant by *rate*. Mutation rate may be expressed per genome replication, per generation, or per unit time. For unicellular microbes and viruses, mutation rates per replication and per generation are equivalent; for multicellular organisms they are not. Further, mutation rates can be expressed as per site, per gene, or per genome. From the perspective of natural selection, the relevant mutation rate is the rate per genome, per generation (U). Natural selection (usually) acts via individuals, whose fitness is integrated over the entire genome, and is manifested by its contribution to the next generation. All else equal, natural selection favors a reduction in U. There are two basic ways to effect a change in U. First, the per-site mutation rate may remain unchanged and the number of sites necessary to build the organism changes. Second, the number of sites may remain unchanged and the per site mutation rate changes. In organisms that undergo multiple rounds of genome replication per generation, the number of rounds of replication between generations may be changed.

Given that mutation rate varies among taxa, are there underlying regularities? The answer appears to be yes. In the early 1990s, Jan Drake observed that the *per nucleotide*, per generation mutation rate in DNA-based microbes (viruses and prokaryotes) varied nearly inversely with genome size across four orders of magnitude, leading to a nearly constant *per genome* mutation rate of about 0.003–0.004, which he and others argued must be due to the existence of a globally optimum mutation rate, presumably due to the existence of a cost of fidelity associated with replication speed.

As data from multicellular eukaryotes accumulated, it became apparent that the per nucleotide, per generation mutation rate varies positively with genome size in cellular organisms (prokaryotes and eukaryotes), although the per nucleotide, per *replication* mutation rate is similar between microbes and multicellular organisms. What could explain the remarkable difference in scaling within microbes, on the one hand, and among cellular organisms (including multicellular eukaryotes), on the other? Michael Lynch has argued that the difference is related to the relationship between genome size, body size, and N_e. Distilled, the argument is as follows: larger organisms have smaller population sizes, reducing the efficiency of natural selection to reduce mutation rate; at some point a further decrease in mutation rate provides such a small fitness advantage that selection cannot overcome drift. Although the absolute cost of fidelity can be the same among different groups (it need not be), groups differ in how close to the optimum (low) mutation rate they can get, based on population size.

7. VARIATION IN MUTATION RATE: WITHIN THE GENOME

Not all parts of the genome evolve at the same rate, nor do they harbor the same amount of genetic variation.

One obvious possibility is that mutation rate varies consistently with particular features of the genome; another is that natural selection does. Three features stand out: level of transcription, chromatin architecture, and local recombination rate. Transcription per se is believed to influence the probability of mutation in two opposing ways. First, DNA is necessarily single stranded during transcription, and single-stranded DNA is more vulnerable to damage (*transcription-induced mutation,* TIM); the effect is more pronounced on the nontranscribed (coding) strand. Second, transcription-coupled repair (TCR) mechanisms repair damage to transcribed DNA, but the repair is primarily on the transcribed (noncoding) strand. Both TIM and TCR predict *strand asymmetry*, in which mutations are more likely on the coding strand; the degree of strand asymmetry has been shown to be positively correlated with transcription level.

Chromatin can be broadly classified as "open" or "closed," based on the degree of compaction. In the human genome, regions of open chromatin are both gene rich and enriched for broadly expressed genes. Interestingly, mutation rates appear higher in closed than in open chromatin. The cause of the distinction is not known; a possibility is that open chromatin is more accessible to the DNA-repair machinery.

It is overwhelmingly clear that within-species polymorphism correlates positively with local recombination rate. There are two possible (nonexclusive) explanations. First, natural selection is more efficient in regions of high recombination (the Hill-Robertson effect). Second, recombination might be mutagenic. The "mutagenic recombination" hypothesis predicts that both polymorphism within species *and* genetic divergence between species should be correlated with local recombination rate, whereas selective hypotheses in general do not predict a relationship between recombination and divergence. Early evidence from *Drosophila melanogaster* found no association between recombination and divergence, which was taken as support for selection underpinning the relationship between polymorphism and recombination. At this writing the evidence must be considered equivocal, although it seems very likely that selection plays the predominant role.

8. THE MUTATIONAL SPECTRUM AND MUTATIONAL BIAS

To say that mutation is a random process with respect to fitness does not mean that all mutations are equally probable: numerous sources of mutational bias are known or suspected. For example, certain kinds of mutations are more likely to occur on the lagging strand during DNA synthesis (others are not), leading to base composition asymmetry over evolutionary time. Base pairs consisting of A:T are more likely to mutate to G:C or C:G base pair than vice versa, and transitions appear more common than transversions. Similarly, gene conversion is often biased such that A:T pairs are more often converted to G:C or C:G than vice versa. Most genomes are too GC rich, given the apparent extent of the mutational bias, suggesting that equilibrium base composition is established by mutation-gene conversion balance.

Genomes, even those of closely related taxa and of the same ploidy level, can vary substantially in size. Changes in genome size have their ultimate cause in mutation; the equilibrium must be established by the balance of insertions and deletions, mediated (maybe) by natural selection. TEs provide an obvious source of (selfish) insertion bias. There appears to be an overall bias toward small deletions in all taxa, although the extent of the bias is stronger in microbes. Natural selection must at some point establish a lower bound on genome size, and presumably an upper bound as well. Understanding the relative influences of mutational bias, genetic drift, and selection on the evolution of genome size is an important unresolved issue.

9. MUTATION, GENOME SIZE, AND GENOMIC COMPLEXITY

In 1971, Manfred Eigen introduced the concept of an *error threshold*, in which the relationship between mutation rate, genome size, and information content of the genome was first formalized. The basic idea is that the error (= mutation) rate puts an upper bound on the length of a biological sequence (e.g., an RNA virus); for a given mutation rate, natural selection will be unable to maintain a (functional) sequence longer than the critical length in the face of mutational loss of information. This finding, referred to as an "error catastrophe," led to an apparent paradox, because mutation rates of RNA viruses, which lack proofreading capacity, appeared too high to allow the evolution of a proofreading enzyme in the first place. Importantly, this theory applies in an infinite population and is therefore a deterministic phenomenon. Various solutions to the paradox have been proposed, including the suggestion that there is no paradox, but the general inverse relationship between genome size and information content, on the one hand, and mutation rate, on the other, seems robust.

The consequences of the relationship between genome size, genome "complexity" (roughly defined as the number of features), mutational target size, and natural selection have been extensively explored by Lynch. In general, increasing the size and/or number of genomic features increases the probability of deleterious mutation. For natural selection to favor increasing the size of a feature, the selective benefit must outweigh the cost of increasing the mutational target size. Since mutations with selective effects $s < 1/2N_e$ are effectively neutral, the ability to grow the genome is more constrained in organisms with large N_e (e.g., microbes). It seems intuitive that complex organisms must somehow require complex genomes, and the genomes of multicellular eukaryotes are arguably more complex than those of microbes. However, the direction of causality is not certain, and (employing a favorite phrase of evolutionary biologists) "theory predicts" that genome complexity should scale inversely with N_e.

10. MUTATION AND EXTINCTION

In 1964, H. J. Muller observed that finite populations of nonrecombining organisms would accumulate mutations by the combined action of mutation and drift. Once the least-loaded genome in the population is lost by drift, it can never be reconstituted (except by back mutation). Thus, mean fitness of finite asexual populations will steadily decay over time by the mechanism known as *Muller's ratchet.* The long-term consequences of the ratchet in an ecological context have been investigated by Lynch, Lande, and others. Once fitness declines below the point at which individuals replace themselves on average, population size begins to decline and the rate of accumulation of deleterious mutations increases as selection becomes progressively less efficient, in a self-reinforcing process culminating in extinction, dubbed a *mutational meltdown.* Although the effect is more pronounced in asexual populations, sexual populations are not immune from the cumulative long-term effects of very slightly deleterious mutations.

11. MUTATION AND EVOLUTION: OTHER LONG-TERM CONSEQUENCES

Deleterious mutations have been implicated as a cause of or leading contributor to a wide variety of evolutionary phenomena that are difficult to explain, including the evolution of: sexual reproduction and recombination, ploidy, mating systems, sex chromosomes, sexual selection, and senescence, among others. Many of these topics are covered in more detail in other chapters; the reader is also encouraged to delve into the further reading suggested below.

FURTHER READING

Drake, J. W., B. Charlesworth, D. Charlesworth, and J. F. Crow. 1998. Rates of spontaneous mutation. Genetics 148: 1667–1686. *Remains the most comprehensive review of the properties of spontaneous mutations across all domains of living organisms, and includes a very readable summary of the theoretical predictions regarding the evolution of mutation rate, circa turn of the century.*

Loewe, L., and W. G. Hill, eds. 2010. *The population genetics of mutations: Good, bad, and indifferent.* Philosophical Transactions of the Royal Society B 365. *A theme issue, on the occasion of Brian Charlesworth's 65th birthday, dedicated to the population genetics of mutations. Filled with fascinating reading.*

Luria, S. E., and M. Delbrück. 1943. Mutations of bacteria from virus sensitivity to virus resistance. Genetics 28: 491–511. *In which it was first shown that mutations are random and not "directed" with respect to future fitness. Also an elegant example of the application of mathematical theory in biology.*

Lynch, M. 2007. The Origins of Genome Architecture. Sunderland, MA: Sinauer. *The grand synthesis of the Lynchian worldview, in which effective population size and slightly deleterious mutation are sufficient to explain many disparate features of gene and genome evolution.*

Lynch, M. 2010. Evolution of the mutation rate. Trends in Genetics 26: 345–352. *In which it is argued that random genetic drift, not a "cost of fidelity," imposes the lower bound on the achievable mutation rate.*

Prendergast, J.G.D., H. Campbell, N. Gilbert, M. G. Dunlop, W. A. Bickmore, and C.A.M. Semple. 2007. Chromatin structure and evolution in the human genome. BMC Evolutionary Biology 7: doi72 10.1186/1471-2148-7-72. *Evidence that open and closed regions of chromatin have characteristically different mutation rates.*

Sturtevant, A. F. 1937. Essays on evolution. 1. On the effects of selection on mutation rate. Quarterly Review of Biology 12: 467–477. *Perhaps the first paper to clearly lay out the basic issues involved in the evolution of mutation rate, it remains extremely influential.*

Wilke, C. O. 2005. Quasispecies theory in the context of population genetics. BMC Evolutionary Biology 5: doi44 10.1186/1471-2148-5-44. *Concisely explains the relationship between the body of theoretical work concerning genome evolution derived from Eigen's pioneering work, on the one hand, and mainstream theoretical population genetics, on the other hand.*

CENOZOIC

Age (Ma)	Period	Epoch	Age
		Holocene (0.0117–0 Ma)	
0	Quaternary	Pleistocene	Calabrian
	Quaternary	Pleistocene	Gelasian
	Neogene	Pliocene	Piacenzian
5	Neogene	Pliocene	Zanclean
	Neogene	Miocene	Messinian
10	Neogene	Miocene	Tortonian
	Neogene	Miocene	Serravallian
15	Neogene	Miocene	Langhian
20	Neogene	Miocene	Burdigalian
	Neogene	Miocene	Aquitanian
25	Paleogene	Oligocene	Chattian
30	Paleogene	Oligocene	Rupelian
35	Paleogene	Eocene	Priabonian
40	Paleogene	Eocene	Bartonian
45	Paleogene	Eocene	Lutetian
50, 55	Paleogene	Eocene	Ypresian
	Paleogene	Paleocene	Thanetian
60	Paleogene	Paleocene	Selandian
65	Paleogene	Paleocene	Danian

MESOZOIC

Age (Ma)	Period	Epoch	Age
70	Cretaceous	Late	Maastrichtian
80	Cretaceous	Late	Campanian
	Cretaceous	Late	Santonian
	Cretaceous	Late	Coniacian
90	Cretaceous	Late	Turonian
	Cretaceous	Late	Cenomanian
100, 110	Cretaceous	Early	Albian
120	Cretaceous	Early	Aptian
	Cretaceous	Early	Barremian
130	Cretaceous	Early	Hauterivian
140	Cretaceous	Early	Valanginian
	Cretaceous	Early	Berriasian
150	Jurassic	Late	Tithonian
	Jurassic	Late	Kimmeridgian
160	Jurassic	Late	Oxfordian
	Jurassic	Middle	Callovian
	Jurassic	Middle	Bathonian
170	Jurassic	Middle	Bajocian
	Jurassic	Middle	Aalenian
180	Jurassic	Early	Toarcian
	Jurassic	Early	Pliensbachian
190	Jurassic	Early	Sinemurian
	Jurassic	Early	Hettangian
200	Triassic	Late	Rhaetian
210	Triassic	Late	Norian
220	Triassic	Late	Carnian
230	Triassic	Middle	Ladinian
240	Triassic	Middle	Anisian
	Triassic	Early	Olenekian
250	Triassic	Early	Induan

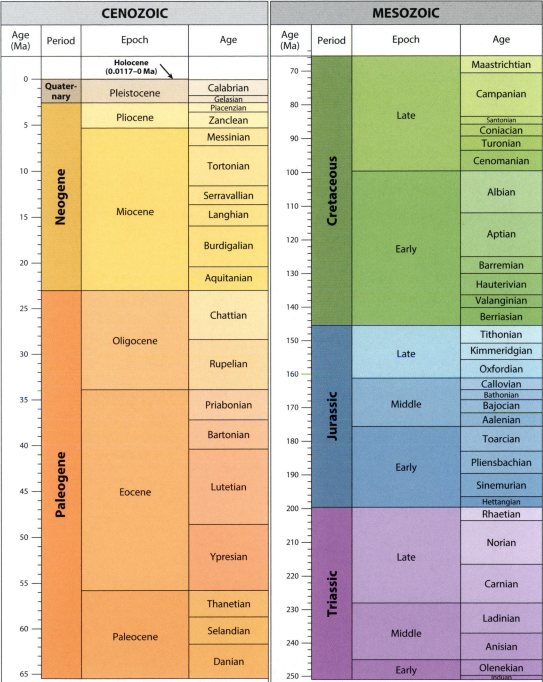

Plate 1. The 2009 geologic timescale accepted by the International Commission on Stratigraphy. The timescale shows eras, periods, epochs, and ages for the Phanerozoic eon and eons, eras, and periods for the Precambrian supereon. The age in millions of years of each time interval is shown at the left. Note that the height of each time interval is scaled linearly in time, but that the Cenozoic, Mesozoic, Paleozoic, and Precambrian columns each have a different linear scale. (*Continued*)

Precambrian

Geologic time (Ma) ~4,600

PALEOZOIC

Age (Ma): 260, 280, 300, 320, 340, 360, 380, 400, 420, 440, 460, 480, 500, 520, 540

Period	Epoch	Age
Permian	Lopingian	Changhsingian
		Wuchiapingian
	Guadalupian	Capitanian
		Wordian
		Roadian
	Cisuralian	Kungurian
		Artinskian
		Sakmarian
		Asselian
Carboniferous	Pennsylvanian	Gzhelian
		Kasimovian
		Moscovian
		Bashkirian
	Mississippian	Serpukhovian
		Visean
		Tournaisian
Devonian	Late	Famennian
		Frasnian
	Middle	Givetian
		Eifelian
	Early	Emsian
		Pragian
		Lochkovian
Silurian	Pridoli	
	Ludlow	Ludfordian
		Gorstian
	Wenlock	Homerian
		Sheinwoodian
	Llandovery	Telychian
		Aeronian
		Rhuddanian
Ordovician	Late	Hirnantian
		Katian
		Sandbian
	Middle	Darriwilian
		Dapingian
	Early	Floian
		Tremadocian
Cambrian	Furongian	Stage 10
		Stage 9
		Paibian
	Series 3	Guzhangian
		Drumian
		Stage 5
	Series 2	Stage 4
		Stage 3
	Terreneuvian	Stage 2
		Fortunian

PRECAMBRIAN

Age (Ma): 600, 850, 1,100, 1,350, 1,600, 1,850, 2,100, 2,350, 2,600, 2,850, 3,100, 3,350, 3,600, 3,850, 4,100, 4,350, ~4,600

Eon	Era	Period
Proterozoic	Neoproterozoic	Ediacaran
		Cryogenian
		Tonian
	Mesoproterozoic	Stenian
		Ectasian
		Calymmian
	Paleoproterozoic	Statherian
		Orosirian
		Rhyacian
		Siderian
Archean	Neoarchean	
	Mesoarchaean	
	Paleoarchaean	
	Eoarchean	
Hadean		

Plate 1. (*Continued*)

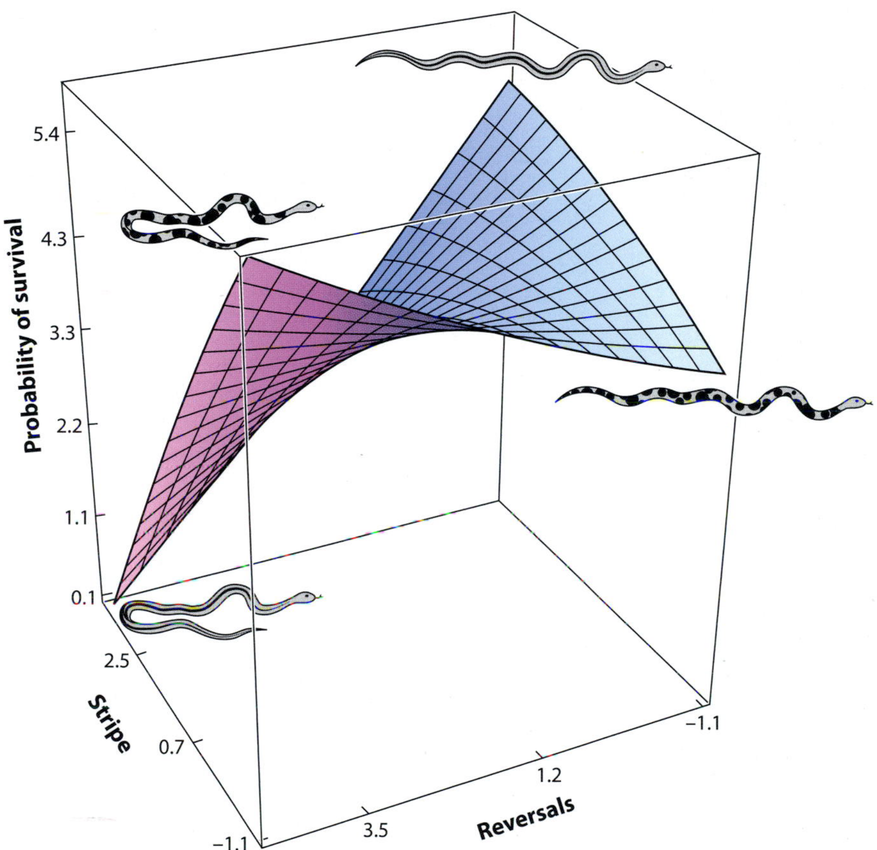

Plate 2. Saddle-shaped correlational selection in garter snakes favors specific combinations of escape behavior and color patterns. The surface shown is estimated from a three-year study of survival of more than 600 individual snakes in the wild. Individuals with striped patterns were most likely to survive if they performed few reversals, and those with more spotted patterns survived if they performed many reversals. The other combinations of color pattern and behavior experienced high mortality.

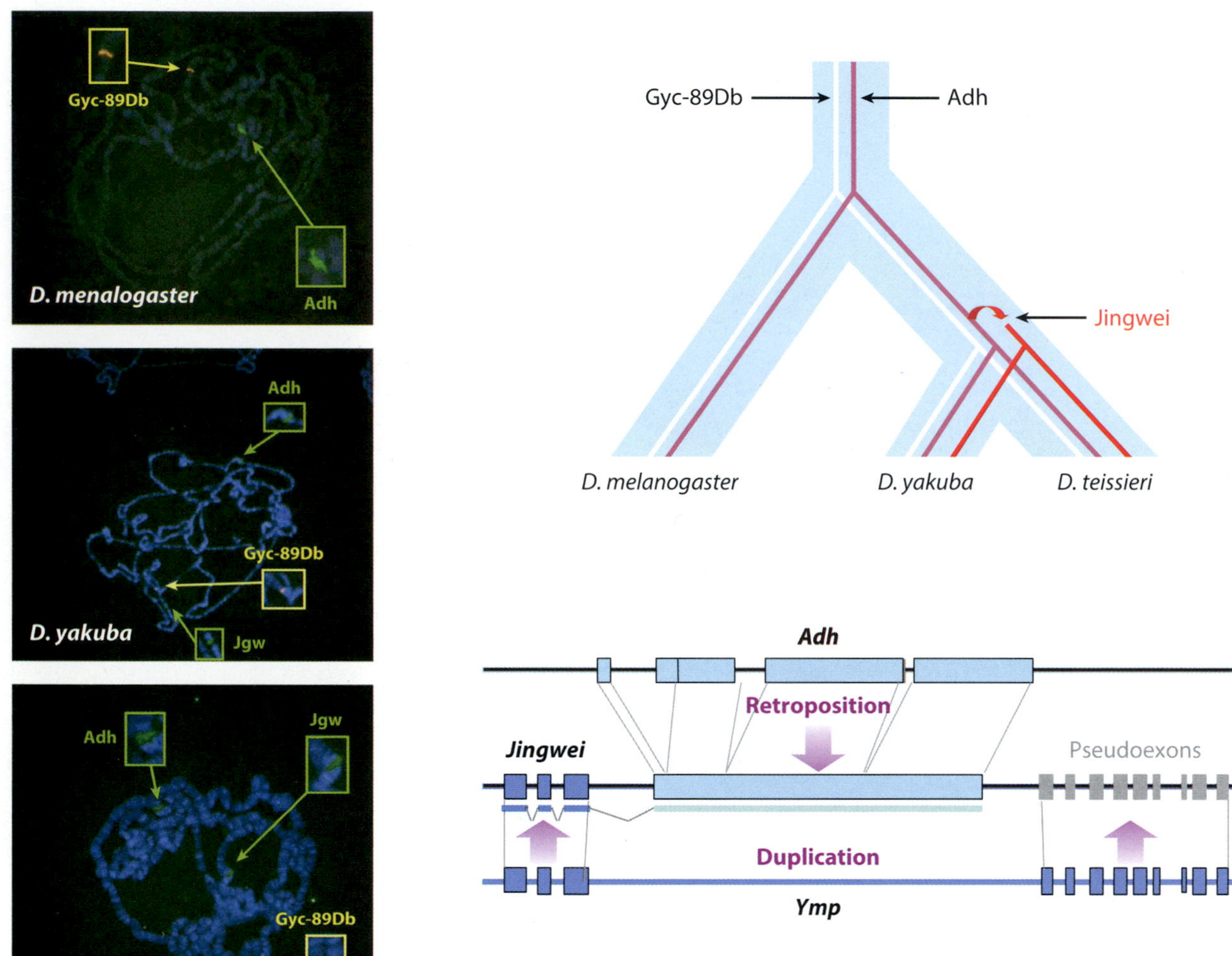

Plate 3. An example of the rapid generation of a chimeric gene: the *Jingwei* gene in *Drosophila*. Based on the distribution of *Jingwei* in related species, the lineage of this gene was inferred in the phylogeny of three species, *D. yakuba* and *D. teissieri*, and *D. melanogaster*. Sequence analyses of genomic DNAs and transcripts reconstructed the process of gene evolution, revealing that retroposition, DNA-based gene duplication, and exon/domain shuffling worked together to create the *Jingwei* gene 3 million years ago in the common ancestor of the *African Drosophila*.

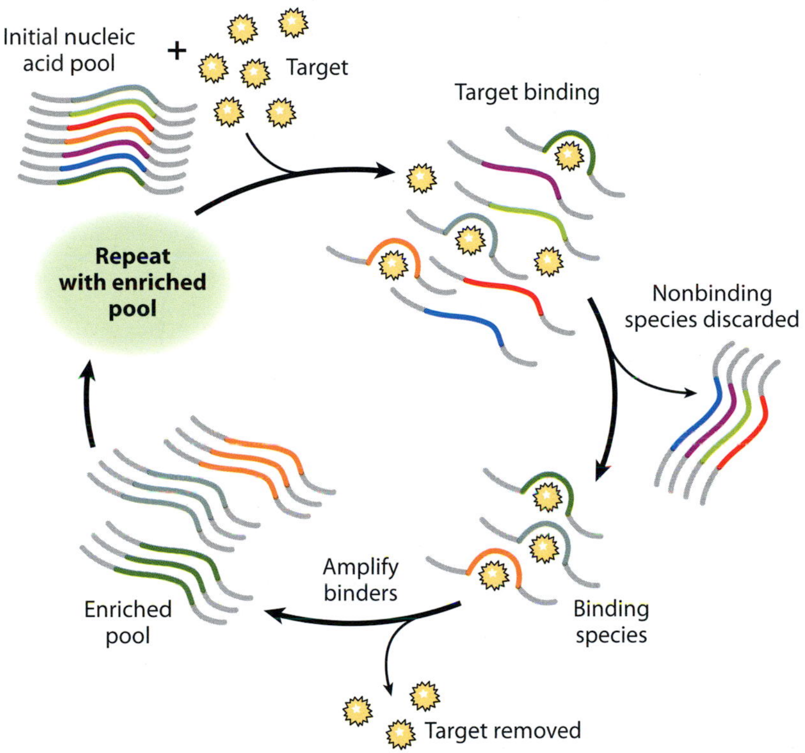

Plate 4. Aptamer selection experiment. (Illustration prepared by Angel Syrett.)

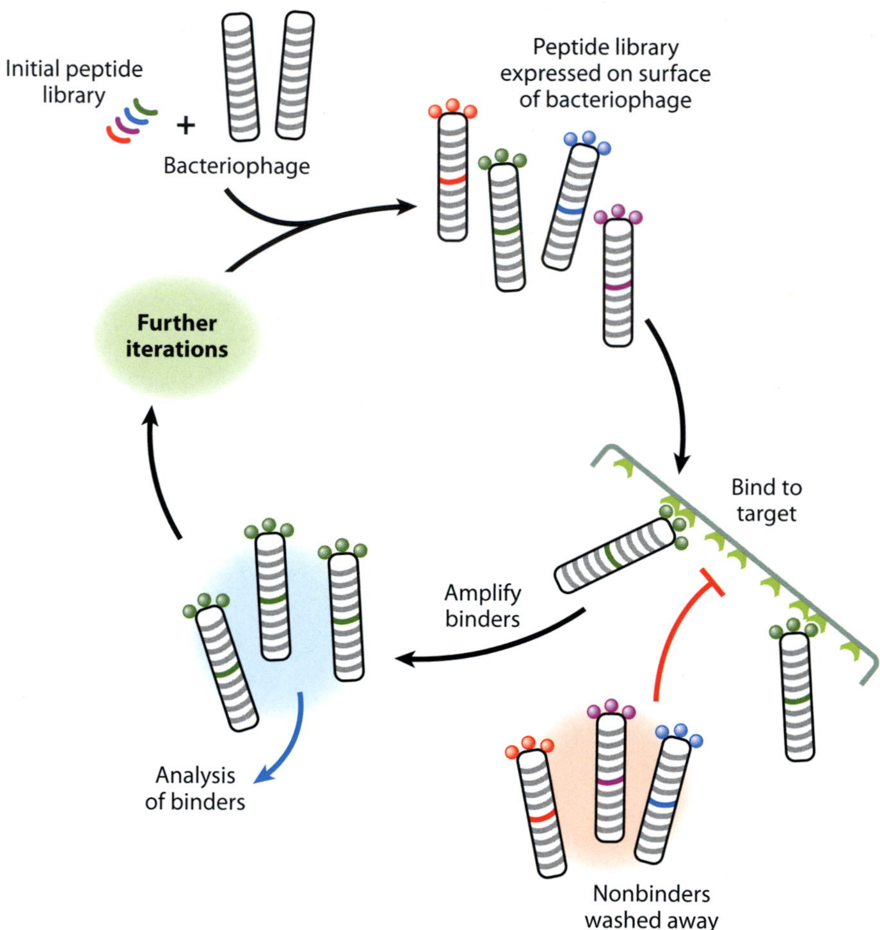

Plate 5. Phage display experiment. (Illustration prepared by Angel Syrett.)

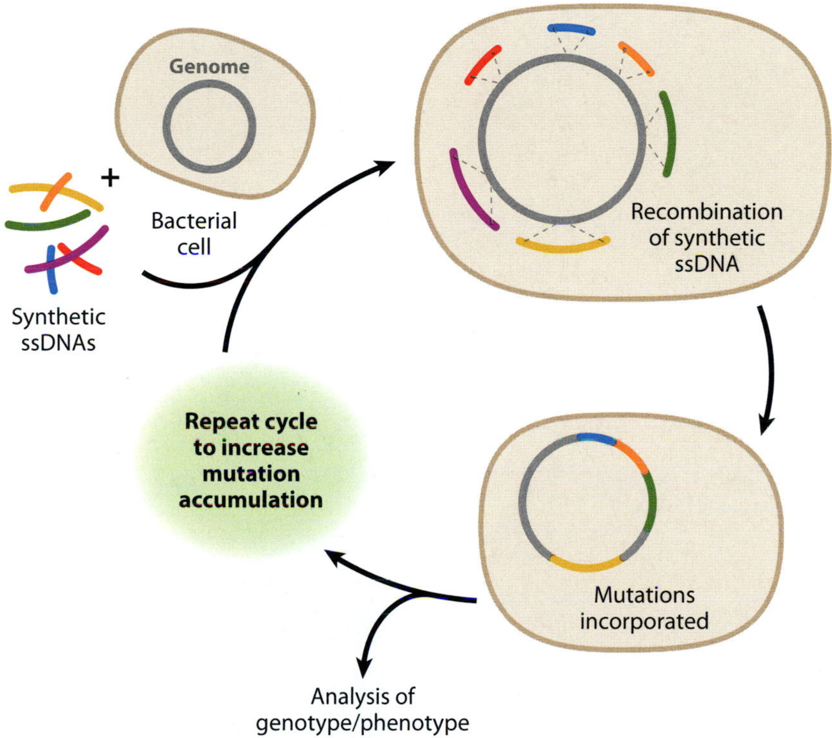

Plate 6. Multiplex Automated Genome Engineering (MAGE). (Illustration prepared by Angel Syrett.)

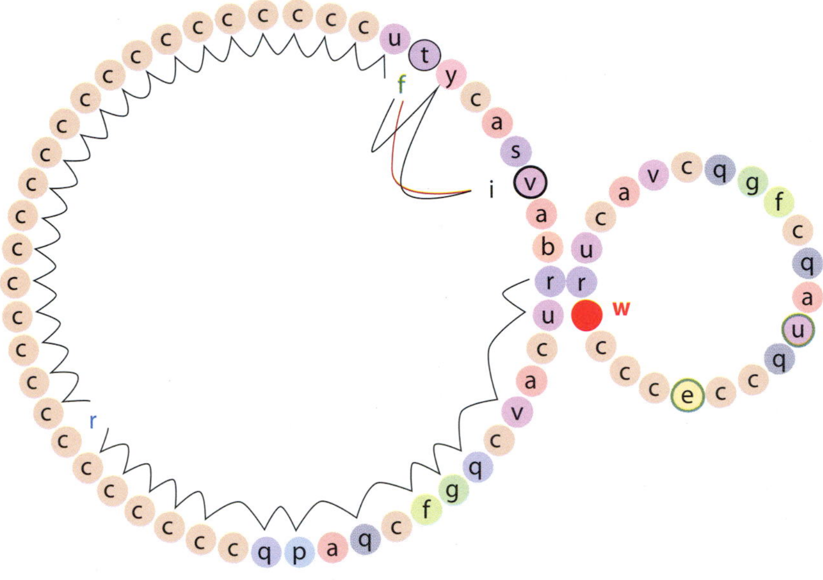

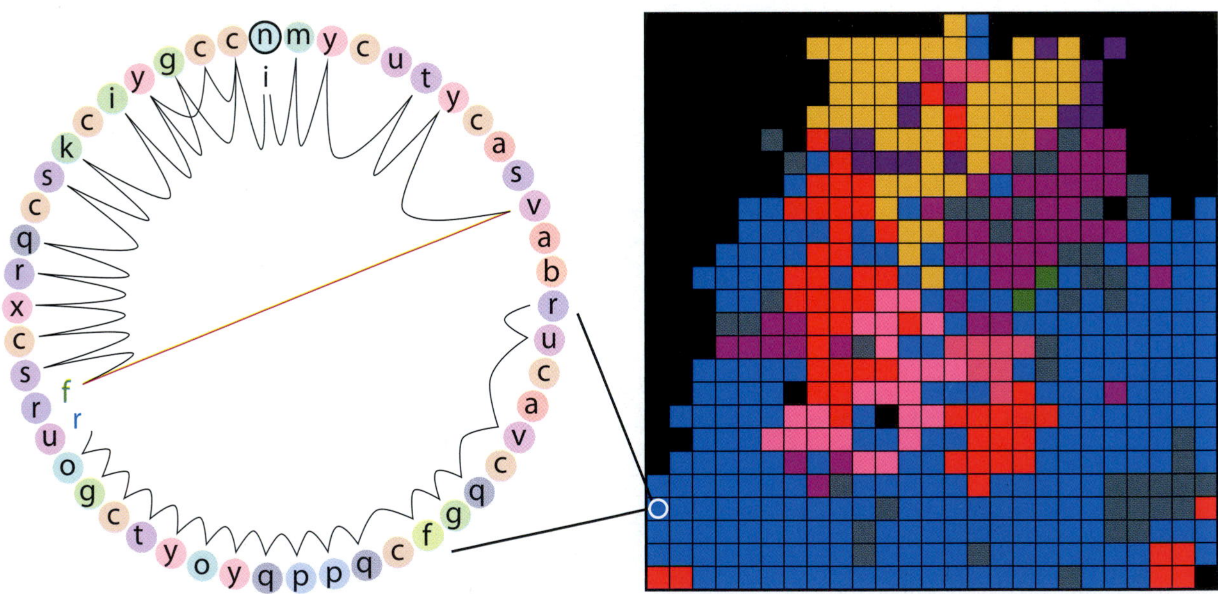

Plate 7. Digital evolution in the Avida-ED system begins with a digital organism with a circular genome consisting of a sequence of instructions that cause its behaviors, including (A) self-replication. Random mutations in the genome of the daughter cell as it buds off from a parent cell can result in new variations that may be neutral, advantageous, deleterious, or lethal. The genome of a descendant cell (B) may thus be significantly different from its ancestors. Natural selection occurs in a population of digital organisms (C) as they compete in a virtual environment that differentially rewards possible traits. (http://avida-ed.msu.edu.)

IV.3

Geographic Variation, Population Structure, and Migration

Ophélie Ronce

OUTLINE

1. The causes of spatial structure in genetic diversity
2. Individuals and their genes move around
3. Gene flow shapes patterns of spatial genetic structure
4. Evolution in spatially structured populations
5. Implications for conservation

Species are rarely genetically homogeneous sets of individuals, and genetic diversity is not distributed randomly through space within their ranges. Spatial patterns in genetic diversity can be observed at many different scales. At a very fine scale, within a continuous population, the genetic similarity between two individuals generally declines with increasing distance between them. For instance, in the annual plant *Medicago truncatula*, two individuals growing within 0.5 m of each other in an old field were found to be on average about 10 times more likely to carry identical variants for some highly polymorphic DNA sequence than two individuals separated by 7 m in the same field. Genetic similarity often varies along environmental gradients. Distribution of genetic diversity can also be patchy. Spatial patterns of genetic variation can be manifest at very broad geographical scales. Both genetic drift and selection can explain the emergence of spatial patterns in genetic diversity. Movement of individuals through space, resulting in gene flow, shapes these patterns. In turn, the nonrandom distribution of genetic diversity through space has multiple consequences for the evolution of mating systems, life histories, and more generally, fitness. Finally, spatial genetic structure and its evolutionary consequences bear many implications for the conservation of biodiversity in the context of global changes.

GLOSSARY

Cline. A gradient of continuous variation through space for a genetic or phenotypic character within a species

Dispersal Kernel. The probability density that an individual initially at coordinates (0,0) is found at coordinates (x,y) after dispersal.

Gene Flow. The partial mixing between different populations.

Genetic Rescue. Increase in a population's mean fitness due to the introduction of genetically divergent individuals.

Gene Swamping. Lack of significant response to selection because gene flow is too high.

Habitat Selection. The nonrandom distribution of individuals after dispersal across habitats.

Heterosis. Higher fitness of progeny born to parents originating from different populations rather than from the same population.

Isolation by Distance. Decreasing genetic similarity with increasing distance; due to shared ancestry when dispersal is limited.

Local Adaptation. Higher fitness of resident genotypes in their native environment relative to that of immigrant genotypes in that same environment.

Metapopulation. A set of discrete populations connected by dispersal.

Migration Load. Decrease in mean fitness of a population because of immigration.

Outbreeding Depression. Lower fitness of progeny born to parents originating from divergent populations than from related populations.

Wahlund Effect. The higher proportion of homozygotes due to local mating compared with that expected in a well-mixed population with the same genetic diversity.

1. THE CAUSES OF SPATIAL STRUCTURE IN GENETIC DIVERSITY

Genetic Drift

At a fine spatial scale, when offspring move a short distance from their parents, the formation of local pedigree structures such that nearby individuals are highly related to each other is a basic explanation for patterns of genetic isolation by distance, as observed in *M. truncatula*. Similar arguments hold at different spatial scales: if exchanges between populations are rare, two individuals found in the same population are more likely to share a common ancestor in the relatively recent past (in that same population) than are two individuals in different populations, whose common ancestor probably dates back to an earlier time. When lineages have been separated for a longer time, the additional time has allowed for mutations to appear and develop differences in gene copies that initially descended from the same ancestor, which can explain the greater divergence between genes sampled in different populations rather than in the same population.

This process is accentuated by small population size. When the number of individuals reproducing locally is not large, it is more likely that two local inhabitants share a common ancestor in a recent past. Many individuals will then carry identical gene copies that have not been altered by mutation since they descended from the same ancestor. Consanguinity then results in decreased genetic diversity and increased homozygosity at the local scale. At the extreme, local genetic diversity can be entirely lost. As this random drift process is blind, it is likely that different genes have spread stochastically to fixation in different populations when the latter are isolated from one another (see chapter IV.1).

The demographic history of populations thus has much potential to shape spatial genetic patterns. Many species have undergone relatively rapid spatial expansion after the retreat of glaciers in the quaternary. Colonization of new areas by a reduced number of founders affects spatial patterns of genetic diversity, because genotypes found in the new part of the range are only a small sample of the diversity found in the original range. The spatial spread of chloroplast variants at the continental scale has therefore been used to infer the various recolonization routes of oak trees from the distinct glacial refuges in southern Europe, with good agreement with the pollen fossil records.

Selection

The other main explanation for spatial genetic structure is that selection favors different genotypes in different locations. In the mosquito *Culex pipiens*, genes conferring resistance to insecticide are obviously strongly selected for in areas near the coastline where pesticides are used. The mutated protein that is the insecticide target, however, functions less well than the original one, so that mosquitoes sensitive to the insecticide outperform the resistant ones in nontreated areas in the north. Coloration in the walkingstick *Timema cristinae* confers differential crypsis depending on the host plant. The unstriped morph is more cryptic on *Ceanothus*, whereas the striped morph is more cryptic on *Adenostoma*. The less cryptic morph on each host decreases in frequency within each generation, as expected if it is subject to higher predation. In other cases, such as the latitudinal cline in wing size in *D. subobscura*, the agent of selection is less clear. *D. subobscura* is of European origin and was introduced into both North and South America. The convergent evolution of such similar genetic clines de novo in both continents after introduction, however, suggests that variable selection (somehow linked to temperature), not drift, is responsible for the formation of the pattern.

When new mutants are obtained by mutagenesis, their relative fitness can be assayed in different environments (e.g., different gene deletions in yeast grown with different sugars). Such experiments have been conducted in a number of rapidly reproducing organisms (e.g., bacteria, viruses, fruit flies) and have revealed that the effect of genetic variation on fitness is in general dependent on the environmental context. Given the existence of adequate genetic variation, selection pressures varying in space should result in the emergence of patterns of local adaptation. Reciprocal transplantations of different genotypes across sites often show that local genotypes outperform foreign genotypes in their environment of origin (e.g., in about 70 percent of the sites in data compiled from 35 transplantation experiments with various plant species).

The fitness of different genotypes may also vary from one locality to the other, not because of extrinsic environmental differences, but because of intrinsic variation in the genetic composition of the local populations. This is the case in particular in presence of genetic incompatibilities, such that crosses between some genotypes are partially sterile. When for historical reasons different incompatible genotypes are more frequent in different locations, rare genotypes incompatible with the locally dominant genotype then suffer from a large fecundity disadvantage, which can maintain strong spatial patterns in genetic diversity.

Habitat Choice

A last cause of spatial genetic structure is when habitat choice is dependent on an individual's genotype. For instance, preference for different host plants is in part genetically determined in some butterflies. Female butterflies with innate preference for some host plant will lay their clutches on patches of that host plant, and the corresponding genotype will be more frequent in such patches, even in the absence of any differences in performance of the different genotypes on that plant. Heavier fledglings of great tits preferentially settle in the less crowded parts of Wytham Wood, while lighter birds settle in denser areas. Phenotype- or genotype-dependent dispersal could thus be a source of phenotypic and genetic divergence among sites.

2. INDIVIDUALS AND THEIR GENES MOVE AROUND

Movement Does Not Amount to Gene Flow

Movement is an essential feature of life. Even sessile organisms such as plants and corals have evolved adaptations like mobile larvae or winged fruits to facilitate movement at some stage in their life cycle. Movement serves many functions, such as foraging for food, finding mates, colonizing new territories, and escaping predators or deteriorating environmental conditions. Not all types of movement affect gene flow: after having spent most of their lives in the sea, salmon undertake long and dangerous journeys to return to mate in the same river where they were born. Despite very large distance movement, there is therefore little genetic mixing among pools of fish from various localities. Conversely, dispersal movements by which individuals leave their natal site to reproduce elsewhere, or to reproduce in different locations in different attempts, are relevant for genetic exchanges across space; moreover, effective gene flow requires that dispersing individuals succeed in spreading their genes in that new location. For instance, colonial naked mole rats are highly xenophobic, and immigrants are often killed when they intrude on a new colony. Depressed or enhanced reproductive success of migrants hence alters patterns of gene flow.

Evolutionary Consequences of Movement beyond Gene Flow

If movement does not necessarily result in gene flow, movement still causes individuals to experience different environmental conditions, and thus different selection pressures. Salmon have, for instance, evolved the capacity to adjust their physiology to different salinity levels. Movement also affects population dynamics, the distribution of population sizes through space, and spread rates, with many consequences on genetic diversity, which are partly distinct from gene flow issues. For instance, experimental populations of the small plant *Cardamine hirsuta* experience frequent crashes and local extinction when the proportion of seed dispersed between patches of plants is either too small or too large. Genotypes found at the very edge of the range in species undergoing rapid spatial expansion (e.g., in invasive species) can benefit from the demographic wave of spread, increasing in frequency even if they have weak negative effects on fitness. In outcrossing plants, seed dispersal and pollen dispersal both contribute to gene flow between localities, but seed dispersal has distinct demographic consequences, allowing in particular the colonization of new areas. Gene flow can therefore be partly uncoupled from the other evolutionary consequences of dispersal.

Heterogeneity in Dispersal and Gene Flow

Despite the general ubiquity of adaptation to facilitate movement, there is a large heterogeneity among species in the spatial extent of resulting gene flow. Even within a set of highly mobile species such as birds, recovery of ringed individuals shows that among 75 species breeding in UK, the mean natal dispersal distance varies from about 2 km in the Dunnock (*Prunella modularis*) to about 70 km in the Grey Heron (*Ardea cinerea*). This heterogeneity is also found within species. For instance, within several insect species, some individuals have fully functional wings while many others carry atrophied wings. The spatial extent of gene flow is often summarized by the mean distance between the location where a parent reproduced and that where its offspring reproduced. Given the highly stochastic nature of movement, dispersal is, however, better described by the entire distribution of distances between offspring and parents. Such dispersal kernels are often highly asymmetrical, with many short-distance dispersal events and a few long-distance events; for instance, viable airborne tree pollen can move over hundreds to thousands of kilometers in some conifers.

Complex Patterns of Gene Flow

It is tempting to describe the movement of living organisms by analogy with the random diffusion of molecules; however, patterns of movement in nature are much more complex. Even for organisms relying on passive dispersal by wind or water, there are often strong patterns of directionality. Most pollen that fertilized seeds in a Swedish population of *Pinus sylvestris* was found to have originated from higher latitudes by 1 to 2 degrees, a finding that might be due to climatic conditions that year. Dispersal has often been found to be influenced by an organism's perception of its own internal condition and of its environment. Even in the small ruderal plant *Crepis sancta*, the proportion of seeds equipped with a parachute-like structure facilitating wind dispersal increases when the mother plant is grown in stressing conditions. Habitat selection is a pervasive feature of many organisms that has the potential to strongly shape the patterns of gene flow through space.

3. GENE FLOW SHAPES PATTERNS OF SPATIAL GENETIC STRUCTURE

Gene Flow Makes Genetic Patterns Vary More Smoothly in Space

The partial mixing of genes across locations tends to blur the spatial genetic structure generated by drift and selection. In particular, genetic similarity among pairs of individuals due to pedigree structure declines more slowly with increasing distance when gene flow occurs on greater spatial scales, and when effective population density is higher. Consistently, genetic similarity generally declines faster with distance in herbaceous plants than in trees, as the latter disperse their genes farther. Patterns of isolation by distance are often used to estimate the spatial scale of gene flow. Such indirect estimates from genetic spatial patterns (e.g., an average distance of 123 m between parent and offspring in the damselfly *Coenagrion mercuriale*) can agree well with more direct demographic estimates of dispersal distance (128 m traveled within a lifetime in that same species). Gene flow makes genetic clines broader than the spatial scale over which selection changes through space. While the transition between pesticide-treated and untreated areas is relatively sharp, for *C. pipiens* the changes in resistance frequency through space are quite smooth. Cline shape results from the tension between divergent selection, which enhances genetic differentiation, and the homogenizing effect of gene flow, which erases it. Variation in cline shape can therefore be used to estimate selection intensity and the spatial extent of gene flow.

Migration Load

In the walkingstick *T. cristinae*, the frequency of the less cryptic morph on a given host plant is much higher when a neighboring population uses the alternate host plant than when such patches of alternate host are found much farther away. Gene flow can thus introduce maladapted individuals in populations and constrain local adaptation. Gene flow can entirely prevent adaptation to local selection pressures in small environmental pockets surrounded by larger areas of a different habitat. Such *gene swamping* occurs when the force of gene flow is much greater than that opposed by selection: local adaptation polymorphism is lost, and genes advantageous in the dominant habitat spread to fixation. Depending on the spatial scale of gene flow, environmental pockets have a critical size below which adaptation to these specific environmental conditions is lost.

Different Genetic Characters Show Different Patterns of Spatial Variation within the Same Species in the Same Set of Localities

Substantial genetic divergence for genes involved in local adaptation can be maintained when the rest of the genome is homogenized by gene flow. In temperate forest trees, for instance, there is very little divergence among most DNA sequences found in different locations, across very large distances, because of long-distance pollen flow and large population sizes; there is, however, much evidence for fine-scale adaptation to local climate in forest tree species, with marked genetic clines along latitude or altitude for frost resistance or the timing of flowering.

Gene Flow at Range Margins

The constraining effect of gene flow on local adaptation is expected to be stronger in small populations at the periphery of the range, where many migrants are received from core populations adapted to different environmental conditions. Asymmetrical gene flow would make marginal populations more genetically similar to core populations that would be optimal in their own environment. When transplanted into various common gardens, populations of *Pinus sylvestris* from marginal locations with extreme climate indeed grow better in milder conditions, closer to the core, than in their original locations. The relative role of evolutionary constraints linked to gene flow, lack of genetic diversity, interspecific competition, and other demographic asymmetries in explaining the evolution of range limits (i.e., failure to adapt to environmental conditions outside the range) is still debated.

Gene Flow Allows the Genetic Cohesion of a Species across Space

By constraining genetic divergence between interbreeding populations, gene flow can be seen as the glue binding the collection of populations constituting a species. In particular, gene flow allows the spread of new favorable mutations across the species range. The delta-32 mutation in the *CCR5* gene in humans confers resistance to infection by HIV and other diseases. This mutation is found in Europe and western Asia, with a high frequency in northern Europe, and is thought to have originated about 3000 years ago. Analysis of spatial patterns for the frequency of the delta-32 mutation suggests that it has spread as a result of rapid dispersal (more than 100 km per generation) and strong selection but has had insufficient time to expand to the entire range of the human species. The constraining effect of gene flow on divergence explains why the speciation process is often initiated in a geographic context that leads to disruption of gene flow (allopatric speciation).

The strength of the normalizing effect of gene flow in maintaining species integrity and preventing speciation can, however, be questioned. Gene flow at the scale of the range is very rare in some species, such as between populations of the snail *Cepea nemoralis* found in different valleys of the Pyrenees Mountains. Stabilizing selection, rather than gene flow, might then be instrumental in maintaining some phenotypic uniformity at the scale of the species range. Conversely, some cases of phenotypic divergence, and of further evolution of reproductive isolation, have been documented in geographic contexts where gene flow was initially not absent (parapatric or sympatric speciation).

Gene Flow Affects Levels of Genetic Diversity within Populations

Dispersal shapes the distribution of genetic diversity through space, by increasing the proportion of total diversity contained within rather than between populations. Accordingly, in selfing plants, which lack pollen dispersal, much diversity is contained between populations rather than within populations. Addition of a relatively moderate number of immigrants each generation (one migrant or more) suffices to maintain within each population a large fraction of the total genetic diversity contained in the whole metapopulation.

The maintenance of high levels of variation for traits closely linked to fitness remains a paradox where stabilizing selection is acting to reduce variation within populations. Genetic diversity for adaptive traits within populations should increase with gene flow from divergent populations, up to the point at which polymorphism is lost because of gene swamping. If gene flow between differently adapted populations is a persistent source of genetic variation, there should be strong correlations between genetic diversity within a population and the amount of heterogeneity in the environment around that population. In a study of 142 populations of lodgepole pine, variation for growth indeed correlated with climatic heterogeneity in the region of origin of populations.

Gene Flow Can Advance Adaptation to Changing Environments

By replenishing genetic variance eroded by drift and selection within populations, gene flow can facilitate adaptation to new environmental conditions. In particular, this is the case when selection varies in both time and space: genetic variation that was favored in some other part of the range may become useful in a new location. The evolutionary arms race between parasites and their hosts provides an example of a case where increased migration actually improves local adaptation. When host populations have evolved resistance against infection by their local parasites, introduction of new genetic variants that are less well recognized by the defense system of the local host could allow the parasite population to overcome host resistance more quickly. Experimental coevolution of viruses with their bacterial host *Pseudomonas fluorescens* in microcosms confirmed such prediction. Local viruses were more infectious on their local host than were foreign viruses, but only when a fraction of viruses were regularly transferred between cultures.

4. EVOLUTION IN SPATIALLY STRUCTURED POPULATIONS

Spatial proximity generally means that individuals are likely to mate, compete, or more generally interact with each other. Genetic resemblance between individuals that are close spatially has therefore many evolutionary consequences.

Different Behaviors and Life History Traits Are Selected For

Ecological interactions with neighbors can greatly affect the fitness of an individual, for example, in competition for the same resource pool, interfering agonistically or, conversely, providing help. Genetic resemblance among neighbors generates some association between the genotype of an individual and the phenotypes of individuals affecting its fitness, with many consequences for the evolution of traits involved in such interactions. For instance, some strains of the yeast *Saccharomyces cerevisiae* consume glucose very quickly but with a low energetic yield (selfish strains), while other strains consume glucose more slowly but with higher yield (prudent strains). In well-mixed cultures, selfish strains readily invade populations of prudent strains. When competing in a spatially structured metapopulation, prudent strains resist the invasion by selfish strains, because patches of prudent strains more efficiently turn resources into population growth. Prudent yeasts do better when they are surrounded by other prudent yeasts, while selfish yeasts do worse when their competitors are also selfish. Spatial genetic structure sets the stage for kin selection to similarly alter the evolution of cooperation, life histories, and dispersal.

Genetic Load can Increase or Decrease with Population Structure

Frequent mating between genetically similar parents results in an increased proportion of gene copies being carried by offspring in homozygous form, compared to what would be expected under random mating (*Wahlund effect*). Many mutations have innocuous or weak effects when present in a single copy in an individual, but are strongly deleterious in homozygous form. In the French-speaking Canadian population of Saguenay–Lac-Saint-Jean in northeastern Quebec, a population with few founders and historically isolated because of religious and linguistic issues, carriers of recessive genetic disorders, very rare or unknown in other human populations, can represent up to 4 percent of the local population. Thus, consanguinity due to spatial structure can in the short term greatly depress the mean fitness of the population when deleterious mutations previously hidden in heterozygous form are expressed. In the long term, however, consanguinity helps purge such mutations, potentially ameliorating the genetic load (see chapter IV.5). Purging occurs only when enough genetic diversity remains for selection to sieve from. Loss of local genetic variation means that selection within populations is inefficient in removing badly adapted genes. Genetic load can therefore either decrease or increase in the long term in spatially structured populations, compared to a reference well-mixed population. This depends on the extent of spatial genetic structure, the distribution of mutation effects on fitness, and the exact details of the life cycle, which govern the ways in which competition within spatial entities contributes to change in gene frequencies across generations.

Crosses between Populations Produce Fitter Individuals

The rare plant *Ranunculus reptans* has a fragmented distribution, with many small isolated populations. Plants whose parents originate from different populations produce more seeds than plants produced by crosses within the same population. This pattern of *heterosis*, or increased vigor of hybrids, is frequently observed and is used in agriculture, for instance, in corn production. In *R. reptans,* crosses between populations produce fitter individuals when the parental populations are small and characterized by low genetic diversity, but not when they are large and diverse. The random fixation of different weakly deleterious recessive mutants in distinct small isolated populations could therefore explain the higher fitness of heterozygous hybrids, in which such mutations are masked.

Crosses between Distant Populations Produce Less Fit Individuals

Conversely, hybrids between distant populations may suffer from outbreeding depression. Outbreeding depression can result from loss of local adaptation, when hybrids have intermediate phenotypes, which make them poorly adapted to their parent environment, or when genetic divergence among populations having evolved in isolation results in genetic incompatibilities. *Arabidopsis thaliana* is a selfing plant with a very large range and very few likely genetic exchanges among distant populations. The gene *HPA*, with essential function for plant development, is duplicated in *A. thaliana*. Plants from different geographical origin (Colombia vs. Cape Verde Islands lines) carry different mutations, disrupting the expression of either one of the duplicates of *HPA*, while the alternate duplicate gene is still functional. In a cross of the two lines, recombinant offspring carrying the two mutated loci in homozygous form have no functional copy of *HPA* and abort as embryos. Similar epistatic interactions are often found to depress the fitness of progeny from crosses between distant populations, with deleterious effects visible after several generations of interbreeding.

5. IMPLICATIONS FOR CONSERVATION

Global changes alter patterns of gene flow, as well as their putative consequences. In particular, gene flow may be critical to helping a species adapt to climate change, when genetic variation already exists somewhere in the range that could foster more rapid adaptation to warming temperature. For many species, habitat loss and fragmentation have resulted in both a drastic reduction in local population sizes and disrupted gene flow between populations. Increased inbreeding in small isolated populations can increase their extinction risk, as was found in the Glanville fritillary butterfly.

Genetic rescue has been proposed as a management option that seeks to increase fitness in a population by introducing unrelated individuals. The Florida panther population had dropped to fewer than 20 individuals in 1995; this small isolated population had accumulated many phenotypic defects, including low sperm quality, suggesting that deleterious mutations had drifted to fixation. Eight females from Texas were translocated in Florida in 1995 and since then the panther population has risen to more than 100, with much reduced frequency of phenotypic defects in panthers of Texas ancestry; however, the success of such *genetic rescue* must be balanced by the potential risks of translocating individuals across large distance, including the spread of diseases, outbreeding depression, and the swamping of local adaptation. Recurrent introduction of one migrant per generation in small populations with high genetic load is considered sufficient to replenish genetic variation and counter the effects of genetic drift, while still allowing the preservation of local adaptation.

See also chapter III.4, chapter III.14, chapter IV.6, chapter VI.3, and chapter VI.6.

FURTHER READING

Clobert, J., E. Danchin, A. A. Dhondt, and J. D. Nichols. 2001. Dispersal. Oxford: Oxford University Press.

Garant, D., S. E. Forde, and A. P. Hendry. 2007. The multifarious effects of dispersal and gene flow on contemporary adaptation. Functional Ecology 21: 434–443.

Lenormand, T. 2002. Gene flow and the limits to natural selection. Trends in Ecology & Evolution 17: 183–189.

Tallmon, D. A., G. Luikart, and R. S. Waples. 2004. The alluring simplicity and complex reality of genetic rescue. Trends in Ecology & Evolution 19: 489–496.

IV.4

Recombination and Sex

N. H. Barton

OUTLINE

1. Molecular recombination
2. Rates of recombination
3. Linkage disequilibrium
4. What generates linkage disequilibria?
5. Recombination facilitates selection

Sex and recombination are among the most striking features of the living world, and they play a crucial role in allowing the evolution of complex adaptation. The sharing of genomes through the sexual union of different individuals requires elaborate behavioral and physiological adaptations. At the molecular level, the alignment of two DNA double helices, followed by their precise cutting and rejoining, is an extraordinary feat. Sex and recombination have diverse—and often surprising—evolutionary consequences: distinct sexes, elaborate mating displays, selfish genetic elements, and so on. Indeed, a substantial fraction of molecular evolution—as measured by the rate of protein evolution—is driven by sex. For example, the most striking changes along the lineage leading from our common ancestor with chimpanzees are in genes expressed in the testis, presumably influencing sexual selection between sperm. Although sex and its consequences are most obvious among eukaryotes, whose genes regularly pass through meiosis, sex is also important—and perhaps, essential—for bacteria and archaea, which often adapt to new environments (and to antibiotics) by bringing in genes from other lineages. The evolution of sex itself is discussed in another chapter (see chapter III.9). Here, I focus on the molecular mechanism of recombination, its effects on the composition of a population, and its interaction with other evolutionary processes—especially, with selection.

GLOSSARY

Allele. A particular form of a gene.

Centimorgan (cM). A distance on the genetic map that corresponds to a rate of recombination of $c = 1\%$.

Epistasis. A state in which the value of a trait is not equal to the sum of effects of the genes that influence it.

Gene Conversion. During meiosis, a DNA heteroduplex forms; repair of mispaired heterozygous sites leads to an excess of one or the other allele.

Hitchhiking. The increase in a neutral allele that happens to be associated with a selectively favorable allele at another locus.

Introgression. Movement of genes from one genetic background to another, as a result of hybridization between individuals from distinct populations.

Linkage. Genes that are carried on the same chromosome are said to be linked.

Linkage Disequilibrium. Nonrandom associations between alleles at two or more genetic loci.

Meiosis. A cellular division process in eukaryotes in which gametes are produced, each with half the number of copies of each chromosome as the parents.

Recombination. The generation of new combinations of genes.

Sex. Production of offspring that are a mixture between two different parental genotypes.

1. MOLECULAR RECOMBINATION

Soon after the rediscovery of Mendel's work in 1900, it was found that alleles at different genes sometimes tend to be inherited together. This phenomenon of linkage could be used to identify the linear order of genes on a chromosome. Consider a diploid parent that is heterozygous at two genes, *A* and *B*: one genome carries alleles *ab*, and the other, *AB*. The fraction of recombinant gametes (*Ab* and *aB*) that are passed on can be measured by crossing to a true-breeding stock. If the genes are on different chromosomes, this fraction is $c_{AB} = 50\%$. If they are closely linked on the same chromosome, then c_{AB} is small, and measures the probability of a crossover between the two genes; by measuring rates of recombination between multiple alleles, one can determine their order on the chromosome; for example, if three genes lie close together, in the order *ABC*, then we expect that $c_{AC} = c_{AB} + c_{BC}$. This relationship is not exact, because there can be multiple crossovers in an interval; an even number will yield a nonrecombinant gamete, while an odd number will produce a recombinant gamete. If crossovers occur independently, then the chance of observing an "effective" recombination between *A* and *C* is the chance of an effective recombination between *A* and *B*, but not between *B* and *C*, and vice versa: thus, $c_{AC} = c_{AB}(1 - c_{BC}) + (1 - c_{AB})c_{BC}$. Thus, genes *A*, *Z* that are far apart on the same chromosome may appear to be unlinked (i.e., $c_{AZ} \sim 50\%$), but can be shown to be linked by mapping the genes between them. Distances on the genetic map are measured in centimorgans, with 1 cM corresponding to a 1 percent probability of crossover; 1 morgan = 100 cM.

The most important finding of classical genetics was that this linear genetic map corresponded to the linear arrangement of the chromosomes. Genetic mapping of model organisms, especially *Drosophila*, became ever more detailed, ultimately identifying the location of mutations within genes. Once the genetics of bacteria and their viruses was established in the 1950s, it was possible to map large numbers of mutations very precisely; by the mid-1960s, the order of mutations in the genetic map was shown to be the same as their order in the protein sequence, thus identifying the physical basis of the abstract alleles that had been mapped by classical genetics.

At the molecular level, the primary function of recombination is to repair double-stranded breaks in the DNA. If both strands of the double helix are broken, accurate repair is possible only if the broken strands can be aligned with an intact homologue, and the missing information copied across. An intermediate structure is formed, which can either be resolved into the two original strands, or lead to a crossover (figure 1). In either case, a small segment is copied from one fragment to the other, leading to gene conversion: any heterozygous sites within the segment will be "converted" into a homozygote. Molecular recombination is crucial for the repair of double-stranded breaks, and remarkably efficient: if human cells in tissue culture are irradiated with ultraviolet light, their chromosomes are broken into many separate fragments, yet such extreme damage can be almost perfectly repaired.

In this article, I focus not on the process of molecular recombination but on its consequences for the evolution of populations. In this context, the terminology is different: *recombination* refers to any process that produces different combinations of genes, and includes the segregation of different chromosomes at meiosis, as well as crossing-over between homologous chromosomes, as described above. More broadly, the transfer of DNA from one bacterium to another is a form of recombination—albeit one that is asymmetrical, and involves only a small part of the genome. The term could even refer to the transfer of genes from the mitochondrial to the nuclear genome that followed the symbiotic union of an alphaproteobacterium with the ancestor of modern eukaryotes (see chapter II.12).

Sex has a slightly different meaning, referring to the coming together of genes from different individuals; the term may also be used broadly, applying to both prokaryotes and eukaryotes. If sexual union is followed by segregation of a single chromosome pair, to produce haploid offspring identical to the parents, then there has been sex but no recombination. Such an alternation between haploid and diploid phases is nevertheless important, since deleterious recessive alleles are masked in the diploid stage (see chapter IV.8).

2. RATES OF RECOMBINATION

The amount of recombination depends on the number of chromosomes, and on the length of the genetic map, summed over the chromosomes. In eukaryotes, these both vary widely; for example, *Drosophila melanogaster* has three chromosomes (plus a tiny nonrecombining chromosome) with a total map length in females of 2.4 morgans, while humans have 23 chromosome pairs, with a map length of about 35 morgans. The number of chromosomes ranges up to many hundreds, while the map length per chromosome is limited by the (usual) requirement that there be at least one crossover per chromosome arm, to ensure proper segregation of the chromosomes at meiosis; however, if the crossover is at the tip of the chromosome, then it may contribute negligible recombination among genes. There are exceptions: one reason *Drosophila* is a convenient model is that no crossing-over occurs at all in males.

Rates of recombination per base pair vary substantially between species, because both the length of the genetic map and the physical length of the genome vary greatly; however, in both humans and *D. melanogaster*, the rate of recombination between adjacent base pairs averages about 10^{-8} per generation (allowing for the

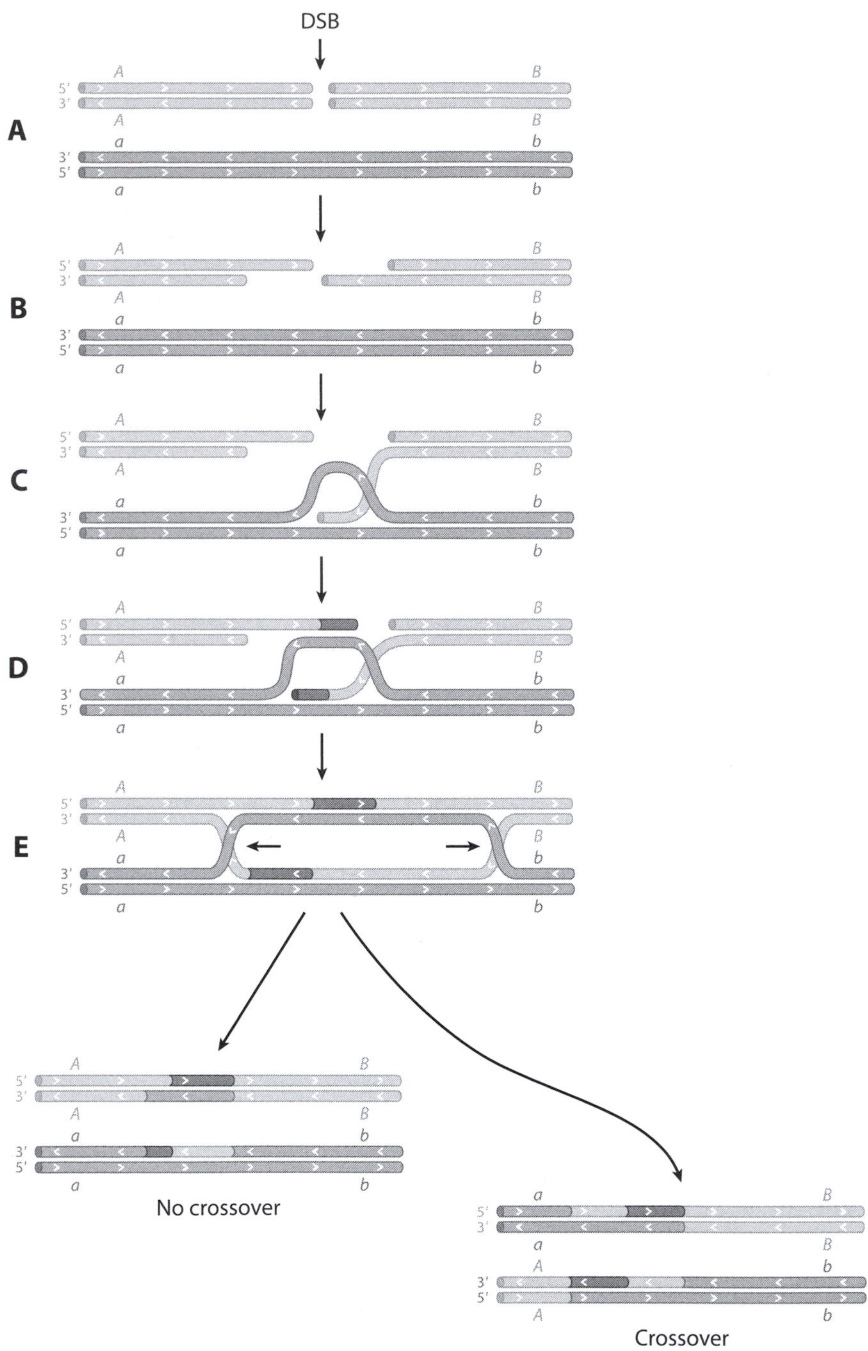

Figure 1. Molecular recombination between two homologous DNA strands is initiated by a double-stranded break (DSB). The outcome can be resolved in two ways: with or without generating a crossover between the two loci, A and B. In either case, a region of the DNA is converted to the homologous allele (orange). (A) Two DNA double helices are aligned, and a double-stranded break is made in one of them. (B) DNA is degraded to make two single-stranded tails. (C) One strand invades the intact double-stranded homologue. (D) New DNA is synthesized (orange), homologous to the invading allele. (E) Strands are rejoined, producing two "Holliday junctions" that can migrate along the DNA. These can each be resolved by breaking and rejoining the strands in two ways. (F) shows the outcome with no crossover, while (G) shows the outcome with a crossover. Note that in both cases, there is gene conversion (orange segments), in which heterozygous sites may become homozygous. (After Watson et al. 2004. Molecular Biology of the Gene. New York: Cold Spring Harbor Laboratory Press.)

absence of crossing-over in males in *Drosophila*). This average figure masks great heterogeneity across the genome; typically, the rate of recombination per base pair is much lower near the centromere. This broad-scale variation in recombination rate provides an opportunity to see the evolutionary effects of recombination. In *Drosophila melanogaster*, genetic diversity is strongly correlated with recombination rate—an observation that has stimulated much work on molecular evolution (see below, and chapter V.1).

On a still-smaller scale, variation in recombination rates can be extreme, with most recombination concentrated in "hot spots." Classical genetics cannot measure such fine-scaled variation in recombination rates; recombination hot spots were first discovered by screening very large numbers of human sperm for recombinants between closely linked genetic markers. They were confirmed by population genetic methods (discussed below) that have allowed detailed maps of hot-spot locations across the human genome. Surveys of single-nucleotide polymorphisms in large pedigrees allow the precise location of recombination events, and they have shown that approximately 60 percent of these occur in hot spots that were estimated by population genetic methods.

The molecular basis of recombination hot spots has recently been determined, and it has interesting evolutionary implications. In mammals, they are initiated by PRDM9, a methyl transferase that marks specific sites on the chromosome; these are then targeted by Spo11, a highly conserved enzyme that initiates double-stranded breaks. The system is puzzling, because any binding site variant that increased the local recombination rate would tend to be eliminated by Spo11, and replaced by the alternative, less active allele—in other words, gene conversion would tend to favor "cold spots." Hot spots are indeed transient, being polymorphic within the human population, and being almost entirely distinct from the hot spots found in chimpanzees. It is possible that a dynamic equilibrium exists between loss of hot spots by gene conversion, and the generation of new binding sites by mutation of the *PRDM9* gene, to recognize different sequences; this latter process could itself be driven by broad-scale selection to maintain the optimal distribution of recombination across the chromosomes. This hypothesis is supported by the conservation of broad recombination patterns across species, despite the rapid evolution of the *PRDM9* gene.

In bacteria, recombination appears to be a side effect of other processes: DNA from other bacteria may be acquired through transfer of plasmids, infection by viruses, or feeding on DNA from outside the cell. It occurs very rarely per cell division, but because bacterial populations are so large, the total number of recombination events can be large, and cause significant evolutionary consequences. Because only small fragments are transferred, they can come from very different lineages, and still function; for example, antibiotic resistance can be acquired from bacteria that are more than 20 percent divergent in sequence. Rates of transfer by direct uptake of DNA do decrease with sequence divergence; nevertheless, selectively favored alleles can be picked up from very distant lineages.

3. LINKAGE DISEQUILIBRIUM

This article concentrates on recombination (including both segregation of different chromosomes, and crossing-over within chromosomes). Recombination produces offspring gametes that carry new combinations of the alleles inherited from either parental genome, and so can generate enormous variability: if the parents differ at 40 sites, recombination can generate $2^{40} \sim 10^{12}$ combinations. However, at the level of the whole population, recombination has no effect on average if the alleles are already randomly combined; further shuffling makes no difference. Thus, recombination can alter the composition of the population only when there are nonrandom associations among alleles. Such associations are technically called *linkage disequilibria*—an unfortunate term, since there can be linkage disequilibria between alleles not physically linked on the same chromosome, and because there may be a steady level of associations at equilibrium. To understand how recombination influences evolution, we must understand how nonrandom associations can be produced.

We measure the strength of associations between alleles simply by the difference between the actual frequency of a particular pair of alleles, p_{AB}, and the frequency expected if they combine at random, $p_A p_B$: $D_{AB} = p_{AB} - p_A p_B$, where p_A and p_B are the frequencies of the *A* and *B* alleles at the two loci. (An equivalent definition is that $D_{AB} = p_{AB}p_{ab} - p_{aB}p_{Ab}$.) Linkage disequilibria are defined for particular sets of alleles, but it is always true that $D_{AB} = D_{ab} = -D_{aB} = -D_{Ab}$. Describing multiple alleles, and more than two genes, is much more complicated: we can define coefficients of association among sets of alleles, but because so many genotypes are possible, a correspondingly large number of coefficients is needed. This fundamental problem makes the population genetics of multiple recombining loci difficult.

In the simple case of two loci with two alleles, recombination at a rate c simply reduces D_{AB} by a factor (1 – c) in every generation; if the loci are unlinked, c = 1/2. Thus, linkage disequilibria will typically persist for approximately 1/c generations: alleles 10 cM apart will remain associated for about 10 generations while alleles 1 kb apart on the human genome recombine at about 10^{-5} per generation, and so stay together for about 100,000 generations—or about halfway back to our common ancestor with chimpanzees.

4. WHAT GENERATES LINKAGE DISEQUILIBRIA?

Recombination changes the composition of a population by breaking up statistical associations between alleles (linkage disequilibria); thus, its effect on evolution depends on what generates these associations. Mutation typically acts independently at different sites, and so breaks down associations.

Selection

Epistatic selection can favor certain combinations of alleles, and so generate linkage disequilibria; recombination will thus reduce fitness by destroying the favorable associations just built by selection (see chapter IV.5). An important example occurs near a sex-determining locus, where alleles that increase fitness in one or other sex will accumulate, leading to strong selection against recombination, and eventually, to sex chromosomes that do not recombine with each other at all (see chapter V.4).

Migration

Migration can also generate strong nonrandom associations. These are seen most strikingly in narrow hybrid zones between genetically distinct populations, in which strong linkage disequilibria are maintained by a balance between mixing and recombination (see chapter VI.6). Because such associations allow selection to act on whole sets of alleles, rather than on each one individually, they can greatly reduce the effective rate of gene exchange by increasing the effectiveness of selection. Migrants bring in sets of alleles that may not be adapted to the local environment or genetic background, and that are eliminated by strong selection. Recombination breaks up these associations, scattering incoming alleles across different native genetic backgrounds and making it harder to eliminate them.

Random Drift

Perhaps the most important, and the most general, cause of linkage disequilibria is random genetic drift (see chapter IV.1). As we look forward in time, there will be random fluctuations in linkage disequilibrium, simply because individuals that carry some combinations of alleles happen by chance to leave more offspring. The variance in linkage disequilibrium between two alleles depends on the number of recombination events between them per generation: $\mathrm{var}(D_{AB})$ is proportional to $1/(1 + 4N_e c)$, where N_e is the effective population size.

Looking back, blocks of genome will share the same genealogical ancestry, to the extent that they have passed intact through meiosis without being broken apart by recombination. This correlation in ancestry is described by an elegant extension to the coalescent process. As we trace the ancestry of a segment of genome back through time, it may encounter a recombination event, such that portions are inherited from different ancestral genomes, and from then on back in time have separate genealogies. Different lineages may coalesce, so the blocks they carry become identical by descent from some ancestral genome —an event in which one parental genome passed a block of genome on to two offspring, both blocks surviving to be found in our present-day sample. In the simplest case of a single well-mixed population, with constant effective size, recombination and coalescence occur at rates that do not change through time. Thus, each present-day genome traces back to many different ancestral genomes, each contributing one or a few small segments.

Typically, blocks of genome of length $c \sim 1/2N_e$ will have the same ancestry. This can be seen directly in the genome sequence: although the proportion of sites that are heterozygous averages $\pi = 4N_e\mu$, this nucleotide diversity varies greatly along the genome, as the genealogy changes abruptly from one block to the next. In the human genome, the boundaries between such haplotype blocks are sharpened by recombination hot spots, but even if recombination rates were uniform, there would be abrupt changes as discrete recombination events occurred in the ancestry of the sample.

The generation of linkage disequilibria by random sampling is seen most strikingly in the spread of a new mutation. This mutation arises on a particular genome and carries a fragment of that one genome with it as it increases in frequency. If the mutation takes T generations to get to its present frequency, then on average, it will carry with it a block of map length $c \sim 1/T$. Thus, the pattern of reduced diversity around such a new allele can give an estimate of the strength of the selection that drove it (see chapter V.14). The same argument applies both to favorable mutations that increase rapidly through selection—the classic process of hitchhiking (see chapter V.14) —and to deleterious mutations that increase through random drift, despite selection against them. These patterns allow us to estimate the age of some alleles. For example, this method has shown that the ΔF508, a deleterious allele that causes cystic fibrosis, arose approximately 3000 years ago.

5. RECOMBINATION FACILITATES SELECTION

In the examples above, of sex chromosomes and of migration with local adaptation, recombination reduced mean fitness. This, together with the obvious costs of sex and recombination, raises the question of why they are so widespread, at least among eukaryotes. At the end of the nineteenth century, August Weissman argued that sexual reproduction provided variation that would allow more efficient adaptation by natural selection, and this intuitive explanation was widely accepted; however, it is not at all easy to show exactly how sex and recombination can generate useful variation, and to show that it gives an advantage that can outweigh the various costs (see chapter III.9).

How might recombination facilitate selection? In principle, selection can be effective on a strictly asexual population, acting simply on the variation generated by mutation. Indeed, if mutations are fixed one at a time, recombination makes no difference, since only two alternative types at most are ever present together; however, if new favorable mutations arise while others are still on their way to fixation, there is strong interference between them—they can be brought together only by recombination. This difficulty can be avoided in very large populations, so large that many copies of every possible mutation arise in every generation. However, in more modestly sized populations ($N < 1/\mu$, say), the rate of adaptation may be limited primarily by the rate of recombination. Another way to look at the issue is to see that recombination randomizes alleles across genetic backgrounds of different quality, allowing selection to disentangle the effects of any particular allele from the effects of the random set of alleles with which it happens to find itself in any one individual.

To see the evolutionary role of recombination in a wider perspective, it is helpful to think of it in relation to speciation. The separation of populations into distinct biological species restricts the field of recombination, and so allows each species to specialize in different ecological niches. Hybrids produced by recombination between species are typically less fit, because they contain new combinations of alleles that have not been favored by selection and may be poorly adapted to the niche of either parent; however, speciation also reduces the size of the gene pool, making it more important to bring together the best combinations of mutations, whose supply is limited by the population size. It may be that regular sex and recombination have made it possible for eukaryotes to adapt to specialized niches, involving large body size and slow reproduction, despite the small population size that such specialization implies.

FURTHER READING

Barton, N. H., D.E.G. Briggs, J. A. Eisen, D. B. Goldstein, and N. Patel. 2007. Evolution. Cold Spring Harbor, NY: Cold Spring Harbor Laboratory Press. *Chapter 15 describes how recombination interacts with random drift to shape the ancestry of genomes; chapter 23 shows how recombination evolves, and its consequences for other aspects of the genetic system.*

Baudat, F., J. Buard, C. Grey, A. Fledel-Alon, C. Ober, M. Przeworski, G. Coop, and B. de Massy. 2010. PRDM9 is a major determinant of meiotic recombination hotspots in humans and mice. Science 327: 836–840. *This landmark paper shows how recombination hot spots move over time, even though the overall pattern of recombination is conserved.*

Burt, A. 2000. Sex, recombination and the efficacy of selection—Was Weissman right? Evolution 54: 337–351. *A review that discusses Weissman's explanation for the function of recombination in modern terms.*

Charlesworth, B., and D. Charlesworth. 2010. Elements of Evolutionary Genetics. Greenwood Village, CO: Roberts & Company. *Chapter 8 discusses the evolutionary role of recombination.*

Coop, G., and M. Przeworski. 2007. An evolutionary view of human recombination. Nature Reviews Genetics 8: 23–34. *A clear and concise review, with a focus on humans.*

Whitehouse, H.L.K. 1973. Towards an Understanding of the Mechanism of Heredity. London: E Arnold. *A classic text that gives an excellent historical account of the development of our understanding of recombination.*

IV.5

Genetic Load

Aneil F. Agrawal

OUTLINE

1. Genetic load
2. Mutation load
3. Other types of load
4. Consequences of load

At a cursory level, evolution by natural selection is simple. Some genotypes are better than others. Those genotypes increase in frequency, wiping out the alternatives. The population then consists of only the best genotype(s). Once this has occurred, the population has achieved its maximum fitness. However, a number of phenomena—especially those processes that generate the variation necessary for adaptation—prevent populations from achieving this selective nirvana in which only the best genotype exists. In other words, nonoptimal genotypes persist at equilibrium. The presence of these nonoptimal genotypes means that the average fitness of individuals within the population is lower than the fitness of the best genotype; this reduction in fitness is *genetic load.*

GLOSSARY

Epistasis. The deviation in fitness of a genotype carrying multiple mutations from its expected fitness based on independent effects of individual mutations.

Genetic Load. The reduction in mean fitness due to the presence of nonoptimal genotypes.

Mean Fitness. In the context of load, mean fitness refers to the average fitness *relative to* the optimal genotype.

Mutation Load. The reduction in mean fitness due to the presence of deleterious alleles maintained by mutation-selection balance; it is one form of genetic load.

1. GENETIC LOAD

For a number of reasons, a population can harbor genotypes that are less fit than the optimal genotype. Such a population carries a genetic load because the average fitness is less than it could be. Here we will consider several types of genetic load, though mutation load will be discussed in greatest detail. Usually, *x load* can be interpreted as meaning how much more fit would the average individual be if one could stop x from happening. For example, a mutation load of 60 percent (L_{mut} = 0.6) means that as a result of new deleterious mutations, the average individual is 60 percent less fit than the best genotype. If mutation could be stopped, selection would eventually eliminate all deleterious alleles, so that all remaining individuals would possess the best genotype, thereby increasing this relative measure of mean fitness by 60 percent.

As discussed in greater detail below, theory tells us that some types of genetic load can be shockingly large, meaning that mean fitness is much lower than it could be; however, even when the load is high, it is unclear whether this "matters." Some have argued that the burden of high genetic loads might severely threaten population persistence. Others have argued that genetic load is no more than a mathematical construct with no meaningful consequences. The truth lies somewhere in between, with the answer depending on the circumstances and also the metric used to assess what it means to "matter." These issues will be discussed in the final section.

2. MUTATION LOAD

The Classic Theory

The vast majority of mutations that affect fitness are deleterious (see chapter IV.2). Because selection should eliminate such alleles, one might naively assume that deleterious mutations are an uninteresting and unimportant part of evolutionary biology; however, more careful consideration of the problem changes that perspective. Although selection pushes deleterious alleles out of populations, new mutations are constantly being introduced. Thus, no population can ever be free of deleterious alleles. For example, recent estimates from the 1000 Genomes project and other sources suggest that the average person carries several hundred (or more) deleterious alleles.

In a classic theory paper, J.B.S. Haldane (1937) considered the equilibrium between the opposing forces of selection and mutation. Consider a single autosomal locus in a diploid organism where the wild-type allele A mutates to the deleterious allele a at rate μ. Given the fitness relationships $W_{AA} = 1$, $W_{Aa} = 1 - hs$, and $W_{aa} = 1 - s$, the mean fitness in the population is $\overline{W} = 1 - 2q(1 - q)hs - q^2s$, where q is the frequency of the deleterious allele.

At equilibrium, the deleterious allele is expected to exist at a low, but nonzero, frequency, $q \approx \mu/hs$ (assuming that $\mu \ll hs$). Because the deleterious allele is rare at equilibrium ($q \ll 1$), mean fitness is well approximated by $\overline{W} \approx 1 - 2qhs \approx 1 - 2\mu$. We can define the mutation load as the reduction in mean fitness when mutation occurs relative to what it would be in the absence of mutation,

$$L_{mut} \equiv \frac{\overline{W}_{NoMut} - \overline{W}_{Mut}}{\overline{W}_{NoMut}}.$$

In the case of our single locus model, this becomes

$$L_{mut} = \frac{1 - (1 - 2\mu)}{1} = 2\mu.$$

This remarkable result says that the mean fitness is reduced by an amount 2μ relative to what it would be in the absence of mutation (when all individuals would be of type AA). Counter to common intuition, the reduction in mean fitness is independent of the strength of selection (hs) against the mutation. This is because if a mutation were to be more strongly selected, it would be rarer at equilibrium ($q \approx \mu/hs$), but each copy present would cause a bigger effect on mean fitness; these two opposing effects of hs cancel out in calculating the mean.

With a realistic per locus mutation rate of $\mu = 10^{-6}$, the mutation load from our single locus model is so small, it seems hardly worth our time to consider it at all; however, we must remember that deleterious mutations are occurring at all n loci in the genome. As a first approximation to calculating the genome-wide mutation load, Haldane made two simplifying assumptions: (1) genes affect fitness independently (no epistasis), and (2) deleterious alleles are randomly distributed across loci (no linkage disequilibrium). Under these assumptions, the mean fitness with respect to the entire genome is simply the product of the mean fitnesses with respect to each individual locus,

$$\overline{W}_{whole\ Genome} = \prod_{i=1}^{n} \overline{W}_i \approx \prod_{i=1}^{n} (1 - 2\mu) \approx \prod_{i=1}^{n} e^{-2\mu}, \\ = e^{-2n\mu} = e^{-U}$$

where $U = 2n\mu$ is the average number of new mutations per diploid genome per generation. This calculation leads to some disturbing conclusions once we start considering real values of U. While a fair bit of uncertainty remains in estimates of U, there is good evidence that many multicellular eukaryotes have deleterious mutation rates in the range of $U = 1$ (see chapter IV.2). By the equation above, we obtain $\overline{W} = e^{-1} = 0.37$, implying a mutation load of more than 60 percent. This major reduction in mean fitness is not due to any one particular deleterious allele being common but rather to the presence of many different rare mutations, perhaps all of very small effect, dispersed across the genome.

Before proceeding, it is worth making a few semantic clarifications. The use of the term *mutation load* here follows the formal mathematical definition above, but readers should use caution when interpreting its usage elsewhere. Sometimes, "mutation load" is used more loosely to describe any scenario where deleterious alleles are present, often in reference to the relative commonness of such alleles, or the ways they affect the genetic variance in fitness, the consequences of inbreeding, or the risk of extinction. The formal definition above pertains precisely to the reduction in "mean fitness," but interpreting this can be tricky. Mutation load is best thought of as the degree to which the average individual would be less fit relative to a mutation-free individual placed in the population. *Mean fitness* as used in the current context is not defined in terms of absolute fitness in a way relating directly to population size or growth, an issue to be discussed later. In addition, beneficial mutations are not considered here, even though they are incredibly important for adaptive evolution and fitness. The purpose of excluding beneficial mutations is to focus on the effects of deleterious mutations. Arguably, a more accurate term for the ideas discussed here would be *deleterious mutation load* rather than just "mutation load," as we are interested in quantifying the impact of deleterious mutations, not all mutations; nonetheless, the traditional terminology is used here.

Extending the Theory

Building from earlier work, Kondrashov and Crow (1988) provided a more general analysis of mutation load that does not depend on any assumptions about gene interaction or linkage disequilibrium. Their analysis shows that load is $L = U(\bar{z} - \bar{y})$, where $\bar{z}$ and $\bar{y}$ are the mean numbers of mutations carried by the winners and losers of selection, respectively. This result can be understood as follows. Each generation, U mutations per individual enter the population. At equilibrium, the same number of mutations must be removed by selection. The mutation load can be thought of as the individuals who die, or otherwise fail to contribute to the next generation, in order to remove these mutations. The difference $\bar{z} - \bar{y}$ is the number of mutations eliminated per *selective death,* when an individual dies or fails to reproduce because of its genotype. (The concept applies more generally than *survival versus death*; for example, a mutant individual who produces 10 percent fewer offspring can be thought of as representing 10 percent of a selective death.) If many mutations can be eliminated at once by killing off a few individuals, the load will be smaller than if it takes many selective deaths to remove the same number of mutations. In practice, it is difficult to determine $\bar{z} - \bar{y}$ for particular scenarios, limiting the application of the elegant general result above. Nonetheless, the general result helps us interpret more specific results.

Perhaps the best-known extension of the mutation load theory pertains to the role of *epistasis* or gene interaction. For example, consider the fitnesses of the following haploid genotypes: $W_{AB} = 1$, $W_{Ab} = 0.9$, and $W_{aB} = 0.9$. If the double mutant has a fitness of $W_{ab} = 0.6$, then its fitness is worse than expected based on each of the single mutation effects, and we would describe this as *synergistic* or *negative epistasis.* If the double mutant had a fitness of $W_{ab} = 0.85$ (i.e., better than expected), we would say there is *positive epistasis.* Compared with Haldane's prediction, mutation load is larger when there is positive epistasis but lower when there is negative epistasis. With strong negative epistasis (especially truncation selection), load can be greatly reduced because many mutations can be eliminated per selective death (i.e., $\bar{z} - \bar{y} > 1$); however, empirical studies have shown that some gene combinations show positive epistasis but others show negative epistasis, with no strong general trend toward one type of epistasis.

Inbreeding can also reduce mutation load, especially if deleterious alleles are strongly recessive. The clustering of alleles through inbreeding (i.e., excess of homozygosity) allows more deleterious alleles to be eliminated with each selective death (see chapter IV.6). Like typical inbreeding, population structure can also create an excess of homozygotes and therefore also reduce mutation load; however, with population structure, the ecology of selection is important, especially if individuals compete primarily against their relatives for resources. If mutants compete against their mutant relatives, rather than against wild types, their deleterious effects may be sheltered from selection, thereby increasing the load.

Under fairly general circumstances, the load of asexual populations is given by Haldane's result of $L = 1 - e^{-U}$. Though epistasis and dominance affect mutation load in sexual populations, they do not affect the load of asexual populations. This discrepancy serves as an important reminder that it is not the form of selection alone that determines the efficiency with which deleterious alleles are removed by selection; the distribution of deleterious alleles with respect to one another—which is affected by sexual processes such as segregation, recombination, and inbreeding—is also important. Because asexual and sexual populations can have different mutation loads under the same circumstances, population geneticists have wondered whether this may contribute to maintenance of sex.

Empirical Evidence

At least with respect to some groups of multicellular eukaryotes, deleterious mutation rates should be large enough (on the order of 1) to substantially reduce mean fitness according to Haldane's prediction; moreover, there is much evidence that deleterious mutations exist in natural populations. In some cases, specific alleles causing particular genetic diseases have been identified. Inbreeding depression is a common observation in many taxa and is largely attributable to the presence of (partially) recessive deleterious mutations. Molecular population geneticists often find an excess of rare variants at nonsynonymous sites, the expected signature of segregating deleterious mutations.

Although there is little doubt that many deleterious alleles segregate in natural populations, and that new ones are entering populations at a high rate, we do not know if their effect on mean fitness is anything close to the magnitude of Haldane's prediction. Despite the potentially massive effects of mutation load, they have never been quantified. The primary reason is that it is impossible to identify a mutation-free genotype to serve as a reference. One possible solution is to identify a high-fitness genotype to serve as reference. Because it can never be known if this is the best possible genotype, any measure of load using this genotype as a reference must be considered a lower bound on the mutation load. For example, with $U = 1$ and $hs = 0.01$, the mean number of mutations per individual is 100 under Haldane's assumptions. The best genotype one would be likely to find (even with considerable effort) would be expected to have at least 80 mutations. Thus, the true load would be $L = 1 - e^{-U} = 0.63$, whereas the measured load using the "best available" genotype as a reference would be 0.18.

It should be possible to gain insight into the problem by manipulating the mutation rate and comparing the average fitness of individuals from populations of high and low mutation rates. Although this would not be true load because no mutation-free genotype would be present, one could obtain a measure of the load of the high mutation rate population *relative to* that of the low mutation rate population. Under Haldane's assumptions, the relative load is given by the difference in mutation rates, $L_{\text{relative}} = 1 - \exp(U_{\text{Low}} - U_{\text{High}})$. Provided the difference in mutation rates between treatments is on the order of 1, we would expect a sizeable relative load, if Haldane was even approximately correct. Bruce Wallace (1991) conducted this type of experiment by maintaining some populations of *Drosophila melanogaster* under normal conditions while other populations were exposed to X-ray radiation each generation. He found differences in fitness of 5–30 percent; however, these results are difficult to interpret for three reasons. It is unknown the degree to which X-ray radiation increased mutation rate. Though reasonable, the fitness assays he performed would not be considered particularly accurate reflections of fitness, compared with more modern assays. Finally, it is possible that the X-ray procedure may have resulted in inadvertent selection for somewhat different phenotypes than the control (e.g., X-ray resistance); nonetheless, these studies by Wallace represent one of the only attempts to examine the ways in which mutation-selection balance affects mean fitness. There is clearly a need for more studies of this nature.

3. OTHER TYPES OF LOAD

Drift Load

Drift load is sometimes viewed casually as a type of mutation load, though others like to keep the concepts separate. *Drift load* usually refers to the reduction in fitness due to the deviation of deleterious alleles away from their deterministic expected value under mutation-selection balance (e.g., $q = \mu/hs$). When selection is weak relative to the inverse of the population size, $s << 1/4N_e$, then the fate of a deleterious allele is more strongly determined by neutral processes than by selection. Deleterious alleles may drift to high frequencies or even reach fixation, thereby reducing fitness. In populations of reasonable size, only very minor deleterious alleles can drift all the way to fixation. Unlike rare segregating deleterious mutations, fixed mutations provide a likely target for compensatory adaptation, especially if a single beneficial mutation can compensate for multiple fixed deleterious effects.

Migration Load

Divergent ecological selection can favor alternative alleles in subpopulations connected by migration. Conceptually, migration load is very similar to mutation load. Just as deleterious alleles are constantly entering a population through mutation, locally maladapted alleles can be maintained within subpopulations through migration. At equilibrium, the frequency of a maladapted allele is $q \approx m/hs$ (assuming $m << hs$), where m is the migration rate. The presence of these alleles reduce mean fitness, creating a single-locus load of $L = 2m$. Although only a small fraction of loci are likely to experience divergent ecological selection, the load per affected locus is likely much larger than that from mutation-selection balance, because migration can be much stronger than mutation, that is, $m >> \mu$. A number of empirical studies provide suggestive evidence that maladapted alleles appear to be maintained by migration-selection balance, but there are no direct estimates of the migration load itself.

Segregation and Recombination Load

Although sexual reproduction does not change allele frequencies, it can change the way alleles are distributed with respect to one another. Because selection is expected to result in an excess of good combinations, rearranging these combinations through segregation and recombination can result in lower fitness. This is the essence of segregation and recombination load.

The simplest scenario to consider is a single locus model with heterozygote advantage: $W_{AA} = 1 - s$, $W_{Aa} = 1$, $W_{aa} = 1 - t$. In the absence of segregation, the population would evolve to consist entirely of heterozygotes and mean fitness would be 1; however, because of sex (specifically, segregation), homozygotes are present at equilibrium. The equilibrium allele frequency is $q = s/(s + t)$, resulting in a mean fitness of $\overline{W} = 1 - st/(s + t)$, which confers a segregation load of $L = st/(s + t)$. Although only a small fraction of loci may be subject to heterozygote advantage, each one can make a substantial contribution to genetic load (relative to a single gene at mutation-selection balance). If we imagine that 50 loci (of the 10,000+ genes in a typical eukaryote) are subject to heterozygote advantage with $s = t = 0.05$, then we expect a segregation load of about 70 percent, assuming independent fitness effects.

Segregation load need not require heterozygote advantage. Consider the case of a recessive lethal: $W_{AA} = W_{Aa} = 1$ and $W_{aa} = 0$. In a facultatively sexual organism where asexual reproduction is common, the allele can accumulate to a reasonably high frequency in the heterozygous state. Whenever sex occurs, segregation will reduce fitness by converting healthy heterozygotes into lethal homozygotes, thereby reducing mean fitness.

Just as segregation disrupts favorable intralocus associations, recombination can break up favorable interlocus associations (e.g., coadapted gene complexes). By converting fit genotypes to less fit ones, recombination can reduce mean fitness, and this is the *recombination load.*

As reviewed by Becks and Agrawal (2011), a number of experimental studies in facultatively sexual organisms (e.g., *Chlamydomonas, Daphnia*, rotifers) have found that sexually derived offspring have substantially lower mean fitness than asexually derived offspring. This "sex load" could be due to segregation load and/or recombination load, as both genetic processes are involved in sex. A study in *Drosophila melanogaster* found that recombinant chromosomes were less fit than nonrecombinant chromosomes, reflecting a recombination load.

Gender Load

Because males and females share largely the same genome (excepting Y and W chromosomes), one or both sexes can be constrained from achieving its optimal phenotype when sex-specific optima differ. Analogous to migration between habitats, biparental inheritance continually moves alleles back and forth between the sexes. Populations may fix alleles that are a selective compromise between the sexes, though suboptimal for both genders. This represents a "fixed" gender load, which is quite hard to detect. Alternatively, balancing selection between the sexes can maintain diversity within populations, resulting in a "segregating" gender load. Such segregating polymorphisms should create genotypes that develop into above-average females but below-average males, or vice versa. This type of negative intrasexual genetic correlation in fitness has been observed in several insect and plant systems. In principle, a more direct measure of the gender load could be obtained by preventing *gene flow* between the sexes. Prasad et al. (2007) used cytogenetic tricks in *Drosophila* so that much of the genome was transmitted patrilineally (e.g., Y-like) and thereby became subject to selection only in males. After 25 generations of being free from selection in females, the resulting genomes produced males that were 15 percent more fit than controls. This represents a lower bound to the segregating gender load.

The gender load described above results from opposing selection on the sexes with respect to the same gene(s), that is, intralocus sexual conflict. A second type of gender load results from interlocus sexual conflict. This occurs when the sexes are in conflict over a "shared" trait, such as remating rate, for example, males typically want to mate more frequently than is optimal for females. A different set of genes (and phenotypes) may underlie a male's ability to influence remating rate from those possessed by a female (e.g., male courtship traits vs. female preference functions). Evolutionary advances by each sex prevent one or both sexes from reaching its optimal value. Because this scenario does not involve opposing selection between the sexes at the gene level, it is more likely to contribute to fixed than segregating gender load.

Substitution Load

Evolution by natural selection is not a magical transformation of poor genotypes into good ones. Rather it is replacement process; good alleles spread to fixation by outcompeting the alternative. The removal of these less fit types was described by Haldane as the "cost of natural selection." The associated term *substitution load* can be confusing, because adaptive substitution hardly seems to involve a load as it obviously increases fitness, rather than decreases it. The substitution load refers to the selective deaths required to make a substitution via replacement relative to an imaginary scenario whereby substitution was by instantaneous (and magical) transformation. Formally, the load resulting from increasing an allele at frequency p to fixation is $L_{sub} = 2\ln(1/p)$. This represents the number of selective deaths per capita but is somewhat different from other loads as it is distributed across many generations. When selection is strong, this load is concentrated into fewer generations. Some have argued that the substitution load may set an upper limit on the number of simultaneous substitutions in a limited time, especially in organisms that do not have large amounts of reproductive excess to "absorb" selective losses. Another interpretation of this result is that assuming the population is initially viable, it is not possible for selection to be very strong at multiple newly substituting loci unless the most fit and least fit genotypes are capable of producing vastly different numbers of offspring. Because the lower limit on reproduction is zero, the maximum difference in fitness is limited by the upper bound on absolute fitness, which is lower for organisms such as elephants than for flies.

4. CONSEQUENCES OF LOAD

Under reasonable conditions, the predicted loads from mutation, migration, segregation, and antagonistic selection between the sexes can be very large. But does it matter? From one perspective, the answer is a definitive yes; individuals are of lower genetic quality than they otherwise would be, and this is inherently interesting. From another perspective, the answer is a more ambiguous maybe. Consider that a recent estimate for genome-wide deleterious mutation rate in humans is $U = 4.2$. Following Haldane, this predicts $\overline{W} = 0.015$, or a load of 98.5 percent. That is a massive mutation load! But what does this mean?

In principle, this means that the average individual from the real population would have a fitness of less than 2 percent if he or she were placed in a population with no deleterious alleles at all. Other potentially misleading interpretations of load are also used; for example, a load of L is sometimes taken to mean that if the average real individual is capable of producing n offspring, then an unloaded individual should be capable of producing $n_{\text{unloaded}} = n/(1 - L)$ offspring. Using human values of L = 98 percent and $n = 10$, then $n_{\text{unloaded}} = 500$.

This value seems ridiculous, so we should consider it carefully. First, we should acknowledge our limited ability to assess the plausibility of this number. It is unlikely that any human, or any other large mammal, has fewer than 100 deleterious mutations, so we have no real basis to evaluate what might be possible in the absence of any deleterious mutations. Second, the calculation uses $n = 10$ based on potential reproductive capacity, but in reality, the average individual produces fewer offspring; in a stable population the average female produces $n = 2$ offspring, reducing maximum reproductive capacity to a lower, but still large, value of $n_{\text{unloaded}} = 100$. Of course, the reason most individuals do not realize their reproductive capacity is because of competition for limited resources. This leads to the third, and probably most important, issue. Load refers to fitness relative to the optimal type, not to absolute fitness. Much of the load may be manifest as reduced competitive ability; if so, it is misleading to attempt to calculate maximum reproductive capacity. An unloaded individual may have no greater reproductive capacity, but rather be much more likely to obtain the resources necessary to reach that potential when in competition with loaded individuals. Conversely, the average individual carrying about 700 deleterious alleles may not seem too unfit if competing against others carrying a similar mutational burden, rather than against a hypothetical mutant-free genotype. From this perspective, one might argue that load does not really "matter" too much with respect to realized absolute fitness.

We can continue this line of thinking with respect to population-level consequences of load. Usually, demography is not a simple linear function of the average "genetic quality" relative to the optimal genotype. Consequently, it is hard to predict whether load will have any meaningful population-level consequences. Consider the scenario above involving a resource-limited population in which much of the load is expressed as reduced competitive ability. Those with more loaded genotypes allow those with less loaded genotypes to succeed. If there were no loaded genotypes, then ecological (nonselective) deaths due to resource limitation would replace genetic (selective) ones. In this case, population size can be largely unaffected by the amount of load. This can explain how populations persist in the face of seemingly "unbearable" loads.

The scenario above, with complete ecological compensation for selective deaths, represents one extreme. Whenever load directly affects birth or death rates or the ability to convert resources into survival or reproduction, load is likely to have demographic consequences, even if resource limitation imposes additional density regulation. Moreover, most discussion of load ignores the importance of interspecific competition. When two closely related species compete, their loads relative to one another will either affect their abundances quantitatively or, more importantly, determine whether there is coexistence or competitive exclusion. This latter notion is implicit in load-based hypotheses for the sexual populations competing against their obligately asexual sister taxa. Although the ecological consequences of load have received little attention in more general contexts, this neglect is unjustified. It remains a major empirical challenge to find useful ways to measure load in terms of the decline in fitness relative to the optimal genotype and also its demographic and ecological consequences.

FURTHER READING

Agrawal, A. F., and M. C. Whitlock. 2012. Mutation load: The fitness of individuals in populations where deleterious alleles are abundant. Annual Review of Ecology and Systematics 43: 115–135. *Provides a more detailed review of mutation load with respect to theoretical results, relevant empirical data, and ecological consequences.*

Becks, L., and A. F. Agrawal. 2011. The effect of sex on the mean and variance in fitness in facultatively sexual rotifers. Journal of Evolutionary Biology 24: 656–664. *Includes a review of empirical data on segregation and recombination load.*

Haldane, J.B.S. 1937. The effect of variation on fitness. American Naturalist 71: 337–349. *Provides the classic derivation of mutation load.*

Kondrashov, A. S., and J. F. Crow. 1988. King's formula for mutation load with epistasis. Genetics 128: 853–856. *Provides a generalized model of mutation load.*

Prasad, N. G., S. Bedhomme, T. Day, and A. K. Chippindale. 2007. An evolutionary cost of separate genders revealed by male-limited evolution. American Naturalist 169: 29–37. *An elegant study demonstrating gender load.*

Wallace, B. 1991. Fifty Years of Genetic Load: An Odyssey. Ithaca, NY: Cornell University Press. *A winding tour through the first half century of theoretical and empirical work on load; not comprehensive but interesting.*

IV.6

Inbreeding

Deborah Charlesworth

OUTLINE

1. Inbreeding
2. Measuring the degree of inbreeding
3. Measuring inbreeding coefficients and rates of self-fertilization and other inbreeding
4. Long- and short-term consequences of inbreeding
5. Consequences of inbreeding for molecular evolution and genome evolution
6. Inbreeding depression, heterosis, and purging

Inbreeding (mating between individuals with recent common ancestors) led to populations of organisms being more homozygous than predicted by the familiar Hardy-Weinberg formula for random-mating populations. Inbreeding is one form of nonrandom mating, and it can occur in populations where numbers of potential mates are limited because of small population size or restricted dispersal, or in species or populations with preferential mating with related individuals, either natural (e.g., in naturally self-fertilizing organisms such as many hermaphrodite plants and animals) or enforced (e.g., in sib-mated "inbred lines" of mice, or when inbred strains or breeds are created by crop breeders or animal breeders). Inbreeding has many consequences: the action of natural selection and the amount of genetic variation in inbreeding populations both differ from the situations in outcrossers. Because so many situations of interest to plant and animal breeders involve inbreeding, and because several "model species" important in modern biology are inbreeders, it is interesting to understand these differences. This chapter also outlines the concepts of inbreeding depression and hybrid vigor.

GLOSSARY

Allele. The "type" of a gene, for example, whether the allele is wild type or a mutant.

Diploid. An organism, or stage in the life cycle of an organism, in which individuals carry alleles from a maternal and a paternal parent (as opposed to *haploid* individuals or stages, with only one allele of each locus).

Genotype Frequencies. The proportions of the different homozygous and heterozygous genotypes at a gene with more than one allele present in a population or sample of individuals. These frequencies depend on the allele frequencies and on the mating system.

Hermaphroditic Organisms. Organisms with both male and female functions that can potentially mate in isolation (as opposed to having separate male and female individuals, which can reproduce only by mating with another individual). Some plants have separate male and female flowers on the same individual; as far as the mating system is concerned, such "monoecious" plants can often be treated as hermaphrodites.

Inbreeding Depression. The lower survival or fertility of progeny of inbred matings compared with progeny produced by outcrossing.

Self-fertilization (Selfing). Mating of a hermaphroditic individual (or a monoecious plant) with its own gametes, or mating between individuals that are genetically the same individual. The most extreme form of selfing occurs in haploid plants, when self-mating produces progeny that are homozygous at all loci. Selfing in other hermaphrodites increases the proportion of homozygous loci, but more slowly.

Sib-mating. Mating between full siblings.

1. INBREEDING

Inbreeding is the occurrence of mating between individuals with common ancestors (figure 1). This can happen in various ways (table 1), and an important division is between two general types of situations, as follows:

- inbreeding caused by lack of opportunities to encounter and mate with unrelated individuals,

for example, when population size is restricted, including populations that have experienced a recent bottleneck
- inbreeding caused by patterns of mating with related individuals, for example, through self-fertilization in hermaphroditic plants or animals (often called *selfing*), or in populations in which individuals regularly or frequently mate with their siblings (e.g., when a preference for sibs as mating partners exists, as in some human cultures)

This chapter deals mostly with the second type of inbreeding, but inbreeding cannot always be neatly categorized into these two types. For example, the situation with animals in which the progeny of females do not disperse and must mate with their siblings, as occurs in social spiders, cannot be categorized as one type or the other. More generally, given that dispersal in real organisms never leads to fully random mating throughout the entire species (i.e., mating with no relationship to the origins of the individuals), some inbreeding must often occur in most species, even if at a very low frequency. Despite its prominence in textbooks, random mating and the familiar Hardy-Weinberg genotype frequencies are not the reality for most species (see chapter I.3). Even populations of large, mobile animals such as humans show evidence of some subdivision, clearly indicating that, at least until recent times, matings have been mostly between individuals from close localities.

Many hermaphroditic animals and plants (including monoecious plants with separate male and female flowers on each individual) have *self-incompatibility* systems preventing inbreeding populations (see chapter IV.8). On the other hand, sib-mating may occur in many randomly mating animals and plants, because dispersal is generally insufficient to guarantee that all potential mates are unrelated, and many organisms have no means of recognizing sibs or avoiding them as mates (this is an example of the first kind of inbreeding listed above).

2. MEASURING THE DEGREE OF INBREEDING

One important measure of inbreeding is the proportion of progeny in a generation or cohort that are generated by the inbreeding. For instance, for hermaphroditic plant populations, we can estimate the selfing rate. Clearly, such selfing rates implicitly assume that the inbreeding occurs according to a regular system (with the same rate for all individuals in the population, remaining the same every generation). It is in principle possible to estimate selfing rates for individuals and

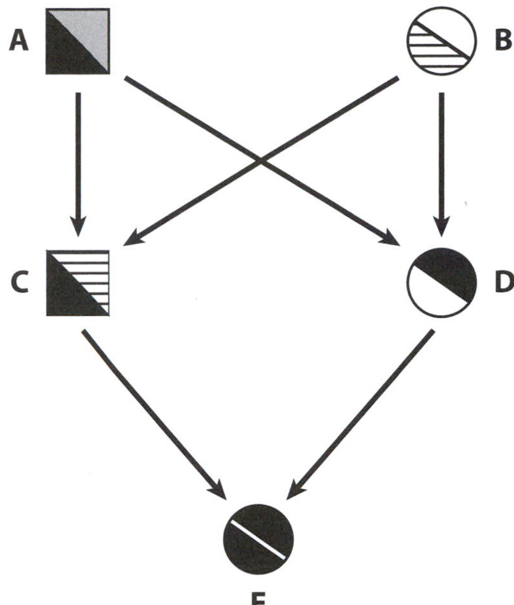

Figure 1. Identity by descent in an inbreeding pedigree (a full sib-mating). Males and females are symbolized by squares and circles, respectively, and the individuals are labeled with letters, and their different alleles are shown with different shadings. Individual E, produced by a full sib-mating, inherits the black allele from her grandfather (individual A) via both her father (C) and her mother (D).

determine whether they are genetically controlled and/or related to particular morphologies (such as larger flowers) or individual characteristics (e.g., the occasional self-compatible individuals in self-incompatible plant species), though this is rarely done. However, it is clear that inbreeding is often temporally variable or context dependent. For example, the selfing rate in a plant or animal population may be higher in years or locations where the density of members of the species is lower, or it can be higher when mating partners are unavailable to snails or other animals than when they are present (e.g., Escobar et al., 2011).

More generally, inbreeding is measured in terms of inbreeding coefficients, which express the probability that both the copies of a gene in a diploid individual were inherited from a copy in a common ancestor—the probability that they are "identical by descent." The increase in the inbreeding coefficient under a regular system of inbreeding can be predicted over the generations under a given mating system. For instance, Mendel's rules of inheritance show that for progeny produced by selfing, this probability is 0.5, while an outcrossed mating produces progeny whose inbreeding coefficient is zero (for the offspring of sib-mating, it is 0.25). Over the generations, a mixture of selfing with some outcrossing therefore leads

Table 1. Some biological situations in which inbreeding occurs

Organisms and situations	*Form of inbreeding*	*Effect on heterozygote frequency and inbreeding coefficient*	*Equilibrium inbreeding coefficient*
Hermaphroditic plants (many weeds) and animals (many parasites)	Self-fertilization, sometimes facultative (e.g., after colonization)	Halved each generation of selfing	<1 if selfing is partial (mixed mating)
Haploid plants	Intragametophyte self-fertilization	Genotypes become completely homozygous in a single generation	As for selfing
Cyclically asexual animals, e.g., aphids, *Daphnia*	Mating between clone mates	Genetically equivalent to self-fertilization	As for selfing
Malaria parasite	Mating between haploids derived from a single diploid infecting parasite	Genetically equivalent to self-fertilization	As for selfing
Many animals	Sib-mating, especially when dispersal is limited, and in zoo populations, but also when sibs are preferred as mates	Decrease with each generation of inbreeding	Can be 1, if exclusive sib-mating occurs (e.g., in inbred lines of mice)

to an inbreeding coefficient that is less than 1 (figure 2 shows this for several selfing rates), whereas slower inbreeding can lead to an inbreeding coefficient of 1 if no outcrossing occurs (e.g., complete sib-mating in figure 2). Pedigrees of much greater complexity than these simple examples include information about the possible lines of descent, and they can also be used to predict the inbreeding coefficient of any member of the pedigree, relative to that of the earliest ancestors in the pedigree (Falconer and Mackay 1996).

Inbreeding coefficients are a very general and useful measure, because they can be applied to any form of inbreeding and compared between different populations or experiments. For inbreeding of the first category above, the inbreeding coefficient can be predicted by modeling the situation in the population of interest, and calculating the time since common ancestry of alleles in any generation (Falconer and Mackay 1996). Methods to do this often use the concept of "inbreeding effective population size," which takes common ancestry into account (see chapter IV.1).

3. MEASURING INBREEDING COEFFICIENTS AND RATES OF SELF-FERTILIZATION AND OTHER INBREEDING

Several reviews have recently been published of procedures for estimating inbreeding rates in natural populations (e.g., Jarne and David 2008). One procedure for estimating inbreeding coefficients is based on the probabilities of identity by descent of the alleles in individuals, which determines the genotype frequencies. This is easily understood for an autosomal locus with two variants, A_1 and A_2, with frequencies p and q. For one of the two alleles of an individual taken from the population, the probability that it is A_1 is simply the A_1 frequency in the population, p, and the chance that the allele is identical by descent (IBD) with the individual's other allele is its inbreeding coefficient, f; ignoring mutation, this allele must also be A_1 because it is identical by descent. The probability that the other allele is non-IBD is $1 - f$, in which case the probability that the individual's genotype is A_1A_1 is p^2, neglecting mutation. Therefore, the total probability that the individual's genotype is A_1A_1 is equal to

$$fp + (1 - f)p^2 = p^2 + fpq.$$

Similarly, the frequency of A_2A_2 is $q^2 + fpq$; by subtraction from 1, the frequency of A_1A_2 is $2(1 - f)pq$. Therefore, heterozygosity is reduced by inbreeding in proportion to f.

Genetic markers allow one to measure the genotype and allele frequencies in a population of interest, giving an estimate of the f value in these equations, and to test whether the frequencies of homozygotes and heterozygotes correspond with random mating ($f = 0$), or whether there is a deficiency of heterozygotes, indicating inbreeding (at least if selection against heterozygotes can be discounted; such selection is implausible, as a factor

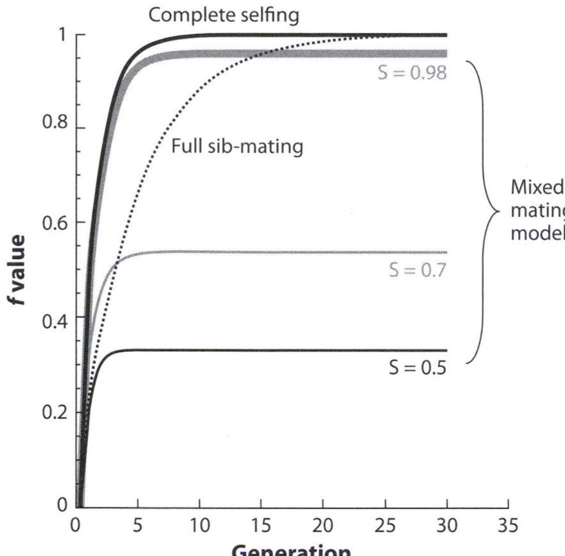

Figure 2. Approach to equilibrium inbreeding coefficients (f) in populations under regular self-fertilization at different self-fertilization rates (denoted by S), and under regular full sib-mating. At equilibrium under the mixed mating model, the expected f value is $S/(2 - S)$.

affecting many loci, and in practice, estimates of mating system usually employ genetic markers such as microsatellites or allozymes that are assumed to be selectively neutral). If so, and if the population's biology suggests that a particular regular system of inbreeding is likely, for example, selfing, the rate of inbreeding can be estimated by assuming that the population has reached equilibrium under that system, and finding the rate that yields the observed genotype frequencies.

Note that inbreeding estimates may not correspond to the values predicted from known pedigrees, because genes may become homozygous more slowly than expected, because of linkage to loci where homozygosity lowers fitness; this is explained in more detail below.

Alternatively, selfing rates can be estimated using family arrays. It may even be possible to estimate the number of generations of inbreeding in the ancestry of individuals. With this kind of approach, individual outcrossing rates can sometimes be estimated using paternity assignment employing suitably informative markers, and it may then be possible to associate selfing rates with characteristics such as, in a plant, different anther-stigma distances in the flowers of different individuals. Both these approaches indicate that, although most populations are found to be either largely outcrossing or largely selfing, intermediate mating systems also exist in both plants and animals (Escobar et al. 2011). Some very important model species in modern biology are among the high inbreeders, notably the plant *Arabidopsis thaliana* and the nematode *Caenorhabditis elegans*.

4. LONG- AND SHORT-TERM CONSEQUENCES OF INBREEDING

Effects on Genotype and Allele Frequencies and on Genetic Variation

Inbreeding increases the frequencies of homozygotes. If inbreeding occurs at a high rate in a population, as in highly selfing species, frequencies of heterozygotes quickly decrease. If a population becomes highly inbred, its individual members can become homozygous at all or most loci. A very important consequence of homozygosity is inbreeding depression, which will be discussed below.

Overall, inbreeding decreases genetic variation within populations and increases isolation between populations, for the following reasons. A longer-term consequence of homozygosity is that the genetic drift process affects homozygous genotypes, rather than individual alleles. Given enough time, populations lose genetic variants through genetic drift, and the difference just mentioned makes this happens faster under inbreeding than in an outcrossing population. Because genetic drift acts independently in different populations, and inbreeding increases isolation between populations by decreasing gene flow, inbreeding leads to increased population differentiation. This can be enhanced by extinction and recolonization processes.

The Effect of Inbreeding on Genetic Recombination, and Differences between Inbreeding and Asexual Reproduction

If a population becomes highly inbred, so that its individual members are homozygous at all or most loci, as in highly selfing species, recombinant genotypes will rarely be formed, even if meiosis and sexual reproduction occur normally, with fertilization of eggs by sperm. This effect occurs simply because double heterozygotes for any pair of loci are rare (rare heterozygotes that are formed when individuals outcross will, of course, yield recombinant progeny), and crossing-over produces recombinant progeny only if the parent is heterozygous at both loci. At the extreme of complete self-fertilization for multiple generations, all individuals will be homozygotes, and reproduction yields progeny whose genotypes are the same as those of their mother, as also occurs with many kinds of asexual reproduction, but the cause is very different. Indeed, asexual species (see chapter IV.4) are often highly heterozygous (such species are often formed by hybridization events).

5. CONSEQUENCES OF INBREEDING FOR MOLECULAR EVOLUTION AND GENOME EVOLUTION

Long-Term Consequences: Hitchhiking

The lowered recombination rate in highly inbreeding populations means that genetic "hitchhiking" events affect larger genome regions than would be affected in an outcrosser. For example, the spread of an advantageous mutation will reduce diversity across a region of genome, and the region will be large in a selfing species. This may make it difficult to detect natural selection in selfers through the low diversity regions caused by selective sweeps.

These hitchhiking effects add to the effect of genetic drift explained above, and can greatly reduce diversity in highly inbreeding populations versus comparable outcrossers (with similar values of the main determinants of diversity, the number of individuals, and the mutation rate). This lower diversity occurs because selective sweeps in selfing species lead to low diversity across large regions of genome, and also because selective elimination of deleterious mutations will cause greater lowering of diversity than under outcrossing. These processes act within populations, so different populations may remain genetically different, and species-wide diversity may not be low. However, in subdivided populations species-wide diversity will generally also be reduced. The net effect on the distribution of diversity within and between populations is generally to greatly reduce within-population diversity in highly inbreeding species, and thus increase the proportion of total diversity that is found between populations.

Haldane's Sieve

Another very interesting consequence of inbreeding is that advantageous mutations will be more likely to spread, even if recessive, than in more outcrossing populations. In a diploid population, advantageous mutations first appear in heterozygotes, and in outcrossing species, homozygotes will remain rare until a high allele frequency is reached. Therefore the selection in heterozygotes determines whether mutations can spread (i.e., the product hs, where s is the selection coefficient and h is the dominance coefficient). Recessive mutations (low h values) therefore have a low chance of fixation. In partially self-fertilizing species, however, less dominant mutations can spread more readily, because inbreeding produces homozygotes even for rare mutations. In contrast, under outcrossing, advantageous mutations are most likely to spread if they are dominant. It is sometimes suggested that inbreeding is advantageous, because it allows the population to take advantage of both dominant and recessive mutations; however, this is a population-level advantage, and (as further discussed below) unlikely to outweigh the strong immediate disadvantages to individuals of inbreeding caused by inbreeding depression.

Chromosome Evolution

Another consequence of high homozygosity produced by inbreeding is that the disadvantages to a new chromosome arrangement are reduced, because these disadvantages often arise through recombination (in translocation heterozygotes) or mis-segregation in heterozygotes. Chromosome rearrangements are indeed observed to be commoner in inbreeding than in outcrossing plants.

Sex Allocation

Hermaphroditic inbreeding populations are predicted to allocate reproductive resources more toward female than male functions. In hermaphroditic selfing organisms, this is indeed found, as detected by lower pollen-ovule ratios in plants, and lower amounts of testis tissue in animals (Petersen and Fischer 1996). In species with separate sexes, inbreeding populations may have female-biased sex ratios.

6. INBREEDING DEPRESSION, HETEROSIS, AND PURGING

Inbreeding Depression and Heterosis

The survival and fertility of offspring of related individuals is usually reduced. Such inbreeding depression effects are well documented in many organisms, including higher frequencies in inbred (consanguineous) families than in outcrosses of major abnormalities such as chlorophyll-deficient albino seedlings in plants and developmental defects in fish, or genetic diseases, as in humans (Bittles 2003). The survival and fertility of individuals in experimentally produced inbred lines is frequently so low that many such lines go extinct. Hybrids made by intercrossing surviving lines often have higher quality than their inbred parents, frequently exceeding the best parent values for several characters. This increased performance of F1 hybrids is called *heterosis*, or *hybrid vigor*.

Models and Empirical Evidence

Inbreeding depression is caused by increased homozygosity of individuals, and its occurrence implies that genetic variation in fitness traits exists within the population. What kind of variation is involved, and why is it present in natural populations? Two distinct kinds of variation can contribute:

- alleles at loci with heterozygote advantage (*overdominance*)
- partially recessive detrimental mutations present in populations at low frequencies due to mutation-selection balance

However, it seems unlikely that hybrid vigor will often be caused by overdominant alleles present in two inbred lines that are intercrossed. This would require that by chance, one of the lines has become homozygous for one allele at the locus, and the other line for the other. In a given cross, heterosis is manifested in many different characteristics, and it becomes implausible that many genes with overdominant alleles will undergo such lucky chances (most inbred lines will become homozygous for the allele whose homozygotes have the highest fitness).

In contrast, the mutation-selection balance hypothesis can readily explain the occurrence of hybrid vigor. Heterosis in an intercross between two populations or genetically uniform strains depends on the existence of genetic differences between them. Although detrimental mutations are expected to be individually rare within large populations, minor-effect mutations can reach high frequencies by genetic drift in small populations, or even become fixed. In different populations, different mutations will reach high frequencies; thus, when two such populations are intercrossed, there is a high chance that the parent genotypes involved will each be homozygous for a different set of deleterious mutations, and so the progeny will be heterozygotes. If the mutations are wholly or partially recessive, even with almost intermediate dominance, the deleterious mutation effects will be wholly or partially masked, leading to heterosis if enough such loci contribute.

The mutation-selection balance hypothesis can also readily explain the occurrence of inbreeding depression. While, as just explained, heterosis will be due largely to mutations with small detrimental effects that can reach high frequencies in populations, rare mutations with moderately large effects predominate in inbreeding depression. For inbreeding within a population to lower fitness, homozygotes for mutations must have much lower fitness than heterozygotes; in other words, the small number of mutations that become homozygous in a given individual must be quite recessive. There is good evidence supporting the involvement of rare recessive alleles (Charlesworth and Willis 2009).

This hypothesis can account quantitatively, without invoking overdominant loci, for the magnitude of inbreeding depression observed in several species where the relevant spontaneous deleterious mutation rates and their effects can be estimated (Charlesworth and Willis 2009). There is also little direct evidence for overdominance from estimates of the magnitude of additive versus dominance genetic variance, which could potentially distinguish whether mutational load or heterozygote advantage contributes most to genetic variation in fitness (overdominance would yield high dominance variance). With few exceptions, characters related to fitness give results consistent with the mutational model, and the exceptions do not suggest a major role of heterozygote advantage (Charlesworth and Willis 2009).

The most obvious approach to determining the genetic basis of the variants causing inbreeding depression or heterosis would seem to be genetically mapping them; however, this is not currently practicable. One problem is that the low resolution of QTL mapping cannot tell us whether a genetic factor that is detected is a single gene versus several genes being involved. Thus, an apparent genetic factor showing heterosis could prove to be a *pseudo-overdominant* situation, in which deleterious mutations at two distinct loci complement one another in heterozygotes. Indeed, fine mapping of particular cases has separated some apparently overdominant factors into situations with deleterious recessive alleles in repulsion. Another difficulty is that, if inbreeding depression is indeed caused by rare mutations, different families should carry different QTLs, so that mapping on one family does not help discover genes important in other families.

The existence of inbreeding depression is probably an important reason why outcrossing has evolved and is often maintained by natural selection. Yet many organisms inbreed by self-fertilization, or exhibit some tendency to prefer matings with relatives. The evolution of inbreeding is discussed in chapter IV.8.

Purging and Failure to Purge Deleterious Mutation

The deleterious mutation hypothesis predicts that inbreeding will expose the mutations to selection in homozygotes, and that inbreeding should thus reduce the load of such mutations in the population; this is called *purging*. Theoretical models of mutations show that major purging effects occur mainly when the mutations greatly lower fitness, and are minor for slightly detrimental mutations. In addition, large purging effects occur only in certain selective situations, and purging is less effective in organisms with large chromosome numbers and high recombination rates.

There is much evidence that deleterious mutations exist in organisms and can fail to be purged during inbreeding. For example, in *Drosophila*, multiple generations of inbreeding lead to lower homozygote frequencies than predicted by the inbreeding coefficient expected from the pedigree, indicating that natural selection against homozygotes for some loci or genome regions is preventing homozygosity. Even in organisms with very high levels of inbreeding in nature, such as *Caenorhabditis elegans*, some genome regions resist becoming homozygous.

A question that has interested many biologists working on mating systems is whether inbreeding leads to lower levels of adaptation than outbreeding, and therefore to a greater chance of extinction. Recall that high levels of inbreeding cause effects similar to those in small populations, and that genetic drift of weakly deleterious mutations is therefore of greater importance than for an outbreeder (see section on heterosis above). Fertility and survival of individuals in inbreeding populations might therefore be low because of high frequencies of such deleterious mutations. On the other hand, the removal of recessive and largely recessive mutations due to exposure to selection in homozygotes (see explanation of purging above) acts in the opposite direction. We have already seen that purging does not completely remove such mutations. Empirical studies are starting to uncover evidence that selfing plants may indeed show maladaptation or *genetic degradation*. The lineage leading to the highly selfing plant *A. thaliana* has been found to have accumulated more substitutions in its coding sequences (i.e., the DNA encoding proteins) at sites that change the amino acid, compared with a related outcrossing lineage (Slotte et al. 2010).

It is less clear whether these effects are strong enough to reduce the long-term survival of inbreeding species. Inbreeders have long been claimed to be "evolutionary dead ends," and it is now becoming possible to use phylogenetic trees based on DNA sequences to infer when changes from outbreeding to inbreeding, and the reverse, have occurred in suitable taxa. These studies have shown that inbreeding probably evolves often, but rarely persists for long evolutionary time, and that there are few large taxa of highly inbreeding organisms (Takebayashi and Morrell 2001).

This, however, does not imply that the failure of inbreeders to generate large taxa of descendant species is because their loss of genetic diversity leads to a lower ability to adapt to changing environments. As explained above, the lower genetic diversity under inbreeding is a long-term consequence, involving genetic drift. It is unlikely to cause a mating system shift to evolve, or to influence such shifts. The short-term effect of homozygosity in causing inbreeding depression is likely to be a much more important cause of failure of inbreeding lineages to evolve, or to persist for long if they do evolve. Self-fertilization is certainly often strongly selectively favored in the short term, through its well-known "transmission advantage," as well as "reproductive assurance" when density is low so that mating opportunities are scarce (or because pollinators are scarce for an animal-pollinated plant), and also when locally adapted genotypes are selected to avoid receiving gametes from different environments. With its many advantages, inbreeding is not a great puzzle—the bigger puzzle is why so many organisms outcross (see chapter IV.8). Possibly, inbreeding often arises in local populations of a subdivided species, but severe inbreeding depression selects for outcrossing with conspecifics from other populations (Schoen and Busch 2008). There is certainly no sign that selfing plants are unable to adapt to diverse environments, as witnessed by many studies of local adaptation in *A. thaliana*.

FURTHER READING

Bittles, A. 2003. Consanguineous marriage and childhood health. Developmental Medicine and Child Neurology 45: 571–576.

Charlesworth, D., and J. H. Willis. 2009. The genetics of inbreeding depression. Nature Reviews Genetics 10: 783–796.

Escobar, J. S., J. R. Auld, A. C. Correa, et al. 2011. Patterns of mating-system evolution in hermaphroditic animals: Correlations among selfing rate, inbreeding depression, and the timing of reproduction. Evolution 65: 1233–1253.

Falconer, D. S, and T.F.C. Mackay. 1996. Introduction to Quantitative Genetics. 4th ed. Harlow, Essex: Longman.

Jarne, P., and P. David. 2008. Quantifying inbreeding in natural populations of hermaphroditic organisms. Heredity 100: 431–439.

Petersen, C. W., and E. A. Fischer. 1996. Intraspecific variation in sex allocation in a simultaneous hermaphrodite: The effect of individual size. Evolution 50: 636–645.

Schoen, D. J., and J. W. Busch. 2008. On the evolution of self-fertilization in a metapopulation. International Journal of Plant Sciences 169: 119–127.

Slotte, T., J. Foxe, K. Hazzouri, and S. I. Wright. 2010. Genome-wide evidence for efficient positive and purifying selection in *Capsella grandiflora*, a plant species with a large effective population size. Molecular Biology and Evolution 27: 1813–1821.

Takebayashi, N., and P. P. Morrell. 2001. Is self-fertilization an evolutionary dead end? Revisiting an old hypothesis with genetic theories and a macroevolutionary approach. American Journal of Botany 88: 1143–1150.

IV.7

Selfish Genetic Elements and Genetic Conflict

Lila Fishman and John Jaenike

OUTLINE

1. What are selfish genetic elements?
2. Diversity of selfish genetic elements
3. Selfish genetic elements and genome evolution
4. Selfish genetic elements and population variation
5. Selfish genetic elements and speciation
6. Applied uses of selfish genetic elements

While all successful genes can be said to be selfish, the term *selfish genetic element* (SGE) refers to heritable units that spread despite their adverse effects on individuals and populations. SGEs are remarkably abundant and diverse, ranging from mitochondrial variants to transposable elements to heritable symbiotic microorganisms. Because the spread of such elements entails a cost to other components of an organism's genome, there can be strong selection to suppress the action of SGEs. SGEs affect several important features of organisms and populations, including genome size and structure, mutational load and mean population fitness, sex ratio, and speciation. SGEs are of applied significance, being important tools in basic research, crop development, and control of infectious diseases.

GLOSSARY

Genetic Conflict. The spread of a selfish genetic element occurs at a cost to other (nonallelic) genetic elements in the same genome or organism.

Genome Parasites. SGEs that colonize and multiply within a genome, such as mobile DNA elements.

Meiotic Drive. Historically, any process causing overtransmission of an SGE to gametes. We restrict this term to chromosomal elements that drive during meiosis rather than during gamete formation or later.

Non-Mendelian Inheritance. In a diploid nuclear locus, transmission of alternative alleles to gametes and/or progeny that deviates from the expected 1:1 ratio. Many reproductive parasites cause non-Mendelian inheritance, also known as segregation distortion or transmission ratio distortion, but other processes, such as inbreeding depression or hybrid incompatibility, can as well.

Reproductive Parasites. An SGE that spreads by altering host reproduction in ways such as meiotic drive, disabling gametes, cytoplasmic male sterility, feminization, parthenogenesis induction (PI), male killing (MK), and cytoplasmic incompatibility (CI).

Selfish Genetic Element (SGE). A heritable unit that can spread despite its adverse effects on an organism's fitness or on other heritable elements carried by those organisms.

Somatic Parasites. SGEs that multiply within individual cells (such as petite mitochondria) or within a body of cells (such as most cancers).

1. WHAT ARE SELFISH GENETIC ELEMENTS?

For more than a century, flies of the genus *Drosophila* have been among the most important model organisms for studies of genetics. As in mammals, the X and Y chromosomes of *Drosophila* males typically obey the Mendelian law of segregation, so that males produce a 1:1 ratio of X- and Y-bearing sperm and thus equal numbers of daughters and sons; however, in some species of *Drosophila*, many of the females brought in from the wild produce only daughters, in clear violation of the standard rules of genetics. Chances are that such females had mated with a male that carried a "sex-ratio" gene on the X chromosome that prevents development of functional Y-bearing sperm. By incapacitating Y-bearing sperm, the sex-ratio gene—and the X chromosome on which it is located—is passed on to all of a male's offspring. In contrast, an X chromosome lacking the sex-ratio gene plays by the rules of Mendelian segregation and is passed on to only half a male's offspring. Therefore, by subverting the process of spermatogenesis, the sex-ratio gene gains a substantial transmission advantage and can rapidly spread through a population, even if it has negative effects on an organism's survival and fertility, that is, its fitness. Such genes are just one example of an extraordinarily widespread, diverse, and influential group of heritable entities referred to a selfish genetic elements (SGEs).

In this example, any genes not linked to sex ratio will segregate normally and thus not experience an enhanced transmission advantage. However, they will experience the reduced fitness of occurring within an individual that carries sex ratio; therefore, most of the genome will be selected to suppress the activity of sex ratio and restore normal Mendelian segregation. Exactly this has been found in some species of *Drosophila*, where males from the wild appear to be genetically normal, producing offspring with 1:1 sex ratios; however, if you cross flies to get individuals carrying the X chromosome from one population and the autosomes from another (or from a closely related species), some of the resulting males will produce all-female progeny. Such a pattern indicates that a population carries a sex-ratio gene that has been suppressed by local autosomal genes, while the substitution of autosomes from another population allows the unfettered expression of the sex-ratio phenotype. This battle between the X-linked sex-ratio gene and the autosomes is an example of genetic conflict. Such conflict and coevolution between SGEs and the rest of the genome have played important roles in shaping eukaryotic genome structure, population biology, and diversification.

The concepts of selfish evolution and genetic conflict have deep roots in theoretical population genetics. J.B.S. Haldane (1932) noted, "A higher plant species is at the mercy of its pollen grains. A gene which greatly accelerates pollen tube growth will spread through a species even if it causes moderately disadvantageous changes in the adult plant." Haldane's pollen example falls into a gray area between sexual selection, selfish evolution, and parent-offspring conflict but demonstrates a very early recognition of the multiplicity of competitive arenas experienced by genes and the conflicts that result. In the same vein, Ronald A. Fisher's (1941) classic model demonstrating the inherent 3:2 transmission advantage of an allele causing complete self-fertilization was an explicit mathematical argument against the popular idea that evolution by natural selection always involves variation in organismal fitness and increases in population mean fitness. Thus, it has long been clear (to theoretical population geneticists, at least!) that selfish genetic elements *should* exist; however, it has taken most of the last century to recognize the wonderfully diverse forms taken by selfish elements, and their pervasiveness across organisms.

The history of selfish element research generally parallels the development of tools for studying genetic transmission, from microscopic observations of chromosomes in the first decade of the 1900s to the detailed analyses of molecular variation now possible. In many cases, selfish elements were described as physical phenomena (e.g., unpaired B chromosomes in bugs, visible *Wolbachia* endosymbionts in insect cells) or through major phenotypic effects in polymorphic populations (e.g., cytoplasmic male sterility, skewed sex ratios) decades before they were recognized as SGEs. The early exceptions were segregation distorters in *Drosophila*, mice, and maize, all of which are unusually tractable genetic model systems. The occurrence of SGEs in many model organisms suggests that they are far more common and diverse than generally thought. In fact, SGEs are present in every cell in our bodies and have been continuously present in our evolutionary lineage for hundreds of millions of years.

2. DIVERSITY OF SELFISH GENETIC ELEMENTS

While there is no generally accepted means by which to classify the tremendous diversity of SGEs, one way to group them is by the arena within which they compete for transmission. *Genome parasites* and *soma and cell parasites* replicate within the genome or body, respectively, of an organism. Thus, they consist of populations of elements that can multiply within individuals, and their effects on individual fitness often reflect their frequency or numbers. In contrast, most "reproductive parasites" compete via transmission to gametes or offspring; thus they act as competing alleles within the organismal population. However, cytoplasmic elements such as mitochondria and, particularly, symbionts operate in both realms, competing via both intraindividual replication ability and manipulation of transmission. We summarize the diversity of SGEs in table 1 and highlight example SGEs from each arena below.

Genome Parasites

Transposable elements (TEs) are segments of DNA that propagate within genomes by various "copy and paste" or "cut and paste" transposition mechanisms, thus multiplying as populations within the habitat of the host genome (figure 1). The act of transposition can result in disruptions to the regulatory or coding sequence of a host gene, thereby resulting in deleterious mutations. Despite this, TEs are extremely abundant in

Table 1. Examples of selfish genetic elements

Category	*What they are*	*Allelic replacement or population growth*	*Where found*	*Mode of action*	*Deleterious aspects*
Segregation Distorters (sperm/spore killers)	Autosomal and sex-linked genes May occur in males or in fungi Spore killer in Neurospora/Podospora	Allelic	Insects, mammals, fungi	Kill spores/sperm bearing alternative alleles → overrepresentation among functional gametes	Biased sex ratio: produce excess of more abundant sex Lower mean individual fitness resulting from (1) spread of deleterious alleles linked to drive locus within inversions; (2) deleterious pleiotropic effects of drive locus itself
Female meiotic drive	Centromeres? Knobs (neocentromeres)	Allelic?	Insects, plants, mammals?	Preferentially segregate to egg pole during asymmetrical female meiosis	Deleterious pleiotropic effects? Linked deleterious loci
B chromosomes	Largely heterochromatic, supernumerary chromosomes	Population	Insects, plants	Multiply independently of other chromosomes	Deleterious pleiotropic effects
Transposable elements		Population	Widespread	Multiply across sites within the genome	Increase in genome size Ectopic recombination Elevated rate of mutation (primarily deleterious)
Cytoplasmic male sterility	Mutant mitochondria	Allelic (mtDNA haplotypes)	Plants	Sterilize male function in hermaphroditic plants, diverting extra resources to seed production	Sex-ratio bias in population Failure to produce pollen and consequent reduction of nuclear gene transmission
Petite	Mutant mitochondria	Population	Yeast	Transmission advantage due to faster replication rate within cell	Reduced capacity for aerobic respiration and colony growth rate
Medea		Allelic	*Tribolium*	Kill descendants that do not inherit the element	Death of 50% of offspring of heterozygotes
Modification restriction system		Allelic	Bacteria	Kill descendants that do not inherit the element	
Homing endonuclease	Endonuclease-encoding sequences	Allelic	Archaea, Bacteria, Eukarya	Get copied into chromosomes lacking such elements	
Endosymbionts	Bacterial or protist maternally transmitted	Population (effects often density dependent)	Insects, arthropods	feminization, CI, mk, PI host dependence	Death of individuals (MK, CI) Biased sex ratio Loss of sexual reproduction and attendant advantages
Cancer	Somatic cells that escape normal regulation of cell proliferation	Population	Multicellular organisms (plants, insects, vertebrates)	Spread and consume resources at expense of the rest of the organism Death of the organism is the end of the line for the cancer lineage	Decreased individual survival Arena in which competition occurs is individual organism, not the organisms population

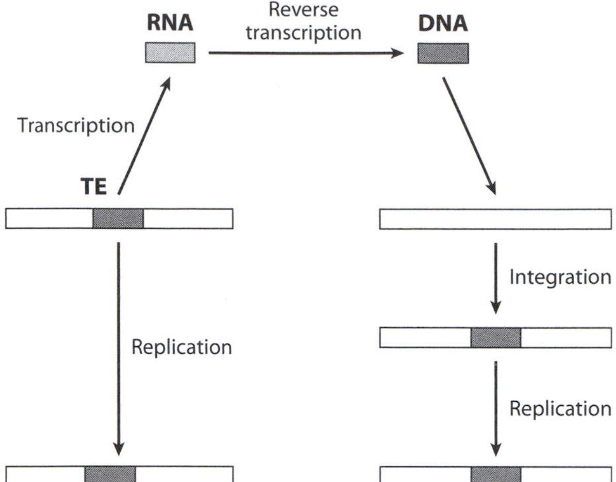

Figure 1. Retrotransposition, one of the means by which transposable elements (TEs) spread. In this case, the TE is transcribed into an RNA intermediate, which then undergoes reverse transcription to DNA, which in turn is integrated into a new position in the genome.

the genomes of plants and animals (~45% in humans, up to 70% in maize), and also occur in fungi, protists, Archaea, and Bacteria. In fact, only a handful of primarily parasitic eukaryotes appear (provisionally) to be TE free. The ubiquity of mobile elements, and their high numbers in most eukaryotes, suggests that the average per element cost is generally small. This allows their greater accumulation in taxa with smaller effective population sizes, in which selection against slightly deleterious genes is less efficient. Because individual TE insertion/deletion events can be a significant source of mutational pressure, and because repetitive tracts of TE DNA can serve as templates for recombination and rearrangement, TEs nonetheless have major impacts on genome structure and function. In addition to these impacts, phylogenetic analyses indicate that one category of TEs—retrotransposons—gave rise to retroviruses, such as HIV.

Soma and Cell Parasites

Somatic mutations can cause cancer by promoting uncontrolled cell division. Cancerous cell lineages spread within individuals, usurping resources and shortening life span, and thus reduce the fitness of all genes within the germ line. Because every cell division has the potential to lead to a cancerous mutation, species with the most total cell divisions—those that are large and/or long lived, like humans—are the most vulnerable. Animals have evolved a variety of mechanisms to effectively detect and control rogue cancerous cells during the reproductive life span of the individual. The rapid propagation of aberrant somatic cells does not generally increase transmission to the next generation, so such lineages cannot spread within populations. Interestingly, cancers are relatively rare in plants, despite their potential for large size and long life span; however, because plant tissues are not segregated into soma and germ lines, there is much greater potential for genetic transmission of a cancer-causing mutation. Thus, selection has probably favored strong anticancer mechanisms in plants.

Petite colonies in yeast are similarly caused by repeated mutation of short-lived selfish elements. Yeast mitochondria with mutations that reduce respiratory activity can replicate more rapidly than normal mitochondria and thus spread to high frequency within a cell; however, this leads to slow growth by the host cells, producing the "petite" phenotype, and low colony fitness. It has been argued that the uniparental inheritance of mitochondria, which has evolved independently in numerous eukaryotic lineages, is in part a strategy to minimize opportunities for interorganelle competition within individuals and forestall the spread of such selfish mutants via sexual reproduction.

Reproductive Parasites

A wide variety of SGEs spread through manipulation of various components of an organism's reproduction to promote their own transmission. These reproductive parasites can be divided into female meiotic drivers, allelic killers, and cytoplasmic distorters.

Female Meiotic Drivers

In most plants and animals, female meiosis is asymmetrical, with only one of the four products becoming the oocyte (or megagametophyte in plants). Thus, any chromosomal element that preferentially segregates to the egg pole gains a transmission advantage. Such a process is termed *meiotic drive*. But such drive can entail significant costs: chromosomal competition can cause errors in segregation resulting in chromosome loss and gamete sterility, and deleterious alleles can accumulate in the driving regions of chromosomes, which tend to be areas of unusually low recombination.

The best-characterized example of meiotic drive is the Ab10/knob neocentromere system in maize, in which an aberrantly long arm of chromosome 10 (Ab10) locks together large heterochromatic regions of repetitive DNA (*knobs*) and a number of genes that cause the knobs to speed to the outer spindle poles ahead of the centromeres. In heterozygotes, this neocentromeric behavior can gain the Ab10 variant up to 70 percent transmission (vs. 50% expected). Ab10 is found at generally low frequencies (<10%) in populations of maize and its wild relative teosinte, indicating that correspondingly strong costs must oppose its spread. In populations with Ab10 present, additional knobs accumulate on the arms of other chromosomes and also preferentially segregate to the egg in Ab10 carriers, acting as secondary parasites on the system.

Paradoxically, the usual mediators of Mendelian chromosomal segregation—centromeres—may themselves act as selfish elements. It has been argued that any centromeric variant that could bias its own segregation toward the egg pole should spread, leading to repeated rounds of centromere turnover and conflict or coevolution between the DNA and protein components of the centromere. Though logically compelling, this model remains controversial in the absence of evidence of driving centromeres.

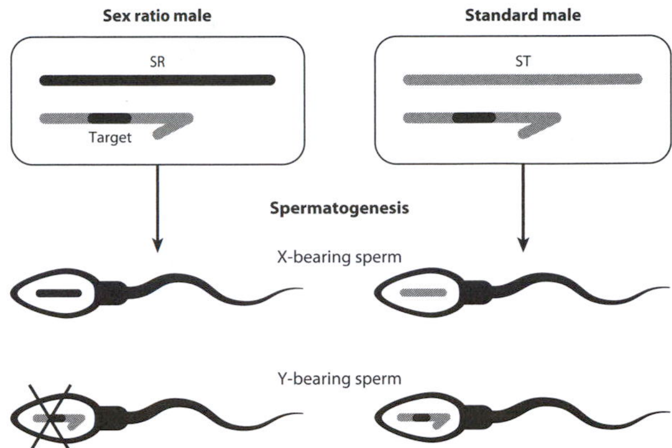

Figure 2. X drive in male flies. Males that carry a sex-ratio X chromosome (SR) produce only X-bearing sperm, whereas males with a standard X chromosome (ST) produce equal numbers of X- and Y-bearing sperm.

Allelic Killers

In contrast to female meiotic drive elements, which subvert equal meiosis to enhance their absolute representation in a heterozygote's gametes, *allelic killers* increase their relative fitness by disabling products of meiosis or offspring that carry alternative alleles. Although such processes are often referred to as meiotic drive or segregation distortion, allelic killers act postmeiotically.

The best-studied autosomal allelic killers are *t* haplotype in mice, Segregation Distorter (SD) in *Drosophila melanogaster*, and Sporekiller in *Neurospora* fungi. Each of these systems comprises, minimally, a killer locus with killer/nonkiller alleles and a target or responder locus with sensitive/insensitive alleles. These loci are locked together in region of low recombination, with the killer/insensitive and nonkiller/sensitive associations predominating, as a killer/sensitive combination would be suicidal. Thus, tight linkage of killer and insensitive alleles is essential for maintaining the observed polymorphisms in sperm/spore killer systems.

When competition between siblings is strong, a parental allele can spread by killing offspring that do not carry it. Examples include maternal effect Medea loci in *Tribolium* beetles, several maternal effect killers in mice, a paternal effect killer in the nematode *Caenorhabditis elegans*, and it has been speculated, restriction-modification systems in bacteria. All of these systems appear to have a killer + target structure similar to that of gamete killers.

Conflicts between gametes or offspring can be especially acute in organisms with chromosomal sex determination, as gametes or offspring carry either the X or the Y from a given parent, and there is essentially no recombination between these chromosomes. Sex chromosome drive is particularly common in flies and mosquitoes, which have XY (male) versus XX (female) sex determination. As discussed above, in males carrying a "sex ratio" X chromosome, sperm carrying the Y fail to differentiate or function properly, resulting in strongly female-biased offspring sex ratios (figure 2). Males with a nondriving X produce equal numbers male and female offspring. Mechanistically, this X chromosome drive entails an interaction between loci on the X and Y chromosomes. In population genetic terms, however, the competition is between the sex-ratio and standard X chromosomes. When females mate with only a single male, the X of a sex-ratio male is passed on to all the offspring, while the X of a standard male is passed on to only half of them; thus, a sex-ratio X chromosome can potentially spread to the point of causing extinction of a population or species.

Cytoplasmic Distorters

Uniparentally inherited organelles and symbionts can spread by increasing the frequency of the carrier sex at the expense of the noncarrier sex. The two best examples of this are mitochondria that cause cytoplasmic male sterility (CMS) in flowering plants and *Wolbachia* (and other maternally transmitted microbial symbionts) in arthropods, both of which are remarkably widespread and have far-reaching effects on organismal variation.

Cytoplasmic male sterility. Maternally transmitted organelles (mitochondria and chloroplasts) have a strong interest in increasing female fertility. Most flowering plants are hermaphroditic, producing both male and

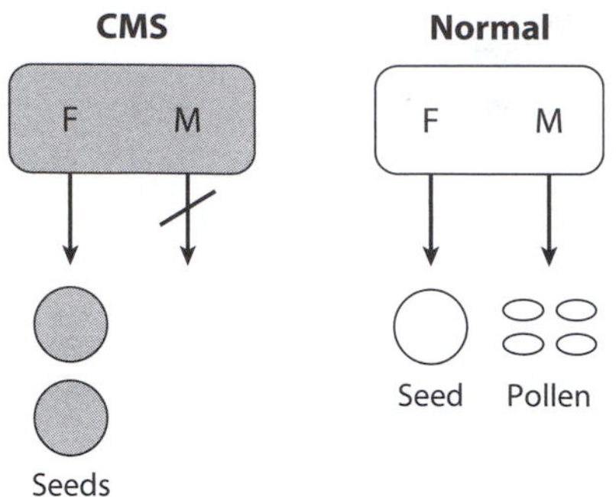

Figure 3. Cytoplasmic male sterility (CMS), found in numerous plants. A specific mitochondrial haplotype results in male sterility and failure to produce pollen, resulting in availability of resources for production of additional eggs, through which mitochondria are transmitted.

female gametes within each flower. This generates a conflict over floral development between organellar (only transmitted in seeds) and nuclear genes (transmitted in both pollen and seeds). (For simplicity, we use the term *seed* to refer to a plant megagametophyte.) Mitochondrial CMS mutants are associated with increased resource allocation to seeds at the expense of pollen production and thus have a significant transmission advantage (figure 3). However, as such a CMS variant spreads, nuclear mutations that restore pollen fertility are favored, both to regain lost male fitness and to take advantage of the increasingly female-biased population sex ratio. This coevolution is predicted, under most conditions, to result in rapid fixation of both the CMS mitochondria and its matched restorer. However, under some conditions, polymorphism of CMS and restorer may be maintained, resulting in a mixture of female and hermaphroditic plants in the population (gynodioecy). About 7 percent of plant species are gynodioecious, suggesting that CMS-restorer interactions are common and contribute to plant mating system diversification. Male sterility due to CMS-restorer mismatch is common in inter- and intraspecific hybrids, indicating that many currently hermaphroditic populations have histories of CMS-restorer coevolution. Rapid evolution under positive selection of the gene family that includes most known restorer alleles also supports the hypothesis of a widespread and ongoing history of cytonuclear conflict in angiosperms.

Wolbachia and other symbionts. Insects and other arthropods are infected with a variety of maternally transmitted microbial symbionts, some of which spread by mechanisms involving manipulation of host reproduction. Such selfish symbionts include bacteria belonging to the genera *Wolbachia* and *Rickettsia* (Alphaproteobacteria), *Arsenophonus* (Gammaproteobacteria), *Spiroplasma* (Firmicutes), and *Cardinium* (Flavobacteria), and microsporidian protozoa. Of these, *Wolbachia* is by far the most common, infecting perhaps two-thirds of all insect species. *Wolbachia* can manipulate host reproduction in a variety of ways, including feminization (developmental conversion of a genetic male into an egg-producing female), male killing (embryonic death of the males, giving their infected sisters access to more resources), parthenogenesis induction (forgoing production of males altogether), and cytoplasmic incompatibility (in which matings between infected males and uninfected females lead to high levels of offspring mortality). Phylogenetic analyses indicate that *Wolbachia* do not infect insect lineages for long evolutionary periods, implying very high rates of colonization of new host species by *Wolbachia* and extinction from currently infected species. Recent studies have shown that some strains of *Wolbachia* that are reproductive parasites can also protect their insect hosts from the adverse effects of RNA virus infection. Thus, *Wolbachia* can function both as a selfish reproductive parasite and as a context-dependent mutualist.

3. SELFISH GENETIC ELEMENTS AND GENOME EVOLUTION

A major fraction of the DNA of many eukaryotes (including humans) consists of transposable elements, making transposable element numbers a primary determinant of genome size. Although TEs make up nearly half the human genome, they contribute relatively little to current mutational variation, as most are dead (missing key components) or otherwise inactivated. In contrast, TEs are extremely abundant and active in many flowering plants. Much of this variation is unexplained, but some undoubtedly reflects aspects of the organismal ecology, such as mating system, that influence the evolutionary dynamics of TEs. For example, small effective population sizes in the highly inbreeding annual plant *Arabidopsis thaliana* may reduce the efficiency of selection against TE-caused mutations and explain its higher load of TEs relative to closely related outcrosser *A. lyrata*. On the other hand, long-term self-fertilization or asexuality reduces opportunities for TE movement among lineages and may even select for TEs with "self-control" by favoring organismal lineages with few or weaker TEs. The latter prediction is borne out by the complete lack of several otherwise-widespread families of transposons in the long-term asexual bdelloid rotifers, despite evidence of TE acquisition by horizontal gene transfer. Under conditions of asexual reproduction or long-term selfing, the interests of all genetic elements within an organism are aligned, thus effectively eliminating conflict and the problems of SGEs.

Mobile elements also affect the physical structure of genomes. TEs are major components of the heterochromatic centromeric and telomeric regions of most chromosomes, contributing to the function of these regions in ways that are only beginning to be understood. Telomeres are repetitive DNA regions at chromosome ends that are eroded during replication and thus must be actively maintained (generally by a telomerase) for cell division to occur properly. In *Drosophila*, end-directed transposition of three types of retrotransposon actually maintains telomere length, an important function for the individual that indicates domestication of a formerly selfish element.

The accumulation of closely related TE sequences throughout a genome provides substrates for recombination between nonhomologous chromosomal regions; it is thus a major source of inversions, translocations, and other rearrangements. Such rearrangements contribute to rapid evolution of chromosome structure in some lineages, particularly flowering plants. If rearrangements vary in their transmission through female meiosis, as has been demonstrated in mice, humans, chickens, and flies, meiotic drive may also contribute to the fixation of such variants. Chromosomal rearrangements are implicated in the evolution of species barriers, both as direct causes of low hybrid fitness and as suppressors of recombination. Thus, the incidence and distribution of TEs across lineages, as well as the opportunities for drive during meiosis, may influence mechanisms and patterns of species diversification.

Other kinds of selfish genetic elements also influence genome- or chromosome-level processes. Most notably, reproductive SGEs may be an important source of selection on recombination rates. A mutation (such as a chromosomal rearrangement) that suppresses recombination in the vicinity of a drive allele will reap the benefits of the linked driver's transmission advantage. Linked mutations that enhance drive will also be favored, favoring low recombination over increasingly larger regions as additional enhancers accumulate. There is abundant evidence for both chromosomal and allelic suppression of recombination around active drive loci. For example, both the maize Ab10/knob system and a centromere-associated female meiotic driver in yellow monkeyflower (*Mimulus guttatus*) encompass vast chromosomal regions locked together by inversions. The rapid spread of selfish nuclear elements along with reduced recombination can result in the spread of linked deleterious alleles. Unlinked loci (that is, most of the genome) should favor greater recombination in the vicinity of a driving element, both to increase the efficiency of selection against deleterious hitchhikers and to break up associations between killer and enhancer/insensitive alleles. For example, the higher recombination rates generally observed near centromeres in female versus male mammals may have evolved to suppress centromere drive in females.

4. SELFISH GENETIC ELEMENTS AND POPULATION VARIATION

SGEs interact with other components of population biology in diverse ways, acting as everything from sources of mutation and expression variation to selective factors in the evolution of mating systems and phenotypic diversification.

As TEs jump around, they often land in functional genes, thus disrupting their expression or function. In taxa with a large percentage of active transposable elements, TEs are a significant source of (primarily deleterious) mutational pressure. For example, in *Drosophila*, TEs are responsible for up to 80 percent of visible newly arising visible mutations. TE insertions into regulatory (nonprotein-coding) regions may affect gene expression in potentially adaptive ways. In *D. melanogaster*, genome scans suggest that a small number of TEs in regulatory regions may have been involved in adaptation to temperate conditions during its recent range expansion out of Africa. Over the longer term, there are numerous individual cases in which TE sequences have been co-opted for other functions. For example, the V(D)J recombination mechanism, which underlies the ability of the vertebrate immune system to recognize an extraordinary diversity of antigens, relies in part on an enzyme derived from a TE transposase gene. These integrations of TEs into the "normal" machinery of an organism do not explain their evolutionary proliferation, but they do beautifully illustrate how evolution by natural selection works with whatever raw materials are at hand.

Reproductive parasites also contribute to standing variation in individual fitness within populations, but through balancing selection on the selfish element (and/or linked loci) rather than mutation. Allelic killers can reduce the fertility of carriers, and segregation distorters can become associated with linked deleterious alleles and thus influence diverse fitness traits. For example, the driving *t* haplotype in mice negatively affects male territorial behavior and female fertility, in addition to its direct effects on male fertility. In *Drosophila melanogaster* in Africa, the driving SD locus has carried along a single linked haplotype representing nearly 40 percent of chromosome 2. Given that both these SGEs were first identified through segregation distortion of genetic markers in thoroughly studied lab crosses, such major effects in wild populations suggest that undiscovered selfish elements may frequently contribute to natural fitness variation in less genetically tractable taxa.

If the spread of an SGE alters the properties of a population, this can bring about selection on morphological, behavioral, or life history traits. Some SGEs bias offspring sex ratios (X and Y drive elements, some endosymbionts) or the sex expression of adult individuals (CMS, some endosymbionts), thus affecting the population-level sex ratio. In sexually dimorphic stalk-eyed flies (*Cyrtodiopsis*), for example, high frequencies of a driving X chromosome lead to extremely female-biased sex ratios. Females that mate with males carrying a nondriving X are favored, as they produce male offspring, which being rare in the population, have exceptionally high mean fitness. Stalk-eyed fly females exhibit a preference for males with long eye stalks, which are genetically associated with the nondriving X. Evidence of selfish elements influencing sexual selection, particularly female choice, has also been documented in guppies, mice, butterflies, and fruit flies. In flowering plants, cytoplasmic male sterility similarly alters selection on reproductive allocation, and females and hermaphrodites in gynodioecious species tend to be dimorphic for a variety of floral characters in addition to pollen production. Given the direct effects of many selfish elements on individual fertility and population sex ratios, they are likely to be generally important factors in sexual selection and mating system evolution in a wide variety of taxa.

5. SELFISH GENETIC ELEMENTS AND SPECIATION

Reproductive parasites and their suppressors tend to evolve rapidly, interact epistatically, and negatively affect fertility when mismatched. Theoretically, joint fixation of driver and suppressor is often more likely than polymorphism, and evidence from X-linked drivers in flies and CMS in plants suggests that hidden drive systems are indeed common. Molecular analyses of hybrid incompatibilities in flies, mice, plants, and yeast indicate that SGEs contribute significantly to the evolution of species barriers.

The endosymbiont *Wolbachia* causes cytoplasmic incompatibility (CI) in a wide variety of insects in a *Wolbachia* strain-specific manner. Therefore, if two populations are fixed for different *Wolbachia* strains, there can be bidirectional CI, a potentially significant source component of postzygotic isolation between populations. While CI may not be sufficient to cause speciation, it probably does contribute to overall reproductive isolation between incipient species and thus contributes to speciation and diversification.

6. APPLIED USES OF SELFISH GENETIC ELEMENTS

Selfish elements have been surprisingly useful to humans, and a variety of novel applications are in development. Both transposable elements and restriction-modification systems are used extensively as molecular genetic tools, allowing targeted manipulation of DNA sequences. Because male sterility facilitates the generation of F1 hybrid seed, CMS is an important tool in agriculture. Research has recently begun on using reproductive parasites to control the insect vectors of human diseases such as malaria and dengue fever. These approaches take advantage of the potentially rapid spread of SGEs to manipulate the genetics or population biology of wild populations in a cost-effective manner. In principle, such approaches could be used either to knock down populations of vectors or to bring about the spread of factors that reduce vector effectiveness (e.g., malarial resistance). For example, experimental introduction of *Wolbachia* into the mosquito *Aedes aegypti* results in the expression of CI; thus, *Wolbachia* has the potential to spread through populations of these mosquitoes. Remarkably, these *Wolbachia* also render the mosquitoes unsuitable as vectors for dengue virus; therefore, as *Wolbachia* spreads through mosquito populations, the incidence of dengue fever in humans is expected to decline. Because such applications involve the release of biologically modified organisms, these approaches must contend with practical, ethical, and social hurdles prior to implementation. Understanding the evolutionary dynamics of naturally occurring SGEs is key to evaluating the costs and benefits of selfish-element–driven vector control.

FURTHER READING

Burt, A. 2003. Site-specific selfish genes as tools for the control and genetic engineering of natural populations. Proceedings of the Royal Society B 270: 921–928.

Burt, A., and R. Trivers. 2006. Genes in Conflict: The Biology of Selfish Genetic Elements. Cambridge, MA: Harvard University Press.

Chase, C. D. 2007. Cytoplasmic male sterility: A window to the world of plant mitochondrial-nuclear interactions. Trends in Genetics 23: 81–90.

Dawkins, R. 1976. The Selfish Gene. Oxford: Oxford University Press.

Doolittle, W. F., and C. Sapienza. 1980. Selfish genes, the phenotype paradigm and genome evolution. Nature 284: 601–603.

Haig, D., and A. Grafen. 1991. Genetic scrambling as a defence against meiotic drive. Journal of Theoretical Biology 153: 531–558.

Henikoff, S., K. Ahmad, and H. S. Malik. 2001. The centromere paradox: Stable inheritance with rapidly evolving DNA. Science 293: 1098–1102.

Hurst, G.D.D., and J. H. Werren. 2001. The role of selfish genetic elements in eukaryotic evolution. Nature Reviews Genetics 2: 597–606.

Hurst, L. D., A. Atlan, and B. O. Bengtsson. 1996. Genetic conflicts. Quarterly Review of Biology 71: 317–364.

Jones, R. N., M. Gonzalez-Sanchez, M. Gonzalez-Garcia, J. M. Vega, and M. J. Puertas. 2008. Chromosomes with a life of their own. Cytogenetic Genome Research 120: 265–280.

Kazazian, H. H. 2004. Mobile elements: Drivers of genome evolution. Science 303: 1626–1632.

Lyttle, T. W. 1991. Segregation distorters. Annual Review of Genetics 25: 511–557.

Merrill, C., L. Bayraktaroglu, A. Kusano, and B. Ganetzky. 1999. Truncated RanGAP encoded by the segregation distorter locus of *Drosophila*. Science 283: 1742–1745.

Naito, T., K. Kusano, and I Kobayashi. 1995. Selfish behavior of restriction-modification systems. Science 267: 897–899.

Orgel, L. E., and Crick, F.H.C. 1980. Selfish DNA: The ultimate parasite. Nature 284: 604–607.

Pardo-Manuel de Villena, F., and C. Sapienza. 2001. Nonrandom segregation during meiosis: The unfairness of females. Mammalian Genome 12: 331–339.

Turelli, M., and A. A. Hoffmann. 1991. Rapid spread of an inherited incompatibility factor in California *Drosophila*. Nature 353: 440–442.

Werren, J. H., L. Baldo, and M. E. Clark. 2008. *Wolbachia*: Master manipulators of invertebrate biology. Nature Reviews Microbiology 6: 741–751.

IV.8

Evolution of Mating Systems: Outcrossing versus Selfing

Spencer C. H. Barrett

OUTLINE

1. Definitions and measurement
2. Variation in mating patterns
3. Evolution of self-fertilization
4. Mechanisms of selection
5. The problem of mixed mating
6. Evolutionary history

Mating systems vary enormously among groups of organisms. This has led to diverse definitions and approaches for investigating their evolution and maintenance. Animal mating systems are characterized by different patterns of parental investment in offspring and variation in the extent to which sexual selection shapes male and female traits (see chapter VII.4). A primary focus of most studies is determining the causes and consequences of variation in mate number for females and males. In contrast, in hermaphrodite organisms, particularly plants, the emphasis is largely on determining the incidence of cross- and self-fertilization and its fitness consequences. Most studies of mating-system evolution have been conducted on plants and largely concern the ecological and genetic mechanisms responsible for evolutionary transitions from outcrossing to predominant self-fertilization, and the extent to which mixed mating can be maintained as a stable strategy. Models of mating-system evolution involve the balance between the transmission advantage of alleles affecting the selfing rate and the reduced fitness of self-fertilized offspring because of inbreeding depression. Selfing provides reproductive assurance whenever outcrossing is limited by the availability of pollinators or mates and thus low density often plays an important role in mating-system transitions. Reconstruction of the evolutionary history of mating systems using phylogenies indicates that transitions to predominant selfing have occurred on numerous occasions, although selfing lineages are often short-lived because of the negative effects of selfing on the genome.

GLOSSARY

Floral Design and Display. The morphological features of flowers and inflorescences that influence pollen dispersal and mating patterns in flowering plants; *floral design* involves characteristics of individual flowers, including their size, structure, color, and the spatial and temporal presentation of female (pistil) and male (stamens) sex organs, and *floral display* concerns the number of open flowers on a plant and their arrangement within and among inflorescences.

Inbreeding Depression. The reduction in viability and/or fertility of inbred offspring in comparison with outbred offspring. It results primarily from the expression of deleterious recessive alleles in homozygous genotypes and is expressed most strongly when inbreeding occurs in primarily outcrossing populations; a key parameter (δ) in models of mating-system evolution.

Mating System. Broadly defined, the mode of transmission of genes from one generation to the next through sexual reproduction; however, in species with separate sexes, it largely concerns the quantity and quality of mates obtained by males and females. In contrast, in hermaphrodites, it is usually defined as the relative frequency of cross-fertilization (*outcrossing*) and self-fertilization (*selfing*) in a population. In this article the last definition is used.

Modes of Self-pollination. The various ways in which self-pollination occurs in flowering plants, distinguished primarily by reproductive expenditure, whether a pollen vector is involved, and timing, relative to cross-pollination. The evolution of selfing depends critically on the particular mode of self-pollination.

Pollen and Seed Discounting. Important reproductive parameters for determining whether selfing will evolve in flowering plants; *pollen discounting* is the loss in outcrossed siring success caused by self-pollination, whereas *seed discounting* is the reduction in outcrossed seed production caused by selfing, either because selfing preempts ovules that could have been outcrossed, or because self-fertilized seeds consume resources that could have been allocated to outcrossed seeds.

Pollen Limitation. The reduction in potential seed production that occurs when ovules remain unfertilized and too few embryos survive genetic death to compete for maternal resources; results from insufficient delivery of pollen quantity and quality to flowers. Persistent pollen limitation can result in the evolution of selfing.

Reproductive Assurance. An increase in seed production caused by selfing when conditions for outcrossing are unfavorable because of an absence of pollinators or mates; requires plants to be self-compatible and generally capable of autonomous self-pollination.

Self-compatible and Self-incompatible. The two conditions that determine whether fertile hermaphrodites have the potential to produce offspring from self-fertilization. In plants, self-compatibility is the ability to produce abundant seed following self-pollination, whereas in a self-incompatible plant, few, if any, seeds are produced by self-pollination. Self-incompatibility is the most common antiselfing mechanism in flowering plants.

1. DEFINITIONS AND MEASUREMENT

The mating system is a key life history trait because it determines the quantity and genetic quality of offspring produced by an individual and thus the individual's reproductive fitness. One of the most striking features of organismal diversity is the existence of numerous mating strategies, even though all serve the same basic function: to promote parental reproductive success. Why do such diverse adaptations exist for an activity crucial for persistence of species of sexual organisms? Organismal variation in mating systems is determined, in part, by the biological features that characterize a particular taxon. For example, whether a group is sedentary or mobile and what type of sexual system it possesses (e.g., hermaphroditism versus separate sexes) influences how and with whom a plant or animal can mate. But although the types of mating systems are closely linked to the distinctive features of particular taxa, it is not uncommon to find quite different mating strategies among closely related species, especially in flowering plants. This indicates that mating systems are evolutionarily labile and can respond to natural selection; therefore, understanding mating-system diversity has been a major theme in evolutionary biology since Charles Darwin's influential work on the topic.

Reproductive biology is replete with terms for what are seemingly similar phenomena. Before we begin, it is important to clarify what is meant by "mating system" and how it relates to other terms associated with sexual reproduction. This is important because the term has various definitions and different usages, particularly in the botanical versus zoological literature. The distinctions depend largely on whether biologists are working on species with separate sexes (dioecy) versus those studying hermaphrodite (cosexual) organisms. Terms that are commonly used in reproductive biology involve the relative importance of sexual versus asexual reproduction (reproductive system) (see chapter III.9), the occurrence of separate sexes versus hermaphroditism (sexual system), and within a population, who mates with whom and how often (mating system) (see chapter VII.4); however, workers studying hermaphroditic organisms usually define *mating system* more specifically as the relative frequency of cross-fertilization and self-fertilization, and this is the definition used here, because the main focus of this article is mating patterns in hermaphrodites. It is also important to note that the term *breeding system* is often used synonymously with mating system in the animal and plant literature but usually more broadly, to include mating behavior and parental care in animals, and the diverse reproductive traits that promote particular mating patterns in plants.

Animal biologists in the field of behavioral ecology classify mating systems largely on the basis of mate number per male and female and are particularly interested in determining the variance in reproductive success of the sexes and thus the scope of sexual selection. Here, the mating system is viewed as a behavioral strategy for obtaining mates, which also encompasses the nature of parental care. Typical classes of animal mating systems include *monogamy*, in which each sex mates with only one partner during its lifetime, *polygyny*, where females mate with a single partner but males are variable in mate number, and *polyandry*, in which the reverse pattern occurs. Information on the types of animal mating systems provides a powerful tool for predicting the degree and direction of sexual dimorphism, as Darwin originally pointed out. These topics are discussed in more detail in Section VII: Evolution of Behavior, Society, and Humans.

Darwin wrote extensively about animal mating systems and sexual selection, but he also published three books on the reproductive biology of plants. This body of work initiated contemporary studies of mating-system evolution in plants. Because of the predominantly hermaphroditic sexual systems of plants, Darwin was particularly interested in how they avoid self-fertilization and its harmful effects on progeny fitness (see chapter IV.6), which he demonstrated in numerous species through controlled pollinations and comparisons of the performance of self- and cross-fertilized progeny. He demonstrated that plants possess numerous structural contrivances (floral designs) that reduce the incidence of self-pollination and promote cross-pollination. Darwin also identified remarkable diversity in floral biology and mating systems among closely related species. This variation has enabled investigation of the ecological and genetic mechanisms responsible for evolutionary transitions in mating systems. In the following sections we will explore this topic further, focusing in particular on plants, about which the most is known, but also drawing parallels where possible with hermaphroditic animal groups.

Early efforts to understand mating patterns in hermaphrodites depended on inferences from observations of behavior and the morphology of reproductive organs. For example, in plants, the relative positions of female and male sexual organs along with observations of pollinator foraging provided clues to the likely occurrence of cross- versus self-fertilization; however, such inferences were often unreliable, and more importantly, they did not provide quantitative estimates of the types of mating events that occur. In the 1970s, this situation changed following demonstration of the utility of genetic markers for measuring mating parameters in populations. Since then, estimates of the proportion of offspring produced by selfing, s, or its complement, the outcrossing rate ($t = 1 - s$), have been obtained for a wide variety of species, providing unprecedented insights into the extent of mating-system variation. Estimates of t and s portray selfing and outcrossing through female function and provide population-level estimates of mating patterns. In addition, the degree of biparental inbreeding and the extent to which progeny are full sibs can also be estimated. More recent developments using hypervariable genetic markers such as microsatellites now enable measurements of mating patterns at the individual level and also estimates of outcrossed mating success through paternal function. The assessment of paternity and male siring success is important because it provides an opportunity to evaluate the importance of sexual selection in plant populations.

2. VARIATION IN MATING PATTERNS

The use of genetic markers for measuring mating patterns in populations has uncovered considerable variation among plant species, and this diversity often provides the template for evolutionary transitions. Most observed variation in outcrossing rate occurs in self-compatible species in which environmental and demographic factors can play an important role in determining the amount of outcrossed pollen delivered to flowers by pollen vectors (animals, wind, and water). Species with strong self-incompatibility systems, or those that are highly selfing (*autogamous*), exhibit much less variation in outcrossing rate; because of this, they have tended to be underrepresented in mating-system surveys. Nevertheless, a picture is emerging of the distribution of outcrossing rates among seed plants, and this has stimulated much interest and controversy because of its relevance to theoretical models of mating-system evolution.

Marker-based estimates of outcrossing rate from 345 species representing 78 families by Goodwillie and colleagues (2005) demonstrated near-continuous variation from highly selfing to obligately outcrossed species; however, among this sample, 42 percent of the species could be classified as exhibiting *mixed mating* (a mixture of both outcrossing and selfing) in which t ranged from 0.2 to 0.8. As we will see later, the occurrence of mixed mating presents an important challenge for theoretical models of mating-system evolution. A particularly striking pattern is the finding that the distribution of outcrossing rate differs significantly between animal-pollinated (biotic) species and those pollinated by wind or water (abiotic). Biotically pollinated species are almost twice as likely to exhibit mixed mating compared to those that are abiotically pollinated. Why this difference occurs is at present unclear but may be associated with stronger selection in abiotically pollinated plants for mechanisms that reduce the incidence of self-pollination owing to the huge quantities of pollen typically produced, particularly in wind-pollinated species. Regardless of the mechanisms responsible, this difference does point to the important influence of pollination systems on mating patterns.

There are much fewer data available on mating-system variation in hermaphroditic animals. It has been estimated that about 5 percent of animal species are hermaphroditic, although this value is much higher, about one-third, if arthropods are excluded. Hermaphroditism occurs in the majority of animal phyla and is especially common in Annelida (ringed worms), Cnidaria (particularly corals), Mollusca (particularly the gastropods: snails and slugs), Nematoda (roundworms), and Playhelminthes (flatworms), where most mating-system work

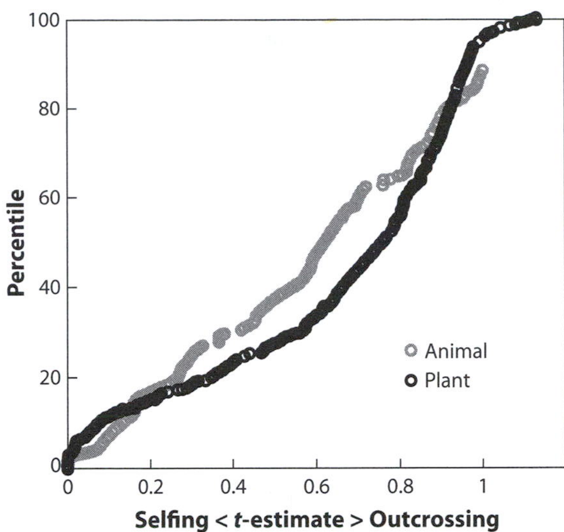

Figure 1. The distribution of outcrossing rates in plants (n = 342 species) and animals (n = 142 species); plant data from Goodwillie et al. (2005), animal data from Jarne and Auld (2006). The outcrossing rate (t) for each species is plotted. (Modified from Jarne and Auld [2006] with permission.)

has been conducted. Estimates of outcrossing rate for animals are based largely on indirect measures from population structure data on the inbreeding coefficient (F_{IS}). This approach assumes that populations are at inbreeding equilibrium and that selfing is the sole of source of inbreeding. Despite this limitation, the available data are not dissimilar to those reported for plants with a near-continuous distribution, from complete outcrossing to predominant selfing, and a significant representation of species with mixed mating.

Species-level surveys of the distribution of outcrossing rates, like those illustrated in figure 1, typically use average values across populations, masking any intraspecific variation that occurs. This variation is important because it provides opportunities to determine the proximate mechanisms governing mating-system variation and can provide important clues to the causes of mating-system transitions. For example, the annual insect-pollinated tropical aquatic plant *Eichhornia paniculata* exhibits a wide range of outcrossing rates, from near zero to complete outcrossing. Populations in Brazil are largely outcrossing, whereas those that have colonized Caribbean islands are highly selfing. This variation results from an evolutionary transition in mating system and is governed by the interaction of stochastic forces (genetic drift and founder effects) and natural selection favoring selfing variants when pollinator service is unreliable. Small population size plays an important role in this mating-system transition. Investigation of intraspecific variation in mating patterns can reveal much about the demographic and genetic factors responsible for the evolution of selfing.

3. EVOLUTION OF SELF-FERTILIZATION

The evolution of predominant self-fertilization (*autogamy*) from high levels of outcrossing is the most frequent mating-system transition in hermaphroditic organisms. No accurate estimates have been made of the total number of origins of selfing, but it is likely that in large groups, such as the angiosperms, this transition has occurred many hundreds of times. The evolution of selfing from outcrossing has attracted the attention of evolutionary biologists because it has diverse biological consequences. First, the effects of selfing on relative fitness through inbreeding depression are well established, and as we shall see, these effects play an important role in determining the dynamics of mating-system evolution. Second, the shift to selfing profoundly influences many aspects of sex allocation and life-history evolution. Third, selfing individuals have the ability to establish colonies at low density or following long-distance dispersal, a phenomenon known as Baker's law (after the plant evolutionist Herbert G. Baker), and this capability has significant ecological, demographic, and biogeographic implications. Finally, high selfing rates have profound genetic and evolutionary consequences, influencing population genetic structure, evolutionary rates, and patterns of evolutionary diversification.

Evolutionary transitions to selfing decrease genome-wide diversity compared with that of outcrossing populations. Homozygosity increases with the selfing rate, causing reductions in effective population size (N_e) up to twofold with complete selfing; moreover, because of higher linkage disequilibrium in selfing populations, other processes in the genome, including selective sweeps and background selection, further reduce genetic variation. These processes can also be influenced by the demographic and life history characteristics of species, especially genetic bottlenecks, when single individuals found colonies. Populations of many selfing species are characterized by frequent colonization-extinction cycles, and these demographic processes can lead to strong population subdivision. Thus selfing species are characterized by limited genetic variation within populations but a high degree of differentiation among populations, a pattern that is the reverse of what is normally found in outcrossing species.

Most groups have striking ecological and life history correlates of mating-system variation, indicating that not all species are likely to experience selection for increased selfing rates. In flowering plants, the distribution of mating systems is correlated with longevity and size; long-lived woody plants are more outcrossing than herbaceous perennials, and the highest selfing rates occur in short-lived annual species. Variation in plant life history is associated with important genetic and reproductive consequences that can explain these correlations. Long-lived species tend to maintain higher genetic loads, perhaps owing to higher per generation mutation rates because of an increased number of somatic mutations, resulting in strong selection against inbred offspring. This limits opportunities for the spread of genetic modifiers influencing the selfing rate and probably explains why there are few reports of highly selfing trees. Long-lived species also tend to be large in size, growing either vertically, as occurs in trees, or laterally, as in highly clonal plants. One consequence of large size is that during reproduction, floral displays also tend to be very large, resulting in extensive pollen transfer between flowers on the same individual, resulting in a mode of self-pollination known as *geitonogamy*, with genetic consequences identical to those for intraflower selfing. Because of strong inbreeding depression, antiselfing mechanisms such as self-incompatibility and dioecy are especially well developed in long-lived trees and clonal plants.

By contrast, the uncertain conditions typical of ephemeral habitats occupied by species with very short life cycles favor a mating system that provides reproductive assurance. This explains why selfing is very common in annual plants, which have only a single opportunity for mating. Annuals commonly possess well-developed powers of seed dispersal and frequently establish colonies at low density, a demographic context that selects against obligate outcrossing and favors individuals with a capacity for self-fertilization. Not surprisingly, many invasive annual plants and hermaphroditic animals are highly selfing.

These associations between mating systems and life history are particularly evident in herbaceous flowering plant families in which perennials tend to be outcrossing and annuals are more likely selfers. These patterns are well exemplified in the Polemoniaceae (Phlox family), which exhibits considerable ecological versatility, particularly in western North America, where species occupy montane, forest, and desert environments. Reconstructions of the evolutionary history of the life cycle (annual versus perennial) and mating system (outcrossing versus selfing) using molecular phylogenies of the family indicate repeated transitions from outcrossing to selfing, either coincidentally with or following transitions from perennial to annual life history. The lability of reproductive mode and life history in many plant families suggests that phylogenetic constraints rarely limit opportunities when ecological conditions require evolutionary shifts in mating system.

Hermaphroditic animals capable of selfing are typically associated with low population densities and a sessile, sluggish, or parasitic lifestyle. Under these demographic conditions, opportunities for encountering mating partners are much more limited than for mobile animals with separate sexes, and this favors self-fertilization. Indeed, comparative evidence has been found among multicellular eukaryotic organisms for the correlated evolution of the sexual system with the mode of locomotion for searching for a mate. Hermaphroditic groups capable of selfing are more likely to have low mate searching efficiency than species with separate sexes. Evidence has also been found among hermaphroditic snails that individuals can respond to variation in mate availability by delaying the time before selfing occurs. This indicates that mate availability is a key target of natural selection on the mating system in animal populations.

4. MECHANISMS OF SELECTION

The most appropriate theoretical framework for understanding the selective mechanisms governing mating-system evolution is to use a population genetics approach that considers the conditions that might favor the spread of a variant capable of self-fertilization in an obligately outcrossing population. The mathematical geneticist Ronald Fisher first pointed out that all else being equal, a gene causing self-fertilization increases in frequency in each generation, because, on average, selfers generally contribute more gene copies to the next generation than outcrossers. In plants this arises because selfers are the maternal and paternal parent to the self-fertilized seed they produce, and their pollen can also participate in outcrossing to other plants; this gene transmission advantage can be seen by considering the scenario shown in table 1.

Not all species are selfing, however; indeed, many more are predominantly outcrossing, indicating that other factors must play a role in determining the direction of mating-system evolution. Today, it is recognized that three principal factors serve to limit the spread of selfing genes in outcrossing populations: inbreeding depression (δ), pollen discounting, and seed discounting, and each factor has the potential to reduce the gene transmission advantage of a selfing variant. *Inbreeding depression* causes reduced fitness of self-fertilized offspring. If they survive and reproduce only half as well as outcrossed offspring, the advantage of selfing disappears. Thus, when $\delta > 0.5$, outcrossing is favored, but when $\delta < 0.5$, selfing should spread. Fisher's model also does not consider whether self-pollination reduces the amount of pollen available for outcrossed siring success. If there is a reduction in pollen available for outcrossing, a selfer's fitness through male function is reduced proportionately (*pollen discounting*). Similarly, *seed discounting*, the reduction in outcrossed seed production caused by selfing,

Table 1. Average Gene Contribution

	Outcrosser	*Selfer*
Ovule parent	1	1
Pollen parent	1	2
Total	2	3

also reduces the transmission advantage of a selfing variant. Current work is attempting to measure these three parameters in an effort to test models and predict the course of mating-system evolution. Measurements of inbreeding depression are relatively straightforward, but pollen and seed discounting are more challenging to estimate, and the extent to which they influence selection on the selfing rate in plant populations is at present unclear.

The other general explanation for the evolution of selfing involves the advantage of selfing individuals over outcrossing individuals when pollinators or mates are scarce, such as commonly occurs under low density. This is referred to as the *reproductive assurance hypothesis* and was originally suggested by Charles Darwin and the German naturalist Hermann Müller. The reproductive assurance hypothesis differs from Fisher's *automatic selection hypothesis* because it predicts increased seed production in selfers compared to outcrossers and requires that self-compatible plants have the ability to self-pollinate autonomously in the absence of a pollinator. Considerable biogeographical evidence indicates that selfing populations occupy range margins, ecologically marginal sites, or areas with reduced pollinator densities where outcrossers are absent, all circumstances predicted by the reproductive assurance hypothesis; however, surprisingly few field studies have provided experimental evidence in support of this hypothesis, despite the widespread occurrence of pollen limitation of seed set in animal-pollinated species. It seems likely that both automatic selection and reproductive assurance will turn out to be involved in many transitions from outcrossing to selfing in flowering plants, with demographic context determining their relative importance.

5. THE PROBLEM OF MIXED MATING

Most genetic models of mating-system evolution, based on the balance between the transmission advantage of selfing and the costs of inbreeding depression, result in populations in which either predominant selfing or outcrossing is an alternative stable state. Thus, theory predicts a bimodal distribution of outcrossing rates resulting from selection for the maintenance of outcrossing in historically large populations with inbreeding depression greater than 0.5, and selection for selfing when demographic factors such as population bottlenecks reduce inbreeding depression to less than 0.5, and partial selfing purges deleterious recessive alleles; however, as we have seen, surveys of outcrossing rates in plants and animals indicate that many species exhibit mixed mating. What accounts for this apparent discrepancy and is there evidence that mixed mating can be a stable mating strategy?

Several explanations help to reconcile the mismatch between theory and empirical evidence on the distribution of mating systems. First, existing estimates of the distribution of outcrossing rates are likely biased in terms of taxonomic representation (many estimates are concentrated in some well-studied taxa, e.g., pines, gum trees, and grasses) and, as discussed earlier, an underrepresentation of the two ends of the distribution, thus inflating the frequency of species with mixed mating. Second, theoretical models predict the equilibrium mating system in a population, and it is probable that some species are not at equilibrium when sampled and, given sufficient time, could be driven to predominant selfing or outcrossing. Third, in many animal-pollinated species, the selfing component of mixed mating is likely a nonadaptive cost that plants pay by having large floral displays to attract pollinators. Geitonogamy occurs because pollinators transfer pollen between flowers on a plant; this provides little benefit to fitness because of strong inbreeding depression and pollen discounting. Fourth, where pollinators exhibit spatial and temporal fluctuations in density, mixed mating can arise by modes of self-pollination that provide reproductive assurance. For example, delayed selfing arises when flowers self-pollinate after opportunities from cross-pollination have passed, and models indicate that this will always be adaptive. Finally, more elaborate theory, integrating more ecological and genetic details, has been able to show stable mixed mating, albeit often under restricted conditions. Nevertheless, the adaptive significance of mixed mating remains problematic for most species, and the search for a general explanation continues to be elusive, probably because the phenomenon has diverse causes.

6. EVOLUTIONARY HISTORY

The recent availability of phylogenetic information for groups with variation in mating systems has enabled the reconstruction of the evolutionary history of outcrossing and selfing, leading to several generalizations: (1) the transition from outcrossing to predominant self-fertilization often occurs repeatedly within lineages and is usually unidirectional; (2) selfing species most often occur at the tips of phylogenies and are therefore relatively young in age; (3) species-rich groups and higher taxa are usually composed of predominantly outcrossing species. These findings support an early view proposed by the plant evolutionist G. Ledyard Stebbins that the evolution of selfing represents an evolutionary dead end. Why should this be so, given that selfing clearly has short-term advantages in some ecological contexts?

High rates of self-fertilization are associated with negative genomic consequences that make it unlikely that selfing species will persist over long timescales. These effects include a reduction in the amount of adaptive genetic variation in populations, limiting their capacity to adapt to environmental change, low recombination and a reduced efficacy of natural selection, lowering opportunities for purging deleterious mutations as well as for fixing adaptive mutations. Thus, weak purifying selection and/or the operation of Muller's ratchet (the irreversible accumulation of deleterious mutations) should lead over time to mutational decay and the ultimate extinction of selfing lineages compared with those that habitually outcross.

Evidence supporting the ephemerality of selfing lineages comes from comparative work on the tomato family, Solanaceae, where the distribution of self-incompatibility (SI) and self-compatibility (SC), traits that directly influence outcrossing and selfing, respectively, are associated with contrasting rates of species diversification. Many independent transitions from SI to SC occur in this group, but significantly, SI is never regained, a pattern commonly observed in other plant families. This raises the question of how SI can be maintained over long timescales if it is being continuously broken down because of selection for selfing. The answer lies in the higher extinction rates of SC lineages compared to SI lineages, counterbalancing the frequent loss of SI. Transitions from outcrossing to selfing are driven ultimately by short-term ecological and genetic mechanisms in local populations; however, they can also have long-term macroevolutionary consequences for patterns of biodiversity, because mating systems influence character evolution and species diversification.

FURTHER READING

Barrett, S.C.H. 2003. Mating strategies in flowering plants: The outcrossing-selfing paradigm and beyond. Philosophical Transactions of the Royal Society B 358: 991–1004. *A review of the distinctive features of plants that influence mating, emphasizing the functional links among flowers, inflorescences, and plant architecture.*

Barrett, S.C.H., and L. D. Harder. 1996. Ecology and evolution of plant mating. Trends in Ecology & Evolution 11: 73–78. *A concise summary of the functional dimensions of plant mating drawing particular attention to the relations between pollen dispersal and mating patterns.*

Charlesworth, D., and B. Charlesworth. 1987. Inbreeding depression and its evolutionary consequences. Annual Review of Ecology and Systematics 18: 237–268. *The classic review on the importance of inbreeding depression to mating-system evolution in plants and animals.*

Goodwillie, C., S. Kalisz, and C. G. Eckert. 2005. The evolutionary enigma of mixed mating systems in plants: Occurrence, theoretical explanations and empirical evidence. Annual Review of Ecology and Systematics 36: 47–79. *A major synthesis of empirical and theoretical work on outcrossing rate variation in seed plants and its relevance to the problem of mixed mating.*

Jarne, P., and Auld, J. R. 2006. Animals mix it up too: The distribution of self-fertilization among hermaphroditic animals. Evolution 60: 1816–1824. *Parallels between mating-system variation in hermaphroditic animals and plants.*

Lloyd, D. G. 1992. Self- and cross-fertilization in plants. II. The selection of self-fertilization. International Journal of Plant Sciences 153: 370–380. *Theoretical models of the selection of selfing from outcrossing, including the key parameters: inbreeding depression, pollen and seed discounting, and reproductive assurance.*

Shuster, S. M., and M. J. Wade. 2003. Mating systems and strategies. Monographs in Behavior and Ecology. Princeton, NJ: Princeton University Press. *A quantitative framework for investigating mating systems and sexual selection in animals with separate sexes.*

Vallejo-Marín, M., M. E. Dorken, and S.C.H. Barrett. 2010. The ecological and evolutionary consequences of clonality for plant mating. Annual Review of Ecology and Systematics 41: 193–213. *The ways in which plant clonal strategies affect mating patterns and the ecological and evolutionary consequences of geitonogamous self-fertilization caused by large floral displays.*

V

Genes, Genomes, Phenotypes

Hopi E. Hoekstra and Catherine L. Peichel

Darwin's 1859 theory of evolution by natural selection has three main tenets: (1) phenotypes vary among individuals, (2) phenotypic differences lead to differential fitness, and (3) these fitness-related phenotypes are heritable. Over decades, Darwin amassed huge amounts of data on natural variation and its effects on organismal fitness. By contrast, he knew almost nothing about heredity. While he knew phenotypes were inherited (i.e., that offspring resemble their parents), he had no knowledge of the mechanism by which this occurred. He acknowledged this missing link in his argument for evolution, and when pushed, devised a theory for the mechanism of inheritance (i.e., pangenesis), which was one of his few major errors. Of course, it was only later, in 1900, that Gregor Mendel's experiments, which elucidated the laws of inheritance, were unearthed. Subsequent discoveries, such as Thomas Hunt Morgan's experiments in *Drosophila* (1915) demonstrating that genes are carried on chromosomes, elucidated the material basis for heredity, and the discovery of the three-dimensional DNA structure by Watson and Crick (1953) showed that DNA is the molecule of inheritance. These seminal findings were central to the rise of molecular biology, and more recently of genomics, as powerful tools in all areas of biological inquiry, including evolution. Most certainly, Darwin couldn't have envisioned the "era of genomics," but he surely would be most pleased that genomic data further supported his theory.

Our ability to comprehensively describe genetic variation by sequencing complete genomes, the basic blueprint of an organism, represents an extraordinary technological advance that is remaking the field of biology in general, and evolutionary biology in particular. The rate at which whole genome sequences are being generated is astonishing. This enterprise started more than 30 years ago, when the modest 5386-base-pair genome of a bacteriophage was decoded. This achievement was quickly followed by the sequencing of several other, larger viral genomes. But sequencing free-living organisms with larger genomes required technological and computational advances, and it took almost another two decades to complete the entire genome sequence of *Haemophilus influenzae* (1.8 million base pairs). Fewer than six years later came the first complete human genome sequence (2.91 billion base pairs) at a cost of nearly $100 million. However, as the acquisition rate of complete genome sequences has increased, costs have decreased by four orders of magnitude, so that sequencing a complete human genome today costs roughly $10,000 and is expected to cost even less in the near future.

Arguably, all species now have the potential to be "genome enabled," and comparisons of these genome sequences—both within and between species—is shedding unprecedented light on evolutionary processes. As the chapters in this section demonstrate, biologists now have the opportunity to observe evolution at a fundamental level; that is, to know which genotypes change over time. And biologists can obtain sequences not only from extant organisms but also from extinct species and historical specimens of existing species, allowing changes in genotype to be observed over time (see chapter V.15). Changes in genotype can occur on a number of levels, affecting the whole genome (see chapters V.2 and V.3), whole chromosomes (see chapter V.4), gene number and content (see chapters V.5 and V.6), gene expression (see chapters V.7 and V.8), and interactions among genes (see chapter V.9). These changes in genotype are translated into changes in phenotype through the process of development, and the new field of evolutionary developmental biology seeks to understand these connections (see chapters V.10 and V.11). Biologists are also concerned with determining which nucleotide differences in genome sequence actually contribute to differences in phenotype (see chapters V.12 and V.13). However, connecting genotype to phenotype is not enough for a complete understanding of evolution; it is also crucial to learn how evolutionary processes such as genetic drift, mutation, and particularly natural selection influence changes in genotype at a specific gene or across the entire genome over time (see chapters V.1 and V.14).

Our ability to sequence the complete genome of (almost) any individual has brought many surprises. Before the human genome sequence was revealed, guesses at the number of protein-coding genes in the genome ranged widely, from as few as 25,000 to as many as 120,000 genes. While a few estimates were close, no one had imagined that humans carry only about 23,000 genes, approximately the same number as most other eukaryotes. Most of the genome, in fact, comprises noncoding DNA (e.g., transposable elements, untranslated regions, and introns), and it is variation in this "other stuff" that is largely responsible for differences in genome size among species. Understanding genome dynamics—the processes responsible for the evolution of genome complexity—remains an exciting area of study (see chapter V.2). And with the sequencing of the human genome came the realization that comparisons of genomes across diverse species would greatly facilitate the identification of genes and regulatory elements. Comparative genomics might also provide an approach to finding regions of the genome important for phenotypic evolution, such as those evolving extremely rapidly, or those that are ultraconserved, which may be indicative of their functional significance (see chapter V.3). Genomics is also providing new insights into an unusual region of the genome: the sex chromosomes, which are inherited differently in the two sexes. How such sex chromosomes, including their gene content and gene expression levels, evolve is an exciting question, especially given the diversity of sex-determining mechanisms identified across species (see chapter V.4).

Despite the fact that change in gene number is not the major driver of genome size evolution, comparative genomics has revealed that gene content can vary by two orders of magnitude across species. Several mechanisms generate variation in gene number, including whole genome or whole chromosome duplication, as well as duplications of individual genes (see chapter V.5). Such gene duplicates are retained at early stages in their evolution if the original functions of the parent gene are divided between the parent gene and the duplicate; the gene duplicates can then evolve new functions at later stages of evolution. Although gene duplication is the primary source of new genes, additional mechanisms for generating new genes do exist, and new genes can even arise de novo (see chapter V.6). New genes arise at a surprisingly high rate, and recent evidence demonstrates that even very young genes have evolved essential functions within species.

However, gene duplications or new genes are not absolutely required for a new function or phenotype to evolve; often, changes in where or when a particular gene is expressed (i.e., "turned on") are involved (see chapter V.7). Heritable genetic changes, either in proteins that regulate gene expression or in the DNA sequences they bind, can lead to the evolution of gene expression differences among species. Although evolutionary biologists are traditionally accustomed to thinking about the contribution of gene expression to evolution, recent research has demonstrated that changes in gene expression can also occur by epigenetic changes (i.e., changes can be stably transmitted across generations in the absence of change to DNA; see chapter V.8). The contribution of epigenetic change to evolutionary change has not yet been fully demonstrated, but investigating how genetic and/or epigenetic changes in gene expression contribute to phenotypic evolution is a promising area of future research.

When considering the contribution of a genetic or epigenetic change to phenotypic evolution, it is imperative to remember that genes do not act in a vacuum but interact with other genes in complex genetic networks (see chapter V.9). Recent technological advances have allowed biologists to investigate entire networks of genes, rather than analyzing only a single gene at a time. A surprising finding of such analyses is that biological networks are not randomly organized, suggesting that constraints imposed by the structure of genetic networks might influence evolutionary trajectories. Consideration of possible genetic and developmental constraints is also an important component of the study of evolutionary developmental biology, or *evo-devo* (see chapter V.10). In particular, organisms in which similar phenotypes have evolved repeatedly in response to similar environments provide powerful experimental systems to investigate the relative importance of natural selection and developmental constraints during phenotypic evolution. Coupled with studies of the molecular pathways and networks that underlie particular phenotypes, new insights are being gained into the ways in which the genome-level changes discussed above (e.g., genome size, gene number, gene expression, gene interactions) are translated through the process of development into phenotypic changes during evolution (see chapter V.11).

Although genome sequencing allows biologists to catalog nearly all the genetic changes that occur between species, a remaining challenge is to determine which of these genetic changes are responsible for the phenotypic changes observed between species. Work over the past decade in a few systems has begun to provide insights into the types of genetic changes that underlie phenotypic evolution, particularly for morphological traits (see chapter V.12). The challenge now is to widen this search to include additional phenotypes, such as behaviors and life history, and to additional organisms.

This will allow evolutionary biologists to determine whether the genetic changes underlying phenotypic changes are indeed predictable, or whether the lessons learned so far are idiosyncrasies of the organism or phenotype studied. New technological and analytical tools will make this challenge easier to meet, particularly in organisms not amenable to genetic studies in the laboratory, and for traits without a simple genetic basis, that is, *complex traits* (see chapter V.13).

The ultimate challenge is to ask whether phenotypic changes observed between species are adaptive, that is, whether selection has played a role in their evolution. Once the genetic changes responsible for a particular phenotypic change are identified, it is possible to use the tools of molecular evolution to determine whether the genetic changes underlying phenotypic evolution are evolving neutrally, or under natural selection (see chapter V.1). This top-down approach starts with the phenotype, then identifies the underlying gene, and finally tests for molecular signatures of selection in the pattern of nucleotide variation in these genes. A complementary approach is to first identify the locations in the genome that appear to be under selection (see chapter V.14) and then work up to phenotype. This bottom-up approach is rapidly being used in a variety of systems because of the relative ease and low cost of sequencing whole genomes; however, these population-genomic studies still must connect the genomic regions under selection with actual phenotypes. While this remains challenging, such studies have already provided important new insights into the effects of natural selection at the level of the genome.

Evolutionary biology is in large part about reconstructing the past. While fossils provide a direct glimpse into the past, the fossil record is largely incomplete. A complementary approach is to compare extant organisms (or their genomes) and then infer ancestral states; however, genomic approaches offer yet another glimpse into the past via the sequencing of ancient DNA (see chapter V.15). Specifically, DNA can be extracted from long-deceased individuals (e.g., ancient humans) or extinct species (e.g., woolly mammoths or Neanderthals), and then sequenced at a candidate gene to gain information on a particular trait, at a large number of noncoding regions to infer demographic history, or increasingly across the entire genome, to more fully elucidate both past phenotype and past demography. Such studies will clearly continue to shed light on phenotypic traits, genetic origins, and biological relationships of now-extinct individuals to present-day populations and species. An increasing number of ancient DNA sequences provide another approach to deducing past events that will stand alongside discoveries from the fossil record.

The "era of genomics" is truly an exciting time for evolutionary biology, because there is now an unprecedented opportunity to directly answer fundamental and long-standing questions about the genetic basis of Darwin's theory of natural selection. Evolutionary biologists can now identify change across the genome over time, determine the phenotypic effects of these genetic changes, and directly assess the role of natural selection in the evolution of genes, genomes, and phenotypes.

V.1

Molecular Evolution

Charles F. Aquadro

OUTLINE

1. What is molecular evolution and why does it occur?
2. Origins of molecular evolution, the molecular clock, and the neutral theory
3. Predictions of the neutral theory for variation within and between species
4. The impact of natural selection on molecular variation and evolution
5. Biological insights from the study of molecular evolution
6. Conclusions

The molecules of life (DNA, RNA, and proteins) change over evolutionary time. Much can be learned about evolutionary process and biological function from the rates and patterns of change in these molecules. The study of these changes is the study of *molecular evolution*. This chapter discusses why these molecules change, what can be learned about pattern and process from these changes, and how the changes in the molecules of life can be used to infer important past evolutionary events.

GLOSSARY

Fixation. The population process in which, either by drift or by natural selection, a new mutation increases in frequency in a population until it replaces all other variants and reaches a frequency of 100 percent.

Molecular Clock. When the time at which organisms last shared a common ancestor is plotted over time (e.g., estimated time from the fossil record), a roughly linear accumulation of genetic changes in DNA and the encoded proteins is frequently observed. In 1965, the rough linearity of this accumulation of change motivated Emile Zuckerkandl and Linus Pauling to propose that these data represent a sort of "molecular clock" by which the amount of molecular divergence could be used to infer the date of a last common ancestor.

Molecular Evolution. Changes in the molecules of life (DNA, RNA, and protein) over generations, for many reasons, including mutation, genetic drift, and natural selection, resulting in different sequences of these molecules in different descendant lineages. The study of molecular evolution is the study of the patterns and process of change that result in these different sequences.

Mutation. Heritable change in genetic material, including base substitutions, insertions, deletions, and rearrangements; the ultimate source of new variation in populations.

Neutral Theory. Short for *neutral mutation–random drift theory of molecular evolution*, proposing that molecular variation is equivalent in function (*selectively neutral*), making genetic drift the main driver of molecular genetic change in populations over time.

Positive Selection. New advantageous mutations, or changing environments, can present opportunities for new, or currently existing, variants to now have a reproductive advantage. They thus relentlessly increase in frequency until they fix in the relevant population. The selective pressure that leads to this fixation is termed *positive selection*.

Purifying Selection. Selection against harmful (i.e., deleterious) mutations "purifies" the population of these harmful variants. Such selection is due to constraint, typically to maintain a specific important biological function.

1. WHAT IS MOLECULAR EVOLUTION AND WHY DOES IT OCCUR?

The molecules of life (DNA, RNA, and proteins) are not static. They change over evolutionary time, hence the term *molecular evolution*. Some molecules evolve rapidly and some only very, very slowly. Their change is due to the interplay between two fundamental evolutionary processes: mutation and fixation. Before discussing these two processes, it is important to note that mutations come in two basic varieties: heritable mutations that occur in the germ line and are passed on in the genome of progeny, and somatic mutations that can occur in the process of cell division during normal growth and development. For example, if the latter affect the control of certain cellular processes, they can lead to uncontrolled cell growth and cancer. Molecular evolutionary studies have traditionally focused only on heritable genetic changes that accumulate within and between organisms.

Inherited mutations occur primarily during DNA replication in the production of gametes and introduce new genetic variants into the population. For animals and plants, the relevant genetic molecule is DNA. For some viruses, the heritable molecule is RNA. Certain segments of DNA or RNA genomes are translated into proteins by the cells, and some changes lead to changes in the encoded protein sequence, leading to molecular evolution at the amino acid level as well. Heritable mutations are largely considered to occur at random in time and space and location across our genome, and at a relatively constant rate, at least for many organisms (see chapter IV.2).

The second process of molecular evolution is fixation, which is fundamentally the "population" phase leading to molecular evolutionary change. Most new mutations are lost from populations as a result of chance (genetic drift; see chapter IV.1), or because they are harmful (deleterious). Mutations remain in populations because of chance or because they increase the reproductive success of offspring carrying them. Drift can lead to rapid changes in frequencies of mutations in very small populations, but is much less influential in large populations. Ultimately, the outcome of drift alone is always the loss or fixation of every new mutation; in other words, every mutation will eventually reach 0 or 100 percent frequency. Fixation means that the new mutation now replaces all previous variants present (segregating) in the population at a particular site in the genome.

Fixation can also be caused, or assisted, by natural selection. Selection acting to directly increase a variant frequency, as the result of an increased relative reproductive success or survival of individuals carrying it, is termed *positive selection*, or often simply *adaptation*. Such selection can cause rapid changes of allele frequencies in populations, over tens of generations, versus millions of generations by genetic drift alone in large populations. While the underlying mutation rate is thought not responsive to selective challenges, the fixation process is strongly influenced by selection.

Harmful or deleterious mutations, or ones that reduce reproductive output or success, often will not be fixed but rather reduced in frequency or eliminated from populations. This is known as *purifying selection*, and it prevents the otherwise-inexorable, but very slow, march of successive neutral mutation fixations over time in all finite populations.

2. ORIGINS OF MOLECULAR EVOLUTION, THE MOLECULAR CLOCK, AND THE NEUTRAL THEORY

Molecular evolution as a field of study originated sort of accidentally, as biologists discovered how to determine the sequence of proteins and started collecting data from diverse organisms in the 1950s and 1960s. Key early data were for hemoglobin and histones, proteins chosen for their biomedical importance. The study of proteins was emphasized because of their clear functional relevance, and because methods were developed first for sequencing these biological molecules. Comparison of sequence divergence among proteins revealed two important patterns: specifically, fibrinopeptides, hemoglobin, and histones from various vertebrates with well-defined fossil records revealed that (1) different proteins evolved at different rates, but (2) each protein seemed to accumulate changes at a surprisingly consistent rate, a pattern that led to the concept of a "molecular clock" (figure 1).

The term *molecular clock* refers to both a mechanism and a tool for evolutionary studies. As a mechanism, the presence of a roughly constant accumulation of change led to the inference that chance, not local adaptation, is the cause of much of the observed molecular change. As a tool, the presence of a clock representing a molecule also meant that if its rate could be calibrated with organisms of known age (e.g., from the fossil record), then observed sequence differences for organisms without a fossil record could also be "dated."

Until the mid-1960s, most evolutionary biologists considered natural selection the primary determinant of evolutionary change. Genetic drift was considered important only in small populations; for example, drift played an important role in Wright's shifting balance theory, but not as a primary driving force in evolution, rather as a source of new combinations of alleles on which selection could act; thus, drift was rarely considered, and most models focused on selection.

Several observations in the mid-1960s and early 1970s challenged the dominance of selection as the driver of evolution. First, high levels of protein polymorphism (i.e., variation within populations) were observed in fruit flies, humans, and bacteria. Could all that variation within populations really be maintained by natural selection? Second, extrapolating from available data,

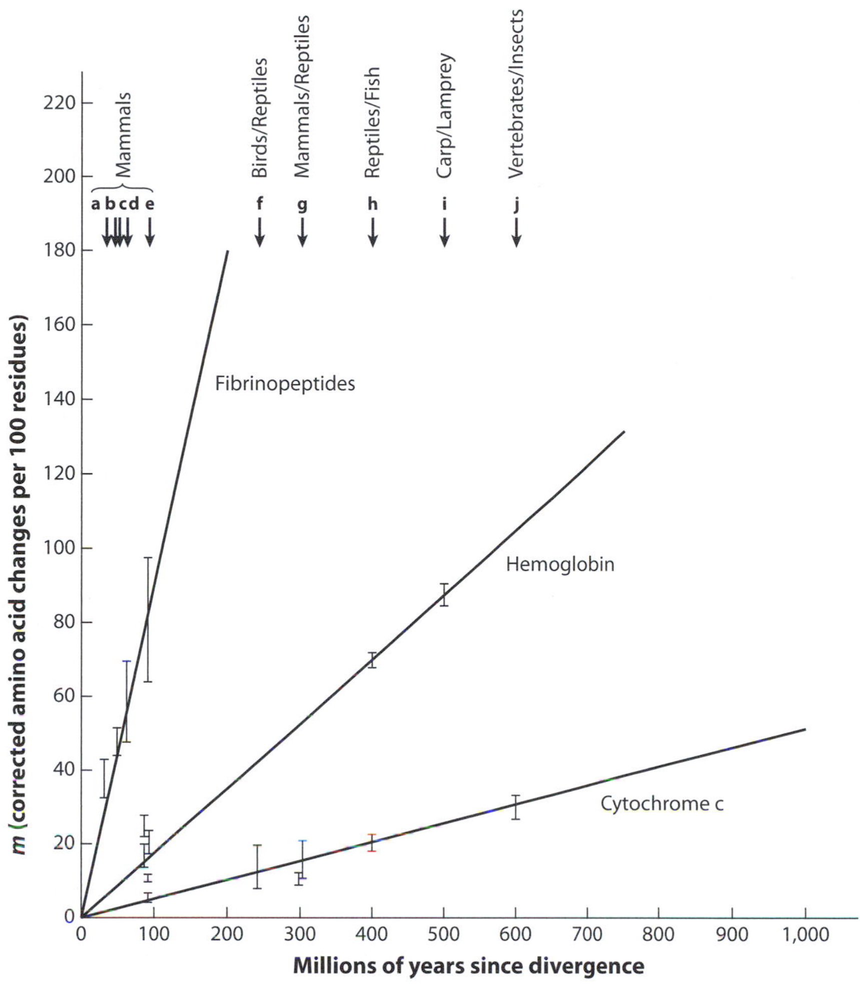

Figure 1. Molecular clocks. Rates of amino acid substitution in three proteins: fibrinopeptides, hemoglobin, and cytochrome *c*. The number of amino acid differences (per 100 residues, and corrected for multiple changes at the same residue) is plotted for comparisons between various mammals, birds and reptiles, mammals and reptiles, reptiles and fish, carp and lamprey, and vertebrates and insects, all lineages of organisms for which fossil data provided estimates of the time since divergence. Note that while some comparisons of the three proteins are from the same pair of organisms and time of divergence, the three proteins are evolving at very different rates (i.e., fibrinopeptide the fastest, and cytochrome *c* the slowest). In addition, the rough linearity of the rate of accumulation of molecular divergence with time illustrates the molecular clock concept. (Modified from Richard E. Dickerson. 1971. The structure of cytochrome *c* and the rates of molecular evolution. *Journal of Molecular Evolution* 1: 26–45).

Motoo Kimura estimated that there were as many as two amino acid replacements per generation across the genome in mammals. Again, could selection really drive that many amino acid replacements to fixation without an intolerable reduction in population fitness? Third, Jack King and Thomas Jukes considered the genetic code with its built-in redundancy (meaning some nucleotide changes were "silent" and didn't change the encoded protein sequence) and the conservative characteristics of many amino acid changes seen between proteins isolated from different organisms. They argued, like Kimura, that much of the variation observed in proteins within and between species did not alter function and therefore accumulated by mutation and genetic drift. Surely adaptation occurred, but Kimura argued that it occurred in only a small proportion of the genome at any one time and that natural selection was unlikely to account for the maintenance of extensive molecular variation observed within species and for the fixation of variation between species. It was also inferred that many mutations were in fact harmful and eliminated from populations, so those regions of genes and the genome that were functionally critical would remain largely invariant. If selection could not reasonably explain these observations, then it followed that there might indeed be a significant role for drift.

This broad concept became known as the neutral allele theory of molecular evolution, and it has formed a conceptual and model framework on which much of the current field has been based. Perhaps most important has been the recognition that all populations of organisms are finite in size, so that the stochastic process of genetic drift forms a background on which all other evolutionary forces act. The term *neutral theory* (as it is often called) can be misleading, as not all variation is selectively neutral; rather, the theory allows that a significant portion of new mutations are strongly deleterious and nearly immediately removed from the population. And the theory does allow for a limited number of adaptive mutations; however, the remaining mutations are selectively equivalent (neutral), and their dynamics are determined solely by genetic drift. Thus, the majority of variation we see within and between species is assumed to have no effect on fitness of organisms.

In the 1970s, the emergence of methods to directly sequence DNA, which were much less laborious than protein sequencing, began an inexorable shift to the study of DNA sequences in the field of molecular evolution. Not only does DNA sequence data provide an estimation of the frequency of variation at individual nucleotides, but the tight genetic linkage of adjacent nucleotides also means that sequences retain more of their evolutionary history. The availability of these correlated evolutionary histories allowed for the development of new statistical and computational approaches for testing models of molecular evolution, particularly the neutral theory. The field was no longer theory rich and data poor, as now the data began pouring in at an astounding rate. The ability to obtain larger sample sizes of DNA sequences has also increased the statistical power to discriminate models and to infer evolutionary history, demography, and the targets and magnitude of selection acting on the genome. The strict neutral theory was clearly an oversimplification but has provided the field with a valuable reminder of the importance of stochastic processes in all populations and a valuable null hypothesis against which to evaluate data.

3. PREDICTIONS OF THE NEUTRAL THEORY FOR VARIATION WITHIN AND BETWEEN SPECIES

The probability that a neutral allele will eventually become fixed is equal to its frequency in the population. And the rate at which new alleles become fixed in a population (the substitution rate) is essentially equal to the "neutral" mutation rate per generation. Thus, if the neutral mutation rate remains constant, so should the rate of evolution. For new neutral mutations destined to be fixed by drift, average time to fixation (in units of generations) is approximately four times the long-term population size. For mutations destined for loss, average time to loss is quite short. Thus, for large populations, we expect long times to fixation, and thus lots of "transient" genetic variation in populations drifting slowly through them. Together, these processes mean that the level of variation within species is a function of population size and mutation rate.

For a stable population, the balance of new mutations and loss or fixation by drift leads to an equilibrium level of variation. It can be shown that this level of variation is such that the probability that an average nucleotide site shows a difference between two randomly chosen chromosomes (or is heterozygous in a randomly chosen diploid sexually reproducing organism) is approximately equal to four times the long-term population size multiplied by the rate of mutation. The amount of divergence between two sequences sampled from two different species will be equal to the mutation rate times twice the time since speciation plus an additional amount equal to the expected number of differences between two randomly chosen chromosomes in the ancestral population. Because variation between species is but an extension of variation within species, and both are ultimately driven by mutation, then strictly neutral variation within species should be positively correlated with strictly neutral variation between species.

4. THE IMPACT OF NATURAL SELECTION ON MOLECULAR VARIATION AND EVOLUTION

Mutations that confer a fitness advantage will increase in frequency in the population because of positive selection. If the variant goes to a frequency of 100 percent, the population has now undergone a substitution of one variant for another (e.g., a new A has replaced the ancestral G nucleotide). Positive selection can lead to very rapid rates of fixation, orders of magnitude faster than the rate of fixation due to genetic drift alone.

Because adjacent nucleotides are tightly linked genetically, selection impacts not only the beneficial mutation but also the region of the genome in which that variant is located. Rapid fixation can therefore fix not only the favored variant but also the surrounding segment of the genome, resulting in a "selective sweep" and consequently a genomic region of initially no or very reduced adjacent neutral variation. Only over time will new mutations introduce variation back into this region. Perhaps surprisingly, the average divergence of linked neutral sites is unchanged from the neutral prediction. Such patterns provide much insight into the frequency and location of adaptive fixations throughout the genome (see chapter V.14).

While the rate of new mutations in the population is unchanged with natural selection, the fixation rate for beneficial mutations is dramatically increased, leading to increased sequence divergence between species for those specific sites under positive selection compared with adjacent neutral variants whose dynamics are determined by stochastic processes of genetic drift alone. The study of protein-coding sequences provides a particularly illustrative, and useful, example of how this contrasting pattern of positively selected, negatively selected, and neutral variation can be used to infer where and how natural selection has acted in the genome. Amino acids are encoded in mRNA in a triplet code of three nucleotides. While there are 61 possible combinations of three nucleotides that encode amino acids (three additional encode protein synthesis "stop" signals), there are only 20 common amino acids. Many amino acids are encoded by more than a single nucleotide triplet. Those triplet codons that encode the same amino acid differ by what is known as *synonymous* or *silent variants*. Those that result in a change in the encoded amino acid are termed *nonsynonymous or replacement variants*.

Since protein function is largely determined by its amino acid sequence, the fitness consequences of nonsynonymous mutations are much greater than those of synonymous mutations. Constraints on protein function result in strong purifying selection on nonsynonymous variants, preventing them from reaching substantial frequencies in populations (*polymorphism*) or from going to fixation (*substitutions*). These proteins evolve at very slow rates (e.g., histones in figure 1). Proteins with very relaxed constraints on amino acid sequence (e.g., fibrinopeptide, which are nonessential "spacer peptides" cleaved from fibrinogen in the clotting of blood, figure 1) evolve near the neutral rate of evolution. Some proteins play key roles in adaptation to new enzyme substrates or respond to biotic or abiotic challenges. Here, positive selection favoring new amino acid variants leads to the accelerated fixation of mutations. Some adaptive responses require repeated changes in protein sequence (such as at the antigen-binding sites of some immunity proteins), resulting in successive accelerated replacements. Somewhat counterintuitively, such positive and negative selection has little to no effect on the rates of substitution of adjacent strictly neutral variants. Thus, contrasting levels of variation and/or divergence at nonsynonymous to synonymous sites can provide estimates of the strength of both positive and purifying selection for a protein-coding gene. The ideal neutral benchmark is found in "dead genes" that no longer function (*pseudogenes*), in which levels of variation and divergence are usually close to those seen at synonymous sites. These contrasts of nonsynonymous and synonymous variation and divergence form the basis of several tests to detect natural selection acting on genomes and uncover the functional targets of that selection (see chapter V.14).

Not all mutations are simply strictly neutral, lethal, or strongly favored; rather, functional and population genetics studies have demonstrated that many mutations affect function only slightly, most in a slightly negative manner but sometimes improving function a bit. In these instances, whether a mutant acts as a neutral variant can be influenced by the population size. Consideration of the relative strength of natural selection and genetic drift reveals that if the difference in reproductive success (fitness) is less than the reciprocal of the long-term population size, then the mutant will behave as a neutral variant, even if it would have a (slight) selective advantage or disadvantage in an infinitely large population.

Data have also revealed just how much of observed DNA variation segregating in a population is nearly neutral, with much of it being slightly deleterious. One impact of this class of variants is that fluctuations in population size among lineages, or even along lineages, leads to fluctuations in the ability of natural selection to "see" these variants. This phenomenon is most clearly observed in the "generation time effect" observed in the molecular clock for some types of variants. For example, germ line mutations are most often caused by DNA replication; therefore, short-generation mammals have more synonymous mutations per year than do long-generation species; however, short-generation mammals also tend to have very large population sizes, so that selection is more efficient and thus could result in more nearly neutral mutations. As predicted then, rates of protein evolution are slower on a per generation basis in short-generation mammals than rates of substitution in long-generation mammals.

5. BIOLOGICAL INSIGHTS FROM THE STUDY OF MOLECULAR EVOLUTION

The comparison of sequences from different organisms, and particularly from short-generation organisms such as microbes and viruses, has provided one of the clearest illustrations of a core principle of evolution: *descent with modification*. For example, these valuable studies have provided real data sets with known phylogenies with which to evaluate the accuracy of statistical methods to estimate the phylogenetic relationships of organisms and their ancestors when we have only sequence data from the extant end points of the evolutionary process for study (figure 2). Experimental microbial and viral evolution studies are also allowing direct tests of evolutionary hypotheses regarding adaptation, including

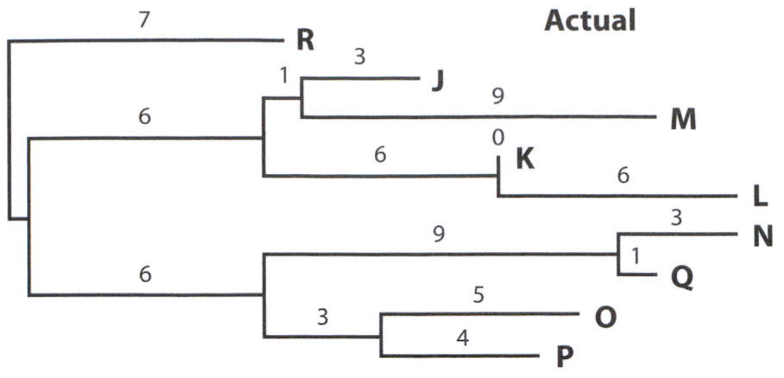

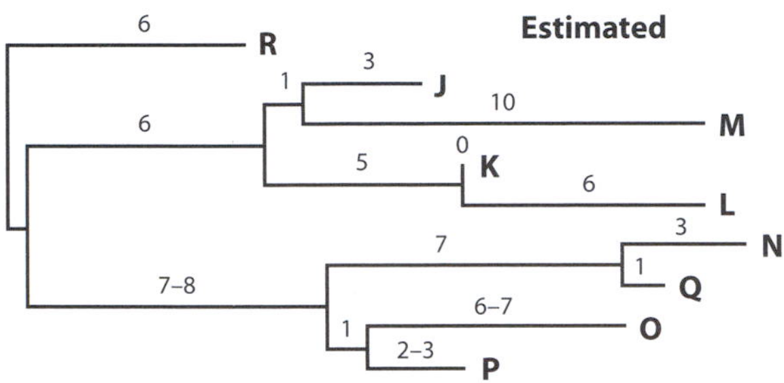

Figure 2. Mutational change in virus cultures demonstrates descent with modification and molecular evolution, and the molecular clock concept. Shown is a comparison of an actual "true" phylogeny of a virus (bacteriophage T7) with an estimated phylogeny constructed using only DNA sequences of the viruses in the experimental population. This population was initiated with one virus and split in binary fashion into several sequential derived lineages, the end points being denoted as the letters on the right side of the phylogeny. Numbers above the branches indicate the actual or estimated number of nucleotide substitutions that occurred along each branch, respectively. Actual substitutions were determined by sequencing 1091 base pairs of the ancestral viruses at each branch point in the tree. (Modified from David M. Hillis, John P. Huelsenbeck, and Clifford W. Cunningham. 1994. Application and accuracy of molecular phylogenetics. *Science* 264: 671–677).

whether recurrent adaptation takes place by the same or novel mechanisms, particularly since samples of every step of the evolutionary process can be saved for future study and even functional reanalysis.

With the introduction of the polymerase chain reaction (PCR) for amplifying specific genomic regions from very small samples (and even ancient bones, scat, and skins; see chapter V.15), coupled with wave after wave of advances in sequencing technology and automation, the field of molecular evolution has gone from being data limited to data overloaded, and no organism is now technically inaccessible. A striking example is the genomic study of microbes inhabiting our guts, our skin, the soil, and the oceans. Many of these microbes were not even known to science, since they could not be cultured in the lab or identified on the basis of morphology alone. Many are now known and identified only via their DNA sequences, and the relationships of these microbes are being revealed by the extent of sequence divergence from other known microbes.

Patterns and rates of molecular evolution can tell us about function in two ways. First, regions of genes and genomes that do not change are likely of critical and unchanging function. This basic principle of molecular evolutionary conservation has pervaded all of biology, and has been a guiding principle underlying the study of function in the exponentially growing number of genome sequences now being completed for virtually every type of plant, animal, microbe, and virus. Conserved regions of genes are what have changed biology from being simply organism focused to drawing on comparative functional data about a gene of interest from all studied forms. Second, regions of molecules that *do* change have turned out to include both those of little or no function, and those for which rapid change is itself adaptive. Distinguishing between adaptive change and relaxed functional constraint can be challenging; however, numerous statistical and computational methods have been developed over the last two decades allowing discrimination between positively selected change and relaxed constraint, and an area of very active research is that of evaluating the functional and fitness consequences of the adaptive fixations.

Another feature of molecular evolution that has emerged in the last decade is how much of the detected positive selection appears associated with conflict (e.g., "arms races" between hosts and pathogens), and not adaptation in the traditional sense (e.g., to new environments or nutrients). Microbial and viral pathogens clearly have driven, and still drive, much of the rapid molecular evolution in genomes, but additional conflicts between males and females, between host genomes and transposable elements, and even between hosts and endosymbionts have emerged as important drivers of rapid molecular evolution, including the evolution of new genes and new gene functions.

The role of noncoding portions of the genome in adaptation has also become strongly apparent in recent years. Variation in regulatory sequences (including enhancers, splicing machinery, and transcription factors) has been demonstrated as key to certain adaptive evolutionary changes in both molecules and phenotype (see chapters V.11 and V.12). Additionally, new and unanticipated types of functional sequences, such as small and long noncoding RNAs of various types (e.g., microRNAs, long noncoding RNAs [lncRNAs], and long intergenic noncoding RNAs [lincRNAs]), have been identified and not only demonstrate remarkable levels of evolutionary conservation and control of key developmental and cellular processes but also could underlie instances of adaptive change. Clearly, much remains to be discovered about genomes and the means and mechanisms by which the molecules of life evolve and adapt.

6. CONCLUSIONS

The study of the patterns and rates of evolution of biological molecules has provided data and results that clarify the relative roles of mutation, genetic drift, and natural selection in populations. The changes in these molecules underlie the evolution of organismal form and function, and the field of molecular evolution is alive with new discoveries about how genomes evolve and how observed molecular changes contribute to the stunning biological diversity of life we observe around us.

FURTHER READING

Graur, D., and W-H. Li. 2000. Fundamentals of Molecular Evolution. 2nd ed. Sunderland, MA: Sinauer. *A solid introduction to many of the core principles of molecular evolution, though it was written in the "pre-genome" era and is thus missing many of the recent discoveries.*

Kimura, M. 1983. The Neutral Theory of Molecular Evolution. Cambridge: Cambridge University Press. *The classic and well-written summary of the neutral theory argued by Motoo Kimura, a central figure in its development.*

Kumar, S. 2005. Molecular clocks: Four decades of evolution. Nature Reviews Genetics 6: 654–662. *A comprehensive review of the development and use of the molecular clock concept from the 1960s through 2004.*

Lynch, M. 2007. The Origins of Genome Architecture. Sunderland, MA: Sinauer. *A recent summary of the patterns and mechanisms of the molecular evolution of genomes.*

Nei, M., and S. Kumar. 2000. Molecular Evolution and Phylogenetics. New York: Oxford University Press. *An accessible review of methods of analysis of DNA and protein-sequence data for molecular evolutionary studies, including the powerful and easy-to-use software MEGA, available for download at www.megasoftware.net/.*

Nielsen, R. 2005. Molecular signatures of natural selection. Annual Reviews of Genetics 39: 197–218. *A clear and concise summary of the statistical inference of natural selection from molecular data from within and between species.*

Page, R.D.M., and E. C. Holmes. 1998. Molecular Evolution: A Phylogenetic Approach. Oxford: Blackwell Science. *An introduction to molecular evolution and molecular population genetics with particularly clear figures and graphs.*

Yang, Z. 2006. Computational Molecular Evolution. New York: Oxford University Press. *A mathematically and statistically rigorous review of methods of analysis of DNA and protein-sequence data for molecular evolutionary studies.*

V.2

Genome Evolution

Sara J. Hanson and John M. Logsdon Jr.

OUTLINE

1. Evolution of genome architecture
2. Genome expansion and restructuring
3. Drivers of genome evolution

The entirety of an organism's DNA content—its *genome*—is a heritable storage system containing all information a cell needs to dictate the organism's growth, development, and phenotypic characteristics. Throughout all forms of life, huge variation exists in the size and content of genomes, demonstrating the highly flexible, dynamic, and complex nature of their evolution. The frequently striking amounts of noncoding DNA present in eukaryotic genomes—largely absent in prokaryotic genomes—is particularly striking. This includes intragenic (introns and untranslated regions) and extragenic (regulatory sequences) as well as transposable elements: features that dominate eukaryotic genomes and usually make up the vast majority of nuclear DNA. Processes including recombination and transposition of mobile genetic elements have been hypothesized as mechanisms for the expansion of eukaryotic nuclear genomes. Both adaptive and neutral processes have been implicated in the origin and evolution of these genomic elements, and understanding the nature of such mechanisms for genome evolution can provide important insights into the evolution of prokaryotic and eukaryotic diversity.

GLOSSARY

Alternative Splicing. The generation of mRNA isoforms through differential use of splice donor and acceptor sites, retention of introns, and/or exon skipping.

Constructive Neutral Evolution. Conditions that decrease the efficacy of selection make it more likely that novel elements such as introns, untranslated regions, and modularity in gene expression will become fixed in a population. As a result, the increased genome size and content in eukaryotes derives from the fact that they have smaller population sizes that stem from increased cell size relative to prokaryotes.

C-Value Paradox. The mass of DNA in a haploid cell—or *C-value*—corresponds to an organism's genome size in base pairs but displays no clear correlation with organismal complexity.

Modularity. In eukaryotic organisms, a gene is expressed under the control of its own promoter and a combination of *trans*-acting factors that interact with other regulatory sequences. This is in contrast to prokaryotes, where a single promoter and set of regulatory sequences and few *trans*-acting factors dictate the coordinated expression of groups of linked genes.

Mutation Bias. Processes that generate unequal outcomes for seemingly reciprocal mutational events. For example, small deletions in genomic DNA occur at higher frequency than small insertions, resulting in smaller genome size over time.

Noncoding DNA. Genomic region that does not encode a protein or functional RNA product. These include introns (intragenic sequences removed following transcription), untranslated regions (transcribed sequences upstream of the translation start codon and downstream of the translation stop codon), and all other intergenic DNA.

Nucleoskeletal Hypothesis. The size of an organism's genome shapes the size of the nucleus required to contain it (the genome serves as a "nucleoskeleton"). Cell size and nucleus size coevolve such that increased cell size corresponds with increased genome size.

Recombination Hot Spot. Genomic regions where crossovers occur at much higher rates than in other regions of the genome.

Selfish DNA Hypothesis. Increased genome size is attributed to proliferation of transposable elements. Transposable elements multiply until they begin to affect (reduce) host fitness, thereby natural selection prevents their further proliferation.

Transposable Element. Mobile DNA segments that are capable of self-proliferation—within and between genomes—through either "cut-and-paste" or "copy-and-paste" mechanisms.

1. EVOLUTION OF GENOME ARCHITECTURE

Before the advent of high-throughput sequencing technologies and the resulting plethora of available genome sequence data, studies of genome evolution concentrated on comparing genome sizes across the tree of life. Such early studies focused on estimates of the mass of DNA within haploid cells, termed the *C-value*, which can be extrapolated to an estimate of the number of base pairs composing an organism's genome. The initial—and seemingly reasonable—hypothesis was that the number of genes contained in an organism's genome would increase with the increasing complexity of organisms. As such, prokaryotes and single-celled eukaryotes would possess fewer genes than multicellular eukaryotes; however, this relationship was not observed, and it was instead found that genome sizes ranged greatly, even within relatively closely related groups of taxa. In 1971, C. A. Thomas Jr. described these perplexing observations as the "C-value paradox," as genome size apparently did not account for increasing organismal complexity.

Although a clear correlation between genome size and organismal complexity was not realized, two generalizable genome configurations are evident. First, in prokaryotes, genomes are small and compact, comprising circular pieces of DNA. Intergenic space in these organisms is limited, and blocks of genes—called *operons*—which largely encode for genes with functions in the same pathway or process, are cotranscribed using the same promoter and regulatory sequence(s). Second, in eukaryotes, genomes are dramatically larger and contained on one or more linear chromosomes. This difference (presumably due to expansion) results from modular gene regulation (in which each gene is transcribed separately, with limited overlapping use of regulatory elements) as well as sometimes-massive amounts of noncoding DNA, including introns, untranslated regions (UTRs), and repetitive elements. Furthermore, the linear nature of eukaryotic chromosomes requires additional elements for proper maintenance and segregation of chromosomes; these include centromeres and telomeres, generally comprising repetitive DNA sequences, needed for segregation during mitosis and meiosis and the maintenance of chromosome ends, respectively.

Genome sequence data have revealed clear patterns relating to variation in genome size, most notably in the characterization of noncoding DNA elements in a wide range of prokaryotes and eukaryotes. The general trend appears to be that genome size increases with genome complexity, which, in turn, is correlated with increasing organismal complexity (although there are notable outliers). This pattern is well illustrated by comparisons among individual genome components, including the observed correlation between genome size and intron and intergenic DNA content. In fact, noncoding DNA is nearly exclusively responsible for differences in eukaryotic genome size: a 10,000-fold difference in the range of genome size exists between prokaryotes and eukaryotes, but only a 100-fold range in the amount of their protein-coding DNA. As shown in figure 1, the relative amount of protein-coding DNA decreases with increasing genome size while other genomic elements, such as transposons, increase.

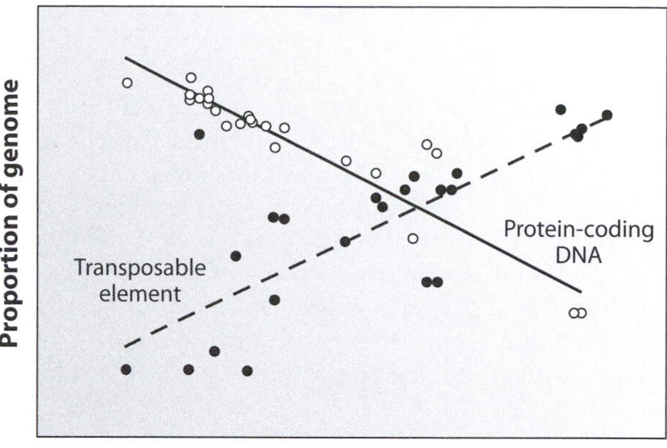

Figure 1. Relative contributions of two components of eukaryotic genomes. As genome size increases, the relative amount of protein-coding DNA decreases (white circles). In contrast, transposable element content increases in larger genomes. Thus, larger genomes contain proportionately fewer genes and more transposable elements than smaller genomes. (Adapted from Gregory 2005b.)

2. GENOME EXPANSION AND RESTRUCTURING

How does genome restructuring occur? What processes result in changes in genome size? Several mechanisms are thought to play a role in large-scale changes in genome architecture, including those that shuffle genotypes and, thus, alter the structure of chromosomes, as well as processes that result in addition or relocation of new DNA sequences in the genome.

Recombination

Recombination, the repair of double-stranded breaks (DSBs) in DNA, has important influences on organismal evolution, including both generating and reducing genetic variation (see chapter IV.4). DSBs may be incurred exogenously through exposure to environmental agents at any point during an organism's life cycle, or endogenously during meiosis in eukaryotes. Repair of these breaks frequently involves using a homologous piece of DNA as template—typically a sister chromatid or homologous chromosome. When a reciprocal exchange of DNA between homologous chromosomes occurs, it is referred to as a *crossover event*. Efficient repair of these breaks is critical, because their presence will disrupt replication and transcription. Errors in recombination can be devastating to an organism, but when they occur in the germ line, they also provide heritable restructuring events in genomes that contribute to genomic evolution in eukaryotes.

Repetitive elements and self-replicating mobile elements are present throughout eukaryotic genomes. Because these elements have multiple homologous templates in the genome, recombination can potentially occur between any two, even if located on different chromosomes. When ectopic recombination occurs between these elements as a result of their sequence similarity, large-scale chromosomal rearrangements can occur, including sequence duplications, deletions, or inversions of large sections of chromosomes, and translocation of a chromosomal section from one chromosome to another. These changes can disrupt protein-coding sequences directly, as well as remove or add regulatory sequences that can result in aberrant expression of genes.

During meiosis, DSBs are induced and repaired in a process mediated by a cell's machinery. Because of the inherent risk associated with the formation of these breaks, it is unsurprising that meiotic recombination appears a tightly regulated and evolutionarily constrained process. More unexpected are the constraints limiting the number of these breaks that result in a crossover event. Furthermore, these crossovers do not occur equally across the genome; rather, they are concentrated at *hot spots*, where rates of recombination are higher by several orders of magnitude than their flanking genome regions, or *cold spots*, in which crossover rates are extremely low. These hot spots are rapidly evolving and dynamic. Organisms as closely related as humans and chimpanzees share no overlap in the genomic locations of hot spots, and intraspecific variation has even been observed within humans. Despite this fast rate of evolution of hot spots, there is mounting evidence that their location is sometimes dictated by specific sequence motifs. In the fission yeast *Schizosaccharomyces pombe*, several discrete sequences seven base pairs in length have been identified at active hot spots. In humans, one degenerate thirteen base-pair motif has been characterized at 41 percent of identified hot spots. Furthermore, in humans, the transcription factor Prdm9 is required for activation of these hot spots, and the amino acids that interact with the thirteen base-pair motif are under strong positive selection. Prdm9 may therefore act as a driver in hot spot evolution, or may be evolving rapidly in response to changes in hot-spot sequence motifs.

This observed specificity in DNA sequence at active hot spots is puzzling. If a specific sequence is required for increased recombination, that sequence should also be lost as a result of the very recombination that it induces. A model was recently proposed to explain this apparent paradox. If a specific sequence is required for hot-spot activation, then a single base-pair change will inactivate it. Conversely, there are many sites in the genome that require a single base-pair change in order to become an activated hot spot; therefore, an evolutionary equilibrium may exist in which hot spots are degraded and introduced through these single base-pair changes. This explanation, coupled with the rapid evolution of hot-spot activators like Prdm9, may explain the dynamic nature of genomic hot-spot locations even in very closely related organisms.

Transposable Elements

An astonishingly large fraction of many eukaryotic genomes is composed of mobile DNA elements. These self-replicating pieces of DNA, which frequently contain their own protein-coding and regulatory sequences, make up about 50 percent of the human genome. Generally, there are two classes of mobile elements characterized primarily by their mode of replication. First there are DNA transposons, which replicate through a "cut-and-paste" mechanism, in which an enzyme (transposase)—which may be encoded by the transposon itself or by a separate transposable element—excises the DNA sequence prior to its insertion into a new genomic location. Proliferation of these elements relies on the horizontal transfer of new elements from one organism to another, such as the transmission of small circular chromosomes containing the elements between prokaryotes. The second class of mobile elements is collectively referred to as *retrotransposons*. These elements replicate by "copy-and-paste" mechanisms, in which an RNA intermediate is produced and reverse transcribed (by a retrotransposon-encoded reverse transcriptase) before insertion. Such elements can proliferate horizontally, as described for DNA transposons, as well as vertically, when they proliferate within cells in the germ line and can then be transmitted to the next generation.

It is easy to see how the replication and insertion of mobile elements in a host genome could be slightly or strongly deleterious. For example, transposon insertion into a protein-coding region would most likely result in a frameshift, premature stop codon, or otherwise-aberrant protein sequence. Potentially, for this reason, a host has mechanisms to defend its genome from such elements.

These include transcriptional silencing of the elements by chromatin modifications and transcription of small interfering RNA molecules that target mRNA produced by the element for destruction, thus depriving the transposable element of the machinery it needs for proliferation. Because successful proliferation of a mobile element depends on the success of a host genome, it is also possible that transposable elements have built-in self-regulatory mechanisms preventing them from uncontrolled proliferation that would drive a host to extinction; however, such mechanisms have not been characterized.

Importantly, as with all forms of mutation, mobile element insertion can on rare occasion give rise to evolutionary novelty. Because mobile elements encode their own machinery, multiple consequences can arise following their insertion into a new location. First, the elements contain protein-coding sequences and thus can introduce new coding regions into the genome (see chapter V.6). Second, these protein-coding sequences in mobile elements frequently have their own regulatory elements that can modify gene expression patterns of sequences, especially when adjacent to the insertion site. For example, the promoter region of a gene in the transposable element may recruit transcriptional machinery to a location near a host gene that has tight temporal or spatial regulation, causing it to be transcribed when it is normally silent. Indeed, it is hypothesized that centromeres and telomeres are often derived from mobile elements, and in some cases (e.g., *Drosophila*), mobile elements provide a mechanism for telomere maintenance. Finally, there is also evidence that mobile elements play a role in DNA double-strand break repair by using double-strand breaks as sites of insertion.

Noncoding Elements

Noncoding DNA sequences are those that do not determine a functional product. This chapter will consider the evolution of two types of noncoding elements, untranslated regions (UTRs) and introns. UTRs are parts of genes that are transcribed but not translated into an amino acid sequence, and are found both preceding the translation initiation site (5′ UTRs) and following the termination of translation (3′ UTRs). The addition of 5′ UTRs to eukaryotic genes is a risky prospect when the potential inclusion of an alternative translation initiation site is considered. Mutation of the 5′ UTR to contain such a site could have dramatic effects on the resulting amino acid sequence, resulting in a nonfunctional product. Because of this increased mutation risk, it is not clear what, if any, advantage eukaryotes gain through the addition of 5′ UTRs, but their presence and length are consistent across eukaryotic diversity. Although the addition of 3′ UTRs to eukaryotic genes does not appear to carry the same risks as 5′ UTRs, these elements are important in several aspects of mRNA regulation. The 3′ UTRs are critical for mRNA stability and nuclear export, and they have important regulatory functions in several aspects of translation. It is likely these features arose subsequent to the evolution of the 3′ UTR itself; therefore they cannot provide an explanation for the addition of this element.

The mechanisms for evolution and origins of introns are much better understood than 5′ UTRs. Despite the similarity in the length and number of protein-coding genes across eukaryotic diversity, there is substantial variation in the amount of intronic DNA. In eukaryotes, introns in nuclear genes (spliceosomal introns) are processed by a nucleoprotein complex—the spliceosome—which is present in all eukaryotes and thus likely present in the most recent eukaryotic ancestor. In humans, an average gene contains 7.7 introns, with an average intron length of 4.66 kilobases (kb). Compared to the average length of a human exon sequence (0.15 kb), it is clear that the total length of a human (and in general any eukaryotic) gene is dominated by introns. This density of introns allows for a large number of potential transcripts per locus through alternative splicing, which in humans is responsible for the average 2.6 transcripts produced per gene. Although the current importance of introns is at least partly understood (alternative splicing, regulatory element content, etc.), the origin and evolutionary mechanisms responsible for the proliferation of introns in eukaryotes remains unclear.

Debate over spliceosomal intron origin has been divided into two camps: those that propose the early evolution of introns prior to the divergence of eukaryotes and prokaryotes, and those that posit a later origin exclusively in eukaryotes. The resolution of this debate rests primarily on the hypothesized relationship of eukaryotic spliceosomal introns with the self-splicing group II introns found in some prokaryotes, which some argue are homologous. Whether spliceosomal introns arose early or late, there has been massive divergence in intron content in eukaryotes, making our understanding of the mechanisms underlying intron gain and loss of great importance.

Both intron loss and gain can be mediated by recombination, with intron loss hypothesized to result from replacement of a genomic gene copy with a reverse transcribed mRNA transcript of that gene (see chapter V.6), while hypotheses for mechanisms of intron gain include ectopic insertion of DNA fragments during an alternative DNA repair mechanism known as *nonhomologous end joining* (NHEJ). During NHEJ, fragments of DNA with very little sequence identity (microhomology) may be joined to repair DSBs, and aberrant insertion of a DNA fragment within a coding region may explain the origin of novel introns. NHEJ may be an intron loss mechanism as well, if microhomology between an intron's splice junctions is used for repair. Consistent with this, species that are intron poor have high conservation of their splice sites, whereas intron-rich species have more degeneracy in their intron splice sites. The hypothesized role of NHEJ in intron gain is supported by the observation that intron-rich species use NHEJ more frequently during DNA repair.

3. DRIVERS OF GENOME EVOLUTION

A challenge that remains for our understanding of genome evolution is explaining how the addition of DNA to the genome and the existence of more complex genomic elements are possible. The presence of these elements is inherently risky, as they provide additional locations at which deleterious mutations can occur. For example, the addition of an intron to a protein-coding region now adds splice junctions, a branch point, and other regulatory elements that are evolutionarily constrained. One could argue that such genomic complexity is necessary for the evolution of organismal complexity; however, the diversity in content of these complex elements suggests otherwise. Adaptive and neutral arguments for the evolution of genomic complexity are described below.

Adaptive Evolution

What evolutionary pressures might be acting on genome size? Some data suggest that the forces may be mutation bias, such that small (< 400 kb) deletions occur more frequently than insertions, resulting in reduction in genome size over time. For example, work performed in *Caenorhabditis elegans* demonstrated that at genomic sites not under selective constraints (i.e., pseudogenes), the rate of deletion was 2.8-fold higher than the rate of insertion. These data offer an explanation for the relatively compact size of the *C. elegans* genome, and suggest a more generalizable trend of deletions outnumbering insertions: selective pressures may favor a smaller genome.

Some other, less generally supported hypotheses suggest selective pressures might underlie the evolution of genome size as a result of the phenotypic consequences of these differences, primarily the effect of genome size on cell size. For example, the *nucleoskeletal hypothesis* proposes that increasing genome size requires an increase in the size of the nucleus, which coevolves with cell size. According to this hypothesis, a larger cell has greater requirements for transcription and translation, and thus requires a larger nucleus and genome to meet its needs; however, the nucleoskeletal hypothesis does not account for accommodation of a larger cell's needs through increased rates of transcript production as opposed to increased DNA content.

Because beneficial outcomes are extremely unlikely for the majority of transposable element insertions, adaptive hypotheses for the existence of these elements can be excluded for the most part. These elements are more frequently thought of as parasitic or selfish because of their lack of dependence on host machinery for replication, and their likely detrimental effects on host fitness. The role of mobile elements in genome evolution is therefore referred to as the *selfish DNA hypothesis*, which posits that genome expansion is mediated by proliferation of mobile elements, and that such elements will spread until the point at which their impact on host fitness is so great that natural selection prohibits their further proliferation. This hypothesis does not account for the role of other elements present in eukaryotic genomes, such as introns, and therefore cannot fully explain the increased genome size in eukaryotes.

There are also several hypotheses for adaptive mechanisms underlying intron evolution in eukaryotes. First, large introns within genes increase the likelihood that incorrect splicing will result in the introduction of a premature stop codon that will be recognized early and will result in the degradation of the mRNA—a process known as nonsense-mediated decay. Second, the presence of one or more introns allows for alternative splicing to occur, in which introns can be excised or retained, exons can be skipped, or exon length can vary depending on the usage of specific splice junctions. This diversity in mRNA products from a single locus greatly increases the number of potential protein products resulting from that locus and allows for increased variation and complexity in molecular pathways (see chapter V.3). Further, the modular nature of genes that result from the inclusion of introns may have allowed for exon shuffling, in which mixing of domains from several different genes gives rise to genes with novel functions (see chapter V.6).

Neutral Evolution

Because eukaryotic genome expansion likely gave rise to sources of vast phenotypic novelty, it is tempting to develop adaptive hypotheses for their origination, such as those described above. However, the main explanation for the existence of these novel features may lie in *neutral evolutionary processes*—those that result from changes that have little or no effect on host fitness, but arise and become fixed in a population through genetic drift. A general framework for understanding the origin and evolution of such genomic novelties was proposed by Arlin Stoltzfus in 1999, which he called "constructive neutral evolution." Expanding on this, Michael Lynch proposed a synthetic hypothesis that posits neutral processes as being largely responsible for the origin of genomic elements that, in turn, gave rise to the expanded genome size observed in eukaryotes. Since eukaryotic cells are typically much larger than prokaryotic cells, which generally result in much smaller population sizes for eukaryotes, the effects of genetic drift are amplified, making it much more likely that neutral or even slightly deleterious mutations—including unusual genetic features—will become fixed in a population (see chapter IV.1).

As described above, incorporation of the features most responsible for increased genome size would have been a very risky prospect for early eukaryotes. In particular, noncoding elements like introns and UTRs dramatically increase the number of sites at which deleterious mutation may occur. Further, the origin of these elements would have been extremely dangerous, as their addition would interrupt protein-coding regions, potentially causing frameshifts, premature stop codons, or alternative translation start sites. The inclusion of these elements therefore would likely have immediate deleterious effects on an organism, or at best would not confer an immediate benefit to be acted on by natural selection.

Instead, neutral processes may account for the initial fixation of these features in early eukaryotic populations. Small eukaryotic populations increased the impact of genetic drift and reduced the efficacy of selection such that these genomic elements could become fixed despite not conferring an advantage on a cell. Any beneficial effects these elements currently have were therefore subsequently acquired and may contribute to their maintenance in a population, but adaptive arguments are unlikely to explain their original fixation in eukaryotes.

Our understanding of the evolution of genome structure and content has substantially improved in the past decade. Fast-moving advances in DNA sequencing technologies have provided unfettered access to complete genomes from across the entire tree of life. Decoding the content of these genomes has been only a first step in understanding their biology. A deeper and more satisfying view of genome biology is emerging in which genomes are not only repositories of genes but also evolving entities with emergent and sometimes-unusual properties that are increasingly explicable within a solid theoretical framework.

FURTHER READING

Denver, D. R., K. Morris, M. Lynch, and W. K. Thomas. 2004. High mutation rate and predominance of insertions in the *Caenorhabditis elegans* nuclear genome. Nature 430: 679–682. *An exemplar study using an experimental approach to examine the evolution of genome size.*

Farlow, A., E. Meduri, and C. Schlotterer. 2011. DNA double-strand break repair and the evolution of intron density. Trends in Genetics 27: 1–6. *Proposed model for role of introns in recombination and double-strand break repair.*

Gregory, T. R, ed. 2005a. The Evolution of the Genome. London: Elsevier Academic Press. *A comprehensive overview of genome diversity and evolution, including the evolution of specific genomic features.*

Gregory, T. R., 2005b. Synergy between sequence and size in large-scale genomics. Nature Reviews Genetics 6: 699–708. *Discussion of the impact of genome sequencing technology on the analysis of genome content at a large scale.*

Kazazian, H. H., Jr. 2004. Mobile elements: Drivers of genome evolution. Science 303: 1626–1632. *This review is a concise introduction to the impact of transposable elements on genomes.*

Lynch, M. 2007. The Origins of Genome Architecture. Sunderland, MA: Sinauer. *Detailed description of the mechanisms of genome evolution in the context of the theory of constructive neutral evolution.*

Roy, S. W., and W. Gilbert. 2006. The evolution of spliceosomal introns: Patterns, puzzles, and progress. Nature Reviews Genetics 7: 211–221. *Overview of the origin and maintenance of introns in eukaryotes.*

Stolzfus, A. 1999. On the possibility of constructive neutral evolution. Journal of Molecular Evolution 49: 169–181. *Initial presentation of constructive neutral evolution theory for the evolution of eukaryotic genomes.*

Thomas, C. A., Jr. 1971. The genetic organization of chromosomes. Annual Review of Genetics 5: 237–256. *Original description of the C-value paradox for chromosome size and organismal complexity.*

Wahls, W. P., and M. K. Davidson. 2011. DNA sequence-mediated, evolutionarily rapid redistribution of meiotic recombination hotspots. Genetics 189: 685–694. *Presentation of a model encompassing the rapid evolution of recombination hot spots and the protein Prdm9.*

Webster, M. T., and L. D. Hurst. 2012. Direct and indirect consequences of meiotic recombination: Implications for genome evolution. Trends in Genetics 28: 101–109. *Summary of the effects of recombination on genome structure and content from the perspective of population genetics.*

V.3

Comparative Genomics

Jason E. Stajich

OUTLINE

1. Comparative genomics and genome evolution
2. Evolution of gene number
3. Identifying regulatory regions
4. Copy number variation
5. Rapidly evolving regions
6. Ultraconserved elements
7. The future of comparative genomics

Genome sequencing now makes it possible to inventory the genetic material of most living and, in some cases, ancient organisms. To date hundreds of genomes have been sequenced from bacteria and single-cell eukaryotes, as well as animals, plants, and fungi. The vast majority of sequences have come from microbes, including Bacteria and Archaea from diverse environments including those living in hot springs, inside animal digestive tracts, in soil, and in disease-causing organisms. The fungi are the most sampled eukaryotes, but dozens of animals and plants, with their large genomes, have also been tackled. The pace of this sequencing is still accelerating, and it is conceivable that many thousands of species will be sequenced in the next decade, with similar numbers of genomes from multiple individuals or populations of species. By comparing the genome sequences of species, we are learning how the genetic inventory changes over time. Those genes or DNA loci that have been maintained among many species often encode important or essential functions for cells, while those that change rapidly may be superfluous or provide a needed function only in select species. The ability to inventory and compare genomes requires computational tools and models of evolutionary change.

GLOSSARY

Alternative Splicing. The production of multiple transcripts from a single gene locus by the alternative inclusion of partial or entire exons through variation in splicing of the RNA.

C-Value Paradox. The idea that genome size scaled with complexity was found to be inconsistent when large genomes were identified for seemingly less complex organisms.

Negative Selection. Sometimes called *purifying selection*, it indicates that selection is purging changes that cause deleterious impacts on the fitness of the host.

Operon. A cluster of genes that function as a single unit under common regulatory control where the proteins often perform multiple steps in a common biochemical pathway.

Orthologue. A gene found in different species that evolved from a gene present in the common ancestor of the species.

Phylogenetic Shadowing. The identification of conserved blocks in sequences by aligning multiple sequences from different species to find regions of high conservation. Also called *phylogenetic footprinting*.

Positive Selection. Often called Darwinian selection, it describes selection where changes improve the fitness of the host.

Posttranscriptional Regulation. Regulation of the gene product after an mRNA is made. Mechanisms include premature degradation and delay or block of translation into protein.

Pseudogene. Gene locus with accumulated mutations and that is no longer functional.

Synteny. Shared gene order, where genes are arranged in the same order and orientation along chromosomes of difference species, indicating they were likely in the same order in the common ancestor.

Transposable Element. DNA segments that can relocate themselves in the genome.

Ultraconserved Elements. Conserved genomic regions stretching over hundreds of bases that are highly similar over millions of years of evolutionary time.

1. COMPARATIVE GENOMICS AND GENOME EVOLUTION

Rates of Change and Molecular Evolution

Studying the process and consequences of genome evolution involves comparing the sequences of modern-day organisms and reconstructing a history of events. The bases for the approaches are rooted in molecular evolution theory, which provides models and tools for the study of changes at the DNA or protein sequence level (see chapter V.1). Comparisons of genomes can identify regions that change, regions that are static or change less, regions that are unique to a specific lineage, or regions that are rapidly evolving. For any given region that is shared among species, one can count the number of differences and similarities over a stretch of DNA sequence and compute a rate of change. An evaluation of the entire genome allows a total average rate to be computed; individual regions can then be classified as faster or slower than an average or background rate. This classification of fast- or slow-evolving regions is useful, because it indicates the types of pressure exerted on the region by natural selection. There are three main classifications: negative selection, indicating slow to no change; neutral evolution, representing the average or background level of change without the influence of selection; and positive selection, indicating faster than the background rate of change (see chapter V.14).

Negative selection limits the mutations that fix in a region, most often because changes cause a decrease in fitness. As a consequence, a genomic region will appear to be evolving more slowly than the genome average because most mutations that occur are purged (e.g., individuals with specific changes in the region die or are less likely to reproduce). For example, a gene region that encodes an essential protein-coding gene, like histones, would be expected to be evolving slowly because any mutation that disrupted its function would render the organism inviable. This classification of negative selection is computed from alignments of a genomic region among multiple organisms. Regions under negative selection will typically have many fewer changes than are observed elsewhere in the genome. Regions with almost no changes between species are deemed *ultraconserved* and are discussed later in the chapter.

Regions that are evolving near the genome-wide average are considered to be evolving neutrally, and they typically do not encode functional elements on which natural selection would act. Examples include pseudogenes, which are inactivated or dead genes, and inactivated transposable elements. Pseudogenes that formed before a species split are particularly useful in trying to date the evolutionary divergence between organisms, because sequences are similar enough to align so the differences can be counted; since they are evolving neutrally, pseudogenes will have accumulated mutations at a rate proportional to the amount of time since the divergence of the species.

Genomic regions that are changing more rapidly than the background rate are considered to be under positive selection. It is not that these regions are experiencing a higher mutation rate than the rest of the genome but that mutations in the region are more likely to persist and be passed to the next generation because of the fitness advantage associated with them. Thus, these regions appear to change at a rate higher than the background rate for the genome. Since not every mutation that occurs will be fixed, those in neutrally evolving regions will be lost by genetic drift at a rate proportional to the effective population size of the species. But if a beneficial mutation occurs in a region and presents a fitness improvement that natural selection can act on, the change will be passed on to the next generation at a higher rate than neutral mutations, thus producing a higher rate of change overall. It is easier to detect and interpret positive selection in protein-coding genes where the impact is seen as higher frequency of amino acids changing mutations relative to mutations that do not change amino acids. Since mutations that change amino acids will alter the sequence and potentially the function of the protein, it is easier to observe and interpret these types of faster rates of change in gene regions. For genomic regions that do not code for proteins, the types of mutations that are beneficial are less defined, but overall faster rates of fixation indicate a relative importance of the region for fitness.

Gene Content Comparison

Sequence comparisons and evidence for selection are not limited to the level of DNA nucleotides. The gene content of an organism can also be important for interpreting its past and present ecological niche and types of competition pressures. An overall inventory of the genes with a shared history from a group of species can be obtained. This is done by first identifying homologous gene sequences using sequence searching tools such as BLAST (Basic Local Alignment Search Tool) to find significantly similar sequences. The genes are clustered by the significance of the similarity to find groups likely to be descendants from a copy that was present in the common ancestor of the species. These groups of shared genes can be automatically constructed from the total gene sets of multiple organisms using sequence similarity and clustering approaches that vary from simple distance methods to graph-theoretic approaches and phylogenetic tree construction. The tools build an overall collection of genes shared among organisms, and the clusters can be compared to see how the content varies among species.

With these collections of genes clustered into orthologous groups, one can investigate how genes are classified into particular phyletic patterns, those found in common among all species, or falling into groupings that represent particular hypotheses. For example, searching for genes found only in common among plant pathogens might provide ideas about genes involved in pathogenesis. The collection of genes shared among all or most species gives an idea about core processes that are essential for survival, since no organism has lost them.

Lineage-specific genes can be markers for specific processes unique to an organism (see chapter V.6); however, one must be cautioned not to overinterpret the set without some validation, as it is possible for some predicted genes to be false-positive gene annotations. Even though false positives are enriched in the lineage-specific gene data set, there will be interesting contents within the group. First, true lineage-specific genes may be rapidly evolving genes that have changed so fast as to lack identifiable sequence similarity with homologues in other species. In addition, novel genes may arise through other processes to create genes de novo even from random mutations. Lineage-specific genes may also be the result of transfer of a gene fragment from a genome parasite like a virus or transposable element.

The order of genes encoded on a chromosome can also contain information that can be compared among species. Shared gene order, or *synteny*, can be used to compare the content of genomes at an even finer scale than simple presence or absence of individual genes. The same gene order in different species typically indicates the genes were in the same order in the ancestor and would give further support to the idea that the genes are orthologous. In bacteria, shared gene order often indicates the related function of genes, since operons typically encode a cluster of genes in a common order to produce a common transcript. In some eukaryotes the existence of shared gene order often represents an ancient order that has not had sufficient time to degrade or indicates a region that is under selection to maintain the physical clustering of the genes. One example is the *Hox* gene cluster, which is a syntenic cluster found in animals that has persisted since the common ancestor of vertebrates and invertebrates.

2. EVOLUTION OF GENE NUMBER

Does having more genes mean a more complex organism? This was certainly the expectation as genome projects in model systems such as yeast, fly, worm, mustard weed, and even human sequencing were under way. A human gene count wager had even been initiated that varied in number from 15,000 to 80,000 genes based on the number of transcripts identified through mRNA sequencing; however, when the human genome was finished and annotated, it was clear that the number is closer to 23,000 genes and not much different from the number in the (seemingly) less complex worm or fly with roughly 20,000 genes. So how does gene number evolve, and how important is the gene count for predicting the capabilities of the organism? In short, the number of genes is probably less important than the diversity of ways that genes can be regulated (e.g., the temporal and spatial expression pattern), the number of interactions of the gene products, and the variation and modifications the transcripts can undergo as part of cellular processes.

Several caveats and definitions are required to fully explore the ideas of the way gene number evolves. First, a gene is a locus of DNA that encodes sequence that will be transcribed by an RNA polymerase. The produced RNA either encodes sequence for a protein, or the RNA itself will fold into a structure that allows it to do some work of the cell. The basis for gene count correlating with complexity originates from the "one gene, one enzyme" hypothesis that led to the Nobel Prize for Beadle and Tatum. The idea was put forth based on experiments in the fungus *Neurospora crassa* and showed that for their studies of enzyme pathways, one gene locus encodes only one functional product. The observed diversity of functional gene products and enzymes found in many organisms suggests that the number of genes would scale with complexity of the organism; however, comparisons from increasingly more genomes, sampling broadly from the diversity of eukaryotes, indicates lack of an overall correlation of gene count and organismal complexity. The idea that complexity did not scale with genome size had already been revealed by the C-value paradox (i.e., genome size does not scale with complexity), but it was surprising that even gene count did not scale as expected with complexity.

The source of the variation comes from the number and types of RNAs that can be made from the gene loci to be translated into protein or folded into regulatory modules. The number of these transcripts is controlled by the gene locus, but also by alternative splicing and post-transcriptional regulation of these products, which injects additional variation into the system. For example, it has been shown that alternative splicing is an important aspect driving diversity of transcripts in organisms classified as complex; nearly 94 percent of the genes in the human genome are alternatively spliced, while fewer than 1 percent are alternatively spliced in the yeast *Saccharomyces cerevisiae*. Additional regulation also takes place at the transcriptional level, where transcription factors and chromatin modifications can precisely regulate when an RNA is made or whether or not it is available for immediate translation or folding into a functional protein (see chapter V.7). Simply put, the complexity comes not from the building blocks (genes) alone but also from the number of ways they can be modified and made to interact, which controls the variable types of structures and behaviors in organisms we consider more complex.

It is worth noting that *complexity* is a loaded term that tends to assume that what we see as larger or more successful at colonizing the earth are in fact the more complex organisms. However, other measures, such as the total number of diverse tissue types, developmental stages, and elaborateness of body systems, are more objective measures that can be applied without anthropomorphic biases.

Despite these caveats, studying gene count differences is still useful for exploring recent changes among close organisms. Because increasing the copy number of a gene in the genome is one route to production of more of the gene product, gene family expansions can be successful routes to adaptation to an environment when more of something is needed. For example, drug resistance can evolve in microbes through the increase in copy number of pumps that shuttle the drug out of the cell. Studies of overall counts and examples of recent duplications can thus provide indications of recent changes that are important for adaptation to a new environment.

3. IDENTIFYING REGULATORY REGIONS

In addition to studies of the regions of protein-coding genes, comparative genomics can be useful for looking at genomic conservation around the coding regions of genes to identify regulatory regions. A challenge is that regulatory regions do not evolve under the same rules as the protein-coding part of a gene locus. The main features of regulatory regions are binding sites for *transcription factors*, which are proteins that can activate or repress transcription of the gene (see chapter V.8). A mechanism for identifying regulatory regions from comparative genomics utilizes the premise that regions that evolve more slowly than the background or neutral evolutionary rate are likely to encode a region that provides a function. Identifying regulatory elements is more difficult than gene regions because it may be that the specific order or orientation of the binding sites is not important, simply the presence of one or more of them.

Orthologous versus Coregulated Genes

An important aspect of the analysis requires choosing an appropriate collection of sequences for comparison. One approach is to compare the regulatory regions of orthologous genes in different species. A reasonable assumption is that when a gene is regulated for liver development in humans, it might have been under the same control in the ancestor of both humans and mice. Thus, there may be common binding sites found among species when comparing the same region in an orthologous gene. Gathering orthologues and searching for motifs found in common among the regulatory regions, or performing alignments on these regions, can identify elements that are in common and have been preserved among the gene copies. This approach may be successful for genes that have not had a change in gene regulation since divergence from a common ancestor. This orthology-focused comparison is useful for studying slowly evolving regulatory patterns that are probably essential or inflexible to change. It is unclear what proportion of genes in any given genome these encompass. There are examples of very conserved processes, such as mating in the yeasts of the *Saccharomyces* group, in which the regulatory modules have completely changed even though the same protein components are all involved.

If gene regulation components have changed, there are additional approaches that can be applied to identify regulatory motifs among genes from a single organism having a common regulatory pattern. The assumption here is that similarly regulated genes share a common collection of regulatory modules. The inference of gene regulation pattern can be made through comparing changes in gene expression over time during developmental processes or in response to various changes in conditions (e.g., temperature, nutrients). By gathering genes with a common gene expression pattern, one can search for shared sequence motifs that may explain their similar regulation. This approach focuses on motif identification in only one species, but if multiple parallel studies are performed in different species, a comparison of types of regulation can be elucidated.

Alignment-Based Regulatory Region Analyses and Phylogenetic Shadowing

To find binding sites or motifs that are in common among sequences, alignment of putative regulatory regions is undertaken to identify islands of conservation, which may be in a different order from the sequences compared. Computational approaches have been developed to identify alignable regions by finding small blocks of sequences without requiring regions to be in the same order or even orientation. Since identifying the general rules for evolution of regulatory elements is still an area of active research, it is not yet clear which heuristics are best, but it does appear that the main regulatory units are blocks of conserved sequences that can be shuffled in order and arrangement. The full extent of the relationship between binding site configuration and gene expression remains an enigma, so identifying "rules" for optimizing alignments of regulatory regions is still an area of active research.

A further extension of the alignment-based approach can be applied to pick out the more slowly evolving regions likely to be functional elements. *Phylogenetic shadowing* scores the rates of change on a per nucleotide basis so that regions that are evolving slower or faster than neutral rate can be identified. This analysis ideally starts with a pairwise or multiple alignment from several species of varying phylogenetic distance. The choice of species will influence the depth of conservation that can be identified. The boundaries of the regulatory elements are found by identifying stretches of sequences where the substitution rate, or average number of changes, is similar. This allows for the identification of boundaries of conserved regions based on the rates of change rather than only the best-scoring alignment blocks, providing more nuanced determination of the likely regulatory regions.

Motif Enrichment and Identification

Binding site motifs and regulatory regions can also be found by searching regulatory regions for short "words" that are overrepresented in genome sequences. Presuming that the binding site sequences are a shared feature of most gene regions regulated by a particular transcription factor, logically there should be one or many copies of the sequence within regulatory regions of genes that have a common expression pattern. Motif-searching tools use this assumption and compare the overrepresentation of a sequence motif of some specified size range (typically 6–12 base pairs long) in the simplest fashion; that is, by counting all the observed motifs and comparing the distribution of observed counts with a theoretical background distribution that takes into account overall sequence bias (such as G + C). An easy way to generate this null background distribution is to also sample motifs from a set of sequences that serve as a negative control group of nonregulatory sequences or, less ideally, simply from a collection of random sequences. Additional modern approaches that use information theory and phylogenetic relatedness can improve the accuracy of identified motifs. Experimental validation of the accuracy of these approaches to identifying truly functional elements is still ongoing research, but it appears that enrichment and phylogenetic conservation are very often strong indicators of true positive functional binding sites.

A twist on the motif enrichment approach is also to search for motif avoidance, where a functional element, if present, would negatively impact the fitness of the organism. For example, if binding sites for transcription factors were found within the protein-coding region of a gene, it would be deleterious as they might allow a factor to bind and prevent efficient transcription by RNA polymerase. As a consequence, there should be a paucity of binding sites in the wrong place. Studies of bacterial genomes provide support for this approach: when searching for the RNA polymerase binding sites, researchers found these motifs significantly underrepresented in the protein-coding regions compared with the upstream or downstream regions of a gene. This approach may additionally validate potential elements by determining whether the occurrence is restricted in some regions of the genome, indicating that the site has functional activity that must be limited to specific sequence contexts.

4. COPY NUMBER VARIATION

The genome is not a static entity; as such, there is not one version of a genome sequence that completely represents a species. Variation among individuals can be in the form of differing nucleotides, insertions, or deletions, and in the total number of copies of genomic regions. Copy number variation (CNV) can increase or decrease copy number of all or part of a chromosome. In humans, CNV has been linked to disease or abnormal phenotypes such as autism and cancer; however, a great deal of variation can be detected that is not associated with disease. Several studies in humans and other species have indicated that CNV can be adaptive. For example, in the diploid yeast species *Candida albicans*, CNV is often observed in response to adaptation to stresses like antifungal drugs; both losses and duplications of parts of chromosomes that contain a drug target or important drug pumps have been observed.

CNV in an individual can be detected with comparative genome hybridization using microarrays constructed with probes designed to be regularly spaced across each chromosome. These arrays can detect copy number by scoring the relative intensity of hybridization of a region in one sample as compared to a control. Because the probes are spaced evenly across a chromosome, the boundaries of the CNV can be mapped with relatively high precision. CNV can also be detected with next-generation sequencing by aligning the short reads back to genome sequence assembly and examining the depth of the reads. Regions with CNV will be higher or lower average coverage than the rest of the genome. As technologies continue to improve, we will gain a deeper understanding of the extent of CNV between individuals, populations, and species as well as its evolutionary implications.

5. RAPIDLY EVOLVING REGIONS

Comparison of rates of change across genomic regions can provide insight into their relative functional importance.

As discussed previously, conserved regions evolving more slowly than expected likely encode functional elements needed for survival of the organism. The identification of regions evolving more quickly than the background rate can suggest changes that may be important in adaptation. Methods that search for these rapidly evolving regions of the genome identify where more changes than expected have accumulated by comparison to background rate of evolution. These methods can be used to test for faster changes across the entire genome, but they are most often applied specifically to protein-coding regions. By identifying fixed sequence differences between species within a protein-coding gene, tests can be performed to compare the number of DNA substitutions that change an amino acid (i.e., nonsynonymous) to the number of silent mutations (i.e., synonymous). Silent mutations occur because of the redundancy of the genetic code, so that several different codons encode the same amino acid residue. These approaches compare the rates of synonymous and nonsynonymous changes in a gene region, where the ratio of these rates can indicate the strength and direction of selection (see chapter V.14). Excess nonsynonymous mutations indicate positive selection, while an excess of silent changes indicates negative selection. A ratio close to 1 indicates a region evolving neutrally and can be considered the background rate, as one would expect for inactivated transposable elements and pseudogenes.

Searching for rapidly evolving protein-coding regions can identify those genes that may be involved in adaptation. Examples of these types of functions include genes involved in host-pathogen interactions, in evolution of drug resistance, and in organisms' changes in lifestyles, food sources, or ecologies. For example, a plant chitinase, which degrades the chitin in an attacking fungal pathogen, shows an excess of amino-acid–changing mutations, suggesting that it has been evolving under positive selection. The changes identified occur in the active site of the enzyme and are interpreted to have improved the plant's ability to further recognize and degrade the fungal chitin in response to avoiding defense strategies deployed by the fungus. Many other studies have identified positive selection, for example, due to sexual competition in cell-surface-recognition genes in plant pollen grains and in lysin, a sperm-recognition receptor protein in marine invertebrates.

Population genetic tests can also be used to find changes occurring among or between populations by examining changes in allele frequencies rather than fixed differences between species (see chapter V.14). For example, population analyses of *Plasmodium falciparum*, the causal agent of malaria, identified recent positive selection in genes that had acquired mutations conferring resistance to antimalarial drugs. Studies of human populations living in higher elevations of the Tibetan plateau revealed alleles under positive selection for transcription factors necessary for the hypoxic, or low oxygen, response. Both these studies found a change in the frequency of alleles between two populations differing in resistance or altitude, respectively; these data suggest that the changes in allele frequencies were driven by natural selection.

Recent efforts have also further identified nonprotein-coding genomic regions that represent RNAs having a functional role in gene regulation. Katherine Pollard and colleagues have identified fast-evolving regions in the human genome based on comparisons with a chimpanzee and a collection of other animal genomes. The study employed methods that estimate the rate of evolution of a sequence region on each branch of the phylogenetic species tree. The rates on each branch were compared with a likelihood ratio test to identify cases where the human branch evolved much faster than the background rate seen in the other species. The likelihood ratio is applied to eliminate cases where the region is rapidly evolving in general to make it possible to identify cases where only the human branch is faster. Using this approach, they found 49 "human-accelerated regions" (or HARS). For one of these regions, a noncoding RNA gene was identified and shown to be expressed in the brain. It has been proposed that this change may contribute to human intelligence; thus, the increased rate of change could be linked to increased fitness as intelligence developed.

6. ULTRACONSERVED ELEMENTS

In contrast to rapidly evolving regions, slowly evolving segments of the genome likely encode regions performing an essential function that permits few changes. Conveniently, these regions can be easy to identify, since they are stretches of DNA that have not changed substantially between organisms of increasingly distant evolutionary distance. To find these, simply applying parameters like sequence alignment windows of at least 100 base pairs (bp) and at least 90 percent identity should reveal loci that have changed little since divergence if the comparisons are between species with an average identity of less than 90 percent. Using this approach, researchers in several groups found long stretches of DNA significantly conserved among humans and other animals. The lengths and degree of conservation were much greater than would be expected given the time since divergence of the species, indicating the regions were under strong selection. One study found 481 *ultraconserved* segments between mouse, human, and rat that were 100 percent identical across at least 200 bp, and more than 5000 that were at least 100 bp long. The finding of nearly unchanged sequence regions for millions of years suggests strong functional significance for ultraconserved elements, but in many cases they did not appear to encode protein and only partially overlapped exons and introns of coding genes.

What functional role do these regions play? Through experimental manipulation and investigation of binding sites from the ultraconserved regions, it has been shown that some are functional regulatory elements; however, deletion of many megabases of DNA containing ultraconserved elements in mice showed no obvious fitness impact, indicating that the importance of these regions, despite evidence that they are under strong negative selection, still remains an enigma. Future experimental work and functional testing of these regions is needed to reveal the larger picture of how and why ultraconserved elements persist in genomes.

7. THE FUTURE OF COMPARATIVE GENOMICS

Comparative genomics utilizes evolutionary theory and computational techniques to provide a powerful tool that allows us to make sense of genome sequences. These approaches provide the means to generate experimentally testable hypotheses about likely functional elements, including the presence of regulatory regions, rapidly evolving regions that may be important for adaptation, and slowly evolving regions that may encode important but often not understood function. The ability to generate a sequence of whole genomes from nearly any organism is now within reach for most biologists, providing even more resources by which comparative genomics can shed light on the evolutionary history of the organisms and its features.

FURTHER READING

Bejerano, G., M. Pheasant, I. Makunin, S. Stephen, W. J. Kent, J. S. Mattick, and D. Haussler. 2004. Ultraconserved elements in the human genome. Science 304: 1321–1325. *One of the first descriptions of ultraconserved elements by comparison of the human genome with genomes of rodents, dogs, chicken, and fish.*

Girirajan, S., C. D. Campbell, and E. E. Eichler. 2011. Human copy number variation and complex genetic disease. Annual Review of Genetics 45: 203–226. *This article is a review of the role of copy number variation in human disease and challenges for the future.*

Margulies, E. H., and E. Birney. 2008. Approaches to comparative sequence analysis: Towards a functional view of vertebrate genomes. Nature Reviews Genetics 9: 303–313. *This article is a review of methods for comparative sequence analyses with an application to vertebrate genomes.*

Mu, J., R. A. Myers, H. Jiang, S. Liu, S. Ricklefs, M. Waisberg, K. Chotivanich, et al. 2010. Plasmodium falciparum genome-wide scans for positive selection, recombination hot spots and resistance to antimalarial drugs. Nature Reviews Genetics 42: 268–271. *This manuscript describes discovery of rapidly evolving regions in the genome of the causal agent of malaria and the ways these are linked to adaptation by drug resistance and creation of a variable cell surface to evade the human host immune system.*

Ohno, S. 1970. Evolution by Gene Duplication. New York: Springer-Verlag. *This book provided some of the first hypotheses concerning the ways in which gene duplication is a major driving force in evolution of novelty.*

Pollard, K. S., S. R. Salama, N. Lambert, M. A. Lambot, S. Coppens, J. S. Pedersen, S. Katzman, et al. 2006. An RNA gene expressed during cortical development evolved rapidly in humans. Nature 443: 167–172. *This paper builds on work to discover rapidly evolving regions in the human genome and demonstrates that one region is a noncoding protein expressed specifically in the brain.*

Wang, E. T., R. Sandberg, S. Luo, I. Khrebtukova, L. Zhang, C. Mayr, S. F. Kingsmore, G. P. Schroth, and C. B. Burge. 2008. Alternative isoform regulation in human tissue transcriptomes. Nature 456: 470–476. *This manuscript describes diversity in alternative splicing of human genes and discovery of sequence motifs through comparative genomics that can be correlated with the patterns of expression in different tissues.*

Yi, X., Y. Liang, E. Huerta-Sanchez, X. Jin, Z. X. Cuo, J. E. Pool, X. Xu, et al. 2010. Sequencing of 50 human exomes reveals adaptation to high altitude. Science 329: 75–78. *This manuscript provides evidence that several genes are under positive selection in Tibetan individuals who have adapted to life at high altitudes.*

V.4

Evolution of Sex Chromosomes

Doris Bachtrog

OUTLINE

1. Origin of sex chromosomes
2. Y (W)-chromosome degeneration
3. Dosage compensation of the X
4. Gene content evolution of sex chromosomes
5. Diversity of sex determination

Sex is universal among most groups of eukaryotes, yet a remarkable diversity of sex-determining (SD) mechanisms exist. The evolution of separate sexes has been accompanied by the acquisition of sex chromosomes many times across fungi, plants, and animals. Despite independent origins, sex chromosomes of many organisms share common features, reflecting similar evolutionary forces acting on them. Sex chromosomes are of particular interest to biologists for different reasons. First, sex chromosomes determine the gender of many species; thus they contain the gene ultimately responsible for sex determination. Second, Y or W chromosomes often lack recombination and undergo chromosome-wide degeneration. Finally, sex chromosomes show sex-biased transmission (that is, they spend different amounts of time in males and females), and they occur in different copy number in the two sexes (for example, in mammals, the X is diploid in females, but haploid in males). These features drive many unusual patterns of genome evolution, and sex chromosomes uniquely contribute to many evolutionary processes, such as speciation and evolutionary conflict.

GLOSSARY

Dosage Compensation. A process that balances expression of sex-linked and autosomal genes in the heterogametic sex.

Environmental Sex Determination (ESD). The process by which sex differentiation is determined by external environmental factors (e.g., temperature or pH) during offspring development.

Female Heterogamety. A sex chromosome system in which males have two identical sex chromosomes (two Z chromosomes) and females have two different sex chromosomes (a Z and a W chromosome).

Genotypic Sex Determination (GSD). The process by which sex differentiation is determined primarily by genetic factors, most commonly on the sex chromosomes.

Haplodiploidy. A sex-determination system in which sex is determined by ploidy level. Males are haploid and develop from unfertilized eggs, whereas females are diploid and develop from fertilized eggs. Females typically have control over fertilization.

Hemizygosity. A state in which only one copy of a gene is functioning in an otherwise-diploid organism (for example, the X chromosome in an XY male).

Heterogametic Sex. The sex with a pair of different sex chromosomes (e.g., male XY in mammals; female ZW in birds). The heterogametic sex produces two different types of gametes, one with one type of sex chromosome and one with the other.

Homogametic Sex. The sex with a pair of identical sex chromosomes (e.g., female XX in mammals; male ZZ in birds), therefore producing only gametes with one type of sex chromosome.

Male Heterogamety. A sex chromosome system in which females have two identical sex chromosomes (two X chromosomes) and males have two different sex chromosomes (an X and a Y chromosome).

Sex Determination. Any of various mechanisms in which the sex of an individual is determined.

Sex-Biased Expression. Expression of genes that show different absolute expression levels in males and females.

Sexually Antagonistic Selection. Selection that differs in direction between males and females; that is, an allele is favored in one sex and unfavored in the other.

1. ORIGIN OF SEX CHROMOSOMES

Male versus Female Heterogamety

In many taxa, including some plants and many animal species, sex is determined by a pair of sex chromosomes. The most familiar sex chromosome system is that of humans, in which females have two identical sex chromosomes, called X chromosomes (i.e., a female is XX), and males have two different chromosomes (an X and a Y chromosome). In some species, including birds, snakes, and butterflies, this pattern is reversed: females carry two different sex chromosomes (termed the Z and W chromosomes), and males carry two identical sex chromosomes (ZZ). The sex with the identical pair of sex chromosomes (the homogametic sex) produces gametes with only one type of sex chromosome, while the sex with different sex chromosomes (the heterogametic sex) produces gametes of two different types (figure 1).

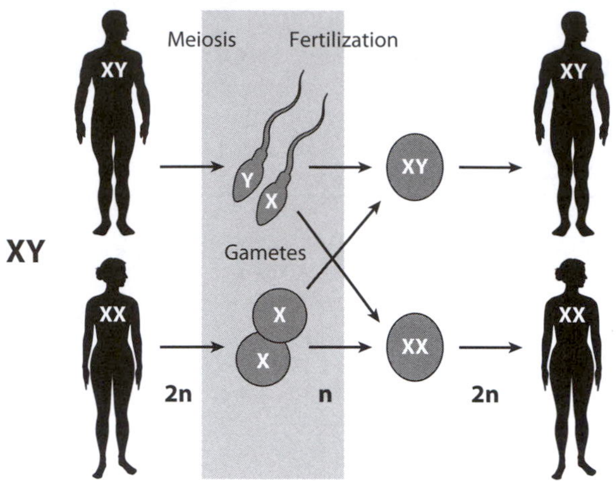

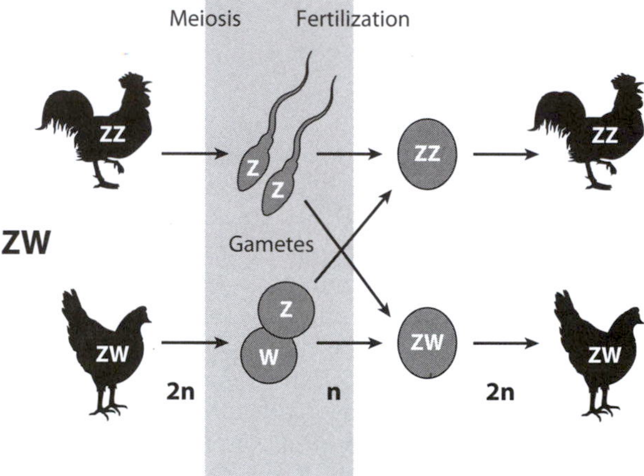

Figure 1. Male versus female heterogamety. In some species, including mammals and *Drosophila*, males have an X and a Y chromosome, while females have two X chromosomes. Here, females produce gametes with one type of sex chromosome (an X), and males produce gametes bearing either an X or a Y chromosome. In female-heterogametic species, such as birds or butterflies, males are the homogametic sex (producing Z-carrying gametes only), while females are heterogametic (producing Z- and W-carrying gametes).

Independent Origins of Sex Chromosomes

Sex chromosomes are phylogenetically widespread and originated many times independently in different organisms, including plants and animals. For example, the XY sex chromosome system shared by all mammals originated about 150 million years ago and is not homologous to the ZW sex chromosome system of birds. Among other vertebrates, ZW sex chromosomes evolved independently in snakes, and both ZW and XY systems have evolved multiple times in other reptiles, amphibians, and fish. Sex chromosomes are also widespread in invertebrates, which contain both male- and female-heterogametic systems. Many Diptera, including the fruit fly *Drosophila*, are XY, while Lepidoptera (moths and butterflies) are ZW. In addition, sex chromosomes arose multiple times in plant species. Examples of independently evolved sex chromosomes include those of papaya (XY), cannabis (XY), wild strawberry (ZW), and white campion (XY). Despite independent origins, sex chromosomes share many similar characteristics. Notably, the chromosome present only in the heterogametic sex (the Y or W chromosome) is often gene poor and has accumulated repetitive DNA. In contrast, the X and Z chromosomes superficially resemble autosomes, apart from their difference in copy number between males and females; however, X chromosomes have often evolved special regulatory mechanisms to compensate for this gene dose difference (see section 3 below). The similar appearance of sex chromosomes in different organisms suggests that similar selective pressures have acted to shape the evolution of sex chromosomes in different taxa. But where do sex chromosomes come from, and what evolutionary processes drive their evolution?

Sex Chromosomes Originate from Autosomes

Sex chromosomes arise from autosomes. The first step in the evolution of sex chromosomes is the acquisition of a sex-determining function on a proto-sex chromosome (genetic sex determination). Genetic sex determination can arise in a species that has no separate sexes (i.e., hermaphrodites, in which individuals carry both female and male reproductive functions), or in a species with separate sexes but where sex is determined by environmental cues (for example, temperature, such as in many

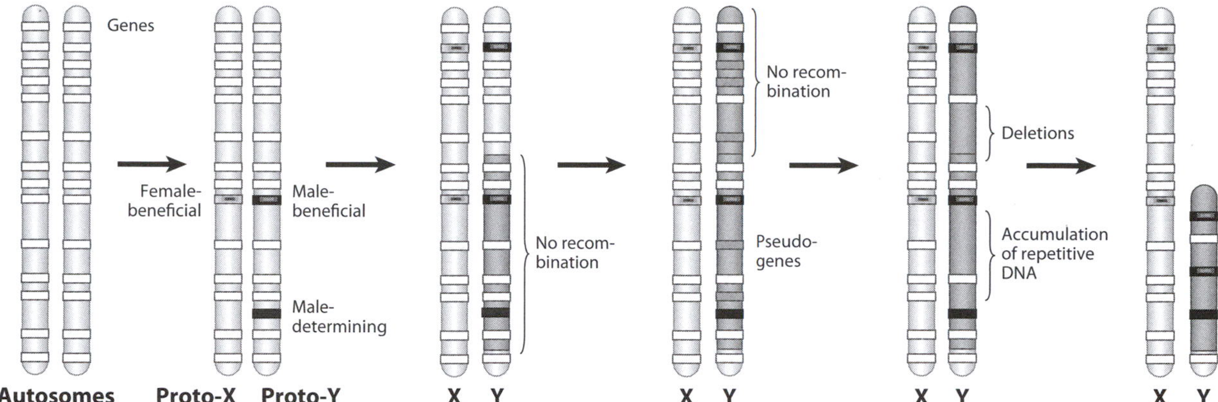

Figure 2. The origin of sex chromosomes (shown is the XY system). Sex chromosomes arise from a pair of initially identical autosomes with the same sets of homologous genes (white boxes). The first step in the evolution of heteromorphic sex chromosomes is the acquisition of a sex-determining gene (in black) on one of the autosomes. The accumulation of sexually antagonistic mutations close to the sex-determining region (shaded boxes) selects for a repression of recombination between nascent sex chromosomes (darker-shaded area along the chromosome). The nonrecombining Y chromosome loses most of its original genes (gray boxes indicate pseudogenes) and degenerates. This results in a pair of heteromorphic sex chromosomes. Different scenarios are possible. If the initial mutation is a female-determining mutation, a ZW system could evolve. In the case where sex determination arises in a hermaphroditic species, evolving separate sexes (and sex chromosomes) would actually require mutations at two loci (a male and a female sterility mutation).

turtles and crocodiles). If sex chromosomes arise in a hermaphroditic species, as is the case for most plants and many animals, a likely path for the evolution of separate sexes and sex chromosomes is that one proto-sex chromosome acquires a male-sterility mutation and the other proto-sex chromosome acquires a female-sterility mutation. Depending on the dominance relationship of these mutations, this would generate a proto-X/proto-Y or proto-Z/proto-W chromosome and the population would transition through a stage in which both hermaphrodites and females, or hermaphrodites and males, are present. There is then strong selection to restrict recombination between the male-sterility and female-sterility mutations on the different chromosomes, since a recombination event could place both mutations on the same chromosome and would generate a sterile individual. In a species with environmental sex determination, a dominant male-determining mutation on a former autosome would create a proto-Y chromosome, while a dominant female-determining mutation would result in the origination of a proto-W chromosome (figure 2).

The *SRY* gene in mammals is such a dominant master sex-determining locus and is found on the Y chromosome; it initiates the differentiation of the developing embryo into a male. Thus, acquisition of the *SRY* gene on an autosome likely triggered the formation of a proto-Y chromosome in mammals, and its former homologue that lacked the *SRY* gene became a proto-X chromosome. On autosomes, recombination homogenizes the gene content between the homologous paternal and maternal chromosomes and shuffles segregating mutations across different chromosomal backgrounds. To allow nascent sex chromosomes to evolve independently, it is necessary that recombination between the proto-Y and proto-X chromosomes becomes suppressed, allowing each chromosome to accumulate independent mutations; the X and the Y can then diverge from each other in sequence and function. But why should recombination become restricted on a pair of proto-sex chromosomes beyond the sex-determining region?

Sexual Antagonism Drives Recombination Suppression

The driving force for the evolution of restricted recombination between proto-sex chromosomes is generally thought to result from sexually antagonistic alleles accumulating close to the sex-determining region. *Sexual antagonism* refers to a situation in which genes cause opposing fitness effects in the two sexes. Males and females in many species differ in their morphology, behavior, and physiology; however, in the absence of sex chromosomes, they share identical sets of genes. It is possible that a large number of genes or mutations may have opposing fitness benefits in the two sexes; specific mutations can be good for one sex, but bad for the other. For example, in guppies, females prefer males with bright, colorful ornaments; despite increased predation risk, such males have a mating advantage over less brightly colored ones. Less colorful females, however, maximize their fitness by avoiding predation. Thus, a mutation that causes a brightly colored spot is selected for in males, but selected against in females. Such a sexually antagonistic mutation can become established in the population only if the benefit to males outweighs its harmful effects in females; however, if this mutation arises in close proximity to the male-determining region, it will find itself more often in males, the sex in which it is beneficial, and it can become established more easily in the population. Thus, sexually antagonistic mutations are expected to accumulate close to sex-determining genes. Indeed, several color genes in guppies are closely linked to a male-determining gene. Sporadic recombination events between the male-beneficial mutations and the male-determining region would transfer these color genes onto the X chromosome, and they would be expressed in females. Thus, there is selection to eliminate recombination between the sexually antagonistic alleles and the sex-determining region, to ensure that such genes are restricted to the favored sex. Recurrent accumulation of sexually antagonistic mutations on the proto-sex chromosomes can select for the repression of recombination over most or all of the length of the proto-sex chromosomes (figure 2). A consequence of the restriction of recombination between the nascent sex chromosomes is that the heterogametic sex chromosome (the Y or W) is completely sheltered from recombination, while the other sex chromosome (X or Z) can still recombine in the homogametic sex. The lack of recombination on the Y or W chromosome results in its degeneration.

2. Y (W)-CHROMOSOME DEGENERATION

A Lack of Recombination Causes Y Degeneration

The most dominant characteristic of heteromorphic sex chromosomes is the lack of functional genes on the Y or W chromosome. Y (or W) chromosomes degenerate because of their lack of recombination. Autosomes and the X in female mammals (or Z in male birds) always exist in two copies, a paternal and a maternal one, and undergo meiotic recombination. As explained below, this enables selection to efficiently purge deleterious mutations, and allows the X or Z to maintain its original gene content. Y (or W) chromosomes, in contrast, completely lack meiotic recombination for most of their length. The efficacy of natural selection is reduced on a nonrecombining chromosome, and is the basis for the degeneration of the Y (or W) chromosome (figure 3). Natural populations are subject to recurrent mutations. Some of these mutations increase the fitness of their carrier, for example, by increasing survivorship, or increasing fertility (beneficial mutations); most mutations, however, are detrimental and reduce the function of a well-adapted gene (deleterious mutations). The general role of natural selection is to incorporate beneficial mutations (i.e., adaptation) and remove deleterious ones (purifying selection; see chapter V.1). On a recombining chromosome, natural selection can act on individual mutations by reshuffling mutations and putting them on different genomic backgrounds. In contrast, in the absence of recombination, new gene combinations cannot be generated and selection must act on the entire chromosome. That is, different selected mutations on a nonrecombining chromosome can interfere with each other, thereby reducing the efficacy of natural selection. This reduction can lead either to an accumulation of deleterious mutations, by a process known as Muller's ratchet or by genetic hitchhiking with beneficial mutations, or to reduced rates of adaptive evolution (the ruby in the rubbish model; see figure 3 for a description of these processes). Under the Muller's ratchet and genetic hitchhiking model, Y-linked genes continuously decrease in fitness relative to their X homologues, as a result of the accumulation of deleterious mutations. The ruby in the rubbish model, in contrast, states that reduced fitness of Y-linked genes relative to the X instead results from a lower rate of incorporation of beneficial mutations. Under both scenarios, dysfunctional Y-linked alleles will eventually become silenced and lost from the degenerating Y, and in the long run, only a few genes remain on old Y chromosomes, if any.

Old Y Chromosomes

The degeneration of Y (or W)-linked functional genes is associated with an accumulation of repetitive DNA, such as transposable elements (TEs) or satellite DNA (see chapter V.2). In parts of the genome that recombine, such insertions can normally be efficiently purged. On the Y (W), however, repetitive DNA can accumulate as a result of the reduced efficacy of natural selection. In addition, as more and more genes degenerate, the chance that a new TE inserts into a functional gene decreases, and TEs can start to accumulate neutrally. Thus, the size of an evolving Y chromosome can increase dramatically because of the accumulation of repetitive DNA; however, as the gene density becomes lower and lower, large deletions can occur, and the Y or W chromosome can shrink in size and carry fewer and fewer genes (figure 2). Eventually, the Y (W) may carry only the sex determining gene and a few other genes beneficial to the heterogametic sex. Ultimately, a species might evolve an alternative sex determination signal, for example, the ratio of X to autosomes can determine sex (as is the case in *Drosophila*). In such a situation, the Y chromosome

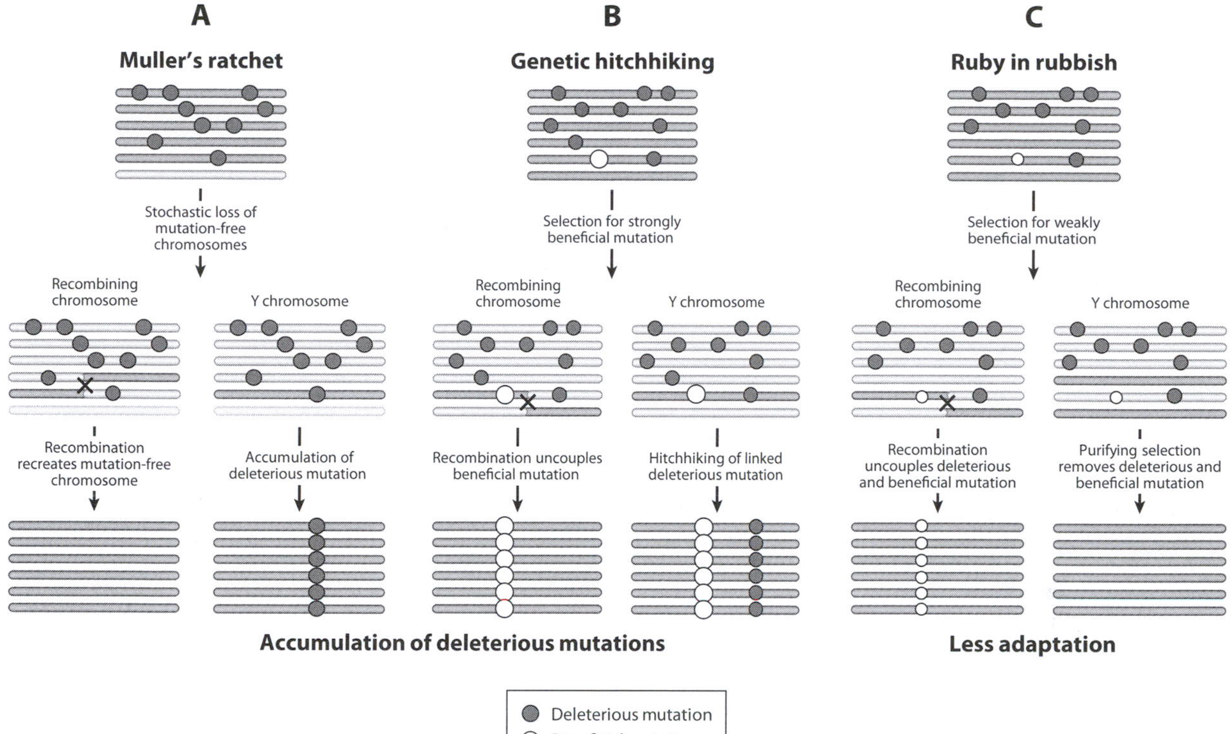

Figure 3. Models of Y degeneration. Y/W chromosomes degenerate since they accumulate deleterious mutations at ancestral genes. Two main processes have been proposed to explain this accumulation: Muller's ratchet, the irreversible accumulation of deleterious mutations (gray circles) in a finite population, and genetic hitchhiking of deleterious mutations together with beneficial alleles (white circles). Y or W chromosomes may also undergo less adaptive evolution, as a result of linkage of beneficial alleles with deleterious mutations (the ruby in the rubbish model). (A) Muller's ratchet. Mutation-free chromosomes can be lost in finite populations as a result of stochastic effects. Recombination allows the re-creation of mutation-free chromosomes, whereas this loss is irreversible on a nonrecombining Y chromosome. (B) Genetic hitchhiking. Newly arising beneficial mutations might occur on a chromosome that also contains deleterious mutations. Recombination enables the beneficial allele to disassociate from the deleterious mutation, while the fixation of the beneficial mutation on a nonrecombining Y chromosome will drag along the deleterious mutation. (C) Ruby in the rubbish. Beneficial mutations of weak effect linked to more strongly deleterious mutations will be eliminated by purifying selection on a nonrecombining Y chromosome, since such chromosomes will have no net fitness advantage.

can be lost entirely, and a species becomes XX (females) and X0 (i.e., only one X chromosome and no Y in males, as has happened in *Caenorhabditis elegans*; see section 5 below).

3. DOSAGE COMPENSATION OF THE X

Gene Dose Deficiency in Heterogametic Sex

The amount of gene product correlates with the number of gene copies for a gene. Y degeneration creates the problem of reduced gene dose in males; genes that degenerate from the Y are expressed at a lower level in males. In many gene networks, however, the dose of genes is important, and gene dose imbalances may have negative fitness consequences. Thus, many organisms have evolved compensatory mechanisms to counterbalance this gene dose deficiency in the heterogametic sex, and different species have found different strategies to achieve dosage compensation (i.e., the balancing of gene product of sex-linked genes in the heterogametic sex). The primary selective pressure driving the evolution of dosage compensation is to balance expression levels between autosomal and sex-linked genes in the heterogametic sex, which has too little gene product for genes that have been lost from the Y chromosome. A by-product of the acquisition of dosage compensation is that expression levels for X-linked genes become similar between the sexes; that is, dosage compensation equalizes expression levels of X-linked genes between the sexes. Note that this is a consequence of selection for dosage compensation in males, and not the primary selective pressure driving it.

Different organisms with independently evolved Y chromosomes have found vastly different evolutionary solutions to achieve dosage compensation.

Different Paths to Dosage Compensation

The most direct way to compensate for a deficiency in X chromosomal gene product is to upregulate X-linked genes specifically in the heterogametic sex. This requires a compensatory mechanism that is male-specific and exclusively recognizes and targets X-linked genes. *Drosophila* has evolved dosage compensation along this path, using a male-specific ribonucleoprotein complex (a complex that consists of RNA and proteins) that specifically targets the X chromosome and results in transcriptional upregulation of the X (figure 4). This mechanism restores the balance between X and autosomal gene product in males, without changing expression levels of the X in females, and it results in similar levels of X-linked gene product in both sexes. A more indirect path to evolve dosage compensation is followed in mammals and *C. elegans*. Here, it is believed that the initial upregulation of X-linked genes was not sex limited. Instead, selection in males to upregulate the X also resulted in increased X expression in females. Thus, while a general upregulation of X-linked genes restores the gene dose problem in males, females now produce too much gene product from their two X chromosomes relative to their autosomal expression; this in turn creates selective pressure in females to evolve a compensatory mechanism to reduce elevated expression levels of the X. In mammals, this is achieved by females completely inactivating one of their two X chromosomes using noncoding RNAs. *C. elegans* XX hermaphrodites, on the other hand, approximately halve expression from each of their two active X chromosomes. Thus, dosage compensation in these organisms is achieved through a two-step process: upregulation of the X in both sexes, followed by downregulation or inactivation of the X in females. Again, this results in balanced X-autosome expression in males (and females), and similar levels of X-linked expression in both sexes.

Lack of Dosage Compensation in Some Systems

Not all taxa appear to have evolved dosage compensation. This could be a consequence of recently evolved sex chromosomes that have not yet had time to have acquired dosage compensation (as in some fish species), or because the X chromosome contains only few genes, none of which is dosage sensitive. However, recent empirical data show that birds, butterflies, and schistostomes (a trematode worm) all lack dosage compensation; these taxa are all considered to have "old" and large sex chromosomes. A common feature of these three groups is that they all have female heterogametic sex determination. W chromosomes closely resemble Y chromosomes; that is, they are degenerate with few active genes. Thus, in ZW systems, females have reduced expression of Z-linked genes, and dosage compensation should function to upregulate Z genes in females, to compensate for their degenerate W chromosome. At present, it is unclear whether the lack of dosage compensation in ZW species is a general feature of these systems, and what the evolutionary explanation for this might be.

4. GENE CONTENT EVOLUTION OF SEX CHROMOSOMES

Differential Accumulation of Sexually Antagonistic Mutations

As discussed above, the driving force in the evolution of restricted recombination between the proto-sex chromosomes is sexually antagonistic selection. Males and females share the same genome, but they often differ in their morphology, behavior, and physiology. This implies that many genes may have different optimal functions in males and females. Sex-biased transmission and hemizygosity shape patterns of gene content evolution of sex chromosomes. In particular, Y chromosomes are limited to males and W chromosomes to females, whereas X chromosomes are transmitted more often through females (females have two X chromosomes, and males, only one; thus, an X chromosome is found two-thirds of the time in a female, and only one-third of the time in a male) and Z chromosomes show male-biased transmission. In general, if a chromosome spends more time in one sex, it should be better adapted to the specific needs of that sex. Autosomes spend equal amounts of time in males and females and are thus exposed to selection equally in the two sexes, while sex-biased transmission of the sex chromosomes implies that they might accumulate sexually antagonistic genes. In addition, X and Z chromosomes are hemizygous in the heterogametic sex, greatly influencing the fixation probability of recessive mutations, especially for mutations favoring the heterogametic sex. Together, these peculiarities uniquely affect the evolutionary dynamics of sex chromosomes, making them a hot spot for the accumulation of sexually antagonistic mutations.

Gene Content of Y Chromosome

Y chromosomes are male limited, and thus selected only in males, whereas W chromosomes are under selection only in females; the Y should therefore accumulate male-beneficial and the W female-beneficial genes. Indeed,

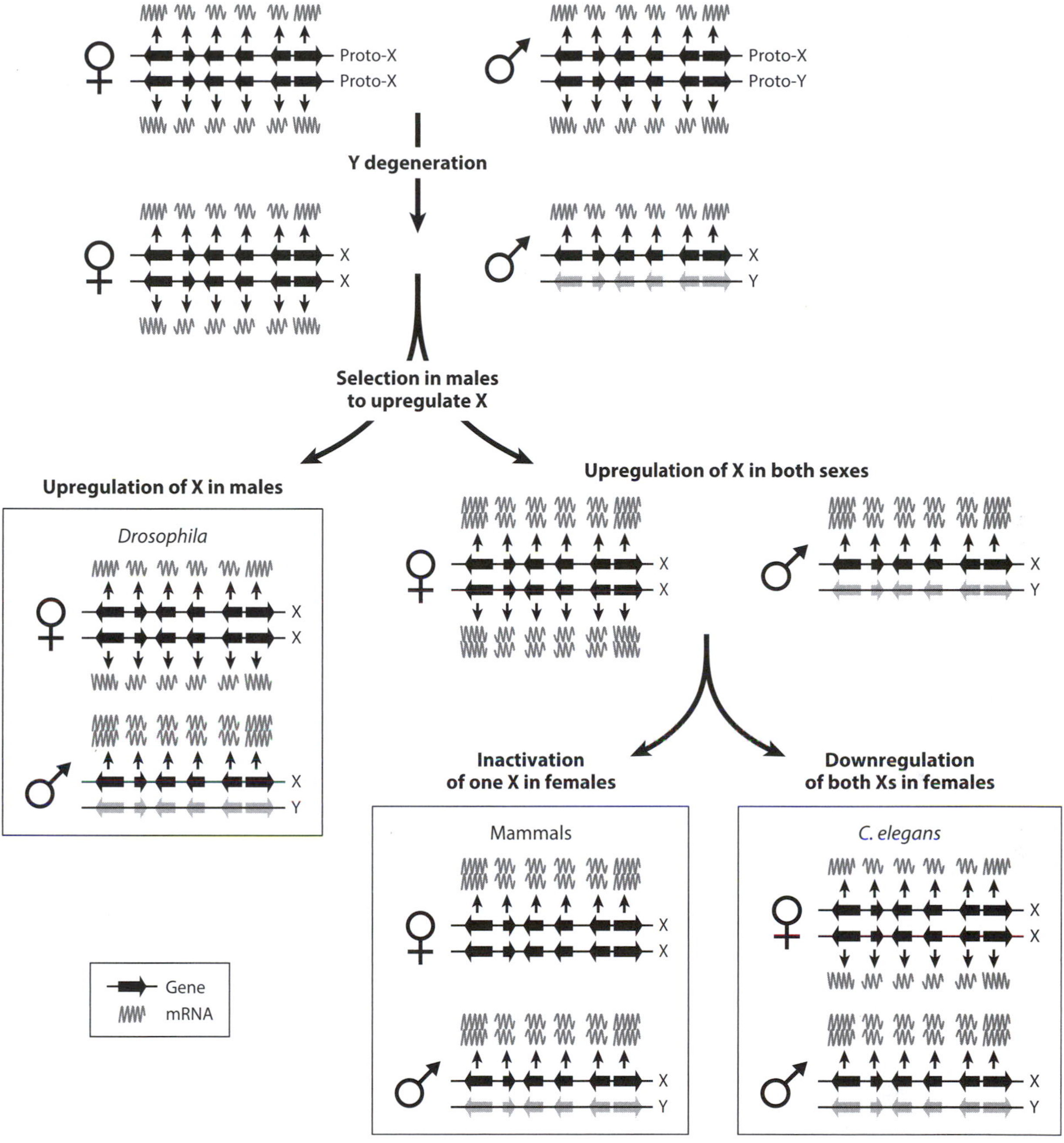

Figure 4. The evolution of dosage compensation. Y degeneration results in reduced gene dose of X-linked genes, selecting for an upregulation of X-linked genes in males. This upregulation may be male specific (as in *Drosophila*). Alternatively, this initial upregulation may not be sex specific, thus resulting in selective pressure to downregulate the X in females. Mammals completely inactivate one of their X chromosomes, whereas C. *elegans* halve expression from both their X chromosomes in hermaphrodites.

most of the genes on the Y chromosome of *Drosophila* and mammals, which are the best-studied sex chromosome systems to date, have testis-specific expression. Also, *Drosophila* males that lack a Y chromosome (which are X0) are viable but sterile. Thus, the *Drosophila* Y contains no genes necessary for viability, but harbors genes required for male fertility. Extensive sequence analysis has revealed a peculiar structure of genes present on the human Y chromosome. In particular, most testis-specific genes on the human Y are contained within ampliconic and palindromic structures (i.e., present in multiple copies in opposite orientations). The gene copies in these palindromes frequently undergo Y-to-Y gene conversion (a recombination-related process), which can be an efficient way to prevent degeneration of Y-linked gene functions. Much less is known about the gene content of W chromosomes, but similar ampliconic gene families have been detected for the chicken W chromosome. This suggests that gene amplification and intrachromosomal gene conversion may be an important way to retard degeneration and preserve gene function on a nonrecombining chromosome.

Sex-Biased Transmission versus Hemizygosity on X and Z

Expectations for gene content evolution are straightforward for Y and W chromosomes, but the situation is more complex for the X and Z, since sex-biased transmission and hemizygosity may work in opposite directions. On one hand, the X shows female-biased transmission, which favors the accumulation of female-beneficial genes on the X, since X-linked genes are more often under selection in females than in males. Conversely, Z chromosomes are more often transmitted through males and male-beneficial genes should accumulate; however, hemizygosity of the X and Z chromosomes may favor an accumulation of mutations beneficial to the hemizygous sex (i.e., male-beneficial mutations on the X, and female-beneficial mutations on the Z), depending on their dominance coefficient. Many beneficial mutations, including those with sexually antagonistic fitness effects, may be recessive. Selection is more effective in incorporating recessive mutations on sex chromosomes. In particular, hemizygosity of the X in males implies that recessive, male-beneficial mutations can become incorporated more easily on the X relative to autosomes, while hemizygosity of the Z in females favors the accumulation of recessive female-beneficial alleles. Thus, X chromosomes may simultaneously accumulate dominant female-beneficial mutations (due to sex-biased transition) and recessive male-beneficial mutations (due to hemizygosity of the X in males), while dominant male-beneficial and recessive female-beneficial mutations may accumulate on Z chromosomes.

Gene Content of X and Z Chromosomes

Empirical patterns of gene content evolution are complex on X or Z chromosomes, as expected given contrasting selective forces. In particular, genome-wide expression studies have shown that genes with sex-biased expression (that is, genes expressed at different levels in males and females) show a nonrandom distribution on sex chromosomes. Different expression levels of genes in the two sexes could be a consequence of the resolution of sexual antagonism. By reducing the expression level of a gene in the sex that suffers a selective disadvantage from the phenotype encoded by that gene, fitness of that sex would increase; thus, genes with sex-biased expression may represent genes that are (or were in the past) under sexually antagonistic selection. Genes with male-biased expression are depleted from the X chromosomes of *Drosophila*, while female-biased genes are more common on the X, relative to autosomes. In mammals, female-biased genes expressed in ovaries and placenta are also overrepresented on the X. This suggests that the X chromosome in these species has become "feminized." Such a pattern is consistent with sexually antagonistic mutations being at least partially dominant, and with female-beneficial genes accumulating on the X, which is more often transmitted through females.

Male-biased genes expressed during early spermatogenesis also appear to be in excess on the X of mammals, which suggests that recessive, male-beneficial genes also accumulate on the X. Studies of the genomic distribution of sex-biased genes in ZW systems found similar patterns. There is a deficit of female-biased genes on the Z chromosome of chickens, and testis-specific genes appear enriched on the Z of *Bombyx*. Again, this is consistent with an accumulation of dominant, male-beneficial mutations and a removal of female-beneficial mutations on the Z chromosome.

Recent research, however, has suggested that the biased distribution of male or female genes on the X also depends on the evolutionary age of the gene, and it is unclear how good sex-biased expression really is as a proxy for identifying past sexual conflict. Thus, while genes with different functions and expression patterns clearly show biased distributions on X chromosomes, the underlying causes for this are not yet clear. It is likely that a complex suite of evolutionary forces, such as the transcriptional inactivation of the X chromosome during spermatogenesis in some species, affects the distribution of sex-biased genes in the genome.

5. DIVERSITY OF SEX DETERMINATION

Mechanisms of Sex Determination

Sex is a universal feature of eukaryotic organisms (see chapter IV.4), and sex determination is a vital biological process: imprecise sex determination leads to the production of faulty, intersexual individuals, and consequently to reproductive impairment. Sex determination must therefore be subject to strong selective pressures; nevertheless, sex-determination mechanisms can undergo rapid evolutionary change, and tremendous variation in sex-determination mechanisms exists, not only within major phylogenetic groups, but also occasionally even within species. Although the number and range of modes of sex determination is large, they can be broadly divided into two major categories: genetic sex determination (GSD) and environmental sex determination (ESD). Sex can be determined based on the genetic makeup of the fertilized zygote (GSD), or sexual development can be under the control of environmental cues (ESD); note, however, that this simple dichotomy is somewhat misleading, and sex determination is always controlled by genes. With ESD, an initial external agent is used to stimulate either male or female development, but the pathway for sexual differentiation of course involves genes. In addition, in some species both mechanisms exist simultaneously, and GSD systems can sometimes be overridden by environmental cues, such as bacterial infections. The principal exception to this evolutionary instability of sex-determining mechanisms is the case of systems with highly evolved, heteromorphic sex chromosomes, since transitions involving a degenerate Y or a dosage-compensated X are difficult. In general, however, the evolution of sex-determination mechanisms presents a conundrum: the trait itself is under strong selective control, yet it can undergo frequent and rapid evolutionary change. The reasons for this diversity of mechanisms and the forces driving transitions between different systems are unclear.

Environmental Sex Determination

With ESD, the primary signal to trigger development into either a male or a female is given by environmental cues. Environmentally derived signals utilized for sex determination include temperature of egg incubation during a critical period, which is employed by several reptiles such as crocodiles and turtles, as well as by some invertebrates, including the gall midge *Heteropeza*, and the fungus gnat *Sciara*. Other environmental clues used for sex determination include photoperiod, as utilized by some amphipods that develop into males early during the season, and into females later. ESD systems also include species that practice sex change: in some snail species such as slipper limpets, young mobile adults are males and later change into sessile females. In some fish species, sex change is induced by social organization. Anemone fish, for example, form social units with a size-based dominance hierarchy composed of a breeding pair and several nonbreeders. If the female of a group—which is the largest individual—dies, the male grows and changes sex to become the breeding female, while the largest nonbreeder grows and becomes the breeding male. In the echiuran worm *Bonellia viridis*, the vast majority of sexually undifferentiated larvae metamorphose into dwarf males that live inside the female when exposed to females, but differentiate into females when developing in the absence of other females. In some arthropods, sex is determined by infection with certain bacteria. *Wolbachia*, for example, infects a high proportion of insect species and can result in feminization of infected males. In this case, an environmental stimulus (*Wolbachia* infection) overrides a GSD system.

Genetic Sex Determination: Sex Chromosomes

The mechanisms of GSD can vary enormously, with clearly distinguishable sex chromosomes being only one possibility. As mentioned above, chromosomal sex determination includes systems in which males are the heterogametic sex (XX/XY sex chromosomes), such as found in mammals, many insects and invertebrates, and several plant species. Alternatively, females can be the heterogametic sex (ZW/ZW), and this system is found in birds, some reptiles and amphibians, some insects, and invertebrates. In some species, the Y chromosome has been entirely lost; here females have two X chromosomes (XX) and males, only a single X (X0), as in a number of insects, including some crickets, grasshoppers, and cockroaches, as well as in *C. elegans*. The mechanism by which sex chromosomes trigger sex determination functions either through a dominant male- or female-determining gene on the Y or W chromosome, or through the ratio of sex chromosomes to autosomes (the X:A ratio). Mammals have a dominant masculinizing gene on the Y (the *SRY* gene), whereas *Drosophila* and X0 species use the X:A ratio for sex determination. Some species contain multiple sex chromosomes instead of a single pair of sex chromosomes; for example, some invertebrates, fish, and mammals contain two X chromosomes (males are X_1X_2Y and females are $X_1X_1X_2X_2$). A very peculiar sex chromosome configuration is found in platypuses, which have ten sex chromosomes. Males have five X and five Y chromosomes ($X_1X_2X_3X_4X_5Y_1Y_2Y_3Y_4Y_5$), while females have ten X chromosomes ($X_1X_1X_2X_2X_3X_3X_4X_4X_5X_5$). In other species, sex is determined by a single locus (genic sex determination), but they have no visually distinguishable sex chromosomes (e.g., phorid flies, several fish species).

Genetic Sex Determination: Other Systems

Another familiar mode of sex determination is *haplodiploidy* (1N-2N). Here, unfertilized eggs develop into haploid individuals that are males, and fertilized eggs develop into diploid females. This mechanism is utilized in hymenoptera (bees, ants, and wasps) and some mites and beetles. Other, rare mechanisms of genetic sex determination include paternal genome elimination, X chromosome elimination, and monogeny. In some species of scale insects, for example, both sexes develop from fertilized eggs and all embryos are initially diploid. However, during early development of the male offspring, the paternal half of the genome is either deactivated through heterochromatinization or completely eliminated (paternal genome elimination; i.e., males are either functionally or actually haploid). In other systems, such as pea aphids, only the X chromosome is eliminated to form X0 males (while females are XX), but the autosomes remain diploid in males (X chromosome elimination). Gall midges (Diptera) reproduce by a mechanism in which all offspring of each individual female are either exclusively male or exclusively female, and is a result of a single maternal effect autosomal gene (monogeny).

Thus, despite the antiquity of the two sexes, a vast diversity of sex-determination mechanisms exist, and these mechanisms evolve rapidly in some lineages. The evolutionary forces that drive rapid change in sex-determination pathways remain largely a mystery but are an area of active research.

FURTHER READING

Bachtrog, D. 2006. A dynamic view of sex chromosome evolution. Current Opinion in Genetics and Development 16(6): 578–585. *Focuses on the dynamics of processes shaping sex chromosomes over evolutionary time.*

Bachtrog, D., M. Kirkpatrick, J. E. Mank, S. F. McDaniel, J. C. Pires, W. Rice, and N. Valenzuela. 2011. Are all sex chromosomes created equal? Trends in Genetics 27(9): 350–357. *Contrasts various sex chromosome systems, including female and male heterogametic systems, and sex chromosomes in haploid organisms.*

Charlesworth, B., and D. Charlesworth. 2000. The degeneration of Y chromosomes. Philosophical Transactions of the Royal Society B 355(1403): 1563–1572. *A comprehensive overview of the evolutionary processes driving Y degeneration.*

Charlesworth, D., and J. E. Mank. 2010. The birds and the bees and the flowers and the trees: Lessons from genetic mapping of sex determination in plants and animals. Genetics 186(1): 9–31. *An overview of the diversity of sex chromosome systems.*

Lahn, B. T., N. M. Pearson, and K. Jegalian. 2001. The human Y chromosome, in the light of evolution. Nature Reviews Genetics 2(3): 207–216. *An overview of the evolution and gene content of the human Y chromosome.*

Marshall Graves, J. A. 2008. Weird animal genomes and the evolution of vertebrate sex and sex chromosomes. Annual Review of Genetics 42: 565–586. *Describes the diversity of sex chromosome systems among vertebrates.*

Ming, R., A. Bendahmane, and S. S. Renner. 2011. Sex chromosomes in land plants. Annual Review of Plant Biology 62: 485–514. *An overview of sex chromosomes in land plants.*

Vicoso, B., and D. Bachtrog. 2009. Progress and prospects toward our understanding of the evolution of dosage compensation. Chromosome Research 7(5): 585–602. *An overview of current understanding of the process of dosage compensation and its evolution in model species.*

Vicoso, B., and B. Charlesworth. 2006. Evolution on the X chromosome: Unusual patterns and processes. Nature Reviews Genetics 7(8): 645–653. *A comprehensive overview of the evolutionary processes operating on X chromosomes.*

V.5

Gene Duplication

Jianzhi Zhang

OUTLINE

1. Mechanisms of gene duplication
2. Fixation of duplicate genes
3. Pseudogenization after duplication
4. Stable retention of duplicate genes
5. Rate of gene duplication
6. Determinants of gene duplicability
7. Functional redundancy among duplicate genes
8. Functional diversification of duplicate genes
9. Future directions in the study of gene duplication

The number of genes in a genome varies by two orders of magnitude across cellular organisms. A primary mechanism underlying this variation is gene duplication, which provides raw genetic materials from which new genes and new gene functions arise. As with other types of genetic mutations, gene duplication first occurs in an individual organism, and its population genetic fate depends on its fitness effect. Even after a duplicate gene is fixed in a population, it will degenerate into a pseudogene unless its presence is beneficial to the organism. Stably retained duplicate genes are quite common in almost all eukaryotic genomes examined. These duplicates form gene families, whose members typically have similar but nonidentical functions or expression patterns. This chapter first describes the processes through which duplicate genes are generated, fixed, and stably retained. It then discusses the rate of gene duplication and factors influencing this rate. Finally, it examines the functional redundancy and divergence among duplicate genes.

GLOSSARY

Alternative Splicing. An RNA splicing process by which the exons of the RNA produced by transcription of a gene are reconnected in multiple ways. Alternative splicing leads to the production of multiple different proteins from a single gene.

Aneuploidy. Loss or gain of one to several chromosomes but not a complete set.

Concerted Evolution. An evolutionary process that explains the observation that individual members of a gene family within one species are more similar to one another than to members of the same gene family in other species, even though these members were generated prior to the divergence of the species. Concerted evolution is usually attributed to frequent gene conversions among gene family members within species.

Functional Redundancy. The condition in which two paralogous genes perform the same function.

Gene Conversion. An event in genetic recombination that converts one DNA sequence to another. It can homogenize the sequences of duplicate genes of the same species.

Neofunctionalization. Acquisition of a new function that may be qualitatively or quantitatively different from the previous function.

Paralogous Genes. Genes that are related through duplication.

Pseudogene. A dysfunctional relative of known genes that has lost its protein-coding ability or is no longer expressed.

Retroposition. Integration into the genomic DNA of a sequence derived from reverse transcription of RNA.

Subfunctionalization. Division of multiple functions of a progenitor gene into its daughter genes such that the total functions of the daughter genes are the same as those of the progenitor gene.

Unequal Crossing-over. Crossing-over between homologous chromosomes that are not precisely paired, resulting in nonreciprocal exchange of material and chromosomes of unequal length.

1. MECHANISMS OF GENE DUPLICATION

Gene duplication refers to the duplication of a segment of DNA that contains one or more genes. In 1936, Calvin Bridges reported the first case of gene duplication, observed in mutant fruit flies (*Drosophila melanogaster*) exhibiting extreme reduction in eye size. Through observation of the polytene chromosomes from the salivary glands of *D. melanogaster* larvae, Bridges showed that the mutant phenotype was caused by the doubling of a small segment of the X chromosome.

Gene duplication occurs by one of the three general mutational mechanisms: unequal crossing-over, retroposition, and chromosomal (or genome) duplication. Unequal crossing-over refers to crossing-over between homologous chromosomes that are not precisely paired, resulting in nonreciprocal exchange of material and chromosomes of unequal length (figure 1A). That is, one of the resultant chromosomes contains an extra copy of a chromosomal segment, while the other loses this segment. This mechanism typically generates tandem gene duplicates that are arrayed next to each other along the chromosome. Depending on the position of crossing-over, the duplicated region may contain part of a gene, an entire gene, or several genes.

Retroposition occurs through a completely different mechanism: a messenger RNA (mRNA) of a gene is retrotranscribed to complementary DNA (cDNA) and is then inserted back into the genome, resulting in an extra gene copy (figure 1B). Genes duplicated by this mechanism are also known as *retroduplicates.* Because they arise from mRNAs, retroduplicates lack introns and regulatory sequences such as the promoter, and often contain poly A tracts at the end. Furthermore, in contrast to genes duplicated by unequal crossing-over, a retroduplicate is unlinked to its mother gene, because the insertion of cDNA into the genome is more or less random. It is also impossible to have blocks of genes duplicated together by retroposition. Because retroposition must occur in the germ line to be heritable, only genes expressed in the germ line are subject to heritable retroposition.

Chromosomal duplication refers to the phenomenon whereby one to several (but not all) chromosomes in a genome are duplicated. It occurs by nondisjunction of homologous chromosomes during meiosis and leads to aneuploidy (figure 1C). By contrast, *genome duplication* refers to the duplication of the entire genome, and is also known as *polyploidization. Autopolyploids* are polyploids with multiple chromosome sets derived from a single species. They arise from a spontaneous, naturally occurring genome doubling (figure 1D), or the fusion of unreduced gametes. In comparison, *allopolyploids* are those with chromosomes originated from different species, as a result of doubling of chromosome number in an F1 hybrid of the two species.

2. FIXATION OF DUPLICATE GENES

The probability that a duplicate gene gets fixed (i.e., is found in every individual) in a population is determined by the fitness effect of duplication, as for other types of mutations. Depending on the type of gene duplication and the genes involved, gene duplication may be deleterious, beneficial, or neutral to the organism in which the duplication occurs (see chapter V.1).

Gene duplication could be deleterious for several reasons. First, because gene transcription and translation costs energy, gene duplication may impose a fitness cost. Second, gene duplication may break the often-sensitive balance in dosage (i.e., the precise amount of RNA or protein relative to other genes) that is required for certain genes. This is why *trisomy*, the presence of three instead of two copies of a chromosome in a diploid individual, is usually deleterious. In humans, all autosomal trisomies are lethal, except for trisomy 21, which causes Down syndrome and, less often, trisomy 13 (Patau syndrome) and 18 (Edwards syndrome). Third, a retroduplicate may be inserted into a gene or a functional element, causing a deleterious effect.

Gene duplication, however, may be beneficial when extra gene product is useful to the organism. For example, cells need a large number of ribosomes for rapid protein synthesis; thus duplication of ribosomal protein and RNA (rRNA) genes is typically beneficial. A recent study found that duplication of the human salivary amylase gene is advantageous in certain human populations with high starch diets, apparently because the increased amount of amylase helps digest starch.

Gene duplication can also be neutral or nearly neutral. For instance, duplication of a gene expressed at low levels imposes very little energy cost and hence virtually no fitness cost unless it has other effects. Most retroduplicates are not expressed because of the lack of promoters; thus they have effectively no fitness effect if they do not interfere with the expression or function of other genes.

As is true for other types of mutations, most duplicate genes do not get fixed in a population; however, among those duplicates that did get fixed and are observed today, a major question remains: were they fixed mostly by positive selection or by random genetic drift? Some authors proposed that positive selection for enhanced gene dosage is the primary mechanism for duplicate gene fixation, but available genomic data do not seem to support this view, although positive selection is clearly involved in a few cases.

A large number of duplicate genes are segregating within populations in their paths to fixation or loss. These

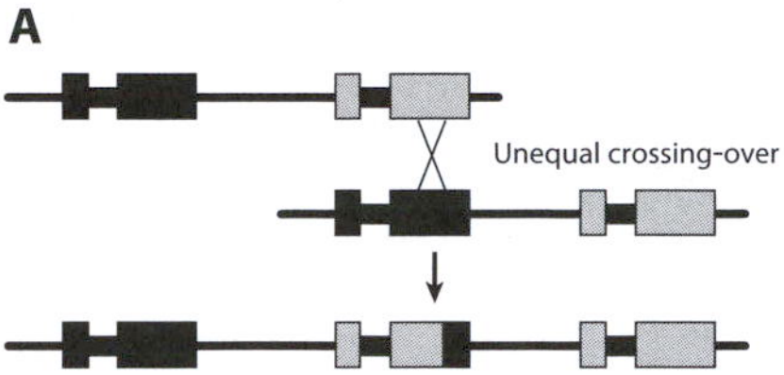

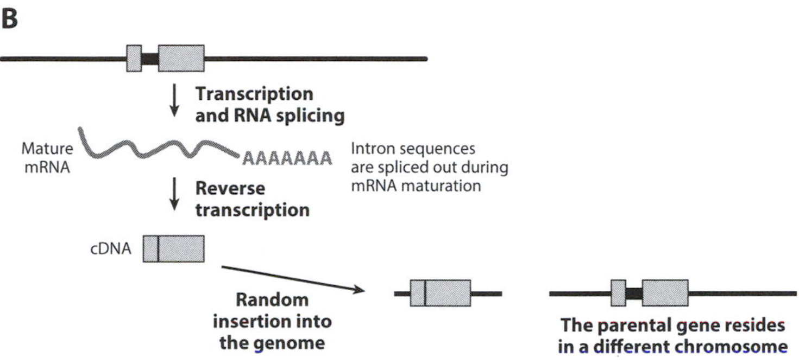

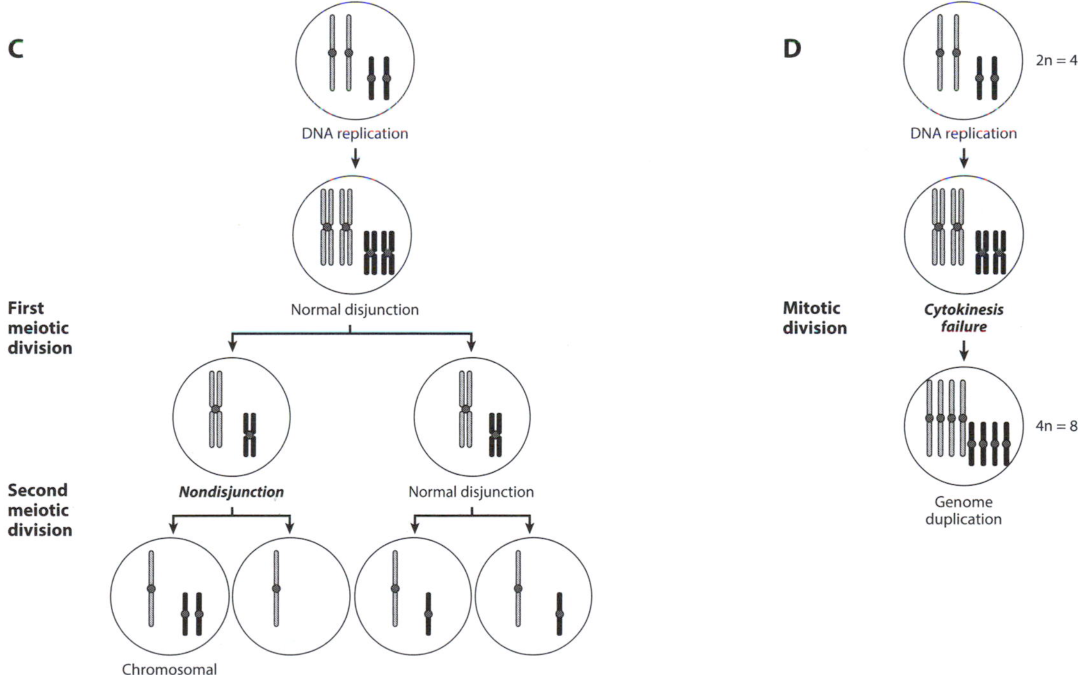

Figure 1. Mechanisms of gene duplication. (A) Unequal crossing-over, which results in a recombination event in which the two recombining sites lie at nonidentical locations in the two parental DNA molecules. (B) Retroposition, a process by which a messenger RNA (mRNA) is retrotranscribed to complementary DNA (cDNA) and then inserted into the genome. In (A) and (B), squares represent exons and bold lines represent introns. (C) Chromosomal duplication via nondisjunction during meiosis. (D) Genome duplication via cytokinesis failure in mitosis.

polymorphisms constitute a form of copy number variation (CNV). Recent studies have shown a surprisingly large number of CNVs in humans. While most CNVs seem to have no visible phenotypic effect, others are associated with human diseases. It is likely that CNVs cause disease by disturbing the dosage of the gene involved or the dosage of the involved gene relative to those of other genes in the genome (i.e., dosage balance).

3. PSEUDOGENIZATION AFTER DUPLICATION

Even after a duplicate gene gets fixed in a population, it may not remain functional and thus be stably retained in the genome for a long time. This is because the daughter gene is usually identical in function to its mother gene; their functional redundancy implies that the loss of one of them has no fitness consequence. In other words, mutations that knock out the expression or function of one of the duplicates can accumulate and the gene gradually becomes a *pseudogene*, defined as a dysfunctional relative of known genes that has lost its function (i.e., its protein-coding ability or its expression). Given enough time, pseudogenes are no longer recognizable because they either diverge too much from their functional relatives or get deleted from the genome. Analysis of the age distribution of duplicate genes in model eukaryotes demonstrated convincingly that pseudogenization is the most common fate of duplicate genes, as the number of (nonpseudogenized) duplicate genes declines sharply with age during the first few million years after gene duplication.

While pseudogenization after duplication seems uninteresting because of its lack of impact on phenotype or fitness, it is important to note that this process may contribute to the formation of reproductive isolation between populations and hence speciation (see chapter VI.8). Let us imagine a duplication event that results in a pair of chromosomally unlinked genes A and B in one species. Shortly after the duplication, the species is split into two populations that are geographically separated. Assume that gene A is pseudogenized in one population, while B is pseudogenized in the other. If the geographic barrier is removed and the two populations merge, the hybrid from a cross between the two populations will have one functional allele and one null allele at locus A as well as at locus B. Thus, a quarter of the gametes produced by the hybrid contain null alleles at both loci (figure 2). If the functions of the genes involved are important, these gametes may malfunction. For example, if the functions of A and B are required for gamete survival, one-quarter of gametes will die. Consequently, the hybrid has a fecundity of $f = 0.75$, relative to an individual from either population. It is easy to see that if n pairs of duplicate genes are reciprocally pseudogenized in the two populations, $f = 0.75^n$, which drops quickly with n. Thus, this model, known as divergent resolution, provides an explanation for a rapid rise in genetic reproductive isolation between geographically isolated populations simply by random degeneration of redundant duplicate genes. This model may be particularly important in lineages that experience whole-genome duplication (WGD), because of the abundance of unlinked duplicate pairs that are subject to divergent resolution.

In addition to those pseudogenes that arise gradually from functional duplicate genes, there are also so-called dead-on-arrival pseudogenes that have never been functional. For instance, retroduplicates lack their own promoters and hence do not have the machinery to have ever been expressed. These pseudogenes, also known as processed pseudogenes for their lack of introns, are highly abundant in many eukaryotic genomes. Because the higher the expression of a gene in the germ line, the greater the probability of retroduplication, one can infer the germ line expression of a gene in ancient times from the numbers of its processed pseudogenes that belong to certain age groups. Furthermore, because retroduplicates arise from mRNAs, processed pseudogenes of different ages provide information on ancient mRNA processing such as alternative splicing that may be absent today. Thus, processed pseudogenes in a genome can be viewed as fossilized ancient transcriptomes that permit an otherwise-impossible glimpse into ancient gene expressions.

4. STABLE RETENTION OF DUPLICATE GENES

For a duplicate gene to be stably retained in a genome, it must be useful to the organism, such that loss of the gene would cause an immediate decrease in fitness too large to spread through the population, because of natural selection. Several mechanisms have been proposed to explain the ways in which duplicates make fitness contributions. First, when gene duplication is immediately beneficial as a result of increased gene dosage, both gene copies can be stably retained because the loss of either gene decreases the dosage and thus fitness (figure 3A). For example, the retention of multiple rRNA genes is easily explained by this mechanism. Duplicate rRNA genes within a genome tend to be highly similar in sequence despite the fact that the duplication may be quite ancient. The high sequence similarity is the result of gene conversion, which is a mutational process homogenizing DNA sequences within a genome. Presumably, highly homogenized rRNA genes are beneficial over heterogeneous rRNA genes, so the product of gene conversion is selectively maintained. This mode of duplicate gene evolution is known as *concerted evolution*.

Second, gene duplication allows one gene to perform the ancestral (and presumably important) function and

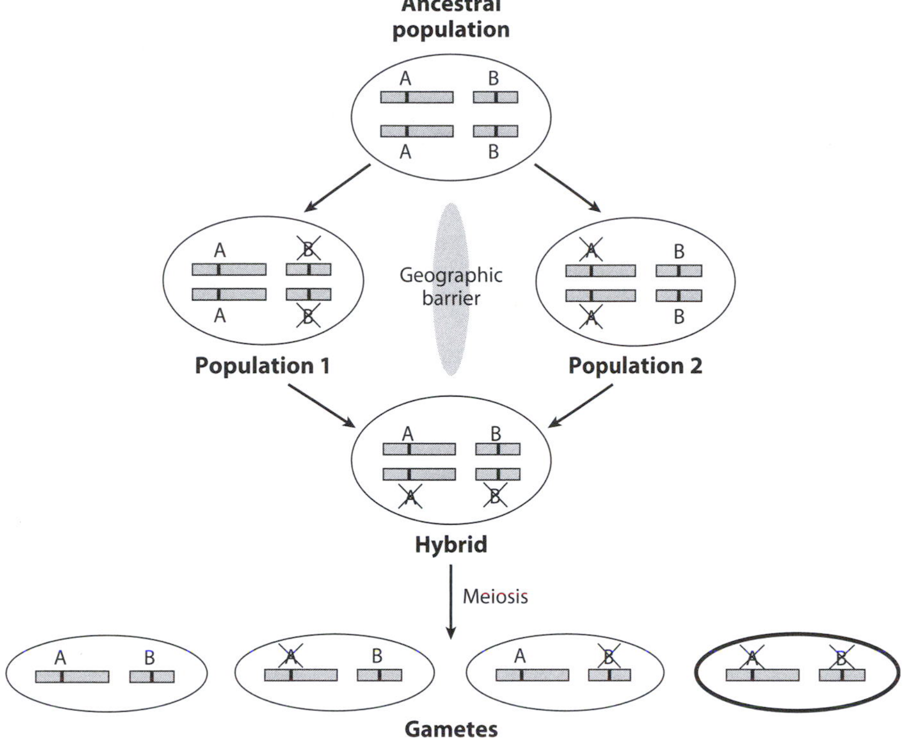

Figure 2. Divergent resolution of duplicate genes can lead to reproductive isolation and speciation. Horizontal boxes represent chromosomes and black bars represent genes. A and B are a pair of functionally redundant duplicate genes. A cross on a gene name indicates pseudogenization. The bold circled gamete has neither functional A nor functional B, and thus is less fit than the other three gametes.

the other to adopt a new function that may be prohibited prior to the duplication because of the impossibility of one gene performing both functions (figure 3B). This route, known as *neofunctionalization*, is generally thought to be the most important contribution of gene duplication to evolution. For example, the primate eosinophil cationic protein (ECP) gene was duplicated from the eosinophil-derived neurotoxin (EDN) gene; in a relatively short evolutionary time after the duplication, ECP acquired an antibacterial activity that is not found in EDN. Rapid sequence evolution driven by positive Darwinian selection (see chapter V.1) was detected in the neofunctionalization of ECP.

Third, an ancestral gene may already possess dual functions; after duplication, each copy may adopt one of the ancestral functions such that they together possess both functions of the ancestral gene (figure 3C). This molecular division of labor, known as *subfunctionalization*, does not increase organismal fitness, but permits each duplicate to make a fitness contribution. For instance, the zebra fish *engrailed-1* gene is expressed in the pectoral appendage bud, while its paralogue *eng1b* is expressed in a specific set of neurons in the hindbrain/spinal cord. This pair of duplicates was generated in teleosts after they diverged from tetrapods. In tetrapods such as mice and chickens, the single-copy *En1* gene is expressed in both expression domains: the developing pectoral appendage bud and specific neurons of the hindbrain and spinal cord.

Fourth, a model that is becoming increasingly popular is called the *escape from adaptive conflict (EAC)*, or the specialization model, which can be viewed as a hybrid of the neofunctionalization and subfunctionalization models (figure 3D). In EAC, the ancestral gene already possesses dual functions, but neither function can be optimized because optimizing one function compromises the other. After duplication, the ancestral functions can be subdivided into the duplicate copies, and the removal of the conflict allows each function to be optimized. Different from the pure neofunctionalization model, EAC asserts that both duplicates will acquire enhanced functions yet no entirely novel function is gained in either copy. EAC is also distinct from the pure subfunctionalization model in that it requires an improvement of ancestral functions. EAC also implies that the improvement of one function in a gene is realized by

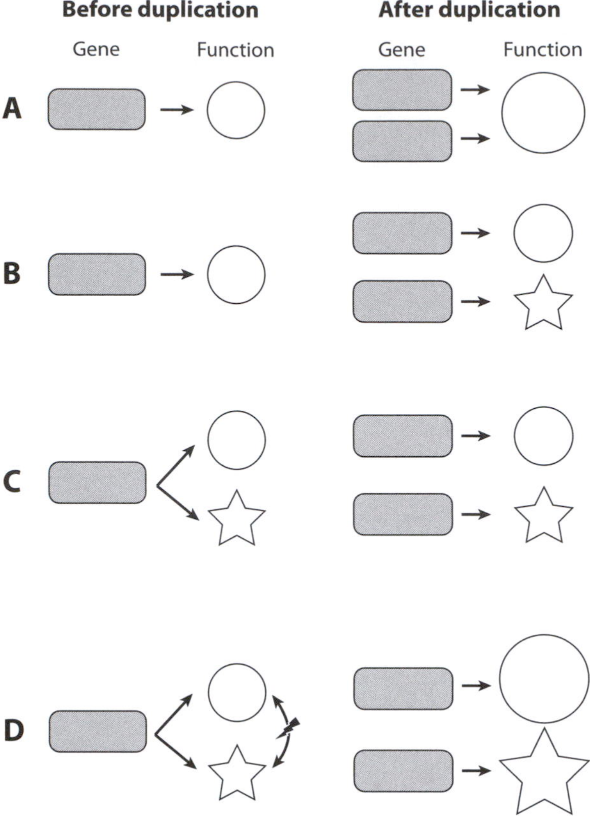

Figure 3. Mechanisms allowing the stable retention of duplicate genes. (A) Duplicate genes maintain sequence and functional similarity, typically by concerted evolution, when having a high concentration of the gene product is beneficial. (B) Neofunctionalization, in which one daughter gene acquires a new function while the other performs the old function. (C) Subfunctionalization, in which each daughter gene inherits one of the ancestral functions. (D) Escape from adaptive conflict, through which each of the two daughter genes inherits one ancestral function and improves it. The improvement is impossible in the progenitor gene because of the adaptive conflict between the two functions, represented by a lightning bolt. Rectangles represent genes, and circles and stars represent different functions. Larger symbols indicate enhanced activities or improved functions.

the degeneration of the other function in the same gene, requiring that these two processes are coupled both in time and in molecular mechanism. The EAC model is best illustrated by the evolution of the duplicated *GAL1* and *GAL3* genes in the yeast galactose use pathway, where the dual functions of their progenitor gene have been divided into the duplicates and further improved. The replacement of *GAL1* with the progenitor gene reduces the fitness, so does the replacement of *GAL3* with the progenitor gene. It should be noted, however, that while the evolutionary patterns of the *GAL* genes fit the EAC model, the initial force that permitted the evolutionary retention of these genes could be pure subfunctionalization.

This last point also emphasizes that the apparent mechanism responsible for the retention of a pair of duplicate genes today may be different from the mechanism underlying the initial retention of the duplicates. Most empirical studies, including the examples provided above, have revealed the mechanisms for current rather than initial retention. Mechanisms for initial retentions are necessarily inferred from comparisons of present-day properties of duplicate genes, and thus should be taken with a grain of salt.

Which of the four models above best explains the initial retention of duplicate genes? To address this question, it is more productive to analyze gene function/expression data than their proxies such as the evolution rate of gene sequences, because the four models are all about gene function or expression. At the genomic scale, yeast duplicate genes exhibit large differences in patterns of expression and protein-protein interaction (PPI). This observation suggests that the model of dosage advantage or concerted evolution cannot be the primary mechanism for duplicate gene retention, because divergences in function and expression are prohibited under this model. Examination of accurately measured genome-wide expression levels of duplicate genes in two yeast species and two mammal species showed that duplicated genes have significantly reduced expression levels compared to their unduplicated progenitor genes. This result further rejects the dosage advantage model and suggests that at least with respect to the amount of gene expression, subfunctionalization has occurred; however, this finding per se is insufficient to establish the role of subfunctionalization in the initial retention of duplicates. In an analysis of tissue expressions of human genes and PPIs of yeast genes at the genomic scale, it has been shown that duplicate genes experience substantial subfunctionalization as well as substantial neofunctionalization, but the former happens quickly after gene duplication, while the latter is a much slower process. This analysis was based on the comparison among groups of duplicate genes that were generated at different time points in the past and found high degrees of subfunctionalization among all age groups of duplicates but high levels of neofunctionalization only among old duplicates. These findings suggest that subfunctionalization is more likely than neofunctionalization to underlie the initial retention of duplicate genes. Because of the lag of neofunctionalization compared to subfunctionalization, this result appears to be inconsistent with the EAC model; nonetheless, in EAC, neofunctionalization is a quantitative improvement rather than a qualitative change in function. EAC could still occur right after duplication, because such quantitative functional improvements are underdetected in the above study of gene expression and PPI, and it is possible that the observed slow neofunctionalization is of a fundamentally different type that is unrelated to EAC.

These empirical findings are generally consistent with theoretical predictions. Specifically, Lynch and colleagues showed that the probability of duplicate gene retention by subfunctionalization is much greater than that by neofunctionalization, especially when the population size is not very large. When the population size gets larger, the chance that a beneficial neofunctionalizing mutation occurs and gets fixed before the occurrence of subfunctionalization increases, and the relative role of neofunctionalization in duplicate gene retention expands.

5. RATE OF GENE DUPLICATION

Examining the numbers of duplicate genes of different ages in a genome, Lynch and Conery (2000) estimated the first genome-wide rate of gene duplication. They estimated from a number of eukaryotic model organisms that fixed duplicate genes arise at a rate of about 0.01 per gene per million years, but the vast majority become pseudogenes within a few million years. Their estimate, however, was only approximate, because of the limited genomic data available at that time and some simplifying assumptions. For example, it was later pointed out that gene conversions between duplicate genes make them look younger than they are, resulting in an overestimation of the duplication rate; also, whole genome duplication (WGD) was not separated from individual gene duplications in the above estimation. WGD is much more frequent in plants than in animals, although a dozen or so WGD events are known in animals.

Another way to estimate the rate of gene duplication is through the examination of mutation accumulation (MA) lines, which are very small populations of organisms maintained in a constant environment for hundreds to thousands of generations. The lack of virtually any natural selection in small populations allows the estimation of the duplication rate per generation at the mutational level. Recently, the genomes of several MA lines of the yeast *Saccharomyces cerevisiae* and nematode *Caenorhabditis elegans* were sequenced. Surprisingly, the rate of the appearance of new duplicates was found to be on the order of 10^{-6} per gene per generation in yeast, about 10^{5} times that measured for duplicate genes that are eventually fixed in a population. Similarly, the rate was on the order of 10^{-7} per gene per generation in *C. elegans*, about two orders of magnitude greater than that measured for fixed duplicate genes. This discrepancy—between the rate of appearance of new duplicates and the rate of fixation of duplicates—strongly suggests that the vast majority of gene duplication events are deleterious and thus do not reach fixation. This conclusion is also supported by recent surveys of CNVs in fruit flies.

6. DETERMINANTS OF GENE DUPLICABILITY

What factors determine gene duplicability, the probability that a gene is duplicated at the mutational level, fixed, and stably retained in the genome? Although three processes (mutation, fixation, and retention) are involved here, it is often difficult to differentiate them in the study of gene duplicability because typically only retained duplicates are observed; nevertheless, under the assumption that duplication rate at the mutational level is not widely different among genes, gene duplicability studies help identify important factors influencing the fixation and retention of duplicate genes.

An important observation of gene duplicability is that compared to other genes, those encoding members of protein complexes have reduced rates of individual gene duplication. Because components of a protein complex need to be balanced in concentration, dosage imbalance brought about by doubling the concentration of one but not other members of the protein complex is likely deleterious. WGD, however, creates an opposite situation. The individual carrying WGD is usually reproductively isolated from other individuals of the same species. In other words, WGD is immediately fixed if the lineage with WGD will survive in evolution; thus, duplicate genes that are present long after WGD tell us what genes tend to be retained after fixation. Interestingly, genes encoding protein complex members tend to be maintained after WGD, because individual gene losses would cause dosage imbalance just as individual gene duplication does.

It has also been reported that the more complex a gene is, the higher its duplicability, where gene complexity is measured by protein length, number of protein domains, and number of *cis*-regulatory motifs in the promoter of the gene. This phenomenon appears to be attributable to higher retention probabilities for more complex genes, presumably because more complex genes are subject to faster subfunctionalization and hence greater likelihood of stable retention. Hence, gene duplication increases both gene number and gene complexity, two factors in the origin of genomic and organismal complexity.

Interestingly, in terms of the duplicability bias among genes of different importance to the fitness of the organism, it has been shown in yeast that less important genes have a greater duplicability than more important ones. There are two apparent reasons for this phenomenon. First, yeast genes encoding members of protein complexes tend to be more important than other genes, and the balance hypothesis explains why the former should have a lower rate of individual gene duplication than the latter. Second, in yeast, gene importance is negatively correlated with the number of *cis*-regulatory motifs in the promoter of the gene, and the gene complexity hypothesis explains why less important genes, which have more *cis*-regulatory motifs, should duplicate more than important genes.

There are several other gene functional properties that have been observed to correlate positively with gene duplicability, although the underlying mechanisms are often unclear. These include functioning as metabolic enzymes, interacting with the external environment, interacting with fewer protein partners, controlling physiological traits (rather than morphological traits), having more phosphorylation sites, and functioning as intermediary proteins (rather than receptors) in signaling pathways.

7. FUNCTIONAL REDUNDANCY AMONG DUPLICATE GENES

Functional redundancy refers to the functional similarity between genes, and is typically demonstrated by measuring the fitness effect of gene deletion. For example, it has been shown at the genomic scale in *S. cerevisiae* and *C. elegans* that deleting a duplicate gene has a significantly smaller fitness effect than deleting a singleton gene. Furthermore, deleting a pair of duplicate genes has a much greater fitness effect than deleting either gene alone. Evolutionary theory predicts that the degree of functional redundancy between a pair of duplicates gradually declines with the time since duplication, as a result of subfunctionalization and/or neofunctionalization; however, it has been reported that some duplicate genes are highly redundant even hundreds of millions of years after duplication. Of course, in the case of ribosomal proteins, rRNAs, histones, tRNAs, and other molecules in high demand in the cell, functional similarity among duplicates is selectively favored and thus requires no other explanation.

For other duplicate genes, several hypotheses have been proposed to explain the unexpectedly long retention of functional redundancy. First, some believe that redundancy is beneficial in itself because it protects the organism from the potential harm of deleterious mutations, much like the backup role of a spare tire for a car. This backup hypothesis, however, cannot on its own be correct, because the benefit of backup cannot be detected by natural selection unless the product of population size and mutation rate is orders of magnitude greater than that observed in cellular organisms. Second, the piggyback hypothesis asserts that paralogous genes have some nonoverlapping functions as well as some overlapping functions, and the existence of the latter is a by-product of the former owing to strong protein structural constraints. Third, reduction of gene expression after duplication, a special form of subfunctionalization, is commonly observed. Suppose 100 protein molecules is the optimal expression level for a gene. On duplication of the gene and subsequent evolution, one daughter gene may produce 60 protein molecules, and the other, 40. In this scenario, both gene copies can be stably retained, yet no functional divergence between them is expected even long after duplication. With respect to fitness, there is no redundancy between the two genes because both copies are required to reach the highest fitness; however, because of the nonlinear relationship between dosage and fitness, deleting both genes would almost always have a much greater fitness effect than deleting either one of them, creating the apparent phenomenon of redundancy. The relative importance of the second and third hypotheses explaining the functional redundancy is unclear and will likely remain a topic of intensive study.

8. FUNCTIONAL DIVERSIFICATION OF DUPLICATE GENES

While the maintenance of functional redundancy among duplicate genes is not unusual, the most common observation among stably retained duplicates is their functional divergence. The degree of functional divergence varies greatly. In many cases, duplicate genes perform similar types of function, but with different activities or specificities, or at different times or locations. For example, isozymes catalyze the same biochemical reaction but usually have different catalytic parameters; they are encoded by duplicate genes that are often expressed in different tissues or at different developmental stages. Duplication also expands the scope of a basic function. For example, odorant receptor (OR) genes form the largest gene family in the vertebrate genome. Each OR is able to recognize only a limited number of odorants. Vertebrates are believed to be able to detect 10,000 or more odorants because of the possession of hundreds of functional OR genes that recognize different ligands.

As mentioned, retroduplicates are usually dead on arrival because of the lack of a promoter; occasionally, however, they may be expressed when they are fortuitously inserted into a genomic region that harbors a promoter, for example, into the intron of another gene. There is now accumulating evidence that retroduplication is also an important source of new genes (see chapter V.6). Probably because of completely different expression patterns and the involvement of gene fusion, functions of retroduplicates can sometimes differ dramatically from those of their mother genes.

9. FUTURE DIRECTIONS IN THE STUDY OF GENE DUPLICATION

Two theoretical questions about the functional divergence of duplicate genes are yet to be resolved. First, is the functional difference between duplicates attributable mainly to subfunctionalization or to neofunctionalization? While some authors stress the former, others emphasize the latter. Still others believe that both happen in each duplicate pair and that their relative contributions depend on the time since duplication. Second, what is the role of natural selection in the functional divergence of duplicate genes? This question is related to the first, because the role of selection is different in different types of functional changes. For instance, subfunctionalization by degenerate mutations does not require positive selection. By contrast, EAC must involve positive selection. It is commonly thought and has been demonstrated in case studies that positive selection is involved in neofunctionalization, but in theory, neofuctionalization can also occur by random fixation of neutral mutations; the utility of the new function may be realized only after its fixation, on an alteration of the genetic background or environment. The role of purifying selection in neofunctionalization is also understudied. Clearly, neofunctionalization in a daughter gene requires at least a partial relaxation of the selective constraint associated with the functions of the progenitor gene. But whether neofunctionalization has a greater chance of occurring in the presence of some functional constraints or in the presence of no constraint is not entirely clear, because in the presence of no constraint, the gene may become a pseudogene before acquiring new function, as has been recently demonstrated experimentally. These uncertainties notwithstanding, it is apparent that gene duplication is the primary source of new genes in evolution and that it has contributed to biodiversity at the genomic, functional, and organismal levels.

FURTHER READING

Conant, G. C., and K. H. Wolfe. 2008. Turning a hobby into a job: How duplicated genes find new functions. Nature Reviews Genetics 9: 938–950. *A recent review on the mechanisms of functional changes after gene duplication.*

Force, A., M. Lynch, F. B. Pickett, A. Amores, Y. L. Yan, and J. Postlethwait. 1999. Preservation of duplicate genes by complementary, degenerative mutations. Genetics 151: 1531–1545. *A classic paper that proposed the subfunctionalization model of duplicate gene evolution.*

Gu, Z., L. M. Steinmetz, X. Gu, C. Scharfe, R. W. Davis, and W. H. Li. 2003. Role of duplicate genes in genetic robustness against null mutations. Nature 421: 63–66. *A genome-wide study that revealed substantial functional redundancy among duplicate genes.*

He, X., and J. Zhang. 2005. Rapid subfunctionalization accompanied by prolonged and substantial neofunctionalization in duplicate gene evolution. Genetics 169: 1157–1164. *Temporal patterns of subfunctionalization and neofunctionalization in duplicate genes were revealed by genome-wide data on protein-protein interaction and gene expression.*

Kaessmann, H., N. Vinckenbosch, and M. Long. 2009. RNA-based gene duplication: Mechanistic and evolutionary insights. Nature Reviews Genetics 10: 19–31. *A recent review on retroduplication.*

Lynch, M., and J. S. Conery. 2000. The evolutionary fate and consequences of duplicate genes. Science 290: 1151–1155. *Most fixed duplicate genes become pseudogenes in a few million years.*

Ohno, S. 1970. Evolution by Gene Duplication. New York: Springer-Verlag. *A classic book that greatly stimulated the study of gene duplication.*

Papp, B., C. Pal, and L. D. Hurst. 2003. Dosage sensitivity and the evolution of gene families in yeast. Nature 424: 194–197. *Gene duplication is deleterious when breaking dosage balance.*

Qian, W., B. Y. Liao, A. Y. Chang, and J. Zhang. 2010. Maintenance of duplicate genes and their functional redundancy by reduced expression. Trends in Genetics 26: 425–430. *Expression reduction after gene duplication allows the long-term retention of functionally redundant duplicate genes in the genome.*

Zhang, J. 2003. Evolution by gene duplication: An update. Trends in Ecology & Evolution 18: 292–298. *A review on the role of gene duplication in genomic and organismal evolution.*

V.6

Evolution of New Genes

Manyuan Long

OUTLINE

1. Mutational mechanisms to generate new genes
2. Rates of new gene origination
3. Patterns of new gene evolution
4. Evolutionary forces acting on new genes
5. Functions and phenotypic effects of new genes

Every gene has its first moment: this is its origination, when a new gene appears in a genome and evolves a distinct or new function(s) that did not previously exist. The genes in extant organisms are of different ages, from ancient to very young. To understand the origination of a gene is to understand the earliest stage of its evolution; however, the origination process cannot be directly observed for most genes because they are ancient. For these ancient genes, there were likely multiple evolutionary events, which may have obliterated the early signature of the gene's origination process. An alternative is to examine the genes that have formed recently, which are called *new genes* or young genes; in these cases, a reconstruction of the origination process is feasible and provides an exciting glimpse into the evolution of new genes.

Several questions are relevant to understanding the origination of new genes. First, what mutational processes generate new genes in a genome? Second, how often do new genes reach 100 percent frequency (i.e., *fixation*) in a species population? Third, if the new genes frequently appear in genomes during evolution, are there any patterns or rules underlying their origination? Fourth, what evolutionary forces are responsible for the fixation of a new gene within a genome? And how did the new gene accumulate mutations to optimize its function? Finally, what are the roles of new genes in phenotypic evolution? Understanding the functions and phenotypic effects of new genes is critical to determine their role in evolution.

Since the first truly young gene, *Jingwei*, was found in *Drosophila* by Manyuan Long and Charles Langley two decades ago, new techniques for molecular and genomic analyses have been invented, and sequence databases have expanded at a previously unimaginable rate. This large amount of data has shed new light on the origination of new genes, such that a general picture is emerging of the general process of genetic and phenotypic evolution.

GLOSSARY

Chimeric Gene. A type of new gene whose domains or encoded exons originated from a combination of different genes.

Copy Number Variation. Newly formed gene duplicates or gene deletions that have not been fixed within a population.

Gene Trafficking. The transfer of gene copies between sex chromosomes and autosomes during evolution.

Neofunctionalization. An evolutionary process in which a new function is acquired by a gene.

New Gene. A gene that appears in a genomic location where it had not previously existed.

Orthologue. A gene found in different species that originated from a common ancestral gene and diverged after speciation.

Paralogue. A gene formed by duplication within a genome.

Retrogene. A gene that originated via retroposition, in which a parental gene is transcribed, the subsequent RNA is reverse transcribed, and the new copy is inserted into a new location in the genome. When retrogenes have no function because of lack of regulatory systems for expression, they are called *retropseudogenes* or *processed pseudogenes*.

Transposable Element. A segment of DNA that is capable of moving within and between genomes.

1. MUTATIONAL MECHANISMS TO GENERATE NEW GENES

What genetic mechanisms underlie the formation of a new gene? Until the early 1990s, three general models, termed the Muller model of duplication (DNA-based duplication), the Gilbert model of exon/domain shuffling, and the Brosius model of retroposition (RNA-based duplication), had been proposed for the origination of new genes (figure 1). Importantly, these three mechanisms are not mutually exclusive and can be used simultaneously to create a new gene. For example, in the ancestor of African fruit flies (*Drosophila yakuba* and *D. teissieri*), a retrogene from the *alcohol dehydrogenase* (*Adh*) gene was inserted into a previously existing duplicate of the *yellow emperor* (*Ymp*) gene, which led to the creation of a chimeric gene (*Jingwei*) that functions in the pheromone metabolism recruitment pathway. Additional mechanisms of new gene formation, via transposable element insertion, lateral gene transfer, and frameshift mutations (see below), have since been discovered, as well as gene fission and fusion mechanisms (figure 1).

In the mid-1990s, it was observed that transposable elements (TEs) could be "domesticated" to create a new coding portion of a nuclear gene. For example, in the human genome, an Alu TE was inserted into the coding portion of the decay-accelerating factor (DAF) and created a new hydrophilic carboxy-terminal region in DAF that later evolved a new function to inhibit DAF from moving into the membrane. As many as 400 human gene families have since been found to be hybrids between a nuclear gene and a TE; these examples represent many different types of TEs. Furthermore, TEs are also found to facilitate recombination, leading to the formation of chimeric genes. For example, in two dozen cases in *Drosophila*, DNAX TEs are associated with new duplicate copies and parental copies in the *melanogaster* subgroup, and Pack-Mule TEs in rice have been involved in the formation of approximately 2000 chimeric genes.

Lateral gene transfer (LGT) was known to happen frequently among prokaryotic organisms, but it was previously thought not to occur in eukaryotes; however, several examples have recently been documented in eukaryotes. For example, genes and genome fragments of the parasitic bacteria *Wolbachia* have been observed in the genomes of *Drosophila*, mosquitoes, and bees; several mitochondrial genes that encode ribosomal and respiratory proteins were subject to horizontal transfer between distantly related species in flowering plants; and the pea aphid has recruited genes encoding multiple enzymes for carotenoid biosynthesis from the genome of an ancestral *Phycomyces* fungus. Thus, the role of LGT in eukaryotic genome evolution was previously underappreciated.

Finally, frameshift mutations, which are usually thought to be deleterious, have been found to contribute to the formation of new genes. In a survey of human genomes, there were about 470 novel protein families that could have been created by a reading frameshift mutation in duplicate copies. Although many of these observations might derive from sequencing and assembling errors, this model is interesting in that it proposed the rapid creation of novel proteins.

In all the aforementioned mechanisms, new genes are derived from previously existing genes; thus, it was surprising to discover that several dozens of new protein-coding genes in *D. melanogaster* appear to have no orthologues in closely related species (even those that diverged only a few million years ago), suggesting that new genes can arise de novo (figure 1). Besides, de novo genes have also been reported in plants, mice, humans, fish, and viruses. One simple interpretation is that previously noncoding or intergenic regions can accumulate enough mutations to create functional open reading frames.

2. RATES OF NEW GENE ORIGINATION

At present, the genomes of thousands of species have been sequenced from almost all major types of organisms, including bacteria, archaea, protozoa, fungi, plants, and animals. Comparative analyses from eukaryotic genomes, especially from the various model organisms, have identified thousands of young genes (see chapter V.3). These observations suggest that the origination of new genes is a common process and that genomes have been modified frequently by adding new genes with new functions. These genomic sequences also provide data to estimate the rate of new gene origination.

The origination rates of new genes created from a few mechanisms have been analyzed. The first rate was estimated for gene duplication (see chapter V.5). Michael Lynch and colleagues extensively analyzed the duplication rates in major model organisms, including humans, mice, *Arabidopsis thaliana*, and *Saccharomyces cerevisiae*, and observed an average duplication rate of 0.01 per gene per million years, although a majority of new duplicates became silenced in a few million years. This duplication rate implied that the fixation of duplicate genes occurs at a high rate: 100 new duplicates per million years per 10,000 genes. This would suggest that one to three new duplicates are fixed in the genomes of *Drosophila* and humans every 10,000 years. These estimates may be impacted by gene conversion and other factors (e.g., insufficient annotation that tends to ignore

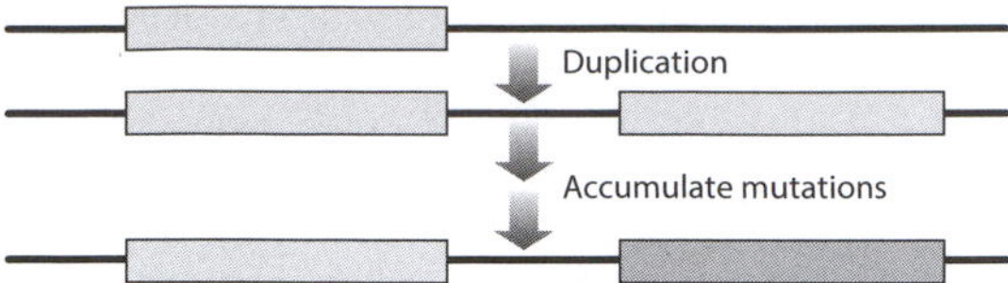

1. The Muller model of duplication

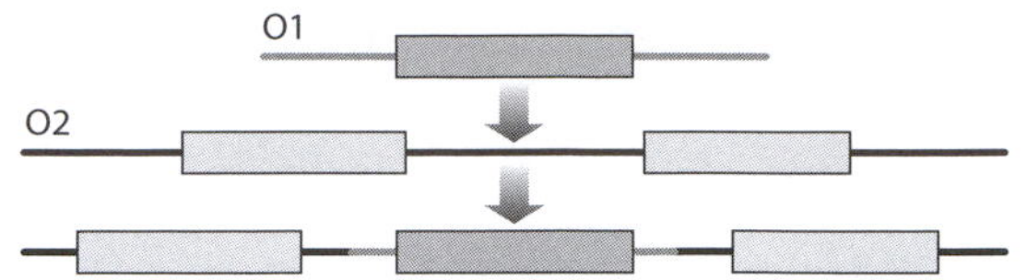

5. Lateral gene transfer between organisms O1 and O2

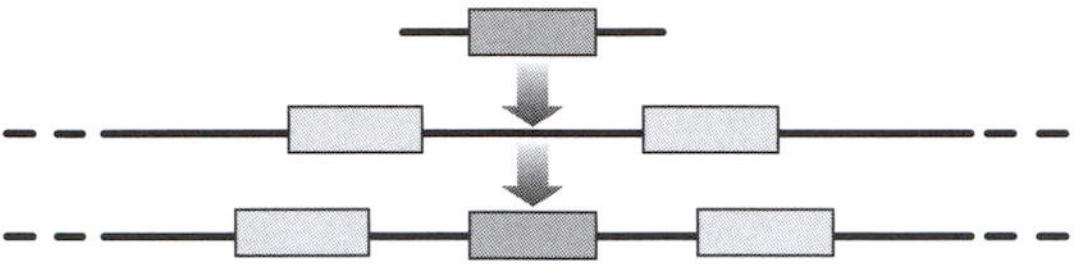

2. The Gilbert model of exon/domain shuffling

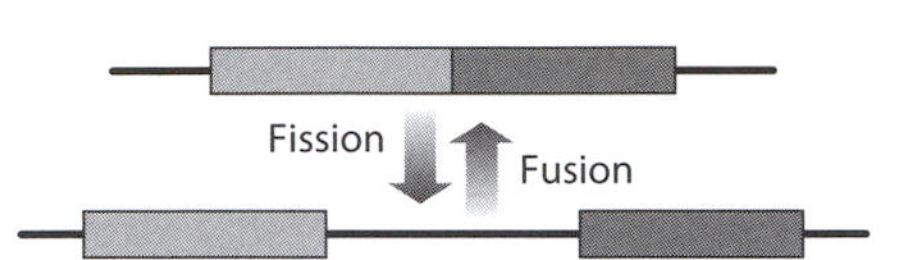

6. Gene fission and gene fusion

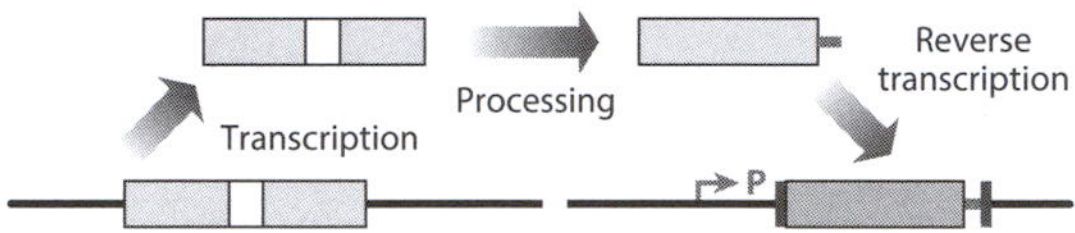

3. The Brosius model of retroposition

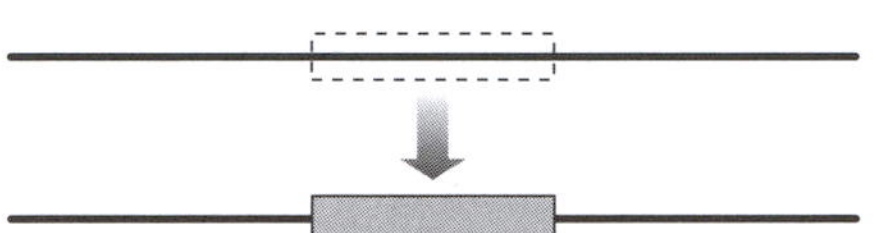

7. De novo gene origination

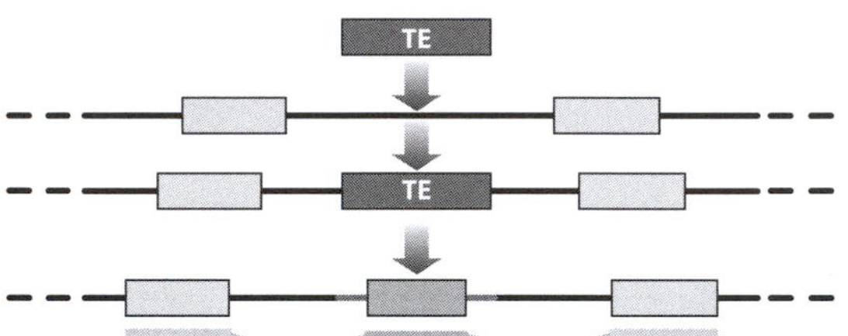

4. Transposable element (TE) domestication

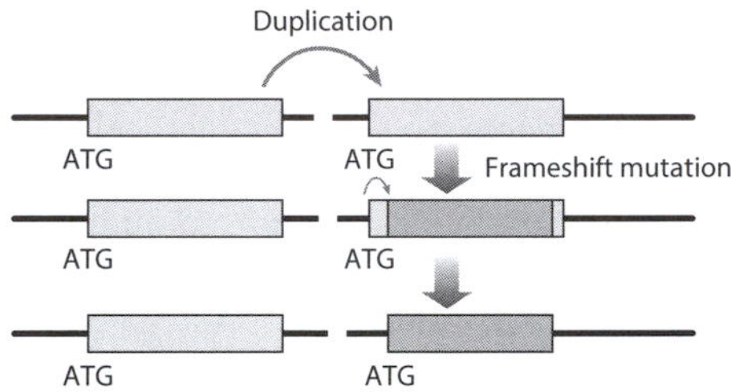

8. Reading frame shift model

Figure 1. Molecular mechanisms of new gene evolution. The first three are general models: the Muller model, Gilbert model, and Brosius model, followed by five mechanisms that are also frequently used in various organisms. In the Brosius retroposition model, the poly(A) tail and short flanking sequences are also labeled as two dark gray bars with the poly(A) tail before the second bar, with fortuitously recruited regulatory system R. Although lateral gene transfer from one species (O1) to the other (O2) is more often observed in prokaryotes, it does contribute to new gene formation in eukaryotes. Frameshift mutations can be caused by insertion or deletion of nonintegers of 3 (e.g., one or two nucleotide insertion or deletion in a codon).

the duplicates that have recently appeared); however, recent estimations based on young duplications in mammalian and *Drosophila* genomes give a rate with the same order of magnitude.

It is still unknown how many of these new duplicates will evolve novel functions, rather than maintain redundant functions. A gene created by a structural change (e.g., a chimeric gene) is likely to have functions that diverge from its parental copy. A change in the temporal or spatial expression of a gene can also allow it to acquire a new function. Extensive surveys of the origination rates of new genes have been conducted in *Drosophila* because of the availability of genomic sequence data in multiple species in the genus. In *Drosophila*, the origination rate of new genes including DNA-based duplication, retrogenes, and de novo origination was approximately 23 per genome per million years, and the majority of these genes evolved via a chimeric structure by recruiting new exons and new untranslated regions (UTRs). These data suggest that new gene functions, rather than redundant functions, evolve frequently, with one functional chimeric gene originating every 50,000 years in *Drosophila*.

In the human genome, a majority of retrogenes are chimeric genes that have recruited new exon regions from the surrounding insertion sites. In primates, it has been estimated that the rates of retroposition, and the formation of chimeric genes by retroposition, are 1 and 0.01 per million years per genome, respectively. In the grass family, the rate of chimeric genes created by retroposition is estimated to be as high as 7 per genome per million years. These observations suggest a surprisingly rapid rate of new gene evolution.

3. PATTERNS OF NEW GENE EVOLUTION

Are there any rules governing the origination of new genes? The large numbers of new genes detected in various organisms have provided an excellent set of data to detect possible patterns or rules underlying the processes of new gene origination. It is important to understand the evolutionary patterns of new genes because these patterns may provide clues for understanding the mechanisms underlying the formation of new genes and, in turn, for formulating theories from which to make predictions. So far, three evolutionary patterns (discussed below) associated with the origination of new genes have been detected; these discoveries have stimulated further interest in the mechanisms responsible for these patterns.

Soon after the genomic sequence of *D. melanogaster* became available, it was realized that a computational identification of retrogenes and their parental copies across the genome would be feasible. It was possible to discriminate the derived retrogene copies from the parental genes because the retrogenes are copied from processed messenger RNA transcripts; newly created retrocopies do not have introns but do have a poly(A) tail and a pair of short duplicate sequences flanking the retrogenes (figure 1). Although the molecular signatures of the poly(A) tail and flanking duplicate sequences might be eroded over substantial evolutionary time (e.g., longer than 10 million years), the loss of introns becomes a permanent feature of retrogenes. Thus, by looking at the paralogous copies from a single species, the ancestral relationship between parental copies and the derived copies can be explicitly characterized.

The fixation of a retrogene and the genomic location of its parental copy are influenced both by mutational events and subsequent evolutionary forces. To determine the relative roles of these processes, the observed genomic distribution of retrogenes can be compared to a hypothetical distribution based on the mutation rate, often the null hypothesis based on the neutral theory of molecular evolution (see chapter V.1). Because retropseudogenes are likely to be evolving neutrally, their distribution can be tested against neutral expectations and used to estimate a distribution of neutral mutations. If they are neutral and the incidence of mutations is not biased among chromosomes, two simple predictions can be derived: (1) the number of parental genes on a chromosome should be proportional to the number of genes on that chromosome, and (2) the number of retrogenes on a chromosome should be proportional to the length of the chromosome. In these analyses, it is important to take the relative population sizes of the X chromosomes and autosomes into account, because they are impacted by additional population genetic factors (e.g., sex ratio) (see chapter V.4). Using data from human genomes, these predictions were tested and confirmed in the functionless retrogenes (processed pseudogenes), suggesting that retroposition is a neutral process.

Trafficking of New Genes between Sex Chromosomes and Autosomes

Computational analysis of *D. melanogaster* genomic sequences was used to characterize the distribution of retrogenes and parental genes from a database containing all possible retrogenes identified in genome sequences. Compared with the neutral expectation, the retrogene data revealed unique patterns: (1) the observed distribution is significantly different from the expectation; (2) a significant excess of X-linked genes are the parental genes of retrogenes; (3) an excess of the retrogenes are found on autosomes; and (4) retroposition events between the two autosomes or from the autosomes to the X were significantly lower than expected. Thus, the *D. melanogaster* genome showed evidence for directional trafficking of retroposed genes from the X to the autosomes. The genome sequences of 11 additional *Drosophila* species revealed the same trend—X-linked genes copied and then pasted in autosomes—suggesting that X-to-autosome gene trafficking is a general process of gene evolution in the *Drosophila* genus. By contrast, similar analyses applied to the human and mouse genomes revealed bidirectional gene trafficking in mammalian genomes: there was a high excess of retroposition both from the X to the autosomes and from the autosomes to the X chromosome.

Association of Sex-Specific Expression with Trafficking of New Genes

Analyzing the expression pattern of retrogenes and parental genes revealed another pattern: the vast majority of the X-derived autosomal retrogenes have evolved a testis-specific expression pattern in both *Drosophila* and mammals. Conversely, very few of the X-linked retrogenes copied from autosomal parental genes are expressed in the testes, but instead often evolved female-specific expression. In addition, the parental genes had significantly lower expression in testis. In *D. pseudoobscura*, gene movement out of the neo-X chromosomes was also observed, with unidirectional gene movement from the X to autosomes and the subsequent evolution of expression in the testes. Similar patterns have been observed in other organisms, including mosquitoes (*Anopheles gambiae),* stalked-eye flies (*Teleopsis*), and mammals.

New Genes Are Preferentially Located in Specific Chromosomal Environments

Gene trafficking between the X and autosomes reflects a preference of new genes for a genomic environment that distinguishes sex chromosomes and autosomes. Is there preference for specific genomic environments within chromosomes? The answer is yes, based on evidence from the particular genomic regions flanking or adjacent to new genes. For example, in the human genome, examination of about 50 functional retrogenes revealed a significant connection between the presence of a functional retrogene and the transcriptional potential of the flanking regions: on average, more expressed genes were identified from the regions surrounding the functional retrogenes in comparison to the regions flanking the retrogenes that are transcriptionally silenced. This observation indicated that retrogenes might take advantage of the regulatory environment formed by nearby genes for their expression.

4. EVOLUTIONARY FORCES ACTING ON NEW GENES

What are the underlying evolutionary forces responsible for the evolution of new genes? Two significant events occur as stages in the evolution of new genes. The first stage involves the fate of a new gene within a species in which the new gene can either be lost from the population or spread to fixation in the population. The second stage involves the further accumulation of mutations in the new gene sequence to further improve its function. After this stage, the new gene is subject to the same processes of molecular evolution as any other gene in the genomes (see chapter V.1); however, the roles of evolutionary forces, particularly natural selection and genetic drift (see chapter IV.1), are interesting and peculiar in these first two stages of the evolution of new genes.

Evolutionary Forces Acting on the Fixation of New Genes

Recent genomic technologies have enabled the investigation of the trajectory of a newly arisen gene toward its final fixation or loss from the population. For example, a population-genetic study was conducted on copy number variation (CNV) in *D. melanogaster.* It was observed that a majority of polymorphic duplicates are found in intergenic regions. This result suggests a role for purifying selection against gene duplication, especially complete gene duplication, consistent with the conjecture that the initial gene duplication is slightly deleterious; however, five recent gene duplicate events, involving genes responding to toxins, were found at high frequency (>70%), suggesting that positive selection is favoring these new gene duplicates.

Daniel Schrider and colleagues recently presented the first study of retrogene polymorphisms, using next-generation sequencing in 37 inbred lines derived from a North Carolina *D. melanogaster* population. By comparing between-species divergence and within-species polymorphism, they found an excess of fixed retrogenes that were copied from X-linked parental genes on autosomes. This recent result reveals a significant role of positive selection in fixation of new retrogenes within species. They also conducted a similar study in humans and detected a positive selection in fixation of retrogenes.

Evolutionary Forces Acting on New Genes Subsequent to Their Fixation

It is conceivable that when a new gene is fixed, further evolutionary modification may be necessary to optimize its function. Such a verbal model predicts a period of rapid sequence evolution in a new gene, which will eventually slow down in later stages. A very young gene, *Jingwei*, that originated 3 million years ago in the common ancestor of the three African *Drosophila* species (*D. yakuba, D. teissieri, D. santomea*) provided data in support of this model (plate 3). While there were no fixed synonymous changes in this new gene, nine amino acid substitutions occurred in the ancestral stage before the first speciation event that led to *D. yakuba* and *D. teissieri*; moreover, after the divergence of *D. yakuba* and *D. teissieri*, there was an excess of amino acid substitutions over synonymous changes, in comparison to within-species polymorphism, suggesting a role of Darwinian positive selection in the evolution of *Jingwei* (see chapter V.14).

The *Adh* gene has been involved in the formation of two additional chimeric genes in two different *Drosophila* species groups: *Adh-Finnegan*, a DNA-based duplication in the *repleta* group, and *Adh-Twain*, an RNA-based duplication (retroposition) in *D. subobscura, D. guanche, and D. madeirensis*. Comparison of all three *Adh*-derived new genes (including *Jingwei*) reveals two interesting patterns. First, there is evidence for convergent evolution (see chapter V.12), because the same amino acid substitutions are fixed in these different genes in different organisms! Second, a recent analysis of a large set of new genes in 12 *Drosophila* species revealed early and rapid substitutions driven by positive selection, with later and slower evolution shaped by purifying selection. These data provide clear evidence that the new genes continue to be under positive selection subsequent to their fixation.

The Targets of Selection

The analysis of duplicate genes described above clearly reveals a role for natural selection and mutational mechanisms in determining the genomic positions of new genes. For example, the testis-expression patterns shown by the new X-derived autosomal retrogenes (described above) have been interpreted as resulting from natural selection. Several explanations for this pattern have been put forward, including sexual antagonism, sexual genomic conflict, degree of dominance, sexual selection, dosage compensation, and male sex-chromosome inactivation.

The classical sexual antagonism model proposes that a mutation with sexually antagonistic effect (i.e., those that are advantageous for males and disadvantageous for females) would be spread from very low frequency to high frequency in a population if its genetic effect is recessive. If a new modifier inhibits its expression in females, or if the gene appears in a small population (in which the effect of genetic drift effect is large), such a sexually antagonistic new gene can be preferably fixed on the X. By contrast, if the new gene is genetically dominant, the fixation of a new antagonistic gene is favored in an autosomal location. It is also likely that dosage compensation can restrict the development of male-biased expression of the X-linked genes, thus favoring the genes that moved to autosomes.

Much interest in recent years has been focused on another aspect of the mechanisms involved in new gene evolution: male sex-chromosome inactivation. In mammals (e.g., humans and mice), it has been observed that when the male germ line cells enter the meiotic stage, the X and Y sex chromosomes are condensed into an *X/Y body,* and genes on these chromosomes are "silenced" (i.e., not expressed). Thus, there should be strong selection for any genes necessary at these stages of spermatogenesis to be located on autosomes rather than on the X chromosome. This prediction of gene trafficking from the X to the autosomes in mammals, also likely in *Drosophila* and *Anopheles*, is confirmed by large-scale analyses of gene expression; however, the mechanisms responsible for the biased genomic distributions of new genes remain to be further elucidated.

5. FUNCTIONS AND PHENOTYPIC EFFECTS OF NEW GENES

The aforementioned observations and analyses, made primarily with data from *Drosophila* and mammals, have revealed that new genes have originated and fixed frequently in the genomes of various organisms. Expression analyses of these new genes suggest that many have reproductive functions (e.g., expressed specifically in testis); however, this tissue-specific expression pattern is just a first step toward understanding the function and phenotypic effects of new genes. Additional information is critical to understanding how new genes have evolved and how their evolution has contributed to organismal evolution. Analysis of sequence evolution of gene duplicates has provided ample information about their functional evolution. The new functions and phenotypic effects of genes that have arisen in recent evolutionary time are particularly informative.

The previously mentioned new gene, *Jingwei*, was extensively investigated for the evolution of its biochemical function. Because of the high sequence similarity in its *Adh*-derived domain, it was initially expected that the new gene might have maintained the functions of its *Adh* parental gene and that new functions might have been added from its N-terminal domains (plate 3). However, the enzymatic activities of the Jingwei protein were assayed by testing its activity on more than 30 different alcohol substrates. It was observed that the gene evolved new metabolic activities: two chemicals, farnesol (involved in the biosynthesis of juvenile hormone) and geraniol (the pheromone for communication among the individuals), became the specific substrates of *Jingwei.* These evolutionary and experimental analyses revealed that the new enzymatic function involving new substrates had evolved in a short evolutionary time.

Additional evidence for the rapid evolution of new functions for new genes comes from a study knocking down the expression of more than 195 new genes that originated within the past 35–3 million years in *Drosophila.* The conventional expectation was that only ancient and conserved genes would be functionally important and that the recently evolved genes would be associated with interesting but dispensable minor functions. Surprisingly, 30 percent of new genes were observed to be lethal when knocked down, which was the same percent of old genes found to be essential. Further assays of the phenotypic effects of 59 of these essential new genes revealed that all affected the development of *D. melanogaster.* These observations together suggest that the genetic program of development contains species-specific or lineage-specific components. An important conceptual connection from this study can be made: the developmental effects of new genes appear to be adaptive, as there was significant evidence for positive selection on these genes. To this end, the relationship between adaptive evolution and evolution of development can be explicitly linked, suggesting that microevolution and evolution of development are not mutually exclusive but combined under the same mechanism found by Darwin: natural selection.

FURTHER READING

Betrán, E., K. Thornton, and M. Long. 2002. Retroposed new genes out of the X in *Drosophila*. Genome Research 12: 1854–1859. *This is the first paper reporting the directional movement of retrogenes from the sex chromosome to autosomes and shows that the vast majority of autosomal retrogenes have evolved male-biased functions. This finding of the dynamic process of gene evolution provides a mechanistic interpretation for the preferential distribution of male-biased genes on autosomes.*

Brosius, J. 1991. Retroposons: Seeds of evolution. Science 251: 753. *This paper presents a model of gene evolution and points out the important role of retroposition (RNA-based duplication) in the origin of new genes.*

Chen, L. B., A. L. Devries, and C.H.C. Cheng. 1997. Evolution of antifreeze glycoprotein gene from a trypsinogen gene in Antarctic notothenioid fish. Proceedings of the National Academy of Sciences, USA 94: 3811–3816. *This report presents evidence of de novo origination of antifreeze protein in Antarctic fish, showing a clear role for environmental factors in driving the evolution of a novel antifreeze protein.*

Chen, S., Y. Zhang, and M. Long. 2010. New genes in *Drosophila* quickly become essential. Science 330: 1682–1685. *This report shows that many young genes have evolved essential functions in development in* Drosophila. *Thus, the developmental program of a species can evolve rapidly with the birth of species-specific and lineage-specific components of the genetic systems underlying development.*

Emerson, J. J., H. Kaessmann, E. Betrán, and M. Long. 2004. Extensive gene traffic on the mammalian X chromosome. Science 303: 537–540. *This report reveals bidirectional gene trafficking between the X and autosomes in the human genome. In one direction, X-linked parental genes were copied onto autosomes and evolved testis-biased expression; in the other direction, an excess of female genes and nonsex-related genes were moved to the X.*

Gilbert, W. 1978. Why genes in pieces? Nature 271: 501. *This essay presents a model of gene evolution by exon shuffling. The evolutionary products derived from exon shuffling, chimeric genes, have since been widely observed.*

Kaessmann, H., N. Vinckenbosch, and M. Long. 2009. RNA-based gene duplication: Mechanistic and evolutionary insights. Nature Reviews Genetics 10: 19–31. *This article provides a comprehensive review of new gene origination by retroposition (RNA-based duplication) from the underlying mechanisms to the generation of new functions.*

Levine, M. T., C. D. Jones, A. D. Kern, H. A. Lindfors, and D. J. Begun. 2006. Novel genes derived from noncoding DNA in *Drosophila melanogaster* are frequently X-linked and exhibit testis-biased expression. Proceedings of the National Academy of Sciences USA 103: 9935–9939. *This research article reports five de novo genes that evolved from noncoding regions, mostly on the X chromosome and all predominantly expressed in the testis in* Drosophila.

Long, M. Forthcoming. Evolution and function of new genes from flies to humans. Annual Review of Genetics 47.

Long, M., E. Betrán, K. Thornton, and W. Wang. 2003. The origin of new genes: Glimpses from the young and old. Nature Reviews Genetics 4: 865–875. *This review presents the first general picture of new gene evolution known at the time. It emphasizes the advantage of using young genes to investigate gene evolution.*

Long, M., and C. H. Langley. 1993. Natural selection and the origin of *Jingwei*, a chimeric processed functional gene in *Drosophila*. Science 260: 91–95. *This report presents the first discovery of the evolution of a new gene. Population genetic and molecular analyses revealed that Darwinian positive selection acted on this newly emerged chimeric gene, which consists of an* Adh*-derived retrogene and a duplicate of an unrelated gene.*

Makalowski, W., G. A. Mitchell, and D. Labuda. 1994. Alu sequences in the coding regions of mRNA: A source of protein variability. Trends in Genetics 10: 188–193. *Transposable elements were conventionally viewed as selfish DNAs. This paper presented the first evidence that Alu transposable elements can contribute to protein diversity.*

Muller, H. J. 1936. Bar duplication. Science 83: 528–530. *In this report of Bar duplication, the model of new gene evolution by DNA-based duplication was explicitly described for the first time.*

Schrider, D. R., K. Stevens, C. M. Cardeño, C. H. Langley, and M. W. Hahn. 2011. Genome-wide analysis of retrogene polymorphisms in *Drosophila melanogaster*. Genome Research 21: 2087–2095. *Using molecular population genetics, this analysis detects a significant role for positive selection in the fixation of retrogenes.*

Vibranovski, M. D., Y. Zhang, and M. Long. 2009. General gene movement off the X chromosome in the *Drosophila* genus. Genome Research 19: 897–903. *This paper shows that retrogene trafficking, first proposed by Betrán et al. (2002), is generally true in the genus* Drosophila. *It further demonstrates that DNA-based duplication has contributed to gene trafficking from the X chromosome to autosomes; thus gene trafficking is not a property of biased mutation patterns but a consequence of natural selection.*

Wang, W., H. Yu, and M. Long 2004. Duplication-degeneration as a mechanism of gene fission and the origin of new genes in *Drosophila* species. Nature Genetics 36: 523–527. *This research report presented the first mechanistic analysis of gene fission, a process whereby one gene splits into two genes, revealing that duplication is an intermediate step.*

V.7

Evolution of Gene Expression

Patricia J. Wittkopp

OUTLINE

1. The importance of regulatory evolution: A historical perspective
2. Finding expression differences within and between species
3. Genomic sources of regulatory evolution
4. Enhancer evolution
5. Evolution of transcription factors and transcription factor binding
6. Evolutionary forces responsible for expression divergence

Genetic changes affecting either the function or regulation of a gene product can contribute to phenotypic evolution. Studies of evolutionary mechanisms have historically focused on changes in protein-coding sequences, but during the last decade, multiple lines of evidence have shown that changes in gene expression are at least equally important. The last few years have brought great progress in understanding the genetic basis of expression differences within and between species. From a growing collection of single-gene case studies and comparative analyses of gene expression on a genomic scale, common themes and patterns in regulatory evolution have begun to emerge.

GLOSSARY

Chromatin. The higher-order complex of DNA, histones, and other proteins that packages nuclear DNA within a eukaryotic cell.

Chromatin Immunoprecipitation. A technique in which transcription factors are cross-linked to DNA, the DNA is sheared, and fragments binding to a specific transcription factor of interest are isolated using an antibody. Identity of the isolated DNA fragments can be assessed by PCR, microarrays, or sequencing.

***Cis*-Regulatory Element.** A DNA sequence (such as an enhancer or promoter) located near the coding region of a gene and that has allele-specific effects on gene expression.

Co-option. Using existing functional parts of a genome for new purposes.

Ectopic Expression. Expression in cells that do not usually express the gene of interest.

Orthologous Genes. Homologous genes that diverged following a speciation event.

Pleiotropy. Occurs when a mutation or gene affects more than one phenotype.

Quantitative Trait Locus (QTL). A region of the genome shown to influence a (quantitative) phenotype of interest.

RNA Interference (RNAi). A technique in which short RNAs are used to interfere with the successful production of proteins for a gene of interest.

Transcription Factor. A protein that binds to DNA in a sequence-specific manner and affects transcription.

1. THE IMPORTANCE OF REGULATORY EVOLUTION: A HISTORICAL PERSPECTIVE

For most of the twentieth century, conventional wisdom among biologists was that, as François Jacob described it, "cows had cow molecules and goats had goat molecules and snakes had snake molecules, and it was because they were made of cow molecules that a cow was a cow." Near the end of the twentieth century, however, it became clear that this was not the case. Species-specific genes exist (see chapter V.6), but they are the exception rather than the rule; much of the biological diversity seen in nature is produced by genes whose functions are highly conserved among species. Discovering this conservation was a boon to the medical genetics community, because it justified the use of model organisms such as fruit flies and mice to investigate human disease, but also presented a paradox: How can divergent traits be constructed using conserved genes?

The answer to this question is, in part, by modifying the regulation of gene expression. Expression of a gene is necessary before it can impact the phenotype of an organism; that is, the DNA sequence encoding a gene product must be transcribed into RNA and then (usually, but not always) translated into a protein before the gene can function in a cell. Each cell expresses only a subset of the genes in its genome, and the specific genes expressed determine a cell's fate (see chapter V.11). In 1969, before the molecular details of gene regulation were known, Roy J. Brittan and Eric H. Davidson proposed a theory for the regulation of gene expression in eukaryotic cells. They viewed gene regulation as integral to evolution and suggested that differences among species could be attributable to changes in the regulation of gene expression. Six years later, Mary-Claire King and A. C. Wilson published a seminal paper showing that the amino acid sequences of homologous human and chimpanzee proteins appeared to be more than 99 percent identical. Based on this result, they argued that the degree of protein divergence was insufficient to account for the extensive morphological, physiological, and behavioral differences between these two species.

Despite these (and similar) predictions more than 35 years ago, the idea that changes in gene expression might be a common source of phenotypic divergence did not gain mainstream acceptance among evolutionary biologists until after the turn of the twenty-first century. Seeds of this acceptance were sown when developmental biologists, using newly developed tools for visualizing gene expression, began comparing expression among species. This approach catalyzed the expansion of evolutionary developmental biology, a field of research known today as *evo-devo*. Within a few years, researchers acquired many examples of cases in which divergent RNA and/or protein expression of genes known to be important for development correlated with morphological divergence between species. Such correlations suggest that the genetic changes responsible for altered gene expression might be the same changes responsible for altered phenotypes. In parallel, quantitative geneticists mapping the mutations responsible for phenotypic differences among individuals of the same species or (less commonly) different species were finding that changes in protein sequence could not always account for the phenotypic effect of a quantitative trait locus (QTL) (see chapter V.12).

2. FINDING EXPRESSION DIFFERENCES WITHIN AND BETWEEN SPECIES

Early comparative studies of gene expression focused on one or a small number of genes within or between species. These low-throughput types of studies were (and still are) critical for establishing links between divergent gene expression and divergence of a particular phenotype; however, they are not suitable for obtaining the genomic measures of expression required to identify global trends in the evolution of gene expression. Rather, *microarrays*, which are short DNA sequences complementary to transcribed sequences from a particular species arrayed onto a filter or a microchip, have been used to quantitatively compare the abundance of RNA from hundreds to thousands of expressed genes in the genome simultaneously. Today, microarrays are largely being replaced by a method known as *RNA-seq* that uses massively parallel sequencing to obtain quantitative measures of gene expression (i.e., RNA abundance). Techniques for measuring protein abundance (which is not always highly correlated with RNA abundance) on a genomic scale are also available (e.g., two-dimensional gel electrophoresis, mass spectrometry), but thus far they have not been used to compare protein expression genome-wide in an evolutionary context.

By contrast, the transcriptome (i.e., the collection of all RNAs expressed in a biological sample) has been analyzed in a wide variety of taxa, including human, mice, fish, flies, yeast, and plants. Comparing transcriptomes has shown that differences in RNA abundance are common both within and between species and that the number of genes showing expression differences between a pair of species is often proportional to their divergence time. For example, in one of the first published transcriptome comparisons between species, microarrays containing sequences complementary to approximately 12,000 human genes were used to measure mRNA abundance in the white blood cells, liver, and brain of humans, chimpanzees, orangutans, and macaques. Comparing expression in samples from three humans, three chimpanzees, and one orangutan showed extensive variation within both humans and chimpanzees. The extent of expression divergence between humans and chimpanzees was smaller than the divergence observed when either of these species was compared to the orangutan, suggesting that expression divergence correlates with phylogenetic distance. In the samples derived from brains, one human was found to differ more from another human than from a chimpanzee, but this type of relationship is rare: polymorphic gene expression within a species is typically less extensive than divergent gene expression between species.

In a slightly different experiment, macaques were used as an out-group, and gene expression in humans was found to have evolved faster in the brain than in the liver or blood. Although it is tempting to speculate that this apparently accelerated evolution of gene expression in the human brain may have contributed to the evolution of human-specific cognitive abilities, a reanalysis of these data that more completely modeled the sources of variance in the experiment found more genes with differential expression in the liver than in the brain between humans and chimpanzees. This example illustrates the potential tremendous impact of statistical analysis methods on the conclusions drawn from this type of work. Particularly problematic in this case (and in other cases where a microarray with sequences from one species is used to compare expression between species) is accounting for the effects of sequence divergence between the microarray probes and the heterologous species. The newer RNA-seq method of quantifying and comparing RNA abundance among species circumvents this problem, but presents its own set of challenges for proper data analysis and interpretation.

3. GENOMIC SOURCES OF REGULATORY EVOLUTION

Heritable differences in the distribution of RNA or protein within or between species often result from changes in the sequence of genomic DNA. To understand the types of sequences in the genome that can be mutated to alter gene expression, one must consider the molecular mechanisms controlling transcriptional and posttranscriptional regulation of gene expression. Within prokaryotes and eukaryotes, these mechanisms are highly conserved, but they differ significantly between the two groups. The remainder of this chapter focuses solely on transcriptional regulation in eukaryotes because it has been studied most extensively in an evolutionary context. Also, the term *gene expression* is used synonymously with *transcription* from this point forward.

When, where, and how much mRNA is produced from a particular gene is determined by its *cis*-regulatory DNA sequences as well as the *trans*-regulatory transcription factor proteins and noncoding RNAs present in a cell. These *cis*-regulatory DNA sequences include the basal promoter that binds to RNA polymerase and its associated cofactors as well as one or more enhancers that encode instructions for spatiotemporal expression and the amount of mRNA to produce (figure 1). Basal promoter sequences are located near the transcriptional start site and are more highly conserved than enhancer sequences, because they bind to transcription factors such as the TATA-binding protein required for transcription of most genes. Enhancer sequences typically comprise a few hundred base pairs, can be located upstream (5'), downstream (3'), or in an intron of the associated gene (figure 1), and are bound by transcription factors that activate expression from the basal promoter in a subset of cells or under a subset of environmental conditions. In multicellular eukaryotes, expression of a gene tends to be controlled by multiple enhancers, each acting independently and controlling expression in a particular place, time, or environment. Because of their more limited effects on an organism (i.e., lower pleiotropy), enhancers are commonly thought to be more likely to harbor mutations that survive in natural populations and give rise to polymorphism and divergence than mutations in basal promoters or coding sequences of transcription factors.

Chromatin can also have *cis*-regulatory effects on gene expression. Like the rest of the genome, *cis*-regulatory sequences are wrapped around histones and packaged into nucleosomes that form chromatin structure. The state of chromatin influences interactions between *cis*-regulatory sequences and *trans*-regulatory factors, thus it is also an important component of transcriptional regulation. Methods suitable for comparing chromatin structure within and between species have recently become available and researchers are investigating how chromatin structure evolves, as well as how this evolution impacts gene expression. Many transcription factors are known to modify chromatin, for example, by acetylating or deacetylating histones, so changes in *cis*-regulatory sequences affecting binding of such transcription factors could be responsible (at least in part) for differences in chromatin structure when they are observed (see chapter V.8).

Determining whether an expression difference between two genotypes is caused by genetic changes in *cis*- or *trans*-regulation can be done using transgenic analysis, allele-specific expression, or genetic mapping. In the first two cases, activity of homologous *cis*-regulatory sequences controlling a divergent expression pattern of interest are compared in the same cellular environment (i.e., when regulated by the same set of *trans*-acting factors). If a difference in the activity of the two *cis*-regulatory sequences is observed, this indicates that there has been functional *cis*-regulatory divergence. This test can be performed by using transgenes to move *cis*-regulatory sequences from species A into the *trans*-acting genetic background of species B (and vice versa) (figure 2A) or by simply crossing the two genotypes and testing for differences in allele-specific expression when the two *cis*-regulatory alleles are in the same heterozygous *trans*-acting genetic background (figure 2B). Putatively *cis*- and *trans*-acting changes can also be inferred from genetic mapping, in which regions of the genome contributing to the expression difference of interest are identified. If such a region is located close to the affected gene, it is assumed to act in *cis*; if such a region is located far from the affected gene, it is assumed to act in *trans*.

As a group, studies using transgenes to investigate regulatory evolution provide evidence for both *cis*- and *trans*-regulatory changes underlying expression divergence, with *cis*-regulatory divergence detected most

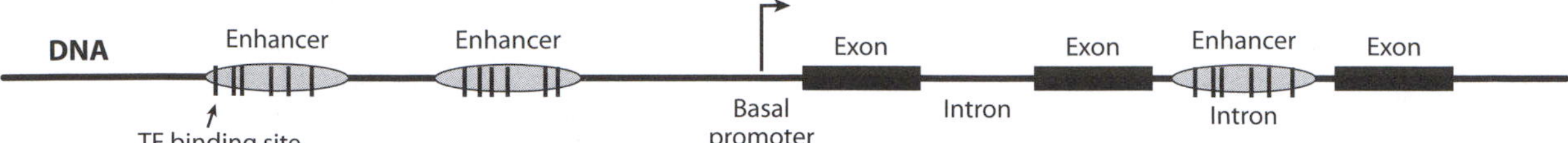

Figure 1. Basic eukaryotic gene structure. *Cis*-regulatory sequences include enhancers and the basal promoter. Most transcription factors (TFs) bind to sequences in enhancers, and transcription factors that compose the RNA polymerase II complex bind to sequences in the basal promoter (neither are shown).

often. Allele-specific expression has been used to examine sources of polymorphic and divergent expression genome-wide in flies, yeast, and plants, and these data suggest that *trans*-acting variation is the predominant source of expression differences among individuals of the same species, whereas *cis*-regulatory changes play a larger role in expression divergence between individuals of different species. Genetic mapping of expression differences has thus far been limited to variation within a species, but it also shows an abundance of variants with apparent *trans*-acting effects on gene expression segregating within a species. Both genetic mapping and allele-specific tests show that although *cis*-regulatory polymorphisms within a species are more rare than *trans*-regulatory polymorphisms, they tend to have larger effects on expression and are less likely to be recessive than *trans*-regulatory variants. The additivity of *cis*-regulatory mutations, combined with the expected lower levels of pleiotropy relative to *trans*-acting mutations, may also make them more likely to contribute to phenotypic evolution.

4. ENHANCER EVOLUTION

As described above, enhancer sequences are an important source of evolutionary change. This class of *cis*-regulatory sequences has been studied in the most detail during the last decade, and these studies have revealed a complex relationship between the DNA sequence and function of an enhancer. The evolution of enhancer sequences that are functionally conserved, functionally divergent, and those that have acquired novel activities are discussed below.

Because enhancer sequences are critical for the proper development and physiology of an organism, most mutations that alter their activity are expected to be deleterious and removed from a population by purifying selection. Consequently, enhancer sequences should be more highly conserved than surrounding nonfunctional DNA. In fact, they are, and this conservation is a helpful tool for finding enhancers within a genome (see chapter V.3). The degree of sequence conservation in an enhancer is typically lower than that of protein-coding sequences, however, because of the structure-function relationship of enhancers: the same enhancer activity can be produced by multiple arrangements of transcription factor binding sites, and most transcription factors binding sites are degenerate, meaning that the same transcription factor can bind to multiple sequences.

This complex relationship between enhancer sequence and function has been nicely illustrated by comparative studies of two *Drosophila* enhancers whose activity appears to be conserved between species. In the first case, the DNA sequence and transcription factor binding sites of an early embryonic enhancer (controlling "stripe 2" expression of the *even-skipped* gene) have been extensively changed between species, yet the function of the elements remains the same. Orthologous *cis*-regulatory elements from *D. melanogaster* and *D. pseudoobscura* had similar activities in transgenic *D. melanogaster,* whereas chimeric enhancers containing the 5' half from one species and the 3′ half from the other showed abnormal activity. Similarly, extensive rearrangement of transcription factor binding sites was found in an enhancer driving conserved expression in the developing eye of *Drosophila.* The *D. melanogaster* allele of this enhancer has been extensively analyzed, allowing predictions to be made about the consequences of some observed changes. These types of analyses provide insight not only into evolutionary processes but also into enhancer architecture in general.

If enhancer sequences can change extensively and still retain their original function, how much does an enhancer need to change to acquire new activities? A number of studies have been published during the last few years, most notably from the laboratories of Sean B. Carroll, David M. Kingsley, and David L. Stern, that are suitable for addressing this question (see chapter V.12). In some cases, as little as a single nucleotide change is sufficient to account for the divergent activity of an enhancer, whereas in others, multiple mutations (on the order of 10 or fewer) are responsible for expression differences. In addition to single nucleotide changes, larger lesions also contribute to divergent activity. For example, in the threespine stickleback, recurrent deletions that disrupt the activity of an enhancer contribute to the repeated loss of pelvic structures in freshwater populations. In *Drosophila*, deletions in an enhancer of the *desatF* gene have been shown to contribute to expression divergence by (surprisingly) creating novel binding sites

A Using transgenes to compare spatiotemporal expression

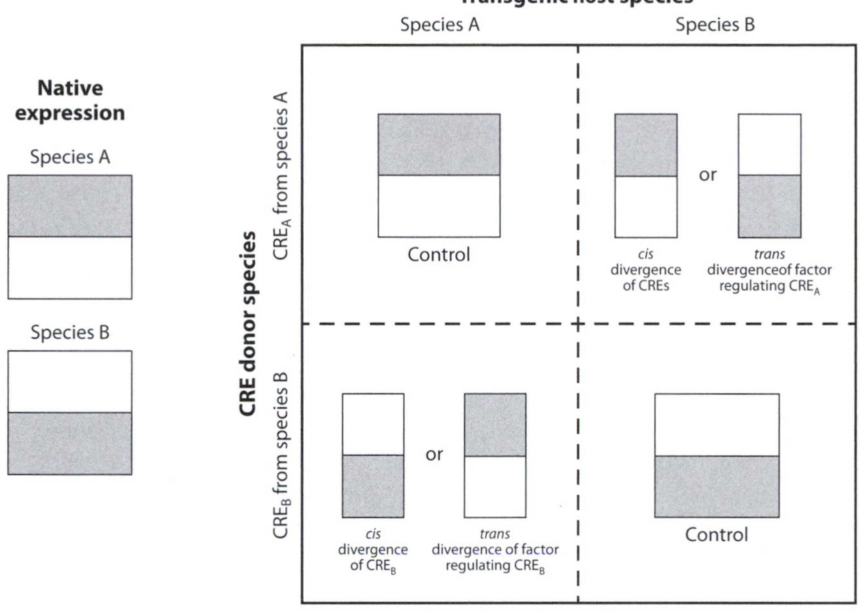

B Using allele-specific expression to compare RNA abundance

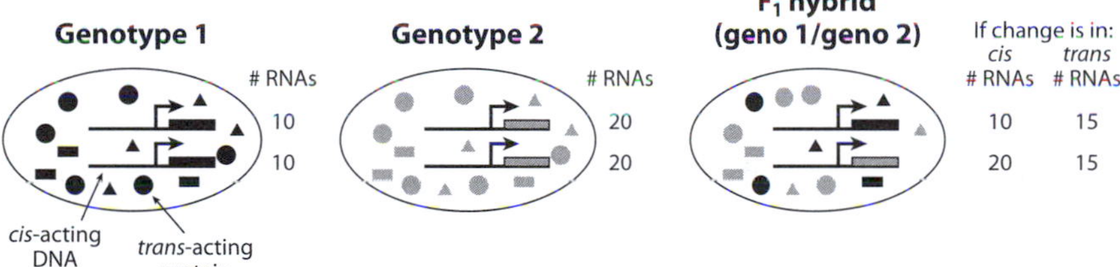

Figure 2. Determining whether divergent expression is due to a change in *cis*- and/or *trans*-regulation. (A) Transgenic analysis can distinguish between *cis*- and *trans*-regulatory divergence by comparing the activity of orthologous *cis*-regulatory elements (CREs) in the presence of the same set of transcription factors. This can be done by creating a pair of artificial genes, each with a CRE controlling expression of a protein that is easy to detect. These so-called reporter genes are then introduced into the genomes of the two species from which the CREs were derived. Different patterns of expression are expected if *cis*- or *trans*-regulatory changes occur between species; a hypothetical example of this is shown in which each box represents a region of tissue, and gray represents either native expression in species A and B or the expression of the CRE tested in each species using a reporter gene. (B) Measures of allele-specific RNA abundance can also be used to distinguish between *cis*- and *trans*-regulatory changes in diploid organisms. Schematic representations of cells from two different (homozygous) genotypes (two different species or two different genotypes from the same species) are shown. A schematic cell from an F1 hybrid produced by crossing genotype 1 and genotype 2 is also shown. In each cell, two copies of a gene are shown with the transcribed region indicated by a gray rectangle and the promoter location indicated by an arrow. The solid black line represents DNA, including the CRE, as indicated. Circles and triangles represent two different transcription factor proteins, each of which is present in multiple copies per cell. Hypothetical numbers of RNA molecules produced by each allele in each cell (# RNAs) is also shown. The F1 hybrid contains a CRE allele and transcription factors from each of its parental genotypes. If the expression difference observed between genotypes 1 and 2 is due solely to *cis*-regulatory changes (i.e., the *trans*-acting transcription factors are equivalent between genotypes), each allele produces the same number of RNA molecules in the F1 hybrid as it did when homozygous in genotype 1 or 2. If, on the other hand, the *cis*-regulatory sequences are functionally equivalent between alleles and the difference in RNA abundance observed between genotypes 1 and 2 results from differences in *trans*-acting factors between genotypes, the two CRE alleles in the F1 hybrid will produce an equal number of RNA molecules, with the precise number (15, in this example) determined by the specific type of *trans*-regulatory divergence. Combinations of *cis*- and *trans*-regulatory changes are also possible, with the *cis*-regulatory difference always reflected in relative expression between the two alleles in the F1 hybrid.

for an unknown transcription factor that activates expression. In the few cases where multiple changes have been implicated in expression divergence and their effects tested individually, the substations have been found to interact in a nonadditive (i.e., epistatic) fashion.

The majority of work on enhancer evolution has focused on cases in which enhancer activity is either conserved or divergent. But what about new enhancers? How do they evolve? Simulations suggest that new point mutations could frequently generate novel transcription factor binding sites and that they could fix over microevolutionary timescales, even in the absence of selection. This suggests that new enhancers driving novel expression patterns might frequently arise de novo. Despite this finding, all the cases of (putatively) novel enhancers characterized to date appear to have evolved using other mechanisms (i.e., duplication and divergence, transposition, or co-option), with co-option (i.e., repurposing) of existing regulatory elements the most common mechanism—in both fruit flies and primates, *cis*-regulatory sequences controlling novel expression patterns have been shown to include sites required for one or more preexisting enhancers.

5. EVOLUTION OF TRANSCRIPTION FACTORS AND TRANSCRIPTION FACTOR BINDING

To function, *cis*-regulatory sequences must be bound by *transcription factors* (TFs), which are proteins that bind to specific DNA sequences and influence (i.e., either activate or repress) transcription. Molecularly, TFs typically contain a DNA binding domain, one or more protein-protein interaction domains, a transcriptional activation or repression domain, and sometimes a chromatin modification domain. As a group, genes encoding TFs are among the most highly conserved in eukaryotic genomes, especially in their DNA binding domains. This high degree of similarity among species is seen not only in terms of protein sequence but also with functional tests. Perhaps the most seminal of these tests showed that the *Drosophila eyeless* and mouse *Pax-6* genes are orthologous genes, and that ectopically expressing either of them in developing *Drosophila* wings or legs was sufficient to transform cells into ectopic eyes. Importantly, both the *Drosophila* and mouse alleles of this gene induced similar morphological transformations, with cell types and organizational structures resembling the normal *Drosophila* eye. This study, and others like it that followed, demonstrated that development is often controlled by highly conserved *master regulatory* proteins.

Conserved master regulatory proteins such as *Pax-6* can create divergent structures by regulating different sets of target genes in different species. Changes in the identity of target genes are mediated by the evolution of TF binding, resulting primarily from changes in *cis*-regulatory sequences. Recently, techniques for monitoring the binding of a particular transcription factor genome-wide have been developed, and comparative studies show that the gain and loss of TF binding sites is very common among species. Between closely related species, changes in the quantitative binding of a TF to a particular site rather than the gain or loss of individual binding sites appears more prevalent. In the next few years, these types of experiments, which rely on chromatin immunoprecipitation, will likely be combined with genomic measures of gene expression and *cis*-regulatory sequence divergence to provide a more complete understanding of how changes in DNA sequence impact TF binding, and how this in turn affects gene expression.

As described above, many of the TFs functionally tested in vivo show conserved functions between species, but this is not always the case—even for highly pleiotropic regulators of development. For example, the function of the HoxA-11 protein has acquired a novel function required for pregnancy in placental mammals, and the Hox genes *fuzi tarazu* and *Ultrabithorax* have diverged between *Drosophila melanogaster* and other insects. In each case, the proteins seem to have retained some ancestral functions while gaining and losing others.

6. EVOLUTIONARY FORCES RESPONSIBLE FOR EXPRESSION DIVERGENCE

With differences in mRNA expression cataloged for a variety of species, researchers are now faced with the daunting task of figuring out what these expression differences mean for organismal phenotypes, especially fitness. Classic genetic mutants and reverse genetic techniques such as RNA interference (RNAi) can be used to assess the function of individual genes, but these techniques are rarely able to predict the consequences of the quantitative changes in expression commonly found in nature. To complicate matters further, mRNA levels do not always correlate with protein abundance, and similar changes in expression of different genes will almost certainly have different effects; for example, a 10 percent change in expression of one gene might have a larger effect on the phenotype than a 1000 percent change in expression of another gene. Connecting changes in gene expression to specific phenotypes is currently best done by studying one gene and one phenotype at a time; however, high-throughput phenotyping strategies currently being developed should soon make it possible to address this question more systematically.

Knowing the impact of a change in gene expression on fitness can help determine the likelihood that the change resulted from natural selection. Assessing the relative roles of neutral and nonneutral processes is a major challenge for evolutionary biology in general.

To date, three main strategies have been used to investigate the role of natural selection in the evolution of gene expression: the comparative method, tests of neutrality, and empirical patterns (see chapter V.14). In the comparative method, evidence of natural selection is inferred when a change in expression is found to correlate with an environmental or other biological factor in a manner that exceeds the correlation expected simply because of shared ancestry. To use tests of neutrality, patterns of regulatory evolution expected from neutral processes must be specified or inferred from the data available. Studies of mutational variance for gene expression provide a starting point for developing these neutral models, but much more remains to be learned about the neutral expectations for regulatory evolution. Finally, empirical patterns, especially comparisons between polymorphism and divergence for expression of a particular gene, capture elements of regulatory variation that cannot easily be incorporated into neutral models. With this approach, one or more representative "baseline" genes assumed to be evolving neutrally are used as references to test for selection, but it is generally not clear which genes should be considered to be evolving neutrally. Presently, there is no consensus about the relative roles of selection and drift in shaping regulatory evolution, although most species show a strong signal of stabilizing selection within a species, indicating that expression levels do matter for fitness.

Regardless of the evolutionary forces underlying the evolution of gene expression, understanding how this important molecular phenotype evolves is a critical component of understanding how the evolutionary process works. The pressing question is no longer whether changes in gene expression contribute to phenotypic evolution but rather when and how they do. The development of many new tools for studying gene expression combined with the recent rapid accumulation of expression and transcription factor binding data suggest that researchers may be able to answer these questions soon.

FURTHER READING

Carroll, S. B. 2005. Evolution at two levels: On genes and form. PLoS Biology 3: e245. *In honor of the thirtieth anniversary of the seminal work of King and Wilson, who provided some of the earliest experimental evidence of the importance of changes in gene expression for evolution, one of the current leaders of the evo-devo field takes a look at the experimental evidence supporting this assertion today.*

Carroll, S. B., J. K. Grenier, and S. D. Weatherbee. 2004. From DNA to Diversity: Molecular Genetics and the Evolution of Animal Design. 2nd ed. New York: Wiley-Blackwell. *This book describes the central role of gene regulation in development and evolution with accessible discussions of specific case studies beautifully illustrated.*

Davidson, E. H. 2006. The Regulatory Genome: Gene Regulatory Networks in Development and Evolution. San Diego, CA: Academic. *This book provides a comprehensive summary of the data and logic behind the assertion that gene regulatory networks are essential for development and play a critical role in evolution.*

Halder, G., P. Callaerts, and W. J. Gehring. 1995. Induction of ectopic eyes by targeted expression of the eyeless gene in *Drosophila*. Science 267: 1788–1792. *This landmark study demonstrated the remarkable conservation of regulatory proteins by showing that homologous genes from species as diverse as fruit flies, mice, and humans all had similar effects on development when introduced into the fruit fly.*

Stern, D. L. and V. Orgogozo. 2008. The loci of evolution: How predictable is genetic evolution? Evolution 62: 2155–2177. *After clearly describing the rationale and alternative models for regulatory evolution, this review provides the most thorough summary to date of studies identifying the types of genetic changes responsible for a divergent phenotype, contrasting the relative frequency of changes attributable to coding and to regulatory mutations.*

Wittkopp, P. J., and G. Kalay. 2012. *Cis*-regulatory elements: Molecular mechanisms and evolutionary processes underlying divergence. Nature Reviews Genetics 13: 59–69. *This review uses data from case studies revealing the genetic and molecular changes responsible for divergent cis regulatory sequences to examine how these types of sequences evolve and to gain insights into more general questions about the mechanisms of evolution.*

Wray, G. A., M. W. Hahn, E. Abouheif, J. P. Balhoff, M. Pizer, M. V. Rockman, and L. A. Ramano. 2003. The evolution of transcriptional regulation in eukaryotes. Molecular Biology and Evolution 20: 1377–1419. *Despite being nearly a decade old, this review remains one of the most complete discussions of the mechanics of gene regulation and its relationship to evolution.*

V.8

Epigenetics

Florian Maderspacher

OUTLINE

1. The concept of epigenetics
2. The history of epigenetics
3. Epigenetics and gene regulation
4. Molecular epigenetics
5. Epigenetic processes
6. Transgenerational epigenetic effects
7. Lamarckism and neo-Lamarckism
8. Epigenetics and evolution
9. Plasticity and assimilation
10. Epilogue

Although coined by Waddington more than 70 years ago, the term *epigenetic* has become widely used only in the past 15 years. A concept, rather than a discipline, epigenetics is being constantly redefined, often controversially. In its broadest sense, epigenetics refers to stable phenotypic changes without a change in genotype. As phenotypes are the result of gene activity, epigenetics is studied by molecular biologists mainly in the context of gene regulation during cellular differentiation. Of particular interest for evolution are epigenetic changes, often induced by the environment, that can be transmitted across generations. At present, it is unclear whether such epigenetic mechanisms have contributed to evolutionary change.

GLOSSARY

(Genetic) Assimilation. A concept put forward by Conrad Hal Waddington, whereby an externally induced phenotype eventually emerges through random genetic mutations.

Chromatin. The material, DNA and protein, of which eukaryotic chromosomes are made. The DNA is wrapped around complexes of histone proteins, forming so-called nucleosomes, in order to accommodate the long DNA strands in the cell nucleus.

Epialleles. States of a gene with the same nucleotide sequence, but different activity that can be transmitted across generations. Known examples in plants and animals are due to differential DNA methylation.

Germ Line. The sequence of genetic material that gets passed on continually from generation to generation. It is distinguished from the soma, which perishes with each individual generation.

Histones. Nuclear proteins that package DNA into chromatin. Histones are chemically modified in many different ways, for instance, by addition of acetyl, methyl, or phosphate groups.

(Genetic) Imprinting. Silencing of alleles of a gene depending on whether they originate from the father or the mother. Imprinting is found in animals and plants and thought to be the consequence of genetic conflicts between parents.

Paramutation. Interaction between two alleles of a gene, whereby one allele permanently silences expression from the other. Examples are known from maize and possibly mice.

Reprogramming. The resetting of the epigenetic state of a cell or chromosome, either naturally or artificially induced. It occurs regularly in the life cycle of many plants and animals and often involves DNA demethylation.

Soft Inheritance. A term coined by Ernst Mayr, referring to the Lamarckian idea that nongenetic changes in phenotype can be transmitted to the next generations.

Transposable Elements. Mobile genetic units that can "jump" between regions of the genome and make up a large fraction of the genome (40% in humans, 85% in maize).

X-Inactivation. A mechanism for balancing gene dosage in some animals in which one sex has more copies of a sex chromosome than the other. In female mammals, one of the two X chromosomes is randomly silenced in somatic tissues.

1. THE CONCEPT OF EPIGENETICS

Epigenetics is not so much a discipline related to a particular object or process of study but rather a concept, the meaning of which has been (and is being) constantly redefined and debated. Today, the word *epigenetic* is used mainly in the fields of molecular and evolutionary biology. Molecular biologists use it to refer to stable changes in gene activity that are not due to a change in the DNA sequence itself, or more often, to loosely refer to chemical modifications of DNA and its associated proteins. Evolutionary biologists use it to describe nongenetic, often induced, phenotypic changes that can be transmitted across generations.

A few common themes underlie the various meanings of *epigenetic:* first, there needs to be a phenotypic effect: a morphological, biochemical, or behavioral alteration in the organism. For evolutionary considerations, it is important whether the phenotypic effect alters the fitness of the organism and whether it is a response to a change in the environment. Second, the effect should be stably transmitted across cell divisions or across generations, and it should in principle be reversible. Third, the effect should not be due to a change in DNA sequence. This last point is the most basic common denominator of epigenetics, as most of its definitions aim at distinguishing epigenetics from genetics.

2. THE HISTORY OF EPIGENETICS

The term *epigenetic* (literally meaning "above" or "after genetics") was coined by Conrad Hal Waddington in 1939 to refer to causal changes by which "the genes of the genotype bring about genetic effects." Waddington was interested in the ways in which genes control embryonic development and chose the term for its similarity to *epigenesis*—the notion that organisms develop de novo, which was historically set against "preformation," the (erroneous) idea that the structures of the organism are already fully contained in the egg or sperm. For Waddington, the "epigenetic constitution" comprised the signals that control genes to "bring the phenotype into being." In 1959, David Nanney, who studied cytoplasmic inheritance in the ciliate protist *Tetrahymena,* described epigenetic systems as "systems regulating the expression of genetic potentialities." Nanney also noted that epigenetic changes should be reversible. Later, Joshua Lederberg suggested the term *epinucleic information* for all information not directly dependent on the DNA sequence itself.

Until about 20 years ago, the adjective *epigenetic* was rarely used, but in the last 15 years its relative usage in the literature has increased 10-fold compared with the previous 15 years. The modern usage is much narrower than Waddington's original definition and essentially equates epigenetics with chemical modifications of DNA and its associated proteins. This usage can be traced back to Robin Holliday, who in the 1970s and 1980s postulated that the methylation of DNA could serve as a heritable epigenetic mechanism of gene expression control. In the late 1990s and early 2000s, histones—proteins that package DNA in the nuclei of eukaryotic cells—were found to be chemically modified in many different ways. These modifications were interpreted by some as a "histone code" that carries epigenetic information about gene regulation. Around the same time, a renewed interest in the evolutionary aspects of epigenetics emerged. In particular, Eva Jablonka has promoted the idea that instances of acquired traits that are inherited across generations call for an overhaul of Darwinian evolution.

3. EPIGENETICS AND GENE REGULATION

What "brings the phenotype into being" are the gene products—RNAs and proteins—expressed by a given organism or cell. A nerve cell is different from a muscle cell, because it expresses different genes, although both cell types share the same DNA sequence; thus, the difference between the two is epigenetic. Whether a given gene is activated or not is decided mainly at the level of transcription; that is, whether or not a messenger RNA is copied from the gene, and in what quantity. This decision is computed by transcription factors that either promote the gene's transcription or repress it and bind in a sequence-specific manner to regulatory DNA elements located around the gene (see chapter V.7). Small RNA molecules also contribute to gene regulation by preventing messenger RNAs from being translated into proteins. Whether a gene is active or not thus depends on the regulatory sequences associated with it, a genetic component, and the presence of the transcriptional and translational regulators, an epigenetic component.

The transcription factors themselves are regulated in response to internal signals, such as hormones or growth factors. Environmental cues, for instance, a change in temperature, can also affect gene activity. During the development of an organism, changes in gene expression are caused by signals that are often transient in nature, yet whose effect on gene expression needs to be stable, such that a muscle cell always remains a muscle cell. Cells thus form a kind of epigenetic memory of gene activity.

The simplest way of maintaining a gene active or inactive in the absence of the initial regulatory signal is through feedback loops of transcription factors that regulate each other's expression. For instance, the mammalian muscle differentiation factor myoD activates the transcription of several other transcription factors that in turn maintain myoD transcription. These factors are passed on during cell divisions and ensure continued activity of the muscle program in daughter cells. Although such mechanisms are found abundantly during embryonic development and fit Waddington's original definition, they are nowadays rarely referred to as "epigenetic." At the same time, the relationship of what is now called "epigenetic" to gene regulatory mechanisms is a matter of active research and debate.

4. MOLECULAR EPIGENETICS

Modification of DNA-Associated Proteins

In eukaryotic cells, the genetic material is stored as *chromosomes*, which consist of *chromatin*—DNA wrapped around protein complexes of histone proteins. Chromatin is packaged to accommodate the length of the DNA strands in the cell nucleus. It is thought that the degree of packaging is closely correlated with the transcriptional activity of a particular DNA region: active regions are loosely packed such that the transcriptional machinery can gain access, while inactive regions are packed tightly. The degree of packaging depends on various chemical modifications attached to the histones at defined sites.

There are dozens of different histone modifications that—depending on their position and chemical nature—can have different effects on chromatin packaging and other biological functions. The association of a given modification with a particular function or activity state of a gene has led to the idea of a "histone code" or "epigenetic code." Acetyl groups on histones, for instance, are generally associated with transcriptional activity, while other modifications, such as methyl groups, can have different effects depending on where they are added. It is thus likely that the function of an individual histone mark depends on the overall context, rather than being defined by a rigid code.

Two important questions concern the possible epigenetic function of histone modifications: First, how exactly do histone modifications relate to gene activity? There is much evidence that histone modifications can be influenced by physiological or pathological changes in the cell as well as by the environment, but to what extent these changes are a cause or an effect of altered gene expression is not clear. Second, are histone modifications stable, and can they be faithfully transmitted across cell divisions? While there are some indications that histone marks can be replicated in the absence of the original inducers, a clear replication mechanism has yet to be found. And even if there is transmission across cell divisions, is it self-perpetuating, or maintained by *trans*-acting transcription factors? Most evidence up to now points toward histone marks as a consequence rather than a cause of changes in gene expression effected by the transcriptional machinery.

A particular class of proteins is associated with preserving a cell's memory of gene activity: the polycomb group (PcG) and trithorax group (trxG) proteins. Broadly speaking, PcG proteins keep silent genes silent, while trxG proteins keep active genes active even after the initial regulators have disappeared. Most of the genes targeted by PcG and trxG proteins encode transcriptional regulators themselves, in particular the *Hox* genes that define the identity of body regions in animals (see chapter V.11). In *Drosophila melanogaster*, the decision about which *Hox* gene is active where is made early in development, but their inducers are present for only a short time. The activity status of a given *Hox* gene is preserved by PcG and trxG proteins, such that the tissue in an adult fly will retain the *Hox* identity of its embryonic precursor

DNA Methylation

Of all the putative carriers of epigenetic information, DNA methylation has received most attention. DNA's cytosine bases can have a methyl group attached, giving rise to 5-methylcytosine. Aspects of the methylation machinery are similar between divergent types of organisms, suggesting that DNA methylation evolved very early in the history of life. Methylated DNA is generally associated with inactivity of genes, such that when methylation is lost, previously silent genes become transcribed. Aberrant DNA methylation is, for instance, associated with irregular gene activity found in some cancers. Methylation serves mainly to keep genes silent in the long term, but it is a consequence rather than a cause of inactivated gene expression. DNA methylation is particularly important for the silencing of transposable elements. Transposable elements can be potentially harmful if mobilized; thus protection against transposable elements and related genetic parasites may be a primary function of DNA methylation. DNA methylation can be inherited across cell divisions through a mechanism closely linked to the replication of the DNA strands themselves; however, the fidelity with which DNA methylation marks are inherited is many orders of magnitude lower than that of DNA replication, and they generally do not persist across generations.

The extent of DNA methylation varies greatly between species: yeast (*Saccharomyces cerevisiae*) and the nematode *Caenorhabditis elegans* show no DNA methylation, *Drosophila* fruit flies have very little, while maize and mammals show extensive methylation. Likewise, the way DNA methylation is distributed across the genome varies greatly: the genomes of mammals are methylated throughout, with the exception of the regulatory regions of many genes. In some plants, like maize, the genome is almost completely methylated, while in the plant model *Arabidopsis thaliana*, methylation is clustered around repetitive DNA and transposable elements. The degree to which DNA methylation patterns vary between individuals and over time is currently an active area of research. In *Arabidopsis*, for instance, DNA methylation patterns vary between plants that are genetically nearly identical. Most importantly, DNA methylation is relatively stable over generations but in some instances switches back and forth. Much like changes in gene expression, methylation patterns can also vary with age and between different environments.

The predominant function of DNA methylation (and to some extent histone modifications) is to keep genes whose functions in a given cell are not needed—or in the case of transposable elements, not wanted—reliably silenced. Molecular epigenetic systems are thus, quite in the sense of Waddington, vital for bringing the phenotype into being, but they can act only on the genetic program encoded in the DNA sequence. And of course, the whole molecular epigenetic machinery is itself encoded in the genes.

5. EPIGENETIC PROCESSES

The best-studied epigenetic processes concern the permanent silencing of alleles or entire chromosomes. Genomic imprinting, for example, is an epigenetic process in animals and plants that leads to the expression of certain genes in the offspring from only one of the two alleles—either the one inherited from the father, or the one from the mother. In mammals about 100 genes are imprinted, most of which are involved in mother-offspring interactions. Imprinting is established in the germ cells according to the sex of the organism, and in mammals is initiated by DNA binding proteins or noncoding RNAs. In both animals and plants, all the epigenetic systems mentioned in the previous section are involved in maintaining the silenced state.

Genomic imprinting has been proposed to be the evolutionary consequence of a "conflict of interest" between the genes of the mother and those of the father. In placental mammals, where the embryo is nourished by the mother, it is in the mother's interest not only to feed the embryo, but also to protect her own resources, while the father's genes benefit only from a thriving embryo. In line with this, most of the paternally imprinted genes promote the growth of the embryo, while those expressed from the maternal allele restrict it.

During *X-inactivation*, one of the two X chromosomes is permanently silenced in the cells of female mammals. This ensures the same amount of X-derived gene product is made as in males, which have only one X chromosome. X-inactivation is initiated by a noncoding RNA (Xist) that marks the chromosome to be silenced, and DNA methylation is necessary to keep the silenced X chromosome silent (see chapter V.4).

Another paradigmatic epigenetic process is *paramutation*, which has been best studied in the maize *b1* gene that promotes purple pigmentation. Two alleles differentially affect *b1* transcription but share an identical DNA sequence: the *B-I* allele shows high *b1* expression, while the *B'* allele shows low expression. When the two alleles are present in the same plant, *B'* will suppress the expression of *b1* from the *B-I* allele, such that the heterozygous plant looks like a *B'* plant. The *B-I* allele is said to be paramutated and will functionally behave like a *B'* allele in subsequent generations; moreover, it will be able to paramutate other *B-I* alleles, an effect that lasts for many generations.

Although epigenetic marks, such as those established during X-inactivation, can be stable in somatic cells, the marks are erased in many organisms before the beginning of a new generation, a process known as *reprogramming*. In the mammalian life cycle, there are two major episodes of reprogramming: in the early embryo after fertilization, and in the primordial germ cells of embryos. Reprogramming is important in various contexts, especially so that imprints on genes can be reset and renewed according to the sex of the embryo, but also for the fertilized eggs to become *pluripotent*; that is, to be able to give rise to all the different cell types of the organism. Scientists have exploited this particular aspect of reprogramming to generate induced pluripotent stem (iPS) cells from differentiated somatic cells of mammals by adding extraneous transcriptional regulators that revert the differentiated state of a cell.

Unlike animals, plants do not show a global reprogramming in germ cells but rather in the tissues that nourish the plant embryo. From there, small RNAs are thought to reinforce silencing of transposable elements in the embryo. A classic example of a reprogrammed epigenetic response in plants is vernalization. Many plants in temperate regions require a prolonged exposure to cold to induce fast flowering and seed development. Individual plants form a long-lasting epigenetic memory of the transient cold experience. In *Arabidopsis*, the vernalization response is associated with silencing of the gene *FLC*, which encodes a repressor of flowering. In seeds, *FLC* silencing is reset, which is why the offspring of vernalized plants do not retain a memory of the vernalization response.

6. TRANSGENERATIONAL EPIGENETIC EFFECTS

Of particular interest to evolutionary questions are epigenetic changes that are heritable across generations, for which there are a handful of examples in plants and animals. These can come in the form of more or less stable epialleles, which involve gene silencing associated with altered DNA methylation, or they can be induced through environmental stimuli.

Plants

Plants differ from animals in two ways that are relevant for epigenetics: first, they do not set aside their germ line early in development, which along with the absence of large-scale reprogramming might make them more prone to carrying epigenetic changes into the next generation. Second, plants cannot move and thus may have a greater need to modulate their phenotype in response to the environment. Because seed dispersal is often limited, offspring will grow up in an environment similar to that of their parents; it thus may be beneficial for them to be prepared for that environment through a kind of transgenerational epigenetic memory.

Three particularly well-characterized examples of transgenerational epigenetic effects come from heritable epialleles in plants. In toadflax (*Linaria vulgaris*), so-called peloric mutants have radially symmetrical flowers instead of asymmetrical ones. In tomatoes (*Solanum lycopersicum*), *colorless nonripening* mutants do not turn red, and their flesh disintegrates instead of softening. In melons (*Cucumis melo*), certain gynoecious lines, unlike other strains, only form female flowers. In all these cases, the phenotype is relatively stable across generations and has been linked to lowered expression at the underlying gene without a detectable difference in nucleotide sequence between mutant and normal plants; instead, all these mutants show excessive DNA methylation near the silenced gene. However, an adaptive value is not apparent in any of these cases.

Plants have evolved a number of phenotypic responses to environmental stress or predators, and in some instances such responses have been shown to be heritable. One example for such an induced transgenerational effect comes from the yellow monkeyflower (*Mimulus guttatus*). In some strains of *Mimulus,* the density of trichomes—hairs that serve as a defense against insect herbivores—increases in response to insect damage. Offspring from mother plants that showed this response will form more trichomes, even if they have experienced no leaf damage themselves; thus the induced phenotype is transmitted from mother to offspring. While a possible adaptive value of this response is conceivable, it is not clear how it is transmitted, nor for how many generations the effect lasts.

In *Arabidopsis,* stressors, such as UV light or bacterial proteins, induce genome instability—a so-called genomic shock. This effect is heritable through both the maternal and paternal germ lines and persists for about four generations in offspring that do not experience the stressors themselves. The precise basis for this effect is not clear, nor is its possible adaptive value, but DNA methylation and small RNAs are necessary for its transmission. A similar transgenerational effect with a possible advantageous function was found for temperature. When *Arabidopsis* plants were exposed to heat, their third-generation offspring, when confronted with a hot environment, produced many more seeds than plants whose ancestors had not experienced heat.

Animals

In animals, the distinction between germ line and soma is made early in development, and while epigenetic mechanisms play a role in gene silencing during somatic differentiation, the extent to which environmentally induced effects can become transmitted through the germ line is not clear. When transgenerational effects in mammals are considered, it is worth bearing in mind that in the mammalian germ line epigenetic marks are widely reset. Moreover, a pregnant female contains not only her own offspring but also the germ cells of that offspring; thus, an environmental effect on the mother can in principle directly affect the two following generations without necessarily implying transgenerational epigenetic inheritance.

The best-studied transgenerational effect in animals comes from an epiallele of the mouse coat color gene *agouti*, which encodes a protein regulating pigment synthesis. Offspring of mothers carrying the *Agouti viable yellow* (*Avy*) allele can range from yellow to nearly normal coat color despite being genetically nearly identical. The range of phenotypes will depend on the phenotype of the mother: darker mothers tend to have more dark offspring, while lighter mothers have more light offspring. The *Avy* allele is caused by a transposable element near the *agouti* gene, and *Avy* alleles differ in their degree of DNA methylation: highly methylated alleles lead to less agouti expression and darker offspring, whereas less methylated alleles lead to yellow offspring. Most likely, the epiallele persists through reprogramming because of the transposable element being targeted by silencing mechanisms. Interestingly, genetic alleles of *agouti* have been implicated in evolutionary adaptation of coat color in beach mice (see chapter V.12).

Three recent examples of induced transgenerational epigenetic effects come from the nematode worm *C. elegans*, a popular model organism. In *C. elegans,* an introduced RNA virus will trigger a silencing response that prevents expression of the virus components through small RNAs that are complementary to viral sequences. This response can persist over several generations, in some individuals indefinitely, even if the virus is removed and reintroduced in subsequent generations. While this response may be of obvious adaptive value to the worms, it is not clear whether it is relevant in the context of naturally occurring viruses, as the virus was introduced artificially. In another example, the offspring of *C. elegans* worms reared in the presence of attractive odors retain an increased preference for these odors even if they have not experienced them before. If raised in the same conditions over more than four generations, offspring can retain the olfactory imprint for at least 40 generations. This transgenerational imprinting has the potential to prime a new generation's preferences for the presumably favorable conditions in which its ancestors grew up. The third example concerns life span. Genetic mutants for several components of the machinery responsible for generating certain histone modifications live longer by about 20–30 percent. When from these mutants offspring are generated that no longer carry the mutant alleles, these worms still live longer than worms whose ancestors did not carry the mutations. What causes this effect, which vanishes after three or four generations, is not known.

A very well-studied transgenerational effect in mammals concerns maternal care behaviors of rats. When female rats are stressed, they dedicate less time to maternal care. Pups that receive less maternal care will in turn provide less maternal care to their own pups. Less maternal care ultimately translates into lower glucocorticoid receptor expression during the first week of a pup's life. This change in gene activity is carried over into adulthood—when the stimulus provided by maternal care is no longer present—possibly through increases in DNA methylation and histone modifications. Interestingly, this effect is seen not only for a mother's natural, genetically related offspring but also when unrelated pups are introduced to the nest; therefore, the effect must be transmitted completely independently of the germ line, through the mother's behavior.

Another transgenerational effect in rats is due to chemicals that interfere with the hormonal regulation of sexual development. Compounds such as the fungicide vinclozolin can interfere with steroid signaling during gonad development. Male rats exposed in utero to vinclozolin show reproductive defects, most notably reduced sperm number and motility. These effects on male fertility are transmitted across several generations in a sex-specific manner: offspring will inherit the effect from their fathers but not from their mothers.

Humans

Unless one chooses to include culturally transmitted traits such as language—which some do—the evidence for transgenerational epigenetic effects in humans is scant. The most popularized cases concern the effects of parental malnutrition on offspring health. For instance, the children of Dutch women who were pregnant during a famine in World War II were found to be more likely to develop type 2 diabetes; however, this particular effect lasted only into the first generation and thus may simply have been a maternal effect of exposure in the uterus. The only evidence for an effect extending into the second generation comes from the Swedish Överkalix population. In the nineteenth century, increased food availability for males between ages 8 and 12 led to a shorter life span in their grandsons as a result of cardiovascular disease and diabetes. A similar relationship was also seen for grandmothers and their granddaughters. However, such effects have so far not been described elsewhere; thus confounding environmental or social factors cannot easily be ruled out.

7. LAMARCKISM AND NEO-LAMARCKISM

The notion that traits that change as a response to the environment could be inherited is associated with Jean Baptiste Lamarck. Based on the well-known malleability of traits—the muscles of the blacksmith grow as he works—Lamarck posited in 1809 that changed environments lead to changed habits in the use of an organ (see chapter I.2). Through inheritance of the acquired state of the organ, the organ will increase or decrease in the offspring. Lamarck's theory initially received little recognition, but similar ideas were widespread, and even Darwin incorporated them into his theory. Only in the last third of the nineteenth century did Lamarckian ideas gain popularity as an alternative to natural selection, which was rejected by many on moral grounds.

At the same time, experimental evidence was mounting against the inheritance of acquired traits. August Weismann conducted a famous experiment in which he cut off the tails of mice in five subsequent generations and in nearly a thousand offspring never observed any tail shortening. Weismann rejected the idea of inheritance of acquired traits because it was incompatible with his idea of germplasm—what we would today call the *genome*—that was propagated independently and separate from the somatic cells. In addition, Theodor Boveri's demonstration that the chromosomes determine the traits of a cell, the rediscovery of Gregor Mendel's laws of inheritance, and finally the chromosome theory of inheritance by Thomas Hunt Morgan established that genetic factors controlled the phenotype. This notion became enshrined in evolutionary thinking through the modern synthesis that unified genetics and Darwinian evolution (see chapter I.2).

Lamarckian ideas became further discredited because of two men, Paul Kammerer and Trofim Lysenko.

Kammerer had claimed that breeding in water induced mating pads in midwife toads. Midwife toads usually lack these pads, as they breed on land, but because he found that the induced pads could be inherited, Kammerer publicized this as an instance of Lamarckian inheritance. Later, he was accused of fraud, and ended his life. Lysenko studied the vernalization response in crops and claimed it too could be inherited, and that in fact all inheritance was due to acquired characteristics—an idea that resonated well with Communist ideology. Lysenko rose quickly in the ranks of Stalin's Russia and was responsible for widespread agricultural malpractice.

With the advances in molecular epigenetics, a renewed interest in Lamarckian inheritance, sometimes called *soft inheritance*, has emerged in some circles. This interest is due in part to claims that DNA sequence variation alone was not able to explain the diversity of life or the notion of "missing heritability" in human diseases—despite many diseases being heritable, the individual DNA sequence variants found associated with a disease contribute statistically very little to an individual's disease risk (see chapter V.13). To some extent, this interest is echoed in the public domain and can be seen as a backlash to overblown claims of genetic determinism at the onset of the genomic age.

8. EPIGENETICS AND EVOLUTION

The question whether and to what extent epigenetic processes contribute to evolution is a matter of debate. Compared with the vast number of genetically transmitted traits, transgenerational epigenetic effects are indeed extremely rare (or very difficult to detect). Many of the best-documented cases of phenotypic change without concomitant genetic change are epialleles. Epialleles change the activity status of a gene, without changing its DNA sequence, but they are most likely independent of environmental inducers. In the cases known so far, the "mutant" epiallele causes an increase in DNA methylation, thus leading to reduced function of the gene. In many cases, epialleles are linked to transposable elements that remain stably methylated, presumably as part of a genomic defense. From an evolutionary point of view, such epialleles behave largely like genetic alleles. Their frequency could in principle increase or decrease in a population as the result of natural selection depending on their effect on fitness; however, epialleles can be more variable than genetic alleles, as the case of *agouti* illustrates. They could thus provide an additional source of variation, which might be advantageous in fluctuating environments; although for most known epialleles, an adaptive value is not readily apparent. Epialleles are also less stable and revert more frequently to the alternative state, which means that in each generation some of the variation may be lost. So far, there is no compelling evidence that epialleles have been the target of natural selection.

By contrast, heritable epigenetic changes that are induced as a response to environmental stimuli behave rather differently from genetic alleles. Genetic mutations arise more or less randomly (though some regions of the genome are more prone to mutations than others) and independent of the selective pressures and phenotypic change they may effect. New mutations will be rare, and even if they have a beneficial effect, they may either be lost from the population by genetic drift, or they can take several generations to become frequent in the population. By contrast, an induced plastic response can in principle affect many members of a population at once. When the change is heritable, it thus has the potential to lead to widespread adaptation much faster than genetic change. Hence, heritable induced epigenetic effects have been postulated to operate on an intermediate timescale, between the short-term phenotypic plasticity that affects only one generation and the long-term genetic adaptation that takes multiple generations to take effect. However, so far no case is known in which an induced heritable response has been shown to underlie an evolutionary change in nature.

9. PLASTICITY AND ASSIMILATION

The notion of the inheritance of acquired traits is closely linked to *phenotypic plasticity*—the ability of a given genotype to generate different phenotypes depending on environmental conditions (see chapter V.11). Phenotypic plasticity is a basic property of organisms and vital for coping with changing environments. If the environmental fluctuations last longer than the generation time of an organism, it may be advantageous if the offspring is already endowed with the adaptive phenotype. One way of carrying over a plastic response into the next generation is through maternal effects, whereby the mother influences the phenotype of her offspring, for example, by adjusting the chemical composition of the egg; this need not necessarily invoke a hereditary mechanism in the strong (neo-)Lamarckian sense.

An acquired trait that is transmitted over several generations (or is induced anew in each generation) might eventually become fixed genetically if mutations that cause the same phenotype as the induced response accumulate in its carriers. This effect was postulated by Waddington as "genetic assimilation" and tested experimentally. While these ideas have been widely popularized, because of their appeal for explaining fast adaptation, there is currently no direct evidence for either genetic assimilation or the inheritance of acquired traits playing a major role in evolutionary change. By contrast, there are by now numerous examples for the genetic basis of evolutionary adaptations. It is also worth keeping in mind that the very ability to respond to an environmental stimulus with a meaningful change in phenotype requires a dedicated machinery that translates the environmental stimulus into a change in gene expression—a machinery that of course must be genetically encoded and evolved by natural selection.

10. EPILOGUE

Epigenetics in its many guises touches on various ideas that exert great attraction to researchers and the public alike: for the public, because the idea that "your genes are your destiny" has some disconcerting implications if interpreted wrongly; for researchers, because there is a temptation to discover new phenomena that challenge or overthrow dominant paradigms, but also because transgenerational epigenetic effects are inherently interesting biological phenomena. This is not so much because they force us to change the way we think about genetics and evolution but because they are the few exceptions to the rule; as the geneticist William Bateson advised biologists, "Treasure your exceptions."

FURTHER READING

Bird, A. 2007. Perceptions of epigenetics. Nature 447: 396–398. *A brief introduction to key questions in epigenetics.*

Bonasio, R., S. Tu, and D. Reinberg. 2010. Molecular signals of epigenetic states. Science 330: 612–616. *An authoritative summary of molecular epigenetics.*

Daxinger, L., and E. Whitelaw. 2010. Transgenerational epigenetic inheritance: More questions than answers. Genome Research 20: 1623–1628. *A discussion of the evidence for transgenerational inheritance.*

Feng, S., S. E. Jacobsen, and W. Reik. 2010. Epigenetic reprogramming in plant and animal development. Science 330: 622–627. *An authoritative summary of reprogramming mechanisms.*

Haig, D. 2007. Weismann rules! OK? Epigenetics and the Lamarckian temptation. Biology and Philosophy 22: 415–428. *A thoughtful introduction to epigenetics and the controversies around it.*

Henikoff, S., and A. Shilatifard. 2011. Histone modification: Cause or cog? Trends in Genetics 27: 389–396. *A discussion of the functions of histone modifications.*

Jablonka, E., and G. Raz. 2009. Transgenerational epigenetic inheritance: Prevalence, mechanisms, and implications for the study of heredity and evolution. Quarterly Review of Biology 84: 131–176. *A tour de force from the most avid proponent of neo-Lamarckism.*

Maderspacher, F. 2010. Lysenko rising. Current Biology 20: R835–R837. *An opinion piece about public perception of epigenetics.*

Moazed, D. 2011. Mechanisms for the inheritance of chromatin states. Cell 146: 510–518. *A review of transmission mechanisms for epigenetic marks.*

Paszkowski, J., and U. Grossniklaus. 2011. Selected aspects of transgenerational epigenetic inheritance and resetting in plants. Current Opinion in Plant Biology 14: 195–203. *A collection of mechanistic case studies of transgenerational effects in plants.*

Ptashne, M. 2007. On the use of the word "epigenetic." Current Biology 17: R233–R236. *An essay on the different meanings of epigenetics.*

Youngson, N. A., and E. Whitelaw. 2008. Transgenerational epigenetic effects. Annual Review of Genomics and Human Genetics 9: 233–257. *A comprehensive review of epigenetic effects in plants and animals.*

V.9

Evolution of Molecular Networks

Mark L. Siegal

OUTLINE

1. Network representations of biological data
2. Global organization of biological networks
3. Evolution of global network organization
4. Local organization and dynamics of biological networks
5. Evolution of local network organization
6. The future of evolutionary systems biology

The importance of interactions between genes has been evident to biologists since the rediscovery of Gregor Mendel's work at the turn of the twentieth century. Indeed, William Bateson coined the term *epistasis* in 1907 to refer to the masking effect of a variant at one locus on a variant at another locus. Notably, this was two years before Wilhelm Johannsen coined the term *gene*. The study of genetic interactions remains central to many fields, from developmental biology to human genetics to evolution. No longer limited to a single pair of genes at a time, scientists are using new technologies to conduct systematic and comprehensive assays of interactions between biomolecules of various kinds. The large-scale data sets produced by such experiments require new methods of analysis. The tools and perspectives of graph theory, in which collections of objects are represented as networks of pairwise interactions, have become particularly important in organizing and analyzing biological data. Studies of model organisms have revealed that biological networks have very different patterns of connectivity than would be expected if their parts were connected at random. This departure from randomness is true both at the global level (considering all interactions of a specified type) and at the local level (considering small subsets of interactions). Understanding the sources of this nonrandomness is a major challenge for evolutionary biologists. Meeting this challenge will not only yield insights into genome organization but also more broadly impact fundamental, long-standing debates in evolutionary biology. These include debates over the existence of developmental constraints, and over the relative importance of adaptive versus nonadaptive processes in evolution.

GLOSSARY

Degree. The number of edges connecting to a node.

Edge. An interaction between two nodes in a network.

Feedforward Loop. A network motif in which two transcription factors jointly regulate a target gene and one of the transcription factors also regulates the other.

Global Network Organization. The pattern of connections of an entire network, as summarized in statistics such as the frequency distribution of node degree.

Homologous Genes. Genes that are related by descent from a common ancestral DNA sequence.

Local Network Organization. The pattern of connections of a subset of nodes in a network.

Modularity. The extent to which biological functions are divided into *modules*, which are defined as sets of interacting components whose functions are relatively independent of the functions of other such sets.

Network Motif. A subset of nodes connected to each other in a particular pattern.

Node. A component of a network that enters into pairwise interactions with other such components.

Posttranscriptional Gene Regulation. Cellular processes acting on an RNA molecule between the time of its transcription and its translation into protein that alter the ultimate abundance through time and space of the encoded protein.

Posttranslational Gene Regulation. Cellular processes acting on a protein molecule that alter the ultimate abundance and activity of the protein through time and space.

Protein Essentiality. The necessity of a protein for viability, usually determined by testing whether organisms can survive deletion of the gene that encodes the protein.

Subcircuit. A subset of interconnected nodes that performs a specific biological function.

Transcription Factor. A sequence-specific DNA-binding protein that activates or represses expression of a gene by, respectively, increasing or decreasing the probability that RNA polymerase will transcribe the gene.

1. NETWORK REPRESENTATIONS OF BIOLOGICAL DATA

All aspects of cell function require controlled interactions between biomolecules. Certain types of interaction have garnered more interest than others because they undergird the organization and logic of cellular processes. For example, protein-protein interactions are especially important because proteins must physically interact with each other to form macromolecular complexes. Such complexes perform essential activities within the cell, such as replicating DNA, transcribing RNA, and transporting cargo from one cellular compartment to another. Protein interactions also mediate communication between cells via signal-transduction pathways. Interactions between proteins and DNA are also critically important. Proteins must interact with DNA to achieve proper packaging of chromosomes and proper cell division, as well as to control when and where transcription happens. Because biological interactions underlie all aspects of cellular structure and information flow, understanding the ways in which these interactions change through time is central to understanding the evolution of form and function.

Many types of interaction can now be systematically investigated by highly parallel experimental platforms. Any collection of interactions can be represented as a network. The first comprehensive set of biomolecular interactions to be represented as a network was that of the enzyme-catalyzed biochemical reactions that compose intermediary metabolism. The familiar wall chart of these biochemical pathways, created by Gerhard Michal in 1968, was not the product of high-throughput experiments but instead a summary of decades of biochemistry research; still, it is entirely modern in form and serves as an excellent illustration of the elements of network representation. A network is defined by its nodes and its edges (figure 1A). The *nodes* are the components that enter into pairwise interactions. In the network of intermediary metabolism, the nodes are metabolites—the substrates and products of enzymatic reactions. *Edges* represent the interactions. In the network of intermediary metabolism, each edge is an enzymatic reaction that converts a substrate into a product.

There are several key points about network representation that are raised by the biochemical pathways wall chart. First, a given network may have either directed or undirected edges (figure 1A). A directed edge, drawn as an arrow from one node to a second node, implies that the entity represented by the first node acts on or is converted into the entity represented by the second node. An undirected edge, drawn as a line without arrowheads, implies symmetry in the interaction, as is the case in a physical association between two proteins (if A touches B, then B touches A). In the biochemical pathways network, edges are directed—substrates are converted into products; note, however, that in the wall chart, many edges run in both directions, because many enzymatic reactions are reversible.

Second, there is not necessarily one "correct" way to represent a set of biological interactions as a network. The choice to represent metabolites as nodes and reactions as edges is just that—a choice. One could instead represent enzymes as nodes and draw edges between enzymes that share substrates or products. Any network is an abstraction that makes salient some features of a system at the expense of others. For example, a large number of enzymatic reactions include ATP as a substrate or product, yet in the wall chart ATP is not depicted as a single node in the same way that other metabolites, such as glucose-6-phosphate, are. Instead, ATP is repeated throughout the network, wherever a reaction consumes or produces it. This change has no effect on the underlying mathematical representation of the network (e.g., one could easily look up the number of reactions that produce ATP if one wanted to), but it does give a very different visual impression than if ATP occupied a single node (with a very large number of edges connecting to it).

This leads to the third key point, which is that networks are simultaneously mathematical and visual objects. Statistical analyses, such as those described in subsequent sections, are performed on the mathematical object, which usually takes the form of a connectivity matrix. While it might often be clear when a visual representation departs from a strict definition of what constitutes a node or an edge (such as when ATP does not occupy a single node), it can be less obvious what decisions went into defining the connectivity matrix. For example, many types of biological data come from experiments that are prone to error. When gazing at a network, or especially when considering statistical analyses of a network, it is imperative to understand the experimental properties and arbitrary decisions that went into determining whether a node or edge was included.

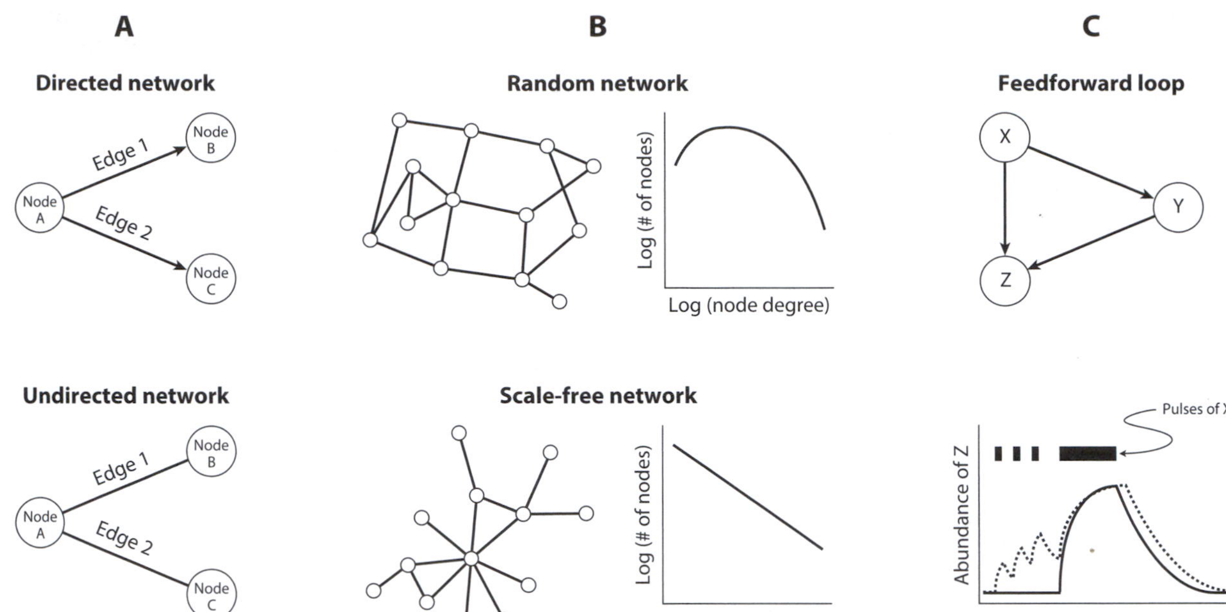

Figure 1. Network terminology and organization. (A) A network comprises nodes (circles) connected by edges. An edge can be directed (arrows, top), when the entity represented by one node acts on or is converted into the entity represented by another node, as in regulatory or metabolic networks. Alternatively, an edge can be undirected (lines, bottom) when the interaction between the two connected entities is symmetrical, as in protein-interaction networks. (B) Global network organization can be described in part by the distribution of node degree (number of interactions per node). In a random network (top), node degree follows a Poisson distribution, whereas in a scale-free network (bottom), node degree follows a power-law distribution with negative exponent. An example of each type of network is shown (left). The characteristic distributions are shown as log-log plots of the number of nodes with each degree (right). (C) A feedforward loop is an example of a network motif. Two transcription factors (X and Y) jointly regulate a target gene (encoding protein Z) and one of the transcription factors (X) also regulates the gene encoding the other (Y) (top). When X and Y are activators, the dynamics of Z expression (bottom) will depend on whether both activators are necessary for target-gene expression (i.e., AND logic applies) or one is sufficient (i.e., OR logic applies). The difference is illustrated by a case in which X is expressed in three brief pulses followed by a sustained pulse (solid bars). If AND logic applies, then Z is expected to be activated only by the sustained pulse of X and to shut down quickly when X is no longer expressed (solid curve). By contrast, if OR logic applies or if the regulation of Z by X is removed to leave a simple linear pathway, then the loop would not filter out short pulses of X and would shut down more slowly (dashed curve).

2. GLOBAL ORGANIZATION OF BIOLOGICAL NETWORKS

Although the first network representation of biomolecular interactions dates back to 1968, the first graph-theoretical analyses of biomolecular networks were reported 32 years later. Perhaps not surprisingly, these too concentrated on metabolism, the system for which we have the most complete knowledge of connections among molecules. In 2000, Albert-Lázsló Barabási and colleagues presented network analyses of intermediary metabolism in 43 organisms, including at least one plant, animal, fungal, eubacterial, and archaeal species (similar analyses were presented at about the same time by Andreas Wagner and David Fell, who concentrated on the metabolic network of one of these species, the bacterium *Escherichia coli*). From the very start, therefore, the statistical analysis of biomolecular networks had a strong evolutionary focus. Intermediary metabolism had not been studied in most of the 43 organisms, so what made the analysis possible was the availability of their complete, or nearly complete, genome sequences. Which enzymes are encoded by the genomes were inferred by standard methods of identifying homologous genes. Thus, edges could be quite confidently drawn between substrates and products despite the lack of any direct biochemical evidence. In other words, the shared evolutionary history among organisms allowed researchers to fill in the gaps when constructing networks in additional species.

The key finding of Barabási and colleagues concerned the frequency distribution of the number of edges per node, otherwise known as the node degree (figure 1B). Consider a network in which the probability of drawing an edge is the same for any two nodes. This is a so-called random network, and its statistical properties have been studied since the seminal work of mathematicians Paul Erdős and Alfréd Rényi in the 1960s (such a network is also called an Erdős-Rényi network). In a random network, node degree follows a Poisson distribution; however, in the metabolic networks studied by Barabási and colleagues, both the number of edges entering a node (the in degree) and the number of edges emanating from a node (the out degree) follow a power-law distribution: the probability of having k edges is proportional to k raised to a negative power. Whereas a Poisson distribution peaks around its mean value, a power-law distribution with a negative exponent always peaks at 1. Such networks therefore have very many nodes with few edges and a few nodes with very many edges. Because there is no peak in the middle of the distribution, and therefore no "typical" node representing the network as a whole, power-law networks have been termed *scale-free*. The power-law distribution gives scale-free networks their characteristic hub-and-spoke appearance, similar to that of an airline route map. Related to this, scale-free networks exhibit the "small world" property, in that the network diameter, or average shortest path between nodes, is very short.

Soon after the analyses of metabolic networks were first reported, analyses of other biological networks followed. Protein-protein interactions have been identified systematically in several organisms, using either yeast two-hybrid assays or affinity purification followed by mass spectrometry. Data sets large enough to examine global network properties exist for model bacterial, fungal, plant, and animal species, as well as for humans. Although there is ongoing debate as to how well these protein-protein interaction data fit power laws, there is no doubt that the degree distribution in each case is a rapidly decreasing function of degree: some proteins have very many interaction partners whereas most proteins have few. The same pattern holds for genetic interaction networks. Genetic interactions are assayed as Bateson would: look for phenotypes resulting from perturbing two genes at once that cannot be explained by adding the effects of perturbing each gene separately. To date, this has been done most comprehensively in the yeast *Saccharomyces cerevisiae*, in which more than 5 million mutant combinations lacking two genes were generated and measured for the ability of cells to grow.

Degree distributions have also been examined for some transcriptional regulatory networks. Systematic identification of the edges in these networks—direct interactions between DNA-binding transcriptional activators or repressors and their target genes—is a difficult and ongoing task. In some model organisms, decades of molecular-genetic research have revealed regulator-target relationships of genes that participate in core cellular functions such as cell-cycle progression, developmental patterning, and stress response; however, especially in multicellular organisms, many transcription factors remain uncharacterized. Currently, the most popular high-throughput method for detecting a transcription factor's direct targets is chromatin immunoprecipitation followed by microarray analysis or deep sequencing of the immunoprecipitated DNA (see chapter V.7). The global network of transcriptional regulation has been best studied in the bacterium *E. coli* and the yeast *S. cerevisiae*, although studies of less complete networks have been conducted in other microbes and in some animal species. In the transcriptional networks that have been analyzed, the distribution of out degree (number of targets of a transcription factor) appears to follow a power law. The distribution of in degree (number of direct regulators of a target gene) is less clear. In some analyses, the in-degree distribution appears to fit an exponential distribution better than a power law, whereas in others the reverse appears true.

It must be kept in mind that the systematic experiments that contribute data to interaction networks have generally been conducted in only one genetic background per species and in one experimental condition. It is virtually certain that the list of interactions would change if different genetic backgrounds or conditions were used. For example, a transcription factor might be posttranslationally modified under certain conditions in a way that affects its binding to DNA. The inferred number and identity of its target genes would therefore depend on the environment used for the chromatin immunoprecipitation experiment. It is unclear whether this limitation in current network data has any effect on the apparent global properties of these networks, such as their hub-and-spoke organization. One cautionary example is the analyses of the in-degree distribution of the yeast transcriptional regulatory network. Early analyses found that the in degree fits an exponential distribution better than a power law, whereas more recent analyses found that it indeed follows a power law. The difference could be caused by the addition of data on more transcription factors, and therefore better discovery of target genes with high in degree. Whatever the reasons for discrepancies between analyses, it should be understood that efforts to explain the evolution of global properties of biological networks must, for some time to come, be continually reevaluated against their best current empirical estimates.

3. EVOLUTION OF GLOBAL NETWORK ORGANIZATION

Making an analogy to nonbiological networks that are scale-free, such as the Internet, Barabási and colleagues stated two related hypotheses about biological networks. First, they proposed that hubs in biological networks should be more important than nonhubs. Second, they proposed that scale-free organization of biological networks makes them robust to random failure. In the Internet, hubs are indeed more important, as measured by their contribution to the efficiency of sending bits of data from one node to another. Disconnecting any desktop computer from the Internet typically has no impact on the ability of other users to send each other information, but disconnecting a major telecommunications center could fragment the network or at the very least cause rerouting delays. It is for this reason that the Internet is robust to random failure: hubs are vastly outnumbered by nonhubs, so if a random node is disconnected it is unlikely to be a hub and therefore unlikely to cause a major disruption.

Note that these two hypotheses about biological networks are, in essence, evolutionary ones. In the first hypothesis, "important" can be translated into biological terms as "making a large contribution to fitness." In the second hypothesis, "robust to random failure" can be translated as "robust to mutation." A priori, it is not obvious that protein-interaction hubs would be important to the extent that Internet hubs are. For example, a "housekeeping" enzyme that is essential for growth might have very few, if any, protein-interaction partners, whereas some large protein complexes are not essential for growth. Nonetheless, the first hypothesis was tested by Barabási and colleagues in 2001, using data from yeast on protein-protein interactions and on protein essentiality (as assayed by gene deletions). A significant positive correlation exists between protein-interaction degree and essentiality: hubs are indeed more likely to be required for viability than nonhubs are (at least under the growth condition assayed). Although this result is striking, there is as yet no consensus as to why the correlation holds. Whereas the edges in the Internet have very consistent meaning (transmission of bits of data), the edges in a protein-interaction network capture a highly heterogeneous set of biological functions, only some of which clearly qualify as information transmission (for example, a kinase phosphorylating its target protein). Understanding the correlation between degree and essentiality will require understanding how the many different types of protein interaction relate to the roles played by proteins in cellular function.

The hypothesis of protein-interaction hub importance has also been tested using data on the rates of evolution of protein-coding sequence. Again, importance has an evolutionary meaning, but this time the assumption is that proteins making a larger contribution to fitness will be constrained to evolve more slowly in amino acid sequence; however, rather than supporting a strong connection between node degree and evolutionary rate, analyses have, surprisingly, challenged the assumption on which they are based. The evolutionary rate/fitness assumption dates back to the landmark 1969 paper titled "Non-Darwinian Evolution" by Jack Lester King and Thomas Jukes and was a cornerstone of Motoo Kimura's neutral theory (see chapter V.1). Nonetheless, recent analyses using comprehensive data from bacteria and yeast on gene dispensability have found absent or very weak correlations with protein evolution rate, especially when the expression level of the gene is controlled for. Controlling for other factors is the principal challenge of this type of analysis, because the relevant factors—such as protein abundance and node degree—tend to be correlated with each other. This problem leads to difficulty in inferring a causative role for any one factor. Consequently, as Eugene Koonin and Yuri Wolf have emphasized, the literature is replete with weak and contradictory claims, a situation symptomatic of, in their words, "a nascent field in turmoil."

The evolution of global network organization has also been examined from the point of view of genome content. In 2003, Erik van Nimwegen showed that the number of genes encoding proteins in a particular functional category scales as a power law with the total number of genes in a genome. For example, across bacterial species representing broad phylogenetic and ecological ranges, the number of genes encoding transcription factors scales with the number of genes in the genome raised to a power of approximately two: for each doubling of genes in the genome, the number encoding transcription factors increases approximately fourfold. The finding of scaling exponents other than one implies that as genomes grow by gene duplication or shrink by gene loss, the probability that a gene will be added or deleted depends on the function of its encoded protein. The evolutionary causes for such departures from equal probability are unknown, but these power laws, like the power laws of node degree, must be explained by any satisfactory model of genome evolution.

It has been pointed out, most forcefully by Michael Lynch and Andreas Wagner, that an adequate null model of genome evolution, as it relates to global network organization, is sorely lacking. Whereas the null models of neutral or nearly neutral evolution have oriented the field of molecular evolution for decades (see chapter V.1), the field of network evolution has been proceeding without such grounding in rigorous population genetics. As Lynch and Wagner have independently noted, most early attempts to explain, for example, the power laws of node degree were strongly adaptationist. Consider the hypothesis that a scale-free network is robust to random failure. This predicted property could be seen merely as a by-product of a possibly nonadaptive process that gave the network its degree distribution. Instead, robustness was presented as a selectively advantageous property and therefore as a cause of the connectivity distribution. Evaluating the validity of such claims will require developing the null models against which to test them.

4. LOCAL ORGANIZATION AND DYNAMICS OF BIOLOGICAL NETWORKS

Local network organization refers to patterns of connection between subsets of nodes. For example, one might ask how the nodes that are connected directly to a particular focal node are connected to each other. In a protein-protein interaction network, such analysis could reveal the organization of proteins into functional modules or complexes. In a regulatory network, local analysis could reveal something less intuitive and therefore potentially more valuable: how network structure relates to network dynamics.

The divide between network structure and network dynamics is especially difficult to bridge because of two major gaps in characterization of regulatory networks, both of which are likely to persist for quite some time. The first major gap is the lack of kinetic rate constants for transcriptional reactions. The rates at which transcription factors bind and release their target DNA sites in vivo are in general difficult to measure, and therefore unknown except in the very rare cases in which advanced methods of single-molecule detection have been used. Rates of mRNA production and degradation have been estimated genome-wide in yeast and some other organisms, but these experiments typically involve nonphysiological conditions or mutational perturbation. Only very recently have methods been developed to observe the dynamics of transcription initiation, elongation, and termination at a single gene in vivo. At present, it is therefore generally not possible to describe the vast majority of regulatory systems with a complete set of coupled differential equations in which all parameter values have been specified from experimental data.

One potential way around the problem of missing rate constants would be to leave the kinetic parameters as variables and to infer their values based on measurements of easily determined quantities, such as mRNA abundance, in experiments where the regulatory system is perturbed away from its steady state. Indeed, this modeling approach to regulatory network inference makes up an extremely active subfield of computational biology. Such inference is extremely challenging, however, both because the number of experiments is often not much larger than the number of parameter values to be estimated, and because it is typically the case that regulatory networks are robust to changes in their parameter values. This robustness is an interesting property in its own right, with important implications for understanding of genetic and phenotypic variation in natural populations (see chapter V.11). What robustness means in practice is that many combinations of parameter values are consistent with the observed data.

The second major gap is the lack of understanding of the ways in which the effects of multiple transcription factors combine to determine target-gene expression. It is commonly accepted that most transcriptional regulation is combinatorial; that is, the effects do not simply add together but instead combine to create what amounts to a logic function. For example, if two activators regulate a given target gene, it might be the case that either one is sufficient to cause transcription (OR logic) or, alternatively, that both are needed (AND logic) (figure 1C). Despite the appreciation that the logic functions can have a large impact on transcriptional dynamics, and therefore that they should be a part of any mathematical model of gene regulation, they are mostly unknown. Ultimately, the logic functions should be reducible to kinetics as well, although the knowledge of cooperative and competitive protein-protein interaction kinetics as they relate to transcription in vivo is even poorer than protein-DNA kinetics. There is no available method for parallel determination of what these logic functions are, and gene-by-gene analyses are time consuming. Moreover, the cases in which the logic is understood might make up a biased subset, because discovering certain forms of regulation is easier than discovering others. For example, if several transcription factors are redundant in their effects on a particular promoter (i.e., OR logic applies), then it is unlikely that any one of them would be discovered by standard mutational analysis.

The severe challenges posed for a network-based understanding of regulatory dynamics, and the evolution of these networks, by the incomplete state of information about transcriptional kinetics and logic are compounded by posttranscriptional and posttranslational gene regulation, which are even less completely characterized; however, some inferences can be made without complete knowledge. Indeed, the robustness of regulatory networks implies that exact values of kinetic parameters do not matter to a large degree. Consider a simple regulatory network consisting of a transcription factor that regulates its own gene's expression. If the transcription factor is a repressor, then over a wide range of parameter values, the system will show predictable behavior. Any increase in expression will lead to more repression, whereas any decrease in expression will lead to less repression. If a sufficient delay exists in the system, then a stable oscillation might be reached, rather than a fixed-point steady state, but in either case the tendency is toward stability. By contrast, if the transcription factor is an activator, then the system will be unstable. Above a certain threshold, an increase in expression will be amplified, producing a switch-like transition of the gene from off to on.

These examples illustrate that in simple cases at least, dynamic properties can be inferred merely from knowing the connections in a regulatory network and the identities of nodes as activators or repressors. This line of reasoning can extend to slightly more complicated cases as well. For example, a regulatory system with two transcription factors that repress each other's expression is expected to behave like a switch. The system can be forced into one of two stable states in which high expression of one factor precludes expression of the other. Small regulatory subnetworks such as these are termed *network motifs*. In 2002, Uri Alon and colleagues introduced the notion of a network motif and investigated whether particular motifs are overrepresented in the transcriptional regulatory networks of *E. coli* and *S. cerevisiae*. They found that particular motifs are indeed overrepresented relative to random expectation. One example is the *feedforward loop*, in which two transcription factors jointly regulate a target gene and one of the transcription factors also regulates the other (figure 1C). The feedforward loop motif is especially relevant because of the link it provides between structure and dynamics. If the two transcription factors are activators, and they are both necessary for target-gene expression (i.e., AND logic applies), then the feedforward loop is expected to act as a noise filter. That is, the target gene will be activated only by a sustained pulse of the upstream activator, because only then will both activators be present simultaneously at sufficiently high levels. This kind of feedforward loop is also expected to shut down more quickly than a simple linear pathway when the upstream regulator is no longer present, because the other regulator is not sufficient to activate the target on its own.

5. EVOLUTION OF LOCAL NETWORK ORGANIZATION

As with global network organization, comparative data can be used to understand the evolution of local network organization. Indeed, Sean Carroll, Eric Davidson, and others have argued that the proper way to understand the evolution of developmental processes is at the local level of regulatory networks (see chapter V.11). For example, Carroll writes of "toolkit" genes, such as those that encode components of signal-transduction cascades, that are deployed for various purposes throughout development and across species. In Davidson's terminology, these would be called "plug-ins"—small subcircuits of genes that perform specific molecular functions but perform potentially many developmental functions. Implicit in the concept of a toolkit or a plug-in is that the regulatory network is modular, comprising groups of genes that function as units. Although modularity also motivates the concept of a network motif (i.e., it makes sense to study motifs to the extent that their behavior is predictable despite their embedding in a larger network), Davidson takes pains to draw a distinction between a subcircuit and a motif, in that the former focuses on biological function whereas the latter focuses on kinetic behavior.

In addition to plug-ins, Davidson defines another type of subcircuit, the "kernel." A kernel also comprises genes that function together, but the distinction is that they function together only to execute a single developmental function. Moreover, a kernel is defined as being evolutionarily conserved and comprising densely interconnected regulatory factors, loss of any one of which leads to developmental failure. The canonical kernel is a set of genes encoding transcription factors that collectively give a developing tissue or organ, such as the heart, its identity.

Davidson hypothesizes that different levels of phylogenetic divergence correspond to divergences of different network elements: deep, phylum-level divergences correspond to the ancient emergence of kernels; intermediate divergences correspond to redeployments of plug-ins; and recent divergences correspond to divergences of terminal-differentiation genes regulated by the kernels and plug-ins. Behind this hypothesis is the argument that particular network structures constitute a form of developmental constraint (see chapter V.10). For example, because a kernel's genes are densely interconnected and essential for normal development, they might be under strong selection not to change their function; however, others have pointed out cases of so-called developmental systems drift, in which the essential output of a regulatory subcircuit remains unchanged despite changes in the subcircuit's membership and interconnections. As pointed out by Lynch and others, an essential regulatory interaction may be lost through an intermediate stage of redundancy with another factor.

Why deep conservation marks some subcircuits, whereas developmental systems drift marks others, is unknown. Likewise, it is unknown why some network motifs, such as the feedforward loop, appear to be overrepresented in regulatory networks. As with global network organization, competing explanations for local organization favor adaptive or nonadaptive processes. Ultimately, these potential explanations of nonrandom features of networks must be measured against rigorous models of genome evolution.

6. THE FUTURE OF EVOLUTIONARY SYSTEMS BIOLOGY

The branch of evolutionary biology dedicated to the understanding of molecular networks has come to be known as evolutionary systems biology. It is difficult to predict the future of this new and contentious field. The points of debate can be subtle but are nonetheless critical. For example, Davidson's distinction between a subcircuit and a motif might seem minor, but it amounts to an argument about the proper research program for evolutionary systems biology. Indeed, Davidson goes so far as to argue that studying kinetics is a distraction, a "siren-like" call to the "mechanistically inclined" to neglect causal regulatory logic in favor of the mere details of its execution. Likewise, Lynch's self-described "contrarian" effort to build a rigorous population-genetic null model of genome evolution is a strong statement on where research effort should be allocated. These are but two of the debates that led Koonin and Wolf to declare evolutionary systems biology to be in turmoil. Despite the uncertain direction of research, what is clear is that these debates intersect with some of the most critical debates in evolutionary biology, including those concerning the role of nonadaptive processes in evolution and the existence of developmental constraints. The coming years will tell whether the approaches and perspectives of evolutionary systems biology can illuminate these long-running debates better than more established ones have.

FURTHER READING

Alon, U. 2007. Network motifs: Theory and experimental approaches. Nature Reviews Genetics 8: 450–461. *A review of the evidence for network motifs in the regulatory networks of diverse species, and of the motifs' functional significance.*

Barabási, A.-L., and Z. N. Oltvai. 2004. Network biology: Understanding the cell's functional organization. Nature Reviews Genetics 5: 101–113. *A review of network terminology as applied to biological networks, and of the putative evolutionary origins and functional importance of scale-free network organization.*

Carroll, S. B., J. K. Grenier, and S. D. Weatherbee. 2001. From DNA to Diversity: Molecular Genetics and the Evolution of Animal Design. Malden, MA: Blackwell Science. *An introduction to the evolution of development with an emphasis on changes in gene regulation and the concept of a genetic "toolkit."*

Davidson, E. 2006. The Regulatory Genome: Gene Regulatory Networks in Development and Evolution. San Diego, CA: Academic/Elsevier. *An introduction to the evolution of development with an emphasis on changes in gene regulation at different levels of network organization.*

Koonin, E. V., and Y. I. Wolf. 2008. Evolutionary systems biology. In M. Pagel and A. Pomiankowski, eds., Evolutionary Genomics and Proteomics. Sunderland, MA: Sinauer. *A review of the impact of systems biology on evolutionary genetics, and of the major unanswered questions in the evolution of networks.*

Lynch, M. 2007. The evolution of genetic networks by nonadaptive processes. Nature Reviews Genetics 8: 803–813. *An argument for the development of appropriate null models of regulatory-network evolution, and an analysis of one such model.*

Wagner, A. 2008. Gene networks and natural selection. In M. Pagel and A. Pomiankowski, eds., Evolutionary Genomics and Proteomics. Sunderland, MA: Sinauer. *A review of the role of natural selection in shaping the architectures of regulatory networks.*

V.10

Evolution and Development: Organisms

Paul M. Brakefield

OUTLINE

1. Evolution of form and function
2. The rise of evolutionary developmental biology
3. Evolutionary constraints and patterns of allometry
4. Patterns of parallel evolution
5. The paradox of morphological stasis
6. Opportunities for future research

The essence of evolution by natural selection is uncomplicated: phenotypic variation among individuals is generated from genetic variation via the processes of development, and this "fuel" is then screened for performance in the ecological and reproductive arena. This chapter examines the extent to which both the generation of fuel and its performance influence the paths of evolution. Paleontologist David Raup showed in the 1960s that the set of all theoretical shapes of snail shells fit a cube, with each axis reflecting a simple mathematical description of one component of shell growth (figure 1). He then observed that only a small proportion of this theoretical morphospace had been occupied in the evolutionary history of gastropod snails. Such findings of substantial areas of potential phenotypic space that have not been explored in evolution are characteristic of this type of study of clades of related species. Furthermore, variation is observed in the density of taxa in the area of morphospace that has been occupied. It is then important to examine the extent to which such patterns reflect how natural selection screens the performance of individuals *and* the production of phenotypic variation. In other words, to what extent is the evolution of optimal phenotypes in response to adaptive landscapes compromised by the numerous potential constraints involved in the generation of phenotypic variation?

GLOSSARY

Allometry. The change in proportion of various parts of an organism as a consequence of growth. Variation in the scaling relationships of parts or organs of organisms typically contributes substantially to evolutionary diversification within lineages.

Artificial Selection. Intentional breeding by humans for certain traits or combination of traits. This contrasts with *natural selection,* in which the environment acts as a sieve to screen variation among individuals for reproductive success.

Constraint. A limit on evolutionary change that slows the rate of adaptive evolution or prevents a population from evolving the optimal value of a trait. Constraints can be considered intrinsic to the organism (e.g., genetic or developmental) and involve the generation of form, or extrinsic and associated with the power of natural selection and the spatial-temporal matching between population structure and the environment.

Developmental Bias. A bias in the production of various phenotypes caused by the structure, composition, or dynamics of the developmental system.

Developmental Drive. Results from developmental processes that facilitate the production of phenotypic variants along certain axes of form and which, therefore, favor evolution in certain directions.

Evolvability. Capacity of a trait, set of traits, or a lineage to adapt to changing conditions.

Genetic Covariance. A summary statistic describing the genetic association or correlation among traits. It quantifies the extent to which the evolutionary response of one trait will be influenced by selection of another. Genetic covariance among traits can yield "axes of least resistance" along which species tend to evolve.

Modularity. Ability of individual parts or modules of an organism, such as repeated limbs or segments, to develop or evolve independently from one another. Different elements within a single module will lack individuality and the ability of independent evolution but show independence from other modules.

Morphospace. A theoretical morphospace represents an N-dimensional geometric hyperspace for the form of organisms produced by systematically varying the parameter values of a geometric or mathematical model for the generation of form. One can then compare observed patterns in the occupancy of morphospace by a group of organisms, or by different groups.

Parallel Evolution. Evolution of similar or identical features independently in related lineages, usually considered to be based on similar modifications of the same developmental pathways.

Stasis. The pronounced morphological stability displayed by many fossil species, often for millions of years. It contrasts sharply with the rapid, often-adaptive, evolutionary change documented in many extant species.

1. EVOLUTION OF FORM AND FUNCTION

Evolution of a trait by natural selection requires three components in any population, namely, phenotypic variation, a consistent relationship between phenotype and fitness, and genetic variation, a resemblance in the trait among related individuals. Thus, if phenotypic variation among individuals is not generated and is not based, at least in part, on some underlying genetic variation, adaptive evolution of the trait will not happen. Variation in the phenotype must exist among individuals so that selection can screen for the performance of individuals with differing phenotypes for some component of viability, survival, and reproductive success. Some genetic basis for the variation among individuals is then necessary for the selection at this phenotypic level to result in an evolutionary change, usually in the mean phenotype, from one generation to the next.

The null hypothesis for natural selection acting in a fully unconstrained manner would be that the observed patterns for the evolution of diversity are the result solely of natural selection at the phenotypic level; however, genetic constraints, such as patterns of genetic variation and especially genetic covariances among traits, can in theory retard, channel, or bias the extent or directions of evolutionary change. Developmental constraints can be considered in a similar framework for the evolution of form. John Maynard Smith made an early effort to integrate studies of evolution and development by bringing together a group of evolutionary biologists and developmental biologists. They established the crucial idea of *developmental bias*: do the properties of the developmental system, and the ways in which it generates variation, bias the course of the evolution of form? The terminology and semantics associated with descriptions of such constraints are extensive and arguably of limited value, but the issue of the mechanisms by which genetic or developmental constraints can influence the trajectories and patterns of evolution alongside natural selection remains a challenge to unravel; however, so-called intrinsic or generative constraints involving the processes by which phenotypic variation is generated are opening up to experimental dissection, and this is in turn leading to a more rounded assessment of the extent of the power of natural selection in the evolution of diversity.

2. THE RISE OF EVOLUTIONARY DEVELOPMENTAL BIOLOGY

Evolutionary developmental biology, hereafter referred to as *evo-devo*, grew out of the rich history of comparative embryology in the nineteenth and early twentieth centuries. It examines how developmental processes evolve, and how knowledge of development can contribute to understanding of evolutionary change. The genotype-to-phenotype map is a major focus of the research program, including an understanding of the concept of *evolvability*, or the potential for morphological traits to exhibit adaptive evolution (see chapter II.6). Although evo-devo is essentially concerned with morphogenesis, other areas of biology such as behavioral biology are embracing the attempt to more fully map phenotypes of functional significance to genotypes.

The field of evo-devo is providing evolutionary biologists with the tools to address in detail the ways in which variation in form is generated through development, and thus to explore the ways in which the developmental processes that build animal or plant forms can be reflected in observed patterns of diversity, whether between widely separate phylogenetic lineages or among closely related species. The mechanisms underlying the evolvability of form are currently being investigated, in the contexts of both fundamental changes in body plan, and adaptive radiations as a lineage of species encounters novel ecological opportunities and explores new morphological and phenotypic spaces. Different modes of key innovations or evolutionary novelties are being explored by research in a wide variety of lineages, as are the processes of elaboration and evolutionary tinkering involved in the exploitation of such novelties.

3. EVOLUTIONARY CONSTRAINTS AND PATTERNS OF ALLOMETRY

When observing many groups of related species of animals or plants, one often sees striking diversity in their shapes. Static allometry refers to the proportional scaling relationship among individuals of a species between one organ and total body size, or between two organs, after growth has ceased or at a single developmental stage. The evolution of static allometry underlies much of the diversity in shape and is therefore a fundamental component

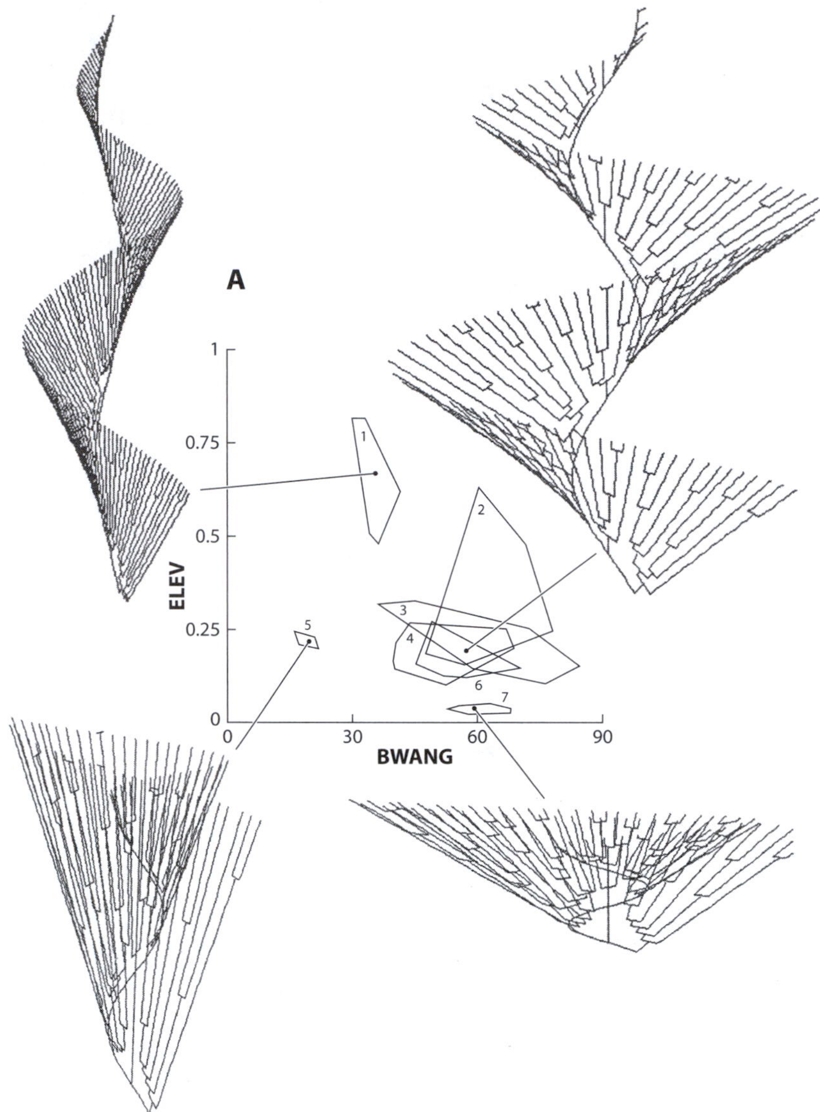

Figure 1. Evolution in morphospace in bryozoa: occupancy of theoretical morphospace and forms that have not evolved. Simulations of helical colony morphologies for living and fossil bryozoa are shown in panel A, while simulations of helical colony morphologies that have never been found are shown in panel B. The graphs are the same in panels A and B, and the polygons illustrate measurements of two geometric parameters of the helical colony model (BWANG and ELAV) taken from living and fossil bryozoans. (George McGhee 2007.)

of an understanding of pattern and process in evolutionary diversification. The evolution of the scaling proportions of various body parts provides an interesting example to discuss in the light of intrinsic versus extrinsic constraints.

Numerous descriptive studies have revealed that patterns of allometry of specific parts of the body within a species are typically remarkably invariant. Thus, each male stalk-eyed fly is characterized by a particular scaling relationship between the width of its eye stalks and its body length. Both these traits can be highly variable, but the relationship between the two is comparatively invariant. While a tight phenotypic correlation is typically observed between traits in individuals of a species, different species of stalk-eyed flies can show remarkably divergent static allometry.

Patterns such as this raise the issue of the mechanisms by which striking differences among species can evolve if the phenotypic variation within species is very limited. This is perhaps especially pertinent where the morphological

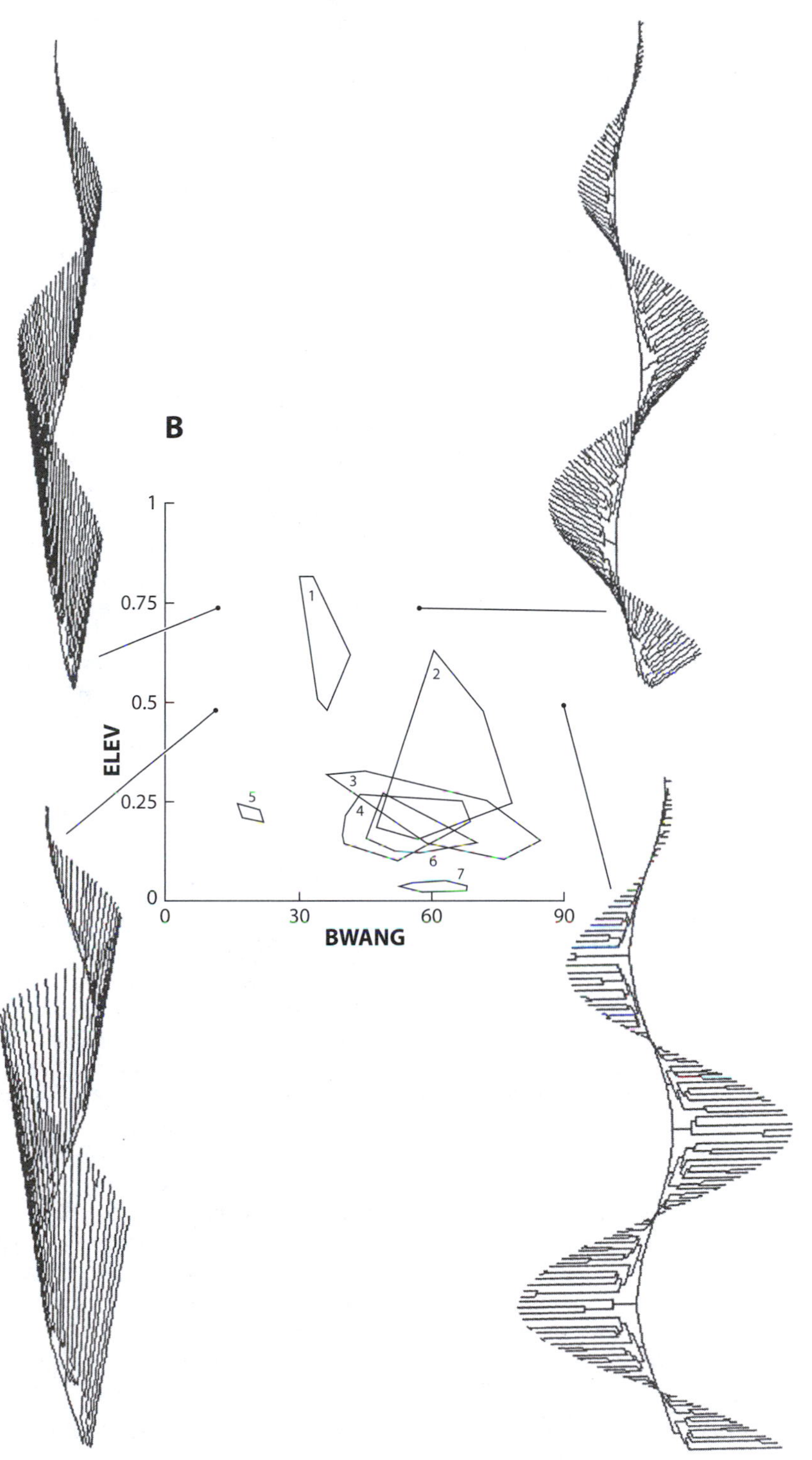

Figure 1. (*cont.*)

traits concerned share developmental processes and show high genetic correlations. An example is the regulation of development of both forewings and hindwings in a butterfly by essentially the same sets of developmental genes and physiological processes. How during evolution do these two sets of wings evolve to gain individuality and become different from each other in divergent patterns of size, shape, and scaling among species? Until recently, there was little knowledge of the developmental mechanisms that regulate such scaling relationships, but this is beginning to change, especially through work on organ development in *Drosophila melanogaster*. Feedback processes among growing organs can modulate their final sizes in relation to one another and to overall body size. In *Drosophila*, organ autonomous changes in expression of a gene called *FOXO* are one component that can alter the extent to which individual organs adjust their size, for example, in response to nutrition during growth.

One way to examine the role of generative constraints is to employ artificial selection. Recent experiments of this type with butterflies have explored the more general issue of the contribution to patterns of diversity in static allometry made by processes intrinsic or extrinsic to the organism; can the mechanisms involved in generating phenotypic variation act as a brake on the evolution in allometry, or is natural selection sufficiently powerful to drive the independent evolution of body parts when this is favored?

Tony Frankino applied artificial selection in lines established from an outbred laboratory stock of the tropical butterfly *Bicyclus anynana* for the scaling of the size of the forewing relative to the hindwing. These two traits show a positive genetic covariance with a genetic correlation not significantly different from unity, and yet selection was able to produce short-term evolution of the scaling relationship and novel phenotypes in each direction (figure 2). Additional experiments with free-flying butterflies in a spacious greenhouse used hybrids obtained from crosses of pairs of opposing selection lines to examine the fitness consequences of the change in phenotype. The results showed that the mating success of male butterflies with a divergent forewing to hindwing allometry was substantially lower than that of those with an intermediate wing scaling. This indicates a strongly stabilizing pattern of selection in favor of the original wild type and against either of the divergent morphologies (figure 2). Thus, the evolution of allometries in this insect is not limited by any developmental constraint, and the scaling relationships are shaped by strong natural selection. Comparable experiments have been performed with *B. anynana* with closely similar results for scaling of the forewing relative to body size.

Artificial selection experiments have also been used to examine the flexibility of evolution in different directions of morphospace for two wing eyespots in this species of butterfly. Each eyespot pattern element consists of concentric rings of epithelial cells containing different-colored pigments. They are often arranged in an anterior to posterior column along the wing margins, and may then function in misdirecting the attacks of predators hunting by sight, away from the vulnerable body. A series of eyespots on the same wing surface behave as a module in which each eyespot is based on a central signal-diffusion system (or "organizer") during development, and they all express the same set of genes during pattern determination. Thus, the eyespots represent a set of repeated elements exhibiting strongly positive genetic correlations for both eyespot size and color composition such that selection targeting either trait for a specific eyespot in *B. anynana* yields corresponding responses in the other eyespots (note that the two eyespot traits show little, if any, genetic covariance). A series of artificial selection experiments performed by Patrícia Beldade examined whether repeated eyespots can evolve independently. By simultaneously targeting two different eyespots on the same wing, the experiments explored the extent to which the pattern of eyespots, in terms of relative size or color composition, is flexible in evolutionary terms; for example, with respect to size, can two different eyespots be selected to both either increase or decrease in size, *and* also for one to become larger and the other smaller? A high flexibility in the evolution of extreme, novel forms toward all four corners of theoretical morphospace for this pattern was observed for the pattern of relative eyespot size over 25 generations of selection. This result is consistent with neither intrinsic constraints nor any strong bias in potential responses to natural selection. This flexibility of response in this experimental system appears to be reflected in a full exploration of morphospace for the same pattern when variation in the species-rich tribe of butterflies is considered. In contrast, experiments on eyespot color yielded very little response and no novel phenotypes when two eyespots were selected in opposing directions. The different responses of the two traits may be traceable to mechanisms of developmental regulation; whereas the size of each eyespot depends on the strength of the organizing "signaling" source at its center, the regulation of eyespot color involves threshold responses to the signals and occurs at the level of the whole wing epidermis.

4. PATTERNS OF PARALLEL EVOLUTION

Parallel evolution describes the development of similar traits or forms in related but distinct species descending from the same ancestor. Such patterns of occupancy in morphospace are of particular interest in considering

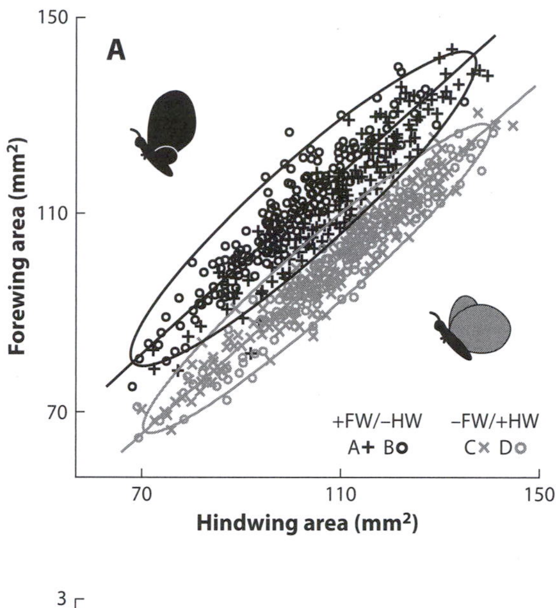

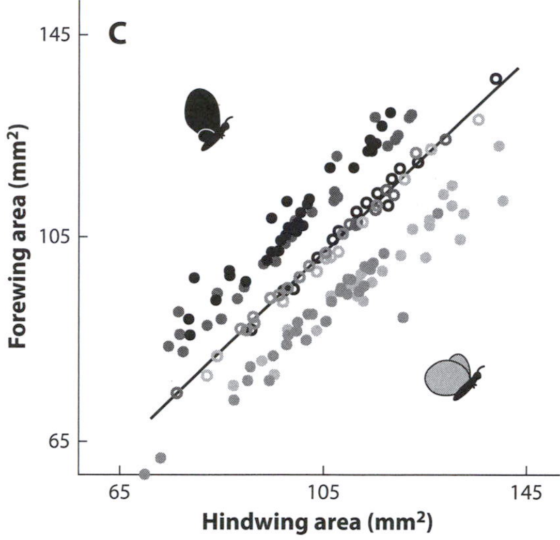

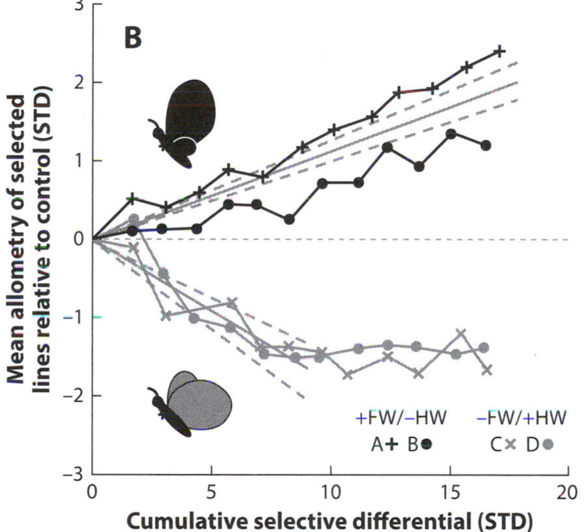

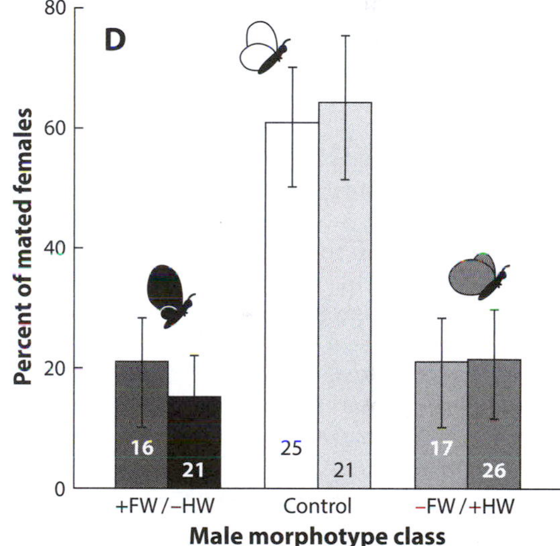

Figure 2. Forewing:hindwing allometry in butterflies: positive responses to artificial selection versus stabilizing sexual selection in favor of wild-type butterflies. (Frankino et al. 2007.)

how evolvability and constraints may influence the trajectories of adaptive evolution. Related clades that have colonized similar environments may show parallel patterns of diversification because the environments provide similar unexploited ecological niches and opportunities for adaptation. This could occur in a largely unconstrained way and be driven by natural selection and ecological performance. On the other hand, related clades may diversify in similar ways in part because they share developmental and genetic systems, and therefore have similar propensities to evolve along certain trajectories or axes of change; in other words, the patterns of parallel evolution may reflect a comparable process of developmental drive. It will become more feasible to examine the role of the evolvability of traits in shaping the ways in which organisms evolve through phenotypic space as research on model species within particular groups of related organisms more effectively maps patterns of phenotypic variation to genetic variation via developmental and physiological processes (see chapter V.12).

One of the most striking examples of parallel evolution is provided by the species flocks of cichlid fish in the Rift Valley lakes of Africa. Each lake was colonized independently by haplochromine cichlids, presumably via rivers entering the lakes. Although the lakes differ in age, topology, and history, each has harbored spectacular radiations of cichlid taxa established over remarkably short periods of time. In particular, the dramatic increases in morphological diversity have been associated with the evolution of clusters of taxa with readily identifiable *ecomorphs* or trophic morphologies. The ecomorph morphologies are remarkably similar across lakes, even though the taxa within each lake are substantially more similar to each other at a genetic level than they are to any taxon from one of the other lakes. Each ecomorph is associated with suites of life history and behavioral traits associated with exploitation of particular ecological environments within the lake. Such ecomorphs are also characteristic of the island faunas of Anoles in the Caribbean, and feeding morphologies are a major component of radiations in other animal groups, including Darwin's finches (see chapter VI.10). The patterns of radiation that have evolved in these cichlid faunas may be driven primarily by natural selection, but shared genetic and developmental systems inherited from the haplochromine ancestors could also have facilitated evolution in certain directions of change, and not in others. Thus, although natural selection may, at least superficially, appear all-powerful, the properties in terms of the evolvability of traits and the generation of phenotypic variation could have biased observed patterns in occupancy of morphospace and contributed to the parallel nature of evolution in independent radiations.

Work by Peter and Rosemary Grant on Darwin's finches of the Galápagos Island archipelago has yielded one of the most complete accounts of how natural selection can yield an adaptive radiation. They have demonstrated very clearly the ways in which bill morphology is a crucial target of natural selection. The diversity in bill morphology can then be compared in a 2-D morphospace by plotting an index of bill depth and width against length. Evo-devo research is now uncovering two developmental genetic pathways that have contributed to the phenotypic diversification in the finch bills. This work includes functional tests of the role of the pathways by employing the chick model for genetic manipulations in embryonic development. This provides promise that future evo-devo work will be able to compare the mechanisms by which diversity has evolved in independent radiations, whether of birds or of cichlid fishes, thus revealing more about the contributions of the properties of developmental systems to patterns of evolution (see chapter V.11).

5. THE PARADOX OF MORPHOLOGICAL STASIS

One of the most striking contrasts observed in patterns of evolution is apparent when studies of microevolutionary change over short periods of time are compared with those of rates of change in morphology over periods of geologic time. Estimates of natural selection obtained from the numerous studies of local populations conducted in recent decades collectively demonstrate a high intensity of natural selection (see chapter III.7). Such observations in nature are also matched by the generality of rapid responses to artificial selection, although there are important differences between the way selection can be made to target phenotypic variation by the human experimenter and the way natural selection interacts with population dynamics in the wild (see chapter III.6). In contrast to these short-term patterns of dynamic evolutionary change, paleontologists studying fossil remains frequently record very little change in the form of animals over geologic time (see chapter II.9). A dramatic disparity can be seen between the observed rates of evolution over short periods of time and those over long periods.

One of the most detailed descriptions of variation in animal form over geologic time is that of cheilostome bryozoans. Cheilostomes are small, clonal marine organisms that form colonies. They are abundant in recent seas and in the fossil record. The results of studies of their biology in present-day populations enable a high degree of confidence in interpretation of the fossil record, for example, with respect to the relationships between variability in morphology and the species-level taxonomy, ecology, and life history evolution of the organisms. A study of two tropical American genera of 11 species found that with both genera, all species tended to remain essentially unchanged in morphology over geologic time. In contrast, new species can appear comparatively abruptly, that is, within the limits of stratographic resolution of sampling of some 150,000 years. Observations from studies of extant species suggest that this pattern extends to stasis in life history traits.

Morphological stasis seems a dominant pattern in the fossil record, thus demanding an explanation in terms of the dynamic behavior of local populations in contemporary analyses. Numerous works have discussed several ways in which this apparent discrepancy could be reconciled. These are not mutually exclusive, and it is indeed likely that there is no single explanation. First, natural selection may tend to oscillate in direction over time on a scale that ensures that patterns of progressive change are seldom observed at the scale of sampling typical analyses of the fossil record. Examples of evolutionary reversals have been observed in studies at a more intermediate timescale, for example, in Darwin's finches and in stickleback fish. A second class of explanations involves considering patterns of genetic pleiotropy and genetic covariances, and the dimensionality of combinations of traits that are the targets of natural selection. Such interactions among traits, both genetic and developmental, can theoretically constrain or restrain evolutionary change, as outlined earlier in this chapter. Third, and on a more spatial scale, several evolutionary biologists have emphasized the difference between observing dynamic evolutionary change in one, or a group, of local populations, and observing the spread of such change throughout the range of a (widely distributed) species. This has been referred to as the ephemeral divergence hypothesis, pointing out the likelihood of interactions with speciation processes. A later acquisition of reproductive isolation by a group of populations will enable the evolution of a phenotypic shift within them to be retained but not be detected as a change within a species in the fossil record. In contrast, stasis is likely to be observed unless speciation intervenes to capture such a pattern of phenotypic change within a subgroup of populations.

6. OPPORTUNITIES FOR FUTURE RESEARCH

There remain many challenges to expanding the analysis of ways in which intrinsic or generational constraints can restrain, bias, or limit evolutionary change; however, there are also increasing opportunities for using comparative and experimental approaches to exploring the extent to which such constraints can contribute alongside the power of natural selection to evolutionary diversification. Research in evo-devo and evolutionary genetics will increasingly refine our understanding of the evolvability of single traits and more complex character sets, and thus describe the potential pathways of evolutionary change within a lineage. This endeavor should increasingly reveal whether the way in which development works can indeed both facilitate certain trajectories of evolution in morphology and make others less likely to occur. We will then be in a better position to disentangle the contributions of different kinds of intrinsic and extrinsic constraints (including selection and gene flow) to the ways in which animal and plant life explore morphospace in evolution.

FURTHER READING

Allen, C., P. Beldade, B. J. Zwaan, and P. M. Brakefield. 2008. Differences in the selection response of serially repeated color pattern characters: Standing variation, development, and evolution. BMC Evolutionary Biology 8: 94. *A description of an experimental analysis of the potential for developmental bias to become reflected in patterns of evolutionary divergence in the eyespots on butterfly wings.*

Brakefield, P. M. 2006. Evo-devo and constraints on selection. Trends in Ecology & Evolution 21: 362–368. *A review of evolutionary constraints that emphasizes how studies of evo-devo are beginning to open up the analysis of intrinsic constraints.*

Eldridge, N., J. N. Thompson, P. M. Brakefield, S. Gavrilets, D. Jablonski, J.B.C. Jackson, R. E. Lenski, et al. 2005. The dynamics of evolutionary stasis. Paleobiology 31: 133–145. *A review of ideas about the evolutionary dynamics of morphological stasis.*

Frankino, W. A., D. S. Stern, and P. M. Brakefield. 2007. Internal and external constraints in the evolution of a forewing-hindwing allometry. Evolution 61: 2958–2970. *A description of experiments to explore how constraints may influence the evolution of scaling of forewing:hind wing size in butterflies.*

Futuyma, D. J. 2010. Evolutionary constraints and ecological consequences. Evolution 64: 1865–1884. *A detailed review of current thinking about evolutionary constraints, especially in an ecological context.*

Jackson, J.B.C., and A. H. Cheetham. 1994. Phylogeny reconstruction and the tempo of speciation in cheilostome *bryozoa*. Paleobiology 20: 407–423. *This paper is part of one of the most thorough analyses of morphological stasis in the fossil record.*

Kirkpatrick, M. 2010. Rates of adaptation: Why is Darwin's machine so slow? In M. A. Bell et al., eds., Evolution since Darwin: The First 150 Years. Sunderland, MA: Sinauer. *A discussion of the contrast between rates of selection observed in extant populations and those apparent from studies of the fossil record.*

Maynard Smith J., R. Burian, S. Kauffman, P. Alberch, J. Campbell, B. Goodwin, R. Lande, D. Raup, and L. Wolpert. 1985. Developmental constraints and evolution. Quarterly Review of Biology 60: 265–287. *An early attempt to facilitate a constructive dialogue between evolutionary biologists and developmental biologists. It yielded a key definition of developmental bias and a discussion of its potential relevance to evolution.*

McGhee, G. 2007. The Geometry of Evolution: Adaptive Landscapes and Theoretical Morphospaces. Cambridge: Cambridge University Press. *A recent book about analyses of theoretical morphospace in the context of adaptive landscapes and ecological performance. Several key examples of analyses are taken from the fossil record.*

Raup, D. M. 1966. Geometric analysis of shell coiling: General problems. Journal of Paleontology. 40:1178–1190. *The classic early study of the occupancy of a theoretical morphospace.*

Shingleton, A. 2011. The regulation and evolution of growth and body size. In T. Flatt and A. Heyland, eds., Mechanisms of Life History Evolution. Oxford: Oxford University Press. *A review about allometry that includes a discussion of recent work on the regulation of growth and body size in* Drosophila melanogaster.

Wagner, G. P., and V. J. Lynch. 2010. Evolutionary novelties. Current Biology 20: R48–R52. *This is a comprehensive primer on current views of evolutionary novelties.*

V.11

Evolution and Development: Molecules

Antónia Monteiro

OUTLINE

1. The goals of molecular studies in evolutionary developmental biology
2. Mapping genotype to phenotype during development
3. Mapping genotype to phenotype during evolution
4. The evolution of novel traits and their underlying gene regulatory networks
5. Future areas of research in evolutionary developmental biology

In this chapter, major research themes and approaches in evolutionary developmental biology, commonly referred to as *evo-devo*, are presented from a molecular perspective. The field is concerned primarily with connecting changes at the DNA level to changes in developmental pathways and gene regulatory networks that lead to the evolution of morphology, physiology, and behavior. Researchers in the field are interested in identifying whether mutations in DNA are altering the regulation or the function of proteins, and describing how these changes alter the output of larger gene regulatory networks and ultimately the adult phenotype. In addition, interest is mounting in understanding how novel gene regulatory networks originate and evolve, and how the environment interacts with these gene regulatory networks to promote either robustness or adaptive phenotypic plasticity in organismal form.

GLOSSARY

Candidate Gene. A gene that is suspected of playing a role in the evolution of a trait typically because its expression domain or temporal pattern of expression is associated with the development of that trait.

Enhancer or *Cis*-Regulatory Element. A sequence of DNA that regulates the temporal and spatial expression of flanking protein-coding genes when bound by specific transcription factors.

Homologous Trait. A trait found in two lineages is homologous if it derives from the same trait present in the common ancestor. For example, pectoral fins in fish and arms in humans are homologous traits.

Modular Gene Regulatory Network. An interacting group of genes that are activated together in response to simple inputs, and in a largely context-independent manner during development. For example, a gene regulatory network specific to the fruit fly eye can be activated in multiple places in the body (e.g., wings and antennae) in response to the expression of specific transcription factors.

Phenotypic Plasticity. The ability of some organisms to modify their phenotype in response to their rearing environment.

Quantitative Trait Loci (QTL) Mapping. A method that involves discovering the genomic position and relative effect of loci responsible for producing phenotypic differences between two individuals that can be crossed.

Selector Gene. A regulatory gene that specifies cell, tissue, organ, or regional identity in animals.

Serial Homologous Traits. Repeated traits within the same body that use a similar gene regulatory network during their development. Examples of serial homologous traits include arthropod segments, teeth, vertebrae, and pelvic and pectoral fins (or arms and legs).

Transcription Factor (TF). A protein that binds to DNA to effect changes in the transcription of flanking protein-coding genes.

Transgenic Organism. An organism that has been genetically modified to carry additional genes. Typically, transgenic organisms are used to test the function of particular protein-coding sequences during development. Transgenics can also be used to test the function of candidate enhancer sequences by attaching them to reporter genes such as green fluorescent protein (GFP), and monitoring GFP expression during development.

1. THE GOALS OF MOLECULAR STUDIES IN EVOLUTIONARY DEVELOPMENTAL BIOLOGY

Over the course of life, organisms have evolved into myriad sizes, shapes, and forms. They also evolved different physiologies, life histories, and behaviors. Most of this diversity is encoded in the molecule that unites all of life, DNA, and a challenge for biologists lies in understanding how variation in this molecule actually produces organismal diversity.

Connecting variation at the DNA level with organismal diversity can be broken into two separate challenges. One involves understanding how DNA sequences in any one organism lead to the development of the traits of that organism. This endeavor is also dubbed "identifying the genotype-phenotype map" for each species. The other challenge involves identifying the relevant changes in the genotype-phenotype map that cause different organisms to evolve different traits. The first challenge falls in the domain of developmental biology, and the second challenge in the domain of evolutionary developmental biology, or *evo-devo*.

In molecular studies of evo-devo (compared with organismal level studies; see chapter V.10), the goal is to understand how the process of genomic evolution, including changes in gene number, gene structure, and gene regulation, is translated via developmental mechanisms into the evolution of morphology, behavior, or physiology. Because phenotypes such as behavior are only just beginning to be studied, this chapter will mostly highlight the various ways that biologists are probing the developmental mechanisms that underlie the evolution of morphology. There have been two main routes to this type of work; one involves investigating the entire genome, whereas the other investigates candidate genes to identify DNA sequences or loci responsible for morphological evolution. The chapter also addresses the role of developmental modules and of modular genetic architecture in body plan evolution and the evolution of novel complex traits, and concludes with some of the unexplored aspects of molecular developmental evolution, including the molecular basis of plasticity.

2. MAPPING GENOTYPE TO PHENOTYPE DURING DEVELOPMENT

While developmental biology aims to understand how genes are involved in building organismal traits via the developmental process, evolutionary developmental biology focuses on the subset of genes and developmental mechanisms responsible for the evolution of morphological variation within and between species. Variation at the DNA level impacts developmental mechanisms and ultimately phenotypes in multiple ways. This section briefly illustrates the fundamental steps connecting DNA to morphology via development, or the unfurling of the genotype-phenotype map.

Mapping genotype to phenotype involves the process of reading the DNA molecule through the course of development. All multicellular organisms start off as single cells that subsequently divide and differentiate into multiple cell types, tissues, and organs. Cell division and differentiation involve complex orchestrations of gene regulation. Genes are inactive in most cells of early embryos, but as development progresses, different genes are activated in different cells of the embryo, producing asymmetries in regulatory states (on/off). These asymmetries in gene expression across the body later translate into visible phenotypic differences. Certain cells will become differentiated to produce pigments that give rise to color patterns, other cells will become muscle cells, and yet others will secrete crystalline proteins that agglomerate to give rise to the lens of an eye.

Asymmetries in regulatory states of genes inside cells are produced by asymmetries in the distribution of important regulatory proteins, *transcription factors* (TFs). TFs induce (or repress) gene transcription by binding to specific regulatory sequences flanking a protein-coding sequence, also called enhancers or *cis*-regulatory elements (see chapter V.7). Binding of TFs to enhancers (usually more than one TF is involved) leads to the recruitment of the RNA polymerase enzyme to the promoters of those genes, and transcription is initiated. Enhancers contain information about when and where a gene will be expressed during development because they contain clusters of TF binding sites that will lead to gene activation (or repression) only when bound by the respective TFs. If a gene contains more than one enhancer, it can be expressed in very different developmental contexts, depending on the sequence of each of its enhancers. The earliest stages of development usually start with TFs that are asymmetrically distributed in the cytoplasm of the single-celled egg, and are responsible for beginning the process of cell differentiation. If a certain cocktail of TFs is present in the right combination and concentration in one part of the embryo but absent from another, then genes responsive to that exact complement and concentration of TFs (i.e., with binding sites for those TFs in their enhancer sequences) will be turned on only in those cells of the embryo. TFs can also induce the expression of signaling molecules that can diffuse some distance within the embryo. These molecules can, in turn, activate a novel set of TFs in the surrounding cells. The process of development is essentially a process of subdividing a growing and uniform field of cells into separate domains expressing unique combinations of TFs at unique concentrations. These TFs then control downstream target genes that build different traits in different parts of the body. Intermediate levels of gene regulation occur after a gene is transcribed, and before traits are built, for instance, by posttranscriptional modifications to proteins, but so far, not much work in evo-devo has explored evolution at this level.

This section has established, in broad brushstrokes, the mechanisms by which genomic information is translated into a phenotype during the course of development; the next section will focus on the methods used by researchers to identify the alterations to developmental programs that lead to distinct morphologies that are characteristic of different species.

3. MAPPING GENOTYPE TO PHENOTYPE DURING EVOLUTION

Two main approaches are used to investigate the genomic loci and/or developmental mechanisms that have been altered to produce morphological change across species: (1) the candidate gene approach and (2) the quantitative trait locus (QTL) mapping approach. These two approaches have different strengths and limitations. The candidate gene approach can be undertaken to investigate morphological change in any set of species, whereas the QTL mapping approach is limited to species that can be crossed in the lab or that cross naturally in the field. In addition, the candidate gene approach can highlight differences in developmental programs that are characteristic of different genera, family, or even phyla, and that have been established deep in the tree of life, whereas the QTL mapping approach usually addresses more recent divergence in developmental programs that result in species-level differences. The main distinction between these approaches is that while the candidate gene approach can identify how developmental programs have changed across species, it rarely can pinpoint the causative mutations that lead to these changes. The QTL approach, on the other hand, can zoom in on the exact genomic loci that have mutated and are responsible for alterations in developmental programs across closely related species (see chapter V.13). Examples of both these approaches are provided below.

Often researchers target candidate genes for their role in causing differences in development across species because of prior knowledge that these genes are expressed during the development of the trait of interest. Genes known to be involved in building a homologous trait in a different species also make good candidate genes. Candidate gene approaches were used to implicate two genes, *Bone morphogenetic protein 4* (*Bmp4*) and *Calmodulin*, in the generation of differently shaped beaks in Galápagos finches. These two genes were differentially expressed in finches with deep and broad or long beaks, respectively. When chickens with artificially modified levels of these genes were produced in the lab, they also showed significant changes in the depth/width and length of their beaks. Taken together, these data suggest that changes in the expression of *Bmp4* and *Calmodulin* during the course of evolution caused changes in beak shape in these finches; however, these data do not necessarily suggest that these genes were themselves modified during the course of evolution to alter the beaks of these finches. Alterations to a gene's expression can occur via alterations in the *cis*-regulatory elements of the gene itself, or by alterations to the cocktail of TFs that bind to these elements (the *trans* regulators) and regulate gene expression. So while the candidate gene approach identifies changes in developmental mechanisms —changes to amounts of Bmp4 or Calmodulin mRNA and protein present in beaks at particular times in development—it cannot always identify the locus that mutated to produce these differences. To further dissect where these differences lie, a reciprocal locus transplantation experiment using transgenics is needed (discussed below).

QTL approaches have also been used to identify genes responsible for morphological evolution across species (see chapter V.12). For example, *Drosophila melanogaster* as well as several other closely related species are covered in small hairs, or trichomes, on the dorsal part of their bodies when they are larvae; however, *D. sechellia* has few trichomes on its body. By performing QTL mapping in laboratory crosses between *D. melanogaster* and *D. sechellia,* the position of the causative locus that explained most of the variation in larval trichome patterns was mapped to the *shavenbaby-ovo (svb)* locus. Modifications to the sequence of at least three different *cis*-regulatory elements of *svb,* each driving expression of *svb* in different sections of the larval body, were responsible for trichome loss in *D. sechellia*. This gene, when overexpressed in epidermal cells without trichomes, was shown to be necessary and sufficient to initiate the developmental program that builds trichomes, so shutting it down by deletion of its multiple epidermal enhancers, is an effective and direct way to eliminate trichome development in *D. sechellia.*

In the *Drosophila* case, unlike the case of the finches above, the ability to cross the two species with differing morphologies enabled the researchers to determine that *cis*-regulatory changes rather than changes to the *trans*-acting factors were responsible for the morphological changes that were observed between the species; however, when genetic crosses between species are not feasible because of reproductive incompatibilities, researchers

Species A and B differ in the stripe expression of the *black* gene. Why? Is this difference due to changes in the *cis*-regulatory region of *black*, or in the *trans*-regulatory environment?

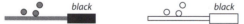

Reciprocal transplants of *cis*-regulatory elements attached to a reporter gene (e.g., GFP) into the different *trans*-regulatory environments can elucidate where the differences lie.

1) Changes in *cis*-regulatory elements but no changes in the *trans*-regulatory environment:

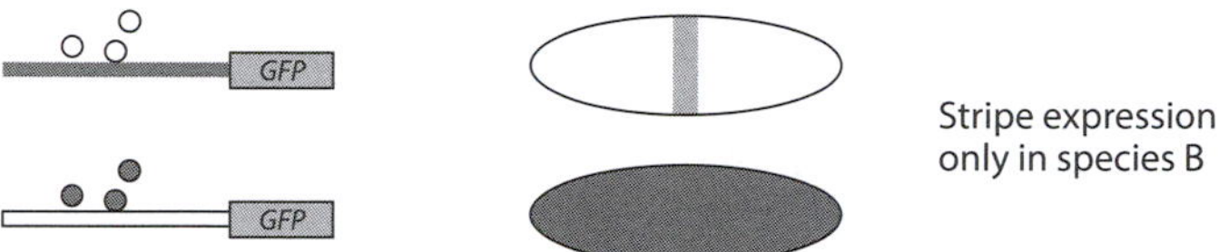

2) Changes in *trans* but no changes in *cis*-regulatory elements:

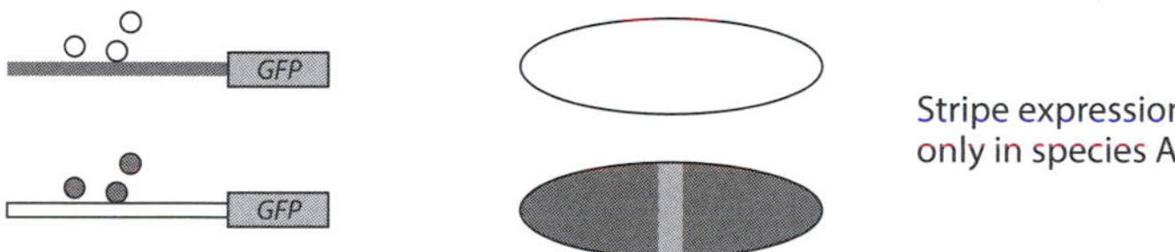

3) Changes in *cis*-regulatory elements and in *trans*:

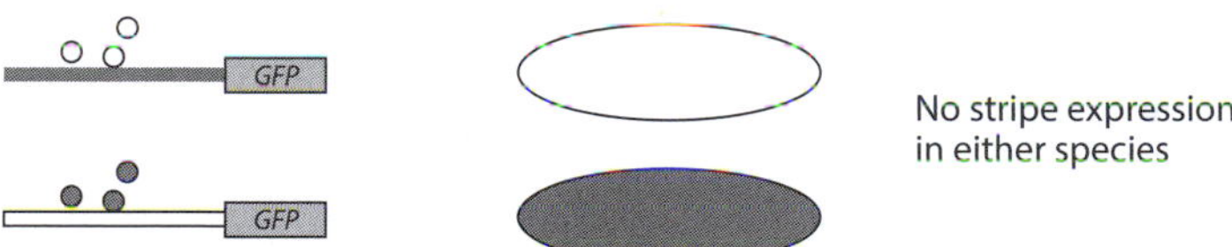

If transgenesis can only be performed in species A, then scenarios 1 and 3 cannot be distinguished, and if transgenesis can only be performed in species B, then scenarios 2 and 3 cannot be distinguished. If no stripe expression is observed in either species, then in the first case, changes in *trans* could also be playing a role, and in the second case, changes in *cis* could also be playing a role. If stripes are observed in species A, we conclude changes in *trans* alone. If in species B, then we conclude changes in *cis* alone.

Figure 1. Schematic of reciprocal genetic transplantation experiments that test whether changes in the *cis*-regulatory elements of a gene or the *trans*-regulatory factors that bind those elements are responsible for the expression differences observed between two species (A and B). In this case, expression differences correspond to the presence or absence of a stripe of *black* gene expression along the body (ellipse). Boxes correspond to protein-coding sequences. Black/white boxes: alleles of *black* candidate gene; gray box: reporter gene (*GFP*). Lines connected to boxes represent *cis*-regulatory sequences.

can turn to transplantation experiments using transgenic tools. The rationale behind these experiments involves taking the candidate gene of one species and introducing it into the *trans*-regulatory environment of the other species, and then performing the reciprocal experiment with the orthologous gene from the second species (see figure 1). These transplantation experiments are commonly performed in only one direction, often because of limitations in transgenic technology in one of the two test species, but an example of a complete reciprocal transplantation experiment was performed with the *lin-48 ovo* gene in *Caenorhabditis elegans* and *C. briggsae*. Researchers hypothesized that differences in the expression pattern of this gene observed between species could be due either to changes in the *cis*-regulatory region of the gene or to changes in the *trans*-acting factors. By performing a complete set of transplantation experiments in which they took the regulatory regions of each gene attached to a reporter gene to monitor expression activity and transplanted them (transgenically) to the *trans*-regulatory environment of the other species, they were able to conclude that changes in both the *cis*-regulatory sequences and the *trans*-acting factors that mediate *lin-48* expression contributed to the species-specific differences.

This type of transplantation experiment can also be done with the complete locus (*cis*-regulatory elements plus protein coding sequence) if alterations at the amino acid level are also suspected of contributing to particular phenotypic differences between species.

4. THE EVOLUTION OF NOVEL TRAITS AND THEIR UNDERLYING GENE REGULATORY NETWORKS

The examples discussed above monitor and dissect the evolution of developmental mechanisms from the perspective of individual genes. Mutations to single developmental genes, however, often modify the expression of many downstream targets and have a large impact on an organism's final phenotype. The group of affected genes depends on the topology of the regulatory network, that is, how many targets are downstream of the mutated gene, including both direct and indirect targets (see chapter V.9).

Some gene-regulatory networks are modular in their effects and may be quite important in body plan evolution. For instance, the *Distal-less* and *Pax6* TFs are important early regulators of limb and eye development, respectively, throughout the Metazoa. These genes, when ectopically expressed in several other parts of the body of a fly, are able to promote limb duplications and ectopic eyes; that is, they control the initiation of gene regulatory networks that lead to limb and eye differentiation. These networks have modular qualities in that they can be initiated in a context-independent manner at multiple locations in the body, somewhat independently of the cocktail of other TFs present at those locations.

The deployment and co-option of these modular networks into novel places in the body, and their recruitment to create repeated or serial homologous traits, and potentially also novel traits, is an active area of research in evo-devo. The idea is that the origination of novel traits may proceed by the co-option and the mixing and matching of modular networks, in novel combinations and at novel places in the body, rather than by the elaboration of preexisting networks one gene at a time (see figure 2). Evolution of novel traits would proceed via the genetic tinkering of modules of interacting genes by modification of the *cis*-regulatory regions of only a small set of individual genes regulating the initiation of each of these modules. The above-mentioned *Distal-less* (*Dll*) gene in the context of the evolution of appendages provides a nice example of the way these modular gene networks may originate.

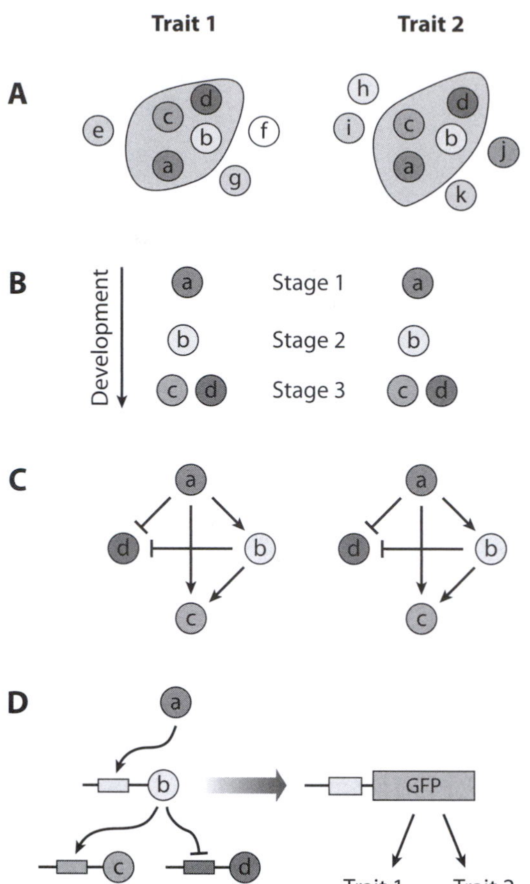

Figure 2. Different types of experiments aimed to test whether four genes (*a*, *b*, *c*, and *d*) expressed in two traits (1 and 2) are part of the same gene regulatory network that functions in the development of both traits. (A) A common set of genes (circled) is expressed during the development of the two traits. (B) The genes are expressed in a similar temporal order. (C) The genes display the same type of regulatory interactions (*a* represses *d*, *b* activates *c*, etc. Note that the regulatory interactions inferred may be direct or indirect). (D) Genes internal to the shared set (expressed at developmental stages 2 or 3, but not at stage 1) may contain unique *cis*-regulatory elements that drive gene expression in the two different developmental contexts. This is depicted by the isolation of the *cis*-regulatory element of the *b* gene, attaching it to a reporter gene (GFP), transforming the genome of the organism with this construct, and observing GFP expression in the tissue precursors of the two traits. (Modified from Monteiro 2012, *Bioessays* 34: 181–186.)

It is possible that in early metazoans, *Dll* became expressed in a novel cluster of cells as a result of evolution of novel positional information in its *cis*-regulatory region, "marking" these cells in a unique way. Other genes, by evolving binding sites for the Dll protein in their *cis*-regulatory regions, would be co-opted for expression in the same cluster of cells. Additional genes would have been gradually added to this basic gene network by developing binding sites either for *Dll*, or for any of the other gene products activated downstream of *Dll*. Perhaps, early in the process of building this network, a small outgrowth emerged from the body wall. If these outgrowths were useful in some way, the genomic information coding for the novel network would be retained. Later, with further network elaboration, the small outgrowths could become proper appendages. Such a network, scaffolded on *Dll* expression, is modular and context insensitive, i.e., when *Dll* is recruited to novel positions in the body, it is often able to direct the complete set of downstream targets and produce a novel outgrowth at these novel locations.

Many classic examples of the evolution of body plans involve changes to modular gene regulatory networks. These include examples where modular networks are modified by the action of region-specific TFs, named selector genes, or are duplicated, repressed, or co-opted into novel locations in the body to create serial homologous traits, or novel body plans. An example of network modification includes the evolution of arthropod appendages into a variety of different shapes and sizes by the action of *Hox* genes, selector genes that are differentially expressed along the anterior-posterior axis of the body and give each region of the body a unique identity. In crustaceans, limbs that develop in regions of the body where anterior *Hox* genes are expressed become feeding appendages, whereas limbs that develop in regions of the body where posterior *Hox* genes are expressed become walking legs. The *Hox* genes appear to bind to the *cis*-regulatory regions of many different genes within a limb network in order to modify their expression, and thus, the final limb phenotype. In addition, *Hox* genes expressed in the abdominal region directly bind to the early limb enhancer of *Dll*, thereby shutting down the limb network in the abdomen of flies, and perhaps most other insects.

Similar to the role of *Hox* genes in specifying the identity of modules along the anterior-posterior axis, modifications to other types of selector gene also underlie modifications to modular gene networks that are repeated in the body. For instance, the *Pitx1* gene controls the identity and the development of the pelvic fins in stickleback fish, but *Pitx1* has no role in pectoral fin development. Multiple independent deletions of a *cis*-regulatory element upstream of *Pitx1* have occurred in different stickleback populations, resulting in the loss of *Pitx1* expression in pelvic fins, and therefore, loss of the pelvic fin structure in these fish. Changes to *Pitx1*, because of its unique pelvic fin expression, are among the few places in the fin gene regulatory network that would allow a complete fin to be lost without impairing the development of the other serial homologue (the pectoral fin).

Modular gene regulatory networks, such as the limb or the eye network, may have also been co-opted into different regions in the body to give rise to novel body parts, or serial homologous traits. For instance, the appearance of horns in the heads of beetles may have originated via the co-option of the insect limb network to the head, as many of the genes found in limbs are also expressed in horns. The evolution of multiple eyes along the mantle of scallops is probably due to the co-option of an early expressed gene from the eye gene regulatory network to the mantel's edge. And the evolution of the most posterior set of fins/limbs in vertebrates is due to the co-option of the vertebrate limb network, initially deployed only in the pectoral fin region of primitive fish, to a more posterior position along the anterior-posterior body axis, thus creating the vertebrate paired appendages.

Co-option of the modular gene regulatory networks mentioned above to the novel locations would involve the evolution of novel positional information for the expression of the network's top regulatory gene. This positional information would be in the form of a novel enhancer sequence where binding sites for one or more TFs expressed at the novel body location would evolve and allow the top regulatory gene to be turned on at that location. It remains possible that completely novel and parallel gene networks were created de novo at these novel body locations; however, this is unlikely, as such networks would take a much longer period of time to evolve and would probably not be fully functional until complete. Many aspects of modular gene regulatory networks are still unclear, such as their frequency in developmental systems, their size distribution (e.g., how many genes are involved), and their evolution, but this information will likely become available as research progresses in this field.

Thinking of development as the temporal stringing together of modular gene regulatory networks also helps explain why there are sometimes dramatic differences between species at the early stages of development, while later stages of development are conserved. Early network modules can evolve as long as the connections to later modules are kept intact. An example involves the earliest steps in embryonic development in *Drosophila:* the determination of where the head is going to lie. This is achieved by a gradient of Bicoid protein that is set up by the mother before the egg is laid. She deposits and attaches *Bicoid* mRNA molecules to the anterior end of the egg. On translation, a gradient of protein is established, and high levels of protein activate a downstream target gene in the anterior half of the embryo, *hunchback*, which defines the head region of the fly. In the beetle *Tribolium*, Bicoid protein is not responsible for head patterning in the early embryo, but the function and expression of *hunchback* is still conserved. Another example of such modularity is the sex-determination pathway in animals where the upstream factors that determine the sex of the animal are very diverse, ranging from a sex chromosome to temperature induction (see chapter V.4), but the downstream effectors are very conserved and usually involve the gene *doublesex* and its homologues. So, gene regulatory networks can evolve in their very earliest steps while downstream components and the final phenotype remain unchanged.

5. FUTURE AREAS OF RESEARCH IN EVOLUTIONARY DEVELOPMENTAL BIOLOGY

The Molecular Basis of Phenotypic Plasticity

While much is beginning to be known about the molecular details of morphological evolution, an area that is still lagging behind concerns investigating the molecular basis of the integration of environmental factors into regulatory gene networks to induce distinct phenotypes. Phenotypic plasticity, or the ability of the same genome to give rise to very different morphological, physiological, or behavioral traits depending on rearing environment, is still poorly understood at the molecular level. A variety of environmental factors such as temperature, light, pressure, food availability, and certain chemicals are known to induce alternative developmental pathways, but the molecular details of the mechanisms by which these factors influence gene regulatory networks are poorly understood.

The evolution of adaptive phenotypic plasticity usually involves changes to gene-regulatory networks that better adapt the organism to different and predictable environments. In many cases, hormones appear to play important roles in coordinating plastic development as they circulate among all the tissues in the body, and are thus able to coordinate changes in multiple modular gene regulatory networks underlying the development of various traits. But how these hormonal signaling systems evolve to interact with specific gene networks and how hormonal systems themselves become sensitive to the environment are still areas of active investigation.

Robustness

The flip side of plasticity is robustness, where developmental networks have evolved extreme insensitivity to environmental and/or genetic perturbations. At the molecular level, robustness is achieved by evolution of regulatory wiring that leads to gene expression homeostasis, by gene duplications, or even by *cis*-regulatory element duplications that lead to more robust patterns of gene expression in the face of perturbation. Robust gene networks can potentially accumulate many mutations that are buffered from affecting network output (creating cryptic genetic variation) by the architecture of the developmental gene network.

Understanding how these two types of gene networks, plastic and robust, bias or channel further evolutionary change is an important area of future research. In particular, the roles of natural and sexual selection are believed by many to be all-powerful in shaping the behavior and morphology of organisms, but these forces can exert change in systems only if these systems produce sufficient phenotypic variation for selection to act on (see chapter V.10). Plastic networks will readily produce variation in response to environmental variation, whereas robust networks will not. On the other hand, selection on phenotypes derived from plastic networks will not lead to evolutionary change, since the variation is not based in genetics but environmentally induced, whereas selection on phenotypes derived from robust networks will produce minimal change, because phenotypes will essentially be the same. Novel environments may favor evolutionary change, and this can lead to novel patterns of selection and changes to network topology in the case of plastic networks, and to the release of accumulated cryptic genetic variation in the case of robust networks, if these networks are altered beyond their natural buffering capacity.

Conclusion

In summary, molecular evo-devo has the ability to explain both micro- as well as more macro evolutionary changes in developmental programs and phenotypes, the evolution of novel traits, and the role played by the environment in modifying development to create plastic phenotypes. Future empirical work with additional species and traits, as well as modeling work, should eventually aim to produce a theory of morphological evolution based on gene networks, and gene interactions, that fully updates the modern synthesis.

FURTHER READING

Carroll, S. B., J. K. Grenier, and S. D. Weatherbee. 2005. From DNA to Diversity: Molecular Mechanisms and the Evolution of Animal Design. 2nd ed. Malden, MA: Blackwell. *A comprehensive text in the field of evo-devo.*

Erwin, D. H., and E. H. Davidson. 2009. The evolution of hierarchical gene regulatory networks. Nature Reviews

Genetics 10: 140–148. *A recent review of several of the major topics discussed in this chapter by leaders in the field.*

Fielenback, N., and A. Antebi. 2008. *C. elegans* dauer formation and the molecular basis of plasticity. Genes and Development 22: 2149–2165. *A nice review article that discusses the molecular basis of plasticity in a model organism.*

Gilbert, S. F., and D. Epel. 2009. Ecological Developmental Biology: Integrating Epigenetics, Medicine, and Evolution. Sunderland, MA: Sinauer. *An engaging and clear exposition of the ways in which the environment affects developmental programs, with many examples at the molecular level.*

Nowick, K., and L. Stubbs. 2010. Lineage-specific transcription factors and the evolution of gene regulatory networks. Briefings in Functional Genomics 9: 65–78. *A review of the ways in which evolution of TF sequence and TF duplications impact gene regulatory networks.*

Stern, D. L. 2010. Evolution, Development, and the Predictable Genome. Greenwood, CO: Roberts & Company. *A very readable account that discusses the reasons certain genes in networks become hot spots of morphological evolution.*

V.12

Genetics of Phenotypic Evolution

Catherine L. Peichel

OUTLINE

1. Genetic architecture of phenotypic evolution
2. Molecular basis of phenotypic evolution
3. Using genotypes to test whether phenotypes are adaptive
4. Genetic basis of repeated phenotypic evolution
5. Prospects for future research

The incredible diversity of life on earth is most easily evidenced at the level of the phenotype, which is any characteristic of an organism that can be observed or measured. Thus, the term *phenotype* encompasses morphological, behavioral, and physiological traits. For evolution of a phenotype to occur, it must have a genetic basis (i.e., be heritable). It is therefore necessary to understand the genetic underpinnings of phenotypic traits to understand the process of phenotypic evolution. This chapter will focus on the genetic and molecular basis of phenotypes that are adaptive; that is, phenotypes that contribute to fitness in a given environment. Although the genetics of adaptation has a long history of study, experimental progress was somewhat limited for much of the last century; however, recent technological advances in genetics and genomics have enabled the identification of genes and mutations that underlie phenotypic evolution in both plants and animals. These initial studies have begun to address long-standing questions about the number and effect sizes of the genetic changes that underlie adaptation, the types of genetic changes involved, the evolutionary history of the genetic changes, and whether the same genetic changes are used when similar phenotypes evolve in independent populations; however, more work needs to be done across a number of systems to gain a complete picture of the genetic basis for phenotypic evolution and adaptation. Fortunately, rapid progress in genome sequencing technologies and the development of new experimental systems are providing an unprecedented opportunity to understand the link between the environmental agents of selection and the phenotypic and genotypic targets of selection.

GLOSSARY

Adaptation. The process by which a population evolves to have higher fitness in a given environment.

Enhancer. A DNA sequence that controls the expression of a gene; enhancers are also referred to as *cis*-regulatory elements.

Inversion. A chromosomal rearrangement in which the sequence of DNA is reversed. When present in a heterozygous state, recombination is suppressed within the inversion.

Phenotypic Effect Size. The amount of variation in a phenotype that can be explained by a particular genetic change.

Pleiotropy. The same gene or mutation affects multiple phenotypes.

Quantitative Trait Locus (QTL) Mapping. A genetic linkage mapping approach that seeks to identify associations between genotype and phenotype on a genome-wide scale. This approach provides information about the genomic location, number, and effect sizes of the genetic loci that underlie a given phenotype. It requires the ability to cross individuals with different phenotypes.

Repeated Evolution. The appearance of similar phenotypes in independent evolutionary lineages that experience similar environments; also referred to as *parallel or convergent evolution.*

1. GENETIC ARCHITECTURE OF PHENOTYPIC EVOLUTION

A fundamental question in evolutionary biology is whether phenotypic changes occur via an infinite number of genetic changes, each with an extremely small effect on phenotype, or via relatively few genetic changes, each with a large effect on phenotype. This question has been hotly debated over the last 150 years, starting with Darwin's view that evolution must occur through many small changes with slight phenotypic effects ("micromutationism") because of the observation that gradual and continuous phenotypic changes are found in nature. The rediscovery of Mendel's laws of inheritance challenged this view because early geneticists began to find mutations that caused large and discontinuous phenotypic effects, at least in the laboratory. This led to a vigorous debate between the Darwinian gradualists and the Mendelian geneticists around the turn of the twentieth century. The successful fusion of micromutationism with Mendelism by the founders of population genetics, particularly Ronald A. Fisher, during the "evolutionary synthesis," seemed to resolve the debate (see chapter I.2). Using mathematical arguments, Fisher demonstrated that an infinite number of small genetic changes could underlie continuous phenotypic variation. Fisher then compared the genetic process of phenotypic adaptation to the process of focusing a microscope. When focusing a microscope, a large adjustment has a much smaller probability of improving the focus than a small adjustment. In the same way, Fisher considered it very unlikely that mutations of large effect would be beneficial and thus concluded that only mutations with extremely small effects would be beneficial and contribute to phenotypic adaptation (the "infinitesimal model"). While satisfactory, for much of the twentieth century this theoretical argument prevented any further empirical work on the question because it seemed pointless to look for infinitesimally small genetic changes.

Fisher's initial theory was later revisited by Motoo Kimura, who realized that while mutations with large phenotypic effects were more likely to be deleterious, those that were beneficial were less likely to be lost due to drift, particularly in small populations. Extending this work, H. Allen Orr showed that mutations of large effect that are beneficial are more likely to be fixed, particularly early in the process of adaptation. Taken together, this theoretical framework predicts that the genetic architecture of phenotypic evolution will involve a few genetic changes with large effects and many genetic changes of smaller effects (the "geometric model").

Although empirical work on this question was limited by the almost-universal acceptance of the infinitesimal model for much of the last century, recent advances in technology have enabled experimental approaches to identify the genetic architecture of phenotypic evolution. In particular, quantitative trait locus (QTL) mapping (see chapter V.13) has been used to identify the genomic location, number, and effect sizes of genetic changes that underlie phenotypic differences among natural populations of plants and animals. Many studies have now conclusively demonstrated that genetic changes of both large and small effect contribute to phenotypic evolution; however, very few studies have had the experimental power to address the relative contributions of these changes during adaptation. More data are needed to determine whether the geometric model will generally hold true, but the recent emergence of new genome sequencing technologies will make it feasible to investigate the number and effect sizes of the genetic changes underlying the evolution of a wide variety of phenotypes across numerous taxonomic groups.

Genetic mapping studies such as QTL mapping also enable researchers to address a second question about the genetic architecture of phenotypic evolution: Are the genetic changes that underlie phenotypic evolution and adaptation found in particular regions of the genome or are they distributed across the genome? This question is of particular interest when an organism adapts to a new environment, because multiple phenotypic changes are usually required for adaptation. For example, a plant living in a cool, wet environment and a plant living in a hot, dry environment will differ in many respects, including morphology, physiology, and life history. Thus, adaptation to a new environment might be facilitated if the same genetic changes give rise to multiple phenotypic differences (*pleiotropy*), or if independent genetic changes are each responsible for a single phenotype, but multiple phenotypes are inherited together (*linkage*).

The evidence for pleiotropy is certainly good after decades of genetic studies in model laboratory organisms; mutations in a single gene often affect multiple phenotypes. Although there is also evidence that multiple phenotypes map to the same locus in genetic studies of natural populations, most of these studies have insufficient resolution to determine whether multiple phenotypes are controlled by the same gene or by tightly linked genes. Classic examples of such *supergene* complexes are found in Müllerian mimicry in *Heliconius* butterflies. Multiple aspects of wing color patterns (e.g., the type and distribution of pigments) map to a single locus, but it is still not known whether these are due to mutations in different enhancer elements of the same gene, or to mutations in multiple, closely linked genes. To distinguish these possibilities, it will be necessary to identify the actual sequence changes that give rise to the color pattern phenotypes (see next section).

In either case, recombination within the supergene complex would create individuals that are not perfect mimics and thus would be less likely to survive. Interestingly, one of the supergene complexes in *Heliconius* is found within a chromosomal rearrangement that suppresses recombination. It has long been thought that such chromosomal rearrangements, particularly inversions, might be hot spots for genes involved in phenotypic evolution and adaptation, as well as speciation (see chapter VI.9). Many closely related species differ by one or more chromosomal rearrangements, and differences in the frequency of chromosomal inversions within species have been correlated with environmental clines. These data suggest that inversions might be important for adaptation, but only recently has this hypothesis been directly tested using yellow monkeyflowers (*Mimulus guttatus*). Two forms of this species are found along the west coast of the United States: a perennial form adapted to the cool and wet coastal region, and an annual form adapted to the hot and dry inland region. Many of the phenotypic differences between the forms, including flowering time and a number of morphological traits, map to an inversion. By a clever crossing scheme, the perennial inversion was placed into annual plants and vice versa. Then, plants were placed in the two habitats; remarkably, the inversion altered the phenotypes and conferred increased survival and fitness in the appropriate environment. Although this study provides good evidence that genes involved in phenotypic evolution and adaptation can be clustered within chromosomal rearrangements and thus coinherited, there are still many cases where the genes that underlie multiple phenotypes required for adaptation to a particular environment map to distinct genomic locations. Comprehensive genetic mapping studies of morphological, behavioral, and physiological traits across a number of systems are required to determine the extent to which pleiotropy and linkage contribute to the genetic architecture of phenotypic evolution and adaptation.

2. MOLECULAR BASIS OF PHENOTYPIC EVOLUTION

An ultimate goal of modern evolutionary genetics is to identify the actual genes and specific mutations that underlie phenotypic evolution. Using classical genetic mapping and candidate gene approaches in combination with modern molecular tools such as transgenic technologies (see chapters V.11 and V.13), it has been possible to pinpoint the specific genes and in some cases the mutations that underlie a wide variety of phenotypic differences among natural populations of both plants and animals. Interestingly, most of these cases involve morphological or physiological traits, while only a handful of genes have been identified underlying the evolution of behavioral phenotypes. This is a fruitful area for future research (see below).

There has been some vigorous debate about whether the genetic changes responsible for phenotypic evolution would more likely occur in coding sequences of genes (i.e., the protein itself) or in the *cis*-regulatory regions of genes (i.e., the regions that control the time and place of gene expression). Thus far, ample evidence has been presented that both coding and regulatory changes can and do contribute to phenotypic evolution; however, the data collected thus far do not represent a completely unbiased set, so it is not yet possible to determine the relative roles of these different types of mutations.

Although finding the actual mutations responsible for phenotypic changes can be challenging, particularly when the mutations are in regulatory regions, this information enables a fine-grained view of the process of phenotypic evolution. As discussed in the previous section, there has been a long debate over whether a few mutations of large effect or many mutations of small effect contribute to phenotypic evolution and adaptation. To date, many studies have found that a single genetic locus can have a relatively large effect on a given phenotype. An interesting question then follows of whether a single genetic locus comprises many mutations, each with small effect, or a single mutation of large phenotypic effect. In the case of coding mutations, there is abundant evidence that a single nucleotide mutation leading to a single amino acid change can be sufficient to cause a large phenotypic effect, as is exemplified by mutations in the *Mc1r* gene that lead to dramatic differences in pigmentation (see chapter I.4). Although there is good evidence that regulatory mutations can have large effects on phenotype, only a few studies to date have identified the actual enhancers and mutations responsible. In two different cases in fruit flies, multiple single base-pair mutations within a single enhancer are required to create a large phenotype effect. By contrast, in stickleback fish, a single mutational event in a specific enhancer is sufficient to create a large phenotypic effect. With such a limited data set, it is too early to speculate on the reasons these differences may exist. Although conducting such studies to identify mutations in regulatory elements is difficult and long term, they can provide unprecedented insights into the process of adaptation, as discussed in the next section.

3. USING GENOTYPES TO TEST WHETHER PHENOTYPES ARE ADAPTIVE

It has been known for a long time that phenotypes do evolve and thus have a genetic basis; thus, some evolutionary biologists have wondered what new insights can be gained by identifying the actual genes and mutations that underlie phenotypic evolution. In fact, there are many long-standing questions about the genetic basis of phenotypic evolution and adaptation that can be addressed with such information.

First, once a gene that underlies a particular phenotype is known, it is possible to investigate whether the trait evolved in response to selection or as a result of neutral processes like genetic drift. Such an investigation can be conducted by looking for molecular signatures of selection, as has been widely done across genomes or at specific genes (see chapter V.14); however, most of these studies have been genotype focused and suggest only that a locus has experienced selection, but they do not reveal the phenotypes associated with those genotypes or the selective agent involved. Thus, selection is only inferred from the molecular data. By first identifying genes that underlie phenotypes known to be under selection, it is possible to determine whether molecular tests for selection do in fact identify loci associated with phenotypes under selection and thereby inform genotype-based approaches. Furthermore, in studies in which it is not known whether a phenotype is under selection, identifying the genes responsible enables tests for selection, as demonstrated by a pair of recent studies in stickleback fish and fruit flies. In both cases, the actual mutations responsible for phenotypic changes were identified: deletion of an enhancer for *Pitx1* gene contributes to loss of pelvic spines in sticklebacks, and mutations in an enhancer for the *ebony* gene contribute to changes in body coloration in fruit flies. Although these phenotypes were predicted to be adaptive, evidence for selection on these phenotypes was obtained by identifying the molecular changes involved and then looking for molecular signatures of selection; thus, these two studies provide compelling examples of the ability to make a connection between selection, phenotype, and genotype.

Even once such connections are made, it is still difficult to discern the selective agents responsible for the evolution of the phenotype. A second advantage of identifying the genes responsible for a phenotype is that it can enable studies to identify specific agents of selection. For example, the fitness effects of alternative alleles at a particular locus can be assessed in the field or in controlled, seminatural habitats. To date, very few studies have done this, but the ones that have are classics. For example, a single locus of major effect, called *YUP*, is responsible for the difference in color between a pink, bumble-bee-pollinated species of monkeyflower (*Mimulus lewisii*) and its red, hummingbird-pollinated sister species (*M. cardinalis*). By swapping the *YUP* locus between these two species and planting them in the field, a reversal in pollinator preference can be observed. These data demonstrate that the change in flower color is adaptive and that pollinator preference is the selective agent at work.

A third advantage of identifying the genetic basis of a phenotype is the ability to learn about the evolutionary history of an adaptive allele. For example, it is possible to determine whether adaptation to a new environment involves new mutations or selection on existing genetic variation. There is now clear evidence that both contribute to phenotypic evolution, sometimes at the same genetic locus. In the case of fruit fly body coloration described above, both existing variation and new mutation at the *ebony* gene played a role in the evolution of darker pigmentation at higher altitudes. It is further possible to use sequence data to determine when adaptive alleles arose within populations, as demonstrated by studies of an allele of the *agouti* pigmentation gene that arose by new mutation in deer mice with light coloration adapted to living on lighter-colored soil in the Sand Hills of Nebraska. Such studies are important because they inform us about the speed of adaptation to new environments, which is particularly relevant when thinking about how quickly organisms might adapt to global changes in climate in the future.

4. GENETIC BASIS OF REPEATED PHENOTYPIC EVOLUTION

The repeated evolution of similar phenotypes in response to similar environmental conditions is generally taken as strong evidence of a role for natural selection in the evolution of that phenotype. Such repeated evolution is referred to as *parallel evolution* when similar phenotypes evolve in closely related lineages and as *convergent evolution* when similar phenotypes evolve in distantly related lineages; however, the distinction between parallel and convergent evolution can be contentious and is further complicated by identification of the molecular basis of these traits. For example, the same genes, and sometimes even the same mutations, underlie the repeated evolution of the same phenotype in independent lineages, as exemplified by the finding that mutations in the *Mc1r* gene underlie differences in coloration among multiple species of fish, birds, snakes, lizards, and mammals. Strikingly, the same mutation in *Mc1r* can be found in some melanic birds and mammals; however, not all pigmentation differences are controlled by *Mc1r*, and even closely related subspecies of mice do not share the same genetic basis for similar pigmentation patterns. Thus, phenotypic parallelism or convergence may not be mirrored at the genetic level, creating an issue of whether these terms should be defined at the level of phenotype or genotype. Despite this semantic debate, an important and interesting question remains: When selection favors the evolution of similar traits, are the same genes and mutations used, or can different genes and mutations create similar phenotypic changes?

Although the available data are far from comprehensive or unbiased, the answer so far appears to be yes to both. Repeated evolution of phenotypes as diverse as flowering time in plants, insecticide resistance in insects, and pigmentation in vertebrates and invertebrates does occur via changes in the same genes, but not always. However, when particular phenotypes appear to evolve using a limited number of genes, it begs the question of why the same genes might be used repeatedly. For some traits, it might be that only a few genes can alter a given phenotype. For example, changes in color vision almost necessarily occur through changes in opsin genes. However, it is likely that for most phenotypes, mutations in any of hundreds of genes might give rise to a particular phenotype. For example, more than 80 genes can regulate flowering time, but only a few of these genes appear to be used during the evolution of differences in flowering time. Though the question of why particular genes are reused during evolution is far from resolved, there are several possible reasons, each supported by empirical evidence.

First, there may be a difference in mutational bias between genes; that is, some genes are larger targets for mutation or found in regions of the genome with higher mutation rates. For example, an enhancer in the *Pitx1* gene has been deleted at least nine independent times in stickleback populations that have lost the pelvic spines. This enhancer is located in a region of the stickleback genome that has features associated with DNA fragility, which may predispose it to deletion. However, not all repeated use of the same genes can be explained by differences in mutation rate and might reflect either selective constraints or historical contingency; these possibilities are discussed next.

A second reason for the repeated use of particular genes is related to pleiotropy. As discussed above, many genes are known to affect more than one phenotype. If a mutation in a particular gene causes one phenotypic change that is beneficial, but also causes additional phenotypic changes that are detrimental and outweigh the benefits, there could be selection against that particular mutation. Thus, genes that are expressed only in a specific tissue might be more likely to be used repeatedly. Consistent with this hypothesis, the *Mc1r* gene has been repeatedly implicated in the evolution of pigmentation across vertebrates and seems to have relatively few pleiotropic effects. Importantly, even though a gene might have an effect on multiple phenotypes, it is possible to identify mutations in the gene that have no pleiotropic effects. For example, mutations in an enhancer that drives expression in a particular anatomical location might be less likely to confer deleterious pleiotropic effects than a mutation in the coding region or in a general promoter region of a protein expressed in multiple tissues or with multiple functions. The *Pitx1* gene mentioned above provides a nice example; this gene is expressed in a number of tissues, and laboratory mice with a deletion of *Pitx1* die shortly after birth. In sticklebacks, the deletion of an enhancer that drives expression of the *Pitx1* gene only in the developing pelvic region is therefore a mutation with no pleiotropic effects, despite the fact that the gene itself exhibits pleiotropy.

Third, the use of a particular genetic change for phenotypic evolution might be dependent on the historical background of genetic variation already present within a population. If a large ancestral population repeatedly colonizes similar environments, there may be standing genetic variation that can be repeatedly selected in the new environment, leading to reuse of the same genetic change to evolve the same phenotype. A now classic example occurs in stickleback fish; marine ancestors have colonized freshwater environments and consequently evolved reduction of body armor in stickleback populations across the Northern Hemisphere. Repeated evolution of this phenotype is due to repeated selection of a particular allele of the *Ectodysplasin* gene present at low frequency in the ancestral marine population. In other cases, it is possible that the ancestral genetic background has no phenotypic effect on its own, but rather influences that new mutations have a phenotypic effect and thus are selected. Studies of the *shavenbaby* gene in fruit flies demonstrated that the phenotypic effect of any single mutation in an enhancer was dependent on which other mutations were also present in the enhancer. Not only are interactions within a locus important, but interactions between two or more loci might also influence the types of mutations that are selected. Reduced pigmentation in beach mice that live on light sand dunes in Florida is due to mutations in two previously discussed pigmentation genes, *Mc1r* and *agouti*. In this case, the phenotypic effects of the *Mc1r* mutation can be observed only when the *agouti* mutation is present. Thus, the selective advantage for the *Mc1r* mutation occurs only on a specific genetic background.

As more and more genes are identified that contribute to the repeated evolution of similar phenotypes across a number of different plant and animal groups, a more complete picture will begin to emerge of the relative roles of mutational bias, selective constraints such as pleiotropy, and historical contingencies such as genetic background. Such data will allow us to determine whether there are any "rules" for phenotypic evolution and adaptation.

5. PROSPECTS FOR FUTURE RESEARCH

This is an exciting time to be an evolutionary biologist. Along with a renaissance of interest in understanding the genetics of phenotypic evolution and adaptation, genome sequencing technologies are becoming widely available. Thus, it is now feasible to develop genetic tools and genomic resources for many systems with interesting phenotypic differences, rather than focusing on just a few model systems or a limited number of phenotypes. In particular, very little is known about the genetic changes that contribute to the origin of novel phenotypes or the evolution of behavior. The recent successes in identifying genetic changes that underlie morphological and physiological evolution suggest that by using appropriate systems, progress will be made on these types of traits as well. And continued efforts to identify the genetic underpinnings of many phenotypes in diverse plant and animal systems will surely provide a comprehensive picture of the relative contributions of small- versus large-effect mutations, coding versus regulatory mutations, and new mutations versus standing variation to phenotypic evolution and adaptation.

With these genetic changes in hand, it will also be possible to begin to connect the genotypic and phenotypic targets of selection with the environmental agents of selection. In particular, once a genetic change has been identified that contributes to a phenotype, that genetic change can be made on controlled genetic background. The phenotypic and fitness effects of the genetic change can then be measured in controlled environmental conditions or in seminatural habitats. These types of experiments will ultimately enable identification of the particular environmental conditions that select for a specific genetic and phenotype change. Such studies will provide unprecedented insights into the process by which phenotypes evolve and organisms adapt to their environments.

FURTHER READING

Barrett, R.D.H., and D. Schluter. 2008. Adaptation from standing genetic variation. Trends in Ecology & Evolution 23: 38–44. *Compares the evolutionary consequences of phenotypic evolution through standing genetic variation versus new mutation, and describes approaches to distinguishing between them.*

Bradshaw, H. D., and D. W. Schemske. 2003. Allele substitution at a flower colour locus produces a pollinator shift in monkeyflowers. Nature 426: 176–178. *One of a few studies to directly demonstrate that switching a single genetic locus has a major effect on phenotype and to provide evidence that the resulting phenotypic change is adaptive in the wild.*

Chan, Y. F., M. E. Marks, F. C. Jones, G. Villarreal Jr., M. D. Shapiro, S. D. Brady, A. M. Southwick, et al. 2010. Adaptive evolution of pelvic reduction in sticklebacks by recurrent deletion of a *Pitx1* enhancer. Science 327: 302–305. *This study directly links genotype and phenotype to selection. In this case, new mutations have independently occurred in the same regulatory element to cause repeated phenotypic evolution, and these mutations show molecular evidence of selection.*

Conte, G. L., M. E. Arnegard, C. L. Peichel, and D. Schluter. 2012. The probability of genetic parallelism and convergence in natural populations. Proceedings of the Royal Society B 279: 5039–5047. *Reviews the genetic and molecular basis of repeated evolution.*

Frankel, N., D. F. Erezyilmaz, A. P. McGregor, S. Wang, F. Payre, and D. L. Stern. 2011. Morphological evolution caused by many subtle-effect substitutions in regulatory DNA. Nature 474: 598–603. *Illustrates how a single locus of large effect can consist of many mutations, each with relatively small effect, in a regulatory element.*

Gompel, N., and B. Prud'homme. 2009. The causes of repeated genetic evolution. Developmental Biology 332: 36–47. *Provides an overview of the factors that might contribute to the repeated use of the same genes during phenotypic evolution.*

Joron, M., L. Frezal, R. T. Jones, N. L. Chamberlain, S. F. Lee, C. R. Haag, A. Whibley, et al. 2011. Chromosomal rearrangements maintain a polymorphic supergene controlling butterfly mimicry. Nature 477: 203–206. *Provides evidence that supergenes can be maintained by chromosomal inversions.*

Kirkpatrick, M., and N. Barton. 2006. Chromosome inversions, local adaptation and speciation. Genetics 173: 419–434. *Summarizes the main mechanisms proposed to explain the spread of chromosomal inversions in populations, and proposes an elegant and testable mechanism for the role of inversions in adaptation and speciation.*

Linnen, C. R., and H. E. Hoekstra. 2009. Measuring natural selection on genotypes and phenotypes in the wild. Cold Spring Harbor Symposia on Quantitative Biology 74: 155–168. *Provides a road map for experimental approaches to connect the targets of selection (genotype and phenotype) with the agents of selection (environment).*

Linnen, C. R., E. P. Kingsley, J. D. Jensen, and H. E. Hoekstra. 2009. On the origin and spread of an adaptive allele in deer mice. Science 325: 1095–1098. *Identifies a genetic locus that underlies phenotypic evolution and uses molecular data to infer the age of the mutation as well as the strength of selection on the locus.*

Lowry, D. B., and J. H. Willis. 2010. A widespread chromosomal inversion polymorphism contributes to a major life-history transition, local adaptation, and reproductive isolation. PLoS Biology 8: e1000500. *A direct empirical demonstration that a chromosomal inversion harbors multiple phenotypic traits and plays a role in adaptation to divergent environments.*

Orr, H. A. 2005. The genetic theory of adaptation: A brief history. Nature Reviews Genetics 6: 119–127. *A comprehensive and clear review of theory and data relevant to the debate over the number and effect size of genetic changes that contribute to adaptation.*

Rebeiz, M., J. E. Pool, V. A. Kassner, C. F. Aquadro, and S. B. Carroll. 2009. Stepwise modification of a modular enhancer underlies adaptation in a *Drosophila* population. Science 326: 1663–1667. *Demonstrates that a combination of standing genetic variation and new mutations within a single regulatory element contribute to phenotypic evolution. These mutations show molecular evidence of selection, providing a link between genotype, phenotype, and selection.*

Stern, D. L., and V. Orgogozo. 2008. The loci of evolution: How predictable is genetic evolution? Evolution 62: 2155–2177. *Provides a comprehensive list of genetic changes so far known to contribute to phenotypic evolution in plants and animals.*

V.13

Dissection of Complex Trait Evolution

Bret A. Payseur

OUTLINE

1. Genetic variation in complex traits
2. Using laboratory crosses to map the mutations responsible for phenotypic evolution
3. Using association testing to map the mutations responsible for phenotypic evolution
4. Current challenges and prospects for future research

This chapter describes approaches that use naturally occurring variation to dissect the genetic basis of phenotypic evolution. Emphasis is given to *complex traits*—those phenotypes controlled by multiple genetic and environmental factors. The use of laboratory crosses to locate mutations that affect trait variation is briefly reviewed. Then the strategy of comparing phenotypes and genotypes in population samples of unrelated individuals is explained. The factors that affect success in association testing when conducted on the genome-wide scale are discussed, along with some lessons from studies in humans.

GLOSSARY

Alleles. Different versions of a gene or mutation carried by individuals in a population.

Complex Traits. Phenotypes to which multiple genetic and environmental factors contribute.

Genome-Wide Association Testing. A strategy that compares phenotypes with genotypes at markers throughout the genomes of unrelated individuals from a population to identify mutations associated with complex trait variation.

Haplotype. The particular combination of alleles at a series of polymorphisms that is present on one chromosome in an individual.

Heritability. The fraction of phenotypic variation in a population that is caused by genetic differences.

Linkage Disequilibrium. The statistical correlation between alleles at different loci in populations.

Missing Heritability. The common observation that the combined phenotypic effects of variants discovered through genome-wide association testing leave much of the genetic variation in a trait unexplained.

Quantitative Trait Locus (QTL). A genomic region that contributes to complex trait variation.

Recombination. Exchange of genetic material during meiosis that produces new combinations of mutations in a population.

Single Nucleotide Polymorphism (SNP). A single site in the genome that varies among individuals in a population.

1. GENETIC VARIATION IN COMPLEX TRAITS

Organisms vary in anatomy, physiology, and behavior. For most traits, this variation reflects the combination of multiple environmental and genetic factors. Although understanding the environmental component of trait variation is an important goal, evolutionary biologists have long been obsessed with the genetic piece. This fixation arises because evolution specifically targets inherited variation. Without genetic differences between organisms, evolution stops (see chapter I.4).

The field principally concerned with measuring and interpreting genetic variation in complex traits is *quantitative genetics*, which uses statistical models to partition different sources of trait variation in a population, including environmental and genetic variance. *Genetic variance* is a population quantity that jointly reflects the frequencies and phenotypic effects of all mutations that shape a trait. Phenotypic resemblances among relatives can be expressed as functions of genetic variances, enabling the estimation of these quantities from phenotypic data. Nevertheless, because genetic variances are statistical composites of effects across many loci, the contributions of specific genes and mutations cannot be obtained from them without inspecting DNA variation.

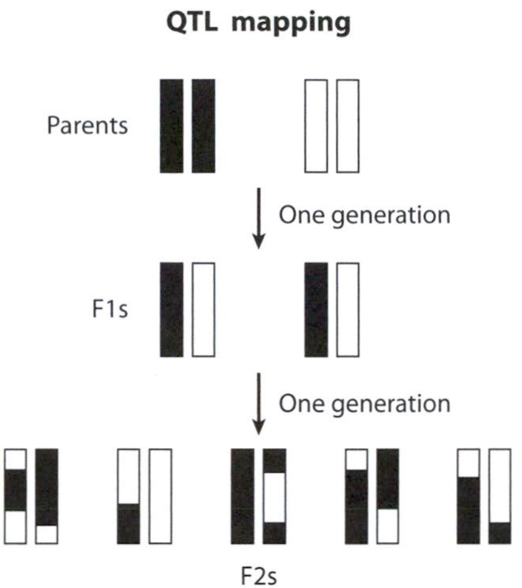

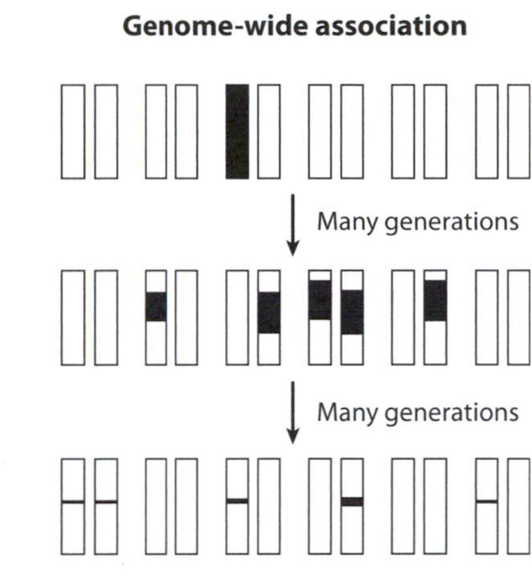

Figure 1. The difference in mapping resolution between QTL mapping and genome-wide association testing. The left panel shows chromosomes in individuals sampled from an intercross between two inbred parental strains. Informative recombination is restricted to one generation, leaving a causative mutation associated with large chromosomal blocks. The right panel shows chromosomes in individuals sampled from a natural population. At the top, a causative mutation (shaded) arises on a single chromosome and is initially associated with the particular haplotype on that chromosome. Over many generations, recombination reduces the sizes of chromosomal blocks (linkage disequilibrium) associated with the causative mutation, dramatically increasing mapping resolution. Only chromosomal blocks harboring the causative mutation are shown (the large diversity of haplotypes is not shown).

The union of quantitative genetics with molecular approaches that allow rapid DNA genotyping throughout the genome has made it possible to map mutations responsible for complex trait variation.

2. USING LABORATORY CROSSES TO MAP THE MUTATIONS RESPONSIBLE FOR PHENOTYPIC EVOLUTION

A powerful approach to mapping mutations that shape complex trait evolution is to look for the cosegregation of phenotypes and DNA genotypes in the offspring from a laboratory cross. A little more than two decades after the necessary statistical machinery was developed, the evolution of a large variety of complex traits in a wide array of species has been mapped to specific genomic regions, commonly referred to as *quantitative trait loci* (QTL).

The simplest and most commonly used QTL design begins with two inbred lines that differ in the trait of interest. In evolutionary studies, these lines are often created by conducting repeated rounds of self-fertilization or brother-sister mating, beginning with individuals sampled from nature. Using two inbred lines reduces the maximum number of possible *alleles* at each locus to two, simplifying genetic analysis (though the procedure for mapping QTL using outbred parents is similar). Individuals from the two (parental) inbred lines are crossed to create F1 hybrids, which inherit half of their genome from each line and are genetically identical to one another along the autosomes. The F1s are mated to a parental strain (*backcross*) or to one another (*intercross*) to yield the next generation of offspring. Importantly, meiotic *recombination* in the F1s scrambles chromosomal pieces so that each offspring inherits a unique combination of alleles from the parental lines. This randomization of the genome is a key element that makes it possible to map QTL.

Phenotypes and genotypes are collected for each individual. Typically, the parental lines harbor different alleles at loci throughout the genome, allowing the source of each chromosomal segment in each hybrid to be determined. Standard backcross or intercross designs feature limited opportunity for useful recombination (only one generation), so that each chromosome is broken into several large pieces (figure 1). As a result, only a small subset of informative loci needs to be surveyed. These molecular markers—usually single nucleotide polymorphisms (SNPs) or short tandem repeat polymorphisms (microsatellites)—are typically chosen to be equally spaced on a genetic or physical map. Markers can be genotyped using a variety of methods, such as polymerase chain reaction, DNA microarrays, or DNA sequencing.

The test for linkage between genetic marker and QTL is straightforward. In genomic regions that do not harbor mutations affecting the measured phenotype, marker alleles and trait values will segregate independently. In genomic regions that contain QTL, individuals with different marker genotypes will differ in phenotype.

The most widely applied method—interval mapping—uses data from the two markers flanking a genomic position to estimate the conditional probabilities of genotypes at that location. Inference proceeds by comparing the likelihood of the data given a QTL at a position to the likelihood of the data given no QTL at that same position. This comparison is usually made by calculating an LOD score, the difference between the $\log_{10}$-transformed likelihoods (similar to the F-statistic from an ANOVA). In the most common procedure, an LOD score is calculated separately for each genomic position and the collection of scores for a chromosome is plotted to identify peaks, where evidence for the presence of a QTL is strongest. Permutation tests are used to determine whether a particular LOD score exceeds that expected by chance when tests are conducted across the genome. In addition to yielding QTL position, this analysis can estimate the QTL effect, including the percentage of phenotypic variance it explains and whether it is additive or dominant. More complex models allow simultaneous detection of multiple QTL and testing for interactions among QTL.

The logic of QTL mapping can be extended to generate a variety of additional genetic resource populations that can help to identify the causative mutations. Collections of inbred lines can be formed using F2s or later-generation intercross progeny as parents. The additional rounds of recombination that occur during the creation of these recombinant inbred lines (RILs) provide finer mapping resolution. RILs also allow the measurement of multiple, genetically identical individuals (in contrast to standard backcross and intercross designs, in which each individual is unique), increasing the accuracy of phenotypic characterization. Researchers can estimate the phenotypic effect of a single QTL region in isolation from the remainder of the genome by constructing introgression lines (sometimes called "nearly isogenic lines" or "congenic lines"). This is accomplished by backcrossing to one parental strain, choosing (heterozygous) offspring that harbor the QTL region from the other parental strain (by genotyping markers in the region), and continuing this procedure for several generations. The result is an inbred strain carrying the QTL region from one parent on the fixed genomic background of the other parent.

The large number of QTL mapping studies completed by evolutionary biologists collectively point to a few general patterns. First, multiple QTL contribute to complex trait evolution, as expected from quantitative genetic models. Second, QTL vary in both effect size and mode of action (e.g., dominance). Third, the expression of a QTL often depends on other factors, including sex, genomic background, and environment. Finally, larger crosses yield more QTL, suggesting that many QTL with modest effects go undetected with the sample sizes typically employed. These conclusions apply to both variation between populations and variation within populations.

A list of QTL for a particular trait helps researchers evaluate models of phenotypic evolution, including hypotheses about the genetic architecture of adaptation (see chapter V.12) and speciation (see chapter VI.8). Importantly, QTL mapping also provides an entrée into determination of the specific DNA variants responsible for phenotypic evolution. QTL mapping aided the identification of genes or mutations that underpin the evolution of bony armor in fish, coat color in mice, domestication traits in corn, and flower color in plants; unfortunately, most QTL studies have not achieved this level of success. The pedigree of an intercross or backcross features two generations of mating, and only one of those generations includes recombination that is useful for genetic mapping. This restriction limits the resolution of genotype-phenotype linkage to large chromosomal regions that contain many genes. Even QTL regions narrowed by mapping in recombinant inbred lines or introgression lines typically comprise multiple genes. The desire to drastically and rapidly improve mapping resolution by sampling many more recombination events is a primary motivation for a different approach: association testing in natural populations.

3. USING ASSOCIATION TESTING TO MAP THE MUTATIONS RESPONSIBLE FOR PHENOTYPIC EVOLUTION

Like organisms in a laboratory cross, individuals in a population are connected by a pedigree, but independent assortment and recombination during meiosis uncouple the pedigrees of different genomic regions. Each stretch of DNA has its own genealogical history, comprising a series of DNA replication events in ancestors that ultimately gave rise to the current generation. The chance that a genomic region has experienced recombination increases with time, so longer genealogies offer higher mapping resolution (figure 1). Although the structures of genealogies are difficult to resolve, two simple criteria can be used to sample longer genealogies and maximize the opportunity for recombination. First, unrelated individuals can be chosen. Whereas a stretch of DNA found in a pair of close relatives traces back to a common ancestor that lived recently, the same region found in two unrelated individuals may trace back to a common ancestor many generations ago. Second, a large number of individuals can be surveyed. Each new individual increases the length of the sampled genealogy. Partly for these reasons, association testing generally involves taking large samples of unrelated individuals from natural populations.

Like QTL mapping, association testing requires the collection of two forms of data. First, phenotypes are measured for each individual in the sample. Next, individuals are genotyped at DNA sites that vary in the population. Broadly speaking, there are two ways to choose mutations to survey. Investigators may focus on a subset of candidate genes with known or inferred functions that imply their involvement in the phenotype (see chapter V.11). A gene could be specifically targeted because previous studies revealed that mutations in this gene affect the phenotype of interest, either in the same species or in a different one. Less directly, a gene whose product is expressed at the right time or in the right tissue could be prioritized. Although this candidate gene approach has produced notable success stories in evolutionary biology, the ability to survey DNA variants from across the genome using high-throughput genotyping, along with growing realizations about the genetic complexity of most traits, has spurred increasing interest in the alternative strategy: *genome-wide association testing*. The reasoning behind this approach is that the mutations responsible for phenotypic variation may lie anywhere in the genome. Investigators genotype a large number of sites across the genome, often focusing on those that were previously identified as variable in the population of interest or in another population. The number of sites required to achieve adequate coverage depends on genome size and the correlations between mutations in the population (see "Linkage Disequilibrium," below).

Association testing proceeds by comparing phenotypes and genotypes across the population sample, one DNA variant at a time. In all tests of association, genotypes are categorical. For an SNP with two alleles, there are three possible genotypes. If the trait has two categories, say, light and dark coloration, associations can be detected using a Fisher's exact test of the 2×3 (phenotype $\times$ genotype) matrix. The linear regression of the frequency of light individuals on genotypic category (recoded as 0, 1, or 2) provides an alternative test of association. When phenotypes take on more than two (unordered) states, multinomial regression can be used.

For traits that vary continuously, analysis of variance (ANOVA) or linear regression can be used to examine genotype-phenotype associations. ANOVA is the more general test in that phenotypic differences between genotypes may assume any form (analogous to Fisher's exact test for discrete traits).

These basic tests of association can be extended in multiple ways. Biological variables that could explain additional variation in the phenotype of interest, such as environmental factors or sex, can be included as cofactors to improve power. Multiple DNA variants can be jointly tested, which can help detect associations with the phenotype that derive from functional interactions among sites (i.e., epistasis). Finally, previous knowledge can be used to prioritize variants for testing. For example, nonsynonymous mutations, those that alter the amino acid sequence of a protein, might be expected to affect phenotype more frequently than synonymous mutations. The incorporation of outside information is perhaps accomplished most naturally through Bayesian association testing, in which variants are weighted by prior probabilities of association with the phenotype.

Several biological and statistical factors influence the ability of investigators to map genetic variants that contribute to complex trait evolution through association testing. Five of the most important variables are the phenotypic effects of causative mutations, the correlations between alleles at different loci (linkage disequilibrium), the frequencies of causative and marker mutations, the homogeneity of the population sample, and the very large number of tests.

Phenotypic Effects of Causative Mutations

The power to detect an association between genotype and phenotype is driven principally by the phenotypic effect of the causative mutation. For a continuous trait, this effect can be measured by comparing phenotypic means between genotypic classes. Half the difference in average trait values between the two homozygotes estimates the effect of substituting one allele, or the *additive effect*. The difference between the mean value for heterozygotes and the mean of the two homozygote values measures the deviation from additivity, or the *dominance effect*. The genetic variance contributed by a mutation is a sum of the squared additive and dominance effects, each weighted by different products of allele frequencies. Additive mutations with large effects are the easiest to detect. Effect sizes also may be reported as the percentage of phenotypic variance explained by marker genotype. For discrete traits with two categories, phenotypic effects are often measured using odds ratios. These ratios compare the odds that an individual exhibits one phenotype versus the other for two genotypic categories (e.g., whether or not the individual harbors a particular allele).

Genetic complexity that dilutes the phenotypic effects of mutations reduces the power of association studies. When trait values are determined by a large number of loci, the contribution of an individual mutation will be small. Phenotypic effects that are contingent on genotypes at other loci (*epistasis*) or the environment (*genotype by environment interaction*) are especially difficult to detect through association testing. Finding mutations with modest effects requires large sample sizes that may be prohibitive for a typical evolutionary study.

Linkage Disequilibrium

In most cases, a correlation between phenotype and genotype detected in a genome-wide association study is not caused directly by the tested polymorphism itself; instead, these associations arise because the tested polymorphism is located near a mutation that affects the trait of interest. The polymorphism is a marker for the causative mutation, providing its genomic address.

Randomly chosen polymorphisms can serve as markers because nearby mutations are often statistically correlated in populations. The force primarily responsible for the magnitude of these correlations is meiotic recombination (see chapter IV.4). When polymorphisms are separated by great physical distances along a chromosome, there is a high probability that recombination events have occurred between them during the history of the sample. These recombination events create new combinations of polymorphisms. Comparatively, the probability that recombination has occurred between closely spaced polymorphisms is low. Without recombination, sets of mutations that have been together in the population stay together. As a result, one can predict the allele carried by an individual at one polymorphism from knowledge of the allele at another polymorphism. This statistical association is called *linkage disequilibrium*. When recombination has shuffled combinations of polymorphisms, such that the frequency of a multisite combination is predicted from the product of its constituent allele frequencies, the population is said to be at linkage equilibrium.

Linkage disequilibrium is straightforward to measure. When the string of alleles on each chromosomal copy carried by an individual—the *haplotype*—is known or can be reconstructed, the population frequencies of haplotypes are compared to their predicted frequencies assuming free recombination. If the observed and predicted frequencies differ, there is linkage disequilibrium. Several summary statistics are available that use this basic principle.

The physical scale of linkage disequilibrium is a primary determinant of marker density in genome-wide association studies. When linkage disequilibrium decays rapidly, more markers must be genotyped to ensure that causative mutations are "tagged." When linkage disequilibrium decays slowly, fewer markers are needed.

Linkage disequilibrium has been measured in a variety of species, but the most exhaustive analyses on the genomic scale come from the International Human Haplotype Map (HapMap) project. During the three phases of this effort, millions of SNPs were genotyped in large samples from populations representing primary axes of human genetic diversity. Four major patterns were observed. First, nearby SNPs often show strong linkage disequilibrium. Second, linkage disequilibrium decays on the scale of tens of kilobases (kb). Third, linkage disequilibrium varies among genomic regions, with lower values in regions with higher recombination rates. Fourth, linkage disequilibrium varies among populations, with lower values in populations of larger size. These patterns are all predicted by population genetic theory. Similar results have been observed on smaller scales in other species.

The observed patterns of linkage disequilibrium have implications for the design of genome-wide association studies. The fact that nearby mutations are often correlated suggests good prospects for association testing, as long as enough markers are genotyped. The required density of markers depends on the local recombination rate and the population size. Because these biological factors vary among species, different marker densities are needed to achieve the same level of coverage in different species. For example, population sizes in *Drosophila melanogaster* are much larger than in humans; as a result, linkage disequilibrium decays faster in fruit flies. Therefore, genome-wide association studies in fruit flies require a higher density of markers to obtain the same level of genomic coverage.

Frequencies of Causative and Marker Mutations

Each mutation that contributes to evolutionary differences in a complex trait takes on a particular frequency in the population. The frequency is determined by a combination of evolutionary factors. Natural selection increases the frequency of beneficial mutations and decreases the frequency of deleterious mutations. The magnitude of selection "felt" by a particular mutation is determined by both the strength of selection on the phenotype and the genetic architecture of the trait. For example, selection acting on trait variation controlled by a large number of loci may not generate substantial changes in the frequencies of individual mutations, whereas selection targeting a phenotype affected by a few loci will likely change mutation frequencies.

As with all DNA variants, the frequencies of causative mutations change randomly as a result of finite population size (see chapter IV.1). Genetic drift can cause mutations to increase or decrease in frequency. When population size is small or differences in genotypic fitness are limited, frequency changes due to genetic drift can overwhelm those caused by selection. Mutation (when rates are high) and migration also affect allele frequency.

Population genetic models predict that mutations contributing to complex trait variation will span the frequency spectrum from rare to common. General statements about the frequency spectrum underlying traits of evolutionary interest are not yet possible, but it is clear that the frequencies of causative mutations affect the success of association testing. For example, holding phenotypic effect size constant and assuming no dominance, mutations with frequencies of 0.5 contribute the maximum amount of genetic variance to the population; genetic variance declines as frequencies move away from 0.5. This frequency effect may be balanced by the expectation that rare alleles exhibit larger phenotypic effects, a biologically realistic idea for deleterious mutations.

Whereas the frequencies of causative mutations are usually unknown, the frequencies of marker mutations are directly measured in an association study. The power to detect trait-genotype associations is maximized when marker and causative mutations have identical frequencies. Even when linkage disequilibrium values are the same, frequency-matched alleles are easier to detect than those with divergent frequencies. As a result, the frequencies of marker mutations chosen for genotyping directly influence the classes of causative mutations that can be discovered through association testing. The common strategy of surveying only common mutations skews detected associations toward those generated by common alleles. Because the frequency spectrum of causative mutations is unknown, this bias should motivate choosing markers to span the frequency range or association testing of all variants in the genome (through whole-genome sequencing).

Homogeneity of Population Sample

Associations can be generated by factors other than a direct mechanistic connection between DNA and phenotypic variation. One confounding variable that has received much attention is the presence of structure within the population sample. Unintentionally combining groups that differ in both trait values and allele frequencies can produce false-positive associations at random loci scattered throughout the genome. Several strategies are available to detect and to account for the effects of population structure in genome-wide association studies. Quantile-quantile plots of association *P* values across loci can reveal genome-wide departures that might be attributable to population structure. Population genetic methods can be applied to genotypes at markers throughout the genome to identify groups in the study sample. Investigators can then split the sample into homogeneous groups for association testing or they can include population membership as a cofactor in association tests.

Not all population structure is an impediment to detecting genotype-phenotype associations. When populations result from gene flow between genetically divergent lineages, phenotypic differences among these admixed individuals may be attributed to allelic differences among their source populations. This "admixture mapping" strategy is particularly useful in hybrid zones.

Multiple Testing

Each comparison between phenotypic values and marker genotypes has some chance of producing a false-positive association. To achieve adequate coverage of the genome, association studies typically require tens of thousands to millions of tests, substantially raising the risk that false positives will be included in the list of identified associations. The simplest approach to control the genome-wide false-positive rate is to simply divide the single-marker critical value by the number of tested markers, but linkage disequilibrium between markers makes this method overly conservative. Instead, researchers can use a permutation procedure that randomizes phenotypes but leaves the linkage disequilibrium structure of marker genotypes intact. Under this method, the corrected significance threshold is obtained by collecting extreme *P* values from many rounds of genome-wide association testing in which the connection between phenotype and genotype has been erased. An alternative approach is to control the *false-discovery rate* (FDR), which is the proportion of false-positive associations among all positive tests. If none of the tested markers are linked to variants that affect the phenotype, the genomic distribution of *P* values should be approximately uniform between 0 and 1. FDR asks whether the observed distribution is instead a mixture between this uniform distribution and one skewed toward lower *P* values.

General Patterns from Genome-Wide Association Studies

Human populations have been the target of most genome-wide association studies to date. As of September 30, 2011, the National Human Genome Research Institute Catalog of Published Genome-Wide Association Studies listed 5103 SNPs showing good evidence of association with at least one phenotype (from 1032 publications). Inspection of these results reveals some general patterns. First, like QTL mapping, genome-wide association studies often point to multiple genomic locations that control a complex trait. Second, a subset of these associations replicates in other human populations or involves loci previously known to affect the phenotype (or both). This congruence suggests that some identified associations reflect mechanistic connections between genotype and phenotype. Third, most identified loci were not previously known to affect the phenotype of interest, indicating that genome-wide association testing is a powerful approach for discovering new variants that contribute to evolution. Fourth, the vast majority of loci exert small phenotypic effects, a finding that has generated considerable attention. Only a small fraction of the genetic variance in the trait that is suspected to exist in the examined populations (estimated from comparisons among relatives) can be explained by summing the effects of detected associations from across the genome. This "missing heritability" problem is probably caused by limited power to detect many mutations with small effects or low frequencies using reasonable sample sizes and common marker alleles. A practical consequence of this challenge is that individual phenotypic values cannot be accurately predicted from genotypes at the loci statistically associated with the trait. An alternative approach to phenotypic prediction that works better in some contexts ("genomic selection") fits the relationship between the trait and all genotyped variants into a single statistical model. For evolutionary studies, the standard genome-wide association strategy is still preferable because it points to specific genes and pathways responsible for phenotypic evolution. Although the vast majority of human studies target disease phenotypes, nondisease traits show similar properties. For example, genetic differences in height are determined by a large number of mutations with small phenotypic effects.

Looking beyond humans, published genome-wide association studies are currently biased toward domesticated plants and animals. Recent and intense selection by humans has produced striking phenotypic divergence within or between these species, increasing the power of association testing. In some cases, loci with large phenotypic effects on complex traits have been identified and a larger fraction of genetic variation has been explained than in human studies. For example, association testing revealed a major role for the insulin-like growth factor 1 (*IGF1*) gene in body size evolution in dogs. Determining whether the genetic basis of trait variation in natural populations of evolutionary interest more closely resembles that in humans or that in domesticated organisms will require genome-wide association studies across a broad array of species. Variation in population history among species will generate differences in linkage disequilibrium, allele frequencies, phenotypic effects of mutations, and population structure, suggesting caution in comparisons across groups.

4. CURRENT CHALLENGES AND PROSPECTS FOR FUTURE RESEARCH

If genome-wide association testing can successfully find variants that contribute to phenotypic evolution, why hasn't this strategy been applied more widely in evolutionary biology? One reason may be that association testing is designed for application to within-population variation; QTL mapping is a better strategy for dissecting the genetic basis of trait differences between populations or species. Other barriers are practical. Simulation studies and results from humans suggest that large sample sizes are typically required. Additionally, surveying molecular variation across genomes is a challenging task in organisms without genomic tools.

These challenges can be overcome. Amassing large numbers of unrelated individuals from a population is an achievable goal, especially for organisms that are easy to collect such as invertebrate animals, plants, and microbes. Although surveying thousands of individuals may not be realistic, sampling hundreds may suffice for finding the loci that contribute disproportionately to trait variation. The genomic resources required for genome-wide association testing can now be developed for a broader variety of organisms. One approach is to sequence the genomes of a handful of individuals using next-generation sequencing technology, compare the sequences to identify (common) SNPs, and develop an array to genotype these SNPs in a larger sample from the population. Although this procedure is not cheap, rapid advances in sequencing and genotyping are bringing the cost within reach of evolutionary biologists studying organisms that are not traditional genetic models. Results from the candidate gene approach should provide further motivation for genome-wide association studies in natural populations. This strategy has identified a variety of genes that contribute to variation in evolutionarily interesting phenotypes in natural populations (see chapter V.12). Although these links have been biased toward genetically simple phenotypes controlled by small numbers of genes, some traits have been more complex, including behavior.

A more fundamental challenge is how to proceed after a genotype-phenotype association is identified. The approaches described above locate the genetic origins of phenotypic differences in genomic regions or even specific genes, but a causative mutation is unlikely to be pinpointed from QTL mapping or associating testing alone, because multiple mutations in linkage disequilibrium will typically be found in associated intervals. Laboratory crosses and association studies do not evaluate the biological mechanism that connects genotype with phenotype. In the future, sequencing genomes and exhaustively testing for associations at all variants (rather than relying on markers) may ameliorate this problem, but the resolution may still be limited by linkage disequilibrium (and the multiple testing burden will only worsen). Although variants can be prioritized by location (e.g., coding vs. noncoding), levels of conservation across divergent species, or predicted biochemical effects, functional studies are ultimately needed to identify the causative mutation. Creative methods for whittling down the number of candidate variants will reach a premium as the ability to sequence whole genomes expands.

FURTHER READING

Balding, D. J. 2006. A tutorial on statistical methods for population association studies. Nature Reviews Genetics 7: 781–791. *This review clearly articulates the issues faced by investigators conducting genome-wide association studies.*

Flint, J., and T.F.C. Mackay. 2009. Genetic architecture of quantitative traits in mice, flies, and humans. Genome Research 19: 723–733. *This review summarizes knowledge about the genetic basis of complex traits in three species and compares the strategies of QTL mapping and genome-wide association testing.*

Hamblin, M. T., E. S. Buckler, and J.-L. Jannink. 2011. Population genetics of genomics-based crop improvement methods. Trends in Genetics 27: 98–106. *This review correctly argues that population history is an important determinant of the success of genome-wide association studies and raises the idea that the inferred genetic architectures of complex traits may differ between agricultural crops and humans.*

Hindorff, L. A., J. MacArthur, A. Wise, H. A. Junkins, P. N. Hall, A. K. Klemm, and T. A. Manolio. A catalog of published genome-wide association studies. Downloadable from www.genome.gov/gwastudiesgwastudies. Accessed Sept. 30, 2011. *This database from the National Human Genome Research Institute compiles genome-wide association study results and is regularly updated.*

International Human HapMap Consortium. 2005. A haplotype map of the human genome. Nature 437: 1299–1320. *This paper describes the scale and magnitude of linkage disequilibrium involving common variants across the human genome. The results have guided the design of genome-wide association studies.*

Lynch, M., and B. Walsh. 1998. Genetics and Analysis of Quantitative Traits. Sunderland, MA: Sinauer. *This book provides a comprehensive review of genetic methods for analyzing complex traits, with an emphasis on laboratory crosses.*

V.14

Searching for Adaptation in the Genome

Dmitri A. Petrov

OUTLINE

1. Evolution as mutation and change in allele frequencies
2. The neutral theory of molecular evolution
3. The McDonald-Kreitman test
4. Population genomics approaches for detecting and quantifying adaptation
5. Remaining challenges

The study of adaptation lies at the heart of evolutionary biology; despite 150 years of intense study, however, many foundational questions about the mode and tempo of adaptation remain unanswered. The development of population genetics theory and the rise of genomics are bringing a promise of new types of data that are able to provide insight into these long-standing issues. This chapter discusses some of the key conceptual underpinnings of methods that use population and comparative genomics data to study adaptation, and it underscores some of the remaining difficulties and challenges.

GLOSSARY

Adaptation. The process by which a population evolves to having greater fitness in a given environment.

Effective Population Size. One common attempt to simplify the modeling of the evolutionary process is to assert that evolutionary dynamics in a real population $N(t)$ (where t stands for time in generations) can be modeled faithfully in an idealized population of a different, constant effective size N_e. N_e is generally smaller than the census population size. Rapid adaptation should be sensitive to fluctuations in N only over short periods of time (short-term N_e) while slower neutral processes should be sensitive to the fluctuations over much longer periods of time (long-term N_e). Short- and long-term N_e can be different by many orders of magnitude from each other.

Fixation. The process in which a new mutation that is present in some individuals in a species becomes present in all individuals in a species.

Hitchhiking. The process of change of the allele frequency of a particular polymorphism allele as a result of its linkage to selected alleles in the genomic vicinity.

Indel. Insertion/deletion mutation, polymorphism, or substitution.

Mutation. A change in the DNA sequence in the genome of the individual. Can range from a change of a single nucleotide to much larger structural changes such as insertions, deletions, translocations, and inversions all the way to whole genome polyploidization.

Polymorphism. A genetic variant (an allele) that is present in some but not all individuals in a species.

Replacement. Mutation, polymorphism, or substitution in the protein-coding sequence that does change the amino acid sequence of the encoded protein. Also often called *nonsynonymous*.

SNP. Single nucleotide polymorphism.

Substitution. The outcome of fixation of new mutation. It is often observed by comparing DNA sequences from different species in which the majority of differences are due to fixations during the long-term evolution of independent lineages.

Synonymous. Mutation, polymorphism, or substitution in the protein-coding sequence that does not change the amino acid sequence of the encoded protein.

Adaptation is the primary process in evolution, yet after a century and a half of intensive study, most key questions about the mode and tempo of adaptation remain largely unanswered. This is troubling but also understandable given the extreme difficulty of (1) identifying individual adaptive genetic changes in a convincing way, (2) understanding their phenotypic effects, and (3) identifying the adaptive nature of these phenotypes in ecological and functional contexts. One approach is to start at the level of phenotype and work down to genes (see chapters V.12

and V.13). Another is to look for the signatures left by adaptation in the patterns of genetic (genomic) variation and then to work from candidate genetic regions to their phenotypic and possibly adaptive effects. Neither approach is foolproof, but much promise rests on the genomes-first, phenotype-second approach. This is because recent technological developments are allowing researchers to ever more quickly and efficiently document genetic variation on a genome-wide scale in multiple organisms and whole populations. The great promise of these approaches is that they can be applied to most organisms in a way that is virtually agnostic to their specific biology and in a way that is not biased by assumptions of which phenotypes are adaptive and which are not.

This chapter will focus only on the study of adaptation using population genomic data. The chapter will not be able to cover all or even a sizable fraction of the methods that have been developed—the ambition is to elucidate the main logic of several key approaches, highlight a few classical, popular, or most logically transparent methods, discuss the key caveats, and present some of the main insights into the adaptive process that have been gathered using population genomics to date.

1. EVOLUTION AS MUTATION AND CHANGE IN ALLELE FREQUENCIES

From the point of view of a population geneticist, evolution can be separated into two key phases: (1) origination of new alleles by mutation and (2) change of allele frequencies within populations. The first phase can involve mutations of varying magnitude, from single nucleotide changes all the way to chromosomal rearrangements and even whole genome duplications (see chapter IV.2); however, no matter how small or large these mutations are in physical scope or phenotypic and fitness effects, they always appear in a single individual or at most in a few siblings at once. For these new genetic variants to become established as differences between species (substitutions) or even be detectable as genetic variants (polymorphisms) within species, they need to increase in frequency very substantially. Hence the central importance to evolution of the second phase: the change in allele frequencies.

The process of allele frequency change is also the place where natural selection acts. Natural selection acts against new deleterious alleles, purging them from the population. Such natural selection is called *purifying* as it keeps the population "pure," preserving the ancestral state and slowing down evolutionary change. This is likely the most common form of natural selection; in contrast, natural selection promotes increase in frequency and even eventual fixation of advantageous alleles. Such "positive" selection speeds up evolution and divergence between species.

The way natural selection affects allele frequencies is at times counterintuitive (see chapters III.1 and III.3). For example, even though the naive expectation is that all advantageous mutations should be fixed (i.e., reach 100% frequency) by natural selection, the reality is that the vast majority are lost almost immediately after they are generated by mutation. The probability that a strongly beneficial mutation escapes loss is roughly equal to its selective advantage, such that only 5 percent of mutations with 5 percent selective benefit are expected to fix, and 95 percent of them are expected to be lost. The main intuition here is that even adaptive mutations are very vulnerable to loss when they are extremely rare, because even a small fluctuation in frequency resulting from random events can remove them from the population and because natural selection is inefficient when allele frequencies are low (see chapter IV.1). Similarly, even the selectively neutral (i.e., natural selection does not favor one over any other) alleles can change in frequency, and one can become fixed as a result of purely stochastic fluctuations that are inevitable in finite populations. These stochastic fluctuations at neutral sites are further exacerbated by selection, both purifying and positive, at sites that are located nearby on a chromosome to a site with a neutral polymorphism. This latter phenomenon is known as *hitchhiking* because neutral or weakly selected polymorphisms can hitchhike on more selectively substantial mutations to which they are physically linked. Hitchhiking has recently received much attention and will be described in greater detail later in this chapter. These stochastic fluctuations also mean that not all deleterious mutations are lost as one might expect. Although the probability of loss of an even moderately deleterious mutation is much higher than loss of a neutral or advantageous mutation, it is not a certainty and some deleterious mutations can even fix in populations.

These considerations make it clear that the mere observation of an allele that reached high frequencies or of a substitution between species is not sufficient to argue that positive selection was involved. Indeed, the mutation in question could have been neutral or even weakly deleterious. Similarly, just because an allele was lost does not mean it was deleterious, because the vast majority of *all* new mutations, be they deleterious, neutral, or adaptive, are lost and lost quickly.

2. THE NEUTRAL THEORY OF MOLECULAR EVOLUTION

To detect adaptation with confidence, much more specific, quantitative expectations under the null model of no adaptation must be generated. The neutral theory, most commonly associated with Motoo Kimura, provides a good example of such expectations (see chapter V.1). The neutral theory postulates that practically all mutations

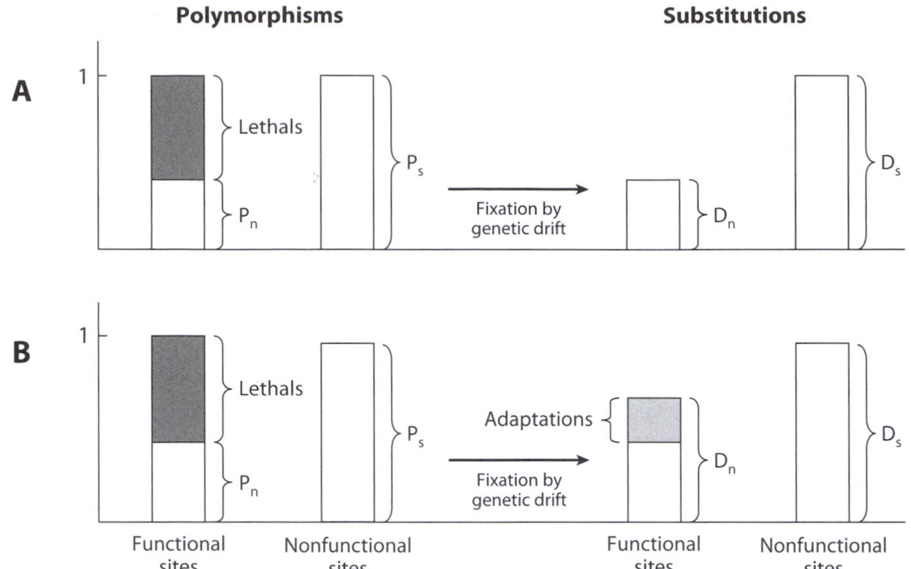

Figure 1. Expected patterns of polymorphism (P) and divergence (D) and functional (subscript n) and neutral (subscript s) sites. (A) The expectation under the neutral theory. Mutations at functional sites come in two classes, either *lethal* or *neutral*. Lethals are eliminated right away; the remaining neutral mutations, at both functional and neutral sites, being identical in fitness, fix in the population and turn into divergent sites at identical rates. As a result, the ratio of the number of polymorphisms at functional and neutral sites (Pn/Ps) is expected to be equal to the ratio of the same ratio at divergent sites (Dn/Ds). (B) If some mutations are adaptive and happen rarely, the approximate expectation is that they are never seen as polymorphisms but do accumulate at a high rate as divergent sites. They reveal themselves as the excess of divergent sites at functional sites (Dn) relative to the neutral expectation (i.e., Dn > Ds × [Pn/Ps]).

are either deleterious or neutral and that practically all detectable polymorphisms and all substitutions are due to neutral mutations. Note that the second postulate is much more restrictive than the first. Even if the adaptive mutations are vanishingly rare compared with neutral or deleterious mutations, they still could easily contribute to the majority of substitutions. This is because the probability of fixation of a new neutral mutation is the reciprocal of the population size (N) (technically of the long-term effective population size N_e), while the probability of fixation of a strongly advantageous mutation is roughly equal to its selective benefit (as mentioned above) and the latter can be much, much larger. For instance, in *Drosophila melanogaster*, where the (long-term effective) population size is roughly 1 million, a mutation that provides 1 percent benefit has a 10,000 times greater chance of fixation than a neutral mutation. This implies that if adaptive mutations of 1 percent advantage were even 1000 times less frequent than neutral ones, they would still correspond to about 90 percent of all substitutions. The neutral theory thus claims that the increased chance of fixation of adaptive mutations does not compensate for their relative rarity.

Whether or not one agrees with the neutral theory as a description of empirical reality—and as a disclaimer, this author does not—it is indisputable that the neutral theory provides a useful null model for the study of adaptation. The next sections discuss some of the key approaches that employ the neutral theory as a null model for the detection and quantification of adaptive evolution.

3. THE MCDONALD-KREITMAN TEST

Consider the process of molecular evolution as envisioned by the neutral theory (figure 1A). Imagine you could classify all positions in the genome into functional (some mutations at these positions alter the functioning of the gene) and nonfunctional (all mutations at such positions have no effect on function at all). For simplicity, consider only protein-coding regions and assume that all synonymous mutations have no functional significance and that some nonsynonymous mutations have functional effects. Synonymous mutations contribute to the synonymous polymorphism (P_s) within species and synonymous divergence (D_s) between species, whereas nonsynonymous mutations contribute to the nonsynonymous polymorphism (P_n) within species and nonsynonymous divergence (D_n) between species (figure 1A).

Imagine that you then take the same number of both types of sites and carefully ensure that mutation rates are the same at both types of positions. Some mutations at functional sites will be deleterious, and according to the neutral theory, they are immediately eliminated from the population. In fact, they are eliminated so quickly that they can never reach frequencies at which they can be detected even as rare polymorphisms in samples of reasonable size. This assumption is equivalent to saying that all deleterious mutations are very strongly deleterious. Furthermore, because the neutral theory postulates that there are no adaptive mutations, the remaining mutations are all neutral at both functional and nonfunctional sites. Thus all observable polymorphisms at both types of sites should be due to neutral mutations, with the number of polymorphisms at functional sites reduced by the immediate elimination of the deleterious mutations. The next insight is that because all polymorphisms are neutral, polymorphisms at both types of sites should be fixing at the same rate such that the ratio of the numbers of functional and nonfunctional polymorphisms should be equal to the ratio of functional and nonfunctional substitutions (figure 1A). This equality is one of the key predictions of the neutral theory.

Now consider a modification of the neutral theory in which adaptive mutations are allowed to occur at some rate (figure 1B), but slightly deleterious mutations are still not permitted. How would this change the picture? The adaptive mutations must have a functional effect and so can take place only at functional sites. They have a much higher probability of fixation and fix much, much more quickly than neutral mutations; indeed, population-genetic theory suggests that the number of generations that a neutral mutation destined for fixation spends as a polymorphism is on the order of the (long-term effective) population size. For instance, in *D. melanogaster*, with the long-term effective population size of about 1 million, it takes on the order of 1 million generations for a neutral mutation to reach fixation. In contrast, an adaptive mutation of 1 percent advantage takes about 1000 generations—approximately 1000-fold less time. Because adaptive mutations fix so fast and the time they spent as polymorphisms is so fleeting, it is hard to detect them as segregating polymorphisms in populations. The adaptive mutations should thus contribute disproportionately to the number of substitutions between species and less so to the polymorphism within species.

These considerations immediately suggest a way both to test the neutral theory and to estimate the number of adaptive substitutions. Figure 1B illustrates that adaptive substitutions should constitute simply the excess of divergence at functional sites compared with what you would expect given the number of substitutions at nonfunctional sites (Ds) and the amounts of polymorphism at functional (Pn) and nonfunctional sites (Ps). Specifically, given that Pn/Ps is the proportion of the mutations at functional sites that are neutral, Ds $\times$ (Pn/Ps) becomes the estimate of the number of neutral substitutions at functional sites, and Dn – Ds $\times$ (Pn/Ps) is the estimate of the excess of substitutions at functional sites compared to the neutral theory expectations and thus also an estimate of the number of adaptive substitutions.

The test of the equality of Pn/Ps and Dn/Ds ratios can be formalized as the test of the neutral theory. This test was proposed in the seminal 1991 paper of McDonald and Kreitman and is now known as the McDonald-Kreitman (MK) test. They used their test with the polymorphisms and substitutions in a single gene (*Adh* in *D. melanogaster* and *D. yakuba*) and showed that the number of nonsynonymous substitutions was much larger than expected under the neutral theory, with practically all substitutions appearing to have been driven by positive selection.

Since its inception, the MK test has proved extremely popular, and it has been extended to other types of functional sites and applied to many individual genes and also genome-wide in a number of species, from yeast to humans and to nonprotein-coding functional sites. The results have been surprisingly mixed. At one extreme, all tested *Drosophila* species showed extremely high rates of adaptation, with about 50 percent of all amino acid substitutions (and similar rates at noncoding sites) in the genome appearing to be driven by positive selection. Similarly, substitutions in putative regulatory regions show high rates of adaptation. High rates of positive selection were also detected in some vertebrates such as house mouse (*Mus musculus*; ~50%) and chicken (*Gallus gallus*; ~25%), but not in humans (0–10%) or some plants (*Helianthus*, *Capsella*, and *Populus*). No evidence of positive selection at the protein level could be detected in *Saccharomyces* and in many species of plants (from *Arabidopsis* to *Oryza* to *Zea*). At first glance, there is no obvious population size or phylogenetic or ecological pattern to which organisms show high and low levels of adaptation according to the MK test.

One possibility is that the rate of positive selection really does vary substantially across different lineages and that there is now an opportunity to understand why. The other, less interesting but important possibility is that the MK test can be substantially biased in some situations, and that the type and direction of the bias varies in different organisms.

Let's consider how the MK test can be biased. What are the fragile points? First, are the counts of the polymorphisms and substitutions that go into the MK test likely to be correct? MK tests generally employ closely related organisms; thus, not many errors are expected to result from multiple events happening at the same position and obscuring each other. The principal error, if it does occur, must come from incorrectly estimating the number of nonadaptive (neutral or slightly deleterious), nonsynonymous substitutions that are subtracted from the total number of nonsynonymous substitutions to arrive at the number of specifically adaptive ones (figure 1B).

A number of ways in which this can happen have been proposed; most of these depend on the existence of slightly deleterious mutations. Slightly deleterious mutations are a class ignored by the neutral theory (they were later considered by the nearly neutral theory), but they might turn out to be of great consequence both for the evolutionary process and also for the ability to estimate the rate of adaptive evolution. Slightly deleterious mutations have such a small deleterious effect that they are not efficiently removed from the population by purifying selection and can segregate in the population at appreciable frequencies. These polymorphisms, however, do feel the effect of purifying natural selection that pushes their frequency down; thus, they have substantially lower probabilities of fixation compared with the similarly frequent truly neutral polymorphisms. If the slightly deleterious nonsynonymous (or other functional) polymorphisms are mistakenly considered neutral, the rate of neutral amino acid (functional) substitutions will be overestimated, and the rate of adaptive evolution will be underestimated. On the other hand, slightly deleterious polymorphisms can also segregate at synonymous sites—by treating these as neutral polymorphisms, estimates of the rate of adaptive evolution will be too high.

To complicate these matters further, the fate of slightly deleterious mutations depends strongly on the population size variation and on the rate and strength of adaptive evolution. The same mutations should behave as strongly deleterious mutations during population booms and as virtually neutral mutations during population busts. This is because the effect of natural selection must be compared to the amount of stochastic noise generated by random genetic drift that in turn is stronger in small populations; thus, selection strong enough to significantly affect the fate of a polymorphism in a large population would have virtually no discernable effect in smaller populations. If the population size during evolution of the two species since their divergence was systematically different from the recent times over which the sampled polymorphisms have been segregating, then the effect that slightly deleterious polymorphisms have on divergence and polymorphism might be over- or underestimated. In addition, some deleterious mutations might happen to be linked to new adaptive mutations and reach unusually high frequencies through this linkage (see below for the description of hitchhiking). These moderately frequent polymorphisms can be easily mistaken for neutral ones even though they have virtually no chance of fixation and thus no chance to contribute to functional divergence.

Overall, it is likely that slightly deleterious mutations are more likely to mask effects of adaptation because most slightly deleterious mutations are found at nonsynonymous sites. A number of approaches have been proposed to deal with this problem, but none of them are likely to be foolproof. These approaches generally involve a way to estimate the proportion and selective effect of segregating slightly deleterious polymorphisms and then statistically adjusting the estimate of the neutral and slightly deleterious functional substitutions. It is still not known whether such approaches are reliable in practice, especially because most of them assume that all sites are evolving independently from each other (although it is possible that interactions between adaptive substitutions and slightly deleterious polymorphisms can be highly consequential) and thus that the fates of different polymorphisms really cannot be modeled independently of each other.

The uncertainty over the veracity of MK tests spurred search for additional signatures of adaptation that would complement MK tests, and that even if not foolproof themselves, would hopefully be immune to the problems that potentially plague the MK tests and thus provide an independent way to verify MK estimates. Approaches that rely on the search for patterns of hitchhiking and selective sweeps are described below.

4. POPULATION GENOMICS APPROACHES FOR DETECTING AND QUANTIFYING ADAPTATION

Selective Sweeps

Some of the most promising approaches to the detection and quantification of adaptation focus on the effects of adaptive alleles on linked neutral variation. Consider a new strongly adaptive mutation that is destined to become fixed in the population. When it has just arisen, it is present on a particular chromosome (figure 2A). Because it is destined to reach fixation, it will increase quickly in frequency.

What about the neutral variants initially present on the same chromosome? At the beginning, all of them start increasing in frequency as well, as a result of the physical linkage with the adaptive variant. With time, recombination will separate neutral variants located along the chromosome far away from the adaptive mutation frequently enough that the dynamics of the neutral variants will not be affected by the dynamics at the adaptive locus. By contrast, neutral variants located close to the adaptive substitution would not be separated by recombination from the adaptive mutation frequently, and thus might be driven to high frequency or even fixation along with the adaptive mutation itself.

This process (termed *adaptive hitchhiking* by Maynard-Smith and Haigh) should perturb the pattern of neutral

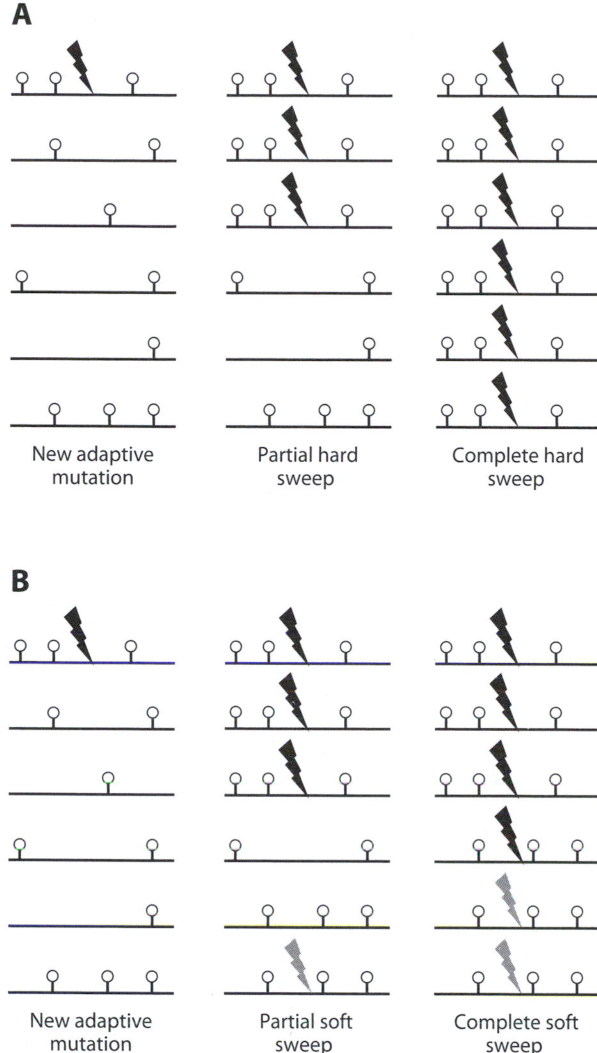

Figure 2. A cartoon representation of expected patterns of selective sweeps driven by adaptation generated by de novo mutation. (A) Pattern expected in the mutation-limited regime. A new adaptive mutation appears (lightning); if it is not quickly lost through stochastic fluctuations near the absorbing boundary and can reach establishment frequency, it quickly starts increasing in frequency and drags with it linked neutral polymorphisms. At the end, a single haplotype reaches high frequencies and ultimately fixation. Not shown are mutations and recombination events taking place during the rise to high frequency. These should generate relatively rare additional haplotypes. (B) Adaptation in the non-mutation-limited regime can proceed differently. Here a second adaptive mutation at the same locus takes place and reaches establishment frequency before the first one reaches very high frequencies. At the end, multiple independent adaptive mutations rise to high frequencies and drag multiple haplotypes with them, generating the signature of a soft sweep.

polymorphism right next to the adaptive mutation that is increasing in frequency and ultimately around the adaptive substitution. The key expectation is that there should be a dearth of polymorphism next to the adaptive substitution because the allele linked to the new adaptive mutation should fix and the rest should be lost. This pattern of variation reduction by an adaptive mutation was termed a *selective sweep*.

The key expectations of selective sweeps are outlined in figure 2. First, during the sweep itself, the sweeping adaptive mutation and the linked region next to it are generating a pattern of a partial sweep (figure 2A), with the pattern of unusually long and unusually frequent haplotypes being its key feature. Note that a partial sweep does not lead to a substantial reduction of the overall levels of polymorphism. Only when the sweep reaches completion or near completion are regions of reduced variability observed. The range over which partial or complete selective sweeps extend are proportional to the strength of selection acting on the adaptive mutation and inversely proportional to the rate of local recombination. The rule of thumb is that the distance over which a sweep can be detected is about 0.1 s/r, such that an adaptive substitution of 1 percent advantage in the region with the recombination rate of 1 cM/Mb (cM or centimorgan is a unit of recombination such that two markers located 1 cm apart from each other will generate on average 1% of recombinant progeny) will generate sweeps of about 100 kb. With time, selective sweeps dissipate as new mutations in the swept region rise in frequency and the level of polymorphism eventually increases to its background level. Individual sweeps become almost impossible to detect within the length of time equal to the (long-term) N_e generations. For some statistics, the time is even shorter if the generated patterns are rapidly broken up by recombination.

Common Statistics for the Detection of Sweeps

Different statistical approaches have been designed to detect partial and complete selective sweeps. This section briefly outlines some of the most instructive and popular approaches. One of the most commonly used and powerful methods for the detection of partial sweep was proposed by Pardis Sabeti and colleagues. The original statistic (that has since been refined) is called EHH (extended haplotype homozygosity), and it is based on the comparison of the lengths of haplotypes that are linked to the two alternative states of a single nucelotide polymorphism, or SNP. SNPs that are linked to unusually long haplotypes might be associated with a sweeping adaptive mutation. A comparison of the length of the unbroken haplotypes linked to the alternative versions of the same SNP is very powerful as it allows one to control for the variation of recombination rate in the genome.

The statistics designed to detect complete sweeps, in contrast, cannot rely on the haplotype structure and must instead use other signatures. As mentioned above, the first and most obvious signature of a complete sweep is the local loss of variation. Although this signature is used commonly, it suffers from the possibility that detected regions are devoid of polymorphism simply because of regionally low mutation rates. To avoid this possibility, several other approaches have been developed that look for additional deviations from the expectations under the neutral theory.

Consider a complete sweep that removes all variation from the population in a region. After some time, additional neutral mutations will arise and start increasing in frequency. For a period after the sweep, their frequencies will tend to be low as it takes some time (on the order of long-term N_e generations) for the neutral alleles to drift to intermediate frequencies. This means that selective sweeps can be detected for a period of time by the paucity of intermediate frequency variants compared to the low-frequency ones. One of the earliest and most popular such tests was developed by Fumio Tajima. Tajima's D, as it is known, is negative if the pattern of polymorphisms shows a bias in favor of rare variants compared with the expectations of the neutral theory. Thus, a lack of polymorphism in the region combined with negative values of Tajima's D (or similar such tests) is often treated as a hallmark of a selective sweep.

Unfortunately, negative values of Tajima's D are also expected under many demographic scenarios, and specifically in populations experiencing recent population growth. Given that many organisms of interest such as humans, *D. melanogaster*, and house mice have experienced sharp recent population growth, this presents a serious problem. One solution is to use genome-average values of Tajima's D and look for regions both devoid of variation and having more negative values of Tajima's D than shown by the genome on average.

Many other statistics have been developed that summarize the allele frequency distribution in different ways and attempt to be sensitive to specific perturbations of the spectrum expected under adaptation. All of them suffer from the same problems: they can be strongly affected by demographic scenarios, fluctuations in recombination rate, and other phenomena that cannot be easily ascertained. Assessing the expected values under refined demographic models and neutrality and defining empirical cutoffs based on the genome-wide assessments of these statistics are both common though imperfect means of dealing with these difficulties.

Soft Sweeps: Single and Multiple-Origin Soft Sweeps

The discussion of sweeps in the previous section represents the classical view of a selective sweep with a single de novo mutation rapidly increasing in frequency. Such classical sweeps are also known as "hard sweeps" (figure 2A). However, sweeps driven by multiple adaptive mutations on multiple haplotypes simultaneously rising in frequency might be even more common than hard sweeps. Such multiple adaptive allele sweeps were first systematically discussed by Hermisson and Pennings, who termed them "soft sweeps" (figure 2B).

What are the scenarios under which adaptive mutations on multiple haplotypes should be present at the same time in the population? There are two key possibilities: adaptation from standing variation and adaptation from de novo mutation in large populations. The first possibility is one in which a neutral or even slightly deleterious allele becomes adaptive as a result of a change in the environment. Because such an allele has been present in the population for a while, it should be present on multiple haplotypes. The second possibility is that multiple independently generated adaptive mutations should be generated roughly at the same time and roughly at the same chromosomal location site; thus multiple-origin de novo adaptive mutations on multiple haplotypes should be spreading through the population at the same time.

The notion of standing variation being the source of adaptation is a very natural one. What about the second, multiple-origin scenario? At first, this scenario appears far-fetched unless the mutational target is very large; indeed, most organisms have been assumed to have effective population size at most in the millions, while the mutation rate per site is on the order of one in a billion. This suggests that unless the same adaptation can be generated by mutations at multiple sites (for instance, a gene loss can be brought about by multiple stop codon generating point mutations and indels), most single-site adaptations generated by de novo mutations should generate hard sweeps. Surprisingly, however, recent analysis of dynamics of adaptation in *D. melanogaster* to pesticide resistance at the *Ace* locus revealed that even point mutations generating specific individual amino acid changes occur multiple times per generation in the population. This suggests that the relevant, short-term effective population size in *D. melanogaster* is more than 100-fold larger than previously thought. In many ways, this makes sense as the short-term N_e relevant to adaptation should be closer to the nominal population size, which is often going to be much, much larger than the long-term N_e. This is because the long-term but not the short-term N_e is very sensitive to any significant population decline occurring at any time over the past hundreds of thousands or even millions of generations. The same pattern can be seen even in humans, where population sizes have increased to an extent that adaptations such as lactase persistence arose in large enough populations to produce soft sweeps.

One can see immediately that soft sweeps should generate very different signatures compared with those generated by hard sweeps (figure 2). For instance, a complete soft sweep is not expected to lead to the complete loss of variation. Because multiple haplotypes increase in frequency simultaneously, the polymorphisms that distinguish these haplotypes from each other would not be eliminated; however, this does not mean that soft sweeps are indistinguishable from neutrality. Even though soft sweeps do not strongly perturb the total amounts of variability per site, they do generate very unusual haplotype patterns that can be detected as regions of high linkage disequilibrium. Tests based on these signatures are only now being designed; thus it is not yet known how prevalent soft sweeps are in comparison to hard sweeps.

5. REMAINING CHALLENGES

The genome-first approach discussed in this chapter clearly holds much promise, especially now that the ability to generate genome-level polymorphism and divergence data is growing by leaps and bounds, and experiments unthinkable even a few years ago are suddenly within reach of even individual investigators. While 10 years ago, barely enough data existed to reject the simplest and most clearly incorrect null hypotheses, it is now conceivable to estimate demography, distribution of fitness, and heterozygous effects of new mutations, and assess the rate of adaptation in different regions of the genome, genes, and pathways. At the same time, much remains to be done.

One of the most difficult questions in the study of adaptation is that of assessment of the rate of generation and the selective advantage of new adaptive mutations. Even more difficult is to assess the way in which the selective advantage of a new allele may vary through time and space and as a function of the genetic background (epistasis). One key question is how often adaptive alleles are advantageous only when rare (frequency-dependent selection or fitness overdominance leading to balancing selection) and how often they are unconditionally advantageous. This is a big question, partly because the answer will determine whether adaptation will increase genetic variation by driving adaptive alleles into the populations but not fixing them. Another important question is the prevalence of multistep adaptation and the number of steps that single adaptive bouts take should they be common. Such multistep adaptation will generate correlations in the fixation of individual adaptive events in space, time, and genomic location and thus could in principle be detectable. Failure to acknowledge this possibility could lead to incorrect inference.

Answers to these questions are not yet known, but the ease with which data can now be obtained should allow time-series studies of populations (natural and experimental) that will provide information about trajectories of adaptive mutations. These time-series data are not commonly available yet but should prove instrumental in further understanding of the adaptive process.

Finally, it is important to reemphasize that however powerful these approaches are, they will never be sufficient by themselves. The hard work of understanding the action of natural selection at the ecological, physiological, and molecular levels will remain. Population genetics can provide candidate loci and estimates of timing and strength of positive selection, but full understanding of the adaptive process must come from comprehensive inquiry that combines population genomics with all the biological levels that lie above the genotype.

FURTHER READING

Andolfatto, P. 2005. Adaptive evolution of non-coding DNA in *Drosophila*. Nature 437: 1149–1152. *Extended application of the McDonald-Kreitman tests to noncoding regions and an argument that the overall rate of adaptation at noncoding sites exceeds that at coding sites.*

Fay, J. C. 2011. Weighing the evidence for adaptation at the molecular level. Trends in Genetics 27: 343–349. *Review of evidence of adaptation in multiple organisms. Argues that some of the commonly accepted methods for the study of adaptation might be substantially biased and must be taken with a grain of salt.*

Hermisson, J., and P. S. Pennings. 2005. Soft sweeps: Molecular population genetics of adaptation from standing genetic variation. Genetics 169: 2335–2352. *A seminal paper that introduced the concept of soft sweeps and argued that such soft sweeps might be more common than the classic hard sweeps that most methods attempt to detect in the population genomic data.*

Karasov, T., P. W. Messer, and D. A. Petrov. 2010. Evidence that adaptation in *Drosophila* is not limited by mutations at single sites. PLoS Genetics 5: e1000924. *A study of rapid recent adaptation at the* Ace *locus, which is the target of most commonly used insecticides. Variation at this locus underlies much of the evolved pesticide resistance in insects in general and* Drosophila *more specifically. The paper argues that adaptation in* D. melanogaster *is currently not mutation limited and that consequently the relevant short-term*

N_e *must be hundreds or even thousands of times larger than the accepted value of* N_e *for* D. melanogaster *(10^6).*

Maynard-Smith, J., and J. Haigh. 1974. The hitch-hiking effect of a favorable gene. Genetics Research 23: 23–35. *A seminal paper that initiated research into correlated evolutionary histories of linked sites and provided a key metaphor of neutral polymorphisms hitchhiking on linked adaptive mutations as they rise to high frequencies.*

McDonald, J. H., and M. Kreitman. 1991. Adaptive protein evolution at the *Adh* locus in *Drosophila*. Nature 351: 652–654. *The seminal paper that described a simple test of the neutral theory using polymorphism and divergence at interdigitated synonymous and nonsynonymous sites. Conceptually it was an extension of the Hudson-Aguade-Kreitman (HKA) test, but its use of interdigitated sites allowed for a natural control of variation in underlying mutation rate and coalescence times across the genome. It also established a simple framework for the estimation of the rate of adaptive evolution.*

Sabeti, P. C., et al. 2002. Detecting recent positive selection in the human genome from haplotype structure. Nature 419: 832–837. *One of the first population genomic papers attempting to identify regions that experienced recent, strong, incomplete selective sweeps. Introduced the idea of using of extended haplotype statistics.*

Sella, G., D. Petrov, M. Przeworski, and P. Andolfatto. 2009. Pervasive natural selection in the *Drosophila* genome? PLoS Genetics 5: e1000495. *A useful review of the various pieces of evidence suggesting that rate of adaptation in* Drosophila *is high and involves some adaptive mutations of large effect.*

Tajima, F. 1989. Statistical method for testing the neutral mutation hypothesis by DNA polymorphism. Genetics 123: 585–595. *One of the first papers that used perturbations in the allele frequency spectrum expected under selection to devise a statistical test for detection of natural selection in genetic sequences. Remains popular in modern population genetics.*

V.15

Ancient DNA

Beth Shapiro

OUTLINE

1. Beginnings
2. The importance of being clean
3. Name that bone: Inserting extinct species into molecular phylogenies
4. Ancient population genetics and phylogeography
5. Ancient genomics
6. The future of ancient DNA

Ancient DNA is a field of molecular evolutionary biology that uses DNA sequence data recovered from poorly preserved organisms, usually deceased for hundreds to hundreds of thousands of years. Ancient DNA data can provide unique snapshots in time to better understand how populations and species evolve. The field was born in the early 1980s, when the first ancient DNA sequences were recovered from preserved muscle of a quagga, a relative of the zebra, which had been extinct for nearly 100 years. Although the early days of ancient DNA were marked by a few spectacular but flawed results, the field has matured into a robust, internally rigorous scientific pursuit with the potential to provide real insight into the mechanisms of evolution at both the species and the population level. Ancient DNA has benefited in particular from recent advances in high-throughput sequencing technologies and from the development of analytical techniques that take advantage of the evolutionary information gained by sampling genetic data over both space and time.

GLOSSARY

Ancient DNA. A field of biology that involves extracting and manipulating sequence data from samples that are old and decayed in some way.

Contaminating DNA. DNA introduced into an experiment from the preservation environment, from excavation, sample handling, or sample processing, or during the experiment itself.

Coprolite. Preserved feces.

Draft Genome. Genomes of ancient DNA published before being considered sufficient in quality to be called "complete."

Mitochondrial DNA (mtDNA). A separate DNA genome of the *mitochondria*, which are maternally inherited organelles found within every cell.

Polymerase Chain Reaction (PCR). An enzymatic technique for amplifying from one to a few copies of DNA by several orders of magnitude.

Postmortem Decay. The DNA damage that accumulates after an organism's death.

1. BEGINNINGS

In 1984, a team of researchers based mostly in Allan Wilson's laboratory at the University of California, Berkeley, cloned two short fragments of mitochondrial DNA (mtDNA) from dried muscle taken from a 140-year-old museum specimen of a quagga (*Equus quagga*), a relative of the zebra that had been extinct since 1883. This work was the first to describe DNA preserved in nonliving tissues in a mainstream scientific journal. It came three years after a Chinese-language publication reported sequences cloned from a mummified human liver, and at the same time a German-language publication described the recovery of DNA from several Egyptian mummies. The quagga work confirmed that preserved tissues contained amplifiable DNA sequences. The results captured international attention, heralding great enthusiasm for this new source of DNA and a race to sequence the oldest, most exciting extinct organism. Crucially, the quagga study also noted what remains the most pervasive problem in the field of ancient DNA: that very little DNA survives postmortem.

An early leader in the field and widely considered "the father of ancient DNA," Svante Pääbo began his work with the aim of genetically characterizing the evolutionary history of Egyptian mummies. The process of rapid desiccation to which the bodies had been subjected immediately after death should have left the DNA molecules in a relatively intact form, making them ideal for ancient DNA analysis. In 1985, he recovered two members of the *Alu* family of human repetitive DNA sequences from a 2400-year-old Egyptian mummy. Although DNA could be recovered from only one of the 23 mummies he tested, close inspection of the data led Pääbo to conclude that few changes had occurred in the DNA postmortem.

A few years later, the polymerase chain reaction (PCR) was invented. This reaction makes millions of copies from only one or a few starting molecules of DNA; it also allows specific DNA sequences to be targeted, providing the means for focused evolutionary research. For ancient DNA, another advantage of PCR was the capability of a more thorough assessment of the ancient sequences. The enzyme used in the PCR to copy DNA was thought to read through undamaged molecules only, so that when errors were encountered, the reaction would simply end. In contrast, the enzymes that had been used during molecular cloning maintained the ability to repair damaged DNA, and this repair process could potentially introduce errors into the ancient sequences. When the same fragment of quagga DNA was amplified using the PCR, the two differences found between the quagga and its closest relative, the plains zebra, turned out to be no different at all. Damage, it seemed, was going to be a problem.

In 1989, Pääbo used the PCR to assess DNA survival in differently aged remains collected from a variety of locations. These results were instrumental in securing a place for ancient DNA as a credible scientific endeavor while warning future practitioners of the specific challenges associated with working with ancient material. He showed that ancient DNA sequences contain chemical modifications, including strand breaks, DNA crosslinks, and modified bases, that make their recovery challenging. He proposed an inverse relationship between fragment length and the number of surviving molecules of that length. He noted that DNA preservation is not determined by specimen age but by the environment in which the specimen was preserved. And crucially, he pointed out that contamination by modern DNA is likely to be the most serious challenge of working with ancient specimens. All these observations remain relevant to ancient DNA research today.

2. THE IMPORTANCE OF BEING CLEAN

DNA damage and contamination are the two biggest problems facing ancient DNA researchers. Initially, degradation occurs through the action of endogenous nucleases. In some circumstances, including rapid desiccation or deposition in very cold, dry, or salty environments, these enzymes will themselves be degraded before they can destroy all the DNA; however, even in ideal circumstances, environmental processes such as exposure to oxygen and water will slowly but steadily break down the surviving DNA until what remains is too damaged or fragmentary to be useful. Eventually, the continuous breakdown of DNA will result in only a few surviving, nonfragmented molecules per sample. The most common form of hydrolytic damage in ancient DNA specimens is deamination, in particular the conversion of cytosine to uracil. This results in the template DNA being read as a thymine, rather than cytosine, and in the erroneous incorporation of an adenine in the complementary strand. Although the exact numbers are still a matter of debate, this form of DNA damage is thought to account for nearly all misincorporated bases observed in amplified ancient DNA sequences. In addition to base misincorporations, double-strand breaks and DNA crosslinks are both common in ancient DNA samples; both lead to the amplification of only very short fragments of DNA. Expectations about DNA damage, and in particular the observation of cytosine deamination, are now used to distinguish authentic ancient DNA from contaminating DNA, and new phylogenetic models use information about damage to estimate the probability that certain mutations are due to decay rather than to evolution.

A variety of experiential protocols have been suggested to minimize the impact of DNA damage and contamination. These range from common laboratory sense, including wearing protective clothing and sterilizing components and work surfaces, to experimentally rigorous procedures, such as using multiple negative controls, performing independent PCRs to generate consensus sequences, and cloning PCR products to detect damage and contamination. Most laboratories comprise two separate, geographically isolated facilities: one in which the DNA extraction is performed and PCR is set up, and another for downstream (post-PCR) molecular biology work. This, in combination with a streamlined daily workflow in which researchers never move from the modern to the ancient lab, helps to ensure that amplified ancient DNA does not itself become a contaminant. These protocols have been modified over the years as technologies advance and as more is learned about how DNA degrades. For example, a requirement that was widely adopted in 2000, that each ancient DNA sequence be independently replicated in a separate laboratory, has been largely abandoned as high-throughput sequencing and population sampling have become more common, and hence contamination easier to identify. There is no doubt, however, that ancient DNA is sensitive to postmortem damage and contamination, and that care should be taken to ensure that published results are authentic.

Before these protocols were put in place and their importance made clear, the race to publish the oldest

DNA produced several results subsequently shown to be false. This period during the 1990s can be considered the "dark days" of ancient DNA. First, a 790-base-pair (bp) fragment of chloroplast DNA from a 16-million-year-old magnolia leaf was published. These data were met initially with skepticism, as it was unclear that DNA should survive for that long, even in the best possible conditions. Not surprisingly, these results were soon shown to be derived from contaminating bacteria. Next, bacterial DNA sequences and DNA from insects were reported from pieces of amber 25–120 million years old. Despite repeated attempts, these results could not be reproduced, but popular culture was already inspired, spawning movies such as *Jurassic Park* in 1993. In 1994, the first dinosaur DNA was published, purportedly isolated from a fossilized 80-million-year-old bone. To achieve this, researchers performed 2880 PCRs on two DNA extracts from the same bone, resulting in amplification of nine 170-bp fragments of mitochondrial DNA. Reanalysis of these fragments by different groups showed them to be mammalian in origin, and most likely a human sequence that, at the time, was not available in a public database for comparison.

These early missteps have not stopped ancient DNA researchers from pushing the technique's temporal limit. During and since the dark days, several hundred-million-year-old bacterial sequences from rocks excavated from salt mines have been published, as have Miocene plant DNA, and protein sequences from two different dinosaur bones. None of these results have been independently replicated, and many remain skeptical about their authenticity; nonetheless, DNA sequences have been recovered and authenticated that are very old, including sequences from 100,000-year-old bones preserved in arctic permafrost and temperate caves. The oldest authenticated DNA published so far are bacterial sequences recovered from permafrost ice cores between 450,000 and 800,000 years old. Importantly, these very old sequences have all been recovered from geographic regions where the preservation conditions favor long-term DNA survival.

3. NAME THAT BONE: INSERTING EXTINCT SPECIES INTO MOLECULAR PHYLOGENIES

At the same time these spectacular mistakes were captivating public and scientific attention, progress was being made in the extraction and investigation of authentic ancient DNA. For example, amplification and sequencing of two mitochondrial fragments of the recently extinct Tasmanian wolf confirmed that the Tasmanian wolf was more closely related to Australian marsupial carnivores than to more similar-looking marsupial carnivores from South America, indicating that their shared morphological features must have evolved independently. Thus began a period of genetically characterizing extinct organisms that continues today.

Ancient DNA has been used to place many extinct species in molecular phylogenies (see Section II: Phylogenetics and the History of Life). One of the earliest results was to reveal that mammoths are, rather unsurprisingly, closely related to elephants. Complete mitochondrial genomes of mammoths and mastodons later resolved this relationship further, revealing that mammoths are more closely related to Asian elephants than to African elephants (figure 1). Ancient DNA isolated from the Oxford dodo showed that this international emblem of extinction was a type of pigeon, rather than in its own evolutionary lineage as previously believed. In another revisionary discovery, ancient nuclear DNA recovered from the remains of several extinct New Zealand moa revealed that the three described species were in fact only two, and that the vast size difference used to distinguish the species from each other was actually due to pronounced sexual dimorphism.

Ancient DNA can be recovered from any element that has been shown to contain DNA, commonly bone, teeth, hair, seeds, muscle, or eggshells. In addition to these individual-specific tissues, DNA can also be recovered from mixed materials such as coprolites and soil. Despite being exposed to more damage-inducing influences than DNA preserved within bones, coprolite-recovered sequences are as reliable as those produced from bone. More interestingly, DNA extracted from coprolites provides both genetic information about the defecator and a genetic survey of that individual's last few meals. Sedimentary DNA, likely from a combination of shed cells and decaying plant material, provides the means to characterize ancient communities in the absence of macrofossil remains, circumventing potential problems with differential survival of representative members of the extinct community, wherein certain species may not leave large numbers of fossils, or certain environments may not be amenable to the long-term preservation of DNA. The DNA present in soil is generally "naked," not bound to anything and therefore not protected from bombardment by damage-inducing environmental events. This makes it difficult to distinguish damage lesions from phylogenetically informative mutations in the recovered sequences. Nonetheless, sedimentary DNA does make it possible to identify when and where species were present, potentially extending the range of locations and species that can be studied using ancient DNA.

4. ANCIENT POPULATION GENETICS AND PHYLOGEOGRAPHY

Most of the studies mentioned above use material recovered either from caves, where the ambient moisture

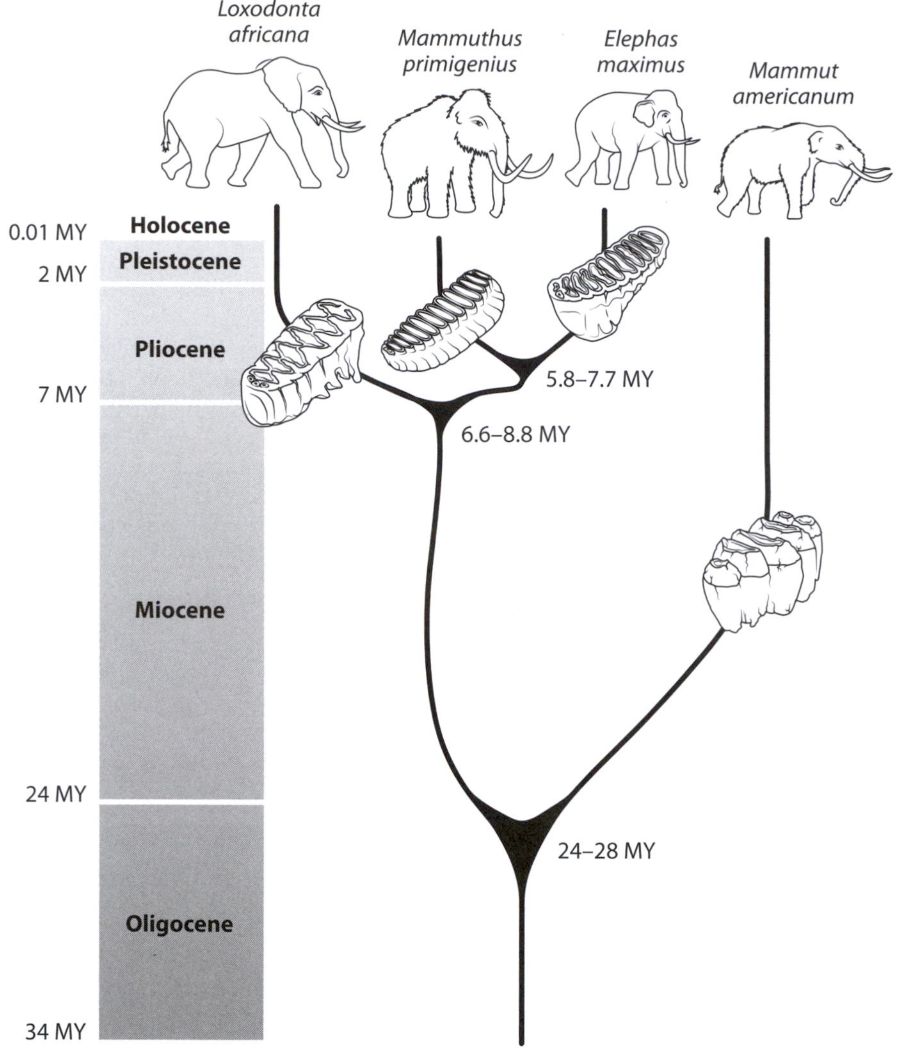

Figure 1. Complete mitochondrial genomes have now been sequenced for the extinct mammoth and mastodon, as well as for both living elephant species. These data have been used to infer both the branching order of the elephantid phylogeny (revealing that mammoths are more closely related to Asian elephants than to African elephants) and the timing of diversification between the different lineages. Elephants are only one example of lineages for which the addition of ancient DNA data has provided significant resolving power for long-standing phylogenetic and taxonomic questions. Other lineages include pigeons, ratites, cow, and even humans. (Reproduced from Rohland et al. 2007.)

and temperature tend to remain constant in both the short and long term, or from the Arctic, where remains are preserved in permanently frozen soil (permafrost). While authentic ancient DNA has been recovered from warmer climates (e.g., Florida, the Caribbean), consistently cold places (e.g., Siberia) are by far the richest source of material for ancient DNA analysis. This perhaps explains why, as the field moves toward analyzing populations of individuals rather than single individuals representing an extinct species, the focus has mostly been on Arctic and cave-dwelling species.

The first analyses of changes in genetic diversity within populations through time, however, took advantage of younger, museum-preserved skins that were more likely better preserved than ice age bones. One of the first population-level analyses amplified mtDNA from skins of three geographically isolated populations of the Panamint kangaroo rat in California collected early in the twentieth century. These data skins were compared with data collected from the same localities but in 1988; surprisingly, the populations had remained genetically isolated from each other throughout the period spanning the sample ages. Soon after, it was shown that pocket gophers from Yellowstone National Park have been genetically isolated from nearby populations for at least 2400 years. Later, mtDNA isolated from rabbit remains from across Europe and North Africa showed that these populations had maintained genetic stability and strong population structure for at least 11,000 years.

This pattern was not found for all species, however. Sequencing of mtDNA isolated from seven permafrost-preserved, Alaskan brown bear bones showed that the existing geographic isolation between brown bear mitochondrial lineages was established 15,000 years ago, and that prior to this time a different geographic pattern prevailed. This work was later expanded to include 36 brown bears ranging from 2000 to more than 50,000 years old. The new results supported the original findings and identified four distinct temporal periods during which the geographic distribution of brown bear mitochondrial lineages had remained stable, with rapid changes occurring between these. Climate change, in particular that linked to the last ice age, was implicated as the driver of most of these demographic changes.

The analysis of population genetic data sampled over time became more common as statistical tools capable of taking advantage of this kind of data were developed. Two complementary approaches were introduced approximately simultaneously, and both have been used to test the role of environmental change in driving changes in population genetic diversity. A major question within the reach of ancient DNA is, What caused the extinction of the ice age megafauna 8000 years ago? New genetic analysis methods allow a full-probabilistic estimation of the demographic history of a set of sampled DNA sequences, enabling the first attempt to answer to this question. The first large, ancient DNA population data set contained more than 600 sequences from North American bison, ranging from only 100 to more than 55,000 years old. These data showed clear evidence for a peak in bison diversity around 35,000 years ago, followed by a rapid decline toward extinction. The timing of the beginning of this decline was surprising, as it predated both the peak of the last ice age and the first appearance of large numbers of humans in North America, two competing hypotheses about the cause of the mass extinction. Later, more sophisticated demographic models further resolved the bison demographic history, revealing that around 13,000 years ago, bison narrowly escaped extinction; this bottleneck was followed by rapid recovery of the genetic diversity that persists today.

Work on this question continues, and population data sets now exist for six herbivores and several carnivores. All of these seem to respond to changes in climate differently, depending on their particular habitat requirements. Horses, for example, peak in genetic diversity slightly after bison peak in North America, probably because they were better able to survive once the steppe grasslands began to disappear at the onset of the last ice age. Although the jury is out regarding the ultimate cause of these extinctions, it is clear that climate change played a major role.

The second approach to analyzing ancient population genetic data takes advantage of the approximate Bayesian computation (ABC) framework. In this framework, genetic data are simulated under proposed models of population evolution and compared to those estimated from real data to identify the most likely demographic scenario. A major breakthrough came with another program that allowed simulated data sets to mimic ancient DNA data sets, in that samples could be drawn from an evolving population at different points in time. This approach was used to show that 3000 years ago, the Argentinean colonial tuco-tuco, a subterranean rodent, suffered a severe population bottleneck in which it lost around 99.7 percent of its mitochondrial genetic diversity.

5. ANCIENT GENOMICS

The recent technological advances collectively known as "next-generation sequencing" have been embraced by the ancient DNA community. These technologies allow millions of sequencing reactions to happen in parallel by creating microreactors and/or attaching DNA molecules to solid surfaces or beads prior to sequencing. These technologies provide a means to explore more fully the amount and quality of DNA preserved in ancient specimens. They also make it feasible to obtain larger amounts of ancient data in a much less time-consuming and often less expensive way than using traditional approaches.

The first complete ancient genomes were published in 2001, long before next-generation sequencing was state of the art. Two teams working independently both published mitochondrial genomes from two species of moa. Each 17,000-bp mitochondrial genome was painstakingly pieced together from overlapping 350–600 bp fragments amplified via PCR. These genomes were proof that ancient genomics is feasible, and could provide useful evolutionary information. The moa genomes were used to estimate the timing of the divergence between ratite birds and provide a temporal framework for the breakup of Gondwana into smaller continental fragments (these eventually became most of the landmasses found in today's Southern Hemisphere, as well as a few landmasses that migrated further north).

Five years later, two complete mitochondrial genomes of the mammoth were published using different techniques. One group pieced together the mammoth mitochondrial genome by targeting only longer, intact fragments, between 1200 and 1600 base pairs in length. A second developed a multiplex PCR approach to coamplify nonoverlapping fragments of mammoth mitochondrial DNA in a single PCR, greatly speeding up the process of data generation and significantly reducing the amount of sample required to perform the experiment. In the same year, a third group took mammoth mitochondrial-genome sequencing into next generation. They used the Roche 454 technology to shotgun sequence a permafrost-preserved mammoth bone. Of the 13 million base pairs of mammoth DNA they recovered, 222 reads, each around 89 bp long, mapped to the mammoth mitochondrial genome.

In a shotgun-sequencing approach, all the DNA extracted from a particular specimen is made into a library, and that DNA library is then sequenced. As a result, sequences are generated not only from the target specimen but also from any bacteria or other organisms that may have colonized the sample during its preservation history, and any DNA that may have contaminated the sample during processing. The sample used in this first study was remarkably well preserved: 45.4 percent of the sequences from the genomic library were identified as mammoth DNA, the remainder likely coming from organisms colonizing the sample after its deposition. In contrast, the libraries that were later used to sequence the complete nuclear genome of the Neanderthal contained only 1–5 percent Neanderthal DNA. In this case, enzymes targeting specific sequences present in bacterial DNA (and absent from Neanderthal DNA) were used to chop up bacterial DNA in the DNA libraries, thereby increasing the ratio of Neanderthal to contaminating DNA. Draft ancient genomes have now been published for a mammoth, a 4000-year-old Paleoeskimo from Greenland, a Neanderthal, and a previously unknown hominin from Denisova Cave in Siberia.

6. THE FUTURE OF ANCIENT DNA

Although the field of ancient DNA is now more than 25 years old, its potential is only beginning to be realized. After the excitement of simply generating complete ancient genomes fades, a new era of ancient DNA research is likely to emerge, in which the unique perspective allowed by ancient DNA is fully recognized. While it is impossible to know what the next discovery will be, two questions stand out.

First, what makes species unique? With the publication of the Neanderthal genome, we now have much more information about precisely which mutations distinguish us from our closest living relative, the chimpanzee (see chapter II.18). Prior to 2010, Neanderthals and humans were known to share a derived allele at the FOXP2 locus, which is involved in speech and language, suggesting that a selective sweep (see chapter V.14) occurred in this region prior to the divergence between Neanderthals and humans. The draft Neanderthal genome revealed large genomic regions that have been under positive selection *since* our divergence from Neanderthals. These regions include genes associated with human-specific maladies, including autism spectrum disorder and type 2 diabetes. Learning more about these genomic regions may reveal much about what it means to be human. Methods to target and capture specific regions of DNA provide a promising route to refining these observations and improving our understanding of what makes species look and act the way they do.

Second, what is the role of environmental change in the maintenance and distribution of genetic diversity? Shotgun sequencing hundreds of individuals for population genomic analyses is still too expensive; however, approaches are in development to capture specific fragments of DNA from DNA libraries. These captured fragments can then be bar-coded, pooled, and sequenced together on a next-generation platform. This approach allows hundreds or thousands of loci to be sequenced simultaneously from hundreds of individuals. It provides a solution to the matrilineal bias of using only mitochondrial DNA, and much more power to detect changes in genetic diversity associated with either particular environmental events or episodes of natural selection.

The results of analyses incorporating ancient DNA data have ranged from obvious (that a mammoth is closely related to an elephant) to surprising (that all non-African humans still contain some Neanderthal DNA). Regardless of what happens in the next 25 years, it is clear that the perspective gained from these data has benefited many aspects of evolutionary research. We know much more about the evolution of life on earth, about how populations respond to climate change, and about our own, recent evolutionary history than we could have known without ancient DNA.

FURTHER READING

Bunce, M., T. H. Worthy, T. Ford, W. Hoppitt, E. Willerslev, A. Drummond, and A. Cooper. 2003. Extreme reversed sexual size dimorphism in the extinct New Zealand moa *Dinornis*. Nature 425: 172–175. *This paper shows that ancient DNA can be used to resolve long-standing taxonomic issues. What were once thought three species of Dinornis are actually only one, with very large females and smaller males.*

Green, R. E., J. Krause, A. W. Briggs, T. Maricic, U. Stenzel, M. Kircher, N. Patterson, et al. 2010. A draft sequence of the Neandertal genome. Science 328: 710–722. *This paper provide details of the draft genome sequence of a female Neanderthal and provides evidence of interbreeding between Neanderthals and anatomically modern* Homo sapiens *after* H. sapiens *left Africa.*

Lorenzen, E. D., D. Nogués-Bravo, L. Orlando, J. Weinstock, J. Binladen, K. A. Marske, A. Ugan, et al. 2011. Species specific responses of Late Quaternary megafauna to climate and humans. Nature 479: 359–364. *This paper provides a detailed assessment of changes in distribution and abundance of six large herbivores (mammoth, bison, horse, caribou, musk ox, and woolly rhino) over the last 50,000 years, including both full-probabilistic and ABC-based inference of changes in population size and an analysis of changes in the distribution of appropriate habitat for each species over this time frame.*

Pääbo, S., H. Poinar, D. Serre, V. Jaenicke-Després, J. Hebler, N. Rohland, M. Kuch, et al. 2004. Genetic analyses from ancient DNA. Annual Review of Genetics 38: 645–678. *This paper contains a concise review of the history and challenges associated with working with ancient DNA specimens and data.*

Poinar, H, C. Schwarz, J. Qi, B. Shapiro, R.D.E. MacPhee, B. Buigues, A. Tikhonov, et al. 2006. Metagenomics to paleogenomics: Large-scale sequencing of mammoth DNA. Science 311: 392–394. *This manuscript provides the first example of use of next-generation sequencing technology to sequence an ancient specimen.*

Rohland, N., A.-S. Malaspinas, J. L. Pollack, M. Slatkin, P. Matheus, and M. Hofreiter. 2007. Proboscidean mitogenomics: Chronology and mode of elephant evolution using mastodon as out-group. PLoS Biology 5: e207. *This manuscript uses ancient DNA data to infer the phylogeny of the elephants, and demonstrates conclusively that the addition of ancient DNA data can considerably improve knowledge of the rate and timing of species divergence.*

Willerslev, E., A. J. Hansen, J. Binladen, T. B. Brand, M.T.P. Gilbert, B. Shapiro, M. Bunce, et al. 2003. Diverse plant and animal genetic records from Holocene and Pleistocene sediments. Science 300: 791–795. *This manuscript describes the amplification of both plant and animal ancient DNA directly from Alaskan permafrost sediments.*

VI

Speciation and Macroevolution

Dolph Schluter

Since the Big Bang, not much has happened in the universe more interesting than the diversification of life on earth. Most of life's current diversity is wrapped up in the genetic and phenotypic differences between species, between the communities of species they form, and between the higher taxonomic groups that species make up, such as families and phyla. For this reason the study of the origin of species—speciation—and its consequences tells us a great deal about how the extraordinary diversity of life arose, how it is distributed across the globe, how it is presently maintained, and how it has changed through billions of years of earth's history.

To Charles Darwin, how new species evolve was the "mystery of mysteries." In his 1859 masterpiece, *On the Origin of Species by Means of Natural Selection*, he took the first big steps to demystifying the process. Darwin recognized that in nature there was no sharp discontinuity between the differences one sees among populations within species and the differences observed between closely related species: "I look at the term species as one arbitrarily given, for the sake of convenience, to a set of individuals closely resembling each other, and that it does not essentially differ from the term variety, which is given to less distinct and more fluctuating forms." The origin of a species is not a sudden instant in the history of life but one that (usually) results from a steady accumulation of differences. Those differences, he explained, are the product of gradual evolution by natural selection of variation present in populations.

We now recognize that Darwin's solution was incomplete. One reason is that our concept of species has evolved. In Darwin's day, species were designated according to the magnitude of morphological differences: "the amount of difference is one very important criterion in settling whether two forms should be ranked as species or varieties." This means that except for the magnitudes involved, morphological criteria grouped populations into species just as species were grouped into genera, and genera into families. Under this concept, *speciation* is the evolution of differences in ordinary phenotypic traits sufficiently large to warrant the taxonomic designation *species*. Darwin appreciated that matings between different species often produced inviable or sterile offspring, but he decided that this was not universal and was less reliable than morphological differences for classifying species.

The focus of speciation study changed with the development of the *biological species concept* (see chapter VI.1). In 1937, Theodosius Dobzhansky defined speciation as the evolution of "isolating mechanisms," traits that reduce gene flow between populations. Subsequently, Ernst Mayr defined species as "groups of actually or potentially interbreeding natural populations, which are reproductively isolated from other such groups." Reproductive isolation doesn't just mean hybrid sterility and inviability—it includes *any* genetically based difference that acts as a barrier to the movement of genes between populations. The mechanical, chemical, behavioral, and ecological traits that reduce interbreeding all contribute to reproductive isolation. Reproductive isolation also includes traits that inhibit fertilization after mating and any evolved behavioral, ecological, and genetic factor causing hybrids to be relatively unsuccessful. Speciation research today is focused on answering the question, How does reproductive isolation evolve?

In the hierarchy of categories that we use to classify life's diversity, the evolution of reproductive isolation is a feature unique to the species category. No equivalent process takes place during the evolution of a new genus or a new family. Speciation therefore has special significance in the evolution of diversity.

Another reason for its importance is that speciation occupies a juncture between the scales of processes—small and large—that have produced the patterns of diversity we see today. Speciation is undoubtedly a microevolutionary process. It is the outcome of accumulated genetic divergence between populations, and all intermediate stages are represented among contemporary populations in nature. Speciation is a macroevolutionary event as well, because species are the basic unit for measuring large-scale changes in life's diversity over long spans of time. This dual interpretation of speciation makes sense because once gene flow is sufficiently restricted, new species can evolve independently of others. The completion of a speciation event changes patterns of biodiversity in many ways. A new species can coexist with its closest relatives without their collapsing to a hybrid swarm. Thus speciation affects the numbers of species coexisting in ecological communities. A new species in its defined geographic range also incrementally affects patterns of species diversity across the globe by influencing the number of species present in each region. Each new species also contributes to the diversity of the larger clade of species—all related by descent—to which it belongs. A speciation event may thus affect the probability of long-term persistence of its whole lineage. To the extent that traits of a new species are shared by its close relatives, and tend to be passed on from ancestor to descendant species (with some modification), the evolution of a new species alters the frequency distribution of traits represented in earth's biota. A speciation event is thus the seed of long-term patterns in the history and future evolution of life.

For all these reasons, speciation has played a large role in many of the processes that brought about the modern diversity of life and its distribution over the face of the earth. But what do we really know about these processes? What drives the origin of species, and what determines their geographic distribution? What is the connection between speciation and the evolution of ordinary phenotypic traits? Are the composition of species communities, and long-term trends in local and regional species numbers, affected by the mechanism of speciation? Why do some lineages speciate more often than others? What determines the overall rates of speciation and extinction? Are the factors that determine the success of some lineages over the long term, and the decline of others, the same as those that drive microevolution and the origin of species? In this section, we review what is known about the causes and global consequences of speciation.

SECTION THEMES

Species Diversity Patterns

Measuring species numbers over time and their distribution over the earth requires that we know what a "species" is. Speciation researchers today focus on the biological species concept, where (nearly complete) reproductive isolation is key. However, this is not always a practical criterion for classifying organisms to species. For example, how does one decide whether two populations are "actually or potentially interbreeding," and thus belong to the same biological species when their ranges do not overlap, or the populations are known only from fossils, or their individuals reproduce asexually? Chapter VI.1 discusses these issues and some alternative species criteria that have also been proposed, their connections to one another, and their implications. Debates over species concepts are largely resolved by recognizing that different species criteria emphasize different stages and features along the continuum of changes that take place during the origin of species.

Centuries of exploration and survey have established that species are extremely unevenly distributed across the face of the earth and through time. The causes of this unevenness continue to challenge biologists. The most prominent spatial pattern is the latitudinal diversity gradient, whereby more species occur in the tropics than in the temperate zone. This pattern is seen in both sexual and asexual species, suggesting similar underlying causes. Chapters VI.2 and VI.3 describe these and other patterns of species diversity in space and time, the possible mechanisms that produce them, and their consequences.

Gene Flow

Divergence of populations is a tug-of-war between the forces that generate differences (mutation, genetic drift, natural and sexual selection) and the main process that erodes differences: gene flow. Speciation marks the point at which barriers to gene flow are strong enough to resist the effects of gene flow. Yet, if there is gene flow, it is difficult for reproductive isolation to evolve in the first place. The amount of gene flow between two populations is strongly influenced by their geographic distributions (see chapter VI.3), being least when the populations are fully separated by a geographic barrier to movement (allopatric), and most when they overlap in distribution (sympatric).

How does gene flow retard speciation? First, it slows or prevents genetic divergence between populations when divergence is not directly favored by selection, such as when it occurs by genetic drift, or when separate populations adapting to similar selection pressures by chance experience and fix different advantageous mutations. In such cases, gene flow moves alleles among populations, eroding genetic differentiation. Gene flow is less destructive when selection is divergent, favoring different alleles in different populations, because an allele that flows from a population in which it is favored to another in which it is not favored will eventually be removed by selection, provided selection is sufficiently strong. Second, even if selection is divergent, gene flow slows speciation by breaking up associations between alleles at different genes. For speciation to proceed, genes responsible for reproductive isolation, such as those that influence mate preferences, must become associated with the genes under divergent selection. The individuals in one population must prefer to mate with other individuals from the same population, rather than with individuals from another population. Gene flow will break up these associations, bringing mate preference genes from one population to the other, breaking down reproductive isolation between them. Consideration of this problem has led theorists to make predictions about the specific circumstances under which speciation with gene flow can nevertheless occur (see chapter VI.3). When it does occur in the face of gene flow, it should leave detectable marks on the types of genetic differences that evolve, on the strength of different kinds of evolved barriers to gene flow (see chapter VI.4), on genome-wide patterns of genetic differentiation (see chapter VI.9), and on spatial distributions of species (see chapters VI.2 and VI.3).

Paradoxically, episodes of gene flow between already well-differentiated species can sometimes be a creative rather than a homogenizing process, resulting in new genetic combinations that have novel phenotypes and represent brand-new "hybrid" species (see chapters VI.6 and VI.9)

Mechanisms of Speciation

What are the forces that generate differences and bring about the evolution of reproductive isolation during speciation? For a long time the answers to this question came mainly from theory and laboratory experiments, which evaluated the plausibility of speciation by genetic drift, founder events, divergent natural selection, and other processes. Only recently have we been able to test these ideas in nature and say with confidence how real species in nature have formed.

Since Darwin the role of natural selection has been of great interest, and we know more about its role than that of any other process. For example, natural selection on ordinary phenotypic traits may incidentally build reproductive isolation between populations as a by-product (see chapter VI.4). Such isolation can occur when separate populations adapt to different environments (*ecological speciation*) as different alleles favored in one environment but not the other gradually accumulate between populations. Alternatively, selection may build genetic differences among populations that experience similar selection pressures if the populations by chance experience and accumulate different sets of advantageous mutations (*mutation-order speciation*). Finally, there has been a long-standing interest in the role of genetic drift—speciation without any natural selection at all—but we still don't know much about its importance.

When speciation involves natural selection, what are its mechanisms? As described in Section III: Natural Selection and Adaptation, adaptation to abiotic factors in the environment such as soil and climate is one possibility, and there are now good examples of this process. Biotic interactions with other species, including predator-prey, host-parasite, competition and mutualism, are also a major (perhaps *the* major) source of selection on populations. Furthermore, reciprocal evolutionary changes between interacting species—coevolution—might bring about rapid divergence between populations within each of the interacting species (see chapter VI.7; see additional examples in chapters VI.10 and VI.16). Strong natural selection can also result from internal genomic conflict. For example, reproductive isolation might evolve as a by-product of conflict resolution between different genetic elements within individuals (*intragenomic conflict*) or between the sexes (*intergenomic conflict*) (see chapters VI.4 and VI.8).

Natural selection might also contribute to speciation when it directly favors stronger prezygotic reproductive isolation between incipient species when their hybrid offspring have reduced fitness (see chapter VI.4). This process, called *reinforcement*, represents the only known circumstance in which natural selection directly favors the evolution of stronger reproductive isolation. Otherwise, as described earlier, the role of selection is indirect—reproductive isolation evolves as an incidental consequence of adaptation.

It has often been pointed out that the most conspicuous differences between closely related species are often in secondary sexual traits, such as in courtship or body coloration of males, rather than in ecological traits (see chapter VI.5). Since the exaggeration of such traits is caused by sexual selection, it seems likely that sexual selection is also frequently involved in speciation. Any such role would likely involve natural selection, too, because it is the process that leads to divergence of mate preferences or that favors the evolution of traits that ameliorate intergenomic conflict.

The genetic changes that underlie the evolution of reproductive isolation (*speciation genes*) are finally being discovered, and the hunt for them represents one of the most exciting directions in modern speciation research (see chapter VI.8). Sometimes, mutations of large effects on reproductive isolation are found, whereas reproductive isolation often results from the accumulated effects of many small-effect mutations. Genetic "signatures" of selection detected on speciation genes provide some of the best evidence that natural and/or sexual selection have been responsible for driving the mutations to high frequency, and hence for the evolution of reproductive isolation. We are beginning to learn why speciation genes are often clustered rather than dispersed within the genome and how the genome evolves collectively during the evolution of reproductive isolation (see chapter VI.9).

A surprisingly common mechanism of sudden speciation is via the evolution of *polyploidy*. Individuals that have more than two sets of chromosomes are occasionally formed (see chapter VI.9), and as a result, they instantly possess some degree of reproductive isolation from their diploid ancestors. This makes polyploidization the fastest mode of speciation known. Often, the polyploids are hybrids between two species. The process is most common in plants, but several examples from animals have recently been discovered.

Adaptive Radiation

When a group of organisms experiences a flurry of speciation events in association with adaptation of nascent species to different ecological niches, the result is *adaptive radiation* (see chapter VI.10). Classic examples include the finch radiations on the Galápagos and Hawaiian Islands. In both cases the species have evolved a wide diversity of beak sizes and shapes that enhance the ability of individuals to exploit particular resources, such as hard seeds, nectar from long-tubed flowers, or insects under bark. In the few adaptive radiations that have been studied intensively, it is clear that the same natural selection pressures that adapt populations to distinct niches also indirectly contribute to the buildup of reproductive isolation between populations. In these few studied cases, at least, there is a close connection between rapid speciation and adaptive evolution.

Adaptive radiations are particularly prevalent where ample resources are available, and few competing lineages take full advantage of them (*ecological opportunity*). Even under such conditions, however, some lineages diversify more readily than others, as though they have intrinsic differences that affect their abilities to speciate rapidly, or to adapt to and usurp, novel resources. One reason might be that the fortunate lineages possess key traits that increase their evolvability or their propensity to speciate (*key evolutionary innovations*; see chapter VI.15). For example, it has been proposed that the huge diversity of angiosperm plants is attributable to the evolution of the flower. Adaptation of flower structures to different suites of pollinators in different environments might speed the evolution of premating reproductive isolation. Another hypothesis is that the evolution of traits permitting certain insects to consume plant tissue is behind the astonishing diversity of phytophagous insects, such as herbivorous beetles, found today. Plants are incredibly abundant and diverse in their leaf structures, chemistries, and life histories, which favors niche specialization and diversification by insects that exploit them. Such hypotheses are challenging to test, but great strides are being made.

Evolutionary Rates

Adaptive radiations represent episodes of particularly fast evolution and speciation. In contrast, study of patterns of evolution in the fossil record and in phylogenetic trees has found that evolution is often slow. Lineages frequently undergo long periods in which little evolution seems to take place—at least in easily identified morphological traits (perhaps rates are not so slow in other aspects, such as at genes involved in fighting disease). The hypothesis of *punctuated equilibria* was an extreme statement about rates of evolution in nature: that evolution hardly ever occurs except in the relatively brief periods during which speciation also takes place. The rest of the time, so the hypothesis goes, species exhibit *stasis*, changing little. This conjecture prompted a great deal of research that continues to examine the true relationships among speciation, time, and trait evolution (see chapters VI.11 and VI.12). A key question is whether the punctuated equilibrium is a caricature of evolutionary patterns in the fossil record. Sustained directional changes in traits might indeed occur infrequently and episodically, but the rest of the time evolution might be better described as oscillating rather than static, or at least not sustained and directional, with fluctuations of varying amplitude taking place through time. Chapter VI.11 describes additional patterns, including the paradoxical observation that measured rates of evolution appear faster the shorter the time interval over which change is measured, and the observation that rates of phenotypic evolution appear to be highest early in a clade's history.

Macroevolutionary Trends

The traits that a species possesses—such as the mean body size of its individuals or its geographic range size—can influence the rate at which it subsequently produces new species and the probability that it will go extinct (see chapter VI.12). If a relationship between a trait possessed by species and speciation or extinction rates holds consistently across multiple lineages and over time, the result will be a large-scale increase in the prevalence of that particular trait in nature. We can think of this process as *species selection*—the macroevolutionary analogous of natural selection on individuals within populations. The notion of species selection has been controversial, and many researchers regard it as a weak force compared with ordinary natural selection within species. For example, a large-scale trend toward larger body size in the fossil record appears to be mainly the result of ordinary natural selection within species accumulated over a long time span (see chapter VI.12). However, species selection need not oppose ordinary natural selection, and it may indeed generate trends in the absence of any net direction to evolution within species. Current research focuses on the evidence for species selection driving macroevolutionary trends, and we now have good examples of the process (see chapters VI.12 and VI.14).

One of the most striking discoveries from the fossil record is that long-term success and failures of lineages are not necessarily determined by the same factors that drive evolution within populations (see chapter VI.13). Dinosaurs possessed exquisite adaptations to the many environments in which they occurred, yet they were wiped out en masse by catastrophic environmental changes never before experienced during their many millions of years of history. Indeed, it does not often happen that the lineages that dominated the earth prior to mass extinctions recover and reassume their dominant positions afterward. More typically, previously minor components of life's ensemble proliferate subsequently and become the new dominants, which in turn results in wholesale changes to the frequency distribution of different kinds of traits represented in nature. These changes may often be due more to chance (*species drift*) than selection, though some species selection seems to occur during mass extinctions, consistently favoring lineages with certain traits such as a broad diet and a large geographic range (see chapter VI.13).

The resolution of these microevolutionary and macroevolutionary forces has left its mark on the composition of life on earth and on the communities of species seen today (see chapter VI.16). The macroevolutionary processes of speciation and extinction, of species selection and species drift, acting over long spans of time created the biodiversity that has assembled into ecological communities and continues to influence how local assemblages change through time. Natural selection on variation within species has produced the myriad adaptations of species to one another and to the abiotic environments they encounter across their geographic ranges. As species adapt to one another, the strength of their interactions changes and the flow of energy and materials is altered, producing consequent changes in the properties and dynamics of the surrounding ecosystem. Thus, microevolution on species within communities generates new species and modifies the traits that species possess and so provides the material that drives macroevolutionary changes. These changes, in turn, will affect the course of future evolution.

VI.1

Species and Speciation

Richard G. Harrison

OUTLINE

1. Species concepts and definitions
2. Speciation as the evolution of intrinsic barriers to gene exchange
3. Classifying barriers to gene exchange
4. Studying speciation

Species are the fundamental units of biodiversity, but the definition of a species remains a subject of debate within evolutionary biology. One resolution of this debate views alternative species definitions as different stages in the process of *speciation*, in which conspecific populations diverge, accumulate intrinsic barriers to gene exchange, and ultimately become exclusive or reciprocally monophyletic groups. Most studies of speciation have focused on the evolution of reproductive isolation or intrinsic barriers to gene exchange. Such barriers may result from a variety of trait differences, some of which are simply a by-product of divergence in allopatry. Barriers may prevent individuals from meeting or mating, they may compromise gamete interactions, or they may reduce the viability and/or fertility of hybrid offspring.

GLOSSARY

Allopatric. Occupying different geographic regions; geographically isolated.

Gene Flow. Movement or incorporation of alleles from one population into one or more different populations.

Monophyletic. A group of organisms (taxa) that all share a most recent common ancestor not shared by any other organisms (taxa).

Sympatric. Occupying the same geographic area, with the opportunity for gene flow.

Zygote. The (usually diploid) cell formed by the union of two (usually haploid) gametes (e.g., sperm and egg).

The diversity of life comprises relatively discrete entities we call *species*. Like cells or individual organisms, species are widely viewed as fundamental units of biological organization. However, the defining qualities of species, the nature of the boundary between species, and even the reality of species remain matters of dispute. It is ironic that the concept or definition of species, so central to the studies of evolution, ecology, and conservation biology, has engendered so much confusion and debate. In contrast, the rules for naming species (for animals embodied in the International Code of Zoological Nomenclature) are very clearly described and widely accepted.

The last two decades have witnessed gradual acceptance of the view that there is not one "right" species concept or definition. One approach to resolving past disagreements is to recognize the difference between a species *concept* and a species *definition*. The former is what is meant by the word *species*; the latter involves defining the criteria used to delimit species. K. de Queiroz has suggested that most would agree that the defining property for a species is "a separately evolving metapopulation lineage," in which a lineage is an ancestor-descendant series. Because species are defined over time as well as space, and because speciation is a process not an event, differences among species definitions may then reflect different landmarks along the path from conspecific populations to separate species.

1. SPECIES CONCEPTS AND DEFINITIONS

The evolutionary biology literature presents a bewildering array of different species concepts or definitions (as many as 24 have been identified). Some of the concepts are subtle variations on basic themes, but at least seven major concepts can be differentiated along a number of axes, including whether they are retrospective (species as products of history) or prospective (species as lineages extending into the future), whether they are relational or nonrelational, whether they are based on pattern or process, and the extent to which they are operational,

Table 1. Major species concepts or definitions

Biological Species Concept
"Groups of actually or potentially interbreeding natural populations which are reproductively isolated from other such groups." (Mayr 1963)
"Systems of populations; the gene exchange between these systems is limited or prevented in nature by a reproductive isolating mechanism or by a combination of such mechanisms." (Dobzhansky 1970)

Recognition Species Concept
"The most inclusive population of individual biparental organisms which share a common fertilization system." (Paterson 1985)

Isolation Species Concept
"The most inclusive population of individuals having the potential for phenotypic cohesion through intrinsic cohesion mechanisms (genetic and/or demographic exchangeability)." (Templeton 1989)

Character-Based Phylogenetic Species Concept
"The smallest aggregation of populations (sexual) or lineages (asexual) diagnosable by a unique combination of character states in comparable individuals." (Nixon and Wheeler 1990)

Genealogical Species Concept
"Exclusive groups of organisms, where an exclusive group is one whose member are all more closely related to each other than to any organisms outside the group." (Baum and Shaw 1995)

Evolutionary Species Concept
"A single lineage of ancestor-descendant populations which maintains its identity from other such lineages and which has its own evolutionary tendencies and historical fate." (Wiley 1978)

Genotypic Cluster Definition
"Genetically distinguishable groups of individuals that have few or no intermediates when in contact." (Mallet 1995)

including whether they can be applied to allopatric (geographically isolated) populations and to asexual lineages. The major concepts are summarized in table 1.

Character-Based Concepts

Perhaps the simplest and most intuitive concepts of species are those that are "character based." Accordingly, the fundamental criterion for defining species is "diagnosability": species are groups of organisms/populations that are diagnosably distinct, that is, groups exhibit fixed differences in states for at least one (but often more than one) character. Such a definition is unambiguous and easy to apply; it relies only on the ability to define and compare sets of characters (morphological, DNA sequence, etc.) in groups of organisms. Of course, care must be taken to ensure that phenotypically distinct groups do not simply represent differences between males and females or differences among life history stages.

Although easy to apply, character-based definitions view speciation as equivalent to divergence. If isolated populations diverged as a result of natural selection that led to local adaptation, or if allele frequencies drifted to fixation for different alleles at a gene locus, these populations would be viewed as distinct species. Strict application of such character-based concepts might result in a tremendous proliferation of species and the elevation of many current subspecies or races to species status.

Darwin clearly recognized that evolution (including the origin of species) is a process and that sampling diversity at any one point in time should reveal populations at all stages in the process. Indeed, Darwin saw the continuum of differences between populations as direct evidence for the evolutionary process. He wrote:

> Certainly no clear line of demarcation has as yet been drawn between species and sub-species....; or again between sub-species and well-marked varieties, or between lesser varieties and individual differences. These differences blend into each other in an insensible series; and a series impresses the mind with the idea of an actual passage.

Thus, species boundaries may be fuzzy and develop gradually, with few rules about how much difference needs to accumulate before lineages should be recognized as different species. However, for allopatric populations, reliance on amount (or quality) of difference seems to be the only possibility. Darwin commented that in cases "in which intermediate links [between populations] have not been found ... naturalists are compelled to come to a determination by the amount of difference between them." When differentiated populations co-occur, a direct test of character-based definitions is whether the populations remain distinct in sympatry. The approach has been formalized in an updated version of Darwin's views, in which species are recognized when there are few/no intermediates in a zone of overlap. This genotypic cluster definition is also character based, but by examining character differences in sympatry, it attempts to assess whether the existing differences persist.

The Biological Species Concept (or Isolation Concept)

For many years, the prevailing species concept, promoted by two famous twentieth-century evolutionary biologists, Ernst Mayr and Theodosius Dobzhansky, has been the biological species concept (BSC). In contrast with character-based definitions, the BSC defines species in terms of gene flow or gene exchange between populations. Species are characterized by interbreeding within a population and reproductive isolation among groups of populations. Reproductive isolation is a consequence of intrinsic barriers to gene exchange, barriers due to the properties of the organisms themselves and not simply geographic separation.

The great strength of the BSC is that it is clearly based on an appreciation for the evolutionary process; it attempts to define species using a criterion (gene exchange) that has important implications for the future of the lineages and is not simply a product of history. The presence of intrinsic barriers to gene exchange suggests that two species will persist, although ecological differentiation may also be a prerequisite for coexistence.

The BSC has a number of obvious limitations, many of which are shared by other definitions of species. First, it is difficult to estimate amount of gene flow in natural populations; therefore, absence of gene flow is often inferred from patterns of differentiation for genotypic or phenotypic markers. Second, the BSC cannot be applied to allopatric populations; if it is, amounts of gene flow cannot be estimated. Mayr's definition introduces the notion of "actually or potentially interbreeding" populations; use of "potentially interbreeding" presumably implies that we should be able to infer whether populations would interbreed should they come into contact. A third limitation is that the BSC is relevant only for organisms that reproduce sexually; interbreeding clearly has no meaning for obligately asexual lineages. Finally, the BSC must confront the issue of the "fuzzy" species boundary. Patterns of gene exchange vary not only in space and time but also across the genome. In some regions of the genome there may be no gene exchange, whereas in other regions, hybridization results in the flow of alleles between the two populations/species. In the extreme, the BSC may need to be defined for individual genes or gene regions. The bright side is that observed patterns of differential gene flow (introgression) can be used to gain insight into the genetic architecture of speciation.

Phylogenetic or Genealogical Concepts

Descent with modification produces a nested hierarchy of traits and taxa. The structure of this nested hierarchy can be revealed by phylogenetic analysis, which documents the pattern of branching events (forward in time) or coalescent events (backward in time) that define extant individuals, populations, or species. A phylogenetic or genealogical perspective suggests that species should be considered to be monophyletic or exclusive groups. An exclusive group is one whose members are all more closely related to one another than to any individual outside the group. The genealogical species concept of Baum and Shaw argues that exclusivity is an important criterion for species status. It seems quite reasonable that all members of a species should be closely related, but because of ancestral polymorphism and incomplete lineage sorting and/or ongoing hybridization, individuals within a species may, in fact, be more closely related to individuals of other species. Relationships depend on which gene or gene region is sampled, and exclusivity may characterize some genome regions and not others. It is not clear what proportion of the genome must be "exclusive" before a group is considered a genealogical species. According to a strict definition, many entities now viewed as independent evolutionary lineages would be considered conspecific, because genealogical speciation requires very long periods of time.

Evolutionary Species Concept

For most of evolutionary history, the only data we have about species and speciation come from the fossil record. Thus, the data are purely phenotypic (usually morphological), and neither the nature nor the quantity of the data allow direct assessment of gene exchange or exclusivity. Fixed differences can be defined, and character-based species definitions apply.

A very different and more general view that has been applied to fossil data is the evolutionary species concept, originating with the paleontologist G. G. Simpson and updated and modified by others. This concept defines species as populations through time (which can be followed in the fossil record) that maintain their separate identity (in the presence of other lineages) and exhibit an independent evolutionary trajectory.

Cohesion and Recognition as the Basis for Defining Species

All the species concepts discussed thus far rely on comparison of two (or more) lineages—they are *relational.* Fixed differences, reproductive isolation, exclusivity, and separate identities are all patterns or characteristics that must be defined in terms of differences between individuals and populations. A number of evolutionary biologists have argued that species concepts should be nonrelational, defined in terms of what is shared, rather than what is different.

The recognition concept defines species as populations of sexual organisms that share a common fertilization system or specific mate recognition system. The emphasis is on defining an interbreeding unit, a "field for recombination," a group of organisms held together by "genetic cohesion." In many ways the recognition concept is a reaction to the mention of "isolation" and "isolating mechanisms" in versions of the BSC. In practice, defining cohesion mechanisms is virtually equivalent to defining reproductive isolation. In either case, a group of individuals has to be partitioned into two or more subgroups, each of which shares fertilization and mate recognition systems within the subgroup but differs in these respects from other subgroups.

Most species concepts focus on genetic cohesion or isolation. But if two species are to persist in sympatry, the competitive exclusion principle from ecology says that they must be ecologically distinct. Ecology has not played much of a role in the development of species concepts. One exception is the cohesion species concept, which defines species as "the most inclusive population of individuals having the potential for phenotypic cohesion," which is mediated by both genetic and demographic exchangeability. The former is essentially equivalent to interbreeding or gene flow. However, demographic exchangeability emphasizes ecological interactions. Groups of organisms that are demographically exchangeable are ecological equivalents. This concept is of particular use in sympatric asexual lineages, in which even in the absence of any gene exchange, demographic exchangeability implies that lineages belong to the same species.

2. SPECIATION AS THE EVOLUTION OF INTRINSIC BARRIERS TO GENE EXCHANGE

In the simplest model of speciation, allopatric populations diverge in the absence of gene flow as a result of the fixation (due to natural selection or genetic drift) of new mutations and/or different ancestral alleles. It is useful to think of the process of divergence as the life history of a species. Fixation of alternative alleles or phenotypes in the two populations results in their being considered character-based phylogenetic species. Eventually, some of the fixed differences affect the ecology, behavior, physiology, or reproductive biology of the diverging lineages. As a by-product of this divergence, the populations accumulate differences that affect the probability of their interbreeding or the success of their progeny should an individual mate with a member of the "other" population. At this point, the populations become biological (or isolation) species. However, these populations are exclusive groups only at some (perhaps relatively few) regions of the genome. Over time, an increasing proportion of the genome diverges, and each of the diverging populations becomes an exclusive group across the genome. At this point, the populations are genealogical species.

The critical event in this life history is the evolution of reproductive isolation, the appearance of intrinsic barriers to gene exchange. This transition alters the outcome if or when secondary contact occurs between the diverging populations. Should the populations become sympatric, intrinsic barriers will prevent gene flow and the erosion of genetic differences. Similarly, in models of sympatric speciation, the evolution of reproductive barriers leads to the cessation of gene flow and enables further divergence of two subpopulations.

Evolutionary biologists who study speciation therefore focus on the nature of intrinsic barriers to gene exchange: what they are, when they act in the life cycle of the organism, to what extent they reduce gene flow, and when in the history of divergence they arose. Answers to these questions emerge from comparisons of closely related (recently diverged) species and "incipient" species." Such comparisons are particularly informative when the diverging populations occur in sympatry, and important contributions have come from the study of hybrid zones where individuals from distinct lineages meet and mate, producing some offspring of mixed ancestry (see chapter VI.6).

3. CLASSIFYING BARRIERS TO GENE EXCHANGE

Many phenotypic differences between diverging lineages can result in barriers to gene exchange. The traditional approach to classifying such barriers is to organize them with respect to whether they act before mating and/or zygote formation or whether they are a consequence of the reduced fitness (viability, fertility) of hybrid offspring. Most classifications recognize three distinct sorts of barriers: (1) premating, (2) postmating but prezygotic, and (3) postzygotic.

Premating Barriers

Premating barriers are those that result from trait differences that prevent hybridization between distinct species.

Barriers to gene exchange will exist if potential mates do not meet (temporal and habitat or ecogeographic isolation), if potential mates meet but do not mate (behavioral isolation), or if attempted copulation does not result in sperm transfer (mechanical isolation). Premating barriers to gene exchange have been studied in many different animal and plant systems, revealing a host of different mechanisms whereby gene exchange is limited or prevented.

Temporal isolation reflects seasonal or diurnal differences in the times at which adults are present or sexually active. Thus, flowering time differences in plants, and major life cycle differences in animals (e.g., different overwintering life stages or different rates of development in insects) lead to partial or complete seasonal isolation. In many marine invertebrates, mating (in the narrow sense) does not occur, and eggs and sperm are simply broadcast in the water column. Spawning times can be determined by lunar cycles, so that in some corals, eggs and sperm from closely related species are unlikely to encounter each other in the water column. Similar patterns are seen in some moths, in which sexual activity is limited to a relatively narrow window in the diurnal cycle and may be displaced from the corresponding window for a sympatric close relative.

Habitat or resource isolation results from the association of particular populations or lineages with specific habitats or resources. Observed associations can be the result of differential adaptation to habitats or differential preference for habitats. Many examples come from the insect–host plant literature, in which insect lineages have apparently diversified by adapting to new host plants, resulting in reproductive isolation between the derived forms. There are also numerous examples of plants with different habitat needs or requirements; some of the best studied involve adaptation to different soils (e.g., serpentine soils or soils contaminated by heavy metals). Geographic isolation can result from local adaptation if habitats or resources are geographically separate; this phenomenon has been termed "ecogeographic isolation." This type of isolation is a special form of ecological or habitat isolation, because unlike most barriers, which are studied in sympatry (where they prevent individuals from meeting or mating), ecogeographic isolation is an intrinsic barrier that characterizes allopatric taxa. A related concept is that of "immigrant inviability," which refers to the "reduced survival of immigrants on reaching foreign habitats that are ecologically divergent from their native habitat" (see chapter VI.4).

If ecological factors do not prevent individuals from meeting, then behavioral differences may well prevent them from mating. Many examples have been documented of species-specific communication systems that function in mate finding and mate recognition. These include visual communication (e.g., plumage coloration in birds, color patterns in fish, flashing patterns of "fireflies," attraction of pollinators to flowers), acoustical communication (e.g., songs of [mostly male] birds, frogs, and insects and the corresponding preference functions of females), and chemical communication (e.g., sex pheromones in insects). Behavioral barriers have been well studied in a diversity of animal systems, and sexual selection is often invoked to explain patterns of divergence. A primary focus has been on sexual selection by female preference. Depending on the nature of female preferences, the outcome of sexual selection may be "arbitrary"—that is, it will not have any "adaptive value" or relationship to environment (e.g., in runaway sexual selection). Prezygotic barriers can then arise as a result of different outcomes of sexual selection in isolated populations. Differences in female preferences (and in corresponding male traits) will act as barriers to gene exchange should populations come into secondary contact (see chapter VI.5).

Finally, if mating is attempted, successful transfer of sperm may not occur. Genitalic mismatch (the "lock-and-key hypothesis") is an oft-cited reason for this failure, but the role of mismatch as a barrier to gene flow is not entirely clear. One well-documented case of mechanical isolation involves snails that show a dimorphism for coiling: some individuals exhibit dextral coiling, and others, sinistral coiling. Mating within coiling types presents no problems, but in matings between types the genital openings fail to match up.

Postmating, Prezygotic Barriers

Mating (or spawning) and zygote formation are often separated in time, and therefore it is important to recognize postmating but prezygotic barriers. These include sperm or pollen competition, cryptic female choice, and gametic incompatibility. Sperm and pollen competition demonstrate that male-male competition can continue after mating: when females are multiply inseminated, production of offspring sired only by conspecific males may occur even when sperm are successfully transferred from both conspecific and heterospecific males, and when heterospecific pollen or sperm are known to be able to combine with eggs to form viable zygotes. This phenomenon, documented in insects, vertebrates, and plants, is known as *conspecific sperm and pollen precedence.* The outcome of competition may be mediated by the female, in which case it is referred to as *cryptic female choice.*

Postmating, prezygotic barriers can also result from reduced gamete compatibility. Sperm-egg (pollen-ovule) interactions are often mediated by specific proteins, and changes in either of the gamete recognition proteins may affect the rate and/or ultimate success of the interactions. In abalone, the sperm protein lysin interacts with an egg receptor, VERL. Both these proteins are rapidly evolving and subject to directional selection, and differences in lysin-VERL interactions in closely related species may be responsible for barriers to gene exchange.

Postzygotic Barriers

The reduced viability and fertility of "hybrids" (often meaning the F_1 offspring of a cross between two distinct parental types, but also more broadly used to refer to any individual of mixed ancestry) have been documented in a wide variety of taxa. Traditionally, postzygotic isolation has referred to developmental defects in hybrids that lead to full or partial inviability and/or infertility. The origin of such barriers presents a challenge (noted by Darwin), because if isolation is due to heterozygote disadvantage at a single locus, then the origin of such a barrier requires passing through a less fit intermediate state (and therefore would be opposed by selection). An alternative model imagines that incompatibilities arise because of fixation of mutations at two different (but interacting) loci, with each mutation arising uniquely in one allopatric population. If the two gene products do not "work well" together, then secondary contact between the divergent populations will lead to less fit hybrids (because they will carry both mutations). This model (the Dobzhansky-Muller model) is now widely accepted, and this scenario has now been supported in a number of different systems (see chapter VI.8).

There is an extensive literature on the genetics of postzygotic barriers, particularly in *Drosophila*. Part of that literature has been motivated by *Haldane's rule*, the observation that "when in the offspring of two different animal races one sex is absent, rare, or sterile, that sex is the heterozygous sex." By "heterozygous" Haldane meant heterogametic, that is, having two different sex chromosomes (individuals that are XY or ZW). Haldane's rule appears to apply in diverse animal taxa, including species in which males are the heterogametic sex (flies and mammals) and species in which females are the heterogametic sex (birds and *Lepidoptera*) (see chapter VI.8).

The postzygotic barriers discussed are "intrinsic" barriers, in the sense that they result from developmental problems, apparently independent of the environment. More recently, it has been recognized that postzygotic barriers can be "extrinsic." In these cases, the reduced fitness of hybrids results because they "fall between the niches" (are less fit than either parental type in the environments in which the species live) or because they are less successful at obtaining mates. For example, two forms of lake sticklebacks (benthic and limnetic) are distinct in morphology and feeding ecology. Hybrids are intermediate in morphology and are less successful than either of the parental types in the habitats in which the parents feed (see chapter VI.4).

4. STUDYING SPECIATION

If speciation is defined as the origin of barriers to gene exchange between diverging lineages, then evolutionary biologists interested in revealing the details of the process must confront the following questions: (1) What are the barriers that are currently operating between species, and to what extent does each of these barriers currently reduce gene flow? (2) When did each of these barriers arise in the history of the diverging lineages, and therefore which of the barriers was initially responsible for a reduction in gene flow? (3) In what geographic context did the barriers arise (and in particular, did divergence occur in the face of some gene flow)? (4) What roles have natural selection, sexual selection, and genetic drift played in driving the divergence? (5) What is the genetic architecture of reproductive isolation? Genetic architecture refers to the number, effect (major or minor), and chromosomal distribution (clustered or distributed across the genome) of the genes that determine speciation phenotypes.

The first question has been examined in a wide variety of animal and plant systems, although comprehensive analyses of barriers and their effects are still relatively rare. A focus on the order in which barriers act (during the life history of the organism) emphasizes the importance of early-acting barriers (those that prevent potential mates from meeting, e.g., ecogeographic barriers, differences in habitat or timing of reproduction) and distinguishes between the absolute strength of the barrier and its proportional contribution to total isolation. Because each barrier can reduce gene flow that persists only after earlier-acting barriers have had their effect, later-acting barriers inevitably make a smaller proportional contribution. However, it is not appropriate to extrapolate from the current contribution of individual barriers to their role in the speciation process, because it is the order in which barriers arose during the history of the lineages, rather than their proportional contribution now, that is most relevant to understanding the speciation process.

Debate about the geographic context in which speciation occurs has been a central issue in the speciation literature for many years (see chapter VI.3) and, as with so many issues in evolutionary biology, can be traced to Darwin. The central debate has been whether sympatric speciation (speciation in the face of substantial gene flow) can occur. Theoretical models support the notion that sympatric speciation is possible, and a few model systems can be cited in evidence, but the consensus remains that sympatric speciation is relatively rare. However, there is also increasing evidence that selection can drive sympatric divergence in the face of some gene flow.

The relative importance of drift and selection in causing divergence that leads to speciation has also been a central theme in the speciation literature. Genetic drift can alter allele frequencies in small populations, and founder events may therefore be an important route to speciation. In certain situations (colonization of oceanic islands, fixation of different chromosome rearrangements), drift may well determine outcomes, but a wider role for genetic drift in speciation is not well supported.

In contrast, selection can play many possible roles in the speciation process, including the following: (1) Populations diverge in allopatry as a result of different selection pressures in two regions or habitats, and barriers arise as a simple by-product of the allopatric divergence. (2) In zones of secondary contact, hybridization results in less fit progeny, and selection favors those individuals that do not or cannot hybridize. (3) Disruptive ecological selection (adaptation to two local habitats or resources) within a single population leads directly to sympatric species. (4) Sexual selection follows different trajectories in allopatric populations, leading to divergence in mate recognition and/or fertilization systems. In many examples, probably two or more of these scenarios may apply.

Finally, because the genomes of sexually reproducing organisms are mosaics of different histories, the pattern of the mosaic can reveal the recent history of selection across the genome. Furthermore, when diverging populations are in contact, genome regions that include genes that contribute to reproductive barriers tend to remain distinct, whereas gene flow (introgression) at neutral loci may erase patterns of differentiation. Recent genome or transcriptome scans have provided insights into genomic patterns of differentiation, which in turn are revealing candidate gene regions worthy of further investigation. These regions are candidates for harboring "barrier genes"—genes that contribute to reproductive isolation.

FURTHER READING

Coyne, J. A., and H. A. Orr. 2004. Speciation. Sunderland, MA: Sinauer.

de Queiroz, K. 2007. Species concepts and species delimitation. Systematic Biology 56: 879–886.

Harrison, R. G. 1998. Linking evolutionary pattern and process. In D. J. Howard and S. H. Berlocher, eds., Endless Forms. New York: Oxford University Press 19–31.

Mayr, E. 1963. Animal Species and Evolution. Cambridge, MA: Harvard University Press.

Schluter, D. 2001. Ecology and the origin of species. Trends in Ecology & Evolution 16: 372–380.

VI.2

Speciation Patterns

Timothy G. Barraclough

OUTLINE

1. Testing the nature of species
2. Speciation patterns in sexual eukaryotes
3. Speciation patterns in asexuals
4. Speciation patterns in prokaryotes
5. Speciation and global diversity patterns
6. Linking patterns with process

Speciation refers to the splitting of a single ancestral species into two or more descendant species. The way in which speciation occurs can affect many aspects of biodiversity, such as phenotypic variation among organisms, the geographic distributions of related species, and the number of species in a geographic region or clade. All these aspects of biodiversity, if studied in relation to speciation, are referred to as *speciation patterns*. These patterns provide vital clues into the causes of speciation and their variation among different organisms. By studying patterns we can understand processes occurring over many human life spans and over large geographic scales that are difficult to study experimentally. Fundamental questions such as the importance of reproductive isolation and sex in speciation can be addressed by comparing speciation patterns among organisms with different lifestyles. Documentation of speciation patterns is also a major goal for understanding the distribution of biodiversity across the planet and among major groups of organisms.

GLOSSARY

Allopatric. Describing species with nonoverlapping geographic ranges.

Clade. A group of organisms descended from a single common ancestor. The pattern of diversity within a clade is determined in part by the speciation events that generated the separate species.

Crossing Experiment. An experiment that tests species boundaries in sexual organisms by attempting to interbreed individuals believed to belong to different species.

Eukaryote. An organism that has a cell nucleus and other organelles enclosed by membranes, such as mitochondria. Eukaryotes include single-celled protists and multicellular animals, plants, and fungi.

Prokaryote. An organism that lacks a cell nucleus or other organelles surrounded by membranes. The two major groups are the true bacteria and the Archaea.

Recombination. The bringing together of genes from different individuals into a single genome. In sexual eukaryotes, it occurs by crossing-over events during gamete formation. In bacteria, it occurs through a variety of mechanisms.

Reproductive Isolation. The result of prevention of interbreeding between two populations or species of sexual organisms, either by prezygotic or postzygotic isolation.

Sexual Reproduction. The formation of an offspring by the combination of genetic material from two parents. There are two steps: the formation of gametes such as sperm and eggs that contain half the genetic material of the parent followed by fusion of two gametes from different parents (called fertilization).

Speciation. The splitting of a single ancestral species into two or more descendant species.

Sympatric. Describing species with overlapping geographic ranges.

Taxonomy. The science of classifying organisms and giving them names.

1. TESTING THE NATURE OF SPECIES

Most biologists agree that species represent a real and fundamental unit of biodiversity. In sexually reproducing organisms, species arise because of reproductive isolation caused by barriers to gene exchange. Evolutionary processes of gene flow, selection, and drift maintain coherence within species but cause divergence between isolated species over time. Repeated speciation events should therefore lead to a discrete pattern of phenotypic and genetic variation within clades and communities.

Perhaps the fundamental question about speciation patterns is, Does diversity really fall into discrete species units, rather than fitting some alternative model? For example, there might instead be a continuum of variation among individuals without the phenotypic gaps predicted by the concept of species. This question is hard to answer using traditional methods for describing biodiversity. Taxonomists have long cataloged the diversity of life and applied names to species taxa. However, although informed by biological understanding, traditional alpha-taxonomy is often subjective and based on expert opinion rather than formal analyses of measured variation. Also, in most cases, it is assumed a priori that organisms fall into distinct species. Consequently, taxonomic species may not reflect real evolutionary entities but instead simply provide convenient labels for categorizing variation.

The solution to this problem is to collect genetic and phenotypic data and to analyze the pattern of variation against alternative hypotheses for the occurrence of diversification. A classic example was the work of G. Evelyn Hutchinson, who measured the morphology of planktonic rotifer species in an attempt to understand the importance of sexual reproduction for generating a pattern of distinct species. By taking repeated measurements for large samples of individuals from a pond, he was able to show that individuals' morphology fell into distinct clusters, indicative of species. The hypothesis that diversity fell into a continuum of variation was rejected. Similar studies have been applied in problematic groups such as asexuals and prokaryotes, as described in later sections.

In a large-scale study to test the reality of species in sexual eukaryotes, Loren Rieseberg and colleagues investigated the correspondence between phenotypic clusters (based on morphological measurements) and the results of crossing experiments across 400 plant and animal genera. Phenotypic clusters were found in more than 80 percent of genera, and 75 percent of those were confirmed to be reproductively isolated from crossing experiments. Although the possibility that an alternative pattern of diversity might better explain the observed variation in these genera was not ruled out, the results did establish the presence of phenotypically distinct groups of individuals that tend to be reproductively isolated. The study also refuted the widely held belief that plant species are less distinct than in animals, because of their assumed greater propensity for hybridization, asexuality, and other evolutionary modes that are believed to blur species boundaries: in fact, distinct phenotypic clusters were just as frequent in plant genera as in animal genera, and plant species were more likely to correspond to reproductively isolated groups than were animal species.

Surprisingly few studies have adopted a quantitative approach to delimiting species, and most species in multicellular eukaryotes are still defined by taxonomic opinion. As an illustration, there are 13,000 genera of flowering plants alone, compared with the 400 genera of plants and animals for which quantitative data were available for the Rieseberg study. In some groups, such as fungi, in which useful morphological characters are either lacking or often deceptive, it has been routine to use formal crossing experiments to identify reproductively isolated species. However, in most organisms this is impractical—imagine the effort required to delimit more than 40,000 species of weevils by pairwise crossing experiments. A recent solution is to use DNA sequence data to test species boundaries. Using a single genetic marker, such as the mitochondrial gene cytochrome oxidase I now widely used for DNA identification, called *DNA bar coding*, it is possible to identify genetic clusters of individuals and check for correspondence between morphological characters and genetic divergence. With multiple genetic markers (i.e., several nuclear genes), it is possible to estimate the level of gene flow between putative species. Finally, in the age of genomics, it is becoming possible to identify genes underlying key ecological differences between species and to test detailed scenarios of divergence of species' genomes.

2. SPECIATION PATTERNS IN SEXUAL EUKARYOTES

Theories of speciation were largely first developed with animals and plants in mind, and speciation patterns played an important role in shaping ideas about which mechanisms of speciation were most prevalent in the real world. Classic work by Ernst Mayr and others used taxonomic information of animal and plant groups to infer likely causes of speciation based on the characteristics of closely related species. Widely held views such as that most speciation involves geographic isolation, and that hybrid and polyploid speciation is common in plants, derived largely from observations: closely related species tend to occupy different geographic areas, and closely related plant species often differ in chromosome number.

Perhaps the biggest question concerning speciation in sexual animals and plants is, How do reproductive isolating mechanisms evolve? (See chapter VI.1.) Given that most reproductive and ecological traits important for the coexistence of species are polygenic, how can these traits evolve from a starting point of gene flow in the common ancestor? Also, what types of barriers to gene flow are most important in initiating speciation? *Prezygotic isolation* refers to barriers to gene flow caused by mating preferences or incompatibility of gametes prior to the formation of a zygote. *Postzygotic isolation* refers to genetic incompatibilities occurring within the zygote, such as poor growth or survival of hybrid offspring. In broad terms, the former emphasizes divergence in mating preferences and sexual characters between populations, whereas the latter emphasizes genetic incompatibility in hybrid offspring.

Important evidence for testing alternative theories has come from analyses of patterns of isolation across pairs of recently diverged species. By focusing on recently diverged species, it is hoped that patterns reflect early stages of speciation rather than subsequent evolutionary changes accumulating in the species. The classic study, which stimulated much of the comparative work on speciation patterns, was by Jerry Coyne and H. Allen Orr in the 1980s. They compiled data on levels of reproductive isolation from published crossing experiments between *Drosophila* species, and asked two questions. First, does prezygotic or postzygotic isolation evolve faster between species pairs? Second, does the rate at which isolation evolves differ in crosses between sympatric pairs (which might have experienced a history of gene flow) versus between allopatric pairs (which are assumed to have diverged without experiencing gene flow)? The finding was striking: prezygotic isolation evolves faster than postzygotic isolation (figure 1A versus figure 1B), but only between sympatric species pairs. The conclusions were far-reaching: first, natural selection is acting to increase reproductive isolation (which is also known as *reinforcement*; see chapter VI.4), as only when species were in contact did strong prezygotic isolation evolve; second, although postzygotic isolation remains a phenomenon of considerable interest for understanding genetic systems, it appears to play a minor role in solving the conundrum of how sexual species are able to coexist. In sympatric *Drosophila* at least, most species are already completely isolated by prezygotic mechanisms by the time postzygotic isolation evolves to a sufficient level to reduce interbreeding.

Since then, other studies have looked for correlates of reproductive isolation in a wide range of groups. For example, Mikael Le Gac and Tatiana Giraud found the same pattern of greater premating isolation between sympatric than allopatric species of mushroom-forming fungi as had been found in *Drosophila*. Subsequent studies have sought to identify the correlates driving the evolution of reproductive isolation. Comparison of patterns of reproductive isolation and ecological differences among sister species pairs have shown that ecological shifts play an important role in driving the evolution of reproductive isolation. For example, in closely related stickleback species diverging between different habitats, such as between freshwater and marine populations, reproductive isolation is caused by changes in breeding habitat and body size associated with adaptation to the local habitat, and by poor survival of hybrid offspring in either habitat, rather than by genetic incompatibilities.

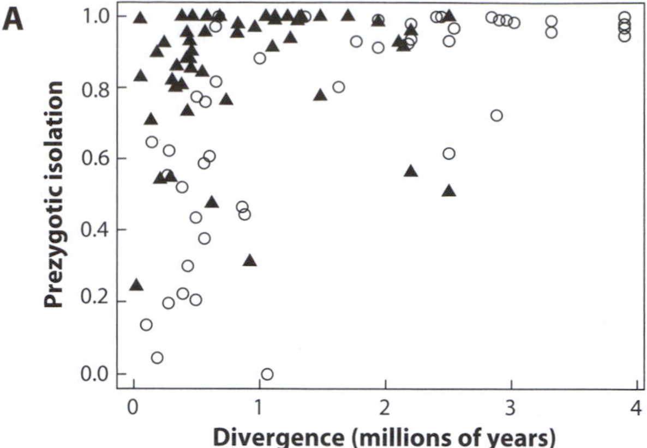

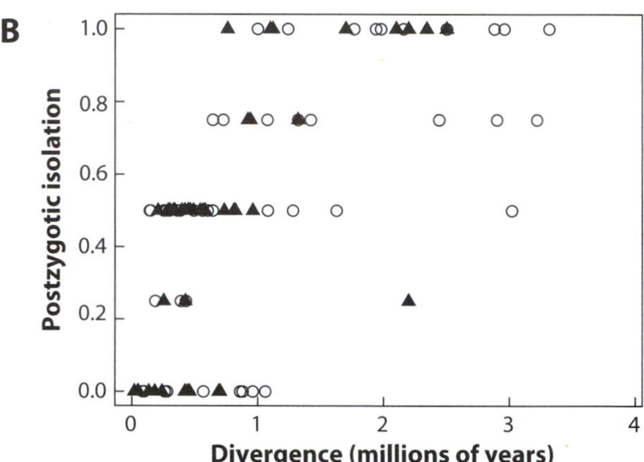

Figure 1. The pattern of (A) prezygotic and (B) postzygotic reproductive isolation between pairs of *Drosophila* species in relation to the divergence time between them. Divergence was calculated from genetic distances calibrated using a rough molecular clock. Black triangles are sympatric species pairs, and white circles are allopatric species pairs. Prezygotic isolation evolves faster than postzygotic isolation between recently diverged species when sympatric but not when allopatric. (Redrawn using data from Coyne and Orr 1997.)

3. SPECIATION PATTERNS IN ASEXUALS

Our knowledge of speciation has been strongly shaped by a few taxa: birds, *Drosophila*, fish, a few plant taxa. It has long been realized that some ideas formulated for these taxa become problematic when applied to the charming variety of eukaryotes that are less well understood. Because of the importance of reproductive isolation for understanding speciation in sexual eukaryotes, perhaps the biggest challenge is posed by speciation in

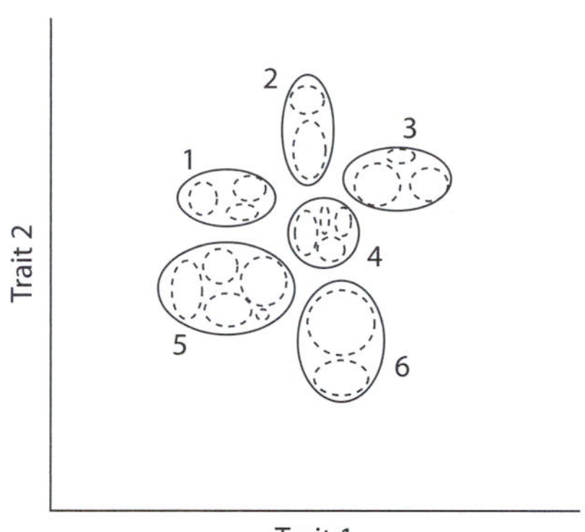

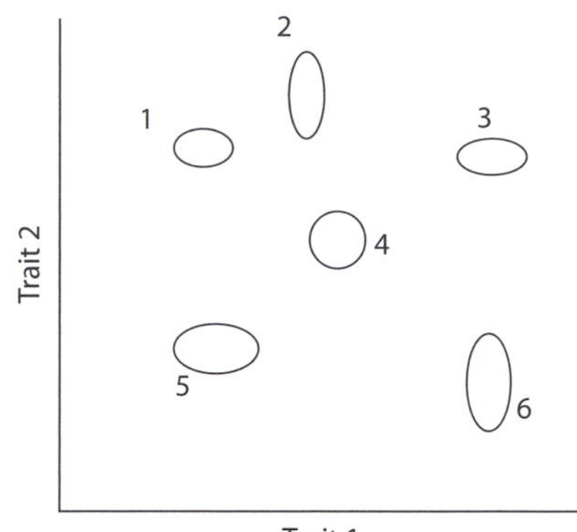

Figure 2. Hypothetical example of variation within an asexual and a sexual clade, both of which have diversified into six independently evolving species. The axes represent two independent measures of trait variation: they could represent morphological traits such as body length and wing length, or a summary measure of genetic variation among individuals. In both cases, the six independently evolving groups that form distinct clusters are indicated by solid ellipses. Processes of selection and drift maintain low variation among individuals within a cluster (a feature known as *coherence* or *cohesion*), but independent evolution allows divergence between clusters. The asexual clade also displays discrete variation within clusters, indicated by the dashed ellipses: each dashed ellipse represents a group of individuals sharing a relatively recent maternal common ancestor. In contrast, variation within sexual clusters should be homogeneous, as long as there is not further population subdivision within clusters, because the traits are determined by many genes, each with a different ancestry in the population. In the example shown, clusters are more divergent and distinct in the sexual clade, which might be expected if faster rates of adaptation in sexuals allow their species to adapt more efficiently to different niches. (Redrawn using data from Barraclough, Birky, and Burt 2003.)

asexuals. A discreet but abundant set of microscopic eukaryotes reproduce, as far as we currently know, purely by asexual reproduction, in which offspring derive all their genetic material from a single parent individual. Clearly, the biological species concept based on reproductive isolation does not usefully apply, as none of their individuals interbreed: reproductively isolated units coincide, redundantly, with the individual level. This has led some authors to argue that species are a property of sexual organisms: asexuals do not "do" species. While this conclusion is true if we subscribe to a strict biological species concept, it is at odds with population genetic theories of speciation. If interbreeding is the very process that makes speciation difficult in sexuals, because it tends to erode genetic differences arising between populations (see chapter VI.3), then removing the ability to interbreed should make divergence easier.

Population genetic theories show that the same conditions promoting divergence and the evolution of reproductive isolation in sexuals, namely, geographic isolation or adaptation to distinct ecological niches, should also cause divergence in asexuals. Distinct genetic and phenotypic clusters are expected to emerge that evolve independently from one another and are maintained coherent by ongoing drift and selection within each cluster (figure 2). The pattern of variation *within* clusters should differ between sexuals and asexuals. Sexuals inherit genes from many different ancestors, and variation in traits such as body size should display normal distributions owing to averaging of multiple genes. Asexuals inherit all their genes from a single individual (i.e., their mother) in each preceding generation, and traits such as body size should exhibit a hierarchical pattern of variation reflecting maternal ancestry in the population. The pattern of variation *between* clusters, however, should be qualitatively similar in sexual and asexual clades if they encounter the same conditions favoring divergence. Whether sexuals or asexuals show a stronger pattern of genotype clustering will depend on the balance of two contrasting effects: asexuals might show stronger patterns of genotypic clustering because they do not have the limiting factor of interbreeding that limits speciation in sexuals, but they might show weaker patterns if they adapt more slowly to new niches than do sexuals (see chapter IV.4).

The main reason that this question has been hard to answer is the difficulty of interpreting observed patterns. Traditional taxonomic approaches for delimiting species cannot help, because species tend to be defined differently in sexual and asexual taxa and so are not comparable. Even with quantitative measurements of phenotypic or genetic clustering, comparisons are limited by complications in how asexuals originate. Most asexual organisms tend to be recently derived from sexual ancestors, which means that variation in part reflects variation in their sexual ancestor and also that there has likely been insufficient time for conditions favoring speciation to have arisen and to have produced their effects on the population. A robust comparison of speciation in sexuals and asexuals requires lineages of asexuals that have survived over long enough timescales (many millions of years) without sexual reproduction.

Diego Fontaneto and colleagues evaluated the pattern of divergence in a celebrated clade of long-lived asexuals, the bdelloid rotifers. Bdelloids are microscopic animals living in permanent or temporary freshwater habitats. By comparing patterns of genetic and morphological variation with null models of variation in a single asexual population, these researchers showed that bdelloids have diversified into distinct and independently evolving clusters broadly equivalent to sexual species. The extent of morphological variation relative to genetic variation revealed that divergent selection on feeding morphology was one cause of their diversification. Whether bdelloids display a stronger or weaker pattern of clustering than equivalent sexual organisms, such as the facultatively sexual monogonont rotifers, remains to be answered. Also, although there has been no evidence for sexual reproduction, Gladyshev and colleagues showed that bdelloids can take up bacterial, plant, fungal, and protist genes from their environment. In principle, bdelloids could use this mechanism to acquire genes that help them to adapt to new environments and in part compensate for the reduced adaptive abilities expected of strict asexuals. Whether this unusual (for eukaryotes) means of acquiring genetic variation has affected their diversification is an interesting question for future work.

Many other groups of eukaryotes display features departing from those of the classical sexual eukaryote "models" for speciation. As well as asexuality, many microscopic eukaryotes have very high dispersal potential, which could affect the relative importance of geographic isolation in driving speciation. Genetic mechanisms might also differ between animals and plants in ways affecting the nature or magnitude of diversification. For example, many mushroom-forming fungi have complex life cycles in which true mating is preceded by a period of "cohabitation" of nuclei belonging to each mating partner within the same cells (these cohabiting fungi are referred to as *heterokaryons*). Cohabitation sometimes ends with each partner going its separate way rather than in the formation of sexual spores. In this way, genetic compatibility of the mating partners can be "tried out" before taking the irreversible step of combining their genes in the same nuclei. For example, poor growth in the heterokaryon resulting from the combination of genes from the partners might be used as a signal to end the relationship. DNA sequencing technology has opened up the opportunity to characterize diversity in the many eukaryotes not amenable to classical biological investigation (i.e., difficult to culture and with few morphological characters). Our view of how speciation occurs is likely to broaden substantially as these organisms are studied in increasing detail.

4. SPECIATION PATTERNS IN PROKARYOTES

Prokaryotes present a big challenge to theories of species and speciation developed for eukaryotes. The dominant concept of species as reproductively isolated groups assumes that individuals engage in sexual reproduction, but prokaryotes do not. Many prokaryotes are predominantly clonal and reproduce by cell division, but they also exchange genes by a variety of mechanisms. So-called homologous recombination occurs when naked DNA is taken up from the environment and incorporated into the genome. Such recombination mostly involves DNA from closely related cells: the probability of homologous recombination declines exponentially with increasing sequence divergence between the recipient and the DNA being incorporated. Alternatively, genes can also be transferred by independently replicating viruses or DNA molecules called *plasmids*. Plasmids often contain a set of functionally related genes, such as those conferring antibiotic resistance, and sometimes cross between distantly related taxa. The frequency of these different mechanisms varies widely among taxa and in all cases departs from the dominant mechanism of recombination in sexual eukaryotes. Speciation patterns therefore might differ between prokaryotes and eukaryotes and among prokaryote taxa.

The complexity of mechanisms of recombination in prokaryotes, and their failure to conform to the assumptions of the biological species concept, has led many biologists to an agnostic approach to prokaryote species: they use the term "species" for convenience but often follow statements on bacterial species with "whatever they are" or claim that bacteria do not have species. DNA sequence data provide the framework for prokaryote taxonomy, but species are usually classified using simple thresholds in pairwise genetic divergence (originally calibrated against "known" species units from traditional culture-based methods), without reference to an evolutionary definition for what those species units represent. For example, current efforts to catalog bacterial diversity in seawater, soil, and other habitats typically amplify and sequence the 16S ribosomal RNA gene and calculate species richness by counting the numbers of groups differing by more than 1 percent in their 16S sequences.

A more satisfactory view of prokaryote species comes from applying population genetic theories to predict patterns of diversity. Irrespective of how genetic inheritance and transfer occurs, the same forces that cause speciation in sexual eukaryotes can also cause diversification in bacterial populations. Frederick Cohan and colleagues pioneered evolutionary definitions of bacterial species by showing how specialization to different ecological niches causes the emergence of independently evolving genetic clusters of individuals. Mutations that are useful for survival and reproduction in one niche spread within the genetic cluster adapted to that niche but not into genetic clusters adapted to different niches. Simulation models of these processes have been used to demonstrate multiple distinct genetic clusters, which they call *ecotypes*, in *Bacillus simplex* and *B. subtilis–B. licheniformis* isolated from dry canyons in Israel. These models apply especially to bacteria with rare recombination, because the spread of new beneficial genes during adaptation to the distinct niches creates a pattern of genetic clustering at neutral DNA markers only if recombination is rare, unless a barrier to recombination evolves in association with ecological divergence.

Christophe Fraser, William Hanage, and Brian Spratt showed how speciation occurs in bacteria with high rates of recombination. When the recombination rate is lower than the mutation rate per gene, bacterial populations behave equivalently to populations without recombination, and their expected speciation patterns are similar to those described for asexual eukaryotes, and by Cohan and colleagues for bacteria with rare recombination. When recombination equals or exceeds the mutation rate, bacterial populations behave effectively as "sexual" organisms with the result that recombination prevents individuals within a population from becoming too genetically divergent from one another. Distinct species can evolve because of barriers to gene exchange caused by the decline in homologous recombination with increasing genetic distance between them, similar to the process of reproductive isolation in sexual eukaryotes. For example, in the *mitis* group of *Streptococcus*, there are four distinct genotype clusters based on variation in six genes involved in basic metabolic functions within the cell. These clusters conform to four traditional taxonomic species named on the basis of phenotypic traits.

Prokaryote diversity can therefore be understood in the same framework as developed for eukaryotes. A survey by Michiel Vos and Xavier Didelot found that 27 of 48 bacterial species had recombination rates above the threshold identified by Fraser and colleagues, and therefore are expected to evolve more like "sexuals" than "asexuals"; 13 fell in a gray area between clear-cut "sexual" or "asexual" evolution, and 8 had effectively "asexual" evolution. There remain important differences, however, from eukaryotes even in bacteria with high recombination rates: in the *Streptococcus* example described, a seventh gene linked to a locus involved in penicillin resistance is exchanged between these otherwise-distinct species. The question therefore remains whether prokaryote diversity is best described by simple units that could be called "species" or whether this model of diversity should be replaced with something more complex.

Our current knowledge of evolutionary processes in bacteria derives from species that can be cultured in the lab or that cause disease. Most prokaryotes cannot be cultured, and their diversity is only just being uncovered through large-scale DNA sequencing. Surveys of marker genes such as the 16S ribosomal RNA gene show that genetic variation falls into discrete clusters, as it does in surveys of single gene markers in both sexual and asexual eukaryotes. At least at neutral loci, it seems that most bacteria display the same pattern of discrete, independently evolving sets of genotypes that eukaryotes do. Metagenome sequencing—in which all DNA in a sample of soil or seawater is sequenced, and not just marker genes—is beginning to uncover ecological variation in bacterial communities, by identifying sets of genes required by the bacterial community at large to function in particular environments. However, piecing together the whole genomes needed for evolutionary and population genetic analyses from the "soup" of genetic information resulting from metagenome sequencing remains difficult.

5. SPECIATION AND GLOBAL DIVERSITY PATTERNS

The number of species in a group of organisms depends on the balance between the origin of new species by speciation and the loss of species by extinction, on the age of the group, and on the availability of space and resources to sustain species within the region occupied by the group (as space or resources diminish, the speciation rate will either decrease, or the extinction rate will increase). Although it is not the only process affecting diversity, speciation is likely to play an important role in shaping many different aspects of diversity (see chapter VI.16). For example, whether closely related species tend to share similar ecological characteristics will depend on the role of ecological divergence in speciation. If speciation tends to involve ecological shifts, then close relatives will tend to differ in ecological traits, whereas if it tends to occur through geographic isolation in identical habitats, then close relatives will tend to have similar ecological traits. These alternatives will also affect whether local communities are made up of distantly related species or of closely related species.

Perhaps the biggest question is whether speciation patterns can explain patterns of species richness among different organisms and in different geographic regions. Key parameters important in speciation models, such as dispersal rate and the strength of mating preferences, have been shown to correlate with species richness among clades. It is difficult, however, to separate effects of speciation from those of extinction, because current diversity is the net product of both processes. By including information on species relationships or species age distributions, it can be possible to tease them apart using statistical models. For example, Jason Weir and Dolph Schluter showed that recent speciation rates in northern latitudes exceeded those in more tropical latitudes, perhaps because of exposure of new ecological opportunities following the retreat of glaciers. Extinction can be detected directly from the fossil record, but few taxa display sufficient taxonomic and temporal resolution to document speciation, extinction, and morphological evolution along lineages. Analyses of the fossil record of single-celled marine eukaryotes called planktonic foraminifera by Thomas Ezard and colleagues showed that speciation rates depend more on the number of species present at the time (and decline as the number of species increases) than by changing physical environments, whereas extinction was driven largely by climate. Growing evidence from these and other studies indicates that speciation rates are strongly influenced by feedback from geographic and ecological limits on available resources within a region. As the number of species within a region and using a particular set of ecological resources increases, then the opportunity for new species to originate and persist declines.

The main frontier for understanding global diversity patterns is in hyperdiverse or previously poorly studied organisms such as tropical insects, tiny marine animals, and especially prokaryotes. To date, our knowledge of global diversity is heavily skewed toward macroscopic eukaryotes, especially vertebrates. Efforts to quantify how many species there are on earth are strongly dependent on being able to define comparable species units among disparate taxa and then to identify those units operationally. In the past, much attention has focused on tropical insects because of their known high diversity from museum collections, but generating a species list for the planet has seemed a distant goal based on the rate at which new species can be described. Extrapolations to estimate the likely total number of species from smaller samples have used assumptions about the turnover of diversity between different regions, about the level of specialization to tree species and the number of tree species in different regions, and the rates of description of new species over time. High-throughput DNA sequencing techniques could speed up the rate of discovery of new species, by reducing the need for the slow process of matching specimens morphologically to museum collections. Even with these methods, the turnover of hyperdiverse groups between regions means there will always be gaps in our sampling. However, by quantifying turnover and habitat specificity at different spatial scales for key groups, it will be possible to generate robust models for extrapolation to estimate global totals.

Prokaryotes present the biggest challenge. Even with the latest high-throughput sequencing techniques and using single marker genes, only a tiny fraction of the true diversity in the environment can be quantified in a given sample. Estimates of true diversity rely on estimating the total abundance of bacteria and the abundance of common species, and then extrapolating to estimate the total number of species likely to be present. This extrapolation requires an assumption of how species tend to vary in their abundances. One approach is to use *neutral theory*, which is an ecological theory that predicts the abundance of different species from mathematical equations based on the simplifying assumption that species do not differ in their ecological characteristics. Graham Bell used this approach to estimate that the estimated 10^{29} soil bacterial cells worldwide belong to perhaps 10^{14} species. Documenting and understanding diversity patterns of this many bacteria in the detail achieved for, say, vertebrates will keep researchers busy for decades, but the work has begun. For example, several studies have shown that pH is a major factor determining species richness and community similarity in soil bacteria.

Strangely, as noted by Graham Bell, the estimated diversity of single-celled eukaryotes is low (only around 50,000 species of protists have been described) relative to the astronomical numbers for bacteria and even compared with multicellular eukaryotes. DNA sequence data are revealing considerable cryptic diversity within protist species defined based on traditional taxonomy, which might increase their estimated diversity by one or two orders of magnitude, but this still leaves them depauperate compared with expectations from their global abundance. The reason for lower diversity in single-celled compared with larger eukaryotes could be that their higher dispersal rates (by wind, rain, or fowl) reduce speciation rates by limiting the opportunities for geographic isolation that arise abundantly in larger organisms. Another reason for lower diversity than bacteria could be the constraining influence of sex described earlier: if the average recombination rate were much higher for protists than for bacteria, and if the main effect of recombination were to limit genetic divergence, then this might explain the lower diversity of protists. Final resolution of this paradox awaits species units to be defined using comparable methods and estimation of rates of recombination in both bacteria and protists, which are both answerable with multilocus genetic techniques.

6. LINKING PATTERNS WITH PROCESS

Together with mathematical models and laboratory crossing experiments, studies of speciation patterns in nature have been at the heart of developing ideas about speciation (see chapter VI.3). Future work will entail establishing even stronger links between patterns and process. Population genetic theories of speciation are well developed, and the mechanisms of genetic divergence are increasingly well known from case studies of pairs of recently diverged species. It has been hard, however, to collect detailed genetic data for broader scales necessary to document speciation patterns across clades and regions. Traditional work focused on easily measurable characteristics such as geographic range and secondary sexual trait differences, and relied on taxonomic lists of species. The next generation of studies will use high-throughput genetic techniques to perform systematic evaluation of population genetic process such as gene flow and selection across broad samples of species. This will allow the nature of species units to be compared across organisms and the causes behind their formation to be revealed. One sticky issue will be where to draw the line for species versus higher levels in the hierarchy of nature, such as genera, or lower levels, such as populations. All biologists know that diversity is often more complex than can be encapsulated in simple units—species are not like chemical elements—and as quantitative approaches are increasingly used, we will be able to investigate the processes promoting diversification without needing to make simplifying assumptions a priori. At the broader scale of global diversity patterns, dynamic models are needed to connect the population genetic processes occurring in populations with the broader environmental context of how conditions favoring speciation and coexistence come and go (see chapter VI.16). This remains a big challenge, as it depends on complex interactions between organisms and their environment over elusive medium timescales: too long for direct observation but too short for most fossil records (with the notable exception of planktonic foraminifera).

FURTHER READING

Acinas, S. G., V. Klepac-Ceraj, D. E. Hunt, C. Pharino, I. Ceraj, D. L. Distel, and M. F. Polz. 2004. Fine-scale phylogenetic architecture of a complex bacterial community. Nature 430: 551–554. *Study that used genetic data to demonstrate the existence of clusters indicative of species in prokaryotes.*

Barraclough, T. G., C. W. Birky Jr., and A. Burt. 2003. Diversification in sexual and asexual organisms. Evolution 57: 2166–2172. *Verbal theory comparing expected patterns of speciation in sexuals and asexuals.*

Bell, G. 2009. The poverty of the protists. In R. Butlin, J. Bridle, and D. Schluter, eds., Speciation and Patterns of Diversity. Cambridge: Cambridge University Press. *A thought-provoking account of differences in species numbers among the major forms of life.*

Coyne, J. A., and H. A. Orr. 1997. Patterns of speciation in *Drosophila* revisited. Evolution 51: 295–303. *Revisit of the classic paper on speciation patterns with some new data added.*

Coyne, J. A., and H. A. Orr. 2006. Speciation. Sunderland, MA: Sinauer. *The best single source of information and insight on all forms of speciation research.*

Fontaneto, D., E. A. Herniou, C. Boschetti, M. Caprioli, G. Melone, C. Ricci, and T. G. Barraclough. 2007. Independently evolving species in asexual bdelloid rotifers. PLoS Biology 5: e87. *A study using quantitative methods to test whether asexuals diversify into species.*

Fraser C., W. P. Hanage, and B. G. Spratt. 2007. Recombination and the nature of bacterial speciation. Science 315: 476–480. *Theory demonstrating how recombination shapes diversity patterns in bacteria.*

McKinnon, J. S., and H. D. Rundle. 2002. Speciation in nature: The threespine stickleback model systems. Trends in Ecology & Evolution 17: 480–488. *A good overview of evidence for speciation patterns reviewed in a well-studied group of fish.*

Rieseberg, L. H., Wood, T. E., and E. J. Baack. 2006. The nature of plant species. Nature 440: 524–527. *A critical investigation of the reality of species—more work like this is greatly needed to answer questions raised in this chapter.*

Weir, J. T., and D. Schluter. 2007. The latitudinal gradient in recent speciation and extinction rates of birds and mammals. Science 315: 1574–1576. *An example of how speciation patterns affect diversity patterns, and how models can be used to separate the effects of speciation and extinction.*

VI.3

Geography, Range Evolution, and Speciation

Albert Phillimore

OUTLINE

1. Geographic patterns of species and speciation
2. The geography of speciation
3. Island patterns and their implications
4. Speciation and area
5. Geographic and geological triggers of speciation
6. Challenges and prospects

The distribution of biodiversity is very uneven across the earth, a testament to geographic variation in the net contribution made by speciation, shifting geographic ranges, and extinction. One of the most obvious diversity patterns on the planet is the tendency for species richness to be greatest close to the equator and to decline toward the poles. However, identifying the mechanisms whereby speciation, geographic range dynamics, and extinction contribute to this global latitudinal diversity gradient continues to confound ecologists and evolutionary biologists alike. Geography features prominently in speciation research—so much so, that until recently the geographic distributions of species undergoing speciation formed the basis for classifying speciation into three main modes: allopatric, parapatric, and sympatric. Although population-genetic-based classifications of speciation are now preferred, the geographic distributions of populations reveals something about the potential for gene flow, which in turn affects the ease with which reproductive isolation and speciation can arise. Unfortunately, because species ranges can expand, contract, and move substantially through time, little can usually be inferred about the geography of speciation from species' present distributions. Island-dwelling species may be the exception; their isolation serves to limit postspeciation range movements. This has made studying the geography of speciation on islands especially informative. Islands offer a further attraction to evolutionary biologists, in that teasing apart the mechanisms whereby immigration, extinction, and speciation contribute to their diversity patterns has proven more tractable than for continental patterns.

GLOSSARY

Allopatric Speciation. Geographically separated populations among which reproductive isolation accumulates in the absence of gene flow.

Parapatric Speciation. Speciation of populations in the face of gene flow at a level that is greater than zero (allopatric speciation) and less than complete mixing (sympatric speciation). Populations may have abutting or geographically separated distributions.

Peripatric Speciation. A mode of allopatric speciation arising when some individuals disperse outside the current range and beyond a preexisting barrier to establish a new population in a remote location. After establishment, the rarity of long-distance dispersal means that little or no gene flow occurs.

Species Richness. The total number of species in a clade or inhabiting a specified area.

Sympatric Speciation. Under a population genetic definition, a mode of speciation that results when two reproductively isolated species arise from within a randomly mating ancestral population.

Sister Species. Two species that are each other's closest extant relative, typically identified using a molecular phylogeny.

Vicariant Speciation. A mode of allopatric speciation that involves a physical barrier, such as an ocean channel or mountain range, that subdivides a range and prevents gene flow between the two resulting populations.

1. GEOGRAPHIC PATTERNS OF SPECIES AND SPECIATION

The number of species found in different parts of the planet varies enormously. To visualize this variation, imagine dividing the surface of the earth into equal-sized grid cells, say 100 km by 100 km, by drawing invisible lines across its surface and then counting the species found within each cell. For breeding birds, species richness would vary from as few as zero to one species in Greenland and Antarctica to about 950 species in the Andean foothills of Ecuador. Similarly, mammalian richness would range from zero to 250 species, with the maximum found in the Great Rift Valley of the Democratic Republic of Congo. Estimates of vascular plant diversity are subject to greater uncertainty, but geographic disparity in their richness is even more pronounced, exceeding 6000 species in a single grid cell in the rain forests of Central America.

While some spatial variation in diversity is idiosyncratic to each group of organisms, there are pronounced general trends. For example, terrestrial species richness tends to be high in the foothills of mountain ranges and low on islands, whereas marine species richness tends to be highest in continental shelf and coastal regions. But the global species-richness pattern that receives the most attention is the tendency for species richness to decline from the equator (low latitude) to the poles (high latitudes), a pattern known as the *latitudinal diversity gradient*. If we reconsider the grid cells where birds, mammals, and plants are most diverse, each of these lies near the equator. In fact, the latitudinal diversity gradient holds across most cosmopolitan groups of terrestrial and marine organisms and is observable in the fossil record.

Identification of the processes that give rise to diversity gradients in general, and the latitudinal gradient in particular, has been the goal of innumerable studies. Yet, perhaps surprisingly, a definitive explanation is still lacking. We know that a location can gain species only via immigration and speciation (see chapter VI.2) and lose species via local extinction; therefore, all richness patterns must arise via differential contributions of these three processes over time. More than 100 hypotheses have been put forward to explain the latitudinal gradient, but most of these can be assigned to one of three classes. These classes of explanation identify the major difference between tropical and extratropical regions as one of (1) ecological limit, or, in other words, the overall carrying capacity; (2) diversification rate; and (3) the amount of time that has been available for diversity to accumulate.

We will consider how these explanations might apply to the latitudinal diversity gradient, and the evidence for each. However, we will later see that variations on these explanations may be applied to other instances where diversity varies spatially—the depauperate nature of islands, for example.

Ecological explanations for the latitudinal gradient often boil down to the following chain of reasoning: because the tropics receive the most energy, plants there are more productive, supporting more species in more niches at successively higher trophic levels. Proponents of this hypothesis point to the positive relationship between energy (often measured as potential evapotranspiration) and diversity that is often apparent on a continental to global scale. However, we do not yet know whether the standing diversity at different latitudes has already reached or is even close to its local carrying capacity. In certain locations, such as some Northern Hemisphere temperate forests, there is good reason to believe that diversity has not reached its potential. The fossil record suggests that prior to the last glacial maxima, the European temperate and East Asian forests were comparable in diversity, but now the former possess about a tenth of the species diversity of the latter. It is also possible that carrying capacities are to some extent elastic and that as diversity increases, key innovations provide the opportunity for the accumulation of greater diversity.

Until recently, ecological explanations for the latitudinal diversity gradient received much more attention than evolutionary ones, though this situation is changing. The *diversification rate hypothesis* states that the net contribution of speciation minus extinction decreases toward the poles. A variety of mechanisms leading to faster speciation in the tropics have been proposed (the "cradle" hypothesis). For example, at high latitudes, selection may be driven by abiotic conditions that vary more seasonally than spatially. In comparison, the seasons in tropical areas are less pronounced, a consequence of which may be that geographic variations in biotic conditions come to the fore as drivers of geographic selection and speciation. Perhaps, higher temperatures in the tropics accelerate the mutation rates of ectotherms and plants. Alternatively, extinction may have been slower in the tropics (the "museum" hypothesis) because either tropical areas are more climatically stable or population sizes are larger. The empirical evidence for increased diversification rates in the tropics is currently quite mixed. Phylogenetic and taxonomic studies on representatives of birds, butterflies, marine bivalves, and flowering plants all found that diversification was fastest among low-latitude clades. However, support for a very different model comes from a comparison of the ages of sister species of birds and mammals at different latitudes conducted by Jason Weir and Dolph Schluter. They found that while both speciation and extinction rates were faster at high latitudes, the diversification rate (the speciation rate minus the extinction rate) was consistent across latitudes. It now seems likely that diversification rates themselves may change through time, perhaps owing to a negative feedback of clade-wide ecological limits on the rate of speciation, implying that the ecological limit and diversification rate hypotheses are both at play and interact.

The third major type of explanation, of which Alfred Russel Wallace was a proponent, posits that the time available for diversity to accumulate is key. A mere 20,000 years ago, ice sheets covered North America as far south as New York, as well as much of northern Europe. Pollen recovered from sediment cores reveals the huge effects of the climate and ice sheets on the geographic ranges of temperate and boreal tree species. The distributions of North American oak and spruce trees are now about 1000 km farther north than they were at the last glacial maximum, and the fossil record for North American mammals tells a similar story of postglacial range expansions and shifts. Therefore, we can surmise that postglacial dispersal and range shifts must have played a major role in forming the communities that now inhabit high-latitude terrestrial areas.

During glacial periods, conditions also changed at lower latitudes: many biomes contracted in total area, and arid areas expanded. Species-rich wet tropical biomes appear to have been relatively climatically stable and, at various times in the last 50 million years, have extended much farther north. It is suggested therefore that a combination of greater area and more time have provided the wet tropical biomes with a greater opportunity for diversification. Support for this explanation comes from phylogenetic studies revealing that clades inhabiting higher latitudes are often nested within (i.e., younger than) those inhabiting lower latitudes.

Each of the processes described by these three classes likely contributes to the latitudinal diversity gradient, though they may differ in their importance. Unfortunately, as we have only one earth and no truly independent replication, even if we accumulate additional data and develop more sophisticated statistical models, a full explanation for this fascinating pattern may continue to elude us.

2. THE GEOGRAPHY OF SPECIATION

While speciation is one of the key contributors to geographic diversity patterns, the process of speciation is itself influenced by geography. In fact, the geography of speciation has been the subject of debate for more than a century. Arguments have revolved around both the feasibility and importance of three main geographic modes of speciation: allopatric, parapatric, and sympatric. Often in the past, these modes have been defined on the basis of geography alone, but here we adopt definitions based on gene flow and used by population geneticists (see the glossary).

Gene flow is useful in defining modes of speciation because, for sexual organisms, it plays a key role in determining the ease with which speciation occurs. In the absence of gene flow, any process that leads to genetic divergence, even genetic drift, will eventually result in allopatric speciation. By contrast, the presence of gene flow impedes divergence by homogenizing genetic variation across populations. The splitting of lineages to form two separate species often requires divergence in two or more traits (e.g., one trait that is subject to divergence under selection and another trait used to choose a mate). However, when there is gene flow between two diverging populations, genetic recombination (the process whereby DNA is exchanged between the two strands of a chromosome during meiosis) can cause the genes that underlie these traits to become disassociated (see chapter VI.9).

Theory has shown that speciation in the face of gene flow is easiest when disruptive (or divergent) selection and assortative mating are both present. Selection is described as *disruptive* when the fitness of individuals with extreme phenotypes is greater than that of individuals with intermediate phenotypes, and *divergent* if different phenotypes are favored in different locations. Assortative mating is the nonrandom mating of individuals; that is, members preferentially mate with individuals of similar phenotype. It is relatively easy to envisage how both conditions may arise in a parapatric geographic scenario. For example, consider a species that inhabits two adjacent habitats. There is likely to be divergent selection for adaptation to each habitat, and spatial separation of populations will have the effect of making individuals occupying each habitat more likely to mate with individuals from the same rather than the different habitat. In comparison, sympatric speciation is easiest where a single trait is subject to disruptive selection and assortative mating (or where the genes underlying the traits are situated in proximity on a chromosome), thereby preventing recombination or making it less frequent. The beak of the *Geospiza* genus of Darwin's finches could be just such a trait: the size and shape of the finch's beak is subject to selection for handling seeds that vary in size and hardness, and the dimensions of the bill also affect the song of male birds, on which basis females choose mates. It is therefore possible that reproductive isolation may emerge as a by-product of natural selection as populations adapt to different resources. However, there is no compelling evidence to suggest that any completed speciation in Darwin's finches has actually been sympatric from start to finish.

Allopatric speciation has been classified into two geographic subtypes, vicariant and peripatric. *Vicariant speciation* is inferred if sister species have nonoverlapping geographic ranges, and the age of the split between them, as revealed by molecular phylogenetics, is consistent with the emergence of a known barrier. For instance, when the Panamanian land bridge joined North and South America about 3 million years ago, gene flow between many coastal marine animal populations inhabiting the Caribbean and Pacific sides ceased. A review of 115 comparisons of nucleotide divergence of populations of echinoderms, crustaceans, mollusks, and fish sampled from both the Caribbean and Pacific coasts found that 30 percent had divergence times consistent with the final closure of the isthmus, but the majority of the remainder (63.5 percent of cases) diverged substantially earlier.

The second type of allopatric speciation, *peripatric speciation*, arises when individuals colonize a new locality via long-distance dispersal and become reproductively isolated from the source population. Drawing on examples of endemic species occupying remote islands, biologists have long been aware that peripatric speciation sometimes occurs, though its frequency of occurrence on continents and in the oceans remains to be established. Ernst Mayr, famous for introducing Modern Synthesis ideas on evolution to the study of systematics and biogeography, was interested in the processes responsible for the peripatric speciation of island populations. In a work published in 1954, Mayr observed that there were islands off the coast of New Guinea that differed little in environment from one another or from New Guinea, and yet many of these islands had their own endemic kingfisher. He argued that the similarity of the islands' environments meant that divergent natural selection could not account for this speciation. Therefore, to explain such endemism he proposed the *founder effect* speciation model. According to this model, a small number of individuals, carrying a fraction of the parent population's genetic diversity, colonize a novel area. Thereafter, genetic drift in a massively reduced population causes many alleles to be rapidly fixed or lost. Mayr argued that this process could dramatically alter the genetic background on which selection operates, thereby allowing rapid speciation even in the absence of divergent selection. For some time the founder effect speciation model and variants thereof were considered by many to be an important mode of speciation, especially on islands. However, they have received a great deal of criticism on theoretical grounds. One of the key objections is that even the superficially similar environments on the islands that Mayr discussed actually differ sufficiently in their abiotic and biotic environments for divergent natural selection to operate.

Given that allopatric speciation and sympatric speciation are two extremes of a continuum and that parapatric speciation covers everything in between, it appears likely that parapatric speciation is common. However, demonstrating that two species arose in parapatry has proven especially difficult. We often observe sister species with adjacent and abutting geographic ranges, sometimes with each species adapted to different habitats, exactly as expected under the parapatric model. But on the basis of this evidence alone, we cannot reject the possibility that such distribution patterns arose through allopatric speciation followed by range expansion and finally competition or hybridization along the abutting ranges (i.e., *allo-parapatric speciation*). One of the best examples of parapatric speciation in action comes from a grass species (*Anthoxanthum oderatum*) that has evolved tolerance to high levels of heavy metals in soils brought to the surface by miners. Accompanying the adaptation, some (though not complete) reproductive isolation through divergent flowering times has arisen between the population tolerant to heavy metals and nearby populations on normal soil.

Because theory tells us that the conditions under which allopatric speciation can proceed are much more permissive than those required for sympatric speciation, the most persuasive empirical evidence for sympatric speciation usually comes from sister species for which an allopatric phase can be ruled out or was unlikely. There now exist a handful of cases that meet this criterion. A famous example concerns two *Amphilophus* cichlid sister species confined to a single volcanic crater lake in Nicaragua. While the two species are genetically quite similar, implying that they diverged very recently (sometime in the last 10,000 years), they have divergent morphology, occupy different niches in the water column, and are reproductively isolated. Whether the ancestral population mated randomly, as required under our definition of sympatric speciation, we will perhaps never know. In the crater-lake cichlids, reproductive isolation seems likely to have accumulated as a by-product of disruptive natural selection and in the face of ongoing gene flow, but few cases of putative sympatric speciation fit this description. More often, it appears that there is a rapid shift to nonrandom mating from within an initially panmictic population, followed by a rapid decline in gene flow. For instance, in speciation via host shift, mating of the diverging forms is often assortative with respect to host type. In polyploid speciation (see chapter VI.9), an important mode of speciation in plants, reproductive isolation between polyploid and diploid individuals can be substantial or even complete instantaneously.

If we accept that allopatric, parapatric, and sympatric speciation is each possible in theory and sometimes occurs in nature, how much does each mode contribute to global diversity? As long ago as the turn of the twentieth century, David Starr Jordan and Moritz Wagner observed that most closely related species had nonoverlapping geographic distributions, which has long been held as evidence that allopatric and/or parapatric speciation predominate. More recently, Daniel Bolnick and Benjamin Fitzpatrick (2007) reported that 224 of 309 (72.2 percent) sister species included in recent studies had completely nonoverlapping geographic ranges. A problem with sister species comparisons is that we know from the fossil and pollen records that species geographic ranges sometimes shift substantially over the course of just a few thousand years, meaning that current distributions may not be informative regarding distributions during speciation. One solution to this problem is to consider the geographic distributions of the youngest sister species, for which post-speciation range movements should have been less extensive. Studies that combine information on phylogeny and geographic range reveal that the youngest sister species of many groups—including birds, mammals, and the South African cape flora—often have completely nonoverlapping ranges. In comparison, studies of some groups of herbivorous insects, most famously *Rhagoletis pomonella* (apple maggot fly), reveal substantial range overlap among young sister species or races, consistent with the important role of sympatric divergence in these groups but also explicable in terms of allopatric or parapatric speciation followed by range expansion. Interestingly, molecular analyses of apple maggot fly host races in North America by Jeffrey Feder and colleagues have shown that the phylogeographic history of the genes that underlie the sympatric host shift—from hawthorn to apple trees, which fruit earlier—is much more complex than had earlier been supposed. While the host shift occurred in sympatry, selection for divergence in the timing of emergence acted on genetic variation that arose much earlier in a more southerly allopatric population.

The contributions of different modes of geographic speciation will almost certainly vary among groups depending on their biology. In fact, some of the polarized opinions in the geography-of-speciation debate reflect the worldview of researchers as shaped by their own study taxa. For instance, ornithologists have been especially vocal proponents of allopatric speciation, which is not surprising given that most sister species of birds have nonoverlapping geographic distributions, whereas researchers working on herbivorous insects point to the frequency of host shifts and have been among the strongest advocates of sympatric models.

3. ISLAND PATTERNS AND THEIR IMPLICATIONS

As discussed earlier, the possibility that species' geographic ranges move postspeciation makes it difficult to infer the geography of speciation from the contemporary distributions of closely related species. For species that are restricted to islands, however, it may be reasonable to assume that their geographic distributions have been more static through time, with expansions and shifts limited by physical constraints on dispersal. As a consequence, studies of island taxa have yielded important insights into the geography of speciation, a point to which we will return shortly.

Island systems (note that lakes and mountaintops can be viewed as island systems) also provide the replication required to conduct robust tests of evolutionary and ecological hypotheses; for this reason, islands are sometimes referred to as "natural laboratories." Indeed, identifying the contributions of immigration, speciation, and extinction to the species richness of islands has proven much more tractable than teasing apart the mechanisms whereby they contribute to the latitudinal diversity gradients in continental and marine systems. Islands tend to have fewer species than continental regions of the same size. The community on an island close to a mainland source of colonists is liable to be composed of a sample of species that are present in the source area(s), implying that immigration is the main species input. Robert H. MacArthur and Edward O. Wilson's (1967) *Theory of Island Biogeography* described how the species richness of island communities might represent a dynamic equilibrium between immigration and extinction. They suggested that as the species richness of an island increases, immigration rates decline and extinction rates increase, with equilibrium species richness located where the lines cross.

On more isolated islands, we often find endemic species (i.e., species that are unique to an island). In fact, as island isolation increases, avian species richness declines, but the number of endemic bird species actually increases. Note that isolation is relative to an organism's dispersal ability, such that birds may frequently reach an island that is remote for other taxa such as amphibians. Two types of island endemic species, anagenetic and cladogenetic, are recognized on the basis of the geographic distribution of their relatives. An island endemic species is termed *anagenetic* if its closest relatives are not confined to the island, and *cladogenetic* if its closest relatives are endemic to the same island.

Anagenetic species are encountered most frequently on islands of intermediate isolation, suggesting that anagenetic speciation may be subject to a "Goldilocks effect." On islands that are situated too close to a mainland source of immigrants, gene flow between immigrants and residents impedes speciation, whereas on islands that are too far from the mainland, too few individuals arrive to initiate speciation.

On the most isolated islands, cladogenetic speciation within islands may assume importance as the greatest source of island diversity. Two stunning examples are the radiations of ancestral cichlids into hundreds of species within each of the African Great Lakes—Victoria, Tanganyika, and Malawi—and within-island speciation of *Anolis* lizards to occupy distinct niches in the Greater Antilles (see chapter VI.10 for a fuller discussion of adaptive radiations).

When cladogenetic speciation creates two or more forms on a single island, we are often interested in determining whether speciation has been sympatric, parapatric, or allopatric. As well as offering evolutionary biologists the aforementioned advantage of making post-speciation range shifts less likely, isolated islands are expected to offer conditions that are conducive to sympatric speciation, such as unoccupied niches and increased intraspecific competition, which may promote disruptive selection. For these reasons, a team of researchers studied the geography of speciation of plants on the minute and remote Lord Howe Island (it has an area of less than 16 km^2, is located 600 km to the east of Australia, and is remote from any other island). Using molecular phylogenies, Alex Papadopulos and colleagues identified 11 cases in which the closest relative of a plant species endemic to Lord Howe Island was also endemic to the island (i.e., cladogenetic species). Although these cases are almost certainly sympatric under a biogeographic definition, we do not know whether they would satisfy our population genetic–based definition. In fact, in-depth study of one of these cladogenetic sister species, the *Howea* palms, revealed that, while speciation likely occurred in the face of strong gene flow, it was most likely parapatric under a population genetic definition. From this research into Lord Howe's flora we might infer that within-island speciation in the face of gene flow is not as rare as has been suspected for some groups. However, as we will discuss, research into the relationship between area and speciation on islands reveals a somewhat different story about the relative frequency of different geographic modes of speciation.

4. SPECIATION AND AREA

The species-area relationship, a positive correlation between geographic area and species richness, is one of the most ubiquitous patterns in ecology. Across small geographic units this pattern is believed to arise solely from ecological processes, with communities assembled through immigration and local extinction. But across larger geographic units, speciation within the geographic unit makes an important contribution.

All else being equal, an increase in geographic area is expected to facilitate speciation. For instance, as a species' geographic range size increases, so does the probability that a randomly placed knifelike barrier, such as a mountain range or river, will bisect it and give rise to vicariant speciation. A larger area should also present more opportunities for long-distance dispersal and colonization of new areas, thereby promoting peripatric speciation. Moreover, given that geographic area tends to correlate with habitat diversity, we expect to find that habitat-driven parapatric speciation will generally be more common in larger geographic areas.

Empirical evidence for the relationship between geographic area and speciation relies heavily on insights from island diversity patterns. Barring one as-yet-unproven case of buntings on remote islands in the Tristan da Cunha archipelago, there is no evidence that birds undergo within-island/cladogenetic speciation on any island smaller than Jamaica (11,189 km^2). In comparison, within-island speciation of anoles, a speciose radiation of New World lizards, starts to make a substantial contribution to diversity on Caribbean islands larger than 3000 km^2. The observation that for both birds and anoles within-island speciation appears only on islands that are larger than a certain size implies that speciation is likely to be allopatric or parapatric and that sympatric speciation is unlikely to play a major role in either group.

Yael Kisel and Tim Barraclough (2010) explored the relationship between island area and the probability of observing cladogenetic species across a broad set of taxa, including birds, mammals, snails, and angiosperms, and found that the probability of within-island speciation generally increases with area. Moreover, when they compared the minimum-sized island on which within-island speciation occurs among groups, they found a positive correlation between island size and dispersal ability. In other words, snails are able to speciate cladogenetically on much smaller islands than are birds. This finding implies that the dominant geographic modes of speciation on islands are those that proceed more easily when dispersal (and therefore gene flow) is reduced or absent—namely, parapatric and allopatric speciation—rather than sympatric speciation.

It is becoming clear that the influence of available geographic area on speciation is more complex than outlined thus far. In an important recent paper, Daniel Rabosky and Richard Glor (2010) found that island area determines equilibrium species richness for anole lizards in the Greater Antilles, but not solely via immigration and extinction, as MacArthur and Wilson suggested. Rather, Rabosky and Glor reported that the speciation rate on each island started high and then declined through time, as would be expected if the diversity of anole species on an island induces a negative feedback on the speciation rate. They found that on the two smallest islands (Puerto Rico and Jamaica), speciation rates had started high in the past and rapidly declined to zero. In contrast, on the largest island, Cuba, the decline in speciation rate was much slower, and contemporary speciation rates are still above zero, meaning that the *Anolis* diversity on this island may still be increasing. This result represents some of the most compelling evidence that the processes responsible for diversity patterns involve an interaction between ecological limits (which are, in part, a property of the area available) and diversification rates.

5. GEOGRAPHIC AND GEOLOGICAL TRIGGERS OF SPECIATION

If geographic area places a limit on diversity, via either a negative feedback of diversity on speciation rates or a positive feedback on extinction rates, a corollary is that colonization of a geographic region where there are few competitors should present increased opportunities for diversification. Sometimes, an organism's superior dispersal ability affords it the opportunity to colonize a new region before its competitors. Indeed, the success of anoles in dispersing to Caribbean islands and radiating there may be a case in point.

Alternatively, geological events may present taxa with new opportunities for colonization and diversification. For example, until the Panamanian land bridge joined North and South America, only stronger dispersers, such as some birds and plants, would cross between the two continents, meaning that many of the lineages making up the two biotas evolved in isolation from one another. The mammalian fossil record reveals that following formation of the land bridge about 3 million years ago, poorer dispersers were able to cross it in either direction for the first time. This event is called the Great American Interchange, and while it precipitated some extinction, it also produced some evolutionary winners. For instance, some lineages, such as the rodent subfamily Sigmodontinae from South America, diversified into many species on the new continent.

6. CHALLENGES AND PROSPECTS

After more than a century of research into the geography of speciation, some general principles appear to hold. For example, if we adopt the population genetic definition, sympatric speciation is unusual. On the basis of contemporary distributions it also seems likely that in many groups gene flow is low, though perhaps not absent, during speciation. But as yet, we do not know for sure what a plot of the amount of gene flow on the x-axis and the frequency of speciation on the y-axis would look like. Nor do we know whether this distribution changes at different stages of speciation. For instance, perhaps speciation often begins in the absence of gene flow but is completed in the face of some gene flow. Recent work on speciation has seen a shift in focus away from classifying cases into broad geographic modes and toward characterizing and quantifying the processes involved. Thus, new questions have arisen, for example, How important are biotic versus abiotic drivers of ecological speciation?

Our understanding of how speciation contributes to broad-scale geographic patterns is also in a state of flux. An explosion in the availability of molecular phylogenies for many groups is yielding fascinating insights into the role of speciation, extinction, and time to species diversity patterns. However, findings such as those of Rabosky and Glor for islands in the Greater Antilles highlight complex interactions among geography, ecology, and diversification. Thus, two questions arise: First, do the same general principles that apply to island diversity also apply to broad-scale continental and marine diversity patterns? Second, if diversity limits are an emergent property of a feedback of diversity on diversification, how elastic are these limits? In light of the increasing availability of molecular and paleontological information, there are good prospects for obtaining answers to question such as these in the coming years.

FURTHER READING

Bolnick, D. I., and B. M. Fitzpatrick. 2007. Sympatric speciation: Models and empirical evidence. Annual Review of Ecology and Systematics 38: 459–487.

Coyne, J. A., and H. A. Orr. 2004. Speciation. Sunderland, MA: Sinauer. *Chapters 3 and 4 provide an excellent summary of the theoretical and empirical evidence for allopatric, parapatric, and sympatric speciation. Chapter 4 also provides criteria that need to be satisfied for the case for sympatric speciation to be compelling. Chapter 11 provides a critique of the evidence for founder effect of speciation.*

Fitzpatrick, B. M., J. A. Fordyce, and S. Gavrilets. 2008. What, if anything, is sympatric speciation? Journal of Evolutionary Biology 21: 1452–1459.

Kisel, Y., and T. G. Barraclough. 2010. Speciation has a spatial scale that depends on levels of gene flow. American Naturalist 175: 316–334. *A study of single-island endemic sister species that points toward allopatric and/or parapatric speciation's predominance for a broad range of taxa, including snails, birds, mammals, and angiosperms.*

Lessios, H. A. 2008. The great American schism: Divergence of marine organisms after the rise of the Central American isthmus. Annual Review of Ecology and Systematics 39: 63–91.

MacArthur, R. H., and E. O. Wilson. 1967. The Theory of Island Biogeography. Princeton, NJ: Princeton University Press.

Mittelbach, G. G., D. W. Schemske, H. V. Cornell, A. P. Allen, J. M. Brown, M. B. Bush, S. P. Harrison, et al. 2007. Evolution and the latitudinal diversity gradient: Speciation, extinction and biogeography. Ecology Letters 10: 315–331.

Price, T. 2008. Speciation in Birds. Greenwood Village, CO: Roberts & Company.

Rabosky, D. L., and R. E. Glor. 2010. Equilibrium speciation dynamics in a model adaptive radiation of island lizards. Proceedings of the National Academy of Sciences USA 107: 22178–22183.

Rosenzweig, M. L. 1995. Species Diversity in Space and Time. Cambridge: Cambridge University Press.

VI.4

Speciation and Natural Selection

David B. Lowry and Robin Hopkins

OUTLINE

1. Types of natural selection contributing to reproductive isolation
2. Types of reproductive barriers and the effect of selection on their evolution
3. Considerations when studying natural selection and speciation
4. Reinforcement
5. Future directions

Natural selection is the process whereby heritable genetic variation changes in frequency as a result of its effect on survival and reproduction. The idea that natural selection plays an important role in speciation dates to Charles Darwin. Even so, major advancements in our understanding of how both ecological and reinforcing selection act to drive speciation have occurred since the mid-1990s. Extensive research investigating the role of selection in the process of speciation has revealed the importance of disruptive and directional selection in causing reproductive isolation between diverging groups of organisms. While the idea that ecological adaptation can cause reproductive isolation is basic, the way in which this process occurs is complex and often involves many agents of selection and multiple reproductive isolating barriers.

GLOSSARY

Assortative Mating. The preferential mating among individuals within a group of organisms based on similarity of phenotype.

Directional Selection. Natural selection favoring one end of the phenotypic spectrum over the other end of the spectrum.

Disruptive Selection. Natural selection favoring extreme phenotypes; intermediate phenotypes are the least favored.

Ecotype. A population or group of populations that have evolved a consistent suite of adaptations in response to local environmental conditions.

Natural Selection. The process whereby heritable genetic variation changes in frequency as a result of its effect on the fitness of an organism.

Parallel Speciation. The independent evolution of the same type of reproductive isolating barriers in response to similar agents of selection.

Reinforcement. The process whereby reproductive isolation increases as a response to natural selection against maladapted hybrids.

Reproductive Isolating Barrier. An evolved difference that acts to reduce the exchange of genetic material between populations, ecotypes, or species.

The main thesis of Charles Darwin's *On the Origin of Species* is that natural selection, as opposed to special creation, is the cause of species diversity on planet earth. Although Darwin made this argument, the details of how he envisioned natural selection contributing to speciation are unclear and often contradictory throughout his writings.

Soon after the publication of *The Origin*, other evolutionary biologists, most notably Alfred Russel Wallace, explicitly argued that ecological adaptation plays a role in the formation of new species. However, by the late nineteenth century, research focused on linking adaptation and speciation had waned. Interest was revived during the 1920s when Göte Turesson published a flurry of papers, in which he coined the term *ecotype*. Often overlooked, Turesson's research and theories inspired biologists of the time, particularly botanists, to conduct research investigating the connection between ecological adaptation and species formation.

Theodosius Dobzhansky's *Genetics and the Origin of Species*, published in 1937, and Ernst Mayr's articulation of the *biological species concept* (BSC) in 1942 defined speciation by the evolution of *reproductive isolating barriers*. Examples of reproductive isolating barriers include differences in mating preference between species, hybrid inviability, and hybrid sterility (see chapter VI.1). The BSC was a crucial development, because it directed researchers to focus on the mechanisms underlying the formation of reproductive isolating barriers to understand the process of speciation. With his third edition of *Genetics and the Origin of Species* in 1951, Dobzhansky articulated a list of ways in which natural selection could result in the formation of reproductive isolating barriers. In the same year, the botanist Jens Clausen published the book *Stages in the Evolution of Plant Species*, which assembled extensive evidence supporting the role of natural selection in the origin of many plant species. While Dobzhansky and Clausen described mechanisms by which natural selection could be important for speciation, they also argued that speciation often occurs through the accumulation of multiple reproductive isolating barriers arising from both selection and genetic drift.

After the 1950s, interest in the role of ecological natural selection in speciation diminished substantially. Studies of speciation became focused instead on the geography of speciation, especially the debate over the relative importance of allopatric, parapatric, and sympatric speciation (see chapter VI.3). Throughout the 1980s, it was thought that the major mechanisms involved in speciation were nonecological, often involving neutral genetic drift and chromosomal rearrangements.

Since the mid-1990s, the role of ecological adaptation in the formation of species has again become a major focus of evolutionary biologists. This research has been greatly enhanced by the feasibility of molecular techniques in diverse taxa. With modern tools in hand, researchers can now test important hypotheses about the role of natural selection in speciation.

1. TYPES OF NATURAL SELECTION CONTRIBUTING TO REPRODUCTIVE ISOLATION

Speciation usually results from the gradual accumulation of many reproductive isolating barriers. Natural selection can play a role in speciation when it leads to divergence between populations, and that divergence results in isolating barriers. Under most circumstances, selection does not directly favor an increase in reproductive isolation but rather reproductive isolation evolves indirectly as a by-product of selection. *Reinforcement* is the process whereby selection favors an increase in reproductive isolation to decrease the production of unfit hybrids.

The field of population genetics has defined three major categories of natural selection: disruptive, directional, and stabilizing selection. Both disruptive and directional natural selection are thought to play a major role in the formation of species and are discussed here in detail. *Stabilizing selection* occurs when intermediate phenotypes within a population are favored. The result of stabilizing selection is an increased frequency of individuals with the intermediate phenotype and a decreased trait variance in a population. Since speciation results from the splitting of one species into two, stabilizing selection is not usually thought to be involved in the initiation of speciation. Rather, stabilizing selection can contribute to speciation by maintaining phenotypic differences among populations that have evolved as a result of other forms of selection.

Disruptive Selection

Disruptive selection occurs when extreme phenotypes are favored over intermediate phenotypes within a population. This type of selection increases the variance of a trait and can divide a population into two distinct groups. If the two groups resulting from disruptive selection mate assortatively, then those groups can diverge to form species. Studies of Darwin's finches on the Galápagos Islands by Peter R. Grant, B. Rosemary Grant, and others support the hypothesis that disruptive selection has contributed to speciation. This work has documented strong disruptive selection on beak size and shape caused by competition for similar food resources. Because of competition, selection favors a bimodal distribution of beak shapes as it facilitates the finches' becoming specialized on different food resources. Differences in beak morphology are associated not only with variation in diet but also with variation in mating songs and mate choice. Therefore, it is thought that *assortative mating* is a by-product of disruptive selection on beak morphology in this system.

Directional Selection

Directional selection occurs when one extreme of the phenotypic spectrum is favored while the other extreme is disfavored within a population. The result of directional selection is that the mean phenotype of the population shifts in the direction of the favored phenotype.

Directional selection is most commonly thought to contribute to the formation of species when it operates differentially across habitats. For example, research conducted by David B. Lowry, Megan C. Hall, and John H. Willis found that coastal perennial and inland annual ecotypes of the yellow monkeyflower (*Mimulus guttatus*) are each adapted to their respective habitats in western North America. Inland plants have evolved early flowering and an annual life history to avoid reproducing during the hot summer seasonal drought. In contrast, coastal plants are sheltered from the drought by cooler temperatures and a persistent summer fog. Because these coastal plants have access to water year-round, they have evolved a perennial life history in which later flowering is favored. However, coastal plants must cope with oceanic salt spray, to which they have evolved salt tolerance. The flowering time and salt spray adaptations that differentiate the coastal perennial and inland annual ecotypes result in strong reproductive isolation between the ecotypes. In other words, adaptation to one of the two environments makes it difficult for individuals to survive and reproduce in the other environment, thus reducing gene flow between the ecotypes.

Uniform Directional Selection

The example of coastal perennial and inland annual ecotypes of *M. guttatus* illustrates how contrasting directional selection across habitats can contribute to reproductive isolation. However, different populations that experience uniformly acting directional selection could develop reproductive isolating barriers as a byproduct of the same types of adaptations, because different populations might respond to the same selective regime through different types of genetic changes. If those genetic changes result in hybrid incompatibilities, then uniformly acting directional selection would be an underlying cause of reproductive isolation between the populations.

Forms of Selection in a Geographic Context

The classification of speciation into geographic categories (see chapter VI.3) is a useful framework for beginning to understand how natural selection can influence the evolution of reproductive isolating barriers. *Allopatric speciation* occurs when populations are completely geographically isolated, such that no migration occurs. If populations are separated long enough, the genomes of those populations can diverge to the point at which they remain distinct even if they come into secondary contact in the future. Under allopatric conditions, alleles contributing to reproductive isolation can spread through neutral genetic drift, sexual selection, and natural selection. Directional selection is likely the major form of selection that drives the evolution of reproductive isolating barriers between allopatric populations. Alleles under natural selection or genetically linked to genes under natural selection tend to have a higher substitution rate $4Ns\mu$ (for which N is the population size, s is the coefficient of selection, and μ is the per generation per gene mutation rate) than neutrally evolving genes, for which the substitution rate is simply μ. Thus, adaptive mutations can be substituted at a faster rate than neutral mutations.

When diverging populations experience migration, successful differentiation is dependent on the strength of selection and the rate of gene flow between those populations. Migration between populations resulting in gene flow allows recombination to homogenize differences. It is therefore unlikely that uniform directional selection could result in the evolution of reproductive isolation between populations that exchange genetic material, because any universally adaptive mutation arising in one population would quickly be spread through all populations that exchange migrants. In contrast, divergent directional selection across habitats can result in reproductive isolation between populations even in the face of considerable migration. If selection is strong enough, regions of the genome that are involved in adaptations to different habitats will remain divergent. Conversely, alleles at neutral loci and alleles beneficial in both habitats will move between populations through migration, thus resulting in the homogenization of those regions of the genome.

Sympatric populations and those that exchange large numbers of migrants are more likely to require disruptive selection to evolve reproductive isolation. Theoreticians have explored this scenario extensively in an effort to understand how speciation might occur in sympatry, despite gene flow. However, efforts to find examples of sympatric speciation in nature have yielded only a few compelling examples (see chapter VI.3).

2. TYPES OF REPRODUCTIVE BARRIERS AND THE EFFECT OF SELECTION ON THEIR EVOLUTION

Considering that there are many agents of selection in nature, it is not surprising that natural selection can contribute to the process of speciation in multiple ways simultaneously. As pointed out by Dobzhansky and Clausen, speciation most often involves multiple reproductive isolating barriers driven by a combination of evolutionary forces. Types of barriers are listed in chapter VI.1. The contribution of natural selection to the evolution of those barriers is explained here.

Habitat Isolation

Habitat isolation is a barrier that reduces gene flow owing to adaptations of populations to divergent habitats. When populations become adapted to different habitats, individuals have a greater chance of surviving and mating in their native habitats compared with foreign ones. The reduced viability of immigrant individuals in the foreign habitat leads to a reduced rate of mating between native and foreign individuals. A classic example of habitat isolation occurs in *Timena cristinae* walkingstick insects. In California, populations of walkingsticks have become locally adapted to living on either *Adenostoma fasciculatum* plants or *Ceanothus spinosus* plants. Each of the walkingstick ecotypes has evolved cryptic coloration and morphological differences to blend in with the foliage of their respective host plants, presumably in response to visual predation by birds and lizards. Because of these morphological adaptations, walkingsticks have greater fitness when occurring on their native plant than on the foreign plant. This differential survival leads to more mating among walkingsticks adapted to the same plant species than among individuals adapted to the alternate plant species.

Temporal Isolation

Reproductive isolation that occurs as a result of populations mating at different times is called *temporal isolation*. Various forms of natural selection can lead to changes in the timing of mating and thereby cause reproductive isolation. Flowering-time divergence is commonly cited as a major form of temporal isolation in plants. Flowering-time evolution within plant species is frequently driven by selection imposed by abiotic stresses that cycle throughout the year. There are many examples of plants that have evolved adaptations to avoid flowering during annually recurring periods of environmental stress, such as drought or cold. These shifts in flowering as a result of habitat-mediated directional selection can lead to temporal reproductive isolation between populations adapted to different habitats.

Pollinator Isolation

Pollinator isolation is reproductive isolation resulting from pollinator behavior, such as preference for phenotypically different flowers. Pollinators impose natural selection on flowering plants that depend on them to mate. Shifts between pollinator guilds, such as from bees to hummingbirds, can lead to very strong reproductive isolation between plant taxa. For example, natural selection by different pollinator communities has led to morphological, color, and nectar production differences between bee-pollinated *Mimulus lewisii* and hummingbird-pollinated *M. cardinalis*. As a result, pollen is rarely transferred between these species.

Behavioral Isolation

Adaptation to different habitats can involve behavioral changes, which in turn cause changes in mating preference. For example, there is strong behavioral reproductive isolation between ecotypes of the apple maggot fly (*Rhagoletis pomonella*). Different ecotypes prefer either apple or hawthorn host plants. Behavioral experiments have demonstrated that flies show a strong preference for fruits of their respective host plant and will alter their orientation to go toward the desired odor associated with that plant. This leads to a reduction in mating between the two ecotypes inhabiting different trees and, consequently, assortative mating among individuals of the same ecotype. Behavioral isolation can also result from reinforcement, which we discuss later in this chapter.

Extrinsic Postzygotic Isolation

Extrinsic postzygotic isolation results when hybrids have reduced fitness because of maladaptation to either of the niches of the parental types. Extrinsic postzygotic isolation appears to be strong in a number of systems, including sticklebacks, leaf beetles, and big sagebrush. Stickleback (*Gasterosteus*) fish have evolved different ecotypes in shallow open water (limnetic) and deep (benthic) portions of lakes along the Pacific coast of Canada. Hybrids between benthic and limnetic ecotypes of sticklebacks have characteristics that are ecologically intermediate between the two ecotypes and as a result are maladapted to either of the lake habitats.

Intrinsic Postzygotic Isolation

Intrinsic postzygotic isolation manifests as hybrid lethality, inviability, or sterility and results from genic incompatibilities or chromosomal rearrangements (see chapter VI.8). Alleles contributing to intrinsic genetic incompatibilities can be spread by natural selection if an adaptive mutation directly contributes to an incompatibility or if an incompatibility allele is genetically linked to an adaptive mutation. Indeed, most of the identified genes underlying intrinsic postzygotic isolation appear to show molecular signatures of natural selection (see chapter VI.8). However, very little is known about the mechanisms of selection in these cases. In principle, external ecological factors could lead to the evolution of intrinsic postzygotic isolation.

There is now evidence that adaptation to internal genomic conflict can also drive the spread of isolating barriers. For example, Nitin Phadnis and H. Allen Orr recently showed that the gene *Overdrive* is a selfish genetic element that distorts Mendelian ratios in the gametes of F1 hybrids of two subspecies of *Drosophila pseudoobscura*, such that one allele has a higher probability of being in offspring than alternative alleles. *Overdrive* also causes hybrid sterility and thus contributes to reproductive isolation. Selfish genetic elements, such as *Overdrive*, can act as agents of selection on host genomes, which evolve mechanisms to repress them. The result is a coevolutionary battle in which the selfish element and the host genome both evolve in response to natural selection imposed by the other, increasing reproductive isolation in the process.

The Importance of Natural Selection in the Evolution of Reproductive Isolating Barriers

One way to assess the importance of natural selection in speciation is to compare the relative strengths of different types of reproductive isolating barriers between pairs of species. Some reproductive barriers, such as immigrant inviability and pollinator isolation, are thought to result primarily from natural selection. Other barriers, such as intrinsic reproductive isolation, could be the result of either neutral genetic drift or selection.

The most comprehensive analysis quantifying the strengths of multiple reproductive isolating barriers was carried out by Justin Ramsey, Douglas W. Schemske, H. D. Toby Bradshaw, and others between two species of *Mimulus*, *M. cardinalis* and *M. lewisii*. To date, the strengths of eight reproductive isolating barriers have been measured in this system. When combined, these barriers have a total strength of 0.9999, where 1.0 is complete reproductive isolation. These two species encounter each other infrequently in nature because *M. cardinalis* is adapted to lower elevations, while *M. lewisii* is adapted to higher elevations. As mentioned earlier, different pollinator communities visit these two species, limiting their pollen exchange. Thus, two ecological reproductive isolating barriers—habitat and pollinator isolation—are responsible for near-complete reproductive isolation in this system. Barriers that are less likely to be the result of natural selection, such as gametic isolation and F1 sterility, are weak between *M. cardinalis* and *M. lewisii*.

Some of the best evidence for a role of natural selection in speciation comes from studies that find the same reproductive isolating barriers evolving independently under similar ecological conditions. This repeated evolution of reproductive isolating barriers has been named *parallel speciation*. The walkingsticks provide an excellent example of parallel evolution of reproductive isolation. Studies of *Timena cristinae* walkingsticks by Patrik Nosil, Bernard J. Crespi, and others have identified eight geographically separated locations that contain *Ceanothus*- and *Adenostoma*-adapted ecotypes. Using molecular data, the researchers showed that these divergent reproductively isolated ecotypes have evolved independently. Across *T. cristinae*, individuals show preference for mating with other individuals adapted to the same host plant.

Studies of reproductive isolating barriers in a single system can provide insights into how speciation can occur. However, comparisons in multiple species pairs are necessary to draw broader conclusions about the relative importance of natural selection in the formation of reproductive isolation. Lowry and others (2008) recently compiled the strengths of multiple reproductive isolating barriers for 19 pairs of plant species. This comparative study revealed that prezygotic barriers were on average twice as strong as postzygotic barriers, with ecological barriers alone often accounting for near-complete reproductive isolation (see chapter VI.1). This is likely an underestimate of the importance of natural selection in the formation of reproductive isolation, because intrinsic postzygotic isolation can also be driven by natural selection. Overall, there is increasing evidence that natural selection accounts for the majority of reproductive isolating barriers involved in the process of speciation.

3. CONSIDERATIONS WHEN STUDYING NATURAL SELECTION AND SPECIATION

Because the process of speciation can take a long time, evolutionary biologists are restricted to studying a snapshot of the process for a given evolving pair of species. This limitation raises a number of important considerations when studying the role of natural selection in speciation.

First, the historical importance of particular reproductive isolating barriers cannot easily be determined. It is very difficult, if not impossible, to determine the historical order by which different reproductive isolating barriers accumulate and thus the role that natural selection plays over the entire process of speciation. The reproductive isolating barriers currently keeping species distinct might not have been the ones critical during the initial stages of speciation.

Another very important point to keep in mind when studying speciation is that reproductive isolation driven and maintained by ecological natural selection can collapse when environmental conditions change. A classic example of such species collapse involves cichlid fish in Lake Victoria, Africa—species that occur at different depths. The gradient of light quality correlated with depth has driven evolutionary changes in a light-absorbing opsin gene. It is thought that the local adaptation of female light perception to light at different depths has in turn led to changes in their preference for male coloration. Males of fish in shallow water, where blue wavelengths of light are abundant, have evolved blue coloration. Males occurring in deeper water, where available light is shifted to the red end of the visible light spectrum, are colored red. Under normal clear-water conditions, assortative mating occurs within deep and shallow populations because female cichlids prefer males that have the most visible colors. The recent introduction of anthropogenic pollution to the lake changed the way light passes to different water depths. This altered visibility has led to interbreeding followed by the collapse of closely related species. Eric B. Taylor, Janette W. Boughman, and others recently documented a similar collapse of deep- and shallow-water stickleback ecotypes in Enos Lake, British Columbia. In that case, changes to the lake caused by the introduction of a foreign crayfish apparently led to a breakdown of assortative mating. Thus, while ecological barriers are thought to evolve quickly during the process of speciation, they may not be sufficient to maintain species boundaries into the future.

Finally, it is challenging to infer whether divergent populations and ecotypes will ever become distinct species. It is clear from decades of research that not all adaptations within a species are necessarily involved in speciation. A major goal of speciation research is to determine why some adaptations are important to the process of speciation while other adaptations segregate within species without leading to appreciable reproductive isolation.

4. REINFORCEMENT

Reinforcement is the process whereby reproductive isolation increases as a direct response to natural selection against maladapted hybrids. This process is generally thought to involve three successive steps. First, two populations diverge in allopatry and accumulate partial reproductive isolation. Second, the two divergent populations come into secondary contact such that their ranges overlap partially or completely. Under this scenario, premating reproductive isolation is not complete and the two populations interbreed. There is selection against the interbreeding either because of direct costs of mating (e.g., copulatory damage) or, more commonly, because the hybrids have reduced fecundity, survival, or mating success. This selection against the production of hybrid offspring favors the evolution of greater reproductive isolation between the two diverging groups. Thus, the third step in reinforcement is the evolution of reproductive isolating barriers that reduce interbreeding or the production of maladapted hybrids.

Originally, reinforcement was thought to occur only when postzygotic isolation is very strong. However, recent theoretical models have shown that reinforcement can increase reproductive isolation between sympatric groups that are just beginning to diverge. During the process of reinforcement, natural selection against costly hybridization acts as a selective mechanism to drive the evolution of traits that prevent mating. The increase in reproductive isolation generally involves premating isolating barriers that either increase assortative mating or vary mate choice in a manner that decreases hybridization.

Reinforcement was originally termed the "Wallace effect" after Alfred Russel Wallace, who first articulated the idea that selection against hybrids might favor the evolution of reproductive isolation. Later, Dobzhansky described the process of reinforcement, as we understand it today, and stressed its potential importance in species formation. However, for much of the twentieth century, reinforcement was controversial.

Recombination and Reinforcement

Genetic recombination is at the core of the controversy regarding the occurrence of reinforcement in nature, because recombination between loci underlying the new prezygotic isolating barrier and loci involved in the original reproductive isolating barrier can lead to increased hybridization. Such increases in hybridization will ultimately result in the extinction or the homogenization of diverging populations.

The problem of recombination is best described using a hypothetical example involving two divergent ecotypes, red and blue. When these ecotypes interbreed, the hybrids produced are partially sterile, so selection favors decreased interbreeding between the ecotypes. A novel mutation then arises in the red ecotype, causing strong preference for red individuals over blue individuals. Similarly, a new mutation arises at the same locus causing the blue ecotype to prefer blue individuals. This type of reinforcement is termed a *two-allele mechanism* because two alternative alleles at a single locus result in assortative mating. Both alleles prevent the formation of partially sterile hybrids and are thus favored. The problem with recombination arises when the two ecotypes hybridize. Recombination between the preference locus and loci contributing to sterility in hybrids leads to the production of individuals that mate with the opposite ecotype. As a result, assortative mating can break down, and the ecotypes may fuse into a single population or go extinct.

Despite the problem that recombination can pose for reinforcement, recently developed theoretical models have convincingly demonstrated that selection can overcome the homogenizing effect of recombination and maintain assortative mating under a variety of scenarios. Furthermore, empirical investigations have found evidence for reinforcement in insects, amphibians, birds, mammals, and plants, thus indicating the potential importance of this process in speciation across a wide range of biological diversity.

One way the problem of recombination can be alleviated is for the same allele to increase assortative mating in both diverging taxa. Joseph Felsenstein first discussed this *one-allele mechanism* of reinforcement in 1981. Under this scenario, a novel allele is favored in both ecotypes because it decreases hybridization in either genetic background. This is a one-allele mechanism because the same allele reduces hybridization in both diverging ecotypes. Empiricists were skeptical that this sort of mechanism could actually exist in nature until Daniel Ortiz-Barrientos and Mohamed Noor found strong evidence for a one-allele mechanism of reinforcement in sympatric populations of *Drosophila pseudoobscura* and *D. persimilis*. A single allele causes both the *D. pseudoobscura* females to mate more with *D. pseudoobscura* males, and the *D. persimilis* females to mate more with *D. persimilis* males.

Geographic Patterns of Reinforcement

Instances of reinforcement are often identified when selection causes increased reproductive isolation in the sympatric but not the allopatric parts of the range of two diverging groups. The pattern arises because reinforcing selection occurs only when two diverging groups interact and attempt to mate in overlapping areas of their ranges. Traits favored by reinforcement in sympatry will often be neutral or even disfavored in allopatry and therefore not spread into those areas of the range.

A classic example of phenotypic divergence resulting from reinforcement is flower color variation in species of the genus *Phlox*. *Phlox drummondii* and *P. cuspidata* both have the same light-blue flower color in allopatric regions of their range. In sympatry, *P. drummondii* has dark-red flower color, while *P. cuspidata* retains a light-blue color. Robin Hopkins, Mark D. Rausher, and Donald A. Levin have shown that this flower color change reduces the production of hybrids between these species. Although both species of *Phlox* are pollinated by the same species of butterflies, individual pollinators rarely move pollen between the *Phlox* species if they have different flower colors. Dark-red-flowered *P. drummondii* plants exchange pollen with *P. cuspidata* plants less frequently than light-blue-flowered *P. drummondii* plants exchange pollen with *P. cuspidata* plants. Thus, trait evolution within a species can result in increased reproductive isolation between species.

The first 75 years of research on reinforcement was predominantly concerned with determining whether the process could exist. Now that there are conclusive theoretical and empirical studies supporting its occurrence, research has turned its focus toward determining how, why, and when it occurs. Compared with many areas of speciation research, which rely heavily on empirical studies, theoretical work has shaped our understanding of the process of reinforcement. Of the examples of reinforcement identified in nature, its genetic basis has been identified only in the *Phlox* system, where mutations in two genes involved in the production of anthocyanin floral pigments are responsible for a change in flower color. More work is required to determine the patterns, if any, in the types of mutations; the genetic architecture of traits underlying reinforcement; and the strength of selection acting during the process of reinforcement.

5. FUTURE DIRECTIONS

Many questions remain unresolved regarding natural selection and speciation. While major progress has been made recently in quantifying reproductive isolating barriers, we still have a poor understanding of the role that different barriers play over the entire process of speciation. An important question is, Are the reproductive isolating barriers important during the initiation of speciation different from those that prevent the collapse of species pairs later in the process?

Major progress will be made through the identification of genes involved in reproductive isolation and the mechanisms that caused their spread. Many of the genes discovered to date that underlie reproductive isolation appear to be driven by natural selection (see chapter VI.8). However, the mechanisms by which natural selection has driven the spread of reproductive isolating alleles at those genes are for the most part unknown. Further, most genes identified as underlying reproductive isolation are involved in intrinsic postzygotic isolation. Greater focus should be placed on identifying the genes underlying other types of reproductive isolating barriers that more clearly involve ecological adaptations. Once those genes are identified, follow-up studies should be conducted to determine the geographic distribution of alleles at those genes to better understand how natural selection spreads reproductive isolation during the process of speciation.

See also chapter III.8, chapter VI.2, and chapter VI.7.

FURTHER READING

Clausen J. 1951. Stages in the Evolution of Plant Species. Ithaca, NY: Cornell University Press. *A classic book by a great evolutionary plant biologist on how the process of speciation occurs.*

Coyne, J. A., and H. A. Orr. 2004. Speciation. Sunderland, MA: Sinauer. *The largest tome compiled on speciation research in modern times.*

Dobzhansky T. 1951. Genetics and the Origin of Species. 3rd ed. New York: Columbia University Press. *The classic must-read book by Dobzhansky for anyone interested in studying the process of speciation.*

Hopkins, R., and M. D. Rausher. 2011. Identification of two genes causing reinforcement in the Texas wildflower *Phlox drummondii*. Nature 469: 411–414. *The first study to identify genes involved in reinforcement.*

Lowry, D. B., J. L. Modliszewski, K. M. Wright, C. A. Wu, and J. H. Willis. 2008. The strength and genetic basis of reproductive isolating barriers in flowering plants. Philosophical

Transactions of the Royal Society B 363: 3009–3021. *A multispecies comparative study of the strengths of different types of reproductive isolating barriers.*

Rundle, H. D., and P. Nosil. 2005. Ecological speciation. Ecology Letters 8: 336–352. *A comprehensive review of the ways in which ecological selection can contribute to speciation.*

Schluter, D. 2009. Evidence for ecological speciation and its alternative. Science 323: 737–741. *Excellent recent review on our current understanding of the role of ecology in speciation.*

Seehausen, O., Y. Terai, I. S. Magalhaes, K. L. Carleton, H.D.J. Mrosso, R. Miyagi, I. van der Sluijs, et al. 2008. Speciation through sensory drive in cichlid fish. Nature 455: 620–626. *This excellent study elegantly combined molecular and field research to achieve a deep understanding of how behavioral reproductive isolation can occur as a by-product of natural selection.*

Servedio, M. R, and M. A. F. Noor. 2003. The role of reinforcement in speciation: Theory and data. Annual Reviews of Ecology and Systematics 34: 339–364. *A great review on reinforcement.*

Turesson G. 1922. The species and the variety as ecological units. Hereditas 3: 100–113. *The classic piece in which Turesson's synthesis of his extensive common garden studies resulted in his coining the term* ecotype *and launched interest in the role of ecology in speciation among botanists.*

VI.5

Speciation and Sexual Selection

Janette W. Boughman

OUTLINE

1. Can sexual selection generate diversity?
2. Patterns of speciation by sexual selection
3. The mechanisms of sexual selection that cause speciation
4. Sexual selection and postmating isolation

How does the diversity of life arise? Many of the most striking differences among species occur in traits involved in mating—especially traits that males use to compete with other males or to attract females, or that females use to select mates. Big differences in these same traits also make mating between species unlikely, contributing to reproductive isolation. It seems fairly intuitive, then, that whatever process causes differences in mating traits is involved in the formation of new species. The most likely process is sexual selection, which is defined as variation in mating success among individuals varying in phenotype within a population, caused either by males competing for access to females or females preferentially mating with some males over others. Sexual selection might be able to explain why some groups of organisms have many species—hundreds of thousands of beetles in a single order of insects—while others have very few—only five species of horseshoe crabs represent an entire class. But is sexual selection the only, or even the most important, process creating new species? How do we find out? One approach is to use information in the patterns of diversity of mating traits and species richness. If sexual selection is important in speciation, then groups of organisms that experience sexual selection should have more species than those that do not. Another approach is to explore how sexual selection causes mating traits to diversify and to ask whether differences in mating traits actually keep species from mating with each other. Results from these studies can be compared with those investigating natural selection in speciation to determine their relative roles. These are the lines of evidence considered in this chapter.

GLOSSARY

Allopatry. Geographic distribution in which populations or species are completely separated, with no contact and no opportunity for gene flow.

Antagonistic Coevolution. Coevolution in which the evolutionary interactions between two parties (two sexes or two species) impose costs on each other because of different evolutionary interests. The metaphor of an arms race is often used to describe the process and its outcomes.

Assortative Mating. Mating between individuals that are similar in a trait or set of traits, such as size assortative mating in which large males mate with large females and small males with small females. Also used to indicate preferential mating between individuals of the same species over individuals of other species.

Biological Species Concept. The classification of species as groups of potentially interbreeding natural populations reproductively isolated from other such groups.

Coevolution. The process in which evolutionary change of one species influences the evolution of another species.

Dimorphism. Differences between males and females of a species in size, structure, color, ornament, or other morphological trait(s), not including the sex organs.

Fisherian Runaway Sexual Selection. A model of sexual selection conceived by R. A. Fisher to explain the exaggeration of both male display traits and female preferences in the absence of benefits to females. Genetic correlation between male trait and female preference creates continual evolutionary exaggeration of both until the reproductive benefit of the exaggerated male trait is balanced by the cost of producing it.

Gene Flow. Movement of genes from one population or species to another through mating between individuals of those populations or species.

Good Genes Sexual Selection. A form of sexual selection in which females obtain genetic benefits, or "good genes," from mating with particular males. Female preferences evolve for male traits that indicate genetic benefits.

Local Adaptation. Evolutionary change to increase fitness of organisms in the local environment. Locally adapted individuals or populations have higher fitness in the local environment than the source (ancestral) population or other populations experiencing different conditions.

Mate Preference. Selection of mates based on criterion values of specific trait(s). Preference influences the propensity of individuals to mate with certain phenotypes and consists of two components: preference functions describe the way in which trait values are ranked, and choosiness is the effort an individual invests in making its choice.

Mating Trait. Secondary sexual traits involved in mating. These include display traits and competitive traits for males, and mate search and preference for females.

Reproductive Isolation. Speciation that occurs via the evolution of isolating barriers, which are characteristics of organisms that keep individuals in one population from exchanging genes with other populations. Reproductive isolation can occur by preventing individuals of separate species from mating (premating isolation) or by selecting against hybrids (postmating isolation).

Sensory Drive. A mechanism of sexual selection in which mate preferences evolve as a by-product when communication systems adapt to local conditions. A communication system includes sensory adaptations, preferences, and signaling traits. Although most often studied in the context of mating, sensory drive can occur for communication in any behavioral context.

Sexual Conflict. Occurrence of conflicting evolutionary interests and optimal strategies for reproduction between the two sexes, including aspects of reproduction such as mating rate. Sexual conflict can give rise to antagonistic coevolution of traits in each sex, including both behavior and morphology, that mediate such conflicts.

Sexual Isolation. A form of premating isolation in which the choosy sex of one population or species (usually female) is less likely to accept members of the other population or species as mates.

Sexual Selection. Variation among individuals with different trait values in the number of mates acquired and in overall reproductive success, measured as the number of offspring produced. Intersexual selection involves choosiness by one sex for mates of the other sex based on trait values (often called *female choice*). Intrasexual selection involves competition within a single sex for access to the other sex, frequently through contests (often called *male-male competition*).

Signal. A trait modified by selection to convey information and to influence the behavior of individuals receiving it. Signals can be in various modalities, including visual, auditory, olfactory, and tactile, and can involve specialized structures or be purely behavioral. Male display traits are signals.

Speciation. The process by which one or more species evolves from another via genetic changes and the evolution of mechanisms that restrict gene flow.

Sympatry. Geographic distribution of species in which at least part of the ranges of two species overlap, allowing individuals to encounter one another and making gene flow possible.

1. CAN SEXUAL SELECTION GENERATE DIVERSITY?

This chapter focuses on sexual selection as a force causing evolutionary change and speciation. *Sexual selection* is defined as a process that arises from differences between individuals in the ability to attract mates or in the number of successful matings caused by differences in underlying traits. Males with traits that females prefer mate more often and leave more offspring; these offspring also inherit the traits of their (preferred) father. Over time, populations come to be composed primarily of trait values preferred by females. Because of the high importance of mating success and reproduction to overall fitness, sexual selection is likely to be strong. For reasons explored in this chapter, males of one species are very unlikely to have traits that females of another species prefer. The combination of strong selection and opportunity for mismatch of traits and preferences is one important reason that sexual selection is thought to lead to reproductive isolation.

Reproduction is a key element of an individual's lifetime, and choosing a mating partner can have huge effects on overall fitness. The number of offspring an individual produces is the main measure of fitness, so even if an individual survives, without mating its fitness will be zero. Successful reproduction often involves mating displays typically produced by males, and preferences for those displays typically used by females to guide their choice of mates (see chapter VII.6). Male displays are often exaggerated adornments including patches of color and extravagant structures, or elaborate courtship behavior including dances, songs, and gift giving. Female preferences for one color over another, one song feature, or one smell over another can be quite extreme, leading the female to reject most suitors in favor of a particular male who meets her selective criteria. In many taxa, females have invested heavily in producing eggs and will continue to invest in parental care, so commonly, they do the choosing to find the best mate to father their offspring. Choosy females generate sexual selection on males, leading to evolutionary change in mating traits.

These considerations provide a second reason that sexual selection is thought to be important to speciation. Sexual selection is of fundamental importance to mating, and it causes mating traits to evolve and diverge, thereby influencing premating isolation. This chapter focuses on premating isolation, because sexual selection is especially likely to influence its evolution. Sexual selection can affect some aspects of postmating isolation as well, which is considered briefly at the end. A key form of premating isolation is termed *sexual isolation*, which occurs when mating preferences and/or display traits differ between populations, so that females of one population or species do not find males in the other attractive and hence refuse to mate with them. This *assortative mating* between species, or "like mates with like," can reduce gene flow and enhance reproductive isolation. Sexual isolation is commonly the primary isolating barrier between species in nature. Because sexual selection influences the evolution of mating traits leading to sexual isolation, sexual selection likely drives speciation in nature.

A third reason that sexual selection is thought to be important to speciation is that frequently, the most conspicuous differences between closely related species are traits involved in mating. Examples of this pattern abound. Male birds of paradise have spectacularly elaborated plumage including elongated feathers and brilliant colors, and males use these plumes in odd courtship dances to attract highly choosy females. No two species are alike, and both the elaboration of plumage and courtship as well as the marked differences among species are due to strong sexual selection. When species differ substantially in mating traits but little in ecological traits, sexual selection is implicated as especially important. This characteristic is also seen in lacewing insects: although many species are morphologically and ecologically very similar they produce vibratory mating duets that differ substantially. Males and females who cannot duet properly because they "sing" a different vibratory song do not mate, and this is the key mechanism that isolates the species. Such patterns indicate that sexual selection plays a role in speciation.

Scientists used to think that speciation would take a very long time to occur and that they could understand it only by looking back in time for millions of years (see chapter VI.1). But it turns out that we can witness speciation in action. In some cases, new species arise over tens or hundreds of organismal generations, and within a human lifetime: speciation is happening all around us. Thus we can look through the window this opens on the process of speciation and study what makes it tick. Rapid speciation (speciation in action) is especially likely when sexual selection or natural selection is involved, because these evolutionary processes typically increase the rate of evolution. In one example, Susan Masta and Wayne Maddison studied a group of jumping spiders found in the "sky islands" of Arizona (mountain ranges isolated by intervening desert) that diversified as recently as 10,000 years ago. Males sport intricate color patterns—swaths, stripes, and splotches of red, yellow, blue, black, and white on their face and forelegs—that differ markedly among populations inhabiting different mountains. Males use their facial and foreleg "paint" to attract females, indicating that sexual selection plays a role in the extreme differences among populations in color pattern, which is leading to rapid speciation. Speciation is not yet complete and has proceeded further in some populations than others. Therefore, pairs of populations that are strongly reproductively isolated can be compared with those less isolated to test specific mechanisms causing the evolution of reproductive isolation.

In recent years, research has gone past the initial step of describing differences in mating traits between reproductively isolated species to answering other pressing questions about when and how sexual selection causes reproductive isolation. We now have insight into the forms of sexual selection involved and the way in which sexual selection and natural selection interact to cause divergence. There are two complementary ways to address the role of sexual selection in speciation, covered respectively in the next two sections. First, we can look for patterns of diversity consistent with the action of sexual selection. Second, we can study the process of sexual selection and ask if it generates diversity.

2. PATTERNS OF SPECIATION BY SEXUAL SELECTION

Does sexual selection lead to higher diversity? This question can be addressed by testing for correlations between the presence or degree of sexual dimorphism (or other proxies for sexual selection) and measures of taxonomic diversity, such as the number of extant species. By testing such a correlational relationship in many taxonomic groups, one can assess whether sexual selection is generally associated with higher diversity, as would be expected if it promotes speciation across the tree of life. For example, Nathalie Seddon and colleagues showed that phylogenetic groups of antbirds that are highly dichromatic (male and female plumage colors differ) are more species rich than antbird groups that have little or no dichromatism. Many other comparative studies have found positive associations between measures of sexual dimorphism and species richness, supporting the idea that sexual selection causes speciation. However, not all studies have found this pattern, raising some doubts about the role of sexual selection in driving speciation generally. One reason for the discrepancies among studies may be that proxies for sexual selection like dimorphism are inexact ways to measure sexual selection, or that sexual selection is an important driver of speciation in some groups but not in others. Another possibility is that these studies typically do not take into account species extinctions. Counting the number of existing species in a phylogenetic group reflects not just speciation rate but also extinction rate. If sexual selection influences both, then its role in initiating speciation would be obscured. By addressing these possibilities and evaluating the relationship across many comparative studies, Kraaijeveld and colleagues calculated an overall estimate of how strongly sexual selection correlates with species richness in many groups of animals. The correlation is positive but low (correlation value of 0.07 to 0.14), which suggests that sexual selection does indeed contribute to speciation in many groups of animals, but that it does not act alone. Interestingly, this correlation is similar in magnitude to that found by Daniel Funk and colleagues between measures of ecological divergence and reproductive isolation, which hints that sexual and natural selection may contribute equally to speciation.

Does sexual selection increase the amount of reproductive isolation? To answer this question, one can compare the extent of reproductive isolation between closely related species with the strength of sexual selection between them, and with the degree they have diverged in sexually selected traits. The logic is that strong (and divergent) sexual selection causes a large amount of divergence in sexually selected traits between closely related species, which then results in reproductive isolation. For example, pheasants, whose species differ substantially in the traits males use to attract females, are expected to exhibit more reproductive isolation than parrots, whose species have much smaller differences in sexually selected traits. The relative importance of sexual selection to speciation is evaluated by comparing the amount of divergence in sexually selected traits and female preferences to that in traits thought to have diverged by natural selection instead. These divergence metrics are then correlated with reproductive isolation to suggest which force, natural or sexual selection, is more important. Key findings from such comparisons are that the amount of difference among species in male mating traits and female preferences predicts the extent of reproductive isolation better than the strength of female preferences for the desired male trait within species. These results suggest that sexual selection by itself is not enough to cause reproductive isolation, even if it is strong. Sexual selection needs to be divergent to cause new species to form.

3. THE MECHANISMS OF SEXUAL SELECTION THAT CAUSE SPECIATION

In theory, female preferences can generate sexual selection on traits by several mechanisms, with some more likely than others to cause divergence between populations in those traits, and ultimately to generate sexual isolation. The mechanisms of sexual selection include sensory drive, good genes, Fisherian runaway, and sexual conflict. Which of these are most likely to cause divergence in mating traits? The mechanisms can be grouped into two general classes according to underlying causes: differences between the environments occupied by populations (sensory drive and good genes), and interactions between the sexes (Fisher's runaway and sexual conflict). The environmentally dependent processes (sensory drive and good genes) lead to predictable associations between mating traits and environmental differences. In contrast, interactions between the sexes that generate divergent sexual selection by Fisher's runaway process or by sexual conflict have little to do with environment and generate evolutionary change in arbitrary, unpredictable directions. Whether environmentally dependent mechanisms are more or less likely to cause reproductive isolation than the arbitrary mechanisms is the subject of ongoing research.

Sensory Drive: Local Adaptation of Communication Systems as a Cause of Sexual Isolation

Sensory drive is a process by which some aspect of the sensory world predisposes individuals to attend to and prefer particular features of communication signals. Sexual selection through sensory drive is essentially a hypothesis about the effect of environment on shaping sensory systems, the preferences that depend on senses, and the display traits or signals that are preferred. Animals rely on their sensory systems to acquire information on predators, prey, and mates, and a sensory system that works well in one environment may not be so effective in another. This means that populations in different environments are likely to evolve differences in details of their sensory systems. These differences might include which senses they rely on most (e.g., vision for daytime visual predators, and hearing for nocturnal animals), and how those sensory systems are tuned, for example, what sound frequencies they evolve to hear best, what colors they evolve to see best, what smells they evolve sensitivity to. For example, deep in the ocean the prevailing light is blueshifted because water absorbs red wavelengths. Species of snapper fish that live in deep water have evolved to see blue light much better than closely related snappers that live in shallow estuaries—the eyes of deep-water species are tuned to blue light. This sensory tuning is favored because heightened sensitivity to blue light helps the fish see and discern objects in their environment, whether those objects are prey, predators, refuges, or members of their own species. Sensory tuning can, in turn, affect which mating traits are preferred even when conscious choice is not involved; those that match sensory tuning are likely to be easier to detect, and evolution favors mate choice based on easy-to-detect mating traits. Thus, sensory systems influence mate preferences, which should result in sexual selection on mating traits to match sensory systems.

How well a signal is transmitted through the environment should also influence how well it is detected, owing to the physics of sound, light, or chemical diffusion. Signals that travel easily through one environment may be degraded in another because of the physical interaction of the signal with the environment. Degraded signals are likely to be harder to detect, and degradation may obscure features of the males' signal essential for attracting females. In one example, populations of torrent-eared frogs that live near noisy rivers and waterfalls have evolved to produce louder and higher-pitched calls that can still be heard over the noise. Selection acts on signals to favor those that are well adapted to local transmission conditions.

Local adaptation of communication systems means that populations in different environments are likely to evolve differences in mate preferences and mating traits, which can lead to sexual isolation as a by-product of local adaptation of communication systems. Females in one population prefer trait values that males in the other population do not have and thus do not mate with them. Reduced mating arising from this mismatch between preference and trait among populations slows gene flow and leads to sexual isolation, pushing populations toward becoming distinct species. Sensory drive is thought to be widespread, potentially making it a very important way that sexual selection causes speciation.

Sensory drive causes sexual isolation in lake-dwelling threespine stickleback fish. Populations of two species that live in clear water see red well and prefer it. Males of these species display large patches of bright red color during the breeding season. Red is highly visible in clear water that has full-spectrum light. However, many lakes in the northern latitudes have tannin-stained, redshifted water because organic molecules in the water absorb blue wavelengths. The stickleback species that live in redshifted water do not see red as well, females do not prefer it, and males have evolved to display black color instead of red. In these redshifted habitats, the black males are contrasted against the red background light, making their black mating garb highly visible and easy to detect. Local adaptation of color vision and coloration matters to speciation, because the more two stickleback species differ in color preference and male color, the less likely they are to mate. This characteristic generates sexual isolation between species in different light environments and is an important factor in their rapid speciation. Sticklebacks are an example of speciation in action, as the distinct species arose in lakes that formed after the glaciers receded in the last ice age less than 15,000 years ago.

Additional evidence that sensory drive plays a pivotal role in speciation can be found by identifying the genes involved in sensory adaptation that also underlie reproductive isolation. This approach revealed that speciation in African cichlid fish occurs at least partly by evolution in genes that control color vision, and implicates sensory drive in the evolution of divergent female preferences for male color. Males of different cichlid species display either red or blue bodies to attract females, who prefer color patterns displayed by males of their own species; thus color differences enhance sexual isolation. Color vision depends on visual pigments called *opsins* found in specialized cone cells in eyes. Having different opsins imparts differences in how well individuals see particular colors, which has been shown to influence color preference and the strength of sexual isolation. In an elegant series of experiments, Ole Seehausen, Karen Carleton, and their colleagues (2008) found that differences in opsin genes underlie differences in color vision and color preference in several species of African cichlids from Lake Victoria. Opsins have evolved in response to the light environment, which varies among locations in the lake from clear to murky to redshifted. The fish have evolved to see the dominant water color best. These scientists have also shown that water color is correlated with the body colors males display to attract females—they display bright red in deep water, and blue in shallow water. When color signals and preferences differ among cichlid species, they are sexually isolated. This series of studies identified the sources of natural selection and sexual selection, their contribution to sexual isolation, and the genes that underlie female preferences in different species. Although only one particular set of cichlid species has been studied intensively, similar patterns of divergence in coloration with water clarity and depth are also seen in other groups of African cichlids, implicating sensory drive and color evolution as an important cause of speciation in this group. This is no small feat, as the African cichlids are a textbook case of extraordinary diversification. Hundreds of species have arisen within a very brief period in several of the large African Rift Valley Lakes.

Good Genes: Environmentally Dependent Benefits of Mate Choice and Locally Adapted Males

Sexual selection for good genes comes about because compared with other males, some males in a population carry superior alleles (good genes) that confer high fitness, such as alleles that help resist disease. Females who mate with these males will obtain these "good genes" for their offspring. Offspring of preferred males should have higher fitness than offspring from unpreferred males who do not have "good genes." Mating preferences that help females select these superior males will be favored by natural selection, and sexual selection will favor male traits that indicate genetic quality. Key requirements for good genes sexual selection to occur are variation in male genetic quality coupled with male traits that honestly indicate that quality.

Until recently, good genes sexual selection was not thought to contribute to reproductive isolation, but we now know two ways in which it can play a central role. Both have to do with differences in environment, and both will generate sexual isolation between populations from different environments. In the first way, benefits that females derive from choosing particular males as mates can depend on the environments the females and their offspring inhabit. These are known as *context-dependent benefits*. Because benefits from mate choice guide preference evolution, context-dependent benefits can lead to different preferences for populations in different habitats. Alison Welch found that female gray tree frogs gain benefits from mating with males who produce calls of long duration when they live in low-density habitats but may pay costs for mating with these same males in high-density habitats. In frogs, large size is advantageous, but so is rapid development. Offspring of long-calling males are larger at metamorphosis in low-density habitats (an advantage) but take longer to mature in high-density habitats (a disadvantage). This feature is expected to lead to different preferences by females in those two habitats. Females in low-density habitats should prefer long-calling males so that their offspring will grow large, but females in high-density habitats should avoid mating with long-calling males because their offspring will mature slowly and be at a disadvantage.

In the second way that good genes sexual selection can play a central role, locally adapted males have the particular alleles that make them well suited to local conditions; therefore, they have "locally good genes." Being locally adapted means that males are well suited to feed on local foods, avoid locally abundant predators and parasites, and deal with the local climate; thus they are likely to be in good condition. In many cases, mating traits are bigger or brighter when males are in good condition; this characteristic is known as *condition-dependent* male trait expression. Females can gain benefits by mating with these locally adapted males because their offspring will inherit the locally adapted alleles and thus have high fitness. Moreover, females who prefer male traits that are condition dependent are more likely to choose these locally adapted, high-condition males, primarily because those locally adapted males will display the bigger and brighter traits that females prefer. In contrast, males who have immigrated from other populations will not have the locally beneficial alleles likely to be in poor condition, and as a consequence, to display small or dull traits. Local females will not prefer to mate with them, and female rejection of nonlocal males will create sexual isolation between the local population and more distant ones, especially when environmental conditions differ. Sander van Doorn, Pim Edelaar, and Franz Weissing developed these ideas in a theoretical model in 2009. The model awaits empirical tests, but this scenario may prove to be widespread.

Fisherian Runaway: The Role of Arbitrary Divergence of Mating Traits in Speciation

Male display traits are often bizarre and exaggerated to the extent that it seems they cannot possibly be adaptive. A classic example is the peacock tail. This elaborate display limits male flying ability and increases predation, so is costly to bear. Peahens strongly prefer these showy tails even though this preference does not benefit them, because the showy males provide nothing but sperm to females and are not necessarily genetically superior. R. A. Fisher developed a hypothesis to explain the exaggeration of male display traits and female preferences in the absence of benefits that is known as *Fisher's runaway sexual selection*. The name comes from the "runaway" evolution of extremely elaborate male traits and strong female preferences for them. Fisher's idea was that female preferences generate sexual selection on male display traits, and the male traits evolve more elaboration in response to this selection. The female preferences become genetically correlated with the male traits because the offspring produced from these matings inherit both their father's trait value and their mother's preference, and pass these genetic combinations to their own offspring. This genetic correlation causes the female preferences to evolve along with the male traits and yields the "runaway" exaggeration. As female preferences evolve to become stronger, male traits evolve to become more exaggerated. Because they are genetically correlated, evolutionary elaboration of male traits indirectly causes the evolution of stronger female preference. In turn, the elaboration of female preferences causes stronger sexual selection by females for ever-more-exaggerated male traits. Thus, the male and female mating traits coevolve in response to each other in a positive feedback loop, becoming ever more elaborated as evolution proceeds. The main driver of this cyclic coevolution is not the environment but the interactions between the sexes. Under this hypothesis female preferences are arbitrary (i.e., nonadaptive) in two ways: they confer no fitness advantage or cost to the female and are not influenced by the environment (in contrast with the important influence of the environment under the hypothesis of sensory drive).

Fisher's runaway process has the potential to rapidly amplify differences between populations in both male traits and female preferences and, by doing so, to cause sexual isolation. Because coevolution between male trait and female preference is not dependent on environmental differences, it can occur in a multitude of directions. Allopatric populations are therefore likely to evolve in different directions even if they occur in similar environments. The direction of divergence is determined by chance factors such as the mutations present in the populations or differences in starting allele frequencies. With divergence possible in many arbitrary directions and the possibility for rapid change in mating traits, initially it seemed very likely that Fisher's runaway could lead to sexual isolation between populations. Many theoretical models of sexual selection and speciation are built on Fisher's runaway process, probably because the model is simple, as it does not assume natural selection on the female preference, and because the theoretical structure is elegant. Few empiricists, however, think that Fisher's runaway by itself is likely to be responsible either for the evolution of female preference or for divergence in mating traits sufficient to cause sexual isolation in nature. A primary reason for this doubt is that Fisher's process is not expected to lead to runaway evolution when females experience costs for being choosy. Costs such as increased search time, increased exposure to predators, or increased chance of remaining unmated seem quite likely to exist in nature, although these costs of choice are difficult to measure. When choice is costly and there are no compensating benefits, the expected evolutionary outcome is weak or no preference, and little or no exaggeration of male traits, neither of which is likely to enhance sexual isolation.

Some have suggested that Fisher's runaway process might cause speciation in sympatry, where it is envisioned to cause a single population to split along two paths that evolve in a runaway fashion in different directions. This would occur if there are females in a population that strongly prefer quite different male traits, such as when some females prefer orange color and others prefer black color, or when some females favor complex song and others prefer simple song. Each type of female would select males they prefer, causing sexual selection and evolution in the male trait. Because of the genetic correlation established between trait and preference, the female preference would evolve in concert with the male trait, causing the male traits to evolve ever more exaggeration in two directions at once. Moreover, females who prefer black would be unlikely to mate with orange males, and vice versa, generating reproductive isolation between the diverging subpopulations. In this way the initial population containing both black and orange males with their black-preferring and orange-preferring females would end up as two separate reproductively isolated sympatric populations of all black or all orange. At this point the population will have split into two distinct species. This scenario requires disruptive sexual selection on male traits, which is created by the different female preferences. However, it also requires disruptive natural selection on the female preferences, which may be uncommon in nature. Female reproduction is limited primarily by the ability to acquire sufficient resources to reproduce, and many scientists think this is unlikely to generate disruptive selection on the preferences they have for male traits. Even so, new theoretical work is exploring these possibilities. The special combination of factors needed may be one reason why sympatric speciation via Fisherian sexual selection is probably rare.

The genetic correlation between male traits and female preferences central to the Fisher process is nevertheless likely to occur whenever there is sexual selection by female choice, even under scenarios involving good genes or sensory drive, because the genetic correlation arises inevitably from nonrandom mating, which occurs whenever females prefer some males over others. For example, under a good genes scenario, sexual selection is predicted to lead to a genetic correlation between female preference and male trait. Moreover, the male trait will be correlated with offspring fitness because preferred males also possess good genes; in this case their offspring will inherit a suite of genes: the fitness-related "good" genes, those for the elaborated male trait, and those for the stronger female preference. In practical terms, this means that detecting the mere presence of a genetic correlation between male trait and female preference in a natural population does not imply that the Fisher process is behind the evolution of exaggerated male traits, because every known process of sexual selection is predicted to lead to such a correlation.

Sexual Conflict: Coevolutionary Dynamics, the Mating Dance between the Sexes, and Divergence

Even though successful reproduction increases fitness for both males and females, the sexes achieve high fitness in different ways, which leads to different evolutionary interests of the sexes and can result in conflict over reproductive strategies and outcomes. The best outcome for males is likely to impose costs on females, and the reverse is also true. Termed *sexual conflict*, this feature can lead to rapid and dynamic evolution of male and female reproductive traits. The sexes coevolve, but do so antagonistically. The evolutionary response in one sex ameliorates the costs it experiences, but this adaptation imposes costs on the other sex. This can cause cyclic changes or escalations as adaptations and counteradaptations evolve.

For example, male water striders benefit by mating often and have therefore evolved behavioral strategies and morphological structures to increase mating rate, such as persistent courtship behavior and grasping structures. Female water striders need to mate only once, and when mating rates are high, females experience costs from male harassment, exposure to predators during courtship and copulation, and reduced time spent feeding. The result is selection for traits in females to resist frequent mating, such as high choosiness, evasive behavior, and morphological structures that resist grasping. When females are at their optimum mating rate (the mating rate that yields the most offspring over their lifetime), males are not, and vice versa. This difference generates continual evolutionary change. The antagonistic nature of coevolution also results in negative genetic correlations between male and female mating traits, and between male and female fitness, in contrast with the positive correlations expected with good genes sexual selection.

These interactions between the sexes generate strong sexual and natural selection: they cause male traits to evolve quickly in response to female resistance to persistent males, and females to evolve quickly to counteract these male adaptations. Sexual conflict thus creates dynamic evolutionary change in each sex in response to the other. Males and females are engaged in an evolutionary dance, often termed an *arms race*. Crucially, allopatric populations of the same species are likely to follow different evolutionary trajectories as a consequence of differences in ecology or genetic variation and the strong, dynamic selection imposed by each sex on the other. For example, if a mutation occurs in one population that alters the structures water strider males use to grasp females, but in another population a mutation occurs that increases male aggression, the counteradaptations of females are likely to be different in the two situations (perhaps exaggerated morphology in the first case and exaggerated behavior in the second). The result is that the two populations diverge; they will evolve different adaptations in males to increase mating rate, and different adaptations in females to reduce it. Similar to Fisher's runaway, this coevolution can proceed in a multitude of directions, determined by the particulars of the populations involved. As a by-product of the coevolutionary arms race, sexual isolation can evolve because males and females from different populations are mismatched. However, the alternative outcome is also possible if males in one population are better able to overcome resistance from females in another population because those females have not evolved resistance to their persistence mechanisms. Therefore, whether sexual conflict promotes or inhibits sexual isolation is a matter for debate. This uncertainty is echoed in theoretical and empirical work on the subject.

Theory initially suggested sexual conflict would be likely to cause rapid speciation; however, those early claims have been modulated by later theoretical findings, which found that the conditions required for speciation were restrictive, and sexual isolation was rarely the predicted outcome. Several comparative studies of groups of species have indicated that sexual conflict is indeed associated with increased diversification rates or higher species richness, at least in insects. However, evidence is mixed from studies that experimentally manipulated the presence of sexual conflict in laboratory populations and followed evolutionary change over successive generations. Some experiments found that sexual isolation accumulated over time, but other experiments found no increase in sexual isolation. The jury is still out on whether sexual conflict causes speciation, but this active research area hopes to provide answers soon.

4. SEXUAL SELECTION AND POSTMATING ISOLATION

Sexual Selection against Hybrids

Even though sexual selection is primarily involved in the evolution of premating isolation, it may also be involved in *postmating isolation*, defined as reductions in hybrid fitness after hybrids are formed. For example, postmating isolation occurs if females discriminate against hybrid males. If hybrids are relatively rare, hybrid males will compete with males of the parent species for mating opportunities. The hybrids are likely to have intermediate, mixed or incompletely expressed phenotypes. If females of both parental species consequently find those hybrids unattractive, the hybrid males will have low reproductive success. Thus sexual selection against hybrids will limit gene flow even after hybrids have formed. Hybrid male chorus frogs are an example. The calls of males from the two parental species differ in pulse rate: one species gives calls with a more rapid pulse rate than the other, and females of each species prefer their own males' pulse rate. Hybrid males call at an intermediate pulse rate and are not attractive to females of either parental species. The strength of sexual selection against these hybrids is substantially stronger than natural selection, indicating that sexual selection can, at times, be an important part of postmating reproductive isolation.

Sexual Selection and Genetic Incompatibility

Another possible way in which sexual selection can contribute to postmating isolation arises from genetic changes it causes between populations. Sexual selection causes rapid evolutionary divergence, and the genetic changes that occur in different populations may be incompatible when combined in hybrids. The alleles present in one population work well on that genetic background. If they did not, they would be selected against and not rise in frequency in the first place. However, those alleles have not been tested on the genetic background of other populations and may be incompatible with the genetic background of foreign populations. If this incompatibility occurs, hybrids may not survive or may have diminished reproductive success. This postmating isolation will restrict gene flow, and is termed *Dobzhansky-Muller incompatibility*. The role of sexual selection in this process is unknown at present, and there are no good examples yet from nature; however, given the rapid evolutionary change that sexual selection is expected to generate, it may be a fruitful avenue of future research and deserves more study. This and other genetic mechanisms of reproductive isolation are considered thoroughly in chapter VI.8.

See also chapter IV.8, chapter VI.4, chapter VII.4, chapter VII.5, and chapter VII.7.

FURTHER READING

Andersson, M. B. 1994. Sexual Selection. Princeton, NJ: Princeton University Press. *Still the best book on sexual selection.*

Boughman, J. W. 2001. Divergent sexual selection enhances reproductive isolation in sticklebacks. Nature 411: 944–948. *An empirical study showing that sensory drive caused divergence in mating traits, enhancing reproductive isolation.*

Gavrilets, S. 2000. Rapid evolution of reproductive barriers driven by sexual conflict. Nature 403: 886–889. *An influential model of speciation driven by sexual conflict that spurred follow-up theoretical and empirical work.*

Kirkpatrick, M., and V. Ravigne. 2002. Speciation by natural and sexual selection: Models and experiments. American Naturalist 159: S22–S35. *This review clarified productive lines of inquiry in speciation research for theoreticians and empiricists alike.*

Kraaijeveld, K., F.J.L. Kraaijeveld-Smit, and M. E. Maan. 2011. Sexual selection and speciation: The comparative evidence revisited. Biological Reviews 86: 367–377. *Reviews the comparative evidence, concluding that sexual selection is important primarily in the early stages of speciation.*

Lorch, P. D., S. Proulx, L. Rowe, and T. Day. 2003. Condition-dependent sexual selection can accelerate adaptation. Evolutionary Ecology Research 5: 867–881. *This theoretical paper explores how sexual selection facilitates adaptation by removing maladapted males, and can also generate divergence between populations, leading to speciation.*

Parker, G. A., and L. Partridge. 1998. Sexual conflict and speciation. Philosophical Transactions of the Royal Society B 353: 261–274. *An early and influential review laying out the logic of sexual conflict causing speciation.*

Price, T. D. 2008. Speciation in Birds. Greenwood Village, CO: Roberts & Company. *Considers sexual selection in bird speciation, covering the full gamut of processes.*

Seehausen, O., Y. Terai, I. S. Magalhaes, K. L. Carleton, H.D.J. Mrosso, R. Miyagi, I. van der Sluijs, et al. 2008. Speciation through sensory drive in cichlid fish. Nature 455: 620–623. *An elegant and thorough study of cichlid speciation via sensory drive that considers the role of male color, color vision, and opsin evolution.*

van Doorn, G. S., P. Edelaar, and F. J. Weissing. 2009. On the origin of species by natural and sexual selection. Science 326: 1704–1707. *An elegant theoretical model considering how good genes sexual selection can lead to reproductive isolation without the need for female preference evolution.*

VI.6

Gene Flow, Hybridization, and Speciation

C. Alex Buerkle

OUTLINE

1. Gene flow leads to species cohesion
2. Gene flow and the origin of species
3. Hybridization: A common phenomenon
4. Evolutionary outcomes of hybridization
5. How to think about species in the context of gene flow and hybridization

The extent to which genetic material moves between divergent populations and species is a critical determinant of their evolutionary independence. High gene flow causes homogenization of populations and leads to their evolutionary cohesion, whereas low gene flow is more permissive of evolutionary divergence and independence. When divergent lineages mate or hybridize, there is the potential for genetic material to move between them. Gene flow through hybrids can erode evolved differences and can lead to stable hybrid zones, and to evolutionary novelty, including new species. The genetic, ecological, and evolutionary processes that affect the success of gene flow and hybrids are those that determine the conditions for the maintenance of reproductive isolation between species and for the origin of novel species.

GLOSSARY

Allele. An alternative nucleotide at any site in the genome, whether the locus is within a gene or elsewhere in the genome.

Gene. A region of an organism's genome that codes for the chemical precursor of a protein.

Genome. The entirety of the genetic or hereditary material that is passed between parents and offspring, including chromosomes (in organisms that have them) and all nucleic acids that are inherited.

Hybridization. The process by which progeny are produced from matings between genetically divergent parents, including individuals from different species.

Hybrid Speciation. The formation of an independent evolutionary lineage through hybridization, either through the union of some combination of unreduced gametes, leading to an increase in the number of copies of chromosomes (polyploidy) in the hybrids, or through standard gametes and homoploid progeny.

Hybrid Zone. A geographic region in which two species come into contact and hybridize.

Introgression. Gene flow between species or lineages that moves foreign alleles into the native genetic background.

Reproductive Isolation. The lowered probability that members of different populations will mate with one another when they co-occur, relative to a randomly mating group of all individuals in the populations. Likewise, the lower fitness of progeny from crosses between populations, relative to randomly mated individuals.

1. GENE FLOW LEADS TO SPECIES COHESION

The dispersal of juvenile individuals from their parents potentially leads to the exchange of genetic material among populations, whether this process involves a seed that is transported by an animal or wind, or a juvenile animal that opportunistically settles in a favorable site. The movement of genetic variants, or *alleles*, among geographic locations and populations is referred to as *gene flow*. (Somewhat confusingly, when biologists refer to the movement of genetic variants [gene flow] among populations, they are referring to the movement of any genomic material, not just protein-coding regions. Given that genes constitute a very small fraction [about 1 percent] of the genomes of many eukaryotes, this is an important point.) The net effect of gene flow is to make populations genetically more similar to one another than they would be in the absence of this exchange, because novel alleles that arise by mutation in an individual at one location are passed on to potentially dispersing progeny, rather than being retained only at that location. Gene flow homogenizes differences among populations that arise due to chance fluctuations in allele frequencies (genetic drift) or to the action of natural selection in different populations. In other words, gene flow opposes differences that arise due to any evolutionary processes, by homogenizing allele frequencies among populations. Consequently, populations that are connected by gene flow evolve collectively to some extent. Additionally, gene flow will export adaptive mutations that arise locally and disperse them more broadly across the geographic range of a species. In contrast, populations that do not exchange genetic material, or do so only rarely, have the capacity to evolve independently along different trajectories.

The capacity for dispersal and gene flow varies widely among taxa. For example, the offspring of some plants with heavy fruits only fall passively to the ground beneath their seed-producing parent; in contrast, the seeds of other plants are carried by wind for thousands of kilometers between islands in the Pacific Ocean. Whereas marine turtles and salmonid fishes can move great distances across the globe in their lifetime, in many cases individuals do not disperse very far from their place of birth to breed but instead return to the same beaches and rivers where their parents reproduced. Thus, the movement of individuals during seasonal migratory periods or other life stages is not the same as dispersal for breeding. Furthermore, the capacity for dispersal is not equal to the capacity for gene flow, because gene flow requires not only dispersal but also successful establishment and reproduction in the new location. That is, for gene flow to occur, not only would a seed need to be dispersed to a new location but, additionally, the seed would need to germinate, mature to reproduction, and successfully leave progeny of its own in the new location.

One might think that gene flow would necessarily be closely tied to sexual reproduction, and in many cases it is the union of gametes in sexual reproduction that introduces immigrant alleles into a new population. However, gene flow need not involve sexual reproduction. For example, in the case of asexual organisms, a population often consists of diverse individuals that reproduce clonally. The immigration of clonal lineages could introduce novel alleles into a population and otherwise alter the frequency of alleles, just as with gene flow involving sexual reproduction. For example, despite their showy yellow flowers, most common dandelions (*Taraxacum officinale*) make seeds without the need for sexual reproduction. Dispersal of seeds leads to gene flow among populations and shapes the evolution of dandelion populations, even in the absence of sexual reproduction. Likewise, in the case of bacteria, entirely new genes or allelic variants of an existing gene may be transferred between distantly related lineages by horizontal gene transfer without reproduction. Recent studies have found that horizontal gene transfer is not restricted to bacteria but is evident among eukaryotes, including transfers of genes and divergent alleles between distantly related plants (e.g., unrelated species of grasses, or between a flowering plant and a fern).

For species with small geographic ranges relative to the scale of their dispersal, gene flow is expected to thoroughly homogenize allele frequencies among all populations. Other species have much larger geographic ranges that dwarf the scale of typical single-generation dispersal of individuals. Large geographic ranges are likely to span a broad set of biotic and abiotic conditions that affect the performance of organisms, to encompass potential physical barriers to dispersal, and to include both suitable and unsuitable habitats. For example, consider small invertebrates that are restricted to tidal pools along the Pacific coast of North America, or butterflies that breed in mountain meadows that contain particular host plants on the high peaks of the Rocky Mountains. Each of these organisms possesses particular requirements for survival and successful reproduction, and dispersal of individuals and gene flow among populations requires traversing inhospitable sites.

Given that geographic ranges can be large relative to the scale of dispersal, there is the potential for relatively isolated populations of a species to diverge evolutionarily from one another. In the absence of homogenizing gene flow, mutations that confer fitness advantages in a local habitat will increase in frequency, and local adaptation of populations may occur. For example, novel mutations may lead herbivorous insects to shift to eating a novel host plant in particular locations. The evolutionary fate of the novel mutations will depend on a myriad of processes that affect their frequency, including the potential for gene flow from immigrants to eliminate local, novel alleles or to spread novel alleles among populations and lead to greater evolutionary cohesion among populations.

2. GENE FLOW AND THE ORIGIN OF SPECIES

It follows that if the evolutionary cohesion of populations is enhanced by gene flow, its diminishment allows for diversification of populations and lineages, and ultimately speciation. The origin of species is tied very closely to patterns of gene flow and reproductive isolation between ancestral and derived lineages. Although there are many criteria by which to recognize species (often referred to as species "concepts"), all of them share the idea that new species are formed when lineages evolve traits that reduce gene flow with other populations or lineages and thus become reproductively isolated and evolutionarily independent (see chapter VI.1).

Speciation can usefully be thought of as the accumulation of traits in diverging lineages that contribute to the diminishment, or complete cessation, of gene flow between them. Many evolutionary biologists study these traits because they promote reproductive isolation and increase a lineage's potential to evolve independently and to harbor novelty relative to other species and lineages. The large diversity of biological features and mechanisms that serve to isolate lineages can usefully be divided into those that function prior to the formation of zygotes (*prezygotic* isolating mechanisms; a *zygote* is a fertilized egg) or after zygotes are formed (*postzygotic* isolating mechanisms; see chapter VI.1). These can further be subdivided into (1) prezygotic mechanisms that arise as features of the ecology, behavior, and reproductive biology of the organisms and that reduce the frequency of matings or fertilizations and (2) postzygotic mechanisms associated with the viability or fertility of the hybrid progeny that result from crosses between the lineages. It is important to recognize that in any given pair of evolutionarily independent lineages, a diversity of organismal traits and features of their environment may each contribute to the reproductive isolation of the lineages. For example, two closely related plant species could have a low probability of gene flow because of differences in pollinators, flowering time, the habitats they occupy, greater fertilization success of conspecific pollen, and some inviability and infertility of hybrid progeny.

3. HYBRIDIZATION: A COMMON PHENOMENON

Many species lack complete reproductive isolation from other evolutionarily independent lineages and, instead, hybridize. Hybridization is a common phenomenon across the diversity of life, with hybrids between species known to occur in most familiar groups of organisms. The commonness of hybrids might seem to contradict the concept of species as evolutionary independent lineages. However, hybridization does not necessarily lead to a complete loss of independence; hybrids might be restricted geographically to a hybrid zone along an ecological gradient, or the hybrids might fail to contribute to gene flow between lineages because they are largely inviable or infertile. If one takes the view that complete reproductive isolation and evolutionary independence lie at one end of a gene flow continuum, then progeny from crosses between any two divergent lineages can be considered a type of hybrid (e.g., hybrid corn or maize varieties dominate North American agricultural production and are the result of crosses between divergent lineages of the same domesticated species). Crosses between divergent lineages are likely to occur, or are at least possible, in most organisms, which leads to the conclusion that hybridization is pervasive. Furthermore, isolation and evolutionary independence are not all-or-nothing characteristics but are quantitative attributes that vary by degrees between lineages. Isolation may also vary across the genomes of a pair of species, because exchange of alleles is more effectively counteracted in some regions of the genome than others (see chapters VI.1 and VI.9). Consequently, hybridizing lineages might possess recognizable trait differences that are maintained by divergent natural selection while experiencing substantial gene flow in portions of the genome that do not underlie important trait differences.

There are many examples of naturally occurring hybrids in various taxonomic groups. Some of these have been studied extensively because hybrids play a prominent or notable role in the evolution of these groups. For example, species of sunflowers (*Helianthus*) commonly hybridize, as do ragworts and groundsels (*Senecio*), and researchers have studied the genetics and ecology associated with their hybridization (see further reading). Likewise, tree species in several genera have a high propensity for hybridization, including oaks (*Quercus*), cottonwoods and aspens (*Populus*), and spruce (*Picea*), and have been studied extensively. The commonness of hybridization is by no means restricted to plants. For example, researchers have examined the role of hybridization in the invasion biology of fish (sculpins, suckers, sticklebacks) and salamanders (*Ambystoma*). And it has been estimated that approximately 10 percent of primate species hybridize in the wild. Hybrids between species belonging to the same genus, and even to different genera, are common in birds. The evolutionary and ecological significance of hybrids has been studied in butterflies, ants, corals, mussels, and many other animal groups. Laboratory techniques that allow molecular assays of genetic variation have been instrumental in detecting hybrids and confidently distinguishing them from variants within parental species. As biologists discover and study hybrids in more taxonomic groups, there is a growing appreciation that we can learn about the nature of species boundaries and reproductive isolation by studying instances in which isolation is incomplete and hybrids are formed.

4. EVOLUTIONARY OUTCOMES OF HYBRIDIZATION

The fitness of hybrid progeny can exceed (as in hybrid corn or maize), be equivalent to, or be lower than that of parental forms (e.g., sterile hybrid progeny). This variation can result from intrinsic features of the organism, including genetic and developmental determinants of viability and fertility. Likewise, variation in fitness of hybrids can be shaped by the ecological context in which hybridization occurs. Intrinsic and extrinsic factors interact to affect the outcomes and dynamics of hybridization. For example, the relative abundance of hybrids and parental species, spatial and temporal variation in abiotic and biotic determinants of fitness, and genetic variation among hybrid phenotypes all contribute to determining the fate of hybrids. Biologists can learn a great deal about the ecology and evolution of species by studying the role of different factors in determining the outcomes of hybridization. Three categories of outcomes may be recognized as heuristic points of reference among the complex and dynamic states that occur in nature.

First, it is likely that the most common outcome of hybridization between divergent lineages is their homogenization and the loss of any evolutionary genetic novelty (sets of mutations and trait differences) that had accumulated and been restricted previously to one of the diverging lineages. This must happen frequently when divergent lineages come into contact as a result of geographic range shifts. Given the fluctuations of climate over geological and evolutionary time scales (e.g., glacial advances), shifts in geographic ranges have been common, and divergence that might have accumulated in geographic isolation will have been erased on secondary contact, unless protected by traits that contributed to reproductive isolation. Likewise, in recent history, humans have disturbed natural habitats and caused range shifts that have brought previously isolated lineages into geographic contact. If hybrids have fitness equal to that of their parental lineages, divergent alleles will flow between them, and they will cease to evolve independently. Complete loss of some divergent lineages is expected to be common, but difficult to observe directly. Well-studied examples of loss of divergence due to hybridization include fish species (e.g., sticklebacks, suckers, sculpins, trout, ciscoes, and cichlids) and a large number of flowering plants (including hybridization between crops and their wild relatives). For example, rainbow trout that were transplanted and introduced by humans hybridize with and threaten the persistence of cutthroat trout in many drainages in the western United States.

A second outcome of hybridization is the formation of *hybrid zones*, areas in which a population of hybrids persists adjacent to the parent species. Hybrid zones can occur at geographic range margins—where two species meet—or can be less spatially structured, occurring as a patchwork within the ranges of the parental species (*mosaic hybrid zones*). The dynamics of hybrid zones are affected by the rate of input of parental alleles through ongoing interspecific hybridization and inputs from crosses between hybrids themselves. Likewise, the composition and persistence of hybrid zones is affected by the dispersal of adults and gametes, spatial and temporal variation in ecological conditions, and the contribution of extrinsic and intrinsic factors to fitness variation. Consequently, hybrid zones are inherently dynamic: their composition can change over time, they can move across the landscape, or they can vanish when species cease hybridizing. Some of the best-studied hybridizing species illustrate some of this variation among hybrid zones. Two forms of house mice (*Mus musculus* and *M. domesticus*) come into contact from the east and west in Europe and form a long hybrid zone that stretches 2400 km from Denmark to the South Caucasus. Transects through the hybrid zone have indicated that the area of hybridization extends for more than 50 km, with many agricultural barns occupied by populations of hybrid mice. This extensive, highly geographically structured hybrid zone of mice (and many other similar examples) can be contrasted with many hybrid populations in which hybrids co-occur with both parental species in a single, relatively small area, or different forms occur in a more complex geographic mosaic (e.g., sunflowers, sticklebacks within a single lake, and riparian cottonwoods). In many cases, there are multiple hybrid zones between the same pair of species, so that the outcomes and dynamics of hybridization may be compared. Despite decades of study of hybrid zones, we are only beginning to understand key aspects of the nature of genetic exchange that occurs within them.

By virtue of the incomplete and variable reproductive isolation that can be found in hybrid zones, they have the potential to offer us key insights into the traits and genetics that underlie reproductive isolation between species, which is otherwise difficult to study between lineages that are completely reproductively isolated. Hybrids between divergent lineages will possess novel combinations of alleles (and in some cases novel ploidy levels). These genotype combinations in hybrids that are rare or missing in the parental lineages can lead to inviability or infertility and contribute to reproductive isolation (e.g., Muller-Dobzhansky incompatibilities; see chapter VI.8) or lead to novel phenotypes in hybrids that have high fitness in certain contexts.

Finally, hybrids can become reproductively isolated from both their parental species and form new species. Hybrid speciation comes in two forms. In homoploid hybrid speciation, the genomes of the parental species merge without an increase in the number of chromosomes. This means that the genome of the hybrid species is a mosaic of genetic material from each of the parental species. In allopolyploid speciation, which involves a doubling (or other multiple) of chromosome number (polyploidy), the genomes of both parental species are retained in the novel hybrid species, at least early in its evolutionary history. Over time, the original genome multiplication can become fractionated and evolve toward a diploid number of copies of each genomic region. Both homoploid and polyploid hybrid speciation are examples of speciation with gene flow, since they begin with a hybridization event, but once it is formed, the derived hybrid must itself avoid homogenizing gene flow with the parental lineages if it is to persist and become evolutionarily independent. Thus hybrid speciation is made more likely if the hybrid lineage is somewhat spatially or ecologically isolated from the parents. For example, if the genotypes of hybrids predispose them to breed at a different time than a parental species, or to occupy a habitat in which the parental species cannot survive, this increases the chances of their success as an independent species. This type of ecological shift and resulting isolation has occurred in homoploid hybrid species of sunflowers (*Helianthus*): each of the three known hybrid species occupies an extreme habitat relative to the parental species. Likewise, among butterflies, hybrid species of *Lycaeides* in the Sierra Nevada of California occur at higher altitude and utilize a novel host plant relative to the ancestral lineages. Homoploid hybrid speciation may be more common than once thought and is currently the subject of intense investigation in a variety of taxonomic groups, including *Heliconius* butterflies, sculpins, house sparrows, and pines.

In general, it is important to recognize that hybridization often results in a genetically and phenotypically diverse hybrid progeny. Given this variability and known ecological influences on the outcomes of hybridization, hybridization is likely to lead to a diversity of genetic and evolutionary outcomes. For example, hybridization is common between two annual sunflower species (*Helianthus annuus* and *petiolaris*) wherever their geographic ranges meet. As noted earlier, in at least three instances this hybridization has led to novel, homoploid species that differ from the parental species. More commonly, their hybridization simply leads to a mixed population of various hybrid sunflowers and the parental species. Similarly, hybridization is common among various members of the genus *Senecio* (like *Helianthus*, also plants in the family Asteraceae). Hybridization in this genus has also led to new species, including novel species with the same chromosome number or with double the number of chromosomes of the parental species, but also has resulted in hybrid zones. Understanding the causes of this variation in outcomes of hybridization will continue to be the focus of considerable research and will lead to a better understanding of how species are formed.

5. HOW TO THINK ABOUT SPECIES IN THE CONTEXT OF GENE FLOW AND HYBRIDIZATION

The processes that lead to speciation have been a subject of long-standing interest in evolutionary biology, and recent research has advanced knowledge of the complex interactions of gene flow, hybridization, and speciation. Recognition that hybridizing species can nevertheless evolve independently as a result of traits that contribute to their isolation has clarified the expectation that different portions of their genomes will differ in their degree of divergence and amount of genetic exchange as a result of variation across the genome in mutation rates, effective population size, natural selection, and recombination rates. This means that a pair of hybridizing lineages will appear to exhibit different levels of isolation and divergence depending on which portion of the genome is examined (see chapters VI.1 and VI.9). This is a much more dynamic and realistic view of species, in contrast with the previously held notion that reproductive isolation between species is a genome-wide property. From a practical standpoint, variability in degree of isolation across the genome makes the task of recognizing and naming species more difficult. This challenge is not unique to evolutionary biology but is also encountered in other domains of biology where levels of differentiation may vary depending on which components are measured, such as the transition between juvenile and adult, or the boundary between cell types of different tissues.

Ours is an era of evolutionary biology in which researchers are seeking substantial advances in our understanding of the origin of species and the biological means by which novel lineages escape gene flow from their ancestors. Many researchers are studying traits and genomes in recently diverged lineages and their hybrids, in both natural and experimental settings. It is unlikely that the fundamental concept of species as isolated, independent lineages will be changed by this research, but it is likely that many mechanisms and processes, and ultimately generalities, associated with speciation will be revealed and that these will provide a more complete understanding of how novel species arise.

FURTHER READING

Arnold, M. 1997. Natural Hybridization and Evolution. New York: Oxford University Press.

Bock, R. 2010. The give-and-take of DNA: Horizontal gene transfer in plants. Trends in Plant Science 15: 11–22.

Coyne, J. A., and H. A. Orr. 2004. Speciation. Sunderland, MA: Sinauer.

Endler, J. A. 1977. Geographic Variation, Speciation and Clines. Princeton, NJ: Princeton University Press.

Harrison, R. G. 1993. Hybrid Zones and the Evolutionary Process. New York: Oxford University Press.

Mallet, J. 2005. Hybridization as an invasion of the genome. Trends in Ecology & Evolution 20: 229–237.

Rieseberg, L. H., and J. H. Willis. 2007. Plant speciation. Science 317: 910–914.

Rieseberg, L. H., T. E. Wood, and E. J. Baack. 2006. The nature of plant species. Nature 440: 524–527.

VI.7

Coevolution and Speciation

John N. Thompson

OUTLINE

1. Coevolution and the divergence of species interactions
2. Speciation with character displacement
3. Predators, parasites, and diversification
4. Mutualistic networks and speciation
5. Coevolved symbionts and speciation
6. Escape-and-radiate coevolution
7. Cospeciation
8. Conclusions

The web of life constantly changes as species impose strong natural selection on one another. During the past century alone, there have been dozens of examples of rapidly evolving interactions between parasites and their hosts, predators and their prey, competitors, and mutualists. This process sometimes involves reciprocal evolutionary change in interacting species driven by natural selection, which is called *coevolution*. We know that the coevolutionary process is responsible for many of the adaptations found in species, and it may also be responsible for many instances of speciation and adaptive radiation. This chapter explores the current hypotheses and results regarding coevolution as a driver of speciation, and the possible contributions of the coevolutionary process to the adaptive radiation of species.

GLOSSARY

Character Displacement. Evolutionary shifts of the ecological traits of two of more competing species in environments in which they co-occur.

Coevolution. Reciprocal evolutionary change among interacting species driven by natural selection.

Coevolutionary Hot Spots. Environments in which selection on interactions between species results in reciprocal evolutionary change; in contrast with cold spots, where selection does not favor reciprocal change.

Cospeciation. A macroevolutionary pattern of speciation in which two or more interacting lineages undergo matched speciation events during their phylogenetic history.

Ecological Speciation. Divergence of populations driven by divergent natural selection among environments.

Geographic Mosaic Of Coevolution. Variation among environments in the structure and strength of selection on interspecific interactions and in the genes and gene combinations under selection

Selection Mosaic. Divergent natural selection on interspecific interactions driven by differences in the expression of genes or the ecological outcomes of interactions among environments.

Symbiont-Induced Speciation. Speciation in host species driven by divergence in interactions between host and symbiont populations.

1. COEVOLUTION AND THE DIVERGENCE OF SPECIES INTERACTIONS

Much of the diversity of life is a result of the diversity of interactions among species. Many of the morphological, physiological, behavioral, and life history traits that we use to distinguish species from one another are traits involved in their interactions: the various forms of flowers, the beaks of birds, the running abilities of mammals, and the warning colors of many toxic invertebrates. It is therefore likely that much of the diversification of life may be the result of speciation driven by interactions among species (see also chapters VI.4, VI.10, and VI.16). Nevertheless, it has turned out to be much simpler to demonstrate that species adapt to one another than to demonstrate that interactions among species cause speciation. Part of the problem is time. Although adaptive change can occur in some interactions over a decade or even less, speciation is a much longer process.

The crux of the problem is to understand the extent to which speciation is ecological speciation—that is, a direct result of divergent natural selection acting on populations living in different environments. If speciation is a continuation of the process of adaptation, then studies of divergent adaptation of populations are studies of the process of speciation. The evidence for that view has been increasing steadily. In recent years, multiple studies have provided direct and indirect evidence that natural selection commonly plays a central role in speciation (see chapter VI.4). Many examples involve divergent selection imposed by interactions with other species, such as plant-feeding insect populations adapting to different host plants, or fish populations adapting to environments that differ in competition and predation.

Where speciation is driven by interactions with other species, geographic differences in the structure and strength of coevolution may contribute to the process. There is no reason to assume that all interspecific interactions coevolve, but when coevolution does occur, it may differ among populations living in different environments. Moreover, divergent coevolution among populations may increase the overall rate of adaptive divergence among populations over what would occur if populations were adapting only to their physical environments. When a population adapts to a physical environment, selection can often act to improve adaptation of the population to that particular range of physical conditions. A population is able, over time, to climb toward an adaptive peak. But as a population of one species adapts to a population of another species, the environment (i.e., the other species) often becomes a moving target; the adaptive peaks continue to change. Each adaptation in the first species can produce a counteradaptation in the other species. If populations coevolve in different ways in different places, then the coevolutionary process can rapidly lead to multiple highly divergent populations.

More broadly, coevolution can fuel the divergence of populations through three sources of variation in interspecific interactions: geographic selection mosaics, coevolutionary hot spots, and trait remixing. Any interaction between two or more species can be characterized as a genotype by genotype by environment interaction ($G \times G \times E$). The fitness of any genotype in one species will often depend not only on the distribution of genotypes found in the local population of the other species but also on the particular environment in which the interaction takes place. The expression of genes often differs among environments, making some traits more effective in some environments than in others. In addition, the ecological and evolutionary outcomes of interactions among species are bound to differ among environments, because the surrounding web of life will affect how any two species interact with each other.

The result is a selection mosaic across landscapes as natural selection favors different traits in different environments. An interaction may be antagonistic in all environments, or mutualistic in all environments, but natural selection can favor different defenses, counterdefenses, or mutualistic traits in different environments. Ecologically important selection mosaics have been demonstrated for multiple interactions. These selection mosaics sometimes include coevolutionary hot spots, where reciprocal selection on interacting species is strong, and coevolutionary cold spots, where selection is nonreciprocal. For example, interactions between some woodland star (*Lithophragma* spp.) species and their pollinating *Greya* moths are mutualistic in environments where copollinating bees and bee flies are rare, but commensalistic or even antagonistic in some environments where copollinators are common. In addition, the geographic mosaic of coevolution is further fueled by new beneficial mutations that appear only in some populations, move through gene flow to some populations but not to others, become lost in some populations because of random genetic drift, and continually shift in occurrence and frequency among populations owing to metapopulation dynamics. The combination of traits available for coevolution therefore becomes continually remixed, providing further fuel to coevolutionary change.

The geographic mosaic of coevolution that results from selection mosaics, coevolutionary hot spots, and trait remixing sets the stage for ecological speciation. At the extreme, divergent natural selection may pull mutualistic interactions in one or more directions; antagonistic interactions in multiple other directions; and commensalistic interactions in yet other directions. Multiple interactions have been shown to coevolve as a geographic mosaic. Examples include toxic newts and garter snakes that differ geographically in the levels of tetradotoxin in the newts, and detoxification abilities in the snakes; wild parsnips that differ geographically in the levels of multiple defensive furanocoumarins, and parsnip webworms that differ geographically in the combinations of P450 gut enzymes that detoxify these compounds; and Australian wild flax and its *Melampsora* rusts that differ geographically in the genes involved in defense and counterdefense in these gene-for-gene interactions. Collectively, divergent selection among populations found in multiple studies suggests that the geographic mosaic of coevolution can, in fact, fuel the early stages of ecological speciation.

2. SPECIATION WITH CHARACTER DISPLACEMENT

Competition between populations adapting to different environments has been suggested as a major form of coevolutionary interaction driving ecological speciation. Evolutionary theory predicts that characters with the greatest effect on competition between populations will be displaced more than other characters when populations come into contact. Either one or both populations undergo this displacement in ecological characters in regions where they co-occur, and only the latter is considered coevolutionary displacement.

Character displacement can take multiple spatial and temporal forms, and it not yet known which conditions most often result in coevolutionary displacement rather than evolutionary displacement of just one of the co-occurring populations. At one extreme, populations could undergo some adaptive divergence in allopatry, then meet again in a hybrid zone during a time of range expansion of both populations. Character displacement would occur within the hybrid zone. Alternatively, two or more populations of the same species could colonize a habitat at different points in time and, hence, have different lengths of time to become adapted to the new environment. Sometime after the first colonist population becomes adapted to the habitat, the other colonist population arrives and imposes selection on the resident population while itself being subject to selection by the resident population. A wide range of intermediate situations is possible in the ways in which allopatric populations become sympatric, sometimes forming complex mosaic hybrid zones. Hybridization within the contact zone could itself contribute to the speciation process as genes introgress from one species into the other and come under selection within their new genomes (see chapter VI.6). A major current problem to be solved in coevolutionary biology is which of these ecological situations involving competition and character displacement is most likely to result in speciation.

Character displacement during sequential colonization of a habitat has been shown in detailed studies of threespine sticklebacks in British Columbia. These coastal fish are commonly found in nearshore saltwater environments, but during the Pleistocene some populations colonized nearby freshwater habitats and became trapped. In some lakes, a secondary colonization occurred, and the two populations diverged into a benthic form and a limnetic form. Sympatric populations of benthic and limnetic forms have now been found in five lakes, and each sympatric pair appears to have resulted from a separate sequential set of colonizations followed by character displacement. The ancestral populations are thought to have been pelagic marine forms, which resemble the limnetic form found in the lakes. Limnetic fish have a slender body, extensive body armor, and many gill rakers. In contrast, the benthic form is less slender, has reduced body armor, and fewer gill rakers. The two forms show a strong tendency to mate with others of the same morphological type. Multiple additional studies have shown that these two forms are morphologically, genetically, and ecologically distinct in ways consistent with divergence from a common ancestor, followed by character displacement in the lakes where they have come into contact.

Repeated character displacement can result in an adaptive radiation of species, as has been shown in *Anolis* lizards in the Lesser Antilles. These lizards use a wide range of habitats, and they have diverged in size and shape as they have evolved to live mostly in one of three habitats: on the ground, the lower trunks of trees, or the upper tree canopies, creating what are often called *ecomorphs*. On Cuba, Hispaniola, Jamaica, and Puerto Rico, these lizards have diverged repeatedly into predictable ecomorphs. Each island has between four and six morphs, and phylogenetic analyses have shown that most morphs originated after colonization of each island. The exact sequence of morph divergence on each island is not known, and it is therefore not yet possible to determine the fraction of morphs that originated through reciprocal evolutionary change rather than through evolutionary change in later colonizers adapting to the range of morphs already in place. It is clear, though, that divergence among populations of these species has been driven by selection repeatedly favoring character displacement among coexisting populations on these islands.

3. PREDATORS, PARASITES, AND DIVERSIFICATION

Interactions between trophic levels may be as important a driver of speciation as competition within trophic levels. If divergence in ecologically important traits represents the first stage of most speciation events, then many species show evidence of incipient speciation driven by antagonistic interactions between tropic levels. There are now multiple studies showing that interactions between predators and prey, or parasites and hosts, have resulted in divergent adaptation among populations. Some of the best-studied examples are of plant-feeding insects in which different populations have adapted to different plant species, forming what are commonly called *host races*. In some insect groups, speciation has been attributed directly to shifts of these populations onto new plant species. In other taxa, there is clear evidence that defenses and counterdefenses have escalated more in some populations than in others.

Speciation appears to be driven both by antagonistic trophic interactions and by competition, as has been shown in the interactions among conifers, squirrels, and crossbills on multiple continents. Squirrels selectively harvest conifer cones that are easy for them to handle, and they cache large quantities of these cones. Where squirrels are abundant, multiple conifer species have evolved cone sizes and shape that are difficult for the squirrels to handle and extract seeds. In some regions where squirrels are absent, crossbills are the major seed predators of conifers, and in these regions conifers have evolved cone structures that are difficult for the crossbills to handle. The result has been a geographic mosaic of coevolution in which some populations of some conifer species have coevolved only with squirrels, others only with crossbills, and yet others with squirrels and crossbills together.

Geographic divergence in the crossbills has been even more extreme. The bill of a crossbill is a precision tool that aids the bird in prying apart the scales of closed cones to reach the seeds. Different populations of crossbills have evolved differ bill sizes and shapes to specialize on the cones of different conifer species. Some crossbill populations show evidence of extreme adaptation to the local population of particular conifer species. In some outlying regions of the North American Rocky Mountains where squirrels are absent, local populations of crossbills and conifers have coevolved to such an extent that the crossbills are regarded as separate species. These birds differ from other crossbills in their bill morphology and their songs.

These studies of crossbill speciation have shown that population divergence can be driven by geographic differences in the unique defenses and counterdefenses found in different populations, sometimes mediated by competition imposed by yet other species. Much work remains to be done to determine how antagonistic trophic interactions and competition work together to drive divergence and speciation. The current studies, though, show that geographic differences in the complex web of antagonistic differences may be a powerful force in speciation.

4. MUTUALISTIC NETWORKS AND SPECIATION

Ecologically important mutualisms among free-living species occur in all major ecosystems. Among the most evident are the interactions between plants and their pollinators and seed dispersers in terrestrial environments, and cleaner fish and host fish in the oceans. In many of these interactions, individuals interact with multiple other individuals of the other species during the course of a lifetime. These interactions therefore differ from *symbiotic mutualisms*, in which two mutualistic individuals live in intimate contact for an extended period of time. The distinction between mutualisms involving free-living species and symbiotic mutualisms is important for our understanding of speciation, because symbiotic mutualisms often involve symbionts that become adapted to a single host species, whereas nonsymbiotic mutualisms favor the evolution of multispecies webs. Reciprocal specialization between pairs of species is therefore uncommon in coevolved mutualisms among free-living species.

As mutualisms among free-living species continue to coevolve, they often draw additional species into the interaction. Species converge over time in traits that allow them to exploit the interaction. The process often favors the formation of groups of unrelated species that have all converged to exploit a group of relatively closely related species. Examples include plants from multiple families that have converged in their floral traits to attract hawkmoths or hummingbirds as pollinators. At the same time, these insects and birds continue to evolve to exploit these groups of plants with similar floral traits.

Because these interactions tend to form multispecies networks rather than pairwise interactions, it is often more difficult to study exactly how coevolution has shaped speciation. At the broadest evolutionary scale, however, it seems clear that pollinators are a driving force in the divergence of plant lineages, because these animals are directly responsible for plant reproduction through their movement of pollen among plants. It is equally evident that plants have shaped the diversification of pollinators, as can be seen in the diversification of insect mouthparts and bird bills as tools for extracting nectar.

When reciprocally specialized interactions have evolved between pairs of pollinators and plants, other aspects of the interaction often favor specialization. The most extreme example is pollination by floral parasites: adult pollinators lay their eggs in the flowers of host plants while pollinating the flowers, and the larvae feed on a subset of the developing seeds. Specific examples include the interactions between figs and fig wasps, yuccas and yucca moths, *Lithophragma* plants and *Greya* moths, *Glochidion* plants and their *Epicephala* moths, and globeflower plants and their pollinating flies. Some plant lineages involved in these interactions include hundreds of plant species, each of which is pollinated by one or a few highly specialized pollinating seed parasites. In these cases, specialization in the insects appears to be driven by the parasitic phase of the interaction. The plant has taken advantage of the egg-laying behaviors of these insects and remolded the parasitic interaction into a mutualistic interaction. Because these highly specialized insects are the sole pollinators of their host plants, they completely control the pattern of gene movement among plants and therefore have the potential to control the pattern of divergence among populations of their host plant.

Some other forms of mutualism also can favor population divergence and possibly speciation by favoring local matching of traits among interacting species. The clearest examples include *Müllerian mimicry*, in which co-occurring distasteful species converge on a color pattern that warns predators of their toxicity. Müllerian mimicry complexes have evolved in many taxa, including butterflies in terrestrial environments and sea slugs in the oceans. Within each mimicry complex, all species at a locality evolve to use the same visual cues to warn predators, but different populations of these species sometimes converge on different warning patterns in different regions. Hybridization between populations that are members of different mimicry complexes could therefore foster speciation, because selection could disfavor individuals with a warning pattern that falls outside the usually warning cues used by the local predators. The role of hybridization in the formation of multispecific mimicry complexes remains an active area of research.

5. COEVOLVED SYMBIONTS AND SPECIATION

All major ecosystems are built on a web of coevolved mutualistic symbioses, and we are still at the early stages of understanding how these symbioses affect diversification in the web of life. In terrestrial environments, major mutualistic symbioses include lichens, which are coevolved interactions between fungi and algae; mycorrhizae, which are interactions between fungi and the roots of most terrestrial plants; and rhizobia, which are interactions between nitrogen-fixing bacteria and the roots of some plants. Tropical communities are dominated by termites, which rely on their coevolved gut symbionts to digest cellulose in their diet, and larger plant-feeding animals rely on the symbionts in their gut for many aspects of digestion. In Central and South American forests, leafcutter ants cultivate particular species of coevolved fungi for their food. The richest marine environments are those dominated by corals and the symbiotic zooanthellae that provide nutrients to these corals. And deep-sea vents are built on coevolved interactions between invertebrates and the symbiotic microbes that take over the role of photosynthetic organisms in these lightless environments. Even more broadly, the worldwide proliferation of complex organisms has relied on mitochondria and chloroplasts, which are the products of ancient coevolved interactions. Much of the web of life is therefore a result of the diversification of coevolved interactions that transform inorganic compounds or poorly digestible organic compounds into food and energy.

How these symbiotic interactions affect speciation either in the microbial symbionts or their hosts is one of the questions at the frontier of speciation theory, because it was not possible even to begin exploring genetic divergence in these interactions until the advent of advanced molecular methods in recent decades. So far these methods have shown that the diversity and specificity of microbial symbionts is much greater than previously imagined. All eukaryotic organisms harbor a wide range of mutualistic and potentially pathogenic microbial species. It is easy to understand that hosts offer such different environments that these are likely to be important drivers of speciation in symbionts. It is also likely that symbionts are at least sometimes involved in host speciation, but studies of symbiont-induced speciation are still in their early stages.

One way in which coevolved symbionts might affect speciation is through distortion of sex ratios in their hosts. Bacteria in the genus *Wolbachia* were discovered just several decades ago. Since then, surveys have suggested that they occur in the cells of the majority of insect species and in a wide range of nematodes, interacting with their hosts as antagonists in some cases and as mutualists in others. *Wolbachia*, and some similar symbionts, often cause partial reproductive isolation between *Wolbachia*-harboring host populations and those that either lack these bacteria or have different strains of the bacteria. This partial reproductive isolation could serve as the initial stage of symbiont-induced speciation. It has long been suggested that subsequent coevolution between these symbionts and their hosts could drive speciation among host populations, because the host populations would rapidly diverge in genes favored by natural selection to mitigate the negative effects of the bacteria on host reproduction. Only recently, however, have studies of these interactions begun to analyze how partial reproductive isolation caused by these symbionts could drive speciation among populations during the early stages of population divergence.

At an even broader level, the maintenance of sexual reproduction itself has been attributed in part to natural selection favoring sexual organisms, which are better at keeping pace with coevolving parasites than asexual organisms. An asexual female produces offspring that are genetically identical, except when rare mutations occur. In contrast, a sexual female produces offspring that are each genetically unique. Sexual females are therefore more likely to produce offspring with novel genotypes to which local parasites are not adapted. Much of the process of speciation in eukaryotes is about the development of reproductive isolation among sexual populations, since most eukaryotic species are sexual. Hence, coevolution could lead to speciation not only by driving divergent adaptation among populations but also by favoring the process of sexual reproduction itself. Sexual reproduction then opens opportunities for assortative mating that, in turn, make speciation possible in sexual species. Hence, whether directly or indirectly, coevolution, sexual reproduction, and speciation are linked as driving forces in the diversification of life.

6. ESCAPE-AND-RADIATE COEVOLUTION

Coevolution has the potential to drive even large adaptive radiations by favoring speciation into new adaptive zones, thereby creating macroevolutionary patterns from microevolutionary processes. The leading hypothesis was first proposed by Paul Ehrlich and Peter Raven and is often called *escape-and-radiate coevolution*. It is a three-step process that was initially described for the adaptive radiation of plants and butterflies, but it could readily apply to any two or more lineages of interacting species. The process begins with one or more mutations in a host or prey population that allows individuals to escape attack from enemies. That mutant population then expands its geographic range and undergoes a starburst of speciation in the absence of the interaction as it colonizes a wider range of environments. Eventually, a mutant parasite population overcomes the new host defenses and radiates in species, with each new parasite species specializing on one or more of the many hosts now available to it. The process then repeats itself.

The result of escape-and-radiate coevolution is a temporal series of alternating starbursts of speciation on both sides of the interaction, forming entire clades with new defenses and counterdefenses. This view of coevolution makes clear predictions. First, novel defenses will occur among, rather than within, clades, because defenses accumulate starburst by starburst rather than species by species. Second, parasites will not colonize hosts within each host clade in any systematic fashion, because there is no ancestral-descendant pattern of accumulation of defenses within each starburst of host speciation. Rather, the pattern of escape-and-radiate coevolution appears at higher taxonomic levels, where each starburst of species on one side of the interaction is matched later with a starburst of speciation on the other side. In fact, different taxa could be involved at different points in the radiation of defenses and counterdefenses in this form of coevolution-mediated speciation.

Escape-and-radiate coevolution was the first hypothesis to explain how coevolution could affect not only speciation but also major patterns in the adaptive radiation of entire lineages. It has inspired a great deal of research on how interspecific interactions might drive speciation, and it remains a major framework for thinking about how the process of reciprocal selection might shape the web of life at multiple levels. It is, however, a difficult hypothesis to test for any particular group of interacting lineages, and it is only one of many ways in which reciprocal evolutionary change can drive the diversification of interacting lineages. Perhaps its greatest impact on evolutionary research has been to show that the coevolutionary process will often not result in a macroevolutionary pattern of matched speciation events during the phylogenetic history of interacting species.

7. COSPECIATION

The extreme alternative to escape-and-radiate coevolution is *cospeciation*, or parallel cladogenesis, in which each speciation event on one side of an interspecific interaction is matched with a speciation event on the other side of the interaction (figure 1). This result is impossible with escape-and-radiate coevolution, because parallel cladogenesis at the species level cannot occur during reciprocal starbursts of speciation.

The growing number of studies of the geographic mosaic of coevolution suggests that cospeciation should be uncommon in truly coevolving species. Species coevolve as sets of genetically distinct populations that have their own patterns of adaptation, coadaptation, and rates of divergence. As species coevolve, they can undergo speciation at different rates, and they can differ in the tendency of some populations to shift their interactions to yet other taxa. One local population could coevolve strongly with a particular species, while other populations could coevolve with yet different species. Some populations could coevolve equally with two or more closely related species, and occasionally, a population could even shift its interactions to a taxon phylogenetically far removed from the usual coevolutionary species partners. The long-term pattern of coevolution results from the complex mix of successes and failures of these many different coevolutionary experiments spread among all the populations within those species.

There are, however, three situations that favor cospeciation. One is coevolution of maternally inherited mutualistic symbionts and their hosts. Among the clearest examples are those involving obligate symbionts that provide nutrients required by their host. In these interactions, the mutualistic symbionts are often transmitted directly from mothers to their offspring during host reproduction. As a result, population divergence in the host results directly in population divergence in the symbiont. In some cases, bacterial symbionts have diverged in parallel with their insect hosts over millions or tens of millions of years.

The second situation that favors cospeciation is the tracking of hosts by maternally inherited commensalistic, rather than mutualistic, species. In these interactions, the species have not truly coevolved through reciprocal selection. Instead, the species have simply codiverged as their shared environment has fragmented over time. Each local commensal population diverges with its local host from other geographically separated populations. It is an important process contributing continuity in the structure of species assemblages.

The third situation that favors some degree of cospeciation is the special case of pollinating seed parasites. This specialized form of pollination favors highly host specific insect pollinators. Since these pollinators completely control the pattern of movement of pollen among plants, speciation in the plants follows from the pattern of specialization and speciation in the pollinators. Even

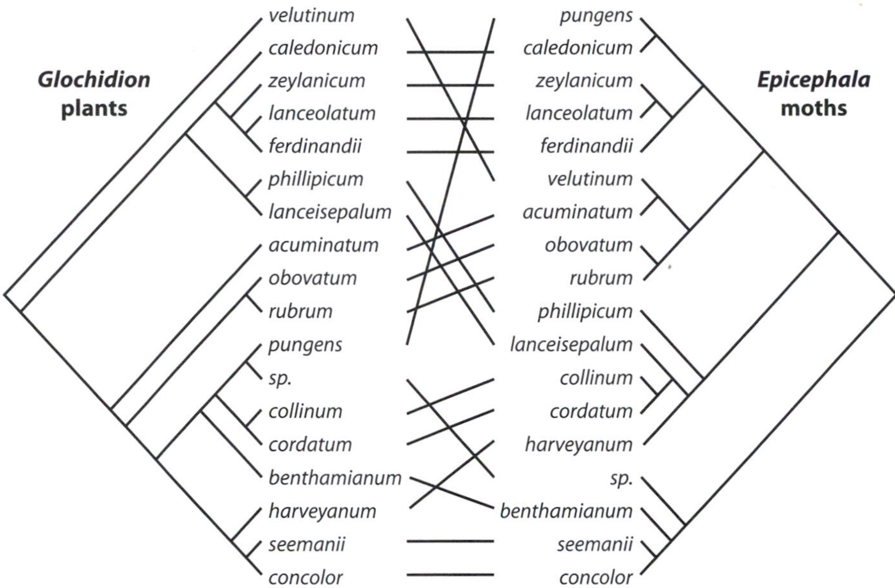

Figure 1. Cospeciation in *Glochidion* plants and their coevolving pollinating floral parasitic moths in the genus *Epicephala*. In perfect cospeciation, each speciation event in either the plants or the moths would be matched with a speciation event in the other lineage. In this case, the plants and moths show an overall pattern of cospeciation with occasional shifts of the moths onto more distantly related plants. (Redrawn from A. Kawakita, A. Takimura, T. Terachi, T. Sota, and M. Kato. 2004. Cospeciation analysis of an obligate pollination mutualism: Have glochidion trees [Euphorbiaceae] and pollinating *Epicephala* moths [Gracillariidae] diversified in parallel? Evolution 58: 2201–2214.)

so, host shifts sometimes occur, leading sometimes to overall cospeciation but with occasional shifts of pollinators to distantly related host plants (figure 1).

In general, few lineages show sustained cospeciation with other lineages. Over the history of a lineage, species often have repeated opportunities to shift their habitats and also their preferences in their interactions with other species. For example, feather lice of birds have switched among avian orders at least twice during their evolutionary history, and the obligate marine worms that live within echinoderms have undergone occasional shifts onto new host lineages during the several hundred million years of their association.

Most codivergence of interacting lineages probably involves a complicated mix of ecological and evolutionary processes. Some species sharing the same habitats will codiverge in some regions and not in others. Some coevolving populations will cospeciate, while other populations of one of the partners will switch their interactions to other species. The result is constantly varying degrees of codivergence of interacting species at different geographic scales and timescales. A well-studied example is the divergence of coevolving leaf-cutter (attine) ants and the fungi they cultivate as food in their fungus gardens. These fungi are directly transmitted by the ants to new colonies generation after generation, creating the opportunity for codiversification of these ants and their symbiotic fungi. Although attine ant and fungal lineages show overall patterns of codiversification, over millions of years there have been multiple instances of acquisition of new fungi by some attine species and occasional shifts of fungal cultivars among attine lineages.

8. CONCLUSIONS

Coevolution is one of the major processes driving the adaptive divergence of populations, shaping interactions between species in different ways in different environments. Coevolution also has the potential to drive population divergence faster than does adaptation only to different physical environments. There is, however, much we still do not know about the genetic and ecological mechanisms by which different forms of coevolutionary selection contribute to speciation and adaptive radiation.

FURTHER READING

Benkman, C. W. 2010. Diversifying coevolution between crossbills and conifers. Evolution: Education and Outreach 3: 47–53. *Summarizes research into one of the best-studied examples of how coevolution beween predators and prey may drive speciation.*

Futuyma, D. J., and A. A. Agrawal. 2009. Macroevolution and the biological diversity of plants and herbivores. Proceedings of the National Academy of Sciences USA 106: 18054–18061. *Reviews the variety of ways in which interactions between plants and herbivores shape large scale patterns in diversification.*

Kay, K. M., and R. D. Sargent. 2009. The role of animal pollination in plant speciation: Integrating ecology, geography, and genetics. Annual Review of Ecology and Systematics 40: 637–656. *Discusses the mechanisms by which pollinators can shape patterns of speciation in plants.*

Losos, J. B. 2010. Adaptive radiation, ecological opportunity, and evolutionary determinism. American Naturalist 175: 623–639. *Provides an overview of how environmental heterogeneity and species interactions can shape adaptive radiation.*

Moran, N. A., J. P. McCutcheon, and A. Nakabachi. 2008. Genomics and evolution of heritable bacterial symbionts. Annual Review of Genetics 42: 165–190. *Describes the pervasiveness of bacterial symbionts in eukaryotes and their potential roles in speciation.*

Pfennig, D. W., and K. S. Pfennig. 2010. Character displacement and the origins of diversity. American Naturalist 176: S26–S44. *Considers how competition between species can drive the process of speciation and the diversification of lineages.*

Schluter, D. 2009. Evidence for ecological speciation and its alternative. Science 323: 737–741. *Summarizes current views on how divergent natural selection among environments, especially in species interactions, can lead to speciation.*

Schluter, D. 2010. Resource competition and coevolution in sticklebacks. Evolution: Education and Outreach 3: 54–61. *Summarizes research into one of the best-studied examples of how character displacement driven by competition and predation may lead to speciation.*

Segraves, K. A. 2010. Branching out with coevolutionary trees. Evolution: Education and Outreach 3: 62–70. *Provides an overview of the processes shaping phylogenetic diversification, including the process of escape-and-radiate coevolution.*

Thompson, J. N. 2005. The Geographic Mosaic of Coevolution. Chicago: University of Chicago Press. *Synthesizes research showing how coevolving interactions diverge among environments, leading to local adaptation and ecological diversification.*

VI.8

Genetics of Speciation

H. Allen Orr and Daniel McNabney

OUTLINE

1. Genetics of prezygotic isolation
2. Genetics of postzygotic isolation
3. Summary

This chapter reviews current understanding of the genetic basis of speciation. Much progress has been made in the past several decades in uncovering how new species arise genetically. These recent studies analyze the barriers that prevent gene flow between closely related populations or species.

GLOSSARY

Dobzhansky-Muller Model. A model that explains how hybrid sterility and inviability can evolve unopposed by natural selection. The model emphasizes interactions among two or more loci.

Epistasis. Nonadditive interaction between loci.

Extrinsic Postzygotic Isolation. Reproductive isolation that results when hybrids are intermediate in phenotype and fall between parental niches.

Intrinsic Postzygotic Isolation. Reproductive isolation that results when hybrids suffer from developmental defects.

Pheromone. A chemical that can affect the behavior of other individuals, typically individuals that belong to the same species.

Most biologists define speciation as the evolution of reproductive isolation between populations (see chapter VI.1). This reproductive isolation is traditionally broken into two types: prezygotic and postzygotic. In prezygotic reproductive isolation, genes do not flow readily between populations or species because of barriers (e.g., courtship differences) that act before the formation of hybrid offspring. In postzygotic reproductive isolation, genes do not flow readily between populations or species because of barriers (e.g., hybrid sterility) that act after formation of hybrid offspring. (Genes cannot move between two populations if all hybrids between these populations are inviable or sterile.) In general, we are most interested in those forms of reproductive isolation—and in the genes giving rise to them—that appear early during the process of speciation. We are less interested, for example, in those forms of isolation that appear after the attainment of complete reproductive isolation. In practice, however, it is sometimes difficult to determine which forms of isolation arose or which genes diverged earliest during speciation.

A good deal is now known about the genetic basis of reproductive isolation, and much of this information has been obtained over the last two decades or so. Indeed, among contemporary evolutionary biologists, much of the progress in and excitement about speciation has focused on developments in the genetics of speciation. Here we summarize some of these developments. As will become clear, we know far more about the genetics of postzygotic than prezygotic isolation.

1. GENETICS OF PREZYGOTIC ISOLATION

Although the genetic basis of prezygotic isolation is less well understood than that of postzygotic isolation, evolutionary biologists have long appreciated its importance in speciation. For instance, Ernst Mayr—a key figure in the Modern Synthesis—argued that "if we were to rank the various isolating mechanisms of animals according to their importance, we would have to place behavioral isolation far ahead of the others." Also, recent empirical studies in taxa including fruit flies and birds reveal that prezygotic isolation can evolve more quickly than postzygotic isolation, especially when populations or species occur in the same geographic area.

A number of theories explain how prezygotic isolation evolves between populations. Many emphasize the role of sexual selection in the evolution of traits. When such evolution occurs independently in two geographically separated populations, it can yield taxa that may no longer find each other attractive once they encounter each other again. (Variants of sexual selection include Fisherian runaway, good genes, sensory drive, and sexual conflict; see chapter VI.5.)

Three main classes of phenotypes can give rise to prezygotic reproductive isolation: ecological, behavioral, and gametic. Although we know something about the genetic basis of each class, some of these barriers are better understood than others. We consider each in turn.

Ecological Isolation

Adaptation to different local environments can lead indirectly to reproductive isolation. One of the best-understood examples of ecological isolation involves two species of monkeyflower (*Mimulus*). These species are adapted to different pollinators, and this differential adaptation gives rise to reproductive isolation. In particular, *M. lewisii* has pink flowers and is pollinated primarily by bumble bees, whereas *M. cardinalis* has red flowers and is pollinated primarily by hummingbirds. Despite overlap in part of their geographic ranges, these two species show essentially complete reproductive isolation.

Genetic analyses of 12 floral traits that differ between these species—differences that likely underlie adaptation to bumble bee versus hummingbird pollinators—reveal that a total of 47 different chromosomal regions are involved. For 9 of these 12 traits, a locus of major effect (defined operationally as a locus that explains at least 25 percent of the species difference) was found, suggesting that these trait differences might have a reasonably simple genetic basis. One of these chromosomal regions includes the *YUP* locus, which regulates the amount of yellow carotenoid pigment incorporated into the petals of flowers. Genetically moving the *YUP* region from each species into the other causes a dramatic shift both in flower color and visitation by bee versus hummingbird pollinators in the field.

Habitat isolation is thought to explain the persistence of a hybrid species of wild sunflower (*Helianthus*). The hybrid species, *H. paradoxus*, which formed between two salt-sensitive species, *H. annuus* and *H. petiolaris*, shows at least a fivefold increase in fitness under high-salt conditions relative to its parental species. Increased salt tolerance has allowed the hybrid species to invade brackish salt marshes in which neither parental species can survive. Filling this novel niche appears to isolate the hybrid species from the two parental types. Genetic analysis of the parental species shows that at least 17 chromosomal regions contribute to the ability to survive in high-salt areas; some evidence suggests that genes that are involved in calcium transport play a role in this salt tolerance.

Temporal isolation, which occurs when two species are reproductively isolated because of differences in the timing of breeding, represents another form of prezygotic isolation often connected to ecology. Unfortunately, rigorous studies of the genetic basis of temporal isolation appear to be lacking.

Sexual Isolation

Differences in courtship signals or rituals can also cause prezygotic isolation between species. During courtship, one sex may present a signal that must be interpreted correctly by the opposite sex. If the signal and/or the preference for the signal diverge between independently evolving populations, reproductive isolation can result. Indeed, in many taxa individuals prefer signals from individuals belonging to the same species relative to those from individuals belonging to other species.

To take a well-known example, differences in pheromones between populations or species can cause sexual isolation. A large body of work has examined this phenomenon in the fruit fly *Drosophila*. Using a simple method to transfer pheromones between species, Jerry Coyne and colleagues showed that pheromonal differences explain several examples of sexual isolation between species of the *D. melanogaster* group (which includes the species *D. melanogaster*, *D. simulans*, *D. mauritiana*, and *D. sechellia*). *D. melanogaster* and *D. sechellia* are both sexually dimorphic species (i.e., males and females differ in their predominant pheromone), whereas *D. mauritiana* and *D. simulans* are sexually monomorphic (i.e., males and females have the same predominant pheromone).

Studies have shown that at least five regions of the third chromosome contribute to differences in female pheromones between species of the *D. melanogaster* group. Genes that contribute to a difference in male pheromones between *D. simulans* and *D. sechellia* map throughout the genome, with at least one gene on each major chromosome arm. In other groups of *Drosophila*, epistatic interactions between genes on the X and second chromosome alter the relative amounts of two pheromones between *D. pseudoobscura* and *D. persimilis*. In the *virilis* group of *Drosophila*, genetic mapping of *D. virilis*, *D. novamexicana*, and *D. lummei* female pheromonal differences has identified several chromosomal regions on the autosomes.

The European corn borer, *Ostrinia nubilalis*, provides another striking example of the role of pheromones in sexual isolation. A difference in pheromone blend isolates two forms of this species. Just two loci account for both the difference in pheromone production and the preference for strain-specific pheromone blend seen in these corn borer populations.

Divergence in color pattern can also cause behavioral isolation. In cichlids found in Lake Victoria, females prefer male color that contrasts with the surrounding light environment. (The brightness and color of light varies with water depth, among other factors.) Female vision has adapted to local light conditions, and in response, males have evolved color patterns that stand out against their local environments. Differences in female vision reflect changes in opsin proteins that allow females to distinguish colors. One of the genes that encode opsin proteins, long-wavelength-sensitive opsin, is the most variable opsin gene in cichlids and shows strong signs of divergent natural selection among populations. (Genetic differentiation at other genes that are thought to be neutral confirms that populations are reproductively isolated.) As human activity has caused the water in Lake Victoria to become cloudy, species diversity has fallen. It appears that increased turbidity prevents females from distinguishing between males belonging to the same versus different species, leading to a collapse of prezygotic reproductive isolation. Taken together, this evidence suggests that female preference for male color is a key isolating barrier in Lake Victoria cichlids.

Another example of sexual isolation that involves color pattern differences occurs among Solomon Island flycatchers. Plumage color pattern acts as a species recognition signal among populations. In particular, Makira Island flycatchers have a chestnut belly, whereas flycatchers on nearby islands have black bellies. Albert Uy and colleagues hypothesized that the *MC1R* gene, which has been shown to explain color differences among a number of animal species, might underlie differences in plumage color. After sequencing *MC1R* from individuals from each population, Uy and colleagues identified a change in a single amino acid between the Makira Island and Santa Ana Island populations that shows perfect association with the color pattern difference, suggesting a role for *MC1R* in sexual isolation between those islands. Interestingly, a black-bellied population that resides on Ugi Island does not show the same association between amino acid change and plumage difference, which suggests that different genes may cause plumage changes in different populations of Solomon Island flycatchers.

Songs often represent an important component of courtship rituals. Courtship songs in animals such as crickets, fruit flies, frogs, and birds can diverge, giving rise to sexual isolation. In *Laupala*, a genus of Hawaiian crickets, differences in courtship song between the species *L. paranigra* and *L. kohalensis* are caused by differences in several chromosomal regions. Each of these chromosomal regions appears to explain a small amount of the total species difference in courtship song, which suggests that differences in song in this system involve many genes of small effect.

Gametic Isolation

Finally, prezygotic isolation can occur after mating/spawning but before formation of a hybrid zygote, so-called postmating prezygotic isolation. One type of postmating prezygotic isolation involves isolation between the gametes of two species. Examples of such gametic isolation occur in abalones and sea urchins, both of which release their gametes into the water . The abalone and sea urchin systems have been thoroughly studied genetically. In abalones, fertilization requires the sperm protein lysin to interact successfully with the egg protein VERL. Lysin shows high levels of divergence between species; both lysin and VERL show little polymorphism within species. The combination of high levels of divergence between species and low levels of polymorphism within species suggests that positive selection has driven divergence of these proteins between species. Similarly, in sea urchins, fertilization requires the bindin protein of sperm to interact successfully with a surface receptor on eggs. The bindin protein has also been shaped by positive selection, particularly between species that are found together geographically.

2. GENETICS OF POSTZYGOTIC ISOLATION

Serious discussion of postzygotic isolation, like so much else in evolutionary biology, began with Darwin, who devoted a chapter of *The Origin* to hybrid sterility. Darwin was primarily concerned with the problem of how something as obviously maladaptive as hybrid sterility could evolve under natural selection (this problem is sometimes called "Darwin's dilemma"). Later, during the Modern Synthesis of the 1930s and 1940s, progress was made on both theoretical and empirical aspects of the genetics of postzygotic isolation. The theoretical progress included a plausible solution to Darwin's dilemma, and the empirical progress included work on the genetic basis of hybrid sterility and inviability, in both animals and plants. Although the genetics of postzygotic isolation was largely neglected in the decades following the Modern Synthesis, a new burst of work began during the 1980s, particularly in model systems like the fruit fly *Drosophila*. This work has continued into the genomics era, with the identification and characterization of individual genes that cause the sterility or inviability of hybrids.

Before summarizing our current understanding of the genetics of postzygotic isolation, it is important to distinguish between the two forms that it can assume: extrinsic and intrinsic. In *extrinsic postzygotic isolation*, hybrids suffer low fitness not because of any inherent defect in development but because they fall between the ecological niches occupied by the parental species. In *intrinsic postzygotic isolation*, hybrids suffer low fitness because they suffer from defects in development.

Extrinsic Postzygotic Isolation

If different populations or closely related species are adapted to different ecological conditions—as they surely are—extrinsic isolation might often appear among their hybrids. Surprisingly, however, we know little about the genetics of this kind of reproductive isolation. Perhaps the best-studied example in animals involves a small fish, the threespine stickleback (*Gasterosteus aculeatus*). Two different forms or "morphs" of these sticklebacks are sometimes found in freshwater lakes along the west coast of North America, e.g., Paxton Lake on Texada Island. The limnetic morph is adapted to open waters and has a narrow morphology; the benthic morph is adapted to the littoral zone and has a broader, deeper morphology. Morphological differences include traits plausibly involved in predation, feeding, and adaptation to different habitats, e.g., jaw morphology and number of gill rakers.

Hybrids between the two morphs occur rarely in nature and can be produced readily in the laboratory. For many morphological characters, hybrids are intermediate between the benthic and limnetic parental forms. Studies by Schluter, Rundle, and colleagues have shown that these hybrids, although intrinsically healthy, suffer low fitness when placed in either the benthic or limnetic habitats. In short, these hybrids, with their intermediate morphologies, appear to fall between the morphologies required to succeed in either parental niche (benthic or limnetic). Genetic analysis reveals that some of the relevant morphological differences reflect divergence at many loci, while others reflect divergence at a modest number of genes.

Other examples of extrinsic postzygotic isolation—some involving intermediate behavior in hybrids (e.g., hybrid birds that migrate in an incorrect direction)—are well known. Unfortunately, most such cases have not yet been rigorously genetically analyzed.

Intrinsic Postzygotic Isolation

Much more is known about the genetics of intrinsic than extrinsic postzygotic isolation. It is clear that a variety of genetic mechanisms can, and sometimes do, cause developmental problems in hybrids. It has long been known, for instance, that polyploidy—the sudden doubling of chromosome number—plays a part in hybrid sterility in many plants (see chapter VI.9). Chromosomal arrangements that differ between species also sometimes cause fertility problems in species hybrids. Some species of mice, for example, feature a number of different chromosome rearrangements; when present in heterozygous form in hybrids, these rearrangement differences disrupt meiosis, thereby lowering fertility. At least in animals, however, intrinsic postzygotic isolation appears often to result from incompatibilities between *genes* in hybrids.

Early in the twentieth century, Bateson, Dobzhansky, and Muller each showed how such genic incompatibilities could evolve unopposed by natural selection, thus resolving Darwin's dilemma. The key to the *Dobzhansky-Muller model* (Bateson's precedent was appreciated only much later) is that new alleles at two or more genes might have beneficial fitness effects (or no fitness effects at all) within species but cause sterility or inviability when brought together in species hybrids. For example, consider a simple two-locus example in which two geographically separated species begin with the genotype *aabb*. In one species, an *A* mutation appears and spreads, yielding *AAbb* individuals; in the other species, a *B* mutation appears and spreads, yielding *aaBB* individuals. Both mutations are fit on their usual species genetic background. But if the two species later meet and cross, there is no guarantee that the resulting *AaBb* hybrids will be fertile and viable. The reason is that the *A* and *B* alleles have never been "tested" together in a common genome. The Dobzhansky-Muller model thus emphasizes the role of epistasis in speciation: genes may interact in unpredictable ways within hybrids, possibly causing inviability or sterility. (It should also be noted that the Dobzhansky-Muller model can cause hybrids to suffer a loss of fitness that is both post- and prezygotic: if two populations independently evolve different mating behaviors, the combination of relevant genes in hybrids may yield individuals that are sexually unattractive and so suffer low mating success.)

A considerable body of genetic data now supports the Dobzhansky-Muller model. During the 1980s and 1990s, geneticists mapped the loci that cause hybrid sterility or inviability in many different pairs of species. These studies nearly always revealed that incompatibilities are the result of between-locus interactions: sterility or inviability arises when some chromosomal region from one species is brought together in hybrids with other chromosomal regions from the other species, as predicted by the model. In some cases, hybrid sterility or inviability appears to involve a single Dobzhansky-Muller incompatibility; that is, hybrid sterility or inviability is caused solely by the interaction in hybrids between two (or perhaps three or four) genes. In other cases, species appear to be separated by many Dobzhansky-Muller incompatibilities; that is, many different combinations of genes from two species independently cause developmental problems and lower hybrid fitness.

Indeed, hybrids between species that have diverged for a long period of time can suffer a kind of "overkill": many different Dobzhansky-Muller incompatibilities may *each* be capable of killing or sterilizing hybrids. The rapid accumulation of genes that cause sterility or inviability of hybrids, leading to this kind of genetic overkill, is expected on theoretical grounds and has been dubbed the "snowball effect."

Surprisingly, genetic analyses during the 1980s and 1990s revealed another pattern: the genes involved in these hybrid incompatibilities are often on the X chromosome. This so-called large-X effect is connected closely to another pattern that characterizes postzygotic isolation, at least in animals: Haldane's rule, which states that when only one hybrid sex is sterile or inviable, it is the "heterogametic" sex, that is, the sex that carries both an X and a Y chromosome. A flurry of genetic studies, mostly in the fruit fly *Drosophila*, showed that Haldane's rule has two likely causes: (1) the alleles involved in Dobzhansky-Muller incompatibilities are mostly recessive in their effects on hybrid fitness (and thus are fully expressed when they reside on the X chromosome of XY hybrids); and (2) the X chromosome has an especially high concentration of genes involved in Dobzhansky-Muller incompatibilities. The reasons for this recessivity and high concentration of postzygotic isolation genes on the X chromosome remain somewhat uncertain and are the focus of much current work.

In the last decade, evolutionary geneticists have devoted much attention to the molecular identification and characterization of the genes that cause intrinsic postzygotic isolation. This effort has proven difficult. The reason is simple: the attempt to identify "speciation genes" is the attempt to do genetics where it is, by definition, nearly impossible to do—between species, that is, between taxa that do not readily exchange genes. Several genetic and molecular techniques have been used to overcome this problem, and as a result, approximately a dozen genes that cause some hybrid sterility or inviability have been identified at the DNA sequence level. (Most of these studies were performed in *Drosophila*, though others were performed in vertebrates.)

Although this sample of genes is small, several patterns have already emerged from these studies. First, the genes that cause intrinsic postzygotic isolation have many different biological functions: some encode DNA-binding proteins, some encode enzymes, and yet others encode structural proteins. Second, comparison of DNA sequences between two species that produce sterile or inviable hybrids shows that the genes that cause these hybrid problems are often rapidly evolving. Third, this rapid evolution is often caused by positive selection. (This can be shown via several molecular population genetic tests—for example, the McDonald-Kreitman test—that use DNA sequence data from the two species that produce sterile or inviable hybrids.) The precise nature of the selection involved remains somewhat unclear, however. Geneticists are currently investigating two possibilities: that rapid evolution reflects (1) adaptation to the external ecological environment, or (2) "genetic conflict," that is, adaptation to the "selfish" effects of other genes in the genome. Some evidence supports this second possibility—for example, some genes involved in hybrid sterility also appear to be involved in forms of genetic conflict. But more data are required before confident conclusions can be drawn about the frequency with which ecological adaptation versus genetic conflict drives the evolution of the genes causing intrinsic postzygotic isolation.

3. SUMMARY

In conclusion, evolutionary biologists now know a good deal about the genetic basis of postzygotic reproductive isolation and a growing amount about the genetic basis of prezygotic reproductive isolation. It seems clear that both forms of isolation typically involve a history of selection, whether natural or sexual. It is also clear that genes that have large effects on reproductive isolation exist, although reproductive isolation sometimes seems to result from the divergence of many genes, each of smaller effect.

FURTHER READING

Coyne, J., and H. A. Orr. 2004. Speciation. Sunderland, MA: Sinauer. *A book-length review of speciation, including the genetics of speciation.*

Noor, M. A., and J. L. Feder. 2006. Speciation genetics: Evolving approaches. Nature Reviews Genetics 7: 851–861. *A review of speciation genetics focusing on the use of modern genetic techniques such as high-throughput molecular techniques and gene manipulation.*

Nosil, P., and D. Schluter 2011. The genes underlying the process of speciation. Trends in Ecology & Evolution 26: 160–167. *A review of speciation genetics focusing on the importance of characterizing genes that play a role in the initial species split.*

Palumbi, S. R. 2009. Speciation and the evolution of gamete recognition genes: Pattern and process. Heredity 102: 66–76. *A review of the biology, including the genetics, of gametic reproductive isolation between species.*

Presgraves, D. C. 2010. The molecular evolutionary basis of species formation. Nature Reviews Genetics 11: 175–180. *An up-to-date review of progress on attempts to characterize the actual genes and proteins that cause reproductive isolation between species; focuses on the genetics of postzygotic isolation.*

Rieseberg, L. H., and J. H. Willis. 2007. Plant speciation. Science 317: 910–914. *A review of reproductive isolation between plant species that includes discussions of prezygotic isolation, postzygotic isolation, and polyploid speciation.*

Uy, J.A.C., R. G. Moyle, C. E. Filardi, and Z. A. Cheviron. 2009. Difference in plumage color used in species recognition between incipient species is linked to a single amino acid substitution in the melanocortin-1 receptor. American Naturalist 174: 244–254. *An example of a recent study identifying genetic changes that contribute to prezygotic isolation.*

VI.9

Speciation and Genome Evolution

Jeffrey Feder, Scott P. Egan, and Patrik Nosil

OUTLINE

1. From beanbags to genomes
2. Geography and gene flow
3. Primary versus secondary geographic contact
4. Selection-recombination antagonism and genomic heterogeneity
5. Empirical data and patterns
6. Chromosomal rearrangements and speciation
7. Polyploidy and speciation
8. Sex chromosomes and speciation
9. Speciation and genomic architecture

Speciation involves the splitting of one group of interbreeding natural populations into two or more reproductively isolated groups. Therefore, to understand speciation one must understand how genetically based barriers to gene flow (i.e., reproductive isolation) evolve between populations. Progress has been made in discerning the importance of different factors, traits, and individual *speciation genes* in generating reproductive isolation. However, we are just beginning to understand how speciation genes are embedded and arrayed within the genome, and thus how genomes evolve collectively during population divergence. Although it is now clear that different regions of the genome often vary in their level of genetic divergence between populations, major questions remain about how genomic architecture facilitates or impedes speciation. This chapter reviews our current understanding of genome-wide patterns of genetic divergence during speciation. The focus is on the processes causing genomic divergence and on their consequences for speciation.

GLOSSARY

Chromosomal Inversion. A rearrangement of a chromosome in which a segment of the chromosome is reversed end to end in its orientation, causing the inverted region of the chromosome to have an inverse linear order of genes compared with the corresponding "collinear" arrangement.

Divergence Hitchhiking. A term used to describe the process by which physical linkage to a divergently selected gene(s) increases genomic divergence for adjacent regions along a chromosome.

Divergent Selection. Selection that acts in different directions between two populations, including the special case in which selection favors two extremes within a single population (i.e., disruptive selection).

Genome Hitchhiking. A term used to describe the process by which genetic divergence across the entire genome is facilitated, even for loci unlinked to those under selection, by a global reduction in average genome-wide gene flow caused by selection.

Genomic Island Of Divergence. A region of the genome, of any size, whose divergence exceeds neutral background expectations based on overall divergence across the genome.

Heterogeneous Genomic Divergence. A term used to describe the highly variable levels of genetic divergence between populations across different regions of the genome.

Isolation By Adaptation (IBA). A pattern of positive correlation between the degree of adaptive phenotypic divergence between populations and their level of molecular genetic differentiation, independent of geographic distance.

Linkage Disequilibrium. The nonrandom association (i.e., correlation) of alleles at two or more loci.

Recombination. In eukaryotes, a process by which a piece of DNA is broken and then joined to a different piece during gamete formation (i.e., meiosis) via chromosomal crossing-over, which leads to offspring that have different combinations of alleles from those of their parents.

Selection-Recombination Antagonism. A term coined by Joseph Felsenstein to describe how recombination breaks up associations between selected loci and loci that cause reproductive isolation, impeding genetic divergence across the genome, and constraining speciation with gene flow.

1. FROM BEANBAGS TO GENOMES

Speciation is a fundamental evolutionary process responsible for creating the great diversity of life on earth. Speciation occurs as one interbreeding population evolves into two or more reproductively isolated groups or taxa (see chapter VI.1 on the concepts of species and speciation). Defining speciation in this manner leads to a basic research question: How do genetically based barriers to gene flow that cause reproductive isolation evolve between populations? Identifying factors promoting population divergence and genetically characterizing traits that cause reproductive isolation are therefore central endeavors for students of speciation (see chapter V1.4 on different reproductive barriers and the role of natural selection in speciation). Another important question concerns how the genes that cause reproductive isolation are positioned relative to one another in the genome. Discerning the physical arrangement of such "speciation genes" is important, because when populations are not fully reproductively isolated and still interbreeding, this genomic architecture may facilitate or impede further divergence. Thus, the question of genome structure links the evolution of individual speciation genes to their collective consequences for speciation.

Our empirical and theoretical understanding of the genetics of speciation has been dominated by what Ernst Mayr described as "beanbag thinking"—a focus on identifying and characterizing individual genes that cause reproductive isolation. However, we are now capable of rapidly scanning large portions of the genome in both model and nonmodel organisms for genetic differentiation during speciation. This ability has enabled researchers to begin studying how speciation genes are embedded and arrayed within the genome and thus how genomes evolve collectively during population divergence. This shift in focus is due, in part, to the rapid technological advances in mass sequencing technologies (i.e., next-generation sequencing platforms) that allow the surveying of many more genomic regions at a fraction of the cost and time. Consequently, the field of evolutionary genomics is moving away from "beanbag" approaches and purely descriptive studies of individual genes and their individual effects toward a more predictive framework that tackles the causes and consequences of genome-wide patterns. In this chapter, we examine what is currently thought about how speciation genes are arrayed in the genome and how genome structure may influence speciation.

2. GEOGRAPHY AND GENE FLOW

The geographic context under which populations diverge is a critical consideration for the role that genome architecture may play in speciation. When populations are completely geographically isolated in allopatry, there is no migration of individuals between populations. Hence, genetic differences can readily accumulate anywhere in the genome between populations by natural selection, sexual selection, and genetic drift. These differences can cause reproductive isolation as an inadvertent by-product if the populations come into contact. In cases where speciation occurs completely in allopatry, genome structure may not be critical to speciation, as the position of genes in the genome may not greatly affect their overall potential to diverge between populations. Thus, given enough time, divergence across the entire genome is inevitable in allopatry. Consequently, although much has been learned about specific reproductive barriers and individual speciation genes from studying allopatric taxa, it is difficult to ascribe any special significance to a particular genetic change or genetic architecture in such systems. Indeed, allopatric *Drosophila* fruit fly taxa, which have been the focus of much study, typically differ by hundreds of genes contributing to sterility and inviability distributed throughout the genome.

In contrast, genomic architecture could be very important for taxa that overlap geographically and exchange migrants and genes during the speciation process, either throughout the process (i.e., sympatric or parapatric) or during secondary contact between formerly allopatric populations before speciation is complete. Here, populations do not evolve completely independently of one another. Factors causing populations to diverge must overcome gene exchange due to migration and hybridization (interbreeding) if speciation is to proceed or, in cases of secondary contact, to continue. Thus when populations overlap and gene flow occurs during the speciation process, aspects of genomic architecture such as the distribution of genes within and among chromosomes (physical linkage relationship), and recombination rates among genes, may be important for enhancing the efficacy of selection and genetic drift in differentiating populations (see chapter VI.6 for additional discussion of gene flow and hybridization).

3. PRIMARY VERSUS SECONDARY GEOGRAPHIC CONTACT

Another important consideration for understanding the role genomic architecture may play in speciation is the timing of geographic overlap and the timing of the evolution of reproductive isolation. "Speciation with gene flow" can include speciation events during primary geographic contact (i.e., sympatric or parapatric speciation), where reproductive isolation evolved initially in the face of gene flow, and speciation events with secondary geographic contact, where reproductive isolation evolved partially in allopatry but then continues in sympatry. The timing of onset of divergence can vary widely among different instances of speciation, and in most cases of incipient speciation with gene flow present, it is very difficult to determine whether gene flow was present throughout divergence or only recently. The important concept here is that regardless of the starting point, when populations along the speciation continuum come into geographic contact and experience gene flow, genomic architecture can play an important role, albeit one dependent on the timing of the onset of gene flow. In cases of primary contact, genomic architecture can be critical to the initial stages of divergence; in cases of secondary contact, genomic architecture can help maintain (partial) reproductive isolation that evolved initially in allopatry.

4. SELECTION-RECOMBINATION ANTAGONISM AND GENOMIC HETEROGENEITY

Classic theoretical work by Joseph Felsenstein regarding the roles that selection, gene flow, and recombination play in shaping patterns of genetic divergence during speciation was described in his paper "Skepticism towards Santa Rosalia, or Why Are There So Few Kinds of Animals?" which clarified the potential importance of genomic architecture in affecting speciation with gene flow. This insightful paper developed the term *selection-recombination antagonism* to describe the tug-of-war between divergent selection—which builds up associations between loci adapting populations to different habitats and loci causing assortative mating—and gene flow and recombination, which break these favorable combinations apart. Gene flow and recombination impede divergence because they cause individuals adapted to a particular habitat to not breed true with other individuals possessing the same suite of locally adapted genes, and thus to generate genetically mixed offspring that do not exclusively possess the locally adapted phenotypes of their parents. The implication of selection-recombination antagonism was therefore that if different genes affect assortative mating and local adaptation, little progress could be made toward speciation unless the genes were fortuitously very tightly linked on a chromosome, such that recombination between them was very limited.

If the limitations imposed by the selection-recombination antagonism are pervasive in nature, then taxa speciating with gene flow should display a highly heterogeneous pattern of genomic divergence. A subset of the genome containing tightly linked genes under disruptive selection will display differentiation, while the homogenizing effects of gene flow and recombination should preclude divergence across the majority of the remaining genome (or in the case of secondary contact, eradicate existing levels of differentiation). Early empirical observations were consistent with this view of a highly heterogeneous genome (see further discussion) and led to the metaphor of genomic islands of speciation. A *genomic island* is any region of the genome that exhibits significantly greater differentiation between speciating taxa than predicted by neutral evolution, in which neutral patterns are based on overall divergence across the genome. The metaphor thus draws parallels between genetic differentiation observed along a chromosome, and the topography of oceanic islands (figure 1). In this case, isolated genomic islands contain putatively selected loci plus closely linked hitchhiking genes that rise above the statistical threshold predicted by neutrality, represented by sea level. In contrast, the ocean floor below sea level represents the majority of the neutrally evolving genome. Factors such as physical proximity between selected and other loci, rates of recombination, and strength of selection affect the height and the size of the genomic islands. Therefore, the metaphor of genomic islands offers a testable hypothesis for taxa speciating in the face of gene flow: if selection-recombination antagonism is common, the few genes under strong selection or those genes physically linked to loci experiencing strong divergent selection will exhibit strong differentiation and will tend to be clustered together in the genome, whereas gene flow will homogenize the remainder of the genome below sea level.

Despite the logic and heuristic appeal of the island metaphor, several aspects of Felsenstein's model were not meant to be biologically realistic but rather were intended to highlight general points about the selection-recombination antagonism. For example, many organisms mate in preferred habitats rather than in a common mating pool, which can help alleviate the selection-recombination antagonism by reducing gene flow among populations with differing habitat preferences. The same is true when assortative mating is based on the phenotypic similarity of organisms rather than on their genotypes at assortative mating loci. Moreover, if selection is strong and migration is not random between populations, an association (i.e., linkage disequilibrium) is established between survival and habitat choice genes even in the absence of physical linkage. These considerations have led to the idea that *multifarious selection*, which is selection affecting multiple loci across the genome, could also kick-start speciation with gene flow. In this case, widespread selection could reduce gene flow sufficiently such that divergence could build up or be maintained across a larger extent of the genome.

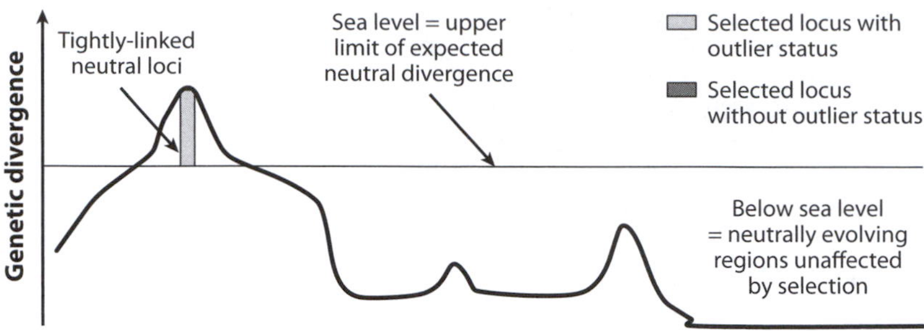

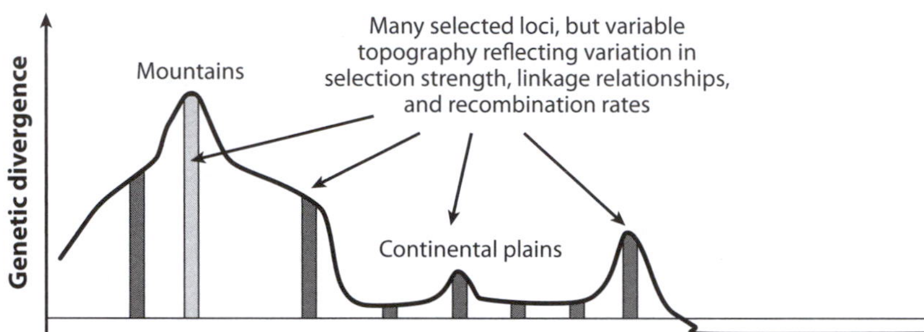

Figure 1. Illustration of the island (A) and continent (B) metaphors for genomic divergence. Note, these views are not mutually exclusive but represent parts of a continuum of genomic divergence among populations. (Figure from Michel et al. 2010 and reprinted with permission of the National Academy of Sciences of the United States of America.)

5. EMPIRICAL DATA AND PATTERNS

Researchers conduct "genome scans" to try to characterize patterns of genetic differentiation between speciating taxa and to identify genomic regions containing candidate speciation genes. For example, a genome scan might compare genome-wide patterns of genetic divergence among populations of an insect that inhabits different host plants (i.e., differing selective environments) and exhibits adaptive divergence and reproductive isolation associated with each environment. A genome scan study typically looks at genetic divergence across dozens to hundreds of thousands of molecular markers, such as amplified fragment length polymorphisms (AFLPs), microsatellites, single nucleotide polymorphisms (SNPs), or DNA sequences themselves, in a large number of individuals per population in each of two or more populations. These studies use the distribution of genetic divergence among markers to distinguish loci whose level of genetic divergence between populations exceeds neutral expectations. Such loci with unusually high levels of genetic divergence are deemed "outlier loci" and are interpreted as putatively being under divergent selection or closely linked to genes that are. The remainder of the genome (i.e., "nonoutlier loci") is deemed to be neutrally evolving (at least statistically) and thus subject to homogenization among populations owing to gene flow. Genome scans thus have the potential to quantify the number of genomic regions under selection and their distribution across the genome (figure 2).

Genome scan studies of populations diverging in the face of gene flow tend to find a few, often physically isolated, outlier loci and thus have generally been interpreted as supporting the island view of genomic divergence. For example, an AFLP-based genome scan performed by Scott Egan and colleagues in 2008 compared *Neochlamisus* leaf beetle populations undergoing speciation associated with adaptations to different host plants. In pairwise comparisons of populations on different host plants, roughly 5 percent of the AFLPs exhibited strong divergence, and these "outliers" appeared to be physically isolated based on linkage disequilibrium analysis. In contrast, evidence for more genomically

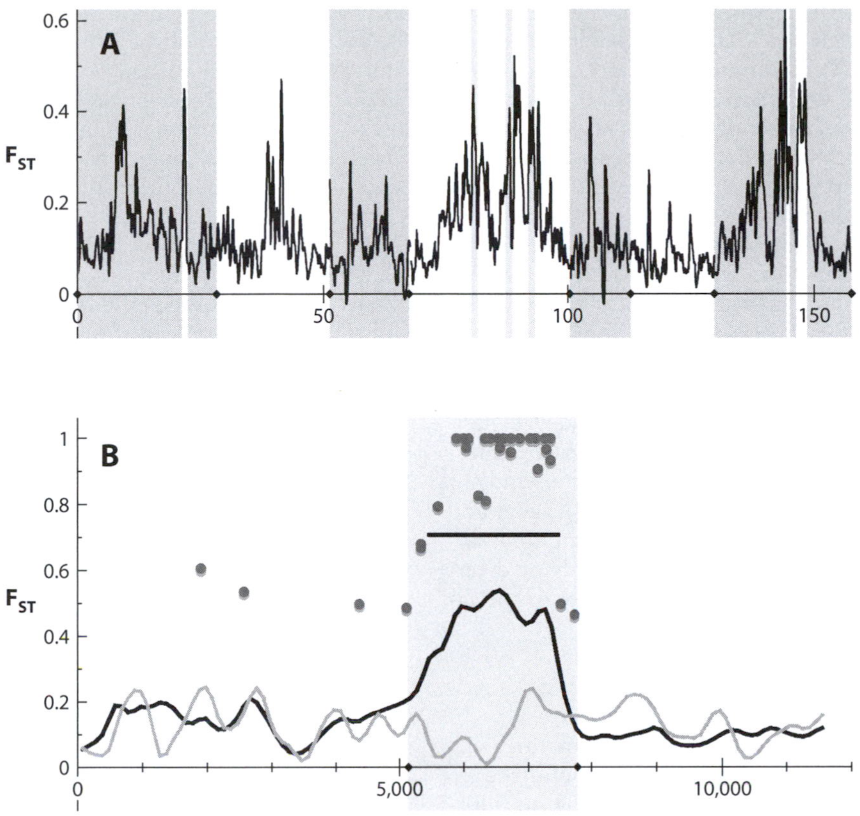

Figure 2. Genome-wide differentiation (F_{ST}) between populations of the threespine stickleback (*Gasterosteus aculeatus*). Panel A shows the distribution of genetic differentiation across 8 of the 21 linkage groups for an overall comparison of oceanic and freshwater populations. Vertical shading indicates boundaries of linkage groups. Panel B highlights a close-up comparison on linkage group 21 of an overall comparison of oceanic and freshwater populations (in black) and among freshwater populations (in gray). The black bar represents bootstrap significant genetic differentiation of a genetic region; dots represent significant point estimates of differentiation from a G-test corrected for multiple testing. (Figures from Hohenlohe et al. 2010 and reprinted with permission of the Public Library of Science.)

widespread divergence, as expected for multifarious selection, is rarer. However, this may stem from limitations in relying on genome scans alone for detecting selection. Genome scans can predestine an island view, because only the most diverged regions will be identified as statistical outliers. Other loci affected by selection, but more weakly, will go unnoticed and be considered part of the mostly "undifferentiated" and neutral genome.

In contrast with genome scans of populations observed in the wild, manipulative experiments exposing individuals to different conditions or factors mirroring those experienced by populations in the wild might allow the detection of weak selection acting across numerous regions of the genome missed by traditional genome scans. For example, field experiments could be conducted in which experimental populations are created on novel and native (i.e., control) environments (e.g., herbivorous insects could be transplanted to their native host or to a host-plant species used by a close relative). Loci that showed consistent and statistically significant allele frequency changes across replicates in the novel environment, but not in the control, would likely be affected by selection in the novel environment, even if such frequency changes were slight. If properly replicated, experiments could thus detect weak selection and distinguish selection from genetic drift and experimental noise to a greater degree than observational genome scans alone.

In 2010, Andrew Michel and colleagues conducted an experimental test of the genomic islands hypothesis in apple- and hawthorn-infesting host-plant races of the fruit fly *Rhagoletis pomonella*, a model for speciation with gene flow. In the experiment, groups of individuals from the apple and hawthorn host races were raised under the overwintering conditions of their native host and the alternative host plant, which differ in their fruiting times and, thus, the prewinter period each host race experiences prior to undergoing a winter diapause. Contrary to expectations of the islands hypothesis, the researchers reported widespread genetic divergence and selection throughout the *Rhagoletis* genome, with the majority of loci (33 microsatellites and six allozymes) displaying within-generation responses to selection in the manipulative overwintering experiment. Additionally, the majority of loci showed associations with latitude, associations with an ecologically relevant trait (adult eclosion time), and significant host differences in nature despite levels of gene flow that are too high to allow divergence via genetic drift. These results were coupled with linkage disequilibrium (LD) analyses, which test for combinations of alleles appearing more often in a population than would be expected based on random combinations of alleles proportional to their frequency in the population. This LD analysis provided experimental evidence that divergence was driven by selection on numerous independent genomic regions and not on a few islands. Based on their findings, Michel and colleagues proposed that "continents" of multiple differentiated loci, rather than isolated islands of divergence, may characterize even the early stages of speciation (figure 1B).

The authors stress, however, that the "island" versus "continent" views of genomic divergence represent ends of a continuum, rather than mutually exclusive hypotheses. Importantly, standard outlier analyses in this same study system detected only two independent outlier gene regions. Thus, experimental data and biological information on gene flow in nature were critical for detecting weaker, yet widespread, divergence across the genome. Additional such studies combining genome scans and selection experiments are needed before general conclusions can be drawn about how selection affects genomic patterns of differentiation.

Divergence Hitchhiking and Physical Linkage

To fully understand patterns of genomic differentiation and their significance for speciation we must not only empirically characterize genome-wide patterns of genetic divergence but also understand the mechanism responsible for generating the patterns. One mechanism potentially linking evolutionary constraints imposed by the selection-recombination antagonism with the concept of speciation islands is called *divergence hitchhiking*. According to this mechanism, a gene under divergent selection creates a localized window of reduced gene flow around it in the genome, because migrants moving into the population will tend to have maladaptive gene combinations for specific loci under selection, but—importantly—other (e.g., neutral) genes will also be linked to the selected loci. If there is not enough time for these associations between selected and other loci to be broken down by recombination, genes of both types may be eliminated by natural selection. Thus, not only will effective gene flow be reduced for the selected locus but also for nearby sequences, facilitating the initiation and buildup of genomic islands, and potentially their subsequent growth, via aiding the establishment and divergence of new mutations between populations. In contrast, unlinked sequences that can freely recombine away from maladapted genes will more easily invade other populations and be subject to higher levels of gene flow. As a result, in the early stages of speciation with gene flow, only a few genomic islands containing a handful of genes under divergent selection may separate taxa. However, the reduction in effective gene flow around these loci may allow the buildup of larger regions of genetic differentiation in surrounding loci subject to divergent selection. Thus, the initial isolated islands may become progressively larger through time.

Genome Hitchhiking and Isolation by Adaptation (IBA)

Divergence hitchhiking has great intuitive appeal as a mechanism ameliorating the selection-recombination antagonism and explaining heterogeneous patterns of genetic divergence. However, the efficacy of divergence hitchhiking is tied to the size of the windows of reduced gene flow for genes surrounding those under divergent selection. In a 2010 theoretical study, Jeff Feder and Patrik Nosil found that under conditions of strong selection, low migration, and small population sizes, this window of reduced gene flow could be relatively pronounced. However, these conditions are rather restrictive, especially for earlier stages of speciation with gene flow, when migration rates are expected to be relatively high and population sizes not small. Thus, critical incipient stages of speciation may require strong multifarious selection acting on several different loci and traits to kick-start the process. Feder and Nosil did find, however, that once a few loci under strong selection become established, effective migration rates can be reduced sufficiently (on average) on a genomically global basis for newer, less well favored mutations to establish throughout the genome. This process of *genome hitchhiking* is therefore distinguished from divergence hitchhiking in that the former does not require tight physical linkage of new mutations with previously selected loci. Genome hitchhiking is consistent with the empirical results of Michel and colleagues in the *Rhagoletis* study discussed earlier and could play an important role in the evolutionary transition of partially reproductively isolated races to more fully differentiated species

Empirically, the reduction in effective gene flow due to genome hitchhiking is predicted to generate a positive association among population pairs between levels of adaptive divergence (a proxy for the strength of selection) and genetic differentiation. This pattern has been referred to as *isolation by adaptation* (IBA) and may occur even for neutral loci unlinked to genes under selection. Evidence for IBA has been found in nature. In a study by John Grahame and colleagues in 2006, intertidal snails adapted to different sympatric shore habitats exhibit partial reproductive isolation. The authors compared hundreds of putatively neutral loci. For a given geographic distance, they observed an increase in neutral genetic differentiation between habitats—relative to within-habitat comparisons—consistent with a general barrier to gene flow across the habitat transition. This study provided a direct example of IBA, since these two habitats are very close to each other, such that some gene flow between them occurs. In a 2009 review by Nosil and colleagues of 22 studies that could test for this pattern of IBA, 68 percent showed evidence for IBA of putatively neutral loci. This study suggests that the IBA may be common in nature, as it was found across many different types of organisms, including lizards, wolves, leaf beetles, stick insects, and flowering plants.

6. CHROMOSOMAL REARRANGEMENTS AND SPECIATION

As described earlier, the antagonism between selection and recombination can impede genomic divergence. It therefore follows that factors that reduce recombination, such as chromosome inversions, can facilitate genomic divergence. Consistent with this prediction, in the 2010 *Rhagoletis* study on apple and hawthorn host races conducted by Michel and colleagues, regions of the genome harboring chromosomal inversions had, on average, twice the genetic divergence of collinear regions. Similar findings have also been reported by groups working on *Anopheles* mosquitoes and *Drosophila* fruit flies. Inversions may therefore help create more elevated oceanic islands and broader, more mountainous continents of divergence between taxa. Similar arguments can be made for other features of the genome that result in reduced recombination, such as, for example, proximity of genes under selection to centromeres, as observed in the European rabbit.

The basic premise for a role of inversions in speciation is therefore that they reduce introgression across the regions of the genome they encompass by protecting favorable genotypic combinations from being broken up by recombination, by reducing crossing-over. Essentially, the favorable genes within the inversions are more favorable together in their natal habitat than they would be individually, and less favorable in the habitats of other populations than they would be alone. Hence, gene flow in the inversion is reduced. Moreover, in addition to preserving blocks of adaptively diverged genes, inversions also provide larger targets in the genome for divergence hitchhiking to work; by suppressing recombination among populations, inversions enlarge the area of the genome in which new favorable mutations could arise linked to already-diverged genes.

An example in which inversions were involved in the evolution of reproductive isolation associated with adaptation to different environments stems from field experiments with ecologically divergent ecotypes of *Mimulus guttatus*. In 2010, David Lowry and John Willis showed that traits involved in local adaptation and reproductive isolation between coastal and inland ecotypes are located within a chromosomal inversion. They then established *Mimulus* lab populations, where they incorporated this inverted gene region into the standard genetic background of the alternative plant ecotype, and used them in a field experiment to demonstrate that the inversion contributes to divergent adaptation and reproductive isolation. In a similar vein, Mohamed Noor and colleagues have shown in a series of studies that many of the genes contributing to the reproductive isolation of two *Drosophila* fruit fly sister species that overlap in the western United States reside in inversions.

In addition to direct experimental and gene mapping data, empirical patterns in nature also provide evidence that inversions can promote speciation. For example, Noor and colleagues reviewed the *Drosophila* speciation literature in 2001 and found that sympatric pairs of species contained more inversions than allopatric species. This pattern was interpreted to indicate that on secondary geographic contact of populations following a period of allopatry, inverted regions of the genome were more resistant to gene flow than collinear regions (i.e., regions with orthologous linear orientation) with free recombination, and thus the inversions remained differentiated while collinear regions exchanged alleles readily and homogenized.

We stress, however, that although inversions may facilitate divergence, they are not required for speciation. Indeed, theoretical work by Feder and Nosil in 2009 showed that there is no reason for the vast majority of loci contributing to reproductive isolation to reside in inversions; they should also be commonly found in collinear regions, particularly when selection is strong relative to migration. Genome scan studies seem to bear out this supposition, as outlier loci presumably associated with adaptive divergence are often widely distributed across the genome. A study by Strasburg and colleagues in 2009 underscores this point. The authors examined divergence between hybridizing sunflower species at 77 loci distributed across the genome and reported that divergence is not accentuated within inversions, except perhaps near chromosomal breakpoints, in contrast with the results from *Rhagoletis*, mosquitoes, and *Drosophila*. Moreover, the authors observed widespread adaptive divergence in collinear regions.

Nevertheless, recent theoretical advances have shown that the presence of inversions themselves can be an outcome of adaptive divergence and an indicator of speciation with gene flow. The recombination suppressing effects of inversions are often not strongly favored in allopatric populations, where directional selection can act unhindered by gene flow and recombination to generate locally adapted sets of genes. Moreover, most new inversions have detrimental consequences during meiosis. Thus, it may be rare for inversions to differentially fix between taxa in allopatry. Instead, it is gene flow between locally adapted populations that favors the recombination suppressing effects of inversions and their spread. This important point was made by Mark Kirkpatrick and Nick Barton in a 2006 paper showing that new inversions that originate in sympatric populations exchanging genes that fortuitously happen to trap a combination of genes all adapted to one habitat versus another can be selectively favored over collinear arrangements, and increase in frequency to differentiate populations. Jeff Feder and colleagues recently demonstrated how such adaptive spread of inversions in the face of gene flow is facilitated if the inversions arise in allopatry, such that they contain the perfect complement of locally adapted alleles, and then rise to high frequency when gene flow ensues on secondary contact. Such a "mixed mode" of geographic divergence, alternating between allopatry and sympatry, allows for the establishment of inversions under a wide range of conditions.

7. POLYPLOIDY AND SPECIATION

Polyploidy describes the phenomenon in which there is a numerical increase in the whole set of chromosomes within an organism; polyploidy has been associated with speciation. Most species of multicellular eukaryotes are *diploid*, meaning that they have two sets of chromosomes—one set inherited from each parent—whereas polyploids have three or more complete sets of chromosomes. Polyploidy can occur via *autopolyploidy*, which describes the increase in ploidy within a species, or *allopolyploidy*, which describes the increase in ploidy resulting from hybridization between two distinct species. Polyploidy can arise via three pathways, regardless of whether it arises via hybridization or within a lineage: somatic doubling, meiotic nonreduction, and polyspermy. *Somatic doubling* occurs when irregularities arise during mitosis, and mitotic products fail to segregate to each pole, thus giving rise to a polyploid. If this doubling occurs in somatic tissue that gives rise to reproductive tissue or very early in development, a polyploid individual will result. *Meiotic nonreduction* occurs when a cell wall fails to form in late meiosis, generating diploid gametes, followed by fusion of two unreduced gametes fusing or one unreduced gamete and one normal gamete to form a triploid. This triploid in turn generates an unreduced triploid gamete, which fuses with a normal gamete, generating a polyploid. *Polyspermy* occurs when two sperm fertilize a single egg, yielding a triploid individual. Any of these mechanisms of polyploidy then relies on the event spawning a population of individuals that can reproduce.

Most eukaryotes have likely gone through a polyploidization event in their evolutionary history; thus polyploidy is a prominent feature in the evolution of eukaryotic life. However, the frequency of polyploidy seems to differ between animals and plants. While polyploidy has been clearly documented in animals, such as salmon and salamanders, it appears to be more common among plants, including ferns and flowering plants. In general, 47 to 70 percent of all plants are the descendants of polyploidy. More specifically, in 2009, Troy Wood and colleagues found that 31 percent of all speciation events in ferns and 15 percent of all speciation events in flowering plants were accompanied by an increase in ploidy. The important implication of polyploidy to speciation is that the polyploidization event itself can instantly generate some degree of reproductive isolation across the whole genome in sympatry, owing to meiotic difficulties experienced by the offspring of the diploid ancestor and the derived polyploid: their uneven number of sets of chromosomes causes varying degrees of sterility or inviability in the hybrids. The reduction in gene flow due to polyploidy is not always complete, and the tempo and magnitude of this reduction can vary among the different modes of polyploidy. However, if the level of gene flow between the parental and derived polyploidy is sufficiently reduced, divergent selection can then act to generate widespread genomic differentiation.

8. SEX CHROMOSOMES AND SPECIATION

Sex-determining chromosomes may also play an important role in the speciation process relative to the rest of the genome, termed *autosomes* (see chapter VI.8 for additional examples of the role sex chromosomes play in the speciation process). In a review of sex-linked speciation genes, Dorothy Pashley Prowell (1998) describes a common pattern in butterflies: a disproportionate association between traits that distinguish closely related butterfly species and the X chromosome. These traits also tend to be associated with prezygotic reproductive barriers, such as habitat preference. In a specific study using the fall army worm (*Spodoptera frugiperda*), Prowell found that of the nine traits that divided two ecologically divergent strains adapted to corn or grass, 33 percent were on the sex chromosomes compared with the 30 (or more) autosomal chromosomes. She posits that this association provides strong evidence that speciation genes are likely to be sex-linked in general across taxa. However, this prediction has yet to receive overwhelming support.

Additional data on the role of sex chromosomes in speciation comes from fish. Among some fish lineages, there has been an uncommon pattern of sex-chromosome turnover, such that many closely related species have different sex-chromosome systems. This finding has led to the idea that sex chromosomes may play an important role in the speciation process. Using threespine sticklebacks (*Gasterosteus aculeatus*) from Japan, Jun Kitano and colleagues (2009) provided strong evidence for the role of sex chromosomes in the speciation process. Here, the common sex chromosome in other populations has fused to a segment of another chromosome, forming a neo–sex chromosome. Using genetic mapping techniques, Kitano found that prezygotic (male courtship displays) and postzygotic (hybrid male sterility) traits map to the neo–sex chromosomes in this system, providing strong evidence for sex-linked genes in speciation. Again, the generality of these findings has yet to be determined.

9. SPECIATION AND GENOMIC ARCHITECTURE

We are just beginning to understand how speciation genes are embedded and arrayed within the genome, and thus how genomes evolve collectively during population divergence. Divergent selection can play multiple roles during the speciation process, such as affecting fitness-associated loci and loci physically linked to these loci, and facilitating genetic drift across the genome by reducing gene flow. Moreover, physical aspects of the genome may facilitate speciation through chromosomal rearrangements, genome duplications, and the evolution of sex chromosomes, often reducing effective recombination between populations. Thus, heterogeneity in genome divergence, polyploidy, chromosomal rearrangements, and the disproportionate role of sex-determining chromosomes in the evolution of reproductive isolation can all have important effects. Major questions therefore still remain about the generalities of genomic architecture during speciation and how this architecture either facilitates or impedes further divergence, especially for populations diverging with gene flow. Since speciation is usually not a discrete event (with the possible exception of polyploidy) but rather a continuous process, studying the genomic architecture along this continuum of divergence—using comparisons that vary in the degree to which speciation has progressed—will allow us to further address the role of the genome in speciation. In this regard, a major focus in the immediate future will be to determine how the selective effects of individual genes themselves, divergence hitchhiking, and genome hitchhiking contribute to speciation with gene flow.

FURTHER READING

Feder, J. L., S. P. Egan, and P. Nosil. 2012. The genomics of speciation-with-gene flow. Trends in Genetics 28: 342–350. *Review of the emerging field of speciation genomics and description of a four-phase model of speciation defined by changes in the relative effectiveness of divergence and genome hitchhiking.*

Feder, J. L., and P. Nosil. 2009. Chromosomal inversions and species differences: When are genes affecting adaptive divergence and reproductive isolation expected to reside within inversions? Evolution 63: 3061–3075. *The results of this theoretical study show that genes contributing to reproductive isolation should be found not only in inverted regions of the genome but elsewhere as well. Inverted regions may retain suites of diverged genes for longer, but double recombination events and gene conversion will eventually break them down here, too.*

Feder, J. L., and P. Nosil. 2010. The efficacy of divergence hitchhiking in generating genomic islands during ecological speciation. Evolution 64: 1729–1747. *The results of this theoretical study show that divergence hitchhiking can generate large regions of neutral differentiation around loci under selection, but under limited conditions. However, they also demonstrate that when many loci are under selection, genome-wide divergence can occur.*

Felsenstein, J. 1981. Skepticism towards Santa Rosalia, or why are there so few kinds of animals? Evolution 35: 124–138. *Classic theoretical work regarding the roles that selection, gene flow, and recombination play in shaping paterns of genetic divergence during speciation.*

Gavrilets, S. 2004. Fitness Landscapes and the Origin of Species. Princeton, NJ: Princeton University Press. *Explores many of the theoretical issues about speciation discussed in this chapter.*

Hohenlohe, P. A., S. Bassham, P. D. Etter, N. Stiffler, E. A. Johnson, and W. A. Cresko. 2010. Population genomics of parallel adaptation in threespine stickleback using sequenced RAD Tags. PLoS Genetics 6: e1000862. *A high-density genome scan of genetic differentiation using >45,000 SNPs between ecologically divergent populations of the threespine stickleback. The results confirm genomic regions previously associated with speciation while also identifying additional genomic regions subject to divergent selection.*

Kirkpatrick, M., and N. Barton. 2006. Chromosome inversions, local adaptation and speciation. Genetics 173: 419–434. *Theoretical paper demonstrating that chromosomal inversions may rise to high frequency within populations owing to natural selection by protecting combinations of well-adapted alleles due to reduced recombination.*

Lowry, D. B., and J. H. Willis. 2010. A widespread chromosomal inversion polymorphism contributes to a major life-history transition, local adaptation, and reproductive isolation. PLoS Biology 8: e1000500. *One of the first experiments to show that chromosomal inversions play an important role in adaptation and speciation in nature, confirming theoretical predictions.*

Michel, A. P., S. Sim, T.H.Q. Powell, M. S. Taylor, P. Nosil, and J. L. Feder. 2010. Widespread genomic divergence during speciation. Proceedings of the National Academy of Sciences USA 107: 9724–9729. *An empirical counterexample to the island model of genomic divergence showing widespread divergence across all chromosomes among ecologically divergent insect host races.*

Noor, M., K. L. Grams, L. A. Bertucci, and J. Reiland. 2001. Chromosomal inversions and the reproductive isolation of species. Proceedings of the National Academy of Sciences USA 98: 12084–12088. *This paper was one of the first to argue for an important role of inversions in speciation. Evidence was presented for traits conferring reproductive isolation between two geographically overlapping species of* Drosophila *fruit flies mapping almost exclusively to two rearranged regions of the genome. A verbal argument was developed for how reduced recombination in inversions could help account for the observed pattern.*

Nosil, P., D. J. Funk, and D. Ortiz-Barrientos. 2009. Divergent selection and heterogeneous genomic divergence. Molecular Ecology 18: 375–402. *Review paper that explains the interplay of evolutionary forces generating reproductive isolation and heterogeneous patterns of genetic divergence across the genome and integrates them with the concept of ecological isolation by adaptation.*

Rieseberg, L. H. 2001. Chromosomal rearrangements and speciation. Trends in Ecology & Evolution 16: 351–358. *An argument is developed that rearrangements reduce gene flow more by suppressing recombination and extending the effects of linked isolation genes than by directly reducing fitness themselves.*

VI.10

Adaptive Radiation

Peter R. Grant

OUTLINE

1. Biodiversity
2. Origin and development of the concept
3. The ecological theory
4. Speciation
5. Ecological opportunity
6. Species interactions
7. Intrinsic factors: Key innovations
8. Hybridization
9. Testing the ideas
10. Future prospects

The world has millions of species, and they display an astonishing variety of size, color, and behavior. Adaptive radiations comprise groups of distinctive yet closely related species that have evolved from a common ancestor in a relatively short time. Studies of these radiations help reveal the causes of their evolution. As a result of natural selection during and after speciation, descendant species differ morphologically or physiologically in the way they exploit different environments. Adaptive differentiation also depends on the absence of constraints from competitor species. The guiding force of natural environments is revealed in the observation that the same evolutionary pathway is often taken by different organisms in the same environment. Taxonomic groups vary in their intrinsic potential to diversify because they possess traits that are key evolutionary innovations or because they readily exchange genes through hybridization. Invasion of an underexploited environment allows species to initially multiply at a high rate, and diversify morphologically and ecologically. The fossil record and reconstructions from molecular phylogenies show that both speciation and diversification rates later decline. Experiments in the laboratory with bacteria replicate the pattern of diversification through observable time. Bacteria respond to ecological opportunity by diversifying into a maximum number of ecologically differentiated types.

GLOSSARY

Allopatry. The occurrence of species that occupy separate environments, such as islands.

Competition. The struggle between two or more individuals or species for a resource in limited supply that they jointly consume (e.g., nutrients for a plant, prey for a predator, or places for prey to avoid being eaten). Competition may take the form of an aggressive interaction such as fighting, or differential depletion of a resource by the competitors.

Disparity. The degree of phenotypic difference among individuals or species in one or more traits.

Ecomorph. A recognizable association between morphology of individuals or species and use of the environment.

Evolvability. The genetic and developmental properties of members of a species that determine the likelihood that it will undergo evolutionary change.

Introgressive Hybridization. The interbreeding of two species or genetically divergent populations and subsequent breeding of the offspring with members of one of the parental populations, resulting in the transfer of genes.

Parapatry. The occurrence of species in adjacent or contiguous distributional ranges.

Sympatry. The occurrence of species that occupy the same area.

1. BIODIVERSITY

For many evolutionary biologists the most important pair of questions that need to be answered are: Why do so many species exist on this planet? And why do they differ so greatly from one another? Species number in the millions, varying in size from viruses to whales and from algae to trees; varying in color from bright butterflies to dull and cryptic moths; varying in behavior from solitary polar bears to highly social honey bees; and varying in numerous other ways in exploiting the environment for food, avoiding their enemies, and reproducing. How is all this variation to be explained?

Adaptive radiations provide rich material for seeking answers to these questions because they comprise groups of distinctive yet closely related species. An *adaptive radiation* is the product of differentiation of an ancestral species into an array of descendant species that differ in the way they exploit the environment. When the differentiation has proceeded rapidly, the evolutionary transitions from one state to another can readily be characterized and strongly interpreted.

Angiosperm plants, dinosaurs, and marsupial mammals are typical examples at high taxonomic levels. Typical examples at lower levels are Darwin's finches on the Galápagos Islands, honeycreeper finches, *Drosophila*, spiders, the silversword alliance of plants in the Hawaiian archipelago, cichlid fish in the Great Lakes of Africa, and *Anolis* lizards in the Caribbean. These examples have the following in common: (1) they comprise several to many species, (2) the species vary morphologically in conspicuous ways, and relatedly, (3) they occupy a diversity of ecological niches. Most of the species were (4) derived from a single ancestor in their current environment, and (5) most diverged relatively rapidly. Cichlid fish in Lake Malawi are an outstanding example. Hundreds of species—the exact number is unknown—were derived from one or a few common ancestors in the last 2 million years, and they have diversified into many trophic forms, including algae-, insect-, snail-, and fish-eating specialists. Their mouth and teeth morphologies reflect their diets, and for this reason the variation is inferred to be adaptive, that is to say, the product of diverse natural selection. One group alone, the rock-dwelling "Mbuna" of the genus *Tropheops*, comprises 230 species. Ole Seehausen has calculated that one new species arose every 46 years!

2. ORIGIN AND DEVELOPMENT OF THE CONCEPT

The term *adaptive radiation* was coined in 1902 by a paleontologist, H. F. Osborn, and the phenomenon it refers to was popularized by another, G. G. Simpson, about 50 years later. Simpson viewed the evolutionary radiation of a major group of animals, such as marsupial mammals, as various lines of descent from a common ancestor arising more or less simultaneously and diverging in different morphological and ecological directions, rather like spokes radiating from the hub of a wheel. This image is powerful yet fails to represent the correct evolutionary pathway of bifurcating branches in a treelike structure. Nonetheless, the term has stuck. The *adaptive* adjective is applied because the products of a radiation are conjectured or known to be adapted to exploiting the environment in different ways.

In the last 20 or 30 years the range of extant organisms that have been studied in detail has increased dramatically, owing largely to the availability of molecular phylogenies for inferring relatedness among species and the pattern and rates of diversification. With these studies has come increasing scrutiny of the term itself, and debate on definitions. Should an unusually high rate of diversification be an essential ingredient of the definition? How is an adaptive radiation to be distinguished from a nonadaptive radiation? These questions become important in comparative studies when generalizations are sought across a broad taxonomic range of organisms. There are no simple answers because there is no clear line of demarcation or break point between adaptive radiations—defined by numbers of species, variety or rates of diversification—and all others. As used in this chapter the term *adaptive radiation* is most usefully applied to those groups that have diversified rapidly and interpretably, such as the ones cited earlier.

3. THE ECOLOGICAL THEORY

By placing the occurrence of fossils within a time frame, paleontologists like Simpson were able to detect a pattern in the history of a radiation. A radiation was seen to begin with rapid multiplication of species as well as diversification of morphological types. As the radiation proceeded, both species proliferation and morphological evolution slowed down. The observed pattern gave rise to an inferred process, as follows. Invasion of an underexploited environment allowed species to multiply at a high rate. At the same time they diversified phenotypically and ecologically. Eventually, both speciation and diversification rates declined as competition increased for a diminishing variety of unexploited or underexploited resources. This is an explicitly ecological interpretation. On one occasion Simpson used T. H. Huxley's phrase "filling the ecological barrel" to capture the essence of a limited environment. The greatest opportunity for occupancy occurs at the beginning, when the barrel is empty, and by implication there is an increasing difficulty for newcomers to fit as the barrel fills.

Dolph Schluter converted a coherent explanation into a theory by identifying all the elements and framing them as hypotheses to be tested by their predictions. The key elements of the theory are (1) phenotypic differentiation caused by natural selection arising from differences among environments, (2) competition for resources, and (3) speciation governed by both processes. The theory is not one to be rejected by the first contrary observation so much as an organizational framework for investigating causes in individual cases. This framework has stimulated a large amount of quantitative analysis of large data sets: literally hundreds of species in the case of cichlid fish.

A few words are warranted here about the domain of applicability of the theory. Some groups diversify without radiating. For example, *Plethodon* salamanders in eastern North America diversified rapidly early in their history, and more slowly later, as in classical adaptive radiations, but speciation was largely allopatric, resulting in little morphological diversity and, as far as is known, relatively little adaptation. The 46 extant species owe their existence to few niches but many habitat fragments and a long history, and not, with some notable exceptions, to competitive interactions. Similarly, speciation may occur repeatedly through diversification of mating signals by sexual selection, such as the color of cichlid fish or petals of flowers (e.g., *Mimulus*), vocalizations of birds, pheromones of moths, and the stridulation of crickets and leafhoppers. In these and suggested examples of diversification through random drift or founder effects it is difficult to rule out the role of ecological factors in speciation, and for this reason the distinction between adaptive and nonadaptive causes of species proliferation is sometimes blurred. Instead, nonadaptive processes are likely to contribute to an adaptive radiation, being complementary rather than a strict alternative to adaptation. This means that a particular radiation may have heterogeneous causes: adaptive and nonadaptive processes, natural selection and sexual selection, competition and predation, and so forth.

4. SPECIATION

In an adaptive radiation one species gives rise to many. Derived species either do not interbreed or interbreed with a limited degree of genetic exchange (see chapter VI.1). It is easy to envisage how reproductive isolation could arise in an archipelago where populations are isolated by physical barriers and undergo independent evolution: they adapt to different environments. Divergence of the signaling and response system deployed in mate choice occurs as a correlated effect of adaptive divergence. A sympatric phase of the speciation process then follows the allopatric phase, when a derived species disperses into the environment of another. Coexistence in the same habitat, as described by the term *sympatry*, depends on ecological and reproductive differences having arisen largely or completely in allopatry.

If this were a universal route to species multiplication, those environments without physical barriers should have fewer species, area for area, than fragmented environments, but this is not always observed. For example, relatively homogeneous lakes contain a very high diversity of fish species. In the case of African Great Lake cichlids, their diversity traces back to one or a very few colonizing species, so they must have evolved in broad-scale geographic sympatry within the lake, though possibly in local parapatry or allopatry. Just how a population splits into two under these circumstances is not clear. The question is, How does reproductive isolation evolve through disruptive selection in the face of gene flow counteracting divergence? (See chapter VI.6.) A solution in plants but rare in animals is polyploidy: autopolyploidy through chromosomal doubling, or allopolyploidy through hybridization.

Speciation and adaptive radiation should not be equated. Adaptive radiation usually implies speciation, whereas speciation implies neither adaptation nor a radiation.

5. ECOLOGICAL OPPORTUNITY

Adaptive differentiation depends on both the availability of ecological resources (niches) to sustain a variety of organisms, and properties of the organisms that facilitate their evolution. Adaptive landscapes provide a way of visualizing opportunities for diversification provided by the environment. The adaptive landscape is an abstraction first developed by Sewall Wright in 1932 to explore possibilities and limits to change of genotype frequencies in a population. Transformed for more general usage by Simpson and others, an adaptive landscape represents variation in fitness in relation to combinations of traits or environmental conditions (figure 1). There are hills and valleys in the landscape. Fitness is at a maximum when a population occupies the top of a hill, be it flat or peaked. Each hill can be equated with ecological opportunity. The spatial distribution of hills reflects another property of environments: some are closer and more within reach of evolving organisms than others, and proximity governs the sequence in which hills can be climbed through the action of natural selection. Neither the environment nor the landscape it represents should be thought of as static. Over time, hills gain or lose height, become more or less peaked, and move closer together or farther apart.

The guiding force of natural environments is revealed by such observations as the same evolutionary pathway being taken by different organisms in the same environment. For example, flowering plants from a diverse array of families have invaded freshwater from the land at least 50 times, with repeated evolution of floating, bladderlike structures in the leaves. *Dioecy*, the presence of two separate sexes in a population, has evolved

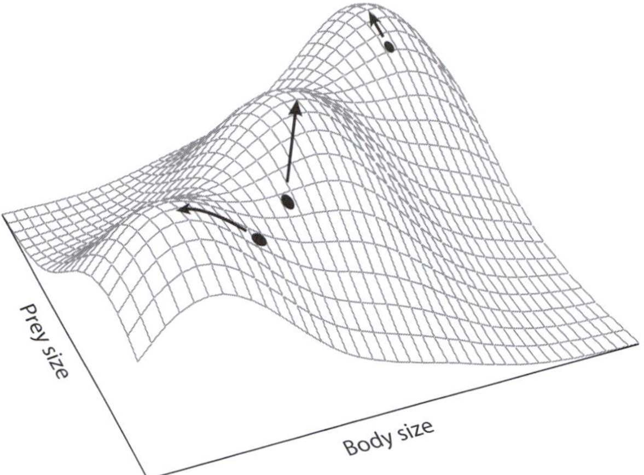

Figure 1. The adaptive landscape. Mean fitness for a phenotype with a given combination of body size and prey size is indicated by the height of the surface. Arrows indicate paths of steepest fitness ascent for three populations in the vicinity of three peaks. (From Schluter 2000.)

repeatedly in different lineages of plants colonizing the Hawaiian archipelago. Algal-scraping, mollusk-crushing, fish-scale-scraping cichlid fish and other ecomorphs seen in Lake Malawi are also seen in Lakes Victoria and Tanganyika, even though their origins differ and their evolution has been independent. Thus organisms diversify in response to ecological opportunities. Hand in hand with the opportunity is a challenge from the environment in many cases. For example, coping with ultraviolet radiation and predation from fish and invertebrates is a challenge faced by planktonic *Daphnia* in the upper waters of lakes. The repeated response of different *Daphnia* lineages has been the evolution of melanism in the face of the first challenge, and head shields and spines when confronted by the second.

Evolution in similar environments is thus predictable to a degree, because ecological opportunities are similar. An example is provided by the adaptive radiation of Caribbean *Anolis* lizards. Their history has been reconstructed from a molecular phylogeny. It shows that the four largest islands of the Greater Antilles (Cuba, Hispaniola, Jamaica, and Puerto Rico) were colonized by different species, which then underwent parallel evolutionary diversification into the same ecotypes occupying the same spatial niches (ground, tree trunk, branches, and twigs). Subsequent evolutionary diversification on a given island took the form of variation on these four themes.

Evolutionary history repeats itself at the higher taxonomic levels, too. As pointed out many years ago by Osborn, the marsupial radiation led to convergence in many phenotypic and ecological traits with eutherian mammals: browsing, burrowing, and gliding habits and associated morphological adaptations evolved independently in the two groups, on separate continents. Parallel radiation within two continents yielded convergence between them. This is strong indirect evidence for the driving force of environmental factors in adaptive radiations.

6. SPECIES INTERACTIONS

An important factor in rapid diversification in a new environment is the absence of competitors—typically, related species. In fact, when the concept of adaptive radiation was first developed, the absence of competitors was emphasized as a key facilitating or predisposing factor. For example, islands were viewed as empty environments when the first colonists arrived, and diversification proceeded until all ecological niches were filled, one species per niche. The Gaussian principle that no two species can occupy the same ecological niche is, in modern language, no two species can occupy the same adaptive peak, for reasons of competitive inequality. Therefore, if all peaks in a landscape are occupied, there is no ecological opportunity for a radiation to occur, or if one has begun, there is no opportunity for it to continue.

The objective reality of adaptive peaks can be demonstrated only with environmental data. For granivorous Darwin's finches in the Galápagos the peaks have been quantified on several islands by measuring food supply, and fitness has been estimated for each island with a given seed size profile by measuring population sizes in relation to beak sizes. The adaptive landscape for these finches is rugged, not smooth. In agreement with expectation, only one species is ever associated with one peak. Interestingly, different species are sometimes associated with the same peak on different islands: they are interchangeable, although ecologically incompatible.

Competitive interactions between species lead either to the exclusion of one species by another or to evolutionary adjustments to each other. Competitive displacement in food-related body size and shape among sticklebacks has been demonstrated experimentally under controlled conditions in ponds. Limnetic species are smaller and more slender than benthic species. A solitary species intermediate in size and shape between them suffered slower growth in the experimental presence of a limnetic species, and as a result, their average body sizes diverged.

Under natural conditions character displacement has been observed in Darwin's finches. During a drought large members of the medium ground finch population died at a disproportionately high rate. They were outcompeted by a larger, more efficient species, the large ground finch, when feeding on a diminishing supply of large and hard seeds; as a result, the average beak size of the medium ground finch population decreased. Beak size is heritable; therefore, the change produced by natural selection gave rise to evolution: average beak size remained low in the generations produced in the following six years.

7. INTRINSIC FACTORS: KEY INNOVATIONS

Ecological opportunity is one side of a coin; organism responsiveness or intrinsic evolvability is the other. The importance of intrinsic factors is hinted at by differences among taxonomic groups in how far and how fast they radiate. For example, speciation in lacustrine fish appears to be faster than speciation in terrestrial birds and arthropods; time required for speciation has been estimated to be 15,000 to 300,000 years for fish, and half a million to1 million years for birds and arthropods on islands. These values are calculated from simply knowing the number of existing species and then estimating time since they were derived from a common ancestor.

Diversification of cichlid fish has been rampant, but most of the other fish taxa in apparently similar circumstances in African lakes have failed to diversify beyond a few species. Since different environmental factors are not clearly implicated, the contrast suggests there are some intrinsic genetic, developmental, or physiological factors that differ among the groups. Intrinsic factors may potentiate evolutionary change, or they may constrain it. All that is known is that the radiating groups of these fish cluster together in the grand African phylogeny, implying a common inheritance of one or more predisposing factors for radiating. In general, these factors are poorly known, if at all. Identifying them is an important challenge for future studies of adaptive radiations.

The most striking evidence of intrinsic potential is correlative and therefore indirect: it is the association between the evolution of a novel trait and a large number of related species that share the trait. The trait is described as a key evolutionary innovation, facilitating a novel and diverse way of exploiting the environment (see chapter VI.15). Prominent among such novel traits are the pharyngeal jaws of cichlid fish. These are plates with toothlike projections in the roof of the mouth that enable their possessors to split the functions of procuring and processing food into oral and pharyngeal regions of the mouth. Fish with pharyngeal jaws have diversified greatly in oral jaws and associated diets, as different as grazing algae and catching other fish. However, pharyngeal jaws are not sufficient to explain cichlid radiation, because groups that have not radiated also possess them.

The evolution of antifreeze glycoproteins by fish in the suborder Notothenioidei is another example of a key innovation that apparently facilitated invasion of Antarctic waters and subsequent diversification in a relatively empty environment. Detailed comparative work is needed to determine whether a key innovation resulted in an enhanced diversification rate. For example, nectar spurs in columbines (*Aquilegia*) vary in shape and color, which affect reproductive isolation by attracting different pollinators, and apparently facilitated speciation, because there are more species in clades with than without spurs. The evolutionary invention of resin canals by some plants, constituting a defense against chewing insects, is another example.

8. HYBRIDIZATION

One possible potentiating factor is *introgressive hybridization*. An exchange of genes can result in enhanced genetic variation and evolutionary potential, and under the right ecological circumstances it can lead to the formation of a new species (see chapter VI.6). The work of Loren Rieseberg on sunflowers (*Helianthus*) and Tom Whitham on poplars (*Populus*) in western North America, and several others elsewhere, has shown that introgression of certain genes is selective according to the nature of the environment (soil, microclimate, herbivores, etc.). Populations of sunflower hybrids in peripheral and ecologically extreme environments undergo large-scale genome reorganization, leading to reproductive isolation from the parental populations. These findings have a bearing on the early stages of major radiations on the assumption that what we observe now reveals what happened in the past at the beginning of those radiations. For example, introgressive hybridization is widespread in many young radiations, including the silverswords of Hawaii, Darwin's finches of the Galápagos, African cichlids, and *Heliconius* butterflies in South and Central America, but is almost absent in older radiations such as Hawaiian honeycreeper finches and Caribbean *Anolis* lizards.

Seehausen has suggested that gene mixing through hybridization does more than accompany a radiation: it creates a hybrid swarm and thereby facilitates a radiation. The evidence of ancient hybridization is a mismatch between phylogenetic reconstructions of the same organisms based on nuclear and cytoplasmic genes. For example, the cichlids of Lake Victoria, with a history of 15,000 years, are far younger than the fish in Lake Malawi and correspondingly have much lower mitochondrial gene diversity, yet they have the same level of nuclear diversity as their Lake Malawi counterparts. Similar evidence of incongruence between nuclear and chloroplast genes is found in one lineage of Hawaiian silverswords. The pattern is even repeated in human history. More modern studies of hybridization like the sunflower research are needed to further explore these ideas on diversity generation.

9. TESTING THE IDEAS

Many adaptive radiations happened over millions of years. Their study is therefore retrospective, and interpretations are necessarily inferences that to be plausible, should be consistent with known biological processes. How can the interpretations be tested? For example, in the absence of fossils, how can we tell whether extinction has been important or trivial? An indirect method (see the following section) is to devise alternative models and to use statistical analyses of data to see which model fits the data best. Taxa comprising large numbers of species are necessary for this method to be effective, and well-resolved molecular phylogenies are desirable. And as described later, direct tests can be performed experimentally with microorganisms in the laboratory.

Timing of Speciation and Diversification

The first major pattern of adaptive radiation is an increase in number of species or higher taxa, at a diminishing rate. The second major pattern is an increase in phenotypic disparity, that is, an increase in morphological differences among the species.

The fossil record shows a pattern of high extinction rates at times of geophysical perturbation, followed by speciation that is initially rapid and subsequently slower (see chapter VI.13). Fossils are missing for most living taxa; therefore, molecular phylogenies have been used in their place to see whether the "early burst" phenomenon with subsequent decline is a consistent element of adaptive radiations. A phylogeny describes how the number of species at any one time has increased from the starting point. Accumulation of species on the y-axis is plotted against time on the x-axis to give a lineage-through-time plot. Since the same pattern of accumulation of species diversity can result from a declining speciation rate, as expected, or an increasing extinction rate, it is necessary to decompose diversity into its speciation and extinction components, via mathematical models. For example, Daniel Rabosky and Richard Glor constructed 12 different ways in which species multiplication might be expected to occur as continuous processes of birth (speciation) and death (extinction). Using a molecular phylogeny of Caribbean *Anolis* lizards to reconstruct the pattern of diversification, they found that one model outperformed others in fitting the data. In this model, speciation rates declined toward a presumed equilibrium on three out of four islands.

However, it is still not known whether the major radiations of the last few million years—which currently are receiving much attention—reflect approximately constant but different speciation and extinction rates, or hidden pulses of each, occurring either synchronously or asynchronously. Agamid lizards in Australia, warblers and Lampropeltine snakes in North America, and pythons in the Australo-Papuan region all declined in rate of diversification. The data fit models with varying speciation rate better than models with varying extinction rate. Conversely, in the most comprehensive analysis to date of many radiations, Jonathan Losos and colleagues found that most of the groups they studied did not fit the early burst pattern. One of several possible reasons is that the broad aspects of the environment do not remain constant, contrary to assumption. For example, Darwin's finch species increased in parallel with an increase in the number of islands of the Galápagos archipelago. In this case, diversification was facilitated, if not actually driven, by expanding spatial and temporal ecological opportunities. Seehausen found a different pattern of diversification in a changing environment. Species diversity of the rapidly evolving cichlid fish in Africa increased in pulses interspersed with periods of relative stasis, as might be expected if the environment is occasionally perturbed in some significant way. Thus speciation and extinction may not vary in a coordinated manner through time; either or both may change systematically or erratically.

With ecological opportunity being maximal at the beginning of a radiation, initially high rates of phenotypic diversification are expected, which then decrease as the environment becomes progressively filled with a large number of ecologically diverse exploiters. This expectation has been tested with the large number of species that make up the Caribbean *Anolis* faunas. The expectation was met: rates of diversification of two ecologically important traits, body size and limb length, decreased with increasing radiation, and decreased as the number of inferred potential competitors increased. These results matched paleontological evidence over larger time spans and taxonomic categories: morphological disparity increased among hard-bodied invertebrates in the so-called Cambrian explosion and in certain groups of organisms following the end-Permian and end-Cretaceous mass extinctions.

To what extent are speciation and morphological change coupled? This question is difficult to answer. Branching points in a phylogeny can be estimated, but morphological (and ecological) change independent of speciation can be dated only with fossils, and then with difficulty. Morphological change is likely to accompany speciation (*cladogenesis*) and be a vital part of it, but change may nonetheless continue without further speciation (*anagenesis*). Anagenesis may have contributed to diversification of Lampropeltine snakes in North America because morphological disparity increased prior to the Pliocene, close to the time when rates of speciation decreased. Selective extinction through competitive interactions is another process that leads to the same enhancement of morphological disparity.

Experimental Adaptive Radiations

Some of the problems involved in inferring history can be circumvented with studies of living organisms. Direct tests of radiation theory can be performed experimentally with microorganisms in the laboratory; bacteria have the enormous advantage of short generations and rapid evolution. Introduced to a plate of agar, an inoculum of the asexual bacterium *Pseudomonas fluorescens* diversifies into three main morphological types. Diversification in a microcosm is an analogue of the radiations of sexually reproducing organisms into many species over millennia or more. For example, in the longest-running experiment performed so far, genetic changes affecting metabolism of *Escherichia coli* have been investigated for more than 50,000 generations. Another advantage of microbes is they can be stored in the freezer, then taken out years later to compare their performance with that of the descendants of the parent population to test the hypothesis that competitive ability evolves. It does.

The *Pseudomonas fluorescens* experiment shows a repeated pattern of evolutionary diversification into three main ecotypes recognized by their distinctive morphology. A colonizing type ("smooth") gives rise to two more: a biofilm-producing "wrinkly spreader" and a "fuzzy spreader." They are specialized on different parts of their environment (spatial niches). Through repeated mutation they give rise to (asexual) clones that are morphological and metabolic variations on these three main themes. The experiments show that the variants within a spatial niche compete for resources and replace one another. The number of morphologically distinct clones (richness) reaches a peak through time and then declines, thus overshooting the long-term carrying capacity of the environment.

The range of morphological variation—the disparity—shows a different pattern: it rises to a maximum and remains there. Thus the bacteria, in the absence of predation and competitor species, respond to ecological opportunity by diversifying into a maximum of ecologically differentiated types. What happens then depends on the environment. Either it remains fixed, as imposed by the investigators, or changes with concomitant changes in the community. Little is known in the laboratory or nature about how a gradually changing environment affects the course of an adaptive radiation, or how adaptive radiation of one group facilitates further radiation through positive feedback from other organisms with which it interacts (e.g., mutualisms). An increase in aridity in the last couple of million years altered the speciation-extinction balance in Caribbean birds. It opened up new niches in the Galápagos, and Darwin's finches responded by evolving seed-eating specialists.

10. FUTURE PROSPECTS

The study of adaptive radiations is becoming increasingly quantitative, experimental, and comparative. The goal is to understand (1) general properties and (2) differences, according to time of occurrence, taxonomic group, and particular environment. Greater understanding of the causes of diverse adaptive radiations will come from a variety of sources. One is the discovery of new systems. Vertebrates and some plant groups have dominated investigations of extant groups so far, although the recent exploitation of microcosms for experimental investigation has revealed an enormous potential residing in microorganisms. Additional experimental potential at the level of ecological communities has scarcely been tapped. A second source is genetics—specifically, gene expression of ecologically important traits during development—for an understanding of comparative evolvability in different lineages. A third is speciation, introgressive hybridization, and the interrelationship of the two. Experimental investigations have a larger role to play in both revealing and testing the causal factors that observations imply. And inferences about how radiations unfold will improve as analytical methods are refined. Eventually, the knowledge obtained from studying adaptive radiations will be integrated with what will be learned from all the rest of evolution, environments, and earth history, to provide a more comprehensive understanding of the richness of the biological world.

FURTHER READING

Gavrilets, S., and J. B. Losos. 2009. Adaptive radiation: Contrasting theory with data. Science 323: 732–737. *Illustrates the interplay of theoretical and empirical studies in the development of understanding how adaptive radiations proceed.*

Gavrilets, S. A., and A. Vose. 2005. Dynamic patterns of adaptive radiation. Proceedings of the National Academy of Sciences USA 102: 18040–18045. *A theoretical analysis of adaptive radiation.*

Givnish, T. J., and K. J. Sytsma, eds. 1997. Molecular Evolution and Adaptive Radiation. Cambridge: Cambridge University Press. *Many chapters provide a molecular genetic basis for reconstructing phylogenies and using them for describing and interpreting patterns of adaptive radiation.*

Grant, P. R., and B. R. Grant. 2008. How and Why Species Multiply: The Radiation of Darwin's Finches. Princeton, NJ: Princeton University Press. *Darwin's finches are used as a model system for understanding how species form in adaptive radiations.*

Losos, J. B. 2009. Lizards in an Evolutionary Tree. Berkeley: University of California Press. *An exceptionally comprehensive account of one of the most impressive and diverse adaptive radiations.*

Meyer, J. R., S. E. Schoustra, J. Lachapelle, and R. Kassen. 2011. Overshooting dynamics in a model adaptive radiation. Proceedings of the Royal Society B 278: 392–398. *A good example of the power of experimental microcosms for testing in the laboratory ideas generated by observations in nature.*

Rabosky, D. L., and R. E. Glor 2010. Equilibrium speciation dynamics in a model adaptive radiation of island lizards. Proceedings of the National Academy of Sciences USA 107: 22178–22183. *A comprehensive modeling approach to questions of how adaptive radiations increase in size and complexity through time.*

Seehausen, O. 2006. African cichlid fish: A model system in adaptive radiation research. Proceedings of the Royal Society B 273: 1987–1998. *A good overview of the most diverse and rapid radiation known.*

Schluter, D. 2000. The Ecology of Adaptive Radiation. Oxford: Oxford University Press. *The standard modern work on adaptive radiation.*

Simpson, G. G. 1953. The Major Features of Evolution. New York: Columbia University Press. *A general discussion of evolutionary radiations that integrates population genetics with paleontology.*

VI.11

Macroevolutionary Rates

Luke J. Harmon

OUTLINE

1. How "fast" is evolution?
2. Rates of speciation and extinction
3. Rates of trait evolution
4. Are there relationships between rates of trait evolution and diversification?

Rates of evolution—both the rate of trait evolution and the rate at which new species form and go extinct—vary tremendously through time and across lineages. Variations in these rates relate to a number of core theories of evolutionary change over long timescales.

GLOSSARY

Adaptive Radiation. The rapid evolution of ecological differences among species in a diversifying lineage.

Birth-Death Model. A mathematical model describing speciation (birth) and extinction (death) of a branching lineage through time, in which the probabilities of speciation or extinction are constant, but the number of species fluctuates by chance.

Clade. A group of lineages that represent all the descendants of some common ancestor

Darwin (Evolutionary Rate). The rate of evolutionary change expressed as the difference in log-transformed trait values divided by the elapsed time in millions of years.

Diversification. The formation of many new species from a single common ancestor.

Diversification Rate. The rate at which new species form, typically in units of lineage^{-1} my^{-1}.

Ecological Opportunity. The presence of unexploited niche space experienced by an evolving lineage.

Haldane (Evolutionary Rate). The rate of evolutionary change expressed in units of within-population standard deviations per generation.

Lineage. A group of organisms that form a line of descent from an ancestor.

Living Fossil. A lineage that experiences little or no trait change over a long period of evolutionary time.

Stasis. A lack of change in a species' traits over evolutionary time.

Taphonomy. The study of the process of fossilization and its variation across lineages and through time.

Tree Balance. A measure of the relative sizes of sister clades in a phylogenetic tree; trees with sister clades of similar size are balanced, while trees with sister clades of dramatically different sizes are unbalanced.

1. HOW "FAST" IS EVOLUTION?

Darwin's finches capture our attention because of their extremely rapid evolution. From one ancestor, these finches have radiated into a great diversity of species, each distinct from the others in size, feeding, song, and many other traits. All this has happened within just a few million years, which is—to evolutionary biologists—the blink of an eye. These species have diverged so rapidly that it is difficult to determine how the species are even related to one another. Reconstructions of their evolutionary history using genetic data show a nearly simultaneous explosion of species. Contrast the rapid radiation of Darwin's finches to the coelacanth, a species of fish that was known only from fossils tens of millions of years old until it was discovered in the deep ocean in the early twentieth century. The coelacanth is often called a *living fossil*, a species that has undergone little or no evolutionary change over a very long period (*stasis*).

These examples seem exceptional, but how do we know whether Darwin's finches and coelacanths are evolving extraordinarily rapidly (or slowly)? The answer is to calculate and compare their rates of evolution. Rates of evolution have been of primary importance since Darwin. Much of the current thinking on evolutionary rates traces to George Gaylord Simpson, who wrote about them in his book *Tempo and Mode in Evolution*. Rates of evolution still play a key role in several important controversies in evolutionary biology. For example, the debate over punctuated equilibrium is really a debate over whether the rate of evolutionary change at speciation is much faster than at other times in the history of lineages. Another example can be found in the debate about the importance (and existence) of adaptive radiations.

This chapter reviews the concept of evolutionary rates. The discussion includes rates of diversification (speciation and extinction) and rates of trait evolution, and how these might relate to one another. Both sections emphasize how measuring rates of evolution has enhanced our understanding of the history of life on earth.

2. RATES OF SPECIATION AND EXTINCTION

During the course of evolution, new species form while others go extinct. The rates of these events determine the number of species expected to form and go extinct in a given time interval. Most often, we are interested in the process of speciation and extinction within evolutionary *clades*—groups of lineages that represent all the descendants of some common ancestor. Thus, rates are typically calculated on a *per lineage* basis, that is, the average rates of speciation and extinction per lineage per million years. These rates vary tremendously across the tree of life.

The *birth-death model* is a useful mathematical model for describing the way lineages speciate or go extinct. In a homogeneous birth-death model, each lineage has a constant probability of either splitting to form a new species or going extinct per unit time. Over the long term, birth-death models predict that the total number of lineages in an evolving clade will increase exponentially (if speciation rate is greater than extinction rate) or decay to zero (if extinction rate is greater than speciation rate). Birth-death models are commonly used in two ways: (1) to estimate the rates of speciation and extinction in evolving clades over time and (2) as a reference point to compare to data. Any deviations from the expected pattern under a birth-death model might reflect biologically interesting phenomena, like higher rates of speciation in certain clades compared with others.

Fitting birth-death models to real data is more complex than one might expect, because direct and complete information about speciation and extinction rarely exists. The most common information about speciation and extinction rates comes from fossil occurrence data—that is, from the appearance and disappearance of lineages in the fossil record; however, in these data, many fossils cannot be identified to the level of individual species, and fossil preservation rates vary across both lineages and geologic time periods. (The study of variation in fossil preservation is called *taphonomy*; see chapter II.9). A variety of statistical approaches have been invented to estimate speciation and extinction rates from incomplete fossil records, and these rates have been estimated for a variety of clades and over a wide range of time intervals.

Studies of speciation and extinction rates have produced three general findings. First, speciation and extinction rates tend to be roughly equal to each other, at least over long periods and over large spatial scales, which suggests that the history of life is characterized by high turnover and that new species continually form and replace species that have gone extinct. Second, the total number of species in individual clades waxes and wanes through time. Thus, some clades, like mammals, that are now common were at one time much rarer; still other clades, like trilobites, were common in the past but have now gone entirely extinct. Finally, both speciation and extinction rates show tremendous variation, both through time and across clades. This variation is especially interesting because even homogeneous birth-death processes predict that clades will, by chance, have substantial variation in lineage diversity. However, real data sets show even more variation than expected from homogeneous birth-death models. For example, the earliest split in lepidosaurs (lizards and their relatives) separates two clades of extraordinarily different diversities: Sphenodontia (the tuatara, either one or two species depending on taxonomy) and Squamata (snakes and lizards, about 9000 species). This difference in diversity of two clades of the same age is even greater than what would be expected from the variability of the homogeneous birth-death process and likely reflects differences in the dynamics of diversification between the two clades. Patterns like this, in which a very diverse clade is closely related to a clade with low diversity, are very common in the tree of life.

Variation in diversification rates is perhaps most apparent at times of mass extinction, when multiple clades experience dramatically increased rates of extinction over a relatively short time interval. There have been several mass extinctions throughout the history of life on earth, the largest of which are known as the "big five" (see chapter VI.13). Interestingly, a number of studies have shown that some clades experience dramatically increased diversification rates following these mass extinctions. This result suggests that mass extinctions create widespread ecological opportunities that can be exploited by surviving lineages.

A second approach to estimating speciation and extinction rates relies on phylogenetic data. Phylogenetic trees show the evolutionary relationships among species, which occur at the tips of the branches of the tree. Typically, phylogenetic trees have branch lengths that reflect the time between splitting events (nodes) in the tree. The challenge of using phylogenetic data to study diversification is that only extant species are available, and the tree is missing data on the number and timing of extinction events. Even without direct information about extinct lineages, the shape and branching pattern of trees capture information about the relative propensity of clades to diversify.

For example, comparison of the number of descendant species (the diversity) among different clades within a tree reveals the relative propensities of these lineages to diversify. Such comparisons typically focus on *sister clades*, clades that are each other's closest relative. Sister clades are by definition the same age, so that any differences in diversity might reflect differences in diversification rates. The earlier example of lepidosaurs reflects this approach. Information about differences in diversity across clades is captured in measures of *tree balance*, which summarize how evenly the tip species are distributed across every node in a phylogenetic tree. Trees can range from perfectly balanced—where every node or branching point on the tree has the same number of descendant species on each side—to perfectly imbalanced, where every node represents a contrast between a single species and a larger clade. Trees generated by a homogeneous birth-death process tend to have a balance intermediate between these two extremes. One strength of tree balance measurements is that the branch lengths of the tree have no effect. In cases where we don't know anything about the timing of diversification, we can still use tree balance to learn about variation in diversification rates across clades.

One of the first generalities to emerge from the study of phylogenetic tree balance was that almost all phylogenies of taxa in nature are more imbalanced than would be expected from a homogeneous birth-death model. This observation is evidence that diversification rates—either speciation or extinction, or both—vary across lineages in phylogenetic trees. That is, lineages within some clades speciate faster than others, and other clades are more likely to go extinct. This result is concordant with results from the fossil record, which show great variation in speciation and extinction rates among lineages.

Often, phylogenetic trees include information about branch lengths, which represent the time intervals between recorded speciation events. Typically, these trees do not include any information about extinct species. Still, we can (at least in theory) use them to estimate rates of both speciation and extinction. Information about extinction is contained in the most recent branches, that is, those connecting the youngest species in the tree. If there is extinction, most young species haven't yet had time to experience its effects, so we expect to see an overabundance of very short branches near the tips of the tree.

In practice, these methods can generate good estimates of net diversification rates (speciation minus extinction) but have difficulty estimating either speciation or extinction rates separately. The method of fitting birth-death models to real phylogenetic trees often yields unrealistic parameter estimates. Many studies using this approach have inferred very low (or even zero) rates of extinction in living clades, which seems to contradict results from the fossil record suggesting that almost all groups experience some extinction. More recent applications to very large trees have uncovered high turnover rates that are more consistent with fossil calculations. It is possible that some estimates are biased because the phylogenetic trees used are incorrect or because the mathematical assumption that constant rates of speciation and extinction apply across a whole tree is incorrect. Perhaps the rates vary through time or across clades, for example. Some authors have suggested that birth-death models are not an accurate description of the diversification of life on earth and that we need to consider models that take into account how species in communities might interact with one another. This is currently an active area of research.

Many phylogenetic studies have compared diversification rates across lineages and through time. Tests using branch lengths have confirmed that rates of speciation and extinction can vary dramatically among closely related groups, mirroring results from measurements of tree balance. For example, across vertebrates, some clades have very high rates of diversification: Boroeutheria (most placental mammals), Percomorpha (perches and perch-like fish), and Neoaves (most birds) all have net diversification rates close to 0.1 per lineage per million years. Such a rate will produce about 22,000 surviving species, on average, from a single ancestor in 100 million years. By contrast, other clades apparently diversify much more slowly. For example, old clades with low diversity like the bowfin, coelacanth, and tuatara might have net diversification rates 100 times slower than the high rates seen in other vertebrate clades.

Rates also appear to vary through time. For instance, phylogenetic trees often seem to show evidence that diversification rates slow as lineages approach the present day. This pattern has been cited as evidence for density dependence in diversification rates; that is, perhaps diversification rates slow as the number of lineages in a clade increases. The prevalence of this pattern across a wide range of organisms might even suggest that as clades diversify, they attain some environmentally determined "carrying capacity" for the number of species that can coexist. However, a number of biases could also explain this pattern. One possibility is that researchers are using biased estimates of phylogenetic trees. When phylogenetic trees are estimated from molecular data, a mathematical model must be applied describing how gene or protein sequences change through time; if the model is incorrect, the resulting tree will be biased. Another potential bias involves sampling—perhaps scientists are failing to include many young species from these trees. Typically, phylogenetic trees include only taxa that are considered "good species" (under various criteria) and leave out younger lineages not considered species but that might provide information about recent speciation events. Leaving out these young lineages might then falsely suggest a slowdown in speciation toward the present. Additionally, these temporal slowdowns seem to contradict ecological studies suggesting that hardly any natural species communities are "saturated" with species, preventing new species from forming and surviving. Instead, new species can almost always be added to natural communities without resulting in extinction. Still, the common observation of slowdowns in diversification through time is an intriguing general result that demands an explanation.

In summary, speciation and extinction rates vary among clades and through time. These patterns have been found in both fossil occurrences and phylogenetic trees, and patterns are more or less concordant across these two types of data. Other intriguing patterns, like density-dependent speciation rates, are still being investigated.

3. RATES OF TRAIT EVOLUTION

Lineages also have traits that evolve and change through time. Scientists estimate the rate of change of these species traits in two ways. First, they can carry out an *allochronic* study by measuring the difference in the mean trait of a lineage at the start and end of a given time interval. Dividing this difference by the amount of time yields the amount of change per unit time, which can be used to estimate the rate of evolution. Alternatively, they can carry out a *synchronic* study by measuring the difference of trait means of two present-day lineages that both diverged from a common ancestor at some known point in time. This difference divided by the time since the common ancestor yields the rate of divergence across these two lineages, which reflects the evolutionary rate. Both these methods yield average rates over time, which may underestimate the amount of change that occurs each generation. For example, if a trait oscillates rapidly over a timescale of a few generations, but measurements are available only every 100,000 generations, the average per generation rate of change of the character will be underestimated. Furthermore, synchronic data can miss important changes when species are all sampled from the same time slice—usually the present day. For example, if there has been a trend in the direction of trait change across multiple lineages, synchronic studies will not be able to distinguish it from present-day differences and, consequently, will underestimate evolutionary rates.

With either allochronic or synchronic rate estimates, the rate of evolution can be quantified in either of two ways. First, the rate in *darwins* can be calculated as

$$r_d = \frac{\ln x_2 - \ln x_1}{t}, \tag{1}$$

where x_1 and x_2 are the mean trait values for two populations or species, t is the elapsed time separating the populations, and ln stands for the natural logarithm. For synchronic studies, t is twice the time since divergence from a common ancestor, because both species have been evolving independently for that period.

Second, the evolutionary rate in *haldanes* can be calculated as

$$r_h = \frac{x_2/s_p - x_1/s_p}{g}, \tag{2}$$

where x_1 and x_2 are the same as in equation (1), g is the elapsed time in generations, and s_p is the pooled standard deviation of traits within populations. Often, data are ln transformed before rates are calculated in haldanes, so that s_p is the pooled standard deviation of ln-transformed traits. This transformation is necessary when populations or species having higher trait means also have higher trait variances, which is usually the case for morphological characters.

Rates in darwins and haldanes are not directly comparable, and each has strengths and weaknesses. Darwins can easily be calculated and do not require extensive sampling within species. However, rates calculated in haldanes are in units of within-population trait variation. This means that haldanes relate more naturally to population genetics models of trait evolution than do darwins and are thus typically more informative about evolutionary processes (see later discussion). If one trait is more variable within species than another, then comparing their rates of evolution in haldanes is more appropriate than using darwins.

It is also possible to estimate rates of evolution from phylogenetic trees and trait data. The most common way to do this is to assume that trait evolution mimics a "random walk" and to model evolution of a quantitative trait (a trait that varies continuously among individuals) using a statistical process called *Brownian motion*. The rate of change under Brownian motion is described by a rate parameter σ^2 that describes how fast the traits "wander" through time. There are statistical techniques for estimating the rate (σ^2) from data that include a phylogenetic tree and trait data for all available species at the tips of that tree. If the branch lengths in the tree are measured in millions of years, and the trait values are ln transformed, the square root of σ^2 is related to rate estimates measured in darwins.

Rates of trait evolution can be used directly to address many questions in evolutionary biology. For example, are traits in some lineages evolving more quickly than in others? Did a given lineage evolve more quickly at one time period than another? Both allochronic and synchronic studies have shown very convincing evidence that rates of evolution can vary tremendously across characters, lineages, and time periods. The highest rates, typically measured over very short timescales, can approach 1 haldane in rapidly changing lineages like Darwin's finches. By contrast, rates measured over longer, paleontological timescales tend to be slower, ranging from 10^{-6} to 10^{-2} haldanes.

If rates of evolution are measured in haldanes, it is possible to compare the rate of evolution in a lineage to what we would expect under specific evolutionary hypotheses. For example, one classic approach, pioneered by Russell Lande and Michael Lynch, is to compare observed rates with the expected rate of trait change through time under the hypothesis of genetic drift. If the observed trait is under directional selection, we expect to see much faster change than that expected under drift. By contrast, stabilizing selection around a fixed optimum predicts that the observed rate will be much slower than that predicted by genetic drift. By comparing the actual rate of evolution with the prediction under neutral (drift) and nonneutral models, we can infer whether processes like selection have long-term effects on patterns of trait evolution. Analyzing data in this way has revealed two main findings. First, when we look at trait change over very short timescales—say, from one generation to the next—rates of evolution are much faster than we would expect under genetic drift. The high rates mentioned earlier (~1 haldane for Darwin's finches) are a good example of rates that are too high to explain using drift. This is perhaps not that surprising, because we know that many traits (especially the ones we prefer to study!) have strong effects on organismal fitness and are under selection. However, long-term data on evolutionary change have revealed a striking pattern in the opposite direction: differences across species are less than those expected under the hypothesis of genetic drift. For example, Lande pointed out that long-term rates of evolution across a variety of mammals show, on average, only about one-tenth the amount of change expected under genetic drift. That is, drift alone predicts that species should have traits that are much more different from one another compared with what we actually see in the tree of life. Another way to say this is that many species undergo stasis, long periods of time with little or no evolutionary change.

In fact, there is a general relationship between observed rates of evolution and the time interval over which that change is measured. We always measure the fastest rates over the shortest time intervals—from one generation to the next in Darwin's finches, for example. This relationship seems at first to be a paradox (sometimes called the "paradox of stasis"). However, detailed analyses of microevolutionary studies suggest a resolution of this paradox. Although traits often show dramatic change from one generation to the next, this trend is rarely sustained over time. Instead, traits tend to change in dramatically different directions from one generation to the next. The result of these "balancing" changes is a slow net rate of evolutionary change. Selection is often strong, but populations tend to experience selection in different directions from one generation to the next. Only rarely do populations experience sustained directional change. This observation explains the difference in the rates of evolution measured over different time intervals.

Both haldanes and darwins measure evolution of quantitative characters, traits that vary continuously within species. However, we are also interested in the rate of evolution of discrete traits, those that can occupy one of a number of states (e.g., black, white) and tend not to vary within populations. Studies of discrete trait evolution have exhibited the same patterns as described for quantitative traits: rates of trait change vary dramatically among characters, among lineages, and through time.

One common pattern is that trait differences among species (*trait disparity*) in a clade are "bottom heavy"—that is, they evolve at peak rates early in a clade's history. This pattern is often attributed to the effects of ecological opportunity in spurring evolutionary innovation. That is, early in a clade's history, competition and an abundance of unused ecological niches (i.e., ecological opportunity) drive the rapid evolution of many species that are very different from one another.

A number of long-standing debates in evolutionary biology relate to rates of trait evolution. For example, the debate over punctuated equilibrium (see chapter VI.12) centers on the relationship between trait change and speciation but also reflects a debate about the constancy of evolutionary rates through time. Steven J. Gould contrasted "gradualism"—a caricatured view that rates of evolution are more or less constant through time—with his favored model of *punctuated equilibrium*, in which traits change only at speciation events and are in stasis the rest of the time. We now know of instances of trait change that are not strictly associated with speciation (see chapter VI.12). We also know that "stasis" isn't really static. Species do not remain constant but instead typically undergo rapid changes over short intervals that may not be sustained over time. Nevertheless, traits of some species may undergo episodic bursts separated by relative stasis, although bursts of change may or may not be associated with speciation events.

In summary, it is clear that rates of trait evolution vary tremendously across different traits, lineages, and time periods. This pattern is concordant with what we have found for rates of speciation and extinction. For trait evolution, one general finding is that the apparent rate of evolution shows a negative relationship with the time interval over which that change is assessed. This pattern likely reflects the fact that evolution involves rapid change accompanied by frequent reversals.

4. ARE THERE RELATIONSHIPS BETWEEN RATES OF TRAIT EVOLUTION AND DIVERSIFICATION?

Elevated rates of both speciation and trait evolution are key components of the concept of adaptive radiations (see chapter VI.10). Although some clades have been shown to evolve faster than others, there has never been a systematic study showing that clades typically called *adaptive radiations* have faster rates of trait evolution or speciation than other clades. There is quite a bit of evidence that many clades follow a pattern whereby diversification rates start out high but slow through time; however, the corresponding pattern for trait evolution seems much less common.

One can also ask whether rates of trait evolution and rates of lineage diversification tend to change together—that is, are times of rapid lineage diversification also accompanied by rapid trait evolution? Such a relationship is postulated by several theories of macroevolution, most notably both punctuated equilibrium and the ecological theory of adaptive radiations; however, several studies have shown that rates of lineage diversification and trait evolution are "decoupled," so that groups with many lineages may or may not harbor exceptional levels of trait disparity. More work is needed in this area.

Rates of evolution play a key role in a number of evolutionary ideas. Rates of speciation, extinction, and trait evolution can be estimated from both fossils and phylogenetic data. The two types of data are complementary in showing tremendous variation in rates, both through time and across lineages. A number of theories have been proposed to account for this variation in rates. The challenge for the future is to use our rich new knowledge of the tree of life to address long-standing questions in macroevolution.

FURTHER READING

Alroy, J., C. R. Marshall, R. K. Bambach, K. Bezusko, M. Foote, F. T. Fursich, T. A. Hansen, et al. 2001. Effects of sampling standardization on estimates of Phanerozoic marine diversification. Proceedings of the National Academy of Sciences USA 98: 6261–6269. *An innovative paper demonstrating how incorporating variation in the likelihood of sampling fossils across clades and through time can affect our view of past diversity.*

Foote, M. 1997. The evolution of morphological diversity. Annual Review of Ecology and Systematics 28: 129–152. *A comprehensive review of patterns of morphological diversity through time, focusing on paleontological approaches.*

Gingerich, P. D. 2009. Rates of evolution. Annual Review of Ecology and Systematics 40: 657–675. *A review of trait evolutionary rates spanning a wide range of timescales.*

Grant, P. R., and B. R. Grant. 2002. Unpredictable evolution in a 30-year study of Darwin's finches. Science 296: 707–711. *Perhaps the best existing study of short-term evolutionary change in the wild.*

Hendry, A. P., and M. T. Kinnison. 1999. The pace of modern life: Measuring rates of contemporary microevolution. Evolution 53: 1637–1653. *Discusses methods for measuring rates and how these rates can be compared with one another.*

Jablonski, D. 1986. Background and mass extinctions: The alternation of macroevolutionary regimes. Science 231: 129–133. *The unique properties of mass extinctions compared with "background" extinctions.*

Lynch, M. 1990. The rate of morphological evolution in mammals from the standpoint of the neutral expectation. American Naturalist 136: 727–741. *An important paper linking long-term patterns of trait change with expectations from quantitative genetic models.*

Nee, S. 2006. Birth-death models in macroevolution. Annual Review of Ecology and Systematics 37: 1–17. *A useful general review of birth-death models and how they can be applied to questions in macroevolution.*

Simpson, G. G. 1944. Tempo and Mode in Evolution. New York: Columbia University Press. *A key book from the Modern Synthesis that unified microevolutionary perspectives from population genetics with observations from paleontology.*

Uyeda, J. C., T. F. Hansen, S. J. Arnold, and J. Pienaar. 2011. The million-year wait for macroevolutionary bursts. Proceedings of the National Academy of Sciences USA 108: 15908–15913. *A provocative recent paper that uses a huge data set of trait changes over a wide range of time spans to argue for the importance of bounded evolution punctuated by rare rapid changes that occur on the order of once per million years.*

VI.12

Macroevolutionary Trends

Gene Hunt

OUTLINE

1. Directionality in evolution
2. The scope of trends
3. Trend mechanisms
4. Examples of trend hypotheses

Trend hypotheses suggest an underlying directionality to evolution in which some changes are more probable than others. Such trends can operate narrowly—within a single species—or so broadly as to encompass all life. By tracking the characteristics of species and clades over time, paleontologists have documented trends in many kinds of traits and across many types of organisms. Selected examples are used here to illustrate aspects of trend hypotheses, including their scope, evidence, and expression in the fossil record. Fundamentally, two kinds of mechanisms can generate trends: (1) biased microevolutionary changes within species and (2) differential proliferation of species with different characteristics (species selection). From these two mechanisms, an enormous number of specific trends have been proposed. Assessing these trend hypotheses is important for understanding what generalities apply to macroevolution, and for understanding the long-term trajectory of earth's biota.

GLOSSARY

Clade. A group consisting of all species that have descended from a particular ancestral species.

Cope's Rule. A specific trend hypothesis that suggests that body size tends to increase in lineages and clades over time.

Macroevolution. Evolution occurring over very long periods of time or across more than one species, in contrast with the short-term microevolutionary changes in populations from one generation to the next.

Punctuated Equilibrium. A macroevolutionary model in which most trait divergence is concentrated into punctuations associated with lineage splitting (speciation).

Species Selection. The differential proliferation of traits resulting from trait-related differences in speciation or extinction rates. Species selection can generate trends if traits have consistent effects on speciation or extinction rates.

Trend. A persistent temporal change in a characteristic of a lineage or clade.

1. DIRECTIONALITY IN EVOLUTION

Systematic exploration of the fossil record began in earnest in the early nineteenth century. Almost as quickly as fossils could be pulled from the ground and described, paleontologists and others began to wonder what lessons this record held about the temporal trajectory of earth and its biota. Central to these discussions was the issue of *directionality*: Does the sequence of life-forms reveal repeated or sustained trends, or is it instead a story of fluctuations and catastrophes? This theme emerges in many different contexts, ranging from the specific (did horses systematically evolve to larger sizes?) to the universal (does life proceed from simple to complex?), and it remains an active area of research spanning paleontology, evolution, and comparative biology.

An evolutionary *trend* can be defined as a persistent temporal change in a characteristic of a lineage or clade. Trends need not be monotonic—reversals in direction are allowed—but they do imply that some kinds of evolutionary change are more probable than others. Given enough time, changes in a preferred direction can accumulate into substantial increases or decreases in the variable of interest. But the defining characteristic of trends is a bias in direction, not the magnitudes or rates of change they yield. Historically, trends were associated with the idea that evolution is progressive—that in some objective sense, evolution improves organisms or species over time. Although attempts have been made to recast progress into terms consistent with modern evolutionary understanding, most workers eschew this term, as it burdens an unambiguous concept—directionality—with a value judgment about whether that directionality is desirable.

Trends can operate over intervals as short as a few generations, but the focus of this chapter is on those trends that span periods long enough to be documented in the fossil record.

2. THE SCOPE OF TRENDS

One can categorize trends according to the biological unit that is traced over time. This unit can be as narrow as a single species or as broad as the entirety of life. The scope at which a trend operates is important, because different kinds of generating mechanisms are relevant when considering evolution at different scales in the biological hierarchy.

Trends within Species

The most focused kind of trend occurs when a species attribute increases or decreases systematically through time. Paleontologists have assessed this kind of trend hypothesis by measuring features of particular lineages through successively younger rock layers. Because the fossil record is dominated by mineralized hard parts (bones, teeth, and shells), paleontological trends usually involve the size and shape of these skeletal elements. Many kinds of organisms have been studied this way, resulting in specific trend hypotheses about directional change in, for example, the sizes of human brains, the shapes of bivalve shells, and the characteristics of mammal teeth, to name just a few.

Charles Darwin thought that natural selection ought to transform species steadily, and he was troubled by the apparent lack of such trends in the known fossil record. Darwin's solution to this conflict was to argue that the geological record was so woefully incomplete that preservation of these gradual species transformations was unlikely. For many years this view was widely held by paleontologists, only to be challenged in 1972 by the *punctuated equilibrium* model of Niles Eldredge and Stephen J. Gould. These authors argued that species-level trends rarely appear in the fossil record not because of incompleteness but because such trends truly are rare, at least at the timescales that paleontologists can normally resolve. Eldredge and Gould suggested that most species exhibit little net change through time and that, instead, changes are concentrated into punctuations associated with the splitting of lineages as they form new species. The association between speciation and phenotypic evolution was not read directly from the fossil record but rather was inferred to be a consequence of how new species were thought to form.

The punctuated equilibrium model was controversial for several reasons. Particularly spirited disagreements stemmed from its potential implications for evolutionary processes and for the relationship between micro- and macroevolution. More fundamentally, some paleontologists questioned whether the model was even correct about pattern. These critics argued instead that gradual species-level trends were, in fact, not uncommon among well-preserved fossil taxa. This conflict over pattern proved very productive in that it motivated a series of studies designed to test the frequency of within-species trends. Early interpretations of these studies varied greatly, but recent overviews conclude that strongly directional trends are quite rare in fossil lineages; instead, most species show fluctuating or meandering changes. Presumably, the directional changes that differentiate closely related species accrue over temporal intervals shorter than paleontological sampling can usually resolve (at least 1000 to 10,000 years in most sedimentary records).

Trends within Groups of Species

Trends can also be considered at a broader genealogical scale, usually in groups of related species called *clades*. Such trends are detected by tracking the mean trait value over the life of a clade, at least for traits measurable on a continuous scale such as size and shape. Trends in qualitative or categorical variables manifest as systematic changes in the frequencies of the different states over time.

Many of the iconic examples of trends in the fossil record have operated within groups of related species. For example, humans and their immediate relatives do not form a single lineage but rather a tree with several major branches and a dozen or more species. The large brains that are so characteristic of modern humans are the result of increases spanning multiple branches of this tree. Thus this trend has occurred at the clade level, albeit at an accelerated rate in humans. The story is similar for that archetype of paleontological trends, horse evolution. Although it is sometimes (wrongly) depicted as a lineal sequence of ancestors and descendants, the net trajectory of horses from small, browsing ancestors to large-bodied grazers plays out over a complex fossil history that spans 55 million years and well over 100 extinct species.

Some trend hypotheses have even broader scope than individual clades. Paleontologists since the 1960s have compiled large databases of where and when different taxa occur, and these data have been analyzed to get a big-picture view of the major features of life over time. One common version of this approach looks over the past half-billion years or so at all durably skeletonized marine invertebrates; this selection of taxa spans many clades that are only distantly related (e.g., mollusks, corals, brachiopods, trilobites), and it excludes subcomponents of these groups (vertebrates, nonmarine, and poorly preserved members of the included groups). There are parallel analyses for terrestrial vertebrates. Earlier, informal versions of these are the source for the familiar statements that link periods of geologic time to taxa common during those intervals. Every child learns that the Mesozoic was the Age of Dinosaurs, and the Cenozoic was the Age of Mammals; other labels involving less charismatic organisms exist but are less often invoked.

In addition to documenting the changing dominance of life's different players, whole-fauna (or flora) analyses have been interpreted to suggest many large-scale trends in the history of life. One that has received intense scrutiny is the trajectory of species richness through time. Animal diversity is almost certainly quite a bit higher now than it was in the early Paleozoic, but the magnitude, timing, and causes of this increase are all actively debated. Other suggested trends or state shifts through this interval include temporal declines in rates of origination and extinction, shifts in the frequency of different modes of life (e.g., sessile filter feeding versus mobile grazing), increases in the intensity of predation, and increases in organismal and ecological complexity.

3. TREND MECHANISMS

Although the biological context of each trend is unique, fundamentally there are only two kinds of mechanisms capable of generating macroevolutionary trends: (1) biased microevolutionary changes within species and (2) the differential proliferation of species with different characteristics, also known as species selection.

Trends as Accumulated Microevolution

The most straightforward mechanism for producing a trend involves a bias in the direction of phenotypic changes within species. Usually the cause of this bias is assumed to be natural selection favoring one direction of change more than others. When this bias is restricted to a single lineage, a species-level trend is produced; when it applies across groups of related species, the result is a trend across an entire clade.

An example of a trend commonly described in these terms is *Cope's rule*, the notion that animal body size preferentially increases within lineages. It has been postulated that larger-bodied individuals have numerous advantages over their smaller conspecifics, including superior physiological buffering, greater ability to subdue prey and defend against predators, and greater success in the competition for mates and resources. While these suggestions are not unreasonable, they can be difficult to test, and quite often the ultimate causes of trends are uncertain. Well-constrained examples often involve climatic drivers because of the abundance of climate-related information preserved in the geological record. Biotic interactions, by contrast, usually leave few detectable traces in the fossil record, and so they are much more difficult to evaluate as putative drivers of trends.

Because paleontological trends typically unfold over millions of years, even the most dramatic trends can be generated by very small biases acting in each generation. One can show, for example, that even the largest evolutionary transitions within single traits in horse evolution can be produced by just a few selective deaths per million individuals per generation. This intensity of selection is very weak—much weaker than genetic drift—and so it is probable that macroevolutionary trends, when driven by natural selection, unfold episodically in response to local and changing conditions, rather than uniformly over time.

In a classic paper, Steven M. Stanley proposed another scenario that can generate a trend via accumulated microevolution. Stanley was interested in explaining Cope's rule, but he did not assume universal advantages for larger bodies. Instead, Stanley suggested that there may be a lower limit to body size below which evolution is unlikely to explore. For endothermic vertebrates, this lower limit might arise from difficulties in maintaining body temperature at very small sizes, but the proximate biological reasons for these limits would depend on the relevant taxon's suite of adaptations and constraints. Moreover, Stanley argued that smaller organisms are, on average, less morphologically specialized than larger organisms, and thus they should preferentially found higher taxa. Starting at a small size near a lower boundary will result in an asymmetrical expansion in body sizes: maximum and mean sizes increase as species diverge over time, but minimum size stays stable because it cannot decrease below the lower limit. Thus, a trend is observed, even though increases and decreases are equally likely except in the neighborhood of a boundary.

This kind of trend has been called "passive" because it arises from diffusion-like evolution in the presence of a boundary; "active" or "driven" trends are those characterized by a uniform bias in the direction of evolutionary change. Passive and active trends can be distinguished empirically by their different effects on maxima and minima, and by comparing the dynamics of species near and far from the putative bound.

Trends from Species Selection

Though it might seem counterintuitive, it is possible to generate clade-level trends even in the absence of directional microevolutionary changes. All that is required is that the trait have a systematic relationship with net rates of species diversification. For example, if large-bodied species proliferate more successfully than small-bodied species, then, over time, that fauna will become increasingly dominated by large-bodied species, and as a consequence, that clade's mean size will increase steadily. This Cope's rule pattern will hold even when body size increases and decreases are equally frequent within species. This hypothetical trend is generated not by biased transformation of species but rather by biases in their proliferation, a process called *species selection* (see chapter VI.14). Just as natural selection follows from variation in reproductive success among individual organisms, so does species selection result from variation among species in their propensities to persist and generate "offspring" lineages through speciation. Accordingly, the traits most likely to be important for species selection are those that influence, directly or indirectly, the rates at which species form or the rates at which they become extinct.

Although species selection is not applicable to trends within a single species, most workers agree that it is a plausible mechanism for clade-wide trends. Its empirical importance, however, is much less clear. The main problem is that assessing a trait's influence on speciation or extinction requires a large set of species of known relationships. But the fossil record is less complete and robust at the species level, and so rates of extinction and origination are usually estimated for genera instead. However, high-quality, species-level data sets are starting to become more common, which should allow for better tests of species selection as a driver of trends. Another way around this obstacle may be afforded by recent methodological advances that permit one to assess the influence of traits on speciation and extinction rates using phylogenies of extant species, with no fossil record required. At present, rather large trees are needed to obtain reliable results, but the explosive growth of molecular phylogenetics has made such trees increasingly available.

At the broadest level, it is worth noting that trends related to the waxing and waning of distinct clades must be driven by differences in diversification among the clades. Sometimes, these diversification differentials are consistent over long periods. For example, the current dominance of flowering plants on land is a consequence of their elevated diversification rates compared with those of other vascular plants. These differences in diversification have persisted since late in the Cretaceous, roughly 100 million years ago. At the other end of the continuum, single extreme events such as mass extinctions can trigger permanent shifts in the dominance of different taxa, as in the shift in terrestrial communities resulting from the extinction of nonavian dinosaurs and other large vertebrates during the end-Cretaceous mass extinction.

4. EXAMPLES OF TREND HYPOTHESES

Cope's Rule in Mammals

Cope's rule has been studied quite broadly, but it has been particularly emphasized in vertebrate paleontology. The enormous sizes achieved by many dinosaur lineages have led to suggestions of Cope's rule for this group, but mammals are the fossil group that has been analyzed most extensively for body size trends. The most comprehensive analysis to date is one by John Alroy, who tracked body mass (estimated from tooth dimensions) in North American mammals over the past 80 million years. By looking at the difference between putative ancestor and descendant species pairs, Alroy was able to demonstrate a bias toward body size increases: on average, descendants were about 9 percent larger than their ancestors. The trend mechanism here is therefore a bias in microevolutionary changes within species, rather than differential sorting among species.

What ultimately explains this preferential size increase within lineages? For a group as diverse and heterogeneous as mammals, it is likely that multiple factors have been important. Fossil horses offer perhaps the best case study because of their richly documented fossil record, especially from North America. The earliest-appearing horses were relatively small, estimated to weigh 30 kg or less, and horses remained at about this size for the next 25 million years or so. Then, average body size among horse lineages increased rapidly following a climate-driven expansion of open, grass-dominated habitats. These body size shifts were associated with changes in other features interpreted as adaptations to grasslands, such as high-crowned teeth (which are better able to withstand high-grit, grassy diets). Thus, the timing of the body size increases, along with their environmental context, is consistent with large-scale shifts in the habitats as the driver of this trend.

Organismal Complexity

Perhaps no trend hypothesis has had a firmer hold on scientists' imagination than the idea that life has progressed from simple to complex. The fossil record has long been marshaled in support of this notion because the first organisms to appear in the fossil record are single-celled prokaryotes, a form of life that presumably exists near a lower boundary for organismal complexity. Moreover, for its first 2 billion years, the fossil record of life consists entirely of structurally simple microscopic organisms. Evidence for animals, possibly sponges, first appears around 800 million years ago, but it is not until the Cambrian period more than 250 million years later that complex bilaterian animals diversify.

The most persistent obstacle to evaluating this trend hypothesis is pinning down what exactly it means for an organism to be complex. Daniel W. McShea has argued that the criteria used to identify complexity have been loose and impressionistic, sometimes reflecting only gross similarity to *Homo sapiens*. According to McShea, complexity comes in different forms, but the type most amenable to testing in the fossil record is related to the number and diversity of subcomponents, or parts, that constitute an organism. The more parts and kinds of parts an organism bears, the more complex it is. This concept is made operational by deciding what most usefully constitutes a part in the system under study.

Perhaps the most general attempts to evaluate animal complexity involve counting as parts the number of distinct cell types in different animal groups. Merging these indicators of complexity with information on the first fossil appearances of those groups suggests an increase in maximum animal complexity over time. Animal taxa with few cell types, such as sponges and cnidarians, appear earlier in the fossil record than groups like birds and mammals that today have many different kinds of cells. These data are sufficiently uncertain, however, that it is difficult to establish with precision the timing and pattern of the increase or even if *average* complexity follows the same trajectory as *maximum* complexity. In part, this uncertainty stems from ambiguities in enumerating distinct cell types across taxa that vary greatly in how well their microanatomy has been studied; with greater investigation, more cell types will be discovered, and finer categorization of cells into types will be possible. Emerging comparative genomic data, when similarly mapped to the fossil record, may permit more unequivocal tracking of the trajectory of animal complexity through time.

Escalation

The hypothesis of *escalation*, as articulated by Geerat J. Vermeij, suggests an overarching ecological directionality to the history of life, in which organisms have increased in their ability to acquire and control resources. This trend is said to be episodic rather than uniform over time, but the net result is a biotic environment that has become more and more perilous to organisms as they confront increasingly dangerous predators, competitors, and prey. This trend is detectable in the fossil record by tracking the origin and frequency of adaptations related to this biological arms race, such as arthropod limb modifications for crushing shelly prey and the complementary features of shelled organisms that defend against such attacks.

The scope of this hypothesized trend is broad, encompassing whole faunas of interacting organisms. Biased microevolution and species selection are thought to act jointly to generate the escalation trend. Microevolutionary transitions are inferred to preferentially favor behaviors and morphologies that are more active and energy intensive. Species selection may also contribute to this trend when escalated species radiate, and poorly defended forms preferentially go extinct.

Definitive tests of the escalation hypothesis are complicated by the diversity of escalation predictions and by uncertainty about the relative energy intensiveness of morphologies and modes of life in extinct taxa. Moreover, because increasingly dangerous modes of life are thought to incur steep energetic demands, there might be temporal and spatial refuges—intervals and environments in which resource limitation slows or reverses the normal tide of escalation. Indeed, empirical studies that address aspects of escalation have produced a variety of results, not all of which are consistent in their timing, coordination, or magnitudes of ecological change. Nevertheless, over suitably long periods, many studies have found net positive trajectories in escalation-related measures, including increasing body size (mostly in the early Paleozoic), increasing dominance of active predators and other mobile lifestyles, diversification of shell-crushing predators (especially in the middle Paleozoic and late Mesozoic), increasing frequency of traces of shell predation (drill holes and repair scars), increasing depth of burrowing (presumably an escape from predation), increasing frequency of morphologies interpreted to defend against predators (e.g., narrow apertures in snails), and a corresponding decrease in susceptible forms.

See also chapter II.9, chapter VI.11, and chapter VI.13.

FURTHER READING

Alroy, J. 1998. Cope's rule and the dynamics of body mass evolution in North American fossil mammals. Science 280: 731–734.

Bush, A. M., and R. K. Bambach. 2011. Paleoecologic megatrends in marine metazoa. Annual Review of Earth and Planetary Sciences 39: 241–269. *The most up-to-date and synthetic treatment of ecological trends over geologic time.*

Eldredge, N., and S. J. Gould. 1972. Punctuated equilibria: An alternative to phyletic gradualism. In T.J.M. Schopf, ed., Models in Paleobiology. San Francisco: Freeman, Cooper.

McNamara, K. J., ed. 1990. Evolutionary Trends. Tucson: University of Arizona Press. *This edited volume has separate chapters covering trends in many major groups of fossils; an overview chapter by Michael McKinney is a useful introduction to trends in the fossil record.*

McShea, D. 1996. Metazoan complexity and evolution: Is there a trend? Evolution 50: 477–492. *A lucid dissection of the concept of complexity and a review of paleontological evidence for a complexity trend in animals.*

Stanley, S. M. 1973. An explanation for Cope's rule. Evolution 27: 1–26. *This seminal paper introduced the idea of a passive trend.*

Vermeij, G. J. 1987. Evolution and Escalation. Princeton, NJ: Princeton University Press.

VI.13

Causes and Consequences of Extinction

Michael J. Benton

OUTLINE

1. Species extinction
2. Some definitions: Extinction styles and magnitudes
3. Mass extinctions
4. Declining extinction risk and resetting the clock
5. Extinction and the drivers of macroevolution

Species extinction is a normal part of evolution, but there have been many times in the earth's history when higher-than-expected numbers of extinctions have occurred. During sudden extinction events, and especially during mass extinctions, major physical environmental crises have wiped out large portions of life. The fact that selectivity during extinction events differs from natural selection suggests that higher-level macroevolutionary processes have continually affected the evolution of life.

GLOSSARY

Court Jester. The model of macroevolution that concentrates on changes in the physical environment as the main drivers (cf. Red Queen).

Ecospace. A combination of habitat and ecological activity at any scale.

Macroevolution. Evolution above the species level.

Mass Extinction. The sudden, worldwide loss of many species of diverse ecologies.

Morphospecies. A species defined on the assumption that all members share the same morphology, and other species show different external form.

Pseudoextinction. "Extinction" of a species when it evolves into another species.

Red Queen. The model of macroevolution that concentrates on biotic interactions as the main drivers (cf. Court Jester).

Taxon. A species or larger division of the tree of life.

1. SPECIES EXTINCTION

Extinction is the disappearance of a species or larger taxon. The geographic scale can be local or global. The concern here is with the latter, corresponding to the complete disappearance of a genetic lineage worldwide, not the local disappearance of a species by emigration or environmental change. In the global case, as has often been said, extinction is forever.

The extinction of species is inevitable. Each species has a duration, which is not predetermined but may be characteristic of the wider taxon. A common assertion, developed by George Gaylord Simpson and Steven Stanley, is that mammals evolve at 10 times the rate of clams, which means they originate and go extinct at rates differing by an order of magnitude. This declaration could be an artifact of how human taxonomists identify morphospecies (see chapter VI.1)—perhaps they subdivide mammalian species 10 times as finely as they do those of bivalves, possibly responding to the evident visual differences among mammalian species while missing the less visible species-specific cues in mollusk shells. Nonetheless, assuming that species of mammals and mollusks are somehow equivalent sections of the tree of life, then there are broad differences in mean species durations through geologic time, and therefore also in species extinction rates (table 1).

Such wide differences in species extinction rates and macroevolutionary rates have clear implications for the interpretation of times of intense extinction, such as extinction events and mass extinctions in the past, and the current biodiversity crisis (see chapter VIII.6): one would expect the fast-evolving species to be more liable to extinction and, indeed, more likely to recover following the crisis than the slowly evolving species. For example, the ammonoids, a long-lived group of mollusks, typically had short species durations but suffered near-complete wipeout during four mass extinction

Table 1. Estimated mean durations of fossil species, taken from various sources

Group	*Mean duration (My)*
Reef corals	25
Bivalves	23
Benthic foraminifera	21
Bryozoans	12
Gastropods	10
Planktonic foraminifera	10
Echinoids	7
Crinoids	6.7
Monocot plants	4
Horses	4
Dicot plants	3
Freshwater fish	3
Birds	2.5
Mammals	1.7
Primates	1
Insects	1.5

Source: Summarized by McKinney 1997.
Note: Marine groups show longer durations (6.7–25 My) than terrestrial groups (1–4 My).

events, yet recovered rather rapidly in comparison with other marine invertebrates.

The prevalence of extinction— its inevitability—for all species is obvious to evolutionary biologists and paleontologists, but perhaps less so to nonscientists, and is germane to wider discussions about the current biodiversity crisis. Clarity is needed on three issues in this context: lineage extinction at some point is the norm, species differ innately in their extinction risk by wider clade membership, and these points are distinct from the immediate risk of extinction of any named species according to current ecological threats.

The aim here is not to discuss extinction as it manifests itself in the context of natural selection or phylogeography (see chapters II.5, III.1, III.6, III.7, and VI.4), nor in terms of its role in rates of evolution and species selection (see chapters VI.11, VI.12, and VI.14) and in the current biodiversity crisis (see chapter VIII.6) but at the macroevolutionary level and in two broad contexts: first, as a part of the debate about biotic and abiotic drivers of evolution, and second, in terms of the role of extinction events in punctuating the history of life.

2. SOME DEFINITIONS: EXTINCTION STYLES AND MAGNITUDES

The extinction of a species may occur according to one of two patterns in phylogenetic terms: the species terminates without leaving any issue, or it evolves sufficiently to be called a new species. In most molecular phylogenetic approaches, species terminate at the present day, and the issue of extinction does not arise. When fossil taxa are incorporated into phylogenies, they are generally treated as discrete entities that terminate with a definitive extinction. In densely sampled fossil records, however, some species apparently evolve directly into others, and the extinction of the older parts of the lineage is termed a *pseudoextinction* because the gene pool of the populations that constitute the original lineage continues into the replacing species. The relative prevalence of such pseudoextinctions is hard to determine: it could be argued that they are in fact rare, and quoted examples are based on nonobjective interpretations of sequences of rather simple fossils through numerous sampling horizons. Conversely, critics of cladistics have claimed that such transitional successions of species are relatively common and represent a challenge to the cladistic method because it can identify only species that arise by splitting.

If species extinction is the end of a lineage, the term *extinction* is also more widely used by evolutionists and paleontologists to denote the end of a clade or paraphyletic group. For example, the "extinction of the dinosaurs" means the end of all nonavian Dinosauria, in other words, the set of clades that includes all animals popularly called dinosaurs but not including the dinosaurian subclade Aves (Avialae), the birds. In this case, the extinction of the dinosaurs does mean the termination of a large number of clades, such as Ornithischia and Sauropodomorpha, and among the theropods, Ceratosauria, Carnosauria, Troodontidae, and Dromaeosauridae. In other cases, however, the term *extinction* is applied to even less cohesive groups that may share some general ecological characteristics, such as body size or geographic region. An example is the end-Pleistocene extinction of large mammals in the Northern Hemisphere, sometimes termed the "extinction of megafauna," meaning some, but not all, large animals, in some, but not all, parts of the world.

Paleontologists divide extinctions into three categories: background extinctions, extinction events, and mass extinctions, each of which is a useful concept in particular contexts, but between which there are no sharp divisions.

Background extinction is the sum of all normal species terminations during a defined time interval (*time bin*). The termination of any particular lineage is not predictable, but the mean rate across a large clade, across a region, or worldwide for all life is predictable. Hence, all things being equal, global extinction is a stochastic process, and its rate should be predictable, dependent on the standing crop of species and their distribution through major clades (each of which has a characteristic mean extinction rate).

Extinction events are times when many species go extinct for a shared reason. Extinction events can be of

Table 2. The "big five" mass extinctions, with principal victims and possible causes

Event	*Ma*	*Victims*	*Possible cause*
End-Ordovician	444	Nautiloids, trilobites, brachiopods, crinoids, bryozoans, corals	Glaciation
Late Devonian	372	Trilobites, brachiopods, bivalves, corals, nautiloids, sponges, crinoids, fishes (ostracoderms*, placoderms*)	LIP, ocean anoxia
End-Permian	252	Brachiopods, blastoids*, trilobites*, crinoids, rugose* and tabulate* corals, pareiasaurs*, synapsids	LIP, ocean anoxia
End-Triassic	201	Bivalves, ammonoids, gastropods, conodonts*, basal archosaurs*	LIP, ocean anoxia
End-Cretaceous	66	Dinosaurs*, pterosaurs*, plesiosaurs*, mosasaurs*, ammonites*, belemenites*, bivalves (rudists*), gastropods, corals, foraminifera, nannoplankton	Meteorite impact; LIP

Note: LIP, large igneous province—basaltic eruptions
*Groups that entirely died out

all magnitudes, but the term is usually reserved for those smaller events that do not qualify as mass extinctions. Under this assumption, extinction events may be regional in scale or may apply to only certain clades or certain ecological guilds. The best-known example is the end-Pleistocene extinction of large mammals in the Northern Hemisphere, but there have been many others over the past 600 million years, such as the early Toarcian ocean anoxic event, 183 million years ago, that killed much of marine life in Europe, or the series of small extinctions in the late Cambrian, about 490 million years ago, each of which marked a major turnover in the trilobite faunas. Causes of these extinction events were varied, but they were generally associated with dramatic changes in the environment that affected many species at least in one or more world regions, such as the retreat of the northern ice 11,000 years ago, the spread of humans and their voracious hunting, an oceanic anoxic event, or major topographical change.

Mass extinctions are the most notable of all, the times of global disappearance of much of life, when many species of wide ecological range died out worldwide, and geologically speaking at least, did so rapidly. Paleontologists have struggled to constrain the terms "much of life" and "rapidly," but without success, because the distribution of extinction event magnitudes is apparently continuous, with no qualitative distinction between small- and large-scale crises. David Raup and Jack Sepkoski famously identified a statistical distinction in which mean familial extinction rates were assessed for 100 time bins through the Phanerozoic, each 5–6 million years in duration, and they found that five of the points stood out as statistical outliers, beyond the 95 percent confidence envelope (figure 1). This result was broadly reasonable, as the five unusually high global extinction rates corresponded to the "big five" mass extinctions, but the method was statistically unreasonable because the error bars included negative extinction rates, which cannot occur.

3. MASS EXTINCTIONS

The identification of what is and is not a mass extinction is variously impossible (because there is a continuum of extinction events of all magnitudes, and so the dividing line between small extinction events and mass extinctions is a matter of choice) and trivial (because there is no category of unique entities called mass extinctions, there is no need to determine which event at the margin is or is not a mass extinction, nor to seek common rules or laws that apply to all). There is, however, a need for paleontologists to engage with the issue, because the subject has achieved wide popular interest and feeds through to concerns about the current biodiversity crisis: Are we living through the sixth mass extinction, as Richard Leakey and Roger Lewin termed it, or not?

The standard list of the big five mass extinctions comprises the end-Ordovician, Late Devonian, end-Permian, end-Triassic, and end-Cretaceous events (table 2). If these are the five, then the current biodiversity crisis can be said to scale with those events of the past, at least in terms of the rate of species loss in the past 500 years, and so it can be termed the "sixth mass extinction." Annoyingly for the headline writers, however, there were earlier extinction events that might merit the term *mass extinction*, including the end of the Ediacaran faunas in the Neoproterozoic, 541 million years ago, and the assembled late Cambrian crises. So, is the present crisis the "eighth mass extinction"? Perhaps that designation is in doubt, as others, including Richard Bambach and colleagues, have argued quite reasonably that there are mass extinctions and mass extinctions: three of the big five were not rapid, single-cause events but summations of pulses of species losses, and perhaps only three of the large events count as mass extinctions: the end-Ordovician, end-Permian, and end-Cretaceous. These three stand out as times of unusually high rates of species loss compared with neighboring time intervals, and the catastrophic losses of biodiversity were caused primarily by high extinction rates. In contrast, during the Late Devonian and end-Triassic events, part of the depletion in biodiversity was caused by unusually low origination rates, and so these appear to have been complex episodes of turnover crisis, rather than simply mass killing.

However they may be defined and counted, there has been much study of the big five mass extinctions (table 2). Today, with thousands of publications each year, it might seem surprising that geologists and paleontologists hardly considered these events until the 1970s—indeed, somehow the "death of the dinosaurs" and earlier crises were ignored or trivialized. It seems that geologists were afraid of being labeled as crazy "catastrophists" at a time, even in the 1950s and 1960s, when it was considered dangerous to admit that the earth had ever been hit by large meteorites. Everything changed after 1980.

The tipping point for geologists occurred with the publication of Luis Alvarez's proposal that the earth had been hit by a 10 km meteorite at the end of the Cretaceous period, that the impact threw dust high into the atmosphere, blacked out the sun, and caused global darkness and freezing for long enough to kill off much of life. This proposal was based on seemingly limited evidence—two locations, in Italy and Denmark, where there was a relatively high concentration of the platinum-group element iridium (the iridium spike) exactly at the Cretaceous-Tertiary (KT) boundary. This, Alvarez reasoned, indicated the arrival on earth of a massive amount of extraterrestrially derived material, transported through the medium of a meteorite or comet, because iridium does not generally occur naturally on the earth's surface. Through a simple calculation, Alvarez and colleagues estimated the volume of dust needed to black out the sun, then the size of the crater required to generate such a dust volume (150 km diameter), and then the size of the colliding rock (10 km diameter).

These proposals were variously met with massive enthusiasm and angry denunciation, but the criticisms diminished as substantial amounts of confirming evidence were identified during the 1980s: the iridium spike was found everywhere at the KT boundary in both marine and terrestrial rocks; additional evidence for impact was identified (high-pressure minerals such as shocked quartz, coesite, and stishovite); and indeed, the crater itself was found, at Chicxulub in Mexico.

The Alvarez hypothesis led to a second consequence, the suggestion that all major extinctions, not just the big five, were triggered by impacts: Raup and Sepkoski presented evidence for periodicity of extinctions during the past 250 million years, noting a statistically prominent 26-million-year period between such events. Only three of the big five mass extinctions occurred within the past 250 million years, but Raup and Sepkoski identified many other medium-sized species extinctions during the Triassic, Jurassic, and Cretaceous. Indeed, the last of their events, in the Miocene, occurred 11 million years ago. The consequences of the periodicity theory were profound: all mass extinctions had a single cause, that cause was almost certainly extraterrestrial and involved impacts, and the next event will occur in 15 million years. The proposals led to massive interest from scientists across many disciplines, with contributions coming from astronomers, mathematicians, geologists, and biologists. The analyses were sophisticated, and some paleontologists and mathematicians are still intrigued by the proposal, yet the raw data are far from convincing: the fossil databases are patchy, revision of geologic timescales casts doubt on the periodic signal and the period length, and most devastatingly, many of the intermediate "smaller" extinctions disappear when inspected closely, as argued by Mike Benton and others.

Key questions about mass extinctions concern the causes, the victims, and the recovery. Here is not the place to present too much detail on the *causes* of mass extinctions—the literature on each event is huge, and the postulated causes, especially if older literature is included, are manifold. Recent work has concentrated on identifying plausible models, especially models that might explain more than one event. Some would still identify a single astronomical model as a driver and so explain all mass extinctions as the result of impact and perhaps a killing model akin to that for the KT event. Most paleontologists, however, are content to accept impact as the sole or major reason for the KT mass extinction, but they seek other explanations for the earlier mass extinctions. The most ubiquitous model appears to be volcanic eruption and its consequences, most notably, massive basaltic eruptions that span several hundred thousand years and form large igneous provinces (LIPs), and that appear to have coincided with at least three of the big five events (table 2), as well as the KT (the Deccan Traps in India). The model for extinction associated with such massive eruptions, as summarized by Paul Wignall, focuses on the huge volumes of carbon dioxide spewed out during the eruptions. This is a greenhouse gas and so causes global warming. In normal circumstances, excess carbon dioxide would be consumed by green plants through photosynthesis, but repeated and continuing large-scale eruptions perhaps swamped the normal feedback processes and caused increasingly severe atmospheric warming. On land, plants and animals succumbed if they could not move to the poles, and in the seas, warming of surface waters caused stagnation as the normal circulation of deep cold waters to the surface was slowed, and so oxygen could not reach the seabed, and life there died.

The *victims* of mass extinctions seem to be a random selection of life of the time. Raup famously contrasted the two assumptions about victims of extinction: they suffer from either "bad genes or bad luck." In normal, Darwinian, evolution the focus is on bad genes; a species dies out because of some aspect of natural selection, perhaps competition with another species, or inability to adapt to changing conditions. However, during mass extinctions, environmental stresses are severe and unpredictable, and so species cannot be selected for their ability to survive such rare events, and those that succumb may be simply unlucky. Nonetheless, there might be biological characteristics that by chance enable species to survive the shock of the extinction crisis or the tough conditions that follow. Among such general characteristics the most important appears to be wide geographic range at the clade level, regardless of the geographic range of individual species. Other useful characters that seem to improve a species' chances of survival are adaptation to a broad diet and broad physiological requirements, and modest body size.

The *recovery* of life after a mass extinction has clear significance for modern conservation concerns. Certainly, it seems that the rapidity of recovery is proportional to the scale of the extinction, but there may be nonscalar components: if a mass extinction removes certain species from ecosystems, the scaffold of the ecosystem may be available after the crisis for new species to slot in. If, however, most components of an ecosystem are removed by a larger extinction event, recovery may involve the construction of entirely new ecosystems, and so perhaps takes longer. Species recover according to their normal evolutionary dynamics, so it is notable, for example, that ammonoids recovered quickly after the end-Permian mass extinction, whereas other groups such as bivalves and echinoderms seem to have taken longer. Further, there may be a major difference between initial and subsequent recovery, meaning the initial rapid filling of ecospace versus the construction of longer-term, more stable ecosystems. So, for example, after the end-Permian mass extinction, species numbers within faunas—and globally—seemed to bounce back within 1–2 million years, but these consisted largely of *disaster taxa*, short-lived lineages that did not contribute to the eventual major clades or to the longer-term structure of the ecosystems. For example, on land, after the end-Permian event, the initial *Lystrosaurus* fauna was unusual in that it was dominated by one species; was associated with many amphibians, but no larger herbivores or carnivores; and was cosmopolitan. It took perhaps 10–15 million years for ecosystems to stabilize with a full range of body sizes and trophic levels—with a balance of the major clades that were to be significant for some time thereafter—and for continent-scale endemicity to become reestablished.

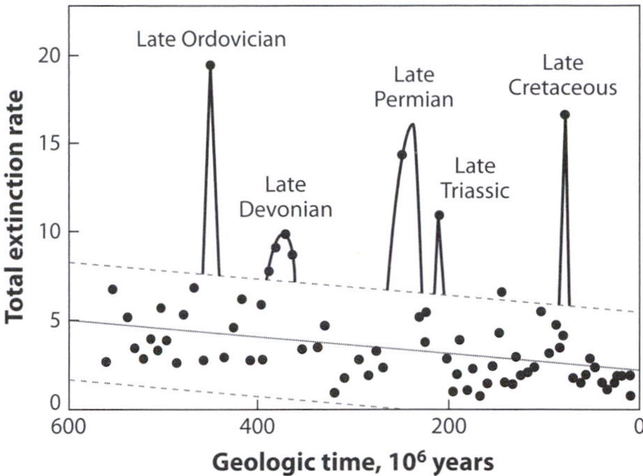

Figure 1. Plot of total extinction rate through time for animal life in the sea. The timescale spans the late Proterozoic and Phanerozoic, the time of relatively abundant large animal life. The total extinction rate is assessed as the mean number of families becoming extinct per million years, in each geological stage (mean duration, 5–6 My). The solid lines indicate the best-fitting regression, and the dashed lines the 95 percent confidence envelope. The plot was interpreted to show declining mean extinction rate through the past 600 million years and to identify six times of unusually high extinction, the named positive outliers. (From Raup and Sepkoski 1982.)

4. DECLINING EXTINCTION RISK AND RESETTING THE CLOCK

One of the key points of Raup and Sepkoski's review of extinction rates (figure 1) was to demonstrate that these rates showed an apparently statistically significant decline through time, which these authors interpreted as a general improvement in the ability of organisms to resist extinction—presumably, as global mean extinction rates fell, mean duration of families of marine invertebrates increased. If this interpretation is correct, it would represent cogent evidence for progress in evolution, a notoriously tricky concept to define and prove.

The evidence has been disputed, for more or less geometric reasons. The fact that the analysis was carried out on families, not genera or species, immediately gives pause for thought: What if the families are largely human constructs, and we simply interpret families differently in older rocks? Further, all other things being equal, species are less likely to be preserved in older rocks than in younger ones, and so "families" in the Cambrian might well include far fewer species than families from younger rocks: with a constant species extinction rate, familial extinction rate will decline as the number of species per family increases. Third, even with a perfect fossil record, it is likely in any case that the number of species per family will increase through time, simply because the geometry of evolution demands lineage splitting and expansion of clades through time. Again, with a constant rate of species extinction, familial extinction rates must decline through time. In the end, then, there is no evidence that the mean of all family-level or genus-level extinction rates at any time can be compared with mean rates in neighboring time bins. Together, this means that there is no evidence for declining rates or improving competitive ability through time.

It has often been said that mass extinctions, or extinction events in general, reset the clock of evolution, cutting across all the existing arms races, coevolutionary species pairs, food webs, and ecosystems, and kick-start an entirely new phase in the history of life. Leigh Van Valen, for example, suggested that the history of life in the sea followed two major evolutionary cycles, one beginning with the origin of animals in the late Neoproterozoic and Cambrian, after which mean per-taxon extinction rates (probabilities of extinction) declined rather steadily to the end of the Permian. The huge end-Permian mass extinction then killed off all but 10 percent of species, which subsequently gave rise in the Triassic to new lineages that at first showed very high mean extinction probabilities, which in turn began a second long-term declining trend toward the present. In this case, he argued that the other mass extinctions had negligible effect on the broad patterns.

It would be wrong to assert that mass extinctions literally "reset" evolution, in the sense of wiping out the preexisting interactions and lineages and opening the world for something entirely unexpected and new. Indeed, many lineages survived even the most severe of mass extinctions, and they became reestablished in similar ecological roles after the crisis, occupying the same positions within ecosystems. These chance survivors may indeed have helped retain the frame of postextinction ecosystems, into which new taxa inserted themselves during the recovery process.

Nonetheless, mass extinctions do reset the pattern of macroevolution in enabling the radiation of clades that might otherwise not have been able to radiate, or not at the same time. For example, in the Early and Middle Triassic seas following the end-Permian mass extinction, several groups of marine reptiles—ichthyosaurs, thalattosaurs, and sauropterygians—became established as entirely new top predators. Likewise, on land the first dinosaurs emerged at this time and, after a further extinction event, took their important role in terrestrial ecosystems. Even better known perhaps are the ascents of modern mammals and birds after the extinction of the dinosaurs 65 million years ago. Had the extinctions not happened, these major groups might not have had their chance to rise to importance.

5. EXTINCTION AND THE DRIVERS OF MACROEVOLUTION

What, then, has been the significance of extinction events in driving evolution? The answer addresses the wider question of the relative roles of biotic and abiotic drivers of macroevolution, characterized as the Red Queen and Court Jester models. The *Red Queen* model, developed by Van Valen, stems from Charles Darwin's work—in which he viewed evolution as primarily a balance of biotic pressures, most notably competition—and was characterized by the Red Queen's statement to Alice in *Through the Looking-Glass* that "it takes all the running you can do, to keep in the same place." In contrast, the *Court Jester* model, presented by Tony Barnosky, is that evolution, speciation, and extinction rarely happen except in response to unpredictable changes in the physical environment, recalling the capricious behavior of the licensed fool of medieval times. Note that neither model was meant to be exclusive, and both Darwin and Van Valen allowed for extrinsic influences on evolution in their primarily biotic, Red Queen views.

Species diversity in a Red Queen world depends primarily on intrinsic factors, such as body size, breadth of physiological tolerance, or adaptability to unusually harsh environmental conditions. In a Court Jester world, species diversity depends on fluctuations in climate, landscape, and food supply. In reality, of course, both worldviews can prevail in different ways and at different times. Traditionally, biologists have tended to think in a Red Queen, Darwinian, intrinsic, biotic factors way, and geologists in a Court Jester, extrinsic, physical factors way.

Much of the divergence between the Red Queen and Court Jester worldviews may depend on the scale of observation. It is evident that biotic interactions drive much of the local-scale success or failure of individuals, populations, and species (Red Queen), but natural selection and the Red Queen also accommodate constantly changing climate and topography at the scale of intergenerationally differing selection pressures. However, perhaps these processes are overwhelmed by large-scale tectonic and climatic processes at timescales above 100,000 years (Court Jester), which may be too drastic for most species lineages to adapt, and they go extinct locally or globally.

The Red Queen and the Court Jester are in opposition in that their consequences differ. Further, the two models could be said to emanate from two different starting points: the Red Queen from considerations and observations of natural selection experiments and evolutionary ecology, the Court Jester from paleobiological and geological studies of global change over longer time spans. The divergence between the two could be interpreted as epistemological, a result of differing methodologies, or ontological, meaning it is real. Evolutionary biologists and paleobiologists are often warned not to scale processes between levels, for example, to assume that large clades act like species in competition and predator-prey interactions, or to assume that geologically instantaneous processes can be ecologically instantaneous also. In this regard, macroevolution is likely pluralistic, with intense biotic interactions shaping ecosystems and species evolution on a daily and yearly basis, and abiotic drivers acting over all timescales, but especially on timescales of centuries to millions of years.

Importantly, no matter whether either the Red Queen or the Court Jester model actually prevails in evolution and how they interact, extinction has a key role in marking the tempo of evolution within clades, and in punctuating the larger-scale, long-term patterns of the history of life.

See also chapter II.9.

FURTHER READING

Alvarez, L. W., W. Alvarez, F. Asaro, and H. V. Michel. 1980. Extraterrestrial cause for the Cretaceous-Tertiary extinction. Science 208: 1095–1108. *The classic presentation of impact as a cause of extinction.*

Bambach, R. K., A. R. Knoll, and S. C. Wang. 2004. Origination, extinction, and mass depletions of marine diversity. Paleobiology 30: 522–542. *A thoughtful consideration of ecological aspects of mass extinctions.*

Barnosky, A. D. 2001. Distinguishing the effects of the Red Queen and Court Jester on Miocene mammal evolution in the northern Rocky Mountains. Journal of Vertebrate Paleontology 21: 172–185.

Benton, M. J. 1995. Diversification and extinction in the history of life. Science 268: 52–58.

Benton, M. J. 2009. The Red Queen and the Court Jester: Species diversity and the role of biotic and abiotic factors through time. Science 323: 728–732. *A review of the relative roles of biotic and abiotic factors on macroevolution.*

Chen, Z. Q., and M. J. Benton. 2012. The timing and pattern of biotic recovery following the end-Permian mass extinction. Nature Geoscience 5: 375–383. *A review of the recovery of life from the most profound mass extinction of all.*

Leakey, R., and R. Lewin.1996. The Sixth Extinction: Patterns of Life and the Future of Humankind. London: Weidenfield and Nicolson.

McKinney, M. L. 1997. Extinction vulnerability and selectivity: Combining ecological and paleontological views. Annual Reviews of Ecology and Systematics 26: 495–516. *An exploration of biological correlates of vulnerability to extinction.*

Raup, D. M. 1992. Extinction: Bad Genes or Bad Luck? New York: W. W. Norton. *A lively presentation by the master of numerical paleobiology, giving a slant to extraterrestrial causation.*

Raup, D. M., and J. J. Sepkoski Jr. 1982. Mass extinctions in the marine fossil record. Science 215: 1501–1503. *The classic attempt to define mass extinctions statistically.*

Raup, D. M., and J. J. Sepkoski Jr. 1984. Periodicity of extinctions in the geologic past. Proceedings of the National Academy of Sciences USA 81: 801–805. *One of the most cited papers in paleobiology, in which evidence for periodicity of mass extinctions is presented.*

Simpson, G. G. 1953. The Major Features of Evolution. New York: Columbia University Press. *One of several books by George Simpson, founder of numerical paleobiology.*

Stanley, S. M. 1979. Macroevolution: Pattern and Process. New York: W. H. Freeman.

Van Valen, L. 1973. A new evolutionary law. Evolutionary Theory 1: 1–30.

Van Valen, L. 1984. A resetting of Phanerozoic community evolution. Nature 307: 50–52.

Wignall, P. B. 2001. Large igneous provinces and mass extinctions. Earth-Science Reviews 53: 1–33. *An excellent overview of an earthbound model for mass extinction arising from massive volcanic eruption.*

VI.14

Species Selection

Emma E. Goldberg

OUTLINE

1. Concepts and consequences
2. History and controversy
3. Empirical tests

The logic of Charles Darwin's view of evolution by natural selection applies not only to individual organisms within populations but also to other levels of the evolutionary hierarchy. Entire species can differ from one another in traits that interact with the environment to affect speciation and extinction, and when those traits are inherited through lineage divergence, species selection occurs. This process has the potential to drive evolution on a large scale, making some clades more species rich than others and determining how commonly particular traits are possessed across groups of species. The precise scope of the definition of species selection and its feasibility as an evolutionary force have been debated for decades. Research now focuses on the empirical question of its prevalence, strength, and consequences in a variety of study systems. Because natural selection may act simultaneously at multiple levels, a great challenge in this endeavor is to separate the contribution of species selection itself from evolution at other levels, especially adaptation within species. Analyses of character evolution and diversification among fossilized or living species help illuminate the significance of species selection in shaping patterns of biological diversity.

GLOSSARY

Aggregate Trait. A characteristic of a species that summarizes a trait present in its individual organisms. Variation within each species in this trait should be smaller than differences in the aggregate trait among species. Body size is one example.

Clade. A group consisting of all the species, living or extinct, descended from a particular ancestral species.

Emergent Fitness. The heritable ability of a species to survive and reproduce; the expected difference between speciation and extinction rates, also called the *net diversification rate*.

Emergent Trait. A characteristic of a species that is not defined by the traits of its individuals. Any particular value of an emergent character may result from many combinations of organismal properties. Geographic range size is one example.

Levels of Evolutionary Hierarchy. Nested units of the complex organization of life. Examples include the species level, the level of populations below it, and the level of clades above it. Selection at any one level may have consequences that appear at other levels.

Species Heritability. The fidelity with which a trait of one species is passed along to its two daughter lineages during a speciation event.

1. CONCEPTS AND CONSEQUENCES

Species-Level Traits, Fitness, and Heritability

There are three basic requirements for evolution by natural selection: variation in the values of some traits, interaction of these traits with the environment to affect fitness (mortality and reproduction), and inheritance of the traits and their fitness consequences from generation to generation. In Darwin's original view, this process played out among individual organisms within populations. Other units of the evolutionary hierarchy, however, may also be viewed as units subject to selection (discussed more generally in chapter III.2), including entire species, with *fitness*, *trait*, and *inheritance* defined at the species level. Selection at the species level may have profound effects on the distributions of species numbers and characteristics across the tree of life.

The fitness of a species is determined by its survival and reproduction—that is, how well it avoids going extinct and how successfully it gives rise to new species. From the standpoint of species selection, speciation and extinction propensities are properties of the species as a whole and together define its *emergent fitness*. The tremendous variation in species richness and characteristics among different groups of organisms is shaped to various degrees by environmental conditions, ecological interactions, and random occurrences. Among these many factors, the defining question for species selection is, What differences in emergent fitness are determined by the interaction of the environment with heritable traits at the species level?

What sorts of traits do species have? Each species exists over some limited portion of earth, so geographic range size is one example. Range size is not the property of a single organism but rather is determined from the group of organisms that make up the species as a whole—it is therefore an *emergent trait* at the species level. Other emergent traits include sex ratio, population density, genetic structure across populations, and social institutions or cultures. Another feature of an emergent trait is that its evolutionary consequences do not depend on how it is determined at lower levels. For example, a species may be narrow ranged because its individuals either lack wings for long-distance dispersal or tolerate only a very particular climate. Geographic range size "screens off" these organismal-level traits, however, if it is simply the small spatial area occupied by the species that puts it at risk of extinction.

In contrast, the body size of an animal is an example of a trait defined at the organismal level. If the variation in body size is lower within species than among them, then it also makes sense to treat average body size as a property of a species. This is an example of an *aggregate species trait*, one defined by a characteristic of individuals within the species. Other aggregate traits include generation time, the degree of ecological specialization, and various modes of reproduction, such as asexuality versus sexuality, monogamy versus polygamy, or wind versus animal pollination.

Many of these traits might reasonably be expected to affect rates of speciation or extinction, although proving so is a more difficult matter. Species with small geographic ranges are especially extinction prone because habitat quality, climate, or predation need turn unfavorable in only one location. The effect of range size on speciation is less clear intuitively and more variable empirically. Larger ranges present more opportunities for geographic barriers to arise and separate existing populations, but small-ranged species may be more sensitive to such barriers and hence more likely to become reproductively subdivided. Empirical results on the relation between body size and diversification are also mixed. Associations among large body size, small population size, and long generation time may make large-bodied species more prone to extinction and slower to complete the speciation process. In contrast with sexually reproducing species, lineages that can reproduce asexually (for example, tulips producing bulbs or lizards developing from unfertilized eggs) are expected to exhibit higher extinction than speciation rates, owing perhaps to low genetic diversity reducing the ability to adapt during environmental changes. In plants, populations with greater pollinator specificity may more rapidly become reproductively isolated and hence have higher speciation rates.

For species selection to occur, not only must species have traits that affect emergent fitness, but those traits must be inherited during the process of speciation. For many aggregate traits, species-level heritability follows naturally from organismal-level inheritance. Unless rapid evolution of the trait in question drives the divergence, as during ecological speciation, each new daughter species will be composed of a lineage of individual organisms drawn from the same pool of parental variability. Emergent traits can also be inherited across speciation events, though this is more difficult to show. The best examples again come from studies of species geographic distributions, which find more similar ranges among more closely related species of mollusks, birds, mammals, and plants.

Consequences of Species Selection

When all the ingredients are present for species selection to occur—heritable variation in traits that affect fitness at the species level—what are the potential consequences of this process? Only if trait variation already exists among species can it drive differences in speciation and extinction rates, so species selection cannot directly cause adaptation within species. Once a species acquires a new trait, however, if that trait increases extinction risk, it can be removed from circulation through species selection. Alternatively, if the trait increases speciation rate, species selection can result in a large clade in which the trait is common. Even characters that do not affect diversification may become more or less prevalent if they tend to be associated with, or "hitchhike" on, another trait subject to species selection. Its effects can also extend beyond simply altering the relative frequencies of characters. Traits that persist over longer timescales are more available for possible further modification, so species selection shapes the background from which new characters evolve.

One of the most celebrated features of species selection is its potential to drive evolution in a different direction than does selection at the organismal level. In extreme cases of cross-level conflict, traits that are advantageous for individuals within populations also increase extinction risk. Evolutionary change that compromises the amount of sexual reproduction or outcrossing, argued George C. Williams, is especially likely to exhibit a balance between two different levels of selection. Within a species, individuals that gain the ability to reproduce without a mate will have a marked advantage and become increasingly common. Furthermore, once traits like the ability to self-fertilize, propagate vegetatively, or develop from unfertilized eggs catch on, they rarely disappear from within a species. Selection at the species level can counter this trend toward reproductive self-sufficiency. Lower extinction rates or higher speciation rates of sexually reproducing species may help explain why most animals require mates and why many plants have intricate adaptations that encourage outcrossing.

Species selection may be more difficult to detect when it works in the same direction as organismal selection. Selection can reduce heritable variation in a trait, and when this happens at both levels simultaneously, trait variation and hence the possibility of evolution will exist over a shorter window of time. Therefore, even if species selection was strong in the past, its signature may not be apparent in living species. The separate contributions of selection at two levels will also be more difficult to measure when they have similar effects. In particular, methods that infer species selection by showing a balance between selection at different levels or by ruling out selection at other levels will not be applicable.

Finally, species selection can also drive trends in the absence of selection at other levels. This is a prominent component of the theory of punctuated equilibrium, discussed later (see also chapter VI.12).

2. HISTORY AND CONTROVERSY

The term *species selection* was first applied to evolution by differential proliferation of lineages by Hugo de Vries in 1905, but it did not come into popular use until 70 years later. It has had a stormy history in the scientific literature. Disagreement over the scope of its definition and collateral damage from debates on related topics have at times muddled its interpretation and thrown its utility into question. A clearer picture of species selection has emerged, however, leading to a firmer grasp on designing and conducting tests of its empirical significance.

Natural Selection, Species Sorting, and Effect Macroevolution

Including species selection under the umbrella of natural selection is not universally accepted. Darwin's original arguments for evolution by natural selection were formulated in terms of individual organisms, but his compelling logic can apply to any level of the biological hierarchy. Illustrious names appear on both sides of the issue, but the prevailing view in multilevel selection theory is that species selection is one form of natural selection.

The broad term *species sorting* subsumes any process in which speciation and extinction rates differ among lineages, without regard to cause. Species selection is one example, with differences in emergent fitness produced by the interactions of species' traits with the environment. *Species drift*, in which speciation and extinction differences are determined by random factors, is another type of species sorting. Sorting on geographic location is a third possibility. For example, if clades diversify rapidly on islands but slowly on the mainland, sorting is driven more by geography than by intrinsic traits.

The discussion so far has considered species selection in the broad sense, allowing a fairly liberal definition of species-level traits. Some prefer to apply the term *species selection* in a stricter sense, limiting it to cases in which the traits that affect emergent fitness are themselves emergent. When species sorting is instead driven by aggregate traits, the term *effect macroevolution* is applied. The reasoning here is that when a trait is expressed at the organismal level, selection on it must ultimately be reducible to processes at that level, making apparent species-level effects an artifact of causation cascading upward.

The more common view, however, is that the defining feature of species selection is the level at which selection occurs. Emergent traits may indeed provide the most compelling examples of species selection, especially because they may be harder to explain by organismal-level evolution. Selection at the species level can also act on aggregate traits, however. The interactions by which a trait affects organismal survival and reproduction may in general be quite different from those by which it affects extinction and speciation. Regardless of terminology, the ultimate goal for understanding multilevel selection is to identify the hierarchical level of the unit that is undergoing selection and to establish how its properties interact with the environment to determine its fitness.

Group Selection

The 1960s' debate over group selection did not touch specifically on species selection, but it affected the general perception of the multilevel framework. Early group selection theories, notably those of V. C. Wynne-Edwards, argued that when an organismal trait is detrimental to the individual but provides an advantage to the group within which it lives, this necessarily implies the action of selection at the group level. Behaviors like giving birth to fewer offspring than is physiologically possible, or deteriorating in health when old, were speculated to evolve "for the good of the group," to regulate population size. This logic was attacked especially by Williams, who argued that such regulatory adaptations often do not really exist. When there are group benefits, he reasoned further, they are better explained by more careful consideration of organismal-level selection, such as accounting for fitness across the whole life span of an individual.

Williams's rebuttal very effectively forced more careful treatments of fitness and its consequences for adaptation. Unfortunately, his view that group-level explanations for traits should be called on only when lower-level explanations fall short caused many to unjustly discard selection at higher levels, including species selection, as even a potentially viable force. Williams did not dispute the basic logic of group-level selection when properly applied, however, and he clearly saw the potential power of selection at the level of species or higher taxa to shape the diversity of earth's biota.

Punctuated Equilibrium

In the wake of the recoil from group selection, Stephen J. Gould was a strong advocate for the importance of multilevel selection, especially in macroevolution. Species selection plays a prominent role in his and Niles Eldredge's original theory of punctuated equilibrium. In this conceptual model, evolutionary change does not accumulate significantly within species, a situation termed *stasis*. Instead, trait variation develops rapidly and in any direction during speciation, when peripheral populations become isolated and diverge. It is then the process of species selection that drives trends by preferentially eliminating much of the new variation while allowing some of it to survive and proliferate.

The punctuated equilibrium theory has been controversial since its presentation in the 1970s. One of its hotly contested claims is that natural selection within populations is primarily stabilizing or constrained, yielding long periods of stasis and thus requiring higher-level selection to produce large-scale evolutionary patterns. Although stasis may make the action of species selection more obvious, by clearly defining species as entities and removing a competing explanation for trends, it is not a prerequisite. Species selection operates equally well on variation among species regardless of whether that variation originated through punctuated bursts or gradual accumulation. Therefore, although punctuated equilibrium brought it into the spotlight, species selection should be judged independently, on its own assumptions and evidence.

3. EMPIRICAL TESTS

Assembling and analyzing data from natural systems to test the action of species selection is not straightforward. Nevertheless, several strong cases for species selection have been built, using a variety of data sources and mathematical tools.

Fossil-Based Tests

Clades and traits that are well preserved in the fossil record provide excellent opportunities for tests of species selection. Speciation and extinction rates can be estimated directly from the dated deposits in which fossils are found. Tying those rates to particular traits is more difficult, however.

Challenges in identifying the target of species selection are well illustrated by three decades of study of geographic range size and larval dispersal mode in marine mollusks. In this system, species are classified by whether they possess a larval stage that swims and feeds on plankton. Such planktotrophs are carried by ocean currents for weeks or months before settling hundreds of miles or more from their parents; they thus disperse much farther than do nonplanktotrophs. The large yolk required by nonplanktotrophs affects shell shape, so larval mode can be inferred for extinct species. Species' geographic ranges are measured from the deposits in which their fossils have been found.

Dispersal ability is expected to affect geographic range size and genetic population structure, and consequently perhaps extinction and speciation rates. Work by Thor A. Hansen, David Jablonski, and colleagues uncovered species selection in several groups of gastropods from the Gulf Coast of North America. They found that planktotrophy was associated with larger geographic range size, longer species durations (lower extinction rate), and a lower rate of speciation. Larval mode is not the sole force behind selection at the species level, however. Within each larval mode, there is still substantial variation in diversification that must be attributed to other factors. Ecological specialization and trophic level are not found to be sufficiently explanatory. Contrast with another group is more illuminating: marine bivalves show a similar correlation between geographic range and species duration, but little association between larval mode and geographic range or extinction. From generalized linear models identifying the factors that best predict survivorship, geographic range indeed emerges as the dominant trait, with little additional predictive power provided by larval mode. This last analysis is a particularly important step in choosing from among correlated traits, even across hierarchical levels, those that best account for emergent fitness. The possibility remains that population genetic structure affects speciation more directly than does range size, but it cannot be determined for extinct species. Finally, the heritability of geographic range is established with regressions and randomization tests that show closely related species to have especially similar range sizes.

One complication sidestepped by this case study is trait evolution within lineages. Leigh Van Valen used the mammal fossil record to present genus-level selection as a force opposing previously documented trends of size increase within lineages. (Sufficient data at the level of species were not available, but genus selection is analogous to species selection, with the defining processes being extinction of all species within a genus and the origination of new genera.) Using a method adapted from the balance between mutation and selection in population genetics, he found a selective disadvantage to large body size in mammals. Because their lineage durations were longer, Van Valen concluded that large-bodied genera have lower origination rates.

A different framework for incorporating within-species trends into species selection analyses is provided by the equation named for George R. Price. Overall changes in traits are separated into two components: the correlation between trait and fitness (attributed to species selection) and other changes across generations (attributed to within-lineage evolution and biased trait inheritance). Carl Simpson used this approach in his analysis of complexity of the calyx, the cuplike portion in crinoids that contains reproductive and digestive organs. He estimated the first component by computing origination and extinction rates over time for genera possessing different values of the calyx complexity trait and then regressing net diversification on the trait. A separate set of calculations based on subclade comparisons estimated the second component. Simpson found that calyx complexity decreased over time both because lineages with simpler calyxes diversify more rapidly and because genera tend to be simpler than their ancestors. The reasons for these tendencies are not known, however. The conceptual application of the Price equation allows both within- and among-lineage selection to be treated on equal footing, rather than attributing to higher levels only what cannot be explained at lower levels.

Phylogeny-Based Tests

Studies based on living species are not limited to organisms and traits that fossilize well. The trade-off is that contemporary data will not directly provide a historical record of trait values, speciation events, and extinctions. Using molecular sequence data to quantify the relationships among species is an alternative means of gaining insights into the past. Such *phylogenetic trees* are rapidly increasing in scope and precision, and the mathematical and computational tools for inferring evolutionary processes from them are likewise advancing.

One popular means of testing whether a trait affects diversification is to compare *sister clades*, which share a common ancestor and hence have the same age and evolved from the same background (see chapter VI.15). Differences between sister clades in a trait and in their numbers of species, and a consistent association between trait and net diversification differences across many pairs of sister clades, can indicate species selection. From numerous applications of this method, characters related to sexual selection have emerged as one class that may influence diversification. These characters include traits associated with female mating preferences, such as showy male colors and elongated fins or feathers, and also reproductive factors that are antagonistic between the sexes, such as the evolution of seminal fluid chemistry to reduce female remating. Because these traits can evolve quickly and in somewhat arbitrary directions, they can drive rapid reproductive isolation between populations and hence increase speciation rates. Sister clade analyses of sexual dichromatism versus monochromatism in birds and fish, and of polyandry versus monandry in insects, do indeed indicate that traits related to sexual selection increase diversification. These analyses cannot distinguish the separate contributions of speciation and extinction, however, and extinction may play a role here if such traits have detrimental effects—for example, by attracting the attention of predators or reducing total fecundity.

Characters that evolve multiple times within a clade are especially valuable for tests of species selection. Correlations with other traits may be broken, and a repeated association with changes in speciation or extinction provides stronger support for a causal connection. The sister clade approach does not deal well with traits interdigitated on the tree, but the last decade has seen significant advances in phylogenetic methods that more powerfully integrate trait changes with diversification. A powerful approach is to fit mathematical functions of rates for trait evolution, speciation, and extinction to a phylogeny. Simultaneously accounting for all these processes is difficult, but recent work by Wayne P. Maddison and colleagues has made it possible under some circumstances.

This procedure was used to study the evolution of self-incompatibility, a genetic mechanism that prevents self-fertilization by causing a plant to reject its own pollen. The ability of individuals to reproduce without a mate is expected to be favored within a species and rarely to disappear from it once it takes hold. Analysis by Boris Igić and colleagues of the alleles involved in self-incompatibility indeed shows that evolutionary transitions to self-compatibility are frequent but that the reverse process has not occurred within the nightshade family Solanaceae. Fitting a model of trait evolution and diversification to a large phylogeny from this family provides estimates of the rate of loss of self-incompatibility within species and the rates of speciation and extinction associated with each breeding system. The results match well with Williams's expectation that species selection can balance organismal selection favoring self-fertilization: the net diversification rate for self-incompatible species was much higher than that for self-compatible species, offsetting the loss of self-incompatibility and causing both states to coexist within the family.

Plants' interactions with pollinators have been hypothesized to influence speciation and extinction, because relying on specialized pollinators may increase both the ease of reproductive isolation and the risk of insufficient reproduction. To look for such effects, phylogenetic models have also been applied to the evolution of floral traits. In a group of tropical vines, presence of a resin reward, which attracts bee pollinators, was not found to affect diversification. An analysis of morning glories, however, showed higher speciation rates for species with pigmented flowers (typically pollinated by bees, butterflies, or hummingbirds) than those with white flowers (typically pollinated by bats or moths).

Although potentially powerful, this framework has not yet been used to separate the effects of correlated characters. For example, self-compatible species tend to be annual rather than perennial, herbaceous rather than woody, rapidly flowering, and found in temperate climates and on islands; any of these traits could also influence diversification rates. Identifying the true targets of species selection is an ongoing challenge that will continue to be attacked with a wide array of data and techniques.

See also chapter III.2, and chapter VI.11.

FURTHER READING

Arnold, A. J., and K. Fristrup. 1982. The theory of evolution by natural selection: A hierarchical expansion. Paleobiology 8: 113–129. *A discussion of natural selection operating at different levels, the Price equation, larval mode, group selection, punctuated equilibrium, and evolutionary constraints.*

Jablonski, D. 2008. Species selection: Theory and data. Annual Review of Ecology and Systematics 39: 501–524. *A lucid summary of conceptual issues, plus extensive tables of traits proposed to affect speciation and extinction.*

Lloyd, E. A., and S. J. Gould. 1993. Species selection on variability. Proceedings of the National Academy of Sciences USA 90: 595–599. *Discussion of broad- and strict-sense species selection and effect macroevolution. The amount of variability within a species may itself be subject to species selection.*

Rabosky, D. L., and A. R. McCune. 2010. Reinventing species selection with molecular phylogenies. Trends in Ecology & Evolution 25: 68–74. *Argues for the widespread importance of species selection, based on modern phylogenetic data and analyses.*

Stanley, S. M. 1975. A theory of evolution above the species level. Proceedings of the National Academy of Sciences USA 72: 646–650. *Species selection is introduced as an analogue to natural selection within populations. The article argues for its importance via punctuated equilibrium. Fossil-based estimates of speciation and extinction rates for mollusks and mammals are given.*

Williams, G. C. 1992. Natural Selection: Domains, Levels, and Challenges. Oxford: Oxford University Press. *Gives requirements for species selection (Chapter 3; the term* clade selection *is used instead of* species selection *to allow for selection at levels higher than the species) and characters it is most likely to affect. Discussion of group selection and punctuated equilibrium, plus a fascinating range of philosophical and empirical topics.*

VI.15

Key Evolutionary Innovations

Michael E. Alfaro

OUTLINE

1. Key innovation concepts in evolutionary biology
2. Where do key evolutionary innovations originate?
3. How do key innovations lead to evolutionary diversity?
4. Testing hypotheses of key innovation
5. Problems with the idea of key innovations

Biologists have long suspected that evolution of traits with strong ecological significance fuels rapid diversification in both species formation and phenotypes. New tools and advances in macroevolutionary theory have helped clarify how key evolutionary changes are expected to affect diversification. Empirical studies often reveal that the relationship between key traits and evolutionary does not conform to simple expectations.

GLOSSARY

Adaptive Zone. A set of closely related niches exploited in a similar manner by a lineage that has evolved a key trait.

Clade. All the descendants of a common ancestor in a phylogenetic tree.

Comparative Method. A statistical method for comparing traits of lineages that incorporates phylogenetic relatedness.

Diversification. An increase in species richness or morphological diversity within a clade.

Ecological Opportunity. A set of niches newly available to a lineage experiencing adaptive radiation.

Exaptation. A trait that arose via natural selection for one function and was then co-opted for a new function by a change in selective pressure.

Key Evolutionary Innovation. A trait or functionally related series of traits of outstanding ecological significance that is thought to have contributed to either the species richness of a lineage or its ecological diversity, or to both.

Lineage. A series of species connected by ancestor-descendant relationships.

Sister Clade. The clade most closely related to (sharing a most recent common ancestor with) a focal clade.

1. KEY INNOVATION CONCEPTS IN EVOLUTIONARY BIOLOGY

With more than 10,000 species distributed across the world in most major habitats, birds are widely considered a story of evolutionary success. Can this success be tied to a key evolutionary feature of the lineage like wings or powered flight? This is the essence of the *key evolutionary innovation hypothesis* in macroevolution, which posits that exceptionally diverse lineages owe their evolutionary success to the evolution of a small number of traits of great ecological or functional significance.

The idea of key innovations has a long history in evolutionary biology. Originally, the term was used to describe the ecological traits believed to be most important in producing higher taxonomic groups. Thus, wings might be proposed as a key innovation for the order Aves (birds), while hardened scales and elongate tongues might be suspected as key innovations explaining the origin of pangolins—a much smaller clade containing only eight species. Such use of the term emphasizes evolutionary distinctiveness rather than species richness. Despite their low species richness, pangolins are phenotypically unique from other lineages of mammals owing to key traits such as an elongate tongue and keratin scales that, presumably, allow them to persist in a unique adaptive zone even if the zone does not permit the same level of species diversification as do wings.

Modern uses of the term *key evolutionary innovations* treat them as traits that confer exceptional evolutionary "success" to a lineage, but biologists differ in their definitions and measures of success. The great disparity in patterns of species richness is perhaps the most pervasive feature of the tree of life, and much of the research on key innovations over the last 20 years has centered on testing whether key traits can explain why some lineages have evolved so many species. Under this conception of evolutionary success, key traits provide a fitness advantage to the lineage itself, leading to higher rates of speciation and/or lower rates of extinction compared with related lineages that lack the trait. Key innovations in this context are a mechanism for driving species selection (see chapter VI.14). With the emergence of a more rigorous theoretical framework for studying adaptive radiations (see chapter VI.10), biologists have recently focused on the expected link between ecological adaptive radiation and the evolution of traits associated with novel niches. As a result, more recent studies of key innovations sometimes define evolutionary success as the degree of morphological and ecological diversity within a lineage. The section "Testing Hypotheses of Key Innovation" explores this idea in greater detail.

The scale at which key evolutionary theory is applied is flexible. Bird wings, mammalian hair, and the amnion of terrestrial vertebrates are all examples of key traits that have been suggested as underlying the success of vast radiations. However, key traits are also used to explain radiations at much smaller scales. The evolution of grinding pharyngeal jaws has been suggested as the key trait underlying a radiation of about 90 species of herbivorous parrot fish, while several traits associated with mouthbrooding have been suggested to be innovations that allowed diversification of lake-dwelling haplochromine cichlids in Africa within the last 8 million years. The important point for modern evolutionary biologists is that the acquisition of key traits at any scale is predicted to alter the tempo of evolutionary diversification.

It is often said that key innovations are easy to propose and hard to test! Interesting key innovations that have been proposed include flowers as the key trait that enabled the astonishing diversification of living angiosperms; the evolution of phytophagy (plant feeding) as underlying success in several insect lineages, including species-rich clades of beetles; powered flight as underlying the success of bats, birds, pterosaurs, and flying insects; and bipedalism as underlying the evolutionary success of hominids. Hundreds more examples can be found within the primary literature on evolutionary biology, although the number of studies that critically test this hypothesis is much smaller. This chapter focuses on the role of key evolutionary innovation in explaining patterns of biodiversity: how evolutionary success is measured, how key innovations are thought to contribute to this success, and what problems are associated with the application of theory to empirical data sets. The concept of key evolutionary innovation has played an important role in shaping the kinds of questions that biologists ask and challenged them to find new ways to test these appealing but often vexing explanations of biodiversity.

2. WHERE DO KEY EVOLUTIONARY INNOVATIONS ORIGINATE?

The concept of key innovations has been applied to both simple and complex changes in a trait. In some cases, a small change in a character can allow a lineage to cross a major functional or ecological threshold. For example, a mutation leading to a single amino acid substitution may confer resistance to a toxin or pathogen and allow a lineage access to a previously unavailable habitat, or a small increase in jaw muscle size may allow a predator to crack shelled prey. Other proposed key innovations are more complicated. Powered flight in birds relies on several proposed key traits, including wings and feathers. Wings themselves are complex structures that comprise heavily modified forelimbs, including the elongation of a reduced number of digits; a specific arrangement of feathers; and physiological and behavioral changes to support flight. This trait really represents a large collection of changes from the ancestral phenotype. When the key innovation represents a suite of functionally related characters, the evolution of the key trait may involve the co-opting of characters that arose under natural selection for a different function. For example, because feathers are found on many species of flightless, nonavian theropod dinosaurs, the origin of feathers cannot be explained as an adaptation for flight. Instead, it is likely that feathers initially evolved for insulation and/or display. The asymmetrical, highly modified feathers found on the wings of modern birds resulted from a shift in selective pressure from this ancestral function to satisfy new demands associated with powered flight. Traits that have experienced functional shifts over their evolutionary history in this way are called *exaptations* (see chapter II.7). Other anatomical changes associated with flight such as modifications to the forelimb and pectoral girdle similarly represent the co-opting of existing structures to novel functional demands. The evolution of a key innovation may thus represent several important evolutionary steps that lead to ever-increasing functional ability. Once a sufficient number of traits have evolved to allow a lineage to fully enter a new adaptive zone, the rate of evolutionary diversification is expected to increase.

3. HOW DO KEY INNOVATIONS LEAD TO EVOLUTIONARY DIVERSITY?

It is readily apparent that certain traits confer large ecological advantages. Wings allow birds access to habitats that are out of reach to most flightless predators and competitors and provide a means of rapid escape as well as a way to quickly reach new habitats if local conditions become unfavorable. How might these ecological advantages translate to evolutionary success? Biologists historically recognize three avenues. Traits might allow a lineage to exploit a new *adaptive zone*—a set of related niches that can be filled only by species possessing an evolutionary novelty such as wings. Second, novel traits might confer a competitive advantage on species possessing it, allowing a lineage to drive competing species to extinction. Fish lineages that evolve the ability to protrude their jaws might enjoy a greatly improved ability to capture prey using suction feeding. On ecological timescales, this trait might result in populations of jaw-protruding species that owing to a competitive advantage in exploiting food resources, are larger than populations of species that lack the trait. Since smaller populations are more vulnerable to extinction, species within lineages lacking the trait might go extinct at a faster rate than those with jaw protrusion. At evolutionary timescales, this sequence could lead to a proliferation of species with the key innovation (see discussion of species selection in chapter VI.14). A third way in which an innovation could produce evolutionary success is by increasing the potential for reproductive isolation and species formation. The evolution of complex mating behaviors within a lineage, for example, might lead to increased potential for speciation between geographically isolated populations and cause the species richness of that lineage to increase.

The ecological theory of adaptive radiation also provides a link between key innovations and evolutionary success. An ecological *adaptive radiation* is the rapid evolution of morphological differences and species richness in a closely related group (see chapter VI.10). Adaptive radiations are spurred when a lineage gains access to *ecological opportunity*—the potential to diversify into new niches along a similar ecological axis. Within the framework of ecological adaptive radiation theory, key innovations can be thought of as traits that grant a lineage ecological opportunity by allowing them to reach a new adaptive zone. As an illustration of these ideas consider the evolution of algal grazing in fish. Algal grazing has evolved in several lineages of marine and freshwater fish including parrot fish, surgeonfish, damselfish, and cichlids, and each of these lineages exhibits conspicuous adaptations of the skull and jaws that are presumably key traits associated with this lifestyle. Niches associated with grazing on algae are likely unavailable to closely related fish lineages that nevertheless lack modified jaws to efficiently scrape algae from rocks or reefs. Once lineages evolve these traits (the key innovations), they gain ecological opportunity and enter a new adaptive zone related to herbivory. Freed from competition with other species for food, the member species of the algal-feeding lineage may initially invade new habitats. Diversification within this lineage may follow an ecological axis as formerly geographically isolated populations that come back into contact evolve reduced competition by specializing on different types of algae or on algae that grow at different depths. Given time, the colonizing species diversifies into a radiation characterized by a unique ecology enabled by a key trait.

One difficulty with many key innovation hypotheses lies in linking the key trait to the process of diversification. Some suggested key innovations play an obvious role in species recognition, and it is relatively straightforward to envision how evolution of the trait would lead to higher rates of speciation. One of the best-documented examples of a key evolutionary trait is the nectar spur of columbines (genus *Aquilegia*). Scott A. Hodges and Michael L. Arnold have statistically demonstrated that the rate of diversification of columbines and other plant lineages with nectar spurs is higher than for those lacking this trait. Additional studies have shown that nectar spur length influences the kinds of pollinators that will visit a flower and that the evolutionary association between pollinators and flowers reflects evolutionary change toward longer spurs as well as shifts to pollinator species with longer tongues. Research on this system supports the hypothesis that nectar spurs have affected both reproductive success and reproductive isolation of the species that possess them, providing a link between the macroevolutionary pattern of high diversity for lineages with the key trait and microevolutionary mechanisms relating directly to the trait that could lead to this diversity For these reasons, the nectar spurs of *Aquilegia* species constitute one of the best-documented examples of a key evolutionary innovation.

Unfortunately, most other suggested key innovations are not so obviously linked to reproductive isolation. In the example of algal grazing in fish, although the ecological significance of scraping jaws for an algae-feeding fish is clear, it is less clear how a trait that is associated with feeding could lead to increased rates of reproductive isolation. One possibility is that key traits not obviously linked to reproductive isolation may still increase species diversification by making either the formation or survival of geographically isolated populations more likely through the ecological advantages they confer. For example, latex and resin canals are defensive structures that have evolved independently in many plant lineages, including conifers, mulberries, and daisies. They are hypothesized to be key innovations that protect plants from pathogens and predators, and Brian D. Farrell and colleagues have shown that lineages with resin canals tend to have a much larger number of species than closely related lineages lacking them. Although this trait is not an integral part of the mating system, as are nectar spurs, they may still turn the engine of species formation by allowing isolated populations to persist at a higher rate than populations without resin canals. A greater frequency of isolated populations, in turn, would be expected to lead to a higher rate of evolution of reproductive isolation within these populations, leading to more species. In support of this idea, Farrell and colleagues have shown at one field site that species with latex and resin canals are more numerically abundant than those without them. This evidence is only suggestive, however, and illustrates that nearly all proposed key evolutionary innovations, including persuasive examples like resin canals, lack an explicit mechanism that demonstrates how the trait gives rise to observed diversity.

4. TESTING HYPOTHESES OF KEY INNOVATION

If key innovations promote evolutionary success, then a simple prediction of key innovation hypotheses is that lineages with the traits should be more successful than closely related lineages lacking them. Success is most commonly measured by the number of species, but more recently, workers have also measured the richness of morphological or ecological diversity contained within a lineage. To assess differences in species richness, biologists often use *sister clade* comparisons, which involve counting the number of species within the lineage that have evolved the trait versus the number of species in the sister clade, the most closely related lineage lacking the purported innovation. One problem with simple counts lies in judging the statistical significance of the difference. If the lineage with the key trait has 20 species and its sister clade contains 5, is the magnitude of the difference sufficiently large to support the key innovation hypothesis? To answer this question, biologists have turned to stochastic models of diversification such as the birth-death model, which treats speciation and extinction as random events controlled by fixed rates of species birth and death (see chapter VI.11). Because the diversification process is probabilistic, sister lineages that have evolved under identical rates of speciation can nevertheless differ in the number of species they contain. These stochastic models form the basis for determining when the difference in species richness between two lineages is so great that it is unlikely they share a single diversification rate. The most compelling tests of key innovation use these methods in conjunction with large-scale phylogenies in which the proposed innovation has evolved multiple times. For example, Farrell has shown that the diversification rate of angiosperm-feeding beetle lineages is significantly higher than that of other phytophagous lineages, which suggests that angiosperm feeding is a key evolutionary innovation for beetles. Hodges and Arnold took a related approach to show that the diversification rate of nectar-spur-bearing plant lineages is significantly higher than that of lineages lacking the trait. The statistical rigor of the comparisons in these studies and the distribution of the key trait over many independent lineages have helped make phytophagy and the latex-resin systems currently two of the most widely accepted examples of key evolutionary innovation in evolutionary biology.

The ecological theory of adaptive radiation has facilitated the testing of key innovation hypotheses by furnishing two main predictions about the tempo of diversification in a lineage that has evolved a key novelty. One is that a lineage should show an increase in the rate of species formation following the acquisition of the key trait. This prediction stems from the idea that key innovations grant a lineage access to new niches. The other is that the rate of evolution of ecological traits related to the innovation should also increase, because diversification in the new adaptive zone is expected to occur along an ecological axis. Both predictions are similar in that they link key innovations to an expected increase in the tempo of evolutionary diversification.

The rise of molecular phylogenetics, which seeks to reconstruct the evolutionary history of taxa using DNA sequence data, combined with recent development of new comparative statistical methods for analyzing data in the context of a phylogeny, has fueled recent work on key innovations. Phylogenetic trees provide at least three important services in the study of key innovations. First, by looking at the distribution of a suspected key trait across all members of a clade in the context of their phylogeny, biologists can infer the time of origin of the trait. This information can be used to assess whether the timing of the appearance of the novelty is consistent with other events that are thought to play a role in allowing the diversification of the lineage. For example, if a key trait is thought to enable a lineage to exploit a new habitat, the age of origin of the trait can be compared with the fossil or paleoclimatic record to determine whether diversification patterns are consistent with the historical availability of that environment.

Second, molecular phylogenies provide a record of the tempo of all the speciation events that gave rise to the present-day members of a lineage. Since key innovations are expected to increase the rate of speciation, a phylogenetic tree can be used to ask directly whether speciation events occur at a faster rate following the acquisition of a key trait. The same is true for the rate of evolution of ecological traits.

Third, phylogenies can be used in conjunction with information about trait diversity within a clade to test whether the tempo of evolution of certain characters changes in a way that is consistent with the key innovation hypothesis. For example, suppose a novel joint between bones in the jaws of fish is thought to underlie the evolution of new and specialized feeding morphologies. If the jaw joint is the key innovation, one would predict that jaw elements related to feeding would be subject to new selective pressures as part of an ecological adaptive radiation spurred by the evolution of the trait. Thus, evolutionary rates of jaw characters should increase shortly after the evolution of the jaw joint. If jaw characters are shown to change more rapidly before the key trait appears, or if the tempo of jaw characters is the same before and after the evolution of the key trait, then the key innovation hypothesis would not be supported. A second prediction might be that the rate of evolution in characters unrelated to the key innovation, perhaps tail shape, should not differ between lineages with and without the presumed innovation. The power of comparative methods for testing historical hypotheses explains why these approaches have emerged as one of the primary ways of evaluating key innovation hypotheses in modern evolutionary biology.

5. PROBLEMS WITH THE IDEA OF KEY INNOVATIONS

Key innovations are sometimes criticized as being little more than evolutionary just-so stories. The idea that wings are the key to the success of birds may seem reasonable, but an expectation of how wings would have shaped the radiation of birds is needed to rigorously evaluate whether the available data support the hypothesis. Phylogenetic hypothesis testing in conjunction with the ecological theory of adaptive radiation has been extremely useful in addressing some of these criticisms. The ecological theory of adaptive radiation generates predictions (i.e., rates of speciation and/or morphological evolution should increase following the acquisition of a key trait), and phylogenetic methods, in conjunction with a phylogeny and data about the distribution of the putative innovation, provide a means for testing these predictions.

A difficulty with evaluating key innovation hypotheses is that the innovations often have evolved only a small number of times. In the case of a single large radiation, a comparative method may reveal that one lineage has a significantly greater number of species than its sister clade. This is the case with living birds, whose species richness and phenotypic diversity dwarfs that of the 24 species of crocodilians, their closest evolutionary cousins. It will generally not be possible, however, to identify one trait as the causal factor of the radiation over another if those traits are codistributed. For example, whereas a functional morphologist might see changes to the jaws as the key innovation, a physiologist might argue that it is the evolution of a novel biochemical pathway to process new foods. If both are inferred to evolve in the same ancestor of a clade, comparative methods alone will not be enough to tease these explanations apart, since they make identical predictions about the timing of a change in the tempo of diversification. When presumed innovations have evolved multiple times within a large phylogenetic tree, the ability to discriminate among competing putative innovations may be much improved. As long as the hypothesized innovations are not identically distributed across the tree, they will make different predictions about when the tempo of diversification will change. The weight of evidentiary data in support of one explanation versus the other can then be assessed with statistical methods. In the case of nectar spurs, Hodges and Arnold showed that the diversification rate of *Aquilegia* is significantly faster than in their sister clade. Although the authors attributed this difference to the nectar spur, one might argue that any other traits that evolved in the common ancestor of *Aquilegia* could have driven the radiation. However, additional comparisons of diversification rate in other plant lineages with nectar spurs with sister taxa lacking them allowed Hodges and Arnold to argue persuasively that the nectar spur itself is the innovation.

Biologists often suggest that key innovations underlie patterns of highly uneven species richness, yet the number of rigorous, phylogeny-based tests of the hypothesis is much smaller than the number of times key innovations have been proposed. Over the last 15 years, the concept of key innovation has been invoked in close to 400 articles within evolutionary biology, yet fewer than 10 percent of these articles tested these ideas with rigorous phylogenetic comparative analysis. And what do these phylogeny-based tests reveal about key innovations? Many studies (including those already discussed examining nectar spurs and angiosperm feeding) are able to quantify a difference in the rate of diversification between lineages that possess a trait and those that do not, providing evidence that key traits are linked to at least some of the major patterns of uneven diversity found on the tree of life. Furthermore, some studies have shown that purported key traits do not, in fact, produce exceptionally diverse groups. N. Ivalú Cacho and colleagues provide one such example, showing that extraflorally derived nectar spurs of some euphorbs have not led to exceptionally diversity within those lineages that possess them. Even when significant changes in diversity are detected, it is also common to find that shifts in the evolutionary rate of speciation or character evolution occurred somewhat later than the origin of proposed innovations. This is especially true for ancient innovations that characterize major taxonomic groups. Duane

McKenna and colleagues found that major episodes of diversification within weevils, one of the most species-rich groups of angiosperm-feeding beetles, occurred 20–30 million years after weevils first colonized flowering plants. Another example concerns the role of genome duplication in fish diversification. Although the evolutionary success of teleosts—which constitute more than 99 percent of living fish—is sometimes linked to a duplication of the entire genome early in teleost history, Francesco Santini and colleagues have shown that most teleost diversity was produced by radiations within lineages that are much younger than the age of this hypothesized key innovation. This result reveals that the present-day species richness of teleosts is more likely to be the result of factors specific to these younger radiations than to the ancient genome duplication event. Phylogenetic approaches have also revealed many instances in which an innovation triggered a radiation in one lineage but not in others. Pharyngeal jaws (a second set of jaws in the throat of some fish) have been proposed as key innovations that underlie the species richness of wrasses (~600 species), damselfish (~360 species), and cichlids (~2000+ species). Yet the surfperches, which have also evolved modified pharyngeal jaws, contain a much smaller number (only ~60 species), suggesting that the trait itself does not always lead to a burst of speciation. Rigorous and explicit tests of the pharyngeal jaws as a key innovation hypothesis using sister clade comparisons or other phylogenetic methods are impeded by the lack of a reliable phylogeny detailing relationships among fish families.

Explaining these apparent exceptions to the predictions of key innovation hypotheses remains an active area of macroevolutionary biology. One possibility is that diversity patterns in large groups have been influenced by more recent events that mask the signal of the initial radiation. Passerines, which include more than 60 percent of extant bird diversity, evolved almost 40 million years after the common ancestor of living birds. The pattern of diversification of this diverse, young group in conjunction with extinction of ancient bird lineages may swamp the signal for the tempo of diversification during the origin of modern birds. Without accurate inference of the pace of early bird diversification, the ability of phylogenetic comparative methods to test key innovation hypotheses will be extremely limited. The "drowning out" of the signal of the tempo of ancient radiations by more recent bursts in the tree may be common for large groups (e.g., beetles, mammals, teleost fish, birds), complicating the study of key innovations. In such cases, the key trait may ultimately be responsible for the initial evolutionary success of the group, but it does not explain some or most of the patterns of diversity that have evolved since the trait was acquired. The ability of simple key innovation hypotheses to explain biodiversity patterns in ancient groups may be severely limited, since richness in these cases will be the outcome of a series of complex historical factors.

A conceptual weakness of key innovation concepts is that they place a great deal of emphasis on the trait itself, whereas there is good reason to expect that the ecological and evolutionary context of the trait is likely to be important as well. Stronger jaws may allow cracking of hard-shelled prey, but evolutionary specialization in sense organs may also be needed to locate prey, or locomotor adaptions may be needed to forage in areas with high abundances of hard-shelled prey. Furthermore, stronger jaws may provide an evolutionary advantage only to lineages that first evolve them within an ecological community. The potential for diversity may also be a function of geography. For example, the geographic area available to an island-dwelling lineage that has evolved a functional innovation may be too small to support more than a few species. In this case the evolutionary potential of the innovation will go unrealized until the lineage is able to colonize a new, larger island or to recolonize the mainland.

To accommodate the idea that the evolutionary response to a proposed key trait may depend on other traits as well as on the ecological and historical context in which the trait is acquired, some biologists suggest that key innovations do not appear all at once in an ancestral species. Rather, key innovations are the outcome of several functional novelties that accumulate over a period that spans multiple ancestral species. Diversification and radiation begin once all the needed traits have evolved and other mitigating conditions are favorable. Testing these conceptions of key innovation is more difficult than testing the simple prediction that diversification rates immediately change once the key trait evolves, because most phylogenetic statistical methods are designed to locate single points on a phylogeny where the diversification rate abruptly shifts. It is still possible to make and test the more general prediction—that the key trait evolves some time before the change in diversification occurs; however, if a clade experiences any pulse of diversification subsequent to the origin of the key trait, it may be difficult to determine whether that pulse represents the end of the lag period and the start of a key trait-fueled radiation or a diversification driven by another, unrelated factor.

Despite these difficulties, the concept of key innovations remains important to macroevolution because it is a theoretical framework that links aspects of ecology to evolutionary patterns of biodiversity. Testing key innovations is not as hard as it once was, but it is apparent that the relationship between diversity and innovation is often complex, while available stochastic models of diversification used for testing are still relatively simple. The trend among researchers toward generating more explicit models of the evolution of key traits and expected patterns of lineage diversification given these models is certain to continue. As more sophisticated methods and better phylogenies for major sections of the tree of life become available, it will become increasingly possible to understand the role that trait innovation plays in generating diversity relative to other macroevolutionary factors like geographic distribution and interactions with other species.

See also chapter VI. 7.

FURTHER READING

Bond, J. E., and B. D. Opell. 1998. Testing adaptive radiation and key innovation hypotheses in spiders. Evolution 52: 403–414.

Cacho, N. I., P. E. Berry, M. E. Olson, V. W. Steinmann, D. A. Baum. 2010. Are spurred cyathia a key innovation? Molecular systematics and trait evolution in the slipper spurges (Pedilanthus clade: Euphorbia, Euphorbiaceae). American Journal of Botany 97: 493–510.

de Queiroz, A. 2002. Contingent predictability in evolution: Key traits and diversification. Systematic Biology 51: 917–929.

Donoghue, M. J. 2005. Key innovations, convergence, and success: Macroevolutionary lessons from plant phylogeny. Paleobiology 31: 77–93. *Makes the argument that the search for a single point of origin of a key innovation may be misguided for complex innovation that arise through a series of significant evolutionary changes to multiple trait systems.*

Farrell, B. D. 1998. "Inordinate fondness" explained: Why are there so many beetles? Science 281: 555–559. *An early example of using quantitative approaches to test for differences in the rate of diversification between lineages with and without a presumptive key innovation.*

Farrell, B. D., D. E. Dussourd, and C. Mitter. 1991. Escalation of plant defense: Do latex and resin canals spur plant diversification? American Naturalist 138: 881–900.

Heard, S. B., and D. L. Hauser. 1995. Key evolutionary innovations and their ecological mechanisms. Historical Biology 10: 151–173.

Hodges, S. A., and M. L. Arnold. 1995. Spurring plant diversification: Are floral nectar spurs a key innovation? Proceedings of the Royal Society B 262: 343–348.

Hunter, J. P. 1998. Key innovations and the ecology of macroevolution. Trends in Ecology & Evolution 13: 31–36.

Mayr, E. 1963. Animal Species and Evolution. Cambridge, MA: Harvard University Press.

Mitter, C., B. Farrell, and B. Wiegmann. 1988. The phylogenetic study of adaptive zones: Has phytophagy promoted insect diversification? American Naturalist 132: 107–128.

Santini, F., L. J. Harmon, G. Carnevale, and M. E. Alfaro. 2009. Did genome duplication drive the origin of teleosts? A comparative study of diversification in ray-finned fishes. BMC Evolutionary Biology 9: 194.

Schluter, D. 2000. The Ecology of Adaptive Radiation. New York: Oxford University Press. *Perhaps the most important recent synthesis of ecological and evolutionary studies relating to adaptive radiation. This book has been foundational in providing biologists with a rigorous and testable framework for studying key innovations and other sources of ecological opportunity.*

Vamosi, J. C., and S. M. Vamosi. 2011. Factors influencing diversification in angiosperms: At the crossroads of intrinsic and extrinsic traits. American Journal of Botany 98: 460.

Yoder, J. B., E. Clancey, S. Des Roches, J. M. Eastman, L. Gentry, W. Godsoe, T. J. Hagey, et al. 2010. Ecological opportunity and the origin of adaptive radiations. Journal of Evolutionary Biology 23: 1581–1596. *A critical review thoughtfully focused on the ecological mechanisms that could allow key innovations and other sources of ecological opportunity to increase speciation and phenotypic diversification.*

VI.16

Evolution of Communities

Mark A. McPeek

OUTLINE

1. What are communities?
2. Microevolutionary change and community evolution
3. Macroevolutionary change and community evolution
4. Geography of speciation and extinction

Changes in abundances of species over time, combinations of species that can and cannot live together, and the number of species that can live together in one place at one time are all influenced by the abilities of each species to deal with the abiotic environment and to interact with the other species they encounter. These abilities are shaped by evolution, and so evolution is the foundational process shaping the properties of biological communities. The genetic diversity of one species can influence the outcomes of species interactions. Also, as species adapt to one another, they alter many aspects of the ecosystem, because of the change in their ability to influence their environment. Speciation and extinction have the most dramatic effects, because these processes introduce new species and eliminate existing species, respectively. In addition, because different modes of speciation generate varying amounts of ecological difference among species, the diverse modes of speciation can result in assorted properties in the resulting communities and can generate a variety of long-term outcomes.

GLOSSARY

Community. A collection of interacting or potentially interacting species on both local and regional spatial scales.

Ecological Speciation. The process of speciation resulting from the adaptation of two or more populations of a species to different ecological environments.

Extinction. The loss of a species from some biological system. The ultimate extinction of a species occurs when the last remaining individual of that species dies.

Functional Group. A group of species within a community that interact with other members of the community in very similar ways. A functional group has a unique ecological role within the community.

Speciation. The process of forming a new reproductively isolated species from preexisting species.

Species Interaction. The mechanism by which one species affects the population growth rate of another species.

Species Richness. The number of species present in a community.

1. WHAT ARE COMMUNITIES?

A biological *community* is a collection of species that live together on both local and regional spatial scales and whose interactions influence one another's distributions and abundances. For example, all the species that live together in a pond form a biological community, and all the species that live in a forest form a community. Communities may be very small—the bacteria, protozoans, and insects inhabiting the pool in a pitcher plant leaf—or they may be very large—all the species living together in Lake Superior.

Interactions among the species within a community influence their distributions and abundances across the regional collection of communities where they are found. Species influence one another's birth and death rates via interactions that can take many forms. For example, some species interact as predator and prey, or pathogen and host. Other species may compete for limiting resources (e.g., light, water, or mineral nutrients) or for biological food, as when two predators compete for a shared prey. Species may mutually benefit one another, as plants and their pollinators do, or they may alter the physical environment in ways that facilitate the performance of others, such as when they recycle inorganic nutrients from detritus into forms that others can then use.

These species interactions shape patterns of species distributions and abundances locally and regionally and also shape the properties of these communities, such as species richness and diversity. Species abundances may change over time owing to their interactions, such as when numbers of lynx and hare cycle relative to each other over multiple years. Also, not all species are capable of living together. For example, different prey species are found in communities depending on the presence or absence of a top predator in the system, because in the former case, prey are killed too quickly.

The performance of species in these interactions is determined by their phenotypes. The success of a prey in avoiding an attacking predator will depend on how its morphology, physiology, and behavior influence how fast it can flee, while the success of a food competitor may depend on how fast it can take up the resource or how well its physiology performs when the resource is scarce, to allow it to survive. As the phenotypes of species evolve as a consequence of these species interactions (see chapter III.15), the properties of the communities in which they live will necessarily also change.

Two fundamental questions regarding communities are: What is the source of all their interacting species? And what are the consequences of their loss from the system? New species enter communities via speciation or immigration from other areas, and they are lost via extinction. Therefore, both micro- and macroevolutionary processes have a profound influence on the properties of communities. Thus development of a fundamental understanding of today's communities must include questions about how past and ongoing evolutionary events endowed species with their current properties.

2. MICROEVOLUTIONARY CHANGE AND COMMUNITY EVOLUTION

The processes of microevolution—mutation, genetic drift, gene flow, and natural and sexual selection—can all change the ecological capabilities of individual species by changing the phenotypes that influence their responses to the environment—including properties like temperature, water, and nutrient availability—as well as their interactions with other species.

The genetic diversity of species can have strong influences on the properties of communities if that genetic diversity is also reflected in phenotypic diversity that is ecologically important. For example, many plant species defend themselves by producing noxious secondary compounds that deter feeding by herbivorous insects. Different genotypes within a species may produce these compounds in varying concentrations and mixtures. Thus, different herbivores will be deterred from feeding on each genotype, and so the herbivorous arthropod assemblages may vary dramatically in species composition and abundance among genotypes of the same plant species. These effects can propagate to higher trophic levels as well if the predators of these herbivores differ in their abilities to exploit the particular herbivore assemblages found on different plant genotypes. Genetic diversity within species has also been shown to affect the likelihood that different plants can live with one another, because the genetic variation in traits of the species determines how they interact.

While mutation, gene flow, and genetic drift can alter the distributions of ecologically important traits in populations over long periods of time, natural selection is the most rapid and powerful evolutionary force shaping the abilities of species to interact with their environment. There is now abundant evidence that natural selection operates in natural populations and that interactions among species are a prime cause of that natural selection (see chapters III.15 and VI.7). Adaptive evolutionary responses in one species can therefore alter species abundances and other community properties throughout the system. For example, as a prey species evolves better defenses against a predator, the abundance of prey should increase because better defenses lower its mortality rate, and the abundance of its predator can be expected to decrease because it then obtains food at a slower rate. Additionally, if a predator evolves in its abilities to capture different types of prey, the abundances of those prey should change as a result. Likewise, if one competitor species evolves increased abilities to take up a limiting resource, its abundance should increase, and the numbers of other competitors for that resource should decrease.

Recent studies have begun to test and confirm such expectations. Studies of protozoan evolution in pitcher plant communities have demonstrated that population growth rates do evolve in response to predators and competitors, and indirect effects propagate throughout the community. For example, the population growth rate of a species of *Colpoda*, a ciliated protozoan, evolved to increase in response to both predation by pitcher plant mosquitoes (*Wyeomyia smithii*) and competitive interactions with other protozoan species.

The adaptation of Trinidadian guppies (*Poecilia reticulata*) to different predation regimes is an exemplar of evolution in the wild. Guppy populations in high predation areas of streams evolve to mature at smaller body sizes and to have higher reproductive effort than those in low predation areas. In artificial stream experiments, invertebrate biomass was higher and algal biomass was lower with guppies from high predation populations than with guppies from low predation populations. These differences mirrored differences in the various guppy diets. Guppies from the two areas also caused differences in various ecosystem properties, including alterations in rates of gross primary productivity, nitrogen flux, leaf decomposition, and standing amounts of benthic organic matter, and these differences mirrored those seen in natural streams.

The impact of alewife (*Alosa pseudoharengus*) on zooplankton assemblages in New England coastal lakes is a prime example of the way a top predator shapes community structure. Alewife are anadromous fish that return to coastal New England lakes to spawn, and as a consequence of their large body size, they feed selectively on large zooplankton, which causes shifts in zooplankton assemblages to favor small-bodied zooplankton species. Dams placed on some rivers by European settlers trapped some alewife populations in the freshwater lakes, and these landlocked populations have evolved much smaller body sizes. These differences in body size, and the concomitant differences in the size of their feeding apparatus, cause predictable differences in zooplankton assemblages between lakes with anadromous and landlocked alewife: small-bodied zooplankton species are found in lakes with anadromous alewife, and larger-bodied zooplankton species live in lakes with landlocked alewife.

Recent laboratory studies have also shown that adaptive evolutionary responses of species to one another can shape short-term population and community dynamics. Predator-prey interactions sometimes produce characteristic cycles in population abundances where predator abundance peaks one-quarter of a cycle behind the peak in prey abundance. For example, such cycles are seen in the historical records of lynx and hare abundances across northern North America. Modeling results show that when the prey and predators are allowed to rapidly evolve in response to each other, the peak abundances of both predator and prey occur farther apart in time and more asymmetrically in time relative to each other; these types of cycles cannot be generated without the evolutionary responses of predator and prey. Recent experiments in laboratory microcosms containing rotifer (*Brachionus calyciflorus*) predators and algal (*Chlorella vulgaris*) prey have demonstrated the validity of these model results. When only one or two genotypes of algae were present, and thus the algae could not evolve, predator and prey abundances rose and fell rapidly in cycles, and prey abundances peaked shortly before predator abundances in each cycle. However, when many genotypes of algae were present, and thus the algal population could evolve, population abundance peaks were significantly farther apart in time, and predator abundance peaked when prey abundance was at its nadir. Similar types of alterations to ecological dynamics caused by evolutionary responses of predators and prey have been demonstrated in laboratory experiments using bacterial prey (*Escherichia coli*) and predatory lytic bacteriophage viruses.

Studies of the evolutionary history of traits also provide windows into the evolution of communities through the changing abilities of species. For example, North American columbines (*Aquilegia*) have long spurs on their flowers for holding nectar. Pollinators must have long tongues to reach the nectar, and in the process of foraging for nectar they will inadvertently transfer pollen they have picked up from other flowers. Evolutionary reconstruction studies suggest that spur length evolves to be longer when a new pollinator with a longer tongue begins feeding on the species. In another compelling example, *Dalechampia* vine and scrub flower traits were shown to have evolved their pollinator reward system of resin secretions first as a defensive mechanism against herbivores; only later did these secretions become a pollinator reward. Conversely, others have hypothesized that some defenses against herbivores may have originated from traits that originally evolved to attract pollinators. These kinds of examples show why past evolutionary events are important for understanding present-day patterns in species interactions and community structure.

3. MACROEVOLUTIONARY CHANGE AND COMMUNITY EVOLUTION

While buildup of ecologically important genetic diversity within species and coevolutionary adaptation among species are certainly important in shaping community properties, the greatest changes to community structure must occur when new species are added to or lost from the community. The macroevolutionary processes of speciation and extinction can fundamentally change species richness and diversity as well as the functional types of species present on both local and regional scales.

Myriad processes can cause speciation (see chapters VI.3, VI.4, and VI.5), some modes of which will have profound effects on the ecological capabilities of the resulting species. Most important among these processes is *ecological speciation*, namely, the reproductive isolation between two or more lineages that results as a byproduct of their ecological differentiation. Ecological population differentiation can occur along many different types of environmental axes and presumably occurs most frequently when lineages adapt to exploit unutilized ecological opportunities. Adaptive radiations are the most spectacular exemplars of ecological speciation (see chapter VI.10), but ecological speciation is a prevalent mechanism creating new species in most clades. Sticklebacks (*Gasterosteus aculineatus*) have been shown to speciate as a result of adapting to different environments (e.g., marine versus freshwater) and prey in different habitats (e.g., prey on the bottom of lakes versus in the water column). *Rhagoletis* flies and *Timema* walking sticks have each diversified to specialize in feeding on many different host plants. *Anolis* lizards have undergone repeated adaptive radiations to utilize different microhabitats of trees on different Caribbean islands. *Enallagma* and *Lestes* damselflies have diversified to live with different top predators (i.e., fish versus large dragonflies) in ponds and lakes across North America. Thus, ecological diversification appears to be a prevalent mode of speciation that is not restricted to any particular type of species interaction.

Most instances of ecological population differentiation probably do not result in new species. As one of the best examples of adaptive differentiation, Trinidadian guppy populations living with different predators show no signs of reproductive differentiation among the populations. However, in the case of ecological speciation, the prime consequence is the introduction of a new functional group to a community. For example, as marine sticklebacks colonized and adapted to living in freshwater lakes as the glaciers retreated 18,000 years ago, they established themselves as a new functional type of predator that fed on both zooplankton and benthic prey in the lakes. Subsequent speciation events via secondary invasions in a small subset of these lakes created two stickleback species: one specialized for feeding on zooplankton in the open water, and the other specialized for feeding on benthic invertebrates. In lakes with only one stickleback species, the dynamics of the benthic and zooplankton prey should be linked by the shared predation of the one stickleback species. Conversely, zooplankton and benthic prey dynamics might be largely decoupled in lakes with two stickleback species, because each has its own predator. These differences also propagate to other features of the ecosystem.

Likewise, ecological speciation in association with recent glacial cycles created three new *Enallagma* damselfly species as they colonized and adapted to ponds and lakes in which large dragonflies were the top predators (but fish predators were lacking). These speciation events introduced a new functional group to the dragonfly-lake community that was much better at avoiding dragonfly predators but poorer at competing for food than other damselfly genera already present. In effect, these ecological speciation events changed this component of the community from a linear food chain to a diamond-shaped food web (i.e., dragonflies as the top predator feeding on two intermediate-level consumer functional groups [the new *Enallagma* species, as well as the damselflies that were already present in the community], and the two consumer functional groups feeding on prey below them in the food web). Because energy and materials flow very differently through a linear as opposed to a diamond-shaped food web, the introduction of the *Enallagma* functional group to the community may have fundamentally altered material flows and the dynamics of community response to perturbations.

One of the central concepts of community ecology is *species coexistence*, namely, ecological differences among species that promote their long-term persistence together. Long-term stable coexistence requires that species be ecologically differentiated from one another such that each has a greater demographic effect on members of its own species than it does on other species; that is, the two species have "different niches." The process of ecological speciation should typically produce species that immediately coexist on either local or regional scales, because the process of ecological differentiation that drives ecological speciation produces species that fill new ecological roles in the community.

Because ecological speciation fills unutilized ecological opportunities (i.e., empty niches), the rate of ecological speciation must also diminish as new species representing new functional groups are added, and so the rate of ecological speciation must depend on the number of species/functional groups already present in a community (see chapter VI.11). Extinction rates may be at some background level when species richness is low but then rapidly increase after all the possible functional groups are present or nearly so, and ecological interactions among species should drive more poorly adapted species extinct via competitive exclusion. As a community is assembled via ecological speciation, ecologically unique species should be added to the community at a diminishing rate until speciation and extinction rates balance, and this balance should exist at a point where nearly all the available niche space is filled. At this macroevolutionary equilibrium, species turnover will continue as species are replaced at existing functional positions in the community, but total species richness should change very little.

In contrast with ecological speciation, other modes of speciation (e.g., chromosomal rearrangements, changes in mate recognition, sexual selection, or sexual conflict) may result in reproductive isolation but little or no ecological differentiation among sister species. For example, evolutionary changes in the traits that males and females use to discriminate conspecific mates from heterospecifics can rapidly generate reproductive isolation among many different lineages simultaneously. However, changes in these traits (e.g., breeding coloration, mate calls, genitalia shapes, biochemical signals for gamete recognition, and compatibility) may have little or no consequences for how these species obtain resources, avoid predators, combat parasites, or interact with mutualists. These speciation modes do not introduce new functional groups into communities but rather add nearly equivalent species to preexisting functional groups of communities where their ranges overlap. Thus, taxa in which these types of speciation modes are most prevalent should display "neutral" community dynamics, in which species relative abundances change slowly and at random over time. Moreover, species richness will greatly exceed the number of available niches in the community. As a result, most species will be on a long, slow sojourn to extinction.

One example of this type of speciation is the recent radiations of the *Enallagma* damselflies. While ecological speciation has played an important role in *Enallagma* diversification to colonize different lake types, the majority of speciation events in the genus appear to have involved changes in the shapes of secondary sexual structures used by males and females to identify potential mates to species, with little to no change in ecologically important traits These speciation events add species to a given lake type (e.g., many new species added to lakes with fish) but do not change the way they interact with other species. As a result, 8 to 12 *Enallagma* species can be found living together in lakes across much of North America. Recent field experiments have shown that these co-occurring species are in fact ecologically indistinguishable. In addition, *Enallagma* relative abundances do not correlate with any of the major environmental gradients (e.g., predator or prey abundances, productivity abiotic conditions) among lakes and appear to vary randomly among lakes: these patterns are also the expected results for ecologically equivalent species.

Different macroevolutionary and macroecological dynamics are thus expected for communities containing taxa like *Enallagma* in which such nonecological modes of speciation dominate. Here, the per lineage speciation rate should be independent of the number of species already present. In addition, extinction rate may be substantially depressed, because the time to competitive exclusion increases nonlinearly as species become ecologically more similar (i.e., extinction rate is inversely proportional to the time to extinction). As a result, the macroevolutionary equilibrium species richness can be quite high. Recall, too, that because these species are ecologically nearly identical with one another, they would represent only one functional group embedded in a larger community; and within this functional group, the species relative abundances should vary randomly through time and change at a rate that is inversely proportional to the total number of individuals in all the equivalent species.

Still other modes of speciation (e.g., hybridization and polyploidization) may produce new species that may be ecologically quite different from their progenitors, but these ecological differences may not coincide with the distribution of available niches in the community, as they do with ecological speciation. For example, hybridization between two plant species will produce a new species that can be phenotypically, and thus ecologically, quite different from both parents, but the phenotype of the new species is produced without respect to the ecological opportunities available in the community. Most new species produced by these mechanisms are probably quickly driven extinct because they have ecologically inferior phenotypes compared with those of their progenitors. However, some will stumble onto superior phenotypes that may allow them to coexist with their progenitors or to invade new ecological conditions or even to replace one of the progenitors.

These speciation modes may produce a third type of macroevolutionary and macroecological dynamic: because species must interact in hybridization and polyploidization modes of speciation, the per lineage speciation rate may actually increase with species richness. But since species will be ecologically fairly different, extinction rate is still expected to increase with species richness. However, species richness is then expected to come to equilibrium at some value above the number of species that can coexist with one another, but the rate of species turnover in the system may be quite high, because speciation and extinction rates equilibrate at high values of each.

4. GEOGRAPHY OF SPECIATION AND EXTINCTION

The geography of speciation and extinction also influences the evolution of community structure (see Section III: Natural Selection and Adaptation). In particular, in addition to the mechanisms that generate reproductive isolation, speciation events are typically classified according to the geographic structure of the differentiating lineages: allopatric, parapatric, peripatric, and sympatric. The geographies of speciation and extinction are important because they define the spatial scale at which new species are added to communities.

Because the differentiating lineages are spatially segregated, allopatric, parapatric, and peripatric speciation events do not add new species to local communities. Each lineage must already be embedded in a local community, and speciation results when these geographically distinct sets of populations differentiate from one another. As with the effects of guppy evolution on different predation regimes, the local evolutionary forces that drive differentiation may consequently alter the local properties of communities. The regional species richness pool will increase, and regional functional diversity will also increase if speciation is driven by ecological differentiation. If one species range (or both) subsequently expands into the other's range, local richness and functional diversity may increase if the two species can live together in the same local community.

Conversely, sympatric speciation events do increase the species richness of a local community. The shift of *Rhagoletis* flies from hawthorn trees to feed and develop on apple trees has increased the number and types of herbivore species in forest and orchard communities of North America in the last 150 years. Similar shifts have presumably occurred to create the diversity of *Rhagoletis* species that are found feeding on trees and shrubs with fleshy fruits across eastern North America. Speciation via hybridization and polyploidy similarly must introduce new species locally, since the daughter species of the process are offspring of parental species individuals.

However, many speciation events do not fit neatly along the classic allopatric to sympatric continuum, particularly those involving ecological speciation caused by habitat shifts. Habitat shifts imply the invasion of a local community by a new functional species type, and so local species richness and functional diversity increases. However, habitats are typically distributed across the landscape in a variegated pattern, and so a speciation event via a habitat shift may occur within the geographic range of the parental species. For example, lakes with fish and lakes with dragonflies as top predators are interspersed across the landscape. In eastern North America, these two lake types have unique assemblages of *Enallagma* species, with each lake constituting a local community. The ancestral lake type for all *Enallagma* was a fish-lake species. Thus, the ecological speciation events within this genus must have been initiated by females of a fish-lake species laying eggs to create a founder population in one or more fishless lakes where large dragonflies were the top predator. Because all these species have broadly overlapping ranges today, these speciation events must have created new species within the ranges of the progenitors (sympatric speciation on a larger spatial scale), but because of their habitat differences, the resulting species are allopatric on a local scale.

Extinction can similarly have a spatial dimension. The presence of one species may drive others extinct on a local scale but have little effect on the much broader regional scales. For example, when fish are introduced to a previously fishless lake, the entire collection of *Enallagma* species in the lake are driven extinct (along with many other invertebrate taxa) and are eventually replaced by the collection of *Enallagma* species that can persist with fish. Such local extinction events occur routinely and are presumably the basis for metapopulation dynamics. Extinction of an entire species then requires that it become locally extinct in all the places where it could formerly support a population. It is a simple mathematical fact that species will become more susceptible to extinction as they become less abundant locally and as they are able to support populations in fewer local communities. Species extinctions are thus much more likely to occur in rare species and to be caused by some factor that influences an entire region.

As the fossil record shows, species have continually been added and lost from biological systems over the history of life on earth (see chapter VI.13). This reality implies an evolutionary dynamism for biological communities that is typically not contemplated. Change in the ecosystem is a result not only of change in the abiotic world but also because of the addition of new species and the loss or alteration of properties of existing species. The evolution of species and higher taxa is a major force for change in ecosystems.

FURTHER READING

Bohannan, B.J.M., and R. E. Lenski. 2000. Linking genetic change to community evolution: Insights from studies of bacteria and bacteriophage. Ecology Letters 3: 362–377. *One of the first experimental community evolution investigations using bacterial and the viruses that attack them.*

Gavrilets, S., and J. B. Losos. 2009. Adaptive radiation: Contrasting theory with data. Science 323: 732–737. *An excellent summary to date of the ideas about how adaptive radiations of clades are sparked, and the types of structures the resulting communities of species may assume.*

Hairston, N. G., Jr., S. P. Ellner, M. A. Geber, T. Yoshida, and J. A. Fox. 2005. Rapid evolution and the convergence of ecological and evolutionary time. Ecology Letters 8: 1114–1127. *A review of the experiments to date on predator-prey interactions in chemostats showing that population cycles are different when predators and their prey can and cannot evolve in response to one another.*

Hughes, A. R., B. D. Inouye, M. T. J. Johnson, N. Underwood, and M. Vellend. 2008 Ecological consequences of genetic diversity. Ecology Letters 11: 609–623. *A review of how genetic diversity in interacting species can alter the types of interactions among species and their outcomes.*

McPeek, M. A. 2008. The ecological dynamics of clade diversification and community assembly. American Naturalist 172: E270–E284. *A comprehensive model of how the mode of speciation in component taxa influences the structure of communities and the phylogenetic patterns of clades evolving in this context.*

Schluter, D. 2000. The Ecology of Adaptive Radiation. Oxford: Oxford University Press. *An exposition of the process of ecological speciation and its consequences for adaptive radiation.*

Strauss, S. Y., J. A. Lau, and S. P. Carroll. 2006. Evolutionary responses of natives to introduced species: What do introductions tell us about natural communities? Ecology Letters 9: 357–374. *A discussion of introductions of nonnative species to explore how species can adapt to one another as a result of interactions between them.*

VII

Evolution of Behavior, Society, and Humans

Allen J. Moore

This section presents the current view of animal behavior and animal societies, and their application and relevance to human evolution, reflecting the return of researchers to the original integration of these themes promoted by Darwin in *The Descent of Man and Selection in Relation to Sex*. Darwin treated the evolution of behavior, society, and humans more extensively and exclusively in this follow-up to *On the Origin of Species* because these traits represented a particular challenge for the theory of natural selection. Why is there so much variation in behavior? How can apparently cognitively complex and advanced behavior (such as mate choice) evolve? What leads to the evolution of diverse mating systems and the associated morphological variations? Why should animals live in societies, cooperate, and help each other? And importantly, given the complexity of humans and human societies, can we find homologous traits between humans and other animals and simpler social systems? *The Descent of Man* provided answers to these questions and showed that social systems and complex behavior exist throughout the animal kingdom. Darwin recognized that behavior can be one of the most complex traits to evolve and can have profound consequences for species, including our own, in driving rapid evolution and promoting great variation both within and among populations and species. Humans may be especially interesting to us, but they aren't special. What began as a challenge to the theory of evolution by natural selection—complex social behavior—has become one of the richest veins of research for testing and validating our current understanding of evolution.

The Descent of Man and Selection in Relation to Sex is arguably now nearly as influential as *On the Origin of Species*, and its title only hints at what was to become a major focus of research in biology in the twentieth century and beyond: investigating the evolutionary mechanisms of sexual selection, and understanding the diversity of mating systems and behavior. Although research and acceptance of sexual selection grew more slowly than research on natural selection, we have come to realize that just as natural selection provides an explanation for much more than the existence of different species, sexual selection explains much more than the descent of humans. In fact, following the Modern Synthesis, research on selection in relation to sex conspicuously ignored humans and focused more on understanding mating and animal behavior in general. Research into sexual selection, mating systems, communication, and social interactions was especially popular but mostly focused on insect, birds, fish, and mammals other than humans. Currently, this research is increasingly being applied to understanding human societies, behavior, and evolution as we find there are many shared aspects of behavioral evolution.

Our focus on Darwin's theory of sexual selection raises a question for modern biologists: Are natural selection and sexual selection conceptually distinct mechanisms of evolution (as Darwin himself believed)? Or is sexual selection better considered as subsumed by natural selection? There is little agreement on this point, and a separation of sexual selection is often implied. Certainly, that is one explanation for this section and its greater focus on ideas developed in the *Descent of Man* over *On the Origin of Species*. Yet from a modern population genetic perspective, the contribution of reproductive success is not distinct from the contribution of survival to *fitness*—defined as the spread of an allele in a population relative to all other alleles in that population. As Futyma states in his introduction to Section III, "Natural selection occurs whenever there is a consistent, average difference in fitness (reproductive success) among sets of 'individuals' that differ in some respect that we may refer to as phenotype." Under these definitions, sexual selection is the subset of natural selection dealing with fitness differences arising from competition for mates. Heuristically, however, it is convenient to treat them as distinct. We (like Darwin) often speak of natural selection as providing the limit to sexual selection. Following Darwin, we use sexual selection to explain the evolution of traits that would appear to be counter to the action of natural selection, that is, those traits that would appear to hinder survival rather than enhance it. So as a means of generating hypotheses, it is useful to consider them separate; however, that is not to suggest that sexual selection in any way contradicts natural selection or somehow provides a challenge or alternative. The expansion of natural selection to concepts that help explain unusual traits—such as kin selection explanations of altruism, or sexual selection explanations for elaborated traits expressed during mating—simply strengthens the logic of natural selection. The same is true for modern extensions of natural and kin selection to multilevel selection.

In this section we explore our current understanding of evolution of complex behavior and social systems—those traits that Darwin suggested might be considered problematic for natural selection—with examples from amoebas to mammals. We start out with proximate influences on behavior and how these relate to evolution. Darwin foresaw much, but he didn't anticipate the discovery of the mechanisms of inheritance, DNA, and molecular genetics! Yet for behavior to evolve, there must be heritable genetic variation, as in any trait that evolves. In chapter VII.1, Yehuda Ben-Shahar provides an overview of genetic influences on behavior and why this matters. We often hear of "genes for" a behavior, but this shorthand is simplistic, misleading, and unfortunate. There is much more to behavior genetics than discovering "the gene for" some behavior. Understanding the influence of genetics on the diversity of behaviors, and when, where, and how this influence matters and varies, is in its infancy. Even less well understood are the details of the genes-brain-behavior relationship for most traits. The natural connection between these different levels is likely to center around hormones, and in chapter VII.2, Ellen Ketterson, Jonathan Atwell, and Joel McGlothlin explore the ways in which hormones and behavior are linked. Ultimately, our understanding of these causal mechanisms of behavior strongly shapes our understanding of the evolution of behavior and interactions.

One of the most powerful associations between understanding human behavior and exploring evolution occurred when George Price recognized that game theory, developed to explain human behavior associated with economics, could be applied to animal behavior and evolutionary problems in the 1970s. In chapter VII.3 John McNamara provides an overview of the remarkable advances in this area since Price's paper appeared (coauthored with one of the major figures in subsequent theoretical developments in the field, John Maynard Smith). Game theory continues to be one of the most powerful paradigms for understanding the evolution of adaptive behavior.

Chapters VII.4, VII.5, and VII.6 cover the details of evolution by sexual selection. Rhonda Snook describes mating system evolution in general in chapter VII.4. She shows how understanding the mating systems of organisms provides insights into the evolutionary potential of populations. (The diversity of mating systems is in large part what inspired the theory of sexual selection by Darwin.) In chapter VII.5, Christine Miller explores how male-male competition results in the evolution of elaborate traits, while in chapter VII.6 Michael Jennions and Hanna Kokko explore how mate choice may do the same. There are clear connections between these chapters, but Darwin recognized two distinct mechanisms of sexual selection: male-male competition and female mate choice. In recent years there has been a focus on female mate choice, originally the more controversial of the two mechanisms. Now that evidence for female mate choice is overwhelming, we are seeing a return to an interest in male-male competition and how it influences sexual selection. Of course, total sexual selection reflects a combination of all mechanisms. Moreover, it is not always the females that choose and the males that compete; the sex roles may well be reversed, depending on the mating system.

Chapter VII.7 moves to communication, which is a fundamental part of most animal social systems yet presents special difficulties for understanding its evolution. Communication is one of those traits (like social dominance) that make little sense as a property of an individual by itself, as by definition communication involves interaction between at least two individuals. Michael Greenfield provides our current understanding of how communication evolves, and covers the various modalities involved in communication.

One of Darwin's main concerns in *The Descent of Man* was the evolution of society. The simplest social group is a parent and offspring, and in chapter VII.8, Mathias Kölliker, Per Smiseth, and Nick Royle provide an overview of the evolution of parental care, and the importance of cross-generational effects for evolution. Chapter VII.9, by Joan Strassmann and David Queller, considers the theoretical basis for the roles of cooperation and conflict in structuring interactions, from cells (single cells to multicellularity) to organisms. Michael Cant follows in chapter VII.10 with a consideration of cooperative breeding, using insects and vertebrates as examples. Thus, the chapters on parental care and cooperative breeding represent a continuum of social interactions, linked by our current understanding of how cooperation and conflict evolve.

The preceding chapters focus mostly on animals and diversity, but as Darwin suggested, sexual selection is a driving force in structuring the evolution of humans and human society. The prolonged association of parents and children is key, and drives a competition for mating partners because of the extreme investment in offspring by humans. Virpi Lummaa in chapter VII.11 expands on this theme by describing current understanding of human behavioral ecology, which is very much influenced by parent-offspring and even grandparent-grandoffspring interactions. Human behavior presents special challenges for researchers, given the limit on the types of studies that are feasible. It also yields special rewards, as there is additional complexity to consider. No more complex is the consideration of "moral facilities" Darwin discusses in *The Descent of Man*. Robert Richardson, in chapter VII.12, presents the modern attempt to apply evolutionary approaches to understanding human psychology. As he shows, this effort has been only somewhat successful thus far, partly because the social interactions in humans are highly developed. This is not the complete justification, however, as shown when we return to animals and consider the most highly developed societies in eusocial organisms such as ants, bees, and wasps. Laurent Keller and Michel Chapuisat discuss the evolution of eusociality in chapter VII.13. Finally, Marc Hauser concludes a consideration of the trait we typically consider to be quintessentially human by examining cognition in chapter VII.14.

One of the dangers of evolutionary biology is simply to assume everything we see is adaptive and evolved "for" a specific purpose. Evolution, however, can be a powerful hypothesis for explaining why animals behave as they do, as demonstrated most clearly when we consider how evolutionary reasoning allows us to explore traits that on the surface, do not appear adaptive. In chapter VII.15 Nathan Bailey discusses traits such as behavior to attract predators, cannibalism, and same-sex sexual behavior as examples that allow us to explore the power of evolutionary approaches to understanding behavior. Finally, Jacob Moorad and Daniel Promislow explore aging and menopause in chapter VII.16, traits that are the exceptions that prove the rule for evolution. On the face of it, forgoing reproduction, aging, and dying are the embodiment of nonadaptive behaviors, yet evolutionary tests provide powerful insights for why these traits exist.

Some of the richest research themes in evolutionary biology have explored behavior. The diversity of behavior in the natural world is breathtaking, as are the often-elaborate traits associated with behavior. Our own species is characterized by complex social interactions, prolonged periods of parental care, extended family interactions, and competition for mates. Darwin recognized this and in his 1871 book, *The Descent of Man and Selection in Relation to Sex*, gave us a framework for exploring this diversity in nature. Behavior fascinates us, perhaps because it is both familiar and mysterious. It also provides a window on how evolution structures biodiversity and reveals the commonalities among organisms, from microbes to humans.

VII.1

Genes, Brains, and Behavior

Yehuda Ben-Shahar

OUTLINE

1. Genes and behavior
2. "Nature versus nurture"
3. What is a "behavioral gene"?
4. Analyzing behavior: Natural variations versus mutations
5. Genomes and systems genetics
6. The future of behavioral genetics: The behavioral epigenome

Behavior is defined as the directed action of an animal in response to a stimulus. In multicellular animals, behavior is the product of the nervous system. Behavioral phenotypes are often conserved across distant taxa and are heritable. Yet, the role of genetics and evolution in determining behavior has been controversial for much of the first half of the twentieth century, often paraphrased as the "nature versus nurture" debate. While heredity clearly plays a role in behavior, linking "behavioral" genes to behaviors has not been easy. Two primary approaches have been used over the years to identify causal loci: studies of natural variations in wild-type populations and mutagenesis-dependent forward genetic screens. Both approaches have strengths and limitations; while mutation analysis has been immensely successful in identifying many causal genes, it is blind to the evolutionary and population levels forces that shaped behaviors. In contrast, studies of natural behavioral variations have often failed to identify causal relationships between specific genetic polymorphisms and the studied phenotype. However, the methodological dichotomy is fast disappearing owing to the exponential pace of technical and theoretical advances in molecular biology, resulting in an improved understanding of how the interactions among genes, brains, and the environment lead to specific behaviors.

GLOSSARY

Central Dogma. The concept, proposed by Francis Crick in 1970, that all cellular proteins are produced via a linear and nonreversible process in which "gene"-specific information encoded in DNA is transcribed into a messenger RNA (mRNA), which subsequently gets translated into a protein.

Circadian Rhythms. Internally driven, circa 24-hour cycle in biological systems.

Epigenetics. In general, an often-transient heritable change in gene expression caused by factors other than changes in the underlying DNA sequence (mutation). In this chapter this term is used to describe the role of chemical modification of DNA and histone proteins by methylation and/or acetylation as a mechanism for regulating gene function in a tissue-specific manner. Under certain circumstances, such modifications can be heritable without any changes to the primary DNA sequence and hence constitute a mechanism for non-genomic inheritance of quantitative traits. It is important to note that epigenetics means different things to different people in different biological fields.

Ethology. The study of behavior of animals in their natural environments. Konrad Lorenz, Karl von Frisch, and Nikolaas Tinbergen won the Nobel Prize for Medicine in 1973 based on their work in this field.

Eugenics. A popular social movement in the early twentieth century that advocated the use of selective breeding for the improvement of hereditary traits of a specific race, often applied in the context of humans.

Experimental Psychology. The study of animal behavior under controlled laboratory conditions. The psychologists Ivan Pavlov and B. F. Skinner were seminal in developing animal models for cognition, although neither studied genetics.

Forward Genetics. An approach for identifying genes underlying a specific phenotype without any a priori assumptions in regard to their identity or biochemical functions.

Genomics. Studies of whole-genome architecture, including DNA sequences, analyses of whole-genome transcriptional regulation, and the role of genome evolution in biological processes such as speciation or heritable diseases.

Geotaxis. The behavioral response of organisms to the vector of gravity, which can have a negative or positive value—that is, move up (negative) or down (positive).

Phototaxis. The behavioral response of organisms to light, which can have a negative or positive value—that is, move away from (negative) or toward (positive).

Pleiotropy. The effects of the function of a single gene on multiple, independent phenotypes.

Quantitative Trait Loci (QTL). Sequence(s) of underlying DNA associated with complex, non-Mendelian, polygenic traits. Each contributing QTL is independently variable in the studied population and hence is responsible for a defined proportion of the overall observed phenotypic variability.

Reverse Genetics. A "candidate gene approach," which implicates previously characterized genes with a novel phenotype.

Systems Biology. An emerging biological framework that promotes a holistic approach to understanding complex biological systems, based on the idea that complex biological systems have irreducible emergent properties that cannot be understood by studies of simpler individual elements.

Systems Genetics. Genetic studies using systems biology, especially in the context of non-Mendelian complex phenotypes, based on the principle that understanding complex phenotypes such as behavior depends on understanding interrelationships among genotypes and phenotypes at the organismal level.

1. GENES AND BEHAVIOR

Behavior is one of the characteristic traits of animals and can be defined as a reaction in response to specific changes in an individual's environment. Despite this ubiquity, finding a simple and common description of behavior is elusive, perhaps in part because the term *behavior* covers so many different actions. This staggering behavioral diversity across the animal kingdom and the inherent phenotypic range associated with behavior have often led to a common belief that behaviors are unique phenotypes that cannot be explained within the general biological framework. Yet many behaviors are stereotypic and seem to be driven by common molecular pathways across diverse and distant animal taxa. Examples of commonality among many animals include aggression over territories (see chapter VII.5) and mates (see chapter VII.6), foraging for food when hungry, circadian rhythms, and the avoidance of harmful chemicals or extreme temperatures. Even the most plastic of traits—learning and memory and cognition (see chapter VII.14)—have genetic homology across organisms.

The conservation of specific behaviors across generations of the same species—independent of cultural and learned phenotypic transmissions—and its evolution over time can mean only one thing: behavior is the product of information encoded in DNA. Yet the acceptance of behaviors as genetically inherited phenotypes shaped by natural selection has long been controversial. This chapter explains and discusses some of the confusions associated with behavioral genetics. It also examines the role behavior has played in animal evolution, and the impact of evolutionary thinking on studies of animal behavior.

Behavioral genetics emerged during the first half of the twentieth century as a multidisciplinary field that built on foundations drawn from important advances in several areas of the modern biological sciences, including statistics, ethology, and experimental psychology. Study in these fields led to the realization that behaviors are often conserved across distant species, that general behavioral principles can be learned by observing animals, and that behavioral traits can be shaped by both natural and artificial selection. The second important contribution to behavioral genetics thus came from the work of Theodosius Dobzhansky and his colleagues, and led to the Modern Synthesis that linked evolutionary theory with genetics. This synthetic framework also included the work by J. B. Haldane, Sewall Wright, and R. A. Fisher, which laid the foundation for the development of population genetics, a discipline that uses quantitative approaches and empirical studies to understand the role of allelic frequencies in populations and the effect of these genetic variations on observed phenotypic diversities.

The third important contribution to behavioral genetics was the emergence of molecular genetics in the second half of the twentieth century, which was fueled by the discoveries of the double helix model for the DNA structure and the development of the *central dogma* in molecular biology (DNA to RNA to protein). These discoveries, in combination with rapid technological advancements, resulted in the ability of geneticists and evolutionary biologists to obtain DNA sequences of specific genes from individuals, revolutionizing the fields of evolution and genetics. For the first time, biologists were able to associate phenotypic variations with physical DNA polymorphisms. These advancements enabled biologists to interpret function and evolution of morphological, physiological, and behavioral traits by studying the DNA sequence and the genetic architectures at the complete genome level.

The fourth important scientific development was the emergence of modern neuroscience. This field served as the foundation for understanding how multicellular animals can integrate and process stimuli, and translate them into an action, a "behavior." Better understanding of the biology of neurons enabled the placement of behavioral genetic findings in the context of the primary organ that drives behavior, the nervous system.

The rest of this chapter presents the theories and empirical data that support the role of genes and evolution in behavior and explains the synergistic role behavioral genetics plays in modern evolutionary thought.

2. "NATURE VERSUS NURTURE"

First coined by Galton in his book *English Men of Science: Their Nature and Their Nurture* (1874)—and despite his perspective that both nature and nurture are important in the development of behavior—the expression "nature versus nurture" became synonymous with the controversies associated with the field of behavioral genetics. The introduction of heredity, and later genetics, to the fields of animal behavior and evolution met with much resistance, likely stemming from earlier views by philosophers such as John Locke, who suggested that all people are born as a *tabula rasa*, or blank slate. The *tabula rasa postulate* asserted that all humans are born equal in terms of their cognitive and behavioral capacities, and it is only their experiences that shape who they become. In contrast, many of the early geneticists in the United States and Europe promoted the hereditarian (nature) view of behavior, which was dominant for most of the first half of the twentieth century. This interpretation resulted in the application of Mendelian genetics, sometimes in the most absurd ways, to many human and animal behavioral traits. These deterministic views changed rapidly during the 1950s as *behaviorism* became a popular philosophical view of human and animal behavior. Guided by the writings of leading experimental psychologists such as J. B. Watson and B. F. Skinner, behaviorists dismissed altogether the role of heredity in determining behavior and advocated that only life experiences matter in shaping an individual's behavior (nurture).

Most modern-day researchers who study human or animal behavior would agree that the nature versus nurture dichotomy is, in fact, oversimplified and archaic, although it is still prevalent in the scientific literature. No organism develops without genes (nature), and no organism develops in the absence of an environment (nurture). Furthermore, as the field of molecular genetics matured, it became clear that many behavioral phenotypes are complex and do not follow simple single-gene Mendelian rules. While the mounting evidence supports the hypothesis that heredity and genetics have an influence on behaviors, most behaviors show a continuous distribution of values (i.e., are *quantitative traits*) rather than a collection of discrete phenotypes. Hence, a refined view of the nature versus nurture debate is that information stored in the DNA determines behavioral phenotypic boundaries rather than specific phenotypic values. This model suggests that a specific behavioral phenotype can be stretched in multiple directions dependent on the strength of the various internal and external stimuli relevant to the phenotype in a species-specific context. Moreover, as is discussed toward the end of this chapter, certain plastic phenotypic alternatives can become fixed across generations via nongenomic, epigenetic mechanisms.

3. WHAT IS A "BEHAVIORAL GENE"?

Despite the tendency of people (and newspapers) to speak of "genes for behavior," specific genes might not necessarily encode directly for a specific behavior but rather are shaped by evolution to set physiological and physical constraints on the expression of behavioral phenotypes. Unfortunately, and perhaps confusingly, geneticists frequently assign specific "functions" for their genes, in ways that simplify their long-term research goals but that lead to the perception that specific genes cause specific behaviors. For example, the scientific literature contains descriptions of genes classified as a "developmental gene" or a "cancer gene." The field of behavioral genetics, which is strongly influenced by developmental biology, followed the categorization of genes with the identification of various "behavioral" genes. Examples included "learning and memory genes," "social genes," and "sex genes." However, genes do not encode for phenotypes such as cancer, development, or specific behaviors. Rather, genes encode for RNAs and proteins. It is the biochemical function of these macromolecules, and the cellular processes they fuel, that drive processes such as the occurrence of cancer, normal development, or the expression of a specific behavioral phenotype (Robinson et al. 2008).

Modern neuroscience teaches that in all multicellular animals, a behavior is the product of the nervous system, a complex and highly specialized organ made of many individual neurons organized in stereotypical neuronal circuits. Therefore, "behavioral" genes and genetic variations are likely to affect behaviors indirectly by determining the development of specific neuronal circuits, their interactions with other organs and cell types, and the capacity of neuronal circuits to respond physiologically to changes in the internal and external environments of individuals. Behavior is typically flexible, quantitative in expression, and difficult to study in genetic terms because of these multiple inputs. Any and all of these levels can be genetically influenced. Nevertheless, none of the levels is purely environmental.

4. ANALYZING BEHAVIOR: NATURAL VARIATIONS VERSUS MUTATIONS

As mentioned in the previous sections, studies of behavior and cognition played a major role in the early days of genetics. In contrast with the scientific fallacies that were associated with eugenics, many positive influences on modern biology were associated with the emergence of the field of genetics, in the work of Arthur Darbishire, Robert Yerkes, E. C. Tolman, and others. These pioneers took advantage of various animal models to investigate the role of heredity in specific behaviors. Their studies were not confined to natural behaviors but also tried to understand the genetics and inheritance of "abnormal" behaviors and the insights they might bring to elucidating how nervous systems drive specific behaviors. Illustrative examples are the studies on heredity in the "Japanese waltzing mouse," a mouse breed that exhibits a tendency to run in circles, and selection studies by E. C. Tolman used to generate rats that were either "dulls" or "brights" in learning how to navigate a maze for a food reward. The latter studies indicated a strong genetic basis for cognitive abilities in rats.

The negative connotations associated with the eugenics movement, and the related atrocities of World War II, led to the disenchantment of many human and vertebrate behaviorists with genetics. But other model systems emerged to support the role of genetics in behavior. Some of the strongest support for the function of heredity in behavior came from studies of the genetic workhorse, the fruit fly *Drosophila melanogaster*, driven by the seminal works of T. H. Morgan and his students. The fly turned out to be an excellent genetic model owing to its small genome, short generation time, adaptability to laboratory conditions, and rapid response to selection and mutagenesis. One of the first to use the fly purely as a model for behavioral studies was Jerry Hirsch, who trained with Tolman, and was influenced by Dobzhansky's views on population genetics. Hirsch (1963) used flies and artificial selection to decipher the "genetic architecture" underlying natural behavioral variation. Hirsch asserted that the only way to understand how behaviors evolve, and what role genetic variation plays in specific behavioral phenotypes, is to use the tools of population and quantitative genetics to quantify the relative contributions of specific *quantitative trait loci* (QTLs) to the overall behavioral variations within a population. An underlying assumption in his studies was that multiple independent genes contribute to behavioral phenotypes and that allelic variations in each locus contribute a defined fraction of the overall behavioral variations in a population. One of his studies that best illustrated this approach was a long-term selection study of flies that showed a strong positive or negative response to gravity (geotaxis). Wild-type flies tend to be somewhat negatively geotactic (run "up" when disturbed). Hirsch selected for genetically homogeneous *Drosophila* strains that showed either extreme positive or negative geotaxis behaviors. He then used chromosomal mapping techniques to estimate the quantitative contribution of each of the fly's three main chromosomes to the genetic differences in behavior between the two extremes. His conclusions were that even for a relatively simple behavior such as geotaxis, many genes contribute to the genetic divergence between the two strains and the responsible genes are likely distributed across all three chromosomes. Hence, his selection led to the allelic stabilization of multiple independent genes, rather than changes in a single major gene. His studies fell short of identifying the specific loci responsible for the behavioral differences between the lines.

While Hirsch's approach to behavioral genetics was strongly influenced by experimental psychology and quantitative genetics, other approaches to studying the genetics of behavior emerged in the early 1960s. The most influential work came from the laboratory of Seymour Benzer, who was strongly influenced by the emergence of the use of mutagenesis to study molecular genetics and the transformative effect it had on the fields of embryology and developmental biology; he saw an overlap with behavior. This profound insight was best captured in his first publication on studies of behavioral genetics in *Drosophila*:

> Complex as it is, much of the vast network of cellular functions has been successfully dissected, on a microscopic scale, by the use of mutants in which one element is altered at a time. A similar approach may be fruitful in tackling the complex structures and events underlying behavior, using behavioral mutations to indicate modifications of the nervous system. (Benzer 1967)

To succeed, Benzer's approach had to rely on the premise that although many genes might contribute to a specific behavioral phenotype, mutations in a single gene could still have measurable effects on the studied behavior relative to wild-type animals of otherwise-identical genetic background. Like Hirsch, Benzer chose to apply his approach first to a relatively simple behavioral phenotype, phototaxis (attraction to light). He devised a clever assay to measure the phototactic response of flies, in which close to 100 percent of wild-type flies showed positive attraction to light. He then used chemical mutagenesis and screened hundreds of animals for any deviations from the expected wild-type behavior. The approach turned out to be immensely successful. Benzer isolated many different phototaxis mutations in individual genes. In contrast with Hirsch's approach, Benzer's studies rapidly identified causal genetic effectors associated with different behavioral and neurophysiological phenotypes such as the molecular identity of the circadian clock, learning and memory, and sexual behaviors (Weiner 1999).

Benzer's successful forward genetics approach, however, comes with a scientific cost. While several animal species are highly amenable to mutagenesis screens, and hence became the darlings of the behavioral genetics community, many others, including people, are not. As a result, much of what we know about the roles of specific genes in neurogenetics and behavior comes from very few model organisms, primarily the fly, mouse, and roundworm. Furthermore, the dominance of mutation analysis studies at the expense of quantitative population genetics led to a major gap in our understanding of behavior in the context of natural selection and evolution.

Many of the early behavioral population geneticists often concluded their studies by saying that the associations between specific genes and behaviors are too complex to allow identification of causal relationships. Yet in cases where a single major polymorphic gene was involved, it was possible to do so. An example of a success story involves the *foraging* gene. *Drosophila* larvae exhibit a natural polymorphism in foraging behavior; when placed on a yeast lawn, some larvae tend to move rapidly while consuming food ("rovers"), while others seem to slow down significantly ("sitters") (Osborne et al. 1997). Genetic analyses indicated that this behavioral polymorphism is mediated by variations in a single major gene. When the gene was finally cloned, it turned out that it encodes a cGMP-dependent protein kinase (PKG), a protein present in all cells and important for activating or inactivating other proteins by mediating phosphorylation. In the case of foraging, high levels of enzyme activity were associated with "rovers," while lower activity was associated with "sitters." Subsequently, it has been found that the role of PKG in regulating feeding behaviors is highly conserved across different animal species. For example, studies suggested that a homologous PKG gene in honey bees and other social hymenopterans (bees, wasps, and ants) is regulated in association with the division of labor among workers (Ben-Shahar 2002). These studies indicated that changes in the foraging gene activity are associated with feeding behavior plasticity in different species, albeit on different timescales: an evolutionary timescale in flies, and a developmental timescale in social insects. The foraging gene story illustrates that complex natural behaviors, undoubtedly influenced by many genes, can still be studied from the standpoint of the contribution of a single gene.

5. GENOMES AND SYSTEMS GENETICS

The early success of Benzer's single-gene-mutations approach to behavioral genetics, and despite a handful of examples such as the *foraging gene* account regarding single genes and their influence on behavior and evolution, suggested to molecular biologists that the approach of Hirsch and some of the other early evolutionary geneticists who studied behavior would disappear from the scientific literature because of the difficulty in identifying the actual molecules, genes, and genetic networks that underlie natural variations in specific behavioral traits. Fortunately, this has not happened, and indeed, the trend is toward studies of more species and more natural variation in behavior. The sequencing of the human genome and the plethora of genome projects that followed led to the reevaluation of studies of natural genetic variations underlying the biology of complex traits, including behavior. This reevaluation was also fueled by the need for a better understanding of the mechanisms underlying complex human behavioral traits, and the rapid transition of evolutionary biology into a molecular biology field.

The exponential growth in biological data acquisition led to the emergence of a "new" biological framework, often termed *systems biology*. The idea behind this approach is that to understand how biological systems work, one has to investigate the emerging properties of the system as a whole rather than looking at its parts individually. *Systems genetics* is a branch of this framework based on the assertion that all organizational levels of biology are interconnected in a complex network that includes both genetic and phenotypic elements, and it is the network that determines the biological characteristics of an individual, or even a group of individuals (Mackay et al. 2009). This approach is in contrast with Mendelian genetics, which looks at each gene as an independent genetic factor.

What is the impact of systems genetics on behavioral genetics? One view argues that there is nothing new in the systems genetics approach, that it is, rather, a rediscovery of the work of Hirsch and others. Nevertheless, the application of new molecular, genetic, and statistical techniques makes systems genetics an exciting field, with its increasing focus on multiple levels, genes, and natural variation. A great example comes from a study that revisited Hirsch's geotaxis studies. In spite of Hirsch's original assertion that the genetic architecture underlying geotaxis behavior is too complex to allow identification of specific contributing loci, Ralph Greenspan and his colleagues used gene expression microarrays, which can be used to simultaneously examine relative expression levels of thousands of genes from a single source, to identify genes that were differentially expressed in the heads of Hirsch's high and low geotactic fly strains. This effort yielded several candidate "geotaxis genes," which were further confirmed by single gene mutations as playing a major role in the geotaxis response of individual flies. Thus, more than 40 years after Hirsch's publication of his geotaxis studies, modern genetics was finally able to merge the approaches of Hirsch and Benzer to decipher the genetic architecture underlining this ecologically and naturally varying relevant behavior.

The efforts to combine genome-level information with natural genetic variations to identify loci responsible for complex behaviors are ongoing. In spite of early difficulties in pinpointing such traits (as in Hirsch's early geotaxis studies), several recent advances suggest that the identification of loci responsible for quantitative behavioral traits is an attainable goal, especially in genetic model organisms such as the fruit fly and the nematode (roundworm). One of the best examples of the use of recent advances in DNA sequencing for creating modern tools for systems genetics comes from the *Drosophila* Genome Reference Panel (DGRP)—a collection of 192 naturally derived inbred lines with fully sequenced genomes generated by Trudi Mackay and her colleagues (2009). The lines can be screened for any number of complex behavioral traits that can then be mapped to specific variable regions in the fly genome. Furthermore, the DGRP is a community resource, which allows many different research groups to perform behavioral screens using the exact same fly populations. Consequently, phenotypic data generated by the research community could be used to identify genetic variations that affect multiple different behaviors—indicating a pleiotropy between traits—as well as to obtain precise estimates of the contributions of genotype by environmental interactions to specific behaviors. Recent similar approaches taken by C. Bargmann, L. Kruglyak, and colleagues successfully identified several quantitative trait loci associated with complex variable behaviors such as "social feeding" decisions, and the response to specific sensory stimuli in the roundworm. In both cases, at least one major gene was identified as responsible for the quantitative behavioral differences between different wild-type individuals. While the tools mentioned are currently available for very few genetically tractable model organisms, it is likely that as DNA sequencing techniques become more economical and more readily available, such tools will be increasingly useful in studies of behavioral genetics and the evolution of behavior in other animals as well.

6. THE FUTURE OF BEHAVIORAL GENETICS: THE BEHAVIORAL EPIGENOME

Despite the successes in identifying genes, QTLs, and transcriptional differences associated with various behaviors, the flexibility of behavior remains something of a mystery. Why can it change rapidly in response to changes in the environment? This gene-by-environment interaction and its role in producing variable behavioral phenotypes is still a major unresolved issue. Identifying the interactions among genes, environments, and flexibility fit will have many important implications for basic biology and in clinical studies of human behavior. The problem is simple even if the answers are complex or hidden: How do we reconcile the slow change of genetics (over generations) with the rapid change of behavior (over hours or minutes)?

One fascinating area in which significant progress has been made in this regard is in *behavioral epigenetics*. Studies of epigenetics focus on specific chemical modifications to DNA and nucleosomes during development. These modifications affect gene transcription in a tissue-specific manner. Recently, principles of epigenetics have been applied to explain how some specific life experiences might lead to differential behavioral outcomes in a group of animals with otherwise-identical sequences of DNA. It is beyond the scope of this chapter to describe in detail the current state of knowledge in the field of epigenetics given that it is relatively new and changing rapidly. Instead, we will focus on one well-established example of the important role epigenetics is playing in the heredity of specific human and animal behaviors.

The idea that environmental changes during development could affect the behavior of adult animals is as old as the nature versus nurture debate, yet the actual mechanisms underlying such functional relationships are still poorly understood. One such convincing example is the epigenetic inheritance of maternal behaviors in rats. In 2004, Michael Meaney and his colleagues showed that laboratory rats had two mothering styles. One mother type exhibited high level of licking and grooming behavior, while the other type showed very low levels of these behaviors. These alternative phenotypes seemed to follow simple Mendelian rules: adult females always exhibited the same behavior they experienced as pups. Surprisingly, cross-fostering experiments in which pups were switched at birth and so had an adoptive (environmental) mother as well as a biological (genetic) one suggested that the maternal care environment of female pups, not the genotype of the biological moms, determined their behavior as adults. This puzzling case of nongenomic inheritance was found to be driven by the high-grooming environment experienced by pups, which resulted in epigenetic modifications of the DNA that encodes the glucocorticoid receptor, a protein important for the regulation of stress-related behaviors in brains of vertebrates. Such modifications led to profound spatial and temporal changes in the brain expression patterns of this protein. These studies indicated that the rat genome potentially encodes two alternative maternal behaviors, and the one that is expressed is plastically dependent on a critical developmental window. The trait is then subsequently transferred across generations as a fixed trait as long as the maternal environment does not change (for example, by being switched at birth by an experimenter). Later investigations by the same group suggested that similar processes affect human behavior as well: similar DNA chemical modifications of the human version of the glucocorticoid receptor in the brain in response to early childhood abuse increased the likelihood of anxiety and suicidal behavior in these individuals as adults. These studies indicated that epigenetics is one of the key molecular processes that link genetic variations to environmental changes in the context of behavior (Zhang and Meaney 2010). This connection is further supported by many recent studies that have demonstrated a role for epigenetic processes in diverse behavioral phenotypes including learning, long-term memory, drug addiction, personality, and various neuropsychiatric disorders.

The fields of behavioral and evolutionary genetics are gaining momentum as a result of the technical and theoretical advancements in molecular genetics. This exciting progress should lead to a better understanding of how flexibility of behavior is maintained, how specific behaviors evolve, what role behavior plays in speciation, which genes are essential for normal neuronal functions, and what are the molecular bases for human behavioral pathologies.

FURTHER READING

Ben-Shahar Y., A. Robichon, M. B. Sokolowski, and G. E. Robinson. 2002. Influence of gene action across different time scales on behavior. Science 296: 741–744. *One of the first studies to show that the role of specific genes in affecting behavior can be studied in nontraditional genetic model organisms such as the honey bee. Furthermore, it showed that specific molecular signaling pathways can affect analogous behavior across distant species.*

Benzer S. 1967. Behavioral mutants of *Drosophila* isolated by countercurrent distribution. Proceedings of the National Academy of Sciences USA 58: 1112–1119. *The first successful genetic screen to identify mutations in single major genes that affect a specific behavioral phenotype.*

Hirsch J. 1963. Behavior genetics and individuality understood. Science 142: 1436–1442. *A well-explained approach to behavioral genetics.*

Mackay, T. F., E. A. Stone, and J. F. Ayroles. 2009. The genetics of quantitative traits: Challenges and prospects. Nature Reviews Genetics 10: 565–577. *A review article describing the modern approaches to understanding quantitative traits by using molecular genetics tools.*

Osborne K. A., A. Robichon, E. Burgess, S. Butland, R. A. Shaw, A. Coulthard, H. S. Pereira, R. J. Greenspan, and M. B. Sokolowski.1997. Natural behavior polymorphism due to a cGMP-dependent protein kinase of *Drosophila*. Science 277: 834–836. *One of the first studies to show that natural variations in a single major gene can have dramatic heritable effects on a specific naturally polymorphic behavioral trait.*

Robinson G. E., R. D. Fernald, and D. F. Clayton. 2008. Genes and social behavior. Science 322: 896–900. *An excellent review of the modern approach to understanding the role of genes and genetic variations in complex social behaviors.*

Weiner J. 1999. Time, Love, Memory: A Great Biologist and His Quest for the Origins of Behavior. New York: Knopf. *An award-winning book that describes the early and exciting days of behavioral genetics in the laboratory of Seymour Benzer at the California Institute of Technology.*

Zhang T. Y., and M. J. Meaney. 2010. Epigenetics and the environmental regulation of the genome and its function. Annual Review of Psychology 61: 439–466. *A review that explains well the role of epigenetics as a mechanism for the integration of environmental and genetic factors in natural behavioral variations across individuals.*

VII.2

Evolution of Hormones and Behavior

Ellen D. Ketterson, Jonathan W. Atwell, and Joel W. McGlothlin

OUTLINE

1. Hormonal mechanisms and phenotypic variation
2. Hormones and phenotypic integration
3. Hormones and microevolution
4. Hormones and macroevolution
5. Summary and future directions

Early evolutionary biologists often focused on either genes or visible phenotypes while neglecting the myriad developmental and physiological mechanisms that link them. Recently, evolutionary biologists have become more interested in these mechanisms and have come to appreciate the role they play in the evolutionary process. This chapter focuses on a major class of physiological mechanisms—hormones—and discusses a few of the many ways that understanding hormonal mechanisms can enrich our understanding of evolution. Although the discussion is biased toward vertebrate animals and those mechanisms that mediate behavior, the same principles apply to other taxa and to other complex phenotypes.

GLOSSARY

Activational Effect. The effect of a hormone that fluctuates with circulating hormone levels and is often reversible (i.e., not permanent, as with organizational effects).

Challenge Hypothesis. The concept that the arrival of an aggressive intruder induces a hormonal response, often an elevation in testosterone, in the animal on which it intrudes. Similar to an immune response to a pathogen, which can prepare an organism for subsequent encounters with the pathogen, the hormonal response to an aggressive challenge is hypothesized to induce physiological preparation for future intrusions.

Correlational Selection. Selection that arises when traits interact in their effects on fitness; may act over time to assemble groups of traits that work well together, including hormone-mediated suites.

Hormonal Pleiotropy. Coordination of a suite of correlated or co-occurring traits by a common underlying hormonal mechanism.

Hormone. A chemical messenger molecule that is released from specialized glands or cells into the circulation and regulates a biological response at target cells or tissues.

Hormone-Mediated Suite. A group of traits that are correlated owing to the influence of a hormone.

Hormone-Mediated Trait. Phenotype affected by the action of a hormone.

Organizational Effect. Developmental effect of a hormone occurring early in life, including prenatal; usually irreversible and often involves a specific sensitive period.

Phenotypic Engineering. An experimental approach used to assess the adaptive value of a trait or traits by manipulating individual phenotypes (e.g., experimentally altering hormone levels and measuring behavioral, performance, and/or fitness consequences).

Phenotypic Integration. Patterns of correlation or interdependence among different parts of the phenotype; can be mediated by common underlying hormonal mechanisms of trait expression or development.

Phenotypic Plasticity. The capability to express more than one phenotype for a given genotype, often mediated by hormonal mechanisms. This may occur at many timescales.

Receptor. A protein on the surface or interior of a cell that binds to a hormone, leading to modulation of cellular functioning (e.g., activation/inhibition of gene transcription or second messenger networks).

Target. Tissues or cells whose function is influenced by the action of a hormone owing to the presence of hormone receptor proteins.

Trade-off. The situation that occurs when two traits (often components of fitness, e.g., mating effort versus parental effort) cannot be simultaneously maximized because the expression of each comes at the expense of the other.

1. HORMONAL MECHANISMS AND PHENOTYPIC VARIATION

What Are Hormones?

Hormones are classically defined as chemical signals produced in specialized glands and carried throughout the rest of the body by the circulatory system. They function by binding to receptor proteins located within or on the surface of the cells of target tissues, leading to altered cell physiology or changes in gene expression. The resulting hormone-mediated changes in target cells and tissues underlie variation in a diverse array of phenotypes, including physiology, morphology, and behavior.

Many kinds of molecules can act as hormones. In animals, two of the largest and best-studied classes are peptide (or protein) and steroid hormones. *Peptides* are composed of chains of amino acids and usually interact with receptors on the cell membrane, typically activating second messenger systems that trigger cascading reactions within the cell. *Steroid hormones* are synthesized from cholesterol and tend to bind to intracellular receptors, which then directly or indirectly regulate gene expression. Examples of protein hormones relevant to behavior include prolactin, arginine vasopressin, and melanocortin, and examples of relevant steroid hormones include estrogens, androgens (e.g., testosterone), and glucocorticoids (e.g., cortisol and corticosterone).

Neurotransmitters, such as dopamine or serotonin, usually act at neural synapses to enable neural transmission and are thus often considered as a separate class of chemical messengers. However, recent studies blur the distinction between hormones and neurotransmitters. Classical hormones such as estrogen can be synthesized in the brain and act locally and instantaneously to modulate behavior, much like a neurotransmitter. Similarly, epinephrine (adrenaline) is produced by the adrenal glands and can influence behavior both as a neurotransmitter and as a classical hormone in non-neural tissues. Further, tissues not traditionally considered endocrine glands are known to secrete hormones; for example, the liver produces insulin-like growth factor (IGF-1), a peptide hormone linked to variation in vertebrate life history and behavior.

Organizational and Activational Effects of Hormones

Hormonal influences that occur as part of developmental processes are referred to as *organizational effects* and often persist throughout the life of the animal. For example, although male and female genomes are nearly identical, minor differences (i.e., a single gene or chromosome determining sex) can give rise to striking dissimilarities in sexual phenotypes because of the organizational action of hormones during early development. Within-sex polymorphisms can also be associated with organizational hormonal differences (e.g., "sneaker" males in fish, and plumage color polymorphisms in songbirds). The hormonal environment that an embryo experiences during sensitive phases of development is often influenced by its mother's hormonal state. Female birds deposit hormones in yolk that affect embryonic development and the behavior of hatchlings; female mammals provide a uterine hormonal environment that can influence the behavioral development of offspring before birth.

Hormonal influences on phenotype that fluctuate with hormone levels are known as *activational effects*. Activational effects of hormones often underlie reversible changes, such as behavioral flexibility in response to changes in the physical or social environment. Some within-sex polymorphisms in insects, lizards, and other groups are activational in nature, allowing, for example, males to alternate between territorial and satellite strategies depending on the situation.

Sites and Modes of Hormone Action

For behavioral traits, the target tissues of hormone action are often located in the central nervous system. These targets include the brain and spinal cord, which are sensitive to both organizational and activational effects of hormones. Other target tissues relevant to behavior include structures related to social interactions (e.g., a cock's comb, which responds to androgens) or locomotion. For physiological traits, hormones may affect a wide variety of target tissues to influence metabolism, biological rhythms, and immune function.

Importantly, variation in the density or location of hormone receptors among the cells of target tissues can determine the phenotypic response to a hormone. Receptor location and density are known to vary dynamically during development, between sexes, between seasons, and over short timescales in adulthood. An example involving sex role reversal can be seen in the African black coucal, a bird species in which "traditional" sex roles are reversed. In the black coucal, females sing, fight, and compete with one another for mates, while males incubate eggs and care for the young. Cornelia Voigt and colleagues have shown that although male black coucals have much higher testosterone levels during breeding, the density of androgen receptors in key brain regions is much lower in males than in females, apparently facilitating the observed "reversal" in sexual behavior.

Hormonal Cascades

Hormones rarely act alone. Many exist as part of a hormonal *axis* or *cascade*, in which the synthesis or release of one hormone is itself modulated by another hormone, usually one produced in another part of the body. The hypothalamic-pituitary-gonadal (HPG) endocrine axis is a classic example. The HPG axis is initiated by a peptide hormone, gonadotropin-releasing hormone (GnRH), that is released from the hypothalamus in the brain and acts on the pituitary gland situated directly below the brain, prompting it to release luteinizing hormone (LH) and follicle stimulating hormone (FSH). LH subsequently modulates the release of steroid hormones (androgens or estrogens) by the gonads (the testes or ovaries). Gonadal hormones may in turn cause negative feedback, downregulating the release of hormones earlier in the pathway to suppress the activity of the axis. Events that regulate the initial release of GnRH from the hypothalamus, which may involve other hormones or neuropeptides, are less completely understood and are the subject of active research.

As demonstrated by the HPG axis, the circulating concentration of any given hormone signal may be regulated at multiple points in a hormonal cascade, resulting in highly flexible hormonal systems. Adding even more complexity, hormonal axes often interact with one another. For example, the hypothalamic-pituitary-adrenal (HPA) axis, which functions in response to acute and chronic stress, interacts in complex ways with the HPG axis. Interactions between these two axes allow animals not only to modulate reproductive effort in response to stressors but also to modulate their sensitivity to stressors in response to reproductive state. As an example, responsiveness of the HPA axis to stressors can decline with age, allowing older individuals to invest relatively more in reproduction at a potential cost to investment in self-maintenance (e.g., growth or immunity).

Sources of Variation in Hormone-Mediated Phenotypes

Hormone-mediated behaviors often show noticeable variation within and among natural populations. Variation in hormone signal concentrations or hormone receptor distributions in target tissues are the most obvious determinants of this variation, but additional layers of complexity are almost always present. The activity of certain hormones is influenced by the presence of hormone-binding proteins, which may facilitate or inhibit the delivery of the hormone to target tissues, depending on the relative binding affinities of the hormone for the binding protein and the target hormone receptor. Similarly, metabolizing enzymes can act to alter hormone signal concentrations over time by affecting their half-life in the circulation or their conversion to related hormones that are active in different tissues. Testosterone, for example, is often converted to a form of estrogen in the brain, where it then binds to estrogen receptors to alter gene transcription.

Adding to the complexity of characterizing hormonal regulation of behavior is that this regulation is not unidirectional: behavioral experiences can also alter levels of hormones and their receptors. Thus, hormones influence behavior, but behavior also influences hormones. One example of behavioral effects on hormone levels is the *challenge hypothesis*. First proposed by John Wingfield, this hypothesis predicts that animals will respond to an aggressive challenge with an increase in testosterone. This prediction has been borne out in a wide array of vertebrate taxa and has even been applied to insects that respond to an aggressive challenge with an increase in juvenile hormone.

Of particular interest to evolutionary biologists is the possibility that individual variation in hormone-mediated phenotypes has a genetic basis and may thus serve as the raw material of evolution. Although studies quantifying heritable genetic variation in hormone levels or hormone-mediated traits are rare, it is clear from both artificial selection and pedigree studies that such variation does exist. Just as often, however, individual variation in hormones and hormonally mediated phenotypes may reflect environmental variation. Cues such as temperature, photoperiod, and food availability can alter phenotypes through the action of hormones. In these cases, hormones serve to mediate *phenotypic plasticity*—that is, to create different phenotypes from a single genotype. Birds fatten before they migrate, and become active at night rather than during the day, hamsters hibernate in winter, and many temperate-zone organisms breed when the days are long. These seasonal changes in phenotypic expression are examples of phenotypic plasticity modulated by the environment, and all have been shown to have a hormonal basis that responds to day length. Hormone-mediated plasticity can also facilitate shorter-term changes in behavior, for example, when a spike in adrenaline or glucocorticoids modulates "fight-or-flight" behavior in stressful situations. The role of hormones in facilitating plastic responses to variable environments is also likely to have a genetic basis that can evolve in response to selection. Thus, studying hormonal mechanisms may provide particular insight into how populations adapt to environmental change.

2. HORMONES AND PHENOTYPIC INTEGRATION

Hormonal Pleiotropy and Phenotypic Integration

Another reason that hormones are of interest to evolutionary biologists is their ability to influence the expression of suites of correlated traits. This property means that hormones are well suited to act as a physiological mechanism underlying *phenotypic integration*, the pattern of correlation among an organism's different traits. By analogy to *genetic pleiotropy*, the term used to describe the effect of an individual gene on multiple traits, an individual hormone's influences on multiple traits may also be thought of as a form of pleiotropy. The classic example of hormonal pleiotropy is testosterone, which has been experimentally and observationally linked to many behavioral, morphological, and physiological processes, including aggression, parental care, ornament expression, and immune function. Like genetic pleiotropy, this shared mechanism may result in trade-offs (e.g., between aggression and parental care).

Phenotypic integration interests evolutionary biologists for two reasons. First, when traits are genetically correlated with one another, they cannot evolve independently; selection acting on one trait will lead to evolutionary change in correlated traits as a by-product. Second, patterns of phenotypic integration often represent adaptive solutions to problems posed by an organism's environment. Ultimately, traits may be correlated because they work well together, and selection has favored mechanisms that cause their expression to be coordinated.

There are many examples of suites of traits that foster survival and reproduction more effectively when they are expressed in a coordinated fashion. Cryptic appearance, for example, is more likely to reduce predation if it is coupled with cryptic behavior. Bright colors may be more attractive to potential mates when associated with loud or complex and engaging displays. Infected animals may be more likely to survive an infection if they run a fever and become lethargic and anorexic. Each of these examples demonstrates how coexpression of behavior, physiology, and morphology can achieve adaptive solutions to problems such as avoiding predation, attracting a mate, or recovering from infection, and each may involve regulation by hormonal mechanisms.

Examples of associations among traits orchestrated by hormones to produce complex phenotypes come from a wide array of organisms, including beetles, butterflies, crickets, lizards, rodents, and songbirds. One theme that emerges is that hormones often act as mechanistic links in trade-offs in life histories. For example, allocation of time and energy to parenting can be in conflict with time and energy allocated to mating, reproduction, and self-maintenance. A common regulatory mechanism like a hormone may allow organisms to plastically adjust multiple traits in a coordinated fashion, allowing changes in allocation over short periods within a lifetime or over environments that vary in space and time.

Hormones and Correlated Evolution

As both integrators of multiple traits and environmentally sensitive links between genotype and phenotype, hormonal mechanisms are uniquely situated to play a particularly important role in shaping the evolution of complex traits. For example, selection of one trait in a hormone-mediated suite (e.g., aggressive behavior) could lead to correlated responses in other traits (e.g., parental behavior) owing to an altered profile of a shared hormonal mechanism (e.g., testosterone). Because of this potential to cause evolutionary change in traits that may not be directly favored by natural selection, hormones are often considered to act as evolutionary constraints. In extreme cases, such as when a hormone-mediated trait is under very strong selection, it is theoretically possible for maladaptive changes to occur in correlated traits. At the other extreme, however, hormone-mediated correlations can cause adaptation to proceed more quickly if selection favors simultaneous changes in many traits.

From another perspective, hormone-mediated suites of traits may be seen as adaptations that have been assembled by natural selection. When traits function together as a group, they are often subject to a type of natural selection called *correlational selection*. Such selection occurs when traits interact in their effects on fitness, as, for example, when the probability of surviving or reproducing depends less on the value of a single trait than on whether the values of two traits match. Thus, sexual ornaments may be advantageous only when coexpressed with large body size, and mismatches of ornament and body size could be detrimental. Correlational selection is likely to be common in nature, and although perhaps understudied in relation to other forms of selection, a few demonstrations in the wild have been reported, including a study of antipredator behavior and color patterns in garter snakes by Edmund Brodie III, and our own work involving body size and color patterns in dark-eyed juncos.

Evidence suggests that such hormone-mediated correlations may facilitate adaptation and diversification, allowing simultaneous shifts for groups of traits that are favored together in a new or changing environment. Comparative studies reveal that hormonal mechanisms and hormone-mediated phenotypic integration are highly conserved across taxa and over long evolutionary timescales. For example, the HPG axis described earlier is incredibly similar in most vertebrates, as are the types of traits regulated by the androgens it produces. However, studies of closely related species and populations also reveal that certain phenotypes can sometimes become dissociated from the systemic effects of hormone signal levels (e.g., via changes in receptor densities in a target tissue). For example, although many species of songbirds show decreases in parental care in response to testosterone, closely related species sometimes show insensitivity to testosterone. This insensitivity seems to be an evolved response to other aspects of the species' life history or environment, such as breeding season length. Such evolutionary change in the makeup of hormone-mediated suites is likely driven by shifts in the strength or direction of correlational selection, which may be caused by environmental change or colonization of a new habitat.

Ongoing research aims to more precisely understand the role that hormonal mechanisms play in both evolution within populations (i.e., microevolution) and large-scale patterns of evolutionary history (i.e., macroevolution). We discuss each in turn in the following sections.

3. HORMONES AND MICROEVOLUTION

Of deep interest to biologists studying microevolution are the relationships among phenotypes, environmental conditions, and fitness, including the mechanisms that generate and maintain variation within populations and lead to phenotypic changes within populations and among closely related populations. In recent decades, both experimental and observational studies have begun to characterize the evolutionary dynamics of hormonal mechanisms and hormone-mediated phenotypes in natural systems.

Phenotypic Engineering

One approach to establishing how selection operates on hormonal variation is to employ *phenotypic engineering*. By manipulating hormone levels or receptor-binding properties with pharmacological agents, it is possible to modify phenotypes and measure the resultant fitness consequences in field studies of free-living animals. This approach has been employed successfully to investigate the fitness consequences of elevated or reduced hormone levels in several natural systems.

Our own work (involving many collaborators over several decades) on dark-eyed juncos, a species of songbird, is an example of this approach. In these studies, testosterone implants were used to elevate hormone levels of male juncos to naturally high levels throughout the breeding season to document behavioral and physiological responses as well as to measure survival and reproductive success. Relative to controls, high-testosterone males sang more often and had higher mating success. The benefits of higher testosterone were accompanied by costs, however. Testosterone led to higher levels of stress hormones (corticosterone), suppressed immune activity, and decreased survival. These experimental results demonstrated that opposing selective forces (survival costs versus reproductive benefits) may act on testosterone and its associated traits. Similar phenotypic engineering approaches have been employed in several other systems, including amphibians, reptiles, and other birds.

Natural Variation and Fitness in the Wild

Despite the power of hormonal manipulations to experimentally demonstrate hormonal mechanisms and trade-offs, one limitation is that manipulations may obscure natural variation in hormones and hormone-mediated phenotypes as exhibited in nature. Natural variation among individuals, which may occur in hormone levels, receptor densities, or other aspects of hormone pathways, is required for selection to operate on suites of hormone-mediated traits.

Several recent studies of natural populations have found correlations between natural variation in hormone levels and both phenotypes and components of fitness. Examples of such covariation measured in wild populations include our own work examining testosterone, aggression, and parental behavior in juncos, as well as Maria Thaker's work on corticosterone, a glucocorticoid stress hormone, and predator escape behaviors in tree lizards. Several recent studies have shown relationships between corticosterone and survival or reproductive success in the wild. Taken together, these results indicate that hormone-mediated traits are often the targets of selection and that understanding the causes and consequences of natural variation in hormone pathways can be informative for understanding the evolution of natural populations.

Artificial Selection

Artificial selection studies conducted in the laboratory have also provided insight into the evolutionary importance of coordinating hormonal mechanisms. Animal breeders have known for centuries that selecting particular behavioral or morphological traits brings along suites of correlated phenotypes. Contemporary scientists have selected for specific behavioral traits or hormone levels and quantified the observed changes. For example, Kees van Oers and colleagues selected for "fast" and "slow" exploratory boldness behavior in male great tits (songbirds) and found correlated responses in testosterone and immune function in just a handful of generations. Similarly, when Suzanne Mills and colleagues selected on immune function in voles, they found it led to correlated responses in testosterone levels.

Such artificial selection studies reveal substantial heritable genetic variation in hormonal mechanisms, which could provide the raw material for selection to shape hormone-mediated phenotypes over short-term evolutionary timescales. Artificial selection studies also demonstrate the types of integrated evolutionary responses that may be expected for hormone-mediated suites in the wild.

Population Divergence in Hormone-Mediated Suites

Another approach to characterizing the microevolutionary dynamics of hormone-mediated traits is to examine variation in hormonal mechanisms among closely related populations. Comparing tropical versus temperate populations of the same species, for example, has revealed that hormonal profiles may differ dramatically within a species across environments, accompanied by correlated changes in life history and behavior. One such example comes from Brent Horton and colleagues, who recently showed that orange-crowned warblers breeding in Alaska exhibit higher early-peak and much more seasonally variable testosterone levels across the breeding season when compared with a population of the same species breeding off the coast of Southern California. Similarly, studies of urban versus wildland populations of the same species revealed that phenotypic divergence in hormone levels and hormone-mediated suites can occur over just a few generations in response to novel or changing environments. In some cases, major behavioral shifts may be required for population persistence in a novel or altered environment, and hormonal responses may accommodate and coordinate such behavioral change. "Common garden" studies of hormonal, behavioral, morphological, and physiological variation—in which individuals from different populations are raised under identical environmental conditions—have indicated that both phenotypic plasticity and genetic differences may underlie hormonally mediated variation among recently diverged populations. As examples, two distinct studies of urban versus wildland songbirds by Jesko Partecke (European blackbirds in Germany) and our research group (dark-eyed juncos in Southern California) have found that short-term elevation of stress hormones (i.e., corticosterone) in response to capture and handling was reduced in recently established urban songbird populations, and these differences persisted, in part, in subsequent common garden studies—indicating that both rapid evolution and phenotypic plasticity likely play a role in hormonal and behavioral divergence in these systems.

4. HORMONES AND MACROEVOLUTION

To study the long-term evolutionary significance of hormonal mechanisms, evolutionary biologists use the comparative method, studying patterns of diversity across taxa to make inferences about the origins and rates of diversification of endocrine systems and hormone-mediated traits. Such studies have revealed that molecules that act as hormones are relatively few in number and phylogenetically ancient. For example, several families of hormones, including vasopressin-oxytocin family peptides, gonadotropin-releasing peptides, and steroids such as androgens and estrogens are present in and serve similar regulatory functions across nearly all studied vertebrate taxa with only minor modifications in their molecular structure. We detail just two of the many examples of how the comparative approach has informed us about the role of hormones in macroevolutionary patterns.

Evolution of Hormones and Their Receptors

Because hormone-mediated phenotypes emerge from interactions between hormones and receptors, one particularly interesting question is, How do hormone-receptor molecules evolve to serve new functions? An example can be found in work by Joseph Thornton and colleagues on two classes of steroid hormones, mineralocorticoids and glucocorticoids. Each class of hormones plays key roles in electrolyte balance and in metabolic responses to stressors, respectively, and each has its own class of receptor, the mineralocorticoid receptor (MR) and the glucocorticoid receptor (GR). Both receptors act as transcription factors that bind to different regions of DNA. Comparative evidence by Thornton's group indicates that the genes that code for MR and GR arose through gene duplication. By comparing the sequences for MR and GR along different fish and tetrapod lineages, scientists have succeeded in re-creating the sequence of the ancestral corticoid receptor that preceded present-day MR and GR by 450 million years. By synthesizing the ancestral receptor in the lab and studying its ability to bind with extant hormones, they were able to show how sequential changes in the amino acid sequence of both classes of receptors affected their ability to bind with both ancient and more modern forms of their hormones. This unique experimental twist on the comparative approach has revealed not only how strongly form, function, and mechanisms of action of hormones are conserved, but also how evolutionary innovation has proceeded across distant taxa over vast timescales.

Comparative Studies of Hormones, Behavior, and Life History

The diversity of behavioral traits and complex phenotypes among closely related species has also been the subject of studies aiming to understand the physiological mechanisms underlying more recent evolution. Several such studies have revealed that relatively minor variations in neuroendocrine mechanisms underlie major behavioral and life history differences among closely related species. For example, a series of studies by Larry Young and colleagues on prairie voles, which are among the few percent of socially monogamous mammals, showed that vasopressin receptor gene expression in the brain's reward circuitry is greater in prairie voles versus nonmonogamous vole species. Furthermore, Young's group found that males of nonmonogamous vole species could be induced to express monogamous-like behavior simply by overexpressing the vasopressin receptor gene in this same brain area. Similarly, studies of closely related songbird species by James Goodson and colleagues showed that different patterns of neuropeptide activation of social brain nuclei underlie striking differences in social systems (e.g., territorial versus gregarious).

5. SUMMARY AND FUTURE DIRECTIONS

Hormonal mechanisms regulate the coordinated expression of multiple behavioral, physiological, and morphological phenotypes in ways that often seem to optimize individual fitness in the face of conflicting demands. Hormones are of particular interest to evolutionary biologists because they have pleiotropic effects and can modulate both genetic and plastic responses to environmental change for suites of traits. Although hormonal mechanisms themselves are quite conserved across taxa, specific traits and trade-offs mediated by these mechanisms show wide diversity across the animal kingdom. Artificial selection experiments and studies of evolution within and among natural populations have revealed the potential for hormone-mediated suites to evolve rapidly.

Despite a growing understanding of the role of hormones in the evolution of behavior and other complex phenotypes, there is still much to learn. For example, we know little about how readily hormone-mediated correlations may be assembled and dismantled by natural selection. Similarly, it is unclear whether hormonal mechanisms are more likely to limit or facilitate adaptation. Emerging experimental approaches in behavioral neuroendocrinology as well as continued development of genomic and bioinformatic tools should lead to a better understanding of the mechanisms by which hormones influence evolutionary change and phenotypic plasticity, and serve to produce a clearer picture of the evolutionary history of endocrine systems.

The principles described here may have practical applications. A few areas in which the intersection between hormones and evolution may provide insight include adaptation of natural populations to anthropogenic environmental change, host-pathogen dynamics including human disease, and the effects of endocrine-disrupting chemicals in the environment.

A key message of this chapter is that both the plastic responses of organisms and the genetic evolution of populations are not likely to involve single traits in isolation but rather correlated suites of traits that are often under hormonal regulation. Such hormonal mechanisms have likely evolved to successfully coordinate multiple behavioral, morphological, and physiological traits, but once established, may influence the evolutionary trajectory of populations. Thus, hormonal mechanisms have the potential to play a fundamental role in limiting or facilitating patterns of diversification of behavior and other complex traits. Understanding this role can provide unique insights into both contemporary microevolutionary processes and historical macroevolutionary patterns.

FURTHER READING

Adkins-Regan, E. 2005. Hormones and Animal Social Behavior. Princeton, NJ: Princeton University Press. *Provides a comprehensive and synthetic overview of the relationships among hormones, behavior, ecology, and evolution.*

Dean, A. M., and J. W. Thornton. 2007. Mechanistic approaches to the study of evolution: The functional synthesis. Nature Reviews Genetics 8: 675–688. *Advocates a synthesis of evolutionary biology and mechanistic molecular biology, reviewing work on hormone receptor evolution by Thornton and colleagues.*

Denver, R. J., P. M. Hopkins, S. D. McCormick, C. R. Propper, L. Riddiford, S. A. Sower, and J. C. Wingfield. 2009. Comparative endocrinology in the 21st century. Integrative and Comparative Biology 49: 339–348. *Emphasizes the role of endocrinology in understanding how organisms respond to global change.*

Goodson, J. L. 2005. The vertebrate social behavior network: Evolutionary themes and variations. Hormones and Behavior 48: 11–22. *Reviews work showing that subtle changes in deeply conserved brain networks can underlie evolutionary divergence of social systems.*

Hau, M., and J. C. Wingfield. 2011. Hormonally regulated trade-offs: Evolutionary variability and phenotypic plasticity in testosterone signaling pathways. In T. Heyland and A. Flatt, eds., Molecular Mechanisms of Life History Evolution. Oxford: Oxford University Press. *Presents a detailed analysis of organismal and evolutionary mechanisms related to testosterone-mediated trade-offs in life history and behavior.*

Ketterson, E. D., J. W. Atwell, and J. W. McGlothlin. 2009. Phenotypic integration and independence: Hormones, performance, and response to environmental change. Integrative and Comparative Biology 49: 365–379. *Examines how suites of traits mediated by hormonal mechanisms may respond developmentally or evolutionarily as integrated or independent units, with many examples from vertebrate systems.*

Ketterson, E. D., and V. Nolan Jr. 1999. Adaptation, exaptation, and constraint: A hormonal perspective. American Naturalist 153: S4–S25. *Refines theory and summarizes a series of landmark field experiments examining how hormonal pleiotropy may constrain (or facilitate) trait evolution.*

McGlothlin, J. W., and E. D. Ketterson. Hormone-mediated suites as adaptations and evolutionary constraints. Philosophical Transactions of the Royal Society B 363: 1161–1620. *Uses a quantitative genetic perspective to conceptualize how correlated traits mediated by common hormonal signals are predicted to evolve.*

Nelson, R. J. 2011. An Introduction to Behavioral Endocrinology. 4th ed. Sunderland, MA: Sinauer. *Provides a textbook overview of the discipline, including a primer on hormonal mechanism, as well as details and examples from several hormone-behavior systems including sex behavior and sex differences.*

West-Eberhard, M. J. 2003. Developmental Plasticity and Evolution. Oxford: Oxford University Press. *The first volume to comprehensively examine the interplay between developmental biology (including phenotypic plasticity) and mechanisms of evolutionary (genetic) change.*

Williams, T. D. 2012. Physiological Adaptations for Breeding in Birds. Princeton, NJ: Princeton University Press. *Provides a review and prospectus of the roles of hormones in avian reproduction, with a strong grounding in life history evolution.*

Zera, A. J., L. G. Harshman, and T. D. Williams. 2007. Evolutionary endocrinology: The developing synthesis between endocrinology and evolutionary genetics. Annual Review of Ecology and Systematics 38: 793–817. *Presents a past, present, and future view of how studies of variation in hormonal systems are informing our understanding of microevolutionary processes including many nice examples from insect systems.*

VII.3

Game Theory and Behavior

John M. McNamara

OUTLINE

1. The basic ideas
2. Examples
3. Issues for consideration
4. Applications
5. Future directions

GLOSSARY

Best Response. A strategy that has the highest fitness given the resident strategy.

Convergence Stable. A strategy that is a local attractor under evolutionary dynamics; that is, the population strategy will evolve to this strategy if it is initially close to it.

Evolutionarily Stable Strategy (ESS). A resident strategy such that any mutant strategy (different from that of the resident) is selected against while rare and so cannot increase in frequency.

Frequency Dependence. The dependence of fitness of a given strategy on the frequency of other strategies in a population.

Mixed Strategy. A strategy in which actions are chosen probabilistically.

Monomorphic Population. A population in which only one (genetically determined) strategy is present.

Mutant Strategy. Any strategy that can arise (as a mutation); usually thought of as rare compared with the resident strategy.

Payoff Matrix. A table specifying how the increment to fitness as a result of playing a game depends on the action of the individual and that of its opponent(s).

Polymorphic Population. A population in which more than one (genetically determined) strategy is present.

Resident Strategy. A strategy that is used by almost all population members.

Symmetric Game. A game in which all players have the same set of possible actions and the same payoff matrix, and there are no role asymmetries.

1. THE BASIC IDEAS

The two main motivating ideas of evolutionary game theory are described in this section. These ideas are, however, just the basics; other necessary ingredients are discussed in section 3.

Frequency Dependence

Suppose that the environment experienced by a species is constant over many generations. Then, natural selection can be envisaged as a hill-climbing process in which species members become fitter over evolutionary time. The end point of this process is that species members will behave so as to approximately maximize fitness within this constant environment. This idea of fitness maximization is useful, since it allows us to predict and understand end points of the process of evolution without considering details of the dynamic process that led to the "top of the hill."

However, the fitness of members of a population typically depends on the behavioral strategies of other members of the population as well as on their own behavior. For example, suppose that a foraging animal contests a food item with another population member. Then, it will be worth the animal's being aggressive and attacking its rival if the rival is not aggressive and liable to run away if attacked; but if the rival is also liable to be aggressive, it may be better to be nonaggressive. The best strategy for the focal animal thus depends on the frequency of aggressive individuals in the population; that is, fitness, and hence the action of selection, is *frequency dependent.*

Even if the physical environment remains constant, when selection is frequency dependent, the "environment" experienced by members of a species changes over evolutionary time as the strategies of members of the species evolve. Thus as natural selection tends to hill climb, the shape of the hill and the position and height of its summit change. *Evolutionary game* theory is concerned with the outcome of evolution by natural selection in this more complex setting.

Evolutionary End Points

It is not clear that the evolutionary process will settle down to some end point when there is frequency-dependent selection; for example, strategies might cycle in their distribution of occurrence. Most of evolutionary game theory is, however, concerned with situations in which stable end points are achieved. The theory attempts to characterize the properties of possible end points.

To analyze outcomes when there is frequency dependence we use the following notation. Suppose that almost all members of a population follow a genetically determined strategy π. We then refer to π as the *resident strategy*. Let π' be some possibly different strategy that is rare in occurrence compared with the resident strategy. We denote the fitness of this mutant strategy by $W(\pi', \pi)$. This formalism thus explicitly expresses the fitness of an individual as a function of the resident strategy as well as its own strategy.

We expect any end point to be *uninvadable*; that is, no *mutant* strategy can invade the population under the action of natural selection. To formalize this concept, suppose that the resident strategy is π^*. Then, for π^* to be uninvadable it is necessary that

$$W(\pi, \pi^*) \leq W(\pi^*, \pi^*) \text{ for all possible mutants } \pi. \quad (1)$$

That is, when the resident strategy is π^*, the fitness of residents is at least as great as that of any mutants. As we explain later, although this condition is necessary for uninvadability, it is not sufficient and needs to be strengthened.

Game theory has its roots in economics, where property (1) is the defining condition for π^* to be a *Nash equilibrium*; the theory developed somewhat independently in biology. The concept of uninvadability is present in the early work of William D. Hamilton on stable sex ratios. Later, John Maynard Smith and George Price (1973) formalized the concept mathematically. Nowadays there is much mutual interplay between the fields of economic and evolutionary game theory.

Table 1. Payoff matrix for the Hawk-Dove game in which two population members contest a resource of value V

Payoff (fitness increment) to focal individual	*Opponent chooses hawk*	*Opponent chooses dove*
Focal chooses hawk	$\frac{1}{2}V - \frac{1}{2}C$	V
Focal chooses dove	0	$\frac{1}{2}V$

2. EXAMPLES

The following are classic games that illustrate the basic principles of evolutionary game theory.

The Hawk-Dove Game

In their seminal paper Maynard Smith and Price modeled the behavior of two individuals that contest a resource such as a food item or mate. It is assumed that gaining the resource increases fitness by V. The model sought to explain why contests are often settled by ritualized display rather than all-out fighting. In their interaction each contestant can take one of two actions, hawk or dove. Hawks attack their opponent; doves display to their opponent and run away once attacked. The possible outcomes are as follows. When two doves interact, they share the resource. If a hawk interacts with a dove, the hawk gets the resource, since the hawk attacks and the dove runs away. If two hawks interact, they fight, and each wins the fight with probability 1/2; the winner obtains the resource, and the loser pays a fitness cost C representing the loss of fitness as a result of injuries incurred in the fight. Table 1 summarizes these fitness consequences to one (focal) individual. The opponent has exactly the same *payoff matrix*. We assume that the cost of injury exceeds the value of the resource ($V < C$).

For this game we may take a strategy π to specify the probability of playing hawk; that is, an individual following strategy π plays hawk with probability π and dove with probability $1 - \pi$. Consider the payoff to a mutant that uses strategy π' when the resident strategy is π. There are four possibilities in a contest between the mutant and a resident: (1) mutant plays hawk and resident plays hawk (a combination of actions that occurs with probability $\pi'\pi$), (2) mutant plays hawk and resident plays dove (occurs with probability $\pi'(1 - \pi)$), (3) mutant plays dove and resident plays hawk (occurs with probability $(1 - \pi')\pi$), and (4) mutant plays dove and resident plays dove (occurs with probability $(1 - \pi')(1 - \pi)$). Averaging over the possibilities and taking the associated payoffs into consideration (table 1), we see that the mean payoff to the mutant is

$$W(\pi', \pi) = \pi'\pi(\tfrac{1}{2}V - \tfrac{1}{2}C) + \pi'(1-\pi)V + (1-\pi')\pi \times 0 + (1-\pi')(1-\pi)\tfrac{1}{2}V.$$

Rearranging terms, we can rewrite the preceding equation as

$$W(\pi', \pi) = (1-\pi)\tfrac{1}{2}V + \frac{C}{2}\left[\frac{V}{C} - \pi\right]\pi'.$$

The difference between the payoff to the mutant and that to a resident is thus

$$W(\pi', \pi) - W(\pi, \pi) = \frac{C}{2}\left[\frac{V}{C} - \pi\right](\pi' - \pi). \quad (2)$$

We are now in a position to find the Nash equilibrium for this game. If $\pi < V/C$, then the right-hand side of expression 2 is positive provided $\pi' > \pi$. Thus any mutant that is more aggressive than the resident strategy can invade the population. If $\pi > V/C$, then expression 2 is positive provided $\pi' < \pi$. Thus any mutant that is less aggressive than the resident strategy can invade. Finally, if $\pi = V/C$, then expression 2 is zero for all π. Thus when the resident strategy is to play hawk with probability $\pi^* = V/C$, all possible mutants are equally fit and have the same fitness as the residents. Thus equation 1 is satisfied, and the strategy $\pi^* = V/C$ is a Nash equilibrium for this game. This is the only Nash equilibrium.

A Sex-Ratio Game

Suppose that female members of a population can control the sex of their offspring; should they produce sons or daughters? To consider this question in its simplest setting we consider a large population with discrete nonoverlapping generations: individuals born in one year reach maturity the next year, attempt to breed, and then die. During breeding each female mates with a single male chosen at random from the males present. The strategy of a female specifies the proportion of offspring that are sons; a female with strategy π produces a proportion π sons and a proportion $1 - \pi$ daughters. Again, for simplicity, we suppose that the number of offspring that survive to maturity does not depend on their sex.

With these assumptions, the sex of offspring does not affect the number that survive to maturity but can affect the total numbers of matings obtained by these offspring and, hence, affects the number of grandchildren produced. We take the payoff to a female to be the number of matings obtained by her offspring.

If the resident population strategy is to produce more sons than daughters ($\pi^* > 1/2$), then the breeding population is male biased. Thus every male in the breeding population gets less than one mating on average. Since breeding females get exactly one mating, breeding females get more matings than males. It follows that a mutant female that produces all daughters will get a bigger payoff than resident females, so the population is invadable. If the resident strategy is to produce more daughters than sons ($\pi^* < 1/2$), then each breeding male gets more than one mating on average. Thus a mutant female that produces all sons will get a higher payoff than resident females, and the population is again invadable. Finally, suppose that the resident strategy is to produce equal numbers of sons and daughters ($\pi^* = 1/2$). Then, it can be seen that any mutant female that does otherwise will do exactly as well as resident females. Thus the strategy of producing equal number of sons and daughters satisfies condition 1 and is hence a Nash equilibrium.

3. ISSUES FOR CONSIDERATION

The preceding simple examples introduced some basics of evolutionary game theory, but many details and subtleties were passed over in describing them. Some of these issues are discussed next.

Strategies

Strategies are rules for choosing actions. As the Hawk-Dove game illustrates, we may allow actions to be chosen probabilistically. In the Hawk-Dove game a strategy is then simply a rule specifying the probability with which each action is chosen. In this game the Nash equilibrium is a *mixed strategy* in that both hawk and dove are chosen with positive probability. If instead we had constrained population members to adopt deterministic strategies, so that each either always played hawk or always played dove, then we would have predicted that the population would evolve to be *polymorphic*, in which a proportion $1 - (V/C)$ of individuals follow the strategy of always playing dove, and a proportion V/C follow the strategy of always playing hawk.

A strategy specifies the action chosen in every possible circumstance. For example, in a generalization of the Hawk-Dove game we might allow individuals to differ in fighting ability; an individual knows its own ability but not that of an opponent. A strategy for this game would specify how the probability of playing hawk depended on fighting ability. We would expect a single Nash equilibrium for this game to be of the following form: if ability exceeds a given threshold, then play hawk; otherwise play dove. Thus the existence of an underlying state (fighting ability) has changed the Nash equilibrium from a mixed strategy to one in which there are deterministic but contingent decisions. In nature most decisions are likely to be based on state rather being determined probabilistically, so the mixed-strategy Nash equilibria of much of game theory are really idealized simplifications.

Payoffs

The *payoffs* in a game are intended to represent the changes in fitness as a result of playing the game. Thus if a contestant receives the resource in the Hawk-Dove game, its fitness is increased by V. As this game illustrates, the payoff an individual receives can vary; for example, it can depend on the action chosen by the opponent. In an analysis of a game the *mean* (average) payoff is the relevant factor (see the discussion of fitness).

In considering a game-theoretical situation it is usual to specify payoffs in advance; however, this approach isolates a single play of a particular game from other parts of an individual's life. This tactic is often artificial and can be misleading. For example, consider a game in which parents are caring for common young and each must decide whether to desert the young, leaving the partner to care for them. Suppose that the male benefits from desertion because he can attempt to find a single female and initiate another breeding attempt with her. However, his chances of finding another single female depend on how many have become available by deserting their young, and on the competition from other males that have deserted. Thus the payoff depends on the outcome of the games played by other population members. In this scenario the payoff of desertion for the male cannot be specified in advance; instead, it emerges from a holistic view of what is happening during the breeding season.

Fitness

It is usually envisaged that strategies are genetically determined and so are inherited. Natural selection acts to change the frequencies with which strategies occur. Also, the population is seen as large, with many copies of each possible genotype, so that it is possible to average; that is, the change in the frequency of strategies is determined by a suitable measure of their *average* reproductive success. It is this measure that is referred to as *fitness*. Thus, although payoffs are usually assigned to combinations of actions, in evolutionary terms what matters is the average payoff under each strategy.

In the simplest setting the fitness of a strategy is the mean number of surviving offspring produced over the lifetime of an individual that follows the strategy. In some situations, however, this measure is not adequate. For example, in the sex-ratio game outlined earlier, all females produce the same number of surviving offspring but may differ in terms of the numbers of matings obtained by these offspring and, hence, differ in terms of the numbers of their grandchildren. The mean number of surviving grandchildren is thus an appropriate fitness measure. If females differ in their phenotypic quality (for example, in dominance status), and offspring tend to inherit the quality of their mother, then even this fitness measure may not be adequate; fitness may need to be based on the mean number of great-grandchildren or beyond. The preceding fitness measures are all based on mean descendant numbers. In other situations the variance in descendant numbers may also be important.

Invasion of Mutants

The Nash equilibrium condition (inequality 1) is necessary for evolutionary stability but does not guarantee it. This condition states that if resident population members follow the Nash equilibrium strategy, then no rare mutant has higher fitness than that of the residents; but this condition does not specify what happens when a mutant has fitness equal to that of the residents. For example, the Hawk-Dove game and the sex-ratio game are both examples in which all possible rare mutants do as well as the residents at the Nash equilibrium. If rare mutants do as well as the residents, the proportion of mutants could increase by genetic drift. Furthermore, once mutants start to become common, the fitness of all population members may change, and mutants may become fitter than residents.

Motivated by these considerations Maynard Smith defined π^* to be an *evolutionarily stable strategy* (ESS) if for every mutant strategy $\pi \neq \pi^*$ either

$$(\mathrm{ES1})\, W(\pi^*, \pi^*)$$

or

$$(\mathrm{ES2})(\mathrm{i})\; W(\pi, \pi^*) = W(\pi^*, \pi^*)$$

and

$$(\mathrm{ii})\; W(\pi, \pi) < W(\pi^*, \pi).$$

This definition of an ESS applies to a two-player game but must be modified to deal with "games against the field" such as the sex-ratio game.

For two-player games the idea behind the definition is as follows. A mutant is certainly selected against when rare if it has a strictly lower payoff against residents than residents do when playing against one another (condition ES1). Suppose, however, that a mutant does equally well when rare [condition ES2(i)]. Then, if mutant numbers start to increase by drift, they will play against one another as well as against residents. Mutants will then be selected against if they do less well than residents in interactions with other mutants [condition ES2(ii)].

In the Hawk-Dove game the Nash equilibrium strategy is to play hawk with probability V/C. This strategy is also the unique ESS for this game.

Adaptive Dynamics and Convergence Stability

The ESS condition is concerned with determining whether a population adopting a particular resident strategy π^* can be invaded but says nothing about whether strategy π^* will evolve in the first place. To analyze this latter question we first have to specify how the resident strategy changes as the population evolves; in other words, we have to specify the evolutionary dynamics. The exact dynamics depend on assumptions about the underlying genetic system; however, one especially simple approach known as *adaptive dynamics* ignores genetic detail and assumes that selection acts to locally hill climb. Given this dynamic we may consider whether a population that initially has some resident strategy π will evolve so that the resident strategy becomes π^*. If this occurs for every π close to π^*, then we can think of strategy π^* as *convergence stable.*

The ESS for the Hawk-Dove game is convergence stable, but many games have at least one ESS that is not. Conversely, a strategy π^* may be convergence stable but not an ESS. In such cases evolution leads toward π^*, but at π^* the resident strategy is at a local fitness minimum. Mutants with both smaller and greater trait values than the resident strategy can then invade, producing disruptive selection, with the possibility of evolutionary branching.

Population Fitness and Evolutionary Stability

In the Hawk-Dove game there are two contestants per contest and one resource of value V. Thus whatever strategies are used, the average value of resources obtained per individual is ½V; however, the average payoff may be lower than this owing to the cost of fighting. If all individuals play dove with probability 1, then there is no cost of fighting, and the average payoff is ½V. If instead all individuals play the ESS strategy, then there is fighting, and the average payoff is less than ½V. Thus the average payoff per individual is not maximized at the ESS.

Game theory provides many such examples in which fitness is not maximized at evolutionary stability: individuals are all doing the best for themselves, but this is not best for the population as a whole. In economic game theory an analogous idea is captured in the example of the "tragedy of the common."

Role Asymmetries

The basic Hawk-Dove game is *symmetric*—there are no a priori differences in the two contestants. In contrast, consider a version of this game in which the resource contested is ownership of a territory; one individual is the current territory owner and the opponent is an intruder, with both contestants aware of their roles. Otherwise the specification of actions and payoffs are as before. For this version of the game a strategy is a rule specifying two probabilities: the probability of playing hawk when a territory owner and the probability of playing hawk when an intruder. In the original game the unique Nash equilibrium was to play hawk with probability V/C. This strategy is an ESS. In the new game there are other possibilities. For example, if all owners play hawk, then the *best response* in the role of intruder is to play dove. Conversely, if all intruders play dove, the best response in the role of owner is to play hawk. Thus the strategy always to play hawk when owner and always to play dove when intruder is a Nash equilibrium strategy. There are actually three Nash equilibria: (1) play hawk with probability V/C regardless of role; (2) play hawk with probability 1 when owner, and dove with probability 1 when intruder; and (3) play hawk with probability 1 when intruder, and dove with probability 1 when owner. Of these Nash equilibria the first is not an ESS; the other two are ESSs.

The preceding example illustrates a general property of games: the mere presence of asymmetry, even if it has no effect on payoffs (e.g., "owner" or "intruder"), can completely alter the predicted evolutionary outcome.

The Importance of Process

The predicted outcome of a game depends not just on the payoff structure but also on the process by which decisions are reached. For example, consider a game between parents over the care of their common young (see chapter VII.8). Each parent decides whether to care for the young. Suppose that the payoffs are as shown in table 2. These payoffs are motivated by the idea that the young do better if cared for by both parents rather than by one (and die if they receive no care), although there are costs of care. In addition, the male can get benefits from deserting (e.g., by obtaining extra-pair copulations). Note that for these payoffs the best action of the female is to care for the young whatever the action of the male.

Consider first a version of the game in which both parents decided on their action without knowing the action of their partner. Suppose that the resident strategy is for the male to always desert and the female always to care. Then, no individual can do as well if it changes its

Table 2. Payoff matrices for a game in which each of two parents decides whether to care for or to desert their common young

Payoff to the male	*Female cares*	*Female deserts*
Male cares	6	3
Male deserts	7	2
Payoff to the female	*Male cares*	*Male deserts*
Female cares	6	3
Female deserts	5	0

action. Thus the resident strategy is a Nash equilibrium (it is also an ESS). Furthermore, it is easy to see that it is the only Nash equilibrium. Thus we predict female-only care.

In contrast, suppose that the female decides on her action first and then the male decides on his action knowing the female's action. For this game a strategy for the female just specifies her choice of action. A strategy for the male is a contingent rule specifying what action to take if the female is caring and what action to take if she has deserted. Consider a resident population in which all females desert, and all males adopt the strategy to care if the female deserts, and desert if the female cares. In this population there is male-only care. It can be seen that this population is at a Nash equilibrium: resident males are doing the best given whatever action is taken by the female; resident females get a payoff of 5 (since males care) but would get a lower payoff of 3 if they cared (since then males would desert).

This example illustrates that the decision process is crucial to the outcome predicted. When parents are ignorant of each other's actions, there is female-only care, and the male gets a payoff of 7. When the female chooses first, there is male-only care, and the male gets a payoff of 3. Thus knowing the action of the female lowers the male's predicted payoff. This is an unexpected feature of behavior that game theory has uncovered—having information may put an individual at a disadvantage.

4. APPLICATIONS

Frequency dependence occurs in many aspects of an organism's life. It is therefore not surprising that evolutionary game theory has been applied to a huge range of problems, including territorial behavior, the distribution of animals across resources, antipredator behavior, and foraging games such as those between producers and scroungers. Some further applications are described next in more detail.

Mate Choice

Suppose that males differ in some aspects of their quality, and females are choosy about their partner (see chapter VII.6). If males are no longer available to other females once chosen, there will be competition for the best males. Partner search by females will then be frequency dependent: the best strategy for a female will depend on the search and choice strategies adopted by other females.

If males are also choosy, partner choice has another game-theoretical aspect. How choosy an individual should be about his or her partner will depend on the choosiness of members of the opposite sex, as well as on his or her own quality. At a game-theoretical equilibrium the choice of strategy for males will be the best given the strategy of females, and vice versa. At this equilibrium there is liable to be assortative pairing in the population.

Parental Effort

In species with biparental care of offspring there are costs and benefits associated with parental effort. Increased care by either parent increases the prospects of the young and hence contributes to the fitness of both parents; however, an increase in level of parental care is usually costly to that parent. One obvious cost to males is the loss of opportunity to have extra-pair matings with other females. Care can also have mortality costs. For example, obtaining food for the young may incur an extra risk of predation or may require the parent to work so hard its condition suffers, and it becomes more prone to dying of disease. Since there is a joint benefit (the offspring), but costs are paid individually, there is a conflict of interest: each parent is better off if the other parent does most of the care.

The predicted level of care can be analyzed using game theory because the optimal level of care by one parent depends on the level of care provided by the other parent. Predictions depend on the assumptions made. In the simplest setting we might assume that the level of care is genetically determined, and individuals do not respond to the amount of care provided by the partner or to the state of the offspring. If instead we allow more flexible strategies, with individuals responding to circumstances, predicted levels of care are liable to be very different.

Alternative Male Mating Strategies

In some species, males in a population adopt different mating strategies (see chapter VII.4). In the Pacific salmon, males can mature as a "hook" or as a "jack." Hooks are large individuals that defend territories and attempt to attract females to their territories to mate. Jacks are smaller, mobile individuals that attempt to sneak matings with females in the territories of hooks. The advantage of being a jack is frequency dependent; if there are many jacks, they compete with one another, and there are fewer territories. Evolutionary game theory thus provides a suitable tool for analyzing this situation.

The three examples considered all concern reproductive behavior, an area that provides the motivation for many games. Furthermore, there is often interaction among these games, so a holistic approach is required. For example, whether females should prefer a particular male may depend on the parental effort he will provide. How much effort a male should provide depends on his chances of attracting a new mate should he cease care. Thus mate choice games and parental effort games are linked.

The Evolution of Cooperation

The action of natural selection produces population members that do the best for themselves (or their genes), so it is not immediately apparent why we see cooperation between unrelated individuals in natural populations (see chapters VII.9 and VII.10). Work in this area has focused on the issue of why an individual might help another at a cost to itself, and the Prisoner's Dilemma game has often been used to capture the underlying conundrum. In this symmetric two-player game each player can either cooperate or defect. In the simplest version of this game cooperation benefits the opponent by an amount b but costs the cooperator an amount c, where $b > c$. The payoff matrix is given in table 3. As can be seen, the best response of each player is to defect regardless of the action chosen by the other player. Thus the only Nash equilibrium is for both players to defect. Consequently, both receive a payoff of 0, whereas both could have received the larger payoff $b - c$ had they both cooperated. Thus cooperation is mutually advantageous in this game but will not evolve, since a population with cooperation as the resident strategy can be invaded by an uncooperative mutant.

Various mechanisms that promote cooperation have been proposed. The mechanism of *direct reciprocity* involves population members helping those who have previously helped them (and possibly punishing those who have not given help). This mechanism has often been modeled using the Prisoner's Dilemma game; population members pair up and play repeated rounds of this game against one another. In this interaction the action of an individual is allowed to depend on the actions chosen by itself and its opponent in previous rounds.

Table 3. The payoff matrix for a version of the Prisoner's Dilemma game

Payoff to player 1	*Player 2 cooperates*	*Player 2 defects*
Player 1 cooperates	$b-c$	$-c$
Player 1 defects	b	0

The mechanism of *indirect reciprocity* involves helping those who have helped others in the past. The mechanism of *generalized reciprocity* involves helping another if help has been received in the past. A common feature of explanations of the evolution of cooperative behavior is that the mechanism tends to produce an assortment in which cooperators tend to interact with other cooperators.

5. FUTURE DIRECTIONS

Evolutionary game theory will continue to be applied to a range of phenomena in biology. In past applications games were largely schematic and simple, which allowed biologists to capture and understand some aspects of the world. It is increasingly being recognized, however, that important aspects that could radically alter predictions were ignored. Some of these features requiring further investigation are as follows: (1) A holistic view needs to be taken in which payoffs are not just imposed but arise as a result of embedding the game in an ecological context. (2) Game theory has largely ignored the existence of variability in natural populations and has effectively treated variability as noise. The type and amount of variability can, however, be crucial in determining the predictions of game theory. (3) Most games have assumed contestants choose a single action without knowing how the opponent(s) chose. However, the process by which individuals interact and hence reach decisions is critical to predictions. Game theory needs to pay more attention to this process. (4) Individuals are not completely flexible in their behavior; instead, they have behavioral consistencies both within and across contexts. In this sense a range of animal personalities exist within natural populations. Researchers need to better understand how much of this variation can be explained using game theory. They also need to explore the consequences of the lack of flexibility for the predictions of game theory. (5) Game theorists need to pay more attention to the decision-making mechanisms employed by animals. For example, emotions such as anger and mental states such as trust are important in human decision making, and it could be profitable to incorporate these mental states more explicitly into models.

FURTHER READING

Diekmann, O. 2004. A beginner's guide to adaptive dynamics. Banach Center Publications 63: 47–86. *Adaptive dynamics is now an important tool in game theory.*

Gintis, H. 2000. Game Theory Evolving: A Problem-Centered Introduction to Modeling Strategic Interaction. Princeton, NJ: Princeton University Press. *Some nice examples of simple games, from biology and other areas.*

Houston, A. I., and J. M. McNamara. 1999. Models of Adaptive Behaviour: An Approach Based on State. Cambridge: Cambridge University Press. *Chapter 7 provides examples of games in which players take a sequence of actions that depend on state and time.*

Maynard Smith, J. 1982. Evolution and the Theory of Games. Cambridge: Cambridge University Press. *A classic general introduction, although incomplete, as it does not contain more recent developments such as adaptive dynamics.*

McNamara, J. M., and O. Leimar. 2010. Variation and the response to variation as a basis of successful cooperation. Philosophical Transactions of the Royal Society B 365: 2627–2633. *Emphasizes how between-individual variation in population members can completely change the predictions of game theory.*

McNamara, J. M., and F. J. Weissing. 2010. Evolutionary Game Theory. In T. Szekely, A. J. Moore, and J. Komdeur, eds., Social Behaviour: Genes, Ecology and Evolution. Cambridge: Cambridge University Press. *A general overview that goes into more depth on some issues than this chapter does.*

Oliveira, R., M. Taborsky, and H. J. Brockmann. 2008. Alternative Reproductive Tactics: An Integrative Approach. Cambridge: Cambridge University Press.

Reeve, H. K., and L. A. Dugatkin, eds. 1998. Game Theory and Animal Behavior. Oxford: Oxford University Press. *Examples of the application of game theory to a variety of biological phenomena.*

Sinervo, B., and R. Calsbeek. 2006. The developmental, physiological, neural, and genetical causes and consequences of frequency-dependent selection in the wild. Annual Review of Ecology and Systematics 37: 581–610.

Smith, J. M., and G. Price. 1973. The logic of animal conflict. Nature 246: 15–18. *The original article that formalized game theory in biology.*

van Doorn, G. S., G. M. Hengeveld, and F. J. Weissing. 2003. The evolution of social dominance. Behaviors 140: 1305–1358.

VII.4

Sexual Selection and Its Impact on Mating Systems

Rhonda R. Snook

OUTLINE

1. What are mating systems and why are they important?
2. Measures of mating systems
3. Plastic, continuous mating systems and the evolution of behavior
4. Mating systems and evolutionary potential
5. Applied relevance of the study of the evolution of mating systems

In sexually reproducing animals, males and females interact for mating, and the way in which they interact defines an organism's mating system. Mating systems reflect the action of sexual selection, including sexual conflict, and can be quantified. Mating systems also determine a population's evolutionary potential and perhaps its ability to respond to anthropogenic changes.

GLOSSARY

Effective Population Size (N_e). The number of individuals that can contribute genes equally to the next generation. This may not equal the census population size.

Environmental Intensity of Sexual Selection. A measure of the degree of competition for mates, which is dependent on the potential reproductive rate (PRR) of each sex and which influences the operational sex ratio (OSR).

Operational Sex Ratio (OSR). Average ratio of fertilizable females (or more generally, the limiting sex) to sexually active males (or more generally, the competing sex) at any given time.

Opportunity for Sexual Selection. Frequently used as a synonym for the intensity of sexual selection but more correctly defined by an equation in which it is equal to the total variance in reproductive success divided by the square of the mean reproductive success, that is, I_{mates}.

Potential for Polygamy (EPP). The degree to which multiple mates, or resources critical to gaining multiple mates, are economically defendable.

Potential Reproductive Rate (PRR). Has multiple definitions; the maximum number of independent offspring that parents can produce per unit time when mating partners are freely available but all other constraints in terms of environmental factors (such as food, number and sizes of nest sites, temperature) remain. The sexual difference in PRR in the population then influences the OSR and has to be estimated experimentally for a sample of the population.

Strength of Sexual Selection. Frequently used as a synonym for intensity of sexual selection.

1. WHAT ARE MATING SYSTEMS AND WHY ARE THEY IMPORTANT?

The simplest definition of a *mating system* for either plants or animals is a description of who mates with whom and when. Beyond this general description, however, discussions of animal and plant mating systems diverge. In plants, genetic relationships define the mating system and capture the degree of outcrossing (see chapter IV.8). In animals, mating system descriptions include the general behavioral strategy employed in obtaining mates, encompassing not only the number of mates acquired and when but other features such as the manner of mate acquisition and the patterns of parental care provided by each sex. Despite such a simple general definition, understanding the evolution of mating systems is complex. This complexity arises because mating system descriptions (e.g., monandry, polygyny; table 1) are idealized, and their definitions are often inaccurate; social and genetic mating systems are not necessarily exclusively the same, and the same species can exhibit plasticity of mating systems across or even within populations. Moreover, methods for quantifying the nature of selection that contributes to the evolution of mating systems are contentious.

The study of mating systems, for both animals and plants, began—like much of the study of evolution—with Charles Darwin. In this context, Darwin contributed his second idea about how evolution works, that of sexual selection. Darwin recognized that the marked sexual dimorphism—one of the most striking and easily observed of natural phenomena—seen in many animals could be contrary to the action of natural selection and thus would undermine his theory of evolutionary change. To resolve this conundrum, he suggested that individuals struggle not only to exist but also to acquire mates, and this selection results in sex-limited traits that are used to either compete for or choose mates. Developments in the 1970s and 1990s extended this arena of competition to sperm-egg interactions, so the modern definition of sexual selection includes that arising owing to competition not only for access to mates but also for fertilization opportunities (see chapters VII.5 and VII.6). The inherent differences between the sexes can lead to conflicts of interest between males and females in evolutionary outcomes; this sexual conflict is now embedded in sexual selection and the evolution of mating systems.

Darwin described animal mating systems using the now-familiar descriptions of monogamy and polygamy. He also intuitively realized that polygamous species, relative to monogamous species, should experience stronger sexual selection because there is a greater opportunity for increased variation in mating success among males in polygamous species. Yet he also believed that most animals exhibited monogamy. Darwin's prejudiced notion leads to an illogical conclusion using his ideas. If the intensity of sexual selection is positively related to the extent of sexual dimorphism, and sexual dimorphism is so conspicuous, then why would most taxa be monogamous? Biologists now recognize that Darwin was wrong about the extent of monogamy; through better observation and molecular parentage tools—and perhaps a more accepting society—most animal systems are now known to be polygamous to some degree. However, this fact raises another question. Since we now generally have a catalog of the mating systems of taxa, what is the point of modern studies of the evolution of mating systems? What is left to understand?

There are several research questions that are fundamental, but unresolved, regarding the evolution of mating systems; we focus on three. First, while it is generally agreed that the intensity of sexual selection influences the mating system, how to measure sexual selection and therefore how to quantify and predict which mating system will be observed is controversial. Second, mating systems can affect the evolutionary potential of a population through a variety of mechanisms. However, these effects are predominantly theoretical and tested across a limited set of taxa. Third, anthropogenic changes via spatiotemporal changes in resource availability may alter mating systems. Such changes in the mating system can alter genetic variability, potentially affecting the ability of a population to respond to environmental change, but few studies have assessed the relationship among these factors.

2. MEASURES OF MATING SYSTEMS

The modern focus on mating systems is couched in understanding how sexual selection acts on a population. Sexual selection arises from differential access to reproductive opportunities. Such selection occurs in two ways: among individuals of one sex for access to the other sex (intrasexual selection; competition; see chapter VII.5) and between the sexes for choice of mate (intersexual selection; mate choice; see chapter VII.6). The strength of sexual selection can be measured using different, highly debated methods, but in essence the larger the variance in mating success experienced by one sex, the stronger the sexual selection on that sex. Thus, theory predicts that sexual selection should be a much stronger evolutionary force in taxa in which some individuals of one sex are successful at both mating and preventing other individuals of the same sex from reproducing. Such a pattern occurs in polygamous compared with monogamous mating systems.

In 1977 the first attempt at a unified evolutionary hypothesis to explain variation in animal mating systems was put forward by Stephen Emlen and Lewis Oring. Their model, herein called the *E and O model*, is descriptive, based on the spatiotemporal distribution of receptive mates and/or the resources used to monopolize mates. The model arises from Darwin, and its subsequent development uses economic cost-benefit analyses, applied to ecology. Darwin wrote that competition for access to mates occurs because one sex is a limiting factor for the other; one sex competes among its members for a limiting resource, leading to variance in mating success for the nonlimiting sex. Thus, understanding why different taxa exhibit different mating systems is, in essence, a process of determining why in certain species one sex is less of a limiting resource than in other species. Robert Triver's parental investment theory, based on A. J. Bateman's work in *Drosophila*, suggested that because females invest more in parental care (including the initial cost of larger eggs—anisogamy), females are the limiting resource, and therefore males compete for them.

The E and O model suggests that male variation in mating success, the extent of which influences the intensity of sexual selection, reflects the ability of males to control (or not) access of others to potential mates. This monopolization may be over females themselves or over resources that females may require. Resources vary in space and time, and thus ecological constraints on the sexes determine the intensity of sexual selection and the mating system. The original intention of the E and O model was to allow for predictions regarding how the environment, through ecological constraints, results in the evolution of animal mating systems. To make the verbal model predictive, they introduced two terms that allowed mating systems to be identified: *operational sex ratio* (OSR) and *environmental potential for polygamy* (EPP). Emlen and Oring argued that the OSR represents an empirical measure of the potential for monopolization. If the OSR is close to unity, then monopolization is unlikely, whereas if the OSR is skewed toward one sex, then that sex has the opportunity to monopolize the other sex. The OSR may be predicted by the *potential reproductive rate* (PRR) of each sex. However, EPP has no quantitative definition and therefore may have limited use in the ability to predict mating systems.

Using the concepts of OSR and EPP, the E and O model classifies mating systems into four simple categories: monogamy, polygyny, rapid multiple clutch polygamy, and polyandry (table 1). Both polygyny and polyandry contain subcategories of mating systems (table 1). These categories represent a description of the degree to which a sex can monopolize resources. Thus in monogamy, neither sex has the opportunity for monopolization, whereas in resource defense polygyny, males have the opportunity to control access to critical resources required by females and thereby indirectly control access to females. In the former category, there will be little sexual selection because the OSR is not skewed toward one sex, but in the latter the intensity of sexual selection will reflect the extent to which the OSR is skewed—the greater the skew, the greater the variation in mating success, so the stronger the sexual selection. The influence of the OSR on variance in mating success depends on the economics of the resource defense. If a resource becomes too expensive to defend, then the mating system may move away from polygyny and toward monogamy.

Much of the E and O model is intuitive, and it enjoyed considerable success, especially early on. However, despite the model's intuitive appeal, there are concerns with it that have been extensively discussed elsewhere. Generally, there are two main problems: the OSR is an incomplete predictor of mating systems, and it takes into account only a subset of the population. Use of only the OSR is insufficient to predict the mating system because EPP is not quantitative; in other words, the relationship between OSR and whatever environmental factors affect monopolization of females is not always positive. For example, in the European bitterling fish, larger males are territorial, and most spawning involves breeding pairs. However, if male density is high, the EPP is reduced because the economical defendability of large territories is untenable. Thus, the mating system changes from one of resource defense polygyny to that of scramble (or explosive breeding assemblage) competition, in which large males abandon territoriality, and group spawning occurs. Counterintuitively, the relatively less male-biased OSR of the population during resource defense polygyny has a larger variance in male mating success than when the population has a more male-biased OSR, when a scramble-based mating system is used and there is less variance in male mating success. While the E and O model predicts this pattern, the dual issues are that the verbal model provides no way to quantify when this switch will happen because EPP is not an empirical measure, and the relationship between OSR and variance in male mating success is not always positive. Therefore, the OSR itself cannot necessarily accurately predict the strength of sexual selection and therefore cannot predict the mating system.

The second problem with using OSR as a predictor of mating systems is that it takes into account only breeding individuals at particular times. Nonreproductive males are not included, which is fundamental to the theory, but this omission has two consequences for understanding a population's response to sexual selection. First, this exclusion causes errors in the calculation of the strength of sexual selection by overestimating population fitness and underestimating variance in fitness. Under strong sexual selection, more individuals are left out of the equation, and these errors become larger. Thus, the predicted mating system, which is based on variance in mating success of one sex, will be underestimated. Second, OSR is an instantaneous measure, and it can change across the breeding season; the effect of this change depends on its magnitude and whether such changes occur relatively early or late in the breeding season.

In an effort to make sexual selection quantifiable, and therefore mating system evolution more predictable, several attempts have been made to move away from the proxy measurement of the OSR to "direct" measures of selection explicitly linked to evolutionary theory (that is, to fitness). Of the several indexes suggested, two have proven most popular: *Bateman's gradient* (β_{ss}) and the *opportunity for sexual selection* (I_s). Both these measures are consistent with quantitative genetic theory and measurement of selection, and are independent of phenotypic traits. Moreover, they are more general

Table 1. Major categories of mating systems using three different classification schemes

Main category	*Subcategory*	*Definition based on mate number*	*Definition based on monopolizability*	*Definition based on opportunity for sexual selection*
Monogamy		1 male and 1 female have an exclusive relationship.	Neither sex has the ability to monopolize additional members of the opposite sex.	Each sex mates once; variance in mate number for both males and females = 0.
Polygyny		1 male has an exclusive relationship with 2+ females.	Individual males control or gain access to multiple females.	Females mate once, whereas males are variable in mate number; variance in mate number for females = 0; variance in mate number of males = ++.
	Resource defense		Males indirectly control access to females by monopolizing critical resources.	
	Female (harem) defense		Males directly control access to females because females group together.	
	Male dominance—two forms: Explosive breeding and assemblages Leks		Mates and resources are not defendable, so males aggregate during breeding season, and females select mates from these aggregations. In explosive breeding assemblages breeding is highly synchronous and individuals short-lived (OSR = 1, so there is little sexual selection). Leks occur when females are highly asynchronous; males compete for territories or dominance (OSR highly male biased).	
Polyandry		1 female has an exclusive relationship with 2+ males.	Individual females control or gain access to multiple males.	Males mate once, whereas females are variable in mate number; variance in female mate number = ++; variance in male mate number = 0.
	Resource defense		Males indirectly control access to females by monopolizing critical resources.	
	Female access		Resembles explosive breeding assemblage, but the OSR is female biased because females limit access to males.	
Polygynandry		2 or more males have an exclusive relationship with 2 or more females.		Both sexes have variable mate numbers, but male mating is more variable than that of females. Thus, male variance = ++, and female variance = +.
Polyandrogyny				Both sexes have variable mate numbers, but male mating is less variable than that of females. Thus, male variance = +, and female variance = ++.
Polygamy		Includes polygyny, polyandry, and polygynandry.		Both sexes have variable mate numbers but they do not differ between the sexes. Thus male variance = +, and female variance = +.

measures of the strength of sexual selection than indexes that require knowledge of the phenotypic trait putatively under selection (e.g., selection gradients, β).

Bateman's gradient is a selection differential, describing the covariance between values of a particular trait and fitness. In the case of understanding sexual selection and the evolution of mating systems, the particular trait of interest is the number of mates, assuming that higher mating success results in more offspring (i.e., greater reproductive success). If the regression of fitness on mating success gives a gradient of zero, then sexual selection is typically nonexistent. In contrast, if the gradient is steep, then sexual selection will act strongly on some trait correlated with mating success. Thus, it has been suggested that β_{ss} provides an estimate of the strength of selection acting on mating success. However, subsequently understanding why either populations or species have different Bateman gradients requires uncovering the cause both of differences among individuals in their ability to acquire mates and of nonzero Bateman gradients.

The opportunity for sexual selection (I_s) measures the standardized variance in mating success and is an upper limit of the strength of sexual selection. One proposed advantage to this measure is that it may predict the level of sexual dimorphism as an outcome of sexual selection. While described by some as being a direct measure of selection, it is actually still a proxy, because although I_s represents the upper limit of selection, there is little knowledge regarding the relationship between it and the actual strength of selection. Thus, the extent to which I_s can predict the mating system is unclear.

Both these measures require knowing mating success. In cases where this parameter is not known, Stephen Shuster and Michael Wade have suggested that calculating how critical resources, particularly females, are clumped in space and time may serve as a measure of the strength of sexual selection, and they use the concept of *mean crowding* to quantify these variables as m^* and t^*, respectively. Essentially, these measures enumerate EPP, making the E and O model quantitative. High values of m^* reflect spatially clumped resources, whereas low values indicate overdispersion; high values of t^* reflect temporal invariability (e.g., synchronous breeding), whereas low values indicate temporal variation. Thus high values of m^* and low values of t^* represent conditions in which one sex would have the potential to monopolize resources, the opportunity for sexual selection would be great, and the mating system would reflect one based on resource or defense. The contrasting values would represent little ability for one sex to monopolize resources, a low opportunity for sexual selection, and a mating system tending toward monogamy.

These different measures of sexual selection, along with other less used suggestions not discussed here, are contentious. Some measures may be more suitable than others under different conditions. However, studies directly comparing these indexes have reached different conclusions about the congruency between them. Seemingly, the only idea with which everyone agrees is that no current specific measure or combination of measures used to quantify sexual selection, and thus mating systems, satisfies everyone.

3. PLASTIC, CONTINUOUS MATING SYSTEMS AND THE EVOLUTION OF BEHAVIOR

There are a number of classification frameworks for mating systems, although the predominant focus is on the number of mates of each sex for each sex (table 1). These frameworks often lack a common terminology, but mating system descriptions also can overlap, be redundant, represent inadequate descriptions, or lend themselves to incorrect usage relative to formal definitions. For example, *monogamy* is frequently defined as the condition in which each sex has only one partner. However, this is an inadequate description because this could be the situation for an organism's entire life or for one season, or an individual might raise offspring with only one mate regardless of whether the offspring are sired solely by that mate (as seen in many passerine birds).

Another problem is multiple definitions of the same word. For example, under the general E and O framework (table 1), *polyandry* is a mating system in which females have variable numbers of mates, while males mate with a single female. In trying to understand female multiple mating, many researchers have asked, What is the evolutionary benefit of polyandry? But in this context, most researchers are following the definition of polyandry provided by Thornhill and Alcock in which polyandry simply means female multiple mating with no conditions regarding the number of mates for males. In most systems, both males and females mate multiply. Perhaps such a system may be described as *promiscuous*, in which both males and females mate with multiple partners. Yet the term is loaded with connotations about indiscriminate mating, which is the direct opposite of what is being asked by researchers studying the functional significance of polyandry, which is primarily focused on the evolutionary rationale for female choice.

In an attempt to solve these descriptive problems, Shuster and Wade use their framework to further delineate mating systems. While quantitative (although the variables necessary to determine mating system [i.e., m^*, t^*, female distribution in brood number, sperm competition level] are largely unavailable), their system is rather unwieldy, resulting in 41 different mating systems under 12 different general categories.

So, how many mating system categories are necessary? Categorizing phenomena is something humans do, but doing so implies rigidness to those phenomena. If the E and O model is mostly correct, then spatiotemporal resource distribution is the driving factor in the evolution of mating systems. Given that resources are frequently ephemeral across space and time, mating systems are likely dynamic rather than static. Intraspecific mating system variability has been associated with changes in such factors as population density, as discussed earlier regarding the European bitterling; predation; food availability; and climate change. Thus, a species "mating system" is likely much more flexible than mating system terminology suggests. Problems associated with correct use of mating system terminology may be more than mere nuisances, as these terms give the impression that mating systems are discrete rather than continuous. The Shuster and Wade model most approximates a continuous distribution of mating systems, but whether it is feasible to implement remains debated.

Sexual selection results in the evolution of behavioral, morphological, and physiological traits. Many of these traits can be plastic and exhibit continuous variation. The ability to predict how these traits change in response to underlying environmental variables that control mating systems is critical. The OSR (and sometimes associated changes in density) has frequently been shown to predictably influence male behavior, such as aggression, territoriality and alternative reproductive tactics, mate guarding, copulation duration, and sperm release. Changes in the OSR can also vary female mate choice. These changes in behavior are in response to the ability of males to monopolize females and to the costs and benefits of mating for females.

The ability to monopolize mates at both pre- and postcopulatory stages is one of the driving forces of sexual conflict and sperm competition. A recent meta-analysis asked whether changes in OSR could predict the outcome on different types of intrasexual selection behavior associated with mating systems: direct mechanisms of monopolization via contest and sperm competition, and indirect mechanisms via courtship behavior, copulation duration, and mate guarding. Overall the study found that behaviors change in response to increasing OSR. In particular, contest competition exhibited a humped distribution, initially increasing but then decreasing as the OSR became more male biased. Sperm competition follows this trend. In mating systems that are predominantly determined through direct competition, males may conserve energy when competition gets extreme and the economic defendability of mates is unsustainable—as predicted by the E and O model. In contrast, behaviors that function indirectly to monopolize mates did not exhibit the same response, either to one another or to direct competition. As the OSR became more male biased, courtship rate decreased, whereas mate guarding and copulation duration/attempts increased. Mate guarding and copulation duration (if it serves as mate guarding) may increase, as it benefits a male to ensure his fertilization success when there are many competitors around rather than to seek additional mates that may already have a partner. If males do not have partners, then increasing copulation attempts is the only option for securing a mate. Additionally, per capita courtship rate decreased because males were competing for the same female. One long-term outcome of increasing male OSR is the evolution of alternative mating tactics.

Female mate choice is also predicted to change when the OSR changes. Females should be more selective when the OSR becomes more male biased because there are more males from which to choose. However, as (receptive) males become more prevalent in the population, sexual coercion by males may increase, and sexual conflict can result. Sexual conflict occurs when the reproductive interests of males and females are incongruent; because of sex differences in PRR, the benefit to males of mating with more females is generally linear, whereas the benefit of multiple mating for females is asymptotic. Sexual conflict is thought to be particularly strong over mating decisions because of this sex disparity in optimal mating rate. Thus, under high male density, when males may be persistent for mating opportunities, the cost to females of resisting these attempts may outweigh the benefits of not mating. In this case, females may become more receptive and acquiesce to superfluous matings (i.e., "convenience polyandry").

These two hypotheses provide contrasting predictions regarding female behavior when the OSR becomes more male biased. If females become more selective, via affecting the proportion of male displays that a female accepts, then female sexual responsiveness should decrease under more male-biased OSRs. However, if female mating rate is determined by male harassment, then female mating rate should increase under greater male-biased OSR. Female water strider mating activity is positively related to male harassment rate, which supports the convenience polyandry hypothesis. The positive association between male harassment rate and female mating was not found in other taxa, however, which suggests that in these taxa the costs of unnecessary matings is greater than the cost of avoiding harassment.

As emphasized earlier, male density and resources frequently change across the breeding season, so a population's mating system can be variable rather than rigid. This flexibility has potential consequences for such behaviors as male harassment and female mating decisions. This flexibility also may affect the evolutionary trajectory of populations, as discussed in the next section.

4. MATING SYSTEMS AND EVOLUTIONARY POTENTIAL

The two hypotheses about the relationship between female mating rate and male coercion have different effects on the opportunity for sexual selection because this value is partly determined by female choosiness, which in turn is dependent on the costs of choice. Convenience polyandry has been predicted to decrease the opportunity for and strength of sexual selection because females mate indiscriminately to avoid harassment. If male harassment changes across a breeding season, then the strength of sexual selection also changes temporally. In fact, the strength of and opportunity for sexual selection may frequently vary temporally if the mating system is as dynamic as studies suggest. For example, mating systems are known to oscillate between contest and scramble competition. This oscillation will have an impact on the rate at which traits evolve owing to changes in the intensity of sexual selection and perhaps also in the direction of selection. Access to mates via contests usually predicts the evolution of costly exaggerated traits, such as male body size and armaments. Thus in populations experiencing contest competitions, these traits will be directionally selected. However, if male density changes across a breeding season, subsequently altering the mating system to one of scramble polygyny, then these costly traits may not be beneficial and subsequently may be selected against.

Two types of constraints may complicate the ability to predict evolutionary trajectories of selected traits in different mating systems: indirect genetic effects and between-sex genetic correlations. Moreover, if sexual conflict is operating in a mating system, then the nature of that conflict may have different effects on its evolution.

Indirect genetic effects (IGEs) occur when the phenotype of one individual is affected by genes expressed in interacting partners. In the case of sexual selection, the evolution of traits influencing male mating success is a consequence of the male's own selective history and the environmental constraint posed by the genes of the interacting female. Thus, males and females have interacting phenotypes, and the evolutionary trajectories of these traits are different from those of traits expressed regardless of social context. The effect is twofold: the evolutionary response may not be constrained by depleted genetic variation in one sex, and the strength of the interaction translates into the relative rate of evolutionary change. Both of these relate to the mating system, since more intense sexual selection is predicted to deplete genetic variation (although condition dependence of such traits may mitigate loss of genetic diversity) but also to result in greater phenotypic change owing to stronger interactions between the sexes. While theory on IGEs is well developed, the empirical application to mating system evolution is nascent.

A separate constraint arises because males and females share a common genome. If the homologous trait under selection is controlled by genes expressed in both sexes, then a between-sex genetic correlation occurs in which the evolutionary change in a trait of one sex is dependent on the magnitude of the genetic correlation with, and patterns of selection acting on, the trait in the other sex. Thus, between-sex genetic correlations are expected to influence the evolution of sex differences given the shared genetic basis and evolutionary history of the trait. In particular, if there is sexual conflict, then these correlations can act as a constraint.

However, these potential evolutionary limitations can be mitigated by, for example, the evolution of sex-specific alleles, which is predicted to occur relatively rapidly under intralocus sexual conflict. This conflict arises when each sex has a different fitness optimum for the expression of a shared trait, which can generate sexual antagonistic evolution. Such antagonism can be resolved through the evolution of sex-specific alleles that putatively can engender the evolution of sexual dimorphism, the pattern that Darwin explained through his ideas on sexual selection. Intralocus sexual selection may also alter the evolution of gene expression to resolve sexual antagonism. Many organisms have been shown to have sex-biased gene expression (that is, genes that are sexually dimorphic in their expression), and these genes are rapidly evolving and nonrandomly distributed across the genome.

Interlocus sexual conflict can also affect the evolution of populations with different mating systems. Such conflict occurs when the sexes differ in the fitness outcome of male-female interactions, and traits associated with this interaction are genetically encoded by independent genetic effects. Because males and females can evolve independently, each sex is predicted to evolve traits that enhance its own fitness at the cost of its mating partner's fitness, which can generate a coevolutionary arms race. Conflict can occur over sex-specific life history traits related to the mating system, such as mating frequency, relative parental effort, reproductive rate, and clutch size. Moreover, costs and benefits of particular responses are environmentally dependent. Thus, sexual conflict can affect the mating system and vice versa. Interlocus sexual conflict is predicted to accelerate adaptive evolution owing to sexually antagonistic coevolution.

Theoretical and empirical results indicate that the sex chromosomes play a strong role in the evolution of sex-specific traits that have been putatively linked to sexual selection. The X chromosome has a smaller effective population size (N_e) than autosomes because males carry only one copy (that is, they are hemizygous) but two autosomes. Under conditions of random mating and equal fitness, the X/A ratio is 0.75. However, this ratio can increase if variance in male reproductive success is skewed, such as in strong sexual selection. This increase occurs because the fewer males that contribute to the gene pool have a greater influence on autosomes than on X chromosomes, since males account for half of all autosomes but only one-third of X chromosomes. A dominant allele that benefits females but is unfavorable to males can be maintained on the X chromosome more easily than on autosomes. In contrast, a recessive X-linked allele that benefits males will immediately be exposed to selection because it is hemizygous, and such alleles should therefore colonize the X chromosome more readily than the autosomes. The nature of X chromosome transmission and the differential effects of different types of alleles in the sexes will facilitate the evolution of X-linked genes related to sexual antagonism and sexual selection. Such patterns have been found in *Drosophila*.

This section outlined the influence of mating systems on the evolutionary potential of a population through a variety of mechanisms. However, much of the current understanding is theoretical, with relatively few empirical tests in a limited set of taxa, and some of it is controversial. Additionally, different mechanisms for impeding or accelerating evolution within a mating system, let alone across mating systems, are generally not considered simultaneously.

5. APPLIED RELEVANCE OF THE STUDY OF THE EVOLUTION OF MATING SYSTEMS

Mating systems may affect population viability, because the sex ratio influences pair number, which can alter population dynamics. Male-biased sex ratios can result in harm to females and decreased reproductive output. Female-biased sex ratios, either naturally occurring or due to selective harvesting, can lead to mate limitation and extinction. Additionally, as a population becomes small, N_e will diminish, and this result will be compounded in species with intense sexual selection in which only a proportion of the population reproduces. Both theory and experiments have shown that because male armaments and ornaments are expensive, sexual selection carries a genetic load that negatively affects mean population fitness. However, sexual selection could prevent extinction for a variety of reasons, one being that it accelerates the rate of adaptive evolutionary change. Yet experiments designed to test whether this acceleration occurs in the face of a changing environment have found little support. Overall, whether polygamous or monogamous systems are more at risk depends on the parameters of the models and the analyses performed. For example, comparative analyses addressing whether monogamous or polygamous taxa may experience a greater threat of extinction have been equivocal. In one study of birds, greater extinction probability was associated with taxa having greater post- but not precopulatory sexual selection. Other studies have failed to find a similar relationship.

Mating systems may change as a consequence of anthropogenic environmental modifications and affect population viability. Anthropogenic influences that result in climate change, habitat fragmentation, or pollution—and even selective harvesting—may be potent mediators of mating system evolution by altering predictors of mating system variation (i.e., m^*, t^*; Lane et al. 2011) and the genetic variation underlying sexually selected traits. For example, one of the first biological alterations in response to climate change is phenological; that is, the temporal distribution of key life history traits is altered. This change may affect mating systems given that it is variation in the life histories of males and females that determines PRR and subsequently the OSR. Habitat fragmentation shifts spatial resource distributions, including females, which may alter mating systems. Pollution can decrease the effectiveness of sexual signals, even to the point of losing species distinctions. Selective harvesting (e.g., trophy hunting) typically removes the largest individuals in a population, which can deplete phenotypic, and the underlying genetic, variation in size and sexually selected traits associated with condition.

The benefits of using a quantitative approach to studying the evolution of mating systems, such as the one Shuster and Wade advocate, is that m^* and t^* can be used to estimate how anthropological effects may shift these variables and therefore shift the mating system. For example, habitat fragmentation through processes of urbanization and resource extraction leads to clumping of resources and an adjustment in m^*. The effect of changes in m^* on the mating system depends on the scale of female clumping relative to the organism's range size. For example, assume a bird species' mating system is polygynous. If female range size is historically large, but fragmentation results in the loss of nesting sites, then these small areas will support fewer females. This situation will decrease a male's ability to monopolize many females, thereby reducing the intensity of sexual selection and potentially converting the mating system in that fragmented area to monogamy. Whether this fragmentation influences population viability depends on several factors, including the extent to which genetic factors are also altered as a consequence of a change in mating system and a possible difference in the mean reproductive rate due to limited ability to find mates. If enough is known about a species' mating system, then proposed changes in habitat use (or pollution or climate change) could be modeled, and the effect on the mating system could be predicted.

Such predictions may be too simplistic, however, as the intensity of sexual selection on a particular trait associated with mating success may depend on an environmental cue not included in current mating system approaches. For example, in the seed beetle (*Stator limbatus*), larger males are preferred by females, but under conditions mimicking scramble competition smaller males have a searching advantage over larger males. This advantage, however, occurs only in cooler temperatures, because smaller males can heat up more rapidly and thus begin searching earlier. The effect of climate change on this system will depend on whether populations experience either cooler or warmer temperatures and the extent to which scramble competition occurs naturally. Quantitative mating system measures might be able to predict environmental conditions that foster scramble competition as a consequence of changing resource distribution, but the current mating system framework would not be able to predict the evolution of a particular trait, such as body size, in relation to changes in both the mating system and the abiotic environment.

In conclusion, there have been tantalizing, but frequently opposing, findings on the interaction of mating systems, environmental change, and population viability. Understanding the effect of the mating system on population viability and the feedback loops between mating systems and environmental change should be a priority for future research on mating system evolution given its applied relevance. Integrating genetic studies that address the influence of mating systems on the evolutionary potential of populations is vital to this endeavor. Appreciating that mating systems, however quantified, are a dynamic continuum, rather than static and fixed, will help in this understanding and will facilitate progress in predicting the evolutionary trajectories of behavioral traits under selection.

FURTHER READING

Arnold, S. J., and D. Duvall. 1994. Animal mating systems: A synthesis based on selection theory. American Naturalist 143: 317–348. *Develops the theory of sexual selection gradients to provide a conceptual framework to predict mating systems. Quantifies Bateman's principles and provides a direct link to theory for selection on quantitative traits.*

Arnqvist, G., and L. Rowe. 2005. Sexual Conflict. Princeton, NJ: Princeton University Press. *The first source accumulating the evidence for the role of sexual conflict in influencing evolutionary trajectories of populations.*

Bateman, A. J. 1948. Intra-sexual selection in *Drosophila*. Heredity 2: 349–368. *The famous paper that empirically demonstrated that males benefited more from multiple mating than females. While aspects of Bateman's experimental design and interpretation have been justly criticized, subsequent studies across a variety of species have demonstrated sex-specific variation in the benefits of multiple mating.*

Darwin, C. 1871. The Descent of Man and Selection in Relation to Sex. London: John Murray. *Darwin's answer to some of the logical problems (e.g., marked sexual dimorphism) of natural selection.*

Emlen, S. T., and L. W. Oring. 1977. Ecology, sexual selection, and the evolution of mating systems. Science 197: 215–223. *One of the most influential papers in behavioral ecology, which proposed that mating systems can be classified by their species' ecology, in particular the spatiotemporal distribution of resources.*

Lane, J. E., M.N.K. Forrest, and C.K.R. Willis. 2011. Anthropogenic influences on natural animal mating systems. Animal Behaviour 81: 909–917. *A novel and thought-provoking view of how the study of mating systems can be brought into applied relevance.*

Mills, S. C., and J. D. Reynolds. 2003. Operational sex ratio and alternative reproductive behaviours in the European Bitterling, *Rhodeus sericeus*. Behavioral Ecology and Sociobiology 54: 98–104. *A classic example of the difficulties of using OSR as a measure of the mating system.*

Moya-Laraño, J., M.E.T. El-Sayyid, and C. W. Fox. 2007. Smaller beetles are better scramble competitors at cooler temperatures. Biology Letters 3: 475–478. *An empirical illustration of the potential benefits of and difficulties with attempting to demonstrate the influence of climate change on mating systems.*

Shuster, S. M., and M. J. Wade. 2003. Mating Systems and Strategies. Princeton, NJ: Princeton University Press. *A book devoted to quantifying the opportunity for sexual selection (I) and using it to study and classify animal mating systems.*

Thornhill, R., and J. Alcock. 1983. The Evolution of Insect Mating Systems. Cambridge, MA: Harvard University Press. *First widely referred use of polyandry as female multiple mating without conditions for the number of mates/males—currently the predominantly used definition for polyandry.*

Trivers, R. L. 1972. Parental investment and sexual selection. In B. Campbell, ed., Sexual Selection and the Descent of Man, 1871–1971. London: Heinemann. *Based on Bateman's work, this revolutionary contribution suggested that because females invest more in parental care (including the initial cost of larger eggs—anisogamy), females are the limiting resource, and therefore males compete for them.*

VII.5

Sexual Selection: Male-Male Competition

Christine W. Miller

It is certain that amongst almost all animals there is a struggle between the males for the possession of the female. This fact is so notorious that it would be superfluous to give examples. —Charles Darwin

OUTLINE

1. Why are males most often the competing sex and females the choosy sex?
2. The processes of sexual selection
3. Male-male competition in the big and small
4. Weapon evolution
5. Additional forms of male-male competition
6. Male-male competition in plants
7. Total sexual selection
8. Sexual selection and ecological context

Males commonly compete for access to potential mates. This chapter addresses these competitive interactions among males, including alternative mating strategies and sperm competition.

GLOSSARY

Aggression. Offensive action, particularly in regard to defending resources. Aggression in defending females and other resources should be favored by selection when there are fewer resources than competitors and when an individual can boost fitness by forcibly removing another individual.

Intersexual Selection. Sexual selection due to interactions between the sexes. Mate choice by males or females is intersexual selection.

Intrasexual Selection. Sexual selection due to interactions within one sex. Male-male competition for access to mates is the major form of intrasexual selection. In some cases females may also compete for mates, especially in sex-role reversed species.

Polygyny. Mating systems characterized by high variance in male reproductive success; a few males mate with many females, and many males mate with few or no females.

Sperm Competition. Competition between the ejaculates of different males to fertilize the ova of a particular female.

Males in many species compete with one another for access to potential mates, a phenomenon called *male-male competition*. Sexual selection, or selection due to variation in reproductive success, is traditionally divided into the processes of *intra*sexual selection—that is, selection between members of one sex for reproductive access to the other sex—and *inter*sexual selection, or mate choice, in which selection of mates is nonrandom. Both sexes can directly compete for matings and exert mate choice, but male-male competition is more apparent than female-female competition, and females are often choosier in selecting mates than are males. Indeed, across species, males are usually the sex with the more elaborated weapons (used in intrasexual competition) and ornaments (used to attract females). When females do compete with one another for males, it is usually in situations where males offer resources such as food, paternal care, or a suitable location to raise offspring.

1. WHY ARE MALES MOST OFTEN THE COMPETING SEX AND FEMALES THE CHOOSY SEX?

The asymmetry between males and females is due to the factors that limit reproductive success for each sex. Individual male reproductive success increases with the

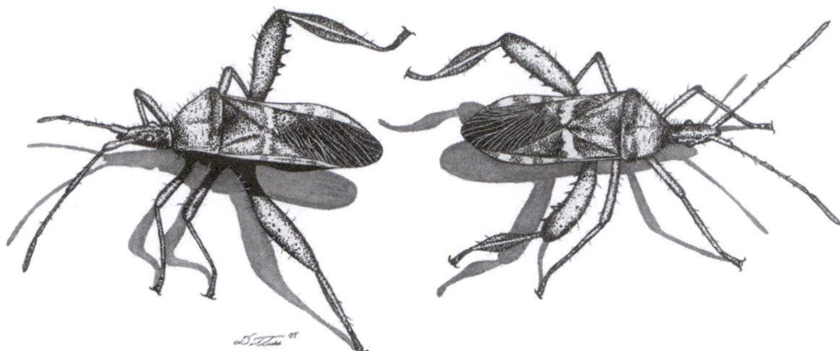

Figure 1. In an escalated competition male leaf-footed cactus bugs line up end to end, wrap their hindlegs around one another, and press their femur spines into the abdomen of their opponent. (Drawing by David Tuss.)

number of mating events. Thus, a male can achieve greater reproductive success by competing with other males to access as many females as possible. Females, however, have an upper limit to their potential reproductive success owing to their greater investment in each individual offspring. Thus, it benefits females to choose mates carefully, to ensure that the offspring in which they invest are high quality. When males invest relatively more in caring for offspring than do females, such as is the case in jacana birds, seahorses, giant water bugs, and other sex-role reversed species, females then become the competing sex and males the choosy sex. Furthermore, in monogamous species in which both parents care for offspring and are limited in reproductive potential, both male-male competition and female-female competitions may exist, as well as mate choice by both sexes. See also chapter VII.6.

2. THE PROCESSES OF SEXUAL SELECTION

In *The Descent of Man and Selection in Relation to Sex* (1874) Charles Darwin noted that "if each male secures two or more females, many males cannot pair." This observation highlights why sexual selection causes the divergence of male and female traits and explains why male-male competition for reproduction is not unusual. To achieve reproductive opportunities, males must (1) fight other males for access to females or guard the resources females need for survival and reproduction and/or (2) attract females with sexual ornaments and/or direct benefits, such as food. The first situation involves intrasexual selection among males. The second situation describes intersexual selection, involving female choice for indirect (genetic) and direct benefits (see chapter VII.6). These mechanisms of sexual selection rarely operate independently, and thus understanding the entire process of sexual selection requires simultaneous consideration of all the agents that result in differential mating success. Furthermore, because mating success does not perfectly translate into reproductive success, attention to those factors resulting in differential fertilization success (sperm competition and cryptic female choice) is also necessary. In the past, intrasexual selection and intersexual selection have garnered varying amounts of attention each, even though both play central roles in the process of sexual selection. Here, we consider these forces of sexual selection separately to better highlight the important aspects of each to the bigger picture.

3. MALE-MALE COMPETITION IN THE BIG AND SMALL

Male-male competition is most commonly observed in polygynous mating systems and where males compete over females (female defense polygyny) or the resources that females need (resource defense polygyny). Examples of male-male competition include the dramatic head butting by male bighorn sheep and fights between roosters. Male elk use elaborate antlers to lock and push one another during the mating season, vying for access to females. But antlers are not limited to ungulates; some male flies also have antlers and competitive behaviors that bear a striking resemblance to those of elk. Another group of flies, the stalk-eyed flies, has eyes that teeter out on the end of long stalks. Males with the longest eyestalks are more likely to win in competitions with other males, and they are also more attractive to females. Leaf-footed cactus bug males engage in wrestling matches by turning around end to end (figure 1), wrapping their elaborated hindlegs around their opponent's body, and squeezing. These are only a few examples of the wide array of male-male competitive behaviors and associated morphologies found in the natural world.

4. WEAPON EVOLUTION

Male-male competition has resulted in the evolution of a tremendous diversity of weapons across taxa, including spurs, tusks, antlers, horns, and mandibles. In nearly all

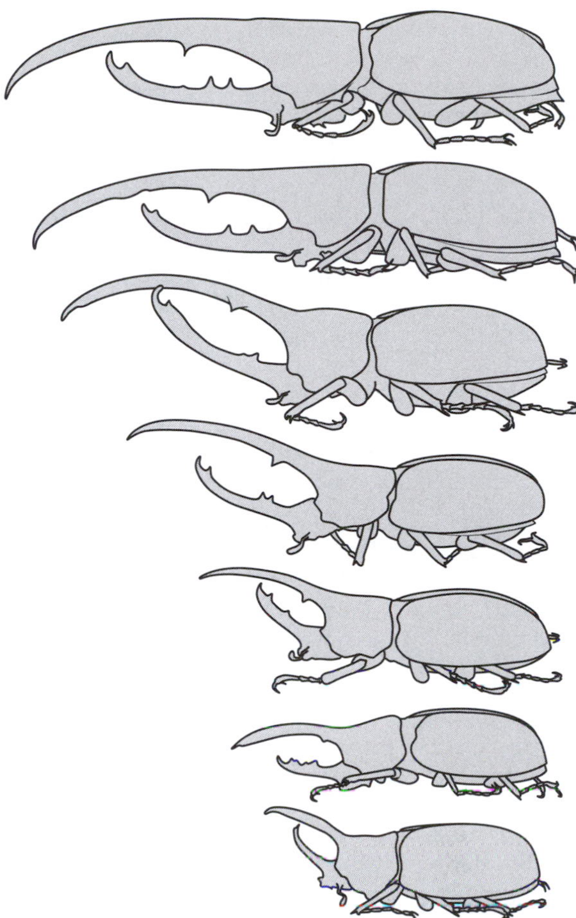

Figure 2. Weapon size and shape are often extremely variable within populations. Pictured here: variation in male *Dynastes hercules* horn size. (Adapted from C. Champy. 1924. Les Caractéres Sexuels Considérés comme Phénomènes de Développement et dans Leurs Rapports avec l'Hormone Sexuelle. Paris: Gaston Doin.)

species that possess sexually selected weapons, essential resources are concentrated in space or time and are thus economically defensible. The costs of investing in weapons can presumably be offset by the benefits of increased reproductive opportunities.

Weapon size can be one of the most variable morphological traits among individuals of the same population (figure 2). Large males or those in good condition generally invest in the largest or most complex weapons, while smaller males often have disproportionately reduced traits. There may be a high cost for a small male to develop a large weapon, both in terms of energetic investment and in terms of severe consequences during escalated competitions. Therefore, these traits are thought to serve as honest indicators of male quality, both prior to and during physical encounters.

Species with complex weapons and other indicators of male quality may actually have a reduced, not increased, danger of injury from physical combat. Theoretical models suggest that males should assess the potential of their rivals using visual, chemical, and audible signals and avoid competing with unbeatable opponents. Weapons may serve a dual function as visual signals: the more complex the weapon, the more clearly it may signal a male's competitive ability. In fact, reconstructions of weapon evolution within clades is beginning to reveal that weapons initially evolve as small, dangerous traits that later evolve to serve more as signals of quality.

Complex weapons may also facilitate male-male contests that are highly ritualized with low risk of lethality. In animals such as elk, dynastid beetles, and even triceratops dinosaurs, male weapons connect in a specific way, allowing for protracted pushing contests that assess strength with little chance of physical damage. For example, baradine weevils have deep pockets in their thorax that accommodate the paired horns of opponent males during contests.

Weapons sometimes exist in both sexes. For example, female cape buffalo have prominent horns, and female tusked wasps have branched facial outgrowths. In most cases, female weapons are smaller than those of their male counterparts and may serve as defense against predators or in competitions with other females over nest sites and food. Existing evidence suggests that most weapons initially evolve owing to natural selection, and male weapons are further elaborated by sexual selection. Evolution commonly proceeds in such a manner—existing features are co-opted for new uses over evolutionary time.

5. ADDITIONAL FORMS OF MALE-MALE COMPETITION

Alternative Mating Strategies

Male-male competition is not limited to obvious direct confrontations. Within a population, some males are able to access females by evading aggressive, dominant males. Males may employ various alternative techniques to come into contact with females, such as superficially resembling females or employing sneaking behaviors. Some of the better-studied examples of species using alternative mating strategies include isopods, ruffs (a wading bird), dung beetles, and Pacific salmon. Males employing alternative mating strategies often also have complementary male morphologies that support the behavioral differences.

When should alternative mating strategies evolve? The answer can be found by examining the average and variance in male mating success for a population. If some males are able to achieve disproportionate access to females, then the average number of mates per male *and* the variance in male mating success will increase. Sexual selection in this scenario will be much stronger than in a monogamous population, because some males will leave many descendants, while many will leave none. Thus, polygyny can lead to a "mating niche" for males employing *unconventional* mating behaviors. Males who avoid direct confrontations with other males may still be able to find and mate with females using alternative means.

Males employing alternative mating strategies and achieving surreptitious matings may yield only a fraction of the fertilization success gained by the conventional males, but if they manage to sire even a few offspring, then they will still have greater mating success than conventional males who are unable to secure mates. The average mating success of conventional and unconventional males is often similar, though the variance is often quite different.

Three genetically distinct male types coexist in the marine isopod *Paracerceis sculpta*. Large alpha males defend females within intertidal sponges, beta males mimic females, and tiny gamma males hide within large harems. All male types have equal mating success, with the beta males and gamma males sneaking copulations. While this mating system is female-defense polygyny, only one of the three male types attempts to defend females. Thus, understanding sexual selection in this species necessitates recognition of alternative mating strategies.

Sperm Competition

Darwin distinguished two contexts for sexual selection: male-male competition and mate choice. He apparently viewed sexual selection as occurring only prior to copulation and that a male's success could be measured in his ability to obtain copulations. Since Geoff Parker's work (1970) it has become clear that this view is incomplete. Females commonly mate with more than one male during a reproductive cycle, thus setting the stage for the postcopulatory equivalent of male-male competition, *sperm competition*, and the postcopulatory equivalent of mate choice, termed *cryptic female choice*. Sperm competition is the competition between the ejaculates of different males to fertilize the ova of an individual female. Cryptic female choice occurs when females can discriminate in their reproductive tract among the sperm of different males, and at least 21 possible mechanisms for this form of discrimination have been described (see chapter VII.6). Sperm competition and cryptic female choice create powerful selective pressures and have shaped many life-history characteristics including body size, reproductive morphology, physiology, and behavior.

Male Adaptations to Sperm Competition

The single most important factor determining the fertilization success of a male is the number of sperm a male inseminates. Thus, across taxa, males with the potential for more intense sperm competition tend to have larger testes and larger sperm storage organs. Some also have larger accessory glands, which produce the seminal fluid in which the sperm are transported. Accessory glands may produce seminal substances that form copulatory plugs that impede further insemination by other males. Seminal fluids can also contain cocktails of chemicals that increase male fertilization success. In the fruit fly *Drosophila*, seminal products have been shown to poison previously inseminated sperm and elevate female egg production. These chemicals raise the likelihood of male paternity but come at a cost (to females) of earlier mortality.

Mate guarding is a common and straightforward means of preventing or minimizing sperm competition. Males may guard females before or after mating, or through prolonged copulation. For example, in golden dung flies (*Sepsis cynipsea*), males locate female on dung pats, guard individual females as they lay eggs, and then attempt to copulate with females after they leave the dung. In squash bugs (*Anasa tristis*), copulations can last for days, during which the large females pull the small males around by their genitalia. Many males suffer injuries owing to these prolonged copulations, but if greater fertilization success is the result, injury and early death may be well worth the investment.

6. MALE-MALE COMPETITION IN PLANTS

Sexual selection, including both mate choice and male-male competition, is not limited to animals. Plants also compete for mating opportunities. First, male plants must have effective means for moving their pollen to the stigma of a female flower, and they do so using wind, water, or animals. Greater pollen production can help plants disperse their pollen and access ovules. In animal-pollinated plants, male-male competition for the attraction of effective pollinators has played an important role in the evolution of flowers.

Once pollen has reached a stigma, it also must then compete with other pollen for fertilization success, akin to sperm competition in animals. Not all pollen has an equal likelihood of siring seeds. For example, in wild radish (*Raphanus sativus*), use of genetic markers has shown that unequal siring of seeds by pollen donors is very common. This differential fertilization success may be due to the speed and effectiveness of pollen in reaching ovules, but it may also reflect properties of the female plant that discriminate among potential mates.

7. TOTAL SEXUAL SELECTION

Male-male competition is often studied separately from female mate choice. Experimental isolation of the individual agents of sexual selection is convenient and can provide keen insights into how each component functions, but these studies provide only limited insights into the whole picture. A complete understanding of sexual selection requires simultaneous consideration of male-male competition, female mate choice, and the other agents that affect differential fertilization success in populations, including sperm competition, cryptic female choice, and alternative mating strategies. Mate choice is dealt with in chapter VII.6 but is considered here in relation to its effects on competition for mates.

An example of a dynamic interplay between male-male competition and female mate choice can be seen in the pronghorn antelope, a species with a female defense polygyny mating system. Male pronghorn compete with other males for groups of females, called *harems*. However, females are not passively herded into harems. Instead, females invest considerable energy sampling many males before choosing a male with which to reproduce. As a result, harem composition fluctuates, and male mating success is associated with the ability to maintain large harems across rut. Sexual selection on males is strong, with only a small subset of males in a population able to achieve mating success in a given year.

As seen in pronghorn, male-male competition and female mate choice commonly select linearly and in the same direction on male traits, such as defensive ability, body size, and weapon size. In some instances, male-male competition and female mate choice may have identical outcomes. For example, females may directly observe male-male competition and mate with the winner of the competitions. However, the agents of sexual selection can also be partially or completely in opposition. For example, in the cockroach *Nauphoeta cinerea*, a male pheromone is used both in establishing male dominance and in attracting females. However, the blend of compounds most useful for establishing dominance is not the same as the blend found most attractive to females. Thus, there are opposing selection pressures on the composition of a male trait used both in male-male competition and in female mate choice. Such opposing selection pressures in this and other species may lead to the maintenance of genetic variation in sexually selected traits.

What determines whether male-male competition and female mate choice are reinforcing, in opposition, or somewhere in between? One factor appears to be the mating system of a species. For example, in social species, such as primates and cockroaches, males may form dominance hierarchies through male-male competition. Dominant males may have only limited control over female mating decisions, and females may, at times, be able to mate with whomever they prefer. Conversely, in resource defense mating systems, females may be able to reproduce only with the subset of males successful in guarding food and nest sites or those successful in sneaking access to females. Thus, in some species, females may not have the opportunity to fully exercise their mating preferences.

The preceding discussion focused primarily on those situations in which male-male competition and female mate choice act on the same traits in males. Body size is an example of a male trait that is often under selection for both male-male competition and female mate choice. However, the agents of sexual selection may also select for distinct traits. Females may be rather unconcerned about male weapons such as male horns, and males vying for status may ignore ornaments such as colorful plumage. In this case, ornaments and weapons may evolve somewhat independently, albeit with potential trade-offs in investment.

8. SEXUAL SELECTION AND ECOLOGICAL CONTEXT

Theory on mating systems has long recognized that ecological variables—in particular, the concentration of resources—shape the form, strength, and even direction of sexual selection for entire populations (see chapter VII.4). However, this perspective has not fully permeated to considerations of the level of the day-to-day interactions among individuals. The availability and quality of resources and other environmental factors can change quickly, and as they do, individual animals may adaptively change their mate choice and male-male competitive behaviors. For example, in times of resource limitation, otherwise-dominant *N. cinerea* cockroaches reduce their aggression and dominance behaviors, while subordinate cockroaches become more aggressive. Thus, the structure of dominance hierarchies, access to females, and the resulting sexual selection pressures may vary according to the amount of resources available.

An ecological view of sexual selection necessitates, first and foremost, a keen understanding of the natural history of study organisms, best gained through careful observations in nature. The vast majority of investigations on mate choice and male-male competition have been conducted in only one or a small range of environmental contexts, and often these contexts are wholly artificial. Such experiments allow for fine-tuned analysis of how sexual selection *can* operate, but they do not provide an ecologically relevant picture of how mate choice and male-male competition function, and fluctuate, in real-world settings. Studies are beginning to demonstrate that sexual selection is indeed variable over time and space. Over the next decades, it will be exciting to learn how environmental variability alters the processes, and outcomes, of sexual selection.

FURTHER READING

Andersson, M. 1994. Sexual Selection. Princeton, NJ: Princeton University Press. *A wide-ranging review of sexual selection and mating systems.*

Emlen, D. J. 2008. The evolution of animal weapons. Annual Review of Ecology and Systematics 39: 387–413. *An extremely readable and fascinating description of animal weapons, including those used by dinosaurs.*

Hunt, J., C. J. Breuker, J. A. Sadowski, and A. J. Moore. 2009. Male-male competition, female mate choice and their interaction: Determining total sexual selection. Journal of Evolutionary Biology 22: 12–26. *An important reminder that male-male competition and female mate choice may exert opposing or reinforcing selection pressures on male sexually selected traits.*

Parker, G. A. 1970. Sperm competition and its evolutionary consequences in the insects. Biological Reviews 45: 525–567. *Seminal work by the founder and developer of modern sperm competition theory.*

Shuster, S. M., and M. J. Wade. 2003. Mating Systems and Strategies. Princeton, NJ: Princeton University Press. *A modern and quantitative synthesis of mating systems and alternative mating strategies.*

VII.6

Sexual Selection: Mate Choice

Michael D. Jennions and Hanna Kokko

OUTLINE

1. Why does mate choice fascinate evolutionary biologists?
2. What counts as mate choice?
3. Choosiness lowers the breeding rate, and there are other costs
4. The rewards of being choosy
5. Why do the sexes differ in choosiness?

The evolution of many extravagant traits and bizarre behaviors is attributed to sexual selection arising from mate choice. Mate choice occurs when individuals' traits or behaviors (*preferences*) make them less likely to produce offspring when they encounter certain individuals of the opposite sex. It involves either some form of rejection of mating or fertilization opportunities, or mating multiply and selectively using the gametes from these mating partners to produce offspring. Such behaviors are expected to increase the interval between successive breeding events. All else being equal, slowing down breeding is costly; thus one expects some compensatory benefits of being choosy to explain its evolution. Choosiness can increase the mean number of offspring produced per breeding event and/or increase offspring quality (i.e., their reproductive value). The latter requires that potential mates vary in their effect on offspring fitness, either because of the genes they transfer or because of the direct benefits (e.g., parental effort) they offer. Females are generally choosier than males, although males often exhibit subtle forms of choice. Large differences in the relative availability of eggs and sperm and of sexually receptive males to females is key to understanding why the sexes differ in whether and how they engage in mate choice.

GLOSSARY

Direct Benefit. An increase in offspring number because preferred mates provide more resources than average, which are converted into additional offspring.

Direct Selection. Selection on traits that increase the bearer's lifetime reproductive output (e.g., mating preferences that provide direct material benefits to the chooser).

Genetic Benefits. An increase in mean offspring fitness because their genotypes differ from those expected under random mating and fertilization.

Heterozygosity. The state in which each parent transfers a different version (allele) of a gene to offspring.

Homozygosity. The state in which both parents transfer the same version (allele) of a gene to offspring.

Indirect Selection. Selection on traits that increase the fitness of relatives rather than the number of offspring the bearer produces (e.g., selection on mating preferences that increase offspring quality through genetic benefits; offspring are relatives).

Operational Sex Ratio. The current ratio of males to females in the mating pool (i.e., adults that are ready to mate).

Polyandry. The propensity for females to mate with more than one male in a single reproductive cycle.

Sexual Conflict. The divergence in evolutionary interests of a male and a female. Conflict is removed only in strict lifetime monogamy.

Sperm Competition. Postejaculatory male-male competition to sire offspring when females mate polyandrously and where sperm compete to fertilize eggs; can cryptically be influenced by females.

1. WHY DOES MATE CHOICE FASCINATE EVOLUTIONARY BIOLOGISTS?

There is often an exquisite fit between organisms and their environment. The manner in which katytid crickets mimic leaves to camouflage themselves from insectivorous birds is just one example of the astonishing machinery organisms use to survive in a hostile world. Individuals detect predators or prey, acquire food, and maintain their bodies with such proficiency that it creates the illusion that they were intelligently designed for survival in a hostile world. Ever since Darwin, natural selection has offered a biological explanation for such economical design.

Yet some characteristics of animals appear positively wasteful. Why, for example, do many birds and fish have gaudy colors and extravagant courtship displays that draw the attention of predators? Why do male fiddler crabs consume energy waving their massive claw to attract a female rather than simply waiting for one to approach? And why does mating involve features that reek of sexual conflict rather than harmonious union, like chemicals in the seminal fluid of fruit flies that manipulate the reproductive behavior of females, and the sharply barbed penises of certain beetles that damage females and shorten their life span?

The existence of characteristics that reduce longevity can be explained by evoking a trade-off between natural selection, which favors increased longevity, and sexual selection for characters that increase reproductive success. Evolutionary success ultimately depends not on how long an organism lives but on how many offspring it produces and, in turn, how successfully they breed. Sexual selection implies competition between members of one sex for access to gametes of the opposite sex. This is partly why females are usually choosier: there are many more sperm than unfertilized eggs available at any given time. The costs of choosing are usually larger for males, because the wait until the next available fertilization opportunity is longer.

2. WHAT COUNTS AS MATE CHOICE?

Sexual selection is traditionally divided into two categories: contests among members of the same sex (*intrasexual*) and interactions between the sexes (*intersexual*). In this classification, mate choice belongs to the latter category. However, the importance of this distinction is easily overstated, and it is often difficult to consider one mechanism of mate choice without considering the influence of the other as well. Dividing sexual selection into purely intersexual and intrasexual mechanisms obscures the shared underlying structure when there is competition within a sex *and* the outcome of this contest is ultimately rewarded by access to gametes of the opposite sex. Victory in competition for mates is rewarded only because the eventual contest winner is more likely to be accepted as a mate (e.g., by virtue of being allowed to stay in the area). With mate choice there is likewise competition among rivals as individuals produce sounds, visual signals, chemicals, and tactile movements that increase the chances that they can induce a member of the opposite sex to mate and use their gametes. There is either a real war or a war contested by more diplomatic means. Nevertheless, to highlight the importance of these two topics to understanding the evolution and evolutionary consequences of behavior, competition between individuals of the same sex is dealt with in chapter VII.5. Here we consider intrasexual competition only in terms of its influence on mate choice.

While the dividing line between inter- and intrasexual selection is relatively irrelevant, it is important to note that sexual selection can occur without involving either mate choice or direct physical contact between rivals. Consider a situation in which individuals occur at low density, and members of one sex actively search for mates. Differences in the ability to locate the opposite sex—for example, variation in olfactory abilities required to find females—will create variation in mating success, even if no mating opportunity is ever rejected, every encounter leads immediately to fertilized eggs, and solitary individuals live at such low densities that no mating involves sperm competition. There is still competition among rivals analogous to that among treasure hunters: located trove is no longer available for rivals to find. Thus sexual selection can increase olfactory sensitivity (or any other trait that improves mate encounter rates, e.g., locomotory ability) beyond its naturally selected optimum, even if no mate choice occurs.

Reaching a clean definition of *mate choice* is challenging. First, and least controversially, conventional (precopulatory) mate choice occurs when individuals are more likely to mate with certain members of the opposite sex when they are encountered. Despite being uncontroversial, the definition strictly speaking states only that choosiness occurs—in practice it is surprisingly difficult to infer solely from an observed distribution of matings who is choosy. A male with a modest ornament may be accepted by some females but not by others. He is thus more likely to mate with some females than others, but this clearly does not imply that he himself was choosy. The correct interpretation relies on additional behavioral insight into which individuals "would have been willing" to mate. In practice, the definition is used in one of the following two ways. (1) When two or more potential mates are simultaneously encountered, a choosy individual is more likely to mate with one type than another. For example, female frogs generally prefer males with lower-pitched calls. (2) When mates are sequentially encountered, there is the additional requirement that some current mating opportunities will be rejected, even though the choosy individual has eggs or sperm available to produce offspring.

As a second category of mate choice, the potential for *cryptic female choice* occurs when females mate with more males than are required merely to fertilize all their eggs. Here, a female creates a situation in which there is competition not among males for a mating but among their sperm to fertilize her eggs. Such sperm competition is often treated as analogous to males' fighting for access to mates. But as already noted, the distinction between intrasexual and intersexual selection is blurry: How should one take into account the various ways that a female's response to competing males (or their sperm) affects the males' eventual reproductive success? Sexual selection clearly occurs, but is it based on mate choice?

Some researchers argue that cryptic female choice occurs only when females actively engage in sperm selection, using the sperm of certain males disproportionately (more often than expected based purely on the relative numbers of sperm inseminated per male). For example, in field crickets, interactions between sperm and the female reproductive tract result in lower fertilization success for sperm from more closely related males. Similarly, in the beautiful Gouldian finches there are two common head color morphs: red and black. If a female mates with males of both types, she is more likely to produce offspring sired by a male of her own color morph. This is an adaptive response from a female perspective, because offspring from genetically mismatched (different colored) parents are less viable. Females can sometimes also influence how many sperm are transferred by different males. Indeed, comparative studies have provided compelling evidence that the amazing diversity in both male genitalia (e.g., that seen in otherwise similar-looking snakes) and behavior during and after copulation have partly evolved because these traits affect a female's propensity to accept, store, and use a male's sperm.

Other researchers argue for a broader interpretation: female choice occurs whenever females set the "rules of engagement" that determine which males are more likely to gain fertilizations. For example, if females of internal fertilizers have a reproductive tract that increases the fertilization success of the last male to mate, this could be considered a cryptic form of choice that biases paternity toward more recent mates. Likewise, if a female's reproductive physiology makes male fertilization success directly proportional to the relative numbers of sperm inseminated, why not describe this as a mechanism of female choice favoring males that produce more sperm? Such forms of mate choice are rather "passive" compared with behaviors such as physically expelling or destroying sperm. The evolutionary outcome is, however, fundamentally the same: certain males gain more paternity owing to interaction between females and copulating males or the sperm they transfer. The case of direct proportionality could, however, also be viewed as the appropriate null model of reproductive tract design, so that it might be difficult to argue it evolved as a reproductive adaptation: From what simpler ancestral mechanism would it have evolved?

A third category of mate choice is often called *cryptic male choice*, though this refers to processes that are not necessarily specific to males. Even if an individual does not benefit from rejecting a mating, it might still benefit from adjusting its mating effort (courtship or gamete expenditure) depending on the types of opposite-sex individual it has encountered. For example, in Gouldian finches both sexes preferentially court mates of the same color morph. More generally, in many species males show a propensity to expend greater effort courting and to inseminate more sperm when they encounter larger, more fecund females. These behaviors qualify as choice, even if males never actually refused to mate with any female, because by courting one female more intensely than another males change the likelihood that they will produce offspring when they encounter different types of females. Likewise, when there is sperm competition, males that produce a larger ejaculate usually have a greater likelihood of siring offspring. To qualify as choice, however, variation in ejaculate size is not sufficient; a male must adjust ejaculate size depending on female traits.

The description of some forms of mate choice as "cryptic" has its origin in cases in which the detection of choice requires information about sperm number and fertilization success inside a reproductive tract. As is clear from the preceding examples, however, not all "cryptic" choice is truly hidden from view; the terminology comes with some historical baggage. "Strategic adjustment influencing whom one has offspring with" is the common theme in all forms of choice.

Male versus Female Cryptic Choice

Male and female cryptic choice follows somewhat different evolutionary rules because the former depends crucially on the competitive behavior of rivals. The success of an ejaculate or the effectiveness of courtship depends on the intensity of sperm or mating competition for different females. Male-male competition means that sperm has to be produced in large numbers, or courtship has to be energetic, for a male to gain paternity. There is an upper limit to how much males can invest, so to maximize their returns they should strategically allocate their reproductive effort across the different females they encounter, depending on the value of each female as a potential mate (which includes considerations of how many of her eggs are likely to be fertilized by rivals). Thus, males engaging in sperm competition often exert choice through strategic allocation of sperm.

To be worthwhile, however, strategic allocation always requires that the savings accrued can be used fruitfully. This consideration points to a factor intrinsic to male-male competition that works *against* cryptic male choice. If female availability is low (e.g., if females reject most mating attempts and/or are usually unavailable to mate because they are providing maternal care), male-male competition is intense, but males are also likely to have replenished their sperm supplies or reenergized themselves for courtship before they next encounter a female. There is then no reason not to invest maximally in each female. So, the lower the rate at which sexually receptive females are encountered, the less likely it is that males will exhibit cryptic or, indeed, conventional choice.

Cryptic female choice is more likely to be driven by variation in the genetic benefits of being fertilized by certain males. As with cryptic male choice, there is strategic investment: a female "spends more eggs" with some males than with others (exactly symmetrically to the situation in which males spend more sperm with some females than with others). But this is not because the female could not otherwise gain access to enough of the chosen male's sperm. Unlike eggs for males, sperm is only rarely limiting for females, so the presence and behavior of other females is less likely to be a significant factor in shaping egg expenditure. The decisive factor is again mate availability: strategic adjustment of egg expenditure per potential mate becomes more beneficial when rejecting one potential mate does not imply a long wait for the next mate. Similarly, for males, strategic adjustment of sperm is more likely in situations with a short wait for the next female.

Previously, we noted that within-sex competition for opposite-sex gametes generally selects against strategic allocation (cryptic choice) of any kind: competition means opportunities are scarce, so existing ones should be neither rejected nor exploited only partially. There are two exceptions, however. First, strong competition can create a situation in which some individuals benefit by pursuing options that at face value offer lower fitness benefits—for example, preferring females with poor fecundity—but become better options by virtue of being ignored by others. This option has been called *prudent choice* and can happen in male-male competition (e.g., in some fish, individuals of low competitive ability preferentially court small females) as well as in the relatively rare situations where females are sperm limited. For example, attractive male ungulates can become temporarily sperm depleted, which can lead to direct female-female competition for access to preferred males. Some females are more likely to win contests for access to these males, so less competitive females consequently decide to mate elsewhere.

Second, intense sperm competition can create conditions in which mating opportunities are abundant for males, but each opportunity offers meager fitness gains. To see why male-male competition can now select for (cryptic) male choice, recall that sperm competition can reach intense levels only if females mate multiply. Consider an extreme case in which all females in the local population mate with all males. Mating opportunities are no longer scarce for males, and males should use sperm relatively sparingly: each mating brings about only modest paternity, and future mating opportunities are likely to occur soon. Even though future mating opportunities, too, yield meager benefits, it remains true that saving sperm now is likely to pay off in the near future, because opportunities arise regularly. In this setting, male-male competition can select for strategic sperm allocation because it does not coincide with the low availability of mating opportunities. To qualify as choice, however, variation in male responses to different females (e.g., based on her fecundity or mating status—virgin or otherwise) is still required.

3. CHOOSINESS LOWERS THE BREEDING RATE, AND THERE ARE OTHER COSTS

The fundamental cost of conventional mate choice is that all else being equal, it lowers an individual's lifetime reproductive output by reducing how often it breeds. This cost can, however, become negligible when many potential mates are simultaneously available. In general, the cost rises when potential mates are encountered rarely.

The simple time delay caused by searching or waiting for an acceptable mate is not necessarily the most significant cost of being choosy. Spending time in other activities before breeding commences can also increase the risk of dying. This risk is present when simply waiting for potential mates, but if mate sampling involves greater mobility, then risks and energetic costs (including less time left to forage) can become even more pronounced, again strengthening the expectation that the lifetime number of offspring produced will decline. Another commonly incurred cost, most relevant in the context of conventional female choice, is that attempts to reject a mate can be costly. Such costs can be substantial despite being subtle. For example, female guppies choose to forage in suboptimal areas to avoid the sexual attention of males. Or costs might be of a more dramatic nature: females that decline to mate run the risk of being injured or even killed by sexually coercive males. Lack of female choosiness in such cases is described as *convenience polyandry*. By contrast, there are very few species in which males that try to reject a mating are damaged by females.

The concept of convenience polyandry predicts that multiple mating is sometimes the least costly option for a female when there is sexual conflict over mating rates (it minimizes the costs of resisting). However, because multiple mating is a prerequisite for cryptic female choice, it is also worth recalling the costs of multiple mating that would be minimized with monogamy. The most obvious costs include acquiring sexually transmitted diseases, being damaged by the male during copulation, being killed by a predator while copulating, and losing male contributions to parental care (in cases in which only one social mate helps with feeding the young; this can be reversed to become a benefit of multiple mating if several male mates are willing to feed).

Multiple mating can also intensify sexual conflict when selection favors male traits that elevate a female's immediate rate of offspring production, which is not always congruent with maximizing her lifetime reproductive output. For example, male seminal fluid can contain chemicals that elevate female egg production at the price of decreasing her life span. Males are selected to speed up their mates' reproduction because a male's likelihood of paternity is highest shortly after the mating: later, his sperm might have died, and the female will more likely have had time to remate.

4. THE REWARDS OF BEING CHOOSY

All the preceding factors create a baseline expectation that choosiness has a negative effect on fitness, albeit with some significant complications. If, for instance, multiple mating is costly, then conventional choice can have a positive effect on female fitness by minimizing such costs. However, if conventional choice leads to monogamy, then the prospects for cryptic female choice are erased. By and large, though, the described costs confirm that choosiness should offer compensatory benefits to overcome its negative expected effect on the rate of offspring production.

What form do these rewards take? A common requirement for both sexes is that choosiness is beneficial only when potential mates vary in the benefits they provide. Although some benefits apply to both sexes, it is easiest to first consider females.

Direct Selection for Material Benefits

The least controversial explanation for female preference for certain males is that this increases the number of offspring produced per breeding event or, more generally, over a female's lifetime. Although rarely explicitly discussed, a female might benefit by choosing a male who helps with parental duties or offers some other kind of material resources to the extent that she can now work less hard to raise young; offspring production per brood might not be elevated, but the female now lives longer. All such positive effects on female fitness result in direct selection on female mating preferences. Males do not have to provide material resources actively (e.g., in the form of parental care or nuptial gifts) for direct benefits to occur. They can also elevate a female's fitness by providing her with access to food or other essential resources. This benefit is common in species in which males defend breeding territories that also contain food. Alternatively, females can use mate choice to avoid males that reduce their life span, such as those infected with transmissible diseases.

Although direct benefits are typically considered to be uncontroversial, the evolution of preferences for them offers intriguing challenges. One caveat is that to benefit from material contributions, females must somehow detect males that offer superior resources. This is straightforward when females can reliably assess the quality of the resources before they mate. It is more difficult to gauge how much parental care a male will provide in the future. Although some sexual traits do appear to predict how much care a male will provide, it is unclear what maintains the honesty of these signals. What prevents a male from promising to be a good father and then reneging on the deal? Early theory about parental care and mate choice asked whether "coyness" of females could evolve as a means to make sure a mate is committed enough to stay, but this idea has scarcely been followed up. On the male side of the equation, the fact that material benefits can rarely be divided among several females without reducing the benefit that each gains means that there are interesting courtship and allocation questions still awaiting study.

Indirect Selection for Genetic Benefits

Females can compensate for costly choosiness if it increases offspring fitness. This is a quality-quantity tradeoff, so that even if a choosy female has fewer offspring than a randomly mating female, she could still end up with more grandoffspring.

There are two main ways that offspring fitness can be elevated. First, as noted earlier, females might choose males that provide more resources, and these are used to nourish each offspring better instead of increasing the total number of offspring produced. Better-nourished juveniles tend to become more fecund adult females or sexually competitive males. This is not a genetic benefit of mate choice. Second, and far more controversially, preferred males might transfer genes that elevate offspring fitness above that of randomly mating females. If this process is strong enough, it could account for both conventional and cryptic female choice. For the process to work, females have to identify males with desirable genes and then preferentially use their sperm, whether by precopulatory (conventional) or postcopulatory (cryptic) choice.

The quality-quantity trade-off raises a major problem with mate choice for genetic benefits. Females often show an open-ended (directional) preference for certain male traits such as brighter colors or more complex songs. Indeed, directional female choice (not to be confused with direct benefits!) provides the main explanation for selection on males to produce ever more elaborate sexual traits to increase their mating success. Male success comes not from being, say, bright; it comes from being bright*er* than rivals. If females choose males based on the expression of traits that do not simply signal information about material resources, this implies that these traits signal heritable variation in fitness ("good genes"). But to overcome the costs of choice, females must identify males who are genetically superior to a randomly chosen male. This must become an increasingly difficult task over time, given that previous female generations have already been selecting males for these same traits. A randomly chosen male of the current generation is expected to have already inherited "good genes." Relentless selection for genes that increase sexual trait expression should eliminate the very variation in "good genes" needed to maintain female choice.

This is a specific example of a general problem in evolutionary genetics: What maintains heritable variation in traits that strongly affect fitness and, ultimately, in fitness itself? A process unique to the maintenance of heritable genetic variation for preferred sexual traits is that these are assumed to reflect the net effect of all the genes that influence a male's ability to assimilate energy and nutrients and maintain his "condition." This measure of condition is an abstraction that equates to fitness prior to the effect of sexual selection on reproductive success. If sexual signals are condition dependent, then they are more likely to signal heritable differences in fitness. Because innumerable genes affect condition, and because condition can also depend on local adaptation (genotype-by-environment interactions), any mutations or minor temporal or spatial changes in the environment will generate variation in genes that improve condition.

Even with condition dependence and genotype-by-environment interactions, the genetic differences between "average" and "good" males might remain slight. This factor predicts, simply, that female choice for indirect benefits is most commonly found where choice is relatively cheap. For example, when male frogs gather to lek in a pond, and males in the best condition have the highest stamina, then a female that simply arrives when it suits her and prefers the most persistently calling male is effectively choosing for male stamina. The female pays a minimal cost (a randomly choosing female would behave nearly identically), but her preference for calls has a great effect on male trait evolution and on which genes are passed on to future generations.

Importantly, some genetic benefits of mate choice can persist without the need for "good genes." Females might benefit by mating with males with more compatible genes, so that the specific genes maximizing fitness vary among females. An obvious example of mate choice for genetic compatibility is inbreeding avoidance. Rejecting mating with close relatives reduces the level of homozygosity in offspring (because relatives more often share the same versions of genes). This choice tends to increase offspring fitness because heterozygotes are better than homozygotes in a range of activities such as immune defense and physiological performance. Given that inbreeding also has a significant beneficial side (it elevates the number of gene copies identical by descent that are transmitted to future generations), inbreeding avoidance implies that genetic compatibility is a very important fitness consideration. It is, however, harder to explain how this kind of preference can lead to directional selection for elaborate mate traits, simply because it is difficult for a male to signal his compatibility to the majority of females in a population unless his heterozygosity, leading to good condition, implies that he is the son of a recent immigrant to the local population (if this is the case, then most females will find that breeding with this male will lead to avoidance of inbreeding).

What about the Benefits of Male Choice, and How Do the Sexes Differ?

Conventional male choice can be favored for the same reasons as female choice. Some females accrue more resources and end up being more fecund than others or produce higher-quality offspring that have been better provisioned. There can, therefore, be direct benefits of male mate choice. To see why conventional mate choice can be rare, though, consider a polygynous species. A choosy male can increase the number of eggs he fertilizes when he mates if rejecting other females allows him to have a maximally large ejaculate on encountering a highly fecund one. This strategy clearly assumes, however, that mating with the other females would have compromised his sperm stores to a significant degree. But if mating opportunities are rare, the likelihood of becoming sperm depleted is probably too small for a male to benefit by rejecting any mating opportunities that arise.

Other potential benefits of male choice follow similar rules to those for females. Some females will have better or more compatible genes than others, so that males can choose mates for genetic benefits, and indirect selection can favor the evolution of male choice, or even cryptic male choice. In reality, cryptic male choice seems to have evolved to maximize the mean rate of offspring production for a given level of male reproductive investment. Effectively, males strategically allocate limited resources (sperm or courtship effort) to maximize the total number of fertilization across all mating encounters.

Whether one considers cryptic or conventional mate choice, male choice for female traits is expected to be strongest when mate availability (for males) is high, and each mating and any allied consequences require a large investment. The potential exception is prudent choice that can evolve as a response to a highly competitive situation in which some males opt out and ignore the most highly competed-for females. The prediction that choosiness occurs when mate availability is high follows from the fact that a male should reduce his mating effort (or actually reject a female) only if savings in the form of time, energy, or sperm are likely to be useful in the near future. This clearly requires that future potential mates are readily available.

5. WHY DO THE SEXES DIFFER IN CHOOSINESS?

Both sexes can be choosy, but females tend to be choosier. This conclusion is most obvious when surveying the occurrence of elaborate male and female sexual traits that have evolved owing to mate choice. There are few female equivalents to the peacock's train or the male nightingale's song. However, this is primarily a statement about conventional mate choice. If we take into account cryptic choice, there is abundant evidence that, for example, males in many species strategically adjust the size of the ejaculates they transfer to different females.

One potential explanation for greater female choosiness is that males may be more variable than females in terms of the benefits they provide as potential mates. Data do not, however, support this idea. Females can differ greatly in fecundity, while in many species without male parental care the only benefit of female choice is variation in heritable male fitness, which is likely to be modest in scope.

The most general explanation for the greater prevalence of strong female than male choice in nature is based not on the benefits of choice but its costs. Males, in general, pay a larger cost if they are choosy. To see why, we must consider the total investment (tallying up the number and size of gametes, the effort expended on courtship, and the subsequent investment in caring for offspring) per mating. This investment is often much greater for females than for males. The largest source of this asymmetry can usually be found in parental care, although in external fertilizers lacking care it can be a consequence solely of eggs being larger than sperm. The net result is that after each mating, females take longer to return to a state in which they can again mate ("recovery time"), or in some cases they may be at risk of dying while performing costly care, so they never return to the pool of potential mates. In either case the result is often a male-biased *operational sex ratio*, reflecting a situation in which there are many more males than females seeking to mate at any given time (see chapter VII.4).

This asymmetry has considerable consequences for mate availability. When the rate at which a female encounters males is far higher than the rate at which a male encounters females, the delay caused by rejecting a potential mate has very different outcomes for each sex. If a female rejects a male, the next potential mate is never far away. (A Helsinki saying is that one should never run for a tram or a man.) In contrast, a male may have to wait a long time until he encounters another female. Consequently, a male should not reject a less profitable than average mate encounter based on the (rare enough to be irrelevant) prospect of taking advantage of a better one soon. This effect is magnified if females have embarked on an evolutionary trajectory toward being choosy. This means that the relevant mate encounter rates for males drop further: ignoring coercive copulations, only females that accept a particular male usefully qualify as potential mates, so mate availability from a male perspective is now even lower.

Of course, there are situations in which males can choose between two or more simultaneously available females. If the situation excludes mating with both, males effectively have to choose: males will reject one for the other, and nonrandom choice is favored. Similarly, conventional male choice is more likely in systems with biparental care or when males otherwise make a substantial postzygotic investment and cannot care for a limitless number of young. Cryptic mate choice can be even more widespread. Multiple mating by females tends to shorten the interval between matings for males, so that prudent sperm usage can become favored as future mating opportunities become more likely for males. Intriguingly, the complexity of male strategic allocation decisions argues against the stereotype that male mating behavior is indiscriminate. Nevertheless, a large asymmetry between the sexes remains: cryptic male choice only rarely leaves females without offspring, whereas female choice (cryptic or otherwise) often creates a large subset of males without genetic descendants. This imbalance has obvious consequences for the overall level of mating effort by each sex seen in nature.

FURTHER READING

Andersson, M. B. 1994. Sexual Selection. Princeton, NJ: Princeton University Press. *This book, while slightly dated, remains the best starting point for a comprehensive overview of the evolution of mating preferences and general sexual selection theory.*

Bonduriansky, R. 2002. The evolution of male mate choice in insects: A synthesis of ideas and evidence. Biological Reviews 76: 305–339.

Kelly, C. D., and M. D. Jennions. 2011. Sexual selection and sperm quantity: Meta-analyses of strategic ejaculation. Biological Reviews 86: 863–884. *The empirical evidence that males adjust ejaculate size in response to the presence and number of rivals, female "quality," and female mating status is quantitatively summarized.*

Kokko, H., M. D. Jennions, and R. C. Brooks. 2006. Unifying and testing models of sexual selection. Annual Reviews of Ecology and Systematics 37: 43–66. *This review attempts to categorize the numerous theoretical models for the evolution of mate choice so that links between seemingly disparate models are made apparent.*

Pryke, S. R., L. A. Rollins, and S. C. Griffith. 2010. Females use multiple mating and genetically loaded sperm competition to target compatible genes. Science 329: 964–967. *An impressive empirical demonstration of biased sperm usage by female Gouldian finches that offers robust evidence for cryptic female choice.*

Rowell, J. T., and M. R. Servedio. 2009. Gentlemen prefer blondes: The evolution of mate preference among strategically allocated males. American Naturalist 173: 12–25. *A theoretical study exploring the effect of male-male competition on the value of males' investment in courting different types of females.*

Simmons, L. W. 2005. The evolution of polyandry: Sperm competition, sperm selection and offspring viability. Annual Reviews of Ecology and Systematics 36: 125–146. *A highly readable, primarily narrative, review of the reasons why females might mate multiply.*

Slatyer, R. A., B. S. Mautz, P.R.Y. Backwell, and M. D. Jennions. 2011. Estimating genetics benefits of polyandry from experimental studies: A meta-analysis. Biological Reviews 87: 1–33. *The empirical evidence from experimental studies that females mate multiply because they can bias fertilization toward males that elevate offspring performance (fitness) (i.e., for genetic benefits) is quantitatively summarized.*

VII.7

Evolution of Communication

Michael D. Greenfield

OUTLINE

1. Elements of animal communication
2. What communication is
3. How does communication originate and how does it evolve?
4. Evolutionary trajectories: Four examples
5. On the reliability of animal communication

The evolution of animal communication remains one of the more fascinating questions in evolutionary biology, but it is also presents us with some of the more complex problems. Owing to the diverse and often elaborate ways in which animals send, receive, and evaluate messages, animal communication attracts considerable attention from a wide range of professional scientists and the lay public. The conspicuousness of animal communication to the human observer and the expectation that an understanding of communication in nonhuman animals may shed light on our own behavior—the origin of human language in particular—are additional factors that draw our interest. Nonetheless, understanding how animal communication evolves has proven to be particularly challenging, for several reasons. First, communication invariably comprises markedly distinct traits: sending messages and receiving and interpreting them. These separate traits may be subject to very different selection pressures, as when males are the message senders and females are the receivers, or vice versa, but at the same time such disparate traits cannot evolve independently of one another. Second, as with most behavioral traits, animal communication seldom leaves direct fossil evidence, and attempts to reconstruct the evolution of communication normally resort to indirect, comparative (phylogenetic) methods and other means of logical inference.

GLOSSARY

Pheromone. A chemical substance, comprising one or several compounds, that an animal emits outside its body and that influences the behavior or physiology of another individual of the same species.

Phylogenetic Analysis. Use of an evolutionary model such as a tree or other diagram that depicts evolutionary relationships among species, to infer the evolutionary trajectory of a particular trait. Importantly, the model used must be based on traits independent of the trait of interest.

Sensory Bias. A perceptual trait that is widespread in a group of related species and that evolved in a context other than communication. At a later point in evolution, communication signals to which this ancestral perceptual trait is particularly sensitive may arise in one or more of these species, in which case the signals are described as having evolved by "exploiting" a sensory bias.

Signal Intensity. The energy transmitted by a mechanical (sound or vibration) or visual signal or the amount of matter disseminated by a chemical signal, as measured at a particular time and location in reference to the signal source (e.g., molecules per cubic centimeter for a pheromone dispersed in air or water, or watts per square meter for a sound signal in air or water).

Signal-to-Noise Ratio. The intensity of a signal, such as an animal's communication message, divided by the intensity of that signal's modality (e.g., sound, light, vibration), as measured in the local environment at a time when the signal is not being transmitted; the relative conspicuousness of a signal against the background.

1. ELEMENTS OF ANIMAL COMMUNICATION

Animal communication comprises four elements: (1) the individual who plays the role of a signaler, (2) the signal that this individual sends, (3) the individual(s) who plays the role of a receiver of the signal, and (4) the channel along which the signal travels from the signaler to the receiver. Signals include chemical, mechanical (sound and vibration), visual, and electrostatic messages transmitted along an environmental channel that may traverse air, water, or the substrate en route from the signaler to the receiver. Not everything that can be perceived can also be a signal. Whereas some animals are capable of perceiving infrared radiation or the earth's magnetic field, no evidence indicates that they can transmit signals and communicate in these modalities. A feature shared by communication signals in all modalities is some modification of the physical or chemical environment that stands out against the background noise and may be perceived by a receiver.

Various constraints imposed by the physical and biotic environment may favor the evolution of one signaling modality over another in a particular species and for a given aspect of its communication. For example, nocturnal species cannot rely on visual signals unless they can generate bioluminescence, as in fireflies and various marine crustaceans. Similarly, signals intended as alarms that warn conspecifics of impending danger may need to be sent very quickly, a prerequisite that could preclude the use of chemical messages where neighbors are separated by relatively long distances: except in the case of odors transmitted by direct contact between individuals or that serve as territorial markers, chemical signals travel largely by convection in air or water, that is, wind or current. Diffusion of odors would be too slow a process for all but the shortest interneighbor distances, and convection would not allow signalers sufficient control over the direction in which messages are sent. Thus, in many species alarm signals tend to be visual, acoustic, or vibrational.

The need for fine control over message transmission may further constrain signaling modalities. A signaler normally has intended receivers (e.g., potential mates or conspecific neighbors), as well as unwanted eavesdroppers (e.g., male rivals or approaching predators), and signals may need to satisfy conflicting demands: the message may be expected to reach a certain intended receiver or as many of them as possible, but it may also be expected to remain hidden from unintended receivers. These opposing pressures may select for signals that are transmitted along a "private channel," one where the signaler can retain control over the dissemination of the message. For example, we might expect a signaler in an acoustic species to forgo its sound messages, which disperse across long distances and more or less in all directions, and use substrate vibration instead when predators are present. Whereas vibrational messages may be limited by distance and require a particular substrate, for example, vegetation having certain mechanical properties, it is this discrete nature of vibration that may prevent unintended receivers, save those few who happen to be situated on the same substrate, from perceiving the messages. On an evolutionary scale, this shift from airborne sound to substrate vibration is observed in various clades of acoustic insects: katydid species found in regions subject to high levels of predation by insectivorous bats that localize the sounds of their prey tend to rely more heavily on vibration. Similarly, males in various insect species that broadcast intense calling songs attractive to females over relatively long distances tend to replace sound with vibration and tactile signals in the courtship that ensues once an attracted female is proximate.

In other cases signals may be primarily selected to reach as many receivers as possible, perhaps with relatively little regard to eavesdroppers. Territorial markers may fall into this category, as such signals are normally intended for conspecific rivals. Here, signals whose broadcast continues even in the absence of the signaler may be favored to provide a more or less permanent message for potential usurpers of the site. Signals that mark territories tend to be chemicals that have been applied to the substrate in key locations, often along the region's boundary. The only other type of signal that could satisfy this demand of permanence would be the special visual message provided by architectural construction. While such construction may play a role in the courtship communication of some vertebrate and invertebrate species—for example, Australian bowerbirds or fiddler crabs whose elevated burrow entrances influence female visitation—it does not appear to be prominent in communication about territories.

A very general constraint on signaling modality, largely overlooked, is the body size of the signaler—and of the receiver. Size poses special physical constraints for both acoustic and visual signaling: animals below certain minimum dimensions are generally unable to radiate sound waves into the surrounding medium—air or water—and they are equally unable to bear pressure-sensitive organs (tympana) sensitive to far-field sound waves. Thus, the smallest insects, like ants, do not tend to rely on acoustic communication for courtship or other functions. Moreover, species at the lower end of the size range at which acoustic communication is feasible are typically restricted to using high sound frequencies, which can be more efficiently radiated and perceived by small organs. Vision, too, is subject to a size constraint, as eyes, either compound or single lens, capable of forming images must exceed a minimum diameter. Again, the smallest insects appear not to use visual communication, or they extract only very crude information such as patterns of movement from a signaler. However, chemical communication is ubiquitous among animals, and it is also found among single-celled organisms. There are apparently no fundamental size constraints on the production and perception of odor, and the chemical modality may have been the first one to evolve in communication by living organisms.

A general property of communication is the transfer of "information" from the signaler to the receiver. Information in this context may be considered as a reduction in uncertainty about the identity, quality, ability, or intentions of the signaler. A signaler may also transfer information about its environment, such as presence of predators or value of local resources. Because effective information transfer is often subject to strong selection pressure, certain signaling modalities may be favored more than others. For example, reliable information about a signaler's individual identity may demand considerable detail so that he or she can be distinguished from other members of the local population. Constraints on both the production and perception of signals may preclude some modalities because the information content of signals would be too low: either the signaler may be unable to fine-tune its broadcast—for example, only a single compound might be available as a chemical message—or the receiver may be unable to discern subtle variants of the message, such as different levels of concentration of an odor or minor changes in its chemical composition. Consequently, acoustic or visual communication might be used, at least for individual recognition.

2. WHAT COMMUNICATION IS

One view, by no means universal, is that communication occurs when a signaler transfers information and the receiver then modifies its behavior and/or physiology such that both individuals benefit (have a net increase in expected fitness). This definition would include many examples of messages sent between different species, including mutualistic animal-plant interactions such as pollination. Cases in which information is inadvertently provided by prey to potential predators, however, would not be considered communication. At the intraspecific level, traditionally the level at which communication is considered to occur, situations arise in which the stipulation of mutual benefits accrued by both parties might be questioned. In sexual communication, a male whose "quality" is relatively inferior may nonetheless encounter a female and mate, possibly by remaining as a "satellite" in the vicinity of a male who was broadcasting a superior signal. Thus, the benefit that the female accrued by perceiving and evaluating that superior signal is unclear. However, if satellites are rare and a female would normally pair with a high-quality mate by virtue of perception and evaluation of the signals, the stipulation of mutual net benefit would be upheld.

Communication may be further delimited as a process in which the information transferred by the signaler is perceived by the receiver's nervous system, often to be stored in its memory. It is the retrieval of this stored information at a later time that results in the change in behavior and/or physiology noted earlier. According to this specification, communication would not include events in which one animal exerts physical force on a second animal, who then withdraws as a direct reaction to that force. Similarly, one animal might transfer a resource, such as a food item, to another, who then ingests the item and achieves improved growth and development as a result. Unless the second individual shows evidence of a behavioral response to the chemical, mechanical, or visual stimulus introduced by the food item, the interchange is simply nutritional.

The most difficult issue concerning the designation of an event as communication is the distinction between *signals* and *cues*. This distinction is most critical in the context of the present chapter because it is directly involved in the evolution of communication. Animals nearly always emit various inadvertent stimuli that may be perceived by another individual. In many cases such perception may benefit that individual, as well as the individual who emitted the stimulus, however inadvertently. For example, arthropods generally incorporate certain hydrocarbon substances in or on their cuticle that may function in protection against water loss or in other physiological processes, but these substances may also reveal the presence of the animal to other individuals. Thus, a male may localize and identify a conspecific female by detecting her cuticular substances, and even determine that she is mature and receptive. Here, both parties benefit from the message provided by the chemical stimulus, but does the message represent a communication signal? According to one perspective, it does not if the chemical stimulus shows no evidence of having undergone any specific evolution in the context of information transfer between individuals. Such evidence might include (1) energy expended specifically during the production and release of the substance that exceeds the expense necessary for the primary or original physiological function of the substance, (2) specialized structures that improve the release of the substance to the outside environment so that it can be more readily perceived by other individuals, and (3) a specific daily schedule for release of the substance that coincides with the periodicity of mating behavior and male receptivity rather than physiological needs, such as prevention of desiccation. Unless some of these indications are present, the stimulus is simply a cue, albeit one to which a male receiver responds. Should the system of chemical production and release evolve, however, toward a process in which any of the described features appear, one may consider the stimulus to be a specialized communication signal. In fact, it is not usual for cuticular hydrocarbons to evolve from ancestral, inadvertent cues to representing components of a sex pheromone.

3. HOW DOES COMMUNICATION ORIGINATE AND HOW DOES IT EVOLVE?

Animal communication requires an audience, either a specific receiver(s) that the signaler has perceived or expects to be in the surrounding area, or some unknown number of potential receivers who are likely to be present within broadcast range of the signaler. Thus, it is imagined that communication evolves in one of two ways, either by means of reciprocal modifications in the behavior of signalers and receivers that proceed in an alternating fashion or by means of modifications in the behavior of signalers in response to a preexisting preference or perceptual capability in receivers. Within the domain of sexual selection, wherein one typically focuses on the origin of mating signals, these two processes correspond with the well-known coevolutionary or Fisherian mechanism for the evolution of traits and preferences, or the evolution of traits by "exploitation" of sensory biases. In other words, the evolutionary origin and continued modification of signaler behavior in the context of communication does not seem likely without some corresponding evolutionary response in receiver behavior. A further clarification is that the descriptions of the two processes of behavior modification refer to events at their origin. Under the Fisherian (or coevolutionary) mechanism, evolutionary changes are expected in both parties, whereas under the sensory bias (or exploitation) process evolutionary change occurs only in signalers. However, the sensory bias process does not exclude the possibility that receiver traits may also undergo modification at a later point during evolutionary time. Rather, it is quite likely that a preexisting perceptual ability eventually experiences some fine-tuning in response to the newly evolved signal. Thus, the primary difference between the two processes is the relative timing of evolutionary change in signaler and receiver traits.

The need for an audience was recognized early in studies of animal communication, and one hypothesis proposed that communication could evolve provided the same or closely linked genetic elements controlled both signaler and receiver traits. This hypothesis, originally known as *genetic coupling*, was invoked on various occasions to explain how mating communication might change during the process of speciation: without such coupling, any change in mating signals would result in removal of the new variants from the population because no females would pay attention to them. The basic hypothesis continues to receive attention from evolutionary biologists studying species recognition in the speciation process as well as mate evaluation in sexual selection. Some limited support for the notion that common genetic factors pleiotropically influence both signaler and receiver traits has been found at the level of species recognition in some animal groups and signaling modalities, for example, visual communication in fish (Medaka) and *Heliconias* butterflies, acoustic communication in Hawaiian crickets (*Laupala*), and chemical communication in *Drosophila*. These findings invite the question, How might a motor trait and a perceptual trait share a common basis at mechanistic (physiological) and genetic levels?

4. EVOLUTIONARY TRAJECTORIES: FOUR EXAMPLES

A classical approach to the evolution of animal communication described a general process of *ritualization*, wherein a postural movement or displacement activity, perhaps initially serving a physiological function such as thermoregulation, becomes incorporated in a message. Over the course of evolution, the various components of the activity gradually become standardized. One may recognize elements of this so-called ritualization process in the sensory bias mechanism noted earlier, as well as in the general transition of cues to signals. The following four examples present possible trajectories by which different forms of animal communication may have evolved.

Oviposition Marker Pheromones

The egg-laying behavior of female tephritid flies (true fruit flies) includes locating a fruit of the host species that is in an appropriate state of maturation and then depositing an egg if another female did not do so previously. The focal female determines whether she has been preceded by an earlier female via detecting an odor left on the fruit surface during oviposition. By sensing and responding to this odor, an oviposition marker pheromone, the focal female avoids having her offspring compete with a larva that is probably in a more advanced state of development and thus likely to win the resource in the event of a contest. By leaving an odor on the fruit surface, the first female to arrive prevents her offspring from having to compete for the resource, a desirable result even though her offspring would probably win. A parsimonious hypothesis for the evolution of oviposition marker pheromones is that they originated as a chemical cue left inadvertently on the fruit surface during deposition of an egg. Because a subsequent female would be under particularly strong selection pressure to shield her offspring from competition, any perceptual ability allowing her to detect the cues of a prior oviposition would have been favored. Once such detection had evolved to a certain level of sensitivity, selection would then have favored some modification of the cue rendering it more conspicuous and thereby ensuring that the offspring of the first female would avoid a costly encounter. Thus, a cue left during oviposition would have evolved into a specialized marker pheromone. Eventually, some refinement of the avoidance responses to the specific marker pheromone would also be anticipated. Similar signals may have evolved in parasitoid wasps that deposit one egg or larva on a host insect. In general, one might expect the evolution of this mode of communication in cases where a resource patch can accommodate one and only one offspring; while sufficient resource patches exist it will pay a second female to leave the patch and search for a previously undiscovered one.

Dance Language of Eusocial Bees

The communication by which Western honey bees (*Apis mellifera*) recruit numerous foragers to a distant food source has been largely deciphered owing to a series of painstaking observations and ingenious experiments performed over the past century by Karl von Frisch and his colleagues and students: having found a valuable patch of flowers, a worker honey bee returns to the colony and repeatedly performs a figure eight movement on a vertical surface in the darkness of the hive interior. Other workers sense the orientation and duration of the figure eight movement via substrate vibration, tactile stimuli, and near-field sound emanating from the dancer, and they use this information to infer the azimuth, that is, the horizontal angle between the sun and the floral patch as seen at the hive entrance, and approximate distance to the patch. Odor cues—scent from the discovered flowers—on the dancer may fine-tune the navigation of the recruited worker once she arrives in the vicinity of the resource patch.

Studies by Claudia Dreller and Wolfgang Kirchner on other members of the genus *Apis* found in Asia shed some light on how the complex recruitment communication in *A. mellifera* may have evolved. Various Asian honey bee species, such as *A. dorsata*, *A. florea*, and *A. laboriosa*, do not nest within enclosed cavities, and successful foragers in these species perform recruitment dances in the open while moving on upper, horizontal surfaces of the colony. Notably, dance signals in these species do not include sound. *A. florea* is believed to be the extant species of *Apis* closest to the root of the genus, suggesting that this dance style may retain some aspects of the ancestral form. One Asian species, *A. cerana*, is a cavity nester like *A. mellifera* and, like *A. mellifera*, performs its dances on vertical surfaces in the dark interior of the colony, and the dances do include sound. We can infer that foragers in the ancestral *Apis* species made certain displacement movements on the exposed hive surface on returning to the colony and that these movements reflected the distance they had just flown and the direction in which they might depart on their next trip. Thus, colony members would have been under some selection to pay attention to these cues because their own foraging would then be more efficient. That is, the colony would have served as an "information center" at which workers could learn the local distribution of floral resources. But because the colony is more than just a center at which information might be passively acquired, each colony member would benefit from an increased foraging efficiency by other members. Consequently, it has been proposed that selection favored the modification of inadvertent cues into specialized signals, that is, a dance that conveyed more accurate information about those resources. The switch to nesting inside dark cavities in the more derived species of *Apis* would have then demanded two further modifications of dance signals. First, these honey bee species could no longer rely on vision for evaluating information in the dance, but the incorporation of near-field sound signals appears to have retained high information content. Second, in the absence of an exposed horizontal surface on the colony, workers could not directly indicate the direction toward floral resources by means of dance movements. However, the development of an indirect mechanism in which the angle between the straight segment of the figure eight and a vertical line represents the horizontal angle between the sun and the floral patch seems to have solved this difficulty. It is this abstract representation of direction that has led these honey bee dances to be commonly referred to as a "language."

Courtship Pheromones in the Lepidoptera

Long-distance communication during pair formation in the Lepidoptera typically involves visual signals (butterflies) or advertisement pheromones emitted by females (moths), but once the male and female have established contact, male courtship pheromones often mediate the final outcome of the encounter. Chemical analyses in some species indicate that these male courtship pheromones may be derived from substances acquired during feeding by larvae in some cases, and by adults in others. In the arctiine moth *Utetheisa ornatrix*, a species studied intensively in the laboratory of Thomas Eisner at Cornell University over many years, the courtship pheromone is a volatile substance that the male produces by converting a chemical, a pyrollizidine alkaloid (PA) acquired from the host plant—several species of legumes in the genus *Crotalaria*— during larval feeding and sequestered in the body. Males also transfer a small quantity of this unconverted host-plant chemical

(PA) to the female along with sperm at the time of mating, and the toxic properties of the PA appear to confer some protection against natural enemies to the female, as well as protecting her eggs. Thus, transfer of the PA represents a "nuptial gift," and the courtship pheromone serves to signal the presence, and possibly the quantity, of this gift to a discriminating female. Because both male and female *U. ornatrix* perceive and respond positively to various chemical cues from their host plant, one may propose that the male courtship pheromone evolved from an ancestral cue, the PA in this case. Evolution of the pheromone may have taken place for any of several reasons. First, the PA in its original form is probably not sufficiently volatile to serve in communication, even at close range. Second, by converting the PA to another substance, the male may achieve an improved signal-to-noise ratio when courtship occurs on or near the host plant: How would a female discern a courting male from the background odor of the plant were he to use an unconverted host-plant chemical as a pheromone? Finally, the primary behavior that an ancestral cue such as the PA might elicit in *U. ornatrix* females may be oviposition, not mating. However, by using a different but chemically related substance as a courtship signal, a male ensures that a female will still perceive and respond but that her response will occur in a different (nonoviposition) context, such as courtship. This specific courtship response assumes that the female receiver, like the male signaler, has undergone some modification during the course of evolution.

Acoustic Communication in the Lepidoptera

Whereas the vast majority of pair formation in moths is accomplished via long-range advertisement pheromones emitted by females and short-range courtship pheromones presented by males, a small but critical percentage of species use sound in mating communication. How did these exceptions originate, and what general lessons about the evolution of communication can we learn by studying them? We begin by noting that the perception of sound is actually quite common in moths, being prevalent in three major superfamilies (Pyraloidea, Geometroidea, and Noctuoidea) and having evolved independently between 7 and 10 times. Hearing in moths occurs largely in the domain of ultrasonic (> 20 kHz) frequencies and appears to have evolved and to function in the context of avoiding predatory bats: phylogenetic and biogeographic analyses suggest that moth hearing arose coincident with the origin of echolocation signals in insectivorous bats, about 55 million years ago. This inference is bolstered by the observation that hearing has been lost secondarily in various moth species found in regions lacking insectivorous bats or that fly at times when bats do not hunt. That is, the primary function of moth hearing is the detection of ultrasonic echolocation signals emitted by bats, which then allows the moth to make appropriate flight movements and evade predation.

Against the evolutionary background of widespread hearing, some moth species also emit sound, in most cases in the ultrasonic frequency range. In some of these species the sounds function, like hearing, in defensive behavior: they may "jam" the bat's echolocation system, or they may serve as an aposematic (warning) signal in some chemically protected moth species. But in a few phylogenetically unrelated species of pyraloid and noctuoid moths the sounds are produced only by males and serve either in close-range courtship (similar to the use of the male pheromone by *U. ornatrix*, described earlier) or as a long-range advertisement to females. In these latter species the sounds are essentially the equivalent of male calling songs found in more familiar acoustic species such as crickets or frogs.

Given the relative prevalence of hearing and rarity of courtship and calling songs, it is most likely that acoustic communication in moths evolved via "exploiting" the ancestral auditory perception. Beyond this basic inference, however, various questions remain unsolved. As in *U. ornatrix*, a female moth would not be expected to respond to a male sound as if it were a courtship message but rather in an antipredator manner, that is, by undertaking a negative, evasive movement. Thus, the origin of acoustic communication in this group is not completely clear. One possible explanation is that moths situated on the substrate, as courting females normally are, may evade bats by remaining stationary, a response that eliminates inadvertent sound that can reveal the moth's presence to a bat searching for prey on vegetation or on the ground. Consequently, a male who emits an ultrasonic song may render a nearby female immobile and therefore more readily courted. But in some moth species that broadcast long-range male calling songs, the male call is delivered with a rhythm that is fundamentally different from that found in most bat species. Here, females distinguish between bat echolocation signals and male calls, and they actively run toward the latter. Evidently, the female response, as well as the male signal, has experienced some fine-tuning subsequent to the initial appearance of male song.

5. ON THE RELIABILITY OF ANIMAL COMMUNICATION

While examples of cheating and deception among nonhuman animals always draw considerable interest from human onlookers, it is becoming increasingly evident that animal communication is by and large an "honest" affair. Honesty in this sense refers to the communication signal's being a reliable indication of the signaler's identity, quality, physical ability, intention, information, and so forth. The expectation that signals are reliable is based on economic grounds: should receivers pay attention to signals that do not reflect the quality of the signaler, for example, they will ultimately suffer reduced fitness. Under such circumstances, receivers are likely to diminish and eventually cease their evaluation of these signals, which, in turn, should influence signalers to discontinue such broadcasts. Production of ignored signals would not be worth the inevitable costs in time, energy, and risk.

It is argued that the need for signal reliability has played a major role in shaping the evolution of animal communication wherever the potential for conflict exists between signalers and receivers. Importantly, some degree of conflict is expected in most interactions, including those between members of the same species. For example, a male and a female may have a common goal of producing offspring, but they also have very different specific objectives related to achieving this goal. A male might be less discriminating and be expected to mate with most females encountered, whereas a female may forgo mating with a particular male if he is of low quality, and future mating possibilities exist. Thus, any signal that pretends to represent male quality would have to be reliable in that it could be produced only by a male who actually bears that quality. The male courtship pheromone of the moth *U. ornatrix* shows how signal reliability might function. The chemical precursor of this pheromone is a substance in the host plant that forms part of a nuptial gift transferred at mating, and the only way the male can acquire the precursor is by feeding on the host plant and sequestering it. Thus, on detecting the pheromone, a female is at least assured that the courting male has successfully foraged on and accumulated the substance—behaviors that might be inherited by her offspring. Additionally, the male can potentially transfer a nuptial gift, the size of which may be commensurate with the concentration of the pheromone, although it is yet possible that the male could withhold his offering at the last moment. That is, he might conserve his gift materials for a future mating opportunity, particularly in the case where he has evaluated the female to be of low reproductive potential.

Another common way in which signal reliability can function is via intensity. If the broadcast of an intense signal demands considerable energy, and available energy is an indication of male quality, only high-quality individuals are expected to signal strongly. Should a low-quality individual imitate a high-quality one by broadcasting an incongruously intense signal, he would deplete so much of his energy store that he might not be able to avail himself of the rewards normally accruing to a strong signaler. Thus, signal strength is expected to be proportional to quality.

Signal reliability may be much less critical in communication in highly social species, particularly those in which most individuals tend to be close genetic relatives. Interindividual conflict is expected to be relatively weak here, and there are few a priori reasons for individuals to broadcast signals that misrepresent themselves or their information. For example, the dance language in honey bees, while not perfectly accurate, is a reasonably good indication of the location of floral resources, and there are no apparent mechanisms by which this accuracy is safeguarded against returning foragers who might deliberately mislead recruited individuals.

The preceding arguments for honesty notwithstanding, dishonest communication is sometimes observed in animals, and it merits our consideration. Some cases of unreliable signals can be explained as communication that on average affords a net benefit to the parties concerned. That is, signals may misrepresent quality or intention until a certain point, beyond which their deception would be selected against. But perhaps a more common source of signal unreliability is the dynamic nature of evolution itself. Populations are continually subject to environmental change, as well as to the arrival of migrants from neighboring populations, in which traits may be somewhat different. Our observations of animal communication are only snapshots of signaler and receiver traits, which may sometimes be less than fully reliable: for example, females may evaluate male signals according to criteria that promised quality under previous environmental conditions but not necessarily under current ones. Thus, the level of reliability in animal communication may reflect the extent to which populations have attained an equilibrium state in which traits are fully adapted to an environment that is, for the moment, stable.

FURTHER READING

Boake, C.R.B. 1991. Coevolution of senders and receivers of sexual signals: Genetic coupling and genetic correlations. Trends in Ecology & Evolution 6: 225–227. *A thorough discussion of the possibility that signal and response traits are controlled by the same genes and the potential importance of this shared control in the evolution of mating communication.*

Dussourd, D. E., C. A. Harvis, J. Meinwald, and T. Eisner. 1989. Pheromonal advertisement of a nuptial gift by a male moth (*Utetheisa ornatrix*). Proceedings of the National Academy of Sciences USA 88: 9224–9227.

Dyer, F. C. 2002. The biology of the dance language. Annual Review of Entomology 47: 917–949.

Endler, J. A., and A. L. Basolo. 1998. Sensory ecology, receiver biases, and sexual selection. Trends in Ecology & Evolution 13: 415–420. *A concise description of the sensory bias mechanism of signal evolution and the criteria for confirming its presence.*

Greenfield, M. D. 2002. Signalers and Receivers: Mechanisms and Evolution of Arthropod Communication. Oxford: Oxford University Press. *The introduction discusses definitions of animal communication with emphasis on distinguishing signals and cues.*

Nakano, R., T. Takanashi, N. Skals, A. Surlykke, and Y. Ishikawa. 2010. To females of a noctuid moth, male courtship songs are nothing more than bat echolocation calls. Biology Letters 6: 582–584.

Nufio, C. R., and D. R. Papaj. 2001. Host-marking behavior in phytophagous insects and parasitoids. Entomologia Experimentalis et Applicata 99: 273–293.

Searcy, W., and S. Nowicki. 2005. The Evolution of Animal Communication: Reliability and Deception in Signaling Systems. Princeton, NJ: Princeton University Press. *A thorough treatment of the expectation of "honesty" in animal communication; examples are mostly taken from bird plumage, bird and frog vocalizations, and the leg-waving displays of crabs.*

Shaw, K. L., and S. C. Lesnick. 2009. Genomic linkage of male song and female acoustic preference QTL underlying a rapid species radiation. Proceedings of the National Academy of Sciences USA 106: 9737–9742.

Steiger, S., T. Schmitt, and H. M. Schaefer. 2011. The origin and dynamic evolution of chemical information transfer. Proceedings of the Royal Society B 278: 970–979. *A particularly novel perspective on the evolution of pheromones and chemical communication based on broad consideration of the chemical ecology of animals as well as plants.*

Sueur, J., D. Mackie, and J.F.C. Windmill. 2011. So small, so loud: Extremely high sound pressure level from a pygmy aquatic insect (Corixidae: Micronectinae). PLoS ONE 6: e21089. *Description of an exceptional insect that despite its small size, is able to broadcast an intense advertisement song. Cases such as this one do not refute the expectation of signal reliability, which predicts that signal characteristics and the "quality" of individuals are correlated within a population.*

VII.8

Evolution of Parental Care

Mathias Kölliker, Per T. Smiseth, and Nick J. Royle

OUTLINE

1. Natural diversity in forms of parental care
2. Origin and evolution of parental care
3. Evolutionary maintenance of parental care
4. Genetics and epigenetics of parental care
5. Sociality beyond family

Across the animal kingdom there are many species in which parents enhance their offspring's fitness by providing various forms of care. In some animal taxa, such as birds and mammals, almost all species have parental care, and parental care is complex and necessary for offspring survival. In other taxa, such as fish, reptiles, amphibians, or invertebrates, parental care occurs more sporadically, is more variable, is often less complex, and is not always obligate. The diversity in the forms of parental care is vast, ranging from the choice of oviposition sites to providing food, shelter, and protection to the young (table 1). These different forms of care are adaptations to one or more ecological challenges and form part of an animal's life history and reproductive strategy. The ecology and life history of a species determine the benefits and costs associated with parental care and, hence, the likelihood that care evolves and is maintained. Evolutionary conflicts between parents and offspring, among siblings, and between male and female parents underlie the evolutionary maintenance of parental care. These conflicts generate novel selection pressures on parental care: how much is provided, who provides it, and how it is allocated among offspring. The evolution of parental care is a coevolutionary process because parental and offspring fitness is determined not only by their respective phenotypic and genotypic characteristics but also by the way in which parents and offspring interact with one another. As a result, evolutionary trajectories of parental care traits can be diverse and complex and play an important role in the evolution of other forms of social behavior, such as eusociality.

GLOSSARY

Correlational Selection. Selection favoring particular combinations of parent and offspring traits rather than individual traits in isolation.

Inclusive Fitness. An individual's own fitness measured through its own survival and reproduction (direct fitness) in addition to the survival and reproduction of related individuals weighed by genetic relatedness (indirect fitness).

Parental Care. Any parental trait—behavioral or other—that increases the fitness of a parent's offspring and that is likely to have originated and/or is currently maintained for this function. Measurement: parental behavior/phenotype.

Parental Effect. The causal effects that the parent's phenotype has on the offspring's phenotype, including its growth and survival, over and above direct effects due to genes inherited from parents. Measurement: change in offspring phenotype due to parental care.

Parental Effort. The combined fitness costs—in terms of reduced mating effort and/or somatic effort—that parents incur owing to the production and care of all offspring in a given biologically relevant period, such as a breeding attempt; ultimate measure of cost. Measurement: reduction in mating and/or survival prospects due to parental care.

Parental Expenditure. The expenditure of parental resources on parental care of one or more offspring; proximate measure of cost. Measurement: time, energy, food spent on parental care.

Parental Investment. Any investment by the parent in an individual offspring that increases the offspring's chance of survival (i.e., offspring fitness) at the cost of the parent's ability to invest in other offspring (Trivers 1972).

Parent-Offspring Coadaptation. The coevolutionary process and outcome of adaptation by parents to variation in offspring traits, and adaptation by offspring to variation in parental care.

Parent-Offspring Conflict. The difference between the optima for parental investment (in terms of inclusive fitness) from the perspective of a gene expressed in the caring parent and a gene expressed in its offspring. Offspring are usually selected to demand more resources than the parent should provide.

Parental care is of interest in evolutionary biology because it is a prime example of an altruistic trait; that is, the recipients of care (offspring) gain a fitness benefit, while the donors (parents) pay an evolutionary cost. Understanding the circumstances under which this occurs is the key to explaining the origins and diversity of forms of care, and sociality. However, although parental care involves altruistic behavior toward offspring, there is also scope for conflict. Asymmetries in relatedness among family members lead to genetic conflicts over the amount and duration of parental care. Furthermore, the genetic bases of traits expressed in families are complex owing to the transgenerational effects on phenotypic expression and the coevolutionary dynamics of parental and offspring traits. This means that parental care is not just a target for selection but generates novel genetic and phenotypic variation that can contribute to the evolution of more complex forms of sociality, such as eusociality (see chapters VII.9, VII.10, and VII.13).

1. NATURAL DIVERSITY IN FORMS OF PARENTAL CARE

Parental care is traditionally studied in animals displaying conspicuous and highly elaborate forms of care, such as bird species where both parents undertake hundreds of foraging trips a day to feed their offspring; or in mammals, where females feed their offspring with highly nutritious and metabolically costly milk. These forms of care are evolutionarily derived and the norm only among birds and mammals. Nevertheless, elaborate parental care is not confined to mammals and birds. Clutton-Brock (1991) provided one of the first comprehensive compilations that detail the diversity of forms and taxonomic distribution of care. For example, various reptiles, fish, insects, arachnids, mollusks, brachiopods, and bryozoans nourish developing embryos via a placenta-like structure similar to that found in mammals, and there are even a small number of amphibians, fish, insects, arachnids, crustaceans, and leeches among which parents provide food for their offspring after hatching or birth, as in birds. Simpler forms of care are more widespread across the animal kingdom, and research on such basal forms of parental care provides important insights into the evolutionary origins of, and later modifications to, parental care. Relatively simple forms of parental care continue to be discovered, as illustrated by recent reports of a deepwater squid (*Gonatus onyx*) whose females tend their egg mass by holding it in their tentacles, and a caecilian amphibian (*Boulengerula taitanus*) whose offspring feed by peeling off and eating the outer layers of their mother's nutrient-rich skin.

There is a vast natural diversity in the forms of care provided by parents that enhance offspring fitness (table 1). These forms of care can be understood as adaptations for dealing with one or more ecological hazards, such as predation, parasites, and food shortages. The most basal and widespread form of care is the provisioning of gametes (eggs) with extra nutrients, such as proteins and lipids, beyond the minimum required for successful fertilization. By definition, this form of care is provided by females, as females are defined as the sex producing the larger gametes. Nevertheless, males of some species may contribute to gamete provisioning by offering nuptial gifts to females. Oviposition- and nest-site selection, as well as nest building, count as parental care provided this behavior enhances the fitness of offspring (usually survival) rather than that of the parent (the parent's fecundity). The construction of nests and burrows is a widespread form of care in both vertebrates and invertebrates. Nests and burrows provide protection from predators or infanticidal conspecifics but also promotes interactions between parents and offspring owing to spatial aggregation, which is an important factor in the evolution of parental care (see further discussion). More advanced forms of parental care include attendance of eggs or offspring, and food provisioning. Egg or offspring attendance/brooding are very diverse forms of care that include remaining with eggs or offspring in a fixed location for protection against natural enemies, or carriage of eggs or offspring by parents. The various forms whereby parents provide food to their young, whether in the form of mass provisioning of food prior to hatching or progressive provisioning of food after hatching, represent some of the most highly derived forms of care.

This diversity of form and broad taxonomic distribution of parental care requires an understanding of how and why it evolved. Many organisms do not have parental care. So a key question is, What are the conditions that promote the evolutionary origin of parental care and its maintenance and later modification?

2. ORIGIN AND EVOLUTION OF PARENTAL CARE

By definition parental care enhances offspring fitness. At the same time, it usually comes at a cost to the parents. Parents expend time, energy, and resources delivering care that can impair the parent's ability to raise other offspring. This evolutionary cost forces parents to balance how much care they direct to an individual

Table 1. The main forms of parental care

Form of parental care	*Definition and taxonomic distribution*
Provisioning of gametes	Eggs are provided with yolk proteins and lipids beyond the minimum required for successful fertilization; in virtually all animal groups, including birds, reptiles, amphibians, spiders or insects, and other invertebrates; to a lesser extent in mammals.
Oviposition- and nest-site selection	Parents choose sites with a suitable microclimate for offspring development/survival and/or that is safe from predation; in virtually all animal groups, including birds, mammals, and insects.
Burrowing and nest building	Cavities are burrowed in a substrate (e.g., soil, wood); many insects, fish, some mammals, reptiles, and birds. A nest is built from materials found in the environment, such as mud and plant materials, from processed plant materials, such as paper or materials produced by the parents themselves, such as silk and mucus; many birds and mammals, and some amphibians, fish, spiders, and insects.
Egg attendance	Parents remain with the eggs at a fixed location after oviposition, usually the oviposition site and often in a burrow/nest, after egg laying; relatively common in amphibians, fish, and invertebrates. Sometimes associated with parental thermoregulation of eggs (e.g., incubation in birds).
Egg brooding	Eggs are carried after oviposition externally or internally in specialized pouches; found in some amphibians, mouthbrooding fishes (e.g., cichlids), insects, spiders, crustaceans, and other invertebrates.
Viviparity	Fertilized eggs are retained within the female reproductive tract during embryonic development; ubiquitous in marsupial and eutherian mammals, more sporadic in squamate reptiles, fish, insects, onychophorans, mollusks, tunicates, echinoderms, arachnids, and bryozoans.
Offspring attendance	Parents remain with offspring after hatching/birth either at a fixed location or by following the offspring as they move around; ubiquitous in mammals, birds, and fish, but also found in insects and spiders.
Offspring brooding	Hatched/born young are carried externally or internally in specialized pouches. External carrying occurs in some frogs, primates, scorpions, and a range of marine invertebrates. Internal carrying occurs in marsupials and mouthbrooding fish.
Offspring food provisioning	Ranges from the transfer of nutrients through a placenta and mass provisioning of brood chambers prior to birth and hatching to progressive provisioning of prey items or specialized secretions (milk) after hatching or birth; ubiquitous in mammals and birds, but also found in some fish, amphibians, insects, spiders, and other invertebrates.
Care after nutritional independence	Offspring are cared for after they have reached the age of nutritional independence, for example, to help offspring in competition with conspecifics or to protect them against natural enemies; found in some birds (cygnets), mammals (hyenas, red squirrels), and insects (burying beetle, earwigs).
Care for mature offspring	Requires long life span and overlapping generations and is correspondingly limited to few taxa. For example, grandparental care in humans: assistance to offspring that have become parents. Other examples include some primates, elephants, and hyenas.

offspring against the number of additional offspring they could produce now or in the future. It is only because offspring are genetically related to parents that such behavior may be worthwhile and makes it possible for parental care to evolve. But even when parents and offspring share genes in common, parental care can evolve only if the benefits of care to offspring outweigh the costs to the parent. Why this is so can be illustrated by the following hypothetical example. Imagine a mutation in a female that leads her to stay with her offspring and protect them against predators. Will this mutation increase in frequency in the next generation and spread through the population over time? By being protected from predators the offspring will, on average, have an increased probability of surviving. Half the female's offspring will carry this mutation, but no other offspring in the population will, and the mutation can spread only if the female directs her care exclusively toward her own offspring. However, even if the female does so, selection does not necessarily favor the spread of the mutation, as it depends on the costs to the female. For example, the female herself may be more exposed to predators by remaining with her offspring, or she may delay or impair her future breeding owing to the time and resources spent on the defense of her current offspring. Thus, answering questions about the origin of parental care requires an understanding of the factors that affect parent-offspring relatedness, the fitness benefits to offspring, and the fitness costs to parents.

Temporary spatial aggregation of parents and their offspring is an important condition that promotes the evolution of parental care for two main reasons. First, it ensures that in an ancestral population where parental care originates, those few individuals that carry a mutation for parental care will pass the benefits of care on to their own offspring, which are also likely to be carriers of the mutation. Second, it increases the probability that parental care can be provided effectively and will improve offspring fitness. If offspring were widely dispersed, it would be more difficult, risky, and time consuming for parents to provide care for their offspring, thus increasing the costs of care to parents. Selection therefore favors the origin of parental care in species that already produce their offspring in clutches or litters and have offspring that remain near the parent for the duration of care.

Variation in life histories of species is another important condition that influences the evolution of parental care. Some organisms develop very quickly, while others have very slow developmental times. Some reproduce only once during their lifetime and produce many offspring in a single clutch or litter (i.e., are *semelparous*), while others produce their offspring in batches spread over their lifetime (i.e., are *iteroparous*). If parental care enhances offspring development and thereby the fitness prospects of these offspring, it should evolve more readily in a species with relatively slow development, because the beneficial effect of care can accumulate over a longer period of time. Parental care may also be more likely to evolve when the prospects for future reproduction of parents are low, and the value of the current offspring is correspondingly high.

Irrespective of an organism's life history, the ecological conditions or hazards faced by parents and their offspring are important determinants of the benefits and costs of parental care. In particular, food availability and natural enemies such as predators or parasites are thought to have played a central role in the origin of parental care. However, it is important to specify precisely how ecological hazards affect parents and offspring. A generally harsh environment will not necessarily favor the evolution of parental care. If harsh environments have negative effects on both adult and juvenile life stages, any potential benefit of care to the offspring may be offset by a high cost to parents. But parental care is much more likely to evolve if the harshness of the environment has a stronger effect on juveniles than on parents. For example, if a main source of mortality is predation by a specialized egg predator that poses no threat to the parent, parental egg attendance might be a very beneficial strategy that should easily spread. Likewise, cannibalism on eggs or offspring by adults from the same species, which is a common phenomenon in many invertebrates, poses typically higher threats on egg and juvenile stages and may favor the origin of egg attendance in a similar way.

There is no single ubiquitous factor that explains the evolutionary origin of parental care across all systems or in a given species. For example, parental protection is not the only possible evolutionary answer to reducing predator-induced offspring mortality. Natural selection can also favor adaptations in the offspring themselves, such as camouflage, that deal with the same hazards. The key factor that promotes the evolution of parental care as opposed to adaptations in offspring is how environmental hazards differentially affect the fitness of parents versus offspring.

3. EVOLUTIONARY MAINTENANCE OF PARENTAL CARE

Once evolved, parental care has a number of important implications for the continuing evolution of a species. In particular, parental care not only is a *target* of selection but also generates variation in offspring phenotypes and survival. It is therefore also an *agent* of selection. In this situation, traits expressed in parents and offspring will tend to coevolve with one another owing to selection imposed by the environmental effects of parental and offspring traits on each other's fitness. For example, the current benefit of food provisioning in many species reflects that juvenile survival of offspring is completely dependent on food provided by parents. It is unlikely that such a dependency would characterize an ancestral population in which food provisioning evolved for the first time, and offspring still retained the ability to forage independently of their parents. Thus, offspring dependency must have evolved secondarily, by coevolving with parental food provisioning and thereby enhancing the adaptive value of provisioning. Conversely, once parental care evolved, conflicts and socially parasitic strategies could have partly undermined the original adaptive value of parental care. As a consequence, studies on the current adaptive value of parental care provide little insight into how parental food provisioning increased offspring fitness in the ancestral state.

The key question then is, What are the main forces that maintain parental care and shape the coevolution of offspring and parental traits? Parents that care for offspring pay a cost by doing so, as they normally cannot remate at the same time. Moreover, the resources parents provide to their offspring cannot also be used to enhance the parents' own reproduction and survival. These attributes of parental care characterize parental care as an altruistic trait and modify selection during the mating period (sexual selection) as well as selection on life-history traits, such as longevity. Furthermore, following the origin of parental care, the social environment in which offspring develop is partly provided by the parent, which has evolved to enhance offspring development. These forms of social evolution and contingency lead to the coevolution of parent and offspring traits. Finally, because parents and offspring are not genetically identical in sexually reproducing organisms, parental care can induce the scope for genetic conflicts of interest that, in turn, may shape adaptations to family life.

In sexually reproducing species, three social dimensions of within-family conflicts have to be considered: sexual conflict, parent-offspring conflict, and sibling conflict. All offspring have two parents, and the benefits of parental care depend on the combined amount of care provided by the two parents. However, because the costs of care to each parent are determined by the amount of care it provides, each parent would do better if the other paid the costs of care (unless there is lifelong monogamy, without divorce or remating following the partner's death). This situation leads to sexual conflict over which parent should provide parental care and, in species in which both parents provide care, conflict over the amount of care that should be contributed by each parent.

Sex differences in the provision of parental care are termed *sex roles*. The most common sex role is female-only parental care, as in most mammals. Male-only parental care also occurs, most notably in some amphibians and some fish; and biparental care, in which both males and females care for offspring, is most common in birds. However, even in species with biparental care the sexes rarely provide parental care in exactly the same way. For example, in many species of birds only females incubate eggs, although both parents may feed nestlings. The two most important factors favoring a divergence in sex roles are sexual selection and certainty of parentage (see chapter VII.3 for a game theory model). The different size of male and female gametes (*anisogamy*) means that there are fewer female gametes (eggs) than male gametes (sperm). This difference leads to sexual selection in males to locate unfertilized eggs, increasing their benefits of mating effort at the expense of parental effort. Sperm competition as a result of multiple males competing for and mating with females lowers the average relatedness of males to young compared with that of females, further decreasing the benefits of paternal care. This may be especially relevant for males that are successful in mating, who will have mating opportunities elsewhere. However, selection favors male parental care when the population of individuals in the mating pool is very male biased, making the probability of success in mating very low. In these circumstances it is better, on average, for males to invest in offspring that already exist (parental effort) rather than to invest in future offspring (mating effort).

For species in which caring parents interact with their offspring, and especially when such interactions occur over long periods, there is ample scope for offspring to influence the care provided by parents. For example, the mammalian fetus is intimately linked to the mother's blood circulation through the placenta, and parent birds and some insects make repeated foraging trips to provide food to their offspring. In terms of genetic relatedness, each offspring is of equal importance to parents, but individual offspring are expected to value their own survival and reproduction more highly than that of their siblings. Each offspring is therefore under selection to demand more resources for itself than the parent is under selection to provide, leading to sibling competition and parent-offspring conflict, which are therefore tightly linked. Whenever offspring are produced in clutches or litters, there is opportunity for sibling competition over the limited resources provided by parents. Because parents often initially produce more offspring than they can rear—as insurance against unpredictability of environmental resource availability or hatching failure—demand typically exceeds supply, thus intensifying the conflict. While close genetic relatedness often leads to the evolution of altruism and cooperation, the mismatch between parental supply and offspring demand tends to override any benefits of sibling cooperation. Thus, sibling competition can be extremely severe and involve lethal aggression, as in cattle egrets (*Bubulcus ibis*), in which older chicks frequently kill their younger siblings.

A large body of theory has shown that parent-offspring conflict is expected to favor the evolution of elaborate forms of communication, such as begging by nestling birds, or high levels of aggression among siblings. Imagine a species in which parental care has recently originated. If parents are sensitive to offspring demands for resources, the genetic conflict favors offspring that exaggerate their demands to manipulate the parents into providing more resources. In a coevolutionary arms race, this form of offspring manipulation generates selection on parents to become less sensitive to these offspring traits, which in turn selects for offspring to increase demand, and so forth, until the conflict reaches an evolutionary stable outcome or resolution. Resolution does not imply that there is no more conflict, just that there are no further opportunities for the manipulation of parents by offspring, and vice versa.

Theoretical models of parent-offspring conflict have shown that evolutionarily stable resolutions of parent-offspring conflict usually require that offspring begging be costly to offspring, thus preventing further evolutionary escalation. Stability is also determined by whether parents or the offspring control the allocation of resources. If parents gain control, and there is selection for offspring to provide costly and honest information about their need or quality, then parents can use that information to allocate resources in a way that optimizes their own fitness. An alternative evolutionary route to resolving conflict is for offspring to gain control over who is being fed by the parent. In that case, the parent has no direct control over the information provided by offspring and allocates resources passively in a way that primarily serves the evolutionary interests of the offspring.

Parental care generates a social environment that is favorable for the growth, development, and survival of offspring. These benefits are intended for the parent's own genetic offspring but may be exploited by any offspring capable of gaining access to the resources provided by parents, be it from the same or a different species. Parental care generates a social niche for parasitic adaptations that exploit parental behaviors.

Socially parasitic strategies are observed both within and between species. For example, it is well documented that females of some birds such as starlings (*Sturnus vulgaris*) lay eggs in foreign nests, a behavior called *egg dumping*. Egg dumping is often used by subordinate females that are unable to breed on their own. By dumping an egg into the nest of a breeding conspecific, they parasitize the caring behavior of the breeder to gain some reproductive success despite not breeding on their own. Social parasitism is not limited to higher vertebrates and can also take place after hatching, as in the European earwig (*Forficula auricularia*). In this species, nymphs are relatively mobile soon after hatching and can disperse to join other earwig broods, thereby parasitizing the care provided by an unrelated female. While female earwigs tolerate foreign offspring, the nymphs can discriminate unrelated nymphs from siblings and often kill and cannibalize unrelated nest mates. A well-known example of brood parasitism between species is the common cuckoo (*Cuculus canorus*), which never cares for offspring itself and parasitizes the parental care of a wide range of passerine host species. As obligate brood parasites, cuckoos exhibit a range of highly specialized adaptations to exploit the parental behaviors of their hosts. Female cuckoos lay an egg into the nest of a host species, timed very precisely to avoid detection and rejection by the host parents. The cuckoo chick hatches early and immediately sets about evicting its competitors—the host's own eggs or chicks. It then uses effective acoustic trickery that makes it sound like a whole brood of host chicks to stimulate its foster parents into providing as much food as they normally provide to a whole brood. The parasitized parents, although typically being much smaller than the cuckoo they feed, provide the rapidly growing brood parasite with all the food it needs to reach independence.

Social parasitism provides a good example of how the original benefits of parental care can be partially undermined once evolved. With increasing frequency of socially parasitic strategies in the population, the evolutionary benefit of parental care is reduced, thereby generating negative frequency-dependent selection on the parasitic strategy and/or selection for defense mechanisms in the hosts (e.g., kin or species recognition). Social parasitism provides a particularly clear example of the important role that social interactions play in driving the coevolutionary processes that result in the evolution and diversification of parental care and associated traits.

4. GENETICS AND EPIGENETICS OF PARENTAL CARE

Parental care must have a heritable basis to evolve, like any other target of selection. But parental care is also an environmental effect that shapes the conditions offspring experience during their development. These transgenerational parental effects can have lasting consequences for trait expression, including the possibility of epigenetic modifications in offspring behavior that are heritable and transmitted to future generations. The complexity of genetic bases of traits expressed in families has a number of interesting consequences for parent-offspring coevolution.

Why do animal families (including humans) typically show considerable variation in the level and duration of parental care and in the intensity with which offspring demand care from the parent and compete with siblings? One explanation is provided by environmental variability. If resources are plentiful, parents can provide all the necessary food for their offspring at relatively low costs. In this case, parental provisioning rate will be high, and as a consequence, offspring resource demand will be low. In contrast, if resources are scarce, provisioning rate will be low—because food is difficult to find—and insufficient for optimal offspring growth, which will lead to high offspring demand.

Recent experimental research shows that variation in parental care may also reflect genetic variation between families in how parents and offspring interact. Why does natural selection not eliminate all the heritable variation and maintain only those parent and offspring genotypes that combine to reflect a unique state of optimally resolved conflict? The main reason is that when parents and offspring interact, parental and offspring traits are determined not only by their own genotype but also by the genotype of the other individual(s) involved. This socially mediated interaction leads to correlational selection favoring particular combinations of parent and offspring traits. Because different combinations are selected, no single parent or offspring genotype will do best in all possible combinations, so genetic variation will be maintained in both parent and offspring traits.

As a consequence of correlational selection, parent and offspring traits become evolutionarily linked, and these traits are therefore also coinherited. For example in great tits (*Parus major*), burying beetles (*Nicrophorus vespilloides*), and canaries (*Serinus canaria*), parents that are good food providers produce offspring that beg more intensely for food. Although this statistical association is robust, the molecular genetic basis for such coinheritance is poorly understood. An exception has been found in studies on laboratory mice based on gene-knockout mutations, which showed that single genes often contribute to some of the variation in the traits of both mothers and pups. One example is the gene *Peg3*, which affects both milk letdown in females and the suckling efficiency of pups.

From a genetic perspective parental care leads to selection for coadapted (or matching) parent and offspring genotypes that cosegregate within genomes. All parents began their lives as offspring, and all surviving offspring must in turn become parents to gain reproductive success. As a result, genes affecting offspring and parental traits will be located in the same genome, and selection through coadaptation favors physical or statistical association of combinations that are important determinants of how competitive an individual is as offspring and how altruistic it is as parent.

Parental care also plays an important role in the evolution of epigenetic modifications to patterns of gene expression. The best-studied examples involve genes in which only the allele that is inherited from one parent (in some cases the father, and in other cases the mother) is expressed. Such parent-of-origin specific gene expression is termed *genomic imprinting*. To illustrate how parental care matters in the evolution of genomic imprinting, consider a hypothetical species in which only females provide care, and females care for successive single-offspring litters fathered by different males. The maternally inherited allele should favor a level of care that balances the benefits to the offspring against the costs to the mother's survival and future reproduction to maximize its transmission to future generations, because the allele will also be transmitted through future half siblings by the same female. In contrast, the paternally inherited allele should favor a level of care that maximizes the benefits to the offspring, because it will not be passed on to future generations through offspring produced by the same female. Theory predicts that genes coding for factors enhancing offspring growth and demand for maternal resources should be expressed when inherited from the father to exploit maternal investment, and silenced when inherited from the mother to limit maternal investment. Evidence consistent with these predictions has been described for a range of genes expressed in the mammalian placenta. An example is the insulin-like growth factor 2 gene (*Igf2*) expressed in the mammalian placenta, which enhances placental and fetal growth and for which only the paternal gene copy is expressed.

In summary, both the coinheritance of parent and offspring traits favored by coadaptation, and the epigenetic inheritance mechanisms favored by asymmetries in how selection operates on genes inherited from the two parents, reflect how parental care can modify genetic trait architecture, which may feed back to further affect the coevolutionary dynamics of parent and offspring traits.

5. SOCIALITY BEYOND FAMILY

Parental care provides an important stepping-stone toward the evolution of greater social complexity. For example, in some species, the amount of care received by the offspring is determined by the efforts of their parents as well as helpers, that is, other adults in the group that do not themselves breed (see chapters VII.9 and VII.10). The most derived form of sociality is *eusociality*, which is relatively common among bees, wasps, ants, termites, and—as the only vertebrate systems in which eusociality has been invoked—possibly occurs also in the naked mole rat and the Damaraland mole rat (see chapter VII.13). In these species, the suppression of reproduction of helpers has evolved to such a level that they forgo the opportunity to reproduce entirely.

Close genetic relatedness among individuals is required for the evolution of more complex social groups, including eusociality, and the close association between parents and offspring in many species with parental care is a key stage in this process (see chapter VII.13). More complex forms of sociality can evolve from parental care when selection favors the adult offspring of the parent to stay and help raising siblings (cooperative breeding) rather than dispersing and breeding independently. This evolutionary step requires that care be expressed not only in parents (mated adults pursuing their own reproduction) but also in nonbreeding adults (helpers or workers). This is possible only for forms of care that nonbreeding individuals can actually provide (e.g., progressive provisioning) and if caring is decoupled from mating and reproduction. If the ecological conditions that favor helping persist long enough, helpers may eventually lose the ability to reproduce and become specialized caregivers (often termed *workers*). Thus, the evolution of parental care represents an important transition that can promote the evolution of ever more complex forms of sociality, such as eusociality.

We have provided a brief overview of the ultimate causes promoting the evolution of parental care (i.e., life history and ecology) and the consequences of that evolution for other phenomena such as evolutionary conflicts, trait inheritance, and sociality. Coevolutionary feedback between life history and parental care traits, mediated by (social) environmental variation and genetic conflicts, makes parental care an evolutionary engine of biodiversity.

FURTHER READING

Clutton-Brock, T. H. 1991. The Evolution of Parental Care. Princeton, NJ: Princeton University Press. *The first comprehensive overview of the diversity of forms of care, and a synthesis for the principles governing the evolution of parental care.*

Forbes, S. 2005. A Natural History of Families. Princeton, NJ: Princeton University Press. *A review of reproductive decisions by parents and family conflicts for a general readership.*

Mock, D., and G. A. Parker 1997. The Evolution of Sibling Rivalry. Oxford: Oxford University Press. *A synthesis and review of principles governing the evolution of family conflicts.*

Mousseau, T. A., and C. W. Fox, eds. 1998. Maternal Effects as Adaptations. Oxford: Oxford University Press. *A synthesis of mechanisms and principles in the evolution of parental effects across taxa.*

Royle, N. J., P. T. Smiseth, and M. Kölliker, eds. 2012. The Evolution of Parental Care. Oxford: Oxford University Press. *A new synthesis of the diversity of forms of parental care, and the conceptual framework for understanding their evolution.*

Trivers, R. L. 1972. Parental investment and sexual selection. In B. Campbell, ed., Sexual Selection and the Descent of Man, 1871–1971. Chicago: Aldine. *A classic paper that discusses the reciprocal effects of parental care and sexual selection.*

Trivers, R. L. 1974. Parent-offspring conflict. American Zoologist 14: 249–264. *The first formal demonstration that the evolutionary interests of parents and their offspring are not always congruent.*

Wilson, E. O. 1975. Sociobiology: The New Synthesis. Cambridge, MA: Harvard University Press. *A broad synthesis of the evolution of social behavior that includes a first discussion of the role of ecology in the evolution of parental care.*

VII.9

Cooperation and Conflict: Microbes to Humans

Joan E. Strassmann and David C. Queller

OUTLINE

1. What is cooperation and why is it so important?
2. Fraternal and egalitarian cooperation
3. Fraternal cooperation is explained by kin selection
4. Egalitarian cooperation requires direct benefits
5. Conflict and control of conflict in fraternal cooperative systems
6. Conflict and control of conflict in egalitarian cooperative systems
7. Organismality results from high cooperation and low conflict

Cooperative interactions characterize all life, giving us spectacular multicellular organisms like kelp and kangaroos; complex societies like army ants, and hyenas; and extensive cooperative networks, like pollinators and their plants. Fraternal cooperation among related, like entities explains multicellularity and social insects. Egalitarian cooperation among different entities explains pollination, cleaning stations, and bacteria-insect symbioses. For cooperation to flourish, exploitation must be controlled; when it is, organismality results.

GLOSSARY

Cheating. Behavior that benefits oneself at a cost to others under circumstances within an otherwise-cooperative framework.

Cooperation. An interaction that benefits the recipient and the actor either directly or indirectly.

Direct Benefit. A benefit that accrues to the actor's personal fitness.

Egalitarian Cooperation. Cooperation between unrelated individuals of the same or different species; the payoff to the actor must be direct.

Fraternal Cooperation. Cooperation between relatives; payoff to the actor can be direct or indirect.

Hamilton's Rule: $rB - C > 0$, where r is the relatedness between the altruist and the beneficiary, B is the increased fitness benefit to the beneficiary, and C is the cost in lost fitness to the altruist.

Inclusive Fitness. The fitness of an individual that includes all ways it has increased its genetic representation in the next generation, from rearing progeny to the share it contributed to nondescendant kin.

Indirect Benefit. A benefit of actions of the actor toward nondescendant kin, who tend to share the actor's genes.

Mutualism. Cooperation between nonrelatives of the same or different species; often, different goods or services are exchanged.

Organism. A living unit with high cooperation and very low conflict among its parts; an adapted unit that is not much disrupted by conflict at lower levels, nor subsumed into adaptation at higher levels.

1. WHAT IS COOPERATION AND WHY IS IT SO IMPORTANT?

Cooperation in an evolutionary sense is defined as an action performed by an actor that benefits its recipient. For cooperation to be evolutionarily stable (see chapter VII.3 on game theory), it should also benefit the genes that underlie the action, as by benefiting the actor directly or by benefiting relatives of the actor. A ground squirrel that sounds an alarm when a coyote is spotted benefits relatives in hearing range, causing them to run for shelter. A social amoeba joins with others to form a motile slug that can travel farther and lift off the ground before forming spores, though in the process some cells die by becoming the supporting stalk and therefore never reproduce. A sterile worker bee helps the queen rear sisters and brothers. That same worker bee takes nectar and pollen from flowers, in return pollinating the ova. A bacterium produces light inside a squid, so the squid casts no shadow in the moonlight when seen from below. Termites are able to eat cellulose only with the aid of their bacterial gut inhabitants. These are just a few examples of evolved cooperative interactions.

Cooperation may be the most underestimated process in the evolution of life, one that affects nearly every topic in this book, from speciation to phylogenetics to adaptation. One reason for its importance is that it is often easier to acquire a new capability by allying with another that already has the capability, as compared with the slow and less reliable process of accumulating mutations that provide the trait de novo, as Nancy Moran has shown so elegantly in the mutualism between sap-sucking insects like aphids and sharpshooters and their bacterial symbionts. The bobtail squid can more easily evolve the ability to house and feed luminescent bacteria than become luminescent itself. Termites cannot easily evolve the ability to digest cellulose that appears to be so easy for their spirochete gut bacteria.

Partly for this reason, cooperation can be enormously successful. The cooperative engulfing of a blue-green alga by a eukaryotic cell produced the green plants on which so much of life depends. Cooperation is the basis of the phenomenal success of the social insects (see chapter VII.13 on eusociality) and is at the heart of the major transitions in life, as highlighted by John Maynard Smith and Eörs Sathmáry. Cooperation is the source of eukaryote cells, multicellularity, and organisms themselves.

Cooperation has been fixed in highly successful alliances, but the lack of current variability in many species makes these less than optimal for studying how cooperation came to be. The most fruitful study thus involves organisms where cooperative alliances are still plastic, subject to measurable pushes and pulls as the conflicting interests of different parties surge and are then quelled. Such organisms include social insects (see chapter VII.13) and vertebrates (see chapter VII.10), which were important in the early development of the theory, and extend to the powerful experimental systems of social microbes and within-genome alliances studied more recently.

Here we divide cooperation into two natural kinds. We then discuss benefits, costs, conflict, and how conflict is controlled. We hope the curious reader will push back the darkness in some new corners of our cooperative world.

2. FRATERNAL AND EGALITARIAN COOPERATION

Fungus-growing ants provide an illustration of the two major types of cooperation. *Atta* ant colonies contain millions of workers that methodically strip the tropical forest for leaves to feed their fungus. The ant workers occur in several fixed forms, or castes, all working to rear the offspring of a single large queen. At the same time, the ants cooperate with a fungus, a little package of which is carried by a new queen when she starts her colony. In addition to dispersing the fungus, the ants feed it the leaves they harvest and, in turn, eat parts of the fungus.

Cooperation can be either *fraternal* or *egalitarian*, according to whether the cooperators are the same kind of entity or are different entities, according to David Queller. Cooperation among the ants themselves is fraternal cooperation, based on shared genes between cooperators. Cooperation between the ants and the fungus they grow is egalitarian cooperation that requires that each party benefit. These terms are taken from the last two components of the French Revolution cry "Liberté, égalité, fraternité!" (the first, liberté, could be viewed as the noncooperative, solitary option). Fraternal cooperation results in alliances of like individuals, including the same molecules in compartments, organelles in cells, cells in multicellular individuals, and individuals in colonies or societies. Egalitarian cooperation results in alliances of different kinds, including different molecules in compartments, different genes in chromosomes, different organelles in cells, and cooperation among different species.

Fraternal cooperation involves cooperation among relatives and so need not pay back directly to the actor's phenotype. Thus, it is the only kind of cooperation that extends to true altruism. The genes causing the action proliferate because they are also present in the recipient. This indirect benefit is explained by William D. Hamilton's *kin selection theory*. Such cooperation is exhibited by social insects, birds with helpers, wolf packs, and multicellularity; many other kinds of cooperation are found among entities that share genes.

Egalitarian cooperation involves cooperative acts in which both parties benefit directly and includes all cooperation between different species, including cleaner fish and their clients, leaches and their blood-digesting bacteria, plants and fungal mycorrhizae, and plants with their pollinators. Unrelated individuals of the same species also cooperate only under conditions favorable to each, so the evolution of this kind of cooperation has more in common with egalitarian than fraternal cooperation, so we treat it under that heading. This category includes males and females in sexual relationships.

3. FRATERNAL COOPERATION IS EXPLAINED BY KIN SELECTION

Like any other evolved adaptation, cooperation must increase the frequency of the genes that cause it. Unlike other evolved adaptations, this may seem to be a challenge, because cooperation involves an action that costs the actor and benefits another. Genes for fraternal cooperation can spread because of their benefits to identical genes occurring in kin. Fraternal cooperation is favored under conditions specified by Hamilton's rule, $rB - C > 0$, where r is genetic relatedness between donor and beneficiary, B is benefit to beneficiary in terms of increased progeny, and C is cost to donor in terms of lost progeny. Genetic relatedness measures shared genes above average frequencies and is usually due to pedigree connections (giving help to partners with average allele frequencies will not change frequencies). This genetic relatedness among cooperators is often estimated using variable Mendelian markers, like DNA microsatellites.

Fraternal cooperation includes true *altruism*, which is behavior that reduces the actor's direct fitness, because nondescendant kin can pass on the genes that underlie the actions. Many of the cooperators we think of first fall into this category. Social insect colonies are based on families. Helpers at the nest in mammals and birds are usually older progeny, related to the brood they rear.

Costs and benefits can be measured in terms of progeny lost and nondescendant kin gained by following a particular strategy. These values are often measured by comparing different strategies. For example, a wasp foundress that nests alone and produces a certain number of progeny could be viewed as a stand-in for a wasp foundress that helps a sister instead of reproducing herself, to provide an estimate of what she might have produced. Comparing the two could give an estimate of the benefit of cooperation.

An important advantage in social insects is division of labor, which means that different participants do different things and may even have different forms. For example, some ant species have workers that are adapted as foragers, and others as soldiers, the latter being larger, with large mandibles for biting. However, the initial advantages were not based on these derived forms.

Predators are a strong initial selective force for sociality. The two main lifestyles in social insects—life insurance and fortress defense—are thought to have evolved to protect against predators and their effects. Life insurance means that adult cooperators can take over half-raised young and finish the job if some adults die. This lifestyle is most likely to apply in species in which adult lifetimes are short, and offspring dependency is long, as in ants, bees, and wasps. Fortress defense means cooperators can take advantage of safe, defensible places for nesting, particularly inside edible resources. Termites, naked mole rats, and social shrimp may be eusocial because of fortress defense.

When there are advantages, the actual trigger for cooperation can be very simple. The sweat bee *Halictus rubicundus* that Jeremy Field and colleagues study in the United Kingdom is solitary in Belfast but has females that remain to help their mother in Sussex, where the growing season is long enough for the firstborn to rear subsequent young. When Field moved females from the more northern, solitary population to the warmer climate in Sussex, some of the females remained with their mother in a cooperative alliance, making the fraternal transition from solitary to social breeding.

Another kind of fraternal cooperation is multicellularity, which has arisen many times and seems to be an easy evolutionary step, if David Kirk's evidence from the volvocine algae is any indication. Multicellular organisms usually have a single-cell bottleneck at the beginning of development, which is crucial, because this bottleneck causes the cells in the multicellular organism to be genetically identical. This means that cells specializing into somatic functions are not in conflict with the gonad cells.

A major advantage of multicellularity is the division of labor among many different cell types, but when the first cells came together they were probably not already specialized, so some other advantages might have been involved. Studies of the early stages of multicellularity in the volvocine algae have begun to examine these advantages. *Chlamydomonas reinhardtii* is a single-celled alga with two flagella that allow it to move. In its multicellular relative *Volvox carteri,* somatic cells numbering around 2000 have flagella, and the 16 or so gonadal cells do not, a division of labor that allows it to swim and reproduce at the same time. But the initial advantages of multicellularity might be sought in intermediate forms, like *Gonium* species, with relatively few largely undifferentiated cells in the multicellular body. Still, it is not entirely clear what initial advantages multicellularity confers. High on the list of candidates are increased motility and reduced predation, both advantages of larger size, but the continued presence of both single-celled and multicellular species argues against a single best strategy.

4. EGALITARIAN COOPERATION REQUIRES DIRECT BENEFITS

Cooperation among unrelated individuals must provide a direct benefit to the actor if it is to evolve. Exactly how this process works is the subject of mutualism theory, discussed in the section on conflict. For example, a female meerkat might join a troop of unrelated individuals and help rear the babies if she has some chance of reproducing in the group in the future. A male and a female northern mockingbird cooperate to rear their progeny, as do many other organisms with biparental care. Cleaner fish eat the parasites and dead cells off the larger fish that come to the cleaning station.

Division of labor often provides advantages from the very beginning of egalitarian cooperation. Since the partners are typically different in egalitarian cooperation, as with males and females in sexual reproduction, it is easy to imagine that they have different talents that their partners can use. Egalitarian cooperation easily provides benefits to each partner, because what is easy for one may be difficult or impossible for the other. Flowering plants provide a reward to their pollinators, who, usually incidentally, carry pollen to waiting, stationary ovules. Animals rely on bacteria for food digestion, providing an environment for growth in return. When the food is highly specialized (e.g., plant sap, blood, or cellulose), only one or two bacteria species may take on the digestive task. The symbionts of some animals, like corals, make their food from sunlight. Plants rely on cooperative relationships to extract nutrients from the soil, as with mycorrhizae. Some bacteria fix nitrogen in exchange for plant-produced carbon. In each of these examples, each partner provides something that is relatively easy for it to manufacture, in exchange for something that would be difficult or impossible for it to do. Among the most spectacular examples of these sorts of bargains are those that form eukaryote cells, with mitochondria providing energy conversion and, in plants, chloroplasts converting carbon dioxide to sugar.

In a powerful demonstration of how easily mutualism can evolve, William Harcombe took a strain of *Escherichia coli* that could not synthesize methionine and mixed it in a lactose environment with a strain of *Salmonella enterica* he had engineered with the capability of producing methionine as a waste product. At the beginning of the experiment neither could grow, because the *Salmonella* did not produce enough methionine for the *E. coli* to grow, so it did not produce enough sugar for the *Salmonella.* But in a structured plate environment, cooperation evolved that benefited both species. Increased methionine production was costly for *Salmonella,* so it evolved only in a private interaction with *E. coli* under a fixed-surface environment that ensured that benefiting neighbors of the other species would create additional benefits to self. In a liquid environment, where the benefits were dispersed more globally, cooperation did not evolve.

5. CONFLICT AND CONTROL OF CONFLICT IN FRATERNAL COOPERATIVE SYSTEMS

Fraternal cooperation requires genetic relatedness among participants, but this does not mean that all interactions among kin are cooperative. The problem of conflict for fraternal cooperators may seem small, since they are related. When the parties share genes, actions that excessively reduce the fitness of one individual will reduce that of the actor also. But if relatedness is less than one, or complete clonality, the interests of two parties are not identical, so there can be conflicts. Even within families, there are many conflicts of interest, and how these play out make up some of the richest stories of fraternal cooperation.

How is fraternal conflict controlled? This is a major question in the study of cooperation within families. When relatedness alone does not eliminate conflicts, outcomes are decided in large part by the relative power of the parties. For example, even when helping pays, it is still generally better to receive altruism than to give it, so there can be conflict over who helps and who receives help. In some cases this potential conflict is reduced because of asymmetries. Age is an important asymmetry, as in the case of a parent helping a child. Parents are often in a position to provide aid to offspring, while offspring may be incapable, at some ages, of helping parents. In other cases, power can determine who helps whom, with the stronger party forcing the weaker into the helping role. In *Polistes* wasps, one foundress usually takes the queen role and lays all the eggs and her sisters help her, taking on the risky foraging tasks. The one taking the queen role is usually the first to begin the nest in the springtime of temperate latitudes. She is typically the largest, which may be why she emerges from hibernation earlier and begins nesting activities sooner.

Once who helps whom is decided, fraternal cooperators may further disagree on the amount of aid given by the altruist. For example, a baby may demand more from its parent than that parent is selected to give if this imposes costs on the parent's other progeny, because a parent is typically related to all its babies by one-half, while the baby is related to itself by one and to its full siblings by one-half. Thus, the baby will favor more investment in itself and less in its siblings. The famous parent-offspring conflict first described by Robert Trivers ensues.

Parent-offspring conflict does not have to involve how much the parent gives but can also involve sex ratios. In social ants, bees, and wasps, workers are more related to their sisters than to their brothers in colonies with a single once-mated queen because of haplodiploid sex determination. Consequently, workers will favor a more female-biased sex ratio among the brood they rear than the queen will favor. This interesting example is covered in more detail in the eusociality chapter (VII.13). The resolution of these and many other within-family conflicts of interest have provided some of the best tests and supports of kin selection theory.

Whereas some conflicts are resolved through relative individual power, others involve the power of a collective. Francis Ratnieks called the group enforcement of common good *policing.* He demonstrated that in honey bees, workers suppress other workers from laying eggs, at least in the presence of the queen. Policing has been demonstrated to be important in controlling many kinds of conflict.

6. CONFLICT AND CONTROL OF CONFLICT IN EGALITARIAN COOPERATIVE SYSTEMS

Conflict is potentially even stronger in egalitarian cooperation, in which neither partner has a genetic stake in the other; there can be a persistent evolutionary push and pull between cooperating parties. We begin with the topic of sexual reproduction. Before a new zygote is formed, in most eukaryotes, the diploid genome is divided in half in a normally cooperative meiosis. The fairness of meiosis is egalitarian cooperation, even though it occurs in a single individual, because some genes do not end up in the egg. This fairness of meiosis is sometimes defeated by *meiotic drive*, the name for the process that causes one allele to always make it into the progeny.

In sexual reproduction, the gametes from male and female fuse in another egalitarian process, at least for nuclear genes, because half come from the father and half from the mother, in most organisms. But mitochondria (and chloroplasts in plants) are usually inherited entirely through the female. Nuclear and mitochondrial genes cooperate in the adaptive function of the eukaryote cell, but the conflicts arising from differences in inheritance are not entirely resolved. For example, mitochondrial genes may cause male sterility in plants, in order to produce more seeds that transmit mitochondria, while nuclear genes act to restore male fertility. Similarly, cytoplasmic parasites, like *Wolbachia,* can bias sexual reproduction toward females who transmit the *Wolbachia*. But genes in the same individual are largely cooperative, and live or die with the individual.

After sexual reproduction, most organisms release the progeny into the world to fend for themselves, as seeds or eggs. But in some organisms, parents greatly increase the chance of progeny survival by caring for them (see chapter VII.8), which creates another arena for egalitarian cooperation. Though the parents are both related to the progeny, they are unrelated to each other and so will disagree on how much each should give the young. In some groups, like mammals, one sex has evolved special abilities for caring (milk production in females). In others, either parent can care, which generates a rich area of research into the specifics of such care. Confidence that one is actually the parent is a factor affecting which individual gives more care. This is usually the female, since her confidence of being the mother is greater than the male's confidence of being the father, at least in organisms with internal fertilization.

There are many other fascinating examples of conflict in egalitarian systems. Toby Kiers and collaborators showed that when soybean rhizobia were prevented from fixing nitrogen by being isolated in a nitrogen-free atmosphere, the plants cut the amount of carbon they allocated to those nodules. Figs have evolved a complex relationship with their pollinating wasps, which enter the fig, lay their eggs, and either actively or passively pollinate the flowers within the fig. In the more basal species with passive pollination, the wasps simply encounter abundant pollen in their natal fig and transport the pollen by chance. In the more derived species, the wasps seek out the pollen-producing flowers in their natal fig, carry the pollen with them, and actively pollinate the flowers in the fig they choose for their eggs. Clearly, the latter form is a tighter mutualism, for the fig is dependent on an act the wasp would not necessarily perform. Jander and Herre found that the actively pollinated species had sanctions against wasps that did not pollinate sufficient flowers: those fruit were simply dropped from the tree and not allowed to ripen, killing the wasps inside.

Control of cheating in egalitarian relationships like those just described is based on how partners are kept to their end of the bargain. These controls take two general forms, called, somewhat confusingly, *partner choice* and *partner fidelity feedback*. Under partner choice, underperformers can be punished. The legume-rhizobium and the fig-wasp examples are examples. The plants reject poorly performing bacterial nodules, or wasps that do not provide sufficient benefits.

Under partner fidelity feedback, the fates of the partners can be so completely commingled that sanctions are rare, for they would hurt both partners. The eukaryotic cell is such a case; with rare exceptions any harm that either mitochondrion or host cell does to the other feeds back as harm to itself. Many phloem-feeding insects rely on bacteria to digest their sugary food and to produce essential vitamins. The aphids and their vertically transmitted *Buchnera* bacteria reproduce through the same pathway and are utterly dependent on each other; neither can do much to gain at the other's expense. In general, cotransmission makes partner fidelity feedback strong.

7. ORGANISMALITY RESULTS FROM HIGH COOPERATION AND LOW CONFLICT

Fraternal and egalitarian cooperation alike can bring formerly separate entities together into alliances of varying degrees. Most of the alliances that are highly cooperative, with conflict at lower levels thoroughly controlled, are called *organisms*. This is the level at which adaptations are most common, and these adapted bundles compete with other adapted bundles. In an earlier paper (Queller and Strassmann 2009) we explored the consequences of taking high cooperation and low conflict as the definition of organismality and argued that other definitions of the organism cannot be consistently applied. Under our cooperation-based definition, widely recognized organisms such as whales and sequoia trees retain their organismality. To the organism list we might add the aphid-*Buchnera* symbiosis, *Dictyostelium* fruiting bodies, honey bee colonies, anglerfish mates, lichens, and some fig-wasp symbioses, to name a few. As with any other definition of organism, there may be gray areas, but this can be an advantage, because there is much to be learned about pattern and process of cooperative alliances at the borders of organismality. Cooperation is not just an activity engaged in by a few special organisms— it is how all organisms came to be in the first place.

FURTHER READING

Bourke, A.F.G. 2011. Principles of Social Evolution. Oxford: Oxford University Press.

Burt, A., and R. Trivers. 2006. Genes in Conflict: The Biology of Selfish Genetic Elements. Cambridge, MA: Belknap Press of Harvard University Press.

Moran, N. 2007. Symbiosis as an adaptive process and source of phenotypic complexity. Proceedings of the National Academy of Sciences USA 104: 8627–8633.

Queller, D. C., and J. E. Strassmann. 2009. Beyond society: The evolution of organismality. Philosophical Transactions of the Royal Society B 364: 3143–3155.

Sachs, J. L., U. G. Mueller, T. P. Wilcox, and J. J. Bull. 2004. The evolution of cooperation. Quarterly Review of Biology 79: 135–160.

West, S. A., S. P. Diggle, A. Buckling, A. Gardner, and A. S. Griffins. 2007. The social lives of microbes. Annual Review of Ecology and Systematics 38: 53–77.

VII.10

Cooperative Breeding

Michael A. Cant

OUTLINE

1. Ecology and evolution of cooperative breeding
2. The evolution of helping
3. Individual differences in helping behavior
4. Reproductive conflict

Cooperative breeding is a relatively rare but taxonomically widespread social system in which adult helpers work to rear offspring that are not their own. Approximately 10 percent of birds and 2 percent of mammals are cooperative breeders; examples are also seen in insects, spiders, crustaceans, and some fish. These diverse systems present an opportunity to investigate how cooperation and helping can be favored in the face of natural selection, which is expected to work for self-interest. Cooperative breeding animals present concrete examples of altruism and helping, together with the possibility of measuring the lifetime fitness consequences of helping decisions. Research on cooperative breeding can also help elucidate the evolution of the unusual human life history, because humans evolved in cooperatively breeding groups in which grandparents, siblings, and other family members contributed to rearing offspring. Some of the main questions posed by cooperatively breeding animal societies are considered here: What ecological conditions favor cooperative breeding? Why do helpers help? Why do some individuals work much harder than others? How do competing individuals resolve conflicts of interest so that the cooperative team can function?

GLOSSARY

Altruism. Acts or behaviors that result in a lifetime direct fitness increase for other individuals, at a lifetime direct fitness cost to the actor.

Cooperation. A social interaction in which individuals enhance each other's inclusive fitness.

Direct Fitness. The number of copies of alleles that an individual contributes to the next generation through offspring.

Harming. Acts or behaviors that reduce the number of offspring of other breeders that are raised to independence.

Helping. Acts or behaviors that increase the number of offspring of other breeders that are raised to independence.

Inclusive Fitness. Direct plus indirect fitness.

Indirect Fitness. The number of copies of alleles that an individual contributes to the next generation by helping nondescendant kin.

Reproductive Division of Labor. The partitioning of tasks involved in reproduction within animal societies among different individuals. Typically, socially dominant individuals produce offspring, while subordinate, nonbreeding individuals help provision or rear young.

Reproductive Skew. A measure of the evenness with which reproduction is distributed among the members of a cooperative group.

1. ECOLOGY AND EVOLUTION OF COOPERATIVE BREEDING

Cooperative breeding is a type of social system in which some adults (known as *helpers*) routinely assist in the raising of offspring that are not their own, even though they have the ability to produce offspring themselves currently or in the future. This broad definition includes a range of species, from social insects such as paper wasps, hover wasps, and halictid (sweat) bees to "helper-at-the-nest" bird systems (e.g., western bluebirds, white-fronted bee eaters), in which offspring delay dispersal and help their parents with the next clutch; and larger bird and mammal societies (e.g., acorn woodpeckers, banded mongooses) with multiple male and female breeders and helpers per group. These systems have proven to be excellent models for studying the evolution of sociality and cooperation because they provide concrete examples of *altruism*, that is, behavior that boosts the lifetime fitness of others at a lifetime fitness cost to the actor (see also chapters VII.9 and VII.13). For example, in paper wasps, groups of overwintered females (called *foundresses*) emerge from hibernation in spring and form groups that cooperate to build a nest. In each group a single dominant female lays most of the eggs and remains safely on the nest while the other females forage for prey to feed the offspring and collect nest material to expand the nest. These helper foundresses are all mated and fully fertile, with the option of building their own nest or attempting to supplant the dominant, so why do they accept a nonbreeding position and risk their lives to help the dominant instead? Studying cooperative breeders can help us understand how cooperation can be favored by natural selection and how cooperative groups remain stable despite conflict over reproduction and social rank. This is also a topic that is relevant to human evolution: many of the puzzling and unusual features of human life history (early reproductive cessation followed by menopause, a long period of offspring dependency, sequential production of multiple dependent young) appear to reflect an evolutionary history of cooperative breeding.

Cooperative societies, while very diverse in terms of social structure and basic biology, share some important features. Populations with cooperative breeders are usually made up of closely knit extended family groups, formed when offspring delay dispersal to remain in their natal groups. Within these groups there is usually (but not always) a reproductive division of labor in which older or socially dominant individuals breed, and lower-ranked or younger individuals provide most of the help. Because helpers retain the ability to reproduce, their behavior reflects a trade-off between their current and future fitness, and between direct and indirect components of inclusive fitness. In this way cooperative breeders differ from *eusocial* species (such as ants, honey bees, termites, some aphids, and naked mole rats) in which there are distinct reproductive and worker castes, and helpers remain functionally or morphologically sterile throughout their lives.

Ecological factors play a central role in both the evolution and maintenance of cooperative breeding. In birds, for example, comparative analyses show that the evolution of cooperative breeding is associated with high adult survival and intense competition among adults for breeding territories. Within species, offspring remain on their natal territory and serve as helpers if no suitable breeding habitat or territory is available but rapidly disperse to breed independently if ecological constraints are relaxed or vacant territories appear. In the Seychelles warbler (*Acrocephalus sechellensis*), for example, adult males and females were transplanted from a saturated island in the Seychelles group (Cousin) to two adjacent uninhabited islands (Aride and Cousine), whereupon they formed breeding pairs and produced offspring who in turn went off to breed (Komdeur et al. 1995). As the vacant islands filled up and the best-quality territories were taken, offspring (particularly those born on high-quality territories) began to delay dispersal and to remain on their natal territory to help. This and other experimental studies of birds and fish (such as the cooperative cichlid *Neolamprologus pulcher*) suggest that ecological constraints on dispersal and high fitness benefits of remaining at home (known as the *benefits of philopatry*) together promote the formation of cooperative breeding groups. Similar constraints on dispersal promote delayed dispersal of helpers in cooperatively breeding mammals (such as lions, African wild dogs, meerkats, and banded mongooses), but here it is often aggressive territorial defense, rather than a lack of available habitat, that constrains immigration into existing groups (Clutton-Brock 2009).

In most cooperatively breeding insects, dispersal is not constrained by a lack of breeding habitat or territory, because nests can be constructed on a range of vegetation types or substrates. There are, however, often severe constraints on independent breeding because mothers have a high probability of dying in the extended period during which offspring are dependent on their care. Offspring that stay to help their mother or join the nests of same-generation females can provide insurance against the failure of the nest due to the death of the breeding female. Experiments have shown that these benefits are substantial and favor staying to help even when relatedness is low and breeding is monopolized by a single female (meaning that helpers stand to gain little indirect or direct fitness benefits from help). In cooperative insects such as paper wasps and tropical hover-flies, subordinates form a social queue and can inherit breeding status on the death of the dominant, which further increases the benefits of remaining in the natal group. Helping in these social queues ensures that nests quickly get through the vulnerable founding phase (i.e., the period before workers emerge), thus safeguarding the potential future fitness benefits that subordinates might gain through inheritance.

2. THE EVOLUTION OF HELPING

Ecological constraints on dispersal may set the stage for delayed dispersal and cooperative breeding, but on their own these constraints do not explain why helpers work to rear the young of breeders, rather than just waiting for breeding vacancies to appear. A general theoretical framework for understanding the evolution of helping behavior (or indeed any trait that affects the fitness of social partners) was provided by William D. Hamilton's *inclusive fitness theory*, set out in a seminal 1964 paper. Hamilton showed that selection favors social traits that satisfy the following inequality (now known as *Hamilton's rule*): $rB - C > 0$, where B is the lifetime direct fitness benefit to the recipient of a social act (e.g., the recipient of help), C is the lifetime direct fitness cost to the actor or possessor of the social trait, and r is the coefficient of relatedness, a measure of genetic similarity between social partners relative to the average "background" genetic similarity in the wider population.

Ecological conditions can affect the magnitude of all three terms in Hamilton's rule: the fitness benefits (B) to recipients of a given unit of help (for example, offspring may benefit more from help when food is scarce); the fitness cost (C) of investing in help (for example, helping may be more costly when food is scarce); and even relatedness (r), given that severe ecological constraints on dispersal mean that most interactions will occur between kin. Relatedness is also dependent on patterns of mating: other things being equal, monogamy is predicted to be more conducive to the evolution of altruism than is polygyny, since relatedness among family members is higher under the former than the latter. Recent phylogenetic analyses support this prediction and show that most cooperatively breeding insects, birds, and mammals arose from monogamous ancestors. It appears therefore that both ecology and family genetic structure exert important influences on the evolution of cooperative breeding.

Within the general framework of inclusive fitness theory, four main evolutionary mechanisms have been proposed to explain how helping behavior can evolve. The first two of these were proposed by Hamilton himself and are usually grouped under the term *kin selection*: (1) indiscriminate helping may be favored if dispersal is limited, so that the recipients of help are on average more closely related than the population at large; (2) individuals may recognize kin and preferentially direct care toward them; (3) there may be immediate or delayed direct fitness benefits that outweigh the immediate fitness costs; (4) helping may be enforced by social punishment, so that the alternative, not helping, results in even greater fitness costs. It is important to recognize that none of these explanations are mutually exclusive: for example, helping may involve both indirect fitness benefits (mechanisms 1 and 2) and direct fitness benefits (mechanisms 3 and 4).

Indiscriminate Helping

The first mechanism based on dispersal constraints, or *population viscosity*, has been the subject of controversy, because constraints on dispersal lead to both high relatedness and high local competition between relatives. There is little to be gained from raising extra offspring if these offspring compete with one another for the same limited number of breeding places. In fact, the first theoretical analysis of this problem by Peter J. Taylor in the early 1990s suggested that the costs of competition arising from dispersal constraints exactly cancel the benefits of increased relatedness for the evolution of altruism. According to this model, therefore, Hamilton's first mechanism for the evolution of altruism should not work. However, subsequent theory has shown that adding in biological features such as overlapping generations and sex differences in dispersal typically recovers Hamilton's prediction that increasing viscosity promotes the evolution of indiscriminate helping.

Extensions of Hamilton's theory can be used to predict harming as well as helping behavior, that is, acts or traits that *reduce* the fecundity of local group members, and can be applied to understand any social life history traits that have an impact on the direct fitness of fellow group members, such as reproduction and the rate at which individuals get old and die. For example, Rufus Johnstone and I have modeled how patterns of dispersal and mating may have predisposed humans and some cetaceans to the evolution of menopause and late life helping. Testing of these models is at an early stage, but proposals such as this with associated tests should lead to a better understanding of demographic influences on life history and helping behavior.

Discriminate Helping

Hamilton's rule is easier to satisfy if helpers can direct care toward more closely related group members, since in this case relatedness is by definition higher than the average relatedness to all potential recipients. The ability to preferentially aid kin increases the inclusive fitness benefits of costly helping, so we might expect animals to evolve mechanisms to recognize close kin. In cooperatively breeding birds and mammals, kin recognition does occur and is typically based on cues that are learned during development, not on direct recognition of genetic similarity. In the long-tailed tit, for example, cross fostering experiments show that offspring preferentially help those individuals with which they were reared, rather than their genetic relatives. In fact, direct recognition of genetic similarity appears to be uncommon in social vertebrates, although there is evidence from laboratory mice and humans that individuals can detect similarity at some very variable genetic regions of the genome, such as the major histocompatibility complex (MHC) of genes that are involved in immune function.

In cooperatively breeding insects, helpers discriminate nest mates and non–nest mates but generally do not discriminate kin from nonkin within groups. In the paper wasp *Polistes dominulus,* for example, 20 to 30 percent of foundress helpers are nonrelatives, but there is no difference between related and unrelated helpers in foraging effort, nest defense, aggression, or inheritance rank. However, recent studies have also shown that these unrelated helpers have measurably different hydrocarbon profiles (volatile chemicals in the cuticle that in other insects are involved in signaling and kin recognition). Thus, although cues to discriminate kin exist, they are not used by wasps in helping decisions.

Direct Fitness Benefits

Examples of hardworking, unrelated helpers (for example, in paper wasps and meerkats) suggest that helping may also yield direct fitness benefits, and that in some cases these direct benefits alone are sufficient to outweigh the costs of helping. Several direct fitness benefits of helping have been proposed. The *skills hypothesis* suggests that the experience of helping allows helpers to pick up parenting or foraging skills, which increases their reproductive success when they become breeders themselves. The *group augmentation hypothesis* suggests that helping can be favored if, as a consequence, helpers inherit a larger, more productive group within which to breed in the future. Finally, the *prestige hypothesis* suggests that helping results in elevated social status and an increased probability of inheriting breeding status in the future.

The skills hypothesis has been tested in several cooperatively breeding bird species by examining the correlation between helping effort and later reproductive success. These studies have found little evidence to support the hypothesis, and correlational tests of this kind are problematic, because any correlation may reflect differences in quality, state, or age of helpers, rather than a causal relationship between help and breeding success. Group augmentation seems a plausible idea, because larger groups are usually more productive in vertebrate cooperative breeders, but the key assumptions of the hypothesis have not been tested, namely, (1) that helping leads to increased recruitment and larger future group size, and (2) that larger group size is beneficial to the direct fitness of helpers. In paper wasps and hover wasps group augmentation benefits do not appear to be a major determinant of helper effort: helpers reduce their helping effort as they get closer to inheriting the breeding position, a pattern that is the opposite of that predicted by the group augmentation hypothesis. Finally, tests of the prestige hypothesis have also yielded little support, although there are intriguing observations of Arabian babblers competing with each other to help, which are consistent with the hypothesis. Overall, evidence for direct fitness benefits of helping is thin on the ground, but more experiments are needed to manipulate helping effort and establish the causal consequences for later breeding success.

Enforced Fitness Benefits

A great deal of theoretical interest in evolutionary biology has focused on the use of punishment and threats to induce cooperation and helping (Cant 2011, 3530). In the context of cooperative breeding, the *pay-to-stay hypothesis* suggests that dominants can use the threat of eviction from the group to induce subordinates to help or to pay "rent" to be allowed to stay. Alternatively, breeders could use acts of aggression to punish lazy helpers, rather than the threat of eviction. These two explanations are different because the pay-to-stay hypothesis is based on the use of a threat (namely, of eviction), whereas aggression represents a form of punishment. Threats differ from punishments because, if effective, a threat rarely needs to be exercised; punishments, however, require overt actions to be effective. There is evidence from cooperative insects and vertebrates that growth and behavior are influenced by "hidden" threats that are triggered only when the social rules they enforce are broken, but detecting these hidden threats requires experiments to break these rules, for example, by preventing helpers from helping or reducing their effort.

The key prediction of the pay-to-stay hypothesis is that experimental reduction of helper effort should lead to eviction from the group. In cooperative cichlid fish and splendid fairy wrens, helpers that were temporarily removed from the group or were prevented from helping were subjected to aggression from dominants, but were never evicted from the group. In naked mole rats dominant queens use aggressive "shoving" to activate lazy workers. Kern Reeve and his colleagues showed that in the paper wasp *Polistes fuscatus* the removal or inactivation of dominant foundresses (by cooling them) led to reduced helper effort, while wing clipping of subordinate helpers led to increased aggression from dominants, as expected if aggression is used to enforce help. There is no evidence from wasps or other cooperative insects that lazy helpers are evicted from the group. Thus, when enforcement does occur in cooperative societies, it appears to be achieved through the use of punishment, not the threat of eviction.

3. INDIVIDUAL DIFFERENCES IN HELPING BEHAVIOR

In most cooperatively breeding insects and vertebrates some helpers work hard to rear offspring, while other individuals in the same group do very little. Hamilton's rule suggests that these individual differences could be attributed to variation in relatedness between helpers and offspring, or to variation in the individual fitness costs of helping or of the benefits to recipients. The evidence for an effect of relatedness on helping effort is mixed. In birds and mammals, helping effort is positively associated with relatedness in some species but not others, while in social insects unrelated helpers typically work just as hard as more related nest mates. There is much stronger evidence that variation in the costs of helping underlies individual differences in helping effort. In meerkats and Arabian babblers, for example, experimental feeding of helpers results in increased helping effort. In paper wasps and banded mongooses, helpers with high expected future direct fitness (i.e., those that have most to lose) work less hard than those with little future direct fitness. Jeremy Field and colleagues tested the impact of future fitness experimentally on the Malaysian hover wasp, in which helpers form a strict age-based queue to inherit the position of breeder. In some groups they removed wasps from the bottom of the queue, which left the inheritance ranks of the remaining wasps unchanged. In other groups they removed wasps from the middle or upper part of the queue, which resulted in a promotion for all the wasps below the removed individual. As predicted, wasps that were promoted reduced their helping effort compared with wasps that did not ascend in rank. This experiment showed that in this social insect, helpers adjust their helping effort according to their expected future direct fitness.

There is also evidence of consistent individual differences or "personalities" within cooperatively breeding groups. In meerkats and banded mongooses, for example, there are consistent differences among helpers in their contributions to pup feeding even when controlling for age, sex, and social status. Similar consistent differences in helping and other forms of social behavior have been found in cooperatively breeding cichlids. A plausible explanation for these differences comes from research on phenotypic plasticity that shows that early life conditions interact with genotype and exert a profound influence on an animal's phenotype. In eusocial insects, variation in provisioning during the larval period triggers genetic switches that alter the developmental trajectory and result in distinct morphological and behavioral castes, even among individuals of the same genotype. Little is known about whether similar developmental effects underlie consistent individual differences in cooperative behavior in cooperatively breeding vertebrates and insects.

4. REPRODUCTIVE CONFLICT

Cooperatively breeding groups can together raise many more young than can solitary breeders, but within groups the role of breeder is much more profitable, in terms of inclusive fitness, than is the role of helper. This difference leads to intense competition over breeding status that can threaten the productivity and stability of cooperative groups. Much research over the last 20 years has focused on how this reproductive conflict is resolved and why there is so much variation among societies in the level of *reproductive skew*—a measure of the evenness with which reproduction is shared among group members. In high-skew societies reproduction is monopolized by one or a few dominant individuals; in low-skew societies all or most adults breed. Reproductive skew varies widely between species and between groups in the same species, and may be different for males and females in mixed-sex groups.

Two main types of model have been proposed to explain variation in skew within and between species. *Transactional* models assume that the stable distribution of reproduction is determined by threats to exercise *outside options*, such as leaving the group or evicting other group members. For example, if dominant individuals fully control reproduction, subordinates can use the threat of departure to extract a reproductive *concession* from dominants. If dominants have no control over subordinate reproduction, then the amount claimed by subordinates will be limited only by the threat of being evicted from the group. In both cases the stable level of reproductive skew is assumed to depend on the value of outside options, which are set by ecological constraint. In contrast with these models, *incomplete control* models assume that all group members can invest effort to exert partial, costly control over reproductive shares. In these models the stable outcome depends on the relative efficiency or strength of the players, and ecological constraints play no role. A key way to distinguish the models, therefore, is to test whether skew and group stability is sensitive to changes in ecological constraints.

Current data do not support the assumption of transactional models that skew is sensitive to outside options or ecological constraints on dispersal. Two studies (on cooperative cichlids, and a social bee) manipulated outside options experimentally and found no effect on reproductive skew. Observations of banded mongooses show that dominant females use eviction to limit reproductive competition, but subordinates do not forego breeding when the chance of being evicted is high, as would be expected if reproduction was limited by the threat of eviction. Finally, experiments to reduce the share of paternity of subordinate males in a cooperatively breeding group have never led to the breakup of the group, as one would expect if subordinates used the threat of departure to extract a share of reproduction from dominants. These lines of evidence suggest that reproductive skew is not influenced by threats to leave or evict other group members, although more experiments are needed.

In light of these results, the focus of research has shifted to understanding the evolution of conflict strategies in cooperative groups: how animals suppress one another's breeding attempts, how conflicts are settled on a behavioral timescale, and why the outcome of reproductive conflict is so variable. Analogous questions can be asked about the resolution of conflict at other levels of biological organization, for example, between genes, cells, and groups. The evolution of biological complexity, from replicating molecules to animal and human societies, has occurred via repeated cooperative transitions whereby individual subunits have come together to form cooperative teams. These transitions require that individual subunits find ways to repress selfishness and resolve conflicts of interest over direct fitness, just as cooperative breeders must resolve conflicts if they are to reap the rewards of teamwork. Theory and experiments that help elucidate conflict resolution in cooperative breeders may therefore also shed light on the fundamental question of how biological complexity arose.

FURTHER READING

Cant, M. A. 2011. The role of threats in animal cooperation. Proceedings of the Royal Society B 278: 170–178

Cant, M. A., and R. A. Johnstone. 2008. Reproductive conflict and the evolutionary separation of reproductive generations in humans. Proceedings of the National Academy of Sciences USA 105: 5332–5336. *Highlights the importance of cooperative breeding for human life history evolution, and in particular, the role of reproductive competition in the evolution of menopause.*

Clutton-Brock, T. H. 2009. Structure and function in mammalian societies. Philosophical Transactions of the Royal Society B 364: 3229–3242.

Dickinson, J. L., and B. J. Hatchwell. 2004. Fitness consequences of helping. In W. D. Koenig and J. L. Dickinson, eds., Ecology and Evolution of Cooperative Breeding in Birds. Cambridge: Cambridge University Press. *An accessible overview of the major findings and outstanding questions in avian cooperative breeding research.*

Field, J., and M. A. Cant. 2009. Social stability and helping in small animal societies. Philosophical Transactions of the Royal Society B 364: 3181–3189.

Gardner, A., G. Wild, and S. A. West. 2011. The genetical theory of kin selection. Journal of Evolutionary Biology 24: 1020–1043. *A review of the current state of inclusive fitness theory from some of the leading theoreticians in the field.*

Hatchwell, B. 2009. The evolution of cooperative breeding in birds: Kinship, dispersal and life history. Philosophical Transactions of the Royal Society B 364: 3217–3227.

Komdeur, J., A. Hufstadt, W. Prast, G. Castle, R. Mileto, and J. Wattel. 1995. Transfer experiments of Seychelles warblers to new islands: Changes in dispersal and helping behaviour. Animal Behaviour 49: 695–708.

Ratnieks, F.L.W., and T. Wenseleers. 2006. Altruism in insect societies and beyond: Voluntary or enforced? Trends in Ecology & Evolution 23: 45–52.

VII.11

Human Behavioral Ecology

Virpi Lummaa

OUTLINE

1. Development of human behavioral ecology
2. Problems and criticism
3. New focus on evolution in the modern societies
4. What can human behavioral ecology contribute to the general study of evolution?

Human behavioral ecology applies the general theories and mathematical models developed for understanding variation in traits across species to test similar questions in humans. The focus is on studying the consequences of particular traits or behavioral strategies for an individual's success at passing on its genes to the following generations, given the ecological and social environment of that individual. Humans experience a wide global range of living conditions and lifestyles, from traditional communities to extreme urbanization, and human behavioral ecologists today use a range of study designs and data sources to investigate all these populations from an evolutionary perspective. The type of data available on humans makes it possible to investigate the details of many central questions in evolutionary biology.

GLOSSARY

Cohort Studies. Longitudinal study designs commonly used, for example, in medical and social science research, and increasingly also in human behavioral ecology. Such studies record the life events of a group (cohort) of individuals sharing a common characteristic or experience (e.g., born during the same year or exposed to a famine in utero) and compare these individuals with other cohorts or the general population.

Demographic Transition. The transition from high birth and death rates to low birth and death rates as a country develops from a preindustrial to an industrialized economic system.

(Historical) Population Records. Registers of births, deaths, marriages, and migrations that have been maintained in many countries over long periods of time (e.g., by the church or governmental departments) and that are now a frequent source of data in human behavioral ecology.

Hunter-gatherer. Ancestral subsistence mode of *Homo* in which most or all food was obtained from wild plants and animals, in contrast with agriculture, which relies on domesticated species. All humans were hunter-gatherers at least until approximately 10,000 years ago.

Intervention Studies. Procedures used to test a cause-and-effect relation in epidemiological studies by modifying the suspected causal factor(s) affecting health outcomes (e.g., by supplementary feeding of a group of subjects or treating them with a given medicine) and recording their future life events in comparison with those of subjects not receiving the treatment.

Microevolution. A change in gene frequency within a population over time.

Optimality Models. Simulations that weigh the costs and benefits of a given trait or behavior compared with another trait or behavior for maximizing fitness.

Pleistocene. A time period 2,588,000 to 12,000 years before the present when key events in human evolution took place.

Twin Registers. A type of data often used in human behavioral genetics recording various traits of up to thousands of twin pairs from a given country or cohort. Such data sets are most commonly used to estimate the relative importance of environmental and genetic influences on particular traits and behaviors in humans by comparing individuals in identical and fraternal twin pairs.

Human behavioral ecology is an evolutionary approach to studying human behavior that applies methods virtually identical with those used by behavioral ecologists studying other species. The focus is on studying the consequences of particular traits or behavioral strategies for an individual's success at passing on its genes to the following generations. The most successful behavior from the viewpoint of evolutionary fitness may vary among individuals depending on attributes such as their wealth, age, living environment, family support available, or set of genes. Empirical studies in human behavioral ecology use data from different human populations to test predictions produced by the general theories and mathematical models developed for understanding variation in traits across species. One of the most widely studied questions is whether variation among individuals in partner choice and reproductive patterns in humans is adaptive: Does mate choice capitalize on reproductive prospects in the future? Does age at first reproduction reflect the "best age" for the given man or woman to start a family to maximize his or her overall number of children reared over a lifetime? Or is there an adaptive explanation for women going through menopause before the end of their life span? For example, it is postulated that women living in an environment with a high mortality hazard benefit from giving birth at a young age to ensure reproducing before dying, despite the risks to both maternal and baby survival associated with early motherhood. In contrast, a woman living in a more stable environment might maximize her overall number of surviving offspring by delaying the onset of motherhood until she has finished growing and maturing.

Application of evolutionary theory to understanding human behavior has grown increasingly popular since the publication of *Sociobiology* by Edward O. Wilson (1975), often considered as "giving birth" to the field. An evolutionary approach to explaining variation among individuals in traits such as mate preferences, marriage patterns, and childbearing—or even differences in hunting patterns, diet, language, diseases, and personality—has gained popularity in disciplines besides biology, such as anthropology, psychology, and more recently, medicine. This approach has also been applied in economics, where—much as in evolutionary thinking—maximization and self-interest are central concepts. In contrast, sociologists, for example, have traditionally been slower at integrating evolutionary theory into their approach to explaining human behavior. Consequently, scientists applying evolutionary theory to understanding human behavior have backgrounds and training in an extraordinary diverse range of disciplines. They often disagree about how evolutionary theory can be applied to understanding human behavior and how such attempts should incorporate any influence of culture, modernity, inheritance of wealth, and other factors often considered particularly relevant in humans as compared with other species.

This chapter focuses on discussing the success of the behavioral ecological approach in explaining variation among humans. The first part introduces the key approaches and assumptions traditionally used in the study of human behavioral ecology, and lists the main areas of research and their findings. The second part discusses the difficulties and criticism faced by such studies. The third part highlights the recent developments in the field that arose in response to such criticism, and points out the areas in need of further investigation. Finally, although studies on humans suffer from many unavoidable methodological difficulties, the last section highlights the particular benefits that working with humans offers for advancing our understanding of evolutionary processes in general.

1. DEVELOPMENT OF HUMAN BEHAVIORAL ECOLOGY

Human behavioral ecology began by testing predictions formulated largely from optimality theory. *Optimality models* weigh the costs and benefits of alternative traits or behaviors for maximizing fitness and have been successfully used to further our understanding of behavioral variation in other animals (see chapter VII.3). In humans, short birth intervals, for example, could be associated with the benefit of producing many offspring over the limited reproductive life span, but such benefits must be weighed against the costs of short birth intervals to both mother and child in terms of mortality risk. The best (optimal) strategy thus involves a trade-off between such factors to maximize the overall possible number of offspring raised in a lifetime. The approach typically considers human behavior to be highly plastic and likely to produce adaptive outcomes in different environmental settings. Such a black box approach assumes that there is a link between genes and behavior, but the existence of this linkage was for a long time not studied in detail (see chapter VII.1).

In humans, most quantitative data to test the models have been collected studying contemporary "traditional" societies, such as extant hunter-gatherer, agropastoral, or horticultural groups, for example, in southern Africa (!Kung San), Kenya (Kipsigis), Amazonia (Yanomamö; Tsimane), and Tanzania (Hadza) (see Hawkes et al. 1997). Only the hunter-gatherer lifestyle (e.g., that of the traditional !Kung San) is usually, strictly speaking, expected to be similar to that during Pleistocene, when human evolution is thought to have been rapid; however, because of the current rarity of such groups, research has expanded to other populations little influenced by globalization and with "natural" mortality and fertility rates, with the idea that studying such tribal groups is close to studying our ancestors. Thus, the traits that increase reproductive success among the currently living traditional populations have also done so in the past and can inform us about selection pressures operating in past environments. Because of the desire to correlate given traits or behaviors with measures of reproductive success, such as the number of living children or grandchildren, the data analyzed on these populations have largely been correlational in nature; that is, they have involved collection of anthropometric, behavioral, and demographic data on individuals without the possibility—available for other shorter-lived organisms—to conduct experiments.

One of the first areas of focus was research on foraging behavior to show that, on the whole, human foragers select food sources that maximize nutrient acquisition, as predicted by optimal foraging theory. Further research has applied the optimal theory framework to investigating mating patterns (e.g., to test whether females may gain higher fitness by mating with a male who already has a mate), life history variation (e.g., age at maturation and first reproduction, birth spacing, and senescence), and parental investment according to the prevailing social and environmental conditions. Overall, although these studies cannot necessarily show that the traits in question are the products of past selection, they have proven that applying the same framework as scientists working on similar questions in other species can indeed produce convincing support for the tested hypothesis and provide insight into how natural selection maintains variation in the trait.

For example, one of the greatest mysteries in human life history has been the existence of female menopause, a complete and irreversible physiological shutdown of reproductive potential, well before the commonly achieved overall life span in all human populations. This phenomenon is evolutionarily puzzling, because all organisms are predicted to seek to maximize their genes in the following generations, a goal that is normally achieved by breeding throughout life. The problem is that adaptive benefits of menopause are difficult to test empirically, because all women experience it; we will never know whether in our evolutionary past, women experiencing menopause produced significantly more and/or superior offspring than women who continued to reproduce until death. What is better understood, however, is that whatever the cause for menopause itself, the extended life span after menopause gives an evolutionary advantage to women. A woman with genes for living beyond her decline in fertility produces more grandchildren (and hence forwards more genes to the following generation) than a woman who dies at menopause, because postreproductive women can have positive effects on their offspring's reproductive success—they help rear their own grandchildren. Among the Hadza of Tanzania, child weight is positively correlated with grandmother's foraging time (see Hawkes et al. 1998 for details). The presence of a grandmother has also been linked to increases in grandchild survival chances in many contemporary traditional as well as historical populations around the world. Finally, research using data available for farming/fishing communities of eighteenth- and nineteenth-century Finnish and Canadian people has shown that mothers indeed gained extra grandchildren by surviving beyond menopause until their mid-seventies. These data show that life span can be under positive selection at least until this age. This effect arose because offspring in the presence of their living postreproductive mothers bred earlier, more frequently, for longer, and more successfully. Such benefits were not present if the mother was alive but lived farther apart from her adult offspring, which suggests that the findings are not a mere artifact of better overall survival of both grandmothers and grandchildren in some families (see Lahdenperät et al. 2004 for details). An additional discussion of the evolution of menopause in humans can be found in chapter VII.16.

Another main interest in human behavior ecology has been to investigate the effect of environmental conditions on the fitness benefits of different traits. For example, costs of reproduction to females need to be analyzed in relation to the energy budget of the woman: high costs of reproduction do not have the same effects on women who have good diets and low levels of physical activity compared with women in poor energetic condition. Such physiological consequences of reproduction for women with differing food access are well documented in humans. Further evidence that resource availability may affect selection on life history traits in humans comes from studies showing a negative relationship between number of offspring and postmenopausal life span among poor landless women, whereas for wealthier women, the relationship between fecundity and postmenopausal life span is often positive. A negative relationship between fecundity and longevity may therefore be expected in women who owing to multiple pregnancies and breastfeeding pay high costs of reproduction that cannot easily be compensated for by increases in dietary intake and reduction in physical activity. In contrast, wealthier women can more easily "afford" both large family size and long life span. Comparable differences in the costs of reproduction could also be created, for example, by differing amounts of help available from other individuals with raising the offspring, such as partners, grandparents, or other helpers in the nest, that affect the level of investment made by the mother, but few studies have investigated such effects.

2. PROBLEMS AND CRITICISM

The downside of the original focus on traditional populations with high fertility and mortality rates is that sample sizes tend to be limited; groups are rapidly disappearing or are affected by globalization; collection of multigenerational data often essential for addressing evolutionary questions is time consuming or impossible; and ages are merely estimates. Focusing preferentially on hunter-gatherers also ignores the fact that human evolution has been most rapid, in terms of generation-to-generation changes in gene frequencies, since the invention of agriculture. Investigating modern populations is equally interesting, because differences in reproductive and survival rates among individuals still lead to selection favoring certain heritable traits over others, albeit that the alleles being favored might also be influenced by culture (see chapter VIII.10), in particular modern medical care. Moreover, modern populations lend themselves to current genomic and population genetic analyses.

First, recent analyses of the human genome have revealed that human genetic makeup has responded to the domestication of plants and animals and the spread of agriculture; numerous genes have experienced recent positive selection, and overall considerable selection has occurred in the past 10,000 years (see chapter VIII.12 for more details and examples). These results are at odds with the claims that natural selection affecting humans stopped with the spread of agriculture or at least with the recent modernization, and investigating only those humans exhibiting lifestyles comparable to those practiced during the Pleistocene is relevant for understanding human evolution. Clearly, agriculture has been a powerful selection force whose effects should be more rigorously investigated, and the continued evolution of humans should be better documented.

Second, analyses of the human genome have also revealed that significant genetic differences both among and, in particular, within human populations have arisen from recent selection events. Many scientists who apply natural selection to understand human behavior have traditionally been uncomfortable with assigning any role for genes in explaining variation among individuals or populations, perhaps because of social Darwinism and racially discriminatory perspectives on human evolution put forward during the early half of the 1900s (see also chapter VIII.11). In contrast, a modern approach to investigating the role of genes in human behavior should focus on studying the effects of mating and reproductive patterns on genetic variation, and genetic constraints on trait evolvability in different populations, as well as on how the documented selection on traits together with their underlying genetic architecture predict responses to such selection.

Third, early attempts to apply evolutionary framework to contemporary Western populations sparked criticism on the ground that some aspects of the modern industrialized world are too novel, and humans may be responding nonadaptively to them, making studies on adaptive traits in such populations pointless. This view ignores the fact that in both industrialized human societies with easy access to modern contraception and medical care and traditional societies there is a large variance in the reproductive success of both sexes. In other words, although survival to old age is high among all individuals, not everyone has the same family size, and many individuals even forego reproduction altogether. Such a variance provides material to natural selection that will capitalize on any heritable trait variation linked with higher reproductive success. Thus, even if many behaviors in novel modern environments turn out to be maladaptive, the large opportunity for selection coupled with heritable traits linked with differences in reproductive output of individuals might lead to rapid changes in the genetic makeup of the population over generations, and selection against any traits genetically linked to maladaptive behavior, because any genetically variable traits associated with the variance in reproductive success will experience selection and evolution regardless of the mechanism by which reproductive variance is affected. Consequently, while social Darwinism should not be tolerated, the reality that humans can continue to evolve should not be negated. Yet because of the trend in human behavioral ecology to focus on the past, and the previous criticism for using other than hunter-gatherers (or to some extent horticulturalists, agropastoralists, or farmers with high mortality and fertility rates) as model populations, only recently have scientists started investigating the behavior of people living in industrialized societies from an adaptationist viewpoint. Even fewer studies have been undertaken to examine how the modern environment itself continues to fuel evolution by favoring or disfavoring certain alleles of the genes, and how the drastic demographic shifts in many populations to low birth and death rates during the recent centuries has affected the overall opportunity for selection or specific trait selection.

Human behavioral ecologists are also criticized for seeking adaptive explanations for behaviors even when such explanations are unlikely. Such criticism applies to all behavioral ecology, but pointing out flaws and factors not correctly considered in the evolutionary models of behavior is obviously easier when the study subject is our own species. It should, however, be stressed that human behavioral ecologists investigate not only how human behavior "fits" the given environment with adaptive benefits but also how environmental conditions constrain individual success. For example, poor early environmental conditions for developing individuals, such as unfavorable month or season of birth, reduce longevity and reproductive performance, yet women commonly reproduce during such times. Social norms, cultural practices, and traditions often lead to reproductive outcomes that are not necessarily beneficial in terms of evolutionary fitness—the study of cultural evolution represents an entire field of research investigating such topics but is not discussed further in this chapter (see chapter VIII.10 for details). Furthermore, poor dietary intake during gestation that leads to reduced birth weight of babies has been shown to be associated with their subsequent risk of adverse health, age at sexual maturation, ovarian function, and life span, which suggests that poor early-life conditions influence development and produce adverse effects later in life. The implications of such effects for evolutionary processes should be considered in more detail.

3. NEW FOCUS ON EVOLUTION IN THE MODERN SOCIETIES

Recent methodological improvements in the ability to measure selection, heritability, and response to selection in natural populations of animals have inspired many human behavioral ecologists. The central focus of human behavioral ecology has recently begun to shift from asking how the behavior of modern humans reflects our species' historical response to natural selection, to measuring current selection in contemporary populations as well as investigating how that might (or might not) cause evolution. Calculations that incorporate a measure of selection and heritable variation in traits allow us to predict how traits under selection could change over time. Such evolutionary changes in human populations are likely, because natural selection operates on several morphological, physiological, and life history traits in modern societies through differential reproduction or survival, and variation in many of these traits has a heritable genetic basis. This change of focus has led to several important changes in methods and approaches used in the field.

First, the type of information that can be analyzed has become more diverse, allowing researchers to take full advantage of the exceptional data available only for humans. Historical demographers, population geneticists, and evolutionary biologists are making increasingly better use of (*historical*) *population records* of agricultural or industrialized populations. Such data sets have the benefit of large multigenerational samples, although the type of data available is usually limited to demographic information such as births, marriages, reproductive events, and deaths. There have been recent promising attempts to make better use of extremely large and versatile *cohort studies* and *twin registers* collected by epidemiologists and social scientists on representative samples of people living in contemporary Europe, the United States, and Australia. *Medical intervention studies* that have collected long-term data on their subjects (e.g., after supplementing mothers' diet during pregnancy) offer a much-needed experimental framework for human behavioral ecologists. These data sets are only now making their way into evolutionary studies. Many scientists are also beginning to use noninvasive manipulations, especially in questions related to sexual selection and mate choice, but also when studying life history strategies. For example, subjects can be exposed to images ("environment") associated with high versus low mortality risk and then asked questions about reproductive investment intentions and preferences. Primatologists have conducted between-species comparisons across primates to draw conclusions on human patterns, and worldwide ethnographies and encyclopedias provide an opportunity to perform similar tests among the large variety of human societies, too. All in all, humans experience the widest global range of living conditions and lifestyles, from traditional communities to extreme urbanization, and human behavioral ecologists today ought to use a wide selection of study designs and data sources to investigate all these populations from an evolutionary perspective.

Second, the focus on studying microevolution in contemporary populations has made it necessary to reexamine the old assumption among behavioral ecologists that the details of trait inheritance do not seriously constrain adaptive responses to ecological variation. Estimating heritability of human traits is often considered problematic: an estimation of heritabilities and genetic correlations requires large multigenerational samples and sample sizes often not available in traditional anthropological studies. Furthermore, effects of a common environment shared by close relatives, and cultural transmission, can inflate estimates of heritability. Nevertheless, a review by Stephen Stearns and colleagues (2010) of studies investigating heritability of life history and health traits in humans suggested that although the heritability levels vary considerably among traits and among study populations, many human traits, such as age at first and last reproduction, cardiovascular function, blood phenotypes, weight, and height have measurable heritability and will respond to selection if they are not constrained by genetic correlations with other traits. Fewer studies have investigated such genetic correlations between traits (caused, for example, by the same gene affecting variation in several traits), but there is some suggestion that such effects can set genetic constraints on trait evolution in humans. For example, a study using the historical pedigree records available on rural Finnish people showed significant negative genetic correlations between reproductive traits and longevity (see Pettay et al. 2005). The existence of this genetic variation and covariation implies that females who reproduced at faster rates also had genes for relatively shorter life span, supporting the hypothesis that rate of reproduction should trade off with longevity. Overall, investigation of genes underlying behavioral differences is only beginning in humans, but studies so far suggest that detailed knowledge of the genetic architecture and its dynamics with environmental conditions can provide helpful information on the current evolutionary processes.

Third, an increasing number of studies show that both the opportunity for selection (variation among individuals in fitness) and selection on particular traits can be strong in contemporary populations (see, e.g., Courtiol et al. 2012). The important question is, Do these results predict any phenotypic changes taking place in the mean trait values or the genetic makeup of the population over generations? Understanding such responses to selection reveals how the rapidly changing culture, such as medical care, is changing the biology of humans. A recent study by Sean Byars and his colleagues (2009) measured the strength of selection, estimated genetic variation and covariation, and predicted the response to selection for life history and health traits in the current US population. Natural selection appears to be causing a gradual evolutionary change in many traits: the descendants of the study women were predicted to be on average slightly shorter and stouter, to have lower total cholesterol levels and systolic blood pressure, to have their first child earlier, and to reach menopause later than they would in the absence of evolution. A similar study on a preindustrial French-Canadian population found natural selection to favor an earlier age at first reproduction among women, a trait that was also highly heritable and genetically correlated to fitness in this population. Age at first reproduction declined over a 140-year period and also showed a substantial change in the breeding value (part of the deviation of an individual phenotype from the population mean due to the additive effects of alleles), suggesting that the change occurred largely at the genetic level. These studies demonstrate that microevolution might be detectable over relatively few generations in humans. It must, however, also be borne in mind that phenotypic changes may not always provide robust evidence of evolution, as they may not reflect underlying genetic trends. Many traits such as height, weight, mortality, age at first reproduction, and family size have shown strong secular changes during a *demographic transition* (the change from high birth and death rates to low ones as a country develops from a preindustrial to an industrialized economic system) that may mostly be associated with rapid changes in diet, medicine, and contraception availability. Further studies focusing on how selection interacts with changing early and later-life environment of individuals and is associated with changes in specific sections of the genome are thus needed.

4. WHAT CAN HUMAN BEHAVIORAL ECOLOGY CONTRIBUTE TO THE GENERAL STUDY OF EVOLUTION?

Evolutionary studies on humans are said to suffer from many drawbacks compared with investigations on model animals, because the data are "correlational" given the difficulty in conducting experiments, and the study objects are exceptionally long-lived, which complicates the collection of lifelong data in the field. Nevertheless, humans make it feasible to investigate the details of many central questions in evolutionary biology.

Only in humans is it possible to work on databases that contain the lifetime vital records, medical history, and a range of physical and psychological details for up to millions of recognizable individuals that can in some cases be traced back for several generations. Such data sets allow researchers to investigate selection on and evolutionary change in physiological and health-related traits that could never be feasibly collected for any other animal in natural conditions. Moreover, such data sets also allow studies in selection on personality and cognitive abilities, which have become popular among behavioral ecologists working on animals, but in humans these can be explored in greater detail than in other species and can be linked to lifetime reproductive success. Furthermore, huge investments in documenting the human genome shadow those available for most other species, and genetic data are sometimes available alongside historical pedigree data. In addition, ongoing large research programs to unravel developmental origins of health and disease in humans should offer excellent opportunities to investigate the evolutionary implications of interplays between developmental conditions and genetics in a much longer lived species than those studied so far.

Data available on humans also allow investigations of fitness in a more reliable way than is often possible in similarly long-lived other species, or even in short-lived species in the wild. Many registers allow accurately determining the numbers of grandchildren for each individual, and these provide a far better measure of fitness than simply the number of offspring born, given the considerable trade-offs detected between offspring quantity and quality in humans (and likely in many other species, too, in which large parental investment improves offspring survival and mating success). Importantly, population-based registers allow inclusion of those individuals who never reproduce into the calculations of variance in fitness, which appears crucial given that in the past as well as present human populations, a large fraction of each birth cohort fail to contribute their genes to next generation, and selection is often strongest through recruitment differences rather than differences in the family size among those who do reproduce.

Many "natural experiments" also offer opportunities to investigate evolutionary questions. Such events involve well-documented famines such as the Dutch Hunger Winter during the Second World War (see, e.g., Roseboom et al. 2001 for details), sex-ratio biases created by wars, documented long-term year-to-year variation in crop success and local ecology linked with individual fitness data, or large-scale changes in the demographic parameters of the population that have occurred repeatedly across the world but at different periods in different countries.

Given that humans exhibit all mating systems documented in the animal kingdom (monogamy, polygyny, polyandry, and even promiscuous mating; see chapter VII.4), they also offer interesting opportunities for investigating how changes in mating system affect selection. For example, over the reproductive lifetimes of Utahans born between 1830 and 1894, socially induced reductions in the rate and degree of polygamy corresponded to a 58 percent reduction in the strength of sexual selection, illustrating the potency of sexual selection in polygynous human populations and the dramatic influence that short-term societal changes can have on evolutionary processes.

Finally, humans are also exceptional in that it is possible to reliably study individual variation in complex cognitive traits. Researchers have used methodology relying on simple experimental settings to collect quantitative data on traits such as mating preferences, cooperativeness, and personality. Similar studies are virtually impossible to conduct on animals because the methods involve a certain degree of abstraction. For example, the same individuals can be asked to choose between large numbers of fictive alternatives, such as hypothetical partners. These preferences for mate characteristics can then be further compared with real-life partner characteristics, and the ecological and individual causes and fitness consequences of the degree of mismatch between preferences and actual pairings can be examined.

FURTHER READING

Alvergne, A., and V. Lummaa. 2010. Does the contraceptive pill alter mate choice in humans? Trends in Ecology & Evolution 25: 171–179.

Byars, S. G., D. Ewbank, D. R. Govindaraju, and S. C. Stearns. 2009. Natural selection in a contemporary human population. Proceedings of the National Academy of Sciences USA 107: 1787–1792.

Courtiol, A., J. Pettay, M. Jokela, A. Rotkirch, and V. Lummaa. 2012. Natural and sexual selection in a monogamous historical human population. Proceedings of the National Academy of Sciences USA 109: 8044–8049

Dunbar, R., L. Barrett, and J. Lycett. 2005. Evolutionary Psychology: A Beginner's Guide. Oxford: OneWorld.

Hawkes, K., J. F. O'Connell, N. G. Blurton Jones, H. Alvarez, and E. L. Charnov. 1998. Grandmothering, menopause, and the evolution of human life histories. Proceedings of the National Academy of Sciences USA 95: 1336–1339.

Hawkes, K., J. F. O'Connell, and L. Rogers. 1997. The behavioral ecology of modern hunter-gatherers, and human evolution. Trends in Ecology & Evolution 12: 29–32.

Lahdenperä, M., V. Lummaa, S. Helle, M. Tremblay, and A. F. Russell. 2004. Fitness benefits of prolonged post-reproductive lifespan in women. Nature 428: 178–181.

Laland, K. N., and G. R. Brown. 2011. Sense and Nonsense: Evolutionary Perspectives on Human Behavior. 2nd ed. Oxford: Oxford University Press.

Moorad, J. A., D.E.L. Promislow, K. R. Smith, and M. J. Wade. 2011. Mating system change reduces the strength of sexual selection in an American frontier population of the 19th century. Evolution and Human Behavior 32: 147–155.

Pettay, J. E., L.E.B. Kruuk, J. Jokela, and V. Lummaa. 2005. Heritability and genetic constraints of life-history trait evolution in preindustrial humans. Proceedings of the National Academy of Sciences USA 102: 2838–2843.

Roseboom, T., J. van der Meulen, C. Osmond, D. Barker, A. Ravelli, and O. Bleker. 2001. Adult survival after prenatal exposure to the Dutch famine 1944–45. Paediatric and Perinatal Epidemiology 15: 220–225.

Stearns, S. C., S. G. Byars, D. R. Govindaraju, and D. Ewbank. 2010. Measuring selection in contemporary human populations. Nature Reviews Genetics 11: 611–22.

VII.12

Evolutionary Psychology

Robert C. Richardson

OUTLINE

1. The Darwinian background for evolutionary psychology
2. The modern-day program of evolutionary psychology
3. Psychological evidence
4. The application of evolutionary models in evolutionary psychology
5. Evolutionary alternatives

Evolutionary psychology is an approach to cognitive psychology that aims to inform work in psychology with evolutionary ideas and to reform cognitive science by placing it in an evolutionary context, that is, by focusing on how psychological traits such as aggression, mate selection, and social reasoning were adaptive in ancestral environments. This methodology involves a variety of psychological and behavioral evidence that is relatively independent but may be interpretable in evolutionary terms; in other cases, it involves psychological models that depend on evolutionary models. One such example is incest aversion, which can be interpreted in terms of kin selection or inclusive fitness. There are problems in integrating the two domains. More specifically, the evolutionary interpretations often lack empirical evidence. In general, it seems evolutionary psychology could benefit from a more inclusive and contemporary infusion of evolutionary theory.

GLOSSARY

Computational Mechanisms. Algorithms that compute determinate input-output functions, dependent only on the structure of representations involved; sometimes called *Turing computability*.

Ecological Rationality. A hypothesis characteristic of evolutionary psychology that what counts as a rational procedure is relative to the ecological context in which it is applied, and cannot be determined without knowing the context in which it is applied.

Environment Of Evolutionary Adaptedness (EEA). The environment characteristic of human evolution, both physical and social; sometimes the EEA is thought of as the environment of Pleistocene ancestors in the African savanna; sometimes it is treated as a statistical composite of ancestral environments.

Incest Taboos. General social prohibitions against sexual relations among more closely related individuals; among humans, the paradigm is the prohibition of sexual activity between individuals closer than second cousins. This is often held to be a human universal.

Modules. Cognitive mechanisms that operate in relative independence from other mechanisms that govern other domains (e.g., face recognition is a capacity in humans that is relatively independent of other capacities).

Social Exchange. Any exchange of value among individuals, with costs and benefits attached; in more interesting cases, these involve iterated exchanges in which reciprocal altruism can be effective.

Evolutionary psychology (EP) is a field of research that seeks to rely on evolutionary biology in the development and elaboration of specifically human psychological hypotheses or psychological mechanisms; more generally, EP looks to the integration of cognitive psychology with evolutionary biology in explaining and interpreting human behavior. EP developed from sociobiology, with the perspective that because humans, like all other organisms, have evolved, and the principles of evolutionary biology are universal, evolutionary theories will help us understand our own origins and features (see chapters II.18 and VII.11). Yet the application of principles of evolution to humans still engenders debate and skepticism (see chapters VIII.11–VIII.15), as it did when the idea of evolution was first introduced.

The primary range of the cognitive models in EP includes such issues as attraction to mates, patterns of jealousy, reasoning applied to social exchange, probabilistic reasoning, and incest taboos; it also includes more controversial topics such as the evolution of rape, differences between male and female aggression, and the patterns of child abuse. Psychological mechanisms are assumed to be subject to shaping by natural and sexual selection, and as a result, current behavioral patterns can be understood in terms of human evolutionary history. EP assumes that current psychological mechanisms are adaptations to ancestral environments and not to contemporary environments. Dietary preferences, for example, that may have been adaptive in ancestral environments, such as a preference for sweet food, are not adaptive in our current environment. The same disparity should apply to other psychological patterns and mechanisms.

Interpreting psychological hypotheses and mechanisms in terms of evolutionary principles is not a simple matter. In one view, evolutionary theory may be used primarily as a heuristic for defining and elaborating psychological hypotheses. Some cases seem to fit in this category, such as gender differences in the sorts of traits that are attractive in potential mates. This use seems relatively unproblematic but makes no substantive use of evolutionary theory.

In contrast, evolutionary models may be integrated into the evidence for the psychological hypotheses, supposedly contributing to their evidential credentials. For example, incest taboos are argued to have evolved by natural selection and function to avoid inbreeding. Examples of this sort are the most controversial. Here, the specific evolutionary models are often not supported by adequate evidence. When this is so, the psychological hypotheses are correspondingly uncertain, or at least no more certain than otherwise warranted by the psychological evidence. For example, human language is plausibly an evolutionary adaptation: given the complexity of the underlying structure and function of the mechanisms, the incorporation of recursive grammars, and the complex patterns in the acquisition of children's languages, this is the sort of complex mechanism we should expect to be an adaptation. In this portrayal, human languages are *adaptations* for human communication, which is no doubt true given the importance of language; but we are left in the dark about the specific features of human languages, such as their recursive structure, that likely make them adaptations. Communication, for example, is adaptive, but the connection of a general appeal to communication to recursive structures is not clear. In such cases, there is a disconnection between the supposed adaptation and the adaptive model.

1. THE DARWINIAN BACKGROUND FOR EVOLUTIONARY PSYCHOLOGY

Attempts to unify evolution and psychology date to Darwin, but contrary to the modern pursuit of identifying the mechanisms of evolution such as kin selection or natural selection that shape human psychology, Darwin's focus was more on arguing that human psychology has evolved than on exploring or suggesting how psychology evolves. There are few mentions of human evolution in *On the Origin of Species*, but Charles Darwin did write in the final chapter that *The Origin* would "open fields for far more important researches. Psychology will be based on a new foundation, that of the necessary acquirement of each mental power and capacity by gradation. Light will be thrown on the origin of man and his history" (1859). Out of context, this appears to suggest that interpreting psychology in the light of *natural selection* would put psychology on a "new foundation." However, this passage appears in a section that lists topics that Darwin felt would be transformed by acknowledging evolution. He did not discuss the mechanism of evolution in this section, just its existence. Thus, while he tackled head-on the most contentious area of all—what would seem to differentiate us from all other animals—human psychology, he did so without reference to the mechanism of natural selection.

Darwin developed his ideas on human evolution more fully in *The Descent of Man and Selection in Relation to Sex* (1871). In the opening passage of *The Descent* he writes:

> He who wishes to decide whether man is the modified descendant of some pre-existing form, would probably first enquire *whether man varies*, however slightly, in bodily structure and in mental faculties; and if so, *whether the variations are transmitted* to his offspring in accordance with the laws which prevail with the lower animals; such as that of the transmission of characters to the same age or sex. Again, are the variations the result, as far as our ignorance permits us to judge, of the same general causes, and are they governed by the same general laws, as in the case of other organisms? (Darwin 1871, 9; italics added.)

In both *The Origin* and in this passage from *The Descent*, Darwin is discussing evolution, or common descent, and not specifically natural or sexual selection. There is appeal to variations and to inheritance, and to the "laws" governing each of them, but there is not a hint of competition or the "struggle for existence," much less of natural selection. Here in *The Descent*, he initially recapitulates the argument for common descent from *The Origin*, extending it to what he calls the "mental faculties" of man, saying at the outset that his object "is solely to shew that there is no fundamental difference between man and the higher mammals in their mental faculties" (1871). Darwin is clear that this commitment to evolution is meant to include what he calls the "moral sense." This was crucial for Darwin. It meant, among other things, that our capacities for social interaction and our psychological propensities were meant to be within the purview of his evolutionary theory.

Darwin was neither the first nor the last to bring evolutionary insights to the discussion of our social sentiments and reasoning. In the nineteenth century, Herbert Spencer had placed his discussion of psychology in an explicitly evolutionary setting before the publication of *The Origin*; William James's psychology was inspired by Darwinian insights, as were other important psychologists at the turn of the century. In the twentieth century, there were other ventures into the evolution of human psychology, some of which are in retrospect less well regarded, such as Desmond Morris's *The Naked Ape*. With the elaboration of models designed to capture social behavior in the middle of the last century, *Sociobiology* by E. O. Wilson dealt with the task of capturing animal behavior in evolutionary terms, and almost as an appendix extended that project to the domain of human social behavior. From Wilson's book, the field of sociobiology was born and thrived in the 1980s, with various attempts to extend sociobiology to encompass the human case. This brings us to the most recent approach, evolutionary psychology, which takes up the Darwinian idea that evolution should shed light on human psychology, and which has usurped sociobiology.

2. THE MODERN-DAY PROGRAM OF EVOLUTIONARY PSYCHOLOGY

Contemporary evolutionary psychology is not a homogeneous collection of views, even with respect to its evolutionary commitments, though it is possible to articulate a loose set of claims that are broadly endorsed, and typical of contemporary adherents. In large part, these are commitments consistent with evolutionary theory as it was articulated during the "evolutionary synthesis" years in the first half of the twentieth century (see chapter I.2), updated by evolutionary models from the 1960s. Not every advocate of EP is committed to precisely the same set of claims, but it is possible to provide a kind of portrait. The following are some characteristic commitments:

- Psychological mechanisms are the result of natural selection and sexual selection. While it is generally acknowledged that evolution has some outcomes that are owing to chance or are by-products of selection for other traits, the focus of EP is on traits presumed to be adaptations and therefore that reflect evolution from selection—traits centered on problems such as finding a mate, cooperative activities like hunting, or the raising of offspring. The assumption is that natural selection will tend to "solve" problems like this with considerable efficiency. Possible alternatives to selection are rarely considered, nor are alternative selectionist regimes.
- Psychological mechanisms can be thought of as computational mechanisms. Among such mechanisms are included cognitive processes (e.g., probabilistic reasoning or problem solving) as well as emotional responses (e.g., jealousy or fear). The idea that psychological mechanisms are computational is a common assumption among a range of cognitive scientists, though its prevalence has faded considerably in the last decade or so. Alternatively, these computational mechanisms can be thought of as exhibited in and causing behavioral *strategies* for responding to environmental challenges, where the strategies are genetically specified.
- Psychological mechanisms evolved in response to relatively stable features of *ancestral* environments, often collectively referred to as the *environment of evolutionary adaptedness* (EEA). EP asserts that because most of human evolution took part during the Pleistocene (roughly 2.6 million to 12,000 years ago), and presumably in the later Pleistocene, what is seen today in terms of psychology evolved to be adaptive in this hypothetical EEA. Often, the EEA is identified with the savanna of the African Pleistocene, with a hunter-gatherer lifestyle. The EEA can also be identified with a kind of statistical aggregate of the total range of ancestral environments.
- Because psychological mechanisms are adaptations to ancestral environments, there is no assumption that they are adaptive in contemporary circumstances. Social environments are a significant part of the environment and are obviously crucial to human evolution. If we assume with EP that our ancestral social environment consisted of small, nomadic bands of relatives, then the difference between that and our contemporary culture suggests that whatever strategies were adaptive for our ancestors, may not be so for us. Likewise, if we assume that our distant ancestors lived in a sugar-deprived environment, then our fondness for sweets might be "natural" though no longer adaptive. In general, EP assumes that evolutionary responses are too slow to

have had any significant effect in the last 12,000 years or so, the earlier advent of agriculture and sedentary life.

- The human mind is a kind of mosaic of mechanisms, each with some specific adaptive function, rather than merely a general-purpose learning machine. Different adaptive problems will require different solutions and different strategies for dealing with them. So, for example, a mechanism for mate selection is unlikely to be of much use in foraging. At least some of this machinery must be domain specific, specialized for particular tasks, and some of these mechanisms may count intuitively as *instincts*. Several advocates of EP treat these mechanisms as *modules*, though others insist that all that is required is distinct domain-specific mechanisms.

3. PSYCHOLOGICAL EVIDENCE

Evolutionary psychologists make use of an array of techniques to evaluate their psychological models, most of which do not specifically depend on the evolutionary assumptions. These methods include, among others, the use of questionnaires, controlled experiments, observational methods, and brain imaging (functional magnetic resonance imaging [fMRI] and positron emission tomography [PET]). Some also make use of a variety of less standard techniques, including ethnographic records, paleontological information, and life history data. Evolutionary assumptions do come into play in advancing and formulating hypotheses, as suggestive of psychological hypotheses to test. Whether they are more than *merely* heuristic is sometimes not clear.

Evolutionary psychologists have articulated and tested a wide array of psychological hypotheses inspired by evolutionary thinking. These include human propensities for such matters as cooperation and cheater detection, differences in spatial memory, and short-term mating preferences. Some simple examples may be sufficient to illustrate the method. Assume that human memory is sensitive to items that affected fitness among our ancestors, such as food items, shelter, or possible mates. Using standard experimental memory probes within psychology that are concerned with recall and recognition for lists of words, researchers found that recall for survival-oriented terms was significantly better than recall for more neutral words. Similarly, theories of parental investment suggest that given monogamous coupling, females will tend to prefer mates that are more likely to invest in offspring. From an EP perspective, this also suggests that males and females will differ in the patterns of jealousy, with females on average more sensitive to emotional infidelity (as a risk of abandonment) and males more sensitive to actual sexual infidelity (as a risk to paternity). These predictions have been supported by straightforward evaluations of preferences using questionnaires, spontaneous recall, and fMRI.

4. THE APPLICATION OF EVOLUTIONARY MODELS IN EVOLUTIONARY PSYCHOLOGY

Relying on work in paleoanthropology and ethnography relating especially to contemporary hunter-gatherers, evolutionary psychologists have elaborated a portrait of ancestral social life. While their description is plausible, it is also controversial among anthropologists. EP assumes ancestral hominids lived in relatively compact kin-based groups of no more than 100 members. It is presumed that a sexual division of labor existed, with males more engaged in hunting and females more engaged with gathering, and that stable male-female bonds existed, as well as long periods of biparental care. Within each kin group, there was cooperative foraging. Much more is known about the biotic and abiotic environment that existed. For example, it is known that during this time, humans were subject to a variety of predators and pathogens and considerable variance in the availability of resources.

Assuming this broad portrait of early human social life and abiotic influences allows evolutionary psychologists to construct a variety of evolutionary scenarios. Depending on the case, they use a variety of resources from evolutionary biology, including theoretical models concerning reciprocal altruism (see chapter VII.9), parental investment (see chapter VII.8), kin selection (see chapters VII.10 and VII.12), and evolutionary game theory (see chapter VII.3). Beginning with the relevant dimensions assumed to be typical in the EEA, evolutionary psychologists construct an account of the adaptive functions that must be satisfied, rather like a design specification. The problem for EP is then to reverse engineer a solution to the adaptive problem and test it in modern populations.

Reverse engineering is a powerful theoretical tool, but it can lead to difficulties and criticism, especially if the "adaptive problem" is not clear. If it is poorly articulated, then it is not clear whether the evolutionary solution is the right one. To use a nonhuman example, if the ecological "problem" is how insects can walk on water, knowing the adaptive "solution" depends on the surface tension of water, and that in turn depends on knowing the saline content; absent the determinate content of the problem, there is no general solution to crucial issues such as foot structure, though the specific solutions are solved readily. In the human case, language is certainly involved in communication, but this fact offers no explanation for the peculiarities of human language—its recursive structure, for example—which are plausibly adaptations.

EP also raises issues about connecting the psychological hypotheses and the evolutionary interpretations. Consider a Darwinian theory of the evolution of incest avoidance. Incest is a very interesting case of a social prohibition, since psychological studies show that disapproval of it survives even the recognition that it will produce no actual harm. It has been a very significant issue, first for sociobiologists and now for evolutionary psychologists. There is a straightforward case against incest from an evolutionary perspective based on inbreeding. Inbreeding can result in reduced fitness, termed *inbreeding depression* (see chapter IV.6). Where inbreeding depression exists, there should be an evolutionary pressure against inbreeding.

The importance of inbreeding depression leads EP advocates to suggest that there is a natural tendency—sometimes a psychological "module"—for incest avoidance. Debra Lieberman, together with Cosmides and Tooby, has suggested that humans have a specialized kin recognition system (there are such mechanisms in other animals). They observe that incest avoidance could facilitate an avoidance of any deleterious consequences associated with inbreeding depression and suggest that this leads to selection for incest avoidance.

The *Westermark hypothesis* posits a mechanism of the sort Lieberman, Cosmides, and Tooby predicted, suggesting that children raised together develop a sexual disinterest, or even a sexual aversion, to one another. The proposed function is to avoid incest, since those who are raised together are most often closely related. Lieberman assumes, reasonably, that coresidence during periods of high parental investment should be a reliable indicator of kinship or would have been a reliable indicator in the EEA with small kinship bands. Together with Cosmides and Tooby, Lieberman shows considerable support for the conclusion that duration of coresidence is *psychologically* predictive of sexual aversion.

The evolutionary interpretation that incest avoidance evolved because of selection against inbreeding is plausible on the surface but nonetheless problematic. The association cannot be directly tested in ancestral populations, but it does fit the patterns of some contemporary "hunter-gatherer" populations (which may be considered a proxy for ancestral populations), though not all. It is, of course, true that siblings would typically be associated with one another during childhood, but the proper question is whether in ancestral groups the set of people that an individual may have selected from when choosing a mate included the siblings with which one interacted as a child.

A more straightforward and general problem with EP is assuming a single EEA. We know that our Pleistocene ancestors did not have simply one lifestyle in one region but lived on the African savanna, in deserts, next to rivers, by oceans, in forests, and even in the Arctic, employing very different foraging methods and living off diverse diets, with technologies ranging from the simple chopping tools of *Homo habilis* to the rich and sophisticated stone, bone, and antler toolbox of late Pleistocene *Homo sapiens*. There is little reason to think that there was a single form of social structure associated with the full range of human physical environments, much less that contemporary "hunter-gatherer" populations are typical of ancestral groups. For the hypothesis of incest avoidance as a mechanism to avoid inbreeding depression, it is hard to know whether association will be limited to siblings, or more likely to be with siblings, absent a fairly specific account of social organization, including the relative viscosity of the groups and issues such as group size.

In many animal species, there is a tendency for animals to disperse prior to mating; they move away from their familial unit. This clearly has the effect of reducing inbreeding, though there doesn't seem to be any consensus on whether incest avoidance is particularly significant in supporting dispersal. Among chimpanzees, when males come of reproductive age, they tend to emigrate from the ancestral clan. There is no need for incest aversion, since they move away from their siblings. To know how to apply inbreeding avoidance models to ancestral human groups, it would be necessary to know whether both males and females remained with the ancestral groups or emigrated. There is some evidence that among early *Homo*, the males tended to move out of their ancestral groups once they were reproductive, as with chimpanzees, but it doesn't matter whether this is correct. The important point is that absent such information, the relevance of the evolutionary models of selection to incest aversion is not clear. If reproductives tend to emigrate, then there is little need for incest aversion, and none for incest taboos.

5. EVOLUTIONARY ALTERNATIVES

The preceding general assumptions that form the backdrop for EP are characteristic of only a selected subsample of work in evolutionary biology. More generally, evolutionary biology incorporates a wide variety of disciplinary perspectives that do not feature in the work within EP. Many of the assumptions characteristic of EP may be problematic because they do not incorporate more recent advances in our understanding of how evolution works. There are many recent developments in genetics, in evolutionary biology, and in developmental biology that might improve EP considerably, bringing it more in line with more recent evolutionary thinking. Several are briefly noted here.

- Natural selection and sexual selection are doubtless potent evolutionary forces. There are alternative evolutionary factors that can, and do, affect evolutionary trajectories. Evolutionary psychologists acknowledge such factors as genetic drift (though it plays no role in their scenarios, and they do not address it as an alternative) but do not incorporate phylogeny or comparative biology. As a result, our primate kin do not typically feature in EP explanations. One salutary change would be to take account systematically of our relatedness to our primate kin and the possibility that features exhibited by humans are inherited from a common ancestor. At the very least, this would provide expectations as to social patterns that feature so prominently in EP, and give testable alternative hypotheses. In particular, it would downplay the commitment to natural selection acting on specifically human social capacities.
- EP typically assumes that the relevant selection forces are relatively ancient and that recent changes would be insignificant. From the perspective of EP, modern humans are Pleistocene relics. Yet we know that there have been substantial changes in the human genome over even the last 10,000 years and that these changes are ongoing (see chapter VIII.12). Many of the evolutionary changes reflect the adoption of agriculture and the domestication of animals, environmental changes that surely impose selection.
- The environment of the Pleistocene is known to have been highly spatiotemporally variable. The environment of the early Pleistocene was very different from, say, the late Pliocene. Moreover, humans came to be widely dispersed, occupying a variety of distinctive environments. Given what we know, it would be reasonable as well to think that social structures would be different in different physical environments—for example, some would be more conducive to sedentary lifestyles, and others to more mobile ones. Though humans are not as genetically diverse as many other animals, there is sufficient genetic variation to support genetic changes in relatively short amounts of time.
- Human behavior is both adaptive and malleable. EP tends to assume, by contrast, that evolved computational programs are species specific and species universal. When there is within-species variation, the assumption is that the strategies are conditional, evoked in different conditions. The validity of this is questionable. Evolutionary biologists have found that the rate of evolution can be much faster than EP tends to assume (see chapters III.7 and III.8). Advances in our understanding of epigenetics and developmental plasticity provide alternatives to conditionality that are typically not incorporated in EP. It is not that EP assumes some form of genetic determinism; rather, the point is that the kind of interplay seen among genetic factors, epigenetic influences, and learning makes universals less likely.
- There are significant alternatives to the typical emphasis of EP on individual- and gene-centered models of evolution. This is an issue beyond the problems of applying EP's preferred modes of analysis. Gene-culture coevolution may be an important source of evolutionary changes (see chapter VIII.10). This is becoming a well-developed alternative, emphasizing the role of cultural practices in modifying the human brain. In general, gene-culture dynamics can enhance and accelerate rates of evolution, even if we do not yet know how important these are. Multilevel selection models are also being developed. With distinctive, genetically isolated groups that compete as groups, it is possible to develop models for the evolution of social behavior that do not assume the typically individual- and gene-oriented perspective of EP.

There are alternatives that could enrich the work within EP but that typically remain beyond its purview. Darwin was right to think that evolution should reshape our understanding of human psychology. There are many avenues yet to explore in seeing how an evolutionary perspective can contribute to our understanding of human psychology. Most of these avenues are ahead of us.

FURTHER READING

Barkow, J., L. Cosmides, and J. Tooby, eds. 1992. The Adapted Mind: Evolutionary Psychology and the Generation of Culture. New York: Oxford University Press. *A collection that defined the case for evolutionary psychology.*

Bolhuis, J. J., G. R. Brown, R. C. Richardson, and K. N. Laland. 2011. Darwin in mind: New opportunities for evolutionary psychology. PLoS Biology 9: e1001109, doi:10.1371/journal.pbio.1001109.

Buller, D. J. 2005. Adapting Minds: Evolutionary Psychology and the Persistent Quest for Human Nature. Cambridge, MA: MIT Press. *A critique focused on the psychological case for evolutionary psychology.*

Buss, D. M. 1995. Evolutionary psychology: A new paradigm for psychological science. Psychological Inquiry

6: 1–49. *A very important and useful overview from an advocate.*

Buss, D. M., ed. 2005. The Handbook of Evolutionary Psychology. New York: John Wiley & Sons. *A more recent comprehensive collection.*

Laland, K. N., and G. R. Brown. 2011. Sense and Nonsense: Evolutionary Perspectives on Human Behaviour. 2nd ed. Oxford: Oxford University Press. *A criticism of EP from within evolutionary biology and recent psychology.*

Richardson, R. C. 2007. Evolutionary Psychology as Maladapted Psychology. Cambridge, MA: MIT Press. *A philosopher of biology skeptical of the evolutionary credentials of EP.*

VII.13

Evolution of Eusociality

Laurent Keller and Michel Chapuisat

OUTLINE

1. Eusociality: A highly integrated form of social organization
2. What drives eusociality?
3. Working together
4. Intragroup conflicts and their resolution

Animal societies can reach very high levels of coordination and integration. In ants, bees, termites, and naked mole rats, hundreds of permanently nonreproducing workers help rear the offspring of a few fertile individuals, the queens and males. Societies with such a reproductive division of labor are called *eusocial*. The evolution of eusociality puzzled Darwin: How could workers pass on their characteristics to the next generation if they did not reproduce? W. D. Hamilton provided the answer in the 1960s with the concept of *kin selection*, the indirect transmission of genes through relatives, which occurs in stable associations of related individuals jointly exploiting and defending common resources. Despite the high level of cooperation characterizing eusocial societies, conflicts among individuals are still common, and sophisticated social mechanisms often contribute to maintaining social cohesion. Many eusocial species are extremely successful, because the coordinated and cooperative work of many individuals allows them to efficiently use and transform their environment while being robust to perturbations.

GLOSSARY

Cooperation. A collective action that benefits two or more individuals.

Eusociality. A social organization that includes reproductive division of labor, cooperative brood care, and overlap of generations.

Haplodiploidy. A sex determination system in which males are derived from unfertilized (haploid) eggs, and females from fertilized (diploid) eggs.

Kin Selection. Selection on genes for social traits that affect the fitness of relatives.

Philopatry. The tendency to stay close or return to the natal site.

Relatedness. The probability that individuals share identical alleles inherited from recent common ancestors.

Reproductive Altruism. An action by which an individual decreases its own reproduction to help one or several other individuals reproduce.

Self-organization. A spontaneous organization arising from local interactions among individuals, without central control.

1. EUSOCIALITY: A HIGHLY INTEGRATED FORM OF SOCIAL ORGANIZATION

Insect societies have long fascinated human beings, as illustrated by the Bible verse "Go to the ant, thou sluggard; consider her ways, and be wise" (Proverbs 6:6), which acknowledges the industrious nature of insect workers. The major organizing principle of insect societies is reproductive division of labor, whereby one or a few individuals (the queens) specialize in reproduction, whereas the others (the workers) participate in cooperative tasks such as building the nest, collecting food, rearing the young, and defending the colony. This division of labor may lead to amazing specializations, such as ants with a thickened and enlarged head that they use as an armored door to block the nest entrance, or "kamikaze" termites that explode to glue their opponents in toxic secretions.

The term *eusociality* refers to animal societies with a marked and permanent reproductive division of labor. In most of these societies, only one or very few individuals reproduce, although the number of potentially reproductive adults may number in the hundreds or thousands. In general, these societies also exhibit cooperative brood care and have overlapping generations. Eusociality was traditionally thought to occur only in insects; all species of ants and termites, as well as part of the bees and wasps, are eusocial. Colonies of termites are headed by one or a few queens and kings, and workers can be either males or females. In contrast, ants, wasps, and bees form matriarchal colonies in which all workers are females. In these colonies, there may be one or more queens, which usually mate once in early adulthood and store the sperm that they use throughout their lives.

Recently, eusociality was discovered in a small number of species belonging to diverse taxa of invertebrates, including parasitic flatworms, snapping shrimps, ambrosia beetles, gall-forming thrips, and gall-forming aphids. Moreover, two species of mammals were recognized to be eusocial, the naked mole rat and the Damaraland mole rat. The discovery of several taxa with social organizations similar to those of some social insects, and the realization that there is a continuum in the extent to which individuals abstain from reproducing, have led to the conclusion that it is somewhat difficult to classify species as being eusocial or not. For example, in many cooperatively breeding birds and mammals, some individuals forgo their own reproduction to help raise the offspring of others, but generally, this reproductive altruism is reversible and may cease if a breeding opportunity becomes available. This reversibility occurs in some wasps as well.

Despite some uncertainties about the transience or permanence of the reproductive division of labor, the number of truly eusocial species is relatively small, accounting for approximately 1 percent of the described animal species. However, eusocial species are ecologically very successful, being found in almost every type of terrestrial environment and making up a considerable proportion of the animal biomass of the earth—up to 50 percent in some tropical habitats. In many ecosystems, eusocial insects, ants in particular, are dominant organisms that play a crucial role as predators and pollinators, and in soil formation. This tremendous ecological success of eusocial insects is undoubtedly due to their social organization, based on large numbers of individuals cooperating in social groups, which provides multiple competitive advantages over solitary species.

2. WHAT DRIVES EUSOCIALITY?

An Apparent Paradox

The ecological success of eusocial species is based on an organization that seems to contradict the principle of natural selection. In the process of natural selection, genes conferring greater survival and reproduction increase in frequency over generations in a population, since individuals carrying these genes leave more descendants. This feedback loop explains the adaptation of organisms to their environment as alleles underlying more successful traits spread and increase in a population. Yet among many eusocial species, workers have particular morphology and physiology preventing them from reproducing. The paradox of why, in some animal societies, a proportion of the individuals in a population forgo reproduction to assist other group members did not escape the attention of Charles Darwin. In *The Origin* he noted that sterile workers of eusocial insects embodied "one special difficulty, which at first appeared to me insuperable, and actually fatal to my whole theory" (1859). Although inheritance mechanisms were not known at this time, Darwin drafted a solution to this apparent paradox, namely, that selection may operate not only at the level of the individual but also at the level of the family.

Kin Selection

In the mid-1960s, as a graduate student, William D. Hamilton resolved the paradox of reproductive altruism by showing that individuals can transmit copies of their own genes not only through their own reproduction but also by favoring the reproduction of kin, such as siblings or cousins. Kin share identical copies of genes inherited from their common ancestors in the same way a child possesses copies of paternal and maternal genes. Thus by helping their mother produce numerous fertile offspring (the males and the future queens), sterile workers have an excellent way of transmitting copies of their own genes to the next generation. This indirect selection on genes due to their effect on the fitness of relatives has been called *kin selection*.

Hamilton's approach is based on the Darwinian principle of natural selection on genes, combined with Mendel's genetics. The fundamental idea is to consider not only the direct reproductive output, or fitness, of individuals but their *inclusive fitness*, which includes the indirect effects of genes on the fitness of other individuals carrying copies of the same genes (see chapter III.4). A specific application of the general theory of inclusive fitness is to examine the conditions favoring the evolution of reproductive altruism through the mechanism of kin selection. *Hamilton's rule* delineates when an individual transmits more copies of its own genes by behaving altruistically, that is, decreasing its own chances of survival and reproduction to help other individuals reproduce. In short, reproductive altruism is favored when the loss in the altruistic actor's personal fitness is smaller than the gain in the personal fitness of the recipient of the altruistic act multiplied by the relatedness between the actor and the recipient. Personal fitness is an estimate of the number of descendants of an individual, relative to other individuals in the population, while the relatedness is the probability that the recipient carries copies of the actor's genes, inherited from recent common ancestors.

A general description of Hamilton's rule is that altruistic acts are more likely to be selected for when individuals are closely related and when the decrease in the actor's personal fitness is relatively small compared with the increase in the recipient's fitness. However, it is important to note that reproductive altruism does not require exceptionally high degrees of relatedness. It can evolve among distantly related individuals if the benefits for the recipient are high and the costs to the actor low.

The following simple example illustrates Hamilton's rule. Imagine a gene that programs an individual to die so as to save relatives' lives. One copy of the gene will be lost if the altruist dies, but the gene will increase in frequency in the population if, on average, the altruistic act saves the lives of more than two siblings (relatedness = 0.5), more than four nephews or nieces (relatedness = 0.25), or more than eight cousins (relatedness = 0.125). J.B.S. Haldane fully apprehended kin selection theory and Hamilton's rule when he announced, having done some calculations on an envelope in a pub, that he would be ready to give his life to save two brothers or eight cousins.

There is unambiguous theoretical and empirical evidence that kin selection has been central to the evolution of eusociality and reproductive altruism by workers. First, for the problem to be resolved, all models accounting for the evolution of altruism explicitly or implicitly assume that altruistic individuals and recipients of help are related. Second, eusociality invariably evolves within groups of highly related individuals, such as one mother and her offspring. Well-marked reproductive division of labor is rare in societies where individuals are distantly related. There are a few ant species in which the relatedness between nest mates is indistinguishable from zero, but this low relatedness stems from an increase in queen number that occurred long after the evolution of morphological castes and reproductive division of labor. Workers of these ants generally have only vestigial (shrunken) ovaries, and it has been suggested that obligate sterility of workers prevented such societies from collapsing after the drop in relatedness. However, societies with very large numbers of queens are expected to be unstable in the long term, as there is no more selection for workers' altruism.

Ecological and Life History Factors Favoring the Evolution of Eusociality

The evolution of eusociality depends on a combination of ecological, genetic, social, and life history factors that jointly determine whether the best option for a young individual is to stay in the group and sacrifice part or all of its direct reproduction to help others, or to leave the group to breed independently (see chapter VII.10). The payoffs of each strategy are determined by the benefits of helping in terms of group productivity, the genetic relatedness among group members, and the expected success of a young individual that attempts to reproduce solitarily. In turn, these three elements depend on ecological conditions—in particular, on the degree of ecological constraints on independent breeding—as well as on the ecology, life history, breeding system, and family structure of a species. For example, in some cooperatively breeding birds, the decision about whether to become a helper depends on territory availability. In other taxa, such as paper wasps, suitable nest sites are not limited, and females seem to associate because of the benefits conferred by sociality, particularly lower breeding failure.

An ecological factor that favors the evolution of eusociality is the coincidence of shelter and food. Thus, eusocial thrips, aphids, beetles, shrimps, termites, and mole rats live in cavities or burrows in which they obtain their food. Similarly, eusocial flatworms form colonies within their molluskan hosts. Such ways of life may promote sociality for several reasons. First, the high value of a habitat combining food and shelter may favor altruistic self-sacrifice for colony defense, leading to the evolution of a soldier caste. Second, this type of valuable habitat may select for philopatry and helping because the colony can be inherited by offspring, and some helpers may have a chance to replace breeders. Third, living in a confined habitat helps keep relatives in physical proximity and thereby creates opportunities for kin-selected reproductive altruism. Finally, because juveniles in such habitats are frequently self-sufficient with regard to food, they can devote themselves more directly, and at a younger age, to helping raise younger siblings.

High risks of mortality during the period of brood rearing may also promote eusociality, because sociality provides life insurance: if a cooperative breeder or a helper dies early, its investment is not lost, because other group members will finish rearing related brood. In contrast, a solitary breeder that dies before having raised its brood will have zero fitness. Hence, long development time and high mortality risk should favor helping and eusociality. These conditions occur in many species of Hymenoptera, such as ants, wasps, and bees, which have extended parental care and search for food outside nests.

The Role of the Family Structure

One factor that played a crucial role in facilitating the evolution of reproductive altruism is the type of family structure. Mother-daughter associations not only provide an opportunity for the offspring to help while they are still juveniles but also generate a genetic structure that favors the evolution of reproductive altruism.

Forty years ago, there was considerable discussion about whether eusociality evolved within groups composed of a mother and her offspring, or within groups composed of related individuals of the same generation, such as sisters. The asymmetry in relatedness occurring in mother-daughter associations should favor the evolution of eusociality. In such associations, daughters are on average as related to their mother's offspring as to their own descendants (i.e., they share half their genes by recent common ancestry, $r = 0.5$), so they lose nothing by giving up direct reproduction in favor of the mother. By contrast, mothers are twice more related to their own offspring ($r = 0.5$) than to their daughters' offspring ($r = 0.25$); they thus benefit from monopolizing reproduction. Therefore, a pronounced reproductive division of labor and monopolization of reproduction by mothers is expected to evolve and be stable in mother-daughter associations. In contrast, siblings are always more related to their own offspring ($r = 0.5$) than to those of their sisters ($r = 0.25$) and therefore should not easily forgo their reproduction to help their sisters.

The level of relatedness among the queen's daughters also depends on the number of fathers—it decreases when the mother queen mates with multiple males. In a comparative study of 267 species of eusocial Hymenoptera, William O. H. Hughes and colleagues showed that monoandry (a queen mating with a single male) was ancestral on each of the occasions when eusociality evolved and that multiple mating evolved only after workers had lost reproductive totipotency. In fact, the breeding system of most eusocial species is conducive to lifetime monogamy. In ants, bees, and wasps, for example, the queens mate only during the mating flight, hence ensuring high relatedness of their offspring throughout their life. Similarly, termite colonies are typically initiated by a royal couple, which jointly produce all the colony offspring. Thus, the breeding system of species that became permanently eusocial was conducive to the formation of simple families in which offspring were highly related and thus more likely to forgo reproduction if this increased colony survival and productivity.

Caste Differentiation

Colonies of many eusocial insects (e.g., the honey bee, vespine wasps, and most ants and termites) contain distinct morphological castes: the queens are morphologically and physiologically specialized for reproduction, and the workers for other tasks such as foraging and brood care. The degree of polymorphism varies greatly. In some species such as allodapine bees, hover wasps, polistes wasps, and sweat bees, there is little morphological difference between queens and workers; the specialization is mostly behavioral. By contrast, many ants and termites are characterized by striking differences between queens and workers, with the latter sometimes having completely lost their ovaries and developed morphologies adapted for special tasks.

A broad comparison among social insects reveals an association between the queen/worker dimorphism and colony size. Morphological differences between queens and workers are generally absent or small in species forming small colonies, whereas the differences are well marked in species forming large colonies. Such an association can be explained by a relationship between colony size and the probability of workers to become replacement reproductives. This probability drastically decreases in larger colonies, with the effect that there is lower selection to retain reproductive ability. The only vertebrate species in which morphological castes have evolved is the naked mole rat. In this species, the lumbar vertebrae of the breeding female elongate after the onset of reproduction. Of all the vertebrate species, the naked mole rat forms the largest societies, with up to 300 individuals.

3. WORKING TOGETHER

Division of labor among workers plays a major role in the great ecological success of eusocial species. Analogously to somatic tissues in a multicellular organism, workers can specialize in various tasks (division of labor). Moreover, tasks can be divided in sequential actions performed by more than one individual (task partitioning). Hence, work can be performed collectively, with concurrent operations and synergistic interactions generating large benefits to the whole colony. In a feedback loop, the benefits of collective work further promote the evolution of reproductive altruism and contribute to stabilize eusociality. Over evolutionary time, selection has favored societies in which the work is organized in an efficient, robust, and flexible manner.

Efficiency, Robustness, and Flexibility

There are several ways by which division of labor can increase colony performance. First, the capacity to perform tasks concurrently often provides large advantages. In solitary species, a single individual conducts one task at a time, often in a specific sequence, and must complete a set of tasks to reproduce successfully. For example, a solitary sphecid wasp has to excavate a nest, find a prey item, sting it, bring it back to the nest, and then lay eggs. In contrast, eusocial species can conduct many tasks at the same time, seizing opportunities as they arise.

The efficiency of eusocial species is further increased by the collective performance of tasks that would be out of reach of single individuals. For example, colonies of naked mole rats excavate burrows that can be more than 4 km in cumulative length. Six small ants can immobilize a large insect by seizing one leg each, scouts can recruit foragers to a rich food source, or nest temperature can be accurately controlled at all times. Finally, by repeating the same task in one area of the colony territory, for example, collecting food, feeding the brood, or guarding the nest entrance, workers can learn and become more efficient. They also minimize costs associated with traveling between tasks, and time lost in task switching.

Another feature of concurrent systems is robustness. The failure of one individual at one task does not compromise the whole enterprise. The redundancy of the system, with many individuals performing the same task and many concurrent production lines, makes it resistant to perturbations or catastrophic events. Finally, it is important to stress that workers do not usually work in a fixed and rigid way. They show behavioral flexibility, so that the number of workers engaged in each task can vary over time to match the needs of the colony and the changes in the environment.

Mechanisms Regulating the Division of Labor

Colonies face the complex challenge of dynamically allocating the correct number of workers to each task. Early researchers on division of labor considered that workers were rigidly programmed to perform only one task over long periods of their life, with task performance being determined by internal factors such as age, size, or morphology. Indeed, there is often a correlation between age and task in the social insects. Young individuals usually perform tasks within the colony, such as brood care or nest maintenance, while older individuals engage in outside, more risky jobs, such as foraging or colony defense. However, a fixed partitioning of tasks according to age or other internal factors gives little flexibility.

Despite physiological or age-related predispositions for certain tasks, workers are usually able to switch tasks according to needs. For example, if one behavioral caste is experimentally removed, nurses become foragers, or foragers switch to guards. However, task switching is likely to be costly and should occur only when necessary. More recently, researchers have considered that the colony is a self-organizing system in which a flexible division of labor arises from the independent actions and decisions of workers, without any central or hierarchical control. Several models in which division of labor emerges by self-organization have been proposed, based on spatial location, task encounter, or physiological threshold. An important class of models is based on response thresholds, with workers performing a task when a specific stimulus for this task exceeds their individual threshold. In the response threshold model, the task and stimulus are linked in a negative feedback loop that regulates the system, as performing the task decreases the stimulus for this particular task. Variation in response thresholds among individuals results in worker specialization; however, the system retains flexibility and self-adjusts to needs. For example, honey bee workers will start to fan to cool the hive if the temperature exceeds a given threshold. The ones with the lowest response threshold will start to fan first. By so doing they decrease the temperature, which may not reach the threshold of other workers under normal conditions. However, if the temperature continues to rise, other workers with higher response thresholds will start to fan. Hence, a subset of workers become task specialists because of small differences in threshold response, but all workers are able to perform the task if needed. Variation in response threshold can come from many sources, including genotypic differences or experience. A division of labor may emerge spontaneously when individuals with different response thresholds group together. Selection can then favor response threshold distributions that ensure the most efficient allocation of workers to tasks.

4. INTRAGROUP CONFLICTS AND THEIR RESOLUTION

Despite high levels of cooperation and apparent harmony, potential conflicts persist in colonies of eusocial species. Potential conflicts arise because, in contrast with cells of an organism, colony mates are not genetically identical (see chapter VII.9). Hence, kin selection predicts that individuals with partially divergent genetic interests may attempt to favor the propagation of their own genes, possibly to the detriment of their nest mates. Colony members can compete over direct reproduction or over allocation of colony resources to various relatives, and the potential conflict may translate into actual conflict or may remain unexpressed.

Conflict over who reproduces is common in many eusocial species. For example, dominance behavior and linear hierarchies frequently occur within small colonies of wasps, bees, and ants. Some potential conflicts are specific to the social Hymenoptera, which are male-haploid, female-diploid. Queens and workers may compete over the production of males and over the allocation of colony resources to males and females, respectively. These potential conflicts sometimes degenerate into open conflicts. In some ant species the queens and workers both try to influence the relative investment in females versus males. In these species, workers kill brothers to favor their more related sisters, while queens influence colony sex allocation by biasing the sex ratio of their eggs toward males.

Conflict Resolution

Within animal societies, the resolution of potential conflicts still results in a wide range of outcomes. The expression of conflict can range from high levels of actual conflict to its complete absence. Understanding how potential conflicts among individuals are resolved is important to comprehending the emergence of cooperation in social groups, the evolutionary transition toward eusociality, and the further increase in complexity of societies.

Several types of factors and mechanisms contribute to align the divergent interests of colony members, thereby favoring peaceful cooperation in cohesive social groups. A major factor is genetic homogeneity, which results in high and symmetrical degrees of relatedness among group members, thus reducing the area and magnitude of potential conflicts. Other important elements are the multiple benefits of group living, as compared with solitary breeding, as well as the costs of behaving selfishly. In short, solitary or selfish behaviors are most likely to be selected against when cooperation and division of labor provide large synergistic fitness benefits and when open conflicts decrease colony productivity.

Finally, multiple socially mediated mechanisms may contribute to restrain within-group selfishness. These social mechanisms may be based on pacific "social contracts," such as leaving enough reproduction for each breeder to stay peacefully in the group. Social cohesion can also be enforced individually or collectively by direct actions against individuals that behave selfishly, in the form of aggression, coercion, or punishment. Proximately, reproductive altruism within eusocial groups may be socially enforced and may thus reach higher levels than the subordinates' optimum. Power asymmetries, or unequal access to information, may tip the balance in favor of one party or another. In naked mole rats, the breeding female frequently attacks subordinate females and by so doing suppresses their reproductive attempts. In ants, bees, and wasps the workers often police one another: they suppress male-destined eggs laid by other workers, or ally against a sister that tries to overturn their mother. Overall, social processes such as coercion and policing appear to play a major role in preventing outbursts of conflicts within social groups and may thus be very important for the evolution and maintenance of eusociality.

FURTHER READING

Bourke, A.F.G. 2011. Principles of Social Evolution. Oxford: Oxford University Press. *A recent synthesis of the principles governing the major transitions of life, including eusociality.*

Crozier, R. H., and P. Pamilo. 1996. Evolution of Social Insect Colonies: Sex Allocation and Kin Selection. Oxford: Oxford University Press. *A thorough account of the evolutionary genetics of social insects, with a clear presentation of its mathematical foundation.*

Hölldobler, B., and E. O. Wilson. 1990. The Ants. Berlin: Springer-Verlag. *A comprehensive reference on the natural history of ants.*

Keller, L., and M. Chapuisat. 1999. Cooperation among selfish individuals in insect societies. BioScience 49: 899–909. *A concise review of intracolony conflicts and their resolution.*

Keller, L., and E. Gordon. 2009. The Lives of Ants. Oxford: Oxford University Press. *A general-audience overview on the biology of ants and the research they inspired.*

Wilson, E. O. 1971. The Insect Societies. Cambridge, MA: Harvard University Press. *A rich synthesis on the natural history of social insects that was foundational to insect sociobiology.*

VII.14

Cognition: Phylogeny, Adaptation, and By-Products

Marc D. Hauser

OUTLINE

1. What are we measuring?
2. The space of possibilities
3. Novel possibilities and unanticipated outcomes
4. Evolving limitless options

The mind consists of feelings, decisions, plans, and memories generated by the brain. To study how minds evolve, a comparative approach is necessary, one that seeks evidence of phylogenetic similarities and differences, together with evidence of adaptive function. This chapter describes a set of challenges associated with exploring mental evolution, together with a framework for exploring a corner of this problem, focused on the patterns and processes that led to the evolution of human minds. This is a story of phylogeny, adaptation, and by-products. The chapter examines the hypothesis that despite some similarities between human and nonhuman animal minds, there are far greater discontinuities. The uniqueness of human cognitive capacity is due to a suite of changes in brain function that generate the signature of both human universals and cultural variation in language, music, mathematics, technology, and morality.

GLOSSARY

Abstract Thoughts. The ability to represent or think in ways that are detached from the primary sensory and perceptual inputs. This capacity allows individuals to think about things that are beyond their direct experiences with the world, including concepts such as *infinity* and objects that are possible but do not exist, such as hippos with green fur and pink antlers.

Cognitive Decomposition Approach. Dissecting cognitive capacities into their component parts to study the evolutionary history of the components. Students pursuing this approach decompose language, music, mathematics, technology, and morality into a suite of capacities and explore the phylogenies of these components, as well as the evolutionary processes that may have favored them.

Cognitive Promiscuity. The capacity to combine thoughts and feelings from different domains of understanding to create novel solutions. When cognitive promiscuity is in play, capacities that evolved to solve a highly specialized problem are also used to solve problems in other, disparate situations.

Combinatorial Operations. Mixing or combining discrete elements to create novel combinations, and thus new meanings or functions.

Recursive Operations. Computable functions that generate a potentially infinite set of hierarchically organized expressions.

Symbolic Expression. The capacity to represent and express a thought or emotion with a discrete symbol, including spoken, signed, or written words, as well as nonlinguistic representations (e.g., McDonald's golden arches, scales of justice). This capacity allows individuals to reduce memory load through compact storage of discrete and easily retrieved symbols while facilitating lossless information transfer.

The anthropologist Martin Muller has an arresting photo of his hands, each palm up, each holding a part of a recently killed chimpanzee from Uganda. In one hand is the chimpanzee's brain; in the other, one of its testicles. Both are virtually the same size, about the dimensions of a tangerine. Both have distinctive functions; both are targets of selection, with ancient evolutionary histories. All these comparative points are true of humans as well, except one: a human brain would completely cover and outsize a human hand, whereas a human testicle would sit easily within the palm. For any healthy human adult, Muller would be holding a melon in one hand and a diminutive grape in the other.

If we look beyond brain size and male reproductive organs, as well as many other features of comparative anatomy, the differences between humans and chimpanzees are even more striking. Chimpanzees have shown minuscule changes in the 6–7 million years of their evolution. Not one chimpanzee has ever ventured out of Africa to another continent. Not one chimpanzee has moved out of its own country, say, from Tanzania to Kenya. Not one chimpanzee has moved out of its habitat of origin in the tropical forests to life in the mountains or on the beach or in the desert. It appears that chimpanzees use the same communication today as they did millions of years ago. They also have the same mating system and social organization. And although they have made a few clever technological innovations, their material culture is largely the same today as it was in the past.

If you believe these comments belittle chimpanzees, you are wrong. What chimpanzees can do is impressive. Moreover, they have persisted for 6–7 million years, so they are doing something right. Finally, the observation of cognitive and behavioral stasis is true of virtually every other animal—virtually all but one: *Homo sapiens*.

In the 6–7 million years of our evolution, humans have marched out of Africa, inhabited every continent and virtually every habitat offered on earth, and ventured to novel environments on other planets. Our communication today preserves elements of our past but goes far beyond it, quantitatively and qualitatively. Our unique faculties of language and mathematics allow for a limitless variety of expressions, including massive compression of ideas into words and patterns of 1s and 0s. Our material culture today would be incomprehensible to our human ancestors. We have gone from chipped stone hand axes to air-powered hammers, cell phones, airplanes, and scud missiles. All these developments have fueled our nomadic travels and capacity to inhabit a bewildering diversity of environments. We also have transitioned from universal polygyny to a diversity of mating systems that includes polygyny, lifelong monogamy, and serial monogamy, and within each of these systems, opportunities to engage heterosexually, homosexually, or bisexually.

This comparative summary raises several profound evolutionary questions: What accounts for the differences we observe between modern humans and chimpanzees? What changes in the brain allowed us, but not our closest living relatives, to change our material cultures, mating systems, living environments, and systems of communication and thought? What evolutionary processes led to the differences between humans and chimpanzees?

These questions keep scholars of cognitive evolution up at night. Our goal is to provide a few answers to these questions, perhaps allowing for greater rest; then again, perhaps they will stimulate even more restlessness.

A few caveats before starting. There are two reasons to focus primarily on primates, and especially the contrast between humans and chimpanzees. First, there is no other comparative contrast that yields such a striking cognitive gap, especially among such closely related species. To put it starkly, though the molecular evidence shows that humans are more closely related to chimpanzees than chimpanzees are to gorillas, the cognitive evidence shows few differences between chimpanzees and gorillas but massive differences between both apes and humans. This is a delicious problem. It shows both the difficulty of moving from genetic to phenotypic differences, and the challenges of creating phylogenies based on different types of evidence. Second, because of the richness of our understanding of human cognition, including its neural underpinnings, we are well equipped to ask deep questions about comparative differences and similarities. Thus, for example, when we ask whether other animals imitate, experience empathy, perceive musical patterns, or deceive *like* humans, we can rely on a wealth of empirical evidence from cognitive science, neurobiology, and most recently, molecular neuroscience. *Homo sapiens* has joined the ranks of *C. elegans* and *Drosophila* as a model species.

To make sure there is no confusion, none of these initial comments are meant to diminish the elegant comparative evidence for other taxonomic assemblages, including experimental investigations of spatial navigation in insects, fish, birds, and other mammals, and the proximal mechanisms and selective pressures on caching in birds and rodents. In a short chapter, choices are necessary. I hope mine provide a reasonable introduction to some of the most pressing challenges and exciting developments while acknowledging the particular theoretical biases I hold.

1. WHAT ARE WE MEASURING?

Unlike measuring testicles or the brain, asking questions about cognitive evolution poses a measurement problem. What *can* we measure so that we may both make phylogenetic comparisons and devise experimental methods to test for selective pressures and the adaptations they generate? One approach, common to this day, has been to seek comparative evidence of behaviors that are unambiguous indicators of human intelligence, including language, mathematics, culture, music, and cooperation. But this approach is flawed. None of these traits are clearly isolated phenotypes like testicles and brains, readily quantifiable for entry into a phylogenetic analysis or experimental test of adaptation. All these traits depend on a variegated set of mechanisms, some shared in common with other animals, and some uniquely human. Each mechanism, in turn, generates a set of signature behaviors that can be quantitatively measured. To productively explore questions of cognitive evolution, we must decompose these complex phenotypes into their component parts. This is called the *cognitive decomposition* approach.

The good news is that more recent work in cognitive evolution has recognized that phylogenetic and functional analyses require more precise and narrowly specified traits. For example, a number of scholars, such as Derek Bickerton, Terry Deacon, and Tecumseh Fitch, have recognized the need to decompose language into separate mechanisms, including computations for structuring words into sentences, mapping concepts to words, and articulating words with sounds or visual signs. Those interested in the evolution of music, such as Ray Jackendoff, Fred Lehrdahl, and Ani Patel, have appreciated the need to decompose this system into auditory perception, planning, memory, pattern analysis, and combinatorial computations, to name a few. Those interested in the evolution of mathematics, such as Susan Carey, Stanislas Dehaene, Randy Gallistel, and Elizabeth Spelke, have looked at different mechanisms of quantification, including those that rely on discrete and explicit symbols such as the integers, as well as those that do not; and the ways in which quantification relies on more general mechanisms of categorization, memory, and attention; and at how nonexplicit symbol systems for quantification evolve and develop into explicit ones. Students interested in cultural evolution, such as Robert Boyd and Michael Tomasello, have recognized the importance of looking at different mechanisms of transmission, including teaching, imitation, and observational learning, together with the importance of innovation, conformity, and social organization. From such decomposition, students of cognitive evolution have made great strides over the past 15 to 20 years.

There are two important points to note about the cognitive decomposition approach. First, it starts with questions concerning the nature of a particular mechanism and then follows with questions related to adaptive function, including the socioecological conditions that favored its original expression. Consider, for example, the capacity for teaching, an ability that is observed in every human culture, in a wide variety of contexts, and is dependent on different cognitive processes. When humans teach, it is recognized that some individuals are ignorant and may want to learn. To teach requires demonstrating, breaking problems down into simpler components, recruiting the learner's attention, monitoring their progress, revising the pedagogical approach, and so forth. In other animals, teaching appears highly limited in taxonomic scope, and among those animals exhibiting some form of teaching, it appears limited to a single context. Thus, meerkats engage in a form of functional teaching in which adults help prepare pups to develop the skills to kill scorpion prey. It is a form of teaching in that the adults recognize a deficiency in young individuals and then break down the mature form of the skill into components to facilitate learning. Unlike humans involved in teaching, however, meerkats engage in this kind of pedagogy in only one context: predation on scorpions. At present, we lack a coherent account of why even this form of teaching does not occur in other functionally significant contexts among meerkats and why other species with similar social and ecological pressures lack teaching altogether.

Second, the decompositional approach seeks to understand whether the mechanism in question evolved to solve a suite of general problems or a highly specific one while recognizing that a specialized mechanism can be co-opted for more general purposes. Three examples highlight the significance of this point:

- The temporal lobe of the primate cortex evolved for object recognition, but in humans it has been co-opted for the added task of recognizing written word forms.
- A region of the parietal lobe operates in nonhuman and human primates for approximate number estimation, but in humans it is also recruited for precise number computation with explicit symbols.
- A region at the juncture between the temporal and occipital lobes is involved in face recognition in both monkeys and humans—a highly specialized function—but in humans it is also used for other within-category discriminations (e.g., cars) that are unique to our species.

These three cases illustrate why, for any given cognitive function in an organism, it is important to ask for what it evolved, for what it is presently used, for what it could be used, and whether each of these functions is general or specific to a particular problem. The comparative evidence available suggests that most cognitive functions observed in nonhuman animals have evolved for highly specialized problems and are restricted to use in this context. In many cases, we have inherited these highly adaptive capacities, but owing to evolutionary changes in the brain, coupled with the particular environments we inhabit, these specialized functions have been liberated, allowing us to tackle a much broader range of problems. In fact, it appears that many of our most revered cognitive capacities are by-products of a few specific changes in neural function, capacities that are nonetheless highly adaptive today.

2. THE SPACE OF POSSIBILITIES

The fact that traits can evolve for one function and subsequently be used for another raises a challenge for students of cognitive evolution, a problem that finds parallels within the field of functional and theoretical morphology. In particular, as students of evolution, we typically study what is observable or what has been observed in the past if we have access to a fossil record. What is less often studied, except perhaps by students of artificial life and evolution, is what is potentially observable. What we observe reveals information only about the options that species have explored over their history. It does not reveal what could have occurred had they confronted different ecological or social pressures. This is a problem of potential, of possible phenotypic outcomes. In the case of anatomy, say, the coiling patterns of ammonite mollusks, it is a question of the possible shapes they might take under different conditions. In the case of behavior, say, courtship displays in birds, it is a question of the possible movements of the body and vocal tract that might evolve, either to accommodate shifts in perception of the choosy sex or to accommodate changes in the physical environment. In the case of cognition, say, pattern recognition, it is a question of possible ways of computing and classifying patterns that might change as animals develop different systems of communication or social relationships. The idea that for every organism there exists a space of possible anatomical, behavioral, and cognitive forms that have yet to be realized changes how evolutionarily oriented research programs are carried out. It is no longer sufficient to study what animals do or have done. It is important to study what they *might* do when confronted with different situations. In the case of cognition, it becomes a matter of challenging nonhuman and human animals with novel problems. The process of domestication provides an elegant illustration of this problem, and of the unanticipated consequences that can emerge when the selective regime changes in a radical way.

3. NOVEL POSSIBILITIES AND UNANTICIPATED OUTCOMES

Artificial selection (or experimental evolution; see chapter III.6) provides one way to explore the space of possible outcomes. Artificial selection provides a way of tapping into human creativity, allowing us to use our imagination to select for what nature may never imagine. Could we create a square tomato or a hen that lays square eggs, both optimized for packing? Could we select for hairless pets, reducing the mess at home and allowing those with allergies to enjoy the company of other animals? Some of these questions have already been asked and explored, generating a list of successes and failures. The successes show the power of our imagination and selection. The failures reveal either the poverty of our imagination or the hidden constraints that limit the power of selection. Both successes and failures inform our understanding of evolution. Recent work on mammalian domestication provides an elegant illustration, especially with respect to the role of selection in both behavioral and cognitive evolution.

Molecular evidence shows that the domestic dog first differentiated itself from its ancestor, the wolf, about 100,000 years ago. Dogs begin to appear in cave art, and the archaeological record more generally, between 14,000 and 16,000 years ago. Though there is little explicit information about the process of domesticating dogs, most scholars believe that as human populations adopted a more sedentary lifestyle, which included the introduction of agriculture, wolves started scavenging for food. Over time, this process initiated a cascade of morphological, physiological, behavioral, and cognitive changes that led to the first dog breeds. This first wave of change was followed by a second in which humans played an increasingly more directed role in creating new breeds, selecting for differences in size, coat coloration, temperament, and behavioral skills. Some of these differences were selected for pure aesthetics (e.g., shorter snouts, pint-sized bodies, floppy ears), others for functional differences linked to human lifestyle (e.g., herding dogs for those with livestock).

Behavioral research in the last 10 years suggests that the process of domestication in dogs resulted in fundamental changes in cognitive ability—differences that are striking when contrasted with those of their close relatives, the wolves, and their more distant relatives, the nonhuman primates, including chimpanzees. There are two fundamental questions here: What critical differences in cognitive ability are specifically due to domestication? And did these capacities evolve as a result of selection or are they by-products of other cognitive changes? At least part of the answer to these questions comes from an elegant series of studies on dogs and wolves, focusing on their capacity to understand the communicative gestures of humans, especially in relationship to nonhuman primates. Another part of the answer comes from a more targeted project involving the domestication of the silver fox.

When humans point, or turn their heads in a particular direction, those watching immediately understand that the space of interest has shifted. People point and turn their heads to indicate a change in focus, and often, to cause others to change their focus accordingly. Early in development, without any training, human infants recognize that pointing is goal directed, designed to capture another's attention. Infants point to indicate an object or event of interest, and to obtain help in grabbing something that is out of reach. If an observant infant knows that a desirable toy has been hidden in one of two boxes but doesn't know which one, it will readily find the toy by following an adult's pointing gesture. Strikingly, chimpanzees presented with the same situation largely fail to find the hidden object; many individual chimpanzees will even fail after repeated presentations. This is a robust and striking failure. In contrast, dogs presented with the same hide-and-point task readily find the hidden object, and often do so on the first presentation.

The comparative evidence on human infants, dogs, and chimpanzees led to the hypothesis, championed by Brian Hare, that perhaps domestication was responsible for the convergence in cognitive ability between humans and dogs. If domestication is responsible for the dog's cognitive prowess, then wolves should behave like chimpanzees in this task, and fail to understand human pointing, even after repeated exposure. This assumption was borne out in the first series of studies, providing stronger evidence for the role of domestication in dog cognition.

One problem with this comparative evidence is that not only are dogs domesticated whereas chimpanzees and wolves are not but dogs are reared from birth to adulthood in a human environment by humans. Perhaps dogs succeed on the pointing task because of their experience in a human environment, one rich in human pointing. That is, dogs learn about the goal of pointing through either passive exposure and associative mechanisms or through training. To tease apart these factors, wolf pups were raised by humans and then tested on the same task. These individuals readily followed the pointing gesture to the target box. This suggests that the human environment makes a direct contribution to the cognitive abilities of dogs. Further support for this claim comes from the fact that stray dogs, with either no human input or minimal experience, perform like wild wolves. Thus even domesticated dogs, lacking significant exposure to a human environment—and thus functionally feral—have poor comprehension of human communicative gestures.

Where domestication appears to play a unique role in dog cognition is in the relative sophistication of their capacity to read human gestures, a point supported by the work of Hare as well as Adam Miklosi. When dogs reared by humans are compared with wolves reared by humans, dogs perform far better when more subtle forms of pointing are presented, including foot pointing, momentary pointing at a distance, and reading the direction of eye gaze. What these results suggest is that a combination of selective pressure and early experience has shaped the cognitive ability of dogs. This is a lovely illustration of the interplay between evolutionary and developmental processes, or *evo-devo*. What this work is unable to demonstrate is whether the cognitive capacity of dogs (at least with respect to understanding communicative gestures) was selected for in its origins or is the accidental by-product of selection for some other capacity. To address this problem, we turn to work on silver foxes.

The Russian biologist Dmitri Belyaev decided to gain a deeper understanding of the process of domestication by systematically starting the process and controlling it over time. He started with a large population of silver foxes and divided them into several groups. Once they were settled, a human experimenter approached a group and offered some food. Those individuals who stayed near were selected and bred. This process was iterated over 30 generations. The outcome: a tame fox, completely unafraid of humans. This is precisely what Belyaev was hoping for. What he didn't expect, and certainly didn't select for, was floppy ears, a piebald coat, a curly tail, smaller brain, higher levels of serotonin (a neurochemical that regulates self-control), and much greater social cleverness. When tested in the hide-and-point task, domesticated silver foxes outcompeted their wild relatives. They performed at the level of domesticated dogs.

The fox studies led to the conclusion that selection for tameness resulted in a suite of by-products, including social cleverness. Although this is a reasonable conclusion, there is an alternative: those foxes paying attention to the experimenter's eyes, and recognizing the lack of threat, were most likely to stay close. These were socially clever foxes, and they were selected. Though Belyaev thought he was selecting for tameness, he was actually selecting for those individuals most likely to correctly interpret human gestures, including especially the direction of their eye gaze. Belyaev actually selected for cognitive capacity. To rule out this alternative, it would have been necessary for him to approach the foxes with covered faces.

Together, these studies of dogs, wolves, and foxes show both the power of selection and developmental processes to modify phenotypes, and the fact that even targeted selection can result in unforeseen consequences. What has evolved doesn't completely inform what is ultimately possible. Hidden within every organism is a suite of unrealized possibilities, waiting for novel opportunities, either naturally occurring or artificially imposed.

4. EVOLVING LIMITLESS OPTIONS

I stipulated in the introduction that the discontinuity between human and nonhuman minds is massive. I also stipulated that the cognitive chasm separating these two taxonomic groups is much larger than any other comparative contrast. For example, though many have claimed that the monkeys and apes far outperform birds in most cognitive tasks, recent work by Nicola Clayton and Nathan Emery on corvids shows that the gap is actually small, and in some domains such as creative tool use, these birds have the upper hand. Some scholars, such as Marc Bekoff and Frans de Waal, fundamentally disagree with the thesis that there is significant discontinuity in the cognitive achievements of humans and other animals, seeing far greater continuity. Thus, for example, Klaus Zuberbuhler has suggested that monkeys have vocalizations with a semantic richness that is akin to human words, and in some cases they combine these vocalizations in ways that approximate human sentences and their syntactic structure. These claims are controversial. Most linguists conclude that the origins of language, especially its semantics and syntax, cannot be traced to any of the communicative systems observed in other animals. None of these systems have the generative capacity of human language, including the young child's ability to spontaneously acquire a massive vocabulary and to combine and organize these words into novel expressions—that is, to tap the generative power of our syntax to create a limitless number of meaningful sentences. The same claim holds for music, mathematics, technology, and morality. Some components of these systems show continuity with other animals, but the most powerful components show no parallel at all. Why?

To answer this question, we need to consider how the human brain generates variation in behavior, and in particular, a flexible and creative set of responses to novel situations. Humans uniquely evolved four distinctive mechanisms that solve this problem: generative computation, cognitive promiscuity, abstract thought, and symbolic expression. Generative computations consist of recursive and combinatorial operations. *Recursive operations* are computable functions that generate a potentially infinite set of hierarchically organized expressions. *Combinatorial operations* entail mixing discrete elements to create novel combinations, and thus new meanings or functions. *Cognitive promiscuity* allows wildly different ideas to couple, blending thoughts and feelings from different domains of understanding. *Abstract thoughts* are detached from the primary sensory and perceptual inputs, allowing us to think about things that are beyond our direct experiences with the world. *Symbolic expressions* are discrete representations of thoughts, ideas, or emotions, such as spoken, signed, or written words, as well as graphic icons.

The generative computations provide the engines of variation. Cognitive promiscuity enables us to go beyond the often specialized and myopic mechanisms that evolved to solve problems in one context or domain, such as foraging, tool use, cooperation, or communication. Abstract thoughts allow us imagine possibilities that extend far beyond what we feel, see, smell, hear, and touch. Symbols allow us to reduce memory load through compact storage of discrete and easily retrieved symbols while facilitating lossless information transfer. The following makes this less abstract.

A wide variety of animals use tools to solve particular environmental problems. In this sense, tool use is most definitely not unique to humans. This is, however, where the convergence ends. Unlike animal tool use, including that by chimpanzees and New Caledonian crows, human tool use makes use of the four variation-generating mechanisms just noted. For almost any tool that one can think of today, materials are combined to create objects with multiple functions—think pencil or Swiss army knife. The design for almost any tool that one can think of derives from cognitive promiscuity—pencils are designed to write, thereby physically transforming blank pages into marked-up ones while conveying our thoughts and feelings. Here physics, language, and social psychology are combined. In the case of many tools, their inventors went beyond direct experience to create an object with a particular function—no one ever experienced a writing device until someone imagined that marking up stuff, as in the earliest cave art, was important, both for telling stories and conveying information linked to survival. For many tools, the object not only serves one or more functions but is a symbol itself—consider the scales of justice. Animal tools serve one function, never consist of more than one material, are created from direct experience, and do not function as symbols. Animal tools lack the signature of generativity, promiscuity, abstraction, and symbolism.

Though we do not know when, how, or why these four uniquely human ingredients of cognition evolved, what we can say is that they enabled almost everything that we think of as characteristically human: language, mathematics, technology, music, religion, and morality. Each of these systems of knowledge is capable of expressing a virtually limitless range of possibilities. Each of these systems is highly generative, capable of changing as a function of new environmental challenges. In this sense, though each system may have originally evolved as a by-product of combining distinctive and highly adaptive components, selection may act on such by-products to produce highly adaptive outcomes. For example, religion consists of cognitive processes that evolved for nonreligious functions, such as attributing intentions to hidden powers or inanimate entities, and holding beliefs that unify in-group solidarity and inspire out-group hatred. These cognitive processes, and others, played a significant role in the evolution of religion and, especially, the capacity of religious organizations to solve the problem of large-scale cooperation among unrelated strangers.

CONCLUDING REMARKS

This chapter has largely focused on a small branch of the phylogenetic bush, specifically, the comparison between human and nonhuman primates, and especially chimpanzees. This focus was intentional. No other comparative pairing among living organisms presents such small genetic distances accompanied by massive phenotypic differences. Chimpanzees have hair all over their body and are knuckle walkers. Humans are virtually bald and walk on two feet. Chimpanzees have never left the forests of Africa. Humans long ago left the forests of Africa to inhabit mountains, deserts, ice caps, and oceans within and outside Africa, and to explore other planets. Chimpanzees make tools and communicate through sounds and gestures, but each of these expressions is far too primitive to count as a precursor for human technology and language.

Though it has commonly been stated that human intellectual prowess stems from the evolution of language, or the capacity to create culture, or to invent new technologies, these explanations fail on two counts. First, they treat each of these causal factors as monolithic. Language, culture, and technology each rely on a suite of mechanisms, some unique to humans and others shared. To understand the evolution of language, culture, or technology—or any of the other distinctively human expressions—requires decomposition, studying each mechanism's unique evolutionary history. Second, decomposition shows that language, culture, and technology each rely on the uniquely human ingredients of generative computation, promiscuity, abstraction, and symbolic expression. These capacities may each have evolved to solve a particular problem, but today they serve multiple functions. These capacities represent the engine of creativity, a device that delivers a limitless number of solutions to novel problems. These capacities allowed our species uniquely to go where no species has gone before.

FURTHER READING

Dehaene, S. 2005. Evolution of human cortical circuits for reading and arithmetic: The neuronal recycling hypothesis. In S. Dehaene, J.-R. Duhamel, M. D. Hauser, and G. Rizzolatti, eds., From Monkey Brain to Human Brain. Cambridge, MA: MIT Press.

Emery, N. J., and N. S. Clayton. 2004. The mentality of crows: Convergent evolution of intelligence in corvids and apes. Science 306 (5703): 1903–1907.

Fitch, W. T., L. Huber, and T. Bugnyar. 2010. Social cognition and the evolution of language: Constructing cognitive phylogenies. Neuron 65 (6): 795–814.

Hare, B., I. Plyusnina, N. Ignacio, O. Schepina, A. Stepika, R. W. Wrangham, and L. N. Trut. 2005. Social cognitive evolution in captive foxes is a correlated by-product of experimental domestication. Current Biology 15: 226–230

Hauser, M. D. 2009. The possibility of impossible cultures. Nature 460: 190–196.

McGhee, G. R. 1999. Theoretical Morphology: The Concept and Its Application. New York: Columbia University Press.

Penn, D. C., K. J. Holyoak, and D. J. Povinelli. 2008. Darwin's mistake: Explaining the discontinuity between human and nonhuman minds. Behavioral and Brain Sciences 31: 109–178.

Premack, D. 2007. Human and animal cognition: Continuity and discontinuity. Proceedings of the National Academy of Sciences USA 104: 13861–13867.

Ryan, M. 1998. Sexual selection, receiver biases, and the evolution of sex differences. Science 281: 1999–2002.

VII.15

Evolution of Apparently Nonadaptive Behavior

Nathan W. Bailey

OUTLINE

1. What is apparently nonadaptive behavior?
2. Behavior as a transaction
3. Random mutation versus adaptation: Cannibalism
4. Manipulation: Imposter birds and zombie snails
5. Evolution does not equal perfection: Sexual cannibalism
6. Same-sex sexual behavior: A case study
7. Insights from apparently nonadaptive behavior

Behaviors appear to be *nonadaptive* when their costs, in terms of reproductive fitness, appear to outweigh their benefits. As long as there is genetic variation, selection should eliminate such behaviors under those conditions. However, apparently nonadaptive behaviors are much more common than might be expected. Examples are numerous and include counterintuitive responses to infection by pathogens or parasites, sexual cannibalism, and same-sex sexual behavior. This chapter explores how an evolutionary framework can be used to understand how and why such behaviors evolve, and what causes them to be maintained within populations despite what appear to be fitness disadvantages.

GLOSSARY

Altricial. The condition of offspring in which they are unable to feed or fend for themselves without considerable assistance from parents for a period of time.

Altruism. A behavior that benefits a recipient individual at the expense of the fitness of the actor (contributor).

Apparently Nonadaptive Behavior. Behavior that appears to impose a fitness cost on the actor with no compensating benefit.

Arms Race. Cycles of adaptations and counteradaptations leading to the evolution of novel morphological or behavioral traits, occurring between two or more interacting species, or in some cases, between the sexes.

Brood Parasite. In birds, a species that exploits the reproductive behavior of another species by laying an egg in its nest. The chick is then fed by the host species at the expense of the host's offspring.

Filial Cannibalism. The eating of offspring by their parents.

Overdominance. Heterozygote advantage; when alleles in the heterozygous condition increase fitness but decrease it when in the homozygous condition.

Parasite Manipulation. Behavioral, physiological, or morphological changes induced in a host by a parasite that increase the likelihood of transmission of the parasite.

Phenotypic Plasticity. The production of different phenotypes by a single genotype, depending on the environment in which it is expressed.

Reproductive Fitness. An individual's contribution of offspring in the next generation, relative to other individuals in the population.

Same-Sex Sexual Behavior. Sexual behavior, such as courtship, mounting, or copulation, occurring between individuals of the same sex that also occurs between individuals of the opposite sex.

Sexual Antagonism. The expression of genes (or a phenotype) in one sex that confer a fitness advantage, but when expressed in the other sex cause a fitness disadvantage.

Sexual Cannibalism. The killing and consumption of a mate during or after copulation.

1. WHAT IS APPARENTLY NONADAPTIVE BEHAVIOR?

Evolutionary theory provides a robust framework for explaining the diversity of the natural world. Behavior contributes considerably to that diversity, but understanding how and why animals behave the way they do poses challenges, because many behaviors appear to be costly and nonadaptive. The power of evolutionary theory lies in its ability to generate testable hypotheses to explain the reasons behind even the most bewildering traits, and there are a large number of bewildering behaviors from which to choose.

Wire-tailed manakins (*Pipra filicauda*) appear to go out of their way to attract predators and physically exhaust themselves during courtship. Males have a conspicuous yellow breast, a bright red cap, and long black filamentous tail feathers, and they engage in spectacular behavioral displays to elicit attention from females. They fluff up the feathers on their back, wiggle their tail, bob their head up and down and back and forth, quiver their outstretched wings, and leap and bounce rapidly from side to side along horizontal branches, all the while emitting a series of buzzing and whistling vocalizations. Any biologist observing these extraordinary behaviors could be forgiven for wondering how such a flamboyant repertoire could ever evolve, when it clearly requires an enormous energetic expenditure and appears to invite predation. How could this behavior possibly be adaptive for a manakin?

Closer examination of the natural history of many animal species will reveal similar examples of behaviors that appear to be nonadaptive—in other words, behaviors that appear to decrease rather than increase fitness. The purpose of this chapter is to highlight how the framework of evolutionary theory can be used to study such behaviors and to illustrate how historical studies of puzzling behaviors have provided powerful, sometimes surprising, evolutionary insights. It concludes with an analysis of same-sex sexual behavior as a topical case study in apparently nonadaptive behavior.

As with all good theories, the apparent exceptions provide the best test of a theory's robustness and generality. Behaviors that appear to defy evolutionary logic have historically provided some of the most decisive verification of the basic principles of evolution, and research in the field continues to focus on identifying the adaptive value of counterintuitive behaviors, understanding their genetic basis, and working out how they evolve. This evolutionary approach has a long history. The Nobel Prize–winning ethologist Konrad Lorentz wrote in the 1960s that aggression was more likely to occur between individuals of the same species than between individuals of different species. Lorentz puzzled over why animals of the same species behave aggressively toward one another. Aggression is risky and costly. Predation is likely to arrive in the form of a different species, and disease and parasitism present yet other challenges, so if an animal is to expend energy being aggressive, why waste it on a conspecific? Lorentz's logic was that the fiercest competition for resources should occur among members of the same species, whether the resources are materials such as food, water, or nesting substances, or reproductive such as access to mates. In 1973 John Maynard Smith and George Price formalized the evolutionary dynamics of animal conflict using game-theoretical models to clarify the individual benefits of aggressive behavior (see chapter VII.3). There are many other scientists whose theoretical advances have been motivated by puzzling animal behavior. Around the same time as Lorentz, William D. Hamilton developed the theory of kin selection to help explain why animals sometimes behave altruistically, conferring benefits on others at great cost to their own survival or reproduction. The scientific study of apparently nonadaptive behaviors was catalyzed by Charles Darwin, who proposed sexual selection as a mechanism underlying the evolution of conspicuous, costly traits, such as the displaying behavior of the manakins. But are these mechanisms sufficient to explain the persistence of all apparently nonadaptive behaviors?

2. BEHAVIOR AS A TRANSACTION

All animals behave. And all behaviors carry costs and benefits. The currency of this transaction can be thought of as reproductive fitness, or the share of the genes in the next generation contributed by a parent. More genes, through the production of more offspring relative to other parents in the population, equate to higher fitness. Some behaviors obviously increase the fitness of the actor. For example, a guppy initiating an escape response to a predator clearly increases its fitness. But how is this behavior evaluated in terms of costs and benefits? In the case of the fleeing guppy, the costs are not terribly high—they might include a bout of energetic expenditure and time taken away from eating or reproducing. But the benefit is obvious: not dying. Unfortunately, the costs and benefits of other behaviors are not always so obvious, and in many cases the costs appear to exceed the benefits.

There are a number of evolutionary explanations for the existence of behaviors that give the appearance of imposing a fitness cost that is not compensated for by some other benefit, and they can be divided into two broad categories. The first is behavior that appears to decrease fitness but in reality is evolutionarily maintained within populations, whereas the second is behavior that appears to decrease fitness and is in fact selected against. It is important to note, however, that just because they *appear* to decrease the fitness of the individual engaging in them does not mean such behaviors always *do* decrease fitness. Sometimes the fitness benefit is "hidden" and can be revealed through experimentation. Nevertheless, in many cases behaviors that appear to be nonadaptive actually decrease the actor's fitness but are still evolutionarily maintained (figure 1).

	Eliminated by selection	**Maintained by selection**
Behaviors that increase fitness of the actor	1A	1B Sexual cannibalism
Behaviors that decrease fitness of the actor	2A Maladaptive mutations	2B Parasite manipulation Antagonistic coevolution Altruism

Figure 1. An evolutionary framework for studying apparently nonadaptive behavior. The horizontal rows indicate fitness effects of the behavior *on the actor*, and the columns indicate how selection acts *on the behavior*. Examples are given in each quadrant. The top row indicates apparently nonadaptive behaviors that have a "hidden" fitness benefit; in other words, the behavior yields a benefit that is not immediately apparent to scientific observers. Fitness-increasing traits can be counterselected (1A), but this is unlikely to happen as a result of direct selection on the trait; for example, the behavior might be lost because it is genetically correlated with another fitness-reducing trait. Apparently nonadaptive behaviors that increase fitness are likely to be evolutionarily maintained (1B), as in the case of sexual cannibalism in spiders. Some behaviors that appear nonadaptive truly do lower individual fitness (row 2) and are therefore counterselected (2A). However, most apparently nonadaptive behaviors of interest fall into the category of behaviors that incur a fitness cost to the individual, yet are nonetheless evolutionarily maintained in populations (2B). These include behaviors resulting from manipulation by other agents, behaviors arising because of selection at a level different from the individual, or correlated responses to selection such as sexually antagonistic coevolution.

3. RANDOM MUTATION VERSUS ADAPTATION: CANNIBALISM

Behaviors that are not evolutionarily maintained are not ordinarily observed or studied, because, by definition, selection has eliminated them. These include behaviors that arise as a result of random genetic mutation and confer a fitness disadvantage. However, lab studies frequently use mutant strains of *Drosophila* or mice containing genes that have been partially or wholly inactivated, and these are useful for illustrating how genetic mutations can affect the adaptive value of a behavior.

A mutant strain of lab mice called "Tokyo" provides a clear example of the genetic basis for a nonadaptive behavior. The Tokyo strain expresses a mutant vitamin D receptor, VDR. One of VDR's ordinary functions is to act as a transcription factor than binds both DNA and vitamin D, but in the Tokyo strain this function is diminished because the mutant protein lacks a DNA-binding domain. Vitamin D is a hormone that regulates a variety of neurological processes underlying behaviors, and Allan Kelueff and his colleagues found that among other behavioral abnormalities, all female Tokyo mutants cannibalized all their pups, whereas females that did not express the mutant gene did not cannibalize any of their young. The mice in this study were kept in conditions that would not normally provoke offspring killing (*filial infanticide*), and it is reasonable to predict that a comparable mutation arising in a wild population of mice would not be evolutionarily maintained.

The killing of one's own offspring seems so patently maladaptive that it would be easy to dismiss the phenomenon as unusual, pathological, and therefore of little interest. However, offspring killing is surprisingly common. Infanticide has been documented in many mammal species, from rodents to primates, in addition to birds, insects, fish, and rotifers, and it has different manifestations. There are many appealing adaptive explanations for killing the offspring of an unrelated conspecific, but it is less clear how it could be beneficial for a parent to kill its own young. Nevertheless, viewing filial infanticide through the lens of evolutionary theory and identifying the fitness costs and benefits can help reveal reasons for its persistence.

Sometimes it is advantageous for a female to kill her young if she gains resources from them, for example, by eating them. In the burying beetle *Nicrophorus vespilloides*, food for developing larvae is limited to a single liquefying animal carcass that the parents have prepared in advance. If the carcass is too small to support all the offspring that hatch and begin to develop, the parents have a strategy to cut their losses and ensure that at least some of them survive, rather than face the complete loss of the brood. That strategy is to kill and eat some—but not all—of their offspring. Wiping out a portion of the brood ensures that the remaining larvae have enough food to survive, and it has the added advantage of providing the parents with a meal. In other species, a female may sacrifice her young if the cost to her future reproductive success of rearing those young is greater than the cost of simply killing them and waiting to reproduce at a later date. The costs of rearing young are not insignificant in altricial species and particularly in mammals with long, energetically expensive periods of lactation, so the cost-to-benefit ratio of filial infanticide can shift dramatically during times of low resource availability or increased stress. For example, female mice (*Mus musculus*) can reduce their litter size when food is scarce. Sometimes this reduction is accompanied by filial cannibalism. In studies that examine this behavior in mice, it can be difficult to distinguish between active infanticide (i.e., physically killing pups) and simply allowing pups to die from starvation and then eating them. Intriguingly, however, female mice have the capacity to resorb developing fetuses, and they are more likely to do so when conditions are not favorable for rearing offspring.

4. MANIPULATION: IMPOSTER BIRDS AND ZOMBIE SNAILS

Animals do not always behave in their own best interests, and frequently that is because one individual is being manipulated to serve another's needs. Some parasitic relationships resemble a marionette being controlled by an invisible puppeteer: parasites and pathogens can invade the body of their host and induce it to behave in a way that benefits the parasite at the expense of the host. In other cases, the manipulation is accomplished from afar. Either way, fascinating nonadaptive behaviors can result. A classic example is manipulation by cuckoos (*Cuculus canorus*). Female cuckoos lay mimetic eggs in host nests, and newly hatched cuckoo chicks, without any instruction or provocation, quickly set about pushing the host parent's eggs up the wall of the nest on their tiny back, and then hurling them over the edge to their demise. Despite the obvious phenotypic differences between a cuckoo chick and what a host's chick ought to look like, host parents appear to overlook the destructive behavior of the invader and instead expend a remarkable amount of time and energy feeding and protecting the developing cuckoo.

But how and where is this behavior nonadaptive? Cuckoo-feeding behavior is maladaptive for the host, because it reduces its own reproduction, but of great benefit to the brood parasitic bird by increasing the production of offspring, and this imbalance creates interesting evolutionary dynamics. What is adaptive for one species in the interaction is not adaptive for the other, but selection acts on both. Selection on hosts should result in the evolution of traits that decrease the risk of being deceived, while selection on the cuckoos should be expected to increase the ability to deceive hosts. These countervailing pressures can result in an evolutionary arms race wherein host species and parasite species evolve adaptations and counteradaptations, and many cuckoo studies support predictions made by the arms-race hypothesis. For example, because host species should evolve parasite egg discrimination, one would expect the brood parasites to evolve ever-better egg mimicry. Such evolution might in turn select for hosts that can distinguish foreign nestlings after they have hatched, and that is exactly what has been found in the superb fairywren (*Malurus cyaneus*), which is parasitized by two species of bronze-cuckoos (*Chrysococcyx* spp.). However, a theoretical study by Maria Servedio and Russell Lande showed that it is also possible for host egg discrimination and cuckoo egg mimicry to evolve to a stable point at which there is imperfect egg mimicry, weak host discrimination, and a slight tendency of hosts to reject cuckoo eggs.

Parasitic microorganisms that infect a host can also elicit nonadaptive behaviors by hijacking their host's central nervous system to induce behaviors that increase their likelihood of transmission. Many neurophysiological mechanisms for manipulating host behavior have been identified. Pathogens can infect host neurons and disrupt their function, cause inflammation of the central nervous system by triggering host immune responses, induce neuronal apoptosis, and affect neurotransmitter and hormone levels in discrete structures of the host's brain.

A fascinating example is that of snails infected with the trematode parasite *Leucochloridium paradoxum*. For this trematode flatworm to reproduce, it must migrate from its intermediate snail host to its definitive host, a bird. The trematode starts out as an egg within a snail. The egg hatches into a larva called a *miracidium*, and then the miracidium matures into a sporocyst. Multiple sporocysts migrate to the snail's eye tentacles and form swollen, colorful brood sacs that throb rhythmically and resemble caterpillars. Infected pulsating snails further compound their doom by exhibiting positive phototaxis, which means they uncharacteristically move toward sunlight instead of away from it, thereby increasing their vulnerability to bird predators. Once inside a bird, *L. paradoxum* continues its life cycle. A similar locomotory manipulation occurs in insects infected with nematomorph worms. These extremely long, thin worms reproduce only in water, and during their infectious stage they impel the ordinarily land-bound insect host within which they are tightly packed to find water, jump in, and drown so the worms can emerge to swim away and mate.

Although host manipulation is clearly an adaptation for the parasite, these behaviors are not adaptive for the host—that is, the benefits to the host clearly do not outweigh the costs. A good deal of debate surrounds the question of whether selection has acted on parasites to be able to manipulate their hosts, or whether these abnormal behaviors are just by-products of physiological responses hosts mount when they are invaded. However, they differ from other behaviors that hosts mount during a pathological infection to kill pathogens or mitigate the negative effects of infection, such as the behavioral fever some insects induce by moving to a warmer area. As Richard Dawkins articulated in the early 1980s, nonadaptive behaviors often represent the influence of adaptations in other individuals, via the extended phenotypic effects of their genes. Viewing behavior as more than just a linear output starting from neurological processes operating within one individual's body allows a much richer understanding of the selective pressures that maintain apparently nonadaptive behaviors. Behavior is by definition especially responsive to external influences, either from the physical environment or from social interactions with other individuals. This property makes it susceptible to the manipulative effects of genes carried in other individuals. However, the malleability of behavior can also be a major asset to an animal facing an uncertain environment.

5. EVOLUTION DOES NOT EQUAL PERFECTION: SEXUAL CANNIBALISM

No organism is perfectly adapted to its environment, because the environment consists of continually changing, interacting forces. These can be physical, such as ambient temperature or the pH of an aquatic habitat, or they can be biotic, for example, the extent of canopy cover in a rain forest or the abundance of a keystone species in an ecosystem. Social interactions among conspecifics can be particularly variable depending on the behavioral, morphological, or physiological traits expressed by interacting individuals. In such a constantly changing world, it often pays to be flexible.

Phenotypic plasticity (see chapter III.10 on reaction norms) describes the capacity for a single genotype to produce different trait values depending on the environmental conditions in which that genotype is expressed. Phenotypic plasticity is responsible for much of the behavioral variation that evolutionary biologists and behavioral ecologists observe and study. For example, in many amphibians and insects that use acoustic signals to attract mates, males can sing for females, or they can adopt a silent satellite strategy in which they position themselves close to singing males and then intercept females responding to the caller. Which strategy they adopt depends on a number of factors such as the population density, risk of predation, and their own body condition. Similarly, many female birds develop song preferences early in life. These preferences depend on nuances of the male songs to which they are exposed during a critical period of imprinting, which means that whom they choose to mate with as an adult depends greatly on the males they encountered as juveniles.

In an ideal world, selection would favor individuals with the ability to perfectly shape morphology, physiology, and behavior to whatever physical or social environment they happen to encounter, but this does not happen. Plasticity, like any other trait, has costs and limitations. Organisms must possess the necessary sensory organs and central nervous system circuitry to detect environmental variation, and then they must produce an appropriate response. The environment can also fluctuate a great deal, and in some cases environmental cues can be inaccurate.

Many other factors impede perfect adaptation. The physics of surface-to-area relationships limits how large or small animals can grow at a given temperature. The laws of chemistry and electricity dictate how fast nerve impulses can travel. Protein biochemistry determines the flexibility, elasticity, permeability, and durability of biological substances. The work of Geoff Parker and John Maynard Smith was instrumental in describing how selection may drive organismal evolution *toward* an optimal phenotype, but fluctuating environments compounded with other constraints such as these will constantly undermine the process.

Consider sexual cannibalism. Male redback spiders (*Latrodectus hasselti*) are tiny compared with females. They copulate by depositing sperm onto the end of one of their pedipalps, which are paired sensory appendages projecting from either side of the mouth, and inserting it into a female's genital opening. Shortly after initiating this maneuver, the male flips his entire body directly into the female's mandibles using his still-inserted palp as a hinge. Sometimes the female pierces him, injects digestive juices, and then sucks out his liquefied tissues (figure 2). The male thus sacrifices any opportunity for additional mating by becoming a snack for the female.

Maydianne Andrade and her colleagues have studied the behavioral and evolutionary dynamics of the stereotyped flipping behavior males engage in while mating, and they found that perhaps not surprisingly, it increases the likelihood that the female will eat the male. Males are especially prone to being eaten when the female is hungry. However, Andrade's lab has also found that males that get eaten sire a greater proportion of the resulting offspring compared with males that escape, and this result is likely driven by the fact that females are less inclined to remate after eating a male.

At first glance it seems easy to explain why males commit nuptial suicide: males that do it sire more offspring. But surely there is a better way to sire more offspring that doesn't involve killing yourself and thereby removing all possibility of reproducing in the future? A key question is, Why hasn't evolution favored *L. hasselti* males that escape and go on to inseminate other females, perhaps donating a different, nonlethal, type of food item, as happens in many insect species? The answer lies in the fact that *L. hasselti* males almost never remate. In fact, they usually are not able to. The sperm-holding organs on their palps snap off after mating, and their chances of surviving long enough to even find another female are negligible because they are so tiny (figure 2). The evolution of self-sacrificing behavior makes much more sense in light of these vital constraints: if there is no chance of

Figure 2. Female redback spider (*Latrodectus hasselti*) with cannibalized former mate hanging in the web above her fangs. (Photo by Ken Jones; copyright M.C.B. Andrade 2002.)

fathering more offspring in the future, and males do not provide direct care or resources to their young, then there is no fitness advantage to staying alive.

6. SAME-SEX SEXUAL BEHAVIOR: A CASE STUDY

The final portion of this chapter is devoted to a behavior that appears to be nonadaptive and has attracted increasing academic scrutiny over the last several decades: same-sex sexual behavior (SSB). The aim here is to demonstrate how evolutionary causes of a seemingly puzzling behavior, SSB, can be digested into different hypotheses to yield predictions that can then be tested. This process provides an example of the deductive power of evolutionary reasoning for exploring the ultimate reasons for the existence of apparently nonadaptive behaviors, by suggesting potential benefits to outweigh suspected costs.

Same-sex sexual behavior is present in many species, can occur in both sexes, and takes on a variety of forms. In general, it is defined as sexual behavior between individuals of the same sex that is (usually more commonly) expressed between individuals of opposite sex. It includes behaviors such as courtship, mounting, and copulation. Studies of same-sex interactions in nonhuman animals typically focus on same-sex behavior or same-sex preferences, rather than same-sex orientation, which is a more enduring preference and therefore more difficult to ascertain. SSB has been recorded at least since the time of Aristotle:

> And some male birds have been seen to be so effeminate from their birth, that they neither crowed, nor desired sexual intercourse, and would submit themselves to any males that desired them. (Aristotle, *Historia Animalium*, in Cresswell [1891].)

In some species, the proportion of individuals in a population that exhibit SSB can be high; for example, more than 30 percent of pair-bonded birds in a Hawaiian population of Laysan albatross (*Phoebastria immutabilis*) were found to be female-female pairs, and in a study of the deep-sea squid *Octopoteuthis deletron*, researchers observed that about half of all sexual encounters were between males. The manner in which SSB manifests is as varied as the number of species in which it is found. To name just several, male toads (*Bufo bufo*) sometimes amplect other males, female Japanese macaques (*Macaca fuscata*) mount one another and engage in genital contact, and male flour beetles (*Tribolium castaneum*) will mount and attempt to deposit sperm on other males.

The pervasive assumption that SSB is never adaptive has been gradually but steadily replaced by a wide range of hypotheses for evolutionary mechanisms that maintain the behavior, and in many respects the human and primate literature has led this transition. There is no universal reason for the existence of SSB, as it appears to have different functions and origins in different animal species. However, SSB provides a vehicle for studying evolutionary mechanisms underlying many apparently nonadaptive behaviors, for example, sexually antagonistic selection, kin selection, intrasexual competition, cooperative breeding, and others. A recent theoretical analysis published by Sergey Gavrilets and William Rice is illustrative. Gavrilets and Rice examined the feasibility of two major genetic models for the evolutionary maintenance of SSB: overdominance and sexual antagonism. *Overdominance* describes a heterozygote advantage; genes that confer a fitness advantage when in the heterozygous condition might induce the expression of SSB when in the homozygous condition. Given a large enough heterozygote advantage, SSB could be maintained in a population even if it decreases the fitness of individuals that exhibit it. Under *sexual antagonism*, genes that provide a fitness advantage in one sex might antagonistically cause the expression of SSB in the other. Gavrilets and Rice were able to derive genetic and phenotypic conditions under which either scenario could maintain SSB within a population. Empirical evidence from humans supports predictions made by the sexual antagonism model: female relatives of male homosexuals produce more offspring than female relatives of male heterosexuals, and male homosexuality tends to run in maternal lineages.

Additional hypotheses have been advanced to explain the evolutionary maintenance of SSB in other species. SSB in primates, cetaceans, and other mammals has been examined to test whether it provides social cohesion, acting as a form of currency in interactions among group members. The evidence is mixed. In some cases, such as in bottlenose dolphins, male-male SSB represents approximately half of all sexual encounters and appears to facilitate the formation of social alliances within small groups of males. In other systems such as Japanese macaques, the social functions of SSB remain unclear.

In contrast with mammal studies, research in insect systems has focused on ways in which SSB might mediate intrasexual competition. Sara Lewis and her colleagues studied male-male mounting and copulation in the flour beetle *T. castaneum*. Mounting males sometimes transfer sperm onto the males they have mounted. Lewis's group hypothesized that when the mounted males later mate with a female, some of the previously deposited sperm could fertilize females' eggs "by proxy," thus providing an adaptive benefit for male-male mounting. However, while fertilization by proxy can happen, it appears unlikely to occur frequently enough to provide an adequate fitness benefit to explain its maintenance.

The evolutionary reasons for SSB remain unresolved for most species. However, as exemplified by the studies described, research that tests hypotheses for the existence of an apparently nonadaptive behavior demonstrates how it can be studied within an evolutionary framework. These studies provide insights not only into the dynamics of a particular behavior but also into more general evolutionary questions, such as how cooperation or competition manifest in different species, what evolutionary forces shape behavior, and why the genetic architecture of behavior is important.

7. INSIGHTS FROM APPARENTLY NONADAPTIVE BEHAVIOR

Behaviors that appear to be nonadaptive are common. In some respects, labeling a behavior as nonadaptive is a matter of perspective. Behaviors that seem to be detrimental to the fitness of an individual are often actually adaptive on closer inspection, such as in the case of redback spiders. Other behaviors that appear detrimental really are fitness reducing, representing new mutations that have yet to be eliminated from the population, as in the case of genetic mutations seen in lab studies. However, there is a vast gray area between these two extremes describing behaviors that are evolutionarily maintained in populations despite what we perceive to be heavy costs. These include sexual cannibalism, brood parasitism, parasite manipulation, and same-sex sexual behavior, all of which have been empirically and theoretically studied to try to understand the evolutionary mechanisms causing their persistence. Same-sex sexual behavior has been utilized to explore the evolutionary mechanisms that maintain many nonadaptive behaviors, such as sexually antagonistic selection, kin selection, and intrasexual competition. By employing a toolkit of evolutionary hypotheses, future research on SSB has the potential to illuminate surprising and intriguing reasons for its existence that inform other areas of evolutionary biology.

FURTHER READING

Andrade, M.C.B. 1996. Sexual selection for male sacrifice in the Australian redback spider. Science 271: 70–72. *A study examining the fitness benefits to males of self-sacrificing courtship behavior.*

Bailey, N. W., and M. Zuk. 2009. Same-sex sexual behavior and evolution. Trends in Ecology & Evolution 24: 439–446. *The most recent general review of same-sex sexual behavior that examines its causes and consequences across a wide range of species.*

Cresswell, R. 1891. Aristotle's History of Animals. London: George Bell and Sons. *A comprehensive translation of Aristotle's observational writings on animal natural history, behavior, and morphology.*

Dawkins, R. 1982. The Extended Phenotype. Oxford: Oxford University Press. *One of Dawkins's most influential works, this book outlines his observation that genes in one individual can influence the development of traits in other individuals, through "extended" effects.*

Gavrilets, S., and W. R. Rice. 2006. Genetic models of homosexuality: Generating testable predictions. Proceedings of the Royal Society B 273: 3031–3038. *One of very few formal theoretical models of same-sex sexual behavior, this study examined genetic conditions under which the behavior might be expected to persist in populations.*

Klein, S. L. 2003. Parasite manipulation of the proximate mechanisms that mediate social behavior in vertebrates. Physiology and Behavior 79: 441–449. *A comprehensive review article on the evidence for parasite manipulation of behavior, presented from a mechanistic perspective.*

Parker, G. A., and J. M. Smith. 1990. Optimality theory in evolutionary biology. Nature 348: 27–33. *A seminal exposition of the rationale behind optimization models, and their implications for the power of selection to craft adaptations.*

Servedio, M. R., and R. Lande. 2003. Coevolution of an avian host and its parasitic cuckoo. Evolution 57: 1164–1175. *A useful discussion of the arms race hypothesis in systems of brood parasitism, plus a contrasting model for the evolutionary dynamics of such systems.*

Tigreros, N., A. South, T. Fedina, and S. Lewis. 2009. Does fertilization by proxy occur in *Tribolium* beetles? A replicated study of a novel mechanism of sperm transfer. Animal Behaviour 77: 555–557. *The most recent published study taking a hypothesis-testing approach to assess why male-male copulation occurs in flour beetles.*

VII.16

Aging and Menopause

Jacob A. Moorad and Daniel E. L. Promislow

OUTLINE

1. A natural history of aging
2. Theories for the evolution of aging
3. Menopause
4. Pressing questions on the evolution of aging

Given enough time, organisms lose vigor as they age. Traits that may have once seemed optimized for survival and reproduction degrade, increasing the risk of death and reducing fertility. On the surface, it seems paradoxical that natural selection, which always favors increasing fitness, should permit aging to be nearly ubiquitous. However, evolutionary theory provides simple but powerful hypotheses to explain why humans senesce and die. At its heart, this theory states that fitness depends more on what happens early in life than what happens at old age. In other words, there is more natural selection for early-life function. A basic tenet of aging theory is that if the survival or fertility effects of early-acting and late-acting genes are independent but equally distributed, natural selection will favor the evolution of aging in very predictable ways. But do genes really act this way? We humans age, of course, but our species is unusual in that middle-aged females undergo menopause. What is so special about our species, and given that men die sooner than women, why is reproductive cessation in males neither as abrupt nor as complete as in women?

GLOSSARY

Aging. *See* senescence.

Antagonistic Pleiotropy. A proposed mechanism for the evolution of senescence. Under this model, selection favors alleles with early-acting beneficial effects that have pleiotropic but deleterious effects at late age.

Disposable Soma Theory. The theory that senescence evolves owing to trade-offs between investment in reproduction and investment in somatic (bodily) maintenance and repair. The optimal strategy is one that favors limited investment in maintenance and repair, such that senescence is inevitable.

Gene Regulatory Network. The complex web of interacting genes, some of which regulate themselves and/or downstream target genes, some of which are regulated by upstream regulatory genes.

Genetic Correlation. The statistical dependence between two traits caused by genes that determine the values of both traits.

Genetic Variance. A measure of phenotypic differences among individuals that are caused by genetic differences.

Inbreeding Depression. The loss of fitness that is associated with the mating of relatives.

Iteroparous. Capable of reproducing multiple times throughout life.

Menopause. The late-onset, irreversible cessation of reproductive capability experienced by women, usually at around 50 years of age.

Mutation Accumulation Theory. Theory based on the notion that the strength of selection declines with age; as a result, late-acting germ-line deleterious mutations accumulate over evolutionary time, leading to age-related declines in fitness.

Programmed Death. The idea, generally rejected by most evolutionary biologists as a principle cause of aging, that natural selection favors senescence, such that genes that actually cause death can spread through, causing catastrophic mortality.

Semelparous. Reproducing just once, and then dying; examples of semelparous organisms include spawning salmon and some species of bamboo.

Senescence. An age-related decline in fitness components, including vital rates (age-specific survival or fertility), behavior, physiology, and morphological traits; used interchangeably with *aging*.

1. A NATURAL HISTORY OF AGING

Natural selection is a powerful force. It can shape elaborate developmental pathways that give rise to exquisite morphological characters, it can shape complex behaviors that allow organisms to survive in what to us seem the most inhospitable of environments, and it has even endowed organisms with the ability to heal wounds and to repair themselves. But these characteristics leave us with a puzzle: If selection endows organisms with the ability to repair both genetic and structural damage, why can it not prevent organisms from eventually falling apart as they age?

What Is Aging?

In all human populations, the probability of dying varies across ages, following a bathtub-shaped trajectory (figure 1A). Mortality rates are relatively high in utero and in the months immediately after birth. They then decline, reaching a minimum in late adolescence. From this age onward, the probability of dying increases. In humans, the risk of dying doubles every eight years. In the late twenties, fertility starts to decline, as another manifestation of the aging process (figure 1B).

We measure aging (or *senescence*; here we use the terms interchangeably) as the age-related rate of decline in fitness traits. It is manifested at the level of the population and of the individual. *Demographic senescence* refers to the age-related decrease in the frequency of survival and mean reproductive output of groups of same-age individuals. *Physiological senescence* refers to late-age-related changes in individual phenotypes. Life span per se is not a measure of aging but an outcome of cumulative mortality risks, some of which are constant throughout the life span and some of which vary with age.

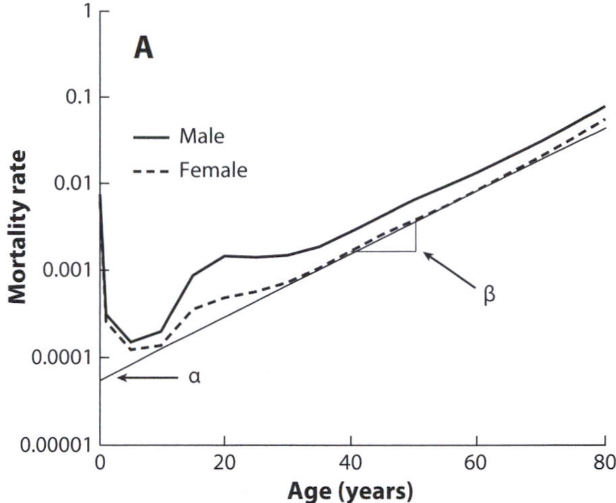

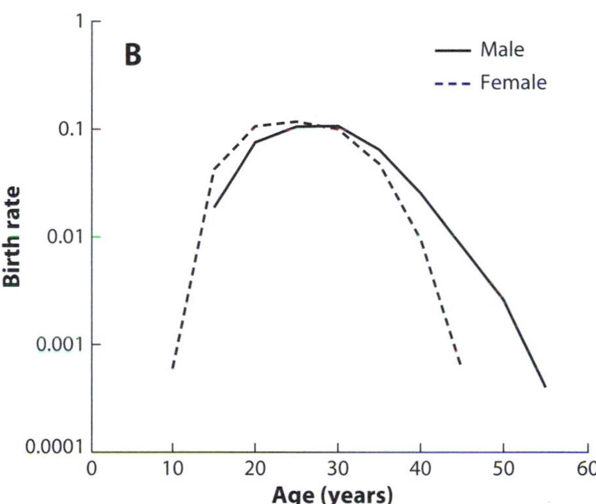

Figure 1. (A) Age-specific mortality in the United States, males and females, 2007 (see http://www.cdc.gov/nchs/fastats/deaths.html). The thin solid line represents a Gompertz curve fitted to adult mortality, $\mu_x = \alpha e^{\beta x}$, where μ_x is instantaneous mortality at age x, α is the intercept of the line, and β, the slope of the line, represents the rate of aging. (B) Age-specific birth rate in the United States, males and females, 2007 (see http://www.census.gov/compendia/statab/cats/births_deaths_marriages_divorces.html).

Aging in Model Systems

Most of what we know about the biology of aging comes from studies of lab-adapted organisms—yeast, nematode worms, fruit flies, mice, and rats. The general patterns of aging in both survival and reproduction seen in humans are much the same in these laboratory populations.

Much has been learned from these lab-adapted species about the way that both environmental and genetic factors can shape longevity. In many species, researchers have been able to enhance age-specific survival and late-age physiological function simply by restricting nutrient intake. In most cases, this effect appears correlated with the cessation of reproductive output. The fact that diet restriction can increase life span in species separated by a billion years of evolution has led some to argue that the ability to survive nutrient stress is an ancient adaptation. However, others have suggested that the response may be a side effect of a very modern adaptation to the lab environment. The response to nutrient restriction is much less clear in wild-caught animals.

In the early 1980s, Michael Rose and his colleagues found that artificial selection in fruit flies could dramatically extend mean life span, demonstrating that there was a large amount of genetic variation for longevity. In subsequent work, researchers have shown that changes in the structure or expression level of single genes in many pathways—most notably, genes associated with insulin/insulin-like growth factor signaling—can greatly enhance life expectancy. These results raise an evolutionary question: If altering a gene can increase life span in a worm or a fly, why has nature not already done the experiment? The answer likely follows from the observation that these life-extending mutations generally *reduce* fitness, such as by decreasing early-age fertility.

What can these lab-based studies reveal about patterns of aging in the wild? Until the 1990s, it was commonly assumed that aging did not, in fact, occur in natural populations. Biologists argued that wild animals would not live long enough to manifest signs of aging. It may be true that wild animals showing *obvious* signs of infirmity are rarely seen, as they are likely to have been killed by predators. But closer demographic analysis—in particular, measures of age-specific mortality and fertility—have shown clear signs of age-specific declines in fitness components in birds, mammals, and even wild insects.

While almost all species show signs of aging, some live notably longer than others. These differences have led to the recognition that there are certain ecological factors associated with long life span. For example, species that fly (bats and gliding mammals, as well as birds) tend to be long-lived relative to terrestrial species of similar mass. Similarly, life span is longer in species that are well armed against predators (like porcupines) and in species that live underground (naked mole rates, queen bees, and ants).

Some species of animals, such as hydra, appear to avoid aging altogether, with a constant risk of mortality throughout life. Other animals simply live what to humans seems like an extraordinarily long time, including Galápagos tortoises (almost 200 years), some species of rockfish (over 200 years), and even some small clams (over 400 years). Whether these long-lived species show signs of senescence is unknown. However, it is clear that aging can vary and is subject to evolution. This observation is borne out by explicit phylogenetic studies as well.

Sex Differences and Senescence

In the temperate rain forest of eastern Australia, males in the marsupial species *Antechinus stuartii* become sexually active during a brief one-week period in the middle of winter. By the end of this brief mating season, all the females in the population are pregnant, and half will survive to breed next year. All the males are dead. Such a dramatic difference between semelparous males and iteroparous females is unusual, but in almost all mammal species, males die sooner than females. In the final section, we consider possible explanations for this widespread pattern.

Aging in Humans

The patterns of aging in humans (and in nonhuman primates) mirror the general pattern seen in model systems, with age-related declines in survival and fertility and higher rates of mortality in males than in females. Human females differ in one important respect in that they show a prolonged period of postreproductive survival (menopause). We discuss the evolutionary explanations for this later (see also chapters VII.10 and VII.11). Perhaps the most striking pattern in human mortality is that which has taken place over the past 250 years: a 90 percent decline in childhood mortality and a 50 percent increase in life expectancy at birth. The percentage of people living past the ages of 60 or 70 is higher now than at any other time in human history. But to fully understand the evolutionary forces that have shaped human aging, it is necessary to know what human demography looked like prior to modern medicine and sanitation.

In nonindustrial indigenous societies, life expectancy at birth can be as low as 35 or 40 years of age. But for those individuals who make it through the riskiest period of infancy, there is a high probability of surviving to age 60 or 70 or beyond. While the immediate risk of dying is higher for adults in these populations than for those living in developed countries, the general pattern of age-specific mortality and fecundity is very similar. Baseline mortality rates have changed dramatically over time, but rates of aging appear to be quite constant among different human populations.

2. THEORIES FOR THE EVOLUTION OF AGING

The natural world offers up limitless examples of the power of natural selection. Aging differs among populations and organisms, and these differences appear to have been shaped by natural selection. But this observation then leaves us with a puzzle: Why does natural selection fail to evolve organisms that can function indefinitely, repairing any damage that arises along the way? In fact, this seems to be the case for hydra. So, why is this not the dominant pattern? Over the past century, three themes have emerged in attempts to understand aging from an evolutionary perspective—aging as adaptation, aging as maladaptation, and aging as constraint.

Aging as Adaptation

The German biologist August Weismann is best known for his *germplasm* theory of inheritance in animals, which recognizes that the germ line and somatic tissue represent two distinct lineages, with inheritance occurring only through the germ line. But Weismann is also credited with formulating the first evolutionary theory of aging in 1881 when he proposed that death in late age is adaptive and that natural selection actively favors the evolution of a death mechanism. This idea has become known as the *programmed death hypothesis*. Weismann emphasized that natural selection acts for the good of the population or species (not the individual) by removing useless individuals from the population and making room for the young.

Researchers have raised many objections to this theory. First, this mechanism requires the preexistence of senescence to explain selection for senescence, because it is predicated on the notion that the old are less fit than the young. Thus, this mechanism certainly cannot explain the *origin* of senescence. Second, the model gives group selection a central role in the evolution of aging but neglects individual-level selection, which should favor longer life span, all else being equal. While recent evolutionary theory considers a role for group selection in the evolution of many kinds of traits (social behaviors, for example), the conditions that allow group-level selection to overwhelm individual-level selection are very restrictive. The evolution of programmed death would require intense group benefits to overwhelm the great loss in individual fitness associated with early death.

Aging as Maladaptation

The first modern explanation for the evolution of aging came from cell biologist Peter Medawar's work in the 1940s and 1950s. Medawar was inspired by evolutionary biologist J.B.S. Haldane's 1941 study of Huntington's disease (HD). Haldane was struck by the fact that HD is a lethal disease caused by mutation at a single, dominant gene. Surely, natural selection should eliminate such a gene from the population, and yet it struck 1 in 18,000 people in England. Haldane explained its prevalence in terms of two forces, mutation and selection (see chapters III.3 and IV.2). While selection works to purge the genome of genes that increase mortality, germ-line mutations that lower survival are constantly being generated. Under mutation-selection balance, natural selection and mutation work in opposition, and an evolutionary equilibrium is met when the two forces are equal in magnitude. Thus, at such an equilibrium, lethal genes can persist in a population. However, the frequency of these alleles should be very low unless selection is very weak.

Consider the case of HD. Affected individuals typically begin to show symptoms around 40 years of age, by which time carriers could have already passed the lethal allele on to their offspring. For most individuals, the lethal consequences of the gene will be seen only after they have finished reproducing. As a consequence, there will be very little natural selection against the allele.

Whereas Haldane saw one particular lethal allele spreading because of its delayed effects, Medawar saw that the same was true for *all* alleles with late-acting deleterious effects. More generally, he recognized that *the later the age at which the effects of a deleterious allele occur, the weaker is the ability of natural selection to eliminate that allele from the population.*

If mutations can have age-specific effects on mortality or fertility (their effect is not manifested at all ages), then at mutation-selection equilibrium, the frequencies of age-specific mutations that decrease fitness will be smallest at early age and greatest at late age. This evolutionary mechanism, usually referred to as the *mutation accumulation,* or simply the *MA hypothesis*, was first suggested by Medawar in 1946. This model has been extremely influential because it is both a very general and very simple evolutionary model that assumes only that selection decreases with age (as William Hamilton, also the father of kin selection, was to show with his mathematical models 20 years later) and that at least some mutations have effects that are confined to specific ages. We can consider these the necessary and sufficient conditions for the evolution of aging.

Aging as Constraint

In 1957, George Williams built on Medawar's insight in one simple but critical respect. Consider a mutation that affects fitness at two different ages, first quite early in life, and then at some much later age. If the direction of the effects at both ages is the same (both beneficial or both deleterious), then natural selection will lead to the mutant allele's fixation or loss, respectively. Neither case is very interesting from the perspective of aging, but suppose that the effects of the mutation are in opposite direction at the two ages. Consider a mutation that decreases survival or fertility early in life but increases it later. Selection might favor its spread and contribute to extended longevity of old individuals, but because selection cares more about changes early in life, the early costs of the allele would tend to outweigh its later benefits. In contrast, mutations with benefits early in life and costs late in life are more likely to be favored by natural selection, and to spread through a population and lead to senescence. Importantly, this *antagonistic pleiotropy* (AP) model does not view aging as an adaptation, since it is not the aging per se that increases fitness, but as a constraint associated with other adaptations that evolve to maximize early life survival and/or fertility.

Twenty years after Williams published his AP model, Tom Kirkwood (1977) suggested a plausible model of AP based on how selection is expected optimize the allocation of limited resources across ages. Kirkwood's *disposable soma* model argued that there is substantial energy demand both from reproductive functions and from functions relating to the maintenance and repair of the individual (the soma). Given that energy is a limited resource, individuals cannot have it all. Selection favors the functions that lead more directly to increased fitness, which in this case, is the production of offspring at the cost of less energy invested in maintaining and repairing old soma. Thus, the equilibrium level of investment in the soma is one that fails to ward off the effects of senescence.

Genetic Variation and Aging Theory

Starting in the mid-1990s and motivated by quantitative genetic models that seemingly provided diagnostic tests of MA versus AP gene action, evolutionary biologists invested considerable effort in trying to describe the standing genetic variation for aging. These studies compared components of genetic variation and inbreeding depression at early and late life with the expectation that they should change if MA causes aging. Work was carried out primarily in fruit flies, but it also included studies of soil nematodes, seed beetles, and even hermaphroditic snails. Most of these studies found putative support for MA. However, Moorad and colleagues have argued recently that quantitative genetic tests of genetic variation and inbreeding depression are not truly diagnostic. From their perspective, the genetic results support the contention that senescence evolved, but they do not favor any one particular model.

Turning to AP, we expect negative genetic correlations between early-age and late-age fitness traits. Research in this area begins with Michael Rose's landmark experiments with fruit flies in the early 1980s. When Rose selectively bred from progressively older individuals, he observed not only a dramatic increase in life span but also a *reduction* in fecundity during early life, a result since replicated in other labs. While these results were consistent with predictions from AP, we now recognize that these results can also be interpreted as evidence for MA, because selection against early-acting fecundity mutations in these experiments is relaxed independently of late-life survival. Researchers also have found evidence for negative genetic correlations between traits at different ages in natural populations of swans and red deer, among other species. However, negative genetic correlations cannot distinguish between early-acting advantageous AP mutations that have caused aging to evolve and new, disadvantageous AP mutations that *reduce* aging. The latter kind of gene may have a negative, but transient, effect on life span before its removal from the population by natural selection.

The most compelling evidence for the AP mechanism comes from single genes that are known to extend life span and occur with high frequency (or are fixed in a population). The best example of this sort of gene is the polymorphic *TP53* gene in humans, which appears to create a genetic trade-off between risk of cancer and risk of aging. One variant of the gene (*R72*) reduces the risk of cancer but decreases longevity. The alternative *P72* allele has the opposite effect.

All these studies address patterns of standing genetic variation. However, existing genetic variation depends not only on how selection acts but also on the nature of the mutations that enter the genome. By examining the distributions of new mutations, it was hoped that a picture of the raw material for evolutionary change could be resolved without the confounding influences of natural selection. These studies revealed four interesting characteristics of new mutations unanticipated by evolutionary theory:

1. New mutations increase mortality more at early age than at late age.
2. Genetic variance caused by new mutation is highest at early ages.
3. New mutations increase mortality at multiple ages (i.e., these generate positive genetic correlations across ages).
4. Effects of new mutations become more age independent (i.e., pattern 3 becomes stronger) as more and more mutations accumulate.

These findings do not support AP (genetic correlations arising from new mutations are positive, not negative). They may also explain two of the aforementioned patterns that are not anticipated by the basic evolutionary models: the occasional reduction in genetic variance with age (2) and mortality deceleration (1).

There is no consensus regarding the primacy of one evolutionary mechanism of aging over the other. Results from quantitative genetic analyses, artificial selection experiments, and mutation accumulation studies are equivocal, which may reflect the fact that MA and AP models are highly idealized and that real aging genes have characteristics of both.

3. MENOPAUSE

Menopause is an unavoidable physiological transition that defines the end of an individual's reproductive capacity. In human females, it lasts between one and three years, occurring at 50 years of age, on average, and is presaged by about 20 years of declining fertility (reproductive senescence). Menopause is marked by a loss of ovarian function (including reduced endocrine production) leading to sterility and a suite of symptoms, including hot flashes, insomnia, mood swings, and increased risk of osteoporosis and coronary heart disease.

Human males also lose reproductive function over time, but men lack a similarly well-defined period of fertility loss. Accordingly, there is no upper limit to the age at which men can reproduce (apart from death); the record for extreme male reproduction appears to be 94 years.

While the existence of female menopause is firmly established in humans, little is known about how widespread menopause is in other animals. Some captive animals, such as rats and rhesus monkeys, appear to exhibit female menopause but do so at advanced ages that are believed to be largely unattainable in the wild. Natural populations of cetacean species, such as short-finned pilot whales and orcas, are observed to have large fractions of females that live beyond the age of reproductive cessation. As one can imagine, there are substantial challenges to collecting data to determine how fertility changes with age in natural populations. Nevertheless, this information is critical to comparative efforts trying to understand the forces of selection that cause menopause to evolve.

Why menopause? is one of the more fascinating (and open) questions of evolutionary demography. If menopause is an adaptation, then we are confronted with the challenge of explaining how natural selection can favor the evolution of a trait that ends one's ability to reproduce and transmit genes to the next generation. One obvious explanation is that menopause is simply a manifestation of reproductive senescence brought on by the age-related decline in the strength of selection. But there is a flaw in this reasoning. This argument assumes that historical human populations were characterized by such high rates of adult mortality that adult women rarely lived into their fifties, such that selection could not act with sufficient strength to avoid the accumulation of late-acting sterility mutations. We see menopause in our postdemographic transition world, this thinking goes, because life spans of modern humans are unnaturally long.

As noted earlier, indigenous human populations have very low life expectancy at birth, but for those who make it through those difficult early years, the probably of surviving well past the age of 50 is quite high. In this light, researchers think that menopause predates modern human societies.

Perhaps the greatest problem with the "menopause as modern artifact" argument is that human males tend to die before females and yet they do not undergo the abrupt reproductive cessation that is observed in females. As mentioned earlier, this pattern of higher male mortality is widespread among mammals. Moreover, evidence from natural populations of baboons and red deer (two species with pronounced elevated male mortality) indicates that the reproductive life spans of males are even more abbreviated than those of females owing to intense male competition for mates (old males do not compete well against their younger counterparts). If menopause were simply reproductive senescence, then we would expect reproductive cessation to be widespread in males and to occur earlier in life than female menopause. This pattern is not observed.

A more tenable hypothesis for female menopause holds that it is an adaptation. However, the fitness benefits of menopause may be realized by the descendants instead of being conferred to the female in menopause (the benefit to a descendant is indirect, while the cost to the female is direct).

There are two common adaptive hypotheses for explaining menopause—the *mother hypothesis* and the *grandmother hypothesis*. In these models, genes that promote menopause are associated with higher fitness in the children of mothers (or grandmothers) in menopause because these children live longer and/or reproduce more than children that are descended from females that do not undergo menopause. At their essence, these models imagine that there is conflict between the fitness interest of the maternal ancestor and her descendants (in this sense the evolution of menopause resembles the evolution of altruism). Menopause is expected to evolve at the age at which the benefits of menopause help the children twice as much as menopause hurts the mother (or helps grandchildren four times as much as it hurts the grandmother). Tests of these hypotheses are currently fertile ground in aging research. The first requirement, that the timing of menopause has a genetic basis, has been met. In humans, the age of menopause in the mother predicts to some degree the age of onset in daughters, suggesting a heritable basis to the age of onset of menopause. The second requirement, that menopause increases descendant fitness, is discussed later. Note that the validity of the mother hypothesis and the grandmother hypothesis are both subject to lively debate, with supporters and detractors on both sides.

The Mother Hypothesis

The mother hypothesis focuses on the observation that the risk of mortality from complications in pregnancy or delivery increases dramatically as women age. For example, rates of gestational diabetes increase fourfold in pregnant women over 40 compared with those under 30. Problems associated with hypertension may be as much as five times more common among older pregnant women. Complications from both of these, as well as an age-related increase in breech births, increase the frequency of operative births in modern societies. Obviously, this was not an option over human evolutionary timescales, and we can safely assume that many of these age-related problems would have resulted in the death of the mother. In the past, these deaths would have denied existing descendants the care or resource provisioning that a mother would otherwise have provided. In this light, menopause might have mitigated a mortality risk in older women. However, recent analyses suggest the increased risk of mortality for existing offspring due to losing a mother, other than for those who have not yet been weaned, is actually minimal. In traditional societies, allocare by a mother's relatives may have greatly reduced the cost of her death.

The Grandmother Hypothesis

The grandmother hypothesis argues that at some age the benefits to women of helping their daughters care for offspring outweigh the benefits of continuing to produce more sons and daughters. At this age, selection will favor a shift to care directed from grandmothers to grandchildren. Studies from preindustrial Western societies as well as from indigenous societies show that in at least some populations, older daughters have more children if their mothers survive and that the grandchildren of living women are larger and have lower mortality than those of dead women. See chapter VII.11 for further details.

Reproductive Competition

A third adaptive argument for the evolution of menopause, recently suggested by Michael Cant and Rufus Johnstone, notes that humans evolved in the context of small groups and that reproduction within these groups came at some expense to the social partners of the mother. Because dispersal among groups seems to have been dominated by young females, females tended to become more closely related to their group as they aged (older females were more likely than young females to have sons in the population). As a result, the social cost of reproduction by older females was borne by her relatives, but the relatives of young females were not affected by her reproduction. Cant and Johnstone reason that the relationship between degree of relatedness and female age caused kin- or group-level selection to favor the cessation of fertility *most* in the older females.

4. PRESSING QUESTIONS ON THE EVOLUTION OF AGING

Why Don't All Species Hit a "Wall of Death"?

In 1966, William Hamilton explained how the force of selection would change with age, given explicit values of age-specific fecundity and survival. Hamilton's model made clear qualitative predictions about the increase of mortality with age. Specifically, there should be three distinct phases. First, age-specific mortality is constant and low from birth until the first age in the population at which reproduction occurs. From then, mortality rates increase with age until that at last reproduction. Beyond this point, selection can no longer act, and postreproductive populations should hit a sudden "wall of death."

Aside from its occurrence in semelparous species like Pacific salmon and *Antechinus*, this wall of death is generally not seen. How can the fact that populations persist beyond the end of reproduction be explained, and that at least in iteroparous plant and animal species, a wall of death is not seen? One reason might be that the effects of individual mutations are less ephemeral than imagined by Medawar or Hamilton, and there are some genes that affect survival at both reproductive and postreproductive ages. As a result, genes that increase postreproductive mortality are not entirely hidden from the purifying effects of natural selection. This sort of gene action may have other, more subtle effects on human mortality, which we discuss in the next section. Another possibility is that even after the age at last reproduction, there will still be selection to survive and help one's offspring reproduce, as discussed earlier.

In many populations, mortality rates level off and can even decline at very late ages. However, scientists have yet to determine definitively whether this pattern of late-age mortality deceleration is a statistical artifact (due to variation in mortality rates among subcohorts within a population) or an evolutionary consequence of selection pressures (or lack thereof) at late ages.

Is There a Limit to Human Life Span?

In a now-famous wager, biologist Steven Austad and demographer S. Jay Olshansky placed a $500 million bet as to whether there will be at least one human who has lived to at least the age of 150, and with mind intact, by the year 2150. Austad and Olshansky don't expect to be around to collect on their wager, so they have each invested $150 and, with careful investment, anticipate that the funds will provide the heirs of the winner half a billion dollars. Austad argues that medical improvements and technological discoveries will lead to 150-year-olds by 2150. Conversely, based on his analysis of existing demographic data, Olshansky argues that humans are already close to the limit of their life span.

Is there an evolutionary response to this question? Consider Rose's experiments with fruit flies, in which selection for late age at reproduction doubled the life span of flies in just a few years. This reasoning can be pursued with the following thought experiment (albeit an extreme one). Imagine a human population in which only men and women over the age of 40 reproduced. All else being equal (that is, considering only the effects on life span and assuming everything else stayed the same), would this give rise to, let's say, in 100 generations, or about 4000 years, a significantly longer-lived population? In the case of Rose's fruit flies, it is thought that selection favored longer life span because only those individuals with genes that promote long life span had a chance to reproduce. In a modern human population, the vast majority of individuals survive to age 40, so there is likely to be little selection on genes affecting survival rates prior to age 40. Rather, a response to an increase in late-age fertility would be expected.

While this thought experiment is an exaggeration, ages at reproduction in industrialized countries are much later now than they were a century ago. If this trend continues, then many centuries from now, our descendants may find that self-imposed selection on fertility has slowed its rate of decline. It remains to be seen whether there are as-yet-unknown deleterious mutations with extreme late-age effects waiting to be discovered once individuals commonly exceed 110 years of age.

Why Do Males Die Earlier Than Females?

There is still no clear answer to why males and females should age differently. However, it is likely that this difference is associated with those traits that define "male" and "female." After all, the fundamental difference that defines two sexes is the size of the gamete that is produced, with female being the sex that produces larger (and typically fewer) gametes. Do sex roles explain why males and females age differently?

Starting with Charles Darwin and Alfred Russell Wallace, evolutionary biologists have been fascinated by the elaborate displays that animals use to attract the opposite sex. Typically, it is the males that bear these traits and/or that compete with one another for access to females. A great deal is now understood about the selective forces that have led to the evolution of secondary sexual traits and mating behavior (see chapters VII.4–VII.6). Less understood are the immediate (proximate) and long-term (evolutionary) costs of these traits.

In thinking about this problem, researchers have created a rich body of literature that falls at the intersection of two conceptually rich fields of study, uniting theories for the evolution of aging with theories of sexual selection and sexual conflict. A 2008 review by Russell Bonduriansky and his colleagues summarized several predictions that have emerged in the literature. These include suggestions that (1) both within and among species, males that invest more heavily in secondary sexual traits should age more quickly; (2) conflicts of interest between the sexes should lead to higher rates of aging, and if one sex has an advantage over the other in this conflict, it should age more slowly than the other; and (3) the influence of sexual selection on sex differences in aging should be mitigated by the degree of genetic correlation between the sexes.

Can We Understand Aging One Gene at a Time?

The evolution of any trait depends on the way in which that trait affects fitness and on the genetic architecture of the trait. The latter component includes the extent to which the focal trait is genetically correlated with other traits that affect fitness, the number of genes that influence the trait, and potential interactions among these genes. With the introduction of high-throughput genomic approaches to the study of aging, researchers are beginning to develop a better understanding of the genetic architecture of aging. At the simplest level, we know that a very large number of genes have the potential to influence life span, and we are just beginning to uncover the much deeper structure that relates the genotype to this particular phenotype. For example, consider the gene regulatory network, which illustrates how genes are connected to one another if they share a common regulatory mechanism. Research on mice has shown that the number of connections among these genes—the overall complexity of the gene regulatory network—declines with age. We are still at a relatively early stage in the study of gene networks and aging. Just how this complex architecture might have influenced, and possibly constrained, the evolution of aging is an important but unanswered question.
See also Chapter III.2.

FURTHER READING

Bonduriansky, R., A. Maklakov, F. Zajitschek, and R. Brooks. 2008. Sexual selection, sexual conflict, and the evolution of ageing and life span. Functional Ecology 22: 443–453.

Cant, M., and R. Johnstone. 2008. Reproductive conflict and the separation of reproductive generations in humans. Proceedings of the National Academy of Sciences USA 105: 5332–5336.

Gurven, M., and H. Kaplan. 2007. Longevity among hunter-gatherers: A cross-cultural examination. Population and Development Review 33: 321–365.

Hamilton, W. D. 1966. The moulding of senescence by natural selection. Journal of Theoretical Biology 12: 12–45. *Developed the key mathematical framework for modeling evolution in age-structured populations.*

Hawkes, K., J. F. O'Connell, N.G.B. Jones, H. Alvarez, and E. L. Charnov. 1998. Grandmothering, menopause, and the evolution of human life histories. Proceedings of the National Academy of Sciences USA 95: 1336–1339.

Kenyon, C. J. 2010. The genetics of ageing. Nature 464: 504–512.

Masoro, E. J. 2005. Overview of caloric restriction and ageing. Mechanisms of Ageing and Development 126: 913–922.

Medawar, P. B. 1946. Old age and natural death. Modern Quarterly 2: 30–49. *Developed the fundamental evolutionary argument that senescence arises owing to an age-related decline in the strength of selection.*

Promislow, D.E.L. 1991. Senescence in natural populations of mammals: A comparative study. Evolution 45: 1869–1887. *Established that senescence, measured as age-related increases in mortality rate, is common in natural populations of mammals.*

Rose, M. 1984. Laboratory evolution of postponed senescence in *Drosophila melanogaster*. Evolution 38: 1004–1010. *The first artificial selection experiment leading to increased longevity and tests of evolutionary theories of aging.*

Sear, R., and R. Mace. 2008. Who keeps children alive? A review of the effects of kin on child survival. Evolution and Human Behavior 29: 1–18.

Williams, G. C. 1957. Pleiotropy, natural selection, and the evolution of senescence. Evolution 11: 398–411. *Established the central role for genetic trade-offs in the evolution of senescence. One of the few papers in the field whose ideas have been adopted by molecular biologists.*

VIII

Evolution and Modern Society

Richard E. Lenski

Many people think of evolution as a fascinating topic, but one with little relevance to modern society. After all, most people first encounter the idea of evolution in museums, where they see the fossilized remnants of organisms that lived long ago. Later exposure to evolution may come in courses that present the basic theory along with evidence from the tree of life and the genetic code shared by all life on earth. For those enamored of wildlife, evolution might also be discussed in shows about exotic organisms in faraway lands, often showing nature "red in tooth and claw." So it is easy to overlook the fact that evolution is relevant to who we are, how we live, and the challenges we face.

The comic strip shown here comes from Garry Trudeau's *Doonesbury* series, and it reminds us that evolution is highly relevant to modern society. In fact, it touches on the four main themes in this final section of the volume: evolution and disease, evolution and technology, evolution and what it means to be human, and evolution in the public sphere. The conversation between the doctor and patient reminds us that despite our efforts to control nature, we remain targets for organisms that have evolved, and continue to evolve, to exploit our bodies for their own propagation. At the same time, the cartoon emphasizes that humans have acquired another mode of response—the use of technology—that allows us to combat diseases far more quickly (and with less suffering) than if we had to rely on a genetically determined evolutionary response. More subtly, the technology, institutions, and language (including humor) that make human societies what they are today all reflect a process of cultural evolution that emerged from, and now often overwhelms, its natural counterpart by virtue of the speed and flexibility of cultural systems. Finally, Trudeau jabs us with the needle of the conflict between evolutionary science and religion that dominates many discussions of evolution in the public sphere, at least in the United States, despite the overwhelming and continually growing body of evidence for evolution.

EVOLUTION AND DISEASE

Diseases are usually studied with a focus on proximate causes. For example, What organ is having problems, and how can it be repaired? What infectious agent (if any) caused the problems, and how can it be eliminated? But one can also ask questions about the evolutionary forces that shape disease, although this is rarely done, and that failure may leave important stones unturned (see chapter VIII.1). For example, why might one group of people be more susceptible to a particular disease than another group? Why are some diseases more prevalent now than in the past, despite improved sanitation and increased access to food? In some cases, we must study the history of our own species to understand the mismatch between our present circumstances and those of our ancestors. In other cases, the infectious agents are the center of attention and inquiry. Why do some pathogens and parasites make us very sick, or even kill us, when closely related microbes are harmless? For many years the "conventional wisdom" was that evolution would favor those parasites and pathogens that were harmless to their hosts. If parasites killed their hosts (so the thinking went), then they would drive their hosts and themselves to extinction. From this perspective, a highly virulent parasite was seen as a transient aberration, perhaps indicative of a pathogen that had recently jumped to a new host—one that would, over time, evolve to become less virulent if it did not burn out first. But this view has been challenged by more rigorous analyses. Even lethal infections do not usually drive their hosts extinct, and the optimum virulence, from the parasite's perspective, likely depends on the balance between within-host growth and between-host transmission (see chapter VIII.2).

The antibiotics that scientific researchers and pharmaceutical companies developed to treat bacterial infections were hailed as a triumph of technology over nature. Only a few decades ago, the most dangerous infections were largely conquered in developed countries. Schools of

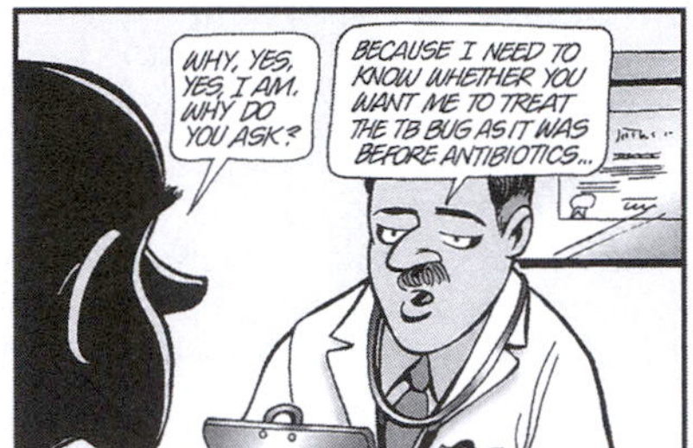

Figure 1. Cartoon strip reminding us that evolution is highly relevant to modern society. (DOONESBURY © 2005 G. B. Trudeau. Reprinted with permission of Universal Uclick. All rights reserved.)

public health shifted their attention from infectious diseases to other threats, while the public looked forward to a cure for the common cold (along with personal jet packs). This benign outlook was shaken, however, in the 1980s by the AIDS epidemic and the discovery that it was caused by a virus. And it continues to be shaken by reports of emerging and reemerging diseases that threaten denizens of even the wealthiest nations, from the SARS virus and bird flu (the H5N1 influenza virus) to multidrug-resistant strains of dangerous bacteria including *Mycobacterium tuberculosis* and *Staphylococcus aureus*. The reemergence of these bacterial pathogens reflects the evolution of varieties resistant to some or all of the antibiotics that were previously used to treat them (see chapter VIII.3). Thousands of tons of antibiotics are used each year, causing intense selection for bacteria that can survive and grow in their presence. As a consequence, pharmaceutical companies must spend vast sums to develop new antimicrobial compounds that will allow us, we hope, to keep up with fast-evolving microbes. Meanwhile, emerging diseases that are new to humankind typically derive from pathogens that infect other animals. The toolbox of molecular evolution and phylogenetic methods is now widely used to determine the source of zoonotic infections and to track a pathogen's transmission through the host population based on mutations that arise as the pathogen continues to evolve during an outbreak. And if these challenges were not enough, some terrorists have deployed pathogens. Investigators must identify the precise source of the microbes deployed in an attack, using evolutionary approaches similar to those used to track natural outbreaks. The "Amerithrax" case, in which spores of *Bacillus anthracis* (the bacterium that causes anthrax) were spread via the US Postal Service, demonstrated the power of new genome sequencing methods to discover tiny genetic differences among samples that may identify relevant sources (see chapter VIII.4).

EVOLUTION AND TECHNOLOGY

Some 10,000 years ago, humans began to harness the power of evolution by selectively breeding various plants and animals for food, clothing materials, and transportation. However, humans were not the first species to invent agriculture. That distinction belongs to ants, some of which began cultivating fungi as food sources millions of years ago (see chapter VIII.5). Some ants even tend other insects, such as aphids. Humans and other farming species change the environment of domesticated species by providing them with shelter, nutrients, and reproductive assistance. Selection in this protective environment reshapes the morphology, physiology, and behavior of the domesticated varieties. While we usually think of the farmer as controlling the domesticated species, their relationship is effectively a mutualism; the farmer, too, may evolve greater dependence on agriculture. For example, humans have evolved an unusual trait among mammals that enables many (but not all) adults to continue to produce lactase, an enzyme that allows milk sugar to be digested.

With the success of agriculture and other technologies, the human population has increased tremendously in size and, moreover, pushed into geographic areas that would otherwise be inhospitable to our species. As a result, humans have altered—by habitat destruction, introduction of nonnative species, and pollution—many environments to which other organisms have adapted and on which they depend, leading to the extinction of some species and threatening many others. Evolutionary biology contributes to conservation efforts in several ways (see chapter VIII.6). For example, phylogenetic analyses are used to quantify branch lengths on the tree of life and determine, in effect, how much unique evolutionary history would be lost by the extinction of one species or another. Given limited resources for conservation efforts, this information can be used to suggest where those resources will have the biggest impact. Also, the mathematical framework of population genetics, which underpins evolutionary biology, is used in the management of endangered populations. In particular, captive-breeding programs and even the physical structure of wildlife preserves can be designed to maximize the preservation of genetic diversity and minimize the effects of inbreeding depression.

Agriculture is the most familiar way in which humans use evolution for practical purposes, but it is not the only way. Over the past few decades, molecular biologists have developed systems that allow populations of molecules to evolve even outside the confines of living cells (see chapter VIII.7). For example, RNA molecules have been selected to perform new functions in vitro, such as binding to targets of interest. After a random library of sequences has been generated, sequences that have bound to the target are separated from those that have not. The former are then amplified (replicated) using biochemical methods that introduce new variants by mutation and recombination. This Darwinian process of replication, variation, and selection is repeated many times, allowing the opportunity for further improvement in binding to the target. Similar approaches allow the directed evolution of proteins, so that today RNA and protein molecules produced by directed evolution are used to treat certain diseases. Perhaps even more remarkably, computer scientists and engineers are harnessing evolution to write code and solve complex problems (see chapter VIII.8). They do so by implementing the processes of biological evolution—replication, variation, and selection—inside a computer. This approach has been used in biology (to test hypotheses that are difficult to study in natural systems) and engineering (where it facilitates the discovery of solutions to complex problems). For example, the design for an antenna used on some NASA satellites was generated not by a team of engineers but in a population of evolving programs (variant codes) that were selected based on the predicted functional properties of the objects encoded in their virtual genomes. Of course, it was then necessary to build the physical objects and test them to see whether they would perform as intended, which they did.

EVOLUTION AND WHAT IT MEANS TO BE HUMAN

How have humans become such an unusual and dominant species on earth? Agriculture, medicine, and other technological innovations are certainly key elements of this story. But the development of technologies depended on the prior emergence of other traits, including the language and culture that allow us to communicate among individuals and across generations, building on prior discoveries and allowing innovations to spread far more quickly than if they had to be hardwired into our genomes. In essence, culture provides a second, and extraordinarily powerful, way of evolving. Other organisms communicate with sounds, chemicals, or visual displays, but human language is unique in its compositional form, which allows an infinite variety of ways for ideas to be combined and expressed (see chapter VIII.9). It is not known when our ancestors evolved language, but studies comparing the morphology and genomes of humans with our relatives (in some cases even extinct species) are providing clues as to how and when the potential for speech evolved. Of course, understanding the capacity for speech does not explain *why* it evolved, but it seems likely that the evolution of human language and social behaviors were tightly connected. In any case, once language emerged, it underwent rapid diversification, with the patterns and processes that govern linguistic evolution suggesting many analogies to biological evolution. Even so, there are important differences between the evolution of culture and that of biology (see chapter VIII.10). Genes encode information about phenotypic solutions to problems that organisms encountered in the past, and that information is transmitted only from parents to offspring. By contrast, cultural information—knowledge, technology, ideas and preferences—can be disseminated broadly, and the information can accumulate within a single generation. Each of us obtains our genetic information in discrete bits from just two individuals (our parents), whereas we can obtain cultural information from many sources and blend that information in myriad ways. Moreover, cultural information offers the potential to plan for the future—for example, by anticipating and ameliorating changes caused by our actions—in ways that biological evolution does not.

Although technological innovation and other cultural influences dominate modern life, we are also the products of biological evolution. As a consequence, we differ from one another by virtue of our genealogical pedigrees as well as our cultural and biological environments, and these differences lead to debates about the contributions of "nature" and "nurture" to vigor, intelligence, and other attributes. The most obvious differences among individuals (in terms of being quickly perceived) are the varied colors of our skin and hair. Based on these superficial differences, people are then categorized into different races. Are these races biologically meaningful, or are they cultural constructs? The word *race* has a particular meaning in biology, corresponding to a genetically distinct lineage within a species (see chapter VIII.11). Indeed, one can quantify the amount and distribution of individual variation within and among populations of the same species. For example, chimpanzees were split into four races, or subspecies, based on morphological differences, whereas genetic analyses indicate there are only three chimpanzee races. Even so, these racial differences account for about 30 percent of the genetic variation in chimpanzees. By contrast, studies of human genetic diversity do not support the existence of biologically defined races. Although differences in ancestry can be detected and associated with geography, "races" account for only a few percent of the variation among humans; and even those small differences indicate a history of genetic admixture as populations spread around the world. Still, humans have adapted to their local environments, and the variation in pigmentation likely reflects compromises between avoiding damage to cells caused by UV radiation and producing vitamin D, which requires UV radiation. However, local adaptation is not equivalent to biological races owing to the few genes involved and the history of population admixture.

And what does the future hold for human evolution (see chapter VIII.12)? Has our biological evolution stopped now that cultural evolution provides technologies, including agriculture, that allow us to control nature and tools, such as medicines, that compensate for inherited differences in our vulnerability to disease? In fact, the opposite may sometimes be true. As agriculture spread across the globe, so, too, did the standing water that mosquitoes need to breed; and with mosquitoes came malaria, a disease that strongly favored individuals with genotypes that conferred resistance. Medicine, too, may promote evolution by relaxing constraints. The large heads that hold the brains that give our species the capacity to communicate and innovate pose a severe risk during childbirth, one that has caused the deaths of countless mothers and infants. Will our species evolve even larger brains and greater intelligence as cesarean births become more and more common? As Yogi Berra said, "It's tough to make predictions, especially about the future." But one prediction seems safe, and that is that the small genetic differences among populations will eventually disappear as a consequence of the increased migration that our technologies allow—that is, unless the colonization of other planets, or some catastrophe here on earth, produces new barriers to migration and gene flow.

EVOLUTION IN THE PUBLIC SPHERE

Evolutionary biology attracts substantial public attention for two reasons. First, many people find it fascinating to understand how humans and other species came into being. Indeed, that question has interested people since the dawn of history, with different cultures and religions providing diverse narratives about the origins of the world and its inhabitants. Second, evolutionary biology often attracts attention because its findings are inconsistent with those narratives. The resulting tension is complex, with many different positions held by scientists and nonscientists alike (see chapter VIII.13). Some people reject religions whose narratives are contradicted by established bodies of scientific evidence. Others emphasize the difference between evidence and faith, viewing them as separate domains of human understanding; these people may retain some religious beliefs and sensibilities while rejecting the literal interpretation of prescientific narratives. Yet other people have suggested that evolution may, in fact, illuminate theology by providing a deeper understanding of nature as it was created.

While some view evolution as an affront to their religion, evolutionary biologists often feel that their field of study is under attack by a unified opposition, especially in the United States, where opposition to teaching evolution is a hot-button issue that generates loud and emotional responses. However, the opposition is far from unified; instead, it is a coalition of creationists expressing views that are inconsistent not only with the scientific evidence but also with the beliefs of other coalition members (see chapter VIII.14). Efforts have been made to unify the creationist coalition and give it credibility by hiding these differences and obscuring their religious basis under the gloss of "intelligent design." However, several US court decisions have recognized the religious nature of the opposition to evolution, and they have disallowed such nonscientific ideas to be presented as scientific alternatives to evolution in public schools.

Fortunately, the exciting discoveries of evolutionary biology also receive considerable media attention (see chapter VIII.15). The field has many gifted writers and communicators among its practitioners, and the excitement draws many reporters and authors, some of whom have immersed themselves in the questions and evidence. Newspapers, books, and television once dominated the media coverage of evolution and usually exerted some quality control; but today websites, blogs, and tweets present a more complex and uneven terrain, even as they provide more opportunities than ever before for the public to explore and examine the discoveries and implications of evolutionary biology.

VIII.1

Evolutionary Medicine

Paul E. Turner

OUTLINE

1. Evolution and medicine
2. Pathogens
3. Defense mechanisms
4. Trade-offs in human traits
5. Mismatches to modernity
6. Implications of evolutionary medicine

Whereas evolutionary biology concerns the *ultimate* origins of trait variation within and among populations, human medicine concerns the *proximate* consequences of individual variation for manifestation of health versus disease. Although knowing *how* disease symptoms arise is essential for practicing medicine, understanding *why* these symptoms appear is additionally crucial. The merger of these two disciplines is *evolutionary medicine*, defined as the use of modern evolutionary methods and theory to better understand human health, with the prospect of improving disease treatment. The central question in evolutionary medicine is, Why has natural selection left our bodies vulnerable to disease? Many possible answers exist, and this chapter focuses on four major ones. First, human evolution is too slow to cope with the coevolutionary arms race involving microbial pathogens. Second, our evolved defense systems against these pathogens may, paradoxically, have harmful effects. Third, there are limitations and constraints on what selection can do, and disease often results from constraints and inevitable trade-offs. Last, human evolution is too slow to cope with novel environments especially of human making, leading to disease resulting from a mismatch to modernity.

GLOSSARY

Coevolutionary Arms Race. The sequence of mutual counteradaptations of two coevolving species, such as a parasite and its host.

Evolutionary Trade-off. A balancing between two traits that occurs when an increase in fitness (reproduction and survival) due to a change in one trait is opposed by a decrease in fitness due to a concomitant change in the second trait.

Germ Theory of Disease. The once-controversial idea that certain diseases are caused by invasion of the body by microbes; research by Louis Pasteur, Joseph Lister, and Robert Koch in the late nineteenth century led to widespread acceptance of the theory.

Hygiene Hypothesis. The idea that a lack of early childhood exposure to infectious agents, symbiotic microorganisms (e.g., gut flora), and parasites increases susceptibility to allergic diseases by modulating development of the immune system.

Mismatch to Modernity. Maladaptation produced by time lags, especially the inability of human adaptation to keep pace with rapid cultural change.

1. EVOLUTION AND MEDICINE

Although disease was not a major focus of Charles Darwin's 1859 book *On the Origin of Species*, it described how intimate species interactions select for special traits of host exploitation in parasites; examples included parasitic wasps whose eggs are laid and larvae develop within the bodies of specific host insects, and cuckoos that parasitize other birds by laying eggs in their nests and relying on them to rear their young. Shortly after the book's publication, Louis Pasteur's experiments in the 1860s improved our understanding of pathogenesis by confirming that diseases could originate from parasitic microbes, which motivated changes in medical hygiene that saved countless human lives. Given these parallel advances in evolution and medicine, it is ironic that the hybrid discipline of evolutionary medicine did not gain traction until the 1990s, when evolutionary biologist George Williams and physician Randolph Nesse were credited for popularizing the notion that the understanding of human illness may be informed through evolutionary thinking.

The historical separation between the two disciplines may be explained by medicine's proximate focus on restoring proper functioning of the human body, without considering how evolution shaped humans while causing us to remain susceptible to diseases and other health problems. This evolutionary consideration helps explain why humans become ill and provides knowledge that can be harnessed to suggest revised or novel methods of treatment. In particular, evolutionary thinking can create key insights and save lives when one is prescribing antibiotics, managing virulent diseases, administering vaccines, treating cancer, advising couples with difficulties in conceiving and carrying offspring, elucidating the origins of recent epidemics of obesity and autoimmune diseases, and answering questions relating to aging. The focus of evolutionary medicine is not to insist that an understanding of evolution is an alternative to current medical training; rather, it is to demonstrate that evolutionary biology is a useful basic science that poses new medical questions, raises possible answers or hypotheses, and thereby contributes to research while also improving medical practice.

The field of evolutionary medicine is relatively new, but the following sections provide examples of evolutionary insights that have informed the understanding of medical issues, and instances in which medical treatments have been modified owing to evolutionary thinking.

2. PATHOGENS

Parasitism is perhaps the most common lifestyle on the planet. Throughout history, humans have suffered extensive morbidity and mortality caused by infectious agents including parasites and pathogens. Some of these infections are evident in archaeological remains, such as ancient Egyptian hieroglyphics that depict humans afflicted with limb deformities characteristic of childhood infection by poliovirus (figure 1). Other evidence comes from mummified human remains showing, for example, the typical scarring due to skin lesions associated with the variola virus infection that causes smallpox. Diseases caused by ancient and novel pathogens constitute a substantial fraction of human mortality, perhaps on the order of 25 percent of all deaths per year. One goal of evolutionary medicine is to promote human health by improving therapies to combat infectious disease agents, based on the knowledge of how pathogens evolve in general, and especially in relation to selection pressures imposed by the human immune system and by current therapies.

The evolution of antibiotic resistance in bacterial pathogens is a popular and important example used to illustrate the process of evolution via natural selection. The age of antibiotics began with the discovery of penicillin in the late 1920s, followed by small-scale attempts to use it to treat patients in the 1930s, and its mass-production and application in the 1940s. Unfortunately, penicillin is no longer effective in combating many pathogens, and the pharmaceutical industry is lagging in the race to develop natural and artificial antibiotics that can replace penicillin and other once-effective antibiotics. The explanation for their failure is that the widespread therapeutic use of antibiotics (and popular use of antibiotics as prophylactics in agriculture) has led to a very large "experiment" in which bacteria have been exposed to vast amounts of antibiotics that select for bacterial mutants with greater resistance to the effects of antibiotics. The large population sizes and short generation times of bacteria virtually assure rapid evolution in response to this strong selection. This unfortunate outcome has also been aided by the

Figure 1. Ancient Egyptian carving showing a priest with a shriveled leg typical of a recovered case of paralytic poliomyelitis (polio). Polio is an infectious disease caused by polioviruses that can permanently damage parts of the nervous system. In some cases, as seen here, it can cause paralysis. This bas-relief was carved around 1500 BCE. © Photo Researchers.

ability of many bacteria to obtain resistance genes via horizontal gene transfer from conspecifics and other species, thereby speeding the process by which resistance proliferates within and among species of bacteria. The emergence of antibiotic resistance is thus easily understood from an evolutionary perspective, and if any puzzle remains, it is why certain bacterial pathogens have not yet evolved resistance against traditional antibiotics.

The realization that misuse and overuse of antibiotics has selected for bacterial resistance has significantly influenced the practice of dispensing these drugs. Physicians are now encouraged to prescribe an antibiotic only when necessary, avoiding dispensation of a drug in the case of viral illness and when trying to combat bacteria against which an antibiotic is widely known to be ineffective. Realizing this to be a global problem, in 2011 the World Health Organization issued an international call for concerted action to halt the spread of antimicrobial resistance, with specific recommendations on government policies in medicine and agriculture to help control the problem. One major contributing factor highlighted in this report was insufficient measures for preventing the spread of resistant bacteria in hospitals and surrounding communities. Evolution of resistance has led to highly resistant superbugs, such as the methicillin-resistant *Staphylococcus aureus* (MRSA) that flourish and spread through hospital-acquired infections. The costs and dangers associated with these superbugs have prompted the development of spatially based evolutionary models that predict the spread of resistant bacteria within hospitals, according to factors such as the number of beds per room and placement of immune-compromised patients. This type of evolutionary thinking is an attempt to help reduce the risk of acquiring a life-threatening pathogen as an unfortunate consequence of a routine hospital stay.

Taking a step back in time, bacteria and fungi had evolved antibiotic production and resistance millions of years before the appearance of the human species, and so it is unsurprising that we are challenged to combat a problem (bacterial resistance) that natural selection figured out long ago. A similar problem is the medical challenge of controlling pathogens that have evolved elaborate traits to uniquely exploit humans. Pathogens that cause human disease often have adaptations to promote their growth within specific target cells inside the body, especially when humans constitute the main host. One example is the protist *Plasmodium falciparum*, which causes malaria. During infection this pathogen enters red blood cells and radically alters the interior of these cells so that it can reproduce. Owing to this cellular intimacy and the huge disease burden in humans caused by malaria, people living in malaria-endemic regions such as sub-Saharan Africa have been strongly selected for resistance, sometimes at an extreme cost. For this reason, malarial parasites have literally shaped the genetic variation of human populations living alongside them. In particular, the high prevalence of malaria in some regions explains why alleles that lead to abnormal hemoglobin and enzyme deficiency have proliferated: they increase resistance to malaria in heterozygotes that carry one copy of the variant allele, despite the burdens of sickle-cell anemia and other disorders suffered by homozygotes that have two copies of the resistance allele. These evolutionary compromises imply that cost-free resistance to malaria parasites is difficult to achieve, and perhaps similar difficulties extend to the largely ineffective efforts to date at generating a malaria vaccine.

Emerging pathogens are defined as those that have recently entered a host population or that are causing disease at increased rates. These pathogens—and future pathogens that fit these criteria—pose a huge medical concern because humans can suffer high rates of mortality when a disease agent jumps into the human population from a nonhuman host, sometimes evolving to become quickly established as a human-specific pathogen. Phylogenetic evidence indicates that simian immunodeficiency viruses (SIV) entered human populations many times in the last century, leading to the emergence and epidemiological spread of human immunodeficiency viruses (HIV) I and II, and high rates of mortality in HIV-I infected individuals. Other emerging pathogens may have a long history of infecting humans, but circumstances may change such that the parasite causes increased mortality. This change may come about owing to pathogen exposure in human populations previously sheltered from the disease and therefore containing little immunity or resistance. For example, European colonization of the Americas brought along the virus that causes smallpox, which contributed to the devastation of Native American populations. Emerging pathogens can thus shape human demographics and alter the course of human history. One goal of evolutionary medicine is to better predict what types of pathogens can emerge in the future and to better prepare the medical community for the inevitable challenges that will be posed by the pathogens of tomorrow that burst on the scene as unexpectedly as the agents for MRSA, HIV, and SARS (severe acute respiratory syndrome) have done in recent memory. At present, the ability to accurately predict which pathogens will next emerge in humans is crude at best. But some patterns seem evident, as disease surveillance data and laboratory evolution studies have shown that generalist parasites that previously evolved to infect multiple hosts appear better able to successfully emerge on novel host species.

Already, evolutionary thinking has revolutionized how we combat influenza virus A, which causes seasonal epidemics in humans and perhaps 500,000 deaths in a typical year. Although flu vaccines can be highly effective in protecting humans against infection, they are strain specific and time-consuming to produce. Thus, a flu vaccine that targets a particular strain must be mass-produced well ahead of the season in which the virus actually circulates in humans. In 1997, Walter Fitch, Robin Bush, and colleagues used evolutionary phylogenetics to analyze influenza virus A isolates, demonstrating that successful lineages of the virus undergo *antigenic drift*; that is, specific changes in their hemagglutinin proteins promote escape from human immune surveillance, allowing these strains to give rise to viruses that dominate subsequent flu seasons. This approach is now generally used to predict which extant strains will most likely give rise to viruses that will dominate in the coming season, such that vaccines against them are mass-produced before they become problematic. However, this attempt to match a vaccine with a major flu variant sometimes fails, especially when a new strain emerges through *antigenic shift* (recombination between viruses that creates a variant novel to the immune system). Goals of evolutionary medicine include more accurate forecasting of seasonal flu variants and better predictions of the emergence of new strains capable of spurring global flu pandemics.

The manifestation of an infection can differ markedly, even between closely related pathogens. This harm caused to the host is called *virulence*. Another goal of evolutionary medicine is to understand how the virulence of a pathogen is shaped by natural selection. If a parasite must rely on direct transmission between hosts, activities of a parasite that increase its opportunity for infectious transmission (e.g., greater within-host production of infectious particles) may greatly weaken the fitness of its current host. These adverse effects may actually reduce the probability of successful transmission; an extreme example would be a parasite that kills its current host faster than a new host is encountered. Thus, if parasite transmission and virulence are tightly coupled, theory predicts that a parasite should evolve an intermediate level of virulence, which balances the costs of harming the current host and the benefits of moving to a new host.

However, virulence is expected to increase when infections move easily between host individuals, as occurs for pathogens spread by insect vectors, via water, or through needles shared by injection-drug users. These situations decouple the success of a pathogen from the viability of its current host, and selection thus pushes the pathogen population to become dominated by more rapidly reproducing variants that exploit the host, leaving it severely debilitated in the process. The situation is often predicted to be similar for a host that is coinfected by multiple pathogen variants, because the advantage goes to the genotype that more quickly replicates, destroying host tissues in the process. Interestingly, the evolution of increased virulence may sometimes be "short-sighted" from the pathogen's perspective, because selection across multiple generations within the host may cause a virulent genotype to be favored, even though this variant may not be the most successful one at being transmitted to a new host individual. This scenario is not the only possible outcome of evolved virulence under coinfection, however. Certain antagonistic interactions among coinfecting genotypes may cause the overall pathogen load to be reduced and lead to lowered virulence through time. Virulence evolution is an active area of research in evolutionary medicine, one in which mathematical models generate predictions that can be subjected to rigorous tests in the laboratory. Such efforts will help inform whether pathogens are expected to evolve increased or decreased virulence through time, such as in response to pressure exerted by the use of new vaccines that may inadvertently exacerbate some diseases by selecting for especially virulent pathogens that evolve to escape the vaccine-induced host response.

3. DEFENSE MECHANISMS

The history of the human species has involved coevolution with various parasites and pathogens, and their short generation times allow them to evolve much more rapidly than we do. To keep pace in this *arms race*, we rely on the protection afforded by mechanisms of innate and adaptive immunity, which in essence provide rapid evolution at the cellular and molecular levels during the lifetime of an individual.

However, a drawback to a complex immune system is that it offers myriad possibilities to malfunction. Diseases in which the immune system attacks the body's own cells, mistaking them for pathogens, are termed *autoimmune diseases*. Similarly, allergy and atopic diseases such as asthma and anaphylaxis represent situations in which the normal evolved processes of defense are inappropriately or excessively activated, thus causing disease.

Genetic differences among individuals affect the potential to develop autoimmune disease, and these differences are usually associated with certain alleles of the human leukocyte antigen (HLA) system, which is the major histocompatibility complex in humans. One example is HLA-B27, an allele that is associated with increased risk of ankylosing spondylitis, a chronic inflammatory arthritis that is more prevalent in men and can cause fusing of the vertebrae in the spine. Because not all males with HLA-B27 develop the disease, this association suggests the involvement of some additional environmental trigger. Some evidence hints that the disease may involve an aberrant response to *Klebsiella* bacteria, which usually are benign inhabitants of the human gut.

Yet other autoimmune diseases may be related to a lack of exposure to certain microbes (figure 2A). Historically, the human species has evolved in environments in which individuals were frequently exposed to severe, persistent infections; in particular, most people carried parasitic worms most of the time. In developed countries, by contrast, humans live in more hygienic environments, such that few people have worms, and few adults die from infection. However, in these same settings the prevalence of asthma, allergies, and chronic inflammatory disorders such as Crohn's disease have increased dramatically (figure 2B). The *hygiene hypothesis* offers a proposed explanation: the vertebrate adaptive immune system coevolved with commensal and pathogenic microbes, and appropriate exposure to this microbiota early in life may be essential for establishing appropriately regulated immunological pathways. In other words, the lack of exposure may cause improper activation of immune responses, which are then manifested as allergic or autoimmune diseases. Helminth worms are especially implicated as stimulators of proper immune function. This new understanding of the evolutionary role of helminths in modulating proper immunity has led to changes in medical treatment. Some doctors now successfully treat autoimmune diseases, such as Crohn's disease, by injecting worm eggs or proteins derived from worms, to activate an inhibitory arm of the immune system that is otherwise suppressed in many modern populations.

Evolutionary medicine studies are also addressing a separate hypothesis to account for escalating allergy rates. Modern environments may contain specific risk factors that were absent from older environments. For example, modern homes are much warmer and drier than the habitations in which our ancestors lived long ago, typically containing fitted carpets, warm bedding, and central heating. This environment is ideal for mites that inhabit house dust and use sloughed-off human skin for food. The feces of these mites are highly allergenic to susceptible people, and they may contribute to escalating allergy rates.

4. TRADE-OFFS IN HUMAN TRAITS

Perhaps the most important generalization that evolution offers to medicine is to consider the body as a collection of trade-offs. No trait is perfect, and when natural selection improves one trait, it might make another worse. If we produced less stomach acid, then we might be less prone to suffer from ulcers, but the trade-off would be greater vulnerability to gastrointestinal infections. This line of thinking is particularly important as we gain increased ability to alter our bodies. Many of us strive to sleep less because we can cram more activities into 24 hours, but natural selection has had the opportunity to adjust the length of our sleep for many thousands of years, and so the adverse health consequences of sleep deprivation may exceed any benefits. Contraceptive use by many women in postindustrial countries causes them to experience an average of 400 menses per lifetime, whereas women in cultures that experience historically typical birth intervals of two and a half years, and breast-feeding during that time, have about 100 menses per lifetime. This perennial

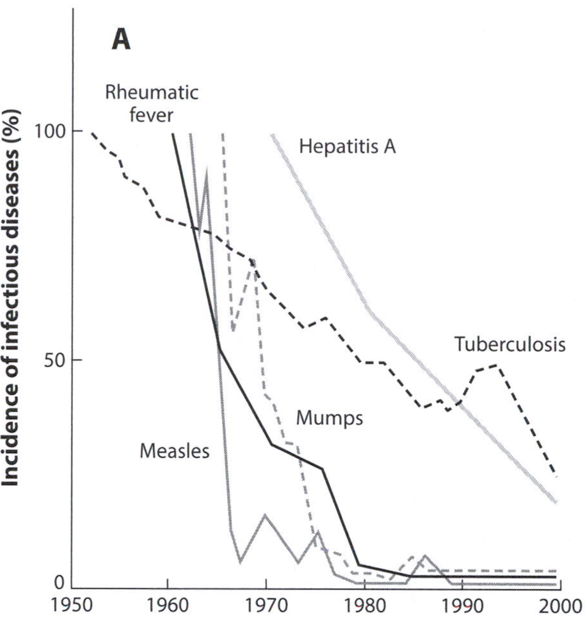

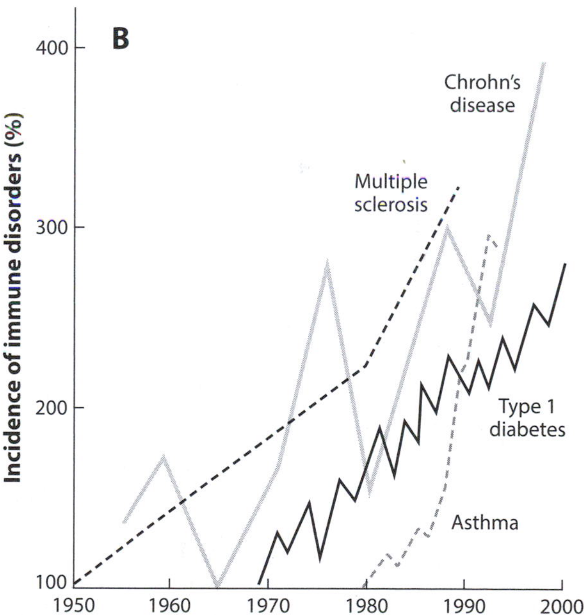

Figure 2. As the incidence of prototypical infectious diseases has fallen in high-income countries (A), the incidence of immune-related disorders has increased (B). (From J. F. Bach. 2002. New England Journal of Medicine 347: 911–920.)

cycling causes prolonged elevated levels of hormones, especially estrogen, which can fuel the growth of cells that cause the tumors often responsible for breast and ovarian cancers. This mechanism may place women at increased risk for certain cancers and might explain why breast cancer rates are often much higher for women in postindustrial societies. Contraceptives need not induce a monthly period, and so perhaps a new solution can be found that allows women to experience a level of estrogen sufficient to maintain bone strength and avoid osteoporosis while avoiding increased risk of cancer.

Every trait can be analyzed in terms of its associated costs and benefits. Trade-offs may limit the extent to which fitness can be improved because, in the limit, an improvement in one trait will compromise some other trait. These compromises can emerge as unpleasant and costly surprises when interventions are made in ignorance of their potential trade-offs. From that perspective, consider that the human life history strategy involves deploying resources toward peak reproductive performance but trading off that investment in reproduction with a reduced investment in reparative functions that might sustain health in the postreproductive years. The life span in most societies has markedly improved owing to large reductions in extrinsic mortality following improvements in public health and, to a lesser extent, in medical care. But one consequence of this longer postreproductive life is an increase in chronic noncommunicable diseases of middle to older age that may reflect insufficient investment in self-repair including, for example, of cellular damage from oxidative stress. The outcomes may include atherosclerosis, arthritis, osteoporosis, cognitive decline, neurodegeneration, and increased susceptibility to infection. An evolutionary perspective might help by developing predictive tests to identify those individuals most at risk of suffering the adverse effects of these trade-offs, which might then allow novel therapies based on a better understanding of this variability.

5. MISMATCHES TO MODERNITY

Evolutionary change is a slow process, especially for organisms with generations as long as our own species. By contrast, our social and physical environments have changed very rapidly owing to our cultural evolution and the impact of technologies on the environments in which we live. Thus, our species has largely evolved under circumstances very different from the present environments that constitute our living conditions. Some of the examples discussed earlier likely reflect the medical consequences of living in an environment that is evolutionarily novel and in which an individual's capacity to adapt physiologically is exceeded. This type of disconnect has been called a *mismatch to modernity*.

It is ironic that mismatches to modernity may result in diseases that stem from humans having access to excess resources. The modern lifestyle allows many people to experience an excess of energy intake relative to energy expenditure (figure 3), especially given technologies that limit the need for physical labor. The body's inability to cope with this metabolic mismatch may well explain the increasing global incidence of obesity, type 2 diabetes, stroke, and cardiovascular disease. Lifestyle changes can combat this problem, but such changes are difficult to achieve because the pleasures of the present often outweigh considerations of the distant future; after all, few of our ancestors lived such long lives that the distant future offset their present needs. Also, our evolutionary history of living with limited resources may have selected for a physiology that benefited from taking on extra nutrients when they were available. Aside from public health approaches that aim to adjust eating behaviors in adults, recent evidence suggests that early-life interventions may be useful. In particular, individuals of lower birth weight are at higher risk of becoming obese and developing metabolic disorders, including hypertension and diabetes. Early nutritional stress is a signal whose evolved response sets the individual on a special developmental course with a physiology effective for conserving energy but ill prepared for abundant food. This finding suggests that in utero cues about nutrition may affect the development of metabolic priorities later in life. In particular, the mismatch between early- and late-life nutritional status may be contributing to increasing obesity rates, rendering those born in poverty and growing into plenty especially vulnerable. A better understanding of the global epidemic of metabolic diseases will require consideration of these early cues and their effects on development, in addition to the interaction between our changing lifestyle and the physiological predilections we inherited from ancestors who lived under very different conditions.

Diseases caused by the mismatch between our bodies and environments can arise from deficiencies as well as excesses. For example, iodine was often routinely consumed by humans who ate seafood and plants grown in iodine-rich soils; however, endemic goiter and cretinism (a specific syndrome of mental retardation) occur when humans live in noncoastal environments deficient in iodine, such as certain mountainous regions. In developed countries, salt is iodized to reduce the risk of both goiter and cretinism. Similarly, the annals of sea exploration contain many stories of scurvy outbreaks caused by a lack of dietary vitamin C, for which fruits and some vegetables are the only natural source. Humans cannot synthesize vitamin C, unlike most other species, and so we are entirely dependent on an environmental source.

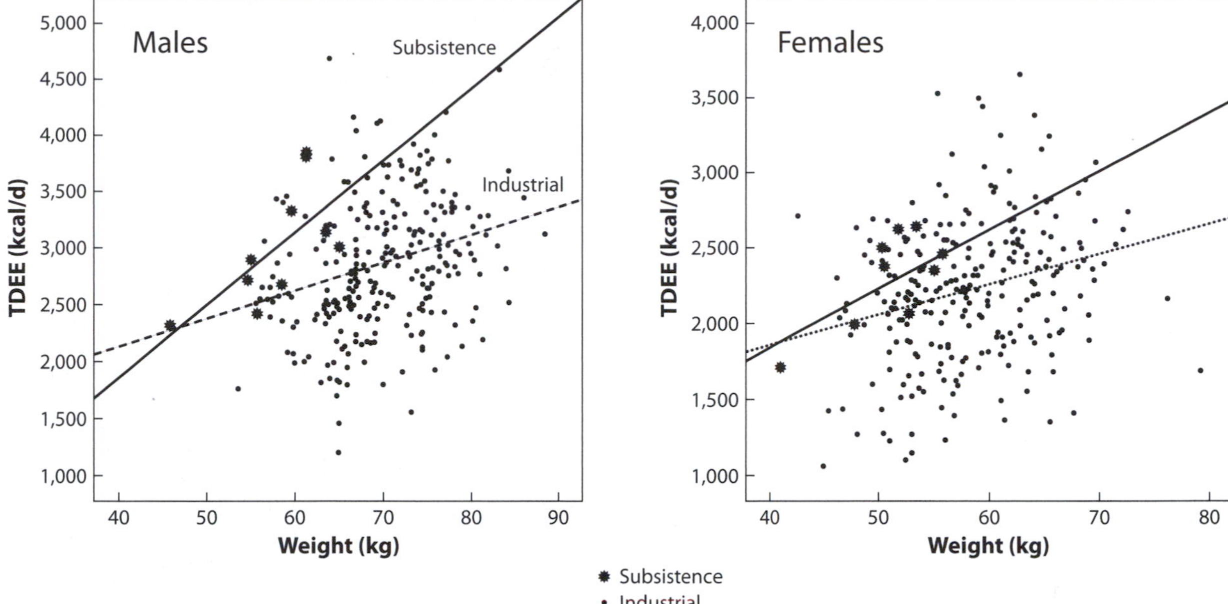

Figure 3. Total daily energy expenditure (TDEE; kcal/d) versus body weight (kg) for adult men and women of industrial and subsistence-level populations. Individuals in subsistence-level groups have systematically higher levels of energy expenditure at a given body weight. (Figure 20.2 from S. C. Stearns and J. C. Koella, eds. Evolution in Health and Disease. 2nd ed. Oxford: Oxford University Press. Reprinted with permission.)

Our ancestors evolved in environments where they had constant access to fruit, and the mutation that caused our inability to synthesize vitamin C was apparently neutral in that environment. But when a person is exposed to a novel environment—say, a long sea voyage—where a dietary source of vitamin C is lacking, scurvy is the result.

6. IMPLICATIONS OF EVOLUTIONARY MEDICINE

Evolutionary medicine should be of interest and use to practicing physicians as well as to biomedical researchers. We continue to be locked into a coevolutionary arms race with pathogens. These professions can therefore benefit from viewing infection from the pathogen's perspective and from anticipating how pathogens may respond evolutionarily to treatments such as antibiotics and vaccines. Evolutionary thinking should also help clinicians and researchers who deal with cancer, reproductive medicine, metabolic disorders, and autoimmune diseases consider how our bodies might be affected by their mismatch with modernity and the inability of human adaptation to keep pace with cultural changes. More generally, medical researchers gain from evolutionary thinking because it brings new perspectives both to posing new questions and addressing old questions in new ways, including tough biomedical problems for which new insights can save lives.

The promise of evolutionary medicine is that this cross disciplinary science will ultimately yield new or improved methods of treatment. However, concrete examples of modified treatments are fairly limited, because the field is relatively young, and it usually takes a long time for basic-research findings to be translated to changes in medical practice. Still, some changes in medical treatment have clearly resulted from an influx of evolutionary thinking. Examples include the aforementioned changes in the ways that antibiotics are used and prescribed, and the use of products derived from helminths to treat certain autoimmune diseases. Also, due to the link between HLA and pathogen resistance, evidence suggests that an HLA mismatch between parents may increase the risk of miscarriage because the embryo is attacked by the immune system; this understanding has informed the practice of sperm donation to reduce the potential risk of spontaneous abortion. Another example is the increasing recognition that the genetic variations in human populations can affect the response of individuals and groups to drug treatments and can increase the likelihood of developing disorders such as alcoholism; this understanding informs the new field of evolutionary pharmacogenomics and may eventually lead to personalized medicine for optimizing a patient's preventive and therapeutic care.

An important educational goal relevant to the growth of evolutionary medicine is to incorporate instruction in evolutionary biology into premedical and medical school education. Currently, few medical schools have evolutionary biologists on their faculties, and none teach evolutionary biology as a basic medical science. Physicians and medical researchers may learn something about evolution before medical school, but few have anywhere near the level of knowledge demanded for other basic sciences. An evolutionary view would correct mistaken notions of the body as a designed machine and would provide physicians with a better sense for the organism and what constitutes disease. A recent change in the Medical College Admissions Test (MCAT) requires competency in evolutionary biology and will probably improve understanding of evolutionary issues among clinicians more than any other potential measure. But in addition to changes in the MCAT, every undergraduate institution could offer courses in evolutionary medicine as part of its premedical curriculum. Specific renovations of the medical curriculum would also infuse evolutionary thinking into medicine. But it is likely that new national policies would be needed to educate physicians, allowing them to make full use of evolution as a crucial basic science for medicine.

Evolutionary medicine is a young field, and as more discoveries demonstrate how evolution-based thinking can improve the understanding and treatment of medical disorders, it should become clear that fundamental knowledge of evolution belongs in the medical toolkit.

See also chapter III.8, chapter III.11, chapter VI.7, chapter VIII.2, chapter VIII.3, chapter VIII.10, and chapter VIII.11.

FURTHER READING

Diamond, J. 1997. Guns, Germs, and Steel: The Fates of Human Societies. New York: W. W. Norton. *A Pulitzer Prize–winning book about Eurasian colonization of other societies, sometimes aided by introducing infectious diseases that decimated native populations.*

Ewald, P. W. 1980. Evolutionary biology and the treatment of signs and symptoms of infectious disease. Journal of Theoretical Biology 86: 169–176. *A perspective on the need to understand whether disease symptoms, such as fever and diarrhea, are beneficial for the host, the pathogen, or neither organism.*

Gluckman, P., A. Beedle, and M. Hanson. 2009. Principles of Evolutionary Medicine. New York: Oxford University Press. *A thorough account of evolutionary medicine geared to a medical audience that includes fundamentals of evolutionary biology.*

Merlo, L. M. F., J. W. Pepper, B. J. Reid, and C. C. Maley. 2006. Cancer as an evolutionary and ecological process. Nature Reviews Cancer 924–935. *A review that uses concepts in ecology and evolution to explain why cancer evolves, and why it is often difficult to treat.*

Stearns, S. C., and J. C. Koella, eds. 2008. Evolution in Health and Disease. 2nd ed. New York: Oxford University Press. *Research advances in evolutionary medicine, mostly from the perspective of evolutionary biologists.*

Trevathan, W. R., E. O. Smith, and J. McKenna. 2007. Evolutionary Medicine and Health: New Perspectives. New York: Oxford University Press. *Collected essays on new treatments of mostly noninfectious diseases that have been informed by the field of evolutionary medicine.*

Turner, P. E., N. M. Morales, B. W. Alto, and S. K. Remold. 2010. Role of evolved host breadth in the initial emergence of an RNA virus. Evolution 64: 3273–3286. *A laboratory test of the general hypothesis that pathogens that historically evolved on multiple hosts should be more likely to infect a novel host.*

Williams, G. C. 1957. Pleiotropy, natural selection, and the evolution of senescence. Evolution 11: 398–411. *A classic paper examining the evolution of aging from an evolutionary perspective.*

Williams, G. C., and R. M. Nesse. 1991. The dawn of Darwinian medicine. Quarterly Review of Biology 66: 1–22. *A review of the power of evolutionary medicine to improve understanding of a wide variety of medical disorders.*

VIII.2

Evolution of Parasite Virulence

Dieter Ebert

OUTLINE

1. Defining virulence
2. The phase model of virulence
3. The trade-off model
4. Vertically transmitted parasites
5. How well do optimality models predict virulence?

Diseases caused by parasite or pathogen infections impair normal functioning in organisms. These impairments can include very diverse symptoms leading to the organism's morbidity and mortality. Studies of the evolution of the virulence of infectious diseases strive to understand the expression of these symptoms as the result of the evolutionary process. This approach is primarily focused on the evolution of parasites (here used to include pathogens), but it may also consider the coevolution of hosts and parasites.

Until about 30 years ago, it was widely accepted that the harmful symptoms of infectious diseases were the side effects of poorly adapted parasites, and that over time virulence would therefore generally evolve toward avirulence. The underlying logic was that a well-adapted parasite should not harm its host, as doing so would deplete its own resource. This view was challenged in the 1980s by Roy Anderson, Robert May, and Paul Ewald, who argued that virulence is often a necessary consequence when parasites exploit their hosts, and depending on the specific conditions, the optimal level of virulence (i.e., the level that maximizes parasite fitness) may range from low to high virulence. Virulence is now understood as a trait whose evolution can be analyzed within the general framework of evolutionary biology, thus considering the roles of history, chance, and natural selection.

The evolution of virulence is not only of academic interest; its conceptual framework also has implications in various applied fields, such as human and veterinary medicine and agriculture. In particular, public health workers can benefit from considering population-biological aspects of the evolution of virulence.

GLOSSARY

Basic Reproductive Number, R_0. The average number of secondary infections resulting from a primary infection in a population of susceptible hosts.

Horizontal Transmission. The passing of a parasite between two hosts that are not related in direct line, for example, vector-borne transmission and sexual transmission.

Kin Selection. Selection on organisms that share genes by common descent. Fitness is said to be inclusive because it considers the fitness of related individuals.

***Myxoma* Virus.** A DNA virus from the family Poxviridae. This virus causes myxomatosis and has been used to control rabbit pests in Australia and Europe. Its virulence evolved rapidly after its release.

Parasite. An infective agent transmitted among hosts and growing or replicating within hosts. Pathogens are included in this definition.

Reproductive Manipulator. Parasites with maternal (vertical) transmission that manipulate host reproduction to increase their representation in the offspring of the next host generation, for example, by killing or feminizing male offspring.

Vertical Transmission. The passing of a parasite from a parent (usually the mother) to an offspring.

Virulence. Parasite-induced morbidity and mortality of a host. More precise definitions, such as parasite-induced host death rate or host fecundity reduction, are used for specific situations.

1. DEFINING VIRULENCE

Virulence is simply defined here as the parasite-induced morbidity and mortality of the host. This definition includes any fitness effect the parasite has on the host, whether that effect is an incidental by-product of the infection or an adaptive trait for the parasite. This definition does not, however, explain how virulence evolves, because it does not specify the link between parasite fitness and virulence. Most attempts to understand the evolution of virulence are based on models of parasite evolution and therefore consider only aspects of virulence that are important for parasite fitness. For an exclusively horizontally transmitted parasite, host mortality is important, as the parasite might die with the host, whereas reduced host fertility or sexual attractiveness—which are important for host fitness—may be of little concern for the parasite. For example, congenital rubella syndrome is a serious illness of babies born to mothers who became infected with the rubella virus during the first trimester of pregnancy. It is a big concern for the human host but is unlikely to have an impact on the evolution of the rubella virus. This picture changes, however, for parasites that rely on vertical transmission. For them, host fecundity becomes an important part of the virulence definition. Thus, understanding the evolution of virulence requires detailed knowledge about the host-parasite system in question. In the application of models of optimal virulence to actual diseases, the key factors to consider are the mode of parasite transmission and the trade-offs among parasite fitness components.

Virulence may be further categorized by distinguishing between effects directly beneficial for the parasite (e.g., when host death is required for parasite transmission, as in many parasites of invertebrates) and those effects that are costly for both the host and the parasite (e.g., host death when infections are transmitted among living hosts). Most models of the evolution of virulence consider the latter scenario. Examples of effects with a direct benefit for the parasite include parasitic castration (which liberates resources for parasite reproduction), impaired host mobility (which may increase access to the host by vectors that transmit the parasite further), and parasite-induced changes in host behavior (which may increase chances of transmission).

2. THE PHASE MODEL OF VIRULENCE

The once-dominant view that only novel diseases are highly virulent and that well-adapted parasites are less virulent is based on two related ideas: first, that virulence is a result of a new interaction between a host and a parasite, and second, that virulence changes as the parasite adapts to the new host. A refined version of this two-stage scenario tries to combine the different aspects of virulence evolution and expression into a unified framework (Ebert and Bull 2008). This model distinguishes three successive phases of disease evolution. Phase 1 is the first contact of a parasite with a host that it usually does not infect, often called *accidental infection*. In phase 2 the parasite has only recently established itself in a new host species, at which point the parasite's virulence is not yet the result of adaptive evolution. In phase 3 the parasite has evolved for some time in a particular host and has adapted to the specific conditions of this host population.

The phase model emphasizes that not all aspects of parasite virulence can be understood as a result of adaptive evolution. Consider, for example, the following diseases of humans. The highly virulent Ebola virus does not circulate long enough in humans to evolve an optimal level of virulence. Transmission chains are short, so it remains in phase 1. The human immune deficiency virus (HIV) entered the human population several decades ago. It is clearly able to persist in humans, but it may not have had time to reach an optimal level of virulence. Thus, it can be considered to be in phase 2. Much older human diseases, such as tuberculosis and leprosy, are likely to be in phase 3.

Phase 1: Accidental Infections

Many terrifying human diseases are caused by accidental infections including, for example, Ebola, bird flu, SARS, anthrax, Lyme disease, Legionnaires' disease, West Nile virus, and echinococcosis. For some of these diseases, untreated infections can approach 100 percent mortality rates. Transmission chains are short, and epidemics do not persist, so the parasites have little opportunity to adapt to their new hosts. At first glance, this case might seem to support the view that novel diseases are highly virulent. However, the most virulent accidental infections are most likely to be recognized, whereas avirulent accidental infections will often go unnoticed, thus producing a strong sampling bias. In fact, avirulent accidental infections may outnumber virulent infections, as experimental transspecies infection trials suggest. Although a few novel infections are highly virulent, most are avirulent. Thus, highly virulent novel diseases are the exception, not the rule, although they may have profound impact on humans and natural populations.

Accidental infections have played and continue to play an important role in applied fields such as medicine and agriculture. In medicine, vaccine development has taken advantage of transspecies host shifts, both by using parasites of closely related species (Jenner's pox vaccine derived from accidental infections with cowpox) and by evolving attenuated parasite lines as vaccines. In agriculture, highly virulent novel infections have been used in pest control, as, for example, to control the very dense populations of European rabbits in Australia and the

United Kingdom with the *Myxoma* virus, which is derived from a virus of a related host species.

Phase 2: Evolution of Virulence Following Successful Invasion

After a parasite infects a novel host species, its transmission success will determine whether it will persist and spread in this species. Initial spread is typically epidemic, and the level of virulence of the nonadapted parasite is unlikely to be close to optimum. Selection typically shapes a parasite's life history and virulence during the epidemic and especially over the following period, as it becomes endemic. Every endemic infectious disease has made the successful transition from phase 1 to 2 at some stage in its evolutionary history, but the number of observed cases is very low, despite many more examples of accidental infections (more than 1000 human diseases, so-called zoonoses). Phase 2 can also help us understand the emergence and spread of novel variants (mutants) of a parasite in an established host-parasite association. These mutants may have extreme effects and may spread rapidly, causing an epidemic and being unaffected by existing host defenses. If they occur frequently, a parasite population may never reach an optimum virulence and thus will remain in phase 2 (Bull and Ebert 2008). In this case, the fitness consequences of having suboptimal virulence are likely to be minor relative to the fitness gains the mutants achieve by evading host immunity and other defenses.

Studying parasites in phase 2 illuminates how the tempo and mode of virulence evolution proceeds in real time. The best example is the *Myxoma* virus. A highly virulent strain of this virus was introduced to Australia to control European rabbits, an invasive species that was causing extensive damage. Within a few years the average virulence of the parasite had changed drastically, attaining a new level far below the virulence of the original strain (Fenner and Kerr 1994). Apparently, the original virulence had been far above the presumed optimum, although not so high that it had prevented the spread of the virus. The *Myxoma* example leaves many questions open, as multiple factors changed simultaneously (e.g., host density, host genetic composition). But it does demonstrate that virulence can evolve rapidly, that virulence evolution does not necessarily lead to complete avirulence, and that virulence coevolves with the host (Fenner and Kerr 1994).

Laboratory experiments that follow parasite evolution after a change of environmental conditions or a host shift are in effect creating phase 2 situations. This situation is particularly true for serial passage experiments, which played an important role in vaccine development, and also for the understanding of virulence evolution. In these experiments parasites are passaged in novel hosts with transmission controlled by the researcher (in some cases, the parasite might go extinct without this intervention). The evolution of the parasite is then monitored over many generations. Evolution typically results in a strong increase in virulence in the host in which the passages take place. At the same time, the parasites display reduced virulence in their former hosts. This attenuation of virulence makes these parasites good candidates for vaccines, as the immune response of the former host still recognizes the parasite, but its low virulence prevents disease or limits its severity. The increased virulence in the novel host is linked to an increase in the parasite's within-host multiplication rate. This increase is likely driven by within-host competition among parasite variants, with the most prolific variants having the highest chance to be transmitted during the next passage.

Phase 3: The Evolution of Optimal Virulence

Parasites that persist for some time in a host are expected to evolve an optimal level of virulence, that is, the level of virulence at which parasite fitness is maximal. It is widely thought that this optimum is characterized by trade-offs among different parasite fitness components. Thus, a key difference between phase 2 and phase 3 is that in phase 2, the parasite is not yet subject to the constraint imposed by the trade-offs. The *Myxoma* virus example was the first to show the existence of a trade-off, in this case between the rate at which rabbits clear the infection (host-induced parasite death) and the rate at which the parasite kills the host (and itself). Highly virulent *Myxoma* strains kill the host too quickly, while strains with low virulence are quickly cleared by the host's immune response. The optimal balance between these two parasite fitness components was shown by Anderson and May (1982) to maximize the parasite's spread in the host population. A mathematical model, latter called the *trade-off model*, and the observed data agreed well.

3. THE TRADE-OFF MODEL

When considering models of optimal virulence, it is essential to define virulence precisely. In the mathematical model first applied to analyze the *Myxoma* data, virulence is defined as the parasite-induced host death rate. Other detrimental effects the parasite may have on the host, such as reduced mating success and fecundity, are not considered because they are not fitness components of the parasite. This simplification is acceptable under the assumption that parasite-induced host death rate is positively correlated with the various expressions of morbidity. These positive correlations also justify the use of surrogate measures of virulence in empirical studies, such as host fatigue, sensitivity to stress, fever, or other physiological parameters. However, it is important to keep in mind that correlations between parasite-induced host death rate and other disease-related traits may be weak or may have a negative sign. For example, many parasites of invertebrates castrate their hosts, which eliminates host fecundity but allows the parasite to keep its host alive for much longer than a noncastrator would be able to. In the following discussion of optimal virulence, parasite-induced host mortality is used as a definition for virulence.

The first and still most used model for the evolution of virulence is the trade-off model (Anderson and May 1982). This model is a powerful starting point for analyzing the evolution of virulence, although its simplicity implies certain assumptions and makes it vulnerable to various criticisms. The model stresses the importance of trade-offs between parasite-induced host death and other parasite fitness components. In its simplest form, parasite fitness is estimated as the number of secondary infections that result from a primary infection, R_0:

$$R_0 = \frac{\beta N}{\alpha + r + \mu},$$

where β is the transmission rate in a susceptible host population of density N, α is virulence (parasite-induced host death rate), r is the rate at which hosts clear infections, and μ is the host background (parasite-independent) mortality rate. Thus, α, r, and μ are all components of the parasite's overall death rate, while βN indicates the production of new infections. Without trade-offs, parasite evolution would maximize β and minimize the total parasite death rate ($\alpha + r + \mu$), thus driving virulence to zero. A positive correlation between β and α, or a negative one between α and r (as in the case of the *Myxoma* example), constrains the evolution of virulence. If the increase in β is leveling off relative to the increase in α (i.e., the relationship is asymptotic), intermediate levels of virulence are predicted. This simple trade-off model relies entirely on the assumption that between-host transmission is the quantity maximized by parasite evolution. Although 30 years old, the hypothesis that between-host transmission is crucial for the evolution of virulence is still not strongly supported by empirical evidence. A critical test will be to show that transmission success has the assumed hump-shaped relationship with virulence. Empirical studies that tested this prediction were conducted with unicellular parasites that infect mice, butterflies, and water fleas (*Daphnia*).

Some of the major limitations of the simple trade-off model are that it ignores the role of multiple infections and within-host evolution of the parasites, kin structure of parasites, and host genetic variation. The model also ignores changes over time in the density of susceptible hosts, including those changes that may occur as the parasite itself evolves.

Multiple Infections, Inclusive Fitness, and Virulence

In the 1990s the trade-off model was extended to include within-host competition. Within-host competition describes scenarios in which different variants of parasites compete within hosts, which strongly influences both their likelihood of transmission to the next host and the level of virulence expressed in the multiply infected hosts. Parasite variants that replicate more quickly are assumed to be superior in within-host competition, even if this reduces host survival and thus shortens the period for transmission to take place. Because higher replication rate is associated with higher virulence, average virulence is expected to be higher in populations with frequent multiple infections than in those with single infections, all else being equal. As a consequence, more virulent parasite variants may dominate although they do not maximize the R_0 as predicted by the simple trade-off model. However, more complex models incorporating additional factors that might occur with multiple infections have been developed that predict the opposite result under specific circumstances. In particular, cooperation among parasites to exploit the host or parasite strategies to exploit one another may lead to the evolution of lower virulence.

Empirical tests on the role of multiple infections have generally supported the key assumption and prediction of the basic multiple infection model. Thus, higher within-host multiplication rates have been found to be associated with both superior competitive ability and higher virulence. Interestingly, it has also been found that some parasites increase their virulence facultatively when the host has been multiply infected, which implies that the parasites are able to sense, either directly or indirectly, the presence of competing parasite variants.

Multiple infection virulence models make a number of other predictions as well. Ecological and demographic factors such as higher host density, longer parasite survival outside the host, longer host life span (e.g., low, parasite-independent mortality), and less spatial structure of the host population (more mixing) are predicted to lead to increased virulence, because these factors should increase the incidence of infections by multiple parasite variants. These predictions have been explained using inclusive fitness theory, which considers the kin structure of the parasite population. Inclusive fitness is a crucial factor in the evolution of virulence, because within-host competition cannot be considered independently from the genetic relatedness of the competing parasites (Frank 1996). More closely related parasites have more common reproductive interests and thus gain less from competition. Using inclusive fitness theory, one can generalize findings that link ecological features with virulence and place the role of transmission mode into a unified context. For example, in well-mixed host populations, multiple infections result mostly from unrelated parasites, which maximizes competition and thus also virulence. In contrast, in viscose host populations, multiple infections more often arise through infections from the same parasite lineage. In these latter cases, lower levels of virulence are expected to evolve because of kin selection in the parasite population. As will be discussed further, kin selection also plays an important role in the evolution of virulence of vertically transmitted parasites.

Host Genetic Variation

Empirical studies have shown that virulence is not only the product of parasite evolution but also the result of the coevolutionary interaction between hosts and the parasites. Some models of virulence evolution have incorporated certain aspects of host genetic variation, but the complexity of the interactions that influence the expression of virulence makes it unlikely that general predictions can be made. Nevertheless, models and empirical data agree that parasite virulence should be lower in genetically diverse host populations. This effect is based on the observation that greater host diversity slows the spread of parasites (parasites spread faster in host monocultures), and this diversity thus reduces multiple infections. Trade-offs in the performance of parasite genotypes across different host genotypes or host species suggest that parasite fitness is compromised (and virulence reduced) in diverse host populations. Finally, host evolution that counters the effects of infections may lead to a reduction in parasite virulence. Incorporating the various effects of host genetic variation into models of disease virulence remains one of the big future challenges for understanding the evolution of virulence.

4. VERTICALLY TRANSMITTED PARASITES

The discussion thus far has focused on those parasites that engage exclusively in *horizontal transmission*, that is, transmission among hosts unrelated in direct line. Many parasites, however, are entirely or partially *vertically transmitted*, usually from mothers to offspring. In this case, the parasite's fitness depends on host reproductive success, and this dependency must be included in models that seek to explain the evolution of virulence. It is best to begin this discussion by focusing on those parasites that are transmitted exclusively from mothers to offspring. Such parasites must either evolve to manipulate host reproduction to their own benefit or evolve to complete avirulence. This necessity can readily be understood when one realizes that a parasite that is transmitted only vertically, and that harms its host's reproductive success without promoting its own transmission, will go extinct as parasite-free hosts outcompete those that are infected.

Some vertically transmitted parasites, such as the bacterium *Wolbachia* and some microsporidians, have evolved mechanisms to manipulate host reproduction in ways that increase their presence in future generations, despite being virulent. These mechanisms include the killing of host sons, feminizing males to become functional females, and inducing cytoplasmic incompatibility. In some of these cases, only kin selection can explain the observed virulence, because the individual parasites that produce the virulent effects (e.g., killing sons, inducing cytoplasmic incompatibility), in essence, commit suicide because they preclude their own propagation. Other individual parasites are genetically identical or nearly so, and they benefit from these behaviors. Parasites that manipulate host reproduction are very rarely transmitted horizontally.

Parasites with exclusive maternal transmission that do not manipulate their hosts must evolve avirulence, because the parasite's reproductive success is perfectly linked to the reproductive success of its host. In these cases, there is no conflict of interest between host and parasite. Some of the best empirical support for virulence theory has come from the experimental evolution of parasites in diverse systems being propagated under contrasting conditions of vertical and horizontal transmission, with the result that the parasites evolve toward avirulence when they are restricted to exclusive vertical transmission. However, complete avirulence no longer fulfills the definition of a parasite, so that the resulting entities might better be described as symbionts.

Many parasites are transmitted both vertically and horizontally. In some cases, the population structure imposes a strict trade-off between the two modes of transmission, such that more frequent horizontal transmission leads to less vertical transmission. Observations and experiments under such conditions have shown that the more horizontal transmission takes place, the more virulent the parasite will be. However, this prediction does not generally hold when the host population is well mixed. In such cases, vertical and horizontal transmission may be positively correlated, making general predictions difficult. Thus, without detailed knowledge about transmission trade-offs and host population structure, the finding that a particular parasite is partially vertically transmitted cannot be used as a predictor of low virulence.

5. HOW WELL DO OPTIMALITY MODELS PREDICT VIRULENCE?

Models of the evolution of virulence are deceptive in their simplicity and power to make testable predictions. Unfortunately, the evidence in support of these models is still rather thin, although many of the key assumptions have been supported experimentally in several systems (e.g., trade-offs have been observed, and multiply infected hosts typically suffer from higher virulence). For example, the field lacks compelling examples of evolutionary changes in virulence associated with trade-offs. Even the frequently cited *Myxoma*–rabbit case leaves open some alternative explanations. Those experimental studies that produced the clearest outcomes also had to use rather extreme conditions (e.g., 100 percent vertical versus 100 percent horizontal transmission) that may limit the ability to generalize their findings. Comparative studies employing data from many different host-parasite systems do not explain much of the variation in virulence, suggesting that other effects may overrule general patterns. For example, the type of host tissue affected by the parasite seems to explain more of the variation in virulence than do either trade-offs or transmission dynamics. Next-generation models that take into account host diversity and multidimensional trade-offs might be able to make more accurate predictions, although they are likely also to suffer in terms of generality.

Despite these limitations, models of optimal virulence are important because they provide a starting point for formulating testable predictions. In those cases in which a single environmental factor changes, it may indeed be possible to predict the associated change in virulence. However, changes in one factor often go hand in hand with changes in other factors, which may exert opposing selection on virulence. For example, the trade-off model predicts that an increase in host life span favors low virulence, but this same change may increase the frequency of multiple infections, which favors higher virulence. Therefore, careful evaluations of epidemiological circumstances and host demographic conditions are likely to be necessary before predictions can be made with confidence. It is currently not possible to make simple and robust recommendations for pest management that will favor the evolution of less virulent parasites. Proposals to manage virulence by changing environmental conditions must therefore be evaluated with appropriate care before they are put into practice.

See also chapter III.4, chapter VI.7, chapter VIII.1, and chapter VIII.4.

FURTHER READING

Alizon, S., A. Hurford, N. Mideo, and M. Van Baalen. 2009. Virulence evolution and the trade-off hypothesis: History, current state of affairs and the future. Journal of Evolutionary Biology 22: 245–259. *A conceptual review of the trade-off model, mainly from a theoretical perspective.*

Anderson, R. M., and R. M. May. 1982. Coevolution of hosts and parasites. Parasitology 85 (pt. 2): 411–426. *The classic reference on this topic and still readable.*

Bull, J. J., and D. Ebert. 2008. Invasion thresholds and the evolution of nonequilibrium virulence. Evolutionary Applications 1: 172–182.

Ebert, D., and J. J. Bull. 2008. The evolution of virulence. In S. C. Stearns and J. K. Koella, eds., Evolution in Health and Disease. 2nd ed. Oxford: Oxford University Press. *The phase model for virulence is here worked out in more detail.*

Fenner, F., and P. J. Kerr. 1994. Evolution of poxviruses, including the coevolution of virus and host in myxomatosis. In S. S. Morse, ed., The Evolutionary Biology of Viruses. New York: Raven. *Much about the biological background for the rabbit–myxoma case.*

Frank, S. A. 1996. Models of parasite virulence. Quarterly Review of Biology 71: 37–78.

VIII.3

Evolution of Antibiotic Resistance

Dan I. Andersson

OUTLINE

1. A medical miracle—and how to ruin it
2. Origins of antibiotics and antibiotic-resistance mechanisms
3. Transmission of resistant bacteria
4. Persistence and reversibility of resistance
5. Can resistance evolution be slowed or even stopped?
6. Will antibiotics become a footnote to medical history?

Antibiotics have revolutionized human and veterinary medicine, and over the last 60 years they have made it possible to treat efficiently most types of bacterial infections. Unfortunately, the extensive use—and frequent misuse—of antibiotics has resulted in the rapid evolution and spread of bacteria that are resistant to antibiotics. Arguably, the global use of antibiotics is one of the largest evolution experiments performed by humans, and the frightening consequence is that we are now at the brink of a postantibiotic era in which antibiotics have lost their miraculous power. This problem originates from the strong selection imposed by the extensive use of antibiotics and the resulting enrichment of resistance mutations and horizontally acquired resistance genes. Together these factors have generated high-level antibiotic resistance in the majority of significant human and veterinary pathogens. Several forces act to stabilize resistance in a population once it becomes established, and resistant bacteria may thus persist for a long time even after use of an antibiotic has been reduced. The development of new classes of antibiotics, coupled with more prudent use of antibiotics, will be required to maintain antibiotics as efficient agents for treating bacterial infections.

GLOSSARY

Antibiotic. An antibacterial compound that may have a natural or synthetic origin.

Biological Fitness Cost. The effect that a resistance mechanism has on bacterial fitness (including growth, persistence, and survival within and outside hosts) in the absence of antibiotic.

Compensatory Evolution. Reduction or elimination of the fitness cost associated with a mutation that has a deleterious side effect (e.g., a resistance mutation) by additional genetic changes (compensatory mutations).

Conjugation. Transfer of genetic material between bacterial cells mediated by direct contact between two cells.

Coselection. Process whereby a nonselected gene indirectly increases in frequency by virtue of its genetic linkage (within a genetic element or a bacterial clone) with a directly selected gene; sometimes also called *genetic hitchhiking*.

Horizontal Gene Transfer. A process in which a recipient organism receives and incorporates genetic material from a donor organism without being the offspring of the donor; sometimes also called *lateral gene transfer*.

Minimum Inhibitory Concentration (MIC). The lowest concentration of an antimicrobial drug that inhibits the growth of a bacterial population.

Nosocomial Infection. Infection contracted during treatment in a hospital or other healthcare facility.

Plasmid. A DNA molecule that is separate from and can replicate independently of the chromosomes in bacteria.

Resistome. A neologism that refers to the set of resistance genes and precursors to resistance genes that are present in all pathogenic and nonpathogenic bacterial species combined.

Transduction. Injection of foreign DNA into bacterial cells by a bacteriophage (i.e., a virus that infects bacteria).

Transformation. Uptake of exogenous DNA into a cell through the cell envelope.

1. A MEDICAL MIRACLE—AND HOW TO RUIN IT

Antibiotics represent one of the most important medical advances in modern times, and since their introduction over 60 years ago they have saved countless lives. Today we often take antibiotics for granted in the developed parts of the world, but we have to go back only to our grandparents to find a generation for which common infections such as pneumonia, meningitis, blood poisoning, and intestinal infections were potentially deadly. Charles Fletcher, a young physician who was involved in early clinical trials of penicillin in the 1940s, describes vividly how the introduction of antibiotics changed modern medicine:

> It is difficult to convey the excitement of actually witnessing the amazing power of penicillin over infections for which there had previously been no effective treatment .... I did glimpse the disappearance of the chambers of horrors which seems to be the best way to describe those old septic wards ... and could see that we should never again have to fear the streptococcus or the more deadly staphylococcus.

In addition to being widely used for the treatment of many common community and *nosocomial* (hospital-acquired) infections, antibiotics are also an essential component in the treatment and prevention of infections associated with advanced medical practices including chemotherapy of cancers, organ transplantation, implantation and replacement of medical devices and prostheses, neonatal care, and invasive surgery.

Unfortunately, the utility of antibiotics is deteriorating at an alarming rate, and the reason for this change is easily understood in the context of Darwinian adaptive evolution. To put it simply, bacteria adapt genetically to the presence of antibiotics by acquiring various types of resistance mechanisms that prevent antibiotics from performing their inhibitory function. These resistance mechanisms allow the bacteria to grow and reproduce in the presence of antibiotics, and evolution thereby nullifies their efficacy in treating infections. The widespread use—and often the overuse—of antibiotics on a global scale (estimated currently to be several hundred thousand tons per year) for human medicine, veterinary medicine, and agriculture is the main reason for the selection and spread of resistance among both human and animal bacterial pathogens. So, although the introduction of antibiotics is often viewed as one of humankind's greatest achievements, we are now at risk of destroying that achievement. At the very least, we are paying a high price for the increased resistance. The overuse of antibiotics reflects several factors, including poor knowledge among prescribers and patients, profits for physicians and pharmacists from the prescription and sale of antibiotics, aggressive marketing from pharmaceutical companies, and the lack of regulations and guidelines for when and how antibiotics should be properly used. Studies have shown correlations between the amount of antibiotics used and the prevalence of resistance at several levels (e.g., country, hospital), as would be expected if antibiotics select for increased bacterial resistance.

As a society, we are paying a high price for the increased levels of bacterial resistance to antibiotics: resistant bacteria limit our ability to efficiently treat bacterial infections, and they also increase the risk of complications and even death. In addition, antibiotic resistance imposes a large economic burden on the healthcare system owing to increased treatment costs as well as the costs of identifying and developing new, alternative compounds. Worldwide there are areas where bacterial infections have become untreatable as the result of antibiotic resistance, and with the recent spread of gram-negative bacteria that produce multidrug-resistant extended-spectrum beta-lactamase (ESBL), the problem is becoming even more acute. This trend toward increasing resistance, combined with diminished research and development of new antibiotics, has led to a dismal situation in which we may face a postantibiotic era.

2. ORIGINS OF ANTIBIOTICS AND ANTIBIOTIC-RESISTANCE MECHANISMS

Antibiotics are compounds that inhibit (bacteriostatic drugs) or kill (bactericidal drugs) bacteria by a specific interaction with some target in the bacterial cell. Some purists limit the definition of antibiotics to only those substances produced by a microorganism, but today all natural, semisynthetic (i.e., a combination of natural and synthetic precursors), and synthetic compounds with antibacterial activity are generally classified as antibiotics. The target for an antibiotic can be an essential enzyme or cellular process such as protein synthesis, cell wall biosynthesis, transcription, or DNA replication. Most medically relevant and industrially produced antibiotics originate in nature and are synthesized by a variety of species, mainly soil-dwelling bacteria (in particular the genus *Streptomyces* among the phylum Actinobacteria) and fungi. The benefits of antibiotics for microbial producers is a matter of debate; antibiotics might be used as ecological weapons to inhibit competitors, but they might have a more benevolent function as signals for cell-to-cell communication in microbial communities. In any case, the synthesis and release of antibiotic compounds in nature means that many bacteria (both producers and bystanders) have long histories of exposure to antibiotics, and as a consequence, many have evolved various resistance mechanisms. These mechanisms likely evolved to protect against self-destruction (in antibiotic producers), to defend against antibiotics produced by other species, to modulate intermicrobe communication, or to perform metabolic functions unrelated to antibiotics. This vast pool of resistance genes, known as the *resistome*, has the potential to be transferred within and between species, and to confer resistance to any antibiotic that might be used against human and animal pathogens. In fact, many of the resistance problems generated by the use of antibiotics since the 1940s are a consequence of the acquisition of preexisting resistance determinants by pathogens via *horizontal gene transfer* (HGT). Transfer mechanisms include conjugative transfer of plasmids and conjugative transposons, transduction via bacteriophages (viruses that infect bacteria), and transformation of naked DNA taken up from the environment by some species. Of these mechanisms, conjugative transfers are the most common mode of acquiring resistance, whereas bacteriophage transfers appear to be rare. Apart from HGT, resistance can also arise by mutations (including point mutations as well as rearrangements and gene amplification) in native resident genes.

For resistance to become a problem, the acquired or mutated resistance genes must be phenotypically expressed in clinically relevant human and animal pathogens. The evolutionary pathways leading to these outcomes are often complex and often not well understood, especially in the case of resistance acquired by HGT. Even when a potential donor of a resistance gene has been identified by genome sequence data (e.g., the CTX-M type of ESBL resistance was likely acquired from *Kluyvera* strains in the environment), several conditions must be fulfilled for resistance to emerge in the case of HGT: (1) a resistance mechanism must be present in a donor bacterium; (2) there must be a genetic mechanism for HGT; (3) there must be an ecological opportunity for transfer between the donor and recipient cells (e.g., in the case of conjugation direct contact is needed); (4) the transferred gene must be stably inherited, adequately expressed, and confer a resistant phenotype in the recipient; and (5) there must be strong enough selection—typically, a sufficient level of antibiotic—to favor the resistant recipient organisms, even though resistance may impose a fitness cost (as discussed in section 4). In the case of resistance that occurs by mutation in resident genes, the process is simpler and requires only a suitable resistance mutation and sufficient selection to favor the resistant mutants. Despite the relative ease by which bacteria can become resistant by mutations, HGT is the predominant route for generation of antibiotic resistance in most human and animal pathogens. The likely explanation is that pathogenic bacteria can acquire high-level resistance to a given antibiotic—and indeed, simultaneous resistance to several antibiotics—by means of a *single* transfer event from the relatively accessible pool of resistance genes in the microbial community's resistome. Of special relevance here are genetic elements called *integrons* that can capture and express arrays of resistance genes; when integrons are transferred on a plasmid, they can convert the recipient strain from being antibiotic susceptible to multidrug resistant. In contrast, mutation-based resistance often produces lower-level resistance and may require several mutational steps to produce high-level resistance, thus requiring a longer evolutionary path to achieve a clinically resistant phenotype. A notable exception, however, is *Mycobacterium tuberculosis,* in which all known resistance mechanisms are the result of mutation rather than HGT, and single mutations sometimes produce high-level resistance (e.g., resistance to streptomycin and rifampicin arises from mutations in ribosomal protein S12 and RNA polymerase subunit β, respectively). Mycobacteria have conjugative plasmids and transducing bacteriophages, and it is unclear why HGT is not associated with resistance evolution in this bacterium.

Horizontally acquired genes and mutations in native genes confer resistance to bacteria by a variety of different mechanisms that either protect the normal cellular target from exposure to the antibiotic or alter the target's structure to prevent the drug from binding. (1) The antibiotic may be enzymatically inactivated by hydrolysis (e.g., resistance to beta-lactam antibiotics conferred by beta-lactamases) or modification (e.g., acetylation, phosphorylation, or adenylylation of aminoglycosides). (2) Uptake of the antibiotic may be reduced by changes in the cell wall (e.g., mutations that confer low-level β-lactam resistance by altering channels called *porins*). (3) Bacteria may express efflux systems that actively pump the antibiotic out of the cell (e.g., efflux pumps that confer β-lactam or aminoglycoside resistance). (4) The target molecule may be modified such that antibiotic is prevented from binding (e.g., mutations in ribosomal proteins or rRNA that inhibit binding of aminoglycosides). (5) Resistance may result from a bypass mechanism whereby the need for the inhibited target is relieved by provision of an alternative target or pathway (e.g., resistance to peptide deformylase inhibitors by inactivation of formyl transferase). Mechanisms 1, 4, and 5 often provide high-level resistance, whereas mechanisms 2 and 3 are typically associated with lower-level resistance.

3. TRANSMISSION OF RESISTANT BACTERIA

Once resistance has evolved in a bacterial pathogen, the extent to which it becomes a medical problem depends on how rapidly and extensively the resistant type is transmitted from its place of origin into the human or animal population and the rate at which it is disseminated among the hosts. Transmission rates of resistant bacteria depend on many factors including host density, patterns of host travel and migration, various hygienic factors (e.g., in hospitals and during food preparation), host immunity (e.g., vaccination), and the intrinsic transmissibility of the resistant pathogens. In principle, humans can influence all these except the last factor.

4. PERSISTENCE AND REVERSIBILITY OF RESISTANCE

Whether antibiotic resistance will persist in a bacterial population after it emerges depends in general on the relative strength of several selective forces. The most obvious of these is the direct advantage to resistant bacteria caused by exposure to concentrations of antibiotics that are lethal or inhibitory to sensitive strains. An opposing force, however, is the fitness cost of resistance, that is, any effect of the resistance mechanism that reduces the ability of the pathogen to grow, persist, or spread in the host population. Such costs will impede the rise of resistant bacteria, and these costs will also affect the likelihood that resistance can be reversed or otherwise eliminated. While these fitness costs offer hope that resistance can be controlled, other forces discussed later can stabilize resistance in a bacterial population, even when the antibiotic is absent or at a low concentration.

Sub-MIC Selection

Selection clearly favors resistant strains when antibiotic concentrations are above the *minimum inhibitory concentration* (MIC) of the susceptible bacteria, but it remains unclear whether levels far below the MIC can also select for resistance. Direct measurements of antibiotic levels in organs and tissues of treated patients and in various natural environments indicate that bacteria are frequently exposed to sub-MIC levels of drugs. In theory, such low antibiotic levels may select for resistance if susceptible bacteria grow even slightly more slowly than resistant strains. Antibiotics can be introduced into the environment in the urine from treated humans and animals, as well as when antibiotics are used in agriculture (for example, on fruit trees). On average, roughly half of all antibiotics (the proportion varies with antibiotic class) consumed by humans and animals enter the sewage system or other environments via urine, and the amount of antibiotics released into the environment is presently on the order of several hundred thousand tons per year. Recent results have shown that the resulting environmental antibiotic concentrations may be important for both the emergence of new resistant strains and the enrichment of existing resistant strains. In competition experiments between susceptible and resistant strains, selection for resistant bacteria can occur at antibiotic concentrations even less than 1 percent of the MIC of the susceptible bacteria; similar antibiotic concentrations can be found in many natural environments.

These findings are important from a public-policy perspective because they suggest that antibiotic releases into the environment through human, veterinary, and agricultural applications contribute significantly to the emergence and persistence of antibiotic resistance. In particular, they indicate the potential benefits of reducing anthropogenically generated antibiotic pollution and avoiding treatment regimens that involve prolonged periods with low levels of antibiotics.

Coselection of Resistance Genes

A resistance gene located on a transmissible element or in a bacterial clone can increase in frequency in a population as a consequence of its genetic linkage to another resistance gene that is under selection. Such linkage and the resulting coselection is a common feature of resistance determinants that have been acquired by HGT, including plasmids, transposons, and integrons, and it can occur more generally in any multidrug-resistant clones. As a consequence, the frequency of resistance to a particular antibiotic can remain stable or even increase in environments where the antibiotic is not currently being used. The linked gene that sustains the unselected resistance gene can be any gene that increases the fitness of the bacterial strain, including another antibiotic resistance gene, a gene that encodes resistance to some heavy metal or disinfectant, or a gene that encodes some virulence-associated function.

Coselection is an important contributor to the long-term maintenance of resistance in bacterial populations, and it may explain why a reduction in the use of a particular antibiotic often has little or no effect on the frequency of resistant bacteria. For example, a recent study reported that an 85 percent reduction in the use of trimethoprim over a two-year period had only a very small effect on trimethoprim resistance in *Escherichia coli*. Similarly minor effects on resistance were recorded in other studies following reductions in use of sulfonamides, macrolides, and penicillin.

Cost-Free Resistance

The fitness cost of any particular resistance gene can vary depending on environmental conditions and the genetic background in which it occurs. For example, some resistance mutations impose no cost under standard laboratory conditions but have large costs in laboratory animals, and vice versa. Also, the cost of a resistance function often depends on the particular bacterial strain in which it occurs as a consequence of epistatic interactions between the resistance gene and other genes. Interestingly, some resistance genes do not appear to have any measurable fitness cost, at least in the environments and strains in which they have been tested. Of course, there may be other conditions under which these resistance genes do impose some costs, and measuring fitness costs under natural conditions is very difficult and rarely done; even in the laboratory, where genetically marked strains can be directly competed, it is difficult to measure fitness differences below about 0.3 percent per generation. It is also difficult to know what costs are relevant with respect to the persistence of an antibiotic resistance gene in a bacterial population. In principle, a fitness cost as small as 0.001 percent per generation would mean that a resistance gene would eventually be purged from the population by natural selection if the use of an antibiotic was stopped, although it might require many decades or even centuries given such small fitness costs.

Fitness-Enhancing Resistance

Although antibiotic resistance often has a fitness cost, in some cases it can actually be advantageous, even in drug-free environments. Interesting examples of such fitness-increasing effects of resistance functions have recently been demonstrated in two bacteria for the fluoroquinolone class of antibiotics. In *E. coli*, fluoroquinolone resistance commonly evolves by a multistep process involving mutations that alter efflux mechanisms and the proteins targeted by the drug. Each resistance mutation alone provides only a small increase in the MIC, so that clinically relevant levels of resistance require the accumulation of several mutations. In laboratory selection experiments, the accumulation of several resistance mutations typically led to reduced fitness in the absence of the antibiotic, but in a few cases an increase in resistance produced higher fitness. *Campylobacter jejuni* provides an interesting example of the background dependence of fitness effects associated with fluoroquinolone resistance. A single mutation in the gene encoding DNA gyrase enhanced the fitness of the resistant strain in a chicken-infection model, but when that same mutation was transferred into a different strain of *C. jejuni*, it imposed a fitness cost. The disturbing implication of these findings is that selection for improved growth may sometimes favor increased resistance even in the absence of drug selection.

Compensatory Evolution That Reduces Fitness Costs

Resistance to an antibiotic may impose a fitness cost because it disrupts the balanced growth of a bacterial cell that has been finely tuned to express genes and functions at levels that maximize fitness. A common process that stabilizes resistance is thus *compensatory evolution*, in which selection favors mutations that restore the cell's balance and thereby reduce or eliminate the cost of resistance, often without any significant loss of resistance. Indeed, several laboratory and animal studies have demonstrated the evolution of mutations that restore fitness and, as a consequence, stabilize resistant populations. Whether adaptation in the absence of antibiotic occurs by compensatory mutations or by reversion (loss of resistance) will depend on several factors particular to any given case, including the mutation rates and fitness effects for compensatory and reversion mutations as well as population size. The genetic mechanisms of compensation vary depending on the particular drug and microbe involved. These mechanisms may include mutations in the resistance gene itself as well as mutations that alter the expression of the resistance gene or other genes in ways that restore the appropriately balanced gene expression.

Plasmid Persistence

Plasmids typically carry genes that are nonessential and beneficial only under specific environmental conditions. Hence, they are often expendable, and their persistence requires either ongoing selection (e.g., for resistance genes) or other mechanisms that assure their continued carriage. The various selective processes discussed earlier can promote the maintenance of both chromosomal and plasmid-encoded resistance functions; there are also several mechanisms that can promote plasmid persistence even without selection for antibiotic resistance. For example, some plasmids enhance bacterial growth even in the absence of antibiotic. Many plasmids encode resolution and partitioning systems that prevent spontaneous plasmid loss during cell division, and some plasmids even have toxin-antitoxin systems that kill cells that lose the plasmid. Also, plasmids can be maintained in bacterial populations by their conjugation-mediated horizontal transfer between cells even if they impose a fitness cost.

5. CAN RESISTANCE EVOLUTION BE SLOWED OR EVEN STOPPED?

A pressing question is whether society can reduce the rate at which antibiotic resistance emerges and spreads. Various approaches have been suggested in the literature, but only a few are known to work. One approach—perhaps the most obvious but still difficult to implement—is to reduce the use of antibiotics, thereby reducing the strength of selection that favors both the emergence and spread of resistant bacteria. The efficacy of this approach follows from basic evolutionary principles and is also supported by numerous studies showing that the frequency of resistance is correlated with the volume of antibiotics used at various levels including individual hospitals, communities, and countries. Global restraint in antibiotic use can be achieved only by concerted action and will require the implementation of several strategies, including (1) avoidance of antibiotic use when none is needed (e.g., when the infection is caused by a virus); (2) discontinuance of the use of antibiotics as growth promoters in animal husbandry; (3) discontinuance of the use of antibiotics in the production of crops and in aquaculture; (4) avoidance of economic situations in which the prescription of antibiotics is profitable for the prescriber; (5) appropriate control and regulation of antibiotic marketing by the pharmaceutical industry (in which prescribers, pharmacists, and consumers are targeted); and (6) prohibition of the sale of antibiotics to the public via the Internet or from pharmacies or other outlets without the need for a prescription.

Also, by increasing use of various hygienic and infection control measures, society can reduce the transmission of pathogenic bacteria and thereby reduce the use of antibiotics. The extent to which these measures will work depends on the pathogen and its mode of transmission among hosts. Pathogens for which hygienic measures have been shown to be particularly successful include various food-borne pathogens (e.g., *Salmonella*) and nosocomial infections such as methicillin-resistant *Staphylococcus aureus* (MRSA). For MRSA infections, screening strategies to track and isolate affected patients, coupled with improved hospital hygiene, have been successful in reducing the transmission of these dangerous bacteria.

Other approaches that have been proposed to reduce the rate at which resistance evolves include changes in dosing regimens and use of antibiotic combinations that reduce selection for resistant mutants without affecting treatment efficacy or safety. The use of drug combinations has been shown to be effective in treating many HIV (the virus that causes AIDS) infections because a mutant that becomes resistant to one drug is nonetheless susceptible to others that are provided at the same time. In addition, drugs and drug targets might be chosen during research and development such that the risk of resistance is minimized. For example, new antibiotics might be developed such that (1) resistance is difficult to acquire by mutation or HGT; (2) the resistance mechanism confers a high fitness cost; and (3) the opportunities for compensatory adaptation are limited. It is interesting to note that no clinical cases of resistance have been reported for certain combinations of drugs and bacteria even after decades of use. For example, penicillin has been used successfully to treat *Streptococcus pyogenes* infections for 60 years. An understanding of the reasons for the lack of resistance evolution might allow more rational choice and design of drugs and drug targets.

In addition to limiting the rates at which resistance emerges and spreads, it might even be possible to reverse the existing problem of resistance by reducing the use of antibiotics. Whether this strategy will be successful depends on the strength of the forces driving reversibility. At the levels of the individual and community, the fitness cost of resistance in the absence of antibiotic is probably the main force pushing toward increased sensitivity, whereas in hospitals the main driving force is probably the continuous influx of patients with susceptible bacteria. In hospitals, mathematical modeling and correlative studies suggest that changes in antibiotic use can cause rapid changes in the frequency of resistance. However, when the fitness cost of resistance drives reversibility, the rate of change is expected to be much slower. The main reasons for this are that in addition to the factors described earlier that can stabilize resistance in bacterial populations, the intrinsic dynamics of reversal is expected to be slow because the strength of selection for sensitivity in the absence of antibiotic is generally much weaker than selection for resistance when antibiotics are used. This inference is supported by clinical intervention studies, performed at both the individual and community levels, in which it has been observed that resistant clones are remarkably stable and persistent even when antibiotic use is reduced.

6. WILL ANTIBIOTICS BECOME A FOOTNOTE TO MEDICAL HISTORY?

How will future generations view our ongoing experiment with antibiotics? Will antibiotics retain their therapeutic value for generations to come? Or will antibiotics be viewed as a failed experiment, one that becomes a mere footnote in the history of medicine? The answers to these questions will depend on many factors, of which two challenges are of particular importance. The first is whether society—including medical practitioners, patients, and the pharmaceutical industry—will have the resolve to use antibiotics in a more restrictive and medically responsible way that will slow the emergence and spread of resistance. Success will require global implementation of changes in healthcare systems and practices that are specifically aimed at reducing the overall use of antibiotics that selectively favors resistant bacteria. Many international resolutions to this effect have been put forward, but so far little has been done to implement any global strategies. What is needed now is leadership and coordination that will allow these recommendations to be put into action. If we fail to implement these recommendations, it is certain that resistance will continue evolving to existing antibiotics as well as to any new ones that are discovered.

The second major challenge is that the pharmaceutical industry has largely abandoned the development of new antibiotics, mainly for economic reasons; as a consequence, few new classes of antibiotics have been introduced for clinical use in recent decades. It is essential that this industry be recommitted to antibiotic discovery and the development of novel drugs. Potential ways forward might include new business models for collaboration between industry and public sectors, including new regulatory rules and funding schemes. Of course, there are real scientific challenges in finding new drugs, including antimicrobials. However, increased knowledge of structural biology, bacterial physiology and metabolism, medicinal chemistry, genomics, and systems biology provide new opportunities for the discovery of novel antibiotics, including ones that might inhibit new targets such that the evolution of resistance is impeded.

See also chapter II.11, chapter III.8, chapter IV.2, and chapter VIII.1.

FURTHER READING

Andersson, D. I., and D. Hughes. 2010. Antibiotic resistance and its costs: Is it possible to reverse resistance? Nature Reviews Microbiology 8: 260–271. *A comprehensive review on the subject of fitness cost of antibiotic resistance and its influence on the emergence, spread, and persistence of resistant bacteria.*

Davies, J., and D. Davies. 2010. Origins and evolution of antibiotic resistance. Microbiology Molecular Biology Reviews 74: 417–433. *From leaders in the field, this insightful review discusses the environmental origin of antibiotics and resistance genes.*

Freire-Moran, L., B. Aronsson, C. Manz, I. C. Gyssens, A. D. So, D. L. Monnet, O. Cars, and the ECDC-EMA Working Group. 2011. Critical shortage of new antibiotics in development against multidrug-resistant bacteria: Time to react is now. Drug Resistance Updates 14: 118–124. *An important paper that demonstrates the serious shortage of new antibiotics in clinical development against multidrug-resistant bacteria and points to the need for the involvement of the public sector into research and development of new antimicrobial drugs.*

Lenski, R. E. 1997. The cost of resistance—from the perspective of a bacterium. In D. J. Chadwick and J. Goode, eds., Antibiotic Resistance: Origins, Evolution, Selection and Spread. Chichester, UK: John Wiley 169. *Uses mathematical models and experiment findings to discuss how the growth, dissemination, and persistence of antibiotic-resistant bacteria might be controlled.*

Martinez, J. L. 2008. Antibiotics and antibiotic resistance genes in natural environments. Science 321: 365–367. *Discusses the potential biological roles antibiotics and resistance genes might have in natural environments.*

Martinez, J. L., F. Baquero, and D. I. Andersson. 2007. Predicting antibiotic resistance. Nature Reviews Microbiology 5: 958–965. *Outlines the methods and knowledge that are needed to allow researchers to predict the emergence of resistance to a new antibiotic.*

Morar, M., and G. D. Wright. 2010. The genomic enzymology of antibiotic resistance. Annual Reviews Genetics 44: 25–51.

White, D. G., M. N. Alekshun, and P. F. McDermott, eds. 2005. Frontiers in Antimicrobial Resistance: A Tribute to Stuart B. Levy. Washington, DC: ASM Press. *This book is a tribute to one of the leaders in the field of antimicrobial resistance that covers many relevant areas, including mechanisms and epidemiology of resistance as well as public policy and public education programs to use antibiotics appropriately.*

zur Wiesch, P. A., R. Kouyos, J. Engelstädter, R. R. Regoes, and S. Bonhoeffer. 2011. Population biological principles of drug-resistance evolution in infectious diseases. Lancet Infectious Diseases 11: 236–247.

VIII.4

Evolution and Microbial Forensics

Paul Keim and Talima Pearson

OUTLINE

1. Evolutionary thinking, molecular epidemiology, and microbial forensics
2. The uses of DNA in human and microbial forensics
3. Genetic technology and the significance of a "match"
4. The Kameido Aum Shinrikyo anthrax release
5. The Ames strain and the 2001 anthrax letters
6. From molecular epidemiology to microbial forensics and back

The tools of molecular biology coupled with the evolutionary methods of phylogenetics have found powerful applications in tracking the origins and spread of infectious diseases. Microbial forensics is a new discipline focused on identifying the source of the infective material involved in a biological crime and it, too, increasingly depends on evolutionary analysis and molecular genetic tools.

GLOSSARY

Clonal Populations. Populations in which members, called *clones*, have diverged without exchanging any genetic material across lineages. Members of such populations (e.g., many recently emerged pathogens) are genetically identical with the exception of variation generated by subsequent mutations.

Homoplasy. A shared genetic (or phenotypic) characteristic produced by convergent evolution or horizontal genetic exchange between lineages, rather than by descent from a common ancestor that shared the same characteristic.

Match. An identical genotypic profile (often called a *DNA fingerprint*) based on a particular technology.

Membership. A phylogenetic concept more useful than a "match" for describing relationships among bacterial isolates. Two isolates can be members of the same phylogenetic group without being absolutely identical in their genome sequences.

Monophyletic. A phylogenetic term referring to all descendants of a common ancestor.

Multiple Loci VNTR Analysis (MLVA). A DNA fingerprinting method widely used to differentiate bacterial types. Here, VNTR stands for variable number of tandem repeats.

Single Nucleotide Polymorphism (SNP). A single base-pair difference between the DNA sequences of two individuals including, for example, two closely related bacterial strains.

The investigation of infectious disease outbreaks has a long history and even predates our understanding of the germ theory of disease, which was formulated by Louis Pasteur in the early 1860s. The classic example, a seminal event in epidemiology, occurred in 1854 when John Snow implicated London's Broad Street water pump as the focus of a cholera outbreak. The correlative association of disease occurrence, potential causative infectious agents, and their sources has grown increasingly sophisticated over the years. Today it is common to examine the genomes of bacteria and viruses to precisely define the pathogen subtype, with the aim of identifying specific case clusters that can reasonably be presumed to be a part of the same outbreak. This approach strengthens any correlative study that aims to identify the disease source by eliminating similar disease cases that did not emanate from the same focus.

These same genomic methods became important after the bioterrorism events of October 2001, when letters laden with *Bacillus anthracis* spores were sent through the US Postal Service, and the investigation that followed sought to identify the source of the letters. Evolutionary theory concerning bacterial populations, mutational processes, and phylogenetic reconstruction were essential for this science-based forensic investigation. The development of the field of microbial forensics was greatly accelerated by the anthrax-letter investigation and it now provides a paradigm for both forensic cases and other public health investigations that involve infectious agents.

1. EVOLUTIONARY THINKING, MOLECULAR EPIDEMIOLOGY, AND MICROBIAL FORENSICS

The fields of molecular epidemiology and microbial forensics are populated by well-educated individuals. Nonetheless, the failure of these fields to employ evolutionary thinking sometimes limits the quality of the evidence and resulting inferences. For example, public health investigations of bacterial diseases have, in recent years, become highly dependent on one particular DNA-based technology called *pulsed-field gel electrophoresis* (PFGE). PFGE has the advantage that it can be applied to any bacteria, but its drawback is that the resulting data preclude more thorough evolutionary analyses. In particular, PFGE generates restriction fragment patterns—often called *DNA fingerprints*—that are analyzed using simple matching algorithms that produce yes/no outcomes, without allowing more sophisticated evolutionary analyses to identify the similarities and differences among the samples of interest. A "match" between fragment patterns is inferred by the analysts based on their experience and the rarity of a particular pattern in large databases. Unfortunately, little effort has been made to understand the evolutionary paths that may connect and explain the varying degrees of similarity among these patterns, and probabilistic models to place confidence estimates on relationships (e.g., a match) are rarely used. Most PFGE practitioners appreciate the validity of evolution, but their use of rigorous evolutionary analysis has been stymied by the difficulty in applying theory to such data and by resistance to making the changes necessary to improve on a widely used method.

In contrast with DNA studies of bacterial diseases, no established uniform technology exists in public health investigations of viral diseases; instead, each pathogen is typically analyzed by sequencing a particular, unique target gene. These sequence data are almost always analyzed using phylogenetic methods, and the analyses frequently include probabilistic models to test alternative hypotheses about the sources of the viruses. These DNA sequence data are in a universal digital format, and evolutionary models of sequence evolution are well developed, allowing for the rapid adoption and application of methodologies from other fields. By contrast, DNA fingerprints are poor substitutes for phylogenetic analyses, and the blind application of phylogenetic algorithms is inappropriate without a better understanding of underlying character state changes. The PFGE-based fragment patterns that constitute the DNA fingerprint can be thought of as complex phenotypes determined by the genotype—but following ill-defined rules—which illustrates the weakness of this approach. However, the lack of evolution-driven approaches in bacterial molecular epidemiology is starting to be overcome as sequence-based methods begin to dominate this discipline, and the costs of sequencing genomes keep dropping. The golden age for the molecular epidemiology of bacterial infectious diseases is arriving with the widespread adoption of whole-genome analysis.

2. THE USES OF DNA IN HUMAN AND MICROBIAL FORENSICS

The utility of DNA fingerprinting for human identification in forensic analysis has had a major impact on society and the legal system: it has led to the exoneration of falsely accused individuals and to the conviction of guilty criminals.The primary methodology is similar in some regards to the PFGE method described for bacteria in the previous section. However, in the case of human forensics, after several years of scientific discussion and debate, the statistical methods used to evaluate matches are firmly grounded on population genetic models and the scientific understanding of human biology, inheritance, and population subdivisions.

But these same statistical models have little utility in microbial forensics owing to the profound differences between bacteria and humans in terms of reproductive biology and modes of genetic inheritance. DNA is the genetic material of both bacteria and humans, of course, but that fact does not mitigate these differences. While DNA analysis in humans and in bacteria may be similar in terms of the molecular methods used, the inferences that can be drawn must reflect their different modes of inheritance and population structures.

It is equally important to realize that in addition to these differences between bacteria and humans, bacterial species—and even populations of the same nominal species—also differ from one another in ways that can influence the interpretation of genetic relationships. One important variable is the relative extent of vertical and horizontal modes of inheritance. Bacteria reproduce asexually, so their inheritance is primarily vertical (mother cell to daughter cell). However, horizontal gene transfer (HGT) between bacterial cells also sometimes occurs, and when it does so, it can move genes not only within but also between different species. HGT can have important consequences, such as the movement of antibiotic-resistance genes between species, and can leave conspicuous genetic evidence when it occurs between distantly related species. However, at a finer scale, many bacterial populations, including many recently emerged pathogens, show little or no detectable HGT. Thus, in many epidemiological and forensic situations, the relevant models and hypotheses are for lineages that are strictly clonal (asexual) in their derivation. In these cases, evolutionary analyses are focused on phylogenetic relationships and mutation rates.

3. GENETIC TECHNOLOGY AND THE SIGNIFICANCE OF A "MATCH"

The idea of a genetic match between two DNA fingerprints is jargon that has entered the scientific lexicon via the fields of human identification and forensics. Because individual humans are almost always the unique product of two unique gametes (identical twins being the exceptions), almost every person can be uniquely identified based on his or her alleles at a relatively few hyperdiverse regions of the human genome characterized by short tandemly repeated sequences. An exact allelic match between DNA samples from two individuals is so unlikely that a "match" has been used as the only physical evidence needed to link an individual to the scene of some crime. Likewise, a "nonmatch" can be used to exonerate a suspect. The idea of unambiguous matches and nonmatches has thus proven to be very powerful in the justice system. Unfortunately, this same terminology is often applied to scenarios in microbial epidemiology and forensics; however, the interpretations may be very different as a consequence of biological differences between humans and microbes.

With microbes, the technological context is also critical to understanding the significance of a "match." A perfect genetic "match" can be lost using methods with greater resolution and discriminatory power. Low-resolution methods, including PFGE and multiple-locus sequence typing, would show that many bacterial isolates have identical alleles, but these methods see only a small portion of the genome. Greater discrimination can be achieved using *multiple-locus VNTR analysis* (MLVA), a technique that involves screening multiple loci with variable numbers of tandem repeats (VNTR), or by sequencing the entire genome of a bacterial isolate. Such whole-genome sequences may seem to be the ultimate standard, but bacterial geneticists have long realized that mutations will generate variation even within a colony of cells separated by only a few generations. Whole-genome sequencing does not detect these mutations because most applications generate a consensus DNA sequence that ignores rare variants; in fact, the accuracy of current technologies is such that rare sequencing errors obscure such rare mutations. In the future, however, new sequencing methods might detect rare variants directly from their individual DNA molecules. Therefore, a seemingly perfect match between two samples can be broken either by increasing the extent of genomic sampling or by searching more thoroughly for variants within the population. When it becomes possible to discriminate even between two colonies derived from the same progenitor strain, the ideal of seeking a perfect genetic match becomes more problematic than useful.

Rather than a match, a microbe's "membership" in a phylogenetic group or clade is a more meaningful concept for epidemiological and forensics work. In a clonal lineage with little or no horizontal gene transfer, one can define phylogenetic relationships based on informative characters with membership in a particular clade based on shared derived states. *Single-nucleotide polymorphisms*, or SNPs, are now commonly employed in this way because they are produced by rare mutation events and thus are usually stable over appropriately long periods. With sufficient data, such as obtained by sequencing whole genomes, this stability can easily be tested by discriminating between convergent (e.g., homoplastic) and vertically inherited matches at the level of each SNP. Moreover, additional SNPs elsewhere in the genome are not problematic for inferring membership, because diversity is hierarchically nested within clades. Thus, additional SNPs produce novel genotypes that are still members of the clade. Even a reversion—a mutation to a prior state—does not change clade membership per se, although it can complicate inferences about membership.

In most cases, multiple point mutations will have occurred along most or all evolutionary branches. However, a single canonical SNP can be used to represent each branch, which can simplify phylogenetic analyses. This paring down of the number of characters is not essential, and it may result in less phylogenetic precision if a sample belongs to some subclade that has not been extensively characterized. In such cases, the failure to include all SNPs along a particular branch may lead to the assignment of that sample to the wrong subclade. This mistake may be caught, of course, by including more SNPs. Thus, the hierarchical redundancy of phylogenetically ordered SNPs creates a safeguard against incorrect assignments of samples to subclades.

4. THE KAMEIDO AUM SHINRIKYO ANTHRAX RELEASE

In the summer of 1993, an attack using a biological weapon was carried out in Kameido, a highly populated suburb of Tokyo, by the Aum Shinrikyo religious cult (currently called Aleph). Although the cult was large, well financed, and had well-educated scientists involved in the planning, the anthrax attack failed to kill or even sicken the targeted population. In fact, it was many years later before scientists realized there had been a failed attack.

In late June 1993, public health officials were notified by Kameido residents of a highly unusual and odiferous mist emanating from the roof of the Aum's facility. Unsure of what was occurring, government health officials collected samples of the spray and submitted them for chemical analysis. The analyses evidently provided no evidence of toxic chemicals, and the cult discontinued the spraying, so no further actions were taken. Two years later, however, the cult carried out a chemical weapons attack by releasing sarin gas in the Tokyo subways. Ten people were killed and hundreds seriously injured. It was only after the arrest of cult members and during their subsequent questioning that the Kameido anthrax attack was discovered for what it was. The mist coming off their building was, they stated, from a culture of *Bacillus anthracis*—the causative agent of anthrax.

Hiroshi Takahashi was the investigating epidemiologist, and in 1997, he discovered a small tube of liquid that had been collected from the Kameido building at the time of the 1993 attack. He transferred this material to the United States, where *B. anthracis* cells were cultured and then genetically analyzed using MLVA, which was the best available technology at that time. Eight variable loci were analyzed, including six on the chromosome and one on each of two extra-chromosomal plasmids that carry virulence factors. Seven of the loci matched a well-known strain of *B. anthracis*, called Sterne, that is used in the production of a vaccine against anthrax. The assay for the eighth locus failed, a result that was also consistent with the Sterne strain because it is missing the pXO2 plasmid that carries this locus. Indeed, the absence of that plasmid is the reason that the Sterne strain is not virulent. Thus, the anthrax attack had failed to kill anyone because the Aum Shinrikyo cult had used a harmless strain of bacteria. The evidence of a vaccine strain raised the question, Why had the cult used a harmless strain? Was it a mistake on the part of the cult? Was it a practice run for a possible later attack? This question remains unanswered today.

B. anthracis is a pathogen with very low genetic diversity, reflecting its recent origin. In the pregenomics era, MLVA was one of the only available methods for distinguishing one *B. anthracis* strain from another. The database at that time contained only 89 distinct genotypes, or fingerprints. Even so, the results of the assays supported several important conclusions: (1) the cult had indeed used *B. anthracis*; (2) several commonly studied and virulent strains (e.g., Ames, Vollum) were excluded as the attack material; (3) the failure of one assay was consistent with the strain's lack of one of two virulence plasmids; and (4) that failure, as well as results from the other seven loci, matched the fingerprint of the widely available vaccine strain Sterne. The first two conclusions were robust. The match to Sterne, however, was less so because other strains share the same seven-locus genotype; the null allele for the plasmid-encoded locus produced additional ambiguity. For the reasons discussed earlier, DNA fingerprinting methods such as MLVA are not well suited for evolutionary inferences. Nonetheless, an important forensic principle is evident—one we will revisit in the next section—in terms of the strength of exclusionary versus inclusionary findings.

5. THE AMES STRAIN AND THE 2001 ANTHRAX LETTERS

Only a few weeks after the September 11, 2001, terrorist attacks had killed thousands of people, the United States faced another shocking incident in October, one that employed a deadly biological weapon. The attacker(s) used the US Postal Service to send at least seven letters containing *B. anthracis* spores. These letters were sent to specific targets, but their routing through the postal system resulted in widespread contamination by spores, which disrupted several mail centers and other government facilities including congressional buildings. Molecular genetics and evolutionary approaches were central to the forensic investigation.

Although whole-genome sequencing methods were eventually brought to bear on this case, the investigation began at a time when that technology was not sufficiently developed to allow it to be used with the immediacy that the circumstances demanded. Public health as well as national security considerations meant that it was critical to identify the likely source—or at least to exclude certain sources—as quickly as possible.

To that end, Paul Keim and colleagues were able to quickly perform an initial analysis of the DNA from the spores in the letters using the same MLVA system used to analyze the *B. anthracis* from the Kameido event, and with an expanded reference database. In 1991, the United Nations Special Commission had discovered weaponized anthrax spores during inspections following the Gulf War. Bacteria were recovered from ordnance, and they were identified as *B. anthracis*, but little other characterization was done at that time. After the 2001 anthrax letter attacks, there was renewed interest in the Iraqi weapon strain, given suspicions of foreign involvement from some quarters. Identifying the Iraqi weapons strain and its relationship to the strain in the anthrax letters was therefore critical.

Within just days of the hospitalization of the first victim in Florida, MLVA showed that the *B. anthracis* isolated from that victim matched the Ames strain at all eight loci. Analyses of samples from the letters also matched the Ames strain. The Ames strain is a virulent one, unlike the Sterne strain that was deployed in the failed attack in Japan. The Ames strain was known to be used in several US government laboratories and, despite its name, it was originally isolated from Texas. The search was then on for the source of the attack strain, with three critical issues at hand. First, was the Iraqi strain also an Ames strain? Second, were other *B. anthracis* strains that could be isolated from nature similar enough to the Ames strain that they would produce a match at all eight MLVA loci? And third, what higher-resolution techniques could be employed to distinguish among sublineages within the clade that contains the Ames strain and its close relatives to trace the attack strain to a specific source?

In December 2001, the Iraqi strain was characterized using the MLVA method, and a match at all eight loci was established to another strain called Vollum. In fact, an Iraqi scientist had purchased the Vollum strain from a culture collection in 1986, indicating the likely source of that strain. Importantly, the Ames and Vollum strains differ at multiple MLVA loci. These differences meant that the *B. anthracis* strain discovered at the Iraqi bioweapons facility could be excluded, with a high degree of confidence, as the source of the spores in the letter attacks.

So, what was the source of the Ames-related material in the letters? Was the material derived from the Ames strain, which had been distributed to various laboratories? Or could it be a different isolate that just happened to match the Ames strain at all eight loci used in the MLVA testing? In fact, the database showed that an isolate obtained in 1997 from a goat in Texas also matched the Ames strain at all eight loci. The circumstances of the attacks made it clear that these anthrax cases were not a natural outbreak. In principle, someone might have reisolated a *B. anthracis* strain from nature that happened to be a close relative of the Ames strain. In any case, it became imperative to employ genetic methods that would allow maximum resolution to determine the source of the attack material.

To that end, whole-genome sequencing was employed to find genetic differences that could be analyzed using phylogenetic methods. SNPs are ideal for this purpose because reversion mutations should be rare in such young lineages as *B. anthracis* and especially the clade containing the Ames strain. Indeed, the extent of homoplasy in species-wide SNP data is only about 0.1 percent across the entire species. This approach identified four SNPs specific to the laboratory Ames strain, which could be used to differentiate it from natural isolates. By screening for these four SNPs, it was possible to exclude other strains including the isolate from the Texas goat (which had matched the Ames strain at all eight loci used in the MLVA test) as well as additional isolates from the same geographic region. Thus, it became possible to determine that a strain was a member of the Ames group of lab-derived isolates with much more confidence than with the fragment-matching approach of MLVA.

Whole-genome sequencing was also employed in other lines of the investigation. With the increasingly strong evidence that the attacks had used spores derived from the Ames strain present in several laboratories, the key genetic issue became one of searching for mutations in the attack materials that might match mutations found in some laboratories but not others. Thus, one line of the investigation involved comparing the genome sequences of the *B. anthracis* isolated from the Florida victim and another Ames-derived strain, called Porton, whose virulence plasmids had been "cured" (eliminated). The Porton strain was used because it was already in the process of being sequenced and analyzed prior to the attacks, thus expediting the investigation. In fact, several mutational differences were discovered between the Florida and Porton derivatives of the Ames strain. However, these differences turned out to be useless for the investigation because all of them were unique to the Porton strain; the mutations probably arose during the mutagenic procedures employed to eliminate the plasmids.

The other line of investigation using genomic sequences proved to be more useful but also quite complicated. In the early stages of the investigation, microbiologists had allowed some of the *B. anthracis* spores taken from the letters to germinate and produce colonies. They observed subtle variation in the appearance of colonies, with one predominant type and several variants at lower frequencies. Thus the differences were heritable, which implied that the differences in colony "morphology" had resulted from mutations. If confirmed, the mutant subpopulations might then provide a signature to distinguish possible sources of the spores used in the attacks. In summary, several clones with variant morphologies were sequenced, and mutations were identified. These mutants had not been seen in previous sequencing attempts because they were rare in the population of spores, and the resulting sequence represented a consensus sequence from the sampled cells. Next, the sampled cells were selected specifically to include these morphological variants. Molecular assays could then be developed to screen for four of these mutations.

In the meantime, the Federal Bureau of Investigation (FBI) had created a repository of more than 1000 samples, all derived from the Ames strain, from about 20 laboratories known to have worked with that strain. These samples were then screened for the four mutations. None of the four mutations were detected in most of the samples, but eight of them gave positive results for all four mutations. (There are many complications related to the sensitivity and specificity of the assays used to detect the mutations, as well as other issues that in the interest of brevity, are not presented here but are discussed in a 2011 report prepared by a committee of experts convened by the National Research Council.) The eight samples were all apparently derived from the same source—a flask of spores identified as RMR-1029—based on information obtained by the FBI. The contents of the flask had been generated by pooling several separately grown batches of spores, to produce a single large stock of material for experiments that would be performed at different times. This manner of preparing the flask of spores might account for the diversity of variant colony types that led to this line of investigation. In any case, these results pointed toward a particular flask and samples taken from that flask as a possible source of the spores placed in the attack letters. The criminal investigation was thus also focused on those individuals who had access to the RMR-1029 flask and its derivatives.

This chapter is focused on the role of evolutionary thinking in microbial forensics; it is not the place to discuss other aspects of the criminal investigation. But for those readers who want to know, very briefly, the outcome of this investigation, the FBI identified a government scientist as the lone suspect of the anthrax letter attacks. Before the US Department of Justice could bring formal charges, that individual committed suicide.

Genomic technologies continue to advance at a rapid pace, and it is possible that spores from the attack letters and from the RMR-1029 flask could be examined even more fully by so-called deep sequencing. That approach could, in principle, expand the analysis of diversity in those samples well beyond the four mutations that were discovered based on the variation in colony morphologies.

6. FROM MOLECULAR EPIDEMIOLOGY TO MICROBIAL FORENSICS AND BACK

Over the course of several decades, increasingly powerful molecular-based methods have been used to identify the source and track the spread of infectious diseases. These methods also served as the starting point for the forensic investigation of the anthrax attacks. Nonetheless, that investigation pointed to the limitation of these methods. The urgency and resulting high levels of funding to investigate the anthrax letters enabled the application of whole-genome sequencing—an approach that molecular epidemiologists had not been able to employ previously owing to its high costs. The genomic methodologies and analytical approaches have now become much less expensive, and so they should be applied much more broadly in molecular epidemiological studies motivated by public health concerns. Both forensic and epidemiological investigations are also well served by using phylogenetic approaches to analyze genomic data for determining relationships among samples, especially as the number of key samples becomes progressively smaller as one hones in on a probable source.

Thus, we predict with confidence that the use of whole-genome sequencing to understand evolutionary relationships will become common in public health. This technological change will bring with it changes in data analysis such that the full power of evolutionary theory, models, and methods can be used to determine infectious sources during natural disease outbreaks.

See also chapter II.1, chapter II.11, and chapter IV.2.

FURTHER READING

Budowle, B., S. E. Schutzer, R. G. Breeze, P. S. Keim, and S. A. Morse, eds. 2011. Microbial Forensics. 2nd ed. New York: Elsevier. *A comprehensive collection of approaches to microbial forensics.*

Committee on Review of the Scientific Approaches Used during the FBI's Investigation of the 2001 *Bacillus anthracis* Mailings. 2011. Review of the scientific approaches used during the FBI's investigation of the 2001 anthrax letters. Washington, DC: National Academies Press. *A hard look at the FBI's investigative methods and results.*

Hillis, D. M. 2009. Evolution Matters. National Institutes of Health, http://videocast.nih.gov/launch.asp?15187. *An overview of the relevance of evolution to emerging diseases and solving certain crimes.*

Jobling, M. A., and P. Gill. 2004. Encoded evidence: DNA in forensic analysis. Nature Reviews Genetics 5: 739–51.

Keim, P., T. Pearson, and R. Okinaka. 2008. Microbial forensics: DNA fingerprinting "anthrax." Analytical Chemistry 80: 4791–4799. *An overview of DNA methods used for investigating the anthrax letter attacks.*

Keim, P., and D. M. Wagner. 2009. Humans, evolutionary and ecologic forces shaped the phylogeography of recently emerged diseases: Anthrax, plague and tularemia. Nature Reviews Microbiology 7: 813–821. *A population model for the emergence and global spread of pathogens.*

Morelli, G., Y. Song, C. J. Mazzoni, M. Eppinger, P. Roumagnac, D. M. Wagner, M. Feldkamp, et al. 2010. Phylogenetic diversity and historical patterns of pandemic spread of *Yersinia pestis*. Nature Genetics 42: 1140–1143. *A detailed evolutionary look at an important clonal pathogen that causes plague.*

Takahashi, H., P. Keim, A. F. Kaufmann, K. L. Smith, C. Keys, K. Taniguchi, S. Inouye, and T. Kurata. 2004. Epidemiological and laboratory investigation of a *Bacillus anthracis* bioterrorism incident, Kameido, Tokyo, 1993. Emerging Infectious Disease 10: 117–120. *The public health report of the Kameido incident.*

VIII.5

Domestication and the Evolution of Agriculture

Amy Cavanaugh and Cameron R. Currie

OUTLINE

1. Domestication
2. Evolution under domestication
3. Agriculture as a mutualism
4. Agriculture in ants
5. Conclusions

Agriculture is an ancient and important factor shaping life on earth. Through the cultivation of food, populations of agriculturalists are able to greatly expand and can even develop a division of labor. This chapter explores the evolution of agriculture, including domestication and selection under domestication, along with the evolutionary events and consequences of farming. It also describes how agricultural associations are perhaps best viewed in the framework of a coevolved mutualism.

GLOSSARY

Artificial Selection. Evolutionary change caused by human breeding in populations of domesticated (or experimental) plants and animals.

Coevolution. Reciprocal evolutionary change between interacting species.

Domestication. Acquisition from the wild of one species by another and breeding it in captivity.

Domestication Syndrome. A suite of traits characteristically found in domesticated species.

Mutualism. An interaction between two species that benefits both.

The domestication of one species by another for food is one of the most significant evolutionary innovations in the history of life on the planet. Indeed, shifting to an agricultural lifestyle, and the concomitant expansion in numbers and range it allows, inexorably alters not only the biology of the species involved but also the ecosystems in which they occur. By establishing a reliable reserve of food, agriculturalists gain an advantage over their hunter-gatherer brethren; the ready source of calories allows the agriculturalist populations to greatly expand and ultimately facilitates the development of a division of labor. Agriculture originated among humans in the Fertile Crescent, but contrary to popular belief, humans were not the world's first farmers. That distinction belongs to a group of ants in the Amazon Basin. These fungus-farming ants maintain specialized gardens of domesticated fungi that serve as the primary nutrient source for the colony. After the origin of agriculture in ants, but still millions of years before humans appeared, other groups of insects also transitioned to farming. In parallel with fungus-growing ants of the New World, some termites farm fungus for food in the Old World. The most diverse farmers are the Ambrosia beetles, represented by more than 3000 species. In all these cases, the utilization of a farmed food source has enabled these insects to expand into a new ecological niche, leading to their diversification and, in some cases, allowing them to become dominant members of their ecosystems. Other insects engage in more rudimentary forms of farming, and some ants even practice animal husbandry by tending aphids and treehoppers. Besides the insects, a marine snail cultivates fungus, and a species of damselfish farms red algae, and recently it has been suggested that even amoebas practice a rudimentary form of bacterial husbandry.

Agriculture most recently originated in our own species approximately 10,000 years ago and has ultimately resulted in our dominating most of the ecosystems on the planet. Humans cultivate around 100 different plant species, which serve primarily as a reliable and more readily stored source of nutrients. Humans have also domesticated a number of animals, obtaining a variety of benefits, including sources of nutrients (e.g., meat and milk), labor (e.g., plowing fields, transporting of goods, and protecting and herding other domesticated animals), and military advancement (e.g., cavalry). Thus, farming provided a reliable source of calories, allowing an increase in human population size, decrease in birth intervals, and specialization of labor leading to stratified societies, while animal husbandry allowed agricultural societies to expand beyond their borders and ultimately to dominate the nonfarming populations with which they came in contact. Based on these advantages it can be argued, as Jared Diamond does in his Pulitzer Prize–winning book *Guns, Germs, and Steel*, that agriculture is the single most important force shaping human history.

Just as agriculture has shaped human society and history, it has also had an important role in the development of evolutionary theory. This influence is evident in Darwin's *The Origin*, which begins with a thorough discussion of domestication and the evolutionary changes caused by human breeding of domesticated plants and animals—an evolutionary force he termed *artificial selection*—even before introducing the tenets of natural selection:

> It is . . . of the highest importance to gain a clear insight into the means of modification and coadaptation. At the commencement of my observations it seemed to me probable that a careful study of domesticated animals and of cultivated plants would offer the best chance of making out this obscure problem. Nor have I been disappointed; in this and in all other perplexing cases I have invariably found that our knowledge, imperfect though it be, of variation under domestication, afforded the best and safest clue.

Of the different domesticated species Darwin investigated to "mak[e] out this obscure problem," the domestic pigeon was the subject of one of his most in depth studies. After thoroughly examining all the breeds of pigeons he could acquire, he determined that more than 20 different characters varied among these breeds. Yet it was believed at the time, and confirmed by his additional studies, that all these diverse breeds had descended from a single wild species, the rock pigeon. Darwin argued that the key factors in creating all this variability among breeds were "man's power of accumulative selection" and use of the large body of literature, both modern and ancient, in which breeders and horticulturalists described in great detail the ways in which they had modified their animals and plants by selectively mating only those individuals with the desired characteristics.

In this chapter, we discuss evolutionary aspects of domestication and selection under domestication in agriculture by humans. We then argue that a useful way to conceptualize the evolution of agriculture is as a mutualism shaped by coevolution. Expanding on this argument, we end with a discussion on the evolution of agriculture in ants, drawing parallels with humans.

1. DOMESTICATION

Domestication is the practice whereby an organism is acquired from the wild and bred in captivity. The population or species that is domesticated can be referred to as the *domesticate*. Domesticates undergo genetic changes during the process of cultivation or breeding that make them more useful to the domesticator and ultimately differentiate them from their wild ancestors.

The first domestication of a plant by humans occurred about 10,000 years ago, when people living in the Middle East (parts of modern Iraq, Iran, Turkey, Syria, and Jordan) began to purposefully plant barley, peas, lentils, chickpeas, muskmelon, flax, and two species of wheat. Not long after agriculture had been established in the Middle East, it arose independently in eastern China. There the available wild species differed, and so the first domesticated crops of Southeast Asia included rice, soybeans, adzuki beans, mung beans, hemp, and two species of millet. Populations within the tropical West African and Sahel regions also appear to have independently begun domesticating species including sorghum, millet, rice, cowpeas, yams, bottle gourds, and cotton. Though the dates are uncertain, people in Ethiopia domesticated coffee, and people in New Guinea domesticated sugarcane and bananas. Although populations in the Americas also independently established themselves as farmers, this transition took place later than those in Eurasia and Africa, most likely owing to the inherent differences in the available wild species. Between 9000 and 3000 years ago, humans began domesticating animals including sheep, goats, cattle, pigs, chickens, and horses in Eurasia and northern Africa. Again, populations in the Americas independently domesticated some animal species, such as the llama and the guinea pig, but they were limited in their efforts because most of the available wild species were unsuitable for domestication.

It might seem that almost any wild species could be domesticated, but history has shown that this is not the case. Although humans have domesticated a number of species, they represent an extremely small proportion of the plants and animals that occur in nature. The wild progenitors of the first crop species were already edible, grew quickly and easily, could be stored, and were self-fertilizing. This last trait is crucial in that self-fertilizing plants will directly pass traits on to their offspring largely unchanged. Species that have never been domesticated fail to meet one or more of the preceding criteria. For example, the oak tree, despite producing nutrient-rich acorns, has never been domesticated, for many reasons. First, the oak is an extremely slow-growing tree, taking more than 10 years to grow from an acorn to a fruit-bearing tree. Second, the bitterness of the acorn is under the influence of many genes, which combined with the long generation time, makes it very difficult to select for mutant, sweet acorns. Finally, acorns are a primary food source for another animal, squirrels. By burying large numbers of acorns, squirrels would undermine any human attempt to plant acorns only from oak trees with desirable traits.

Animals that have been successfully domesticated also share many traits. First, most domesticated animals are herbivores. Owing to the successive loss of energy through each trophic level, it takes much less food to support the growth of a herbivore than a carnivore; therefore, raising herbivores is far more efficient. Although we now eat carnivorous fish, we have only recently begun farming them, and whether this leads to their domestication remains to be seen. Second, as with plants, successfully domesticated animals grow quickly. Extremely large mammals, such as elephants, grow too slowly to be candidates for domestication. Third, domesticated species breed readily in captivity. As Darwin noted, this is a particularly rare trait among animals. Fourth, the animal must have a relatively pleasant disposition. While all large animals, and many small ones, are capable of killing humans, most are much more prone to aggression than the species that have been successfully domesticated. Fifth, they must not be prone to panic, particularly panic that results in the animals' battering themselves to death while trying to escape. This behavioral issue has been a limiting factor in the domestication of many otherwise-suitable herd species, such as gazelles. Finally, many successfully domesticated animals live in herds with well-developed hierarchies and overlapping home ranges; these animals are able to live in proximity to one another and will usually accept a human as the herd leader.

2. EVOLUTION UNDER DOMESTICATION

Although domesticates are species whose wild ancestors possess specific traits suitable for domestication, they are greatly altered by the process of artificial selection imposed by the domesticator. To Darwin, artificial selection was not merely analogous to natural selection but rather represented a clear example of natural selection under a particular set of conditions. The principles are the same, but the environmental conditions in play under artificial selection are those of the human-constructed habitat as opposed to a habitat of nature's making under natural selection. For either selective force to operate there must be variation in the trait under selection, heritability of that trait, and a tendency for individuals with some version of that trait to reproduce, or be bred, more than others with a different version. As people consciously or unconsciously selected the plants and animals that met human needs and preferentially grew and bred them, they were practicing artificial selection. At the same time, people were creating a novel environment for these plant and animal species, and natural selection further increased the frequency of traits that would lead to success in this constructed environment.

Domestication of plants and animals undoubtedly involved the conscious selection of numerous traits. In plants, early protofarmers likely preferentially collected the largest fruits or seeds to consume and to subsequently plant, and likely selected for taste, choosing the least bitter seeds and sweetest fruits. While many plants were selected for their fruit or seeds, others would have been selected for size or fleshiness of other nutritional parts of the plant (e.g., the roots or leaves), their oil content (e.g., olives and sunflowers), or length of fibers (e.g., flax and hemp). Animals likely were consciously selected on the basis of size, for those raised for meat, or reproductive physiology, for those raised for milk or eggs. Sheep and llamas would have been selected for the retention, rather than shedding, of the wool fibers in their coats, while dogs would have been selected for traits such as size, sense of smell, hunting ability, trainability, and herding ability.

Plants and animals were also subjected to a great deal of unconscious selection. For example, the wild progenitors of cereals and legumes typically drop their seeds as a dispersal mechanism. Mutant plants that did not drop seeds would die out quickly because they would leave no offspring. However, such plants would prove beneficial to humans trying to efficiently gather food, as it is much easier to collect a handful of seeds from the top of a stalk than to pick each individual seed from the ground. Once humans began cultivating plants, selection would have also favored plants with faster germination times. After planting, those plants that sprouted first were more likely to be harvested and replanted, compared with those that delayed germination. Finally, while consciously selecting for traits such as size and taste, humans were also unconsciously selecting for plants capable of self-fertilization. In plants that self-fertilize, as most crops do, favorable mutations are maintained, not diluted by recombination with their neighboring wild progenitors.

Humans attempting to breed the largest or best milk-producing variants of a species would also have inadvertently been selecting for animals with the ability to reproduce in captivity. Domestic animals reach sexual maturity earlier than wild animals and have more frequent reproductive cycles. These traits may have been both consciously and unconsciously selected for by humans—consciously by selectively breeding the animals that reached maturity earliest and breeding them as often as they were receptive to it, and unconsciously by eliminating the nutritional constraints that would have limited their reproduction in the wild.

Together, the forces of artificial and natural selection have led to changes in domesticated plants that have come to be known as the *domestication syndrome*. These traits include (1) increased size of reproductive organs (e.g., fruits and seeds); (2) increased tendency for mature seeds to remain on the plant rather than dropping to the ground; (3) faster germination as well as synchronized, predictable germination times; (4) changed allocation of biomass (e.g., larger roots, stems, leaves, or buds); and (5) reduced physical and chemical defenses. Domesticated animals also possess a suite of traits that distinguish them from their wild counterparts. Morphologically, domesticated animals typically exhibit greater variation in overall body size as well as in the size of particular body parts (e.g., length of legs in dogs), as compared with their wild ancestors. Additionally, domestic species have different coloration of fur and feathers than their wild relatives, typically an increase in white or spotted coloration. Although such colors make individual animals more visible and therefore more vulnerable to predation, humans could have inadvertently selected for such individuals because they were easy to see and recover if they wandered away.

3. AGRICULTURE AS A MUTUALISM

Agriculture can be thought of as a *mutualism*—an interaction that benefits both the agriculturalists and the domesticated species. The benefits to humans are obvious, as discussed earlier. But, to some, the benefits of being an "enslaved" plant or animal might not be so clear. However, domesticates do receive numerous benefits, broadly falling into three general categories: (1) protection, (2) increased reproduction, and (3) dispersal. Agriculturalists protect their domesticated crops and animals by significantly reducing interspecific competition, herbivory, and predation. This protection includes growing domesticates in controlled environments and actively weeding, pruning, guarding, and applying chemical treatments. Through the careful planting and cultivation of seeds, farmers increase the probability of seed germination, thus increasing the reproductive rates of domesticated crops. Similarly, domesticated animals have higher reproductive rates, typically owing to shortened interbirth and interlaying intervals. Finally, as agricultural populations spread, they bring their crops and animals with them. By altering the new habitat to be suitable for domesticated species of their homeland, people increase the range of these species. Given the tremendous efforts humans undertake to care for their domesticates and the huge expansion of some plant and animal species following their domestication, Michael Pollan argues in *The Botany of Desire* that it is worth considering the question, Who is domesticating whom?

Even as humans directed the evolution of the species they domesticated, they created new selection pressures on themselves. The transition to an agricultural lifestyle led to changes in both human behavior and physiology. For example, as with domesticated animals, human agriculturalists have increased reproductive rates compared with those of hunter-gatherers. Most likely owing to the increased reliability of a higher calorie diet, interbirth intervals are much shorter in farming societies than in hunter-gatherer societies. In addition, two enzymes, amylase and lactase, show increased expression in members of agricultural societies compared with hunter-gatherers as well as with chimpanzees, our closest nonhuman relatives. In the case of lactase, an enzyme that digests the sugar found in milk, all mammals produce the enzyme as infants but then stop producing it rapidly after weaning. However, in many human populations, a mutation allows the persistent expression of this enzyme into adulthood. The geographic distribution of this mutation is strongly correlated with pastoralism, particularly the raising of animals for milk production. In a case of parallel evolution, two different mutations have been shown to cause lactase persistence in different populations. Both these mutations occur in the promoter region of the lactase gene. Amylase is an enzyme that breaks down starch. In this case, it appears that populations that switched to the starchier agricultural diet evolved extra copies of the gene that produces salivary amylase. These changes in humans, in response to shifting to an agricultural lifestyle, support the view of agriculture as a mutualism. In fact, they suggest that agriculture represents a mutually beneficial association shaped by coevolution, given that both interactors—the farmer and the domesticate—undergo genetic modification in response to the association.

4. AGRICULTURE IN ANTS

Other than agriculture by humans, the best-studied agricultural association is that of fungus-growing ants. Agriculture in ants is ancient, having originated approximately 45 million years ago. As humans have domesticated many species of plants and animals, fungus-growing ants have domesticated multiple species of fungal crops; there are as many as seven different events of free-living fungi being domesticated. Within this agricultural mutualism the ants and their fungal cultivars have coevolved and diversified. Fungus-growing ants include more than 200 species in 13 genera. Likewise, the cultivated fungi are represented by substantial diversity of strains within specific groups of cultivated lineages. At the pinnacle of evolution of agriculture in fungus-growing ants are the charismatic leaf-cutters, which shape neotropical ecosystems through the sheer mass of leaf material that the ants harvest.

The cultivated fungus, maintained in underground garden chambers in most species, serves as the primary food source for workers, larvae, and the queen. The cultivated fungus produces specialized structures called *gongylidia*, which are rich in lipids and carbohydrates. The gongylidia appear to represent an optimized nutrient source for the ants, likely evolved under a form of artificial selection. The ants cannot survive without their fungal crops; without them they literally starve. When establishing new colonies, queens ensure the initial presence of the cultivar by bringing a small ball of fungus collected from her parent colony, effectively transferring the fungus from one generation to the next. Recent genomic studies on leaf-cutters have revealed that fungus-growing ants (like humans) have evolved genetically in response to their dependence on agriculture; in particular, they have lost the ability to synthesize an essential amino acid that they likely obtain from the fungus garden.

Leaf-cutter ants have evolved a complex set of behaviors for cultivating the fungus. Like many human-domesticated species, the ants' fungal crops are unable to survive without the ants. The ants selectively forage for leaf material that promotes the growth of the fungus garden. The garden matrix is thus composed of the fungus and the vegetative substrate that worker ants obtain from outside the nest and then integrate into the fungus garden. Once this leaf material is brought to the colony, the ants lick and chew the material into small pieces. This process breaks down the physical barriers of the leaf that would otherwise prevent the growth of the fungus on the leaf surface. Just as human farmers work manure into the soil, the ants work the leaf pulp into the top layers of the fungus garden. They then bring fungal hyphae from older parts of the garden, plant it onto the surface of the fresh leaf pulp, and continuously add fresh material to the top of the garden.

Besides adding substrate to the garden, the ants also promote the growth of their fungal crop in numerous ways. The ants open and close tunnels to the surface such that they can regulate the temperature and humidity within the growth chambers. There is also evidence that the ants damage the fungus, in a manner akin to pruning, to stimulate increased fungal growth. The fungus produces enzymes that can become disadvantageously concentrated in the garden. When that happens, the ants ingest these enzymes in the areas of high concentration and then defecate them into areas of low concentration, thus creating an equal distribution of the enzymes throughout the garden.

The cultivation of monocultures of clonally propagated crops has led to increased susceptibility to disease. The ants' fungus garden is host to specialized and potentially virulent agriculture pathogens, microfungi in the genus *Escovopsis*. *Escovopsis*—known only from the fungus gardens of these ants, consumes the ants' fungal cultivar and has coevolved with the ants and their fungal crop. The ants engage in meticulous behaviors to deal with the pathogen. They groom out *Escovopsis* by pulling pieces of the fungal cultivar through their mouthparts and collecting the invading microbes in their infrabuccal pocket, a cavity and filtering device within the mouthparts of ants. The ants then deposit this material in the refuse chambers. In cases where the garden has become diseased, the ants remove the affected area in a behavior called *weeding*, which involves ripping out and discarding the infected garden material. Further paralleling human methods for dealing with agriculture pests, the ants employ chemical methods of crop protection. Whereas humans control pests by developing and then spreading chemicals on their crops, the ants form a symbiosis with antibiotic-producing bacteria. These symbionts, Actinobacteria, live on the ants' cuticle and produce antifungal compounds that inhibit the garden pathogen *Escovopsis*.

In summary, agriculture in ants, much like human agriculture, has led to their dominant role in many of the ecosystems in which they occur. Further, they share many of the hallmarks of human agriculture, including multiple domestications of wild species, artificial selection of the domesticates, and cultivation including physical and chemical methods for crop protection. Finally, the recent evidence for agriculturally related genetic changes in both the domesticates and the domesticators in human and ant agriculture suggests they represent coevolved mutualisms.

5. CONCLUSIONS

The ability to cultivate and breed plants and animals represents one of the most important developments in human history, allowing rapid and tremendous population expansion. Today domesticated plants and animals consitute an immense proportion of the global caloric intake by humans. Species that have been successfully domesticated share some important characteristics that predispose them to agriculture, and they have undergone significant genetic modification during domestication. Although the changes in domesticated plants and animals have been recognized for millennia, recent work has shown that humans, too, have undergone evolutionary changes in response to agriculture. These genetic changes in humans have occurred in response to farming and consuming specific plants or animals, and they illustrate the coevolutionary nature of agriculture. These general findings have parallels in agriculture by ants, and they show that agriculture and its evolutionary benefits and processes are not unique to humans.

See also chapter VI.7 and chapter VIII.10.

FURTHER READING

Belyaev, D. K. 1979. Destabilizing selection as a factor in domestication. Journal of Heredity 70: 301–308. *An experimental study of domestication in the silver fox.*

Currie, C. R., J. A. Scott, R. C. Summerbell, and D. Malloch. 1999. Fungus-growing ants use antibiotic-producing bacteria to control garden parasites. Nature 398: 701–704.

Diamond, J. 1999. Guns, Germs, and Steel. New York: W. W. Norton.

Diamond, J. 2002. Evolution, consequences and future of plant and animal domestication. Nature 418: 700–707.

Hölldobler, B., and E. O. Wilson. 2011. The Leafcutter Ants. New York: W. W. Norton.

Pinto-Tomas, A. A., M. A. Anderson, G. Suen, D. M. Stevenson, F.S.T. Chu, W. W. Cleland, P. J. Weimer, and C. R. Currie. 2009. Symbiotic nitrogen fixation in the fungus gardens of leaf-cutter ants. Science 326: 1120–1123. *This paper describes another similarity between human crops and ant crops, the need for symbiotic, nitrogen-fixing bacteria.*

Pollan, M. 2001. The Botany of Desire. New York: Random House.

Suen, G., C. Teiling, L. Li, C. Holt, E. Abouheif, E. Bornberg-Bauer, P. Bouffard, et al. 2011. The genome sequence of the leaf-cutter ant *Atta cephalotes* reveals insights into its obligate symbiotic lifestyle. PLoS Genetics 7: e1002007.

VIII.6

Evolution and Conservation

H. Bradley Shaffer

OUTLINE

1. Evolution, genetics, and conservation
2. Process versus pattern and why both matter
3. The enemies to watch out for
4. What genomics brings to the table
5. Concluding thoughts and prospectus

Traditionally, evolutionary biology has had a distant relationship to conservation compared with ecology and field-based natural history. However, this situation has changed dramatically in the last two decades, particularly as abundant molecular data have become available for at-risk species of conservation concern. As the availability of genome-level data for these species increases, the role of evolutionary biology in conservation management continues to grow to a far greater extent. The combination of these new data from microevolutionary analyses with more traditional input from phylogenetics and systematics has elevated evolutionary biology to a position of primary importance in conservation science.

GLOSSARY

Ecotone. A transition area where two distinct ecological communities meet and integrate.

Endangered Species Act (ESA). The US law that protects critically at-risk species from extinction due to human activities. It was passed into law in 1973 under President Richard Nixon and remains one of the most powerful conservation laws in existence.

Landscape Genetics/Genomics. The fields that integrate population genetics (or genomics) data with features of specific landscapes to study how those features influence the movements of genes and individual. This is a computationally intensive discipline that has become a major part of many conservation programs.

Nongovernmental Organization (NGO). An organization that is independent of any government, and generally has an important advocacy role. Several leading NGOs play a critical role, both locally and globally, in biological conservation.

Phylogenetic Diversity (PD). The amount of character change that evolves along a branch of a phylogenetic tree. PD may evolve along internal or tip branches and may be nonsymmetrical along two branches derived from a common ancestor.

Phylogenetics. The discipline that reconstructs the genealogical relationships of species and lineages.

Systematics. The discipline that names, describes, and infers the evolutionary history of species and lineages.

1. EVOLUTION, GENETICS, AND CONSERVATION

Suppose that you control environmental policy, and you have a choice: you can save either the New Zealand tuatara (*Sphenodon punctatus*) or the western fence lizard (*Sceloporus occidentalis*) from extinction. Whichever you choose, the other will go extinct. The tuatara is a lizard-like animal that is the sole surviving member of a once-diverse but now nearly extinct lineage of vertebrate life. Although that lineage was widespread and globally common 200 million years ago, it is currently down to one (or possibly two, virtually identical) species that occupies a handful of islands off the coasts of both main islands of New Zealand. If any lineage deserves the name "living fossil," it may be the tuatara. The western fence lizard is probably the most common lizard in North America. It is a widespread, abundant, and somewhat-unremarkable member of one of the most diverse and adaptable genera of lizards on earth. As of this writing, 92 species in the fence lizard genus *Sceloporus* are recognized, and new ones are constantly being described and characterized. So, how do you decide?

This kind of "conservation triage" is one of the arenas where evolutionary biology plays a critical role in conservation decision making. Evolutionary biology cannot tell a manager which species is more important, but it can frame the question and provide quantitative insights that

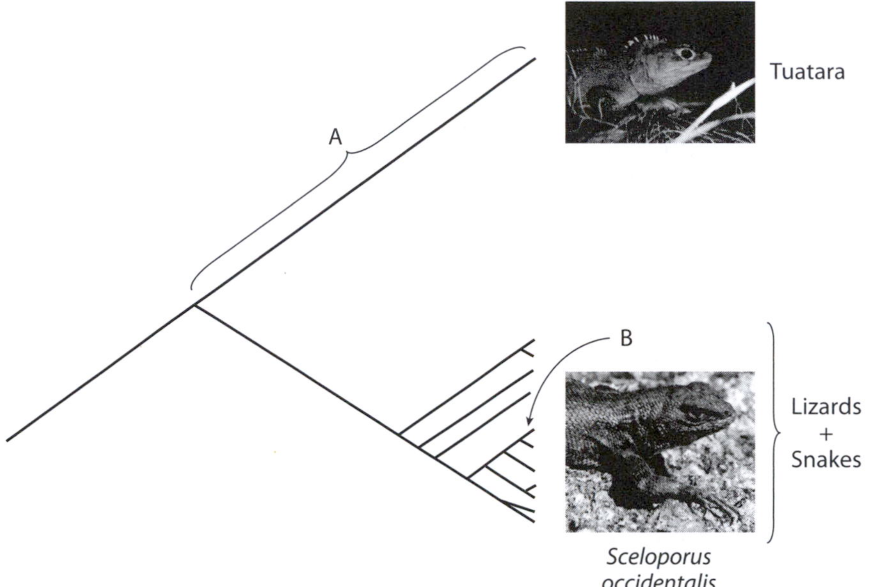

Figure 1. A phylogeny showing the relationships between the tuatara and its closest relatives, the snakes plus lizards. If the tuatara goes extinct, that species plus all the evolutionary history that occurred along the branch leading to it (labeled A in the figure) will be lost forever. If the western fence lizard (*Sceloporus occidentalis*) goes extinct, only that species plus the unique evolution on the much shorter branch B will be lost.

can help guide the decision-making process. In this particular case, virtually all policy-makers would choose the tuatara. The question is, Why? Conserving evolutionary history—that is, long branches of the tree of life that provide a record of the changes that have occurred during the history of life on earth—is a universally recognized component of conservation biology. The logic is that any lineage that took 200 million years to evolve is, in some real sense, more precious than another lineage that has many close relatives with which it shares most aspects of its morphology, ecology, and natural history. The phylogenetic uniqueness of the tuatara, its lack of close relatives, and the incredible length of its branch on the tree of life (figure 1) are all insights that come directly from understanding its evolutionary history and are the primary reasons why it is a global conservation icon. The same is true for many other important conservation targets, including the duck-billed platypus, the remaining rhinoceros species, and the California and Chinese redwoods.

At least three different components of evolutionary biology speak directly and forcefully to problems in conservation biology. The first is *systematics* and the related discipline of *phylogenetics*, and the tuatara is one of the classic examples (see Section II: Phylogenetics and the History of Life for additional detail). Both disciplines now rely heavily on molecular—usually DNA-level—data to make inferences about organisms, and both seek to describe the diversity and interrelationships of life on earth. Given that probably the single most important tenet of conservation biology is that "You cannot protect what you don't recognize," and that one goal of systematics is the delimitation of species and lineages, it seems clear that we require a catalog of life on earth before we can realistically plan for protecting it. For example, until fairly recently it was widely considered that the living tuatara consisted of a single species. However, in 1990, Daugherty and colleagues evaluated the variation found among remnant tuatara populations and hypothesized that the animals on Brothers Island actually constituted a different species, for which they used the name *Sphenodon guntheri*. In so doing, they simultaneously presented the world with one of the rarest species of vertebrates on earth and removed one population from the small catalog of known breeding populations of the critically endangered northern tuatara, *S. punctatus*. However, more recent work, based on additional data and sampling, reversed that decision, instead concluding that the tuatara "is best described as a single species that contains distinctive and important geographic variants." By studying one endangered species in ever-greater detail, this research team has continued to refine our understanding of the evolutionary history of tuataras and thus the populations and potential species in need of conservation actions.

A second, related area in which modern evolutionary biology informs conservation and management is phylogeography. Originally introduced by evolutionary geneticist John Avise in 1987, phylogeography uses genetic data to understand lineage formation and evolutionary diversification within, rather than among, species of organisms (see chapter II.5). As the name implies, a key goal of phylogeographic research is determining the relationship between the geographic location of populations and genetic differentiation among those same populations. In many cases, the recognition of deeply separated lineages within species has led to their independent protection and conservation. For example, recent phylogeographic work from our laboratory on the California tiger salamander (*Ambystoma californiense*) demonstrated that the species consists of at least three genetically independent lineages; two of these are geographic isolates in the south (Santa Barbara County) and the north (Sonoma County), while the third is the larger central group from the Great Central Valley. When the species was protected under the US *Endangered Species Act*, the combination of different levels and types of threats and the phylogeographic recognition of three lineages led to the independent protection of salamanders from Santa Barbara and Sonoma counties as endangered, while the rest of the species' range was separately listed as threatened under the ESA. These different listing levels (threatened versus endangered) actually do matter and could not have been proposed or implemented without this phylogeographic research.

Finally, population genetics has traditionally been the cornerstone of evolutionary biology's contribution to conservation, and this tradition has grown in the last few years. Three decades ago Frankel and Soulé (1981) emphasized the close connections between population genetics and conservation in conceptual areas ranging from minimum viable population sizes to the relationship between inbreeding depression and genetic drift. Frankel and Soulé's book, the first to use the words "conservation" and "evolution" in a single title, was also among the first to explicitly point out the expected relationship between small effective population size (N_e) and population health that is predicted from population genetics theory. Because inbreeding is generally detrimental to most outbreeding populations (see chapter IV.6), Frankel and Soulé argued that small populations would be particularly vulnerable to inbreeding depression and coined their "Basic rule of conservation genetics," which relates the change in inbreeding coefficient, ΔF, to the likelihood that a population will survive into future generations. In particular, they suggested that ΔF greater than about 1 percent constitutes "a threshold rate of inbreeding, above which fitness relentlessly declines" and populations go extinct. A related concept is Frankel's 50/500 rule, which states that on average, populations with a persistent N_e less than 50 may be in immediate danger of extinction, whereas over long time periods, populations with N_e less than 500 may not contain enough genetic variation to adapt to changing conditions (Braude and Low 2010). Although controversial, these "rules" emphasize a key point—when populations become too small and isolated, genetic drift can overcome natural selection, and low-fitness genotypes can rise in frequency by chance alone (see chapter IV.1). If this happens too often, or for too long, extinction may follow.

2. PROCESS VERSUS PATTERN AND WHY BOTH MATTER

A key question in conservation biology is deciding what to conserve and why. The resolution of this question depends on many factors. The country where the action is taking place may have strong conservation laws like the US ESA, or it may have virtually no history or capacity for even the weakest protection of taxa or landscapes. *Nongovernmental organizations* (NGOs) may be prominent partners that have their own opinions and agendas, local jurisdictions may interact in a positive or negative way with national governments and NGOs, and international organizations like the United Nations or World Bank may enter into the conversation.

Regardless of the organization, its politics, or its agenda, anyone who considers human-mediated extinction to be an outcome that should be avoided is really trying to conserve an aspect of evolutionary biology. At the broadest level, one can think about this problem in two ways. First, one can focus on conserving evolutionary history that has already occurred. Alternatively, or in addition, one can attempt to conserve the potential for future evolutionary change. Interestingly, these two approaches sometimes lead to very similar actions, and sometimes to radically different conservation priorities.

Conserving Evolutionary History

The US ESA is one of the most powerful pieces of conservation law in the world. Essentially, it simply states that species (including subspecies) should not be allowed to go extinct and that actions that lead to the further decline of listed species require special permission from the federal government. For evolutionary biologists, this means that the ESA seeks to protect one of the key products of the evolutionary process—species, including incipient species and subspecies. This theme of conserving the outcome of the evolutionary process is at the core of most conservation efforts. It is a very retrospective view of what to conserve—it requires that evolutionary biologists provide a clear picture of how many species, subspecies, and distinct population segments exist in a region, an indication of whether those populations are increasing or decreasing, and a measure of how distinct they are from one another. It is then up to the conservation community to take those results and use them to prioritize species and landscapes and try to preserve as much evolutionary history as possible.

Several specific approaches to conserving evolutionary history above and beyond the basic tenet that species extinction should be avoided are worth discussing in a bit more detail. First is the issue of retaining as much of a phylogenetic tree as possible—the tuatara problem that opened this essay. Two basic approaches have dominated the thinking on this topic. First, one can attempt to conserve the phylogenetic branch length (i.e., the sum of the branches of a phylogeny)—in units of time—that might be lost if an extinction event occurs. In the case of the tuatara, different opinions exist as to when it last shared a common ancestor with its closest relatives, but 271.5 million years seems to be a reasonable estimate. That is, if you lose the tuatara, you lose not only that species but also the 271.5 million years of evolutionary history that it uniquely represents among living organisms.

An alternative approach proposed by Faith (1992) is to conserve *phylogenetic diversity*, or PD. Faith proposed PD as an explicitly character-based approach to identifying taxa to conserve—those that have evolved lots of unique features (characters) contain more important evolutionary history than those that have changed relatively little and therefore remain relatively similar to other taxa. An example might be the human-chimp-gorilla trio of species. Although they all shared a common ancestor about 6 to 8 million years ago, the human lineage has changed considerably more, in a wide variety of biologically important ways, than has the chimp or gorilla from its most recent common ancestor. Thus, in this case, humans would have a far greater PD than the chimpanzee and would be a higher conservation priority if the two species were equally threatened. To the extent that unique features accumulate over time, branch lengths and PD will prioritize species for conservation in the same order. However, evolution does not always proceed in a tidy, time-dependent manner, as the human example points out. In those cases, one must decide what matters most—time or evolutionary novelties—as a target for conservation.

A very different approach is to recognize that the primary reason for human-mediated extinction is habitat alteration and destruction and that certain landscapes or regions tend to accumulate a great number of unique organisms. Certain regions of the world, like New Zealand or the Appalachian Mountains of eastern North America, are rich in endemic taxa found nowhere else on earth. The same can be said, of course, for many parts of the world—the Amazon basin is also very species rich, and most of the species found there are restricted to the Amazon. However, regions like New Zealand contain a disproportionately large number of old, unique lineages, like tuatara, flightless kiwis, and southern beech forests, while the Amazon abounds in species that are often members of widespread tropical genera. Many factors can contribute to these patterns, including the geological age and stability of a landscape, its isolation from other parts of the world, and the extent to which humans have disrupted ecological processes that naturally occur in the area. As phylogenies, and particularly as time-calibrated evolutionary trees, accumulate for the world's fauna and flora, conservation biologists can prioritize those regions that harbor the greatest depth and breadth of the tree of life and try to protect them into the future. In so doing they are saving species, but they are also saving the longest branches in the tree of life.

Conserving the Potential for Future Evolutionary Change

In 1997 Tom Smith and colleagues promoted a very different approach to using insights from evolutionary biology to conserve biodiversity. Smith argued, based on an analysis of a dozen populations of an African bird species, that there is a tremendous level of morphological differentiation between birds in the forest and the same species in the *ecotone* between forest and savanna habitats, and that this variation persists in the face of ongoing movement of individuals and gene flow. While the argument is fascinating in its own right, Smith and colleagues took it one step further, arguing that the ecotone habitat selects for morphologically very different birds from those in the rain forest. Their conclusion was quite radical: if you want to preserve evolutionary *processes* that generate diversity, you should preserve ecotones in addition to pure rain forest.

The importance of preserving habitats critical for the functioning of normal evolutionary processes within species has gained considerable traction in the last decade. The entire discipline of *landscape genetics*, including the emerging subdiscipline of *landscape genomics*, focuses on exactly this issue, and it represents one of the major growth areas in research on the genetics of natural populations. Here, the goal is to use standard population genetics data, in combination with geographic information system (GIS) data layers, to quantify the ways that organisms move across the landscapes they occupy. The approach is directly relevant to landscape management and conservation planning because it takes genetic data from organisms on landscapes and asks whether potential migration corridors and barriers to gene flow function to promote or to disrupt population connectivity. The results can provide unexpected insights into the ways organisms use their environment and can identify those habitat patches and corridors that are most important for maintaining normal evolutionary processes. Such data can take years to collect with traditional mark-recapture methods, but only weeks or months using the insights gained from landscape genetics. Particularly for threatened species, for which decision makers must act quickly, landscape genetics is a powerful conservation tool.

Finally, a very different kind of evolutionary process with enormous conservation consequences has recently been recognized. Given the impact of people on natural landscapes, human activities have the potential to exert strong selection pressures on populations in the wild, and recent studies have demonstrated that organisms are responding to this selection, sometimes in surprising ways. Although the study of rapid evolutionary change due to human activities is still in its infancy, this phenomenon has the potential to profoundly affect conservation outcomes. To take one example, human fisheries generally remove the largest individual fish, which are often both old and female (most fish species have indeterminate growth and continue growing throughout their life). This intense selection on the largest, most fecund females leads to an evolutionary decrease in body size, an earlier age at first reproduction (since having a few babies at a small body size is better than waiting and being caught by a fisherman), and a lower reproductive output for individuals and the species overall. Population models show that this shift from a life history in which individuals wait many years to reproduce, grow to a large size, and have many young, to the alternative strategy of early reproduction at a small size with few offspring leads to lower total biomass, decreased population sustainability, and a greater likelihood for population collapse or extinction. Recent models on the effects of Marine Protected Areas (MPAs) in preventing these conservation disasters have found that some MPAs ameliorate such evolutionary responses to human fishing, but others do not. For example, if there is extensive gene flow between fishing grounds and MPAs, then smaller females will migrate to, and breed in the MPA, and larger, protected females will migrate outside MPAs, where they will be caught and killed; in this case, little is gained from the MPA. Alternatively, if the MPA is very large, or if the fish tend not to migrate extensively, then the larger, protected females will remain in the MPA and provide a constant source of young, genetically unmodified individuals to the outside fishing grounds.

Regardless of how individual case studies play out, humans now constitute a potent force leading to rapid evolutionary change. We clearly need to add ourselves to the list of important processes at the intersection of evolution and conservation.

3. THE ENEMIES TO WATCH OUT FOR

Conservation biology is a complex business that involves equal parts of biology, politics, and economics if real progress is to be made and sustained. An essential element of conservation is to think clearly about what one wants to protect and what one wants to avoid in terms of conservation outcomes. Evolutionary genetics in particular has made very substantial contributions to the identification of these problems and their solutions. To take one of the highest-profile case studies yet conducted, consider the Florida panther, *Puma concolor coryi*. Designated the state animal of Florida in 1982, this endemic subspecies of panther (also known as puma, mountain lion, or cougar in other parts of the species' vast range) was reduced to about 20 individuals in the 1970s, at which time it was showing clear signs of genetic inbreeding depression. As a result, the population was intentionally supplemented with panthers from the adjacent, much larger, and more outbred population in east Texas from the subspecies *Puma concolor stanleyana*. Hybridization occurred, the telltale signs of inbreeding depression disappeared, and Florida's state animal appears to be in a strong phase of population growth and genetic recovery. Except—is it really recovering? There are definitely healthy panthers back in the Florida Everglades, but are they *Puma concolor coryi*? Or has that subspecies been driven extinct by an invasive hybrid panther? The example raises the critical question, What does one want to conserve, and why? Is one protecting native genes, naturally evolved lineages, or ecological roles? Is it better to keep some native Florida panther genes on the Florida landscape than none at all (the hybrid panthers definitely have a lot of native Florida genes), or is hybrid "impurity" worse than extinction? Evolutionary genetics can provide the data, but not the answers, to the moral dilemmas that these questions pose.

Hybridization

Hybridization happens all the time, both because of natural processes and because humans meddle with species and landscapes (see chapter IV.3). Evolutionary genetics can bring great clarity to the status of populations of plants and animals, including precise estimates of the fraction of the genome that is native versus derived from a different species. Most practitioners naturally assume that the primary goal of conservation biology is to preserve pure genetic lineages on the landscapes where they evolved. Thus, in a case that our lab has worked on for several years, human-transported, nonnative Barred tiger salamanders (*Ambystoma tigrinum mavortium*) from Texas and New Mexico have successfully hybridized with native, endangered California tiger salamanders (*A. californiense*) across much of central California, and at least 20 percent of the range of the California tiger salamander is now occupied by hybrids (Fitzpatrick et al. 2010). In this and many other cases, hybrids are viewed as a conservation threat, to be identified, eliminated, and replaced with pure natives if at all possible. Trout, salmon, escaped genes from agricultural plants, and domestic dog genes infiltrating wolf and coyote populations are a few of the better-studied examples of this phenomenon, and the evolutionary analysis of hybridization constitutes the key data on which conservation actions have been based. On the other side of the issue, "genetic rescue," particularly for large mammal populations that have dipped below a genetically sustainable size, remains a viable and occasionally used strategy to augment populations that would otherwise go extinct without genetic intervention—the Florida panther is a classic example. Importantly, these are cases in which evolutionary biology can provide the key insights on expected and realized inbreeding depression, can track the fate of nonnative genes as they move through populations, and can measure the fitness consequences of hybridization. What it cannot do is tell us, as managers of the fate of populations, whether and when we should bring in foreign genes as a last-ditch conservation effort.

Population Bottlenecks, Population Isolation, and Effective Population Size

More books and papers discuss the interface of population genetics and conservation biology than any other aspect of evolutionary conservation biology. There are many reasons for this, but the primary one is that the connection between classical problems in population genetics and conservation biology is both direct and clear. Population geneticists tend to worry about the relationships between genetic drift caused by small effective population sizes, the efficacy of natural selection in shaping variation in the field, and the interplay among mutation, selection, and drift. Conservation biologists spend a great deal of time and energy trying to understand the health of populations in nature, including the effects of small population sizes. Both groups recognize that small populations have a higher chance of becoming inbred and that high levels of standing genetic variation are critical for the current and future health of populations and species. Any good field ecologist knows that populations fluctuate over time and that low numbers are sometimes unavoidable. However, when populations become completely isolated, then migrants from larger populations cannot help those reduced populations recover, either demographically or genetically. The result is small, isolated populations that lose genetic variation over time, become inbred, express deleterious mutations at a higher frequency, and have limited resilience to bounce back from unavoidable population crashes. And given their isolation, when they go locally extinct, they cannot be repopulated—they stay extinct.

One of the holy grails of both population genetics and conservation biology has been to infer past and current demographic parameters using the standing genetic variation that exists in natural populations. At least for genetic markers that are unaffected by strong natural selection (so-called neutral genetic variation), there is a long, rich history of using patterns of variation within and among populations to estimate the amount of migration (or gene flow) among populations. Here, the idea is straightforward—if a mutation arises in one population, and no migrants successfully leave that population and reproduce in a new population, then the mutation will remain exclusively in its site of origin. Such "private" variants will build up over time, such that the longer a population remains in isolation, the greater will be its genetic distinctiveness from other populations. Sewall Wright, one of the pioneers in the field of population genetics, developed a series of statistical methods to quantify this kind of genetic differentiation, and his F-statistics remain the primary way in which such realized gene flow is measured in nature. High values imply little or no gene flow among populations, whereas low values suggest that successful migration and breeding occur regularly. Newer methods can measure the movement of individuals by recognizing that if an occasional migrant moves between somewhat-differentiated populations, that individual can be "assigned" to its population of origin based on its multigene identity. Conducting such assignment tests with confidence requires a great deal of genetic (or genomic) data, but it represents a powerful addition to the conservation biologist's toolkit for measuring how organisms successfully traverse landscapes in nature.

Some of the most compelling and exciting new developments in population genetics allow conservationists to study, with far greater precision, the actual size of a breeding population in nature. Population biologists recognize two different ways to measure population size—N_e, or the effective population size, and N_c, the census population size. The difference is straightforward and absolutely critical: N_c is the number of individuals in a population, whereas N_e reflects the number of individuals who breed and contribute to the genetic variation in the species (note that this represents a simplification of a mathematically complicated concept). For example, if a population has 50 males and 50 females, its census size will be 100. However, if only one of those males and 10 of those females actually breed, its effective size will be close to 10; this latter population will suffer much greater genetic drift and potential inbreeding depression than one in which all 100 individuals breed. Both N_c, and N_e are important, and they measure different aspects of the health of a population. Recent advances in molecular population genetics have provided new tools to measure N_e in nature, sometimes from only a single individual. The math is complex and the requirements for both the number of genetic markers and the proportion of the population sampled may be large, but the results indicate that the effective population size can be estimated, often relatively easily and quickly.

4. WHAT GENOMICS BRINGS TO THE TABLE

Genomics means many things to many people, ranging from data on the full DNA sequence of an organism assembled into complete chromosomes to having "a lot of sequence data." A truly complete genome, in which every base pair has been sequenced and assembled into contiguous chromosomes, has yet to be completed for any vertebrate, although several species, including humans, have essentially complete genomes. However, an increasingly large number of species have had many thousands of genes sequenced, sometimes for multiple individuals and populations. In either case, genomics *always* means having lots of data for each study organism—it may mean billions of nucleotides of sequence data (many vertebrate genomes are around 2–3 billion nucleotides in length), or it may mean thousands, but it is always a lot.

It seems clear that in the next few years, genomic data will dominate population genetics, phylogenetics, and conservation genetics research. Aside from the general truism that more data are always better than fewer data, this onslaught of new information should open several critical avenues of research at the interface of evolutionary and conservation biology (see chapters in Section V: Genes, Genomes, Phenotypes). First, genomic data allow one to study the genetics of functionally important genes as well as neutral ones not affected by natural selection. Presumably, conservation geneticists should focus on genetic variation in the functionally important genes, since they are most important to survival and the ability to adapt to future change. For neutral loci, genetic variation per se is not important to population health, but the standing levels of variation at those loci reflect the past and current effective population size and levels of gene flow or genetic isolation. Both are important to evolutionists and conservationists, but the two are very different. When a large part of the genome is subject to study, the neutral and selected loci can be neatly separated, leading to important insights from both genomic components.

In a similar vein, genomic data help the evolution and conservation community focus much more clearly on exactly what needs to be conserved. For example, the major histocompatibility complex (MHC) is a set of genes involved in the immune response to disease of many vertebrates. Certain diseases, including the fibropapilloma tumors in green sea turtles or the devil facial tumor disease in Tasmanian devils of Australia, may be involved in bringing these endangered taxa to the brink of extinction, and conserving and managing populations for MHC variation may be a way to increase their chances of survival. Genes that allow cold-adapted plants and animals to better cope with human-induced climate change in the next decade are another key class of functional genes that genomics may bring to the conservation table.

The impact of genomic data on evolutionary biology in general, and conservation in particular, is huge, multifaceted, and largely unexplored. It stands as perhaps the most important frontier at the intersection of evolution and conservation biology.

5. CONCLUDING THOUGHTS AND PROSPECTUS

Evolution is all about change—changes in allele frequencies over time, in population size and distributions, and in species composition due to extinction and speciation. Conservation is about managing for change—climate change, invasive species, hybridization, human habitat modifications, and a host of others. As large-scale genetic analyses become increasing available for nonmodel organisms, it seems inevitable that evolutionary genetic analyses will move to center stage in the conservation and management of declining species. Consider, for example, being able to track the reproductive output of captive-reared organisms that are repatriated into the wild, allowing resource managers to measure the impact of their conservation efforts, in the wild, in real time. Or imagine having the data to be able to determine, with very high accuracy, exactly how many individuals have moved between habitat patches historically, and using that information to mimic those patterns with human-assisted migration in fragmented habitats. Or being able to quantify, for any newly proposed protected park, exactly how much of the phylogenetic tree of life is contained in that park—not for specific taxa based on a few genes, but for all life. These are heady ideas, but as genomics, metagenomics, and phylogenomics become affordable and easier to accomplish, they are also very realistic. And they just might help conserve a bit more of our declining biosphere.

FURTHER READING

Allendorf, F. W., and G. Luikart. 2007. Conservation and the genetics of populations. Malden, MA: Blackwell. *A wonderful reference, particularly for the more mathematical aspects of conservation genetics. This book is particularly strong on the interface of population genetics and conservation.*

Braude, S., and B. S. Low, eds. 2010. An Introduction to Methods and Models in Ecology, Evolution, and Conservation Biology. Princeton, NJ: Princeton University Press.

DeWoody, J. A., J. W. Bickham, C. H. Michler, K. M. Nichols, O. E. Rhodes Jr., and K. E. Woeste, eds. 2010. Molecular Approaches in Natural Resource Conservation and Management. New York: Cambridge University Press. *An edited volume, with some very specific chapters that may be of limited general interest, but others of quite broad appeal. A great source of recent examples and case studies using a wide range of molecular genetic tools to inform conservation.*

Faith, D. P. 1992. Conservation evaluation and phylogenetic diversity. Biological Conservation 61: 1010. *The paper that introduced the phylogeny-based concept of conserving character evolution into the mainstream of conservation thinking.*

Fitzpatrick, B. M., J. R. Johnson, D. K. Kump, J. J. Smith, S. R. Voss, and H. B. Shaffer. 2010. Rapid spread of invasive genes into a threatened native species. Proceedings of the National Academy of Sciences USA 107: 3606–3610. *Following a well-documented, human-mediated introduction, this paper shows that some invasive genes can sweep across landscapes at incredibly rates, while other genes are much slower.*

Frankel, O. H., and M. E. Soulé. 1981. Conservation and Evolution. Cambridge: Cambridge University Press. *The original book that brought together the fields of conservation biology and evolutionary genetics—a "must-read."*

Frankham, R., J. D. Ballou, and D. A. Briscoe. 2004. A Primer of Conservation Genetics. Cambridge: Cambridge University Press.

Höglund, J. 2009. Evolutionary Conservation Genetics. Oxford: Oxford University Press.

Schonewald, C. M., S. M. Chambers, B. MacBryde, and W. L. Thomas, eds. 2003. Genetics and Conservation. Caldwell, NJ: Blackburn.

VIII.7

Directed Evolution

Erik Quandt and Andrew D. Ellington

OUTLINE

1. Directed evolution of nucleic acids
2. Directed evolution of proteins
3. Directed evolution of cells
4. The future of directed evolution

Directed evolution is a process in which scientists perform experiments that use selection to push molecular or cellular systems toward some goal or outcome of interest. The objectives of this work include the production of substances of value and improved understanding of the evolutionary process. Elucidating the precise mechanisms by which improvements occur is often of particular interest. In general, directed evolution requires a genetic system in which information is encoded, heritable, and mutable; a means for selecting among variants based on differences in their functional capacities; and the ability to amplify those molecules or organisms that have been selected.

GLOSSARY

Aptamer. A short nucleic acid molecule that binds to a specific target molecule.

Bacteriophage. A virus that infects bacteria.

Esterase. An enzyme that splits esters into an acid and an alcohol in a chemical reaction with water, also known as *hydrolysis*.

Fluorescence-Activated Cell Sorting (FACS). A method for sorting a heterogeneous mixture of cells into two or more containers, one cell at a time, based on the specific light scattering and fluorescent characteristics of each cell.

Lipase. An enzyme that catalyzes the hydrolysis of ester chemical bonds of lipid substrates.

Messenger RNA (mRNA). The product of transcription of a DNA template, which in turn encodes the sequence for the production of a protein.

Peptide. A short polymer of amino acids linked by peptide bonds.

Polymerase Chain Reaction (PCR). Enzymatic reaction in which a small number of DNA molecules can be amplified into many copies.

Protease. A protein capable of hydrolyzing (breaking) a peptide bond.

Quasispecies. A large group of related genotypes that exist in a population that experiences a high mutation rate, in which a large fraction of offspring are expected to contain one or more mutations relative to the parent.

Ribozyme. An RNA molecule with a defined tertiary structure that enables it to catalyze a chemical reaction.

Transfer RNA (tRNA). An RNA molecule linked to an amino acid that is involved in the translation of an mRNA transcript into protein.

Transcription. The process of creating a complementary RNA copy (mRNA) of a sequence of DNA.

Transcription Factor. A protein that binds to specific DNA sequences, thereby affecting the transcription of genetic information from DNA to mRNA.

Translation. The process of decoding an mRNA molecule into a polypeptide chain.

Tumor Necrosis Factor-α (TNF-α). A protein involved in the regulation of certain immune cells that induces inflammation and cell death and thereby inhibits tumorigenesis and viral replication.

Directed evolution involves guiding the natural selection of molecular or cellular systems toward some goal of interest. That goal may be either to enhance basic understanding of natural systems or to produce something of value. In either case, this approach relies on changing the frequency of genotypes over time, with concomitant changes in function and phenotype, just as natural evolution does. However, from the point of view of an "evolutionary engineer" the mechanism by which changes in frequency are obtained is often of particular interest. In general, directed evolution requires a genetic system (a system in which information is encoded, heritable, and mutable), a means for sieving the variants that are present in that system (by differences in either function or fitness), and the ability to amplify those molecules or organisms that pass through the sieve.

The directed evolution of molecules and cells has been carried out for decades, although the methods used for directed evolution have gained in technical sophistication and in the breadth of systems that can be tamed. Indeed, if one includes animal and plant husbandry, then directed evolution has been coincident with the evolution of human society. This chapter first discusses molecular evolution, focusing on the selection of nucleic acids and proteins that have novel functions. The simple rule set for directed evolution described can be satisfied in a surprisingly large variety of ways, including in molecular systems that at first glance appear to have the properties of cells but that are not actually cellular. The chapter then considers how the evolution of cells can be directed and accelerated.

1. DIRECTED EVOLUTION OF NUCLEIC ACIDS

The forefather of the directed evolution of molecules was Sol Spiegelman of the University of Illinois at Urbana-Champaign. Spiegelman and his group studied a small bacteriophage, called Qbeta, that has an RNA genome. In nature, this virus infects host cells of the bacterium *Escherichia coli* to replicate. However, Spiegelman's team found that the protein involved in replicating this bacteriophage, Qbeta replicase, was capable on its own of replicating RNA molecules in a test tube. The only requirements for replication were an initial RNA template, the replicase, nucleoside triphosphates, and appropriate buffer conditions. However, once freed from the confines of a cell, the Qbeta replicase tended to make multiple mutations and deletions in the RNA template, ultimately leading to smaller and smaller RNA molecules. The evolutionary fates of these so-called minimonster variants could be altered depending on the experimental conditions (Saffhill et al. 1970). Spiegelman's demonstration was highly influential and became an icon for the field. However, Qbeta replicase proved too difficult to control, since any successful variants that arose were transient and quickly mutated into a complex and ever-shifting quasispecies. Thus other methods and systems were required to advance the study of directed evolution.

The modern era of directed molecular evolution had to await the development of several technologies, chief among them the chemical synthesis of DNA and the advent of the polymerase chain reaction (PCR). Chemical DNA synthesis allowed a defined, yet random, pool of nucleic acids to be generated, providing the perfect substrate for directed evolution experiments. If constant sequence regions were included at the termini of this pool, then its members could be exponentially amplified by the PCR. All that remained to ensure the selection of functional nucleic acids was to impose some sort of selection that would differentiate the members of the pool from one another, a feat that was performed by Jack Szostak and coworkers (Ellington and Szostak 1990). Each of the different sequences in the pool can fold into a different shape; each of the different shapes therefore potentially has a different function or phenotype. One of the first selections from a random sequence pool involved identifying single-stranded DNA molecules that could bind to a particular molecular dye. The nucleic acid pool was poured down a column containing immobilized dye molecules; some variants stuck to these molecules, while most of the population flowed through. Once the bound variants were eluted (by unfolding the nucleic acids) they were amplified by the PCR, and single strands were prepared from the double-stranded product. Iterative cycles of selection and amplification resulted in the gradual accumulation of those molecules that had high affinity for a given dye (plate 4). One analogy that is often used to describe these experiments is that they are akin to looking for a needle in a haystack; every time you grab a handful of hay that contains a single needle, you then convert that needle into a handful of needles. Given that the hay outnumbered the needles in the experiment described by a factor of almost 10^{10}:1, iterative selection and amplification were essential to the purification of the needle.

The selection of nucleic acids that can bind to ligands became known as *in vitro selection*, and the binding sequences that result from this approach are called *aptamers* (from the Latin *aptus*, "to fit"). In both of the experiments described so far, the visions of the experimenters were driven not so much by specific hypotheses as by the availability of the technology and a desire for unfettered exploration. Spiegelman set off into the unknown to determine whether viral replication in the test tube (outside the host cell) was even possible, whereas Szostak took the great leap that there would be at least a few, previously unknown, shapes in the haystack that could bind to a dye. However, these technological innovations also ended up providing some support for the hypothesis that life may have got its start from self-replicating nucleic acids. Along with the discovery by Tom Cech and Sidney Altman that RNA could act as a catalyst (North 1989), the finding that nucleic-acid-binding variants could be selected from random strings of information implied that nucleic acids might avoid the "chicken-and-egg problem" with respect to the origin of living systems. That is, proteins are the functional machines, while nucleic acids bear information. Proteins are needed to replicate, while nucleic acids must be replicated. It seemed unlikely that both these complex biopolymers arose simultaneously, and this problem was a huge conundrum for thinking about origins. However, once it was clear that nucleic acids were in fact "chicken-eggs" (being both functional machines and information bearing), this conundrum was deftly resolved.

Around the same time that the antidye aptamers were being generated, Larry Gold and Craig Tuerk at the University of Colorado at Boulder showed that protein-binding nucleic acids could be selected from random sequence pools (Tuerk and Gold 1990). The same technology that might contribute to explaining life's origins therefore could also be used to create new drugs. Probably the most famous aptamer produced by directed evolution is known as *Macugen*. It was selected to bind to and thereby inhibit the function of the human protein vascular endothelial growth factor (Ng et al. 2006). This experimentally evolved RNA molecule is now used clinically to combat wet macular degeneration, a common cause of blindness in the elderly.

Catalytic nucleic acids can also be selected from random sequence pools. One fascinating and important ribozyme is known as the *Class I Bartel ligase*, after David Bartel, who discovered it (Bartel and Szostak 1993). This ribozyme has been of seminal importance in understanding the origin and evolution of life. The Bartel ligase seems, on first impression, to be a miracle. The probability of selecting it from a random sequence pool can be calculated, and it turns out that it should have been found once every 10,000 times that Bartel carried out his directed evolution experiment. Although it is possible that Bartel was extraordinarily lucky, the alternative and probably better explanation is that while any particular ligase was unlikely to have evolved, the ligase function itself was much more likely to have arisen. Thus, if the experiment were to be carried out again, a molecule of similar functionality and complexity—but with an entirely different sequence—would emerge and be selected. This finding is extremely important because it means that there are many possible routes from origins to modern, complex systems. As the late paleontologist Steven Jay Gould suggested, if we were to run the tape of life again, we'd probably get a very different answer.

2. DIRECTED EVOLUTION OF PROTEINS

It has also proven possible to direct the evolution of proteins. However, in this instance genotype and phenotype are not embedded in the selfsame molecule—that is, they are not chicken-eggs in the way that some nucleic acids are. Therefore, there must be some other way to connect genotype and phenotype to perform directed evolution on proteins. One of the first ways this was done was by appending a short random library to the gene that encodes a coat protein of a bacteriophage, such that the library of peptides would then be expressed on the surface of the bacteriophage. The displayed peptide variants could then be selected on the basis of their ability to bind a ligand, as with nucleic acids, and amplified not by the PCR but by passage through cells that could be infected by the selected bacteriophage (Smith and Petrenko 1997) (plate 5). Such *phage display* methods have become very popular and are the basis for selecting not just peptides but antibodies and enzymes, as well. For example, Humira, an antibody drug effective in the treatment of rheumatoid arthritis, was selected via phage display. By displaying an antibody library against a protein involved in inflammation response (TNF-α), researchers were able to select and amplify high-affinity antibodies that could bind to the protein.

The same methods were later expanded to cell surfaces, so that individual cells now are the vehicle connecting protein variants on the surface to the genes encoding those proteins. As with the phage, the proteins on the cell surface could be selected for either binding or catalysis. For example, George Georgiou and coworkers showed that libraries of peptide proteases expressed on the surface of *E. coli* could be selected for altered substrate specificity (Varadarajan et al. 2005). Cleavage of a new desired "green" substrate by a given protease variant led to the accumulation of that color on the surface of the bacteria, while there was a parallel opportunity to cleave and accumulate a parental undesired "red" substrate. Both positive and negative selections could thus be applied to tune substrate specificity. The cells expressing protease variants with altered, desired specificities (colored green, but not red) were screened from the protease libraries based on a technique known as *fluorescence-activated cell sorting* (FACS) in which individual cells are sorted based on their fluorescence. The selected variants were further amplified by bacterial growth, and multiple cycles of screening and amplification led to the winnowing of the initial population to those few proteases with the desired new specificities. The ability to tune protease specificities may someday have practical applications, such as destroying undesirable proteins, including viral proteins.

The selection of proteins inside cells is similar to cell-surface display, except that instead of selecting directly for the ability of individual proteins to bind or catalyze reactions, researchers must instead select for the impact of binding or catalysis on cellular phenotypes. For example, some antibiotic resistance genes encode enzymes that modify antibiotics in some way. New resistance functions can be selected by challenging cells with a different antibiotic. For cells to survive, the resistance element must accumulate mutations that change its function. The challenge is often to make sure that selection is focused on a particular enzyme of interest and that the cell does not follow some other evolutionary pathway that leads to survival (i.e., a different cellular enzyme than the one of interest might mutate in a way that leads to antibiotic resistance). To focus selection on a particular enzyme, user-mutagenized libraries of enzymes can be generated, just as they were for the ribozyme and protease selections described earlier. There are various ways to both mutagenize and select a given library. In the 1990s, Pim Stemmer came up with a brilliant technique to speed the evolution of proteins by allowing for in vitro recombination (Stemmer 1994). In this method, known as *DNA shuffling*, different enzyme variants are selected from a library (or may otherwise already be present). By cutting the genes for the enzymes into pieces, and then using PCR to recombine and eventually reassemble them into the full-length gene, many mutations can be brought together in the same gene. This approach is much faster than natural recombination, which generally involves only two gene copies at a time. In either case, the interesting assumption is made that combinations (or at least some combinations) of favorable mutations will themselves be favorable. This assumption has in large measure turned out to be true, perhaps because genes have evolved to evolve. That is, the types of protein sequences and structures amenable to recombination are those that have been successful at responding to changing conditions during the long course of evolution.

The methods so far described all require living cells either to make phage or to express proteins. Other researchers have devised clever ways to carry out protein evolution even without cells, by using in vitro transcription and translation systems that contain all the components necessary to make proteins, including ribosomes and tRNAs. For example, Andreas Pluckthun and his group managed to stall ribosomes in the process of translating mRNAs and thereby could connect the mRNA information being read with the protein function being translated (Hanes and Pluckthun 1997). Similarly, Jack Szostak's group figured out how to use the antibiotic puromycin to covalently couple a protein being translated on the ribosome to the mRNA making that protein (Roberts and Szostak 1997). As with natural selection and the various schemes described earlier, the coupling of genetically encoded information with function is essential for sieving through large sequence libraries to find rare functions. One of the advantages of these ribosome-based methods is that they can be used to look through much larger sequence populations than cell-based methods.

Researchers have also begun to create cell-like bubbles to assist with directed evolution (Griffiths and Tawfik 2006). Water-in-oil emulsions can be created by simply mixing these two components and shaking them (much like making salad dressing). If in vitro transcription and translation components are added to the aqueous component, then each small aqueous bubble in the sea of oil will be capable of making proteins. If only one DNA template or mRNA molecule is captured per bubble, then only one type of protein will be made in that bubble. The problem then becomes how to capture the bubble making the protein variant of interest. One solution is to have the protein feed back on the nucleic acid that produced it, and various methods have been developed that either mark the nucleic acid (for example, by methylation), amplify the nucleic acid (via a translated polymerase), or capture the nucleic acid (via a binding protein). Once the mixture is demulsified, all the nucleic acids are remixed together, but only those that have encoded a functional protein are marked, amplified, or bound, and they can be carried into subsequent rounds of selection and amplification. Another solution to the problem of connecting the appropriate bubble with the phenotype of interest has been to develop methods to capture the bubble itself. Adding a lipid coat to the aqueous bubbles allows them to be stabilized and sorted by FACS. Thus, no direct feedback loop to nucleic acids is required, which simplifies the procedures and greatly expands what kinds of enzymes can be selected. For example, lipases that act on esterases can turn over fluorescent substrates, so that more-active lipase variants will accumulate more fluorescence in their bubbles, which can in turn be sieved from a larger background population (Griffiths and Tawfik 2003). As with all the other techniques described, amplification in vitro provides a selective advantage to the functional lipases by increasing their representation in the next generation.

3. DIRECTED EVOLUTION OF CELLS

The directed evolution of whole cells can yield both the simplest and most complex products. For example, the adaptation of cells to ferment beer (by producing ethanol) and help with baking (by producing carbon dioxide) is almost as old as human society, and researchers from Louis Pasteur onward have bred cells for industrial purposes. In 1928 Alexander Fleming discovered that the antibiotic penicillin was made by the fungus *Penicillium notatum*. While this discovery would in time change the world of medicine, further engineering was necessary. Strong demand for the drug during the Second World War necessitated increased production beyond what the organism naturally produced. To direct the evolution of the organism toward greater antibiotic production, cultures were subjected to X-ray radiation, which was known to mutagenize DNA, thereby accelerating the accumulation of genetic diversity. This mutagenized population was then screened for those cells that produced the most penicillin (Backus, Stauffer, and Johnson 1946), and these selected cultures produced enough penicillin to meet the demand. This use of random mutagenesis followed by screening for a desired phenotype is commonly referred to as *strain improvement* and is a now common practice in many industries in which the production of a particular product is dependent on microbial synthesis.

Experiments that could support these sorts of applications were initiated by Barry Hall, of the University of Rochester. Hall (2003) wondered what would happen if an enzyme of *E. coli* was deleted. This enzyme, β-galactosidase, was responsible for the ability of these cells to grow on the sugar lactose. When the enzyme was deleted, the cells could not grow on lactose, at least not initially. Over time, however, the cells began to grow slowly and, eventually, evolved to grow more rapidly, because another gene in the organism had accumulated mutations that allowed it to break down lactose. By focusing selective pressure on one function, Hall turned the entire organism into a vehicle for finding and improving a suitable enzyme.

While research into the directed evolution of cells led to a better understanding of the source, rate, and type of mutations that could lead to new or modified proteins, it was still difficult to target mutations and selection within a genome, especially for complex phenotypes like antibiotic production that required the adaptation of multiple genes and enzymes in parallel. Going well beyond Pasteur and his contemporaries required the development of tools that can recombine and modify individual sites in a genome, and advances in molecular biology that allow entire genomes to be sequenced. These modern techniques have had a dramatic effect on the time and effort required to generate improved bacterial strains.

Just as protein shuffling was developed to facilitate the accumulation of favorable mutations, other methods have been developed for shuffling entire genomes. Following a single round of classical strain improvement (random mutagenesis and screening or selection), cells are stripped of their cell walls (turned into protoplasts) and then induced to fuse with one another in what is essentially a multiparent mating event. Because a cell can generally accommodate only one genome, the genetic material from fused cells must be resolved into a single unit by the cell's recombination and repair machinery. The recombination process generates mosaic genomes, thereby amplifying the population's genetic diversity. This process can be repeated multiple times, allowing the improved strain to accumulate many additive and even synergistic mutations with respect to the desired phenotype. In a striking example of the power of this method, a group improved the production of the antibiotic tylosin by *Streptomyces fradiae* by about ninefold after only two rounds of genome shuffling that took roughly one year (Zhang et al. 2002). The resulting strains were found to produce as much tylosin as strains that had been independently subjected to 20 rounds of classical strain improvement over 20 years.

Other efforts have focused on reprogramming the cell as a whole by mutating the master regulators that the cell uses to control how and when proteins are made. Changes to these *transcription factors* simultaneously affect the levels of expression for many genes, thereby altering the levels of many proteins and broadly affecting how the cell operates and behaves. The altered regulatory program is presumably not optimal for the organism in its natural context but may be much more productive in an industrial setting. By screening mutant libraries of transcription factors, researchers have isolated strains with increased tolerance to industrial processes and by-products as well as enhanced the production of small molecules.

While genome shuffling and transcription-factor engineering can accelerate evolution, these processes still rely on nondirected changes in sequence or expression as their inputs. To produce a revolution in genome engineering on the scale of that already occurring in protein engineering, it was necessary to direct mutations to particular genes, an achievement that George Church and coworkers published as *multiplex automated genome engineering*, or MAGE (Wang et al. 2009) (plate 6). This technique involves synthesizing single-stranded DNA oligonucleotides corresponding to particular genomic locations. The oligonucleotides can enter the cells, and mutations engineered into the oligonucleotides may then be incorporated into the genome through recombination. By using an automated system to grow cells and deliver the mutant oligonucleotides, the researchers were able to achieve a high rate of mutation at several genomic sites in only a few cycles. This ability to target mutations to specific genomic locations enables researchers to precisely manipulate cells in ways that could be used to optimize the production of proteins or other molecules for industry, or even to rewrite the genetic code at large.

These amazing advances in genomic engineering require concomitantly amazing new analytical tools. With the development of so-called next-generation sequencing platforms that generate massive amounts of sequence data, the directed evolution of a microbial population can now be observed at the level of individual sequence changes within entire genomes. In 1988 Richard Lenski established his "long-term evolution experiment" by inoculating a strain of *E. coli* into minimal media and subsequently transferring a small amount of the previous culture into fresh media each day. Over time, the cells have become adapted to these environmental conditions, and their growth rate has accelerated. The genomes of the evolved organisms from various points through 40,000 generations were sequenced to determine the many underlying mutations responsible for this adaptation (Barrick et al. 2009). As the genomic engineering techniques described earlier are increasingly melded with large-scale acquisition of sequence data, the ability to understand and shape genomes will become commonplace.

4. THE FUTURE OF DIRECTED EVOLUTION

For the pioneers of directed evolution, including Spiegelman and Hall, the experimenter was in charge of directing the selection, whether it was in a population of molecules or organisms. These researchers had to take whatever random mutations the experimental system provided and then study which of these mutations led to changes and functions of interest. This process remains characteristic of the field of directed evolution, since many researchers are still addressing basic questions such as, What outcomes can be produced? What can we make? However, the field is now also beginning to develop models that are predictive, in part based on the paired abilities to direct where mutations occur and to analyze the repertoire of phenotypes associated with vast numbers of mutations. This process will accelerate as protein-structure analysis and prediction provide an understanding of why some mutations work and some do not, and as the field of systems biology begins to more precisely define cellular states. In turn, the shift from phenomenology to quantification and prediction promises to be one of the most exciting aspects of evolutionary biology in the future, and it should ultimately yield a mature discipline of evolutionary engineering.

See also chapter III.3, chapter III.6, chapter VIII.3, chapter VIII.5, and chapter VIII.8.

FURTHER READING

Backus, M. P., J. F. Stauffer, and M. J. Johnson.1946. Penicillin yields from new mold strains. Journal of the American Chemical Society 68: 152.

Barrick, J. E., D. S. Yu, S. H. Yoon, H. Jeong, T. K. Oh, D. Schneider, R. E. Lenski, and J. F. Kim. 2009. Genome evolution and adaptation in a long-term experiment with *Escherichia coli*. Nature 461 (7268): 1243–1247. *A single* E. coli *strain was evolved over thousands of generations. The analysis of this evolutionary path continues to illuminate how bacterial genomes can adapt to new conditions.*

Bartel, D. P., and J. W. Szostak. 1993. Isolation of new ribozymes from a large pool of random sequences. [See comment.] Science 261 (5127): 1411–1418. *The first, surprising isolation of a large and complex ribozyme from a completely random sequence pool. The Class I Bartel ligase is to this day one of the fastest and most complex ribozymes ever discovered.*

Ellington, A. D., and J. W. Szostak. 1990. In vitro selection of RNA molecules that bind specific ligands. Nature 346 (6287): 818–822. *One of the first demonstrations that complex RNA molecules could be isolated from a completely random pool. The selected phenotype, specific dye binding, was unexpected.*

Griffiths, A. D., and D. S. Tawfik. 2003. Directed evolution of an extremely fast phosphotriesterase by in vitro compartmentalization. EMBO Journal 22 (1): 24–35.

Hall, B. G. 2003. The EBG system of *E. coli:* Origin and evolution of a novel beta-galactosidase for the metabolism of lactose. Genetica 118 (2–3): 143–156. *While strain improvement by directed evolution was known, Hall revealed the precise molecular underpinnings of the evolution of lactose utilization. Otherwise cryptic proteins in the* E. coli *genome can evolve novel functions.*

Hanes, J., and A. Pluckthun. 1997. In vitro selection and evolution of functional proteins by using ribosome display. Proceedings of the National Academy of Sciences USA 94 (10): 4937–4942.

Roberts, R. W., and J. W. Szostak. 1997. RNA-peptide fusions for the in vitro selection of peptides and proteins. Proceedings of the National Academy of Sciences USA 94 (23): 12297–12302.

Saffhill, R., H. Schneider-Bernloehr, L. E. Orgel, and S. Spiegelman. 1970. In vitro selection of bacteriophage Q-beta ribonucleic acid variants resistant to ethidium bromide. Journal of Molecular Biology 51 (3): 531–539. *Even before the advent of modern molecular biology techniques such as DNA synthesis and sequencing, Saffhill proved it possible to evolve phage RNAs in vitro for novel phenotypes.*

Smith, G. P., and V. A. Petrenko. 1997. Phage Display. Chemical Reviews 97 (2): 391–410.

Stemmer, W. P. 1994. Rapid evolution of a protein in vitro by DNA shuffling. Nature 370 (6488): 389–391. *Pim Stemmer and coworkers developed a radical new technique for in vitro recombination and consequently demonstrated remarkable enhancements in the speed of directed evolution.*

Tuerk, C., and L. Gold. 1990. Systematic evolution of ligands by exponential enrichment: RNA ligands to bacteriophage T4 DNA polymerase. Science 249 (4968): 505–510.

Wang, H. H., F. J. Isaacs, P. A. Carr, Z. Z. Sun, G. Xu, C. R. Forest, and G. M. Church. 2009. Programming cells by multiplex genome engineering and accelerated evolution. Nature 460 (7257): 894–898. *Until the development of the technique known as MAGE, it was impossible to carry out multiple, site-directed alterations to a bacterial genome in parallel. This technique and others like it should continue to accelerate classic methods for the directed evolution of bacteria.*

Zhang, Y. X., K. Perry, V. A. Vinci, K. Powell, W. P. Stemmer, and S. B. del Cardayré. 2002. Genome shuffling leads to rapid phenotypic improvement in bacteria. Nature 415 (6872): 644–646.

VIII.8

Evolution and Computing

Robert T. Pennock

OUTLINE

1. Unexpected links and shared principles
2. How evolutionary biology joined forces with computer science
3. How evolutionary computation is helping evolutionary biology
4. Evolutionary computation takes off
5. The future of evolution and computing

Shared principles between evolution and computing are opening up fruitful areas for research. This chapter discusses some unexpected connections between evolutionary biology and computer science, such as the core ideas of code, information, and function, and how these are leading to theoretical and practical ways in which each is benefiting the other. The chapter highlights the emerging field of evolutionary computation, giving a brief history and some examples of its utility not only in helping solve basic research problems in biology and computer science but also for generating novel designs in engineering.

GLOSSARY

Digital Evolution. The evolution of digital organisms in a system that instantiates the causal processes of the evolutionary mechanism through random variation, inheritance, and natural selection.

Digital Organism. A model organism, typically with a genome composed of simple instructions, in a computer environment.

Evolutionary Computation. The general term for research and procedures in computer science that take inspiration and utilize insights from evolutionary biology.

Evolutionary Engineering. Use of evolutionary computation approaches for solving design problems in applied engineering contexts, including robotics.

Experimental Evolution. Investigation of evolutionary processes by direct experimental methods, including replications and controls, rather than by indirect comparative methods.

Genetic Algorithm. One form of evolutionary computation; pioneered by John Holland.

What does evolution have to do with computing? What does computing have to do with evolution? At first glance, these fields almost seem to be opposites. On the one hand, evolutionary biology deals with the lush and tangled extravagance that is the living world. Living organisms grow, reproduce, and proliferate in abundant variety and complexity. It was Charles Darwin's genius that began to unravel this complexity and discovered some of the fundamental principles that produce new species and their astounding adaptations. In the century and a half since the publication of *On the Origin of Species*, evolutionary science has become a powerful explanatory framework that illuminates the entire organic world.

Computer science, on the other hand, deals not with organisms but with machines. Machines may get bigger, and computing machines have gotten more powerful, but they don't grow—they are built. The artificiality of computers stands in stark contrast to the naturalness of organisms. Computers are complex, but in quite a different way than living things. One would never confuse the specific patterns of complexity that characterize computing machines designed and built by human beings with the patterns that we find in evolved, living organisms. It is differences of this sort between the natural biological world and the technological world of objects designed and built by human beings that initially made it questionable whether it was even sensible to think that there could be, in Herbert Simon's term, a "science of the artificial" (Simon 1969).

This chapter discusses some of the ways in which evolutionary biologists and computer scientists are discovering, to their mutual benefit, that their fields actually have many concepts in common and that there are significant ways in which they may be united through deep, shared principles. After reviewing some of these principles, we will briefly look at how computer science came to recognize the applicability of evolution to computer science and began to figure out how to incorporate Darwin's findings into its own algorithmic way of thinking. We will see how the mechanism of evolution by natural selection that Darwin discovered can now be not just simulated but actually causally instantiated in a computer, and how this opens the door to surprising new ways for biologists to experimentally investigate evolutionary processes. And finally, we will look at how this new field of evolutionary computation can be applied in practical ways, such as in solving difficult design problems in engineering. As we shall see, evolutionary design has reached the point at which it can equal and sometimes surpass our own problem-solving abilities.

1. UNEXPECTED LINKS AND SHARED PRINCIPLES

The obvious contrasts between living organisms and computing machines hide significant points of commonality between the two fields of study because they focus on *products* rather than on *processes*. Once we begin to compare biology and computing in terms of processes, we find significant and fundamental linkages that were previously overlooked.

One conceptual commonality is the idea that both fields deal at the deepest level with the idea of coded functions. On the computational side, even laypersons understand that computer code runs everything from their notebook or tablet computer to the largest mainframe. It is the coded instructions of the software loaded in one's machine that make it function. On the biological side, everyone knows that organisms similarly depend on their genetic code for their functions. A mistake in the coded program of life can make an organism unable to perform some function as surely as a mistake in a software program can cause a function error. With little exaggeration, one may say that natural organisms are biological machines that run on genetic software that codes for the myriad, complex features that make them work in their native environments. Or that make them fail to work. A severe mistake in the genetic code of an organism can cause it to die, just as a serious coding mistake in some application you are running can bring up the dreaded blue screen of death.

A second, closely related, conceptual commonality between both fields is the idea of information. Here, too, even the language of the two disciplines resonates with deeply shared notions about the significance of information and its flow. For instance, rather than speaking narrowly of computer science, it is becoming more common for many computer scientists to identify their work as *information science*. Again, the term *computer science* makes it seem as though their subject matter is *computers*, whereas they take their real subject matter to be *computing*, which they see as the most basic form of information processing. An information-theoretical approach has not yet been developed nearly as far on the biological side, but here, too, the language of the discipline rings with this idea. An organism's genome is said to code for "biological information," while RNA and DNA are spoken of as "informational molecules."

These and other commonalities have long hovered in the background of scientific investigations in both the biological and computing communities, but as the deep conceptual connections are becoming more appreciated, they are coming to the foreground and being recognized as providing an opportunity for cutting-edge research.

To give just one example, in 2011 the National Science Foundation (NSF) published a letter to researchers calling attention to what it called Biological and Computing Shared Principles (BCSP). Issued jointly by NSF's Biological Sciences (BIO) and Computer & Information Science & Engineering (CISE) directorates, the BCSP letter highlighted a revolutionary transition occurring in the relationship between the fields. There have always been points of mutual influence between biological and computing research, but these are no longer limited to applications of one discipline to the other; the letter highlights "the convergence of central ideas and problems requiring the theoretical, experimental, and methodological competencies of both biology and computing" (US NSF 2011). The reason for the excitement is that shared principles between biology and computing may contribute to *conceptual* advances for both fields.

The BCSP letter identifies a variety of novel areas that are ripe for the identification and investigation of shared principles. Many of these involve specific properties of common interest such as adaptation to unanticipated novel conditions; self-repair and maintenance; coevolution and defense against adaptive adversaries; and general robustness and reliability. Other topics are more abstract, such as knowledge extraction; information flow, processing, and analysis; representations and coding; pattern recognition and pattern generation; network structure, function and dynamics; functions of stochasticity; and theory of biological computation. Although the BCSP letter speaks broadly about computing and "biology," in fact, many of these properties are connected conceptually to *evolutionary* biology.

2. HOW EVOLUTIONARY BIOLOGY JOINED FORCES WITH COMPUTER SCIENCE

Computers are now as much a key instrument in biology today as the microscope was in the nineteeth century. They no longer serve just as fancy calculators that make statistical analysis of lab and field data go faster; their flexibility and power as a universal machine now allows them to also serve a fundamental role in the production of data. One important new role is to allow sophisticated simulations of complex biological entities and processes. To give just one example, recent work by Donahue and Ascoli used computers to model the morphology of neurons and the developmental processes that lead to the elongation, branching, and taper of dendrites. Starting with parameters measured from real cells, modelers can create statistical distributions that can be resampled to form virtual trees and even to simulate somatic repulsive forces thought to be responsible for shaping cells. Such simulations can reveal patterns that may point to important developmental principles. For studying evolution in particular, an even more important advance is that computers now allow scientists to model evolutionary processes directly.

The insight that led to this revolutionary approach was made independently by several researchers (most in the 1960s), who recognized that the mechanism of evolution that Darwin discovered could be instantiated not just in biological systems but in other physical systems as well, including in computers. Probably the most influential of these was University of Michigan computer scientist John Holland, who coined the term *genetic algorithm* for the idea and who implemented it at the level of binary strings of 0s and 1s that could recombine, mutate in a computer, and be subject to selection. In Germany, aeronautical engineer Ingo Rechenberg had a similar idea and developed it with Hans-Paul Schwefel under the name *evolutionary strategies*. A third line of research, dubbed *evolutionary programming*, was begun by electrical engineer Lawrence Fogel. For their pioneering work, Holland, Rechenberg, and Fogel are credited as founders of what now goes by the general term *evolutionary computation*. This is not the place to recount the history of these and other early pioneers, but it is worth mentioning that these initial research streams proceeded separately for over a decade and a half before they discovered one another and began to interact. Today it is recognized that these and other evolutionary computation approaches share the same underlying core principles (De Jong 2006), and a community of researchers has formed around these ideas, spawning a variety of professional societies, conferences (many of which eventually joined together as GECCO, the Genetic and Evolutionary Computation Conference), and journals. *Evolutionary Computation*, the main journal in the field, was introduced in 1993.

This short history returns us to the idea of shared principles between evolution and computing. When evolution is seen as a special sort of algorithmic process, then it becomes possible for a computer to become the evolutionary biologist's lab bench. Properly understood, digital evolution can do more than simulate evolutionary processes, it can instantiate them (Pennock 2007). To see this we need only review the basic elements of the causal principle that Darwin discovered.

Descent with modification, as Darwin defined evolution, occurs whenever three conditions hold. The first is the random production of variations—a diversity of structure, constitution, habits, and so on. The second is that these variations be heritable, meaning that they can be passed on in the process of reproduction to the next generation. Darwin called this the "principle of inheritance." The method of inheritance is not so important as the basic causal principle of heritability itself—the key is *that* the genetic information be copied to the offspring, not the specific mechanism by which that is done. The third condition is that these heritable variations be naturally selected by the environment. If the genome of an individual happens to provide it with some slight variation that gives it any advantage over its competitors in their environment, that individual becomes more likely to survive to the point that it can reproduce, which causes the next generation to have a greater proportion of individuals with its heritable variations than those of its competitors. It is the environment, understood broadly, that naturally selects from among the extant variations, generation after generation. All these causal processes—random variation, heritability, and natural selection—can be instantiated in a computer. (See plate 7.)

Moreover, once one comes to see evolution as a universal causal law—one whose action is not limited to the familiar realm of DNA and flesh—then it becomes easier to see the possibility of sharing other biological concepts with computer science. To give just one example, the concept of a *genome* is not limited to the chromosomes of an organism; it also refers more generally to the complete information-transmitting material of any replicating entity, whether a biological organism or a self-replicating computer program. Indeed, chromosomes are best conceived of as but one possible instance of a genome. Other structures could have been found, and may yet be found, that carry heritable genetic information. Historically, it was not until many decades after the *Origin* that chromosomes were identified as the location of the genetic material. Darwin had simply spoken of "factors" in an abstract causal sense and it was not relevant to his law in *which* material they turned out to be instantiated. In this sense, it is not a metaphor or an analogy to speak of a coded sequence of instructions of a digital organism as its genome, for it is the genetic material of that individual in just the same sense as it is for a biological organism.

There is not the space here to lay out the full argument for this claim, but the rationale for understanding these concepts at this level of abstraction should be clear enough to see the value of bringing them and other shared concepts and principles to the foreground. Recognizing that the evolutionary causal processes can be instantiated in other physical systems, including in a computer, means that digital evolution goes beyond even the utility that a simulation can provide and provides a truly experimental system.

3. HOW EVOLUTIONARY COMPUTATION IS HELPING EVOLUTIONARY BIOLOGY

The late John Maynard Smith, a distinguished evolutionary biologist, was quick to recognize the scientific potential of marrying evolution and computing. In particular, he called attention to how digital evolution provides a way for biologists to escape from the inconvenient limits of our single planet. "So far," he wrote in a 1992 article in *Nature*, "we have been able to study only one evolving system, and we cannot wait for interstellar flight to provide us with a second. If we want to discover generalizations about evolving systems, we will have to look at artificial ones." Since then, others have opened this digital wormhole further.

To illustrate some of the advantages of digital evolution as a model system, we describe a study that used digital evolution to investigate the evolution of complex features (Lenski et al. 2003). Darwin advanced a number of hypotheses about how evolution could produce what he called "organs of extreme perfection and complication." Recognizing that features such as the eye were too complex to have arisen in a single leap, he proposed that their evolution would have involved incremental changes through intermediate forms, including changes of structures from one function to another. Darwin provided indirect evidence for his hypotheses by comparisons across different species but wrote that it would have been ideal if it were possible to precisely trace the details of a single line of descent. With an evolving digital system this is now a reality.

This study used the Avida platform, which is a well-developed model system for digital evolution research. In Avida, the genome of a digital organism (an "Avidian") is composed of simple computer instructions that do little of interest by themselves but when ordered in specific complex sequences can in principle perform any computable function. Such computational functions are the digital organism's phenotype. Among other properties, the genome of an Avidian has the potential ability to self-replicate. In its digital environment, however, the copying process is imperfect, so descendant organisms may have random mutations in their code. They also have to compete for the energy needed to execute their genetic programs. In this system, the digital organisms get energy by performing logical operations that also require specific sequences of instructions to function. Simple functions provide the organism with a small energy boost; more complex functions provide more energy, allowing them to run faster. This process provides an analogue to biological metabolism, but if a digital organism is to perform more complex metabolic functions, it must evolve them, because the ancestor could replicate but not perform any logic functions. As in nature, the digital environment naturally selects those variations that give organisms a competitive advantage. The mutations that arise in the genome are usually deleterious (and some may destroy the Avidian's ability to replicate) or neutral, but a few may improve the organism, resulting in faster replication and thus more offspring with those variations. With all the conditions of the Darwinian mechanism in place, a population of Avidians naturally evolves on its own without any outside assistance.

To investigate how complex features evolve, Lenski and collaborators ran 50 replicate populations, all under identical conditions. Ensuring identical replicates is simple in a digital evolution experiment and more precise than can be done in most natural systems, so statistical replication is simple. The digital system provided other advantages that no natural system could match, in that it allowed the experimenters to track complete lines of descent (from the ancestor across thousands of generations and millions of descendants), record every mutation along the way, and measure whether each was deleterious, neutral, or beneficial. The researchers could thus observe directly as beneficial mutations accumulated to produce one or another logic function, which themselves were later lost or modified for some other, more complex function. And although a given complex function first emerged in a line of descent as the result of just one or two mutations, systematic knockout experiments (removing individual instructions of the genome one at a time) showed that the function always depended on some specific sequence of many instructions that had previously evolved as parts of other functions and that their removal would eliminate the new feature. In an evolving digital system one has the ability to monitor such changes and to analyze them with an extraordinary degree of precision.

Nor are the results of such experiments always predictable. Because this is a real evolving system rather than a numerical simulation, the dynamics of the evolutionary mechanism can yield surprising results, just as in a biological evolving system. One unexpected finding in this particular study was that occasional deleterious mutations were in the line of descent leading to the most complex function in some populations. Some were only slightly deleterious, but a couple reduced fitness by more than 50 percent. Further tests ruled out the possibility that these mutations were just accidental hitchhikers; although the mutations were highly deleterious when they occurred, they became highly beneficial in combination with subsequent mutations.

This study is but one of many that used digital organisms to examine basic questions about evolutionary processes that would have been exceedingly difficult or impossible to perform with biological organisms. In a digital evolution system, one may "replay the tape of life," as Stephen Jay Gould put it, and directly observe, for example, the role that historically contingent events such as mass extinctions can play in the course of evolution. One can devise appropriate controls to test how altruistic behaviors evolve under different selective pressures. One also may investigate the effect of natural selection on different methods of phylogeny reconstruction. Digital evolution researchers have done all this and more.

This is not to say that digital evolution works as a model system for all the kinds of questions evolutionary biologists want to ask. While digital evolution instantiates the core causal processes of the evolutionary mechanism, it does not model, for instance, the defining features of any particular species or the unique properties of particular molecular structures. Thus it will be of little use to someone investigating questions for which such physical structures are salient. But for the biologist interested in questions about the cause-effect relationships of Darwin's law or seeking generalizations that will apply to any evolving system, experimental evolution with digital organisms is a revelation.

Evolution, broadly understood, requires neither DNA nor even living organisms. Evolutionary computation can help biologists understand shared properties such as robustness and nonexpressed code (Foster 2001). The power and flexibility of digital evolution gives researchers unprecedented opportunities to test evolutionary hypotheses, especially those requiring manipulations that are impractical in biological systems or numbers of generations that cannot directly be observed. Equally exciting are the practical applications of these shared principles embodied in evolutionary computation to such fields as engineering.

4. EVOLUTIONARY COMPUTATION TAKES OFF

Asked in 2010 for his judgment about the future course of his field, the president of the National Academy of Engineering, Charles M. Vest, wrote in the *New York Times* that "we're going to see in surprisingly short order that biological inspiration and biological processes will become central to engineering real systems. It's going to lead to a new era in engineering." Vest was no doubt thinking of a range of ways that engineering is beginning to make what we might call "the biological turn," including biomimicry and other uses of the products of evolution, but the use of evolutionary computation is certainly one of the most compelling ways that biological processes are being applied in engineering. As before, let us look at just one example in a bit of detail to show just how far evolutionary engineering has gone—in this case literally into outer space.

In 2004 NASA was preparing for the launch of its Space Technology 5 mission, whose aim was to test technology for measuring the effect of solar activity on the earth's magnetosphere. One of NASA's needs for the mission was a specialized antenna that had to meet a variety of precise specifications. Given certain transmit and receive frequencies, it had to operate within specified ranges for important functional properties. Moreover, the antenna had to fit within a 6 in. cylinder. Members of the Evolvable Systems Group at the NASA Ames Research Center decided to see whether an evolutionary approach could solve the problem.

The research team began by setting up a virtual world with a genetic encoding scheme that could represent the construction of three-dimensional wire forms—the space of possible antenna shapes. They used a tree-structured encoding in which, for example, a branch in the genotype would represent a branch in the wire form. The genotype was allowed to vary at random so as to produce diverse forms with different numbers, lengths, and angles of branches. In an initial population, 200 individuals were evaluated for their fitness for the task that NASA had set—think of this as a competition in which the virtual antennas were tested against each other. The best individuals in a given generation were automatically selected, and most of these were again randomly mutated or recombined to form new variations for the next generation. As this process was iterated over many generations, the shapes of the individuals in the population evolved little by little to better match the functional requirements that NASA had set. The antenna shapes that evolved in these runs were unlike anything that antenna engineers would have come up with themselves. They were very small—not much bigger around than a quarter—and they looked rather like a bunch of randomly twisted paper clips. But at the end of the run, the team built the device that had evolved in the virtual world and found that this misshapen bunch of wires met the required specifications.

The evolved antenna had additional technical benefits, including requiring less power, having more uniform coverage, and not requiring a matching network or a phasing circuit, thus simplifying the design and fabrication. Moreover, all these benefits were achieved with a shorter design cycle: the prototype for the antenna took three person-months to design and fabricate, compared with five person-months for a conventionally designed antenna (Hornby, Lohn, and Linden 2011).

What is especially impressive about this evolved design is that it succeeded where human engineers had failed. Prior to the Evolvable System Group's tackling the problem, NASA had contracted with an antenna engineering group that had produced a prototype design using conventional techniques, after a bidding process among several competing groups. However, the conventional design they produced did not meet the exacting mission requirements, while the evolved design did.

On March 22, 2006, the evolved antenna was launched into space. This was the first time that evolved hardware had reached such heights, both metaphorically and physically, but it was not an isolated instance of the power of evolutionary engineering. Indeed, evolutionary approaches have advanced to the degree that they now routinely equal or surpass human engineers in a variety of design tasks (Koza 2003).

This might seem to a bold claim. By what measures can one say that evolutionary designs can equal or surpass those of human beings? Since 2004, GECCO has held a contest for human-competitive results, and it judges entries using a variety of criteria. To qualify for the competition, an evolved solution must meet at least one of several standards, such as producing a patentable design or a result that is publishable in its own right in a peer-reviewed journal, independent of the fact that is was mechanically created. The evolved antenna shared the Gold Award in 2004. Since then winners have been recognized for human-competitive results in areas as disparate as photonic crystal design, automated software repair, and protein structure prediction. These awards for human-competitive evolved design are appropriately called the "Humies."

5. THE FUTURE OF EVOLUTION AND COMPUTING

One promise of evolutionary approaches to computation and engineering is that they will solve real-world problems. This promise is already being fulfilled: evolutionary computation harnesses the power of evolution, allowing evolutionary processes to work in a digital world just as they work in nature. A further promise of evolutionary computation is that it will help reveal the deeply shared principles between what initially appeared to be quite distinct fields of research. Think of what it means to recognize that it is not a mere metaphor to say that living organisms are running a genetic program. Think of what it means to understand that the functional properties of life—those astounding adaptions—that are coded in the genome were *programmed by evolution.* If this lesson can be learned, the marriage of evolution and computing will have been profound indeed.

See also chapter VIII.7 and chapter VIII.15.

FURTHER READING

Avida-ED Project: Technology for teaching evolution and the nature of science using digital organisms. http://avida-ed.msu.edu/. *With free downloadable digital evolution software as well as background materials and model exercises for undergraduate and AP biology courses, this award-winning project gives students everything they need to perform evolutionary experiments on their own computers.*

Clune, J., H. Goldsby, C. Ofria, and R. T. Pennock. 2010. Selective pressures for accurate altruism targeting: Evidence from digital evolution for difficult-to-test aspects of inclusive fitness theory. Proceedings of the Royal Society B 278: 666–674. *One example of the kind of basic science research that experimental evolution with digital organisms makes possible, this study tested hypotheses about inclusive fitness and the evolution of altruistic behavior.*

De Jong, K. A. 2006. Evolutionary Computation: A Unified Approach. Cambridge, MA: MIT Press. *An authoritative account of the common theoretical underpinnings of different varieties of evolutionary computation.*

Foster, J. A. 2001. Evolutionary computation. Nature Reviews Genetics 2: 428–36. *An excellent review article giving an overview of the field of evolutionary computation for biologists.*

Holland, John H. 1992. Adaptation in Natural and Artificial Systems. Cambridge, MA: MIT Press. *Holland's pioneering book, originally published in 1975, gave a detailed account of how what he called genetic algorithms could model the process of evolutionary adaptation in a computer system.*

Hornby, G. S., J. D. Lohn, and D. S. Linden. 2011. Computer-automated evolution of an X-band antenna for NASA's Space Technology 5 Mission. Evolutionary Computation 19 (1): 1–23. *Scientific account of how a NASA space antenna was produced by evolution engineering techniques.*

Koza, J. R., M. A. Keane, M. J. Streeter, and W. Midlowec. 2003. Genetic Programming IV: Routine Human-Competitive Machine Intelligence. Norwell, MA: Kluwer Academic. *The fourth in a series by John Koza and colleagues about genetic programming, this books lays out the case for how evolutionary techniques have advanced to be able to routinely match human intelligence for a wide variety of problems.*

Lenski, R., C. Ofria, R. T. Pennock, and C. Adami. 2003. The evolutionary origin of complex features. Nature 423: 139–144. *This pioneering study used digital evolution to perform a direct experimental test of some of Darwin's hypotheses about the evolutionary mechanisms that produce complex features.*

Pennock, R. T. 2007. Models, simulations, instantiations and evidence: The case of digital evolution. Journal of Experimental and Theoretical Artificial Intelligence 19 (1): 29–42. *This paper sorts out common confusions about model-based reasoning using digital evolution, explaining why it is a mistake to think of these models as just simulations of evolution and why they are real instances of the causal mechanism that Darwin discovered.*

Simon, H. A. 1969. The Sciences of the Artificial. Cambridge, MA: MIT Press. *Herbert Simon, who was later to win a Nobel Prize, wrote this prescient book about why artificial systems, including computers, could properly be treated as objects for scientific study.*

VIII.9

Linguistics and the Evolution of Human Language

Mark Pagel

OUTLINE

1. What is language?
2. When did language evolve?
3. Why did language evolve?
4. The evolution of human languages
5. Languages adapt to speakers
6. The future of language evolution

This chapter discusses how human language differs from all other forms of animal communication, when and why it evolved, and how elements of language evolve over long periods of time. It closes with a brief account of how languages might evolve in an increasingly globalized world.

GLOSSARY

Cognate. The term in linguistics analogous to *homology* in evolution. Two words are cognate if they derive from a common ancestral word (e.g., the English *water* and the German *wasser* are cognate words, but neither is cognate to the French *eau*).

Homology. A term used in genetics and evolution to identify two genes or two traits thought to derive from a common ancestor. Human and chimpanzee hands and fingers are homologous, as are many human and chimpanzee genes. Bat wings, by comparison, are not homologous to bird wings, even though both are used for flying.

Language. A form of communication unique to humans consisting of discrete words formed into sentences composed of subjects, objects, and verbs.

Phylogeny. A tree diagram like a family tree but conventionally depicting the evolutionary relationships among a group of species. The diagram can be generalized to any evolving objects transmitted from one generation to the next, including languages.

Regular Sound Correspondences. A term used in linguistics to describe statistical regularities in the way sounds change in words over long periods of time. One of the best known of these is the regular replacement of a *p* sound at the beginning of a Latin word with an *f* sound in the Germanic languages, as in the change from Latin *pater* to *father*.

1. WHAT IS LANGUAGE?

We instinctively recognize that human language is unique among all forms of biological communication, but what do we mean by that? Most animals communicate, but humans are the only animals with *language*. Human language is distinct in having the property of being *compositional*: we alone communicate in sentences composed of discrete words that take the roles of subjects, objects, and verbs. This makes human language a digital form of communication as compared with the continuously varying signals that typify the grunts, whistles, barks, chest thumping, bleating, odors, colors, chemical signals, chirruping, or roars of the rest of life. Those familiar sights, sounds, and smells might signal an animal's status or intentions, or indicate its physical prowess; they might tell a predator it has been spotted, or send a message to nearby relatives of an imminent danger. But lacking subjects, verbs, and objects, these acts of communication do not combine and recombine to produce an endless variety of different messages. Thus, your pet dog can tell you it is angry, and even how angry it is, but it cannot recount its life story.

2. WHEN DID LANGUAGE EVOLVE?

No one knows when the capacity to communicate with language evolved, but we can narrow the range of possibilities. Our closest living relatives, the modern chimpanzees, cannot speak, but we can, and so this tells us that language evolved sometime in the 6–7 million years between the time of our common ancestor and the late arrival of modern humans, around 160,000 to 200,000 years ago. For about the last 2.5 million years of that time, the evolutionary lineage of *Homo* or "human" species that would eventually give rise to modern humans, or *Homo sapiens*, was evolving in Africa. One of the best known of these *Homo* ancestors is *Homo erectus*, which probably evolved about 2 million years ago and had a relatively large brain, although at around 700 to 800 cc, it was still only a little more than half the size of the modern human brain. *Homo erectus* skulls reveal the impressions of two slightly protruding regions of the brain—known as Broca's and Wernicke's areas after their discoverers—that neuroscientists have identified as being involved in speech, at least in humans. This finding had led some researchers to suggest that *Homo erectus* had the capacity for speech, even if perhaps a rudimentary one. But Broca's and Wernicke's areas are also enlarged in some apes, so their presence is not by itself a definitive indicator that a species had language. There is also no evidence to suggest *Homo erectus* had anything even remotely close to the complex societies, tools, or other artifacts that we recognize as fully human.

A question that excites great differences of opinion is whether the Neanderthals spoke. This sister species to modern humans last shared a common ancestor with us sometime around 500,000 to 700,000 years ago. Neanderthals had large brains, they could control fire, and they had managed to occupy much of Eurasia by perhaps 300,000 years ago. Speculation that they had language has been fanned by the recent discovery that the Neanderthals had the same variant of a segment of DNA known as *FOXP2*, possessed by modern humans, that has been implicated, among many other effects, in influencing the fine motor control of facial muscles that is required for the production of speech. *FOXP2* is not a gene itself, even though it is often described as one, but rather a short segment of DNA that affects how other genes are expressed. In fact, *FOXP2* affects the human brain by causing about 50 genes to be expressed more and another 50 to be expressed less.

But, just as was true of Broca's and Wernicke's areas in *Homo erectus*, having the same variant of *FOXP2* as modern humans do doesn't tell us that the Neanderthals had language—they might have, but we cannot conclude with certainty that they did on the basis of this short segment of DNA. An analogy explains why. Most people's cars have engines and so do Ferraris. But this doesn't make every car a Ferrari. Closer to our own case, we know the modern human brain differs greatly from those of the Neanderthals in having a more fully developed and highly interconnected neocortex. This is the evolutionarily new and enlarged uppermost layer of the brain that is implicated in symbolic thinking and language. In simple terms, that thick layer of our cortex is why modern humans have foreheads and the Neanderthals did not. Given these differences, there is no compelling reason to conclude that *FOXP2* affected Neanderthals' brains as it does ours.

Further evidence against the Neanderthals' having language comes from the archaeological record, which shows they did not produce the range of tools and symbolic objects that we associate with the cognitive complexity of *Homo sapiens*. For example, by perhaps 60,000 years ago modern humans were producing abstract and realistic art, and jewelry in the form of threaded shell beads, teeth, ivory, and ostrich shells; they used ochre and had tattoos; they had small stone tools in the form of blades and burins; they made artifacts of bone, antler, and ivory, as well as tools for grinding and pounding; and their improved hunting and trapping technology included spear throwers, nets, and possibly even bows. The Neanderthals did not produce any of these things—for instance, there is no evidence they even had needles that would have allowed them to produce sewn clothing. Even where there is some evidence that they might have had jewelry or body decoration, it cannot be ruled out that they acquired these from modern humans who inhabited Europe contemporaneously. And whereas modern humans somehow had the creative capacity to spread out around the entire world, building the technologies that allowed them to inhabit territories from the Arctic to the scorched deserts of Saharan Africa, the Neanderthals never left Eurasia.

This line of reasoning leads to the conclusion that language evolved with the arrival of our own species. Our capacity for language was almost certainly present in our common ancestor, because today all human groups speak, and speak equally well. There are no languages that are superior to others, and no human groups that speak primitive as opposed to advanced languages. If this hypothesis is correct, then language is no more than around 160,000 to 200,000 years old, although some anthropologists think language arose even later, pointing to a sharp increase beginning around 70,000 to 100,000 ago in evidence of symbolic thinking and the complexity of human societies. Even though our species arose prior to 100,000 years ago, it is just possible that all modern humans alive today trace their ancestry to common ancestors who lived around that time, so language evolving at this later date, though unlikely, is possible.

3. WHY DID LANGUAGE EVOLVE?

Evolution does not produce complex adaptations like language in a single moment. Usually, something pulls the trait and its precursors along so that it pays its way by granting some advantage to its bearers. To most people the advantage of having language is obvious—it allows us to communicate. All animals could benefit from being able to communicate as humans do, however, so this does not explain why humans have language and no other species does. Instead, we need to search for ways that having language granted benefits to speakers in our species but not in others. But searching for an advantage to language leads immediately to two evolutionary predicaments. One is that much of what a speaker has to say might benefit someone else, and potentially at a cost to the speaker. Natural selection never promotes naive altruism such as this, and so language would quickly have died a silent death, as selfish people all too willing to listen would have profited but with no intention of returning the speaker's favor. Alternatively, perhaps language evolved to help humans mislead or trick others. It is in fact a fundamental tenet of communication that animal signals evolve to benefit the signalers. This solves the problem of altruism, because language might help speakers benefit at *others'* expense. But this poses the second predicament: if others know that speakers' acts of communication are designed to benefit themselves, surely this will favor people who don't listen.

To understand how language overcame these two predicaments and to grasp why language evolved, we need to appreciate something that is true of human social behavior but that is virtually absent in the rest of the animal kingdom. Members of the human species are the only ones who routinely cooperate with other *unrelated* members of their species. Whereas most cooperation elsewhere in nature is limited almost entirely to helping kin or other relatives—that is, to acts of nepotism—humans routinely exchange favors, goods, and services with people other than those in their immediate families. We have an elaborate division of labor, we engage in task sharing, and we have learned to act in coordinated ways, such as when we go to war or simply combine or coordinate our energies to complete some task. But as powerful as this form of social behavior is, it is risky because in each of these situations we run the risk of being taken advantage of by other members of our group who might hold back, enjoying the spoils of cooperation but without having to pay the costs.

Elsewhere I have suggested that the role of language in this complex social behavior—and the reason it evolved—was to act as the conduit for carrying the information needed to make our form of cooperation work and for keeping people's selfish instincts in check. It is a highly specialized piece of social technology, and so great were the benefits deriving from the cooperation language made possible that it easily paid its way in evolutionary terms. Both evolutionary predicaments are solved because now speakers *and* listeners benefit from having language. We use it to negotiate exchanges, to make plans, to remember lessons from the past, and to coordinate actions. Persons who try to take advantage of this cooperation, perhaps by failing to return someone else's goodwill or by not contributing their fair share, can be exposed as "cheats" and their reputation tarnished. These are all complicated social acts that other animals don't perform, and they require more than grunts, chirrups, odors, and roars. This explains why we and we alone have language: our particular brand of sociality could not exist without it.

4. THE EVOLUTION OF HUMAN LANGUAGES

It is by now widely accepted that modern humans first arose in East or possibly southern Africa, and then later swept "out of Africa" in two waves, one perhaps 120,000 years ago and a second one sometime between 50,000 and 70,000 years ago. Opinions differ about the success of the first wave; some researchers suggest modern humans got no farther than what is known today as the Levant region of the Middle East, where they eventually went extinct. Others believe that people from this first wave survived and spread eastward, establishing the populations that would eventually move into Southeast Asia, Indonesia, Papua New Guinea, and Australia. Whichever scenario is true, by the time of the second wave, modern humans had gone on to establish their presence permanently outside Africa and in a matter of a few tens of thousands of years had occupied nearly every environment on earth.

Why this sudden change in the behavior of a single species? The answer is almost certainly "language." While all other species are largely confined to the environments to which their genes are adapted, having language and the social systems it made possible granted humans the capacity to adapt at the cultural level by producing new technologies, and it was via successive bouts of innovative cultural adaptation that our ancestors came to occupy the world. Language had granted humans the capacity to share ideas. This simple development meant that later generations could benefit from the accumulated wisdom of the past, and new ideas could quickly spread among the members of a population. Once a species has the ability to cooperate with unrelated individuals, a vast store of knowledge and wisdom is unleashed as the society can then draw on a wider range of talents and skills than would be available to a single individual or even to a family. We therefore expect this style of cooperation to be associated with an explosion of complexity, and this is precisely what the archaeological record reveals.

As these newly talkative people moved around the world they evolved different languages. They could do so because, unlike with genes, whose precise sequence of nucleotides determines the proteins they produce, there is no necessary connection—save perhaps for the *onomatopoeic* words—between a word or sound and its meaning. Currently there are approximately 7000 different languages spoken, or 7000 mutually incomprehensible systems of communication around the world. This makes humans unique—and somewhat bizarrely so—among animals in not being able to communicate with other members of their own species. As it happens, the 7000 different languages are not evenly distributed around the world. Instead, *language density*, or number of different languages found in a given geographic area, follows almost exactly the pattern of biological species diversity. Both show strong latitudinal gradients, such that both the number of different languages and number of different species in a given area are small in the northernmost regions of the world, and both become more tightly packed in tropical regions (figure 1). In some tropical regions, languages can be so densely packed that a different language can be spoken every few kilometers! For instance, the Vanuatu island of Gaua in the south Pacific Ocean covers 342 km^2, and like so many of the islands in this region, it is the roughly circular remnant plug of an ancient volcano. Gaua is just 19–21 km in diameter, but this speck of an island supports five distinct languages—*Lakon* or *Vuré, Olrat, Koro, Dorig*, and *Nume*.

It is as if human language groups act like distinct biological species, and this linguistic isolation is clearly evident in the ways that languages evolve. Most of us learn our language from our parents and those immediately around us. This dominant form of language transmission from parent to offspring or from older generation to younger generation means that language evolution shares a number of features with genetic evolution (table 1). Thus genes and words are both discrete elements that evolve by a process of descent with modification as they are passed over many generations: genes can acquire mutations, and words can acquire sound changes. For example, the Old English *brōthor* evolved into the modern English *brother*. Words can also be borrowed from a donor language and incorporated into a recipient language—the English word *beef* is borrowed from the French *boeuf*—just as genes can sometimes be transmitted from one species to another, such as when two closely related species form hybrids.

These and the other parallels between genetic and linguistic evolution mean that when people spread out over an area, and new languages evolve, those languages share many features and form what we recognize as families of languages. Indeed, in *The Descent of Man* (1871), Darwin noted that "the formation of different languages and of distinct species . . . [is] curiously parallel. . . ." This parallel had in fact been recognized in the late eighteenth century, nearly 100 years before Darwin, by an English judge, Sir William Jones, working in colonial India during the rule of the English King George III. To process court papers Sir William found it necessary to learn Sanskrit, and in doing so he became aware of curious similarities among Sanskrit, Latin, and Greek. Jones described these to a meeting of the Asiatic Society in Calcutta in 1786, saying:

> The Sanskrit language . . . [bears] . . . a stronger affinity . . . [to Greek and Latin] . . . both in the roots of verbs and in the forms of grammar, than could possibly have been produced by accident; so strong, indeed, that no philologer could examine them all three, without believing them to have sprung from some common source, which, perhaps, no longer exists."
> Sir William Jones, Calcutta, February 2, 1786

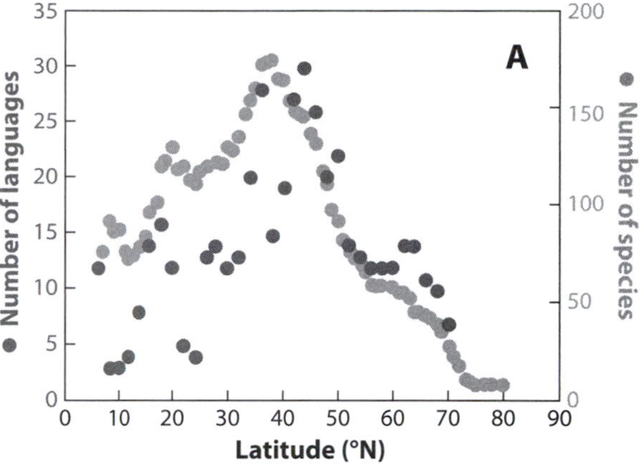

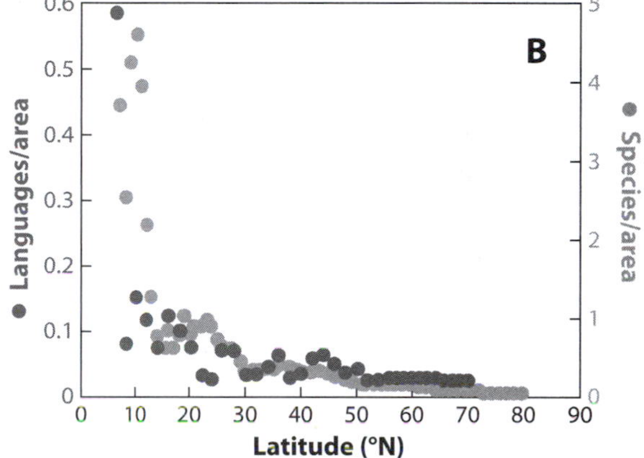

Figure 1. Relationships between languages and species. Numbers of North American human language-cultural groups (before European contact) and mammal species are distributed similarly across degrees of latitude. (A) Numbers of languages and numbers of mammal species at each degree of north latitude in North America. The trends reflect the shape of the continent, being narrow in the south regions and growing wider at higher latitudes. Both trends peak at approximately 40°N, where North America is ~4800 km wide. (B) Densities of languages and mammal species, calculated as the number of each found at the specific latitude divided by the area of the continent for a 1° latitudinal slice at each latitude. (Figure and data are reproduced, with permission, from *Nature Reviews Genetics* [Pagel 2009].)

Table 1. Some analogies between biological and linguistic evolution

Biological evolution	*Language evolution*
Discrete heritable units (e.g., nucleotides, amino acids, genes, morphology, behavior)	Discrete heritable units (e.g., words, syntax, and grammar)
Mechanisms of replication	Teaching, learning, imitation
Mutation—various mechanisms yielding genetic alterations	Innovation (e.g., formant variation, mistakes, sound changes, introduced sounds and words)
Homology (genes that are related by descent from a common ancestral gene and are now found in different species [e.g., ribosomal genes in nearly every species])	Cognates (words that are related by descent from a common ancestral word and are now found in different languages [e.g., *father* and *pater*])
Natural selection	Social selection, trends
Drift	Drift
Cladogenesis (e.g., allopatric speciation [geographic separation] and sympatric speciation [ecological/reproductive separation])	Lineage splits (e.g., geographic separation and social separation)
Anagenesis	Linguistic change without split
Horizontal gene transfer	Borrowing
Hybridization (e.g., horse, zebra, wheat, strawberry)	Language Creoles (e.g., Surinamese)
Correlated genotypes/phenotypes (e.g., allometry, pleiotropy)	Correlated cultural terms (e.g., *five* and *hand*)
Geographic clines	Dialects/dialect chains
Fossils	Ancient texts
Extinction	Language death

Jones had identified what linguists would later recognize as the Indo-European language family, which arose around 8000 to 9000 years ago with the advent of farming in the Fertile Crescent, or what is roughly present-day Turkey and Iraq. Farmers and farming technology then spread out from that region, seeding the languages currently spoken all over Europe, parts of central Asia, and the Indian subcontinent. These include the Germanic languages German, Dutch, and English; Romance languages that derive from Latin including French, Spanish, and Italian; Slavic languages; and even Persian, Hindi, and Punjabi.

By comparing the similarities and differences among a group of languages such as those that form the Indo-European family, it is possible to construct evolutionary or *phylogenetic* trees depicting the probable course of evolution of those languages from their common ancestral or protolanguage. Figure 2 displays just such a tree for a selection of Indo-European languages, including some of its main branches. The most popular and widely used approach to inferring linguistic phylogenies draws on the close parallels between linguistic and genetic evolution such as are listed in table 1. In particular, whereas evolutionary biologists infer phylogenetic trees of species by studying *homologous* genes, evolutionary linguists infer linguistic trees from sets of *cognate* words. Just as slight differences in the sequence of a gene can identify sets of closely related species, slight differences in words can identify sets of closely related languages.

For example, the word *madre* means *mother* in both Spanish and Italian, and we instinctively recognize that both derive from the Latin *mater*. Likewise, we recognize that the English *mother* is similar to the German *mutter*, and again both of these seem to derive from *mater*. In fact, these comparisons identify English as part of the Germanic branch of the Indo-European languages and identify Spanish and Italian as part of the Romance or Latinate branch of this same family (figure 2).

Not all such comparisons are this simple, but linguists make use of what they call *regular sound correspondences* to help them identify whether two words are cognate and how closely related they are. One of the best known of these regular sound changes is the replacement of a *p* sound at the beginning of a word with an *f* sound, as in the Latin *pes* or *ped*, which becomes *foot*, and *pater* becomes *father*. Other regular sound correspondences reveal to linguists that the English *five* is closely related to the German *fünf*, that the French *cinq* is related to the Spanish *cinco*, and less obviously that all four of these

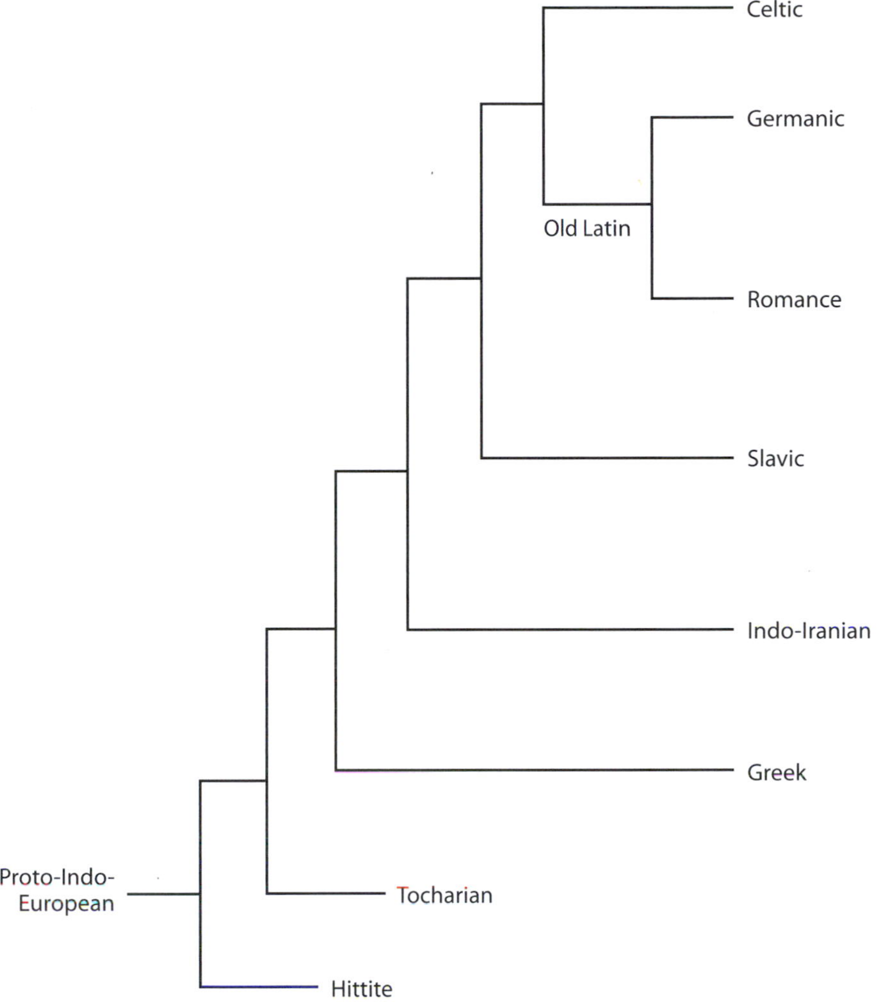

Figure 2. The Indo-European language family. Evolutionary tree showing the relationships among the major branches of this language family. *Celtic* languages include Irish, Breton, and Welsh; *Germanic* languages include German, Dutch, and English; *Romance* or *Latinate* languages include French, Italian, and Spanish; *Slavic* languages include Russian, Czech, and Polish; *Indo-Iranian* languages include Persian, Afghan, Hindi, and Punjabi. Hittite and Tocharian are two extinct languages. The base or root of the tree represents the proto-Indo-European language that might have existed 8000 to 9000 years ago.

words derive from the Latin *quinque*. Of course, not all words are cognate. The Spanish *agua* is cognate to the Italian *acqua*, but neither is cognate to the English *water* or to the German *wasser*. This lack of cognacy serves to reinforce that the Romance languages are a distinct group from the Germanic languages, even though both branches trace their ancestry to a pre-Latin language that might have been spoken 5000 years ago or more.

Once a set of comparisons is made among a group of languages, identifying cognate and noncognate words for a large number of different meanings (a *meaning* being what a word refers to), the resulting data can be used along with formal methods to infer the phylogenetic tree. A variety of different methods are commonly used to infer trees, including parsimony, distance, and likelihood methods, and these are broadly similar whether applied to genetic or linguistic data. *Parsimony methods* seek a tree that minimizes the number of evolutionary events along its branches. Thus, parsimony methods would favor a tree that put English with German, and French with Italian or Spanish, over one, for example, that showed German as more closely related to Spanish than to English. To put German next to Spanish would suggest that the word *water* or *wasser* had somehow evolved twice. *Distance*

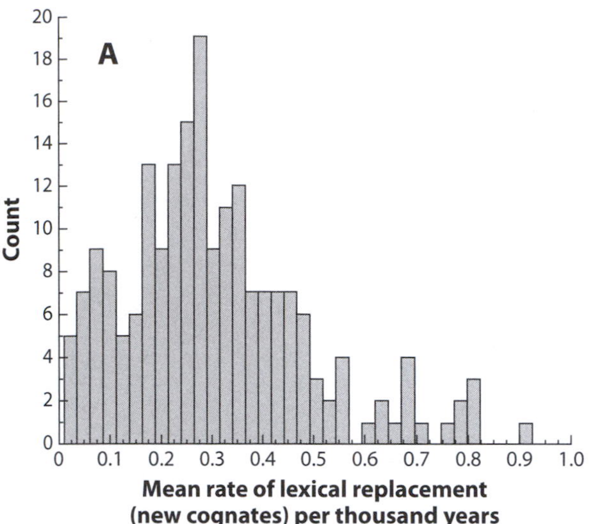

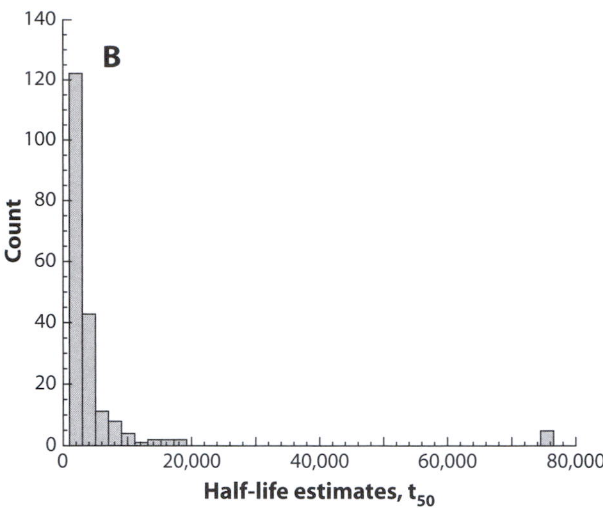

Figure 3. Rates of lexical replacement. (A) Counts of the rate at which a word comes to be replaced by a new unrelated word in the Indo-European languages for 200 common vocabulary words, measured in units of numbers of new words per 1000 years of evolution. Fastest to slowest rate represents more than a 100-fold difference. The average = 0.3 ± 0.18 new words per 1000 years (or roughly one every 3000 years), median = 0.27, range = 0.009 to 0.93. (B) Counts of the word *half-life* estimates as derived from the rates of lexical replacement for the same 200 words. The half-life measures the expected amount of time before a word has a 50 percent chance of being replaced by a new, unrelated word. The average half-life is 5300 years, with a median of 2500 years, and ranges from 750 to a theoretical value of 76,000. (Figure and data are reproduced, with permission, from *Nature Reviews Genetics* [Pagel 2009].)

methods, as their name implies, seek to define a distance between all pairs of languages, and those distances are then used to construct a tree. *Likelihood methods* use formal statistical models to estimate the probability of changes in words through time. A tree is constructed that makes the observed set of changes most probable, given the model of evolution. These methods are all described further in chapters II.1 – II.3.

Trees such as the one depicted in figure 2 have now also been produced for the Austronesian languages, the Bantu languages of Africa, the Arawak languages of South America, the Semitic languages, the Uralic languages of Northern Europe, and some Melanesian languages, and this is an important and growing area of the field of evolutionary linguistics. The existence of sets of languages that comprise families of related languages shows that at least some elements of language evolve slowly enough to preserve signals of their ancestry dating back thousands of years. For example, linguists recognize that the word for *two* of something is probably derived in *all* Indo-European languages from a shared ancestral sound that has been conserved for many thousands of years. Thus, in Spanish the word is *dos*, it is *twee* in Dutch, *deux* in French, *due* (doo-ay) in Italian, *dois* in Portuguese, *duo* (δύο) in Greek, *di* in Albanian, and *do* in Hindi and Punjabi; Julius Caesar would have said *duo*. This conservation leads to the proposal that the original or proto-Indo-European word that was spoken perhaps 9000 years ago was also "two"—as it sounds—and indeed, some scholars suggest it was *duwo* or *duoh*.

A handful of other words, including *three*, *five*, *who*, *I*, and *you*, are also highly conserved. For example, the English word *three* is *tre* in Swedish and Danish, *drei* in German, *tre* in Italian and *tres* in Spanish, *tria* in Greek, *teen* in Hindi and *tin* in Panjabi, and *tri* in Czech, leading to the suggestion that the proto-Indo-European word for *three* might have been *trei*. These conserved words are closely followed by pronouns such as *he* and *she*, and by the *what*, *where*, and *why* words, all of which often show a striking degree of similarity among many Indo-European languages. Other words, however, can vary considerably across these same languages, meaning that they have evolved or changed at far higher rates. The English word *bird*, for example, is *vogel* in German, *oiseau* in French, and *pajaro* in Spanish.

When long lists of words are studied for their frequency of change among the languages of a language family such as Indo-European, it is possible to derive estimates of their rates of change using statistical methods. It turns out that most words can be expected to last somewhere around 1500 to 2500 years, with some lasting far longer and others for shorter amounts of time (figure 3). It is possible from these rates of change to calculate a word's linguistic *half-life*, defined as the amount of time it takes for there to be a 50 percent chance a word will be replaced by some new, unrelated word (figure 3).

This approach shows that some words might be expected to last more than 10,000 years. The linguist Merritt Ruhlen has even proposed a list of "global etymologies" that he thinks are signals leftover from the last common ancestor to all human languages, or our *mother tongue*. Ruhlen's list includes words for *who*, *what*, *two*, *water*, *finger*, and *one*. The mere existence of a mother tongue is controversial, and many linguists dispute the list, but Ruhlen cites evidence that traces of these words are found in many language families from all over the world.

5. LANGUAGES ADAPT TO SPEAKERS

But why should words last as long as they do? Given all the opportunities for neighboring language groups to borrow words from one another, and all the ways that words can change, language changes far more slowly that we might expect; that is, from the standpoint of effective communication, words do not need to last much more than around three generations—the time span covered in a typical population of speakers. If they changed faster than that, then we might not be able to talk to our grandparents. But most words last somewhere around 1500 to 2500 years, so why is there such fidelity in language?

One answer comes from thinking of elements of languages as "replicators" that must adapt to the environment of the human mind. Words are sounds that compete for attention in the mind of speakers. So, to be successful words must get themselves replicated by being transmitted to someone else's mind. It follows that those words that are easiest to remember and say will tend to be those retained. We can see the current competition for space in our mind in common words like *sofa* and *couch*, or *living room*, *sitting room*, *reception room*, and *parlor*. Which of these forms will win? There are no simple answers, but precisely because there is seldom any necessary connection between a sound and its meaning, the competition often focuses on characteristics of the sounds themselves, and thus we can expect the competition to be more intense the more often a word is used. There is a huge disparity in how often different words get used; in fact, so great is the disparity that about 25 percent of all human speech is made up from a mere 25 words. According to the *Oxford English Dictionary*, the English language's top 25 include *the, I, you, he, this, that, have, to be, for,* and *and* (while *they, we, say,* and *she* make it into the top 30). Not surprisingly, perhaps, the words used most frequently have common characteristics: they are short, often monosyllabic, distinct, and easy to pronounce. And this begins to explain why some words last longer than perhaps they need to. It is not that these frequently used words are somehow more important, but that they might be stable over long periods because they have become so highly adapted to our minds that it becomes difficult for a new form to arise and outcompete or dislodge them.

6. THE FUTURE OF LANGUAGE EVOLUTION

Our native language is, perhaps, one of our most intimate traits, being the voice of the "I" or "me" that defines our conscious self. It is the language of our thinking, and it is the code in which many of our memories are stored. Thus it is not surprising that one of the greatest personal losses a people can suffer is the loss of their native language. And yet, currently somewhere about 15 to 30 languages go extinct every year as small traditional societies dwindle in numbers or get overwhelmed by larger neighbors, and younger generations choose to learn the languages of larger and politically dominant societies. Whatever the true numbers of languages going extinct, the loss of languages greatly exceeds the loss of biological species as a proportion of their respective totals.

Some projections say that only a handful of languages will see out this century. This raises the question of which language will win if ever a single language should succeed all others on earth. Currently three languages are spoken by a far greater number of people than any of their competitors. About 1.2 billion people speak Mandarin, followed by around 400 million each for Spanish and English, and these are closely followed by Bengali and Hindi. It is not that these languages are better than their rivals; it is that they have had the fortune of being linked to demographically prosperous cultures. On these counts Mandarin might look like the leader in the race to be the world's language, but this ignores the fact that vastly more people learn English as a second language—including many people in China—than any other. Already it is apparent that if there is a worldwide *lingua franca* it is English.

Still, English itself might be transformed as it is bombarded by the influences of such large numbers of nonnative English speakers who bring along their accents, grammar, and words to English when they speak it. This ability of English to take in so-called foreign words has been the key to its adaptability for at least the millennium since the Norman conquest of the English in 1066 brought an influx of Norman French vocabulary. Just as words must adapt to be competitive in the struggle to gain access to our mind, languages have to adapt as a whole to remain useful to their speakers, and those that do so will be the survivors. Self-appointed human "minders" in the form of reactionary grammarians, sticklers for spelling, or those who deliberately try to exclude some words and phrases will succeed in controlling the rate at which their languages naturally change, but in doing so they might consign these languages to the backwaters of international communication. This might already be happening to French and German, as both governments have ministries devoted to language "purity." The alternative to this control is not the free-for-all that some might fear. If communication is important, languages will never change at rates that imperil the very reason for which they exist.

See also chapter II.1, chapter II.2, and chapter II.3.

FURTHER READING

Cavalli-Sforza, L. L., A. Piazza, P. Menozzi, and J. Mountain. 1988. Reconstruction of human evolution: Bringing together genetic, archaeological, and linguistic data. Proceedings of the National Academy of Sciences USA 85: 6002–6006. *A widely cited early attempt to link genetic and linguistic diversity.*

Enard, W., M. Przeworski, S. E. Fisher, C. S. Lai, V. Wiebe, T. Kitano, A. P. Monaco, and S. Pääbo. 2002. Molecular evolution of *FOXP2*, a gene involved in speech and language. Nature 418: 869–872.

Fitch, W. T. 2010. The Evolution of Language. Cambridge: Cambridge University Press.

Gray, R. D., and Q. D. Atkinson. 2003. Language-tree divergence times support the Anatolian theory of Indo-European origin. Nature 426: 435–439. *This study used language phylogeny to test a historical hypothesis for the timing of the origin of Indo-European languages.*

Pagel, M. 2009. Human language as a culturally transmitted replicator. Nature Reviews Genetics 10: 405–415. *Provides a general overview of language evolution, including a description of methods of phylogenetic inference, and statistical studies of how languages evolve.*

Pagel, M. 2012. Wired for Culture: Origins of the Human Social Mind. New York: W. W. Norton. *One chapter of this book presents many of the arguments given here about why language evolved.*

Pagel, M., and R. Mace. 2004. The cultural wealth of nations. Nature 428: 275–278.

Pagel, M., Q. D. Atkinson, and A. Meade. 2007. Frequency of word use predicts rates of lexical evolution throughout Indo-European history. Nature 449: 717–719. *A general explanation for variation in rates of word evolution.*

Renfrew, C. 1987. Archaeology and Language: The Puzzle of Indo-European Origins. Cambridge: Cambridge University Press.

Wade, N. 2006. Before the Dawn: Recovering the Lost History of Our Ancestors. New York: Penguin.

VIII.10

Cultural Evolution

Elizabeth Hannon and Tim Lewens

OUTLINE

1. What cultural evolution is not
2. Memetics
3. Cultural evolution
4. Nonhuman animal cultural evolution
5. Defining culture

For most of the twentieth century, evolutionary theory focused on phenotypic variation underpinned by inherited genetic variation. Any comprehensive account of the evolution of the human species, and some animal species, must acknowledge that this is at best a simplification of the forces affecting change and stasis in these lineages. Habits, know-how, and technology—what we might consider cultural traits—can also contribute to survival and reproduction. Moreover, these traits are often maintained, in our own species at the very least, by learning from others—that is, they are inherited nongenetically. Further, these traits often show patterns of cumulative improvement as discoveries made in one generation are built on and modified. There also exist subgroups with distinct traits, again often generated and maintained through learning. Theories of cultural evolution start from the observation that humans, and possibly other species, display these important prerequisites for evolution—variation and inheritance of cultural traits—and attempt to build rigorous accounts of cultural change based on this observation. How exactly these accounts should be fashioned, what relationship cultural evolution should have with organic evolution, and how culture itself should be conceived remain open questions. This chapter describes some recent attempts to address these questions.

GLOSSARY

Cultural Evolution. A process of change in the traits manifested within a population that is explained by various forms of social learning among species members.

Horizontal Transmission. Transmission within a generation, sometimes also used to refer to transmission from any nonparent.

Meme. A cultural entity, intended to be analogous to a gene, capable of being replicated and transmitted between individuals.

Replicator. An entity capable of being replicated and capable of influencing its own chances of being replicated through its effects on the world.

Vertical Transmission. Transmission from parent to offspring, usually of genetic material.

1. WHAT CULTURAL EVOLUTION IS NOT

Not all evolutionary approaches that seek to account for cultural phenomena are theories of cultural evolution. Evolutionary psychologists, for example, tend to regard human behavior and culture as the output of cognitive adaptations, and they assume that the most important mechanism producing such cognitive adaptations is natural selection acting on genetically inherited variation (see chapter VII.12). Evolutionary psychologists acknowledge that changes in cultural environments can affect the behavioral outputs of cognitive adaptations, but they tend to downplay the role of nongenetic inheritance. As a consequence, they also tend to be skeptical of the thought that cultural change may affect and underpin the generation of cognitive adaptations themselves.

Nonetheless, cultural inheritance and genetic inheritance are processes that may affect each other in important ways. Cultural changes bring about alterations to the environment, which in turn affect how genes act in development and what selection pressures act on genes. For example, dairy farming, thought to have developed somewhere between 6000 and 8000 years ago, appears to have created a selective environment that facilitated the proliferation of lactose tolerance in those populations where it was practiced. Far from removing humans from the evolutionary fray, our cultural environments may exert selective pressures and thus may be implicated in the ongoing evolution of our physiological natures. These sorts of processes are captured by models of gene-culture coevolution that explore the ways in which cultural changes, adaptive or otherwise, can affect genetic evolution, and vice versa.

The targets of explanation for evolutionary psychology and gene-culture coevolutionary accounts differ from those of theories of cultural evolution. The latter address cultural trends in their own right, rather than as outputs of cognitive adaptations or as selective environments for the natural selection of genes. Further, unlike evolutionary psychology, though in common with gene-culture coevolutionary accounts, theories of cultural evolution allow a significant amount of nongenetic, learning-based inheritance. The remainder of this chapter describes two distinct attempts to build a theory of cultural evolution from this starting point.

2. MEMETICS

A theory of cultural evolution needs some systematic way of modeling the effects of cultural inheritance. One such approach is memetics, which has attracted considerable attention in the popular scientific literature. Initially developed by Richard Dawkins, and following the "gene's-eye view" of natural selection he popularized, memetics takes the view that to explain the sort of transgenerational resemblance needed for cumulative evolutionary change, entities that have the ability to make faithful copies of themselves—so-called replicators—are required. In standard biological models of evolution it is assumed that genes are the relevant replicators; genes make copies of themselves, and (so the story goes) this explains why offspring resemble their parents. For culture to evolve, memeticists argue that replicators—called *memes*—are once again required. Dawkins lists some exemplary memes: "tunes, ideas, catch-phrases, clothes fashions, ways of making pots or of building arches." Note that while it is sometimes assumed that all memes are ideas (and vice versa), Dawkins's list includes other types of things, such as ways of making pots, which are techniques.

Memetics proposes that ideas, skills, practices, and so on, are entities that can be understood to hop from mind to mind, making copies of themselves as they go. For example, you hear a song on the radio as you leave your house in the morning and you sing it at work that day. Your colleague later whistles it as she prepares dinner that evening. Her child hums the song in school the next day and passes it on to his classmates. The meme—in this case, the tune—spreads through its being "catchy." Like genes, memes have differential success in replication: some songs are catchier than others. The rate at which a meme may replicate itself is thought to be dependent on the same factors that determine the rate at which a gene may replicate itself—namely, its effects on the organism it inhabits and on the local environment (partly constituted by the downstream effects of other memes) in which the organism finds itself.

Critics of memetics put pressure on the claim that cultural inheritance is analogous to genetic inheritance. Such criticisms are as likely to originate from those sympathetic to the broad project of developing evolutionary approaches to culture as they are from those outside this project and doubtful of its merits. The remainder of this section outlines some of the key criticisms leveled against memetics.

First, cultural items rarely behave like replicators, and imitation is often very error prone. If you see us dance the tango, and this inspires in you the desire to also dance the tango, we will almost certainly not dance exactly the same steps. Our dances will have been influenced by a wide variety of factors such as who taught us the dance and our own particular physiologies. If this is copying, it is very bad copying indeed.

A second and closely related criticism of memetics draws on the fact that while genetic replication allows us to trace a token copy of a gene back to a single parent, ideas are rarely copied from a single source in a way that allows us to trace clear lineages. Perhaps you learned the tango from several teachers, and your style has been influenced by watching expert dancers. There is no clear single origin for your "tango" meme. Within the realm of biological evolution, an understanding of Mendel's laws has been important in explaining some aspects of evolutionary dynamics. Mendel's laws rely on an understanding of genes as discrete, transmitted units. But if token ideas can appear in an individual by virtue of that individual's exposure to several sources, then it is unlikely that anything close to Mendel's laws will be discovered within cultural evolution. Such an objection need not be fatal for theories of cultural evolution in general, as we shall see, but it does threaten the tight analogy memetics draws between ideas and genes.

Third, memetics seems to demand that we be able to divide culture into discrete units. But it is not clear how this should be done. Ideas stand in logical relation to one another. It is impossible to believe in the theory of relativity without understanding it, and one cannot understand it without holding many additional beliefs relating to physics. It is not at all clear that it makes sense to think that the theory of relativity can be isolated from the rest of physics as an individual meme. One might respond that we have delineated the meme incorrectly here, and those ideas to which the theory of relativity stands in logical relation all form a part of a single more inclusive meme. The worry, now, is that even if we "step back" and consider some broader group of theories, we cannot understand even them without further basic mathematical training, understanding of the operation of measuring apparatus, and so forth. A form of holism looms, according to which single memes will correspond to massive complexes of belief.

These criticisms focus on whether memes exist. Another matter of concern for memetics in particular, and theories of cultural evolution in general, stems from the fact that cultural transmission occurs both vertically and horizontally. Inheritance as understood in mainstream evolutionary biology involves *vertical transmission*, whereby genetic inheritance is taken to occur between parents and their offspring. But in the case of cultural inheritance, traits or memes are acquired from a wide variety of sources. This is *horizontal transmission*, and it can occur at rates much faster than those that result from vertical transmission; a significant proportion of a population might come to possess a meme within a single generation. The same degree of saturation (assuming fairly low fitness differences) can take many generations via vertical transmission. The speed with which memes can spread in a population, potentially ousting other memes along the way, means that they may not have the longevity required for cumulative selection processes. Even if cumulative selection processes are occasionally established, horizontal transmission leaves them extremely vulnerable. So, the criticism runs, if memes or cultural traits cannot, or only very rarely, attain any degree of longevity, we are unlikely to see complex cultural adaptations; and this limits the explanatory power of memetics, or any other theory of cultural evolution that permits horizontal transmission. Thus, since horizontal transmission undermines cumulative natural selection, we are left with no reason to suppose that horizontally transmitted traits will be fitness enhancing. This is a consequence well known to memeticists; Dawkins cites the celibacy of priests as an example of a meme that decreases reproductive potential.

The potential for horizontal transmission in cultural evolution is held to mark a significant break from mainstream evolutionary theory, and a radical overhaul of the latter would be required if theories of cultural evolution were to be brought under its umbrella. It is worth noting, however, that horizontal transmission is not unique to cultural evolution, and it seems an especially important phenomenon among bacteria; horizontal or lateral gene transfer (see chapter II.11) also puts pressure on the strictly vertical notions of inheritance assumed by much mainstream evolutionary theory.

A final worry stems from asking whether, even if memes do exist, the meme concept is of any use. The charge here is that memetics is not particularly enlightening: it only dresses up familiar explanations in a slightly different guise. So, for instance, we might allow that clothes fashions are memes, but even if that is the case, memetics does not explain why fashion memes differentially replicate. To explain why one such meme propagates throughout the population while another one perishes in obscurity, we still require reference to local conditions, consumer psychology, and so on. Any value memetics can bring to the explanation of why one meme is fitter than another is parasitic on conventional work done in psychology. And if individual preferences are subject to change over time, then there may be no general and informative theory of cultural evolution to be had; rather, we will have to settle for local explanations that look to shifting preferences. The upshot of this, the argument goes, is that memetics never gets beyond conventional narrative cultural history and cannot provide us with a new scientific framework for understanding culture.

3. CULTURAL EVOLUTION

Another line of investigation, pioneered by Luigi Luca Cavalli-Sforza, Marcus Feldman, Peter Richerson, and Robert Boyd, concerns the ways in which cultural inheritance can affect evolutionary processes. These models do not assume that cultural inheritance works in the same way as genetic inheritance and thus they differ significantly from memetic approaches. Indeed, they model cultural inheritance in ways that depart quite markedly from genetic inheritance. So it is that many of Robert Boyd and Peter Richerson's models explicitly assume that an individual's cultural makeup is an error-laden blend—synthesized from, and influenced by, many cultural "parents"—rather than a collection of discretely transmitted, self-replicating, gene-like particles. Their work then focuses on the population-level evolutionary consequences of such an inheritance system. Moreover, they tend to concentrate on this form of modeling while remaining noncommitted regarding the precise way in which cultural variants are physically realized.

Such a move can be defended by an appeal to history. Darwin's theory of evolution by natural selection lacked a plausible material theory of inheritance for some time, but this did not prevent Darwin's theory from being useful in the interim. Even without an account of what exactly is inherited in cultural inheritance, work can be done to explain the changes in (cultural) trait frequencies in a population by focusing on the population-level consequences of (cultural) inheritance, selection, mutation, and other forces. So although cultural evolutionary theorists may deny that cultural change should be understood in just the same way as biological change is understood, their approach remains recognizably evolutionary in style.

All the same, one might think that even if cultural change does not require cultural replicators such as memes, cultural replicators are necessary if cultural change is to be adaptive. As discussed in section 1, longevity is required for cumulative selection processes to operate and complex adaptations to arise. One obvious concern in this context stems from the fact that learning is often very error prone. If an individual hits on a fitness-enhancing behavior, that trait may be lost to future generations either because it is miscopied or because it is combined with other, less adaptive traits to produce an averaged, "blended" behavior (recall that a particular version of the tango may be an amalgamation of the influences of several teachers and famous dancers). Again, we might fear that cultural traits will not persist long enough for selection to act on them.

Richerson and Boyd argue that these problems are not fatal, and they have developed a number of valuable models to demonstrate why this is the case. These models assume that individuals will pick up cultural traits from a variety of sources and will frequently make mistakes. They also assume that we possess certain kinds of cognitive biases, and they show how these biases can dampen the spread of error in the population. So even if errors are occasionally made, these isolated errors will tend not to be imitated by others if we possess a *conformist bias*, such that we are more likely to imitate or learn the traits we most commonly encounter. A *prestige bias*, whereby individuals are more likely to imitate a trait possessed by those members of the population who are deemed to be successful, is also thought to keep error in check. It is likely that at least some of the traits possessed by a successful individual will be instrumental in that individual's success. A bias toward imitating successful individuals increases the chances that it is those success-generating traits that are imitated. But if the correct, success-generating trait is not identified, the result is that the individual fails to become successful and so does not become a target for future imitators. Cognitive biases such as these, it is argued, allow for mistakes to be made at an individual level but protect against those mistakes being repeated so widely that they undermine the distribution of cultural traits in a population.

While the existence of such biases can dampen the spread of error, it is far from obvious that they will keep error in check to the extent that cultural inheritance will be robust enough for selection processes to operate. To show that these properties of individual psychology combine to yield population-level inheritance requires some abstract mathematical modeling; much of the novel explanatory payoff of recent work in cultural evolutionary theory comes from the insights gained from this sort of modeling. And the establishment of such population-level consequences is important, for it enables the investigator to revise the constraints one might naively think must bear on cultural inheritance if cumulative cultural evolution is to occur.

This approach allows cultural evolutionists to agree that cumulative evolution requires that fitness-enhancing cultural traits are preserved in the offspring generation as a whole while denying that this requires faithful transmission between individuals. This move also answers one of our earlier criticisms of meme theory: in taking a population-level perspective, cultural evolution offers genuinely novel explanatory resources that go beyond cosmetic redescriptions of what we already know

One may ask why it should be the case that we are able to learn from nonparents at all, given that horizontal transmission enables the spread of maladaptive traits. Cultural evolutionists have defended the thought that the overall adaptive benefits of learning from nonparents outweigh the overall adaptive costs. Determining how best to live in an environment can be difficult, even dangerous, if one attempts to do so without guidance; one may not be able to tell until it is too late which foodstuffs are nutritious and which are poisonous, for example. Similarly, a prestige bias is an achievable solution to a tricky problem: "determining who is a success is much easier than determining how to be a success" (Richerson and Boyd 2005, 124). The contention here is that cognitive biases that incline us toward imitating or learning from certain individuals may not rule out all maladaptive traits spreading through the population, but these biases are nonetheless more adaptive, overall, than available alternatives.

4. NONHUMAN ANIMAL CULTURAL EVOLUTION

There is widespread acceptance of the existence of some degree of culture (sometimes referred to as "tradition") in nonhuman species. For example, distinct dialects exist within the songs of certain species of birds, and tool use in some primates can vary from group to group within a single species. Japanese macaques are a particularly well-studied population, and differences in everything from their grooming behavior to diet have been documented. The macaques of Koshima Island have developed some remarkable behaviors. To attract the macaques to open land so they could be observed, primatologists left sweet potatoes on a beach. This technique was effective but it did leave the potatoes covered in sand. One member of the troop solved this problem by washing the potatoes in a nearby stream. Soon, her peers followed suit. After a while, the group began to use the sea instead of the stream for washing, preferring the taste that the saltwater imparted, and their young took to playing in the sea for the first time. The group also discovered fish discarded by local fishermen and added this to their diet. These changes in their behavioral repertoires have taken place over the course of 50 years and identify them as distinct from other troops of macaques.

This example appears to have some key ingredients for cultural evolution: an individual hits on an innovation, and the innovation spreads throughout the population, which creates behaviorally distinct populations within the species. However, critics argue that this process alone is unlikely to secure cumulative evolution. Not all cultural inheritance involves "observational learning" or imitation, and the worry is that only these forms of social learning will allow the appearance of complex adaptations. We can see why observational learning might be considered crucial with the following example. Certain populations of blue tits learned to remove the foil tops from milk bottles to gain access to the milk inside. The birds' attention was drawn to the milk bottles by the activity of their conspecifics. But it was only through trial-and-error exploration of the milk bottles that each tit worked out how to get to the milk inside. If an individual bird happened on a particularly efficient means of removing the foil top, this technique could not be transmitted to any other bird. Social learning of this sort, which does not rely on observational learning or imitation, means that innovations cannot be combined and built on, and cumulative evolution is unlikely to get off the ground.

Although these sorts of considerations have left some pessimistic about the possibility of significant cultural evolution in nonhuman animals, at worst they merely make complex adaptations as the result of cumulative evolution unlikely. Cultural variations may still play an important role in any evolutionary story of a given species. In the same way that dairy farming led to selection pressure for the ability to digest lactose, cultural changes in nonhuman species may alter the selective environment of those species and instigate a sequence of evolutionary changes. Further, as the macaques of Koshima Island demonstrated, although novel behaviors may not be built on and made more complex, one new innovation can open up previously unexplored parts of the environment and inspire further innovation. At the very least, this sort of example leaves room for the kind of gene-culture coevolutionary models briefly discussed in section 2.

Cultural evolution in nonhumans is an under-researched area, and it remains unclear how widespread observational learning is. Further work will help us establish the significance of culture on the overall evolutionary trajectory of a species, as well as the extent to which we may speak of distinct cultural evolutionary processes in nonhuman species.

5. DEFINING CULTURE

According to Richerson and Boyd (2005), "culture is (mostly) information in brains." Cultural inheritance is then understood as the transmission of this information from one person to the next. So even though Richerson and Boyd deny any strong similarity between genes and cultural variants, they maintain that cultural variants "must be genelike to the extent that they carry cultural information." If we are to understand what is meant by culture and cultural evolution, then we must understand what is meant by *information*. There is no consensus on the meaning of this term, and the definitions that are offered—where they are offered at all—are problematic. For example, in their earlier work, Boyd and Richerson (1985) offered a definition of information as "something which has the property that energetically minor causes have energetically major effects." This is a curious definition; presumably it is meant to evoke intuitive examples whereby small informational "switches" (whether they are literally switches in a designed control system or metaphoric "genetic switches" in developmental pathways) have magnified downstream effects on the systems they influence. However, there are plenty of cases of information-bearing relations in which the energetic inequality is reversed. An instrument's display screen can carry information about solar flares; here, an energetically major cause has an energetically minor effect. Perhaps because of these oddities, their more recent work describes information as "any kind of mental state, conscious or not, that is acquired or modified by social learning and affects behavior" (Richerson and Boyd 2005). They later qualify this description with the concession that in some cases, "cultural information may be stored in artefacts." Alex Mesoudi (2011), on the other hand, does not offer a definition of information in his work but states that it is "intended as a broad term to refer to what social scientists and lay people might call knowledge, beliefs, attitudes, norms, preferences, and skills" while also insisting that culture is "information rather than behaviour." However, because skills involve practiced, embodied behaviors, it is unclear how they can count as a form of cultural information.

Eva Jablonka and Marion Lamb defend the use of "information" on the grounds that it provides us with a term that can free us from worrying about the specifics of modes of transmission. It is taken to cover what is transmitted in genetic material, epigenetic material, environmental structures such as nests, behaviors learned from conspecifics, and the kind of knowledge stored in books. Although this kind of abstraction allows us to formulate hypotheses and theories that bear on all these cases of transmission, grouping them together will highlight what differences exist among them, too, which may encourage us to attend to features of certain types of information and its transmission that we might otherwise overlook. For example, repositories of symbols, most obviously in the form of libraries and computer databases, are vital inheritance systems for humans, allowing the preservation and accumulation of knowledge across generations. Nonsymbolic transmission occurs when some birds inherit their song from adult birds around them. Jablonka and Lamb use the characteristic differences among typical modes of social inheritance in animals and humans to illuminate the impact our own symbolic transmission systems have on human cultural evolution.

In sum, there is some confusion here over what is meant by information and thus how we define culture. The worry is that the term *information* masks some serious issues that any theory of cultural evolution ought to be addressing; we really ought to be clear about what it is we are trying to explain with our theories.

Developing a more fine-grained analysis of cultural inheritance, as Jablonka and Lamb suggest the concept of "information" may allow, can only add to the explanatory power of theories of cultural inheritance, but more work is needed first to clarify some conceptual confusion. While more research exists on human cultural evolution than on nonhuman cultural evolution, both areas are in their infancy. Thus, we should not be surprised to find that we are faced with a paucity of data and concepts in need of some untangling. But although the precise details have yet to be ironed out, the research so far has at least demonstrated that cultural evolution is both possible and plausible.

FURTHER READING

Boyd, R., and P. J. Richerson. 1996. Why culture is common but cultural evolution is rare. Proceedings of the British Academy 88: 73–93. *An overview of nonhuman culture and difficulties for theories of cultural evolution in nonhuman animals.*

Cavalli-Sforza, L., and M. Feldman. 1981. Cultural Transmission and Evolution: A Quantitative Approach. Princeton, NJ: Princeton University Press. *One of the first attempts to construct a theory of cultural evolution.*

Dawkins, R. 1976. The Selfish Gene. Oxford: Oxford University Press. *Chapter 11 contains Dawkins's first discussion of his meme concept.*

Jablonka, E., and M. Lamb. 2005. Evolution in Four Dimensions: Genetic, Epigenetic, Behavioral, and Symbolic Variation in the History of Life. Cambridge, MA: MIT. *Among other things, deals with research into animal culture and defends a coevolutionary model for nonhuman animals. Also includes a discussion of the concept of "information" and its place in evolutionary theory.*

Lewens. T. 2007. Darwin. London: Routledge. *Chapter 7 contains a detailed critique of memetics.*

Odling-Smee, F. J., K. Laland, and M. W. Feldman. 2003. Niche Construction: The Neglected Process in Evolution. Princeton, NJ: Princeton University Press. *One of the most developed accounts of gene-culture coevolution.*

Richerson, P., and R. Boyd. 2005. Not by Genes Alone: How Culture Transformed Human Evolution. Chicago: University of Chicago Press. *An accessible introduction to cultural evolution by two of the founders of the field.*

VIII.11

Evolution and Notions of Human Race

Alan R. Templeton

OUTLINE

1. The biological meaning of race
2. Do biological races exist in chimpanzees?
3. Do biological races exist in humans?
4. Do adaptive traits define human races?
5. Do human races exist: The answer

Races exist in humans in a cultural sense, but it is essential to use biological concepts of race that are applied to other species to see whether human races exist in a manner that avoids cultural biases and anthropocentric thinking. Modern concepts of race can be implemented objectively with molecular genetic data, and genetic data sets are used to see whether biological races exist in humans and in our closest evolutionary relative, the chimpanzee.

GLOSSARY

Admixture. Reproduction between members of two populations that previously had little to no reproductive contact.

Alleles. Alternative forms of homologous genes within a species that constitute the most basic type of genetic diversity.

Evolutionary Lineage. A population that maintains genetic continuity and identity over many generations because of little to no reproductive interchange with other populations.

Evolutionary Tree. A depiction of the ancestral relationships that interconnect a group of biological entities through a diagram in which ancestral nodes can split into two or more descendant types but that does not allow fusion of previously split types.

Gene Flow. Movement of individuals or gametes from the local population of birth to a different local population followed by successful reproduction.

Genetic Differentiation. Differences among populations based on particular alleles they possess, the frequencies of shared alleles, or both.

Haplotype. A specific nucleotide sequence existing among the homologous copies of a defined DNA region, whether a gene or not.

Isolation By Distance. A model of gene flow in which most genetic interchange is between neighboring populations but in which genes can spread to distant populations over many generations because there are no absolute barriers to movement between any pair of neighboring populations.

Local Population. A collection of interbreeding individuals of the same species that live in sufficient proximity that most mates are drawn from this collection of individuals.

Race. A subpopulation within a species, also called a *subspecies*, that has sharp geographic boundaries separating it from the remainder of the species, with the boundaries characterized by a high degree of genetic differentiation defined either through a quantitative threshold or qualitatively as a separate evolutionary lineage.

1. THE BIOLOGICAL MEANING OF RACE

Do human races exist? Many people would answer yes because they have a strong sense of their own racial identity and feel they can classify other people into racial categories. However, the ability to classify oneself and others into races does not mean that races actually exist as a culture-free, biological category. For example, Lao and coworkers (2010) assessed the geographic ancestry of self-declared "whites" and "blacks" in the United States by the use of a panel of genetic markers. It is well known that the frequencies of alleles (different forms of a gene) vary over geographic space in humans. The differences in allele frequencies are generally so modest that any one gene yields only a little information about the geographic origins of one's ancestors. However, with modern DNA technology, it is possible to infer the geographic ancestry of individuals by scoring large numbers of genes. Self-identified "whites" from the United States are primarily of European ancestry, whereas US "blacks" are primarily of African ancestry, with little to no overlap in the amount of African ancestry between US "whites" and "blacks." In contrast, Santos and coworkers (2009) did a similar genetic assessment of Brazilians who self-identified themselves as "whites," "browns," and "blacks" and found extensive overlap in the amount of African ancestry among all these "races." Indeed, many Brazilian "whites" are surprised to learn that they are considered to be "blacks" when they visit the United States, and similarly, some US "blacks" are considered to be "whites" by Brazilians. Obviously, the culturally defined racial categories of "white" and "black" do not have the same genetic meanings in the United States and Brazil. It is clear that an objective, culture-free definition of race is required before the question about the existence of biological races can be answered.

One way of ensuring a culture-free definition of race is to use a definition that is applied to species other than humans. The word *race* is not commonly used in the nonhuman literature; instead, the word *subspecies* is used to indicate the major types or subdivisions within a species. There is no consensus on what constitutes a species (see chapter VI.1), much less a subspecies. Because the US Endangered Species Act mandates the protection of endangered subspecies of vertebrates as well as endangered species, conservation biologists have developed operational definitions of race or subspecies applicable to all vertebrates. We will apply these culture-free definitions to humans to avoid an anthropocentric definition of race.

Biologically, *races* are geographically circumscribed populations within a species that have sharp boundaries that separate them from the remainder of the species. In traditional taxonomic studies, the boundaries were defined by morphological differences, but increasingly these boundaries are defined in terms of genetic differences that can be scored in an objective fashion in all species. Most local populations within a species show some degree of genetic differentiation from other local populations by having either some unique alleles or different frequencies of alleles. If every genetically distinguishable population were elevated to the status of race, then most species would have hundreds to tens of thousands of races. This would make the concept of race nothing more than a synonym for a local population. There is a consensus that race or subspecies should refer to a degree or type of genetic differentiation that is well above the level of genetic differences that exist among local populations. Both quantitative and qualitative criteria are used to define these racial genetic boundaries.

Quantitatively, one commonly used threshold is that two populations with sharp boundaries are considered to be different races if 25 percent or more of the genetic variability that they collectively share is found as between-population differences. One of the oldest measures used to quantify these differences is a statistic known as f_{st}. Consider drawing two homologous genes at random from all the genetic variation collectively shared by both subpopulations. The frequency with which these two randomly drawn genes from the total population are different alleles is designated by H_t, the expected heterozygosity of the total population. Now consider drawing two genes at random from just a single subpopulation. Let H_s be the average frequency with which these randomly drawn genes from the same subpopulation are different alleles. Then, $f_{st} = (H_t - H_s)/H_t$. In many modern genetic studies, the degree of DNA sequence differences between the randomly drawn genes is measured, often with the use of a model of mutation, instead of just determining whether the two genes are the same or different alleles. When this done, the analysis is called an *analysis of molecular variance* (AMOVA). Regardless of the specific measure, the degree of genetic differentiation can be quantified in an objective manner in any species. Hence, human "races" can indeed be studied with exactly the same criteria applied to nonhuman species. The main disadvantage of this definition is the arbitrariness of the threshold value of 25 percent.

A second definition of race defines the genetic differences qualitatively. Sharp boundaries exist in this case because the species is subdivided into two or more evolutionary lineages. An evolutionary lineage is created within a species when an ancestral population is split into two or more subpopulations, often by some sort of geographic barrier, such that there is no or extremely limited genetic interchange after the split. This means that the subpopulations tend to evolve mostly independently of one another, causing the lineages to accumulate genetic differences with increasing time since the split. Immediately after the split, the subpopulations would share most ancestral polymorphisms (gene loci with more than one allele) and would therefore be difficult to diagnose as separate lineages. With increasing time since the split, genetic divergence accumulates, and diagnosing the separate lineages becomes easier. Unlike the f_{st} definition of race, no arbitrary threshold of differentiation is set a priori. A split into separate lineages also means that the genetic differences among the races would define an evolutionary tree analogous to an evolutionary tree of species. Statistical methods exist for testing the null hypothesis that the genetic variation within a species has a treelike structure, and other statistics test the null hypothesis that the entire sample defines a single evolutionary lineage. Therefore, just as with the f_{st} definition, the lineage definition of race can be implemented for all species in an objective fashion using uniform criteria, thereby avoiding a human-specific or cultural definition of race.

2. DO BIOLOGICAL RACES EXIST IN CHIMPANZEES?

Before addressing the existence of human races, we first apply these definitions of race to our closest evolutionary relative, the chimpanzee. In this manner, the definitions can be applied in a context that avoids the emotion and cultural biases that inevitably creep into discussions of human race.

Based on morphological differences, the common chimpanzee (*Pan troglodytes*) has been subdivided into five races or subspecies: *P. t. verus* in the Upper Guinea region of western Africa, *P. t. ellioti* in the Gulf of Guinea region (southern Nigeria and western Cameroon), *P. t. troglodytes* in central Africa, *P. t. schweinfurthii* in the western part of equatorial Africa (mostly southern Cameroon), and *P. t. marungensis* in central and eastern equatorial Africa. Gonder and coworkers (2011) genetically surveyed chimpanzees throughout their range. They discovered sharp genetic differences separating the Upper Guinea and Gulf of Guinea populations from all other populations, but with less sharp genetic boundaries between the equatorial African populations. Table 1 shows the pairwise AMOVA results for these populations. The Upper Guinea and Gulf of Guinea populations are above the 25 percent threshold for contrasts with each other and with all other chimpanzee populations. However, the three regions sampled in equatorial Africa are all well below the 25 percent threshold. Hence, there are three races or subspecies of common chimpanzees using the threshold criterion: *P .t. verus* in the Upper Guinea region, *P. t. ellioti* in the Gulf of Guinea region, and the chimpanzee populations from equatorial Africa.

If chimpanzees are subdivided into separate evolutionary lineages, the genetic differences among lineages should define a treelike structure characterized by splits and isolation. There are genetic differences between different geographic areas (table 1), but such genetic differentiation can also arise when gene flow (genetic interchange associated with individuals who disperse from their birth population) occurs but is restricted by geography. For example, gene flow can be restricted when most dispersal is limited to nearby local populations. Because genes are passed on from generation to generation, a new allele can still spread throughout a species' range over multiple generations by using local populations as "stepping-stones" to reach more distant local populations. Such stepping-stone models yield a pattern of isolation by distance in which the degree of genetic differentiation between two populations increases with increasing geographic distance between them.

Genetic differentiation structured by isolation by distance can be distinguished from genetic differentiation due to lineage splits by testing for constraints on genetic distances. Consider three hypothetical populations (A, B, and C) such that A is closer to B than to C, and B and C are the closest geographic pair. Under isolation by distance, the genetic distance (measured, say, by the f_{st} value between a pair of populations) should increase with increasing geographic distance; that is, the f_{st} between A and B should be less than the f_{st} between A and C. In contrast, suppose populations A, B, and C represent separate evolutionary lineages (races) such that A split from the common ancestral population of B and C in the past, followed by a more recent split between populations B and C. This results in an evolutionary tree of populations such that genetic distance between populations increases with the time since their split from a common ancestral population. In this hypothetical case, the genetic distances between populations A and B and between populations A and C should be the same, since they both involve a split from the same ancestral population. Hence, the expected pattern of genetic distances differs for trees versus isolation by distance, and formal statistical tests exist to determine whether the pattern of genetic differentiation is consistent with the special constraints imposed by an evolutionary tree.

Table 1. Genetic differentiation among populations of chimpanzees as measured by R_{st}

	Upper Guinea	*Gulf of Guinea*	*Southern Cameroon*	*Central Africa*
Gulf of Guinea	0.41			
Southern Cameroon	0.43	0.25		
Central Africa	0.46	0.27	0.07	
Eastern Africa	0.44	0.28	0.05	0.03

Source: Modified from Gonder et al. 2011.

Note: R_{st} is related to f_{st} but incorporates a mutational model for microsatellites.

Another method for testing for a treelike structure is based on finer geographic sampling. As more sites are sampled under an isolation-by-distance model, the geographically intermediate populations should also have intermediate genetic distances. In contrast, when the populations are grouped into a smaller number of evolutionary lineages, genetic distances among populations within a lineage should be relatively small, although they may show an isolation-by-distance pattern within the geographic range occupied by a particular lineage. However, the genetic distances are expected to show a large, sudden increase when the geographic boundary between two lineages is crossed.

When the chimpanzee genetic data are used to estimate an evolutionary tree of populations, the resulting

Table 2. AMOVA of genetic variation in chimpanzees and in humans

			Genetic variance components		
Species	*Number of "races"*	*Number of populations*	*Among individuals within populations*	*Among populations within races*	*Among races*
Chimpanzees	3	5	64.2%	5.7%	30.1%
Humans	5	52	93.2%	2.5%	4.3%

Sources: Chimpanzees—data from Gonder et al. 2011; humans—data from Rosenberg et al. 2002.

tree has the Upper Guinea population splitting off first, followed by the Gulf of Guinea population, and then splits among the equatorial Africa populations. This tree predicts that the Upper Guinea population should be equally distant from all the other populations, and table 1 shows that this prediction is supported when the error in estimating the distances is taken into account. This tree also predicts that the Gulf of Guinea population should be equally distant from all the equatorial African populations but that this distance should be smaller (less time since the split) than the distances involving the Upper Guinea population. Table 1 shows that this prediction is also supported. However, the genetic distances among the three equatorial African populations show the isolation-by-distance pattern on an east-west axis. These three populations are therefore collapsed into a single lineage. Hence, chimpanzees do show a treelike structure of genetic differentiation with three lineages: Upper Guinea, Gulf of Guinea, and the combined equatorial African populations. Hence, races do exist in chimpanzees under the lineage definition, and they correspond exactly to the same three races defined by the quantitative threshold definition of race.

3. DO BIOLOGICAL RACES EXIST IN HUMANS?

Do human races exist according to the same criteria applied to chimpanzees? In 2002, Rosenberg and others performed a genetic survey of 52 human populations. They used a computer program to sort individuals or portions of their genomes into five groups and discovered that the genetic ancestry of most individuals was inferred to come mostly from just one group. Moreover, the groups corresponded to five major geographic populations: (1) sub-Saharan Africans; (2) Europeans, Near and Middle Easterners, and Central Asians; (3) East Asians; (4) Pacific populations; and (5) Amerindians. This paper was the most widely cited article from the journal *Science* in 2002, and many of these citations claimed that this paper supported the idea that races were biologically real in humans. However, Rosenberg and coauthors were more cautious in their interpretation. When they increased the number of groups beyond five, they also obtained an excellent classification into smaller, more regional groups. Hence, they showed that with enough genetic markers, it is possible to discriminate most local populations from one another. Recall that genetic differentiation is *necessary but not sufficient* to define races, so even if there is a consensus that five groups is the right number, genetic discrimination alone does not necessarily mean that these five groups are races.

Assuming for now that the five groups are the meaningful populations, do these groups satisfy the quantitative threshold definition of race? Table 2 shows the AMOVA results for these five groups, along with a comparable analysis of the three races of chimpanzees that satisfy both the threshold and lineage definitions of race. Table 2 shows how the genetic variation is hierarchically partitioned into differences among individuals within the same local population, differences among local populations within the same "race," and among "races." Table 2 confirms the reality of race in chimpanzees using the threshold definition, as 30.1 percent of the genetic variation is found in the among-race component, a result expected from the pairwise analysis shown in table 1. In contrast with chimpanzee races, the five major "races" of humans account for only 4.3 percent of the genetic variation—well below the 25 percent threshold. The genetic variation in our species is overwhelmingly variation among individuals (93.2 percent). According to the threshold definition, there are no races in humans.

As for the lineage definition, a treelike structure of genetic differentiation has been strongly rejected for *every* human data set subjected to testing for the constraints expected from an evolutionary tree of populations. Increased geographic sampling further undermines the idea of separate lineages. When Rosenberg and coworkers published their results in 2002, their geographic sampling was coarse. It is now known that the computer program used in these studies generates well-differentiated populations as an artifact of coarse sampling from species characterized by isolation by distance. Figure 1 shows a plot of the pairwise f_{st} values of

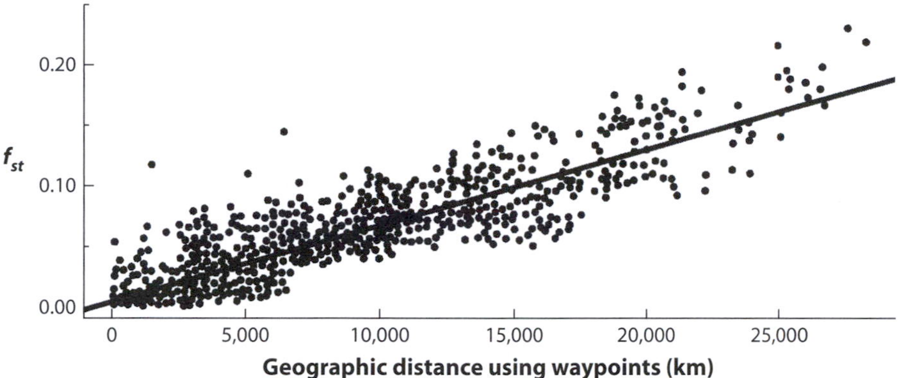

Figure 1. Isolation by distance in human populations. (Modified from Ramachandran et al. 2005. Copyright 2005 National Academy of Sciences, USA.)

humans as a function of geographic distance. The results fit well with the predictions of an isolation-by-distance model. Consequently, it is not surprising that when Behar and coworkers (2010) sampled Old World populations more finely and used the same computer program used in the 2002 study, most individuals showed significant genetic inputs from two or more populations, indicating that most human individuals have mixed ancestries. The "races" so apparent to many who cited Rosenberg and coworkers simply disappeared with better sampling. These results and figure 1 falsify the hypothesis that humans are subdivided into evolutionary lineages.

Another way of testing for distinct lineages is through a technique known as *multilocus nested-clade phylogeographic analysis* (ML-NCPA). Many regions of the human genome experience little to no recombination. The distinct genetic states that exist in such regions (called *haplotypes*) reflect the accumulation of various mutations during evolutionary history. This evolutionary history, called a *haplotype tree*, is the history of the genetic variation in that genomic region and is *not* necessarily the history of the populations that bear this variation. Indeed, if a species has sufficient gene flow, there can be no evolutionary tree of populations because there are no population splits; however, there will still be haplotype trees for each nonrecombining region of the genome. Haplotype distributions can be influenced by population-level history, but the population-level information embedded in a haplotype tree must be extracted carefully. It is never justified to equate a haplotype tree directly to an evolutionary history of populations. ML-NCPA provides a statistically rigorous method for making inferences about population history from haplotype trees. No other technique of phylogeographic inference has been so extensively validated as ML-NCPA by both positive controls (data sets for which outside information exists that indicates a known historical event or process) and computer simulation. These validations show that ML-NCPA is not prone to making false-positive inferences about past splits and is very powerful in detecting separate lineages, even when the split is relatively recent and results in haplotype trees affected by retention and sorting of ancestral haplotypes. Moreover, ML-NCPA can detect lineages even when there is some, but very limited, genetic interchange. ML-NCPA does not require an a priori model of the evolutionary history of a species; rather, the history is inferred directly from the haplotype trees using explicit criteria applicable to all species. Finally, each inference made with ML-NCPA is subject to a statistical test for significance.

Figure 2 shows the inferences from ML-NCPA about human evolution. The oldest inferred event is an out-of-Africa range expansion into Eurasia genetically dated to about 1.9 million years ago—the same time that the fossil evidence indicates that *Homo erectus* spread out of Africa into Eurasia during a major wet period in the Sahara. The paleoclimatic data indicate that the Sahara region experienced repeated minor wet periods such that the Sahara is unlikely to have been a dispersal barrier on a timescale of tens of thousands of years. Consistent with these paleoclimatic data, ML-NCPA infers limited genetic interchange with isolation by distance between sub-Saharan Africa and Eurasia starting no later than 700,000 years ago in the Pleistocene. The null hypothesis of complete genetic isolation during the Pleistocene is decisively rejected. Consequently, even during the Pleistocene, Old World human populations were not subdivided into isolated and independently evolving lineages.

The next major event shown in figure 2 is a second population expansion out of Africa into Eurasia around 700,000 years ago, corresponding to the spread of the Acheulean tool culture out of Africa into Eurasia during the second major Saharan wet period of the Pleistocene. The null hypothesis of no admixture between the expanding population and the Eurasian populations is rejected. Hence, the Acheulean expansion was marked by

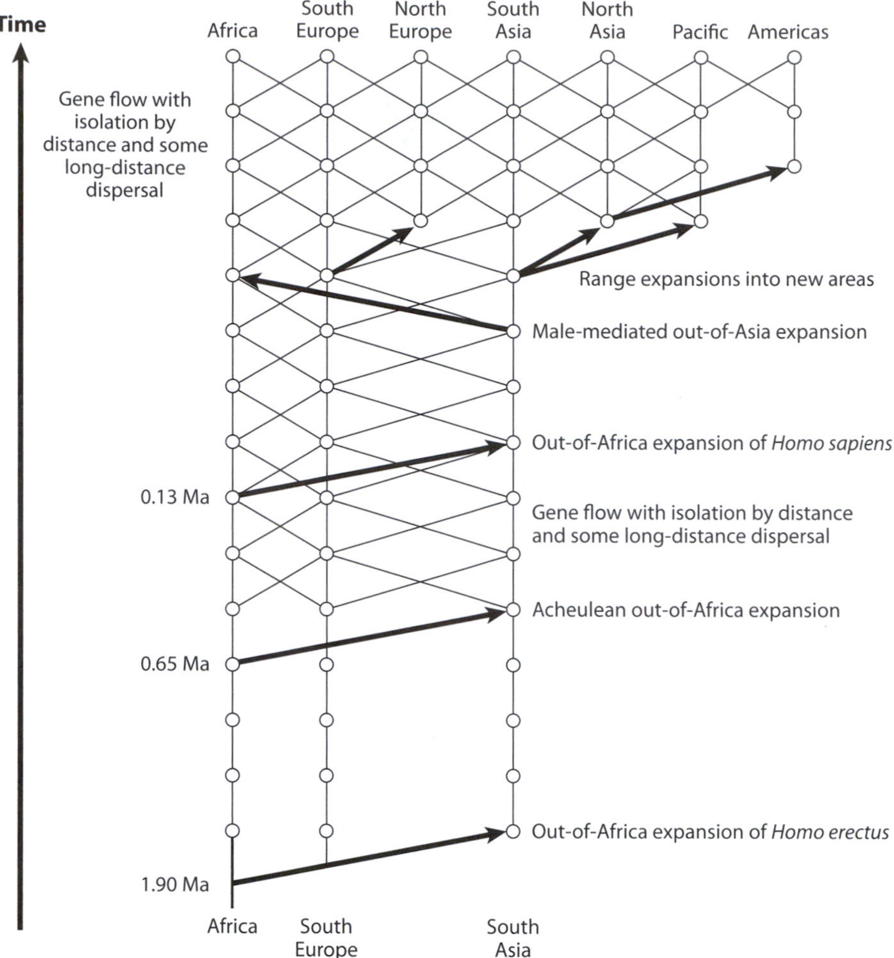

Figure 2. Significant inferences about human evolution from multilocus, nested-clade phylogeographic analysis. Geographic location is indicated on the *x*-axis, and time on the *y*-axis, with the bottom of the figure corresponding to 2 million years ago. Vertical lines indicate genetic descent over time, and diagonal lines indicate gene flow across space and time. Thick arrows indicate statistically significant population range expansions, with the base of the arrow indicating the geographic origin of the expanding population. Lines of descent are not broken, because the population range expansion events were accompanied by statistically significant admixture when they involved expansion into a previously inhabited area. (Modified from Templeton 2005.)

further genetic interchange between African and Eurasian populations, further weakening the hypothesis of isolated Pleistocene lineages of humans. Gene flow then continued until a third major expansion of humans out of Africa into Eurasia occurred around 130,000 years ago, the time of the last major Saharan wet period. The fossil record indicates that modern humans began expanding out of sub-Saharan Africa at 130,000 years ago and reached China no later than 110,000 years ago. The null hypothesis of no admixture is overwhelmingly rejected for this expansion event. This strong rejection of strict replacement without admixing was the most controversial ML-NCPA inference, because the dominant model of human evolution at the time was the out-of-Africa replacement model (see chapter II.18). The inference of admixture has since been supported by studies on fossil DNA of archaic Eurasian populations.

Following the expansion with admixture of modern humans from Africa, there have been additional expansions, mostly into areas not formerly occupied by humans (figure 2). Wherever humans lived, gene flow was established, mostly limited by isolation by distance but with some long-distance dispersal as well. On a timescale of tens of thousands of years, there is not one statistically significant inference of splitting or isolation during the last 700,000 years. Because of gene flow and admixture, humans are a single evolutionary lineage. Hence, there are no races in humans under the lineage definition.

4. DO ADAPTIVE TRAITS DEFINE HUMAN RACES?

Races or subspecies, when they exist, always occupy a subset of the geographic range of their species. Sometimes, environmental factors vary over the geographic range of the species, and some of these environmental factors can induce natural selection that results in local adaptation. Hence, when races exist, they sometimes display local adaptations to the environment associated with their geographic subrange that are not adaptive in the remainder of the species' geographic range. This reasoning leads to the idea that local adaptations can sometimes be biological markers of racial status.

Variation in environmental factors can still induce natural selection that results in local adaptations in species with sufficient gene flow and admixture to prevent race formation. However, in this case, the geographic distributions of the local adaptations reflect the geography of the environmental factors and not racial boundaries. Frequently, different adaptive traits display discordant geographic distributions, thereby indicating that these are simply adaptations of local populations and not markers of higher groupings such as race.

Because humans are not subdivided into races by any of the definitions applied to other species, the locally adaptive traits of humans are not "racial" traits. Skin color is historically the locally adaptive trait most commonly considered a "racial" trait in humans. Skin color is an adaptation to the amount of ultraviolet (UV) radiation in the environment: dark skins are adaptive in high-UV environments to protect cells from radiation damage, and light skins are adaptive in low-UV environments to make sufficient vitamin D, which requires UV radiation. Skin color varies continuously among humans and does not fall into a few discrete "racial" types. Moreover, the geographic distribution of skin color follows the environmental factor of UV incidence and does not reflect overall genetic divergence. For example, the native peoples with the darkest skins live in tropical Africa and Melanesia. The dark skins of Africans and Melanesians are adaptive to the high UV found in these areas. Because Africans and Melanesians live on opposite sides of the world, they are more highly genetically differentiated than many other human populations (figure 1). Europeans, who are geographically close to Africa, are more similar at the molecular genetic level to Africans than Melanesians are to Africans, despite the fact that Europeans have light skins that are adaptive to the low-UV environment of Europe. Hence, skin color is not an indicator of the degree of genetic differentiation, as a true racial trait would be.

Adaptive traits in humans do not define coherent populations. For example, the adaptive trait of dark skin is widespread in sub-Saharan Africa. Another adaptive trait found in Africa is resistance to sleeping sickness, and the responsible gene is found at frequencies up to 80 percent in parts of western Africa where the parasite that causes sleeping sickness is common. However, this adaptive trait is virtually absent in East African populations. Hence, the distribution of sleeping sickness resistance is only a subset of the geographic distribution of dark skin in Africa. Another adaptive trait is resistance to malaria, which is widespread in African populations. However, malaria is also common in some areas outside Africa, and malarial resistance is found in many European and Asian populations as well. Indeed, one of the alleles underlying malarial resistance, the sickle-cell allele, is most frequent in certain populations on the Arabian Peninsula despite often being regarded as a disease of "blacks." Each adaptive trait in humans has its own geographic distribution that reflects the distribution of the underlying environmental factor for which it is adaptive. The discordance in the distributions of adaptive traits in humans makes them useless in defining races.

5. DO HUMAN RACES EXIST: THE ANSWER

Using culture-free, objective definitions of race, the answer to the question whether races exist in humans is clear and unambiguous: no. Human evolutionary history has been dominated by gene flow and admixture that unifies humanity into a single evolutionary lineage. This finding does not mean that all human populations are genetically identical. Isolation by distance ensures that human populations are genetically differentiated from one another, and local adaptation ensures that some of these differences reflect adaptive evolution to the environmental heterogeneity that our globally distributed species experiences. However, most of our genetic variation exists as differences among individuals, with between-population differences being very small. There are no biological races in humans; indeed, despite our global distribution, we are one of the most genetically homogeneous species on this planet.

FURTHER READING

Gonder, M. K., S. Locatelli, L. Ghobrial, M. W. Mitchell, J. T. Kujawski, F. J. Lankester, C.-B. Stewart, and S. A. Tishkoff. 2011. Evidence from Cameroon reveals differences in the genetic structure and histories of chimpanzee populations. Proceedings of the National Academy of Sciences USA 108: 4766–4771.

Lao, O., P. M. Vallone, M. D. Coble, T. M. Diegoli, M. van Oven, K. J. van der Gaag, J. Pijpe, P. de Knijff, and M. Kayser. 2010. Evaluating self-declared ancestry of U.S. Americans with autosomal, Y-chromosomal and mitochondrial DNA. Human Mutation 31: E1875–E1893.

Long, J. C., and R. A. Kittles. 2009. Human genetic diversity and the nonexistence of biological races. Human Biology 81: 777–798. *Formal statistics tests of the hypothesis that human genetic variation is structured in a treelike fashion. They show that the hypothesis of a tree of populations is strongly rejected for humans.*

Ramachandran, S., O. Deshpande, C. C. Roseman, N. A. Rosenberg, M. W. Feldman, and L. L. Cavalli-Sforza. 2005. Proceedings of the National Academy of Sciences USA 102: 15942–15947. *Support from the relationship of genetic and geographic distance in human populations for a serial founder effect originating in Africa*

Relethford, J. H. 2009. Race and global patterns of phenotypic variation. American Journal of Physical Anthropology 139: 16–22. *This paper shows that skin color variation and other morphological traits are clinal and are not well described by discrete racial categories.*

Rosenberg, N. A., J. K. Pritchard, J. L. Weber, H. M. Cann, K. K. Kidd, L. A. Zhivotovsky, and M. W. Feldman. 2002. Genetic structure of human populations. Science 298: 2381–2385.

Santos, R. V., P. H. Fry, S. Monteiro, M. C. Maio, J. C. Rodrigues, L. Bastos-Rodrigues, and S. D. J. Pena. 2009. Color, race, and genomic ancestry in Brazil: Dialogues between anthropology and genetics. Current Anthropology 50: 787–819.

Templeton, A. R. 2003. Human races in the context of recent human evolution: A molecular genetic perspective. In A. H. Goodman, D. Heath, and M. S. Lindee, eds. Genetic Nature/Culture, 234–257. Berkeley: University of California Press. *This paper covers many of the same issues as this chapter, but with older data sets. These older data sets yield the same conclusions found in this chapter.*

Templeton, A. R. 2005. Haplotype trees and modern human origins. Yearbook of Physical Anthropology 48: 33–59. *The results of multilocus nested clade phylogeographic analysis of humans, showing that gene flow and admixture have been such common features of recent human evolution that there is only one evolutionary lineage of humanity.*

Wolpoff, M., and R. Caspari. 1997. Race and Human Evolution. New York: Simon & Schuster. *The authors strongly argue against typological thinking in anthropology that explains human variation in terms of a few types or "races" rather than dealing with the entire range of variation found in living humans. Although the book ignores the nonhuman literature and tends to hyperbole, it remains an excellent introduction to the intertwined topics of racism and models of modern human origins.*

VIII.12

The Future of Human Evolution

Alan R. Templeton

OUTLINE

1. Can we predict how humans will evolve?
2. Has human evolution stopped?
3. Future nonadaptive evolution
4. Future adaptive evolution
5. Eugenics and genetic engineering

How humans will evolve in the future is highly speculative because the process of evolution depends critically on random processes such as mutation, recombination, and genetic drift, and because adaptive evolution is strongly influenced by changing environments. Because the human environment includes culture, which can change quickly, it is difficult to predict future environments and hence future adaptive evolution. Nevertheless, some predictions can be made based on a basic understanding of evolutionary mechanisms.

GLOSSARY

Eugenics. Programs designed to direct evolutionary changes in the human population by controlled breeding, selective abortions, and sterility operations.

Gene Flow. Movement of individuals or gametes from the local population of birth to a different local population followed by successful reproduction.

Gene Pool. The set of genetic variants collectively shared by a reproducing population.

Genetic Drift. The evolutionary force associated with random sampling events that alters the frequencies of genetic variants in the gene pool.

Genetic Engineering. The deliberate modification of characteristics of an organism by manipulating its genetic material.

Heterozygosity. The condition in which the two homologous segments of genetic material inherited from the parents are of a different state.

Mutation. A variety of molecular-level processes by which the genetic material of an organism (usually DNA) undergoes a change.

Neutral Allele. An allele that is functionally equivalent to its ancestral allele in terms of its chances of being replicated and passed on to the next generation.

Recombination. The generation of new combinations of DNA segments that are unlike those that existed in the parents.

1. CAN WE PREDICT HOW HUMANS WILL EVOLVE?

The current biological state of humans is the product of our past evolution (see chapters II.18 and VIII.11). But what of our evolutionary future? We can infer to some extent what our ancestors were like 100,000 or a million years ago, but what will our descendants be like 100,000 or a million years into the future? This question is difficult to answer because the evolutionary process itself is strongly influenced by random factors. There can be no evolution of any sort without genetic variation, and this genetic variation is created by mutation and recombination. Mutation and recombination are molecular-level processes that create new genetic variants before these variants are expressed phenotypically in individuals living and reproducing in an environment. In this sense, the genetic variation that is the raw material for all evolutionary change is random with respect to environmental needs—a fundamental premise of Darwinian evolution. Moreover, once genetic variation is randomly created, sampling error further accentuates the randomness of evolution. For example, suppose a new autosomal mutation occurs such that its bearer would be expected to have 10 percent more offspring than the average for the population. Assuming the population was stable, the average number of offspring per individual would be two for the population as a whole, but for this mutant bearer, the average would be 2.2. This would be regarded as extremely intense natural selection for a highly favored mutation. However, the actual number of offspring of the individual bearing the new mutant is not the same as the expected number. Perhaps this individual died from an accident or a disease unrelated to the mutant effect before he or she could reproduce; perhaps this individual failed to find a mate; perhaps this individual had fewer or more than 2.2 children. It is impossible to predict the exact number of children any particular individual will have even if the average number of children is known. At best what is known is the probability distribution of the offspring number. Suppose this offspring probability distribution is a Poisson distribution (a standard, commonly used probability distribution) with an average of 2.2 offspring. Then, despite the large fitness value this mutant individual has, there is still an 11 percent chance that he or she will have no offspring at all, and hence the mutant will be lost. Moreover, there is the randomness of meiosis. Suppose the mutant individual had three children. Each child would have a 50:50 chance of receiving the mutant, so with a chance of $(1/2)^3 = 1/8$, none of the children would receive the mutant, and the mutant would be lost even though the bearer had more children than the population average. Using a more thorough mathematical analysis, the probability that this strongly favored mutant gene is ultimately lost by chance despite being strongly favored by natural selection is 82 percent! Thus, even strong natural selection cannot completely overcome the randomness inherent in the evolutionary process. These random events affecting the fate of a particular mutant gene are occurring for all genes in the population, and can strongly influence the course of evolution. The evolutionary force associated with these random sampling events that affect existing genetic variation is known as *genetic drift* (see chapter IV.1). Genetic drift, mutation, and recombination ensure that evolution can never be completely predictable.

Another factor that makes future evolution difficult to predict is the environment. Adaptive evolution (see chapter III.1) is always with respect to an environment, and it is difficult to predict what our environment will be like in the distant future. This is especially true for us, because we define much of our environment through our culture (see chapter VIII.10). Culture in turn can change rapidly in unforeseen ways. Therefore, even the evolutionary trajectory of the nonrandom evolutionary force of natural selection is difficult to predict because the raw material on which natural selection operates has a large random component and because our environment, particularly our cultural environment, can change in a manner that is difficult to predict. Nevertheless, some predictions are possible using general evolutionary principles, but specific details must always be regarded as speculative.

2. HAS HUMAN EVOLUTION STOPPED?

One of the more popular predictions about future human evolution is that there is none; that is, human evolution has already stopped. For example, the distinguished evolutionary biologist Stephen Jay Gould (2000) stated:

> There's been no biological change in humans in 40,000 or 50,000 years. Everything we call culture and civilization we've built with the same body and brain.

The basic rationale behind the conclusion that human evolution has stopped is that once the human lineage had achieved a sufficiently large brain and had developed a sufficiently sophisticated culture (sometime around 40,000–50,000 years ago according to Gould, but more commonly placed at 10,000 years ago with the development of agriculture), cultural evolution supplanted biological evolution through natural selection; that is, humans no longer adapt to their environment through natural selection (see chapters III.1, III.3, and III.5) but rather alter the environment to suit human needs through cultural innovations. However, other evolutionary biologists have come to exactly the opposite conclusion. For example, Cochran and Harpending (2009) concluded that "human evolution has accelerated in the past 10,000 years, rather than slowing or stopping, and is now happening about 100 times faster than its long-term average over the 6 million years of our existence."

There are two fundamental flaws in the proposal that human evolution has stopped. First, it ignores the fact that evolution can occur owing to factors other than natural selection. This flaw is discussed in the next section. The second flaw is the premise that cultural evolution eliminates adaptive evolution via natural selection. All organisms adapt to their environment, and we define much of our environment by our culture. Hence, cultural change can actually spur adaptive evolution in humans. Since the development of agriculture, the human population has grown in a roughly superexponential fashion. Agriculture also induced a more sedentary lifestyle. As a result, even early agricultural systems resulted in large increases in local human densities. This in turn created a new demographic environment that was ideal for the spread of infectious diseases. For example, the Malaysian agricultural system, first developed in Southeast Asia, makes extensive use of root and tree crops that are adapted to wet, tropical environments. This tropical agricultural system was introduced to the African mainland about 1500 years ago, and malaria has become a common disease in these new agricultural areas. Because of agriculture, malaria became a major selective agent in African, and other, human populations. The result was that human populations began to adapt to malaria via natural selection. In sub-Saharan Africa, natural selection favored an increase in frequency of the sickle-cell allele at the hemoglobin β-chain locus, which confers resistance to malaria in individuals heterozygous for the sickle-cell allele. Similar selective forces were introduced wherever agriculture created the conditions to allow malaria to become a sustained, epidemic disease, and human populations in turn adapted to malaria by increasing the frequency of a number of alleles at many different loci in addition to the sickle-cell allele. In terms of the numbers of people affected, these antimalarial adaptations alone constitute the vast bulk of the classical Mendelian genetic diseases that afflict current humanity.

Agriculture produced a selective environment that also favored genes associated with risk for common systemic diseases in current human populations, one of the more common of which plaguing humans today is type 2 diabetes. Much of the increased incidence of diabetes is due to environmental changes in diet and lifestyle. However, phenotypes (see chapters I.4 and V.10) arise from the interaction of genes with environment, so a strong environmental component to type 2 diabetes, and many other systemic diseases, does not preclude a genetic component owing to adaptive evolution in recent human history. The idea that genes predisposing an individual to type 2 diabetes could represent recent adaptive evolution was first proposed by James V. Neel in 1962 as the *thrifty genotype hypothesis*. This hypothesis postulates that the same genetic states that predispose one to diabetes also are advantageous when individuals suffer periodically from famines. When food is more plentiful, selection against these genotypes would be mild because the age of onset of the diabetic phenotype is typically after most reproduction has occurred and because the high-sugar, high-calorie diets found in modern societies that help trigger the diabetic phenotype are very recent in human evolutionary history. There is now much evidence for the thrifty genotype hypothesis, including the genomic signatures of strong natural selection at genes shown to increase risk for diabetes in populations with a recent history of exposure to famines or calorie-restricted diets. The thrifty genotype hypothesis has often been portrayed as an example of past adaptation to a Paleolithic lifestyle despite the fact that the populations used to test this hypothesis all suffered from famines in historic times. Hence, the thrifty genotypes present in current human populations are an adaptation to recent events in agricultural systems prone to periodic failures and are *not* a legacy of human evolution having stopped in the Paleolithic.

Agriculture also induced positive selection for humans to adapt to the products of agriculture. For example, with the domestication of cattle and goats, milk and its derivatives became not only a source of nutrition but also a dietary component that protects against nutritional rickets, a common disease associated with high-cereal diets, another by-product of an agricultural environment. The phenotype of adult lactase persistence is determined by a single gene that allows the digestion of milk sugar. This specific allele shows one of the more powerful signatures of strong, recent natural selection in the human genome.

As the preceding examples demonstrate, agriculture—and culture in general—did not stop human evolution via natural selection but rather induced it through its direct and indirect effects on the human environment. Cultural innovations indeed shield some traits from natural selection, but cultural evolution will likely induce further adaptive evolution of many other traits in humans.

3. FUTURE NONADAPTIVE EVOLUTION

Not all evolution is adaptive. Evolution within a species is a change in the type or frequencies of genes or gene combinations in the gene pool over time, with the *gene pool* being the set of genes collectively shared by a reproducing population. Natural selection is a powerful mechanism for altering the frequencies of genes in the gene pool, but developmental constraints (see chapter III.8, patterns of dispersal (see chapter IV.3), system of mating (see chapters IV.6 and IV.8), population size (see chapter IV.1), mutation (see chapter IV.2), recombination (see chapter IV.4), and other factors can also cause alterations in the gene pool. Evolutionary change is determined not by one evolutionary mechanism operating in isolation but rather by several mechanisms operating in concert.

Because evolution emerges from the interaction of multiple evolutionary forces, even a relaxation of natural selection induces further evolution. Many traits are developmentally correlated, so if one trait is made selectively neutral by a cultural innovation, that in turn will alter the evolutionary balance at other, correlated traits, which in turn can induce further nonadaptive evolution via developmental correlations for the neutral trait. For example, most animals adapt to their diet in part through their teeth and jaws, but humans increasingly used tools and fire to prepare their food. These cultural innovations reduced the importance of jaw and tooth evolution as a means of adapting to the dietary environment. Rebecca Ackerman and James Cheverud (2004) tested the hypotheses of selected versus neutral evolution of human teeth and jaws by comparing various hominin fossil measurements with the expected developmental correlations among relative brain size, tooth size, and jaw size as inferred from modern-day humans, chimpanzees, and gorillas. Their analysis indicated the intensity of selection on the face diminished with time in the human lineage, and by 1.5 million years ago there was no longer any detectable selection on human teeth and jaws. This conclusion supports the hypothesis that cultural evolution in the human lineage had indeed eliminated natural selection on these traits. However, this does *not* mean that human teeth and jaws have not evolved over the last 1.5 million years. During the last 1.5 million years, there was a large increase in brain size in the human lineage driven by natural selection, and given the developmental constraints common to humans, chimpanzees, and gorillas, human jaws and teeth continued to evolve as a correlated effect of brain size evolution. In particular, jaws and teeth became relatively smaller for overall human head size as a correlated response to increased brain size, with the jaw becoming relatively smaller more rapidly than the teeth. The result of this correlated evolution is that humans have a small, flat face compared with those of chimpanzees and gorillas, and humans have jaws that tend to be too small for their teeth, leading to tooth crowding in the jaws. This nonadaptive evolution in turn favored the cultural evolution of the profession of orthodontics. One major past selective constraint on brain size has been the difficulty of passing a large-brained baby through the mother's birth canal. With the widespread use of cesarean sections, this selective constraint is being reduced in intensity. If this trend continues and if there is still selection for increased brain size, human jaws will become even smaller relative to the teeth. Therefore, the profession of orthodontics has a secure evolutionary future. As this example shows, the release of traits by culture from natural selection leads to further nonadaptive evolution of these traits—not evolutionary stasis.

As culture makes more mutant alleles effectively neutral with respect to natural selection, then genetic drift and mutation become the evolutionary forces that influence the evolutionary fate of these neutral alleles. As discussed in chapter V.1, the rate of neutral evolution equals the mutation rate to neutral alleles. Hence, to the extent that cultural evolution reduces selective forces in humans, the mutation rate to neutral alleles will increase, which in turn will result in an increase in the rate of neutral evolution in humanity. At first this may seem to be a trivial factor in future human evolution, since by definition this accelerated evolution involves only alleles that have no adaptive significance. However, when a gene has many potential selectively neutral mutations, it is possible for that gene to accumulate many functionally equivalent alleles differing by a series of neutral mutations. In this manner, new forms of the gene can evolve via neutrality that are several mutational steps away from the ancestral gene form. The phenotypic effects of a mutation often depend on other mutations that have occurred previously, so that a mutation that would have been deleterious or neutral on the ancestral allelic background may be selectively favored on the new, derived allelic background. In this manner, neutral evolution can actually increase the adaptive potential of a population and allow for adaptive transitions that would otherwise be unlikely. Hence, cultural evolution that reduces natural selection can increase the long-term adaptive potential of the human species.

Another consequence of cultural evolution is that humans have experienced superexponential growth for the last 10,000 years. The resulting large population size interacts strongly with the random forces of mutation and genetic drift. A small population will have very few new mutations at any given time. For example, suppose a specific nucleotide mutation has a probability of 10^{-9} of occurring per gene per generation at an autosomal locus. In a diploid population of 500, there are 1000 copies of an autosomal gene, so the expected number of new mutations to this specific form in any given generation is 10^{-6}; that is, there is only one chance in a million of this mutation occurring in any given generation. Hence, the randomness of mutation plays a large role in the evolution of this population. The human population size is now at 6.8 billion, so for an autosomal locus we would expect 13.6 occurrences of this specific mutation every generation. The large human population size is causing humans to enter an evolutionary zone that few eukaryotic organisms have ever reached—the zone in which virtually every single-step mutational change occurs in every generation. This in turn greatly reduces the randomness of evolution induced by the mutational process. Recall that the sickle-cell allele became selectively favored in sub-Saharan Africa after the introduction of the Malaysian agricultural complex 1500 years ago. What is more remarkable is that this specific sickle-cell mutation went to high frequency in sub-Saharan African populations from at least four independent mutations of this specific nucleotide. The ability of large populations to produce a huge reservoir of mutational variants means that human populations are more evolutionarily responsive than ever to changes in the environment. As long as the human population size remains large, it will remain in this rare evolutionary zone that increases its adaptive potential.

An expanding population also increases the probability of long-term survival of a new mutant, thereby enhancing the reservoir of mutational variants beyond that of a population of fixed size. For example, consider a mutant with a 10 percent advantage in a stable population in which an individual had an average of two offspring with a Poisson offspring distribution. As indicated earlier, the chances of this highly favorable mutation being lost by chance alone is 82 percent. Now suppose this mutant occurs in a growing population in which the average number of children is three. Then, the probability of loss of the favorable mutant is reduced to 33 percent. However, there is an evolutionary price to be paid for this enhanced survival of favorable mutations. Consider a deleterious mutant that reduces the fitness of its heterozygous bearers by 10 percent. In a constant-size, large population, such a deleterious mutant is eliminated by natural selection with a probability of 1, but in the growing population its chance of elimination is reduced to 53 percent. Hence, beneficial, neutral, and deleterious mutations all accumulate in the human gene pool owing to our unique demographic history. Indeed, recent studies in which the entire DNA sequence of some genes was determined in a sample of nearly 15,000 individuals reveal a large excess of rare variants due to recent mutations in the human gene pool, and many of these recent variants appear to be deleterious.

Exponential population growth cannot be sustained indefinitely in any world with finite resources, so it is inevitable that this phase of human demographic history will end in the future. Indeed, the rate of growth is already dropping. The only question is whether human population will continue to grow to a larger stable size, decrease in size, or fluctuate up and down. The change in demographic environment associated with population size stability or decline will end the era of enhanced survival of mutants, particularly deleterious ones. Indeed, natural selection will in the future start acting to eliminate the reservoir of deleterious variants that have accumulated in the gene pool during the last 10,000 years of population growth. However, as long as our population stabilizes at a large number, the reservoir of genetic variation will remain high, conferring a high degree of adaptability to the human species.

The changing demographic environment will also alter the balance of local genetic drift with gene flow. Although genetic drift causes random fluctuations in allele frequencies, it has some very predictable properties. First, genetic drift causes genetic variation to be lost, and the smaller the population size, the more rapidly genetic variation is eroded (see chapter IV.1). Second, when a species is split into multiple local populations with little genetic interchange between them, genetic drift causes random changes in allele frequencies in all of them. Because the changes are random, they are unlikely to be in the same direction in every local population. Hence, genetic drift leads to genetic differences among local populations, and the smaller the local population sizes, the greater the expected differences among them. *Gene flow* occurs when either individuals or gametes disperse from one local population to another through reproduction. Gene flow can introduce a mutation that arose in one local population into the gene pool of another local population. Hence, gene flow tends to increase the amount of genetic diversity found within local populations. The genetic interchange associated with gene flow also reduces the genetic differences among local populations. Note that genetic drift and gene flow have exactly opposite effects on genetic variation *within* local populations (decreased by drift, increased by gene flow) and genetic differences *among* local populations (increased by drift, decreased by gene flow). As a result, the balance of genetic drift to gene flow is the primary determinant of how a species' genetic variation is distributed within and among its local populations.

There is no doubt that the balance of genetic drift to gene flow has been greatly altered in recent human evolution and continues to change at a rapid pace. The increased human population size associated with the development of agriculture weakens the evolutionary force of genetic drift, and a wide variety of cultural innovations have greatly increased the ability of people to move across the globe and thereby augment gene flow. In addition, our system of mating is changing in response to cultural changes. Currently, about 10.8 percent of human couples on a global basis are related as second cousins or closer, and this subset of human couples is associated with an increased incidence of genetic disease and systemic diseases with a genetic component, as well as increased susceptibility to infectious diseases. Preference for mating with a relative decreases the amount of gene flow, but this preference is rapidly declining with increased urbanization, improved female education, and smaller family sizes. If these cultural trends continue, gene flow and outbreeding will become even stronger in future human populations. All these alterations are increasing the level of genetic variation within local human populations and decreasing the genetic differences among human populations. As long as the ability to disperse over the globe remains high and the trend toward outbreeding continues, much of future human evolution over the next tens of thousands of years will be dominated by decreased local genetic drift and increased gene flow. The result will be increased levels of individual *heterozygosity* (that is, the two copies of an autosomal gene borne by an individual are increasingly likely to be of different allelic states). This rapid and ongoing shift to increased levels of heterozygosity in humans is already having discernible health effects. For example, in studies that control for diet, socioeconomic status, and other factors, several clinical traits have significant beneficial changes with increasing heterozygosity. Similarly, areas of the human genome that lack heterozygosity are associated with diseases with genetic components, such as schizophrenia and late-onset Alzheimer's disease. As heterozygosity levels continue to increase in humans owing to vastly increased abilities to disperse, these beneficial effects are expected to increase even more. This increased heterozygosity will also reduce the deleterious consequences of the many rare, deleterious variants the species has accumulated during its phase of superexponential population growth.

The second effect of this new balance between drift and gene flow will be the eventual fusion of all human local gene pools into a single species-wide gene pool. As described in chapter VIII.11, humans already are one of the most spatially homogeneous species on this planet in a genetic sense. The modest genetic differences observed today among different human populations will be further eroded, and with continual gene flow and large population sizes will eventually be eliminated. The only genetic differences that will be biologically meaningful in the human species will be the differences among individuals, which will be high because of the high levels of genetic variation in the common human gene pool.

One nonadaptive consequence of this genetic fusion of human populations will be the loss of local adaptations. For example, skin color in humans is an adaptation to the local level of ultraviolet radiation and is not a good indicator of "race" or overall genetic differentiation among populations (see chapter VIII.11). The degree of local adaptation reflects the balance of local selective forces favoring genetic differentiation versus gene flow favoring homogenization. If gene flow and outbreeding continue to increase, human populations will display less local adaptation and more genetic homogeneity across the globe.

4. FUTURE ADAPTIVE EVOLUTION

Adaptive evolution is always with respect to an environment, and it is difficult to predict the details of the future human environment. However, much of the past evolution induced by cultural changes has been associated with the alteration of the human demographic environment, and some predictions can be made there. Continued exponential growth is ultimately unsustainable. Two extreme scenarios are possible. The optimistic scenario is that human population size will stabilize, perhaps at a level smaller than today but still quite large, without any major collapse of human civilization. Under this scenario, it is likely that the current trends toward increased dispersal and outbreeding will continue. The level of heterozygosity will increase, improving the overall genetic health of the human species. This demographic environment will also yield a large human population with an immense reservoir of genetic variation of neutral and beneficial mutations but fewer deleterious mutations than at present. The genetic differences among human populations, already small, will become even less significant, and there will be far less local adaptation. However, because of the large reservoir of new mutations and because culture-induced neutrality will allow greater exploration of the mutational state space, the adaptive potential of the human species as a whole will be enhanced. This may be important in adapting to global climate change.

There is a caveat about this greater adaptive potential of future human populations. Although the randomness of genetic drift has been emphasized until now, Sewall Wright, the man most responsible for the development of the theory of genetic drift, emphasized its significance for adaptive evolution. Evolution, including adaptive evolution, arises from the interaction of multiple evolutionary forces, including genetic drift and natural selection. Just as a series of mutationally linked neutral alleles can augment adaptability by allowing a more thorough exploration of the mutational state space, genetic drift can allow a more thorough exploration of the adaptive gene pool state space when there are multiple ways of adapting to the same environment. Multiple adaptive solutions are particularly common when adaptive traits emerge from interactions among multiple genes. Selection in large populations where genetic drift is weak therefore tends to fine-tune a single adaptive solution, whereas populations with stronger genetic drift are more likely to undergo major adaptive innovations. Hence, future humans under this optimistic demographic scenario will have greatly enhanced potential for fine-tuning human adaptations but are unlikely to make major or radical adaptive breakthroughs unless there are also major environmental changes affecting humans at the global level.

The pessimistic demographic scenario is that human population size and civilization will both crash. This will reverse the trends to increasing dispersal and outbreeding, leading to much population subdivision. Because of 10,000 years of population growth, the current human gene pool has a disproportionate number of recent, deleterious mutations. With population fragmentation, some of these globally rare deleterious variants will become locally common, causing a major decline in the overall genetic health of the human species and inducing a period of strong natural selection against deleterious variants after the population crash. Balancing this negative selection, the enhanced reservoir of neutral and beneficial mutations that were also accumulated during the period of population growth when coupled with increased genetic drift makes it likely that some human populations will undergo major adaptive breakthroughs. The nature of these breakthroughs is difficult to predict because of the strong random role that genetic drift will play in this process.

5. EUGENICS AND GENETIC ENGINEERING

The success of agriculture in sustaining 10,000 years of population growth was possible because humans became strong and effective selective agents on crop and livestock species. More recently, the ability to manipulate agricultural species has been augmented with genetic engineering in which humans directly manipulate the genetic material of domesticated species. One possibility for future human evolution is that humans will choose to direct their own evolution by selective breeding (eugenics) and/or genetic engineering.

Eugenic proposals and programs have a long history in human societies. However, this history does not engender much confidence in such an approach to controlling human evolution. For example, the "genetics" used by the American eugenics movement is ludicrous in light of modern genetics, yet this pseudogenetics led to forced sterilizations and major changes in immigration laws, and served as a model for the eugenic excesses of the Nazi regime. When people turn principles of selective breeding and genetic manipulation on themselves, scientific objectivity is frequently lost, and nonscientific social theories and prejudices dominate in shaping eugenic proposals. Moreover, current knowledge of human genetics indicates that the successes attained in plant and animal breeding for agricultural purposes are not likely to be replicated in humans. Phenotypes arise from genes interacting with one another and with the environment. Agricultural breeding is almost always done in stocks or lines that are far more homogeneous genetically than humans. Thus, the effects of any one gene are far more predictable in agricultural breeding and engineering than they would be in humans. The same gene could have dramatically different phenotypic effects on different human genetic backgrounds. Second, phenotypes emerge from genotype by environment interactions (see chapter III.10). In agriculture, humans select and engineer crop and livestock strains specifically for how their genes interact with simple, homogeneous environments. Human environments are not simple or homogeneous, so once again the impact of a single gene can vary tremendously. For example, the single gene locus most predictive of risk for coronary artery disease, the number one killer in the developed world, is the *Apoprotein-E* locus (*ApoE*), which has three common alleles in most human populations: ε2, ε3, and ε4. A retrospective study indicated that individuals bearing the ε4 allele had the highest incidence of coronary artery disease on average. The same study revealed that individuals in the highest tertile for total serum cholesterol level had the highest incidence of coronary artery disease compared with the middle and lower tertiles for cholesterol level. Cholesterol level in turn is affected by many interacting genes (including *ApoE*) and environmental variables such as smoking, diet, and exercise. When genotype and cholesterol levels were combined, the group of people with the highest incidence of coronary artery disease by far were people with high cholesterol levels and the ε2/ε3 genotype. Note that the genotype with the highest absolute incidence of coronary artery disease has the "good" alleles only at the *ApoE* locus. This is the main problem with eugenic and genetic engineering programs for humans: the genetic background and environment is highly heterogeneous in humans, so the consequences of manipulations can never be accurately predicted. Moreover, environments change very rapidly for humans, making eugenic predictions even more prone to error. Unless it is decided to create separate castes of relatively genetically homogeneous human strains and keep them in highly restricted environments, eugenics and genetic engineering is unlikely to play a significant role in future human evolution.

FURTHER READING

Ackermann R. R., and J. M. Cheverud. 2004. Detecting genetic drift versus selection in human evolution. Proceedings of the National Academy of Sciences USA 101: 17947–17951.

Allen, G. E. The misuse of biological hierarchies: The American eugenics movement, 1900–1940. History and Philosophy of the Life Sciences. Section II of Pubblicazioni della Stazione Zoologica di Napoli 5: 105–128. *A brief history of the American eugenics movement and the impact it had on laws and policy in the United States and other countries.*

Bittles, A. H., and M. L. Black. 2010. Consanguinity, human evolution, and complex diseases. Proceedings of the National Academy of Sciences 107: 1779–1786.

Campbell, H., A. D. Carothers, I. Rudan, C. Hayward, Z. Biloglav, L. Barac, M. Pericic, et al. 2007. Effects of genome-wide heterozygosity on a range of biomedically relevant human quantitative traits. Human Molecular Genetics 16: 233–241. *Heterozygosity levels were measured in four different Croatian populations that differed greatly in their degree of gene flow among local populations but that had similar diets, socioeconomic status, and other factors. Several clinical traits were then regressed against relative heterozygosity, and all significant results indicated beneficial effects with increasing heterozygosity.*

Cochran G., and H. Harpending. 2009. The 10,000 Year Explosion: How Civilization Accelerated Human Evolution. New York: Basic Books. *Debunks the idea that human evolution has stopped and instead argues that it has accelerated.*

Coventry, A., L. M. Bull-Otterson, X. Liu, A. G. Clark, T. J. Maxwell, J. Crosby, J. E. Hixson, et al. 2010. Deep resequencing reveals excess rare recent variants consistent with explosive population growth. Nature Communications 1: 131. *One of the first studies to do extensive DNA resequencing, revealing a plethora of recent, rare variants in the human gene pool. Many of these variants are predicted to have deleterious consequences.*

Gould, S. J. 2000. The spice of life. Leader to Leader 15: 19–28.

Ku, C. S., N. Naidoo, S. M. Teo, and Y. Pawitan. 2011. Regions of homozygosity and their impact on complex diseases and traits. Human Genetics 129: 1–15. *This paper shows that areas of the human genome that lack heterozygosity are associated with increased risk for diseases with a genetic component.*

Neel, J. V. 1962. Diabetes mellitus: A "thrifty genotype" rendered detrimental by "progress." American Journal of Human Genetics 14: 353–362. *A classic paper that developed the idea that the genes underlying risk for diabetes could have been adaptive during famine conditions. Much recent work has supported this hypothesis, and variants of the thrifty genotype hypothesis have been proposed for other common diseases in humans.*

Peter, B. M., E. Huerta-Sanchez, and R. Nielsen. 2012. Distinguishing between selective sweeps from standing variation and from a *de novo* mutation. PLoS Genetics 8: e1003011. *Provides evidence for several mutations being favored in recent human evolution, including that for lactase persistence.*

Templeton, A. R. 1998. The complexity of the genotype-phenotype relationship and the limitations of using genetic "markers" at the individual level. Science in Context 11: 373–389. *Discusses why eugenics and genetic engineering should be ineffective in human populations owing to the complex interactions among genes and between genes and environments.*

Templeton, A. R. 2006. Population Genetics and Microevolutionary Theory. Hoboken, NJ: John Wiley & Sons. *This textbook gives the details of many of the examples used in this chapter and also shows how multiple evolutionary forces interact to influence the trajectory of evolution.*

Templeton, A. R. 2010. Has human evolution stopped? Rambam Maimonides Medical Journal 1(1): e0006, doi:10.5041/RMMJ. 10006. *Gives additional details on some the examples used in this chapter. It also argues against the idea that human evolution has stopped, using arguments not found in the book by Cochran and Harpending.*

VIII.13

Evolution and Religion

Francisco J. Ayala

OUTLINE

1. Natural theology and the Bridgewater Treatises
2. Darwin's revolution
3. Evolution and the Bible
4. The problem of evil
5. Evolution: Imperfect design, not intelligent design
6. Evolution and religion: Coda

Theologians and other religious authors have over centuries sought to demonstrate the existence of God by the argument from design, which asserts that organisms have been designed and that only God could account for the design. Its most extensive formulation is William Paley's *Natural Theology* (1802). Darwin's (1859) theory of evolution by natural selection disposed of Paley's arguments: the adaptations of organisms are outcomes of a natural process that causes the gradual accumulation of features beneficial to organisms. There is "design" in the living world, but the design is not intelligent, as expected from an engineer, but imperfect and worse: defects, dysfunctions, oddities, waste, and cruelty pervade the living world. Science and religious faith need not be in contradiction. Science concerns processes that account for the natural world. Religion concerns the meaning and purpose of the world and of human life, the proper relation of humans to their Creator and to one another, and the moral values that inspire and govern people's lives.

GLOSSARY

Evolution. Hereditary change and diversification of organisms through generations.

Intelligent Design. The idea that adaptations of organisms are designed by an intelligent author (= God), rather than resulting from natural processes.

Natural Selection. Differential reproduction of alternative genetic variants.

Problem of Evil. The challenge of explaining the presence of physical evil (e.g., earthquakes that kill millions of people), and biological evil (e.g., the cruelty of predators) if they are designed outcomes of an omnipotent and benevolent Creator.

Religion. Faith in and worship of God or the supernatural.

Religious authors have over the centuries argued that the order, harmony, and design of the universe are incontrovertible evidence that the universe was created by an omniscient and omnipotent Creator. Notable Christian authors include Augustine (353–430 CE), who writes in *The City of God* that the "world itself, by the perfect order of its changes and motions and by the great beauty of all things visible, proclaims . . . that it has been created, and also that it could not have been made other than by a God ineffable and invisible in greatness, and . . . in beauty." Thomas Aquinas (1224–1274), considered by many to be the greatest Christian theologian, advances in his *Summa Theologiae* five ways to demonstrate, by natural reason, that God exists. The fifth way derives from the orderliness and designed purposefulness of the universe, which evince that it has been created by a Supreme Intelligence: "Some intelligent being exists by which all natural things are directed to their end; and this being we call God."

This manner of seeking a natural demonstration of God's existence became later known as the *argument from design*, which is two pronged. The first prong asserts that the universe evinces that it has been designed. The second prong affirms that only God could account for the complexity and perfection of the design. A forceful and elaborate formulation of the argument from design is *The Wisdom of God Manifested in the Works of Creation* (1691) by the English clergyman and naturalist John Ray (1627–1705). Ray regarded as incontrovertible evidence of God's wisdom that all components of the universe—the stars and the planets, as well as all organisms—are so wisely contrived from the beginning and perfect in their operation. The "most convincing argument of the Existence of a Deity," writes Ray, "is the admirable Art and Wisdom that discovers itself in the Make of the Constitution, the Order and Disposition, the Ends and uses of all the parts and members of this stately fabric of Heaven and Earth."

The design argument was advanced, in greater or lesser detail, by a number of authors in the seventeenth and eighteenth centuries. John Ray's contemporary Henry More (1614–1687) saw evidence of God's design in the succession of day and night and of the seasons: "I say that the Phenomena of Day and Night, Winter and Summer, Spring-time and Harvest . . . are signs and tokens unto us that there is a God . . . things are so framed that they naturally imply a Principle of Wisdom and Counsel in the Author of them. And if there be such an Author of external Nature, there is a God." Robert Hooke (1635–1703), a physicist and eventual Secretary of the Royal Society, formulated the watchmaker analogy: God had furnished each plant and animal "with all kinds of contrivances necessary for its own existence and propagation . . . as a Clock-maker might make a Set of Chimes to be a part of a Clock" (Hooke 1665). The clock analogy, among other analogies such as temples, palaces, and ships, was also used by Thomas Burnet (1635–1703) in his *Sacred Theory of the Earth*, and it would become common among natural theologians of the time. The Dutch philosopher and theologian Bernard Nieuwentijdt (1654–1718) developed, at length, the argument from design in his three-volume treatise, *The Religious Philosopher*, where, in the preface, he introduces the watchmaker analogy. Voltaire (1694–1778), like other philosophers of the Enlightenment, accepted the argument from design. Voltaire asserted that in the same way as the existence of a watch proves the existence of a watchmaker, the design and purpose evident in nature prove that the universe was created by a Supreme Intelligence.

1. NATURAL THEOLOGY AND THE BRIDGEWATER TREATISES

William Paley (1743–1805), one of the most influential English authors of his time, formulated in his *Natural Theology* (1802) the argument from design, based on the complex and precise design of organisms. Paley was an influential writer of works on Christian philosophy, ethics, and theology, such as *The Principles of Moral and Political Philosophy* (1785) and *A View of the Evidences of Christianity* (1794). With *Natural Theology*, Paley sought to update Ray's *Wisdom of God* of 1691. But Paley could now carry the argument much further than Ray, by taking advantage of a century of additional biological knowledge.

Paley's keystone claim is that there "cannot be design without a designer; contrivance, without a contriver; order, without choice; . . . means suitable to an end, and executing their office in accomplishing that end, without the end ever having been contemplated." *Natural Theology* is a sustained argument for the existence of God based on the obvious design of humans and their organs, as well as the design of all sorts of organisms, considered by themselves and in their relations to one another and to their environment. Paley's first analogical example in *Natural Theology* is the human eye. He points out that the eye and the telescope "are made upon the same principles; both being adjusted to the laws by which the transmission and refraction of rays of light are regulated." Specifically, there is a precise resemblance between the lenses of a telescope and "the humors of the eye" in their figure, their position, and the ability of converging the rays of light at a precise distance from the lens—on the retina, in the case of the eye.

Natural Theology has chapters dedicated to the human frame, which displays a precise mechanical arrangement of bones, cartilage, and joints; to the circulation of the blood and the disposition of blood vessels; to the comparative anatomy of humans and animals; to the digestive tract, kidneys, urethra, and bladder; to the wings of birds and the fins of fish; and much more. After detailing the precise organization and exquisite functionality of each biological entity, relationship, or process, Paley draws again and again the same conclusion: only an omniscient and omnipotent Deity could account for these marvels of mechanical perfection, purpose, and functionality, and for the enormous diversity of inventions that they entail.

Francis Henry Egerton (1756–1829), the eighth Earl of Bridgewater, bequeathed in 1829 the sum of £8000 with instructions to the Royal Society that it commission eight treatises that would promote natural theology by setting forth "The Power, Wisdom and Goodness of God as manifested in the Creation." Eight treatises were published in the 1830s, several of which artfully incorporated the best science of the time and had considerable influence on the public and among scientists. *The Hand, Its Mechanisms and Vital Endowments as Evincing Design* (1833), by Sir Charles Bell, a distinguished anatomist and surgeon, famous for his neurological discoveries, examines in considerable detail the wondrously useful design of the human hand but also the perfection of design of the forelimb used for different purposes in different animals, serving in each case the particular needs and habits of its owner: the human's arm for handling objects, the dog's leg for running, and the bird's wing for flying. He concludes that "nothing less than the Power, which originally created, is equal to the effecting of those changes on animals, which are to adapt them to their conditions."

William Buckland, Professor of Geology at Oxford University, notes in *Geology and Mineralogy* (1836) the world distribution of coal and mineral ores, and proceeds to point out that they were deposited in remote parts, yet obviously with the forethought of serving the larger human populations that would come about much later. Later, another geologist, Hugh Miller (1858), would formulate in *The Testimony of the Rocks* what may be called the *argument from beauty*, which allows that it is not only the perfection of design but also the beauty of natural structures found in rock formations and in mountains and rivers that manifests the intervention of the Creator.

In the 1990s, a new version of the design argument was formulated in the United States, named intelligent design (ID), which refers to an unidentified Designer who accounts for the order and complexity of the universe, or who intervenes from time to time in the universe so as to design organisms and their parts. The complexity of organisms, it is claimed, cannot be accounted for by natural processes. According to ID proponents, this intelligent designer could be, but need not be, God. The intelligent designer could be an alien from outer space or some other creature, such as a "time-traveling cell biologist," with amazing powers to account for the universe's design. Explicit reference to God is avoided, so that the "theory" of ID can be taught in the public schools as an alternative to the theory of evolution without incurring conflict with the US Constitution, which forbids the endorsement of any religious beliefs in public institutions. The ID movement and the "creationism" claims that preceded it are the subject of chapter VIII.14.

2. DARWIN'S REVOLUTION

Darwin occupies an exalted place in the history of Western thought, deservedly receiving credit for the theory of evolution. In *The Origin*, he laid out the evidence demonstrating the evolution of organisms. However, Darwin accomplished something much more important for intellectual history than demonstrating evolution. Darwin's *Origin of Species* is, first and foremost, a sustained effort to solve the problem of how to account scientifically for the design of organisms. Darwin explains the design of organisms, their complexity, diversity, and marvelous contrivances as the result of natural processes.

There is a version of the history of the ideas that sees a parallel between the Copernican and the Darwinian revolutions. In this view, the Copernican revolution consisted in displacing the earth from its previously accepted locus as the center of the universe, moving it to a subordinate place as just one more planet revolving around the sun. In congruous manner, the Darwinian revolution is viewed as consisting of the displacement of humans from their exalted position as the center of life on earth, with all other species created for the service of humankind. According to this version of intellectual history, Copernicus accomplished his revolution with the heliocentric theory of the solar system; Darwin's achievement emerged from his theory of organic evolution.

Although this version of the two revolutions is correct, it misses what is most important about these two intellectual revolutions, namely, that they ushered in the beginning of science in the modern sense of the word. These two revolutions may jointly be seen as the one Scientific Revolution, with two stages, the Copernican and the Darwinian. The Copernican revolution was launched with the publication in 1543, the year of Nicolaus Copernicus's death, of his *De revolutionibus orbium celestium* (On the Revolutions of Celestial Spheres), and bloomed with the publication in 1687 of Isaac Newton's *Philosophiae naturalis principia mathematica* (Mathematical Principles of Natural Philosophy). The discoveries by Copernicus, Kepler, Galileo, Newton, and others in the sixteenth and seventeenth centuries gradually ushered in a conception of the universe as matter in motion governed by natural laws. These scientists showed that the earth is not the center of the universe but a small planet rotating around an average star; that the universe is immense in space and in time; and that the motions of the planets around the sun can be explained by the same simple laws that account for the motion of physical objects on our planet. These and other discoveries greatly expanded human knowledge. The conceptual revolution they brought about was more fundamental yet: a commitment to the postulate that the universe obeys immanent laws that account for natural phenomena. The workings of the universe were brought into the realm of science: explanation through natural laws.

The advances of physical science brought about by the Copernican revolution drove mankind's conception of the universe to a split-personality state of affairs. Scientific explanations, derived from natural laws, dominated the world of nonliving matter, on the earth as well as in the heavens. However, supernatural explanations, which depended on the unfathomable deeds of the Creator, were accepted as explanations of the origin and configuration of living creatures. Authors such as William Paley argued that the complex design of organisms could not have come about by chance—or by the mechanical laws of physics, chemistry, and astronomy—but was, rather, produced by an omniscient and omnipotent Deity, just as the complexity of a watch, designed to tell time, was fashioned by an intelligent watchmaker. It was Darwin's genius to resolve this conceptual schizophrenia. Darwin completed the Copernican revolution by drawing out for biology the notion of nature as a lawful system of matter in motion that human reason can explain without recourse to supernatural agencies.

The conundrum Darwin faced can hardly be overestimated. The strength of the argument from design to demonstrate the role of the Creator had been forcefully set forth by philosophers and theologians. Wherever there is function or design, we look for its author. It was Darwin's greatest accomplishment to show that the complex organization and functionality of living beings can be explained as the result of a natural process—natural selection—without any need to resort to a Creator or other external agent. The origin and adaptations of organisms in their profusion and wondrous variations were thus brought into the realm of science.

Organisms exhibit complex design, but it is not, in current language, "irreducible complexity," emerging suddenly in full bloom. Rather, according to Darwin's theory of natural selection, the design has arisen gradually and cumulatively, step by step, promoted by the reproductive success of individuals with incrementally more adaptive elaborations.

Natural selection accounts for the "design" of organisms, because adaptive variations tend to increase the probability of survival and reproduction of their carriers at the expense of maladaptive, or less adaptive, variations. The arguments of Paley against the incredible improbability of chance accounts of the adaptations of organisms are well taken as far as they go. But neither Paley, nor any other author before Darwin, was able to discern that there is a natural process (namely, natural selection) that is not random but rather is oriented and able to generate order or "to create." The traits that organisms acquire in their evolutionary histories are not fortuitous but determined by their functional utility to the organisms, "designed," as it were, to serve their life needs.

3. EVOLUTION AND THE BIBLE

To some Christians and other people of faith, the theory of evolution seems to be incompatible with their religious beliefs, because it is inconsistent with the Bible's narrative of creation. The first chapters of the biblical book of Genesis describe God's creation of the world, plants, animals, and human beings. A literal interpretation of Genesis seems incompatible with the gradual evolution of humans and other organisms by natural processes. Even independent of the biblical narrative, the Christian beliefs in the immortality of the soul and in humans as "created in the image of God" have appeared to many as contrary to the evolutionary origin of humans from nonhuman animals.

In 1874, Charles Hodge, an American Protestant theologian, published *What Is Darwinism?*—one of the most articulate assaults on evolutionary theory. Hodge perceived Darwin's theory as "the most thoroughly naturalistic that can be imagined and far more atheistic than that of his predecessor Lamarck." Echoing Paley, Hodge argued that the design of the human eye reveals that "it has been planned by the Creator, like the design of a watch evinces a watchmaker." He concluded that "the denial of design in nature is actually the denial of God."

Some Protestant theologians saw a solution to the apparent contradiction between evolution and creation in the argument that God operates through intermediate causes. The origin and motion of the planets could be explained by the law of gravity and other natural processes without denying God's creation and providence. Similarly, evolution could be seen as the natural process through which God brought living beings into existence and developed them according to his plan. Thus, A. H. Strong, the president of Rochester Theological Seminary in New York State, wrote in his *Systematic Theology* (1885): "We grant the principle of evolution, but we regard it as only the method of divine intelligence." He explains that the brutish ancestry of human beings was not incompatible with their excelling status as creatures in the image of God. Strong drew an analogy with Christ's miraculous conversion of water into wine: "The wine in the miracle was not water because water had been used in the making of it, nor is man a brute because the brute has made some contributions to its creation." Arguments for and against Darwin's theory came from Roman Catholic theologians as well.

Gradually, well into the twentieth century, evolution by natural selection came to be accepted by a majority of Christian writers. Pope Pius XII in his encyclical *Humani generis* (1950, Of the Human Race) acknowledged that biological evolution was compatible with the Christian faith, although he argued that God's intervention was necessary for the creation of the human soul. Pope John Paul II, in an address to the Pontifical Academy of Sciences on October 22, 1996, deplored interpreting the Bible's texts as scientific statements rather than religious teachings. He added: "New scientific knowledge has led us to realize that the theory of evolution is no longer a mere hypothesis. It is indeed remarkable that this theory has been progressively accepted by researchers, following a series of discoveries in various fields of knowledge. The convergence, neither sought nor fabricated, of the results of work that was conducted independently is in itself a significant argument in favor of this theory."

Similar views have been expressed by other mainstream Christian denominations. The General Assembly of the United Presbyterian Church in 1982 adopted a resolution stating that "biblical scholars and theological schools . . . find that the scientific theory of evolution does not conflict with their interpretation of the origins of life found in Biblical literature." The Lutheran World Federation in 1965 affirmed that "evolution's assumptions are as much around us as the air we breathe and no more escapable. At the same time theology's affirmations are being made as responsibly as ever. In this sense both science and religion are here to stay, and . . . need to remain in a healthful tension of respect toward one another."

Similar statements have been advanced by Jewish authorities and leaders of other major religions. In 1984, the 95th Annual Convention of the Central Conference of American Rabbis adopted a resolution stating: "Whereas the principles and concepts of biological evolution are basic to understanding science . . . we call upon science teachers and local school authorities in all states to demand quality textbooks that are based on modern, scientific knowledge and that exclude 'scientific' creationism."

Christian denominations that hold a literal interpretation of the Bible have opposed these views. A succinct expression of this opposition is found in the Statement of Belief of the Creation Research Society, founded in 1963 as a "professional organization of trained scientists and interested laypersons who are firmly committed to scientific special creation": "The Bible is the Written Word of God, and because it is inspired throughout, all of its assertions are historically and scientifically true in the original autographs. To the student of nature this means that the account of origins in Genesis is a factual presentation of simple historical truths."

Many Bible scholars and theologians have long rejected a literal interpretation as untenable, however, because the Bible contains mutually incompatible statements. The very beginning of the book of Genesis presents two different creation narratives. Extending through chapter 1 and the first verses of chapter 2 is the familiar six-day narrative, in which God creates human beings—both "male and female"—in his own image on the sixth day, after creating light, earth, firmament, fish, fowl, and cattle. In verse 4 of chapter 2, a different narrative starts, in which God creates a male human, then plants a garden and creates the animals, and only then proceeds to take a rib from the man to make a woman.

Which one of the two narratives is correct and which one is in error? Neither one contradicts the other if we understand the two narratives as conveying the same message: that the world was created by God and that humans are His creatures. But both narratives cannot be "historically and scientifically true" as postulated in the Statement of Belief of the Creation Research Society.

There are numerous inconsistencies and contradictions in different parts of the Bible, for example, in the description of the return from Egypt to the Promised Land by the chosen people of Israel, not to mention erroneous factual statements about the sun's circling around the earth and the like. Biblical scholars point out that the Bible should be held inerrant with respect to religious truth, not in matters that are of no significance to salvation. Augustine wrote in his *De Genesi ad litteram* (On the Literal Meaning of Genesis): "It is also frequently asked what our belief must be about the form and shape of heaven, according to Sacred Scripture. . . . Such subjects are of no profit for those who seek beatitude. . . . What concern is it of mine whether heaven is like a sphere and earth is enclosed by it and suspended in the middle of the universe, or whether heaven is like a disk and the Earth is above it and hovering to one side." He adds: "In the matter of the shape of heaven, the sacred writers did not wish to teach men facts that could be of no avail for their salvation." Augustine is saying that the book of Genesis is not an elementary book of astronomy. The Bible is about religion, and it is not the purpose of the Bible's religious authors to settle questions about the shape of the universe that are of no relevance whatsoever to how to seek salvation.

In the same vein, Pope John Paul II said in 1981 that the Bible itself "speaks to us of the origins of the universe and its makeup, not in order to provide us with a scientific treatise but in order to state the correct relationships of man with God and with the universe. Sacred Scripture . . . in order to teach this truth, it expresses itself in the terms of the cosmology in use at the time of the writer."

4. THE PROBLEM OF EVIL

Christian scholars for centuries struggled with the problem of evil in the world. The Scottish philosopher David Hume (1711–1776) set the problem succinctly with brutal directness: "Is he [God] willing to prevent evil, but not able? Then he is impotent. Is he able, but not willing? Then, he is malevolent. Is he both able and willing? Whence then evil?" If the reasoning is valid, it would follow that God is not all-powerful or all-good. Christian theology accepts that evil exists but denies the validity of the argument.

Traditional theology distinguishes three kinds of evil: (1) moral evil or sin, the evil originated by human beings; (2) pain and suffering as experienced by human beings; and (3) physical evil, such as floods, tornadoes, earthquakes, and the imperfections of all creatures. Theology has a ready answer for the first two kinds of evil. Sin is a consequence of free will; the converse of sin is virtue, also a consequence of free will. Christian theologians have expounded that if humans are to enter into a genuinely personal relationship with their maker, they must first experience some degree of freedom and autonomy. The eternal reward of heaven calls for a virtuous life, as many Christians see it. Christian theology also provides a good accounting of human pain and suffering. To the extent that pain and suffering are caused by war, injustice, and other forms of human wrongdoing, they are also a consequence of free will; people choose to inflict harm on one another. On the flip side are good deeds by which people choose to alleviate human suffering.

What about earthquakes, storms, floods, droughts, and other physical catastrophes? Enter modern science into the theologian's reasoning. Physical events are built into the structure of the world itself. Since the seventeenth century, humans have known that the processes by which galaxies and stars come into existence, planets are formed, continents move, weather and seasons change, and floods and earthquakes occur are natural processes, not events specifically designed by God for punishing or rewarding humans. The extreme violence of supernova explosions and the chaotic frenzy at galactic centers are outcomes of the laws of physics, not the design of a fearsome Deity. Before Darwin, theologians still encountered a seemingly insurmountable difficulty. If God is the designer of life, whence the lion's cruelty, the snake's poison, and the parasites that secure their existence only by destroying their hosts? Evolution came to the rescue. Jack Haught (1998), a contemporary Roman Catholic theologian, has written of "Darwin's gift to theology." The Protestant theologian Arthur Peacocke has referred to Darwin as the "disguised friend," by quoting the earlier theologian Aubrey Moore, who in 1891 wrote that "Darwinism appeared, and, under the guise of a foe, did the work of a friend" (Peacocke 1998). Haught and Peacocke are acknowledging the irony that the theory of evolution, which at first seemed to remove the need for God in the world, now has convincingly removed the need to explain the world's imperfections as failed outcomes of God's design.

Indeed, a major burden was removed from the shoulders of believers when convincing evidence was advanced that the design of organisms need not be attributed to the immediate agency of the Creator but rather is an outcome of natural processes. If we claim that organisms and their parts have been specifically designed by God, we have to account for the incompetent design of the human jaw, the narrowness of the birth canal, and our poorly designed backbone, less than fittingly suited for walking upright. Imperfections and defects pervade the living world. Consider the human eye. The visual nerve fibers in the eye converge to form the optic nerve, which crosses the retina (to reach the brain) and thus creates a blind spot, a minor imperfection, but an imperfection of design, nevertheless; squids and octopuses do not have this defect. Did the Designer have greater love for squids than for humans and thus exhibit greater care in designing their eyes than ours? It is not only that organisms and their parts are less than perfect but also that deficiencies and dysfunctions are pervasive, evidencing incompetent rather than intelligent design. Consider the human jaw. Humans have too many teeth for the jaw's size, so that wisdom teeth need to be removed and orthodontists can make a decent living straightening the others. Would we want to blame God for this blunder? A human engineer would have done better.

5. EVOLUTION: IMPERFECT DESIGN, NOT INTELLIGENT DESIGN

Evolution gives a good account of this imperfection. Brain size increased over time in human ancestors; the remodeling of the skull to fit the larger brain entailed a reduction of the jaw, so that the head of the newborn would not be too large to pass through the mother's birth canal. The birth canal of women is much too narrow for easy passage of the infant's head, so that thousands on thousands of babies and many mothers die during delivery. Surely we do not want to blame God for this dysfunctional design or for the children's deaths. The theory of evolution makes it understandable: it is a consequence of the evolutionary enlargement of the human brain. Females of other primates do not experience this difficulty. Theologians in the past struggled with the issue of dysfunction because they thought it had to be attributed to God's design. Science, much to the relief of theologians, provides an explanation that convincingly attributes defects, deformities, and dysfunctions to natural causes.

Consider the following. About 20 percent of all recognized human pregnancies end in spontaneous miscarriage during the first two months of pregnancy. This misfortune amounts at present to more than 20 million spontaneous abortions worldwide every year. Do we want to blame God for the deficiencies in the pregnancy process? Many people of faith would rather attribute this monumental mishap to the clumsy ways of the evolutionary process than to the incompetence or deviousness of an intelligent designer.

Evolution makes it possible to attribute these mishaps to natural processes rather than to the direct creation or specific design of the Creator. The response of some critics is that the process of evolution by natural selection does not discharge God's responsibility for the dysfunctions, cruelties, and sadism of the living world, because for people of faith, God is the Creator of the universe and thus would be accountable for its consequences, direct or indirect, immediate or mediated. If God is omnipotent, the argument would say, He could have created a world where such things as cruelty, parasitism, and human miscarriages would not occur.

One possible religious explanation goes along the following lines of reasoning. Consider, first, human beings, who perpetrate all sorts of misdeeds and sins, even perjury, adultery, and murder. People of faith believe that each human being is a creation of God, but this does not entail that God is responsible for human crimes and misdemeanors. As stated earlier, sin is a consequence of free will; the converse of sin is virtue. The critics might say that this account does not excuse God, because God could have created humans without free will (whatever these "humans" might have been called and been like). But one could reasonably argue that "humans" without free will would be a very different kind of creature, a being much less interesting and creative than humans are. Robots are not a good replacement for humans; robots do not perform virtuous deeds.

This line of argumentation can be extended to the catastrophes and other events of the physical world and to the dysfunctions of organisms and the harms caused to them by other organisms and environmental mishaps. However, some authors do not find this extension fully satisfactory as an explanation that exonerates God from moral responsibility. The point made again is that the world was created by God, so God is ultimately responsible. God could have created a world without parasites or dysfunctionalities. But a world of life with evolution is much more exciting; it is a creative world where new species arise, complex ecosystems come about, and humans have evolved. These considerations may provide the beginning of an explanation for many people of faith, as well as for theologians.

The Anglican theologian Keith Ward (2008) has stated the case in even stronger terms, arguing that the creation of a world without suffering and moral evil is not an option even for God: "Could [God] not actualize a world wherein suffering is not a possibility? He could not, if any world complex and diverse enough to include rational and moral agents must necessarily include the possibility of suffering . . . A world with the sorts of success and happiness in it that we occasionally experience is a world that necessarily contains the possibility of failure and misery." The physicist and theologian Robert J. Russell (2007) has gone even further, making the case why there should be natural (physical and biological) evil in the world, "including the pain, suffering, disease, death, and extinction that characterize the evolution of life."

An additional point is that physical or biological (other than human) events that cause harm are not moral evil actions, because they are not caused by moral agents, but are the result of natural processes. If a terrorist blows up a bus with schoolchildren in it, that is moral evil. If an earthquake kills several thousand people in China and destroys their homes and livelihood, there is no subject morally responsible, because the event was not committed by a moral agent but was the result of a natural process. If a mugger uses a vicious dog to brutalize a person, the mugger is morally responsible. But if a coyote attacks a person, there is no moral evil that needs to be accounted for. In the world of physical and biological nature (again, excluding human deeds), no morality is involved. This claim, of course, may or may not satisfy everyone, but it deserves to be explored by theologians and people of faith.

6. EVOLUTION AND RELIGION: CODA

Evolution and religious beliefs need not be in contradiction. Indeed, if science and religion are properly understood, they *cannot* be in contradiction, because they concern different matters. Science and religion are like two different windows for looking at the world. The two windows look at the same world, but they show different aspects of that world. Science concerns the processes that account for the natural world: the movement of planets, the composition of matter and the atmosphere, the origin and adaptations of organisms. Religion concerns the meaning and purpose of the world and of human life, the proper relation of people to the Creator and to one another, the moral values that inspire and govern people's lives. Apparent contradictions emerge only when either the science or the beliefs, or often both, trespass their own boundaries and wrongfully encroach on each other's subject matter.

The scope of science is the world of nature, the reality that is observed, directly or indirectly, by the senses. Science advances explanations concerning the natural world, explanations that are subject to the possibility of corroboration or rejection by observation and experiment. Outside that world, science has no authority, no statements to make, no business whatsoever taking one position or another. Science has nothing decisive to say about values, whether economic, aesthetic, or moral; nothing to say about the meaning of life or its purpose; nothing to say about religious beliefs (except in the case of beliefs that transcend the proper scope of religion and make assertions about the natural world that contradict scientific knowledge; such statements cannot be true).

Science is a way of knowing, but it is not the only way. Knowledge also derives from other sources. Common experience, imaginative literature, art, and history provide valid knowledge about the world; and so do revelation and religion for people of faith. The significance of the world and human life, as well as matters concerning moral or religious values, transcends science. Yet these matters are important; for most of us, they are at least as important as scientific knowledge per se.

The proper relationship between science and religion can be, for people of faith, mutually motivating and inspiring. Science may inspire religious beliefs and religious behavior as we respond with awe to the immensity of the universe, the glorious diversity and wondrous adaptations of organisms, and the marvels of the human brain and the human mind. Religion promotes reverence for the creation, for humankind as well as for the world of life and the environment. Religion often is, for scientists and others, a motivating force and source of inspiration for investigating the marvelous world of the creation and solving the puzzles with which it confronts us.

See also chapter VIII.14.

FURTHER READING

Aquinas, T. 1905. Of God and His Creatures. In J. Rickaby, ed., *Summa contra gentiles.* London: Burns & Oates. *Aquinas is often considered the greatest Christian theologian of all time.*

Augustine. 1998. The City of God, ed. R. Dyson. Cambridge: Cambridge University Press. *An early classic of Christian theology.*

Ayala, F. J. 2007. Darwin's Gift to Science and Religion. Washington, DC: Joseph Henry. *Develops at greater length and in greater depth the ideas of this chapter.*

Ayala, F. J. 2010. Am I a Monkey? Six Big Questions about Evolution. Baltimore: Johns Hopkins University Press. *An easy read about science and religion.*

Haught, J. F. 1998. Darwin's gift to theology. In R. J. Russell, W. R. Stoeger, S. J., and F. J. Ayala, eds., Evolutionary and Molecular Biology: Scientific Perspectives on Divine Action. Vatican City State: Vatican Observatory Press; and Berkeley, CA: Center for Theology and the Natural Sciences. *A modern theologian's view of how modern biology favorably affects Christian faith.*

Miller, K. R. 1999. Finding Darwin's God: A Scientist's Search for Common Ground. New York: HarperCollins. *A scientist's extended argument conciliating Darwinian evolution and Christian theology.*

National Academy of Sciences and Institute of Medicine. 2008. Science, Evolution, and Creationism. New York: National Academy of Sciences Press. *A concise, forceful argument by the most distinguished scientific institution affirming the compatibility of science and religion.*

Paley, W. 1802. Natural theology, or evidences of the existence and attributes of the Deity collected from the appearances of nature. New York: American Tract Society. *A classical treatise expounding the traditional view that the design of the world manifests the existence of the Creator.*

Peacocke, A. R. 1998. Biological evolution: A positive appraisal. In R. J. Russell, W. R. Stoeger, S.J., and F. J. Ayala, eds., Evolutionary and Molecular Biology: Scientific Perspectives on Divine Action. Vatican City State: Vatican Observatory Press; and Berkeley, CA: Center for Theology and the Natural Sciences. *A distinguished Anglican minister and theologian asserts that modern biology provides an enlightened view of creation.*

Russell, R. J. 2007. Physics, cosmology, and the challenge to consequentialist natural theology. In N. Murphy, R. J. Russell, and W. R. Stoeger, S.J., eds., Physics and Cosmology: Scientific Perspectives on the Problem of Natural Evil. Vatican City State: Vatican Observatory Press; and Berkeley, CA: Center for Theology and the Natural Sciences. *Physics and astronomy are shown to be compatible with Christian faith.*

Slack, G. 2008. The Battle over the Meaning of Everything. New York: John Wiley & Sons. *A narrative of the controversies between science and religion.*

Ward, K. 2008. The Big Questions in Science and Religion. West Conshohocken, PA: Templeton Foundation. *A theologian explores a variety of scientific issues that are often seen as contrary to religious faith and asserts that there is no necessary opposition.*

VIII.14

Creationism and Intelligent Design

Eugenie C. Scott

OUTLINE

1. What kind of creationist?
2. The creation-evolution continuum
3. Intelligent design
4. What does the future hold?

Many are unaware that there are several kinds of creationisms, even within the tradition of Christianity. In that tradition, the various creationisms are a function of how the Bible is interpreted, and the differences reflect how much of modern science is accepted. Intelligent design is a more recent form of creationism, but in its particulars it reflects themes similar to other forms of Christian creationism. New forms of creationism may develop in the future, but it is likely that they will reflect the same ideas as their ancestors.

1. WHAT KIND OF CREATIONIST?

There has been a long-standing tension between some religious groups and evolutionary biology, and that tension plays out in schools throughout the United States. At the National Center for Science Education, we monitor the creationism and evolution controversy, and we help parents, teachers, and others cope with challenges to evolution education. All the challenges emanate from people who call themselves—or can be called—"creationists." Often, a student will tell a teacher, "I don't believe in evolution, I'm a creationist." We recommend asking in reply, "What kind of creationist?"

It is a teachable moment: the student has probably never considered that there might be more than one type of creationism, and the teacher has the opportunity not only to expand the student's horizons but also, with luck, to reduce barriers to learning evolution. And yes, it is also an opportunity to help the student understand that scientists don't "believe" in evolution, they *accept* common ancestry as the best explanation for the patterned differences and similarities among living things. That is, the word *belief* evokes positions held with or without evidence; hence, *belief* is at best an ambiguous word to use in the context of science. Scientists don't "believe" in evolution any more—or less—than they "believe" in thermodynamics.

"Belief" in evolution, as it is too frequently termed, occurs at a lower frequency in the United States than in almost any other developed country. Only about 47 percent of Americans accept that all living things have common ancestors, far less than in Western Europe and Japan, where the percentages are above 70 percent and even 80 percent, respectively. Survey research shows a major disconnect between the US public's acceptance of evolution and that of scientists. In one survey of members of the American Association for the Advancement of Science, the world's largest association of scientists, 97 percent accepted the statement that "humans and other living things have evolved over time." High school teachers are more likely than the general public to accept evolution as a scientific concept, but only about 30 percent report that they teach it extensively. Fully 60 percent admit that they either omit evolution or give it short shrift. The reasons include the teachers' apprehension that evolution is a controversial issue, personal religious beliefs, and the feeling that their education did not prepare them to teach the subject well.

Teachers in the United States can expect that their students who describe themselves as creationists will usually base their creationism on some form of Christianity, the religion of most Americans, and almost always on Christianity, Islam, or Judaism. But there are exceptions: some teachers in communities where Native Americans are numerous have also reported pushback on the teaching of evolution. Other forms of creationism based on Hindu and various New Age religious beliefs also occasionally surface in the classroom.

We should therefore speak of creationisms in the plural. This point reveals as problematic the long-standing plea of antievolutionists that teachers should "teach both" evolution and creationism. How should a teacher choose which creationist version to contrast with evolution? Even supposing that there was some reason to privilege Christianity over other religions, there are several distinct versions of Christian creationism, corresponding to the different ways in which scripture is identified and interpreted by various denominations. Mormons revere the Book of Mormon, and Seventh-Day Adventists regard the writings of Ellen Gould White as inspired; which, if either, should a teacher present? Even if only the Bible is considered, whose Bible?—the King James version, the New Jerusalem version (favored by Catholics), the New International Version (favored by Evangelicals), the New Revised Standard Version (favored by mainline Protestants), or one of the scores of texts available? And given a particular version of the Bible, who is to decide which verses are relevant and how they are to be understood? In fact, these complications only scratch the surface, and the following discussion of the varieties of Christian creationism is necessarily abbreviated.

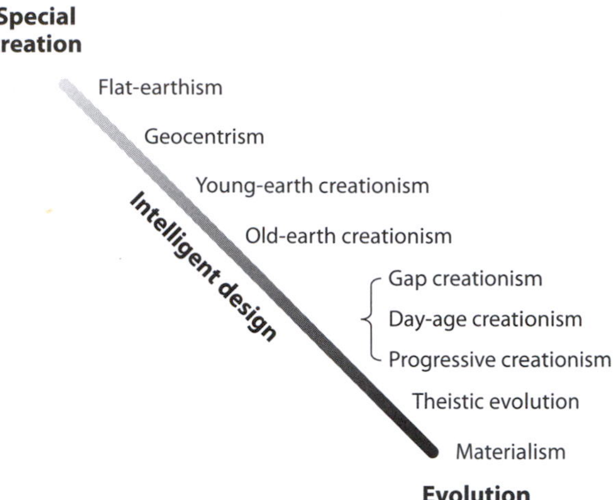

Figure 1. The creation/evolution continuum.

2. THE CREATION-EVOLUTION CONTINUUM

Creationism, as usually encountered in the United States, is based on a "plain" reading of the Bible. Taking the creation narrative in Genesis 1 as authoritative, creationists hold that God specially created the universe, the planet earth, and the living things on it. There are many ways to read the Bible, and the varieties of Christian creationism can be viewed on a continuum reflecting how literally they interpret the words of Genesis and how far their interpretation lies from mainstream science (figure 1).

Flat-Earthism

It is almost comical to believe that *flat-earthers* can exist in the twenty-first century. Nonetheless, until his death in 2001, Charles K. Johnson was president of the International Flat Earth Research Society, a small organization whose interpretation of the Bible is so extreme that passages referring to the "circle of the Earth" (circles are two-dimensional, while spheres are three-dimensional) and the "pillars of heaven" (supports for a metal dome or "firmament" arching over a horizontal planet) are interpreted as stating that the earth is flat. Few Christians take the Bible so literally, but geocentrists are only slightly more liberal in their exegesis.

Geocentrism

Geocentrists believe that the Bible presents earth as the center of the solar system. Passages cited include Joshua 10:12–13, in which God complies with Joshua's plea to stop the sun over the Valley of Ajalon, which requires a stationary earth. Both flat-earthers and geocentrists contend that one cannot pick and choose passages of the Bible to "interpret," so the entire Bible must be accepted as received. If Genesis is not true, they argue, how can we be sure of any of the rest of the Bible, including the New Testament and Revelations, which promise salvation? As Gerardus Bouw, a modern geocentrist author, asked, "If we cannot take God's word as to the rising of the Sun, how can we believe him as to the rising of the Son?"

Geocentrism fought it out with *heliocentrism*, the idea of a sun-centered solar system, during the sixteenth and seventeenth centuries. Heliocentrism eventually would have won because the science was right, but its acceptance was helped by a shift in church doctrine at that time away from the strict biblical literalism of the Middle Ages and early Renaissance periods. As a result, even most creationists accept heliocentrism.

Young-Earth Creationism

Young-earth creationists (YECs) agree on the importance of a plain reading of the Bible, although they do not think that requires belief in a flat earth or geocentrism. YECs are currently the most numerous creationists in the United States. YECs understand Genesis as stating that creation took place over six 24-hour days, during which God created the universe in essentially its present form. Following Archbishop James Ussher, a seventeeth-century Irish cleric, YECs hold that earth was formed only thousands, not billions, of years ago. Animals and plants were created as separate and independent "kinds" and do not share common ancestry. Humans in particular are independent creations, made in God's image.

Most YECs also support the movement known as "creation science," which in its modern form began to be promoted in the 1960s by Henry M. Morris, founder of the Institute for Creation Research (ICR) and its leader until his death in 2006. Morris was highly respected among conservative Christians, and his influence can hardly be overestimated. The movement Morris originated contends that the data and theory of science support the claims of the Bible in all its details. The special creation of all living things by God and the existence of a worldwide Noachian flood (a literal interpretation of Genesis 6–9) are held to be supported not only by faith but also by science.

The science of creation science, however, is decidedly lacking in quality. The logic of creation science is clearly stated by Morris and his followers: evidence *against* evolution is evidence *for* creationism. This approach solves their problem of finding scientific evidence for the sudden appearance of living things in essentially their present form, which special creationism requires. But it also means focusing only on anomalies purporting to disprove evolution and ignoring the massive evidence supporting it. There is ample literature wherein scientists have examined the claims of creation science and found them both factually wrong and theoretically empty. But proponents loudly, if ineffectually, defend their claims that creationism can be made scientific.

Morris firmly believed that a universe measured in thousands of years is foundational to a proper interpretation of the Creation story in Genesis. The only acceptable understanding of creation, then, is young-earth creationism. It is true, of course, that if the earth were young, there would not have been time for much astronomical, geological, or biological evolution. For this reason, YEC institutions including the ICR and Answers in Genesis adhere to Morris's original vision.

YECs insist that just as a young earth is foundational to creationism, so also is the Creation story foundational to Christianity. Christianity's central pillar is the sacrifice of Jesus on the cross to redeem humankind's sins. YECs believe that if Adam and Eve had not been specially created by God, sinned, and punished by being driven from the Garden of Eden, as described in Genesis, then there would not have been a need for a Savior to redeem the sin of Adam. If there was no need for Christ's life, death, and resurrection, as described in the New Testament, then there is no reason to believe the promise of eternal life in the Book of Revelation. The credibility of the entire Bible is thus contingent on the credibility of the special creation of earth and of Adam and Eve. Evolution is therefore unacceptable to YECs.

In Morris's version of young-earth creationism, all sedimentary geological features are the result of Noah's flood, and scientific evidence is sought to support this conclusion. The geological column, it is claimed, only *appears* to present a succession of fossils showing the gradual emergence of present-day forms from earlier forms. Accordingly, the geological succession of fossils resulted from the "hydrodynamic sorting" of the remains of organisms that died in Noah's flood. It is claimed that spherical and smooth organisms such as clams would more likely be found at the bottom of the column because such shapes fall through water more readily than irregular shapes. Jointed organisms with irregular shapes, such as dinosaurs, would be found higher up. And the smarter, more mobile organisms such as mammals would likely have sought higher ground to avoid the floodwaters, explaining their occurrence higher in the geological column. These views are supported by carefully chosen examples—and by ignoring the copious data that refute them.

Old-Earth Creationism

Old-earth creationists (OECs) have perhaps the most variable positions on the continuum. OECs are special creationists, believing that God specially created living things as identifiable "kinds," and thus they reject biological evolution. But they accept the evidence from physical science that our planet and the universe are ancient. Many OECs even accept an earth that is billions of years old. A common view among OECs is to identify the Big Bang as the creative event of Genesis 1. OECs thus consider themselves true to the Bible while accepting the evidence of planetary and cosmic deep time. But among OECs, adherents "interpret" the holy text in various ways to make it compatible with an old earth.

Gap creationism requires the least tinkering with Genesis. Sometimes called "ruin and restoration" theology, gap creationism sees the possibility of two creations in Genesis, with a long period of time between them. The first creation was of a world before Adam and is referenced in the familiar words of Genesis 1:1— "In the beginning, God created the Heaven and the Earth." God then destroyed that creation, a great deal of time passed ("the Earth became without form and void, and darkness was upon the face of the deep"), and then, as stated in Genesis 1:2–31, He created the present world and its inhabitants in six 24-hour days. Gap creationists thus interpret the Bible very literally, though with room for an old earth. It is not surprising that enthusiasm for gap creationism grew in parallel with the rise of modern geological sciences in the late eighteenth and nineteenth centuries.

Day-age creationists, by contrast, believe that the six days of creation were not 24-hour days but instead long periods of indeterminate duration—perhaps hundreds of thousands or even millions of years. They retain reference to a literal Genesis six days but they, too, allow for an old earth. They cite biblical passages such as Psalm 90:4 ("For a thousand years are in your sight like a day") to suggest that the days of creation need not be 24 hours long. Yet another group of OECs downplays the idea of six days, believing instead in an interventionist God who sequentially—and specially—created living things over immense amounts of time. These "progressive creationists" thus accept the geological column as reflecting an accurate history of life on earth but do not believe that the sequences of organisms reflect evolutionary continuity.

All the positions on the continuum discussed thus far are forms of special creationism. But the continuum can be extended to include additional positions on the relationship between the Bible and science. The positions discussed next all accept the mainstream scientific findings of astronomy, geology, and biology—hence, none of them are creationist positions—but they differ from one another on theological or philosophical grounds.

Theistic Evolution

The abandonment of special creationism is clear in the next position on the continuum. *Theistic evolution* (TE) can be described as the belief that evolution has occurred but that God uses evolution to bring about the universe, earth, and living things. Unbeknown to most Americans, TE is mainstream Christian theology, routinely taught in Catholic and Protestant parochial schools. It is considered uncontroversial in many Protestant denominations such as Episcopalians, Presbyterians, United Church of Christ, and in the less conservative branches of Lutherans and Methodists. Thus, when teachers hear students say "I don't believe in evolution, I'm a Catholic," it should be evident that the student is unclear on both science and theology. Embedded in the hallway of a new science building at Catholic Notre Dame University is a large mosaic, 5 ft in diameter, that quotes an aphorism by a famous twentieth-century geneticist, Theodosius Dobzhansky: "Nothing in biology makes sense except in the light of evolution."

In TE, God did not have to create organisms as we see them today: organisms can descend with modification from earlier forms. Believers in TE scorn biblical literalism, being critical of the literalists not only for views incompatible with modern science but also for theological views they consider outmoded and inconsistent. The TE view is that Christian theology must reflect what we know of the world from science if it is to be coherent. Unlike YEC and OEC, the TE view accepts standard scientific interpretations of evidence from geology, physics, chemistry, and biology that indicate that the universe has a long history and that organisms have evolved. Like all theists, adherents to TE believe that the universe was created for a purpose, which science cannot address, although science can address and explain the processes involved in the creation of that universe.

TE believers range along a continuum of their own, varying in how much and in what ways God intervenes over time. Divine intervention is usually conceived as miraculous: with miracles, God violates His created laws, such as by raising Jesus from the dead. However, it is important in TE also for there to be minimal intervention. This "economy of miracles" reflects theological issues not germane to this discussion, such as free will, and the consequences of God "breaking" his own laws. So varieties of TE differ in the degree to which God was "hands-off" in the creation—from one in which God set forth the laws of the universe and allows it to evolve without intervention, to another interpretation in which God also created the first replicating organism (after which evolution proceeded naturally), to yet another in which God intervened also to bring about the evolution of humankind.

The amount of divine action in the creation of the universe is not the only criterion shaping TE views. Like other Christians, believers in TE are also concerned with the degree to which the Deity is personal: an entity who is involved in a meaningful way with the self. One extreme is again a God who created the laws of the universe and is thereafter uninvolved. At the other extreme is the interventionist God to whom one might pray and hope to receive an answer.

The continuum thus far has expressed a greater or lesser reliance on biblical literalism. It has also reflected an inverse acceptance of modern science, with the flat-earthers and geocentrists rejecting some of the most basic facts of modern science, YECs rejecting less familiar but core principles of physics and geology (such as radioisotopic dating) and biology, OECs more or less accepting the physical sciences but rejecting modern biology, and TEs accepting the conclusions of all modern science. All the positions discussed so far have been theistic ones: God exists and is in some way involved in creating the universe in which we live. Next on the continuum are materialists, who reject the concept of a God or higher power.

Materialism

Because this chapter deals with creationism, the nuances and variations of materialism will not be discussed in detail. Briefly, *materialists* believe that matter and energy not only are sufficient to explain the physical universe, as with science, but also are sufficient in a metaphysical sense: there are no gods or supernatural forces or powers. Among materialists there are agnostics, who agree with Thomas Henry Huxley (who coined the term *agnostic*), that one can never know for certain whether there is a God. Agnostics suspend belief. Atheists deny belief in God or gods, and there is a debate among them whether atheism is a philosophical system or merely the denial of the supernatural. Humanism is a nontheistic philosophical system with deep historical roots.

3. INTELLIGENT DESIGN

What about the *intelligent design* (ID) movement? On the diagram of the continuum (figure 1), it is shown straddling OEC and YEC, because ID is, at heart, special creationism, but carefully formulated not to take a stance on the issues that separate OEC and YEC. (In the words of one of the early collections of ID writings, ID espouses "mere creation.") While ID is sometimes erroneously conflated with TE, in practice the ID movement has consistently been antievolutionary in its focus. Also, leaders of the ID community have strongly rejected TE, and the rejection is mutual. Nonetheless, ID has also been criticized by proponents of YEC and, despite some initial enthusiasm, by some leaders in the OEC community.

The reasons for this apparent contradiction lie in the history and content of ID and the strategy its leaders have used to promote their view to the public. The history of ID shows it emerged from a group of OECs (and some YECs) in the mid-1980s. These conservative Christians were dissatisfied with the lack of progress of the YECs in convincing the public to reject evolution or, at least, to accompany its teaching with some form of creationism (such as creation science). At the time, laws promoting equal time for creation science were being tested in the courts, and after a thorough defeat in an Arkansas federal district court, creationists realized that creation science was too obviously tied to Christian religion to survive the Establishment Clause of the United States Constitution. That clause requires public institutions to be religiously neutral. Teaching creation science was judged to be the promotion of religion and thus unconstitutional.

ID emerged as a stripped-down form of creationism out of a series of private meetings (attended by both YECs and OECs) and from the production of a supplemental high school textbook, *Of Pandas and People*, intended to "balance" standard evolution-based textbooks. It ignored creation science favorites, such as the age of the earth and Noah's flood, in favor of the core creationist principle of special creation, although the term *creationism* was (and is) carefully avoided. The ID movement reflected the "argument from design" of William Paley's 1802 book, *Natural Theology*, which compared highly complex biological structures to human-made artifacts. Paley contended that just as a pocket watch could not have assembled itself but required a watchmaker, so, too, a complex biological structure such as the human eye also required a designer and artificer—God. Modern ID examples tend to focus on the complexity of molecular structures. The flagellum of a bacterium is a favorite example of a biological "engine" that is "irreducibly complex" (supposedly too complex to have been produced through natural selection) and thus, it is argued, the product of design by an intelligent agent. Such irreducibly complex structures are called forth in abundance: DNA, the first cell, the body plans of invertebrate phyla of the Cambrian explosion, and so on. Whenever such irreducibly complex structures are discovered, an intelligent designer is invoked, because great complexity is assumed to be unattainable through natural causes.

Who is the intelligent designer responsible for such structures? Proponents of ID are often coy, suggesting that it could be extraterrestrial aliens or time-traveling cell biologists from the far future. However, the more candid among them will acknowledge that they believe the designer to be God, even while agreeing that that is a conclusion unwarranted by science. But when an intelligent agent is invoked at every appearance of an irreducibly complex structure, what is being proposed is actually a form of progressive special creationism. At the grassroots level, ID is understood to be about creationism, with God as the designing agent—even if the leadership of the movement attempts to obscure these identifications to avoid running afoul of the Establishment Clause.

In 1987 the Supreme Court declared in *Edwards v. Aguillard* that teaching creation science in the public schools was unconstitutional. Therefore, when a school board in Dover, Pennsylvania, required teachers to teach ID, lawyers for the plaintiffs in the subsequent 2005 federal district court trial *Kitzmiller v. Dover* sought to demonstrate historical links between creation science and ID. They were successful: such links were crucial in the judge's decision to declare ID a religious rather than a scientific view and that the teaching of ID therefore violated the Establishment Clause.

With ID's roots firmly in creation science, why have the two most prominent YEC organizations, the Institute for Creation Research and Answers in Genesis, attacked ID? Part of the ire of the YECs toward ID arises because of a strategy of ID leaders to omit biblical themes such as the flood of Noah, the special creation of Adam and Eve, and a young age of the earth. OECs tend to outnumber YECs in the leadership of ID and, in fact, ID was largely unknown other than to creationism watchers until 1991, when University of California, Berkeley, law professor Phillip Johnson's book *Darwin on Trial* was published. It is not unusual for antievolution tracts to emerge from creationist institutions or Bible colleges, but such tomes rarely emanate from faculty at major secular universities.

Johnson arguably put ID on the map, as far as the public was concerned, and Johnson's leadership in shaping the legal and philosophical approach of ID was substantial during the 1990s and early 2000s, until ill health required him to take a lower profile. Among other things, Johnson contended that all Christian creationists should unite to attack evolution, setting aside their young-earth versus old-earth squabbles and other differences until they had convinced the public of the scientific and religious shortcomings of evolution. Once evolution was defeated, all the creationists could have a polite discussion over their differences. This strategy may have been appealing to individual creationists, but established creationist organizations were resistant to stepping back from their cherished positions. Eventually they declared that ID—correct in its bashing of evolution—nonetheless was doomed to failure because it would not bring the public to Christianity unless it put the Bible at the center of its mission. And that, of course, was at the heart of the matter. Merely persuading the public that evolution was unsupported by science and inherently atheistic was inadequate: it was necessary to replace evolution with special creation.

4. WHAT DOES THE FUTURE HOLD?

Disagreements among leaders of the creationist movements are only part of the story. What is perhaps more significant is how these movements are viewed by the public. The differences between YECs and OECs are stark, and a choice must be made between an earth that is billions of years old or only a few thousand years old. ID, meanwhile, is commonly considered to be an adjunct to either the YEC or the OEC perspective. Rather than viewing ID as the sophisticated scientific argument dreamed of by its proponents, most members of the public who are familiar with it see it as a generalized form of creationism—which, in fact, it is.

Since the *Kitzmiller v. Dover* trial, the ID star has burned a bit more dimly. Leaders of the movement, affiliated with the Seattle-based Center for Science and Culture at the Discovery Institute, are now encouraging legislation and regulations that would encourage the teaching of "evidence against evolution." Sometimes, evolution is bundled with other "controversial issues" such as global warming and human cloning for special treatment in the curriculum. A common tactic calls for teachers to be given "academic freedom" to bring in "alternative views" to those expressed in the textbook or state standards. In the case of evolution, of course, "alternative views" is a euphemism for creationism. It appears to be a popular strategy for promoting creationism: more than 40 "academic freedom"–style bills were proposed in various state legislatures in the decade 2003–2013, although opponents managed to defeat almost all of them in committee. When such bills reach the floors of their respective chambers, however, they are often difficult for elected officials to publicly oppose. Two bills have passed: one in Louisiana in 2008 and one in Tennessee in 2012.

In contrast, YEC is thriving, although explicit attempts to promote the teaching of creationism in the public schools are rare. Particularly prominent is the Answers in Genesis ministry, which since its founding in 1994 has been remarkably successful at capturing the market for creationism. This success is apparently due in part to its adopting a style heavier on evangelism and lighter on science than the Institute for Creation Research and in part to its use of the latest technology, including a well-crafted website. Answers in Genesis also opened a lavish "Creation Museum" in northern Kentucky in 2007, which may be joined in the future by a Noah's Ark theme park. The Institute for Creation Research has moved to an expanded new campus in Dallas, Texas, and is expecting to rebuild its own museum in that city. Several smaller creationism museums are in the planning stages or have already opened.

All in all, it appears that the creationism movement in the United States is prospering. And given its fragmented history, it is safe to say that even if some constituents fall out of favor, new varieties will emerge somewhere on the continuum.

FURTHER READING

Berkman, M., and E. Plutzer. 2010. Evolution, Creationism, and the Battle to Control America's Classrooms. New York: Cambridge University Press. *Invaluable for its extensive, thoughtful, and fruitful use of survey data, especially its rigorous national survey of high school biology teachers.*

Forrest, B., and P. R. Gross. 2007. Creationism's Trojan Horse: The Wedge of Intelligent Design. Rev. ed. New York: Oxford University Press. *The definitive exposé of intelligent design as a strategy of rebranding creationism, updated with a chapter on events after the* Kitzmiller *trial.*

Larson, E. 2003. Trial and Error. 3rd ed. Oxford: Oxford University Press. *The authoritative history of the legal struggles over the teaching of evolution in the United States, although it stops short of the* Kitzmiller *trial.*

McCalla, A. 2006. The Creationist Debate: The Encounter between the Bible and the Historical Mind. New York: Continuum. *A synoptic history of the creationism/evolution controversy, focusing on the development of the historical sciences and how the ways of interpreting the Bible developed in response.*

Numbers, R. L. 2006. The Creationists: From Scientific Creationism to Intelligent Design. Exp. ed. Cambridge, MA: Harvard University Press. *A monumental work on the history of the creationist movement, newly updated with a chapter on intelligent design.*

Pennock, R. T. 1999. Tower of Babel: The Evidence against the New Creationism. Cambridge, MA: MIT Press. *The first, and still a valuable, examination of the intelligent design movement, by a philosopher who testified at the* Kitzmiller *trial.*

Ruse, M. 2005. The Evolution-Creation Struggle. Cambridge, MA: Harvard University Press. *The distinguished philosopher and historian of science attempts to understand the roots of the creationism/evolution controversy.*

Scott, E. C. 2009. Evolution vs. Creationism: An Introduction. 2nd ed. Berkeley: University of California Press. *A comprehensive history, commentary, and sourcebook on the creationism/evolution controversy.*

VIII.15

Evolution and the Media

Carl Zimmer

OUTLINE

1. Evolution and the birth of modern science communication
2. Evolution and creationism: The dangers of false balance
3. Evolution and the rise of new media
4. The *Darwinius* affair: A cautionary tale
5. Conclusion

On March 28, 1860, the *New York Times* ran a very long article on a newly published book called *On the Origin of Species*. The *Times* explained that the dominant explanation for life's staggering diversity at the time was the independent creation of every species on earth. "Meanwhile," the anonymous author wrote, "Mr. DARWIN, as the fruit of a quarter of a century of patient observation and experiment, throws out, in a book whose title has by this time become familiar to the reading public, a series of arguments and inferences so revolutionary as, if established, to necessitate a radical reconstruction of the fundamental doctrines of natural history."

Today, some 150 years later, evolutionary biologists are continuing to reconstruct natural history, and journalists are still documenting that reconstruction. Each week brings a flood of reports on new research into evolution, ranging from fossil dinosaurs to the emergence of new strains of viruses to evolutionary clues embedded in the human genome. The *New York Times* continues to publish articles about evolution, as do many other newspapers and magazines. But reports on evolution can also take many new forms that were inconceivable in Darwin's day. They can be the subject of television shows, blogs, podcasts, and tweets. This chapter examines the ways in which media has treated evolution over the past four decades, and the rapid changes currently unfolding. There is not space, however, to consider the fascinating relationship of evolution and the media in earlier periods of history (see, e.g., Browne 2001 and Larson 1998).

1. EVOLUTION AND THE BIRTH OF MODERN SCIENCE COMMUNICATION

To understand the relationship between evolution and the media, it helps to take an evolutionary perspective. The journalistic coverage of evolution as we know it today began to take shape in the 1970s. Newspapers, especially in the United States, were growing rapidly at the time and developing new features to attract readers. Many newspapers hired reporters who specialized in science, and many science writers focused much of their attention on evolutionary biology. For example, Boyce Rensberger, a science writer for the *New York Times*, wrote a string of stories about evolution in the 1970s. In one typical Rensberger article (April 12, 1975), titled "East Africa Fossils Suggest That Man Is a Million Years Older Than He Thinks," he described the discovery of a 3-million-year-old fossil of a hitherto-unknown species of hominin, *Australopithecus afarensis*.

Four years later, the *Times* founded a weekly section dedicated to science. It was the first science section ever included in an American newspaper, but in the next few years, many other newspapers followed suit. A number of science magazines were also launched. Old standards like *Scientific American* were joined by start-ups such as *Discover* and *Omni*. All these new publications gave special attention to evolution.

One reason for this focus was that evolutionary biology itself had entered an exciting period of renewal, and so there were many stories for reporters to write about. New fossils like *A. afarensis* provided paleontologists with fresh insights into human evolution. Dinosaurs, which had long been considered sluggish and slumped, received a makeover. During the 1970s, the Yale paleontologist John Ostrom oversaw the reconstruction of dinosaurs as fast-running, warm-blooded creatures—an upgrade from *Godzilla* to *Jurassic Park*.

Geologists were also adding to evolution's cinematic appeal. In the late 1970s Walter Alvarez of the University of California at Berkeley and his colleagues discovered clues that an asteroid smashed into earth 65 million years ago. That collision happened to coincide with the end of the Cretaceous period, a time of mass extinctions that claimed the dinosaurs Ostrom was rehabilitating. Alvarez made a radical connection between the impact and the mass extinctions. Mass extinctions had long been thought to stretch across millions of years, caused by slow-moving processes such as gradual sea level change. Alvarez and his colleagues offered a vision of sudden disaster: the asteroid impact threw dust and rock high into the atmosphere, causing a global environmental catastrophe—darkness for months, acid rain, global warming, and more. In a geological flash, millions of species became extinct.

Alvarez was arguing for a catastrophic mode of evolution. To understand evolution 65 million years ago, we could not simply extrapolate back from the small, incremental changes natural selection produces today from one generation to the next. As a result, some scientists argued, the end-Cretaceous extinctions did not fit into the framework of the Modern Synthesis. The *Modern Synthesis*—an integration of genetics, paleontology, ecology, and other branches of biology—explained life predominantly as the result of natural selection operating on small differences among individuals over vast periods of time.

Challenges to the Modern Synthesis came from studies not just on mass extinctions but on more tranquil periods of the fossil record. Paleontologists Niles Eldredge and Stephen Jay Gould argued that the fossil record revealed a pattern of stasis and change, a pattern they dubbed *punctuated equilibria*: species remained stable for millions of years, and new species rapidly branched off in just thousands of years. Eldredge and Gould argued that this pattern of evolution allowed selection to take place not just between individuals but perhaps also between species.

Science writers chronicled these challenges to the Modern Synthesis, but they also reported on other scientists who were expanding its scope. In 1976 the British zoologist Richard Dawkins, building on the work in the 1960s of George Williams and William Hamilton, published *The Selfish Gene*. Dawkins argued that evolution was best understood from a gene-centered perspective. The Harvard biologist Edward O. Wilson undertook a similar project, interpreting a vast range of behaviors—from the selfless work of sterile worker bees to the bloodshed of human warfare—as strategies for genes to get themselves replicated. In 1975 he unveiled his synthesis in the book *Sociobiology*. Rensberger (May 28, 1975) reported its publication on the front page of the *New York Times* in his article "Sociobiology: Updating Darwin on Behavior."

As evolution was appearing on the front pages of newspapers, science programming was also emerging on television. In 1974, for example, the Public Broadcasting Service developed the *Nova* series. New research on evolution figured prominently in these shows as well. In his 1980 series *Cosmos*, Carl Sagan discussed the basic principles of evolution, along with new ideas about the role of comets and other impacts on the history of life. And in 1981, Walter Cronkite, having just retired from his nightly television news show, hosted a series of science shows called *Cronkite's Universe*. On one episode his guests were Donald Johanson—one of the discoverers of *A. afarensis*—and the paleoanthropologist Richard Leakey. Johanson and Leakey engaged in a heated debate about the place of *A. afarensis* in human evolution. Johanson believed it was on the line that led to *Homo sapiens*, while Leakey considered it a side branch. During the program, Johanson held up a chart showing his version of hominin phylogeny. Next to it was a blank space where he asked Leakey to draw his hypothesis. Instead, Leakey drew an X through Johanson's tree. In its place, he drew a large question mark (Wilford 2011).

Evolutionary biologists debated on television, and they also debated in print. As Dawkins and Wilson garnered attention for their expansion of the Modern Synthesis, Stephen Jay Gould and other scientists launched scathing criticisms, arguing that adaptationists ascribed far too much power to natural selection. They condemned sociobiology as "just-so stories"— plausible-sounding tales of adaptation rather than carefully constructed and tested hypotheses. Most of Gould's attacks took place not in the pages of scientific journals but in popular publications such as *Natural History* and the *New York Review of Books*. Dawkins, Wilson, and others responded in kind, and the debate gave rise to a number of hugely popular books, such as Gould's *Wonderful Life* (1998) and Dawkins's *The Blind Watchmaker* (1996).

2. EVOLUTION AND CREATIONISM: THE DANGERS OF FALSE BALANCE

In December 1981 a number of the top science journalists in the United States converged on Little Rock, Arkansas, to cover a story about evolution. The story did not concern a new fossil, or a new hypothesis about speciation, but a trial. Earlier that year, the Arkansas legislature had passed a law requiring that public school teachers present "creation science" alongside evolution in their biology classes. A group of teachers and religious figures filed a lawsuit challenging the law as an unconstitutional promotion of religion.

Of all the sciences, evolutionary biology attracts an unmatched amount of social controversy. Organized religious opposition to the teaching of evolution in the United States first emerged in the 1920s, leading to the famous Scopes "monkey trial" of 1925. Conflicts over the teaching of evolution have continued to break out in the decades since then. The 1981 case *McLean v. Arkansas* led to the banning of "creation science" from classrooms. But it did not stop the conflict over the teaching of evolution. Journalists have continued to report on the attempts of some state and local school board members to question the validity of evolution and to promote creationism in its various forms. (See chapter VIII.14 for more on the history of creationism in the United States.)

Much of the coverage of evolution found in newspapers, magazines, and television news programs addresses these social conflicts, rather than the science of evolution itself. This bias is an unfortunate result of the nature of modern journalism: editors and journalists seek easily explained conflicts between people. Another weakness in much modern journalism is a craving for *false balance*. If one side in court trial says that evolution is true, then a journalist may feel obligated to unquestioningly quote someone from the other side. This "he said, she said" form of journalism can be legitimate in political reporting, but it is unacceptable in science reporting. It implicitly gives equal credibility to opposing sides, even if one side has no science whatsoever to back up its case. False balance promotes the mistaken impression that evolution is controversial within the scientific community, rather than the foundation of modern biology.

3. EVOLUTION AND THE RISE OF NEW MEDIA

In some ways, the relationship between evolution and the media has changed little since the 1970s. Public television and cable stations periodically air shows dealing with paleontology and human origins. Richard Dawkins and Edward Wilson continue to write best-selling books, and they have been joined by many talented younger evolutionary biologists, such as Steven Pinker, Jared Diamond, Olivia Judson, Sean Carroll, and Neil Shubin. Evolution still inspires abundant journalism in newspapers and magazines. And journalists continue to cover controversies over evolution, including the 2005 *Kitzmiller v. Dover* case and Louisiana's 2008 law protecting creationist science teachers in the name of "academic freedom."

Yet tremendous changes are under way. People are rapidly moving to the Internet to learn about science, including evolution.

Evolution first went online in the 1990s, when a few evolutionary biologists and evolution aficionados began to set up online discussion groups such as the one at www.talk.origins.org. They posted comments about new advances in evolutionary biology and the attempts of creationists to block the teaching of evolution. Later these sites also hosted lists of frequently asked questions about evolution, such as, "If we evolved from monkeys, why are there still monkeys?"

Talk.origins and other evolution discussion groups were founded at a time when few people outside universities had even heard of the Internet. As the number of Internet users grew exponentially, programmers invented more powerful ways to post information online. Blogs allowed people to self-publish their writing; they also made it possible to post podcasts, video, and other media. Today, thanks to the Internet, far more biologists are regularly writing about evolution than ever before (Goldstein 2009).

As blogs have bloomed, the older venues for news on evolution have struggled. A number of science magazines launched in the 1970s and 1980s, such as *Omni* and *Science 80*, eventually folded. Science coverage in newspapers suffered in the 1990s. In 1989, a total of 95 newspapers ran science sections. By 2013 that number had shrunk to just 19. Those shuttered science sections were the victims of an industry-wide blight. Newspapers were being squeezed for greater profits, even as their readerships were declining. They offered their senior staff buyouts to reduce labor costs. A number of the science writers who had been part of the field's first efflorescence left the business.

Many newspapers and magazines now see the Internet as an essential part of their future. The *New York Times*, for example, now has a daily circulation of about 1.5 million readers but receives about 25 million unique visits a month to its website. The news on its site also radiates outward across the World Wide Web as people comment on it in blogs and forums.

These huge changes in readership are changing the way evolution and other branches of science are reported. The print edition of the *New York Times* still includes a science section every Tuesday, but it also offers many untraditional kinds of coverage of evolution. For example, the *New York Times* has published blog posts by evolutionary biologists about their work, and offers podcasts and even short videos about evolution. In 2009 it posted *The Origin* in an online form, with annotations from some of the world's leading scientists.

But the *New York Times* and other publications have to compete with scientists themselves to present evolution to the public. The University of California, Berkeley, has set up a major website called Understanding Evolution (evolution.berkeley.edu), which presents not only the basic concepts of evolution but also new scientific developments. In 2009, Casey Dunn, an evolutionary biologist at Brown University, established a blog called *Creaturecast* about animal evolution (creaturecast.org), where he and his co-bloggers regularly publish innovative videos. One episode explains how single-celled organisms made the evolutionary transition to multicellularity, for example. The film is a stop-action animation of purple modeling clay, which morphs into cells, which then join together into bodies. The video is at once charming and surprisingly enlightening. And most important, it was something no one would have imagined a few years earlier.

4. THE *DARWINIUS* AFFAIR: A CAUTIONARY TALE

Creative efforts such as *Creaturecast* inspire hope for the future of evolution and the media, but they should not inspire a blind optimism. The Internet is also home to a great deal of misinformation about evolution, especially on creationist sites. Some of these sites are relatively obvious, such as Creation Safaris (creationsafaris.com). Other sites cloak their creationism. A site with the harmless-sounding name All About Science (www.allaboutscience.org) has a long page titled "Darwin's Theory of Evolution: A Theory in Crisis." It takes a bit of snooping around to discover that All About Science is produced by a group called AllAboutGod.com.

Too often, journalists for major media provide poor information about evolution online. In fact, the very nature of twenty-first century media fosters bad reporting on evolution. One of the most instructive events took place in May 2009, when journalists reported on the unveiling of a new fossil of a primate dubbed *Darwinius masilae*.

The unveiling was unique in the annals of paleontology. At the American Museum of Natural History, New York City Mayor Michael Bloomberg and other luminaries gazed at the slab preserving a 47-million-year-old specimen (known as Ida, named after the daughter of one of the paleontologists who described the fossil). Minutes before the press conference was to commence, the electronic journal *PLoS ONE* published a paper about the fossil. Some of the paper's authors, speaking at the press conference, described the fossil as both the holy grail of paleontology and the lost ark of archeology.

The scientists were not the only ones to speak that morning. Nancy Dubuc, an executive at the History Channel, said that the fossil "promised to change everything that we thought we understood about the origins of human life" (Pilkington 2009). Why was Dubuc there? Because the unveiling of *Darwinius* was actually a television phenomenon, years in the making.

Television producers had started putting together a big-budget show about *Darwinius* even as the scientists were analyzing the fossil and writing up their results. The documentary's main message was also the chief argument in the *PLoS ONE* paper: *Darwinius* belonged to the lineage that led to monkeys, apes, and humans. As a result, it illuminated how our ancestors diverged from more distantly related primates, such as lemurs. As the air date for the documentary approached, the History Channel cranked up a massive publicity machine. A trade book was rushed into print; ads appeared; YouTube videos spread like viruses. The History Channel set up an elaborate website called Revealing the Link (revealingthelink.com). It featured hyperbolic claims from the scientists, such as "When our results are published, it will be just like an asteroid hitting the Earth."

As press manipulation, the strategy worked well. Newspapers, magazines, and even television news programs ran stories about *Darwinius* on their websites on the day of its grand unveiling. Few of them would have ever considered covering the discovery of an Eocene primate, it's safe to say, without the elaborate publicity. Unfortunately, most reporters simply relayed hyperbolic quotes from their sources. They also demonstrated some deep misunderstandings about evolution. "Fossil is evolution's missing link," announced the *Sun*, falling prey to the common misbelief that paleontologists could ever determine our direct ancestors (Soodin 2009). (In fact, paleontologists compare related species to determine the pattern by which new traits emerged in different lineages.)

Given the upheavals in the media these days, it's not surprising that the press was so swayed by the *Darwinius* publicity machine. The number of skilled science writers who can report a story like this one with the proper skepticism is dwindling. And all media organizations are racing to be the first to get news online.

A few veteran journalists tried to obtain the paper to show it to other experts on fossil primates to get their opinion on its importance, but they were thwarted both by *PLoS ONE* and the authors. Ann Gibbons, a correspondent for *Science*, finally got her hands on the paper the weekend before the press conference, but only after signing a nondisclosure agreement with the television company that produced the *Darwinius* documentary. Gibbons promised not to show the paper to anyone before the press conference (Zimmer 2009).

The first wave of articles about *Darwinius* was based entirely on the press conference and claims from the scientists who had published the paper. Days later, Gibbons and a handful of other science writers published articles that offered a broad look at *Darwinius*, rather than the breathless press conference coverage that dominated the news. Nearly all the other experts reporters contacted thought the fossil was impressive but that the claims of its kinship with humans unjustified. "This hypothesis now lies well outside the scientific mainstream, and the discovery and description of Ida have done little to rehabilitate it," wrote Christopher Beard of the Carnegie Museum of Natural History in the September 2009 issue of *American Scientist*.

In October 2009, five months after the *Darwinius* circus had folded its tents and moved on, paleontologists Erik Seiffert and colleagues published an important new paper on early primate evolution. They described another early primate fossil, called *Afradapis*, and compared its anatomy with that of *Darwinius* and a wide range of other primate fossils. Their analysis placed *Darwinius* on the branch that led to lemurs, not to us. The reaction from the press for this paper was a stark contrast to the pandemonium that greeted *Darwinius* in May. Very few newspapers and other publications even mentioned the new study. Perhaps if there had been a big-budget documentary on *Afradapis*, things might have been different.

5. CONCLUSION

Information about evolution is now available in a staggering range of forms. But readers, listeners, and viewers cannot simply assume that everything they encounter is accurate. People must learn to think critically about what they read, watch, and listen, and should also strive to develop a strong understanding of the basic principles of evolutionary theory. They can also tap into the collective wisdom of the blogosphere. And finally, they should resist the rapid-fire allure of the Internet. After all, the scientific process does not run on a 24-hour-a-day news cycle. It takes years for scientists to gather data and present hypotheses, and for other scientists to test them. Journalism, it is often said, is the first draft of history. In the history of evolutionary biology, that first draft is sometimes wrong.

FURTHER READING

Beard, C. 2009. The weakest link. American Scientist (September–October), http://www.americanscientist.org/bookshelf/pub/the-weakest-link.

Browne J. 2001. Darwin in caricature: A study in the popularisation and dissemination of evolution. Proceedings of the American Philosophical Society 145 (December): 4.

Dawkins, R. 1986.The Blind Watchmaker: Why the Evidence of Evolution Reveals a Universe without Design. New York: W. W. Norton.

Goldstein, A. 2009. Blogging evolution. Evolution: Education and Outreach 2: 548–559

Gould, S. J. 1989. Wonderful Life: The Burgess Shale and the Nature of History. New York: W. W. Norton.

Larson, E. 1998. Summer of the Gods: The Scopes Trial and America's Continuing Debate over Science and Religion. Cambridge, MA: Harvard University Press.

Pilkington, E. 2009. To get a glimpse of the Ida fossil, the media make monkeys of themselves. Guardian, May 19, 2009, http://www.guardian.co.uk/science/2009/may/19/ida-fossil-primate-media-us.

Seiffert, E. R, J.M.G. Perry, E. L. Simons, and D. M. Boyer. 2009. Convergent evolution of anthropoid-like adaptations in Eocene adapiform primates. Nature 461: 7267 (October 22): 1118–1121.

Soodin R. 2009. Fossil is evolution's "missing link." Sun, May 19, 2009, http://www.thesun.co.uk/sol/homepage/news/article2437749.ece.

Wilford, J. 2011. Tracking human lineage through a bramble. New York Times, May 9, 2011.

Zimmer, C. 2009. Science held hostage. Loom, May 21, 2009, http://phenomena.nationalgeographic.com/2009/05/21/science-held-hostage/.

Index

References in *italics* refer to figures.